MyMathLab® = Your Resource for Success

In the lab, at home,...

- Access videos, PowerPoint® slides, and animations.
- Complete assigned homework and quizzes.
- Learn from your own personalized Study Plan.
- Print out the *Student Workbook for Elementary Algebra for College Students* for additional practice.
- Explore even more tools for success.

...and on the go.

Download the free Pearson eText App to access the full eText on your Apple® or Android™ device. MyMathLab log-in required.

Use your Chapter Test as a study tool! Chapter Test Prep Videos show step-by-step solutions to all Chapter Test exercises. Access these videos in MyMathLab or by scanning the code.

Scan the code or go to: www.youtube.com/user/AngelElementaryAlg

Don't Miss Out! Log In Today.

MyMathLab delivers proven results in helping individual students succeed. It provides engaging experiences that personalize, stimulate, and measure learning for each student. And, it comes from a trusted partner with educational expertise and an eye on the future.

To learn more about how MyMathLab combines proven learning applications with powerful assessment, visit **www.mymathlab.com**

VIDEOS • POWERPOINT SLIDES • ANIMATIONS • HOMEWORK • QUIZZES • PERSONALIZED STUDY PLAN • TOOLS FOR SUCCESS

Elementary Algebra

for College Students

Elementary Algebra

for College Students

9e

Allen R. Angel
MONROE COMMUNITY COLLEGE

Dennis C. Runde
STATE COLLEGE OF FLORIDA

PEARSON

Boston Columbus Indianapolis New York San Francisco Upper Saddle River
Amsterdam Cape Town Dubai London Madrid Milan Munich Paris Montréal Toronto
Delhi Mexico City São Paulo Sydney Hong Kong Seoul Singapore Taipei Tokyo

Editorial Director, Mathematics: *Christine Hoag*
Editor in Chief: *Michael Hirsch*
Editorial Project Managers: *Katie DePasquale and Matthew Summers*
Editorial Assistant: *Matthew Summers*
Senior Managing Editor: *Karen Wernholm*
Senior Production Supervisor: *Patty Bergin*
Senior Designer: *Barbara T. Atkinson*
Cover and Interior Designer: *Infiniti Design/Jennifer Willingham*
Digital Assets Manager: *Marianne Groth*
Media Producer: *Jean Choe*
Executive Manager, Course Production: *Peter Silvia*
Director of Content Development: *Rebecca Williams*
Executive Marketing Manager: *Michelle Renda*
Marketing Assistant: *Susan Mai*
Procurement Specialist: *Debbie Rossi*
Production Management and Composition: *Integra*
Art Studios: *Scientific Illustrators* and *Integra*
Cover image: *Margouillat Photo/Shutterstock*

Library of Congress Cataloging-in-Publication Data
Angel, Allen R., 1942-
Elementary algebra for college students/Allen R. Angel, Monroe Community College, Dennis C. Runde, State College of Florida.—9th edition.
p. cm.
ISBN-13: 978-0-321-86806-0 (alk. paper)
1. Algebra—Textbooks. I. Runde, Dennis C. II. Title.
QA152.3.A53 2013
512.9—dc23
2013013673

2 3 4 5 6 7 8 9 10—RRD—17 16 15 14

ISBN-10: 0-321-86806-4
ISBN-13: 978-0-321-86806-0

To my wife, Kathy,
and our sons, Robert and Steven
Allen R. Angel

To my wife, Kristin,
and our sons, Alex, Nick, and Max
Dennis C. Runde

Brief Contents

Contents

Preface

This book was written for students and other adults who have never been exposed to algebra or those who have been exposed but need a refresher course. Our primary goal was to write a book that students can read, understand, and enjoy. To achieve this goal we have used short sentences, clear explanations, and many detailed, worked-out examples. We have tried to make the book relevant to college students by using practical applications of algebra throughout the text.

The many factors that contributed to the success of the previous editions have been retained. In preparing this revision, we considered the suggestions of instructors and students throughout the country. The *Principles and Standards for School Mathematics,* prepared by the National Council of Teachers of Mathematics (NCTM), and *Beyond Crossroads: Implementing Mathematics Standards in the First Two Years of College,* by the American Mathematical Association of Two-Year Colleges (AMATYC), together with advances in technology, influenced the writing of this text.

New to This Edition

One of the most important features of the text is its emphasis on readability. The book is very understandable to students of all reading skill levels. The Ninth Edition retains this emphasis and has been revised with a focus on improving accessibility and addressing the learning needs and styles of today's students. To this end, the following changes have been made:

Content Changes

- Many discussions throughout the text have been thoroughly revised for readability and improved explanation of key concepts. Whenever possible, a visual example or diagram is used to explain concepts and procedures.
- Exercise sets have been carefully written to improve the grading for difficulty. Students can more easily begin with basic concepts and work their way into more difficult problems. Also, the exercises have been paired so that an odd-numbered exercise is followed by a similar even-numbered exercise.
- **Now Try Exercises** are exercises that are closely matched to worked examples to provide students with a framework with which to work key exercises. The **Now Try Exercises** have also been carefully written to better match the worked examples.
- **Warm-Up Exercises** have been revised to better reflect key vocabulary words, concepts, and formulas needed to perform the remainder of the exercise set.
- **Understanding Algebra** boxes draw students' attention quickly to important concepts and facts needed to master algebra. **Understanding Algebra** boxes have been revised to more clearly explain key concepts.
- The pedagogical use of color has been enhanced and continues to support a more visual approach to learning algebra.
- Interval notation is now introduced in Section 2.8.
- Additional material on fractions has been added to help students with concepts such as rational expressions.
- Material on functions has been revised for greater clarity.
- Applications have been updated throughout the book.

Features of the Text

Full-Color Format

Color is used pedagogically to make the text more appealing to, and usable by, students. Important definitions, procedures, and other items stand out via use of color screening and color type. The artwork is enhanced and clarified through the use of multiple colors.

Accuracy

Accuracy in a mathematics text is essential. To ensure accuracy in this book, math teachers from around the country have read the pages carefully for typographical errors and have checked all the answers.

Connections

Many of our students do not thoroughly grasp new concepts the first time they are presented. In this text we encourage students to make connections. That is, we introduce a concept, then later in the text briefly reintroduce it and build upon it. Often an important concept is used in many sections of the text. Important concepts are also reinforced throughout the text in the Cumulative Review Exercises and Cumulative Review Tests.

Chapter Opening Application

Each chapter begins with a real-life application related to the material covered in the chapter. By the time students complete the chapter, they should have the knowledge to work the problem.

Goal of This Chapter

This feature on the chapter opener page gives students a preview of the chapter and also indicates where this material will be used again in other chapters of the book. This material helps students see the connections among various topics in the book and the connection to real-world situations.

The Use of Icons

At the beginning of each exercise set the icons for MathXL®, MathXL®, and for MyMathLab, MyMathLab®, are illustrated to remind students of these homework resources.

Keyed Section Objectives

Each section opens with a list of skills that the student should learn in that section. The objectives are then keyed to the appropriate portions of the sections with blue numbers such as 1.

Problem Solving

Pólya's five-step problem-solving procedure is discussed in Section 1.2. Throughout the book, problem solving and Pólya's problem-solving procedure are emphasized.

Practical Applications

Practical applications of algebra are stressed throughout the text. Students need to learn how to translate application problems into algebraic symbols. The problem-solving approach used throughout this text gives students ample practice in setting up and solving application problems. The use of practical applications motivates students.

Detailed, Worked-Out Examples

A wealth of examples have been worked out in a step-by-step, detailed manner. Important steps are highlighted in color, and no steps are omitted until after the student has seen a sufficient number of similar examples.

Now Try Exercises

In each section, after each example, students are asked to work an exercise that parallels the example given in the text. These Now Try Exercises make the students *active,* rather than passive, learners and they reinforce the concepts as students work the exercises. Through these exercises, students have the opportunity to immediately apply what they have learned. After each example, Now Try Exercises are indicated in purple type such as Now Try Exercise 27. They are also indicated in green type in the exercise sets, such as 27.

Study Skills Section

Students taking this course may benefit from a review of essential study skills. Such study skills are essential for success in mathematics. Section 1.1, the first section of the text, discusses such study skills. This section should be very beneficial for your students and should help them to achieve success in mathematics.

Understanding Algebra

Understanding Algebra boxes appear in the margin throughout the text. Placed at key points, Understanding Algebra boxes help students focus on the important concepts and facts that they need to master.

Helpful Hints

The Helpful Hint boxes offer useful suggestions for problem solving and other varied topics. They are set off in a special manner so that students will be sure to read them.

Avoiding Common Errors

Common student errors are illustrated. Explanations of why the shown procedures are incorrect are given. Explanations of how students may avoid such errors are also presented.

Exercise Sets

The exercise sets are broken into three main categories: Warm-Up Exercises, Practice the Skills, and Problem Solving. Many exercise sets also contain Concept/Writing Exercises, Challenge Problems, and/or Group Activities. Each exercise set is graded in difficulty, and the exercises are paired. The early problems help develop the students' confidence, and then students are eased gradually into the more difficult problems. A sufficient number and variety of examples are given in each section for students to successfully complete even the more difficult exercises. The number of exercises in each section is more than ample for student assignments and practice.

Warm-Up Exercises

The exercise sets begin with Warm-Up Exercises. These fill-in-the-blank exercises include an emphasis on vocabulary. They serve as a great warm-up to the homework exercises or as 5-minute quizzes.

Practice the Skills Exercises

The Practice the Skills exercises reinforce the concepts and procedures discussed in the section. These exercises provide students with practice in working problems similar to the examples given in the text. In many sections the Practice the Skills exercises are the main and most important part of the exercise sets.

Problem-Solving Exercises

These exercises help students become better thinkers and problem solvers. Many of these exercises involve real-life applications of algebra. It is important for students to be able to apply what they learn to real-life situations. Many problem-solving exercises help with this.

Concept/Writing Exercises

Most exercise sets include exercises that require students to write out the answers in words. These exercises improve students' understanding and comprehension of the material. Many of these exercises involve problem solving and conceptualization and help develop better reasoning and critical thinking skills.

Challenge Problems

These exercises, which are part of many exercise sets, provide a variety of problems. Many were written to stimulate student thinking. Others provide additional applications of algebra or present material from future sections of the book so that students can see and learn the material on their own before it is covered in class. Others are more challenging than those in the regular exercise set.

Group Activities

Many exercise sets have Group Activity exercises that lead to interesting group discussions. Many students learn well in a cooperative learning atmosphere, and these exercises will get students talking mathematics to one another.

Group Activities

Many exercise sets have Group Activity exercises that lead to interesting group discussions. Many students learn well in a cooperative learning atmosphere, and these exercises will get students talking mathematics to one another.

Cumulative Review Exercises

All exercise sets (after the first two) contain questions from previous sections in the chapter and from previous chapters. These Cumulative Review Exercises will reinforce topics that were previously covered and help students retain the earlier material while they are learning the new material. For the students' benefit, Cumulative Review Exercises are keyed to the section where the material is covered, using brackets, such as [3.4].

Mid-Chapter Tests

In the middle of each chapter is a Mid-Chapter Test. Students should take each Mid-Chapter Test to make sure they understand the material presented in the chapter up to that point. In the student answers, brackets such as [2.3] are used to indicate the section where the material was first presented.

Chapter Summary

At the end of each chapter is a comprehensive chapter summary that includes important chapter facts and examples illustrating these important facts.

Chapter Review Exercises

At the end of each chapter are review exercises that cover all types of exercises presented in the chapter. The review exercises are keyed using colored numbers and brackets, such as [1.5], to the sections where the material was first introduced.

Chapter Practice Tests

The comprehensive end-of-chapter practice tests enable students to see how well they are prepared for the actual class test. The section where the material was first introduced is indicated in brackets in the student answers.

Cumulative Review Tests

These tests, which appear at the end of each chapter after the first, test the students' knowledge of material from the beginning of the book to the end of that chapter. Students can use these tests for review, as well as for preparation for the final exam. These exams, like the Cumulative Review Exercises, serve to reinforce topics taught earlier. In the answer section, after each answer, the section where that material was covered is given using brackets.

Answers

The *odd-numbered answers* are provided for the exercise sets. *All answers* are provided for the Cumulative Review Exercises, Mid-Chapter Test, Chapter Review Exercises, Chapter Practice Tests, and Cumulative Review Tests. Answers are not provided for the Group Activity exercises because we want students to reach agreement by themselves on the answers to these exercises.

Prerequisite

This text assumes no prior knowledge of algebra. However, a working knowledge of arithmetic skills is important. Fractions are reviewed early in the text, and decimals and percent are reviewed in Appendix A.

Modes of Instruction

The format and readability of this book, and its many resources and supplements, lend it to many different modes of instruction. The constant reinforcement of concepts will result in greater understanding and retention of the material by your students.

The features of the text and its supplements make it suitable for many types of instructional modes, including

- face-to-face courses
- hybrid or blended courses
- emporium-based courses
- online instruction
- self-paced instruction
- inverted classrooms
- cooperative or group study

Student and Instructor Resources

STUDENT RESOURCES

Student Solutions Manual
Provides complete worked-out solutions to
- the odd-numbered section exercises
- all exercises in the Mid-Chapter Tests, Chapter Reviews, Chapter Practice Tests, and Cumulative Review Tests

ISBN: 0-321-92328-6

Student Workbook
- Extra practice exercises for every section of the text with ample space for students to show their work

ISBN: 0-321-92022-8

Section Lecture Videos
- For each section of the text, there are about 20 minutes of lecture covering the concepts from that section with additional examples.
- Captioned in English and Spanish
- Available in MyMathLab®

Video Resources Featuring Chapter Test Prep Videos
- Step-by-step solutions to every exercise in each Chapter Practice Test
- Available in MyMathLab®
- Available on YouTube (www.youtube.com/user/AngelElementaryAlg)

INSTRUCTOR RESOURCES

Annotated Instructor's Edition
Contains all the content found in the student edition, plus the following:
- Answers to exercises on the same text page with graphing answers in the Graphing Answer section at the back of the text
- Instructor Example provided in the margin paired with each student example

Instructor's Resource Manual with Tests and Mini-Lectures
- Mini-lectures for each text section
- Several forms of test per chapter (free response and multiple choice)
- Answers to all items
- Available for download from the IRC and in MyMathLab®

TestGen®
- Enables instructors to build, edit, print, and administer tests using a computerized bank of questions developed to cover all the objectives of the text.
- Algorithmically based, allowing instructors to create multiple but equivalent versions of the same question or test with the click of a button; instructors can also modify test bank questions or add new questions.

Instructor's Solutions Manual
- Provides complete worked-out solutions to all section exercises
- Available for download from the IRC and in MyMathLab®

Online Resources
- MyMathLab® (access code required)
- MathXL® (access code required)

MyMathLab® Online Course (access code required)
MyMathLab from Pearson is the world's leading online resource in mathematics, integrating interactive homework, assessment, and media in a flexible, easy-to-use format. MyMathLab delivers *proven results* in helping individual students succeed. It provides *engaging experiences* that personalize, stimulate, and measure learning for each student. And, it comes from an *experienced partner* with educational expertise and an eye on the future.

To learn more about how MyMathLab combines proven learning applications with powerful assessment, visit www.mymathlab.com or contact your Pearson representative.

Acknowledgments

We thank our spouses, Kathy Angel and Kris Runde, for their support and encouragement throughout the project. We are grateful for their wonderful support and understanding while we worked on the book.

We also thank our children: Robert and Steven Angel and Alex, Nick, and Max Runde. They also gave us support and encouragement and were very understanding when we could not spend as much time with them as we wished because of book deadlines. Special thanks to daughter-in-law, Kathy; mother-in-law, Patricia; and father-in-law, Scott. Without the support and understanding of our families, this book would not be a reality.

Larry Gilligan did not participate in this revision, but we would like to thank him for his contributions to previous editions of this book.

We want to thank Becky Hubiak, Deana Richmond, Paul Lorczak, and John Morin for accuracy reviewing the pages and checking all answers.

Many people at Pearson deserve thanks, including all those listed on the copyright page. In particular, we thank Michael Hirsch, Editor-in-Chief; Katie DePasquale and Matt Summers, Editorial Project Managers; Michelle Renda, Executive Marketing Manager; Patty Bergin, Senior Production Manager; and Barbara Atkinson, Senior Designer.

We would like to thank the following pre-revision reviewers for their thoughtful comments and suggestions:

Elizabeth Bonawitz, *University of Rio Grande, OH*
Connie Buller, *Metropolitan Community College, NE*
Janet Evert, *Erie Community College (South), NY*
Maryann Justinger, *Erie Community College (South), NY*
John Kawai, *Los Angeles Valley College, CA*
Jane Keller, *Metropolitan Community College, NE*
Claire Medve, *State University of New York–Canton, NY*

We would also like to thank the following reviewers and focus group participants of the Eighth Edition:

Darla Aguilar, *Pima Community College, AZ*
Frances Alvarado, *University of Texas–Pan American, TX*
Jose Alvarado, *University of Texas–Pan American, TX*
Ben Anderson, *Darton College, GA*
Mary Lou Baker, *Columbia State Community College, TN*
Sharon Berrian, *Northwest Shoals Community College, AL*
Dianne Bolen, *Northeast Mississippi Community College, MS*
Julie Bonds, *Sonoma State University, CA*
Clark Brown, *Mojave Community College, AZ*
Connie Buller, *Metropolitan Community College, NE*
Marc D. Campbell, *Daytona Beach Community College, FL*
Julie Chesser, *Owens Community College, OH*
Kim Christensen, *Maple Woods Community College, MO*
Barry Cogan, *Macomb Community College, MI*
Pat C. Cook, *Weatherford College, TX*
Lisa DeLong Cuneo, *Pennsylvania State University–Dobois, PA*
Stephan Delong, *Tidewater Community College, VA*
Deborah Doucette, *Erie Community College (North), NY*
William Echols, *Houston Community College, TX*
Dale Felkins, *Arkansas Technical University, AR*
Reginald Fulwood, *Palm Beach State College, FL*
Susan Grody, *Broward College, FL*
Abdollah Hajikandi, *State University of New York–Buffalo, NY*
Olga Cynthia Harrison, *Baton Rouge Community College, LA*
Richard Hobbs, *Mission College, CA*
Joe Howe, *St. Charles Community College, MO*
Laura L. Hoye, *Trident Technical College, SC*
Barbara Hughes, *San Jacinto Community College (Central), TX*
Mary Johnson, *Inver Hills Community College, MN*
Jane Keller, *Metropolitan Community College, NE*
Mike Kirby, *Tidewater Community College, VA*
William Krant, *Palo Alto College, TX*
Gayle L. Krzemine, *Pikes Peak Community College, CO*
Mitchel Levy, *Broward College, FL*
Mitzi Logan, *Pitt Community College, NC*
Jason Mahar, *Monroe Community College, NY*
Kimberley A. Martello, *Monroe Community College, NY*
Constance Meade, *College of Southern Idaho, ID*
Lynnette Meslinsky, *Erie Community College, NY*
Elizabeth Morrison, *Valencia College, FL*
Elsie Newman, *Owens Community College, OH*
Charlotte Newsom, *Tidewater Community College, VA*
Charles Odion, *Houston Community College, TX*
Jean Olsen, *Pikes Peak Community College, CO*
Jearme Pirie, *Erie Community College (North), NY*
Behnaz Rouhani, *Athens Technical College, GA*
Brian Sanders, *Modesto Junior College, CA*
Glenn R. Sandifer, *San Jacinto Community College (Central), TX*
Rebecca Schantz, *Prairle State College, IL*
Cristela Sifuentez, *University of Texas–Pan American, TX*
Fereja Tahir, *Illinois Central College, IL*
Burnette Thompson, Jr., *Houston Community College, TX*
Mary Vachon, *San Joaguin Delta College, CA*
Andrea Vorwark, *Maple Woods Community College, MO*
Ronald Yates, *Community College of Southern Nevada, NY*

Focus Group Participants

Linda Barton, *Ball State, IN*
Karen Egedy, *Baton Rouge Community College, LA*
Daniel Fahringer, *Harrisburg Area Community College, PA*
Sharon Hamsa, *Longview Community College, MO*
Cynthia Harrison, *Baton Rouge Community College, LA*
Judy Kasabian, *El Camino College, CA*
Christopher Yarish, *Harrisburg Area Community College, PA*

To the Student

Algebra is a course that requires active participation. You must read the text and pay attention in class, and, most importantly, you must work the exercises. The more exercises you work, the better.

The text was written with you in mind. Short, clear sentences are used, and many examples are given to illustrate specific points. The text stresses useful applications of algebra. Hopefully, as you progress through the course, you will come to realize that algebra is not just another math course that you are required to take, but a course that offers a wealth of useful information and applications.

This text makes full use of color. The different colors are used to highlight important information. Important procedures, definitions, and formulas are placed within colored boxes.

The boxes marked **Understanding Algebra** should be studied carefully. They emphasize concepts and facts that you need to master to succeed. **Helpful Hints** should be studied carefully, for they stress important information. Be sure to study **Avoiding Common Errors** boxes. These boxes point out common errors and provide the correct procedures for doing these problems.

After each example you will see a Now Try Exercise reference, such as Now Try Exercise 27. The exercise indicated is very similar to the example given in the book. You may wish to try the indicated exercise after you read the example to make sure you truly understand the example. In the exercise set, the Now Try exercises are written in green, such as 27.

Each section is accompanied by a video lecture that covers the concepts discussed in that section, as well as additional example problems. These videos may be accessed through MyMathLab MyMathLab®.

Some questions you should ask your professor early in the course include: What supplements are available for use? Where can help be obtained when the professor is not available? Supplements that may be available include: the Student Solutions Manual; the Lecture Series Videos; the Chapter Test Prep Video; MyMathLab®; and YouTube™. All these items are discussed under the heading of Supplements in Section 1.1 and listed in the Preface.

You may wish to form a study group with other students in your class. Many students find that working in small groups provides an excellent way to learn the material. By discussing and explaining the concepts and exercises to one another, you reinforce your own understanding. Once guidelines and procedures are determined by your group, make sure to follow them.

One of the first things you should do is to read Section 1.1, Study Skills for Success in Mathematics. Read this section slowly and carefully, and pay particular attention to the advice and information given. Occasionally, refer back to this section. This could be the most important section of the book. Pay special attention to the material on doing your homework and on attending class.

At the end of all exercise sets (after the first two) are **Cumulative Review Exercises.** You should work these problems on a regular basis, even if they are not assigned. These problems are from earlier sections and chapters of the text, and they will refresh your memory and reinforce those topics. If you have a problem when working these exercises, read the appropriate section of the text or study your notes that correspond to that material. The section of the text where the Cumulative Review Exercise was introduced is indicated in brackets, [ ], to the left of the exercise. After reviewing the material, if you still have a problem, make an appointment to see your professor. Working the Cumulative Review Exercises throughout the semester will also help prepare you to take your final exam.

Near the middle of each chapter is a **Mid-Chapter Test.** You should take each Mid-Chapter Test to make sure you understand the material up to that point. The section where the material was first introduced is given in brackets after the answer in the answer section of the book.

At the end of each chapter are a **Chapter Summary, Chapter Review Exercises,** a **Chapter Practice Test,** and a **Cumulative Review Test.** Before each examination you should review this material carefully and take the Chapter Practice Test (you may want to review the *Chapter Test Prep Video* also). If you do well on the Chapter Practice Test, you should do well on the class test. The questions in the Review Exercises are marked to indicate the section in which that material was first introduced. If you have a problem with a Review Exercise question, reread the section indicated. You may also wish to take the Cumulative Review Test that appears at the end of every chapter (starting with Chapter 2).

In the back of the text there is an **answer section** that contains the answers to the *odd-numbered* exercises, including the Challenge Problems. Answers to *all* Cumulative Review Exercises, Mid-Chapter Tests, Chapter Review Exercises, Chapter Practice Tests, and Cumulative Review Tests are provided. Answers to the Group Activity exercises are not provided, for we wish students to reach agreement by themselves on answers to these exercises. The answers should be used only to check your work. For the Mid-Chapter Tests, Chapter Practice Tests, and Cumulative Review Tests, after each answer the section number where that type of exercise was covered is provided.

We have tried to make this text as clear and error free as possible. No text is perfect, however. If you find an error in the text, or an example or section that you believe can be improved, we would greatly appreciate hearing from you. If you enjoy the text, we would also appreciate hearing from you. You can submit comments to math@pearson.com, subject for Allen Angel and Dennis Runde.

Allen R. Angel
Dennis C. Runde

1 Real Numbers

A college education is worth money! The amount of average annual income increases dramatically as one's education increases. For example, in 2009 someone with a bachelor's degree earned almost three times as much as a person without a high school diploma. In Exercise 43 on page 18, we will see how to analyze pictorial data to calculate the financial advantages of a college education.

Goals of This Chapter

This chapter will provide you with the foundation that you need in order to succeed in this course and all other mathematics courses you will take. Learning proper study skills is the first step in building this foundation. *Please read Section 1.1 carefully and follow the advice given*. The emphasis of this chapter is to provide you with an understanding of the real number system.

In this chapter, you will learn a five-step problem-solving procedure that will be used throughout the book. Once you have learned the material in this chapter, you will be able to tackle the subsequent chapters in the book with confidence.

1.1 Study Skills for Success in Mathematics

1. Recognize the goals of this text.
2. Learn proper study skills.
3. Prepare for and take exams.
4. Learn to manage time.

This section is extremely important. Take the time to read it carefully and follow the advice given.

Most of you taking this course fall into one of three categories: (1) those who did not take algebra in high school, (2) those who took algebra in high school but did not understand the material, or (3) those who successfully completed algebra in high school but have been out of school for some time and need to take the course again. Whichever the case, you will need to acquire study skills for mathematics courses.

Before we discuss study skills, we will present the goals of this text. These goals may help you realize why certain topics are covered in the text and why they are covered as they are.

1 Recognize the Goals of This Text

The goals of this text include:

1. Presenting traditional algebra topics
2. Preparing you for more advanced mathematics courses
3. Building your confidence in, and your enjoyment of, mathematics
4. Improving your reasoning and critical thinking skills
5. Increasing your understanding of how important mathematics is in solving real-life problems
6. Encouraging you to think analytically, so that you will feel comfortable translating real-life problems into mathematical equations, and then solving the problems.

In addition to teaching you the mathematical content, our goals are to teach you to be more *mathematically literate*, which is also called *quantitatively literate*. We wish to teach you to *communicate mathematically,* to teach you to *understand and interpret data* in a variety of formats, to teach you measurement and geometric concepts, to teach you to *reason more logically,* and to teach you to be able to represent real world applications mathematically, which is called *modeling*. Throughout the book we will strive to increase your mathematical understanding to help you become more successful in mathematics, in your future job, and throughout life.

We also realize that some of you may have some mathematics anxiety. We have written the book to try to help you overcome that anxiety by building your confidence in mathematics.

It is important to realize that this course is the foundation for more advanced mathematics courses. A thorough understanding of algebra will make it easier for you to succeed in later mathematics courses and in life.

2 Learn Proper Study Skills

Have a Positive Attitude You may be thinking to yourself, "I hate math," or "I wish I did not have to take this class." You may have heard of "math anxiety" and feel you fit this category. The first thing to do to be successful in this course is to change your attitude to a more positive one. You must be willing to give this course, and yourself, a fair chance.

Based on past experiences in mathematics, you may feel that this is difficult. However, mathematics is something you need to work at. Many of you are more mature now than when you took previous mathematics courses. Your maturity and desire to learn are extremely important and can make a tremendous difference in your ability to succeed in mathematics. We believe you can be successful in this course, but you also need to believe it.

Prepare for and Attend Class To be prepared for class, you need to do your homework assignments completely. If you have difficulty with the homework, or some of the concepts, write down questions to ask your instructor. If you were given a reading assignment, read the appropriate material carefully before class.

After the material is explained in class, read the corresponding sections of the text slowly and carefully, word by word.

You should plan to attend every class. Generally, the more absences you have, the lower your grade will be. Every time you miss a class, you miss important information. If you must miss a class, contact your instructor ahead of time, and get the reading assignment and homework. If possible, before the next class, try to copy a friend's notes to help you understand the material you missed.

In algebra and other mathematics courses, the material you learn is cumulative. The new material is built on material that was presented previously. You must understand each section before moving on to the next section, and each chapter before moving on to the next chapter. Therefore, do not let yourself fall behind. Seek help as soon as you need it—do not wait! You will greatly increase your chance of success in this course by following the study skills presented in this section.

While in class, pay attention to what your instructor is saying. If you don't understand something, ask your instructor to repeat the material. If you don't ask questions, your instructor will not know that you have a problem understanding the material.

In class, take careful notes. Write numbers and letters clearly, so that you can read them later. Make sure your x's do not look like y's and vice versa. It is not necessary to write down every word your instructor says. Copy the major points and the examples that do not appear in the text. You should not be taking notes so frantically that you lose track of what your instructor is saying.

Read the Text Mathematics textbooks should be read slowly and carefully, word by word. If you do not understand something, reread that material. It is a good idea to read with a pencil in your hand, making notes as you proceed.

Don't panic! As you read the examples, notice that the "flow" is basically downward. It is a challenge but try to understand the reasons for each step. This downward movement is a sequence of steps that takes a problem from statement toward its solution. Each step is important to understand. If you have trouble with the rationale for a step, you should ask your instructor for clarification.

When you come across a new concept or definition, you may wish to underline or highlight it so that it stands out. Then it will be easier to find later. Also, work the **Now Try Exercises** that appear in the text following each example. The Now Try Exercises are designed so that you have the opportunity to immediately apply new ideas. Make notes of things you do not understand to ask your instructor.

There are numerous boxes in the left margin marked **Understanding Algebra.** These boxes give alternative wording and additional illustration of important concepts. You may want to give these special attention as you read and see how they help with topics in the text and examples.

This textbook has other special features to help you. We suggest that you pay particular attention to these highlighted features, including the **Avoiding Common Errors** boxes, the **Helpful Hint** boxes, and important procedures and definitions identified by color. The Avoiding Common Errors boxes point out the most common errors made by students. Read and study this material very carefully and make sure that you understand what is explained. If you avoid making these common errors, your chances of success in this and other mathematics classes will be increased greatly. The Helpful Hints offer many valuable techniques for working certain problems. They may also present some very useful information or show an alternative way to work a problem.

Do the Homework *Two very important commitments that you must make to be successful in this course are attending class and doing your homework regularly.* Your assignments must be worked conscientiously and completely. Do your homework as soon as possible, so the material presented in class will be fresh in your mind. It is through doing homework that you truly learn the material. While working homework you will become aware of the types of problems that you need further help with. If you do not work the assigned exercises, you will not know what questions to ask in class.

When you do your homework, make sure that you write it neatly and carefully. Pay particular attention to copying signs and exponents correctly.

Don't forget to check the answers to your homework assignments. This book contains the answers to the odd-numbered exercises in the back of the book. In addition, the answers to all the Cumulative Review Exercises, Mid-Chapter Tests, Chapter Review Exercises, Chapter Practice Tests, and Cumulative Review Tests are in the back of the book. The section number where the material is first introduced is provided next to the exercises for the Cumulative Review Exercises and Chapter Review Exercises. The section number where the material is first introduced is provided with the answers in the back of the book for the Mid-Chapter Tests, Chapter Practice Tests, and Cumulative Review Tests. Answers to the Group Activity Exercises are not provided because we want you to arrive at the answers as a group.

Ask questions in class about homework problems you don't understand. You should not feel comfortable until you understand all the concepts needed to work every assigned problem successfully.

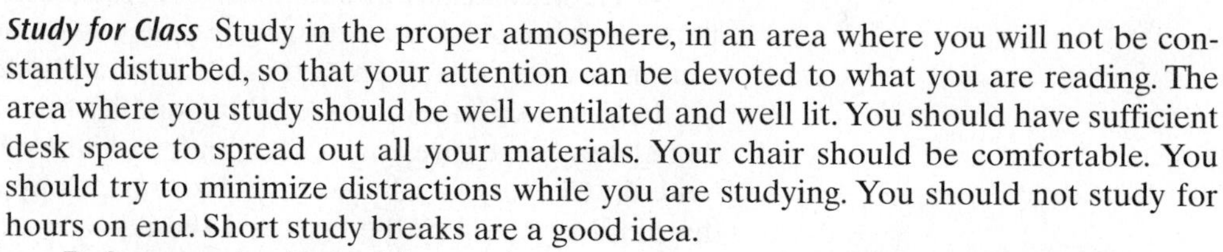

Study for Class Study in the proper atmosphere, in an area where you will not be constantly disturbed, so that your attention can be devoted to what you are reading. The area where you study should be well ventilated and well lit. You should have sufficient desk space to spread out all your materials. Your chair should be comfortable. You should try to minimize distractions while you are studying. You should not study for hours on end. Short study breaks are a good idea.

Before you begin studying, make sure that you have all the materials you need (pencils, markers, calculator, etc.). You may wish to highlight the important points covered in class or in the book.

It is recommended that students study and do homework for at least two hours for each hour of class time. Some students require more time than others. It is important to spread your studying time out over the entire week rather than studying during one large block of time.

When studying, you should not only understand how to work a problem but also know *why* you follow the specific steps you do to work the problem. If you do not have an understanding of why you follow the specific process, you will not be able to transfer the process to solve similar problems.

This book has Mid-Chapter Tests in the middle of each chapter. These exercises reinforce material presented in the first half of the chapter. They will also help you determine if you need to go back and review the topics covered in the first half of the chapter. For any of the Mid-Chapter Test questions that you get incorrect, turn to the section provided with the answers in the back of the book and review that section. This book also has Cumulative Review Exercises at the end of every section after Section 1.2. These exercises reinforce material presented earlier in the course, and you will be less likely to forget the material if you review it repeatedly throughout the course. The exercises will also help prepare you for the final exam. Even if these exercises are not assigned for homework, We urge you to work them as part of your studying process.

3 Prepare for and Take Exams

If you study a little bit each day you should not need to cram the night before an exam. Begin your studying early. If you wait until the last minute, you may not have time to seek the help you may need if you find you cannot work a problem.

To prepare for an exam:

1. Read your class notes.
2. Review your homework assignments.
3. Study formulas, definitions, and procedures you will need for the exam.
4. Read the Avoiding Common Errors boxes and Helpful Hint boxes carefully.

5. Read the summary at the end of each chapter.
6. Work the Chapter Review Exercises at the end of each chapter. If you have difficulties, restudy those sections. If you still have trouble, seek help.
7. Work the Mid-Chapter Test and the Chapter Practice Test.
8. Rework quizzes previously given if the material covered in the quizzes will be included on the test.
9. If your exam is a cumulative exam, work the Cumulative Review Test.
10. Now, if you can arrange it, you may want to consider a session of study with a partner or group from your class. With a partner, you can construct a sample test to take to simulate your actual test and help alleviate test anxiety. Try these steps:

    a) Using three-by-five-inch index cards, go through the text and select representative problems—writing the question on one side of the card and the answer or page reference on the other. Choose questions you think will most likely be asked; don't choose easy problems. Have your study partner do the same thing. Probably 20 to 25 good, representative questions should do it.

    b) Here is the key: *shuffle the cards*. One thing that makes tests more difficult than homework is that homework problems are often of the same type and knowing how to start the problem is not too difficult. But test questions are all mixed up and to simulate that, shuffle the cards.

    c) You take your partner's test—be sure to give yourself the same amount of time your instructor will give you—and your partner takes your test. Try to avoid distractions (music, food, etc.). Grade your partner's test and have your partner grade your test. Then study weak areas and repeat the process if necessary.

Prepare for Midterm and Final Exam When studying for a comprehensive midterm or final exam follow the procedures discussed for preparing for an exam. However, also:

1. Study all your previous tests and quizzes carefully. Make sure that you have learned to work the problems that you may have previously missed.
2. Work the Cumulative Review Test at the end of each chapter. These tests cover the material from the beginning of the book to the end of that chapter.
3. If your instructor has given you a worksheet or practice exam, make sure that you complete it. Ask questions about any problems you do not understand.
4. Begin your studying process early so that you can seek all the help you need in a timely manner.

Take an Exam Make sure you get sufficient sleep the night before the test. Arrive at the exam site early so that you have a few minutes to relax before the exam. If you rush into the exam, you will start out nervous and anxious. After you are given the exam, you should do the following:

1. Carefully write down any formulas or ideas that you want to remember.
2. Look over the entire exam quickly to get an idea of its length. Also make sure that no pages are missing.
3. Read the test directions carefully.
4. Read each question carefully. Show all of your work. Answer each question completely, and make sure that you have answered the specific question asked.
5. Work the questions you understand best first; then go back and work those you are not sure of. Do not spend too much time on any one problem or you may not be able to complete the exam. Be prepared to spend more time on problems worth more points.
6. Attempt each problem. You may get at least partial credit even if you do not obtain the correct answer. If you make no attempt at answering the question, you will lose full credit.

7. Work carefully step by step. Copy all signs and exponents correctly when working from step to step, and make sure to copy the original question from the test correctly.
8. Write clearly so that your instructor can read your work. If your instructor cannot read your work, you may lose credit. When appropriate, make sure that your final answer stands out by placing a box around it.
9. If you have time, check your work and your answers.
10. Do not be concerned if others finish the test before you or if you are the last to finish. Use any extra time to check your work.

Stay calm when taking your test. Do not get upset if you come across a problem you can't figure out right away. Go on to something else and come back to that problem later.

4 Learn to Manage Time

As mentioned earlier, it is recommended that students study and do homework for at least two hours for each hour of class time. Finding the necessary time to study is not always easy. The following are some suggestions that you may find helpful.

1. Plan ahead. Determine when you will study and do your homework. Do not schedule other activities for these periods. Try to space these periods evenly over the week.
2. Be organized, so that you will not have to waste time looking for your books, your pencil, your calculator, or your notes.
3. If you are allowed to use a calculator, use it for tedious calculations.
4. When you stop studying, clearly mark where you stopped in the text.
5. Try not to take on added responsibilities. You must set your priorities. If your education is a top priority, as it should be, you may have to reduce time spent on other activities.
6. If time is a problem, do not overburden yourself with too many courses.

Use Supplements This text comes with a large variety of supplements. Find out from your instructor early in the semester which supplements are available and might be beneficial for you to use. Supplements should not replace reading the text, but should be used to enhance your understanding of the material. If you miss a class, you may want to review the video on the topic you missed before attending the next class.

The supplements available are: the Student Solutions Manual which works out the odd section exercises as well as all the end-of-chapter exercises; The Section Lecture Videos, available in MyMathLab, MyMathLab® which contain about 20 minutes of lecture per section and include additional examples; the Chapter Test Prep Video, which works out every problem in every Chapter Practice Test; MathXL®, a powerful online tutorial and homework system; MyMathLab®, the online course which houses MathXL. The Lecture Series Videos and Chapter Test Prep Videos are available through MyMathLab. The Chapter Test Prep Videos are also available on YouTube (www.youtube.com/user/AngelEA9).

Seek Help Be sure to get help as soon as you need it! Do not wait! In mathematics, one day's material is usually based on the previous day's material. So, if you don't understand the material today, you may not be able to understand the material tomorrow.

Where should you seek help? There are often a number of resources on campus. Try to make a friend in the class with whom you can study. Often, you can help one another. You may wish to form a study group with other students in your class. Discussing the concepts and homework with your peers will reinforce your own understanding of the material.

You should know your instructor's office hours, and you should not hesitate to seek help from your instructor when you need it. Make sure you read the assigned material and attempt the homework before meeting with your instructor. Come prepared with specific questions to ask.

There are often other sources of help available. Many colleges have a mathematics lab or a mathematics learning center where tutors are available. Ask your instructor early in the semester where and when tutoring is available. Arrange for a tutor as soon as you need one.

A Final Word You can be successful at mathematics if you attend class regularly, pay attention in class, study your text carefully, do your homework daily, review regularly, and seek help as soon as you need it. Good luck in your course and remember: *Mathematics is not a spectator sport!*

1.1 Exercise Set MathXL® MyMathLab®

Do you know:

1. your professor's name and office hours?

2. your professor's office location and telephone number?

3. where and when you can obtain help if your professor is not available?

4. the name and phone number of a friend in your class?

5. what supplements are available to assist you in learning?

6. if your instructor is recommending the use of a particular calculator?

7. when you can use your calculator in this course?

8. if your instructor is requiring the use of MyMathLab?

If you do not know the answers to questions 1–8, you should find out as soon as possible.

9. What are your goals for this course?

10. What are your reasons for taking this course?

11. List the things you need to do to prepare properly for class.

12. Are you beginning this course with a positive attitude? It is important that you do!

13. For each hour of class time, how many hours outside of class are recommended for studying and doing homework?

14. Explain how a mathematics text should be read.

15. Two very important commitments that you must make to be successful in this course are **a)** doing homework regularly and completely and **b)** attending class regularly. Explain why these commitments are necessary.

16. When studying, you should not only understand how to work a problem, but also why you follow the specific steps you do. Why is this important?

17. Have you given any thought to studying with a friend or a group of friends? Can you see any advantages in doing so? Can you see any disadvantages in doing so?

18. Write a summary of the steps you should follow when taking an exam.

1.2 Problem Solving

1. **Learn the five-step problem-solving procedure.**
2. **Solve problems involving bar, line, and circle graphs.**
3. **Solve problems involving statistics.**

1 Learn the Five-Step Problem-Solving Procedure

One of the main reasons we study mathematics is to use it to solve real-life problems. To solve most real-life problems mathematically, we need to be able to express the problem in mathematical symbols. We will spend a great deal of time explaining how to express real-life applications mathematically.

You can approach any problem using the general five-step **problem-solving procedure** developed by George Pólya (1887–1985) in his book *How to Solve It.*

Guidelines for Problem Solving

1. **Understand the problem.**
   - Read the problem *carefully* at least twice. In the first reading, get a general overview of the problem. In the second reading, determine (*a*) exactly what you are being asked to find and (*b*) what information the problem provides.
   - Make a list of the given facts. Determine which are pertinent to solving the problem.
   - Determine whether you can substitute smaller or simpler numbers to make the problem more understandable.
   - If it will help you organize the information, list the information in a table.
   - If possible, make a sketch to illustrate the problem. Label the information given.
2. **Translate the problem to mathematical language.**
   - This will generally involve expressing the problem in terms of an algebraic expression or equation. (We will explain how to express application problems as equations in Chapter 3.)
   - Determine whether there is a formula that can be used to solve the problem.
3. **Carry out all necessary calculations.**
4. **Check the answer obtained in step 3.**
   - Ask yourself, "Does the answer make sense?" "Is the answer reasonable?" If the answer is not reasonable, recheck your method for solving the problem and your calculations.
   - Check the solution in the original wording of the problem if possible.
5. **Make sure you have answered the question.**
   - State the answer clearly.

Understanding Algebra

An *expression* is a collection of numbers, letters, grouping symbols, and operations.

In step 2 we use the words *algebraic expression*. An **algebraic expression,** sometimes simply referred to as an **expression,** is a general term for any collection of numbers, letters (called variables), grouping symbols such as parentheses ( ) or brackets [ ], and **operations** (such as addition, subtraction, multiplication, and division). In this section we will not be using variables, so we will discuss their use later.

Examples of Expressions

$$3 + 4, \qquad 6(12 \div 3), \qquad (2)(7)$$

The following examples show how to apply the guidelines for problem solving. In some problems it may not be possible or necessary to list every step in the procedure. If you need to review procedures for adding, subtracting, multiplying, or dividing decimal numbers, or if you need a review of percents, read Appendix A before proceeding.

EXAMPLE 1 **Buying Games** Darla Aguilar is deciding which would be less expensive, buying her son's birthday presents on eBay or buying them at a local toy store. Founded in 1995, eBay is The World's Online Marketplace® for the sale of goods and services by a diverse community of individuals and small businesses. The eBay community includes operations in over 30 countries. The local toy store is only minutes from Darla's house. Therefore, the cost of gasoline for her car will not factor into her decision. On eBay, the three games Darla would like to purchase cost \$5.99, \$9.95, and \$19.95. Shipping costs for the games would total \$11.10. There would be no sales tax on this purchase. At the local toy store, the total cost for the same three games would be \$57.89 plus 8.25% sales tax.

a) Which would be less expensive for Darla, purchasing the games on eBay or at the local toy store?

b) How much would Darla save by making the less expensive purchase?

Solution **a)** Understand the problem A careful reading of the problem shows that the task is to determine if it would be less expensive for Darla to purchase the games on eBay or at the local toy store. Make a list of all the information given and determine which information is needed to solve the problem.

Information Given	Pertinent to Solving the Problem?
eBay includes operations in more than 30 countries	no
\$5.99, \$9.95, and \$19.95 cost of the games on eBay	yes
\$11.10 shipping costs on eBay	yes
no sales tax on the eBay purchase	yes
cost of gasoline not a factor in Darla's decision	yes
\$57.89 cost of the games at the local toy store	yes
8.25% sales tax at the local toy store	yes

To determine whether eBay or a local store would be the better choice for purchasing the games, it is *not* necessary to know that eBay includes operations in more than 30 countries. Solving this problem involves:

- calculating the total cost of games on eBay (including shipping)
- calculating the total cost of games at local store (including sales tax)

To calculate sales tax, you need to determine 8.25% of the cost of the games at the local toy store. When performing calculations, numbers given as percents are changed to decimal numbers. So we will use 0.0825 for 8.25%.

Translate the problem into mathematical language

$$\text{total cost of games on eBay} = \text{cost of each individual game} + \text{shipping costs}$$
$$\text{total cost of games at local toy store} = \text{total cost of games} + 8.25\%\ \text{sales tax}$$

Carry out the calculations

$$\text{total cost of games on eBay} = \$5.99 + \$9.95 + \$19.95 + \$11.10 = \$46.99$$
$$\text{total cost of games at local toy store} = \$57.89 + 0.0825(\$57.89)$$
$$= \$57.89 + \$4.78 = \$62.67$$

Check the answer The total costs of \$46.99 and \$62.67 are reasonable based on the information given.

Answer the question asked It would be less expensive for Darla to purchase the games on eBay.

b) Understand To determine how much Darla would save by making the less expensive purchase, you need to subtract the total cost of the games on eBay from the total cost of the games at the local toy store.

Translate

$$\text{total cost at toy store} - \text{total cost on eBay} = \text{amount Darla would save}$$

Carry Out $\$62.67 - \$46.99 = \$15.68$

Check The answer \$15.68 seems reasonable.

Answer Darla would save \$15.68 by purchasing the games on eBay.

Now Try Exercise 27

EXAMPLE 2 **Processor Speed** Larry Gilligan's iMac G5 computer can perform 2.0 billion operations per second (2.0 billion OPS). How many operations can Larry's computer perform in 0.8 second?

Solution Understand We are given the computer owner's name, the model of the computer, a speed of 2.0 billion (2,000,000,000) operations per second, and 0.8 second. To determine the answer to this problem, the owner's name, Larry Gilligan, and the model of the computer, iMac G5, are not needed.

To obtain the answer, will we multiply or divide? Often a fairly simple problem seems more difficult because of the numbers involved. When very large or very small numbers make the problem confusing, try solving a simpler problem first to determine your problem-solving strategy. Suppose the problem said the computer can perform 10 operations per second. How many operations can the computer perform in 5 seconds? Simply multiply 10×5 to get 50. Since we multiplied when solving the simpler problem, we will multiply when solving the given problem.

Understanding Algebra

When solving word problems, it is often helpful to solve a similar, simpler problem first.

Translate

Number of operations in 0.8 second = $0.8 \times$ the number of operations per second

Carry Out

$$= 0.8 \times 2{,}000{,}000{,}000$$
$$= 1{,}600{,}000{,}000$$

Understanding Algebra

Notice that the answer to the problem in Example 2 is a complete English sentence, not simply a number, and the number has a unit, operations.

Check The answer, 1,600,000,000 operations, is less than the 2,000,000,000 operations per second, which makes sense because the computer is operating for less than a second.

Answer In 0.8 second, the computer can perform 1,600,000,000 operations.

Now Try Exercise 21

EXAMPLE 3 Medical Insurance Sharon Berrian's medical insurance policy is similar to that of many workers. Her policy requires that she pay the first \$100 of medical expenses each calendar year (called a deductible). After the deductible is paid, she pays 20% of the medical expenses (called a co-payment) and the insurance company pays 80%. There is a maximum co-payment of \$600 that she must pay each year. After that, the insurance company pays 100% of the fee schedule. On January 1, Sharon sprained her ankle playing tennis. She went to the doctor's office for an examination and X rays. The total bill of \$325 was sent to the insurance company.

a) How much of the bill will Sharon be responsible for?

b) How much will the insurance company be responsible for?

Solution **a)** Understand First we list all the *relevant* given information.

Given Information

\$100 deductible
20% co-payment after deductible
80% paid by insurance company after deductible
\$325 doctor bill

All the other information is not needed to solve the problem. Sharon will be responsible for the first \$100 and 20% of the remaining balance. The insurance company will be responsible for 80% of the balance after the deductible. Before we can find what Sharon owes, we need to first find the balance of the bill after the deductible. The balance of the bill after the deductible is $\$325 - \$100 = \$225$.

Translate

Sharon's responsibility = deductible + 20% of balance of bill after the deductible

Carry Out

$$\text{Sharon's responsibility} = 100 + 20\%(225)$$
$$= 100 + 0.20(225)$$
$$= 100 + 45$$
$$= 145$$

Check and Answer The answer appears reasonable. Sharon will be responsible for \$145.

b) The insurance company will be responsible for 80% of the balance after the deductible.

$$\begin{aligned}\text{insurance company's responsibility} &= 80\% \text{ of balance after deductible}\\ &= 0.80(225)\\ &= 180\end{aligned}$$

Thus, the insurance company is responsible for \$180. This checks because the sum of Sharon's responsibility and the insurance company's responsibility is equal to the doctor's bill.

$$\$145 + \$180 = \$325$$

We could have also found the answer to part **b)** by subtracting Sharon's responsibility from the total amount of the bill, but to give you more practice with percents we decided to show the solution as we did.

Now Try Exercise 33

Understanding Algebra

Bar graphs, line graphs, circle graphs (or pie charts) are ways of pictorially representing data.

2 Solve Problems Involving Bar, Line, and Circle Graphs

Problem solving often involves understanding and reading graphs and sets of data (or numbers). To work Example 4, you must interpret a bar graph and work with data.

EXAMPLE 4 **Walking It Off** Experts suggest that people walk 10,000 steps daily. Depending on stride length, each mile ranges between 2000 and 2500 steps. **Figure 1.1** is a **bar graph** that shows the number of steps it takes to burn off calories from a garden salad with fat-free dressing, a 12-ounce can of soda, a doughnut, and a cheeseburger.

a) Using the bar graph in **Figure 1.1,** estimate the number of steps it takes to burn off calories from a cheeseburger.

b) If Cliff Jackson can walk a mile in 2000 steps, how many miles will he have to walk in order to burn off the calories from the cheeseburger he ate for lunch?

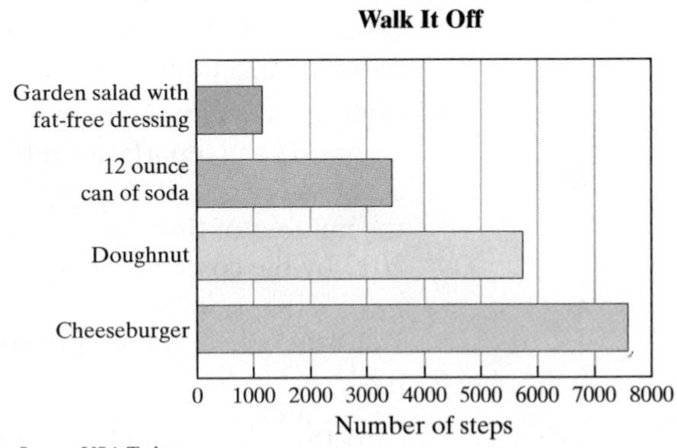

FIGURE 1.1 *Source*: USA Today

Solution

a) Using the bar to the right of Cheeseburger in **Figure 1.1,** we estimate that the number of steps it takes to burn off calories from a cheeseburger is about 7600.

b) Understand Since it takes Cliff 2000 steps to walk a mile, it follows that he would need to take 4000 steps to walk 2 miles, 6000 steps to walk 3 miles, and so on. To determine how many miles Cliff will have to walk in order to burn off the calories from the cheeseburger, we need to divide as follows.

Translate $\text{miles to walk} = \dfrac{\text{number of steps to burn off calories}}{2000}$

Carry Out $\quad$ miles to walk $= \dfrac{7600}{2000} = 3.8$

Check and Answer The answer appears reasonable. Cliff will need to walk 3.8 miles in order to burn off the calories from the cheeseburger that he ate for lunch.

Now Try Exercise 35

Understanding Algebra

The symbol $\approx$ means "is approximately equal to." So we can write $\frac{1}{3} \approx 0.33$, for example.

In Example 5, we will use the symbol $\approx$, which is read "**is approximately equal to.**" If, for example, the answer to a problem is 34.12432, we may write the answer as ≈ 34.1.

EXAMPLE 5 **Super Bowl Ads** The *line graph* in **Figure 1.2** shows the average cost of a 30-second advertisement during Super Bowls from 2004 to 2013. The advertising prices are set by the TV network showing the game.

a) Estimate the cost of 30-second advertisements in 2004 and 2013.

b) How much more was the cost of a 30-second advertisement in 2013 than in 2004?

c) How many times greater was the cost of a 30-second advertisement in 2013 than in 2004?

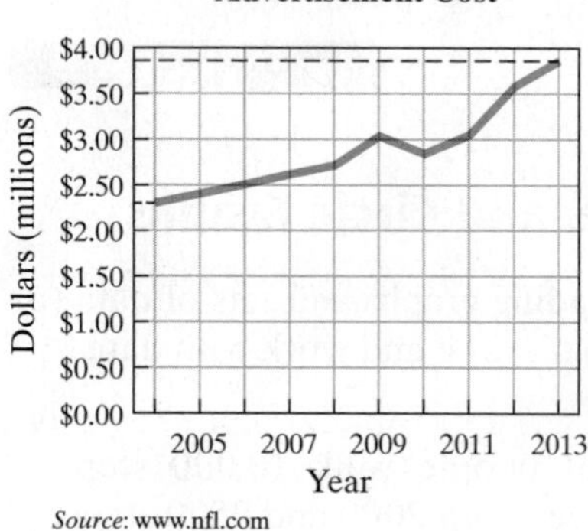

FIGURE 1.2

Solution

a) When reading a line graph where the line has some thickness as in **Figure 1.2,** we will use the center of the line to make our estimate. By observing the graph, we can estimate that the cost of a 30-second advertisement in 2004 was about \$2.3 million (or \$2,300,000). By observing the dashed line on the graph, we can also estimate that the cost of a 30-second advertisement in 2013 was about \$3.8 million (or \$3,800,000).

b) We use the problem-solving procedure to answer the question.

Understand To determine how much more the cost of a 30-second advertisement was in 2013 than in 2004, we need to subtract.

Translate $\quad$ Difference in the cost $=$ cost in 2013 $-$ cost in 2004

Carry Out $\quad = \$3{,}800{,}00 - \$2{,}300{,}000 = \$1{,}500{,}000$

Check and Answer The answer appears reasonable. The cost was \$1,500,000 more in 2013 than in 2004.

c) Understand Although parts **b)** and **c)** may appear to ask the same thing, they do not. The two parts are different in that part **b)** asks "how much more was the cost" whereas part **c)** asks "how many *times* greater was the cost." To determine the number of times greater the cost was in 2013 than in 2004, divide the cost in 2013 by the cost in 2004 as shown below.

Translate $\quad$ number of times greater $= \dfrac{\text{cost in 2013}}{\text{cost in 2004}}$

Carry Out $\quad$ number of times greater $= \dfrac{3{,}800{,}000}{2{,}300{,}000} \approx 1.65$

Check and Answer By observing the graph, we see that the answer is reasonable. The cost of a 30-second advertisement during the Super Bowl in 2013 was about 1.65 times the cost in 2004.

Now Try Exercise 37

EXAMPLE 6 **Stay-at-Home Parents** **Figure 1.3** (on the next page) is a **circle graph** that shows the reasons why married mothers with children under the age of 15 have stayed out of the labor force. Use **Figure 1.3** to determine the number of married mothers with children under the age of 15 who have stayed out of the labor force for the following reasons: to care for home and family, ill/disabled, retired, going to school, could not find work, and other.

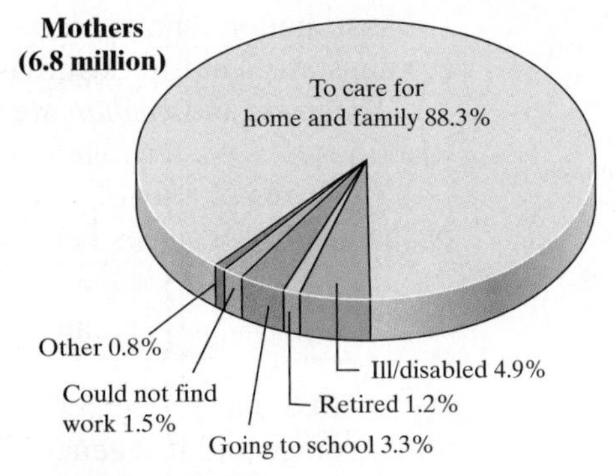

FIGURE 1.3

Solution Understand There were 6.8 million married mothers with children under the age of 15 who were out of the labor force. Of these mothers, 88.3% were out of the labor force to care for home and family. To determine the number of married mothers with children under the age of 15 out of the labor force to care for home and family, we need to find 88.3% of the total number of mothers. To do this, we multiply as follows.

Translate

$$\begin{pmatrix}\text{number to care}\\ \text{for home and}\\ \text{family}\end{pmatrix} = \begin{pmatrix}\text{percent out of the}\\ \text{labor force to care}\\ \text{for home and family}\end{pmatrix}\begin{pmatrix}\text{total number out}\\ \text{of the labor force}\end{pmatrix}$$

Carry Out

$$\begin{aligned}\text{number to care for home and family} &= 0.883(6.8\text{ million})\\ &= 6.0044\text{ million}\end{aligned}$$

Thus, 6.0044 million married mothers with children under the age of 15 were out of the labor force to care for home and family.

To find the number who were out of the labor force for being ill/disabled, we do a similar calculation.

$$\begin{aligned}\text{number of ill/disabled} &= 0.049(6.8\text{ million})\\ &= 0.3332\text{ million}\end{aligned}$$

We do similar calculations to find the number who were retired, going to school, could not find work, or for other reasons were out of the labor force.

$$\begin{aligned}\text{number retired} &= 0.012(6.8\text{ million})\\ &= 0.0816\text{ million}\\ \text{number going to school} &= 0.033(6.8\text{ million})\\ &= 0.2244\text{ million}\\ \text{number who could not find work} &= 0.015(6.8\text{ million})\\ &= 0.102\text{ million}\\ \text{number for other reasons} &= 0.008(6.8\text{ million})\\ &= 0.0544\text{ million}\end{aligned}$$

Check If we add the six amounts, we obtain the 6.8 million total. Therefore, our answer is correct.

$$\begin{aligned}&6.0044\text{ million} + 0.3332\text{ million} + 0.0816\text{ million} + 0.2244\text{ million}\\ &+ 0.102\text{ million} + 0.0544\text{ million} = 6.8\text{ million}\end{aligned}$$

Answer The number of married mothers with children under the age of 15 who were out of the labor force were as follows: 6.0044 million to care for home and family; 0.3332 million were ill/disabled; 0.0816 million were retired; 0.2244 million were going to school; 0.102 million could not find work; for other reasons, 0.0544 million.

Now Try Exercise 39

3 Solve Problems Involving Statistics

Understanding Algebra

The *mean* is found by dividing the sum of the data by the number of data points. The *median* is the middle score of the ranked data.

Because understanding statistics is so important in our society, we will now discuss certain statistical topics and use them in solving problems.

The *mean* and *median* are two **measures of central tendency**, which are also referred to as *averages*. An average is a value that is representative of a set of data (or numbers).

The **mean** of a set of data is determined by adding all the values and dividing the sum by the number of values. For example, to find the mean of 6, 9, 3, 12, 12, we do the following.

$$\text{mean} = \frac{6 + 9 + 3 + 12 + 12}{5} = \frac{42}{5} = 8.4$$

We divided the sum by 5 since there are five values. The mean is the most commonly used average and it is generally what is thought of when we use the word *average*.

Another average is the median. The **median** is the value in the middle of a set of **ranked data.** The data may be ranked from smallest to largest or largest to smallest. To find the median of 6, 9, 3, 12, 12, we can rank the data from smallest to largest as follows.

3, 6, 9, 12, 12

↑
Middle value

The value in the middle of the ranked set of data is 9. Therefore, the median is 9. Note that half the values will be above the median and half will be below the median.

If there is an even number of pieces of data, the median is halfway between the two middle pieces. For example, to find the median of 3, 12, 5, 12, 17, 9, we can rank the data as follows.

3, 5, 9, 12, 12, 17

↑
Middle values

Since there are six pieces of data (an even number), we find the value halfway between the two middle pieces, the 9 and the 12. To find the median, we add these values and divide the sum by 2.

$$\text{median} = \frac{9 + 12}{2} = \frac{21}{2} = 10.5$$

Thus, the median is 10.5. Note that half the values are above and half are below 10.5.

EXAMPLE 7 **The Mean Grade** Jose Alvarado's first six exam grades are 90, 87, 76, 84, 78, and 62.

a) Find the mean for Jose's six grades.

b) If one more exam is to be given, what is the minimum grade that Jose can receive to obtain at least a B average (a mean average of 80 or better)?

c) If there is only one more exam, is it possible for Jose to obtain an A average (90 or better)? Explain.

Solution

a) To obtain the mean, we add the six grades and divide by 6.

$$\text{mean} = \frac{90 + 87 + 76 + 84 + 78 + 62}{6} = \frac{477}{6} = 79.5$$

b) We will show the problem-solving steps for this part of the example.

Understand For the mean average of seven exams to be 80, the total points for the seven exams must be 7(80) or 560. The minimum grade needed can be found by subtracting the sum of the first six grades from 560.

Translate minimum grade needed = 560 − sum of first six exam grades

Carry Out

$$= 560 - (90 + 87 + 76 + 84 + 78 + 62)$$
$$= 560 - 477$$
$$= 83$$

Check We can check to see that a seventh grade of 83 gives a mean of 80 as follows.

$$\text{mean} = \frac{90 + 87 + 76 + 84 + 78 + 62 + 83}{7} = \frac{560}{7} = 80$$

Answer A seventh grade of 83 or higher will result in at least a B average.

c) We can use the same reasoning as in part **b)**. For a 90 average, the total points that Jose will need to attain is $90(7) = 630$. Since his total points are 477, he will need $630 - 477$ or 153 points to obtain an A average. Since the maximum number of points available on most exams is 100, Jose would not be able to obtain an A in the course.

Now Try Exercise 41

1.2 Exercise Set MathXL® MyMathLab®

Warm-Up Exercises

Fill in the blanks with the appropriate word, phrase, or symbol(s) from the following list.

expression	central	mean	median	checking	understanding
equation	approximately	grouping	problem	circle	

1. The ________ of the data 2, 4, 7, 8, 9 is 7.
2. A general collection of numbers, symbols, and operations is called a(n) ________.
3. The symbol $\approx$ means is ________ equal to.
4. The ________ of the data 2, 4, 7, 8, 9 is 6.
5. One of the five important steps in problem solving, seeing if your answer makes sense, is referred to as ________ a problem.
6. The mean and median are types of averages, also called measures of ________ tendency.
7. Graphical representation of data includes bar graphs, line graphs, and ________ graphs.
8. Parentheses and brackets are examples of ________ symbols.
9. In this book we use Pólya's five-step approach for ________ solving.
10. Reading a problem at least twice, making a list of facts, and making a sketch are the problem-solving step called ________ the problem.

Practice the Skills

11. **Test Grades** Jenna Webber's test grades are 78, 97, 59, 74, and 74. For Jenna's grades, determine the **a)** mean and **b)** median.

12. **Bowling Scores** William Krant's bowling scores for five games were 161, 131, 187, 163, and 145. For William's games, determine the **a)** mean and **b)** median.

13. **Electric Bills** The Malones' electric bills for January through June, 2013, were \$96.56, \$108.78, \$87.23, \$85.90, \$79.55, and \$65.88. For these bills, determine the **a)** mean and **b)** median.

14. **Grocery Bills** Antoinette Payne's monthly grocery bills for the first five months of 2013 were \$204.83, \$153.85, \$210.03, \$119.76, and \$128.38. For Antoinette's grocery bills, determine the **a)** mean and **b)** median.

15. **Colonial Massachusetts Population** The bar graph on the right shows the population (in thousands of people) of the Massachusetts colony for the years 1640 through 1780. Use the data shown in the graph to determine the **a)** mean and **b)** median population for these years.

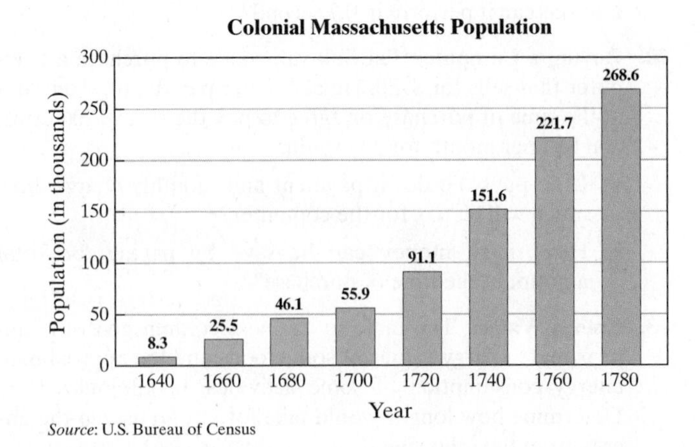

Source: U.S. Bureau of Census

16. **Homes for Sale** Eight homes are for sale in a community. The sale prices are \$124,100, \$175,900, \$142,300, \$164,800, \$146,000, \$210,000, \$112,200, and \$153,600. Determine the **a)** mean and **b)** median sale price of the eight homes.

Problem Solving

17. **Commissions** Barbara Riedell earns a 5% commission on appliances she sells. Her sales last week totaled \$9400. Find her week's earnings.

18. **Empire State Building** May 1, 1931, was the opening day of the Empire State Building. It stands 1454 feet or 443 meters high. Use this information to determine the approximate number of feet in a meter.

19. **Sales Tax** The sales tax in Sarasota County is 7%. Ben Anderson purchases an antique dining room set that costs \$2300 before tax. **a)** What was the sales tax Ben paid on the dining room set? **b)** What was the total cost of the dining room set including tax?

20. **Sales Tax** The sales tax in Luca County is 6.75%. Mary Lou Baker purchases a pair of cross country skis that cost \$300 before tax. **a)** What was the sales tax Mary Lou paid on the skis? **b)** What was the total cost of the skis including tax?

21. **Computer Processor** Suppose a computer processor can perform about 2.3 billion operations per second. How many operations can it perform in 0.7 second?

22. **Buying a Computer** Pat Sullivan wants to purchase a computer that sells for \$950. He can either pay the total amount at the time of purchase or agree to pay the store \$200 down and \$33 per month for 24 months.
    a) If he pays the down payment and monthly charge, how much will he pay for the computer?
    b) How much money can he save by paying the total amount at the time of purchase?

23. **Energy Values** The table in the next column gives the approximate energy values of some foods and the approximate energy consumption of some activities, in kilojoules (kJ). Determine how long it would take for you to use up the energy from the following.
    a) a hamburger by running
    b) a chocolate milkshake by walking
    c) a glass of skim milk by cycling

A green numbered exercise, such as 21, indicates a Now Try Exercise.

Energy Value, Food	(kJ)	Energy Consumption, Activity	(kJ/min)
Chocolate milkshake	2200	Walking	25
Fried egg	460	Cycling	35
Hamburger	1550	Swimming	50
Strawberry shortcake	1440	Running	80
Glass of skim milk	350		

24. **Jet Ski** The rental cost of a jet ski from Don's Ski Rental is \$20 per half-hour, and the rental cost from A. J.'s Ski Rental is \$50 per hour. Suppose you plan to rent a jet ski for 3 hours.
    a) Which is the better deal?
    b) How much will you save?

25. **Gas Mileage** When the odometer in Tribet LaPierre's car reads 16,741.3, he fills his gas tank. The next time he fills his tank it takes 10.5 gallons, and his odometer reads 16,935.4. Determine the number of miles per gallon that his car gets.

26. **Income Taxes** The federal income tax rate schedule for a *joint return* in 2012 is illustrated in the following table.

Adjusted Gross Income	Taxes
\$1–\$17,400	10% of income
\$17,401–\$70,700	\$1740 + 15% in excess of \$17,400
\$70,701–\$142,700	\$9735 + 25% in excess of \$70,700
\$142,701–\$217,450	\$27,735 + 28% in excess of \$142,700
\$217,451–\$388,350	\$48,665 + 33% in excess of \$217,450
over \$388,350	\$105,062 + 35% in excess of \$388,350

Source: U.S. Treasury Department

a) If the Cunninghams' adjusted gross income in 2012 was \$53,298, determine their taxes.
b) If the Kowalskis' adjusted gross income in 2012 was \$156,212, determine their taxes.

27. **Buying Tires** Eric Weiss purchased four tires through the Internet. He paid \$62.30 plus \$6.20 shipping and handling per tire. There was no sales tax on this purchase. When he received the tires, Eric had to pay \$8.00 per tire for mounting and balancing. At a local tire store, his total cost for the four tires with mounting and balancing would have been \$425 plus 8% sales tax. How much did Eric save by purchasing the tires through the Internet?

28. Baseball Salaries In 2012, the highest paid major league baseball player was third baseman Alex Rodriguez of the New York Yankees who was paid \$30,000,000. In the same year the highest paid major league pitcher was Yohan Santana of the New York Mets who was paid \$23,145,011. In 2012 Rodriguez batted 529 times and Santana pitched 117 innings. Determine how much more Santana was paid per inning pitched than Rodriguez was paid per at bat. *Source*: www.mlb.com

29. Balance Consider the figure shown. Assuming the green and red blocks have the same weight, where should a single green block, ■, be placed to make the scale balanced? Explain how you determined your answer.

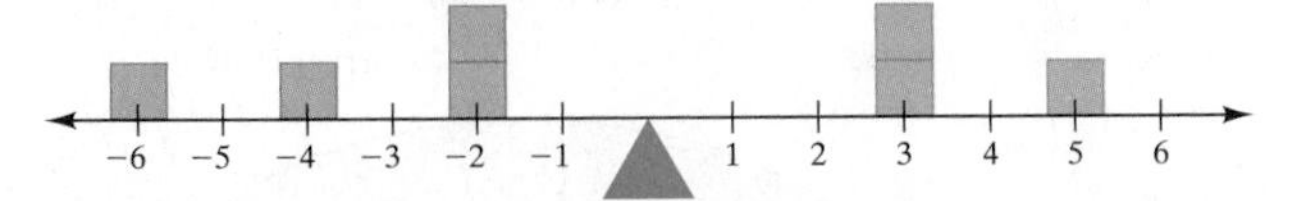

30. Taxi Ride A taxicab charges \$2 upon a customer's entering the taxi, then 30 cents for each $\frac{1}{4}$ mile traveled and 20 cents for each 30 seconds stopped in traffic. David Lopez takes a taxi ride for a distance of 3 miles where the taxi spends 90 seconds stopped in traffic. Determine David's cost of the taxi ride.

31. Leaky Faucet A faucet that leaks 1 ounce of water per minute wastes 11.25 gallons in a day.

a) How many gallons of water are wasted in a (non-leap) year?

b) If water costs \$5.20 per 1000 gallons, how much additional money per year is being spent on the water bill?

32. Conversions a) What is 1 mile per hour equal to in feet per hour? One mile contains 5280 feet.

b) What is 1 mile per hour equal to in feet per second?

c) What is 60 miles per hour equal to in feet per second?

33. Medical Insurance Mel LeBar's medical insurance policy requires that he pay a \$150 deductible each calendar year. After the deductible is paid, he pays 20% of the medical expenses and the insurance company pays 80%. On January 1, Mel's daughter accidentally closed the car door on his finger. He went to the doctor's office for an examination and X rays. The total bill of \$365 was sent to the insurance company. If Mel had not as yet paid any of the deductible,

a) how much of the bill will Mel be responsible for?

b) how much will the insurance company be responsible for?

34. Insurance Drivers under the age of 25 who pass a driver education course generally have their auto insurance premium decreased by 10%. Most insurers will offer this 10% deduction until the driver reaches 25. A particular driver education course costs \$70. Don Beville, who just turned 18, has auto insurance that costs \$630 per year.

a) Excluding the cost of the driver education course, how much would Don save in auto insurance premiums, from the age of 18 until the age of 25, by taking the driver education course?

b) What would be his net savings after the cost of the course?

35. Math Score The bar graph in the next column shows some recent scores from the Program for International Student Assessment, a test for 15-year-olds.

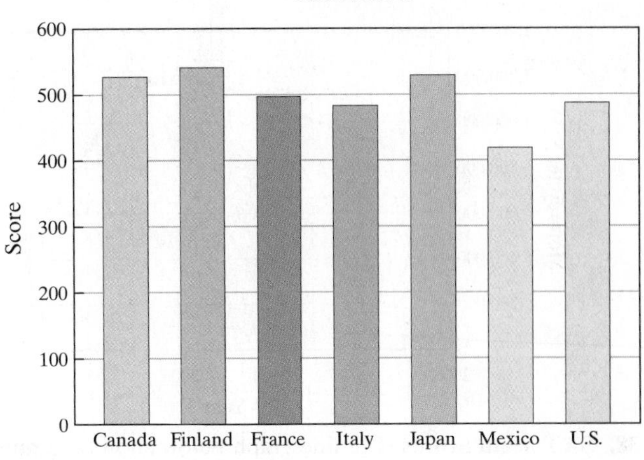

a) Which country had the highest math score shown? Estimate the score.

b) Which country had the lowest math score shown? Estimate the score.

c) Estimate the difference between the scores for Finland and Mexico.

36. Driest States The bar graph below shows the ten driest states and the average annual rain total, in inches of rain, for those states.

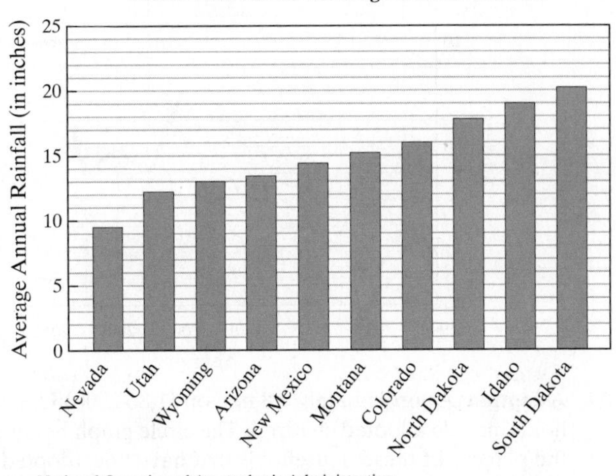

Source: National Oceanic and Atmospheric Administration

a) From the bar graph, estimate the average annual rainfall for South Dakota.

b) From the bar graph, estimate the average annual rainfall for Nevada.

c) How many times greater is the average annual rainfall for South Dakota than the average annual rainfall for Nevada?

37. Motorcycle Sales The line graph at the top of the next page shows the U.S. sales of new motorcycles from 2000 to 2010.

a) Estimate the number of new motorcycles sold in the U.S. in 2006 and 2010.

b) About how many more new motorcycles were sold in the U.S. in 2006 than in 2010?

c) About how many times greater was the number of new motorcycles sold in 2006 than in 2010?

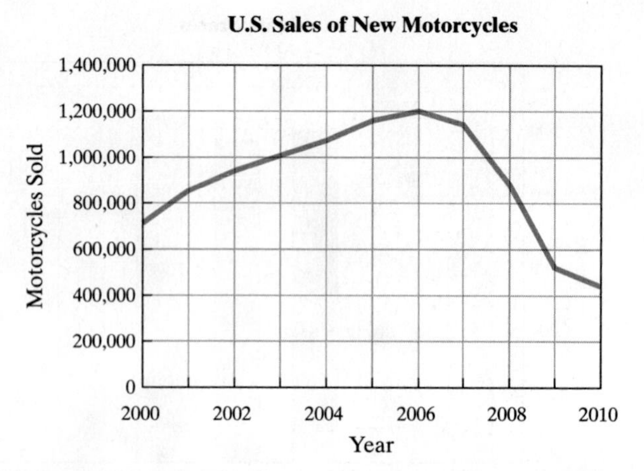

38. SAT Math Scores The line graph below shows the number of freshmen students at Eyre University with high school math SAT scores over 700.

a) During which consecutive years did the number of students remain constant?

During which consecutive years did the number of freshmen with high school math SAT scores over 700

b) increase the most? **c)** decrease the most?

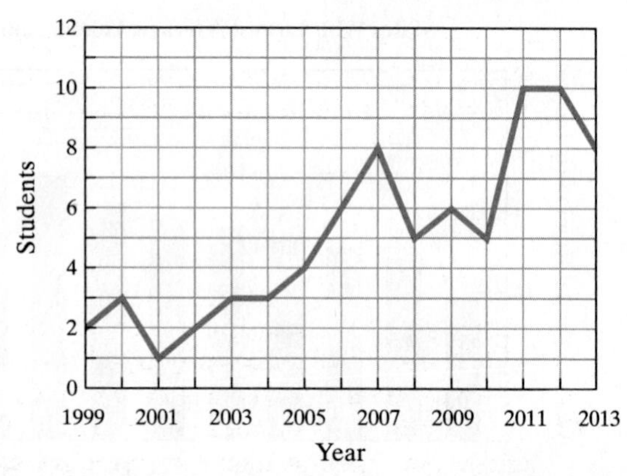

39. Adoption Approximately 1.8 million (1,800,000) U.S. households include adopted children. The circle graph below shows the percent of these households that have one adopted child, two adopted children, and three or more adopted children. Estimate the number of U.S. households that have

a) one adopted child. **b)** two adopted children.

c) three or more adopted children.

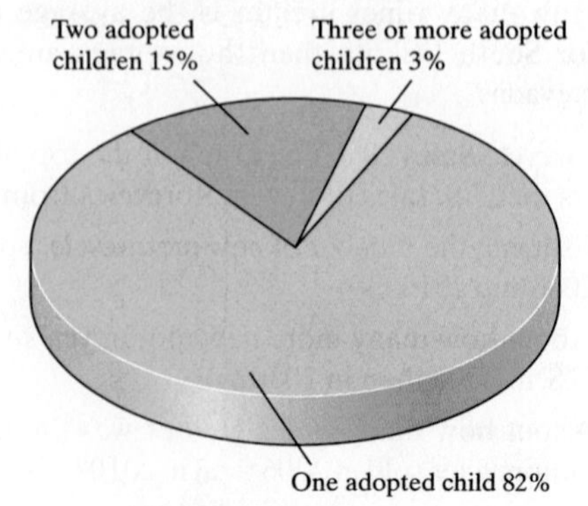

40. Jeopardy! Ken Jennings, a software engineer from Salt Lake City, Utah, is the all-time leading money winner on American television game shows. The circle graph below shows the outcome of the 160 Daily Doubles that Ken attempted on his *Jeopardy*! run. Use the circle graph to determine

a) the number of Daily Doubles Ken answered correctly.

b) the number of Daily Doubles Ken answered incorrectly.

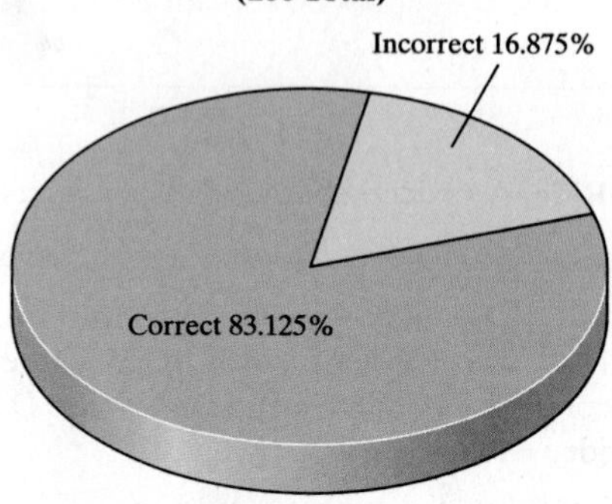

41. Test Grades A mean of 60 on all exams is needed to pass a course. On his first five exams Lamond Paine's grades are 50, 59, 67, 80, and 56.

a) What is the minimum grade that Lamond can receive on the sixth exam to pass the course?

b) An average of 70 is needed to get a C in the course. Is it possible for Lamond to get a C? If so, what is the minimum grade that Lamond can receive on the sixth exam?

42. Test Grades A mean of 80 on all exams is needed to earn a B in a course. On her first four exams, Heather Feldman's grades are 95, 88, 82, and 85.

a) What is the minimum grade that Heather can receive on the fifth exam to earn a B in the course?

b) A mean of 90 is needed to earn an A in the course. Is it possible for Heather to get an A? If so, what is the minimum grade that Heather can receive on the fifth exam?

43. Level of Education The bar graph on the right shows the median annual earnings by level of education in 2009 in the United States.

a) How many times greater are the median annual earnings for those with an associate degree than those with a high school diploma?

b) How many times greater are the median annual earnings for those with a bachelor's degree than those with an associate degree?

c) How many times greater are the median annual earnings for those with a master's degree than those with an bachelor's degree?

44. Exams Mike Ambrosino's mean average on six exams is 78. Find the sum of his scores.

45. Construct Data Construct a set of five pieces of data with a mean of 70 and no two values the same.

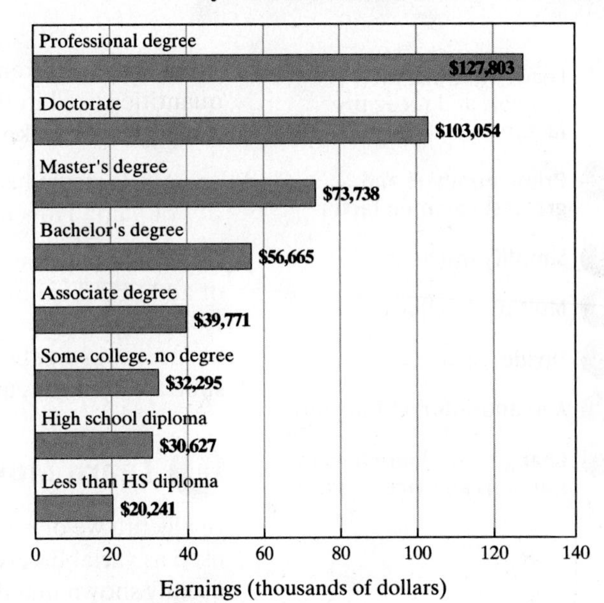

Source: U.S. Census Bureau

Concept/Writing Exercises

46. Suppose a set of data of 10 numbers has a mean of 6. An 11th number is added to the data and it has a value of 5. Will the mean of the new data set increase or decrease? Explain.

47. Consider the set of data 2, 3, 5, 6, 70. Without doing any calculations, can you determine whether the mean or the median is greater? Explain your answer.

Challenge Problem

48. Reading Meters The figure shows how to read an electric meter.

Step 1 Start with the dial on the right. If the arrow falls between two numbers, use the smaller number on the dial (except when the dial is between 9 and 0, then use the 9). Notice the arrows above each dial indicate the direction the dial is moving (clockwise, then counterclockwise).

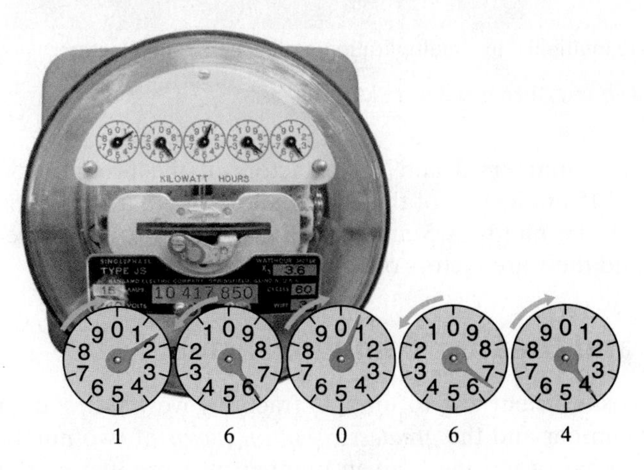

Source: Southern California Edison, Understanding Your Electricity Bill

Step 2 If the pointer is directly on a number, check the dial to the *right* to make sure it has passed 0 and is headed toward 1. If the dial to the right has not passed 0, use the next lower number. The number on the meter on the left is 16064.

Suppose your previous month's reading was as shown on the left, and this month's meter reading is as shown below.

a) Determine this month's meter reading.

b) Determine your electrical cost for this month by first subtracting last month's meter reading from this month's meter reading (measured in kilowatt hours), and then multiplying the difference by the cost per kilowatt hour of electricity. Assume electricity costs 24.3 cents per kilowatt hour.

1.3 Fractions

1. Learn multiplication symbols and recognize factors.
2. Prime numbers and greatest common factor.
3. Simplify fractions.
4. Multiply fractions.
5. Divide fractions.
6. Add and subtract fractions.
7. Change mixed numbers to fractions and vice versa.

What is the difference between arithmetic and algebra? When doing arithmetic, all the quantities used in the calculations are known. In algebra, however, one or more of the quantities are unknown and must be found. Consider the following:

> Mr. Piersma has 1 gallon of paint. In order to paint his bedroom, he needs 3 gallons of paint. How many additional gallons does he need?

This is an example of an algebraic problem. The *unknown* quantity is the number of additional gallons of paint needed. Mr. Piersma needs 2 more gallons of paint.

An understanding of decimal numbers (see Appendix A) and fractions is essential to success in algebra. You will need to know how to simplify a fraction and how to add, subtract, multiply, and divide fractions.

1 Learn Multiplication Symbols and Recognize Factors

In algebra we often use letters called **variables** to represent numbers. Letters commonly used as variables are ***x*, *y*, and *z*,** but other letters can be used as variables. Variables are usually shown in italics. So that we do not confuse the variable x with the multiplication sign $\times$ we often use different notation to indicate multiplication.

Understanding Algebra

$5x$ means 5 times x
and
rs means r times s

Multiplication Symbols

If a and b represent any two mathematical quantities, then each of the following may be used to indicate the product of a and b ("a times b").

$$ab \quad a\cdot b \quad a(b) \quad (a)b \quad (a)(b)$$

Examples

3 times 4 may be written:	3 times x may be written:	x times y may be written:
	$3x$	xy
$3(4)$	$3(x)$	$x(y)$
$(3)4$	$(3)x$	$(x)y$
$(3)(4)$	$(3)(x)$	$(x)(y)$

Now we will introduce the term *factors*, which we shall be using throughout the text.

Understanding Algebra

A *factor* is a number or expression that is multiplied by another number or expression.

Factors

The numbers or variables that are multiplied in a multiplication problem are called **factors.**

If $a\cdot b = c$, then a and b are factors of c.

For example, in $3\cdot 5 = 15$, the numbers 3 and 5 are factors of the product 15. In $2\cdot 15 = 30$, the numbers 2 and 15 are factors of the product 30. Note that 30 has many other factors. Since $5\cdot 6 = 30$, the numbers 5 and 6 are also factors of 30. Since $3x$ means 3 times x, both the 3 and the x are factors of $3x$.

2 Prime Numbers and Greatest Common Factor

In order to understand the most efficient way to simplify fractions, we will first discuss the *prime factorization* of a number and the *greatest common factor* of two numbers. **Prime factorization** is the process of writing a given number as a product of prime numbers. **Prime numbers** are natural numbers, excluding 1, that can be divided by only themselves and 1. The first ten prime numbers are 2, 3, 5, 7, 11, 13, 17, 19, 23, and 29. Can you find the next prime number? If you answered 31, you answered correctly.

To write a number as a product of primes, we can use a *tree diagram*. Begin by selecting any two numbers whose product is the given number. Then continue factoring each of these numbers into prime numbers, as shown in Example 1.

EXAMPLE 1 Determine the prime factorization of the number 120.

Solution We will use three different tree diagrams to illustrate the prime factorization of 120.

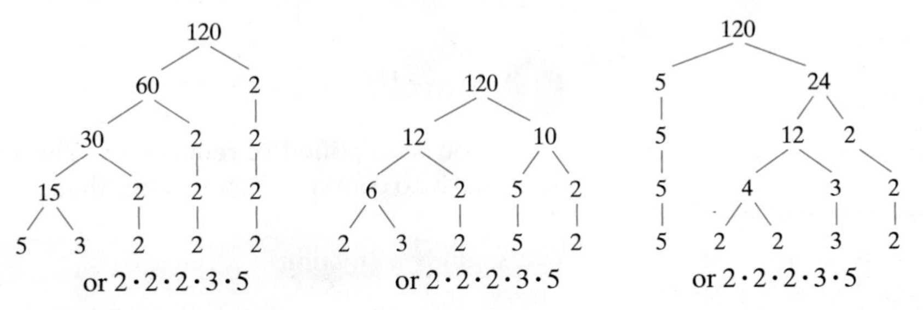

Note that no matter how you start, if you do not make a mistake, you find that the prime factorization of 120 is $2 \cdot 2 \cdot 2 \cdot 3 \cdot 5$. There are other ways 120 can be factored but all will lead to the prime factorization $2 \cdot 2 \cdot 2 \cdot 3 \cdot 5$.

Now Try Exercise 15

Greatest Common Factor

The **greatest common factor (GCF)** of two natural numbers is the greatest integer that is a factor of both numbers. We use the GCF when simplifying fractions.

To Find the Greatest Common Factor of a Given Numerator and Denominator

1. Write both the numerator and the denominator as a product of primes.
2. Determine all the prime factors that are common to both prime factorizations.
3. Multiply the prime factors found in step 2 to obtain the GCF.

EXAMPLE 2 Find the GCF of 108 and 156.

Solution First determine the prime factorizations of both 108 and 156.

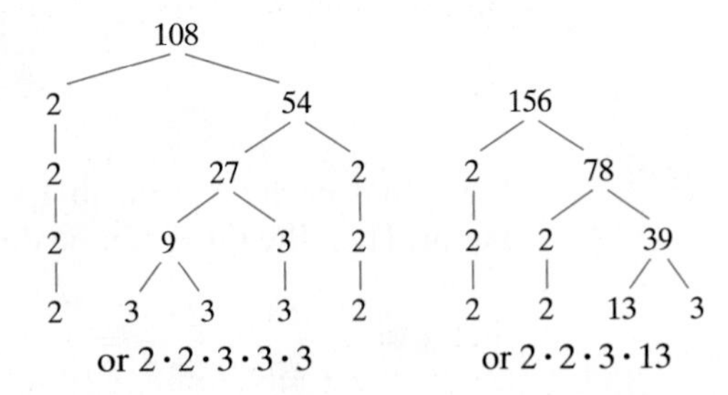

Notice there are two 2s and one 3 common to both prime factorizations; thus

$$\text{GCF} = 2 \cdot 2 \cdot 3 = 12$$

The greatest common factor of 108 and 156 is 12. Twelve is the greatest integer that divides into both 108 and 156.

Now Try Exercise 19

Now we have the necessary information to discuss *fractions*. The top number of a fraction is called the **numerator**, and the bottom number is called the **denominator.** In the fraction $\frac{3}{5}$, the 3 is the numerator and the 5 is the denominator.

Understanding Algebra

The fraction $\frac{3}{5}$ means the same as $3/5$ or $3 \div 5$

HELPFUL HINT

Consider the fraction $\frac{3}{5}$. There are equivalent methods of expressing this fraction, as illustrated below.

$$\frac{3}{5} = 3/5 = 3 \div 5 = 5\overline{)3}$$

In general, $\frac{a}{b} = a/b = a \div b = b\overline{)a}$

3 Simplify Fractions

A fraction is **simplified**, or **reduced to its lowest terms**, when the numerator and denominator have no common factors other than 1. To simplify a fraction, follow these steps.

Understanding Algebra

GCF stands for "greatest common factor." It is the largest number that divides evenly into the two given numbers. The GCF of 12 and 18 is 6.

To Simplify a Fraction

1. Find the **greatest common factor** (GCF) of the numerator and the denominator.
2. Then divide both the numerator and the denominator by the greatest common factor.

EXAMPLE 3 Simplify **a)** $\frac{10}{25}$ **b)** $\frac{6}{18}$ **c)** $\frac{108}{156}$

Solution

a) The GCF of 10 and 25 is 5. Divide both the numerator and the denominator by 5 to simplify the fraction.

$$\frac{10}{25} = \frac{10 \div 5}{25 \div 5} = \frac{2}{5}$$

b) The GCF of 6 and 18 is 6. Divide both the numerator and the denominator by 6.

$$\frac{6}{18} = \frac{6 \div 6}{18 \div 6} = \frac{1}{3}$$

c) In Example 2 we found that the greatest common factor of 108 and 156 is 12. Therefore to simplify $\frac{108}{156}$ we divide both the numerator and denominator by the GCF, 12.

$$\frac{108}{156} = \frac{108 \div 12}{156 \div 12} = \frac{9}{13}$$

Now Try Exercise 29

Note that each fraction in Example 3 could have been written with the GCF as a factor. Then the GCF can be divided out.

$$\frac{10}{25} = \frac{2 \cdot \cancel{5}}{5 \cdot \cancel{5}} = \frac{2}{5}, \quad \frac{6}{18} = \frac{1 \cdot \cancel{6}}{3 \cdot \cancel{6}} = \frac{1}{3}, \quad \frac{108}{156} = \frac{9 \cdot \cancel{12}}{13 \cdot \cancel{12}} = \frac{9}{13}$$

When you work with fractions you should always simplify your answers.

4 Multiply Fractions

To multiply two or more fractions, multiply their numerators together and multiply their denominators together.

To Multiply Fractions

$$\frac{a}{b} \cdot \frac{c}{d} = \frac{ac}{bd}$$

EXAMPLE 4 Multiply $\frac{3}{13}$ by $\frac{5}{11}$.

Solution $\frac{3}{13} \cdot \frac{5}{11} = \frac{3 \cdot 5}{13 \cdot 11} = \frac{15}{143}$

Now Try Exercise 47

Before multiplying fractions, to help avoid having to simplify an answer, we often divide both a numerator and a denominator by a common factor.

EXAMPLE 5 Multiply **a)** $\frac{8}{17} \cdot \frac{5}{16}$ **b)** $\frac{27}{40} \cdot \frac{16}{9}$.

Solution

a) Since the numerator 8 and the denominator 16 can both be divided by the common factor 8, we divide out the 8 first. Then we multiply.

$$\frac{8}{17} \cdot \frac{5}{16} = \frac{\overset{1}{\cancel{8}}}{17} \cdot \frac{5}{\underset{2}{\cancel{16}}} = \frac{1 \cdot 5}{17 \cdot 2} = \frac{5}{34}$$

b) $$\frac{27}{40} \cdot \frac{16}{9} = \frac{\overset{3}{\cancel{27}}}{40} \cdot \frac{16}{\underset{1}{\cancel{9}}} \quad \text{Divide both 27 and 9 by 9.}$$

$$= \frac{\overset{3}{\cancel{27}}}{\underset{5}{\cancel{40}}} \cdot \frac{\overset{2}{\cancel{16}}}{\underset{1}{\cancel{9}}} \quad \text{Divide both 40 and 16 by 8.}$$

$$= \frac{3 \cdot 2}{5 \cdot 1} = \frac{6}{5}$$

Now Try Exercise 49

Understanding Algebra

When we write 1, 2, 3, 4, ..., the three dots, called an *ellipsis,* indicate the pattern continues indefinitely.

The numbers 0, 1, 2, 3, 4, ... are called **whole numbers.** The whole numbers continue indefinitely. Thus, the numbers 468 and 5043 are also whole numbers. To multiply a whole number by a fraction, write the whole number with a denominator of 1 and then multiply.

EXAMPLE 6 **Lawn Mower Engine** Some engines run on a mixture of gas and oil. A particular lawn mower engine requires a mixture of $\frac{5}{64}$ gallon of oil for each gallon of gasoline used. A lawn care company wishes to make a mixture for this engine using 12 gallons of gasoline. How much oil must be used?

Solution We must multiply 12 by $\frac{5}{64}$ to determine the amount of oil that must be used. First we write 12 as $\frac{12}{1}$, then we divide both 12 and 64 by their greatest common factor, 4, as follows.

$$12 \cdot \frac{5}{64} = \frac{12}{1} \cdot \frac{5}{64} = \frac{\overset{3}{\cancel{12}}}{1} \cdot \frac{5}{\underset{16}{\cancel{64}}} = \frac{3 \cdot 5}{1 \cdot 16} = \frac{15}{16}$$

Thus, $\frac{15}{16}$ gallon of oil must be added to the 12 gallons of gasoline to make the proper mixture.

Now Try Exercise 103

5 Divide Fractions

To divide one fraction by another, multiply by the reciprocal of the divisor (the second fraction if written with $\div$).

Understanding Algebra

To divide $\frac{1}{3} \div \frac{5}{7}$

we convert to multiplication:

$$\frac{1}{3}\cdot\frac{7}{5} = \frac{7}{15}$$

To Divide Fractions

$$\frac{a}{b} \div \frac{c}{d} = \frac{a}{b}\cdot\frac{d}{c} = \frac{ad}{bc}$$

Sometimes, rather than being asked to add, subtract, multiply, or divide, you may be asked to *evaluate* an expression. To **evaluate** an expression means to obtain the answer to the problem using the operations given.

EXAMPLE 7 Evaluate **a)** $\frac{3}{5} \div \frac{5}{6}$ **b)** $\frac{3}{8} \div 12$.

Solution

a) $\frac{3}{5} \div \frac{5}{6} = \frac{3}{5}\cdot\frac{6}{5} = \frac{3\cdot 6}{5\cdot 5} = \frac{18}{25}$

b) Write 12 as $\frac{12}{1}$. Then invert the divisor and multiply.

$$\frac{3}{8} \div 12 = \frac{3}{8} \div \frac{12}{1} = \frac{\overset{1}{\cancel{3}}}{8}\cdot\frac{1}{\underset{4}{\cancel{12}}} = \frac{1}{32}$$

Now Try Exercise 55

6 Add and Subtract Fractions

Fractions that have the same (or a *common*) *denominator can be added or subtracted.* To add or subtract fractions with the same denominator, add or subtract the numerators and keep the common denominator.

To Add and Subtract Fractions

$$\frac{a}{c} + \frac{b}{c} = \frac{a+b}{c} \quad \text{or} \quad \frac{a}{c} - \frac{b}{c} = \frac{a-b}{c}$$

EXAMPLE 8 **a)** Add $\frac{6}{15} + \frac{2}{15}$. **b)** Subtract $\frac{8}{13} - \frac{5}{13}$.

Solution

a) $\frac{6}{15} + \frac{2}{15} = \frac{6+2}{15} = \frac{8}{15}$ **b)** $\frac{8}{13} - \frac{5}{13} = \frac{8-5}{13} = \frac{3}{13}$

Now Try Exercise 63

Understanding Algebra

To add or subtract fractions, they need to first have a common denominator. The smallest number that is divisible by two or more denominators is called the *least common denominator* or *LCD*.

To add (or subtract) fractions with unlike denominators, we must first rewrite each fraction with a common denominator. The smallest number that is divisible by two or more denominators is called the **least common denominator** or **LCD.** Sometimes the least common denominator is referred to as the **least common multiple** of the denominators.

To Find the Least Common Denominator of Two or More Fractions

1. Write each denominator as a product of prime numbers.
2. For each prime number, determine the maximum number of times that prime number appears in any of the prime factorizations.
3. Multiply all the prime numbers found in step 2. Include each prime number the maximum number of times it appears in any of the prime factorizations. The product of all these prime numbers will be the LCD.

Example 9 illustrates the procedure to determine the LCD.

EXAMPLE 9 Consider $\frac{1}{6} + \frac{7}{15}$.

a) Determine the least common denominator of the fractions.

b) Add the fractions.

Solution

a) The prime factorization of 6 is $2 \cdot 3$. The prime factorization of 15 is $3 \cdot 5$.

$$6 = 2 \cdot 3 \qquad 15 = 3 \cdot 5$$

We can see that the maximum number of 2s that appear in either prime factorization is one, the maximum number of 3s is one, and the maximum number of 5s is one. We therefore will take the product of one 2, one 3, and one 5.

$$\text{LCD} = 2 \cdot 3 \cdot 5 = 30$$

Thus, the least common denominator is 30. The smallest number that both 6 and 15 will divide into is 30.

b) To add the fractions, we need to write both $\frac{1}{6}$ and $\frac{7}{15}$ as equivalent fractions with a denominator of 30. Since $6 \cdot \mathbf{5} = 30$ we will multiply $\frac{1}{6}$ by $\frac{5}{5}$. Since $15 \cdot \mathbf{2} = 30$ we will multiply $\frac{7}{15}$ by $\frac{2}{2}$.

$$\frac{1}{6} \cdot \frac{5}{5} + \frac{7}{15} \cdot \frac{2}{2} = \frac{5}{30} + \frac{14}{30} = \frac{19}{30}$$

Thus $\frac{1}{6} + \frac{7}{15} = \frac{19}{30}$.

Now Try Exercise 73

EXAMPLE 10 Consider $\frac{7}{108} + \frac{5}{156}$.

a) Determine the least common denominator of the fractions.

b) Add the fractions.

Solution

a) We found in Example 2 on page 21 that

$$108 = 2 \cdot 2 \cdot 3 \cdot 3 \cdot 3 \quad \text{and} \quad 156 = 2 \cdot 2 \cdot 3 \cdot 13$$

We can see that the maximum number of 2s that appear in either prime factorization is two (there are two 2s in both factorizations), the maximum number of 3s is three, and the maximum number of 13s is one. Multiply as follows:

$$\text{LCD} = 2 \cdot 2 \cdot 3 \cdot 3 \cdot 3 \cdot 13 = 1404$$

Thus, the least common denominator is 1404. This is the smallest number that both 108 and 156 divide into.

b) To add the fractions, we need to write both fractions with a common denominator. The best common denominator to use is the LCD. Since $1404 \div 108 = 13$, we will multiply $\frac{7}{108}$ by $\frac{13}{13}$. Since $1404 \div 156 = 9$, we will multiply $\frac{5}{156}$ by $\frac{9}{9}$.

$$\frac{7}{108}\cdot\frac{13}{13} + \frac{5}{156}\cdot\frac{9}{9} = \frac{91}{1404} + \frac{45}{1404} = \frac{136}{1404} = \frac{34}{351}$$

Thus, $\frac{7}{108} + \frac{5}{156} = \frac{34}{351}$.

Now Try Exercise 75

EXAMPLE 11 How much larger is $\frac{3}{4}$ inch than $\frac{2}{3}$ inch?

Solution To find the difference, we need to subtract $\frac{2}{3}$ inch from $\frac{3}{4}$ inch.

$$\frac{3}{4} - \frac{2}{3}$$

The least common denominator is 12. Therefore, we rewrite both fractions with a denominator of 12.

$$\frac{3}{4} = \frac{3}{4}\cdot\frac{3}{3} = \frac{9}{12} \quad \text{and} \quad \frac{2}{3} = \frac{2}{3}\cdot\frac{4}{4} = \frac{8}{12}$$

Now we subtract.

$$\frac{3}{4} - \frac{2}{3} = \frac{9}{12} - \frac{8}{12} = \frac{1}{12}$$

Thus, $\frac{3}{4}$ inch is $\frac{1}{12}$ inch greater than $\frac{2}{3}$ inch.

Now Try Exercise 85

AVOIDING COMMON ERRORS

It is important to remember that dividing out a common factor in the numerator of one fraction and the denominator of a different fraction can be performed only when multiplying fractions. *This process cannot be performed when adding or subtracting fractions.*

CORRECT
MULTIPLICATION PROBLEMS

$$\frac{\overset{1}{\cancel{3}}}{5}\cdot\frac{1}{\underset{1}{\cancel{3}}}$$

$$\frac{\overset{2}{\cancel{8}}\cdot 3}{\underset{1}{\cancel{4}}}$$

INCORRECT
ADDITION PROBLEMS

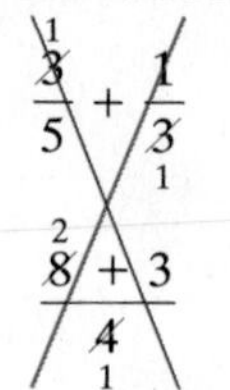

7 Change Mixed Numbers to Fractions and Vice Versa

Understanding Algebra

The mixed number $2\frac{3}{4}$ means $2 + \frac{3}{4}$.

Consider the number $5\frac{2}{3}$. This is an example of a **mixed number.** A mixed number consists of a whole number followed by a fraction. The mixed number $5\frac{2}{3}$ means $5 + \frac{2}{3}$. We can change $5\frac{2}{3}$ to a fraction as follows.

$$5\frac{2}{3} = 5 + \frac{2}{3} = \frac{15}{3} + \frac{2}{3} = \frac{15+2}{3} = \frac{17}{3}$$

Notice that we expressed the whole number, 5, as a fraction with a denominator of 3, then added the fractions.

EXAMPLE 12 Change $7\frac{3}{8}$ to a fraction.

Solution
$$7\frac{3}{8} = 7 + \frac{3}{8} = \frac{56}{8} + \frac{3}{8} = \frac{56 + 3}{8} = \frac{59}{8}$$

Now Try Exercise 37

Now consider the fraction $\frac{17}{3}$. We convert it to a mixed number as follows.

$$\frac{17}{3} = \frac{15}{3} + \frac{2}{3} = 5 + \frac{2}{3} = 5\frac{2}{3}$$

Notice we wrote $\frac{17}{3}$ as a sum of two fractions, each with the denominator of 3. The first fraction being added, $\frac{15}{3}$, is the equivalent of the largest integer that is less than $\frac{17}{3}$.

EXAMPLE 13 Change $\frac{43}{6}$ to a mixed number.

Solution
$$\frac{43}{6} = \frac{42}{6} + \frac{1}{6} = 7 + \frac{1}{6} = 7\frac{1}{6}$$

Now Try Exercise 41

HELPFUL HINT

Notice in Example 13 the fraction $\frac{43}{6}$ is simplified because the greatest common divisor of the numerator and denominator is 1. Do not confuse simplifying a fraction with changing a fraction with a value greater than 1 to a mixed number. The fraction $\frac{43}{6}$ can be converted to the mixed number $7\frac{1}{6}$. However, $\frac{43}{6}$ is a simplified fraction.

Now we will work examples that contain mixed numbers.

EXAMPLE 14 **Plumbing** To repair a plumbing leak, a coupling $\frac{1}{2}$ inch long is glued to a piece of plastic pipe. After the gluing, the piece of plastic pipe showing is $2\frac{9}{16}$ inches, as shown in **Figure 1.4**. How long is the combined length?

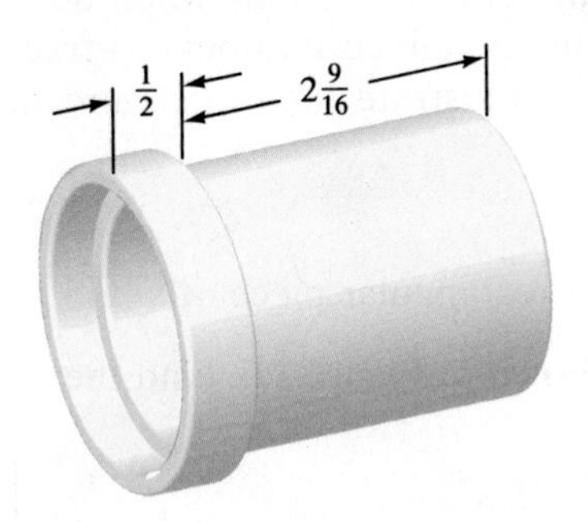

FIGURE 1.4

Solution Understand and Translate We need to add $2\frac{9}{16}$ inches and $\frac{1}{2}$ inch to obtain the combined length. After we write both fractions with a common denominator, we add the numbers.

Carry Out

$$\text{original numbers}\left\{\begin{array}{r} 2\frac{9}{16} \\ +\ \frac{1}{2} \\ \hline \end{array}\right. \rightarrow \left.\begin{array}{r} 2\frac{9}{16} \\ +\ \frac{8}{16} \\ \hline 2\frac{17}{16} \end{array}\right\}\text{fractions rewritten with common denominator of 16}$$

Since $2\frac{17}{16} = 2 + \frac{17}{16} = 2 + 1\frac{1}{16} = 3\frac{1}{16}$, the sum is $3\frac{1}{16}$.

Check and Answer The answer appears reasonable. Thus, the total length is $3\frac{1}{16}$ inches.

Now Try Exercise 107

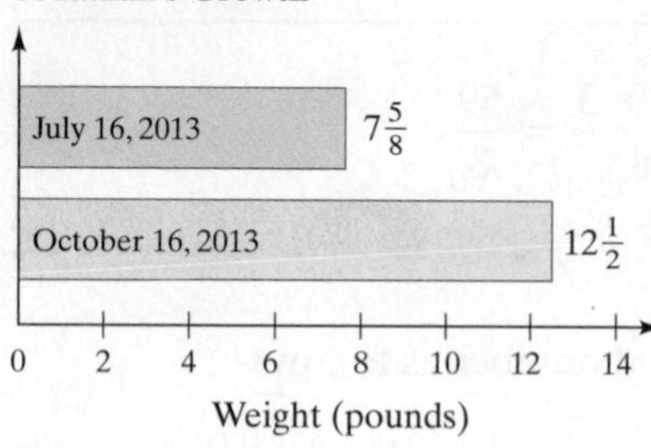

FIGURE 1.5

EXAMPLE 15 Gaining Weight The graph in **Figure 1.5** shows the weight of the Longenbergers' baby boy, Jonathan, on July 16, 2013, and on October 16, 2013. How much weight did Jonathan gain during this time period?

Solution Understand and Translate To find the increase, we need to subtract Jonathan's weight on July 16, 2013, from his weight on October 16, 2013. We will subtract vertically.

Carry Out

$$\begin{array}{r} 12\frac{1}{2} \\ -\ 7\frac{5}{8} \\ \hline \end{array} \quad \rightarrow \quad \begin{array}{r} 12\frac{4}{8} \\ -\ 7\frac{5}{8} \\ \hline \end{array}$$

Since we wish to subtract $\frac{5}{8}$ from $\frac{4}{8}$, and $\frac{5}{8}$ is greater than $\frac{4}{8}$, we write $12\frac{4}{8}$ as $11\frac{12}{8}$. To get $11\frac{12}{8}$, we take 1 unit from the number 12 and write it as $\frac{8}{8}$. This gives $11 + 1 + \frac{4}{8} = 11 + \frac{8}{8} + \frac{4}{8} = 11 + \frac{12}{8} = 11\frac{12}{8}$. Now we subtract as follows.

$$\begin{array}{r} 12\frac{1}{2} \\ -\ 7\frac{5}{8} \\ \hline \end{array} \quad \rightarrow \quad \begin{array}{r} 12\frac{4}{8} \\ -\ 7\frac{5}{8} \\ \hline \end{array} \quad \rightarrow \quad \begin{array}{r} 11\frac{12}{8} \\ -\ 7\frac{5}{8} \\ \hline 4\frac{7}{8} \end{array}$$

Check and Answer By examining the graph, we see that the answer is reasonable. Thus, Jonathan gained $4\frac{7}{8}$ pounds during this time period.

Now Try Exercise 99

Although it is not necessary to change mixed numbers to fractions when adding or subtracting mixed numbers, it is necessary to change mixed numbers to fractions when multiplying or dividing mixed numbers. We illustrate this procedure in Example 16.

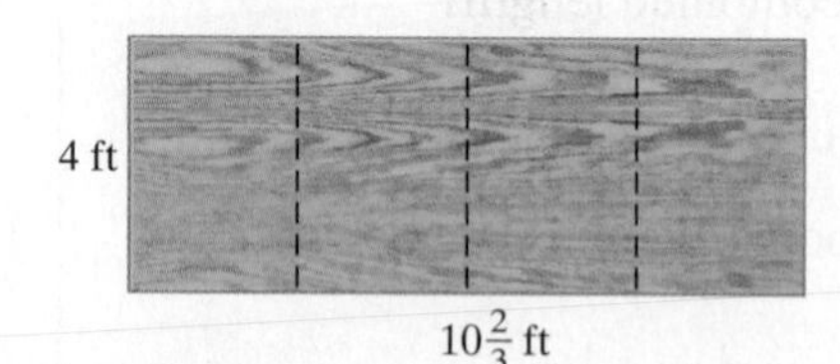

FIGURE 1.6

EXAMPLE 16 Cutting Wood A carpenter is cutting a rectangular piece of wood 4 feet wide by $10\frac{2}{3}$ feet long into four equal strips, as shown in **Figure 1.6**. Find the dimensions of each strip.

Solution Understand and Translate We know from **Figure 1.6** that one side will have a width of 4 feet. To find the length of the strips, we need to divide $10\frac{2}{3}$ by 4.

Carry Out

$$10\frac{2}{3} \div 4 = \frac{32}{3} \div \frac{4}{1} = \frac{\overset{8}{\cancel{32}}}{3} \cdot \frac{1}{\underset{1}{\cancel{4}}} = \frac{8}{3} = 2\frac{2}{3}$$

Check and Answer If you multiply $2\frac{2}{3}$ by 4 you obtain the original length, $10\frac{2}{3}$. Thus, the calculation is correct. The dimensions of each strip will be 4 feet by $2\frac{2}{3}$ feet.

Now Try Exercise 101

1.3 Exercise Set MathXL® MyMathLab®

Warm-Up Exercises

Fill in the blanks with the appropriate word, phrase, or symbol(s) from the following list.

variables	factors	LCD	added	$\frac{2}{3}$	$\frac{3}{2}$
GCF	mixed	ellipsis	$\frac{1}{6}$	denominator	

1. When two fractions are being ________ or subtracted we rewrite them so that they both have the same (common) denominator.

2. $5 + \frac{1}{3}$ is usually written as $5\frac{1}{3}$, which is called a ________ number.

3. Letters that represent numbers are called ________.

4. In the expression 2, 4, 6, 8, . . . , the three dots, called an ________, signify the sequence continues indefinitely.

5. $\frac{1}{3} \div \frac{1}{2} =$ ________.

6. Numbers or variables that are multiplied together are called ________.

7. In the fraction $\frac{3}{4}$, 4 is called the ________.

8. 15 is the ________ of 30 and 75.

9. To perform the division $\frac{4}{7} \div \frac{2}{3}$ we rewrite it using multiplication as $\frac{4}{7} \cdot$ ________.

10. 40 is the ________ of the fractions $\frac{3}{8}$ and $\frac{7}{10}$.

Practice the Skills

Determine the prime factorization of each number.

11. 12 **12.** 18 **13.** 60 **14.** 80

15. 150 **16.** 180

Determine the greatest common factor (GCF) of each pair of numbers.

17. 12 and 18 **18.** 15 and 27 19. 60 and 80 **20.** 45 and 63

21. 150 and 294 **22.** 126 and 162

Simplify each fraction. If a fraction is already simplified, so state.

23. $\frac{8}{10}$ **24.** $\frac{9}{15}$ **25.** $\frac{24}{28}$ **26.** $\frac{24}{42}$

27. $\frac{36}{76}$ **28.** $\frac{16}{72}$ 29. $\frac{18}{42}$ **30.** $\frac{60}{105}$

31. $\frac{18}{49}$ **32.** $\frac{35}{36}$ **33.** $\frac{100}{150}$ **34.** $\frac{112}{144}$

Convert each mixed number to a fraction.

35. $2\frac{13}{15}$ **36.** $15\frac{1}{3}$ 37. $7\frac{2}{3}$ **38.** $14\frac{3}{4}$ **39.** $3\frac{5}{18}$ **40.** $2\frac{2}{9}$

Write each fraction as a mixed number.

41. $\frac{7}{4}$ **42.** $\frac{18}{7}$ **43.** $\frac{13}{4}$ **44.** $\frac{9}{2}$ **45.** $\frac{32}{7}$ **46.** $\frac{110}{20}$

Find each product or quotient. Simplify the answer.

47. $\frac{1}{3} \cdot \frac{4}{5}$ **48.** $\frac{6}{13} \cdot \frac{7}{17}$ 49. $\frac{5}{12} \cdot \frac{4}{15}$ **50.** $\frac{36}{48} \cdot \frac{16}{45}$

51. $\frac{3}{4} \div \frac{1}{2}$ **52.** $\frac{3}{8} \div \frac{3}{4}$ **53.** $\frac{15}{16} \cdot \frac{4}{3}$ **54.** $\frac{3}{8} \cdot \frac{10}{11}$

55. $\frac{10}{3} \div \frac{5}{9}$

56. $\frac{5}{9} \div 30$

57. $\frac{1}{24} \div \frac{3}{16}$

58. $\frac{5}{12} \div \frac{4}{3}$

59. $5\frac{3}{8} \div 1\frac{1}{4}$

60. $4\frac{4}{5} \div \frac{8}{15}$

61. $\frac{28}{13} \cdot \frac{2}{7}$

62. $\left(2\frac{1}{5}\right)\left(\frac{7}{8}\right)$

Add or subtract. Simplify each answer.

63. $\frac{3}{8} + \frac{2}{8}$

64. $\frac{18}{36} + \frac{5}{36}$

65. $\frac{3}{14} - \frac{1}{14}$

66. $\frac{15}{16} - \frac{7}{16}$

67. $\frac{4}{5} + \frac{6}{15}$

68. $\frac{7}{8} + \frac{5}{6}$

69. $\frac{9}{17} + \frac{2}{34}$

70. $\frac{3}{7} + \frac{17}{35}$

71. $\frac{1}{3} + \frac{1}{4}$

72. $\frac{1}{6} + \frac{1}{18}$

73. $\frac{7}{12} - \frac{2}{9}$

74. $\frac{3}{7} - \frac{5}{12}$

75. $\frac{11}{60} + \frac{7}{150}$

76. $\frac{13}{126} + \frac{5}{84}$

77. $6\frac{1}{3} - 3\frac{1}{2}$

78. $5\frac{3}{8} - 3\frac{3}{4}$

79. $9\frac{2}{5} - 6\frac{1}{2}$

80. $4\frac{5}{9} - \frac{7}{8}$

81. $5\frac{9}{10} + 3\frac{1}{3}$

82. $8\frac{2}{7} + 3\frac{1}{3}$

83. How much larger is $\frac{5}{6}$ mile than $\frac{3}{8}$ mile?

84. How much larger is $\frac{1}{5}$ meter than $\frac{1}{7}$ meter?

85. How much larger is $\frac{7}{8}$ centimeter than $\frac{5}{12}$ centimeter?

86. How much larger is $\frac{11}{36}$ yard than $\frac{3}{28}$ yard?

In Exercises 87–90, perform the indicated operation.

87. a) $\frac{3}{4} + \frac{2}{3}$ b) $\frac{3}{4} - \frac{2}{3}$ c) $\frac{3}{4} \cdot \frac{2}{3}$ d) $\frac{3}{4} \div \frac{2}{3}$

88. a) $\frac{5}{6} \div \frac{3}{8}$ b) $\frac{5}{6} + \frac{3}{8}$ c) $\frac{5}{6} - \frac{3}{8}$ d) $\frac{5}{6} \cdot \frac{3}{8}$

89. a) $2\frac{5}{6} \cdot 1\frac{2}{3}$ b) $2\frac{5}{6} + 1\frac{2}{3}$ c) $2\frac{5}{6} \div 1\frac{2}{3}$ d) $2\frac{5}{6} - 1\frac{2}{3}$

90. a) $3\frac{1}{2} - 2\frac{3}{4}$ b) $3\frac{1}{2} \cdot 2\frac{3}{4}$ c) $3\frac{1}{2} \div 2\frac{3}{4}$ d) $3\frac{1}{2} + 2\frac{3}{4}$

Problem Solving

91. **Height Gain** The following graph shows Rebecca Bersagel's height, in inches, on her 8th and 12th birthdays. How much had Rebecca grown in the 4 years?

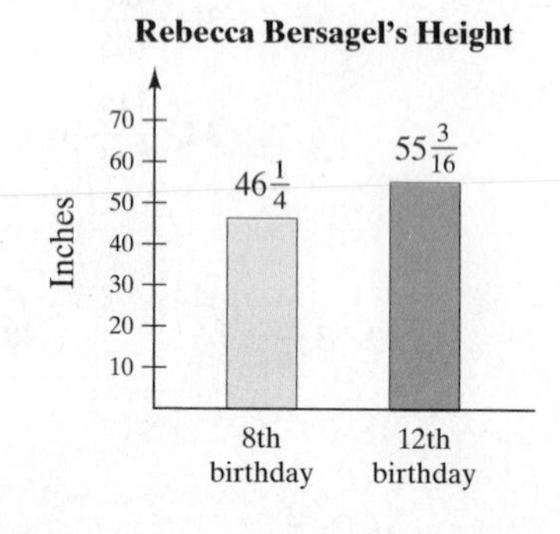

92. **Road Paving** The following graph shows the progress of the Davenport Paving Company in paving the Memorial Highway. How much of the highway was paved from June through August?

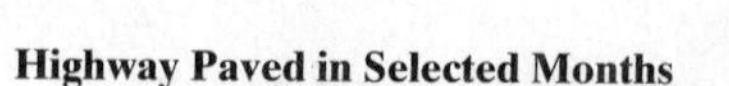

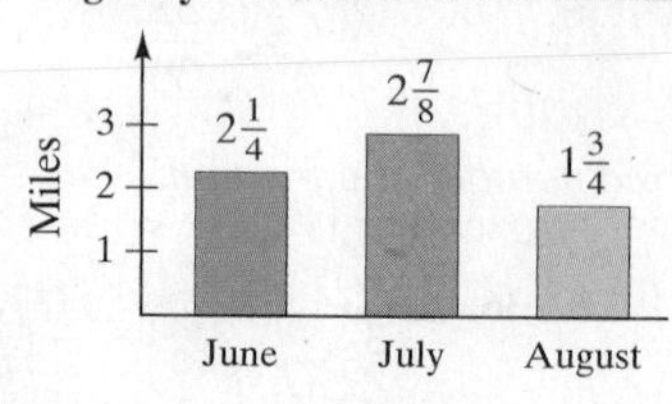

In many problems you will need to subtract a fraction from 1, where 1 represents "the whole" or the "total amount." Exercises 93–96 are answered by subtracting the given fraction from 1.

93. **Putting Success** Lamont is a local golf pro and last year he made $\frac{46}{55}$ of all his putts from within six feet. What fraction of his putts from within six feet did he miss?

94. **Global Warming** The probability that an event does not occur may be found by subtracting the probability that the event does occur from 1. If the probability that global warming is occurring is $\frac{7}{9}$, find the probability that global warming is not occurring.

95. Time to a Degree A recent study at Josephine College found that $\frac{37}{100}$ of first-time freshmen finished their bachelor's degrees in four years. What fraction of freshmen did not finish their bachelor's degree in 4 years?

Freshman Finishing Bachelor's Degree

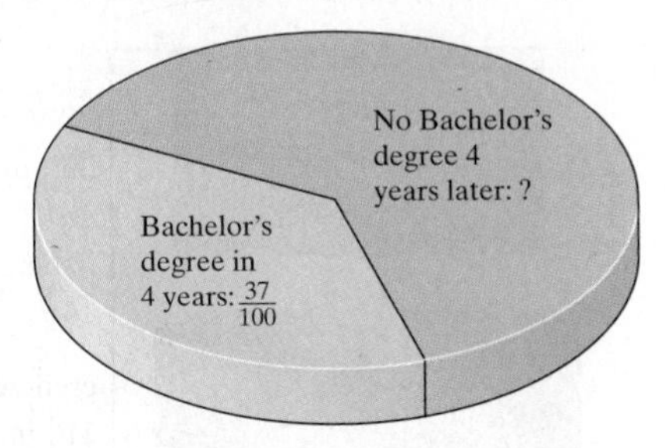

96. Home Heating The following circle graph shows the fraction of U.S. homes built after 1990 that used electricity to heat their homes. Determine the fraction of U.S. homes that did not use electricity.

How Americans Heat Their Homes

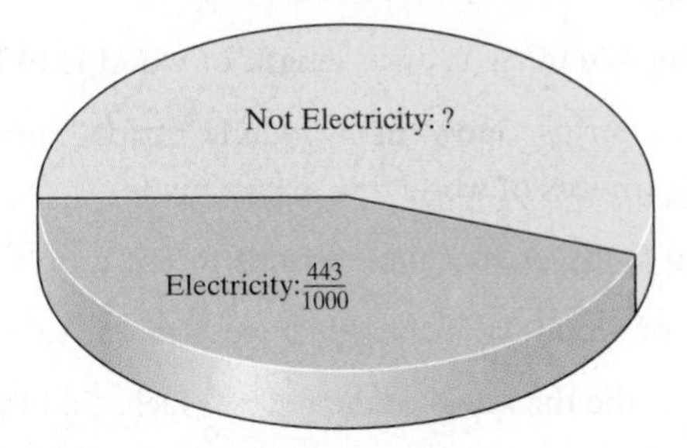

Source: United States Energy Department

In Exercises 97–112, answer the questions asked.

97. Monarch Butterflies The children at Happy Days Daycare Center have two monarch butterflies named Jeremy and April. Jeremy has a wingspan of $8\frac{3}{8}$ centimeters and April has a wingspan of $7\frac{15}{16}$ centimeters. How much larger is Jeremy's wingspan than April's wingspan?

98. Pecan Pie A pecan pie weighs $1\frac{5}{16}$ pounds. If the pie is to be divided equally among 6 people, how much will each person get?

99. Running Denise started a running program in January when she could run a mile in $10\frac{1}{2}$ minutes. After 6 months, Denise could run a mile in $8\frac{1}{5}$ minutes. By how many minutes did Denise improve in 6 months?

100. Baking Turkey The instructions on a turkey indicate that a 12- to 16-pound turkey should bake at 325°F for about 22 minutes per pound. Josephine Nickola is planning to bake a $13\frac{1}{2}$-pound turkey. Approximately how long should the turkey be baked?

101. Wood Cut Debbie Anderson cuts a piece of wood measuring $3\frac{1}{8}$ inches into two equal pieces. How long is each piece?

102. Pants Inseam The inseam on a new pair of pants is 32 inches. If Don O'Neal's inseam is $29\frac{3}{8}$ inches, by how much will the pants need to be shortened?

103. Drug Amount A nurse must give $\frac{1}{16}$ milligram of a drug for each kilogram of patient mass. If Mr. Krisanda has a mass of 80 kilograms, find the amount of the drug Mr. Krisanda should be given.

104. Chopped Onions A recipe for pot roast calls for $\frac{1}{4}$ cup chopped onions for each pound of beef. For $5\frac{1}{2}$ pounds of beef, how many cups of chopped onions are needed?

105. Shampoo A bottle of shampoo contains 15 fluid ounces. If Tierra Bentley uses $\frac{3}{8}$ of an ounce each time she washes her hair, how many times can Tierra wash her hair using this bottle?

106. Fencing Matt Mesaros wants to fence in his backyard as shown. The three sides to be fenced measure $16\frac{2}{3}$ yards, $22\frac{2}{3}$ yards, and $14\frac{1}{8}$ yards.

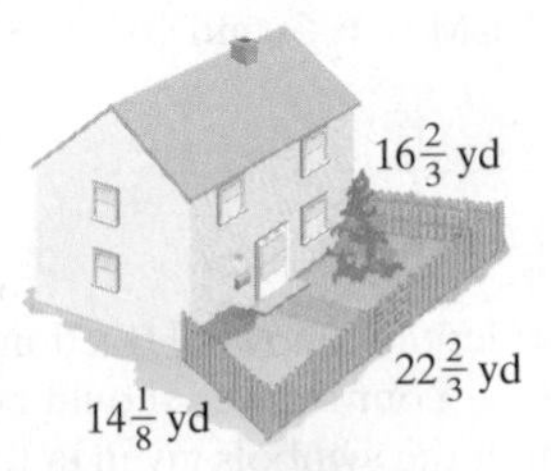

a) How much fence will Matt need?

b) If Matt buys 60 yards of fence, how much will be left over?

107. Windows An insulated window for a house is made up of two pieces of glass, each $\frac{1}{4}$-inch thick, with a 1-inch space between them. What is the total thickness of this window?

108. Truck Weight A flatbed tow truck weighing $4\frac{1}{2}$ tons is carrying two cars. One car weighs $1\frac{1}{6}$ tons, the other weighs $1\frac{3}{4}$ tons. What is the total weight of the tow truck and the two cars?

109. Cutting Wood A 28-inch length of wood is to be cut into $4\frac{2}{3}$ inch strips. How many whole strips can be made? Disregard loss of wood due to cuts made.

110. Fasten Bolts A mechanic wishes to use a bolt to fasten a piece of wood $4\frac{1}{2}$ inches thick to a metal tube $2\frac{1}{3}$ inches thick. If the thickness of the nut is $\frac{1}{8}$ inch, find the length of the shaft of the bolt so that the nut fits flush with the end of the bolt (see the figure below).

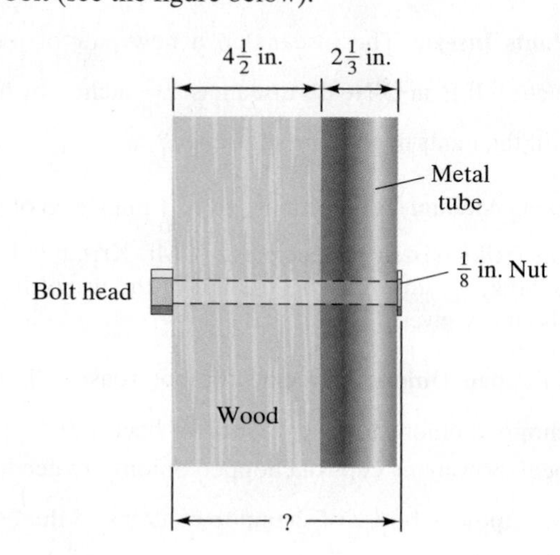

111. Entertainment Center Scott Morningstar just bought a high definition television that will sit atop his entertainment credenza. The TV measures $36\frac{1}{2}$ inches high. Its stand is $14\frac{1}{8}$ inches high and the credenza is $31\frac{3}{4}$ inches high. See figure below.

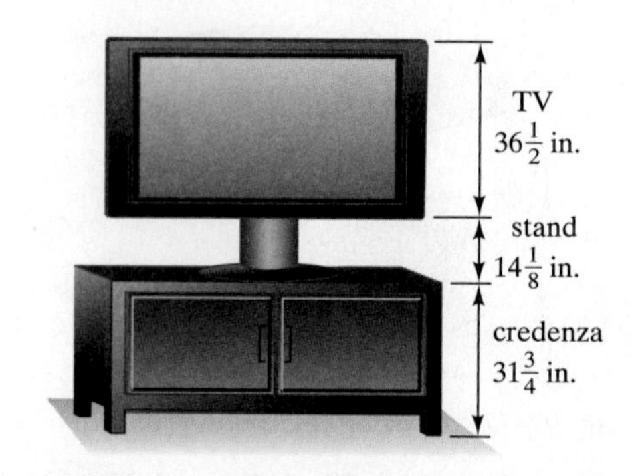

a) If Scott has 8-foot ceilings in his house, will there be sufficient room for this purchase?

b) Find the total height of the TV, the stand, and the credenza.

112. Soda If five 2-liter bottles of soda are split evenly among 30 people, how many liters of soda will each person get?

Concept/Writing Exercises

113. Another way of saying "find the LCD of the 3 fractions $\frac{1}{6}, \frac{2}{3}, \frac{7}{10}$" is to say "find the least common multiple (LCM) of 6, 3, and 10." Write a few sentences to describe how to go about finding the LCM of 6, 3, and 10.

114. Explain how to simplify a fraction.

Challenge Problems

115. Add or subtract the following fractions using the rule discussed in this section. Your answer should be a single fraction, and it should contain the symbols given in the exercise.

a) $\frac{*}{a} + \frac{?}{a}$ **b)** $\frac{\odot}{?} - \frac{\square}{?}$

c) $\frac{\triangle}{\square} + \frac{4}{\square}$ **d)** $\frac{x}{3} - \frac{2}{3}$

e) $\frac{12}{x} - \frac{4}{x}$

116. Multiply the fractions on the right using the rule discussed in this section. Your answer should be a single fraction and it should contain the symbols given in the exercise.

a) $\frac{\triangle}{a} \cdot \frac{\square}{b}$ **b)** $\frac{6}{3} \cdot \frac{\triangle}{\square}$

c) $\frac{x}{a} \cdot \frac{y}{b}$ **d)** $\frac{3}{8} \cdot \frac{4}{y}$

e) $\frac{3}{x} \cdot \frac{x}{y}$

117. Drug Dosage An allopurinol pill comes in 300-milligram doses. Dr. Muechler wants a patient to get 450 milligrams each day by cutting the pills in half and taking one-half pill three times a day. If he wants to prescribe enough pills for a 6-month period (assume 30 days per month), how many pills should he prescribe?

Group Activity

Discuss and answer Exercise 118 as a group.

118. Potatoes The table to the right gives the amount of each ingredient recommended to make 2, 4 and 8 servings of instant mashed potatoes.

Determine the amount of potato flakes and milk needed to make 6 servings by the different methods described. When working with milk, 16 tbsp = 1 cup.

a) Group member 1: Determine the amounts of potato flakes and milk needed to make 6 servings by multiplying the amounts for 2 servings by 3.

b) Group member 2: Determine the amounts by adding the amounts for 2 servings to the amounts for 4 servings.

c) Group member 3: Determine the amounts by finding the average (mean) of 4 and 8 servings.

d) As a group, determine the amounts by subtracting the amounts for 2 servings from the amounts for 8 servings.

e) As a group, compare your answers from parts **a)** through **d)**. Are they all the same? If not, can you explain why?

Servings	2	4	8
Water	$\frac{2}{3}$ cup	$1\frac{1}{3}$ cups	$2\frac{2}{3}$ cups
Milk	2 tbsp	$\frac{1}{3}$ cup	$\frac{2}{3}$ cup
Butter*	1 tbsp	2 tbsp	4 tbsp
Salt†	$\frac{1}{4}$ tsp	$\frac{1}{2}$ tsp	1 tsp
Potato flakes	$\frac{2}{3}$ cup	$1\frac{1}{3}$ cup	$2\frac{2}{3}$ cups

*or margarine

†Less salt can be used if desired.

Cumulative Review Exercises

[1.1] **119.** What is your instructor's name and office hours?

[1.2] **120.** What is the mean of 9, 8, 15, 32, 16?

121. What is the median of 9, 8, 15, 32, 16?

[1.3] **122.** What are variables?

1.4 The Real Number System

1 Identify sets of numbers.

2 Know the structure of the real numbers.

Understanding Algebra

A *set* is a collection of elements—in this book, usually numbers. The set of natural numbers less than 6, for example, is {1, 2, 3, 4, 5}.

This section introduces you to different sets of numbers and to the structure of the real number system.

1 Identify Sets of Numbers

A **set** is a collection of **elements** listed within braces. The set $\{a, b, c, d, e\}$ consists of five elements, namely *a, b, c, d,* and *e*. A set that contains no elements is called an **empty set** (or **null set**). The symbols { } or ∅ are used to represent the empty set.

Two important sets are the natural numbers and the whole numbers. The whole numbers were introduced earlier.

Natural numbers: {1, 2, 3, 4, 5, ...}

Whole numbers: {0, 1, 2, 3, 4, 5, ...}

An aid in understanding sets of numbers is a number line (**Fig. 1.7**).

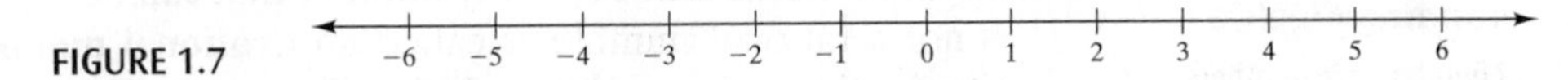

FIGURE 1.7

The number line continues indefinitely in both directions. The numbers to the right of 0 are positive and those to the left of 0 are negative. Zero is neither positive nor negative (**Fig. 1.8**).

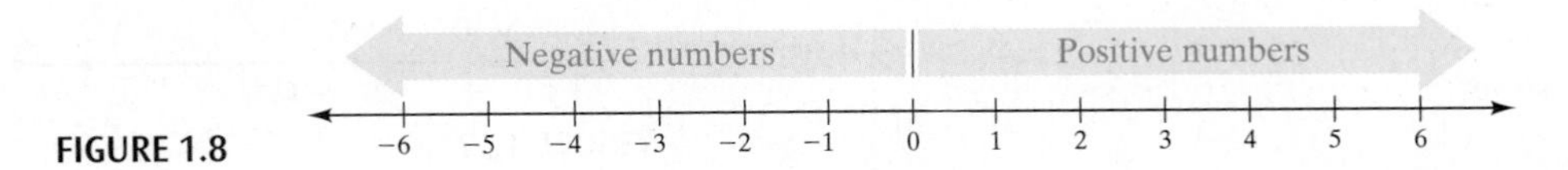

FIGURE 1.8

Figure 1.9 illustrates the natural numbers marked on a number line. The natural numbers are also called the **counting numbers** or the **positive integers.**

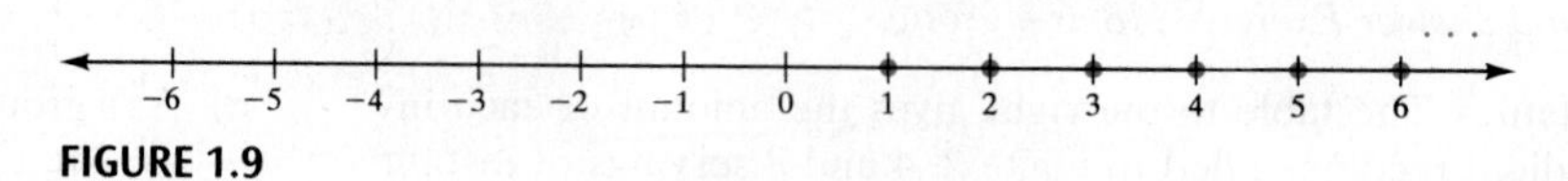

FIGURE 1.9

Another important set of numbers is the integers.

Integers: $\{\ldots, -5, -4, -3, -2, -1, 0, 1, 2, 3, 4, 5, \ldots\}$

Negative integers — Positive integers

The integers consist of the negative integers, 0, and the positive integers. The integers are marked on the number line in **Figure 1.10**.

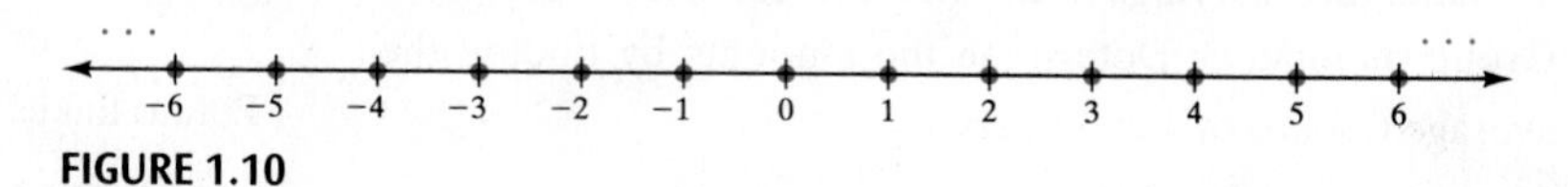

FIGURE 1.10

Fractions and certain decimal numbers do not belong to the set of integers but they do belong to the set of rational numbers. The set of **rational numbers** consists of all the numbers that can be expressed as a quotient (or a ratio) of two integers, with the denominator not 0.

Understanding Algebra

Rational numbers can be expressed as fractions where the numerator and denominator are integers.

or

Rational numbers can be expressed as decimals that either terminate or repeat.

Rational numbers: {quotient of two integers, denominator not 0}

All integers are rational numbers since they can be written with a denominator of 1. For example, $3 = \frac{3}{1}$, $-12 = \frac{-12}{1}$, and $0 = \frac{0}{1}$. All fractions containing integers in the numerator and denominator (with the denominator not 0) are rational numbers. For example, the fraction $\frac{3}{5}$ is a quotient of two integers and is a rational number.

When a fraction that is a ratio of two integers is converted to a decimal number by dividing the numerator by the denominator, the quotient will always be either a *terminating decimal number,* such as 0.3 and 3.25, or a *repeating decimal number* such as 0.3333... and 5.2727... . The three dots at the end of a number indicate that the numbers continue to repeat in the same manner indefinitely. All terminating decimal numbers and all repeating decimal numbers are rational numbers and can be expressed as a quotient of two integers. For example, $0.3 = \frac{3}{10}$, $3.25 = \frac{325}{100}$, $0.3333\ldots = \frac{1}{3}$, and $5.2727\ldots = \frac{58}{11}$. Some rational numbers are illustrated on the number line in **Figure 1.11**.

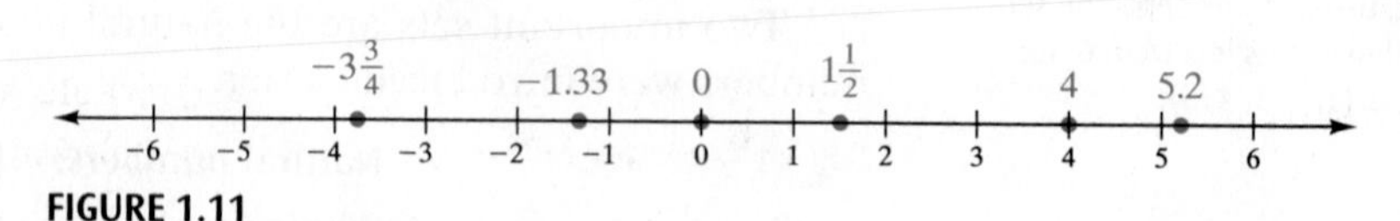

FIGURE 1.11

Understanding Algebra

Irrational numbers are numbers that have infinite, non-repeating decimal representations.

Some numbers are not rational. Numbers such as the square root of 2, written $\sqrt{2}$, are not rational numbers. Any number that can be represented on the number line that is not a rational number is called an **irrational number**. Irrational numbers are nonterminating, non-repeating decimal numbers. For example, $\sqrt{2}$ cannot be expressed exactly as a decimal number. Irrational numbers can only be *approximated* by decimal numbers. $\sqrt{2}$ is *approximately* 1.41. Thus, we may write $\sqrt{2} \approx 1.41$. Some irrational numbers are illustrated on the number line in **Figure 1.12**.

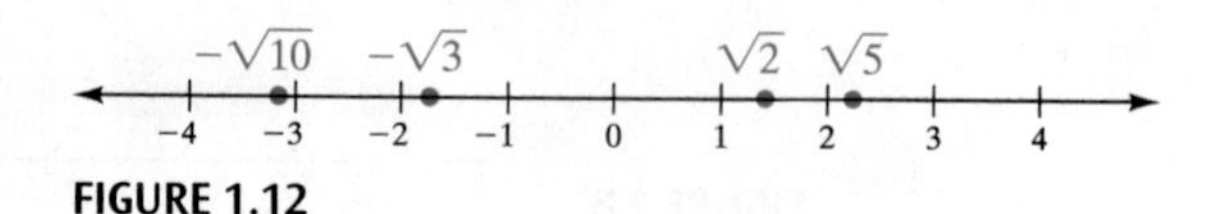

FIGURE 1.12

Understanding Algebra

A *real number* is any number that is either rational or irrational.

The set of all real numbers is denoted by the symbol ℝ

2 Know the Structure of the Real Numbers

Any number that is either rational or irrational is called a **real number** and the number line is often referred to as a **real number line.**

The symbol ℝ is used to represent the set of real numbers. The natural numbers, the whole numbers, the integers, the rational numbers, and the irrational numbers are all real numbers. **Figure 1.13** illustrates the relationships between the various sets of numbers within the set of real numbers.

Understanding Algebra

$\sqrt{2}$ and $\sqrt{5}$ are examples of irrational numbers but $\sqrt{16}$ is rational because it is exactly 4.

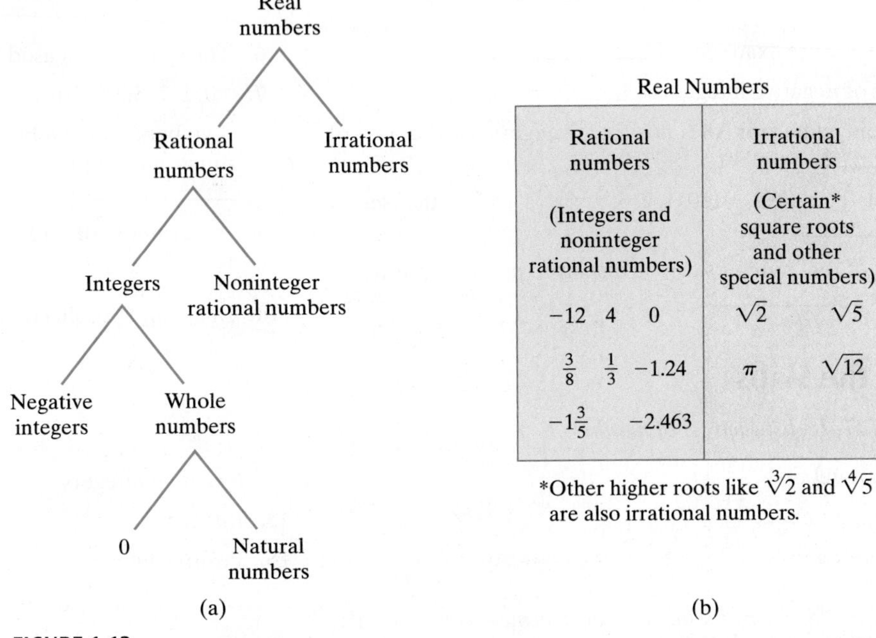

FIGURE 1.13

Consider the natural number 5. If we follow the branches in **Figure 1.13a** upward, we see that the number 5 is also a whole number, an integer, a rational number, and a real number. Now consider the number $\frac{1}{2}$. It belongs to the noninteger rational numbers. If we follow the branches upward, we can see that $\frac{1}{2}$ is also a rational number and a real number.

EXAMPLE 1 Consider the following set of numbers.

$$\left\{-5, -3.5, -\sqrt{3}, -\frac{2}{3}, 0, 0.3333\ldots, \sqrt{5}, 3\frac{1}{4}, 5, 6.4, \sqrt{41}, 9\right\}$$

List the elements of the set that are

a) natural numbers. **b)** whole numbers. **c)** integers.
d) rational numbers. **e)** irrational numbers. **f)** real numbers.

Solution We will list the numbers from left to right as they appear in the set. However, the elements may be listed in any order.

a) $5, 9$

b) $0, 5, 9$

c) $-5, 0, 5, 9$

d) $-5, -3.5, -\frac{2}{3}, 0, 0.3333\ldots, 3\frac{1}{4}, 5, 6.4, 9$

e) $-\sqrt{3}, \sqrt{5}, \sqrt{41}$

f) $-5, -3.5, -\sqrt{3}, -\frac{2}{3}, 0, 0.3333\ldots, \sqrt{5}, 3\frac{1}{4}, 5, 6.4, \sqrt{41}, 9$

Now Try Exercise 51

1.4 Exercise Set MathXL® MyMathLab®

Warm-Up Exercises

Fill in the blanks with the appropriate word, phrase, or symbol(s) from the following list.

set	irrational	rational	empty
$\{1, 2, 3\}$	counting	whole	line
integers	$\{\ldots, -3, -2, -1\}$	$\sqrt{3}$	$\frac{1}{2}$

1. $\sqrt{5}$ and $\sqrt{7}$ are examples of __________ numbers.
2. The set of negative integers is __________.
3. Another name for the positive integers is the set of __________ numbers.
4. The set $\{\ldots, -2, -1, 0, 1, 2, 3, \ldots\}$ is called the set of __________.
5. The set of real numbers can be displayed pictorially as a real number __________.
6. The symbol $\varnothing$ is used to denote the __________ set.
7. $\{0, 1, 2, 3, \ldots\}$ is called the set of __________ numbers.
8. Numbers that can be expressed as a fraction having integer numerator and non-zero integer denominator are called __________ numbers.
9. An example of a real number that is not a rational number is __________.
10. In general, a collection of elements is called a __________.

Practice the Skills

In Exercises 11–16, list each set of numbers.

11. Natural numbers
12. Counting numbers
13. Whole numbers
14. Negative integers
15. Integers
16. Positive integers

In Exercises 17–48, indicate whether each statement is true or false.

17. 5 is a natural number.
18. -5 is a natural number.
19. -4 is a whole number.
20. 4 is a whole number.
21. $\frac{1}{3}$ is an integer.
22. 3 is an integer.
23. 0.57 is a rational number.
24. $\sqrt{3}$ is a rational number.
25. $\sqrt{2}$ is a rational number.
26. 0.666… is a rational number.
27. $-\frac{1}{5}$ is a rational number.
28. $-\frac{2}{3}$ is an irrational number.
29. $\sqrt{5}$ is an irrational number.
30. $-\sqrt{7}$ is an irrational number.
31. Every whole number is a natural number.
32. Every counting number is a rational number.
33. The symbol $\varnothing$ is used to represent the empty set.
34. Every integer is negative.
35. Every real number is a rational number.
36. Every negative integer is a real number.
37. Every rational number is a real number.
38. When zero is added to the set of counting numbers, the set of whole numbers is formed.
39. Some real numbers are not rational numbers.
40. Some irrational numbers are not real numbers.
41. Every negative number is a negative integer.
42. All real numbers can be represented on a number line.
43. The symbol $\mathbb{R}$ is used to represent the set of real numbers.
44. Any number to the left of zero on a number line is a negative number.
45. Every number greater than zero is a positive integer.
46. Irrational numbers cannot be represented on a number line.
47. When the negative integers, the positive integers, and 0 are combined, the integers are formed.
48. The natural numbers, counting numbers, and positive integers are different names for the same set of numbers.
49. **Hotels** Some buildings in Europe have elevators that list negative numbers for floors below the ground level. For example, a floor might be designated as -2. In many countries, floor number 13 is omitted because of superstition. Considering the numbers -2 and 13, list those that are
    a) positive integers.
    b) rational numbers.
    c) real numbers.
    d) whole numbers.
50. **Address** We generally think of house addresses as being integers greater than 0. Have you ever seen a house number

that was not an integer greater than 0? There are quite a few such addresses in cities and towns across America. The house numbers 0 and $2\frac{1}{2}$ appear on Legare Street in Charleston, SC. Considering the numbers 0 and $2\frac{1}{2}$, list those that are

a) integers.

b) rational numbers.

c) real numbers.

51. Consider the following set of numbers.

$$\left\{-\frac{5}{7}, 0, -2, 3, 6\frac{1}{4}, \sqrt{7}, -\sqrt{3}, 1.63, 77\right\}$$

List the numbers that are

a) positive integers.

b) whole numbers.

c) integers.

d) rational numbers.

e) irrational numbers.

f) real numbers.

52. Consider the following set of numbers.

$$\left\{-6, 7, 12.4, -\frac{9}{5}, -2\frac{1}{4}, \sqrt{3}, 0, 9, \sqrt{7}, 0.35, \frac{22}{7}\right\}$$

List the numbers that are

a) positive integers.

b) whole numbers.

c) integers.

d) rational numbers.

e) irrational numbers.

f) real numbers.

Problem Solving

Give three examples of numbers that satisfy the given conditions.

53. An integer but not a negative integer.

54. A real number but not an integer.

55. An irrational number and a negative number.

56. A real number and a rational number.

57. A rational number but not a natural number.

58. An integer and a rational number.

59. A negative integer and a rational number.

60. A negative integer and a real number.

61. A real number but not a positive rational number.

62. A rational number but not a negative number.

63. A real number but not an irrational number.

64. A negative number but not a negative integer.

Three dots inside a set indicate that the set continues in the same manner. For example, $\{1, 2, 3, \ldots, 84\}$ is the set of natural numbers from 1 up to and including 84. In Exercises 65 and 66, determine the number of elements in each set.

65. $\{8, 9, 10, 11, \ldots, 94\}$

66. $\{-4, -3, -2, -1, 0, 1, \ldots, 64\}$

Challenge Problems

The diagrams in Exercises 67 and 68 are called ***Venn diagrams.*** *Venn diagrams are used to illustrate sets. For example, in the diagrams, circle A contains all the elements in set A, and circle B contains all the elements in set B. For each diagram, determine* **a)** *set A,* **b)** *set B,* **c)** *the set of elements that belong to both set A and set B, and* **d)** *the set of elements that belong to either set A or set B.*

67.

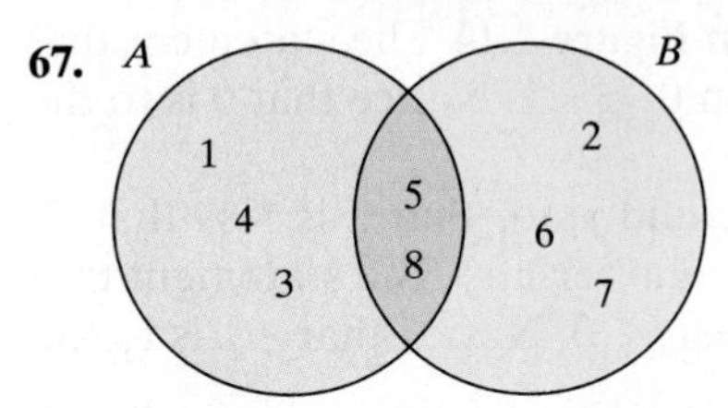

68.

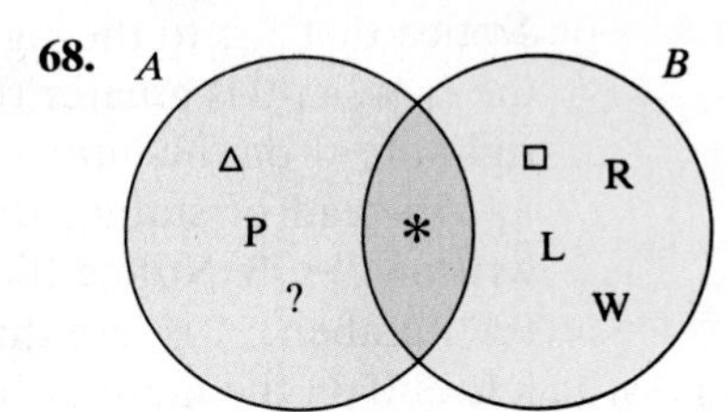

69. Consider the sets $A = \{1, 2, 3, 4\}$ and $B = \{1, 2, 3, 4, \ldots\}$.
 a) Explain the difference between set A and set B.
 b) How many elements are in set A?
 c) How many elements are in set B?
 d) Set A is an example of a *finite set*. Can you guess the name given to a set like set B?

70. How many decimal numbers are there
 a) between 1.0 and 2.0?
 b) between 1.4 and 1.5? Explain your answer.

71. How many fractions are there
 a) between 1 and 2?
 b) between $\frac{1}{3}$ and $\frac{1}{5}$? Explain your answer.

Group Activity

Discuss and answer Exercise 72 as a group.

72. Set A **union** set B, symbolized $A \cup B$, consists of the set of elements that belong to set A or set B (or both sets). Set A **intersection** set B, symbolized $A \cap B$, consists of the set of elements that both set A and set B have in common. Note that the elements that belong to both sets are listed only once in the union of the sets.

Consider the pairs of sets below.

Group member 1: $A = \{2, 3, 4, 6, 8, 9\}$ $B = \{1, 2, 3, 5, 7, 8\}$
Group member 2: $A = \{a, b, c, d, g, i, j\}$ $B = \{b, c, d, h, m, p\}$
Group member 3: $A = \{\text{red, blue, green, yellow}\}$ $B = \{\text{pink, orange, purple}\}$

 a) Group member 1: Find the union and intersection of the sets marked Group member 1.
 b) Group member 2: Find the union and intersection of the sets marked Group member 2.
 c) Group member 3: Find the union and intersection of the sets marked Group member 3.
 d) Now as a group, check each other's work. Correct any mistakes.
 e) As a group, using group member 1's sets, construct a Venn diagram like those shown in Exercises 67 and 68.

Cumulative Review Exercises

[1.3] 73. Convert $5\frac{4}{5}$ to a fraction.

74. Write $\frac{16}{3}$ as a mixed number.

75. Subtract $\frac{7}{8} - \frac{1}{3}$.

76. Divide $\frac{3}{5} \div 6\frac{3}{4}$.

1.5 Inequalities and Absolute Value

1 Determine which is the greater of two numbers.
2 Find the absolute value of a number.

1 Determine Which Is the Greater of Two Numbers

The number line, which shows numbers increasing from left to right, can be used to explain inequalities (see **Fig. 1.14**). When comparing two numbers, *the number to the right on the number line is the greater number, and the number to the left is the lesser number*. The symbol $>$ is used to represent the words "is greater than." The symbol $<$ is used to represent the words "is less than."

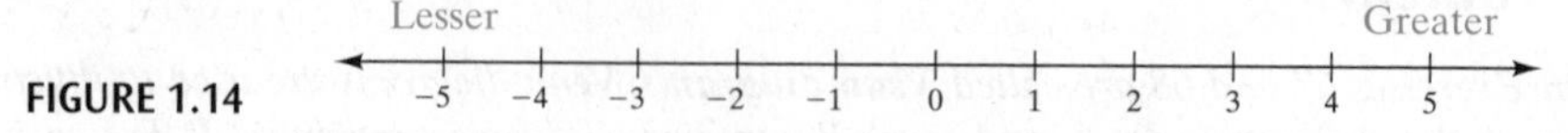

FIGURE 1.14

The statement that the number 3 is greater than the number 2 is written $3 > 2$. Notice that 3 is to the right of 2 on the number line in **Figure 1.14**. The statement that the number 0 is greater than the number -1 is written $0 > -1$. Notice that 0 is to the right of -1 on the number line.

Instead of stating that 3 is greater than 2, we could state that 2 is less than 3, written $2 < 3$. Notice that 2 is to the left of 3 on the number line. The statement that the number -1 is less than the number 0 is written $-1 < 0$. Notice that -1 is to the left of 0 on the number line.

Understanding Algebra

Statement	Expression
"3 is less than 9"	$3 < 9$
"14 is greater than 8"	$14 > 8$
"−6 is less than 5"	$-6 < 5$

EXAMPLE 1 Insert either < or > in the shaded area between each pair of numbers to make a true statement.

a) $2 \quad \square \quad 4$ **b)** $-2 \quad \square \quad -4$ **c)** $\frac{1}{2} \quad \square \quad \frac{1}{4}$ **d)** $-\frac{1}{2} \quad \square \quad -\frac{1}{4}$

Solution The points given are shown on the number line (**Fig. 1.15**).

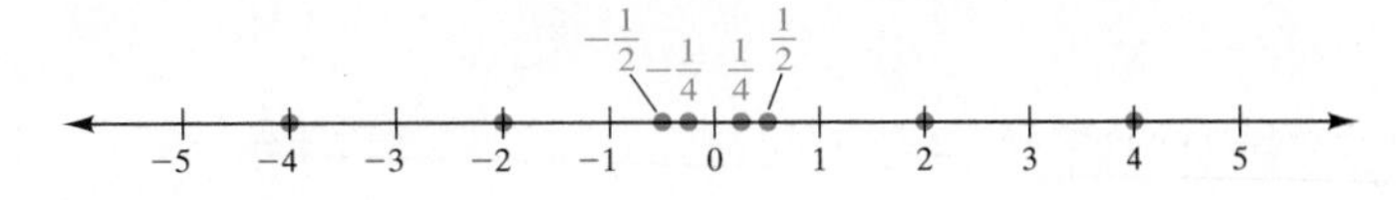

FIGURE 1.15

a) $2 < 4$; notice that 2 is to the left of 4.
b) $-2 > -4$; notice that -2 is to the right of -4.
c) $\frac{1}{2} > \frac{1}{4}$; notice that $\frac{1}{2}$ is to the right of $\frac{1}{4}$.
d) $-\frac{1}{2} < -\frac{1}{4}$; notice that $-\frac{1}{2}$ is to the left of $-\frac{1}{4}$.

Now Try Exercise 27

EXAMPLE 2 Insert either > or < in the shaded area between each pair of numbers to make a true statement.

a) $-3 \quad \square \quad 3$ **b)** $-3 \quad \square \quad -4$ **c)** $-4 \quad \square \quad 0$ **d)** $-1.08 \quad \square \quad -1.8$

Solution The numbers given are shown on the number line (**Fig. 1.16**).

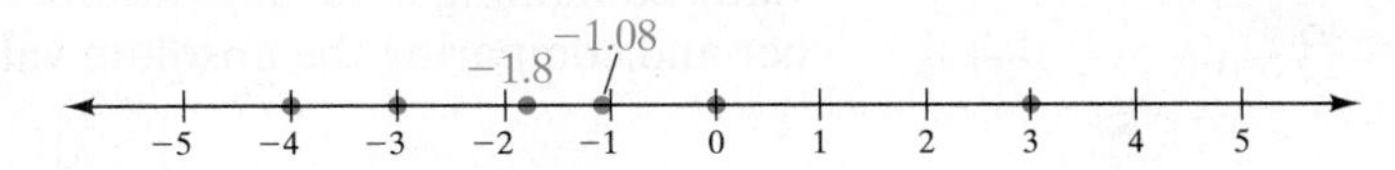

FIGURE 1.16

a) $-3 < 3$; notice that -3 is to the left of 3.
b) $-3 > -4$; notice that -3 is to the right of -4.
c) $-4 < 0$; notice that -4 is to the left of 0.
d) $-1.08 > -1.8$; notice that -1.08 is to the right of -1.8.

Now Try Exercise 43

2 Find the Absolute Value of a Number

The **absolute value** of a number can be considered the distance between the number and 0 on a number line. Thus, the absolute value of 3, written $|3|$, is 3 since it is 3 units from 0 on a number line. Similarly, the absolute value of -3, written $|-3|$, is also 3 since -3 is 3 units from 0. See **Figure 1.17**.

$$|3| = 3 \quad \text{and} \quad |-3| = 3$$

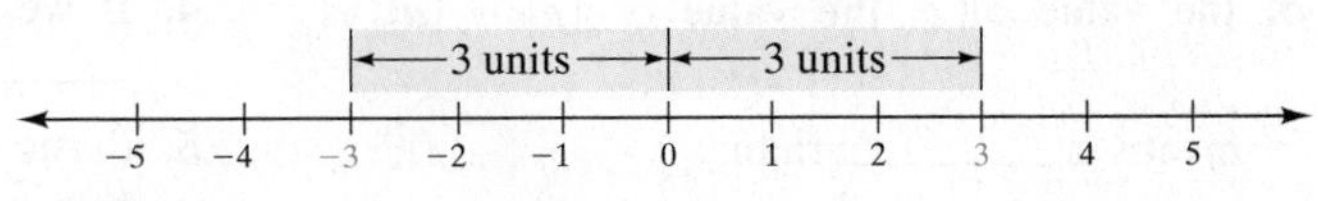

FIGURE 1.17

Since the absolute value of a number measures the distance (without regard to direction) of a number from 0 on the number line, *the absolute value of every number will be either positive or zero.*

Number	Absolute Value of Number
6	$\lvert 6 \rvert = 6$
-6	$\lvert -6 \rvert = 6$
0	$\lvert 0 \rvert = 0$
$-\frac{1}{2}$	$\left\lvert -\frac{1}{2} \right\rvert = \frac{1}{2}$

Understanding Algebra

The absolute value of a real number will always be either zero or positive.

The negative of the absolute value of a nonzero number will always be a negative number. For example,

$$-\lvert 2 \rvert = -(2) = -2 \quad \text{and} \quad -\lvert -3 \rvert = -(3) = -3$$

EXAMPLE 3 Insert either $>$, $<$, or $=$ in each shaded area to make a true statement.

a) $\lvert 3 \rvert \;\square\; 3$ **b)** $\lvert -2 \rvert \;\square\; \lvert 2 \rvert$ **c)** $-2 \;\square\; \lvert -4 \rvert$

d) $\lvert -2 \rvert \;\square\; -\lvert -4 \rvert$ **e)** $\left\lvert -\frac{2}{5} \right\rvert \;\square\; \lvert -0.42 \rvert$

Solution

a) $\lvert 3 \rvert = 3.$

b) $\lvert -2 \rvert = \lvert 2 \rvert$, since both $\lvert -2 \rvert$ and $\lvert 2 \rvert$ equal 2.

c) $-2 < \lvert -4 \rvert$, since $\lvert -4 \rvert = 4$.

d) First note that $\lvert -2 \rvert = 2$. Next, since $\lvert -4 \rvert = 4$, we have $-\lvert -4 \rvert = -4$. Therefore, $\lvert -2 \rvert > -\lvert -4 \rvert$ because $2 > -4$.

e) When an absolute value contains a fraction, we can compare it to an absolute value containing a decimal number by rewriting the fraction as a decimal number and comparing the absolute values of the decimal numbers.

$$\left\lvert -\frac{2}{5} \right\rvert = \lvert -0.40 \rvert = 0.40$$
$$\lvert -0.42 \rvert = 0.42$$

Therefore, $\left\lvert -\frac{2}{5} \right\rvert < \lvert -0.42 \rvert$ because $0.40 < 0.42$.

Now Try Exercise 67

The concept of absolute value is very important in higher-level mathematics courses.

1.5 Exercise Set MathXL® MyMathLab®

Warm-Up Exercises

Fill in the blanks with the appropriate word, phrase, or symbol(s) from the following list.

$\lvert -4 \rvert$	0	$\lvert 6 - (-4) \rvert$	distance	positive	True
greater	-4	less	$\lvert a \rvert$	negative	False

1. Regardless of the value of a, the value of $\lvert a \rvert - \lvert a \rvert$ is ________.
2. The symbol $<$ means is ________ than.
3. The absolute value of the number a is expressed as ________.
4. If we write $x > 0$, alternatively we could say that x is a ________ number.
5. (True or False) If a and b are real numbers and $a < b$, then $b > a$. ________.
6. The symbol $>$ means is ________ than.

7. The distance between 6 and −4 on the number line can be expressed as ________.

8. The distance the number −4 is from zero on the number line can be expressed as ________.

9. The negative of the absolute value of a nonzero number will always be a ________ number.

10. The absolute value of a number represents its ________ from 0 on a real number line.

Practice the Skills

Evaluate.

11. $|7|$ **12.** $|54|$ **13.** $|-15|$ **14.** $|-6|$

15. $|0|$ **16.** $-|0|$ **17.** $-|-5|$ **18.** $-|-34|$

19. $-|26|$ **20.** $-|92|$

Insert either < or > in each shaded area to make a true statement.

21. a) $21 \; \square \; 26$ **b)** $-21 \; \square \; -26$ **22. a)** $31 \; \square \; 29$ **b)** $-31 \; \square \; -29$

23. a) $71 \; \square \; 0$ **b)** $-71 \; \square \; 0$ **24. a)** $-37 \; \square \; 21$ **b)** $37 \; \square \; -21$

25. $\frac{2}{3} \; \square \; \frac{3}{4}$ **26.** $\frac{3}{4} \; \square \; \frac{5}{6}$ **27.** $-\frac{2}{3} \; \square \; -\frac{3}{4}$ **28.** $-\frac{3}{4} \; \square \; -\frac{5}{6}$

29. $\frac{1}{2} \; \square \; -\frac{2}{3}$ **30.** $-\frac{1}{2} \; \square \; \frac{2}{3}$ **31.** $0.1 \; \square \; 0.3$ **32.** $-0.1 \; \square \; -0.3$

33. $-2.1 \; \square \; -2$ **34.** $-1.83 \; \square \; -1.82$ **35.** $0.08 \; \square \; 0.1$ **36.** $-0.08 \; \square \; -0.1$

37. $4.09 \; \square \; 5.3$ **38.** $-4.09 \; \square \; -5.3$ **39.** $0.49 \; \square \; 0.43$ **40.** $-1.0 \; \square \; -0.7$

41. $-0.086 \; \square \; -0.095$ **42.** $0.086 \; \square \; 0.95$ **43.** $0.001 \; \square \; 0.002$ **44.** $-0.006 \; \square \; -0.007$

45. $\frac{5}{8} \; \square \; 0.6$ **46.** $2.7 \; \square \; \frac{10}{3}$ **47.** $-\frac{4}{3} \; \square \; -\frac{2}{3}$ **48.** $\frac{19}{2} \; \square \; \frac{17}{2}$

49. $-0.8 \; \square \; -\frac{3}{5}$ **50.** $-0.7 \; \square \; -0.2$ **51.** $0.3 \; \square \; \frac{1}{3}$ **52.** $\frac{9}{20} \; \square \; 0.42$

53. $-\frac{17}{30} \; \square \; -\frac{16}{20}$ **54.** $\frac{13}{15} \; \square \; \frac{8}{9}$ **55.** $-(-6) \; \square \; -(-5)$ **56.** $-\left(-\frac{12}{13}\right) \; \square \; \frac{7}{8}$

Insert either <, >, or = in each shaded area to make a true statement.

57. $5 \; \square \; |-2|$ **58.** $|-12| \; \square \; |-13|$ **59.** $\frac{3}{4} \; \square \; |-4|$ **60.** $|-4| \; \square \; -3$

61. $|0| \; \square \; |-4|$ **62.** $|-2.1| \; \square \; |-1.8|$ **63.** $4 \; \square \; \left|-\frac{9}{2}\right|$ **64.** $|-5| \; \square \; -|-6|$

65. $\left|-\frac{4}{5}\right| \; \square \; \left|-\frac{5}{4}\right|$ **66.** $\left|\frac{2}{5}\right| \; \square \; |-0.40|$ **67.** $|-4.6| \; \square \; \left|-\frac{23}{5}\right|$ **68.** $\left|-\frac{8}{3}\right| \; \square \; |-3.5|$

Insert either >, <, or = in each shaded area to make a true statement.

69. $\frac{2}{3} + \frac{2}{3} + \frac{2}{3} + \frac{2}{3} \; \square \; 4 \cdot \frac{2}{3}$ **70.** $\frac{3}{4} + \frac{3}{4} \; \square \; \frac{3}{4} \cdot \frac{3}{4}$ **71.** $\frac{1}{2} \cdot \frac{1}{2} \; \square \; \frac{1}{2} \div \frac{1}{2}$

72. $5 \div \frac{2}{3} \; \square \; \frac{2}{3} \div 5$ **73.** $\frac{7}{8} - \frac{1}{2} \; \square \; \frac{7}{8} \div \frac{1}{2}$ **74.** $3\frac{1}{5} + \frac{1}{3} \; \square \; 3\frac{1}{5} \cdot \frac{1}{3}$

Arrange the numbers from smallest to largest.

75. $0.46, \frac{4}{9}, |-5|, -|-1|, \frac{3}{7}$

76. $-\frac{3}{4}, -|0.6|, -\frac{5}{9}, -1.74, |-1.9|$

77. $\frac{2}{3}, 0.6, |-2.6|, \frac{19}{25}, \frac{5}{12}$

78. $-|-5|, |-9|, \left|-\frac{12}{5}\right|, 2.7, \frac{7}{12}$

Problem Solving

79. What numbers are 4 units from 0 on a number line?

80. What numbers are 100 units from 0 on the number line?

In Exercises 81–88, give three real numbers that satisfy all the stated criteria. If no real numbers satisfy the criteria, so state and explain why.

81. less than 4 and greater than 8

82. greater than 4 and less than 6

83. less than -2 and greater than -6

84. greater than -5 and greater than -9

85. greater than -3 and greater than 3

86. less than -3 and less than 3

87. greater than $|-2|$ and less than $|-6|$

88. greater than $|-3|$ and less than $|3|$

89. a) Consider the word *between*. What does this word mean?

b) List three real numbers between 4 and 6

c) Is the number 4 between the numbers 4 and 6? Explain.

d) Is the number 5 between the numbers 4 and 6? Explain.

e) Is it true or false that the real numbers between 4 and 6 are the real numbers that are both greater than 4 and less than 6? Explain.

90. Property Crime Rates The following line graph shows the U.S. property crime rates, in crimes per 100,000 people, for the years 1991 to 2010

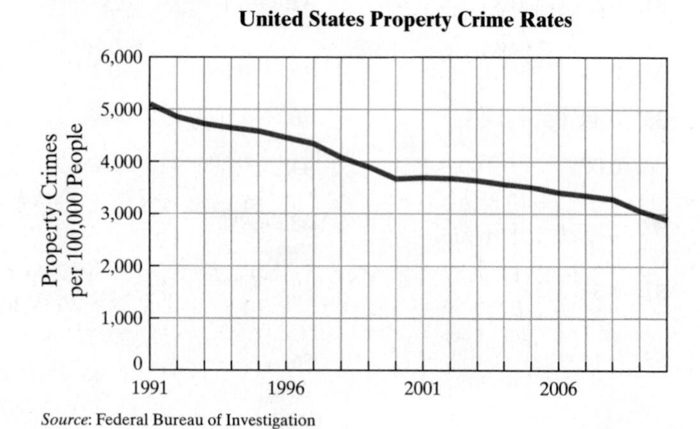

Source: Federal Bureau of Investigation

Estimate the year(s) when the property crime rate was

a) first less than 5000 property crimes per 100,000 people.

b) first less than 4000 property crimes per 100,000 people.

c) greater than 3000 but less than 4000 property crimes per 100,000 people.

91. Peanuts The following bar graph shows the percent of the recommended daily allowance of certain nutrients in 1 ounce of dry-roasted, salted peanuts. Use the bar graph to determine which nutrients provide

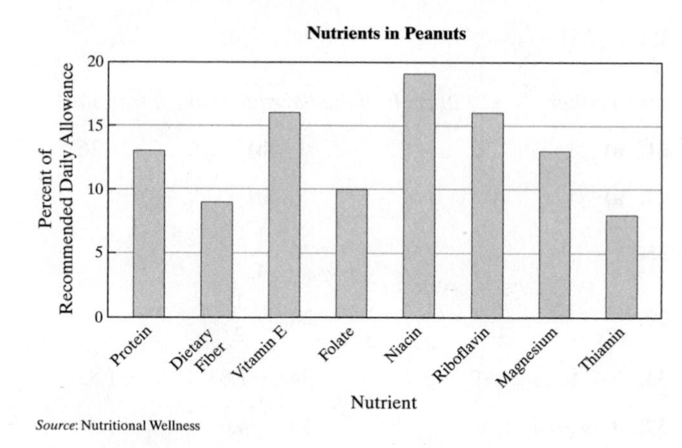

Source: Nutritional Wellness

a) less than 10% of the recommended daily allowance.

b) greater than 15% of the recommended daily allowance.

Concept/Writing Exercises

92. Are there any real numbers whose absolute value is not a positive number? Explain your answer.

93. Will $|a| - |a| = 0$ always be true for any real number a? Explain.

94. Suppose a and b represent any two real numbers. Suppose $a > b$ is true. Will $|a| > |b|$ also be true? Explain and give an example to support your answer.

95. Suppose a and b represent any two real numbers. Suppose $|a| > |b|$ is true. Will $a > b$ also be true? Explain and give an example to support your answer.

Challenge Problems

96. A number greater than 0 and less than 1 (or between 0 and 1) is multiplied by itself. Will the product be less than, equal to, or greater than the original number selected? Explain why this is always true.

97. A number between 0 and 1 is divided by itself. Will the quotient be less than, equal to, or greater than the original number selected? Explain why this is always true.

98. What two numbers can be substituted for x to make $|x| = 3$ a true statement?

99. Are there any values for x that would make $|x| = -5$ a true statement? Explain.

100. a) To what is $|x|$ equal if x represents a real number greater than or equal to 0?

b) To what is $|x|$ equal if x represents a real number less than 0?

c) Fill in the following shaded areas to make a true statement.

$$|x| = \begin{cases} \square, & x \geq 0 \\ \square, & x < 0 \end{cases}$$

Group Activity

Discuss and answer Exercise 101 as a group.

101. a) Group member 1: Draw a number line and mark points on the line to represent the following numbers.

$$|-2|, \quad -|3|, \quad -\left|\frac{1}{3}\right|$$

b) Group member 2: Do the same as in part **a)**, but on your number line mark points for the following numbers.

$$|-4|, \quad -|2|, \quad \left|-\frac{3}{5}\right|$$

c) Group member 3: Do the same as in parts **a)** and **b)**, but mark points for the following numbers.

$$|0|, \quad \left|\frac{16}{5}\right|, \quad -|-3|$$

d) As a group, construct one number line that contains all the points listed in parts **a), b),** and **c)**.

Cumulative Review Exercises

[1.3] **102.** Add $2\frac{3}{5} + 3\frac{1}{3}$.

[1.4] **103.** List the set of integers.

104. List the set of whole numbers.

105. Consider the following set of numbers.

$$\left\{5, -2, 0, \frac{1}{3}, \sqrt{3}, -\frac{5}{9}, 2.3, \pi\right\}$$

List the numbers that are

a) natural numbers.

b) whole numbers.

c) integers.

d) rational numbers.

e) irrational numbers.

f) real numbers.

MID-CHAPTER TEST: 1.1–1.5

To find out how well you understand the chapter material to this point, take this brief test. The answers, and the section where the material was initially discussed, are given in the back of the book. Review any questions that you answered incorrectly.

1. For each hour of class time, how many hours outside of class are recommended for studying and doing homework?

2. Phone Bills Jason Wisely's monthly phone bills for the first six months of 2013 were \$78.83, \$96.57, \$62.23, \$88.79, \$101.75, and \$55.62. For Jason's phone bills, determine the **a)** mean and **b)** median.

3. Checking Account The balance in Elizabeth Mater's checking account is \$652.70. She deposited \$230.75 and then purchased three books at \$19.62 each, including tax. If she pays by check, what is the new balance in her checking account?

4. Boat Rental The rental cost of a boat from Natwora's Boat Rental is \$7.50 per 15 minutes, and the rental cost from Gurney's Boat Rental is \$18 per half hour. Suppose you plan to rent a boat for 4 hours.

a) Which is the better deal?

b) How much will you save?

5. Water Rate The water rate in Livingston County is \$1.85 per 1000 gallons of water used. What is the water bill for a resident of Livingston County who uses 33,700 gallons of water?

Perform the indicated operation.

6. $\frac{3}{7} \cdot \frac{7}{18}$

7. $\frac{9}{16} \div \frac{15}{13}$

8. $\frac{5}{8} + \frac{3}{5}$

9. $6\frac{1}{4} - 3\frac{1}{5}$

10. Garden Aimee Calhoun wants to fence in her rectangular garden to keep out deer. Her garden measures $14\frac{2}{3}$ feet by $12\frac{1}{2}$ feet. How much fence will Aimee need?

In Exercises 11–15, indicate whether each statement is true or false.

11. 0 is a natural number.

12. −8.6 is a real number.

13. $3\frac{2}{3}$ is an irrational number.

14. Every integer is a real number.

15. Every whole number is a counting number.

16. Evaluate $-\left|-\frac{7}{10}\right|$.

Insert either <, >, or = in each shaded area to make a true statement.

17. $-0.005 \; \square \; -0.006$

18. $\frac{7}{8} \; \square \; \frac{5}{6}$

19. $|-9| \; \square \; |-19|$

20. $\left|-\frac{3}{8}\right| \; \square \; |-0.375|$

1.6 Addition of Real Numbers

1. Add real numbers using a number line.
2. Add fractions.
3. Identify opposites.
4. Add using absolute values.

There are many practical uses for negative numbers. A submarine diving below sea level, a bank account that has been overdrawn, a business spending more than it earns, and a temperature below zero are some examples.

The four basic **operations** of arithmetic are addition, subtraction, multiplication, and division. Both positive and negative numbers can be added, subtracted, multiplied, and divided. In this section, we discuss the operation of addition.

1 Add Real Numbers Using a Number Line

Understanding Algebra

With the exception of 0, *any number without a sign in front of it is positive.* For example, 3 means +3 and 5 means +5.

To add numbers, we make use of a number line. Represent the first number to be added (first *addend*) by an arrow starting at 0. The arrow is drawn to the right if the number is positive. If the number is negative, the arrow is drawn to the left. From the tip of the first arrow, draw a second arrow to represent the second addend. The second arrow is drawn to the right or left, as just explained. The sum of the two numbers is found at the tip of the second arrow.

EXAMPLE 1 Evaluate $3 + (-4)$ using a number line.

Solution *Always begin at 0.* Since the first addend, the 3, is positive, the first arrow starts at 0 and is drawn 3 units to the right (**Fig. 1.18**).

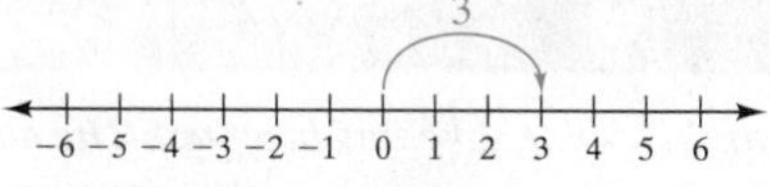

FIGURE 1.18

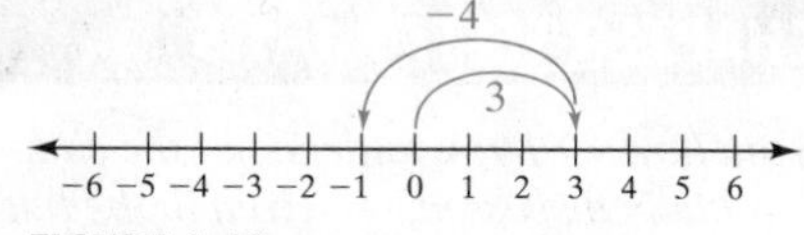

FIGURE 1.19

Understanding Algebra

When adding a positive number and a negative number, the result may be either positive or negative.

$3 + (-4) = -1$

$6 + (-1) = 5$

The second arrow starts at 3 and is drawn 4 units to the left, since the second addend is negative (**Fig. 1.19**). The tip of the second arrow is at -1. Thus,

$$3 + (-4) = -1$$

Now Try Exercise 27

EXAMPLE 2 Evaluate $-4 + 2$ using a number line.

Solution Begin at 0. Since the first addend is negative, -4, the first arrow is drawn 4 units to the left. From there, since 2 is positive, the second arrow is drawn 2 units to the right. The second arrow ends at -2 (**Fig. 1.20**).

FIGURE 1.20

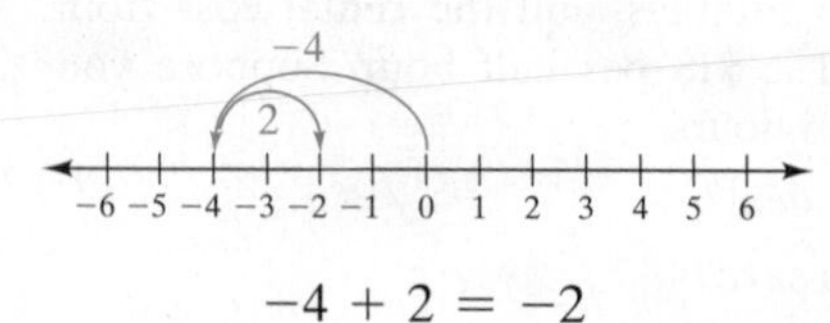

$$-4 + 2 = -2$$

Now Try Exercise 37

EXAMPLE 3 Evaluate $-3 + (-2)$ using a number line.

Solution Start at 0. Since both numbers being added are negative, both arrows will be drawn to the left (**Fig. 1.21**).

FIGURE 1.21

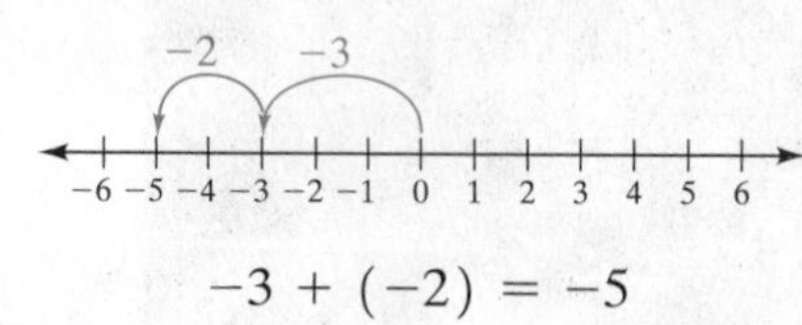

$$-3 + (-2) = -5$$

Now Try Exercise 39

In Example 3, we can think of the expression $-3 + (-2)$ as combining a *loss* of 3 and a *loss* of 2 for a total *loss* of 5, or -5.

EXAMPLE 4 Add $5 + (-5)$ using a number line.

Solution The first arrow starts at 0 and is drawn 5 units to the right. The second arrow starts at 5 and is drawn 5 units to the left. The tip of the second arrow is at 0. Thus, $5 + (-5) = 0$ (**Fig. 1.22**).

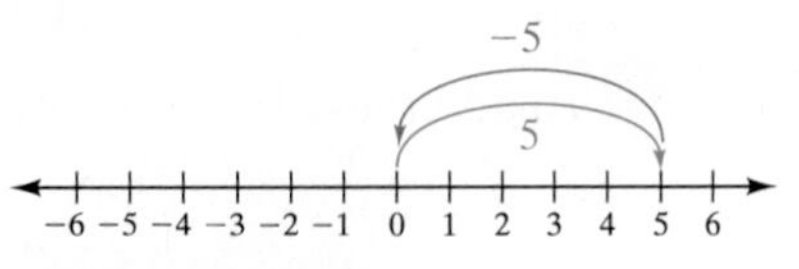

FIGURE 1.22

$$5 + (-5) = 0$$

Now Try Exercise 31

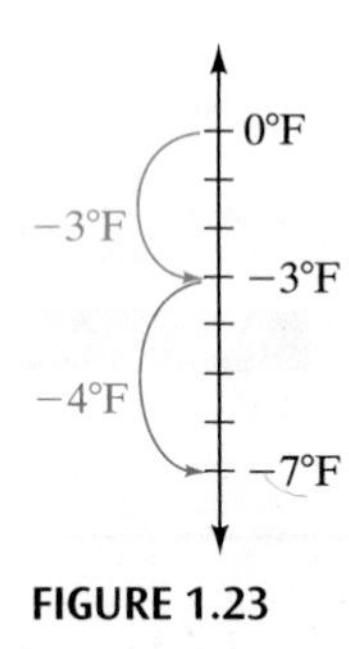

FIGURE 1.23

EXAMPLE 5 **Below-Zero Temperatures** At the beginning of the five o'clock news broadcast on a winter day in Rochester, Minnesota, the chief meteorologist, Alex Runde, reported that the temperature was three degrees below zero Fahrenheit. During the weather segment twenty minutes later, Alex stated that the temperature had dropped four degrees since the beginning of the news broadcast. Find the temperature at 5:20 P.M.

Solution A vertical number line (**Fig. 1.23**) may help you visualize this problem.

$$-3 + (-4) = -7°\text{F}$$

Now Try Exercise 117

2 Add Fractions

To add fractions, where one or more of the fractions is negative, we use the same general procedure discussed in Section 1.3. Whenever the denominators are not the same, we will find the least common denominator (LCD) and then we obtain the answer by adding the numerators while keeping the LCD.

For example, suppose after obtaining a common denominator, we have $-\frac{19}{29} + \frac{13}{29}$. To obtain the numerator of the answer, we may add $-19 + 13$ on the number line to obtain -6. The denominator of the answer is the common denominator, 29. Thus, the answer is $-\frac{6}{29}$.

$$-\frac{19}{29} + \frac{13}{29} = \frac{-19 + 13}{29} = \frac{-6}{29} = -\frac{6}{29}$$

Understanding Algebra

Notice that when you add two negative numbers, the result is negative: $(-3) + (-5) = -8$.

Let's look at one more example. Suppose after obtaining the LCD, we have $-\frac{3}{30} + \left(-\frac{5}{30}\right)$. We add $-3 + (-5)$ on the number line to obtain -8. The denominator of the answer is 30. Thus, the answer before being simplified is $-\frac{8}{30}$. The final answer simplifies to $-\frac{4}{15}$. We show the calculations as follows:

$$-\frac{3}{30} + \left(-\frac{5}{30}\right) = \frac{-3 + (-5)}{30} = \frac{-8}{30} = -\frac{4}{15}.$$

EXAMPLE 6 Add $\frac{7}{16} + \left(-\frac{2}{3}\right)$.

Solution The LCD is 48. Changing each fraction to a fraction with a denominator of 48 yields

$$\frac{7}{16} \cdot \frac{3}{3} + \left(-\frac{2}{3}\right) \cdot \frac{16}{16}$$

$$\text{or} \quad \frac{21}{48} + \left(-\frac{32}{48}\right)$$

To add these fractions, we keep the LCD and add the numerators to get

$$\frac{7}{16} + \left(-\frac{2}{3}\right) = \frac{21}{48} + \left(-\frac{32}{48}\right) = \frac{21 + (-32)}{48}$$

Now we add $21 + (-32)$ on a number line to get the numerator of the fraction, -11; see **Figure 1.24**.

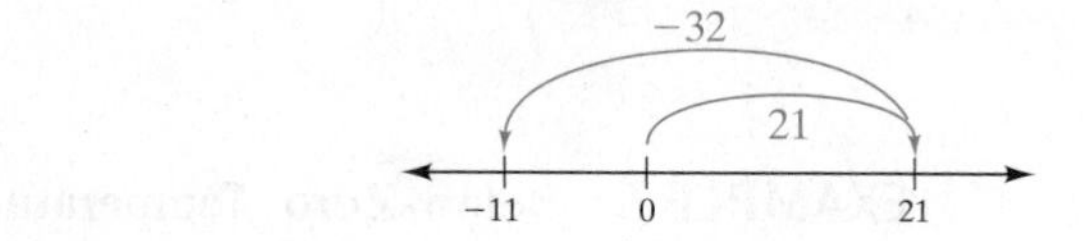

FIGURE 1.24

$$\text{Thus,} \quad \frac{7}{16} + \left(-\frac{2}{3}\right) = \frac{21}{48} + \left(-\frac{32}{48}\right) = \frac{21 + (-32)}{48} = -\frac{11}{48}.$$

Now Try Exercise 77

EXAMPLE 7 Add $-\frac{7}{8} + \left(-\frac{3}{40}\right)$.

Solution The LCD is 40. Rewriting the first fraction with the LCD gives the following.

$$-\frac{7}{8} + \left(-\frac{3}{40}\right) = -\frac{7}{8} \cdot \frac{5}{5} + \left(-\frac{3}{40}\right)$$

$$= -\frac{35}{40} + \left(-\frac{3}{40}\right) = \frac{-35 + (-3)}{40}$$

Now we add $-35 + (-3)$ to get the numerator of the fraction, -38; see **Figure 1.25**

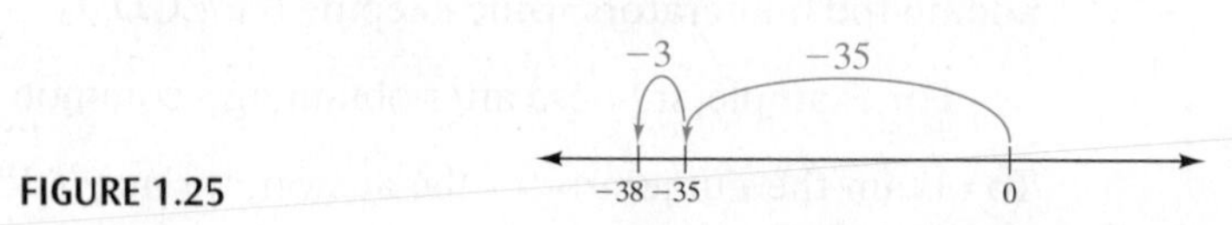

FIGURE 1.25

$$\text{Thus,} \quad -\frac{7}{8} + \left(-\frac{3}{40}\right) = \frac{-35}{40} + \left(\frac{-3}{40}\right) = \frac{-35 + (-3)}{40} = -\frac{38}{40} = -\frac{19}{20}.$$

Now Try Exercise 83

HELPFUL HINT

When the numbers being added are large, you are not expected to actually mark and count the units. For example, when determining $21 + (-32)$ you will not need to count 32 units to the left from 21 to obtain the answer -11.

In objective 4, we will show you how to obtain sums, such as $21 + (-32) = -11$ without having to draw number lines. We present adding on a number line here to help you understand the concept of addition of signed numbers, and to help you in determining, without doing any calculations, whether the sum of two signed numbers will be a positive number, a negative number, or zero.

3 Identify Opposites

Now let's consider **opposites,** or **additive inverses.**

Opposites (or Additive Inverses)

Any two numbers whose sum is zero are said to be **opposites,** or **additive inverses,** of each other. In general, if we let a represent any real number, then its opposite is $-a$ and $a + (-a) = 0$.

In Example 4, the sum of 5 and -5 is 0. Thus, -5 is the opposite of 5 and 5 is the opposite of -5.

EXAMPLE 8 Find the opposite of each number.

a) 3 **b)** -4 **c)** $-\frac{7}{8}$ **d)** 4.03

Solution

a) The opposite of 3 is -3, since $3 + (-3) = 0$.

b) The opposite of -4 is 4, since $-4 + 4 = 0$.

c) The opposite of $-\frac{7}{8}$ is $\frac{7}{8}$, since $-\frac{7}{8} + \frac{7}{8} = 0$.

d) The opposite of 4.03 is -4.03, since $4.03 + (-4.03) = 0$.

Now Try Exercise 15

Understanding Algebra

The *additive inverse* of a number has the property that the sum of the number and its additive inverse is 0. So, (-5) is the additive inverse, or *opposite,* of 5 because $5 + (-5) = 0$.

4 Add Using Absolute Values

Recall that the absolute value of a nonzero number will always be positive.

Adding Real Numbers with the Same Sign

To add real numbers with the same sign (either both positive or both negative), add their absolute values. The sum has the same sign as the numbers being added.

EXAMPLE 9 Add $4 + 8$.

Solution Since both numbers have the same sign, both positive, we add their absolute values: $|4| + |8| = 4 + 8 = 12$. Since both numbers being added are positive, the sum is positive. Thus, $4 + 8 = 12$.

Now Try Exercise 49

EXAMPLE 10 Add $-6 + (-9)$.

Solution Since both numbers have the same sign, both negative, we add their absolute values: $|-6| + |-9| = 6 + 9 = 15$. Since both numbers being added are negative, their sum is negative. Thus, $-6 + (-9) = -15$.

Now Try Exercise 51

Understanding Algebra

The sum of two positive numbers will always be positive and the sum of two negative numbers will always be negative.

Adding Two Signed Numbers with Different Signs

To add two signed numbers with different signs, one positive and the other negative, subtract the smaller absolute value from the larger absolute value. The answer has the sign of the number with the larger absolute value.

EXAMPLE 11 Add $13 + (-4)$.

Solution The two numbers being added have different signs, so we subtract the smaller absolute value from the larger: $|13| - |-4| = 13 - 4 = 9$. Since $|13|$ is greater than $|-4|$ and the sign of 13 is positive, the sum is positive. Thus, $13 + (-4) = 9$.

Now Try Exercise 53

EXAMPLE 12 Add $9 + (-14)$.

Solution The numbers being added have different signs, so we subtract the smaller absolute value from the larger: $|-14| - |9| = 14 - 9 = 5$. Since $|-14|$ is greater than $|9|$ and the sign of -14 is negative, the sum is negative. Thus, $9 + (-14) = -5$.

Now Try Exercise 55

EXAMPLE 13 Add $-35 + 15$.

Solution The two numbers being added have different signs, so we subtract the smaller absolute value from the larger: $|-35| - |15| = 35 - 15 = 20$. Since $|-35|$ is greater than $|15|$ and the sign of -35 is negative, the sum is negative. Thus, $-35 + 15 = -20$.

Now Try Exercise 61

Now let's look at some additional examples that contain fractions and decimal numbers.

EXAMPLE 14 Add $-\frac{3}{5} + \frac{4}{7}$.

Solution We can write each fraction with the least common denominator, 35.

$$-\frac{3}{5} + \frac{4}{7} = -\frac{3}{5} \cdot \frac{7}{7} + \frac{4}{7} \cdot \frac{5}{5}$$

$$= \frac{-21}{35} + \frac{20}{35} = \frac{-21 + 20}{35}$$

Since $|-21|$ is greater than $|20|$, the final answer will be negative. Thus, we can write

$$\frac{-21}{35} + \frac{20}{35} = \frac{-21 + 20}{35} = \frac{-1}{35} = -\frac{1}{35}$$

Thus, $-\frac{3}{5} + \frac{4}{7} = -\frac{1}{35}$.

Now Try Exercise 87

Examples 15 and 16 contain decimal numbers. If you have forgotten how to perform the basic operations of addition, subtraction, multiplication, and division of decimal numbers, read Appendix A now.

EXAMPLE 15 Add $-37.451 + (-26.98)$.

Solution Since both numbers have the same sign, both negative, we add their absolute values: $|-37.451| + |-26.98| = 37.451 + 26.98$. Also recall that $26.98 = 26.980$

$$\begin{array}{r} 37.451 \\ +\ 26.980 \\ \hline 64.431 \end{array}$$

Since two negative numbers are being added, the sum is negative. Therefore, $-37.451 + (-26.98) = -64.431$.

Now Try Exercise 67

EXAMPLE 16 Net Profit or Loss The DuLond Printing Company had a loss of \$4005.69 for the first 6 months of the year and a profit of \$29,645.78 for the second 6 months of the year. Find the net profit or loss for the year.

Solution Understand and Translate This problem can be represented as $-4005.69 + 29{,}645.78$. Since the numbers have different signs, we subtract the smaller absolute value from the larger.

Carry Out $|29{,}645.78| - |-4005.69| = 29{,}645.78 - 4005.69$

$$\begin{array}{r} 29{,}645.78 \\ -\ \ 4005.69 \\ \hline 25{,}640.09 \end{array}$$

Since $|29{,}645.78|$ is greater than $|-4005.69|$ and the sign of 29,645.78 is positive, the sum is positive. Thus, $-4005.69 + 29{,}645.78 = 25{,}640.09$.

Check and Answer The answer is reasonable. Thus, the net profit for the year was \$25,640.09.

Now Try Exercise 121

Understanding Algebra

The sum of two signed numbers with different signs may be either positive or negative. The sign of the sum will be the same as the sign of the number with the larger absolute value.

$13 + (-45) = -32$

$(-17) + 20 = 3$

HELPFUL HINT

Architects often make a scale model of a building before starting construction of the building. This model helps them visualize the project and often helps them avoid problems.

Mathematicians also construct models. A *mathematical model* may be a physical representation of a mathematical concept. It may be as simple as using tiles or chips to represent specific numbers. For example, below we use a model to help explain addition of real numbers. This may help some of you understand the concepts better.

We let a red chip represent $+1$ and a green chip represent -1.

● = +1 ● = −1

If we add $+1$ and -1, or a red and a green chip, we get 0.
Now consider the addition problem $3 + (-5)$. We can represent this as

●●● + ●●●●●
3 −5

If we remove 3 red chips and 3 green chips, or three zeros, we are left with 2 green chips, which represents a sum of -2. Thus, $3 + (-5) = -2$,

●●● + ●●●●●

Now consider the problem $-4 + (-2)$. We can represent this as

●●●● + ●●
−4 −2

Since we end up with 6 green chips, and each green chip represents -1, the sum is -6. Therefore, $-4 + (-2) = -6$.

1.6 Exercise Set MathXL® MyMathLab®

Warm-Up Exercises

Fill in the blanks with the appropriate word, phrase, or symbol(s) from the following list.

opposites	inverse	denominator	addends	sum	positive
absolute	-8	numerator	negative	8	

1. The sum of two negative numbers is always ________.
2. Another name for the opposite of a real number is the additive ________.
3. The expression $|x|$ is read "the ________ value of x."
4. The sum of two positive numbers is always ________.
5. In the statement $(-8) + 5 = -3$, the number -3 is called the ________ of -8 and 5.
6. In the statement $(-8) + 5 = -3$, the numbers -8 and 5 are called ________.

7. $-|-8| =$ ________.

8. $|-8| =$ ________.

9. When adding two fractions with different signs, we first find the least common ________.

10. Two numbers that add up to zero are ________ of each other.

In Exercises 11 and 12, are the calculations shown correct? If not, explain why not.

11. $\frac{-5}{12} + \frac{9}{12} = \frac{-5+9}{12} = \frac{4}{12} = \frac{1}{3}$

12. $\frac{-6}{70} + \left(\frac{-9}{70}\right) = \frac{-6+(-9)}{70} = \frac{-15}{70} = -\frac{3}{14}$

Practice the Skills

Write the opposite of each number.

13. 19 14. 8 15. −28 16. 3

17. 0 18. $-3\frac{1}{2}$ 19. $\frac{5}{3}$ 20. $-\frac{1}{4}$

21. $2\frac{3}{5}$ 22. −1 23. 3.72 24. −0.721

Add.

25. $5 + 16$ 26. $17 + 13$ 27. $4 + (-3)$ 28. $9 + (-12)$

29. $-4 + (-2)$ 30. $-3 + (-5)$ 31. $6 + (-6)$ 32. $-8 + 8$

33. $-4 + 4$ 34. $11 + (-11)$ 35. $-8 + (-2)$ 36. $6 + (-5)$

37. $-7 + 3$ 38. $-6 + 9$ 39. $-8 + (-5)$ 40. $-9 + 13$

41. $0 + 0$ 42. $0 + (-0)$ 43. $-8 + 0$ 44. $0 + (-3)$

45. $-18 + (-9)$ 46. $-7 + 17$ 47. $-33 + (-31)$ 48. $-27 + (-9)$

49. $7 + 9$ 50. $12 + 3$ 51. $-8 + (-4)$ 52. $-25 + (-36)$

53. $6 + (-3)$ 54. $52 + (-25)$ 55. $13 + (-19)$ 56. $34 + (-40)$

57. $180 + (-220)$ 58. $-452 + 312$ 59. $-11 + (-20)$ 60. $-33 + (-92)$

61. $-67 + 28$ 62. $183 + (-183)$ 63. $184 + (-93)$ 64. $-19 + 176$

65. $80.5 + (-90.4)$ 66. $-24.6 + (-13.9)$ 67. $-124.7 + (-19.3)$ 68. $106.3 + (-110.9)$

69. $-12.4 + 16.62$ 70. $13.01 + (-5.1)$ 71. $-97.35 + (-9.8)$ 72. $-73.5 + (-58.68)$

Add.

73. $\frac{3}{5} + \frac{1}{7}$ 74. $\frac{5}{8} + \frac{3}{5}$ 75. $\frac{5}{12} + \frac{6}{7}$ 76. $\frac{2}{9} + \frac{3}{10}$

77. $-\frac{8}{11} + \frac{4}{5}$ 78. $-\frac{4}{9} + \frac{5}{27}$ 79. $-\frac{7}{10} + \frac{11}{90}$ 80. $\frac{8}{9} + \left(-\frac{1}{3}\right)$

81. $-\frac{7}{30} + \left(-\frac{4}{5}\right)$ 82. $-\frac{7}{9} + \left(-\frac{1}{5}\right)$ 83. $-\frac{4}{5} + \left(-\frac{5}{75}\right)$ 84. $-\frac{1}{15} + \left(-\frac{5}{6}\right)$

85. $\frac{9}{25} + \left(-\frac{3}{50}\right)$ 86. $\frac{5}{36} + \left(-\frac{5}{24}\right)$ 87. $-\frac{9}{24} + \frac{5}{7}$ 88. $-\frac{9}{40} + \frac{4}{15}$

89. $-\frac{5}{12} + \left(-\frac{3}{10}\right)$ 90. $\frac{7}{16} + \left(-\frac{5}{24}\right)$ 91. $-\frac{13}{14} + \left(-\frac{7}{42}\right)$ 92. $-\frac{11}{27} + \left(-\frac{7}{18}\right)$

In Exercises 93–108, **a)** *determine by observation whether the sum will be a positive number, zero, or a negative number;* **b)** *find the sum.*

93. $587 + (-197)$ 94. $-140 + (-629)$ 95. $-84 + (-289)$ 96. $-647 + 352$

97. $-947 + 495$ 98. $762 + (-762)$ 99. $-496 + (-804)$ 100. $-354 + 1090$

101. $-375 + 263$

102. $1127 + (-84)$

103. $-1833 + (-2047)$

104. $-426 + 572$

105. $3124 + (-2013)$

106. $-9095 + (-647)$

107. $-1025 + (-1025)$

108. $7513 + (-4361)$

Indicate whether each statement is true or false.

109. The sum of two negative numbers is always a negative number.

110. The sum of a negative number and a positive number is sometimes a negative number.

111. The sum of two positive numbers is never a negative number.

112. The sum of a positive number and a negative number is always a negative number.

113. The sum of a positive number and a negative number is always a positive number.

114. The sum of a number and its opposite is always equal to zero.

Problem Solving

Write an expression that can be used to solve each problem and then solve.

115. Credit Card Clark Brown owed \$94 on his credit card. He charged another item costing \$183. Find the amount that Clark owed.

116. Charge Card Dianne Bolen charged \$142 worth of goods on her charge card. Find her balance after she made a payment of \$87.

117. Football A football team lost 18 yards on one play and then lost 3 yards on the following play. What was the total loss in yardage?

118. Overdrawn Checking Account Julie Chu is unaware that her checking account has been overdrawn by \$56. While shopping, she writes a check for \$162. Find the total amount by which Julie has overdrawn her account.

119. Drilling for Water A company is drilling a well. During the first week they drilled 27 feet, and during the second week they drilled another 34 feet before they struck water. How deep is the well?

120. Coffee Bar The Frenches opened a coffee bar. Their income and expenses for their first three months of operation are shown in the graph to the right above.

a) Find the net profit or loss (the sum of income and expenses) for the first month.

b) Find the net profit or loss for the second month.

c) Find the net profit or loss for the third month.

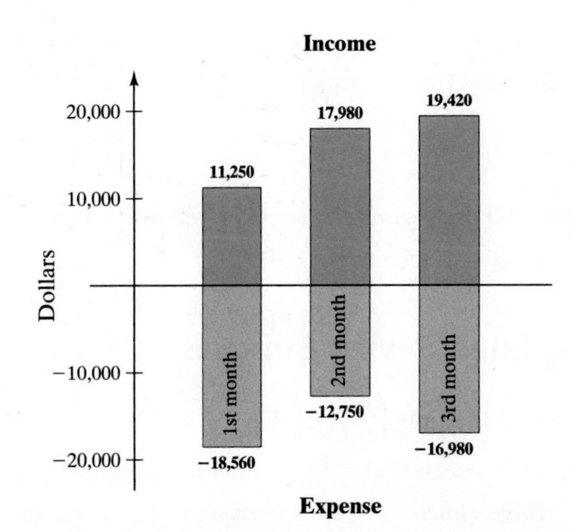

121. High Mountain *Guinness World Records* lists Mauna Kea in Hawaii as the tallest mountain in the world when measured from its base to its peak. The base of Mauna Kea is 19,684 feet below sea level. The total height of the mountain from its base to its peak is 33,480 feet. How high is the peak of Mauna Kea above sea level?

122. Net Profit or Loss The Crafty Scrapbook Company had a loss of \$3000 for the first 4 months of the year and a profit of \$37,400 for the last 8 months of the year. Find the net profit or loss for the year.

123. Surplus and Deficit The graph below shows the surplus or deficit for GKF Publishing Company for the years 2001 through 2013.

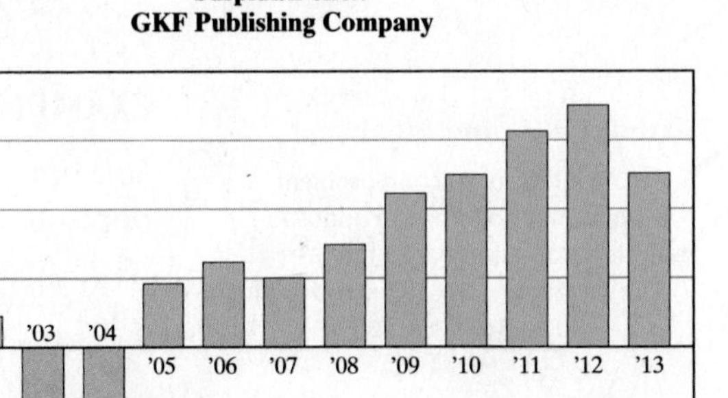

a) Estimate the surplus or deficit for GKF Publishing in 2003.

b) Estimate the surplus or deficit for GKF Publishing for each of the years 2011, 2012, and 2013. Then estimate the total surplus or deficit from 2011 through 2013 by adding these three estimates.

124. Stocks The chart on the right shows percent changes from the first quarter of 2013 through the first quarter of 2014 for the Mutual Canterbury Fund.

Determine the percent change for the fund from the first quarter of 2013 through the first quarter of 2014 by adding the individual percents.

Canterbury Fund	Percent change from previous quarter
1st quarter 2013	4.2%
2nd quarter 2013	5.2%
3rd quarter 2013	0.2%
4th quarter 2013	−13.5%
1st quarter 2014	−3.0%

Challenge Problems

Evaluate each exercise by adding the numbers from left to right. We will discuss problems like this shortly.

125. $(-8) + (-6) + (-12)$

126. $5 + (-7) + (-8)$

127. $29 + (-46) + 37$

128. $4 + (-5) + 6 + (-8)$

129. $(-12) + (-10) + 25 + (-3)$

130. $(-4) + (-2) + (-15) + (-27)$

131. $\frac{1}{2} + \left(-\frac{1}{3}\right) + \frac{1}{5}$

132. $-\frac{3}{8} + \left(-\frac{2}{9}\right) + \left(-\frac{1}{2}\right)$

Find the following sums. Explain how you determined your answer. (Hint: Pair small numbers with large numbers from the ends inward.)

133. $1 + 2 + 3 + \cdots + 10$

134. $1 + 2 + 3 + \cdots + 20$

Cumulative Review Exercises

[1.3] **135.** Multiply $\left(\frac{4}{7}\right)\left(2\frac{3}{8}\right)$.

136. Subtract $3 - \frac{5}{16}$.

[1.4] **137.** True or False: Every number less than zero is a negative integer.

[1.5] *Insert either* $<$, $>$, *or* $=$ *in each shaded area to make a* true statement.

138. $|-3|$ ▭ 2

139. 8 ▭ $|-12|$

1.7 Subtraction of Real Numbers

1. Subtract numbers.
2. Subtract numbers mentally.
3. Evaluate expressions containing more than two numbers.

1 Subtract Numbers

Any subtraction problem can be rewritten as an addition problem using the additive inverse.

To Subtract Real Numbers

In general, if a and b represent any two real numbers, then

$$a - b = a + (-b)$$

Understanding Algebra

We convert a subtraction problem to addition by adding the opposite of the second number to the first number. So, $6 - (-10)$ is rewritten as $6 + 10 = 16$.

EXAMPLE 1 Evaluate $9 - (+4)$.

Solution We are subtracting a positive 4 from 9. To accomplish this, we add the opposite of +4, which is −4, to 9.

$$9 - (+4) = 9 + (-4) = 5$$

(Labels: 9 − Subtract; (+4) Positive 4; + Add; (−4) Negative 4)

We evaluated $9 + (-4)$ using the procedures for *adding* real numbers presented in Section 1.6.

Now Try Exercise 13

Understanding Algebra

The parts of a subtraction problem:

$$\begin{array}{rl} 16 \leftarrow & \text{minuend} \\ -\ \underline{12} \leftarrow & \text{subtrahend} \\ 4 \leftarrow & \text{difference} \end{array}$$

Often in a subtraction problem, when the number being subtracted is a positive number, the + sign preceding the number being subtracted is not shown. For example, in the subtraction $9 - 4$,

$$9 - 4 \text{ means } 9 - (+4)$$

Thus, to evaluate $9 - 4$, we must add the opposite of 4, which is -4, to 9.

$$9 - 4 = 9 + (-4) = 5$$

Subtract Positive 4 Add Negative 4

This procedure is illustrated in Example 2.

EXAMPLE 2 Evaluate $5 - 3$.

Solution We must subtract a positive 3 from 5. To change this problem to an addition problem, we add the opposite of 3, which is -3, to 5.

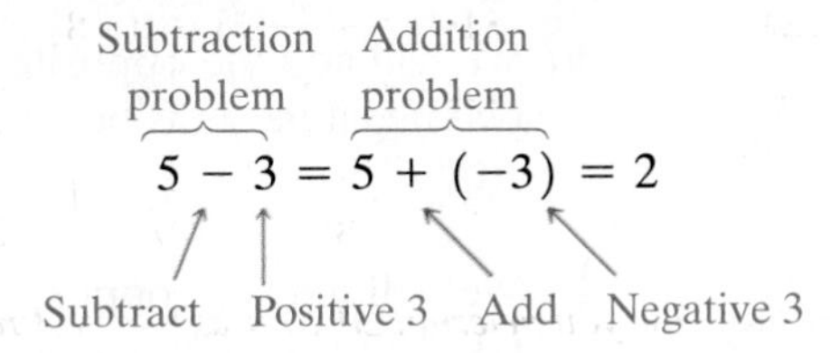

Now Try Exercise 15

EXAMPLE 3 Evaluate.

a) $3 - 10$ **b)** $-6 - 4$

Solution

a) Add the opposite of 10, which is -10, to 3.

$$3 - 10 = 3 + (-10) = -7$$

b) Add the opposite of 4, which is -4, to -6.

$$-6 - 4 = -6 + (-4) = -10$$

Now Try Exercise 17

In Example 4, we subtract numbers that contain decimal points.

EXAMPLE 4 Evaluate $16.32 - 18.75$.

Solution Add the opposite of 18.75, which is -18.75, to 16.32.

$$16.32 - 18.75 = 16.32 + (-18.75) = -2.43$$

Now Try Exercise 53

In Examples 5 and 6, we will show how to subtract a negative number.

EXAMPLE 5 Evaluate $5 - (-3)$.

Solution We are asked to subtract a negative 3 from 5. To do this, we add the opposite of -3, which is 3, to 5.

$$5 - (-3) = 5 + 3 = 8$$

Subtract Negative 3 Add Positive 3

Now Try Exercise 19

Compare the result of Example 5 to the result of Example 2. It is important to see that $5 - 3$ is a different calculation than $5 - (-3)$.

Understanding Algebra

Whenever we subtract a negative number, we always replace the two negative signs with a plus sign.

$$6 - (-9) = 6 + 9$$

HELPFUL HINT

By examining Example 5, we see that

$$5 - (-3) = 5 + 3$$

Two negative signs together → Plus

EXAMPLE 6 Evaluate.

a) $3 - (-10)$ **b)** $-6 - (-4)$

Solution

a) We will add the opposite of -10, which is 10, to 3. Notice that this will result in two negative signs being replaced by a plus sign.

$$3 - (-10) = 3 + 10 = 13$$

b) We will add the opposite of -4, which is 4, to -6. Again, notice that this will result in two negative signs being replaced by a plus sign.

$$-6 - (-4) = -6 + 4 = -2$$

Now Try Exercise 29

Compare the results of Example 6 to those of Example 3. It is important to see the difference in the calculations. Whenever we subtract a negative number, we always replace the two negative signs with a plus sign.

HELPFUL HINT

We will now indicate how we may illustrate subtraction using colored chips. Remember from the preceding section that a red chip represents +1 and a green chip −1.

● = +1 ● = −1

Consider the subtraction problem $2 - 5$. If we change this to an addition problem, we get $2 + (-5)$. We can then add, as was done in the preceding section. The figure below shows that $2 + (-5) = -3$.

●● + ●●●●●

Now consider $-2 - 5$. This means $-2 + (-5)$, which can be represented as follows:

●● + ●●●●●

Thus, $-2 - 5 = -7$.

Now consider the problem $-3 - (-5)$. This can be rewritten as $-3 + 5$, which can be represented as follows:

●●● + ●●●●●

Thus, $-3 - (-5) = 2$.

Some students still have difficulty understanding why when you subtract a negative number you obtain a positive number. Let's look at the problem $3 - (-2)$. This time we will look at it from a slightly different point of view. Let's start with 3:

From this we wish to subtract a negative 2. To the +3 shown above we will add two zeros by adding two $+1 - 1$ combinations. Remember, +1 and −1 sum to 0.

●●● + ●● + ●●

+3 0 0

Now we can subtract or "take away" the two -1's as shown:

●●● + ●~~●~~ + ●~~●~~

From this we see that we are left with $3 + 2$ or 5. Thus, $3 - (-2) = 5$.

EXAMPLE 7 Subtract 12 from 3.

Solution

$$3 - 12 = 3 + (-12) = -9$$

Now Try Exercise 57

HELPFUL HINT

Example 7 asked us to "subtract 12 from 3." The correct way of writing this is $3 - 12$. Notice that the number following the word "from" is our starting point. That is where the calculation begins. For example:

Subtract 2 from 7 means $7 - 2$.	From 7, subtract 2 means $7 - 2$.
Subtract 5 from -1 means $-1 - 5$.	From -1, Subtract 5 means $-1 - 5$.
Subtract -4 from -2 means $-2 - (-4)$.	From -2, Subtract -4 means $-2 - (-4)$.
Subtract -3 from 6 means $6 - (-3)$.	From 6, Subtract -3 means $6 - (-3)$.
Subtract a from b means $b - a$.	From b, Subtract a means $b - a$.

EXAMPLE 8 Subtract 5 from 5.

Solution

$$5 - 5 = 5 + (-5) = 0$$

Now Try Exercise 59

EXAMPLE 9 Subtract -6.481 from 4.25.

Solution

$$4.25 - (-6.481) = 4.250 + 6.481 = 10.731$$

Now Try Exercise 67

Now we will perform subtraction problems that contain fractions.

EXAMPLE 10 Subtract $\frac{5}{9} - \frac{13}{15}$.

Solution Begin by changing the subtraction problem to an addition problem.

$$\frac{5}{9} - \frac{13}{15} = \frac{5}{9} + \left(-\frac{13}{15}\right)$$

Now rewrite the fractions with the LCD, 45, and add the fractions as was done in the last section.

$$\frac{5}{9} + \left(-\frac{13}{15}\right) = \frac{5}{9} \cdot \frac{5}{5} + \left(-\frac{13}{15}\right) \cdot \frac{3}{3}$$

$$= \frac{25}{45} + \left(-\frac{39}{45}\right) = \frac{25 + (-39)}{45} = \frac{-14}{45} = -\frac{14}{45}$$

Thus, $\frac{5}{9} - \frac{13}{15} = -\frac{14}{45}$.

Now Try Exercise 85

EXAMPLE 11 Subtract $-\frac{7}{18}$ from $-\frac{9}{15}$.

Solution This problem is written $-\frac{9}{15} - \left(-\frac{7}{18}\right)$.

We can simplify this as follows.

$$-\frac{9}{15} - \left(-\frac{7}{18}\right) = -\frac{9}{15} + \frac{7}{18}.$$

The LCD of 15 and 18 is 90. Rewriting the fractions with a common denominator gives

$$-\frac{9}{15}\cdot\frac{6}{6} + \frac{7}{18}\cdot\frac{5}{5} = -\frac{54}{90} + \frac{35}{90} = \frac{-54+35}{90}$$

$$= \frac{-19}{90} = -\frac{19}{90}.$$

Now Try Exercise 87

Let us now look at some applications that involve subtraction.

EXAMPLE 12 **Distances Between Cities** When driving north on US Highway 41, Connie Buller drives through the cities of Venice, Sarasota, and Bradenton, FL in that order (see **Figure 1.26**). Venice and Sarasota are 18.6 miles apart. Venice and Bradenton are 31.7 miles apart. How far apart are Sarasota and Bradenton? Assume all three cities are in a straight line.

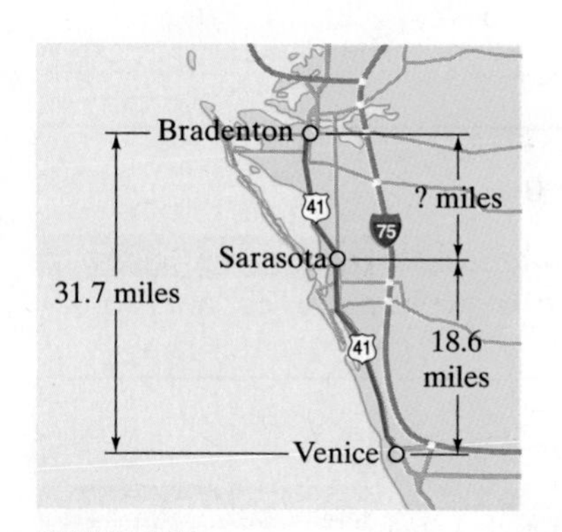

FIGURE 1.26

Solution Understand and Translate By looking at **Figure 1.26**, we can see that the distance from Sarasota to Bradenton can be found by taking the difference between the distance from Venice to Bradenton and the distance from Venice to Sarasota. Here the word *difference* indicates subtraction.

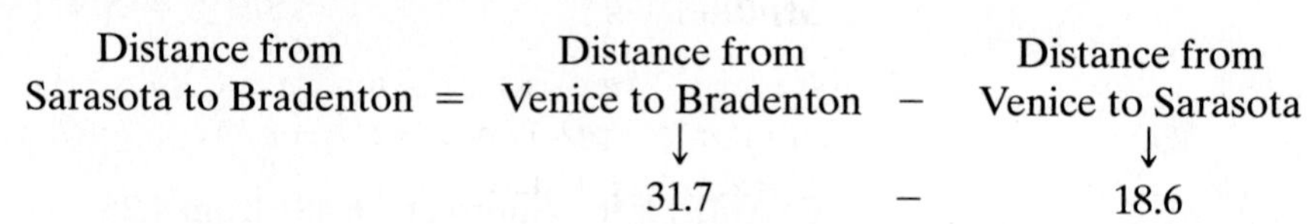

Carry Out $31.7 - 18.6 = 31.7 + (-18.6) = 13.1$

Check and Answer The distance from Sarasota to Bradenton is 13.1 miles.

Now Try Exercise 129

EXAMPLE 13 **Temperature Difference** On January 9, 2011, the high temperature for the day in McAllen, Texas, was 81°F. On the same day, the low temperature in Babbitt, Minnesota, was −29°F. Find the difference in their temperatures.

Solution Understand and Translate The word *difference* in the example title indicates subtraction. We can obtain the difference in their temperatures by subtracting as follows.

Carry Out $81 - (-29) = 81 + 29 = 110$

Check and Answer Therefore, the high temperature in McAllen is 110°F greater than the low temperature in Babbitt.

Now Try Exercise 135

EXAMPLE 14 **Measuring Snow** A kindergarten class in Richfield, Minnesota, has a snow gauge placed outside its window that is left untouched for two days.

Suppose that on the first day $6\frac{3}{8}$ inches of snow falls. On the second day, no snow falls, but $1\frac{1}{2}$ inches of the first day's snowfall melts. How much snow remains after the second day?

Solution Understand and Translate From the first amount, $6\frac{3}{8}$ inches, we must subtract $1\frac{1}{2}$ inches.

Carry Out We begin by changing the subtraction problem to an addition problem. We then change the mixed numbers to fractions, and then rewrite each fraction with the LCD, 8.

$$\begin{aligned} 6\frac{3}{8} - 1\frac{1}{2} &= 6\frac{3}{8} + \left(-1\frac{1}{2}\right) \\ &= \frac{51}{8} + \left(-\frac{3}{2}\right) \\ &= \frac{51}{8} + \left(-\frac{3}{2}\right) \cdot \frac{4}{4} \\ &= \frac{51}{8} + \left(-\frac{12}{8}\right) \\ &= \frac{51 + (-12)}{8} = \frac{39}{8} \quad \text{or} \quad 4\frac{7}{8} \end{aligned}$$

Check and Answer Thus, after the second day there were $4\frac{7}{8}$ inches of snow remaining. Based upon the numbers given in the problem, the answer seems reasonable.

Now Try Exercise 133

EXAMPLE 15 Evaluate.

a) $15 + (-4)$ **b)** $-16 - 3$ **c)** $19 + (-14)$
d) $7 - (-9)$ **e)** $-9 - (-3)$ **f)** $8 - 13$

Solution Parts **a)** and **c)** are addition problems, whereas the other parts are subtraction problems. We can rewrite each subtraction problem as an addition problem to evaluate.

a) $15 + (-4) = 11$
b) $-16 - 3 = -16 + (-3) = -19$
c) $19 + (-14) = 5$
d) $7 - (-9) = 7 + 9 = 16$
e) $-9 - (-3) = -9 + 3 = -6$
f) $8 - 13 = 8 + (-13) = -5$

Now Try Exercise 31

2 Subtract Numbers Mentally

In the previous examples, we rewrote subtraction problems as addition problems. We did this because we know how to add real numbers. After this chapter, when we work out a subtraction problem, we will not show this step. *You need to practice and thoroughly understand how to add and subtract real numbers. When asked to evaluate an expression like* $-4 - 6$, *you need to be able to compute the answer mentally.*

Let's evaluate a few subtraction problems without showing the process of changing the subtraction to addition.

EXAMPLE 16 Evaluate.

a) $-7 - 5$ **b)** $4 - 12$ **c)** $18 - 25$ **d)** $-20 - 12$

Solution

a) $-7 - 5 = -12$ **b)** $4 - 12 = -8$ **c)** $18 - 25 = -7$ **d)** $-20 - 12 = -32$

Now Try Exercise 33

In Example 16 **a)**, we may have reasoned that $-7 - 5$ meant $-7 + (-5)$, which is -12, but we did not need to show it.

EXAMPLE 17 Evaluate $-\frac{3}{5} - \frac{7}{8}$.

Solution Write each fraction with the LCD, 40.

$$-\frac{3}{5} \cdot \frac{8}{8} - \frac{7}{8} \cdot \frac{5}{5} = -\frac{24}{40} - \frac{35}{40} = \frac{-24 - 35}{40} = -\frac{59}{40} = -1\frac{19}{40}$$

Now Try Exercise 77

Notice in Example 17, when we had $-\frac{24}{40} - \frac{35}{40}$, we could have written $\frac{-24 + (-35)}{40}$, but at this time we elected to write it as $\frac{-24 - 35}{40}$. Since $-24 - 35$ is -59, the answer is $-\frac{59}{40}$ or $-1\frac{19}{40}$.

3 Evaluate Expressions Containing More Than Two Numbers

In evaluating expressions involving more than one addition and subtraction, always work from left to right unless parentheses or other grouping symbols appear.

Understanding Algebra

When performing a string of additions and subtractions, always work from left to right.

EXAMPLE 18 Evaluate.

a) $9 - 12 + 3$ **b)** $-7 - 15 - 6$ **c)** $-5 + 1 - 8$

Solution We work from left to right.

a) $\underbrace{9 - 12} + 3$
$= -3 + 3$
$= 0$

b) $\underbrace{-7 - 15} - 6$
$= -22 - 6$
$= -28$

c) $\underbrace{-5 + 1} - 8$
$= -4 - 8$
$= -12$

Now Try Exercise 119

Understanding Algebra

Whenever we see an expression of the form $a + (-b)$, we can write the expression as $a - b$.

$$2 + (-5) = 2 - 5$$

Whenever we see an expression of the form $a - (-b)$, we can rewrite it as $a + b$.

$$6 - (-8) = 6 + 8$$

Rewriting Expressions

In general, for any real numbers a and b,

$$a + (-b) = a - b, \text{ and}$$
$$a - (-b) = a + b$$

Using the given information, the expression $9 + (-12) - (-8)$ may be simplified to $9 - 12 + 8$.

EXAMPLE 19

a) Evaluate $-5 - (-9) + (-12) + (-3)$.

b) Simplify the expression in part **a)**.

c) Evaluate the simplified expression in part **b)**.

Solution

a) We work from left to right. The shading indicates the additions being performed to get to the next step.

$$\begin{aligned} -5 - (-9) + (-12) + (-3) &= -5 + 9 + (-12) + (-3) \\ &= 4 + (-12) + (-3) \\ &= -8 + (-3) \\ &= -11 \end{aligned}$$

b) The expression simplifies as follows:

$$-5 - (-9) + (-12) + (-3) = -5 + 9 - 12 - 3$$

c) Evaluate the simplified expression from left to right. Begin by adding $-5 + 9$ to obtain 4.

$$\begin{aligned} -5 + 9 - 12 - 3 &= 4 - 12 - 3 \\ &= -8 - 3 \\ &= -11 \end{aligned}$$

When you come across an expression like the one in Example 19 **a)**, you should simplify it as we did in part **b)** and then evaluate the simplified expression.

Now Try Exercise 127

1.7 Exercise Set MathXL® MyMathLab®

Warm-Up Exercises

Fill in the blanks with the appropriate word, phrase, or symbol(s) from the following list.

opposite	subtrahend	$-a - b$	$a + b$	difference	$-a + b$
zero	left	minuend	$a + (-b)$	$a - b$	

1. In the equation $4 - 7 = -3$, the 4 is called the ________.
2. In the equation $4 - 7 = -3$, the 7 is called the ________.
3. In the equation $4 - 7 = -3$, the -3 is called the ________.
4. $a - b$ could be rewritten as ________.
5. When subtracting a number, we add its ________.
6. When a number is subtracted from itself, the result is ________.
7. When many numbers are being added and subtracted, we always work from ________ to right.
8. The opposite of the number $a + b$ is ________.
9. $-a - (-b)$ could be rewritten as ________.
10. $a - (-b)$ could be rewritten as ________.

In Exercises 11 and 12, are the following calculations correct? If not, explain why.

11. $\frac{4}{9} - \frac{3}{7} = \frac{28}{63} - \frac{27}{63} = \frac{28 - 27}{63} = \frac{1}{63}$

12. $-\frac{5}{12} - \frac{7}{9} = -\frac{15}{36} - \frac{28}{36} = \frac{-15 - 28}{36} = -\frac{43}{36}$

Practice the Skills

Evaluate.

13. $9 - (+3)$
14. $12 - (+7)$
15. $12 - 5$
16. $9 - 4$
17. $8 - 9$
18. $-6 - 3$
19. $9 - (-3)$
20. $17 - (-5)$
21. $-8 - 8$
22. $-4 - (-3)$
23. $0 - 9$
24. $19 - (-9)$
25. $8 - 8$
26. $10 - 10$
27. $-3 - 1$
28. $-4 - (-4)$
29. $-8 - (-5)$
30. $4 - 9$
31. $6 - (-3)$
32. $6 - 10$
33. $-9 - 11$
34. $37 - 40$
35. $0 - (-9.8)$
36. $-6.3 - 4.7$

37. $-4.8 - (-5.1)$
38. $-5.7 - (-3.1)$
39. $44 - 7$
40. $9 - 9$
41. $-8 - (-12)$
42. $-6 - (-2)$
43. $18 - (-4)$
44. $-25 - 16$
45. $-9 - 2$
46. $-85 - (-8)$
47. $-90.7 - 40.3$
48. $-52.6 - 37.9$
49. $-45 - 39$
50. $-500 - (-400)$
51. $70 - (-70)$
52. $130 - (-90)$
53. $42.3 - 49.7$
54. $81.3 - 92.5$
55. $-3.01 - (-3.1)$
56. $-7.04 - (-7.4)$
57. Subtract 15 from 4.
58. Subtract 7 from 1.
59. Subtract 21 from 21.
60. Subtract 13 from 13.
61. Subtract 24 from 13.
62. Subtract -23 from -23.
63. Subtract -12.4 from -6.3.
64. Subtract 17.3 from -9.8
65. Subtract -7.9 from 10.3.
66. Subtract -11.7 from -5.2.
67. Subtract 8.4 from -3.07
68. Subtract -4.1 from 15.23.

Evaluate.

69. $\frac{2}{3} - \frac{1}{2}$
70. $\frac{3}{5} - \frac{1}{4}$
71. $\frac{2}{15} - \frac{5}{6}$
72. $\frac{5}{12} - \frac{7}{8}$
73. $-\frac{7}{10} - \frac{5}{12}$
74. $-\frac{1}{4} - \frac{2}{3}$
75. $-\frac{4}{15} - \frac{3}{20}$
76. $-\frac{5}{4} - \frac{7}{11}$
77. $-\frac{7}{12} - \frac{5}{40}$
78. $-\frac{5}{6} - \frac{3}{32}$
79. $\frac{5}{8} - \frac{6}{48}$
80. $\frac{17}{18} - \frac{13}{20}$
81. $-\frac{4}{9} - \left(-\frac{3}{5}\right)$
82. $\frac{5}{20} - \left(-\frac{1}{8}\right)$
83. $\frac{3}{16} - \left(-\frac{5}{8}\right)$
84. $-\frac{5}{12} - \left(-\frac{3}{8}\right)$
85. Subtract $\frac{7}{9}$ from $\frac{4}{7}$.
86. Subtract $\frac{7}{15}$ from $\frac{5}{8}$.
87. Subtract $-\frac{3}{10}$ from $-\frac{5}{12}$.
88. Subtract $-\frac{5}{16}$ from $-\frac{9}{10}$.

In Exercises 89–106, **a)** *determine by observation whether the difference will be a positive number, zero, or a negative number;* **b)** *find the difference; and* **c)** *examine your answer to part* **b)** *to see whether it is reasonable and makes sense.*

89. $378 - 279$
90. $483 - 569$
91. $-482 - 137$
92. $178 - (-377)$
93. $843 - (-745)$
94. $864 - (-762)$
95. $-408 - (-604)$
96. $-623 - 111$
97. $-1024 - (-576)$
98. $-104.7 - 27.6$
99. $165.7 - 49.6$
100. $-40.2 - (-12.6)$
101. Subtract 364 from 295.
102. Subtract -433 from -932.
103. Subtract 647 from -1023.
104. Subtract 2432 from -4120.
105. Subtract -7.62 from -7.62.
106. Subtract 36.7 from -103.2.

Evaluate.

107. $7 + 5 - (+8)$
108. $15 - (+9) - (+5)$
109. $-6 + (-6) + 16$
110. $9 - 4 + (-2)$
111. $-13 - (+5) + 3$
112. $7 - (+4) - (-3)$
113. $-9 - (-3) + 4$
114. $15 + (-7) - (-3)$
115. $5 - (-9) + (-1)$
116. $12 + (-5) - (-4)$
117. $17 + (-8) - (+14)$
118. $-7 + 6 - 3$
119. $-36 - 5 + 9$
120. $45 - 3 - 7$
121. $25 - 19 + 3$
122. $-4 - 1 + 5$
123. $-4 - 6 + 5 - 7$
124. $-9 - 3 - (-4) + 5$
125. $17 + (-3) - 9 - (-7)$
126. $32 + 5 - 7 - 12$
127. $-9 + (-7) + (-5) - (-3)$
128. $6 - 9 - (-3) + 12$

Problem Solving

129. **Distances Between Cities** When driving east on US Highway 20, Julie Chesser drives through the cities of Galena, Stockton, and Freeport, IL in that order. Galena and Stockton are 28.1 miles apart. Galena and Freeport are 49.0 miles apart. How far apart are Stockton and Freeport? Assume all three cities are in a straight line.

130. **Distances Between Cities** When driving east on US Highway 224, Kris Mudunuri drives through the cities of Ottawa, Findlay, and Tiffin, OH in that order. Ottawa and Findlay are 22.6 miles apart. Ottawa and Tiffin are 47.7 miles apart. How far apart are Findlay and Tiffin? Assume all three cities are in a straight line.

131. Elevation Differences The highest elevation in California is Mount Whitney at 14,505 feet *above* sea level. The lowest elevation in California is Death Valley at 282 feet *below* sea level. What is the difference in these two elevations?

Mount Whitney, CA

132. Leadville, Co According to *Guinness World Records,* the city with the greatest elevation in the United States is Leadville, Colorado, at 10,152 feet. The city with the lowest elevation in the United States, at 184 feet below sea level, is Calipatria, California. What is the difference in the elevation of these cities?

133. Measuring Rainfall At Kim Christensen's house, a rain gauge is placed in the yard and is left untouched for 2 days. Suppose that on the first day, $2\frac{1}{4}$ inches of rain falls. On the second day, no rain falls, but $\frac{3}{8}$ inch of the first day's rainfall evaporates. How much water remains in the gauge after the second day?

134. Death Valley A medical supply package is dropped into Death Valley, California, from a helicopter 1605.7 feet above sea level. The package lands at a location in Death Valley 267.4 feet below sea level. What vertical distance did the package travel?

135. Temperature Change The greatest change in temperature ever recorded within a 24-hour period occurred at Browning, Montana, on January 23, 1916. The temperature fell from 44°F to −56°F. How much did the temperature drop?

136. Going Home Two college students are driving on an expressway, going home for spring break. Shawntoya travels 58.5 miles in 1 hour. Marcelino travels 67.3 miles in 1 hour.

a) If Shawntoya and Marcelino start at the same parking lot and travel in *opposite* directions, how far apart will they be in 1 hour?

b) If Shawntoya and Marcelino start at the same parking lot and travel in the *same* direction, how far apart will they be in 1 hour?

137. Golf The chart below shows some final scores at the Masters golf tournament, held in Augusta, Georgia, in 2012.

Golfer	Score (above or below par)
Bubba Watson	−10
Louis Oosthuizen	−10
Peter Hanson	−8
Matt Kuchar	−8
Phil Mickelson	−8
Lee Westwood	−8
Rory McIlroy	+5
Tiger Woods	+5
Steve Stricker	+7

Source: www.masters.org

a) If par for the tournament is 288 strokes, determine Bubba Watson's score in 2012.

b) What was the difference in the scores between Steve Stricker and Louis Oosthuizen in 2012?

138. Inseam Christine Henry purchases a new pair of pants whose inseam is $32\frac{1}{2}$ inches. If $2\frac{3}{4}$ inches are cut from the inseam of the pants, what will be the new inseam of the pants?

Concept/Writing Exercises

139. Your friend is having trouble in algebra class distinguishing among these three expressions: $x - y$, $y - x$, and $x - (-y)$. Using 3 for x and 8 for y, explain in writing why all three expressions are different.

140. Simplify $3 - (-9) + (-4)$ by eliminating two signs next to one another and replacing them with a single sign. Explain how you determined your answer and then evaluate your answer.

Challenge Problems

Find each sum.

141. $1 - 2 + 3 - 4 + 5 - 6 + 7 - 8 + 9 - 10$

142. $1 - 2 + 3 - 4 + 5 - 6 + \cdots + 99 - 100$

143. Consider a number line.

a) What is the distance, in units, between −11 and −3?

b) Write a subtraction problem to represent this distance (the distance is to be positive).

144. Stock Amy Tait buys a stock for \$50. Will the stock be worth more if it decreases by 10% and then increases by 10%, or if it increases by 10% and then decreases by 10%, or will the value be the same either way?

145. Rolling Ball A ball rolls off a table and follows the path indicated in the figure on the right. Suppose the maximum height reached by the ball on each bounce is 1 foot less than on the previous bounce.

a) Determine the total vertical distance traveled by the ball.

b) If we consider the ball moving in a downward direction as negative, and the ball moving in an upward direction as positive, what was the net vertical distance traveled (from its starting point) by the ball?

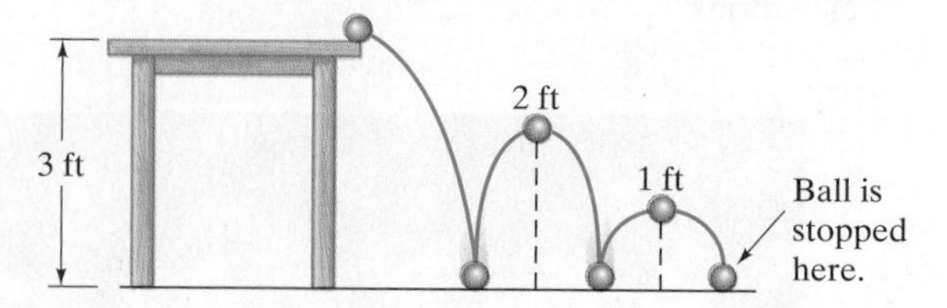

Cumulative Review Exercises

[1.4] **146.** List the set of counting numbers.

147. Explain the relationship between the set of rational numbers, the set of irrational numbers, and the set of real numbers.

[1.5] *Insert either $>$, $<$, or $=$ in each shaded area to make the statement true.*

148. $|-3| \;\square\; -5$

149. $-|-9| \;\square\; -|-5|$

[1.7] **150.** Subtract $\frac{7}{8}$ from $\frac{5}{6}$.

1.8 Multiplication and Division of Real Numbers

1 Multiply numbers.

2 Divide numbers.

3 Remove negative signs from denominators.

4 Evaluate divisions involving 0.

1 Multiply Numbers

The following rules are used in determining the sign of the product when two numbers are multiplied.

The Sign of the Product of Two Real Numbers

1. The product of two numbers with **like** signs is a **positive** number.
2. The product of two numbers with **unlike** signs is a **negative** number.

The product of two positive numbers or two negative numbers will be a positive number. The product of a positive number and a negative number will be a negative number.

Understanding Algebra

Remember, in the expression $(4)(5)$, the numbers 4 and 5 are to be multiplied. When there is no visible operator, the operation is *understood* to be multiplication. So, $(4)(5) = 20$.

EXAMPLE 1 Evaluate.

a) $4(-5)$ **b)** $(-6)(7)$ **c)** $(-9)(-3)$

Solution

a) Since the numbers have unlike signs, the product is negative.

$$4(-5) = -20$$

b) Since the numbers have unlike signs, the product is negative.

$$(-6)(7) = -42$$

c) Since the numbers have like signs, both negative, the product is positive.

$$(-9)(-3) = 27$$

Now Try Exercise 21

Understanding Algebra

In the multiplication

$(4)(5) = 20$

4 and 5 are called *factors* and 20 is called the *product*.

EXAMPLE 2 Evaluate.

a) $(-8)(5)$ **b)** $(-4)(-8)$ **c)** $0(6)$
d) $0(-2)$ **e)** $4.2(-9.7)$ **f)** $(-1.63)(-4.1)$

Solution

a) $(-8)(5) = -40$ **b)** $(-4)(-8) = 32$ **c)** $0(6) = 0$
d) $0(-2) = 0$ **e)** $4.2(-9.7) = -40.74$ **f)** $(-1.63)(-4.1) = 6.683$

Note that zero multiplied by any real number equals zero.

Now Try Exercise 25

Understanding Algebra

The product of two numbers with *different* signs is always negative.

The product of two numbers with the *same* sign is always positive.

HELPFUL HINT

At this point some students begin confusing problems like $-2 - 3$ with $(-2)(-3)$ and problems like $2 - 3$ with $2(-3)$. If you do not understand the difference between problems like $-2 - 3$ and $(-2)(-3)$, make an appointment to see your instructor as soon as possible.

Subtraction Problems	Multiplication Problems
$-2 - 3 = -5$	$(-2)(-3) = 6$
$2 - 3 = -1$	$(2)(-3) = -6$

EXAMPLE 3 Evaluate.

a) $\left(\frac{-1}{8}\right)\left(\frac{-3}{5}\right)$ **b)** $\left(\frac{3}{20}\right)\left(\frac{-3}{10}\right)$

Solution

a) $\left(\frac{-1}{8}\right)\left(\frac{-3}{5}\right) = \frac{(-1)(-3)}{8(5)} = \frac{3}{40}$ **b)** $\left(\frac{3}{20}\right)\left(\frac{-3}{10}\right) = \frac{3(-3)}{20(10)} = -\frac{9}{200}$

Now Try Exercise 41

Understanding Algebra

The product of an *even* number of negative numbers will always be positive.

The product of an *odd* number of negative numbers will always be negative.

Sometimes you may be asked to perform more than one multiplication in a given problem. When this happens, the sign of the final product can be determined by counting the number of *negative* numbers being multiplied. *The product of an even number of negative numbers will always be positive. The product of an odd number of negative numbers will always be negative.*

EXAMPLE 4 Evaluate.

a) $(-5)(-3)(1)(-4)$ **b)** $(-2)(-4)(-1)(3)(-4)$

Solution

a) Since there are an odd number of negative numbers, the product will be negative, as illustrated.

$$\begin{aligned}(-5)(-3)(1)(-4) &= (15)(1)(-4)\\ &= (15)(-4)\\ &= -60\end{aligned}$$

b) Since there are an even number of negative numbers, the product will be positive, as illustrated.

$$\begin{aligned}(-2)(-4)(-1)(3)(-4) &= (8)(-1)(3)(-4)\\ &= (-8)(3)(-4)\\ &= (-24)(-4)\\ &= 96\end{aligned}$$

Now Try Exercise 35

Understanding Algebra

In the division problem

$$24 \div 4 = 6$$

24 is called the *dividend,* 4 is called the *divisor,* and 6 is called the *quotient.*

2 Divide Numbers

The rules for dividing numbers are very similar to those used in multiplying numbers.

The Sign of the Quotient of Two Real Numbers

1. The quotient of two numbers with **like** signs is a **positive** number.
2. The quotient of two numbers with **unlike** signs is a **negative** number.

Therefore, the quotient of two positive numbers or two negative numbers will be a positive number. The quotient of a positive number and a negative number will be a negative number.

EXAMPLE 5 Evaluate.

a) $\frac{10}{-5}$ **b)** $\frac{-45}{5}$ **c)** $\frac{-36}{-6}$

Solution

a) Since the numbers have unlike signs, the quotient is negative.

$$\frac{10}{-5} = -2$$

b) Since the numbers have unlike signs, the quotient is negative.

$$\frac{-45}{5} = -9$$

c) Since the numbers have like signs, both negative, the quotient is positive.

$$\frac{-36}{-6} = 6$$

Now Try Exercise 55

Understanding Algebra

A fraction is really a division problem:

$$24 \div 4 = 6$$

is the same as

$$\frac{24}{4} = 6$$

EXAMPLE 6 Evaluate.

a) $32 \div (-4)$ **b)** $-\frac{3}{7} \div \left(-\frac{1}{2}\right)$

Solution

a) Since the numbers have different signs, the quotient is negative.

$$32 \div (-4) = \frac{32}{-4} = -8$$

b) Since the numbers have like signs, both negative, the quotient is positive. We will multiply $-\frac{3}{7}$ by the reciprocal of $\left(-\frac{1}{2}\right)$.

$$-\frac{3}{7} \div \left(-\frac{1}{2}\right) = -\frac{3}{7} \cdot \left(-\frac{2}{1}\right) = \frac{6}{7}$$

Now Try Exercise 77

EXAMPLE 7 Evaluate. When appropriate, round your answer to the nearest hundredth.

a) $-4.75 \div 1.9$ **b)** $\frac{-7.9}{-1.7}$

Solution

a) Since the numbers have unlike signs, the quotient is negative.

$$-4.75 \div 1.9 = \frac{-4.75}{1.9} = -2.5$$

b) Since the numbers have like signs, both negative, the quotient is positive.

$$\frac{-7.9}{-1.7} \approx 4.65$$

The answer in part **b)** was rounded to two decimal places, or hundredths. If you have forgotten how to round decimal numbers, review Appendix A now.

Now Try Exercise 71

HELPFUL HINT

For multiplication and division of two real numbers:

$$\left.\begin{array}{l}(+)(+) = + \\ (-)(-) = +\end{array}\right\} \quad \left.\begin{array}{l}\dfrac{(+)}{(+)} = + \\ \dfrac{(-)}{(-)} = +\end{array}\right\}$$ Like signs give positive products and quotients.

$$\left.\begin{array}{l}(+)(-) = - \\ (-)(+) = -\end{array}\right\} \quad \left.\begin{array}{l}\dfrac{(+)}{(-)} = - \\ \dfrac{(-)}{(+)} = -\end{array}\right\}$$ Unlike signs give negative products and quotients.

3 Remove Negative Signs from Denominators

We now know that the quotient of a positive number and a negative number is a negative number. The fractions $-\frac{3}{4}$, $\frac{-3}{4}$, and $\frac{3}{-4}$ all represent the same negative number, negative three-fourths.

The Quotient of a Positive Number and a Negative Number

If a and b represent any real numbers, $b \neq 0$, then

$$\frac{a}{-b} = \frac{-a}{b} = -\frac{a}{b}$$

In mathematics we generally do not write a fraction with a negative sign in the denominator. When a negative sign appears in a denominator, we can move it to the numerator or place it in front of the fraction. For example, the fraction $\frac{5}{-7}$ should be written as either $\frac{-5}{7}$ or $-\frac{5}{7}$.

EXAMPLE 8 Evaluate $\frac{3}{7} \div \left(\frac{-12}{35}\right)$.

Solution

$$\frac{3}{7} \div \left(\frac{-12}{35}\right) = \frac{\overset{1}{\cancel{3}}}{\underset{1}{\cancel{7}}} \cdot \left(\frac{\overset{5}{\cancel{35}}}{\underset{-4}{\cancel{-12}}}\right) = \frac{1(5)}{1(-4)} = \frac{5}{-4} = -\frac{5}{4}$$

Now Try Exercise 73

The operations on real numbers are summarized in **Table 1.1** on the next page.

TABLE 1.1 Summary of Operations on Real Numbers

Signs of Numbers	Addition	Subtraction	Multiplication	Division
Both Numbers Are Positive	Sum Is Always Positive	Difference May Be Either Positive or Negative	Product Is Always Positive	Quotient Is Always Positive
Examples				
6 and 2	$6 + 2 = 8$	$6 - 2 = 4$	$6 \cdot 2 = 12$	$6 \div 2 = 3$
2 and 6	$2 + 6 = 8$	$2 - 6 = -4$	$2 \cdot 6 = 12$	$2 \div 6 = \frac{1}{3}$
One Number Is Positive and the Other Number Is Negative	Sum May Be Either Positive or Negative	Difference May Be Either Positive or Negative	Product Is Always Negative	Quotient Is Always Negative
Examples				
6 and −2	$6 + (-2) = 4$	$6 - (-2) = 8$	$6(-2) = -12$	$6 \div (-2) = -3$
−6 and 2	$-6 + 2 = -4$	$-6 - 2 = -8$	$-6(2) = -12$	$-6 \div 2 = -3$
Both Numbers Are Negative	Sum Is Always Negative	Difference May Be Either Positive or Negative	Product Is Always Positive	Quotient Is Always Positive
Examples				
−6 and −2	$-6 + (-2) = -8$	$-6 - (-2) = -4$	$-6(-2) = 12$	$-6 \div (-2) = 3$
−2 and −6	$-2 + (-6) = -8$	$-2 - (-6) = 4$	$-2(-6) = 12$	$-2 \div (-6) = \frac{1}{3}$

4 Evaluate Divisions Involving 0

What is $\frac{0}{1}$ equal to? Note that $\frac{6}{3} = 2$ because $3 \cdot 2 = 6$. We can follow the same procedure to determine the value of $\frac{0}{1}$. Suppose that $\frac{0}{1}$ is equal to some number, which we will designate by ?.

$$\text{If } \frac{0}{1} = \boxed{?} \text{ then } 1 \cdot \boxed{?} = 0$$

Since only $1 \cdot 0 = 0$, the ? must be 0. Thus, $\frac{0}{1} = 0$. Using the same technique, we can show that zero divided by any nonzero number is zero.

Zero Divided by a Nonzero Number

If *a* represents any real number except 0, then

$$0 \div a = \frac{0}{a} = 0$$

Now what is $\frac{1}{0}$ equal to?

$$\text{If } \frac{1}{0} = \boxed{?} \text{ then } 0 \cdot \boxed{?} = 1$$

But since 0 multiplied by any number will be 0, there is no value that can replace ?. We say that $\frac{1}{0}$ is **undefined**. Using the same technique, we can show that any real number, except 0, divided by 0 is undefined.

Division by Zero

If a represents any real number except 0, then

$$a \div 0 \quad \text{or} \quad \frac{a}{0} \quad \text{is undefined}$$

What is $\frac{0}{0}$ equal to?

$$\text{If } \frac{0}{0} = \boxed{?} \text{ then } 0 \cdot \boxed{?} = 0$$

Since the product of any number and 0 is 0, the $\boxed{?}$ can be replaced by any real number. Therefore, the quotient $\frac{0}{0}$ cannot be determined, and so there is no answer. Thus, we will not use it in this textbook.*

Summary of Division Involving 0

If a represents any real number except 0, then

$$\frac{0}{a} = 0 \quad \text{and} \quad \frac{a}{0} \text{ is undefined}$$

EXAMPLE 9 Indicate whether each quotient is 0 or undefined.

a) $\frac{0}{2}$ **b)** $\frac{5}{0}$ **c)** $\frac{0}{-4}$ **d)** $\frac{-2}{0}$

Solution The answer to parts **a)** and **c)** is 0. The answer to parts **b)** and **d)** is undefined.

Now Try Exercise 95

1.8 Exercise Set MathXL® MyMathLab®

Warm-Up Exercises

Fill in the blanks with the appropriate word, phrase, or symbol(s) from the following list.

positive	negative	infinity	product	quotient	-63
zero	63	undefined	$-\frac{a}{b}$	$\frac{a}{b}$	

1. The product of a positive number and a negative number is ________.

2. 16 divided by 0 is ________.

3. 0 divided by 8 is ________.

4. The fraction $\frac{a}{-b}$ may be rewritten as ________.

5. The fraction $-\frac{a}{-b}$ may be rewritten as ________.

6. If x is 9 and y is -7 then the value of xy is ________.

7. The product of two negative numbers is a ________ number.

8. If x is 9 and y is -7 then the value of $x(-y)$ is ________.

9. When two numbers are multiplied, the result is called the ________ of the two numbers.

10. When two real numbers are divided, the result is called the ________.

Determine the sign of each product.

11. $(8)(4)(-5)$

12. $(-9)(-12)(20)$

13. $(-102)(-16)(24)(19)$

14. $(1054)(-92)(-16)(-37)$

15. $(-40)(-16)(30)(50)(-13)$

16. $(-1)(3)(-462)(-196)(-312)$

*At this level, some professors prefer to call $\frac{0}{0}$ *indeterminate* while others prefer to call $\frac{0}{0}$ *undefined*. In higher-level mathematics courses, $\frac{0}{0}$ is sometimes referred to as an *indeterminate form*.

Practice the Skills

Find each product.

17. $8(3)$
18. $7 \cdot 8$
19. $5(-3)$
20. $6(-2)$
21. $(-9)(-6)$
22. $(-6)(-3)$
23. $-7 \cdot 3$
24. $-9 \cdot 5$
25. $-3.2(3)$
26. $-7(5.4)$
27. $-4.67 \cdot 1$
28. $3.29(-1)$
29. $-6.7 \cdot 0$
30. $0(-5)$
31. $(-9)(0)(-6)$
32. $5(-4)(2)$
33. $(21)(-1)(4)$
34. $2(8)(-1)(-3)$
35. $-1(-3)(3)(-8)$
36. $(2)(-4)(-5)(-1)$
37. $(-4)(5)(-7)(10)$
38. $(-3)(2)(5)(3)$
39. $(-1)(3)(0)(-7)$
40. $(-6)(6)(4)(-4)$

Find each product.

41. $\left(\frac{-1}{2}\right)\left(\frac{3}{5}\right)$
42. $\left(\frac{1}{3}\right)\left(\frac{-3}{5}\right)$
43. $\left(\frac{-5}{9}\right)\left(\frac{-7}{15}\right)$
44. $\left(\frac{-9}{10}\right)\left(\frac{7}{-8}\right)$
45. $\left(\frac{6}{-3}\right)\left(\frac{4}{-2}\right)$
46. $\left(\frac{9}{-10}\right)\left(\frac{6}{-7}\right)$
47. $\left(\frac{3}{4}\right)\left(\frac{-2}{15}\right)$
48. $\left(\frac{4}{5}\right)\left(\frac{-3}{10}\right)$

Find each quotient. When appropriate, round you answer to the nearest hundredth.

49. $\frac{-42}{6}$
50. $\frac{-18}{9}$
51. $-16 \div (-4)$
52. $(-25) \div (-5)$
53. $\frac{-36}{-9}$
54. $\frac{-15}{-1}$
55. $\frac{36}{-2}$
56. $\frac{30}{-6}$
57. $\frac{-19.8}{-2}$
58. $-15.6/(-3)$
59. $40/(-4)$
60. $\frac{63}{-7}$
61. $\frac{-66}{2}$
62. $\frac{-25}{-5}$
63. $\frac{48}{-12}$
64. $\frac{-10}{10}$
65. Divide -30 by -5.
66. Divide -36 by -6.
67. Divide 0 by 4.
68. Divide 0 by -13.
69. $-64.8 \div (-4)$
70. $-86.4/(-2)$
71. Divide 30.8 by -5.2.
72. Divide -67.64 by 7.3.

Find each quotient.

73. $\frac{3}{12} \div \left(\frac{-5}{8}\right)$
74. $4 \div \left(\frac{-6}{13}\right)$
75. $\frac{-5}{12} \div (-3)$
76. $\frac{-3}{7} \div (-5)$
77. $\frac{-15}{21} \div \left(\frac{-15}{21}\right)$
78. $\frac{-4}{9} \div \left(\frac{-6}{7}\right)$
79. $(-12) \div \frac{5}{12}$
80. $-16 \div \frac{11}{16}$

Evaluate.

81. $-4(8)$
82. $\frac{-18}{-2}$
83. $\frac{-100}{-5}$
84. $-50 \div (-10)$
85. $-7(2)$
86. $6.4(-8)$
87. $27.9 \div (-3)$
88. Divide 130 by -10.
89. $-100 \div 5$
90. $4(-2)(-1)(-5)$
91. Divide -90 by -90.
92. $(6)(1)(-3)(4)$

Indicate whether each quotient is 0 or undefined.

93. $0 \div 8.6$
94. $\frac{0}{1}$
95. $\frac{5}{0}$
96. $\frac{-2.7}{0}$
97. $0 \div (-7)$
98. $\frac{6}{0}$
99. 8 divided by 0
100. 0 divided by 12

In Exercises 101–116, **a)** *determine by observation whether the product or quotient will be a positive number, zero, a negative number, or undefined;* **b)** *find the product or quotient if it exists;* **c)** *examine your answer in part* **b)** *to see whether it is reasonable and makes sense.*

101. $92(-38)$
102. $-168 \div 42$
103. $-240/15$
104. $0/12$
105. $243 \div (-27)$
106. $(323)(-115)$
107. $(-49)(-126)$
108. $(1530)(0)$

109. $0 \div 5335$

110. $-86.4 \div (-36)$

111. $8.2 \div 0$

112. $-37.74 \div 0$

113. $8 \div (2.5)$

114. $(1.1)(9.72)(6.3)$

115. $(-3.0)(4.2)(-18)$

116. $-288.86/1.43$

Indicate whether each statement is true or false.

117. The product of two negative numbers is a negative number.

118. The product of a positive number and a negative number is a negative number.

119. The quotient of two numbers with unlike signs is a positive number.

120. The quotient of two negative numbers is a positive number.

121. The product of an even number of negative numbers is a positive number.

122. Zero divided by 1 is 1.

123. The product of an odd number of negative numbers is a negative number.

124. Six divided by 0 is 0.

125. Zero divided by 1 is undefined.

126. The product of 0 and any real number is 0.

127. Five divided by 0 is undefined.

128. Division by 0 does not result in a real number.

Problem Solving

129. Football A high school football team is penalized three times, each time with a loss of 15 yards, or -15 yards. Find the total loss due to penalties.

130. Submarine Dive A submarine is at a depth of -160 feet (160 feet below sea level). It dives to 3 times that depth. Find its new depth.

131. Credit Card Leona De Vito's balance on her credit card is $-\$520$ (she owes \$520). She pays back $\frac{1}{5}$ of this balance.

a) How much did she pay back?

b) What is her new balance?

132. Money Owed Brian Philip owes his Dad \$500. After he makes four payments of \$40 each, how much will he still owe?

133. Garage Sale Four sisters made a total of \$775.40 at a garage sale. After they each give their husbands \$50 and split the remaining amount equally, how much will each woman receive?

134. Wind Chill On Monday in Chicago the wind chill temperature was $-30°F$. On Tuesday the wind chill temperature was only $\frac{1}{3}$ of what it was on Monday. What was the wind chill temperature on Tuesday?

135. Test Score Because of incorrect work, Josue Nunez lost 4 points on each of the five questions on his math test.

a) How many points did Josue lose altogether?

b) If the maximum score possible was 100%, what is Josue's test score?

136. Lab Work Jack's job is to monitor the temperature of a superheated piece of metal for ten hours in the lab. He observed that the metal cooled 15° each hour for ten hours.

a) What number represents the total drop in temperature?

b) If the temperature of the metal was originally 678°, what was it at the end of the ten hours?

137. Heart Rate The Johns Hopkins Medical Letter states that to find a person's *target heart rate* in beats per minute, follow this procedure. Subtract the person's age from 220, then multiply this difference by 60% and 75%. The difference multiplied by 60% gives the lower limit and the difference multiplied by 75% gives the upper limit.

a) Find the target heart rate range of a 50-year-old.

b) Find your own target heart rate.

Challenge Problems

We will learn in the next section that $2^3 = 2 \cdot 2 \cdot 2$ *and* $x^n = \underbrace{x \cdot x \cdot x \cdot \cdots \cdot x}_{n \text{ factors of } x}$. *Use this information to evaluate each expression.*

138. 3^4

139. $(-5)^3$

140. $\left(\frac{2}{3}\right)^3$

141. 1^{100}

142. $(-1)^{81}$

143. Will the product of $(-1)(-2)(-3)(-4)\cdots(-10)$ be a positive number or a negative number? Explain how you determined your answer.

144. Will the product $(1)(-2)(3)(-4)(5)(-6)\cdots(33)(-34)$ be a positive number or a negative number? Explain how you determined your answer.

Group Activity

Discuss and answer Exercise 145 as a group, according to the instructions.

145. a) Each member of the group is to do this procedure separately. At this time do not share your number with the other members of your group.

1. Choose a number between 2 and 10.
2. Multiply your number by 9.
3. Add the two digits in the product together.
4. Subtract 5 from the sum.
5. Now choose the corresponding letter of the alphabet that corresponds with the difference found. For example, 1 is a, 2 is b, 3 is c, and so on.
6. Choose a one-word country that starts with that letter.
7. Now choose a one-word animal that starts with the last letter of the country selected.
8. Finally, choose a color that starts with the last letter of the animal chosen.

b) Now share your final answer with the other members of your group. Did you all get the same answer?

c) Most people will obtain the answer *orange*. As a group, write a paragraph or two explaining why.

Cumulative Review Exercises

[1.5] **146.** Insert either <, >, or = in the shaded area to make a true statement.

$$|-3.6| \;\blacksquare\; |-2.7|$$

[1.6] **147.** Add $-\frac{7}{12}+\left(-\frac{1}{10}\right)$.

[1.7] **148.** Subtract -18 from -20.

149. Evaluate $6-3-4-2$.

150. Evaluate $5-(-2)+3-7$.

1.9 Exponents, Parentheses, and the Order of Operations

1. Learn the meaning of exponents.
2. Evaluate expressions containing exponents.
3. Learn the difference between $-x^2$ and $(-x)^2$.
4. Learn the order of operations.
5. Learn the use of parentheses.
6. Evaluate expressions containing variables.

1 Learn the Meaning of Exponents

In the expression 4^2, the 4 is called the **base**, and the 2 is called the **exponent**. The number 4^2 is read "4 squared" or "4 to the second power" and means

$$\underbrace{4\cdot 4}_{\text{2 factors of 4}} = 4^2 \quad (\text{4 is the base, 2 is the exponent})$$

The number 4^3 is read "4 cubed" or "4 to the third power" and means

$$\underbrace{4\cdot 4\cdot 4}_{\text{3 factors of 4}} = 4^3$$

In general, the number b to the nth power, written b^n, means

$$\underbrace{b\cdot b\cdot b\cdot\cdots\cdot b}_{n\text{ factors of }b} = b^n$$

Thus, $b^4 = b\cdot b\cdot b\cdot b$ or $bbbb$ and $x^3 = x\cdot x\cdot x$ or xxx.

Understanding Algebra

Exponents are simply a shorthand notation for repeated multiplication. For example, 3^4 means $3\cdot 3\cdot 3\cdot 3 = 81$.

2 Evaluate Expressions Containing Exponents

Let's evaluate some expressions that contain exponents.

EXAMPLE 1 Evaluate. **a)** 5^2 **b)** 2^5 **c)** 1^6 **d)** $(-9)^2$ **e)** $(-2)^3$ **f)** $\left(\frac{2}{3}\right)^2$

Solution

a) $5^2 = 5 \cdot 5 = 25$
b) $2^5 = 2 \cdot 2 \cdot 2 \cdot 2 \cdot 2 = 32$
c) $1^6 = 1 \cdot 1 \cdot 1 \cdot 1 \cdot 1 \cdot 1 = 1$ (1 raised to any power equals 1; why?)
d) $(-9)^2 = (-9)(-9) = 81$
e) $(-2)^3 = (-2)(-2)(-2) = -8$
f) $\left(\frac{2}{3}\right)^2 = \left(\frac{2}{3}\right)\left(\frac{2}{3}\right) = \frac{4}{9}$

Now Try Exercise 19

Understanding Algebra

If no exponent is written, it is assumed to be 1.

$6 = 6^1$ and $x = x^1$

It is not necessary to write exponents of 1. For example, when writing xxy, we write x^2y and not x^2y^1. *Whenever we see a variable or number without an exponent, we always assume that the variable or number has an exponent of 1.*

Examples of Exponential Notation

a) $xyxx = x^3y$
b) $xyzzy = xy^2z^2$
c) $3aabbb = 3a^2b^3$
d) $5xyyyy = 5xy^4$
e) $4 \cdot 4rrs = 4^2r^2s$
f) $5 \cdot 5 \cdot 5mmn = 5^3m^2n$

Notice in parts **a)** and **b)** that the order of the factors does not matter.

HELPFUL HINT

Note that $x + x + x + x + x + x = 6x$ and $x \cdot x \cdot x \cdot x \cdot x \cdot x = x^6$. Be careful that you do not get addition and multiplication confused.

Understanding Algebra

An exponent refers only to the number or variable that directly precedes it unless parentheses are used to indicate otherwise.

No parentheses:	Parentheses:
-4^2	$(-5)^2$
$-(4)(4) = -16$	$(-5)(-5) = 25$

3 Learn the Difference Between $-x^2$ and $(-x)^2$

An exponent refers only to the number or variable that directly precedes it unless parentheses are used to indicate otherwise. For example, in the expression $3x^2$, only the x is squared. In the expression $-x^2$, only the x is squared. We can write $-x^2$ as $-1x^2$ because any real number may be multiplied by 1 without affecting its value.

$$-x^2 = -1x^2$$

By looking at $-1x^2$ we can see that only the x is squared, not the -1. If the entire expression $-x$ were to be squared, we would need to use parentheses and write $(-x)^2$. Note the difference in the following two examples:

$$-x^2 = -(x)(x)$$
$$(-x)^2 = (-x)(-x)$$

Consider the expressions -3^2 and $(-3)^2$. How do they differ?

$$-3^2 = -(3)(3) = -9$$
$$(-3)^2 = (-3)(-3) = 9$$

HELPFUL HINT

The expression $-x^2$ is read "negative x squared," or "the opposite of x squared." The expression $(-x)^2$ is read "negative x, quantity squared."

EXAMPLE 2 Evaluate. **a)** -5^2 **b)** $(-5)^2$ **c)** -2^3 **d)** $(-2)^3$

Solution

a) $-5^2 = -(5)(5) = -25$ **b)** $(-5)^2 = (-5)(-5) = 25$

c) $-2^3 = -(2)(2)(2) = -8$ **d)** $(-2)^3 = (-2)(-2)(-2) = -8$

Now Try Exercise 23

EXAMPLE 3 Evaluate. **a)** -2^4 **b)** $(-2)^4$

Solution

a) $-2^4 = -(2)(2)(2)(2) = -16$ **b)** $(-2)^4 = (-2)(-2)(-2)(-2) = 16$

Now Try Exercise 25

4 Learn the Order of Operations

Now that we have introduced exponents we can present the *order of operations*. Can you evaluate $2 + 3 \cdot 4$? Is it 20? Or is it 14? To answer this, you must know the order of operations to follow when evaluating a mathematical expression.

Understanding Algebra

Grouping symbols can include combinations of parentheses, brackets, braces, and even fraction bars. For example,

$$\frac{-4(5+2)-8}{19+3(-5)}$$

evaluates to $\frac{-4(7)-8}{19-15}$ or $\frac{-36}{4}$ or simply -9.

Order of Operations: To Evaluate Mathematical Expressions, Use the Following Order

1. First, evaluate the information within **parentheses** ( ), brackets [ ], or braces { }. These are **grouping symbols**, for they group information together. A fraction bar, —, also serves as a grouping symbol. If the expression contains nested grouping symbols (one pair of grouping symbols within another pair), evaluate the information in the innermost grouping symbols first.
2. Next, evaluate all exponents.
3. Next, evaluate all **multiplications** or **divisions** in the order in which they occur, working from left to right.
4. Finally, evaluate all **additions** or **subtractions** in the order in which they occur, working from left to right.

We can now evaluate $2 + 3 \cdot 4$. Since multiplications are performed before additions,

$$2 + 3 \cdot 4 \quad \text{means} \quad 2 + (3 \cdot 4) = 2 + 12 = 14$$

Some students remember the word PEMDAS or the phrase "Please Excuse My Dear Aunt Sally" to help them remember the order of operations. PEMDAS helps them remember the order: **P**arentheses, **E**xponents, **M**ultiplication, **D**ivision, **A**ddition, **S**ubtraction. Remember, this does not imply multiplication before division or addition before subtraction. For example, when evaluating $24 \div 2 \times 3$ we perform the division first and then the multiplication because we work from left to right.

$$24 \div 2 \times 3 = (24 \div 2) \times 3 = 12 \times 3 = 36$$

5 Learn the Use of Parentheses

Grouping symbols may be used (1) to change the order of operations to be followed in evaluating an algebraic expression or (2) to help clarify the understanding of an expression.

To evaluate the expression $2 + 3 \cdot 4$, we would normally perform the multiplication, $3 \cdot 4$, first. If we wished to have the addition performed before the multiplication, we could indicate this by placing parentheses around $2 + 3$:

$$(2 + 3) \cdot 4 = 5 \cdot 4 = 20$$

Sometimes it may be necessary to use more than one set of grouping symbols to indicate the order to be followed when evaluating an expression. When one set of grouping symbols is within another set of grouping symbols, we call these **nested grouping symbols**. Whenever we are given an expression with nested grouping symbols, we always evaluate the numbers in the *innermost grouping symbols first*. Color shading is used in the following examples to indicate the order in which the expression is evaluated.

$$6[2 + 3(4 + 1)] = 6[2 + 3(5)] = 6[2 + 15] = 6[17] = 102$$

$$4[3(6 - 4) \div 6] = 4[3(2) \div 6] = 4[6 \div 6] = 4[1] = 4$$

$$\{2 + [(8 \div 4)^2 - 1]\}^2 = \{2 + [2^2 - 1]\}^2 = \{2 + [4 - 1]\}^2 = \{2 + 3\}^2 = 5^2 = 25$$

HELPFUL HINT

If parentheses are not used to change the order of operations, multiplications and divisions are always performed before additions and subtractions. When a problem has only multiplications and divisions, work from left to right. Similarly, when a problem has only additions and subtractions, work from left to right.

EXAMPLE 4 Evaluate $6 + 3 \cdot 5^2 - 4$.

Solution Colored shading is used to indicate the order in which the expression is to be evaluated.

$6 + 3 \cdot 5^2 - 4$	Exponent
$= 6 + 3 \cdot 25 - 4$	Multiply.
$= 6 + 75 - 4$	Add.
$= 81 - 4$	
$= 77$	

Now Try Exercise 51

EXAMPLE 5 Evaluate $-7 + 2[-6 + (36 \div 3^2)]$.

Solution

$-7 + 2[-6 + (36 \div 3^2)]$	Exponent
$= -7 + 2[-6 + (36 \div 9)]$	Divide.
$= -7 + 2[-6 + 4]$	Add.
$= -7 + 2[-2]$	Multiply.
$= -7 - 4$	
$= -11$	

Now Try Exercise 65

EXAMPLE 6 Evaluate $(8 \div 2) + 7(5 - 2)^2$.

Solution

$(8 \div 2) + 7(5 - 2)^2$	Parentheses
$= 4 + 7(3)^2$	Exponent
$= 4 + 7 \cdot 9$	Multiply.
$= 4 + 63$	Add.
$= 67$	

Now Try Exercise 71

EXAMPLE 7 Evaluate $-8 - 81 \div 9 \cdot 2^2 + 7$.

Solution

$$\begin{aligned} &-8 - 81 \div 9 \cdot 2^2 + 7 && \text{Exponent} \\ &= -8 - 81 \div 9 \cdot 4 + 7 && \text{Divide.} \\ &= -8 - 9 \cdot 4 + 7 && \text{Multiply.} \\ &= -8 - 36 + 7 && \text{Subtract.} \\ &= -44 + 7 && \text{Add.} \\ &= -37 \end{aligned}$$

Now Try Exercise 69

EXAMPLE 8 Evaluate $\frac{3}{8} - \frac{2}{5} \cdot \frac{1}{12}$.

Solution First perform the multiplication.

$$\begin{aligned} &\frac{3}{8} - \left(\frac{\overset{1}{\cancel{2}}}{5} \cdot \frac{1}{\underset{6}{\cancel{12}}}\right) && \text{Multiply.} \\ &= \frac{3}{8} - \frac{1}{30} && \text{Subtract.} \\ &= \frac{45}{120} - \frac{4}{120} \\ &= \frac{41}{120} \end{aligned}$$

Now Try Exercise 81

EXAMPLE 9 Evaluate $\{[(-5 + 1)^2 - 2^3] \div (24 \div 6 \cdot 2)\}^3$.

Solution When working with nested grouping symbols we will evaluate the expressions in the innermost grouping symbols first.

$$\begin{aligned} &\{[(-5 + 1)^2 - 2^3] \div (24 \div 6 \cdot 2)\}^3 && \text{Add.} \\ &\{[(-4)^2 - 2^3] \div (24 \div 6 \cdot 2)\}^3 && \text{Exponents} \\ &\{[16 - 8] \div (24 \div 6 \cdot 2)\}^3 && \text{Divide.} \\ &\{[16 - 8] \div (4 \cdot 2)\}^3 && \text{Multiply.} \\ &\{[16 - 8] \div 8\}^3 && \text{Subtract.} \\ &\{8 \div 8\}^3 && \text{Divide.} \\ &1^3 && \text{Exponent} \\ &1 \end{aligned}$$

Now Try Exercise 93

EXAMPLE 10 Write the following statements as mathematical expressions using parentheses and brackets and then evaluate: Multiply 12 by 3. Add 8 to this product. Subtract 7 from this sum. Divide this difference by 6.

Solution

$$\begin{aligned} &12 \cdot 3 && \text{Multiply 12 by 3.} \\ &(12 \cdot 3) + 8 && \text{Add 8.} \\ &[(12 \cdot 3) + 8] - 7 && \text{Subtract 7.} \\ &\{[(12 \cdot 3) + 8] - 7\} \div 6 && \text{Divide the difference by 6.} \end{aligned}$$

Now evaluate.

$$\begin{aligned} &\{[(12\cdot 3) + 8] - 7\} \div 6 \\ &= \{[36 + 8] - 7\} \div 6 \\ &= \{44 - 7\} \div 6 \\ &= 37 \div 6 \\ &= \frac{37}{6} \end{aligned}$$

Now Try Exercise 125

As shown in Examples 9 and 10, sometimes brackets, [ ], and braces, { }, are used in place of parentheses to help avoid confusion. In Example 10, if only parentheses had been used, the preceding expression would appear as $(((12\cdot 3) + 8) - 7) \div 6$.

6 Evaluate Expressions Containing Variables

Now we will evaluate some expressions for given values of the variables.

EXAMPLE 11 Evaluate $-7x + 9$ when $x = 3$.

Solution Substitute 3 for x in the expression.

$$-7x + 9 = -7(3) + 9 = -21 + 9 = -12$$

Now Try Exercise 105

EXAMPLE 12 Evaluate **a)** x^2 **b)** $-x^2$ and **c)** $(-x)^2$ when $x = 3$.

Solution Substitute 3 for x.

a) $x^2 = 3^2 = (3)(3) = 9$ **b)** $-x^2 = -3^2 = -(3)(3) = -9$

c) $(-x)^2 = (-3)^2 = (-3)(-3) = 9$

Now Try Exercise 95

EXAMPLE 13 Evaluate **a)** y^2 **b)** $-y^2$ and **c)** $(-y)^2$ when $y = -4$.

Solution Substitute -4 for y.

a) $y^2 = (-4)^2 = (-4)(-4) = 16$ **b)** $-y^2 = -(-4)^2 = -(-4)(-4) = -16$

c) $(-y)^2 = [-(-4)]^2 = (4)^2 = 16$

Now Try Exercise 97

Note that $-x^2$ will always be a negative number for any nonzero value of x, and $(-x)^2$ will always be a positive number for any nonzero value of x.

Understanding Algebra

It is easy to confuse the expressions $(-6)^2$ and -6^2. $(-6)^2$ means $(-6)\cdot(-6) = 36$. The expression -6^2 represents the opposite of 6^2. So, $-6^2 = -36$.

AVOIDING COMMON ERRORS

The expression $-x^2$ means $-(x^2)$. When asked to evaluate $-x^2$ for any real number x, many students will incorrectly treat $-x^2$ as $(-x)^2$. For example, to evaluate $-x^2$ when $x = 5$,

CORRECT	INCORRECT
$-5^2 = -(5^2) = -(5)(5)$ $= -25$	~~$-5^2 = (-5)(-5)$ $= 25$~~

EXAMPLE 14 Evaluate $2x^2 + 4x + 1$ when $x = \frac{1}{4}$.

Solution Substitute $\frac{1}{4}$ for each x in the expression, then evaluate using the order of operations.

$$\begin{aligned} 2x^2 + 4x + 1 &= 2\left(\frac{1}{4}\right)^2 + 4\left(\frac{1}{4}\right) + 1 && \text{Substitute.} \\ &= 2\left(\frac{1}{16}\right) + 4\left(\frac{1}{4}\right) + 1 && \text{Exponent} \\ &= \frac{1}{8} + 1 + 1 && \text{Multiply.} \\ &= \frac{1}{8} + 2 && \text{Add.} \\ &= 2\frac{1}{8} \end{aligned}$$

Now Try Exercise 113

EXAMPLE 15 Evaluate $-y^2 + 3(x + 2) - 5$ when $x = -3$ and $y = -2$.

Solution Substitute -3 for each x and -2 for each y, then evaluate using the order of operations.

$$\begin{aligned} -y^2 + 3(x + 2) - 5 &= -(-2)^2 + 3(-3 + 2) - 5 && \text{Substitute.} \\ &= -(-2)^2 + 3(-1) - 5 && \text{Parentheses} \\ &= -(4) + 3(-1) - 5 && \text{Exponent} \\ &= -4 - 3 - 5 && \text{Multiply.} \\ &= -7 - 5 && \text{Subtract, left to right.} \\ &= -12 \end{aligned}$$

Now Try Exercise 121

1.9 Exercise Set MathXL® MyMathLab®

Warm-Up Exercises

Fill in the blanks with the appropriate word, phrase, or symbol(s) from the following list.

base	exponent	grouping	innermost
left	evaluate	middle	outermost

1. When an expression has only additions and subtractions, it is evaluated from __________ to right.
2. After evaluating grouping symbols, the next step in the order of operations is to __________ exponents.
3. Parentheses, brackets, and braces are examples of __________ symbols.
4. In the expression 7^5, the 7 is called the __________.
5. In the expression 7^5, the 5 is called the __________.
6. When grouping symbols are nested, begin evaluating at the __________ group.

Practice the Skills

Evaluate.

7. 2^3
8. 3^2
9. 1^5
10. 1^7
11. 5^1
12. 7^1
13. $(-1)^2$
14. $(-1)^4$
15. $(-1)^3$
16. $(-1)^5$
17. -3^2
18. -4^2
19. $(-3)^2$
20. $(-4)^2$
21. -5^2
22. $(-5)^2$
23. $(-6)^2$
24. -6^2
25. -2^4
26. $(-2)^4$

27. $\left(\frac{3}{4}\right)^2$ **28.** $\left(\frac{4}{5}\right)^2$ **29.** $\left(-\frac{3}{4}\right)^2$ **30.** $\left(-\frac{4}{5}\right)^2$

31. $\left(\frac{3}{4}\right)^3$ **32.** $\left(\frac{4}{5}\right)^3$ **33.** $\left(-\frac{3}{4}\right)^3$ **34.** $\left(-\frac{4}{5}\right)^3$

In Exercises 35–46, **a)** *determine by observation whether the answer should be positive or negative and explain your answer; and* **b)** *evaluate the expression.*

35. -7^2 **36.** -8^2 **37.** $(-7)^2$ **38.** $(-8)^2$

39. $-(-7^2)$ **40.** $-(-8^2)$ **41.** $-(-7)^2$ **42.** $-(-8)^2$

43. $(-1.2)^3$ **44.** $(-2.1)^3$ **45.** $\left(-\frac{5}{8}\right)^2$ **46.** $\left(-\frac{3}{5}\right)^2$

Evaluate.

47. $3 + 3 \cdot 6$ **48.** $7 - 5^2 + 8$ **49.** $6 - 6 + 8$

50. $(8^2 \div 4) - (20 - 4)$ **51.** $-7 + 2 \cdot 6^2 - 8$ **52.** $6 + 2 \cdot 3^2 - 10$

53. $-3^3 + 27$ **54.** $(-2)^3 + 8 \div 4$ **55.** $(4 - 5) \cdot (5 - 1)^2$

56. $-10 - 6 - 3 - 2$ **57.** $3 \cdot 7 + 4 \cdot 2$ **58.** $4^2 - 3 \cdot 4 - 6$

59. $5 - 2(7 + 5)$ **60.** $8 + 3(6 + 4)$ **61.** $-32 - 5(7 - 10)^2$

62. $-40 - 3(4 - 8)^2$ **63.** $\frac{3}{4} + 2\left(\frac{1}{5}\right)^2$ **64.** $-\frac{2}{3} - 3\left(\frac{3}{4}\right)^2$

65. $-4 + 3[-1 + (12 \div 2^2)]$ **66.** $-2 + 4[-3 + (48 \div 4^2)]$ **67.** $(6 \div 3)^3 + 4^2 \div 8$

68. $4 + (4^2 - 13)^4 - 3$ **69.** $-7 - 48 \div 6 \cdot 2^2 + 5$ **70.** $-7 - 56 \div 7 \cdot 2^2 + 4$

71. $(9 \div 3) + 4(7 - 2)^2$ **72.** $(12 \div 4) + 5(6 - 4)^2$ **73.** $[4 + ((5 - 2)^2 \div 3)^2]^2$

74. $(20 \div 5 \cdot 5 \div 5 - 5)^2$ **75.** $(-3)^3 + 8 \div 2$ **76.** $-3^3 + 8 \div 2$

77. $2[1.55 + 5(3.7)] - 3.35$ **78.** $(8.4 + 3.1)^2 - (3.64 - 1.2)$ **79.** $\left(\frac{2}{5} + \frac{3}{8}\right) - \frac{3}{20}$

80. $\left(\frac{5}{6} \cdot \frac{4}{5}\right) + \left(\frac{2}{3} \cdot \frac{5}{8}\right)$ **81.** $\frac{3}{4} - 4 \cdot \frac{5}{40}$ **82.** $\frac{1}{8} - \frac{1}{4} \cdot \frac{3}{2} + \frac{3}{5}$

83. $\frac{4}{5} + \frac{3}{4} \div \frac{1}{2} - \frac{2}{3}$ **84.** $\frac{12 - (4 - 6)^2}{6 + 4^2 \div 2^2}$ **85.** $\frac{-4 - [2(9 \div 3) - 5]}{6^2 - 3^2 \cdot 7}$

86. $\frac{[(7 - 3)^2 - 4]^2}{9 - 16 \div 8 - 4}$ **87.** $\frac{-[4 - (6 - 12)^2]}{[(9 \div 3) + 4]^2 + 2^2}$ **88.** $\frac{[(5 - (3 - 7)) - 2]^2}{2[(16 \div 2^2) - (8 \cdot 4)]}$

89. $\{5 - 2[4 - (6 \div 2)]^2\}^2$ **90.** $\{-6 - [3(16 \div 4^2)^2]\}^2$ **91.** $-\{4 - [-3 - (2 - 5)]^2\}$

92. $3\{4[(3 - 4)^2 - 3]^3 - 1\}$ **93.** $\{4 - 3[2 - (9 \div 3)]^2\}^2$ **94.** $2\{5[(4 - 6)^3 - 1]^2 - 3\}$

Evaluate **a)** x^2, **b)** $-x^2$, *and* **c)** $(-x)^2$ *for the following values of x.*

95. 5 **96.** 8 **97.** -2 **98.** -5

99. 6 **100.** 7 **101.** $-\frac{1}{3}$ **102.** $\frac{3}{4}$

Evaluate each expression for the given value of the variable or variables.

103. $x + 6; x = -2$ **104.** $2x - 4x + 5; x = 3$ **105.** $-7z - 3; z = 6$

106. $3(x - 2); x = 5$ **107.** $a^2 - 6; a = -3$ **108.** $b^2 - 8; b = 5$

109. $3p^2 - 6p - 4; p = 2$ **110.** $2r^2 - 5r + 3; r = 1$ **111.** $-4x^2 - 2x + 1; x = -1$

112. $-t^2 - 4t + 5; t = -4$ **113.** $-x^2 - 2x + 5; x = \frac{1}{2}$ **114.** $2x^2 - 4x - 10; x = \frac{3}{4}$

115. $4(3x + 1)^2 - 6x; x = 5$ **116.** $3n^2(2n - 1) + 5; n = -4$

117. $r^2 - s^2; r = -2, s = -3$ **118.** $p^2 - q^2; p = 5, q = -3$

119. $5(x - 6y) + 3x - 7y; x = 1, y = -5$

120. $4(x + y)^2 + 2(x + y) + 3; x = 2, y = 4$

121. $3(x - 4)^2 - (3y - 4)^2; x = -1, y = -2$

122. $6x^2 + 3xy - y^2; x = 2, y = -3$

Problem Solving

Write the following statements as mathematical expressions using parentheses and brackets, and then evaluate.

123. Multiply 6 by 3. From this product, subtract 4. From this difference, subtract 2.

124. Add 4 to 9. Divide this sum by 2. Add 10 to this quotient.

125. Multiply 10 by 4. Add 9 to this product. Subtract 6 from this sum. Divide this difference by 7.

126. Multiply 6 by 3. To this product, add 27. Divide this sum by 8. Multiply this quotient by 10.

127. Add $\frac{4}{5}$ to $\frac{3}{7}$. Multiply this sum by $\frac{2}{3}$.

128. Multiply $\frac{3}{8}$ by $\frac{4}{5}$. To this product, add $\frac{7}{120}$. From this sum, subtract $\frac{1}{60}$.

129. For what value or values of x does $-(x^2) = -x^2$?

130. For what value or values of x does $x = x^2$?

131. Road Trip If a car travels at 65 miles per hour, the distance it travels in t hours is $65t$. Determine how far a car traveling at 65 miles per hour travels in 2.5 hours.

132. Sales Tax If the sales tax on an item is 8%, the sales tax on an item costing d dollars, before tax, can be found by the expression $0.08d$. Determine the sales tax on a scrapbook that costs $19.99.

133. Baseball Height Anthony Salain throws a baseball upward with an initial velocity of 57 feet per second from a height of 6 feet. The height of the baseball above the ground at any time, t, in seconds, can be found by the expression $-16t^2 + 57t + 6$. Determine the height of the baseball after 1 second.

134. Projectile Height An object is projected upward with an initial velocity of 48 feet per second from the top of a 70-foot building. The height of the object above the ground at any time t, in seconds, can be found by the expression $-16t^2 + 48t + 70$. Determine the height of the object after 2 seconds.

135. New Car Cost If the sales tax on an item is 7%, then the total cost of an item including sales tax can be found by the expression $c + 0.07c$, where c is the cost of the item before tax. Find the total cost of a car if the cost before tax is $21,000.

136. New Motorcycle Cost If the sales tax on an item is 6%, then the total cost of an item including sales tax can be found by the expression $c + 0.06c$, where c is the cost of the item before tax. Find the total cost of a motorcycle if the cost before tax is $13,000.

Challenge Problems

137. Grass Growth The rate of growth of grass in inches per week depends on a number of factors, including rainfall and temperature. For a certain region of the country, the growth per week can be approximated by the expression $0.2R^2 + 0.003RT + 0.0001T^2$, where R is the weekly rainfall, in inches, and T is the average weekly temperature, in degrees Fahrenheit. Find the amount of growth of grass for a week in which the rainfall is 2 inches and the average temperature is 70°F.

See Exercise 137.

Insert one pair of parentheses to make each statement true.

138. $12 - 4 - 6 + 10 = 24$

139. $14 + 6 \div 2 \times 4 = 40$

Group Activity

Discuss and answer Exercises 140–143 as a group, according to the instructions. Each question has four parts. For parts **a)**, **b)**, *and* **c)**, *simplify the expression and write the answer in exponential form. Use the knowledge gained in parts* **a)–c)** *to answer part* **d)**. *(General rules that may be used to solve exercises like these will be discussed in Chapter 4.)*

a) *Group member 1: Do part* **a)** *of each exercise.*

b) *Group member 2: Do part* **b)** *of each exercise.*

c) *Group member 3: Do part* **c)** *of each exercise.*

d) As a group, answer part **d)** of each exercise. You may need to make up other examples like parts **a)–c)** to help you answer part **d)**.

140. a) $2^2 \cdot 2^3$ **b)** $3^2 \cdot 3^3$ **c)** $2^3 \cdot 2^4$ **d)** $x^m \cdot x^n$

141. a) $\frac{2^3}{2^2}$ **b)** $\frac{3^4}{3^2}$ **c)** $\frac{4^5}{4^3}$ **d)** $\frac{x^m}{x^n}$

142. a) $(2^3)^2$ **b)** $(3^3)^2$ **c)** $(4^2)^2$ **d)** $(x^m)^n$

143. a) $(2x)^2$ **b)** $(3x)^2$ **c)** $(4x)^3$ **d)** $(ax)^m$

Cumulative Review Exercises

[1.2] **144. Dogs** The graph shows the number of dogs in various houses selected at random in a neighborhood.

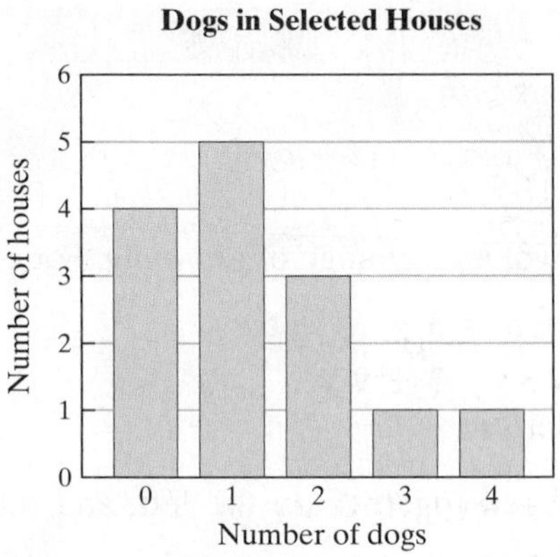

Dogs in Selected Houses

Number of houses (0–6) vs. Number of dogs (0, 1, 2, 3, 4)

a) How many houses have two dogs?

b) Make a chart showing the number of houses that have no dogs, one dog, two dogs, and so on.

c) How many dogs in total are there in all the houses

d) Determine the mean number of dogs in all the houses surveyed.

145. Taxi Cost Yellow Cab charges $2.40 for the first $\frac{1}{2}$ mile plus 20 cents for each additional $\frac{1}{8}$ mile or part thereof. Find the cost of a 3-mile trip.

[1.6] **146.** Add $-\frac{7}{12} + \frac{4}{9}$.

[1.8] **147.** Divide $\left(\frac{-5}{7}\right) \div \left(\frac{-3}{14}\right)$.

1.10 Properties of the Real Number System

1. Learn the commutative property.
2. Learn the associative property.
3. Learn the distributive property.
4. Learn the identity properties.
5. Learn the inverse properties.

Here, we introduce various properties of the real number system.

1 Learn the Commutative Property

The **commutative property of addition** states that the order in which any two real numbers are added does not matter.

Commutative Property of Addition

If a and b represent any two real numbers, then

$$a + b = b + a$$

Notice that the commutative property involves a change in *order*. For example,

$$4 + 3 = 3 + 4$$
$$7 = 7$$

The **commutative property of multiplication** states that the order in which any two real numbers are multiplied does not matter.

Commutative Property of Multiplication

If a and b represent any two real numbers, then

$$a \cdot b = b \cdot a$$

Understanding Algebra
Two numbers can be added in either order or multiplied in either order. Thus, those operations are *commutative*. Neither subtraction nor division are commutative operations.

For example,

$$6 \cdot 3 = 3 \cdot 6$$
$$18 = 18$$

The commutative property does not hold for subtraction or division. For example, $4 - 6 \neq 6 - 4$ and $6 \div 3 \neq 3 \div 6$.

2 Learn the Associative Property

The **associative property of addition** states that, in the addition of three or more numbers, parentheses may be placed around any two adjacent numbers without changing the results.

Associative Property of Addition

If a, b, and c represent any three real numbers, then

$$(a + b) + c = a + (b + c)$$

Notice that the associative property involves a change of *grouping*. For example,

$$(3 + 4) + 5 = 3 + (4 + 5)$$
$$7 + 5 = 3 + 9$$
$$12 = 12$$

In this example, the 3 and 4 are grouped together on the left, and the 4 and 5 are grouped together on the right.

The **associative property of multiplication** states that, in the multiplication of three or more numbers, parentheses may be placed around any two adjacent numbers without changing the results.

Associative Property of Multiplication

If a, b, and c represent any three real numbers, then

$$(a \cdot b) \cdot c = a \cdot (b \cdot c)$$

For example,

$$(6 \cdot 2) \cdot 4 = 6 \cdot (2 \cdot 4)$$
$$12 \cdot 4 = 6 \cdot 8$$
$$48 = 48$$

Since the associative property involves a change of grouping, when the associative property is used, the content within the parentheses changes.

The associative property does not hold for subtraction or division. For example, $(4 - 1) - 3 \neq 4 - (1 - 3)$ and $(8 \div 4) \div 2 \neq 8 \div (4 \div 2)$.

Often when we add numbers we group the numbers so that we can add them easily. For example, when we add $70 + 50 + 30$ we may first add the $70 + 30$ to get 100. We are able to do this because of the commutative and associative properties.

$$\begin{aligned}(70 + 50) + 30 &= 70 + (50 + 30) && \text{Associative property of addition}\\ &= 70 + (30 + 50) && \text{Commutative property of addition}\\ &= (70 + 30) + 50 && \text{Associative property of addition}\\ &= 100 + 50 && \text{Addition facts}\\ &= 150\end{aligned}$$

Notice in the second step that the same numbers remained in parentheses but the order of the numbers changed, $50 + 30$ to $30 + 50$. Since this step involved a change in order (and not grouping), this is the commutative property of addition.

3 Learn the Distributive Property

A very important property of the real numbers is the **distributive property of multiplication over addition.** We often shorten the name to the **distributive property.**

Distributive Property

If a, b, and c represent any three real numbers, then

$$a(b + c) = ab + ac$$

For example, if we let $a = 2$, $b = 3$, and $c = 4$, then

$$2(3 + 4) = (2 \cdot 3) + (2 \cdot 4)$$
$$2 \cdot 7 = 6 + 8$$
$$14 = 14$$

Therefore, we may either add first and then multiply, or multiply first and then add. Another example of the distributive property is

$$2(x + 3) = 2 \cdot x + 2 \cdot 3 = 2x + 6$$

The distributive property can be expanded in the following manner:

$$a(b + c + d + \cdots + n) = ab + ac + ad + \cdots + an$$

For example, $3(x + y + 5) = 3x + 3y + 15$.

HELPFUL HINT

The *commutative property* changes *order*.

The *associative property* changes *grouping*.

The *distributive property* involves *two operations*, usually multiplication and addition.

EXAMPLE 1 Name each property illustrated.

a) $5 + x = x + 5$

b) $5(r + s) = 5 \cdot r + 5 \cdot s = 5r + 5s$

c) $x \cdot y = y \cdot x$

d) $(-12 + 3) + 4 = -12 + (3 + 4)$

Solution

a) Commutative property of addition

b) Distributive property

c) Commutative property of multiplication

d) Associative property of addition

Now Try Exercise 29

HELPFUL HINT

Do not confuse the distributive property with the associative property of multiplication. Make sure you understand the difference.

Distributive Property	Associative Property of Multiplication
$3(4 + x) = 3 \cdot 4 + 3 \cdot x$	$3(4 \cdot x) = (3 \cdot 4)x$
$= 12 + 3x$	$= 12x$

For the distributive property to be used, there must be two *terms* within parentheses, separated by a plus or minus sign as in $3(4 + x)$.

4 Learn the Identity Properties

Now we will discuss the *identity properties*. When the number 0 is added to any real number, the real number is unchanged. For example, $5 + 0 = 5$ and $0 + 5 = 5$. For this reason we call 0 the **identity element of addition** or **additive identity**. When any real number is multiplied by 1, the real number is unchanged. For example, $7 \cdot 1 = 7$ and $1 \cdot 7 = 7$. For this reason we call 1 the **identity element of multiplication** or **multiplicative identity**.

Identity Properties

If a represents any real number, then

$$a + 0 = a \quad \text{and} \quad 0 + a = a \qquad \text{Identity property of addition}$$

and

$$a \cdot 1 = a \quad \text{and} \quad 1 \cdot a = a \qquad \text{Identity property of multiplication}$$

We often use the identity properties without realizing we are using them. For example, when we reduce $\frac{15}{50}$, we may do the following:

$$\frac{15}{50} = \frac{3 \cdot 5}{10 \cdot 5} = \frac{3}{10} \cdot \frac{5}{5} = \frac{3}{10} \cdot 1 = \frac{3}{10}$$

When we showed that $\frac{3}{10} \cdot 1 = \frac{3}{10}$, we used the identity property of multiplication.

5 Learn the Inverse Properties

The last properties we will discuss in this chapter are the **inverse properties**. Numbers like 3 and -3 are *opposites* or *additive inverses* because $3 + (-3) = 0$ and $-3 + 3 = 0$. Any two numbers whose sum is 0 are called *additive inverses* of each other. In general, for any real number a its additive inverse is $-a$.

Numbers like 4 and $\frac{1}{4}$ are *reciprocals* or *multiplicative inverses* because $4 \cdot \frac{1}{4} = 1$ and $\frac{1}{4} \cdot 4 = 1$. Any two numbers whose product is 1 are called *multiplicative inverses* of each other. In general, for any non-zero real number a, its multiplicative inverse is $\frac{1}{a}$. The number 0 does not have a multiplicative inverse. The inverse properties are summarized below.

Inverse Properties

If a represents any real number, then

$$a + (-a) = 0 \quad \text{and} \quad -a + a = 0 \qquad \text{Inverse property of addition}$$

and

$$a \cdot \frac{1}{a} = 1 \quad \text{and} \quad \frac{1}{a} \cdot a = 1 \, (a \neq 0) \qquad \text{Inverse property of multiplication}$$

We often use the inverse properties without realizing we are using them. For example, to evaluate the expression $6x + 2$ when $x = \frac{1}{6}$, we may do the following:

$$6x + 2 = 6\left(\frac{1}{6}\right) + 2 = 1 + 2 = 3$$

When we multiplied $6\left(\frac{1}{6}\right)$ and replaced it with 1, we used the inverse property of multiplication. We will be using both the identity and inverse properties throughout the book, although we may not specifically refer to them by name.

EXAMPLE 2 Name each property illustrated.

a) $2(x + 6) = (2 \cdot x) + (2 \cdot 6) = 2x + 12$

b) $3x \cdot 1 = 3x$

c) $(3 \cdot 6) \cdot 5 = 3 \cdot (6 \cdot 5)$

d) $y \cdot \frac{1}{y} = 1$

e) $2a + (-2a) = 0$

f) $3y + 0 = 3y$

Solution

a) Distributive property
b) Identity property of multiplication
c) Associative property of multiplication
d) Inverse property of multiplication
e) Inverse property of addition
f) Identity property of addition

Now Try Exercise 35

EXAMPLE 3 In parts **a)–f),** the name of a property is given followed by part of an equation. Complete the equation to the right of the equals sign to illustrate the given property.

a) Associative property of multiplication
$(5 \cdot 4) \cdot 7 =$

b) Inverse property of addition
$3c + (-3c) =$

c) Identity property of multiplication
$6y \cdot 1 =$

d) Distributive property
$3(x + 5) =$

e) Identity property of addition
$2a + 0 =$

f) Inverse property of multiplication
$b \cdot \frac{1}{b} =$

Solution

a) $5 \cdot (4 \cdot 7)$ **b)** 0 **c)** $6y$ **d)** $3x + 15$ **e)** $2a$ **f)** 1

Now Try Exercise 55

1.10 Exercise Set MathXL® MyMathLab®

Warm-Up Exercises

Fill in the blanks with the appropriate word, phrase, or symbol(s) from the following list.

commutative	additive	identity	multiplication	distributive
$\frac{1}{5}$	$-\frac{1}{5}$	5	-5	addition
associative	multiplicative	0	1	inverse

1. $(5 + 4) + 6 = 5 + (4 + 6)$ illustrates the ________ property of addition.

2. $14(-3) = -3(14)$ illustrates the ________ property of multiplication.

3. The number ________ does not have a multiplicative inverse.

4. $4 \cdot (25 \cdot 3) = (4 \cdot 25) \cdot 3$ illustrates the associative property of ________.

5. 5 is the additive inverse of ________.

6. $\frac{1}{5}$ is the multiplicative inverse of ________.

7. $-5(x - 3) = -5x + 15$ illustrates the ________ property of multiplication over addition.

8. $x + y = y + x$ illustrates the commutative property of ________.

9. When any real number is multiplied by the number 1, the real number is unchanged. For this reason we call 1 the multiplicative ________.

10. When any real number is added to the number 0, the real number is unchanged. For this reason we call 0 the ________ identity.

Practice the Skills

In Exercises 11–22, for the given expression, determine **a)** *the additive inverse, and* **b)** *the multiplicative inverse.*

11. 6

12. 5

13. -3

14. -7

15. x

16. z

17. 0

18. 1

19. $\frac{1}{5}$

20. $\frac{1}{8}$

21. $-\frac{5}{6}$

22. $-\frac{2}{9}$

In Exercises 23–40, name each property illustrated.

23. $6(x + 7) = 6x + 42$

24. $3 + y = y + 3$

25. $(x + 3) + 5 = x + (3 + 5)$

26. $2(x + 3) = (2)(x) + (2)(3) = 2x + 6$

27. $5 \cdot y = y \cdot 5$

28. $-4x + 4x = 0$

29. $p \cdot (q \cdot r) = (p \cdot q) \cdot r$

30. $1 \cdot x = x$

31. $5w + (-5w) = 0$

32. $3 + (4 + t) = (3 + 4) + t$

33. $3z \cdot 1 = 3z$

34. $0 + 3y = 3y$

35. $2y \cdot \frac{1}{2y} = 1$

36. $x \cdot y = y \cdot x$

37. $3 + x = x + 3$

38. $2 \cdot (5 \cdot 12) = (2 \cdot 5) \cdot 12$

39. $-7x + 0 = -7x$

40. $\frac{1}{3} \cdot 3 = 1$

In Exercises 41–58, the name of a property is given followed by part of an equation. Complete the equation, to the right of the equals sign, to illustrate the given property.

41. associative property of multiplication
$-6 \cdot (4 \cdot 2) =$

42. identity property of multiplication
$(1)\left(-\frac{1}{3}b\right) =$

43. commutative property of multiplication
$x \cdot y =$

44. distributive property
$4(x + 3) =$

45. commutative property of addition
$4x + 3y =$

46. associative property of multiplication
$-9 \cdot (3 \cdot 8) =$

47. inverse property of multiplication
$\frac{1}{3} \cdot 3 =$

48. commutative property of multiplication
$(x + 2)3 =$

49. associative property of addition
$(3x + 4) + 6 =$

50. commutative property of addition
$3(x + y) =$

51. identity property of addition
$-5x + 0 =$

52. associative property of addition
$-5 + (6 + 8) =$

53. distributive property $4(x + y + 3) =$

54. inverse property of addition $(-7a) + 7a =$

55. inverse property of addition $3n + (-3n) =$

56. identity property of addition $0 + 2x =$

57. identity property of multiplication $\left(\frac{5}{2}n\right)(1) =$

58. inverse property of multiplication $\left(\frac{x}{2}\right)\left(\frac{2}{x}\right) =$

Problem Solving

Indicate whether the given processes are commutative. That is, does changing the order in which the actions are done result in the same final outcome? Explain each answer.

59. Putting sugar and then cream in coffee; putting cream and then sugar in coffee.

60. Applying suntan lotion and then sunning yourself; sunning yourself and then applying suntan lotion.

61. Putting your contacts in and then washing your hair; washing your hair and then putting your contacts in.

62. Putting on your sweater and then your coat; putting on your coat and then your sweater.

63. Writing on the chalkboard and then erasing the chalkboard; erasing the chalkboard and then writing on the chalkboard.

64. Getting your hands dirty and then washing your hands; washing your hands and then getting your hands dirty.

In Exercises 65–70, indicate whether the given processes are associative. For a process to be associative, the final outcome must be the same when the first two actions are performed first or when the last two actions are performed first. Explain each answer.

65. Cleaning the kitchen sink, dusting the bedroom furniture, and doing the laundry.

66. In a store, buying cereal, soap, and dog food.

67. Turning on a DVD player, inserting a DVD, and watching the DVD.

68. Putting on a shirt, a tie, and a sweater.

69. Starting a car, moving the shift lever to drive, and then stepping on the gas.

70. Putting cereal, milk, and sugar in a bowl.

71. The commutative property of addition is $a + b = b + a$. Explain why $(3 + 4) + x = x + (3 + 4)$ also illustrates the commutative property of addition.

72. The commutative property of multiplication is $a \cdot b = b \cdot a$. Explain why $(3 + 4) \cdot x = x \cdot (3 + 4)$ also illustrates the commutative property of multiplication.

Challenge Problems

73. Consider $x + (3 + 5) = x + (5 + 3)$. Does this illustrate the commutative property of addition or the associative property of addition? Explain.

74. Consider $x + (3 + 5) = (3 + 5) + x$. Does this illustrate the commutative property of addition or the associative property of addition? Explain.

75. Consider $x + (3 + 5) = (x + 3) + 5$. Does this illustrate the commutative property of addition? Explain.

76. The commutative property of multiplication is $a \cdot b = b \cdot a$. Explain why $(3 + 4) \cdot (5 + 6) = (5 + 6) \cdot (3 + 4)$ also illustrates the commutative property of multiplication.

Cumulative Review Exercises

[1.3] **77.** Add $2\frac{3}{5} + \frac{2}{3}$.

78. Subtract $3\frac{5}{8} - 2\frac{3}{16}$.

[1.6] **79.** Add $102.7 + (-113.9)$.

80. Write the opposite of $\frac{7}{8}$.

Chapter 1 Summary

IMPORTANT FACTS AND CONCEPTS	EXAMPLES

Section 1.2

Guidelines for Problem Solving

1. **Understand the problem.**
2. **Translate the problem to mathematical language.**
3. **Carry out the mathematical calculations necessary to solve the problem.**
4. **Check the answer obtained in step 3.**
5. **Make sure you have answered the question.**

See page 8 for more details on problem solving.

Jacob Thomas can pay either \$725 cash for a desk or pay \$300 down and \$20 a month for 24 months. How much money can he save by paying cash?

Solution: Understand We need to determine how much he would pay if he paid \$300 down and \$20 a month for 24 months. This amount would then be compared to \$725.

Translate

$$\begin{pmatrix}\text{amount if}\\ \text{paid over 24}\\ \text{months}\end{pmatrix} = \$300 + \begin{pmatrix}\text{additional}\\ \text{amount paid}\\ \text{over 24 months}\end{pmatrix}$$

Carry Out

$$\begin{aligned} &= \$300 + \$20(24)\\ &= \$300 + \$480\\ &= \$780\end{aligned}$$

The difference in the amounts is $\$780 - \$725 = \$55$.

Check and Answer

The answer appears reasonable. Jacob can save \$55 by paying the total amount at the time of purchase.

An **algebraic expression** is a general term for any collection of numbers, variables, grouping symbols, and operations.

$$8(24 \div 4),\ 3x - 5,\ a^2 - 6$$

Bar Graph

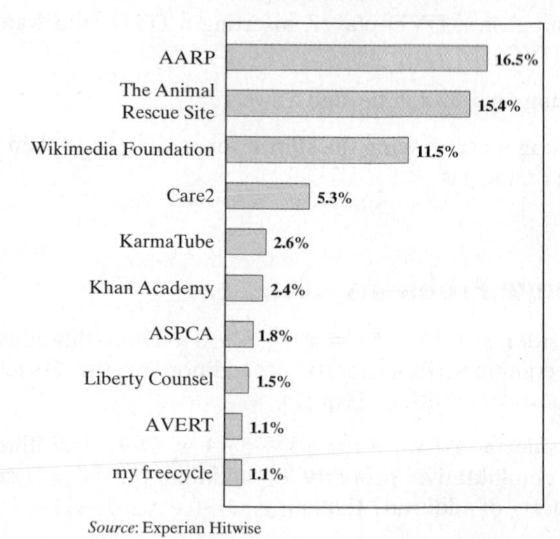

Source: Experian Hitwise

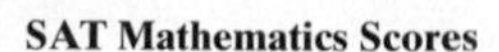

Line Graph

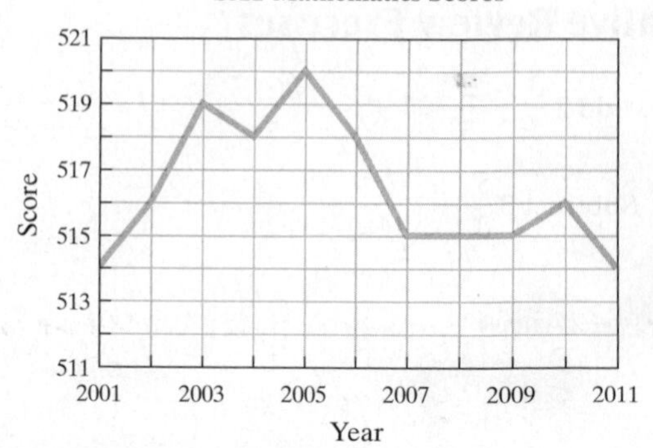

Source: National Center for Education Statistics

IMPORTANT FACTS AND CONCEPTS	EXAMPLES
Section 1.2 (Continued)	
Circle Graph	**Mike's Bikes Sales** Racing 7% Other 15% Mountain 35% Road 32% Beach 11%
The symbol ≈ is read **"is approximately equal to."**	$57.91536 \approx 57.92$
Measures of central tendency or **averages** are values that are representative of a set of data. The **mean** of a set of data is determined by adding all the values and dividing the sum by the number of values. The **median** is the value in the middle of a set of **ranked data.**	The mean and median are averages. The mean of Adam Michael's test grades of 82, 95, 76, 92, and 88 is $\frac{82+95+76+92+88}{5} = \frac{433}{5} = 86.6.$ The median of Adam's five test grades given above is 88 (circled). 76, 82, (88), 92, 95
Section 1.3	
Multiplication Symbols If a and b represent any two mathematical quantities, then each of the following may be used to indicate the product of a and b ("a times b"). $ab \quad a \cdot b \quad a(b) \quad (a)b \quad (a)(b)$	5 times z may be written: $5z, \quad 5 \cdot z, \quad 5(z), \quad (5)z$ or $(5)(z)$
The numbers or variables that are multiplied in a multiplication problem are called **factors.** If $a \cdot b = c$, then a and b are *factors* of c.	In $7 \cdot 9 = 63$, the numbers 7 and 9 are factors of 63.
The numbers 0, 1, 2, 3, 4, . . . are called **whole numbers.**	12, 105, and 0 are whole numbers.
The top number of a fraction is called the **numerator,** and the bottom number is called the **denominator.**	In the fraction $\frac{6}{11}$, the 6 is the numerator and the 11 is the denominator.
To Simplify a Fraction **1.** Find the largest number that will divide (without remainder) both the numerator and the denominator. This number is called the **greatest common factor** (GCF). **2.** Then divide both the numerator and the denominator by the greatest common factor.	The GCF of 36 and 48 is 12. Therefore, $\frac{36}{48} = \frac{36 \div 12}{48 \div 12} = \frac{3}{4}$
To Multiply Fractions $\frac{a}{b} \cdot \frac{c}{d} = \frac{ac}{bd}$	$\frac{3}{4} \cdot \frac{3}{5} = \frac{9}{20}$
To Divide Fractions $\frac{a}{b} \div \frac{c}{d} = \frac{a}{b} \cdot \frac{d}{c} = \frac{ad}{bc}$	$\frac{9}{7} \div \frac{4}{5} = \frac{9}{7} \cdot \frac{5}{4} = \frac{45}{28}$
To Add and Subtract Fractions with Like Denominators $\frac{a}{c} + \frac{b}{c} = \frac{a+b}{c}$ or $\frac{a}{c} - \frac{b}{c} = \frac{a-b}{c}$	**1.** $\frac{3}{13} + \frac{7}{13} = \frac{3+7}{13} = \frac{10}{13}$ **2.** $\frac{4}{5} - \frac{1}{5} = \frac{4-1}{5} = \frac{3}{5}$

IMPORTANT FACTS AND CONCEPTS	EXAMPLES
Section 1.3 (Continued)	
The smallest number that is divisible by two or more denominators is called the **least common denominator** or **LCD**.	In the fractions $\frac{7}{9}$ and $\frac{2}{5}$, 45 is the LCD.
To add (or subtract) fractions with unlike denominators, first rewrite each fraction with a common denominator.	Subtract $\frac{7}{9} - \frac{2}{5}$. $\frac{7}{9} - \frac{2}{5} = \frac{7}{9} \cdot \frac{5}{5} - \frac{2}{5} \cdot \frac{9}{9}$ $= \frac{35}{45} - \frac{18}{45} = \frac{17}{45}$
A **mixed number** consists of a whole number followed by a fraction.	$7\frac{4}{5}$ is a mixed number.
Change a mixed number to a fraction.	$6\frac{2}{3} = 6 + \frac{2}{3} = \frac{18}{3} + \frac{2}{3} = \frac{20}{3}$
Change a fraction to a mixed number.	$\frac{52}{7} = \frac{49}{7} + \frac{3}{7} = 7 + \frac{3}{7} = 7\frac{3}{7}$
Section 1.4	
A **set** is a collection of **elements** listed within braces.	The set $\{2, 4, 6, 8\}$ consists of four elements, namely 2, 4, 6, and 8.
A set that contains no elements is called an **empty set** (or **null set**).	The set of dogs that can fly is an empty set.
Natural numbers: $\{1, 2, 3, 4, 5, \ldots\}$ The natural numbers are also called the **positive integers** or the **counting numbers**.	23, 16, and 1231 are natural numbers.
Whole numbers: $\{0, 1, 2, 3, 4, 5, \ldots\}$	35, 0, and 257 are whole numbers.
Integers: $\{\underbrace{\ldots, -3, -2, -1}_{\text{Negative integers}}, 0, \underbrace{1, 2, 3 \ldots}_{\text{Positive integers}}\}$	-101, 0, and 236 are integers.
Rational numbers: {quotient of two integers, denominator not 0}	$-3, 2.8, 9\frac{1}{2}, 0, 15, -\frac{9}{13}$, and -3.6 are rational numbers.
Irrational numbers: {numbers that can be represented on the number line that are not rational numbers}	$\sqrt{13}$ and $-\sqrt{7}$ are irrational numbers.
Real numbers: {all numbers that can be represented on a number line}	$5.7, -\frac{3}{8}, -16, 0, \sqrt{5}, 3\frac{1}{7}, -2.1, -\sqrt{2}$, and 31 are real numbers.
Section 1.5	
The symbol $>$ is used to represent the words "is greater than." The symbol $<$ is used to represent the words "is less than."	**1.** $3.8 > 3.08$ **2.** $-7 < -1$
The **absolute value** of a number can be considered the distance between the number and 0 on a number line. The absolute value of every number will be either *positive* or *zero*.	**1.** $\lvert 12 \rvert = 12$ **2.** $\lvert 0 \rvert = 0$ **3.** $\lvert -100 \rvert = 100$
Section 1.6	
Add Using a Number Line: Represent the first number to be added (first *addend*) by an arrow starting at 0 on the number line. The arrow is drawn to the right if the number is positive, and to the left if the number is negative. From the tip of the first arrow, draw a second arrow to represent the second addend. The second arrow is drawn to the right or left, as just explained. The sum of the two numbers is found at the tip of the second arrow.	Add $5 + (-7)$. 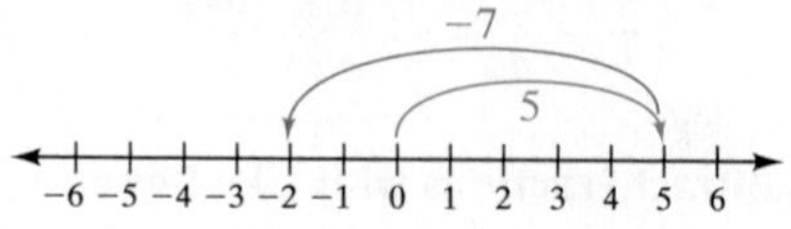 $5 + (-7) = -2$

IMPORTANT FACTS AND CONCEPTS	EXAMPLES
Section 1.6 (Continued)	
Any two numbers whose sum is zero are said to be **opposites** (or **additive inverses**) of each other.	**1.** The opposite of 8 is -8. **2.** The opposite of $-\frac{5}{6}$ is $\frac{5}{6}$.
Add Using Absolute Value: **To add real numbers with the same sign,** add their absolute values. The sum has the same sign as the numbers being added.	Add $-4 + (-9)$. $\|-4\| + \|-9\| = 4 + 9 = 13$ Since both numbers being added are negative, the sum is negative. Thus, $-4 + (-9) = -13$.
To add two signed numbers with different signs, subtract the smaller absolute value from the larger absolute value. The answer has the sign of the number with the larger absolute value.	Add $-25 + 10$. $\|-25\| - \|10\| = 25 - 10 = 15$ Since $\|-25\|$ is greater than $\|10\|$, the sum is negative. Thus, $-25 + 10 = -15$.
Section 1.7	
To Subtract Real Numbers In general, if a and b represent any two real numbers, then $a - b = a + (-b)$ In evaluating expressions involving more than one addition and subtraction, work from left to right unless parentheses or other grouping symbols appear.	**1.** $8 - (+4) = 8 + (-4) = 4$ **2.** $-7 - 5 = -7 + (-5) = -12$ **3.** $-12 - (-6) = -12 + 6 = -6$ **4.** $-8 + 5 - 11 = -3 - 11 = -14$
Section 1.8	
The Sign of the Product of Two Real Numbers **1.** The product of two numbers with **like** signs is a **positive** number. **2.** The product of two numbers with **unlike** signs is a **negative** number.	**1.** $-8(-7) = 56$ **2.** $5(-6) = -30$
The Sign of the Quotient of Two Real Numbers **1.** The quotient of two numbers with **like** signs is a **positive** number. **2.** The quotient of two numbers with **unlike** signs is a **negative** number.	**1.** $\frac{-81}{-9} = 9$ **2.** $\frac{-35}{7} = -5$
The Quotient of a Positive Number and a Negative Number If a and b represent any real numbers, $b \neq 0$, then $\frac{a}{-b} = \frac{-a}{b} = -\frac{a}{b}$	$\frac{3}{-8} = \frac{-3}{8} = -\frac{3}{8}$
Summary of Division Involving 0 If a represents any real number except 0, then $\frac{0}{a} = 0$ $\quad \frac{a}{0}$ is undefined	**1.** $\frac{0}{12} = 0$ **2.** $\frac{5}{0}$ is undefined.
Section 1.9	
In general, the number b to the nth power, written b^n, means $\underbrace{b \cdot b \cdot b \cdot \cdots \cdot b}_{n \text{ factors of } b} = b^n$	$x^5 = x \cdot x \cdot x \cdot x \cdot x$

IMPORTANT FACTS AND CONCEPTS	EXAMPLES
Section 1.9 (Continued)	
Order of Operations **1.** First, evaluate the information within **parentheses** ( ), brackets [ ], or braces { }. These are **grouping symbols**. A fraction bar, –, also serves as a grouping symbol. If the expression contains nested grouping symbols, evaluate the information in the innermost grouping symbols first. **2.** Next, evaluate all **exponents**. **3.** Next, evaluate all **multiplications** or **divisions** in order from left to right. **4.** Finally, evaluate all **additions** or **subtractions** in order from left to right.	$-16 - 3(6 - 8)^2$ $= -16 - 3(-2)^2$ $= -16 - 3(4)$ $= -16 - 12$ $= -28$
Section 1.10	
Commutative Property of Addition If a and b represent any two real numbers, then $a + b = b + a$	$9 + 6 = 6 + 9$ $15 = 15$
Commutative Property of Multiplication If a and b represent any two real numbers, then $a \cdot b = b \cdot a$	$6 \cdot 7 = 7 \cdot 6$ $42 = 42$
Associative Property of Addition If a, b, and c represent any three real numbers, then $(a + b) + c = a + (b + c)$	$(1 + 2) + 3 = 1 + (2 + 3)$ $3 + 3 = 1 + 5$ $6 = 6$
Associative Property of Multiplication If a, b, and c represent any three real numbers, then $(a \cdot b) \cdot c = a \cdot (b \cdot c)$	$(5 \cdot 4) \cdot 3 = 5 \cdot (4 \cdot 3)$ $20 \cdot 3 = 5 \cdot 12$ $60 = 60$
Distributive Property If a, b, and c represent any three real numbers, then $a(b + c) = ab + ac$	$4(y + 9) = 4 \cdot y + 4 \cdot 9 = 4y + 36$
Identity Properties If a represents any real number, then **1.** $a + 0 = a$ and $0 + a = a$ Identity property of addition and **2.** $a \cdot 1 = a$ and $1 \cdot a = a$ Identity property of multiplication	**1.** $15 + 0 = 0 + 15 = 15$ **2.** $8 \cdot 1 = 1 \cdot 8 = 8$
Inverse Properties **1.** If a represents any real number, then $a + (-a) = 0$ and $-a + a = 0$ Inverse property of addition **2.** If a represents any nonzero real number, then $a \cdot \frac{1}{a} = 1$ and $\frac{1}{a} \cdot a = 1$ Inverse property of multiplication	**a)** $9 + (-9) = -9 + 9 = 0$ **b)** $6 \cdot \frac{1}{6} = \frac{1}{6} \cdot 6 = 1$

Chapter 1 Review Exercises

[1.2] *Solve.*

1. **Purchasing Hot Dogs** On Sunday, Frances Alvarado purchases hot dogs to sell in the concession stand during Little League games. Frances purchases 8 packages that contain 30 hot dogs each. The concession stand sells 24 hot dogs on Monday, 31 hot dogs on Tuesday, 17 hot dogs on Wednesday, 49 hot dogs on Thursday, and 53 hot dogs on Friday. Of the hot dogs Frances purchased, how many are left?

2. **Inflation** Assume that the rate of inflation for tuition for a course is 7% per year for the next two years. What will tuition for the course cost in two years if the tuition cost is \$500 today?

3. **Sales Tax** The sales tax in Cattaraugus County is 8.25%.
   a) What was the sales tax that Andrew Jacob paid on a laptop computer that cost \$899.99 before tax?
   b) What is the total cost of the laptop computer including tax?

4. **Plasma TV** Dan Marketos wants to purchase a high definition plasma TV for \$3000. He can either pay the total amount at the time of purchase, or he can agree to pay the store \$400 down and \$225 a month for 12 months. How much money can he save by paying the total amount at the time of purchase?

5. **Test Grades** On Angie Smajstrla's first five exams her grades were 75, 79, 86, 88, and 64. Find the **a)** mean and **b)** median of her grades.

6. **Little League** The number of points scored by a Little League baseball team in its last six games were 21, 3, 17, 10, 9, and 6. Find the **a)** mean and **b)** median of the points.

7. **Commute Times** The bar graph below shows the average commute-to-work times, in minutes, for several cities in the United States. Estimate the average commute-to-work times for the following cities.
   a) San Francisco
   b) Chicago

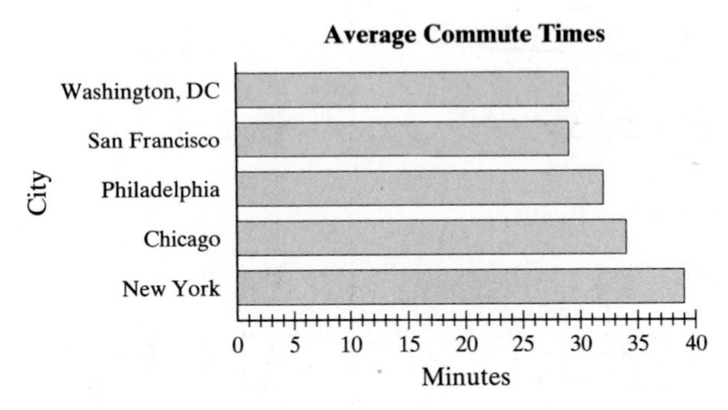

Source: National Public Ratio

8. **Freshman Majors** The graphs below show the majors of the 2010 and the 2013 freshmen classes at North City College.
   a) If there were 900 freshmen in 2010, determine the number of Information Technology majors.
   b) If there were 1100 freshmen in 2013, determine the number of Sports Administration majors.

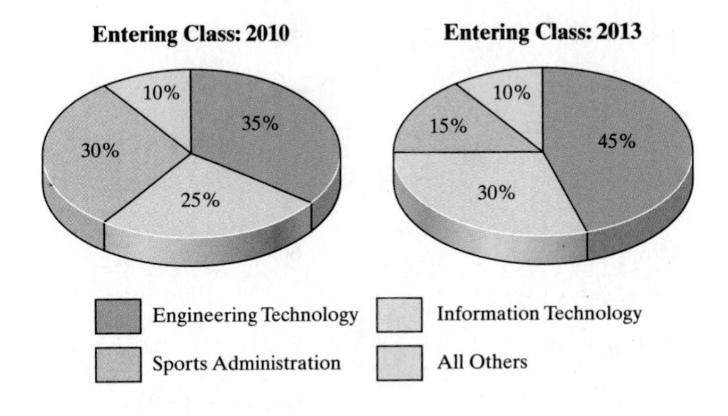

[1.3] *Perform each indicated operation. Simplify your answers.*

9. $\frac{3}{5} \cdot \frac{5}{6}$
10. $3\frac{5}{7} + 2\frac{1}{3}$
11. $\frac{5}{12} \div \frac{3}{5}$
12. $\frac{5}{6} + \frac{1}{3}$
13. $3\frac{1}{6} - 1\frac{1}{4}$
14. $7\frac{3}{8} \div \frac{5}{12}$

[1.4] 15. List the set of natural numbers.

16. List the set of whole numbers.

17. List the set of integers.

18. Describe the set of rational numbers.

19. Consider the following set of numbers.

$$\left\{3, -5, -12, 0, \frac{1}{2}, -0.62, \sqrt{7}, 426, -3\frac{1}{4}\right\}$$

List the numbers that are
   a) positive integers.
   b) whole numbers.
   c) integers.
   d) rational numbers.
   e) irrational numbers.
   f) real numbers.

20. Consider the following set of numbers.

$$\left\{-2.3, -8, -9, 1\frac{1}{2}, \sqrt{2}, -\sqrt{2}, 1, -\frac{3}{17}\right\}$$

List the numbers that are

a) natural numbers.
b) whole numbers.
c) negative integers.
d) integers.
e) rational numbers.
f) irrational numbers.
g) real numbers.

[1.5] *Insert either* $<$, $>$ *or* $=$ *in each shaded area to make a true statement.*

21. $-7 \ \square \ -5$
22. $-2.6 \ \square \ -3.6$
23. $0.50 \ \square \ 0.509$
24. $-\frac{5}{6} \ \square \ -\frac{11}{15}$
25. $-6.3 \ \square \ -6.03$
26. $5 \ \square \ |-3|$
27. $\left|-\frac{9}{2}\right| \ \square \ |-4.5|$
28. $-|-3| \ \square \ -(-3)$

[1.6–1.7] *Evaluate.*

29. $-9 + (-5)$
30. $-6 + 6$
31. $0 + (-3)$
32. $-10 + 4$
33. $-8 - (-2)$
34. $-2 - (-4)$
35. $4 - (-4)$
36. $12 - 12$
37. $2 - 7$
38. $7 - (-7)$
39. $0 - (-4)$
40. $-7 - 5$
41. $\frac{4}{3} - \frac{3}{4}$
42. $\frac{1}{2} + \frac{3}{5}$
43. $\frac{5}{9} - \frac{3}{4}$
44. $-\frac{5}{7} + \frac{3}{8}$
45. $-\frac{5}{12} - \frac{5}{6}$
46. $-\frac{6}{7} + \frac{5}{12}$
47. $\frac{2}{9} - \frac{3}{10}$
48. $\frac{5}{12} - \left(-\frac{3}{5}\right)$

Evaluate.

49. $9 - 4 + 9$
50. $-8 - 9 + 14$
51. $-5 - 4 - 3$
52. $-2 + (-3) - 2$
53. $17 - (+4) - (-3)$
54. $6 - (-2) + 3$

[1.8] *Evaluate.*

55. $7(-9)$
56. $(-8.2)(-3.1)$
57. $(-4)(-5)(-6)$
58. $\left(\frac{3}{5}\right)\left(\frac{-2}{7}\right)$
59. $\left(\frac{10}{11}\right)\left(\frac{3}{-5}\right)$
60. $\left(\frac{-5}{8}\right)\left(\frac{-3}{7}\right)$
61. $0\left(\frac{4}{9}\right)$
62. $(-4)(-6)(-2)(-3)$

Evaluate.

63. $45 \div (-3)$
64. $12 \div (-2)$
65. $-14.72 \div 4.6$
66. $-37.41 \div (-8.7)$
67. $-88 \div (-11)$
68. $-4 \div \left(\frac{-4}{9}\right)$
69. $\frac{28}{-3} \div \left(\frac{9}{-2}\right)$
70. $\frac{14}{3} \div \left(\frac{-6}{5}\right)$

Indicate whether each quotient is 0 *or undefined.*

71. $0 \div 5$
72. $0 \div (-6)$
73. $-12 \div 0$
74. $-4 \div 0$
75. $\frac{8.3}{0}$
76. $\frac{0}{-9.8}$

[1.6–1.8, 1.9] *Evaluate.*

77. $-5(3 - 8)$
78. $2(4 - 8)$
79. $(3 - 6) + 4$
80. $(-4 + 3) - (2 - 6)$
81. $[6 + 3(-2)] - 6$
82. $(-5 - 3)(4)$
83. $[12 + (-4)] + (6 - 8)$
84. $9[3 + (-4)] + 5$
85. $-4(-3) + [4 \div (-2)]$
86. $(-3 \cdot 4) \div (-2 \cdot 6)$
87. $(-3)(-4) + 6 - 3$
88. $[-2(3) + 6] - 4$

[1.9] *Evaluate.*

89. -6^2

90. $(-6)^2$

91. 2^4

92. $(-3)^3$

93. $(-1)^9$

94. $(-2)^5$

95. $\left(\frac{-4}{5}\right)^2$

96. $\left(\frac{2}{5}\right)^3$

97. $5^3 \cdot (-2)^2$

98. $(-2)^4\left(\frac{1}{2}\right)^2$

99. $\left(-\frac{2}{3}\right)^2 \cdot 3^3$

100. $(-4)^3(-2)^2$

Evaluate.

101. $45 \div 15 \cdot 3$

102. $-5 + 7 \cdot 3$

103. $(3.7 - 4.1)^2 + 6.2$

104. $10 - 36 \div 4 \cdot 3$

105. $6 - 3^2 \cdot 5$

106. $[6.9 - (3 \cdot 5)] + 5.8$

107. $\dfrac{6^2 - 4 \cdot 3^2}{-[6 - (3 - 4)]}$

108. $\dfrac{4 + 5^2 \div 5}{6 - (-3 + 2)}$

109. $3[9 - (4^2 + 3)] \cdot 2$

110. $(-3^2 + 4^2) + (3^2 \div 3)$

111. $2^3 \div 4 + 6 \cdot 3$

112. $(4 \div 2)^4 + 4^2 \div 2^2$

113. $(8 - 2^2)^2 - 4 \cdot 3 + 10$

114. $4^3 \div 4^2 - 5(2 - 7) \div 5$

115. $-\{-4[27 \div 3^2 - 2(4 - 2)]\}$

116. $2\{4^3 - 6[4 - (2 - 4)] - 3\}$

Evaluate each expression for the given values.

117. $3x - 7;\ x = 4$

118. $6 - 4x;\ x = -5$

119. $2x^2 - 5x + 3;\ x = 6$

120. $5y^2 + 3y - 2;\ y = -1$

121. $-x^2 + 2x - 3;\ x = -2$

122. $-x^2 + 2x - 3;\ x = 2$

123. $-3x^2 - 5x + 5;\ x = 1$

124. $-x^2 - 8x - 12y;\ x = -3,\ y = -2$

[1.6–1.9] **a)** *Evaluate each expression, and* **b)** *check to see whether your answer is reasonable.*

125. $278 + (-493)$

126. $324 - (-29.6)$

127. $\dfrac{-17.28}{6}$

128. $(-62)(-1.9)$

129. $(-4)^8$

130. $-(4.2)^3$

[1.10] *Name each indicated property.*

131. $(7 + 4) + 9 = 7 + (4 + 9)$

132. $-5(a + 2) = -5a - 10$

133. $6x + 3x = 3x + 6x$

134. $(x + 4)3 = 3(x + 4)$

135. $25 \cdot (4 \cdot 17) = (25 \cdot 4) \cdot 17$

136. $-\frac{1}{5} \cdot (-5) = 1$

137. $8b \cdot 1 = 8b$

138. $-8y + 8y = 0$

139. $0 + 5 = 5$

140. $34 + (6 + 51) = (34 + 6) + 51$

Chapter 1 Practice Test

Chapter Test Prep Videos provide fully worked-out solutions to any of the exercises you want to review. Chapter Test Prep Videos are available via MyMathLab®*, or on* YouTube (www.youtube.com/user/AngelElementaryAlg).

1. Shopping While shopping, Mia Nguyen purchases two half-gallons of milk for $1.30 each, one Boston cream pie for $4.75, and three 2-liter bottles of soda for $1.10 each.

a) What is her total bill before tax?

b) If there is a 7% sales tax on the bottles of soda, how much is the sales tax?

c) How much is her total bill including tax?

d) How much change will she receive from a $50 bill?

2. TV Commercials The line graph below shows the cost for a 30-second commercial during the same time slot on the same channel for 12 consecutive years. How many times greater was the cost for a 30-second commercial in the twelfth year than in the first year?

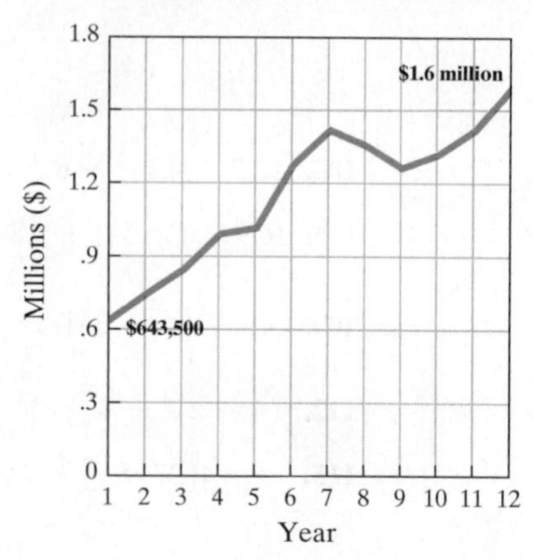

3. Radio The following graph shows the median number of listeners for various radio stations during a specific time.

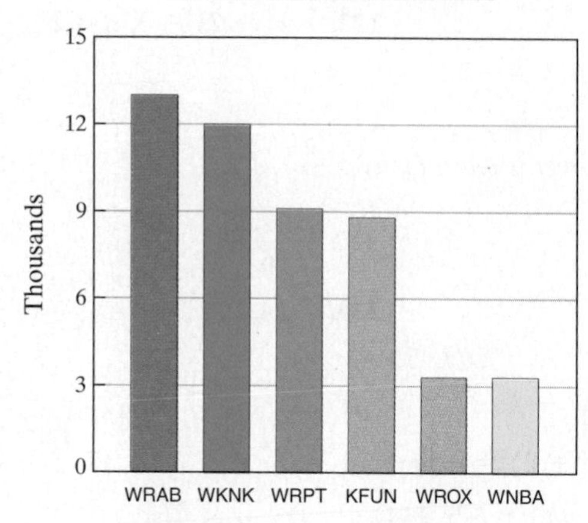

a) Determine the median number of listeners who listened to WRAB during this time.

b) The median number of listeners who listened to KFUN during this time was approximately 8.8 thousand. Explain what this means.

4. Consider the following set of numbers.

$$\left\{-6, 42, -3\frac{1}{2}, 0, 6.52, \sqrt{5}, \frac{5}{9}, -7, -1\right\}$$

List the numbers that are

a) natural numbers.

b) whole numbers.

c) integers.

d) rational numbers.

e) irrational numbers.

f) real numbers.

Insert either $<$, $>$, or $=$ in each shaded area to make a true statement.

5. $-9.9 \quad \square \quad -9.09$

6. $|-3| \quad \square \quad |-2|$

Evaluate.

7. $-7 + (-8)$

8. $-6 - 5$

9. $15 - 12 - 17$

10. $(-4 + 6) - 3(-2)$

11. $(-4)(-3)(2)(-1)$

12. $\left(\frac{-2}{9}\right) \div \left(\frac{-7}{8}\right)$

13. $\left(-18 \cdot \frac{1}{2}\right) \div 3$

14. $-\frac{3}{8} - \frac{4}{7}$

15. $-6(-2 - 3) \div 5 \cdot 2$

16. $\left(-\frac{2}{3}\right)^5$

17. $[6 + ((9 - 3)^2 \div 18)^2]^2$

18. Explain why $-x^2$ will always be a negative value for any nonzero real number selected for x.

Evaluate the expression for the given values.

19. $5x^2 - 8; x = -3$

20. $-x^2 - 6x + 3; x = -2$

21. $6x - 3y^2 + 4; x = 3, y = -2$

22. $-x^2 + xy + y^2; x = 1, y = -2$

Name each indicated property.

23. $x + 3 = 3 + x$

24. $4(x + 9) = 4x + 36$

25. $(2 + x) + 4 = 2 + (x + 4)$

Solving Linear Equations and Inequalities

There are many algebraic formulas that are very useful in everyday life. For example, we can use algebra to determine the amount we owe on a loan, or to convert dollars to other currencies. Algebraic formulas are also useful in sports and recreation. In Example 7 on page 141, we use algebraic formulas to find the area and circumference of a pizza.

Goals of This Chapter

The major emphasis of this chapter is to teach you how to solve linear equations. *To be successful in solving linear equations, you need to have a thorough understanding of adding, subtracting, multiplying, and dividing real numbers.* This material was discussed in Chapter 1.

The first four sections of this chapter will give you the building blocks you will need for solving linear equations. Section 2.5 combines the material previously presented to teach you how to solve a variety of linear equations. In the last few sections of this chapter, you will learn about formulas, ratios, proportions, and solving linear inequalities.

You will be using principles learned in this chapter throughout the book and in real life.

2.1 Combining Like Terms

1. Identify terms.
2. Identify like terms.
3. Combine like terms.
4. Use the distributive property.
5. Remove parentheses when they are preceded by a plus or minus sign.
6. Simplify an expression.

Understanding Algebra

Terms are parts of an expression that get added or subtracted. The expression $2x - 3y - 5$ has 3 terms:

$$\underbrace{\overset{\uparrow}{2x} - \overset{\uparrow}{3y} - \overset{\uparrow}{5}}_{\text{terms}}$$

$2x$ is a term, $-3y$ is a term, and -5 is a term.

Understanding Algebra

Terms are added (or subtracted); *factors* are multiplied.

So, in $4x + 7y$, there are two terms (they are $4x$ and $7y$). The 4 and x are factors of the first term and 7 and y are factors of the second.

Understanding Algebra

A *coefficient* is the numerical factor of a term. If none is shown, it is understood that the coefficient is a "1".

1 Identify Terms

In Section 1.3, we indicated that letters called **variables** are used to represent numbers. A variable can represent a variety of different numbers.

As was indicated in Chapter 1, an **expression,** or **algebraic expression,** is a collection of numbers, variables, grouping symbols, and operation symbols.

Examples of Expressions

$$7, \quad x^2 - 6, \quad 4x - 3, \quad 2(x + 5) + 6, \quad \frac{x + 3}{4}$$

When an algebraic expression consists of several parts, the parts that are *added* are called the **terms** of the expression. Consider the expression $2x - 3y - 5$. The expression can be written as $2x + (-3y) + (-5)$, and so the expression $2x - 3y - 5$ has three terms: $2x$, $-3y$, and -5. The expression $3x^2 + 2xy + 5(x + y)$ also has three terms: $3x^2$, $2xy$, and $5(x + y)$.

When listing the terms of an expression, it is not necessary to list the + sign at the beginning of a term.

Expression	Terms
$-2x + 3y - 8$	$-2x, 3y, -8$
$3y^2 - 2x + \frac{1}{2}$	$3y^2, -2x, \frac{1}{2}$
$7 + x + 4 - 5x$	$7, x, 4, -5x$
$3(x - 1) - 4x + 2$	$3(x - 1), -4x, 2$
$\frac{x + 4}{3} - 5x + 3$	$\frac{x + 4}{3}, -5x, 3$

The numerical part of a term is called its **numerical coefficient** or simply its **coefficient.** In the term $6x$, the 6 is the numerical coefficient. Note that $6x$ means the variable x is multiplied by 6.

Term	Numerical Coefficient
$5x$	5
7	7
$-\frac{1}{2}x$	$-\frac{1}{2}$
$4(x - 3)$	4
$\frac{2x}{3}$	$\frac{2}{3}$, since $\frac{2x}{3}$ means $\frac{2}{3}x$
$\frac{x + 4}{3}$	$\frac{1}{3}$, since $\frac{x + 4}{3}$ means $\frac{1}{3}(x + 4)$

Whenever a term appears without a numerical coefficient, we assume that the numerical coefficient is 1.

Examples

x means $1x$	$-x$ means $-1x$
x^2 means $1x^2$	$-x^2$ means $-1x^2$
xy means $1xy$	$-xy$ means $-1xy$
$(x + 2)$ means $1(x + 2)$	$-(x + 2)$ means $-1(x + 2)$

If an expression has a term that is a number (without a variable), we refer to that number as a **constant term,** or simply a **constant.** In the expression $x^2 + 3x - 4$, the -4 is a constant term, or a constant.

2 Identify Like Terms

Like terms are terms that have the same variables with the same exponents, respectively. Constants, such as 4 and -6, are like terms. Some examples of like terms and unlike terms follow. Note that if two terms are like terms, only their numerical coefficients may differ.

Understanding Algebra

In the expression $2x - 3y - 8x$, we refer to the $2x$ term and the $-8x$ term as *like terms* because their variable parts are identical.

Like Terms	Unlike Terms	
$3x,\ -4x$	$3x,\ 2$ ⟵	(One term has a variable, the other is a constant.)
$4y,\ 6y$	$3x,\ 4y$ ⟵	(Variables differ.)
$5,\ -6$	$x,\ 3$ ⟵	(One term has a variable, the other is a constant.)
$2xy,\ 3xy$	$2x,\ 3xy$ ⟵	(Variables differ.)
$3x^2,\ 4x^2$	$3x,\ 4x^2$ ⟵	(Exponents differ.)
$5ab^2,\ 2ab^2$	$4ab,\ 2ab^2$ ⟵	(Exponents differ.)

EXAMPLE 1 Identify any like terms.

a) $2x + 3x + 4$ **b)** $2x + 3y + 2$ **c)** $x + 3 + y - \frac{1}{2}$ **d)** $x + 3x^2 - 4x^2$

e) $5x - x + 6$ **f)** $3 - 2x + 4x - 6$ **g)** $12 + x^2 - x + 7$

Solution

a) $2x$ and $3x$ are like terms.

b) There are no like terms.

c) 3 and $-\frac{1}{2}$ are like terms.

d) $3x^2$ and $-4x^2$ are like terms.

e) $5x$ and $-x$ (or $-1x$) are like terms.

f) 3 and -6 are like terms; $-2x$ and $4x$ are like terms.

g) 12 and 7 are like terms.

Now Try Exercise 3

3 Combine Like Terms

We often need to simplify expressions that contain like terms. We will do so by using the distributive property (see Section 1.10). For example, in the expression $2a + 3a$ we note that $2a$ and $3a$ are like terms. We rewrite the expression using the distributive property as follows: $2a + 3a = (2 + 3)a = 5a$. To **combine like terms** means to add or subtract the like terms in an expression.

EXAMPLE 2 Combine like terms: $8x + 5x$.

Solution $8x$ and $5x$ are like terms. We can rewrite the expression using the distributive property as follows:

$$8x + 5x = (8 + 5)x = 13x.$$

Now Try Exercise 9

Understanding Algebra

We use the distributive property when combining like terms:

$$5x + 4x = (5 + 4)x = 9x$$

or

$$7y - 3y = (7 - 3)y = 4y$$

The distributive property allows us to combine like terms by using the following procedure.

To Combine Like Terms

1. Determine which terms are like terms.
2. Add or subtract the coefficients of the like terms.
3. Multiply the number found in step 2 by the common variable(s).

Examples 3 through 7 illustrate this procedure.

EXAMPLE 3 Combine like terms: $\frac{3}{5}x - \frac{2}{3}x$.

Solution Since $\frac{3}{5} - \frac{2}{3} = \frac{9}{15} - \frac{10}{15} = -\frac{1}{15}$, then $\frac{3}{5}x - \frac{2}{3}x = -\frac{1}{15}x$.

Now Try Exercise 15

EXAMPLE 4 Combine like terms: $6.47b - 8.39b$.

Solution Since $6.47 - 8.39 = -1.92$, then $6.47b - 8.39b = -1.92b$.

Now Try Exercise 35

EXAMPLE 5 Combine like terms: $3x + x + 5$.

Solution The $3x$ and x are like terms.

$$3x + x + 5 = 3x + 1x + 5 = 4x + 5$$

Now Try Exercise 19

Because of the commutative property of addition, the order of the terms in the answer is not critical. Thus, $5 + 4x$ is also an acceptable answer to Example 5. When writing answers, we generally list the terms containing variables in alphabetical order from left to right, and list the constant term last.

The commutative and associative properties of addition will be used to rearrange the terms in Examples 6 and 7.

EXAMPLE 6 Combine like terms: $3b + 6a - 5 - 2a$.

Solution The only like terms are $6a$ and $-2a$.

$$\begin{aligned} 3b + 6a - 5 - 2a &= 6a - 2a + 3b - 5 && \text{Rearrange terms.} \\ &= 4a + 3b - 5 && \text{Combined like terms.} \end{aligned}$$

Now Try Exercise 21

EXAMPLE 7 Combine like terms: $-2x^2 + 3y - 4x^2 + 3 - y + 5$.

Solution

$-2x^2$ and $-4x^2$ are like terms.

$3y$ and $-y$ are like terms.

3 and 5 are like terms.

Grouping the like terms together gives

$$\begin{aligned} -2x^2 + 3y - 4x^2 + 3 - y + 5 &= -2x^2 - 4x^2 + 3y - y + 3 + 5 \\ &= -6x^2 + 2y + 8 \end{aligned}$$

Now Try Exercise 29

4 Use the Distributive Property

We introduced the distributive property in Section 1.10. Because this property is so important, we will study it again. But before we do, let's briefly review the subtraction of real numbers. Recall from Section 1.7 that

$$6 - 3 = 6 + (-3)$$

In general,

Subtraction of Real Numbers

For any real numbers a and b,

$$a - b = a + (-b)$$

We will use the fact that $a + (-b)$ means $a - b$ in discussing the distributive property.

Distributive Property

For any real numbers a, b, and c,

$$a(b + c) = ab + ac$$

EXAMPLE 8 Use the distributive property to remove parentheses.

a) $2(x + 4)$

b) $-5(p + 3)$

Solution

a) $2(x + 4) = 2x + 2(4) = 2x + 8$

b) $-5(p + 3) = -5p + (-5)(3) = -5p + (-15) = -5p - 15$

Note in part **b)** that, instead of leaving the answer $-5p + (-15)$, we wrote it as $-5p - 15$, which is the preferred form of the answer.

Now Try Exercise 61

EXAMPLE 9 Use the distributive property to remove parentheses.

a) $3(x - 2)$

b) $-2(4x - 3)$

Solution

a) By the definition of subtraction, we write $x - 2$ as $x + (-2)$.

$$\begin{aligned} 3(x - 2) = 3[x + (-2)] &= 3x + 3(-2) \\ &= 3x + (-6) \\ &= 3x - 6 \end{aligned}$$

b) $-2(4x - 3) = -2[4x + (-3)] = -2(4x) + (-2)(-3) = -8x + 6$

Now Try Exercise 63

The distributive property is very important to the study of algebra. Study the Helpful Hint that follows.

HELPFUL HINT

With a little practice, you will be able to eliminate some of the intermediate steps when you use the distributive property. When using the distributive property, there are eight possibilities with regard to signs. Study and understand the eight possibilities that follow.

Positive Coefficient

a) $2(x) = 2x$

$2(x + 3) = 2x + 6$

$2(+3) = +6$

b) $2(x) = 2x$

$2(x - 3) = 2x - 6$

$2(-3) = -6$

c) $2(-x) = -2x$

$2(-x + 3) = -2x + 6$

$2(+3) = +6$

d) $2(-x) = -2x$

$2(-x - 3) = -2x - 6$

$2(-3) = -6$

Negative Coefficient

e) $(-2)(x) = -2x$

$-2(x + 3) = -2x - 6$

$(-2)(+3) = -6$

f) $(-2)(x) = -2x$

$-2(x - 3) = -2x + 6$

$(-2)(-3) = +6$

g) $(-2)(-x) = 2x$

$-2(-x + 3) = 2x - 6$

$(-2)(+3) = -6$

h) $(-2)(-x) = 2x$

$-2(-x - 3) = 2x + 6$

$(-2)(-3) = +6$

Understanding Algebra

Remember, the distributive property is used for multiplying over a sum of 2 or more terms. This is sometimes referred to as *expanding* an expression. So, $2(x - 3y + 9)$ can be rewritten as $2x - 6y + 18$

The distributive property can be expanded as follows:

$$a(b + c + d + \cdots + n) = ab + ac + ad + \cdots + an$$

Examples of the Expanded Distributive Property

$$3(x + y + z) = 3x + 3y + 3z$$
$$2(x + y - 3) = 2x + 2y - 6$$

EXAMPLE 10 Use the distributive property to remove parentheses.

a) $4(x - 3)$ **b)** $-6(5y - 1)$ **c)** $-\frac{1}{2}(4r + 5)$ **d)** $-7(2x + 4y - 9z)$

Solution

a) $4(x - 3) = 4x - 12$

b) $-6(5y - 1) = -30y + 6$

c) $-\frac{1}{2}(4r + 5) = -2r - \frac{5}{2}$

d) $-7(2x + 4y - 9z) = -14x - 28y + 63z$

Now Try Exercise 79

The distributive property can also be used from the right, as in Example 11.

EXAMPLE 11 Use the distributive property to remove parentheses from the expression $(2x - 8y)4$.

Solution We distribute the 4 on the right side of the parentheses over the terms within the parentheses.

$$(2x - 8y)4 = 2x(4) - 8y(4)$$
$$= 8x - 32y$$

Now Try Exercise 83

Example 11 could have been rewritten as $4(2x - 8y)$ by the commutative property of multiplication, and then the 4 could have been distributed from the left to obtain the same answer, $8x - 32y$.

HELPFUL HINT

Students sometimes try to use the distributive property when it cannot be used. For the distributive property to be used, there must be a + or − between the terms *within parentheses* and the terms within parentheses must be *multiplied* by some number or expression. Study the following correct simplifications carefully.

$$4(2xy) = 8xy \qquad \text{(Distributive property is not used.)}$$
$$4(2x + y) = 8x + 4y \qquad \text{(Distributive property is used.)}$$
$$(2x + y) - 4 = 2x + y - 4 \qquad \text{(Distributive property is not used.)}$$
$$(2x + y)(-4) = -8x - 4y \qquad \text{(Distributive property is used.)}$$

5 Remove Parentheses When They Are Preceded by a Plus or Minus Sign

In the expression $(4x + 3)$, how do we remove parentheses? Recall that the coefficient of a term is assumed to be 1 if none is shown. Therefore, we may write

$$\begin{aligned}(4x + 3) &= 1(4x + 3)\\ &= 1(4x) + (1)(3)\\ &= 4x + 3\end{aligned}$$

Other Examples

$$\begin{aligned}(x + 3) &= x + 3\\ (2x - 3) &= 2x - 3\\ +(2x - 5) &= 2x - 5\\ +(x + 2y - 6) &= x + 2y - 6\end{aligned}$$

Understanding Algebra

When no sign or a plus sign precedes parentheses, the parentheses may be removed without changing the expression inside parentheses.

Now consider the expression $-(4x + 3)$. How do we remove parentheses in this expression? Here, the coefficient in front of the parentheses is -1, so each term within the parentheses is multiplied by -1.

$$\begin{aligned}-(4x + 3) &= -1(4x + 3)\\ &= -1(4x) + (-1)(3)\\ &= -4x + (-3)\\ &= -4x - 3\end{aligned}$$

Understanding Algebra

When a minus sign precedes parentheses, the signs of all the terms within the parentheses are changed when the parentheses are removed. For example, when we remove the parentheses in the expression $-(2x - 5)$, we will change the term $2x$ to $-2x$ and the term -5 to $+5$:

$$-(2x - 5) = -2x + 5$$

Thus, $-(4x + 3) = -4x - 3$. *When a minus sign precedes parentheses, the signs of all the terms within the parentheses are changed when the parentheses are removed.*

Examples

$$\begin{aligned}-(x + 4) &= -x - 4\\ -(-2x + 3) &= 2x - 3\\ -(5x - y + 3) &= -5x + y - 3\\ -(-4c - 3d - 5) &= 4c + 3d + 5\end{aligned}$$

6 Simplify an Expression

Combining what we learned in the preceding discussions, we have the following procedure for **simplifying an expression.**

To Simplify an Expression

1. Use the distributive property to remove any parentheses.
2. Combine like terms.

EXAMPLE 12 Simplify $6 + 2(4x + 9)$.

Solution $6 + 2(4x + 9) = 6 + 8x + 18$ Distributive property
$= 8x + 6 + 18$ Commutative property of addition
$= 8x + 24$ Combined like terms.

Note: $8x + 24$ is the same as $24 + 8x$; however, we generally write the term containing the variable first.

Now Try Exercise 89

EXAMPLE 13 Simplify $-\left(\frac{2}{3}x - \frac{1}{4}\right) + 3x$.

Solution $-\left(\frac{2}{3}x - \frac{1}{4}\right) + 3x = -\frac{2}{3}x + \frac{1}{4} + 3x$ Distributive property
$= -\frac{2}{3}x + 3x + \frac{1}{4}$ Rearranged terms.
$= -\frac{2}{3}x + \frac{9}{3}x + \frac{1}{4}$ Wrote x terms with the LCD, 3.
$= \frac{7}{3}x + \frac{1}{4}$ Combined like terms.

Now Try Exercise 101

Notice in Example 13 that $\frac{7}{3}x$ and $\frac{1}{4}$ could not be combined because they are not like terms.

EXAMPLE 14 Simplify $\frac{3}{4}x + \frac{1}{3}(5x - 2)$.

Solution $\frac{3}{4}x + \frac{1}{3}(5x - 2) = \frac{3}{4}x + \frac{1}{3}(5x) + \frac{1}{3}(-2)$ Distributive property
$= \frac{3}{4}x + \frac{5}{3}x - \frac{2}{3}$
$= \frac{9}{12}x + \frac{20}{12}x - \frac{2}{3}$ Wrote x terms with the LCD, 12.
$= \frac{29}{12}x - \frac{2}{3}$ Combined like terms.

Now Try Exercise 103

EXAMPLE 15 Simplify $3(2a - 5) - 3(b - 6) - 4a$.

Solution
$3(2a - 5) - 3(b - 6) - 4a = 6a - 15 - 3b + 18 - 4a$ Distributive property
$= 6a - 4a - 3b - 15 + 18$ Rearranged terms.
$= 2a - 3b + 3$ Combined like terms.

Now Try Exercise 107

HELPFUL HINT

Keep in mind the difference between the concepts of *term* and *factor*. When two or more expressions are *multiplied*, each expression is a **factor** of the product. For example, since $4 \cdot 3 = 12$, the 4 and the 3 are factors of 12. Since $3 \cdot x = 3x$, the 3 and the x are factors of $3x$. Similarly, in the expression $5xyz$, the 5, x, y, and z are all factors.

In an expression, the parts that are *added* are the **terms** of the expression. For example, the expression $2x^2 + 3x - 4$ has three terms, $2x^2$, $3x$, and -4. Note that the terms of an expression may have factors. For example, in the term $2x^2$, the 2 and the x^2 are factors because they are multiplied.

2.1 Exercise Set MathXL® MyMathLab®

Warm-Up Exercises

Fill in the blanks with the appropriate word, phrase, or symbol(s) from the following list.

terms	$4x - 6y - 18$	like	unlike	$-4x + 6y + 18$
constant	factors	coefficient	variable	

1. In the expression $5x - 3y + 17 - 2x$, the 17 is called a ____________ term.

2. When we apply the distributive property to $-2(2x - 3y - 9)$, we obtain ____________.

3. In the expression $5x - 3y + 17 - 2x$, $5x$ and $-2x$ are called ____________ terms.

4. In the expression $5x - 3y + 17 - 2x$, -3 is called the ____________ of the second term.

5. In the expression $5x - 3y + 17 - 2x$, $5x$ and $-3y$ are called ____________ terms.

6. In the expression $12x + 17$, 12 and x are ____________ of the first term.

7. In the expression $-4x^2 + 17x - 90$, $-4x^2$, $17x$, and -90 are called ____________.

8. In the expression $17x$, the x is called a ____________.

Practice the Skills

Combine like terms when possible. If not possible, rewrite the expression as is.

9. $6x + 8x$
10. $4x - 5x$
11. $3x + 6$
12. $4x + 3y$
13. $y + 3 + 4y$
14. $4x - 7x + 4$
15. $\frac{3}{4}a - \frac{6}{11}a$
16. $\frac{3}{4}p - \frac{2}{7}p$
17. $2t - 6x + 5t$
18. $-7 - 4m - 66$
19. $-2w - 3w + 5$
20. $-8y - 4y - 7$
21. $-x + 2 - x - 2$
22. $-3a + 4 + 3a - 13$
23. $3 + 6x - 3 - 6x$
24. $-5y + 7 - 7 + 5y$
25. $5 + 2t - 4t + 16$
26. $7 + d - 13 - 5d$
27. $4p - 6 - 16p - 2$
28. $-6t + 5 + 2t - 9$
29. $3x^2 - 9y^2 + 7x^2 - 5 - y^2 - 2$
30. $-4x^2 - 6y - 3x^2 + 6 - y - 1$
31. $-2x + 4x - 8$
32. $4 - x + 4x - 8$
33. $b + 4 + \frac{3}{5}$
34. $\frac{3}{4}y + 2 + y$
35. $5.1n + 6.42 - 4.3n$
36. $-2.53c + 8.1 - 9.1c$
37. $-\frac{2}{3}x + \frac{5}{9}y + \frac{1}{9}$
38. $\frac{3}{4}p + \frac{1}{7}q + \frac{1}{4}$
39. $13.4x + 1.2x + 8.3$
40. $-4x^2 - 3.1 - 5.2$
41. $-x^2 + 2x^2 + y$
42. $1 + x^2 + 6 - 3x^2$
43. $2x - 7y - 5x + 2y$
44. $3x - 7 - 9 + 4x$
45. $4 - 3n^2 + 9 - 2n$
46. $-5x^2 + 1 - 3x^2 + x$
47. $-19.36 + 40.02x + 12.25 - 18.3x$
48. $-3.4k + 13.01 - 1.09k - 17.3$
49. $\frac{3}{5}x - 3 - \frac{7}{4}x - 2$
50. $\frac{1}{2}y - 4 + \frac{3}{4}x - \frac{1}{5}y$
51. $5w^3 + 2w^2 + w + 3$
52. $3m^3 - 7m^2 + 7m - 2$
53. $2z - 5z^3 - 2z^3 - z^2$
54. $c^3 - 7 + 4c^2 - 2c^2 - 5c^3$
55. $2x^2 + 2x - 5x - 5$
56. $x^2 - 3xy - 2xy + 6$
57. $2a^3 - 6a^2 + 2a + a^2 - 3a + 1$
58. $3b^3 - 3b^2 + 6b + 2b^2 - 2b + 4$

Use the distributive property to remove parentheses.

59. $5(x + 2)$
60. $2(-y + 5)$
61. $5(x + 4)$
62. $-2(y + 8)$
63. $3(x - 6)$
64. $-2(x - 4)$
65. $-\frac{1}{2}(2x - 4)$
66. $-\frac{1}{3}(-6x + 9)$
67. $1(-4 + x)$
68. $-1(5 - x)$
69. $\frac{4}{5}(s - 5)$
70. $-\frac{2}{3}(x - 6)$
71. $-0.3(3x^2 + 5)$
72. $0.4(-3x + 2)$
73. $-\frac{1}{3}(3r - 12)$
74. $-\frac{5}{6}(12x - 18)$
75. $0.7(2x + 0.5)$
76. $-0.3(5x - 0.9)$

77. $-(-x + y)$
78. $-(-p - q)$
79. $-(2x + 4y - 8)$
80. $-3(2a + 3b - 7)$
81. $1.1(3.1x - 5.2y + 2.8)$
82. $2.3(1.6x + 5.1y - 4.1)$
83. $(2x - 9y)5$
84. $(8b - 1)7$
85. $(r + 3s - 19)$
86. $(-p + 2q - 3)$
87. $-3(-x + 2y + 4)$
88. $-4(-2m - 3n + 8)$

Simplify.

89. $5 - (3x + 4)$
90. $7 - (2y - 9)$
91. $-2(3 - x) + 7$
92. $-(3t - 3) + 5$
93. $6x + 2(4x + 9)$
94. $3(x + y) + 2y$
95. $2(x - y) + 2x + 3$
96. $6 + (x - 5) + 3x$
97. $4(2c - 3) - 3(c - 4)$
98. $-5(-3d + 2) + 6(4d - 5)$
99. $8x - (x - 3)$
100. $-(x - 5) - 3x + 4$
101. $-\left(\frac{3}{4}x - \frac{1}{3}\right) + 2x$
102. $-\left(\frac{7}{8}x - \frac{1}{2}\right) - 3x$
103. $\frac{2}{3}x + \frac{1}{2}(5x - 4)$
104. $\frac{4}{5}x + \frac{1}{7}(3x - 1)$
105. $-(3s + 4) - (s + 2)$
106. $6 - 2(2w + 3) + 5w$
107. $4(x - 1) + 2(3 - x) - 4$
108. $4(3b - 2) - 5(c - 4) - 6b$
109. $4(m + 3) - 4m - 12$
110. $-3(a + 2b) + 3(a + 2b)$
111. $0.4 - (y + 5) + 0.6 - 2$
112. $1.03 - (8 - 0.8a) - 0.03 + 1.2a$
113. $4 + (3x - 4) - 5$
114. $2y - 6(y - 2) + 3$
115. $4(x + 2) - 3(x - 4) - 5$
116. $6 - (a - 5) - (2b + 1)$
117. $-0.2(6 - x) - 4(y + 0.4)$
118. $-1.05(4x - 6) + 4(1.5y - 3.05)$
119. $-6x + 7y - (3 + x) + (x + 3)$
120. $3(t - 2) - 2(t + 4) - 6$
121. $\frac{1}{2}(x + 3) + \frac{1}{3}(3x + 6)$
122. $\frac{2}{3}(r - 2) - \frac{1}{2}(r + 4)$

Problem Solving

If □ + □ + □ + ⊙ + ⊙ *can be represented as* 3□ + 2⊙, *write an expression to represent each of the following.*

123. □ + ⊖ + ⊖ + □ + ⊖

124. ⊗ + ☺ + ⊗ + ☺ + ☺ + ☺

125. $x + y + \triangle + \triangle + x + y + y$

126. 2 + x + 2 + ⊖ + ⊖ + 2 + y

In Exercises 127 and 128, consider the following. The positive factors of 6 are 1, 2, 3, and 6 since

$$1 \cdot 6 = 6$$
$$2 \cdot 3 = 6$$
↑ ↑
factors

127. List all the positive factors of 18.

128. List all the positive factors of 24.

Concept/Writing Exercises

129. a) When a minus sign precedes an expression within parentheses, explain how to remove parentheses.

b) Write $-(x - 8)$ without parentheses.

130. a) What are like terms? Determine whether the following are like terms. If not, explain why.

b) $3x, 4y$
c) $7, -2$
d) $5x^2, 5x$
e) $4x, -5xy$

Challenge Problems

Simplify.

131. $4x^2 + 5y^2 + 6(3x^2 - 5y^2) - 4x + 3$

132. $2x^2 - 4x + 8x^2 - 3(x + 2) - x^2 - 2$

133. $2[3 + 4(x - 5)] - [2 - (x - 3)]$

134. $\frac{1}{4}\left[3 - 2(y + 1)\right] - \frac{1}{3}\left[2 - (y - 6)\right]$

Cumulative Review Exercises

[1.5] *Evaluate.*

135. $|-7|$

136. $-|-16|$

[1.7] **137.** Evaluate $-4 - 13 - (-6)$.

[1.9] **138.** Write a paragraph explaining the order of operations.

139. Evaluate $-x^2 + 5x - 6$ when $x = -1$.

2.2 The Addition Property of Equality

1. Identify linear equations.
2. Check solutions to equations.
3. Identify equivalent equations.
4. Use the addition property to solve equations.
5. Solve equations by doing some steps mentally.

1 Identify Linear Equations

A statement that shows two algebraic expressions are equal is called an **equation.** For example, $4x + 3 = 2x - 4$ is an equation. In this chapter, we learn to solve **linear equations** in one variable.

Linear Equation

A **linear equation** in one variable is an equation that can be written in the form

$$ax + b = c$$

where a, b, and c are real numbers, $a \neq 0$.

Examples of Linear Equations

$$x + 4 = 7$$
$$2x - 4 = 6$$

Understanding Algebra

An *equation* is a statement that two expressions are equal.

$$2x - 3 = 7$$
$$y - 4 = 12 \text{ and}$$
$$6t + 14 = 2.3$$

are all *linear equations in one variable* in which the variable is x, y, and t, respectively.

2 Check Solutions to Equations

Solution to an Equation

The **solution to an equation** is the number or numbers that when substituted for the variable or variables make the equation a true statement.

Understanding Algebra

A *solution* to an equation is a number that makes the equation a true statement.

The solution to $x + 4 = 7$ is 3 because when x is 3, the statement is true: 7 equals 7. A solution can be **checked** by substituting your answer for the variable in the original equation. If the result is true, your solution is correct. If the result is false, then you need to go back and find your error. Try to check all your solutions as this will improve your algebra skills.

To check whether 3 is the solution to $x + 4 = 7$, we substitute 3 for each x in the equation.

Understanding Algebra

We use the symbol $\stackrel{?}{=}$ in the process of checking a solution—we are questioning whether a statement is true.

Check

$$\begin{aligned} x &= 3 \\ x + 4 &= 7 \\ 3 + 4 &\stackrel{?}{=} 7 \\ 7 &= 7 \quad \text{True} \end{aligned}$$

Since the check results in a true statement, 3 is a solution.

EXAMPLE 1 Consider the equation $2x - 4 = 6$. Determine whether -3 is a solution.

Solution To determine whether -3 is a solution to the equation, we substitute -3 for x.

Check

$$\begin{aligned} x &= -3 \\ 2x - 4 &= 6 \\ 2(-3) - 4 &\stackrel{?}{=} 6 \\ -6 - 4 &\stackrel{?}{=} 6 \\ -10 &= 6 \quad \text{False} \end{aligned}$$

Since we obtained a false statement, -3 is not a solution.

Now Try Exercise 11

Now check to see if 5 is a solution to the equation in Example 1. Your check should show that 5 is a solution.

EXAMPLE 2 Determine whether 18 is a solution to the following equation.

$$3x - 2(x + 3) = 12$$

Solution To determine whether 18 is a solution, we substitute 18 for each x in the equation.

Check

$$x = 18$$
$$3x - 2(x + 3) = 12$$
$$3(18) - 2(18 + 3) \stackrel{?}{=} 12$$
$$3(18) - 2(21) \stackrel{?}{=} 12$$
$$54 - 42 \stackrel{?}{=} 12$$
$$12 = 12 \quad \text{True}$$

Since we obtained a true statement, 18 is a solution.

Now Try Exercise 15

EXAMPLE 3 Determine whether $-\frac{3}{2}$ is a solution to the following equation.

$$3(n + 3) = 6 + n$$

Solution Substitute $-\frac{3}{2}$ for each n in the equation.

Check

$$n = -\frac{3}{2}$$
$$3(n + 3) = 6 + n$$
$$3\left(-\frac{3}{2} + 3\right) \stackrel{?}{=} 6 + \left(-\frac{3}{2}\right)$$
$$3\left(-\frac{3}{2} + \frac{6}{2}\right) \stackrel{?}{=} \frac{12}{2} - \frac{3}{2}$$
$$3\left(\frac{3}{2}\right) \stackrel{?}{=} \frac{9}{2}$$
$$\frac{9}{2} = \frac{9}{2} \quad \text{True}$$

Thus, $-\frac{3}{2}$ is a solution.

Now Try Exercise 21

3 Identify Equivalent Equations

To **solve an equation** means to find its solution. *To solve an equation, it is necessary to get the variable alone on one side of the equation, that is, on one side of the equals sign. When we get an equation in this form, we say that we isolated the variable.* To isolate the variable we make use of two properties: the addition and multiplication properties of equality. Look first at **Figure 2.1.**

Left side of equation	=	Right side of equation

FIGURE 2.1

Think of an equation as a balanced statement whose left side is balanced by its right side. When solving an equation, we must make sure that the equation remains balanced at all times. *We ensure that an equation always remains balanced by doing the same thing to both sides of the equation.* For example, if we add a number to the left side of the equation, we must add exactly the same number to the right side. If we multiply the right side of the equation by some number, we must multiply the left side by the same number.

Understanding Algebra

The strategy for solving a linear equation is to *isolate* the variable on one side of the equation.

When we add the same number to both sides of an equation or multiply both sides of an equation by the same nonzero number, we do not change the solution to

Understanding Algebra

Think of an equation as a balance scale. When a solution checks, the scale balances:

But when a solution doesn't check, the sides are not in balance:

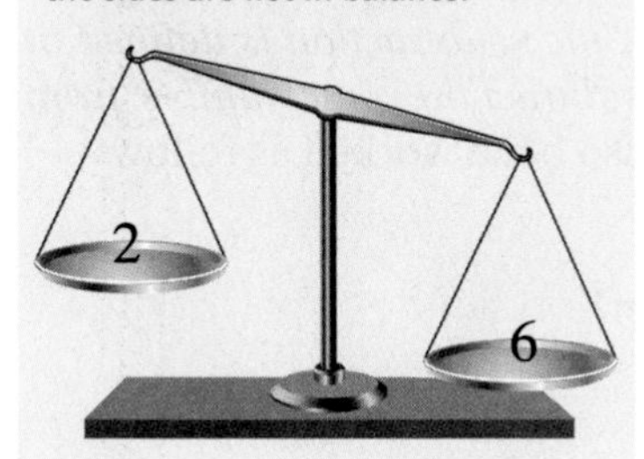

the equation. Two or more equations with the same solution are called **equivalent equations.** The equations $2x - 4 = 2$, $2x = 6$, and $x = 3$ are equivalent, since the solution to each is 3.

Check: $x = 3$

$$\begin{aligned} 2x - 4 &= 2 \\ 2(3) - 4 &\stackrel{?}{=} 2 \\ 6 - 4 &\stackrel{?}{=} 2 \\ 2 &= 2 \quad \text{True} \end{aligned}$$

$$\begin{aligned} 2x &= 6 \\ 2(3) &\stackrel{?}{=} 6 \\ 6 &= 6 \quad \text{True} \end{aligned}$$

$$\begin{aligned} x &= 3 \\ 3 &= 3 \quad \text{True} \end{aligned}$$

When solving an equation, we use the addition and multiplication properties to express a given equation as equivalent equations until we obtain the solution.

4 Use the Addition Property to Solve Equations

Now we are ready to define the **addition property of equality.**

Addition Property of Equality

If $a = b$ then $a + c = b + c$ for any real numbers a, b, and c.

This property means that the same number can be added to both sides of an equation.

To isolate the variable when solving equations of the form $x + a = b$, *we add* $(-a)$ *to both sides of the equation.* This isolates the variable x on the left side of the equation.

Equation	To Solve, Use the Addition Property to Eliminate the Number . . .	. . . By Adding the Opposite to Both Sides
$x - 4 = -3$	-4	4
$x + 5 = 9$	5	-5
$-3 = k + 7$	7	-7
$-5 = x - 4$	-4	4
$-6.25 = y + 12.78$	12.78	-12.78

Now let's work some examples.

Understanding Algebra

Since we can think of an equation as a scale we are trying to keep in balance, adding the same quantity to both sides preserves the balance.

Now, add 4 to both sides: $x - 4 + 4 = -3 + 4$ or simply $x = 1$.

EXAMPLE 4 Solve the equation $x - 4 = -3$.

Solution To isolate the variable, x, we must eliminate the -4 from the left side of the equation. To do this we add 4, the opposite of -4, to *both sides* of the equation.

$$\begin{aligned} x - 4 &= -3 \\ x - 4 + 4 &= -3 + 4 \quad \text{Add 4 to both sides.} \\ x + 0 &= 1 \\ x &= 1 \end{aligned}$$

Note how the process helps to isolate x.

Check

$$\begin{aligned} x - 4 &= -3 \\ 1 - 4 &\stackrel{?}{=} -3 \\ -3 &= -3 \quad \text{True} \end{aligned}$$

Now Try Exercise 27

Space limitations prevent us from showing all checks. However, *you should check all of your answers*.

EXAMPLE 5 Solve the equation $x + 5 = 9$.

Solution To solve this equation, we must eliminate the 5 from the left side of the equation. To do this, we add -5, the opposite of 5, to *both sides* of the equation.

$$x + 5 = 9$$
$$x + 5 + (-5) = 9 + (-5) \quad \text{Add } -5 \text{ to both sides.}$$
$$x + 0 = 4$$
$$x = 4$$

Now Try Exercise 29

Understanding Algebra

"Subtract 5 from both sides" is equivalent to saying "Add (-5) to both sides."

In Example 5, we added -5 to both sides of the equation. We know that $5 + (-5) = 5 - 5$. Thus, adding a negative 5 to both sides of the equation is equivalent to subtracting a 5 from both sides of the equation. *Since subtraction is defined in terms of addition, the addition property also allows us to subtract the same number from both sides of the equation.* Thus, Example 5 could have also been worked as follows:

$$x + 5 = 9$$
$$x + 5 - 5 = 9 - 5 \quad \text{Subtract 5 from both sides.}$$
$$x + 0 = 4$$
$$x = 4$$

Rather than adding a negative number to both sides of the equation, we will subtract a number from both sides of the equation.

EXAMPLE 6 Solve the equation $-3 = k + 7$.

Solution We must isolate the variable, k, which is on the right side of the equation.

$$-3 = k + 7$$
$$-3 - 7 = k + 7 - 7 \quad \text{Subtract 7 from both sides.}$$
$$-10 = k + 0$$
$$-10 = k$$

Check

$$-3 = k + 7$$
$$-3 \stackrel{?}{=} -10 + 7$$
$$-3 = -3 \quad \text{True}$$

Now Try Exercise 25

HELPFUL HINT

Remember that our goal in solving an equation is to get the variable alone on one side of the equation. To do this, we add or subtract *the number on the same side of the equation as the variable* to or from both sides of the equation.

Equation	Must Eliminate	Number to Add (or Subtract) to (or from) Both Sides of the Equation	Correct Results	Solution
$x - 5 = 8$	-5	add 5	$x - 5 + 5 = 8 + 5$	$x = 13$
$x - 3 = -12$	-3	add 3	$x - 3 + 3 = -12 + 3$	$x = -9$
$2 = x - 7$	-7	add 7	$2 + 7 = x - 7 + 7$	$9 = x$ or $x = 9$
$x + 12 = -5$	$+12$	subtract 12	$x + 12 - 12 = -5 - 12$	$x = -17$
$6 = x + 4$	$+4$	subtract 4	$6 - 4 = x + 4 - 4$	$2 = x$ or $x = 2$
$13 = x + 9$	$+9$	subtract 9	$13 - 9 = x + 9 - 9$	$4 = x$ or $x = 4$

Notice that under the *Correct Results* column, when the equation is simplified by combining terms, the x will become isolated because the sum of a number and its opposite is 0, and $x + 0$ equals x.

EXAMPLE 7 Solve the equation $-5 = x - 4$.

Solution The variable, x, is on the right side of the equation. By adding 4 to both sides of the equation, we eliminate -4 from the right and x is isolated on the right side:

$$-5 = x - 4$$
$$-5 + 4 = x - 4 + 4 \quad \text{Add 4 to both sides.}$$
$$-1 = x + 0$$
$$-1 = x$$

Thus, the solution is -1.

Now Try Exercise 35

EXAMPLE 8 Solve the equation $-6.25 = y + 12.78$.

Solution The variable, y, is on the right side of the equation. To isolate the variable, subtract 12.78 from both sides of the equation.

$$-6.25 = y + 12.78$$
$$-6.25 - 12.78 = y + 12.78 - 12.78 \quad \text{Subtract 12.78 from both sides.}$$
$$-19.03 = y + 0$$
$$-19.03 = y$$

Thus, the solution is -19.03.

Now Try Exercise 65

AVOIDING COMMON ERRORS

When solving an equation, our goal is to isolate the variable on one side of the equation. Consider the equation $x + 3 = -4$. How do we solve it?

CORRECT

Remove the 3 from the left side of the equation.

$$x + 3 = -4$$
$$x + 3 - 3 = -4 - 3$$
$$x = -7$$

Variable is now isolated.

INCORRECT

Remove the -4 from the right side of the equation.

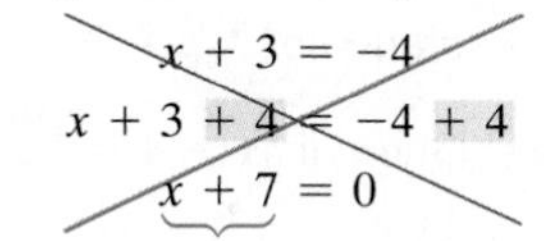

Variable is not isolated.

Remember, use the addition property to *remove the number that is on the same side of the equation as the variable.*

5 Solve Equations by Doing Some Steps Mentally

Consider the following two problems.

a)
$$x - 5 = 12$$
$$x - 5 + 5 = 12 + 5$$
$$x + 0 = 12 + 5$$
$$x = 17$$

b)
$$15 = x + 3$$
$$15 - 3 = x + 3 - 3$$
$$15 - 3 = x + 0$$
$$12 = x$$

When you feel comfortable using the addition property of equality, you may wish to do some of the steps mentally to reduce some of the written work. For example, the preceding two problems may be shortened as follows:

a)
$$x - 5 = 12$$
$$x - 5 + 5 = 12 + 5 \longleftarrow \text{Do this step mentally.}$$
$$x = 12 + 5$$
$$x = 17$$

Shortened Form

$$x - 5 = 12$$
$$x = 12 + 5$$
$$x = 17$$

b)

$$\begin{aligned} 15 &= x + 3 \\ 15 - 3 &= x + 3 - 3 \\ 15 - 3 &= x \\ 12 &= x \end{aligned}$$

⟵ Do this step mentally.

Shortened Form

$$\begin{aligned} 15 &= x + 3 \\ 15 - 3 &= x \\ 12 &= x \end{aligned}$$

2.2 Exercise Set MathXL® MyMathLab®

Warm-Up Exercises

Fill in the blanks with the appropriate word, phrase, or symbol(s) from the following list.

equation	solutions	True	linear	2	opposite
6	isolate	equivalent	False	checking	

1. To solve the equation $6 = x - 2$, first add ________ to both sides of the equation.
2. Numbers which, when substituted for the variable, make the equation a true statement are called ________.
3. The strategy for solving a linear equation in one variable includes a process to ________ the variable.
4. The symbol $\stackrel{?}{=}$ is used when ________ the solution in the given equation.
5. The equation $x - 3 = -19$ is an example of a ________ equation in the variable x.
6. Two equations with the same solutions are called ________ equations.
7. A statement of equality between two expressions, like $x + 7 = 19$, is called an ________.
8. To solve the equation $x - 6 = -2$, first add ________ to both sides of the equation.
9. When solving the equation $x + a = b$ for x, the ________ of a is added to both sides of the equation.
10. (True or False) The equation $x + 19 = -14$ and $x + 60 = 27$ are equivalent equations. ________

Practice the Skills

11. Is $x = 2$ a solution of $4x - 3 = 5$?
12. Is $x = -6$ a solution of $2x + 1 = x - 5$?
13. Is $x = -3$ a solution of $2x - 5 = 5(x + 2)$?
14. Is $x = 1$ a solution of $2(x - 3) = -3(x + 1)$?
15. Is $p = -15$ a solution of $2p - 5(p + 7) = 10$?
16. Is $k = -2$ a solution of $5k - 6(k - 1) = 8$?
17. Is $x = 3.4$ a solution of $3(x + 2) - 3(x - 1) = 9$?
18. Is $x = 5.5$ a solution of $3(x + 3) - 5(x - 4) = 18$?
19. Is $x = \frac{1}{2}$ a solution of $4x - 4 = 2x - 2$?
20. Is $x = \frac{1}{3}$ a solution of $7x + 3 = 2x + 5$?
21. Is $x = \frac{11}{2}$ a solution of $3(x + 2) = 5(x - 1)$?
22. Is $y = -\frac{5}{3}$ a solution of $5(y + 1) - 2(y - 2) = 4$?

Solve each equation and check your solution.

23. $x + 2 = 7$
24. $x + 6 = 14$
25. $-6 = x + 1$
26. $-5 = x + 4$
27. $x - 4 = -8$
28. $p - 5 = -12$
29. $t + 9 = 52$
30. $x + 8 = 17$
31. $-6 + w = 9$
32. $-10 + n = 5$
33. $27 = x + 16$
34. $50 = x - 35$
35. $-18 = x - 14$
36. $-4 = w + 5$
37. $9 + x = 4$
38. $12 + w = 9$
39. $4 + x = -8$
40. $7 + z = -3$
41. $7 + r = -23$
42. $7 + v = -7$
43. $8 = 8 + v$
44. $-5 = -5 + m$
45. $7 + x = -50$
46. $-5 + c = -23$
47. $12 = 16 + x$
48. $62 = z - 15$
49. $15 + y = -50$
50. $-20 = 4 + x$
51. $-15 + x = -15$
52. $-7 + x = -7$
53. $5 = x - 12$
54. $8 = x - 9$
55. $-50 = x - 24$
56. $-37 = x - 12$
57. $43 = 15 + p$
58. $-25 = 74 + x$
59. $40.2 + d = -5.9$
60. $15.8 + b = -4.9$
61. $-37 + x = 9.5$
62. $-29 + x = 7.2$
63. $x - 8.77 = -17$
64. $t - 3.41 = -11$
65. $-5.9 = x + 4.01$
66. $-13.7 = x + 7.28$

Problem Solving

67. Do you think the equation $x + 1 = x + 2$ has a real number as a solution? Explain. (We will discuss equations like this in Section 2.5.)
68. Do you think the equation $x + 4 = x + 4$ has more than one real number as a solution? If so, how many solutions does it have? Explain. (We will discuss equations like this in Section 2.5.)

Concept/Writing Exercises

69. To solve a linear equation, we "isolate the variable." Explain what this means.

70. Explain why these three equations are equivalent:

$$2x + 3 = 5, \quad 2x = 2, \quad x = 1$$

Challenge Problems

We can solve equations that contain unknown symbols. Solve each equation for the symbol indicated by adding or subtracting a symbol to or from both sides of the equation. Explain each answer. (Remember that to solve the equation you want to isolate the symbol you are solving for on one side of the equation.)

71. $x - \triangle = \square$, for x

72. $\square$ + ☺ = $\triangle$, for ☺

73. ☺ = $\square + \triangle$, for $\square$

74. $\square = \triangle$ + ☺, for ☺

Group Activity

Discuss and answer Exercise 75 as a group.

75. Consider the equation $2(x + 3) = 2x + 6$.

a) Group member 1: Determine whether 4 is a solution to the equation.

b) Group member 2: Determine whether -2 is a solution to the equation.

c) Group member 3: Determine whether 0.3 is a solution to the equation.

d) Each group member: Select a number not used in parts **a)–c)** and determine whether that number is a solution to the equation.

e) As a group, write what you think is the solution to the equation $2(x + 3) = 2x + 6$ and write a paragraph explaining your answer.

Cumulative Review Exercises

[1.6] *Add.*

76. $-\frac{7}{15} + \frac{5}{6}$

77. $-\frac{11}{12} + \left(-\frac{3}{8}\right)$

[2.1] *Simplify.*

78. $4x + 3(x - 2) - 5x - 7$

79. $-(2t + 4) + 3(4t - 7) - 3t$

2.3 The Multiplication Property of Equality

1 Identify reciprocals.

2 Use the multiplication property to solve equations.

3 Solve equations of the form $-x = a$.

4 Do some steps mentally when solving equations.

1 Identify Reciprocals

Recall that two numbers are **reciprocals** of each other when their product is 1. Some examples of numbers and their reciprocals follow.

Number	Reciprocal	Product
2	$\frac{1}{2}$	$(2)\left(\frac{1}{2}\right) = 1$
-7	$-\frac{1}{7}$	$(-7)\left(-\frac{1}{7}\right) = 1$
$\frac{1}{5}$	5	$\left(\frac{1}{5}\right)(5) = 1$
$-\frac{2}{3}$	$-\frac{3}{2}$	$\left(-\frac{2}{3}\right)\left(-\frac{3}{2}\right) = 1$
-1	-1	$(-1)(-1) = 1$

The reciprocal of a positive number is a positive number and the reciprocal of a negative number is a negative number. Note that 0 has no reciprocal.

Understanding Algebra

Recall from Section 1.10 that any two numbers whose product is 1 are called *reciprocals* or *multiplicative inverses* of each other. For any nonzero number a the reciprocal is $\frac{1}{a}$ and

$$a \cdot \frac{1}{a} = \frac{1}{a} \cdot a = 1.$$

2 Use the Multiplication Property to Solve Equations

In Section 2.2, we used the addition property of equality to solve equations of the form $x + a = b$, where a and b represent real numbers. In this section, we use the multiplication property of equality to solve equations of the form $ax = b$, where a and b represent real numbers.

HELPFUL HINT

It is important that you recognize the difference between equations like $x + 2 = 8$ and $2x = 8$. In $x + 2 = 8$, the 2 is a *term* that is being added to x, so we use the addition property to solve the equation. In $2x = 8$, the 2 is a *factor* of $2x$. The 2 is the coefficient multiplying the x, so we use the multiplication property to solve the equation.

Understanding Algebra

Since we can think of an equation as a scale we are trying to keep in balance, multiplying the same quantity on both sides preserves the balance.

Now, multiply by $\frac{1}{5}$ on both sides:

$\frac{1}{5}(5x) = \frac{1}{5}(20)$ or simply $x = 4$.

Multiplication Property of Equality

If $a = b$, then $a \cdot c = b \cdot c$ for any real numbers a, b, and c.

The multiplication property of equality allows us to multiply both sides of an equation by the same nonzero number. We will use this property to solve equations of the form $ax = b$. Recall that a is called the *coefficient* of x. We will isolate the variable x by multiplying both sides of the equation by the reciprocal of a, which is $\frac{1}{a}$. Since $\frac{1}{a} \cdot a = 1$, the left side of the equation becomes $1x$ which is the same as x.

Equation	a, the Coefficient of x	To Isolate the Variable x, Multiply Both Sides of the Equation by $\frac{1}{a}$, the Reciprocal of a.
$2x = 12$	2	$\frac{1}{2}$
$-7x = 63$	-7	$-\frac{1}{7}$
$6 = \frac{1}{5}x$	$\frac{1}{5}$	5
$8 = -\frac{2}{3}x$	$-\frac{2}{3}$	$-\frac{3}{2}$

Now let's work some examples.

EXAMPLE 1 Solve the equation $9x = 63$.

Solution The coefficient of x is 9. To isolate the variable, x, we must multiply both sides of the equation by the reciprocal of 9, which is $\frac{1}{9}$.

$$9x = 63$$

$$\frac{1}{9} \cdot 9x = \frac{1}{9} \cdot 63 \quad \text{Multiply both sides by } \frac{1}{9}.$$

$$\frac{1}{\cancel{9}_1} \cdot \overset{1}{\cancel{9}}x = \frac{1}{\cancel{9}_1} \cdot \overset{7}{\cancel{63}} \quad \text{Divide out the common factors.}$$

$$1x = 7$$

$$x = 7$$

Now Try Exercise 9

Notice in Example 1 that $1x$ is replaced by x in the last step. Usually we do this step mentally.

EXAMPLE 2 Solve the equation $\frac{x}{2} = 4$.

Solution Since dividing by 2 is the same as multiplying by $\frac{1}{2}$, the equation $\frac{x}{2} = 4$ is the same as $\frac{1}{2}x = 4$. Therefore, the coefficient of x is $\frac{1}{2}$. We will multiply both sides of the equation by the reciprocal of $\frac{1}{2}$, which is 2.

Understanding Algebra

The equation $\frac{x}{2} = 4$ is the same as $\frac{1}{2}x = 4$. Therefore, the coefficient of x is $\frac{1}{2}$.

$$\frac{x}{2} = 4$$

$$\overset{1}{\cancel{2}}\left(\frac{x}{\underset{1}{\cancel{2}}}\right) = 2 \cdot 4 \quad \text{Multiply both sides by 2.}$$

$$1x = 2 \cdot 4$$

$$x = 8$$

Now Try Exercise 11

EXAMPLE 3 Solve the equation $\frac{2}{3}x = 6$.

Solution The reciprocal of $\frac{2}{3}$ is $\frac{3}{2}$. We multiply both sides of the equation by $\frac{3}{2}$.

$$\frac{2}{3}x = 6$$

$$\frac{3}{2} \cdot \frac{2}{3}x = \frac{3}{2} \cdot 6 \quad \text{Multiply both sides by } \frac{3}{2}.$$

$$1x = 9$$

$$x = 9$$

We will show a check of this solution.

Check

$$\frac{2}{3}x = 6$$

$$\frac{2}{3}(9) \stackrel{?}{=} 6$$

$$6 = 6 \quad \text{True}$$

Now Try Exercise 49

In Example 1, we multiplied both sides of the equation $9x = 63$ by $\frac{1}{9}$ to isolate the variable. We could have also isolated the variable by dividing both sides of the equation by 9, as follows:

$$9x = 63$$

$$\frac{\overset{1}{\cancel{9}}x}{\underset{1}{\cancel{9}}} = \frac{\overset{7}{\cancel{63}}}{\underset{1}{\cancel{9}}} \quad \text{Divide both sides by 9.}$$

$$x = 7$$

We can do this because dividing by 9 is equivalent to multiplying by $\frac{1}{9}$. *Since division can be defined in terms of multiplication* $\left(\frac{a}{b} \text{ means } a \cdot \frac{1}{b}\right)$, *the multiplication property also allows us to divide both sides of an equation by the same nonzero number.*

EXAMPLE 4 Solve the equation $8w = 3$.

Solution To solve the equation we divide both sides of the equation by 8.

$$8w = 3$$

$$\frac{8w}{8} = \frac{3}{8} \quad \text{Divide both sides by 8.}$$

$$w = \frac{3}{8}$$

Now Try Exercise 33

EXAMPLE 5 Solve the equation $-15 = -3z$.

Solution To isolate z, we divide both sides of the equation by -3.

$$-15 = -3z$$

$$\frac{-15}{-3} = \frac{-3z}{-3} \quad \text{Divide both sides by } -3.$$

$$5 = z$$

Now Try Exercise 21

EXAMPLE 6 Solve the equation $0.24x = 1.20$.

Solution Divide both sides of the equation by 0.24 to isolate the variable x.

$$0.24x = 1.20$$

$$\frac{0.24x}{0.24} = \frac{1.20}{0.24} \quad \text{Divide both sides by 0.24.}$$

$$1x = 5$$

$$x = 5$$

Now Try Exercise 35

HELPFUL HINT

When solving an equation of the form $ax = b$, we can isolate the variable by

1. multiplying both sides of the equation by the reciprocal of a, $\frac{1}{a}$, as was done in Examples 1, 2, and 3, or
2. dividing both sides of the equation by a, as was done in Examples 4, 5, and 6.

Either method may be used to isolate the variable. However, if the equation contains a fraction, or fractions, you will arrive at a solution more quickly by multiplying by the reciprocal of a. This is illustrated in Examples 7 and 8.

EXAMPLE 7 Solve the equation $-2x = \frac{3}{5}$.

Solution Since this equation contains a fraction, we will isolate the variable by multiplying both sides of the equation by $-\frac{1}{2}$, which is the reciprocal of -2.

$$-2x = \frac{3}{5}$$

$$\left(-\frac{1}{2}\right)(-2x) = \left(-\frac{1}{2}\right)\left(\frac{3}{5}\right) \quad \text{Multiply both sides by } -\frac{1}{2}.$$

$$1x = \left(-\frac{1}{2}\right)\left(\frac{3}{5}\right)$$

$$x = -\frac{3}{10}$$

Now Try Exercise 39

EXAMPLE 8 Solve the equation $-6 = -\frac{3}{5}x$.

Solution Since this equation contains a fraction, we will isolate the variable by multiplying both sides of the equation by the reciprocal of $-\frac{3}{5}$, which is $-\frac{5}{3}$.

$$-6 = -\frac{3}{5}x$$

$$\left(-\frac{5}{3}\right)(-6) = \left(-\frac{5}{3}\right)\left(-\frac{3}{5}x\right) \quad \text{Multiply both sides by } -\frac{5}{3}.$$

$$10 = 1x$$

$$10 = x$$

Now Try Exercise 57

In Example 8, the equation was written as $-6 = -\frac{3}{5}x$. This equation is equivalent to the equations $-6 = \frac{-3}{5}x$ and $-6 = \frac{3}{-5}x$. Can you explain why? All three equations have the same solution, 10.

3 Solve Equations of the Form $-x = a$

When solving an equation, we may obtain an equation like $-x = 7$. This is not a solution since $-x = 7$ means $-1x = 7$. The solution to an equation is of the form $x =$ some number. When an equation is of the form $-x = 7$, we can solve for x by multiplying both sides of the equation by -1, as illustrated in the following example.

EXAMPLE 9 Solve the equation $-x = 7$.

Solution $-x = 7$ means that $-1x = 7$. We are solving for x, not $-x$. We can multiply both sides of the equation by -1 to isolate x on the left side of the equation.

$$-x = 7$$

$$-1x = 7$$

$$(-1)(-1x) = (-1)(7) \quad \text{Multiply both sides by } -1.$$

$$1x = -7$$

$$x = -7$$

Check

$$-x = 7$$

$$-(-7) \stackrel{?}{=} 7$$

$$7 = 7 \quad \text{True}$$

Thus, the solution is -7.

Now Try Exercise 23

Understanding Algebra

For any real number a, if $-x = a$, then $x = -a$. For example,

if $-x = 7$, then $x = -7$.

If $-x = -2$, then $x = 2$.

Example 9 may also be solved by dividing both sides of the equation by -1. Whenever we have the negative of a variable equal to a quantity, as in Example 9, we can solve for the variable by multiplying or dividing both sides of the equation by -1.

EXAMPLE 10 Solve the equation $-x = -5$.

Solution

$$-x = -5$$

$$-1x = -5$$

$$(-1)(-1x) = (-1)(-5) \quad \text{Multiply both sides by } -1.$$

$$1x = 5$$

$$x = 5$$

Now Try Exercise 25

 Do Some Steps Mentally When Solving Equations

When you feel comfortable using the multiplication property, you may wish to do some of the steps mentally to reduce some of the written work. Now we present two examples worked out in detail, along with their shortened form.

EXAMPLE 11 Solve the equation $-3t = -21$.

Solution

$$-3t = -21$$
$$\frac{-3t}{-3} = \frac{-21}{-3} \quad \longleftarrow \text{Do this step mentally.}$$
$$t = \frac{-21}{-3}$$
$$t = 7$$

SHORTENED FORM

$$-3t = -21$$
$$t = \frac{-21}{-3}$$
$$t = 7$$

Now Try Exercise 61

EXAMPLE 12 Solve the equation $\frac{1}{5}x = 20$.

Solution

$$\frac{1}{5}x = 20$$
$$5\left(\frac{1}{5}x\right) = 5(20) \quad \longleftarrow \text{Do this step mentally.}$$
$$x = 5(20)$$
$$x = 100$$

SHORTENED FORM

$$\frac{1}{5}x = 20$$
$$x = 5(20)$$
$$x = 100$$

Now Try Exercise 63

HELPFUL HINT

The **addition property** is used to solve equations of the form $\boldsymbol{x + a = b}$. The *addition property* is used when a number is *added to or subtracted from* a variable in the original equation.

$$x + 3 = -6 \qquad\qquad x - 5 = -2$$
$$x + 3 - 3 = -6 - 3 \qquad\qquad x - 5 + 5 = -2 + 5$$
$$x = -9 \qquad\qquad x = 3$$

The **multiplication property** is used to solve equations of the form $\boldsymbol{ax = b}$. It is used when a variable is *multiplied or divided by* a number in the original equation.

$$3x = 6 \qquad\qquad \frac{x}{2} = 4 \qquad\qquad \frac{2}{5}x = 12$$
$$\frac{3x}{3} = \frac{6}{3} \qquad\qquad 2\left(\frac{x}{2}\right) = 2(4) \qquad\qquad \left(\frac{5}{2}\right)\left(\frac{2}{5}x\right) = \left(\frac{5}{2}\right)(12)$$
$$x = 2 \qquad\qquad x = 8 \qquad\qquad x = 30$$

2.3 Exercise Set MathXL® MyMathLab®

Warm-Up Exercises

Fill in the blanks with the appropriate word, phrase, or symbol(s) from the following list.

$\frac{x}{3}$	$\frac{3}{x}$	$\frac{1}{3}$	-1	checking
1	multiplicative	3	isolate	$-\frac{7}{x}$

1. The expression $\frac{1}{3}x$ can be rewritten as __________.

2. To solve the equation $3x = 5$, we multiply both sides of the equation by __________.

3. To solve the equation $\frac{x}{3} = 7$, we multiply both sides of the equation by ________.

4. The strategy for solving a linear equation in one variable includes a process to ________ the variable.

5. The symbol $\stackrel{?}{=}$ is used when ________ the solution in the given equation.

6. Another name for the reciprocal of a number is the ________ inverse.

7. The reciprocal of $-\frac{x}{7}$ is ________.

8. Two numbers equal to their own reciprocal are ________ and ________.

Practice the Skills

Solve each equation and check your solution.

9. $5x = 20$

10. $5x = 50$

11. $\frac{x}{3} = 7$

12. $\frac{y}{5} = 3$

13. $-4x = 12$

14. $-6x = 30$

15. $\frac{x}{4} = -20$

16. $\frac{x}{3} = -3$

17. $\frac{x}{5} = 4$

18. $\frac{x}{8} = -3$

19. $-27n = 81$

20. $-7t = 49$

21. $-7 = 3r$

22. $16 = -4y$

23. $-x = 13$

24. $-x = 9$

25. $-x = -8$

26. $-x = -15$

27. $-\frac{w}{3} = -10$

28. $-4 = \frac{c}{7}$

29. $4 = -12x$

30. $7 = -28x$

31. $-\frac{x}{3} = -2$

32. $-\frac{a}{8} = -7$

33. $43t = 26$

34. $-24x = -18$

35. $-4.2x = -8.4$

36. $-3.88 = 1.94y$

37. $3x = \frac{3}{5}$

38. $7x = -\frac{7}{16}$

39. $5x = -\frac{3}{8}$

40. $-2b = -\frac{4}{5}$

41. $16 = -\frac{x}{4}$

42. $25 = -\frac{a}{5}$

43. $-\frac{b}{4} = -60$

44. $-\frac{c}{6} = 30$

45. $\frac{x}{5} = -9$

46. $\frac{d}{7} = -8$

47. $5 = \frac{x}{4}$

48. $-8 = \frac{x}{-5}$

49. $\frac{3}{5}d = -30$

50. $\frac{2}{7}x = 7$

51. $9x = 0$

52. $-7x = 0$

53. $\frac{-7}{8}w = 0$

54. $-\frac{9}{13}z = 0$

55. $\frac{1}{5}x = 4.5$

56. $-\frac{1}{5}x = 3.7$

57. $-4 = -\frac{2}{3}z$

58. $-9 = \frac{-5}{3}n$

59. $-1.4x = 28.28$

60. $-0.42x = -2.142$

Solve each equation by doing some steps mentally. Check your solution.

61. $-8x = -56$

62. $-9x = -45$

63. $\frac{2}{3}x = 6$

64. $\frac{3}{4}k = 9$

Concept/Writing Exercises

65. a) Explain the difference between $5 + x = 10$ and $5x = 10$.

b) Solve $5 + x = 10$.

c) Solve $5x = 10$.

66. a) Explain the difference between $3 + x = 6$ and $3x = 6$.

b) Solve $3 + x = 6$.

c) Solve $3x = 6$.

67. Consider the equation $\frac{2}{3}x = 4$. This equation could be solved by multiplying both sides of the equation by $\frac{3}{2}$, the reciprocal of $\frac{2}{3}$, or by dividing both sides of the equation by $\frac{2}{3}$. Which method do you feel would be easier? Explain your answer. Find the solution to the equation.

68. Consider the equation $4x = \frac{3}{5}$. Would it be easier to solve this equation by dividing both sides of the equation by 4 or by multiplying both sides of the equation by $\frac{1}{4}$, the reciprocal of 4? Explain your answer. Find the solution to the problem.

69. Consider the equation $\frac{3}{7}x = \frac{4}{5}$. Would it be easier to solve this equation by dividing both sides of the equation by $\frac{3}{7}$ or by multiplying both sides of the equation by $\frac{7}{3}$, the reciprocal of $\frac{3}{7}$? Explain your answer. Find the solution to the equation.

Challenge Problems

70. Consider the equation □⊙ = △.

a) To solve for ⊙, what symbol do we need to isolate?

b) How would you isolate the symbol you specified in part **a)**?

c) Solve the equation for ⊙.

71. Consider the equation ☺ = △⊡.

a) To solve for ⊡, what symbol do we need to isolate?

b) How would you isolate the symbol you specified in part **a)**?

c) Solve the equation for ⊡.

72. Consider the equation # = $\frac{☺}{△}$.

a) To solve for ☺, what symbol do we need to isolate?

b) How would you isolate the symbol you specified in part **a)**?

c) Solve the equation for ☺.

Cumulative Review Exercises

[1.7] **73.** Subtract -4 from -8.

[1.8] **74.** Evaluate $(-3)(-2)(5)(-1)$.

[1.9] **75.** Evaluate $4^2 - 2^3 \cdot 6 \div 3 + 6$.

[1.10] **76.** Name the property illustrated.
$2 + (4 + y) = (2 + 4) + y$

[2.2] **77.** Solve the equation $-48 = x + 9$.

2.4 Solving Linear Equations with a Variable on Only One Side of the Equation

1 Solve linear equations with a variable on only one side of the equation.

2 Solve equations containing decimal numbers or fractions.

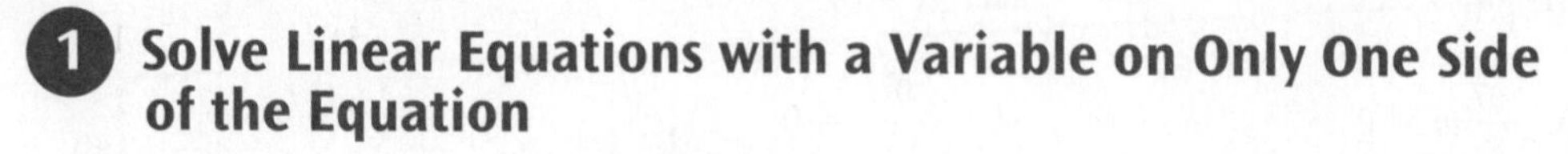

1 Solve Linear Equations with a Variable on Only One Side of the Equation

In this section, we will use *both* the addition and multiplication properties of equality to solve linear equations in which the variable appears on only one side of the equation.

No one method is the "best" to solve all linear equations. But the following general procedure can be used to solve linear equations when the variable appears on only one side of the equation.

To Solve Linear Equations with a Variable on Only One Side of the Equation

1. If the equation contains fractions, multiply *both* sides of the equation by the least common denominator (LCD). This will eliminate the fractions from the equation.
2. Use the distributive property to remove parentheses.
3. Combine like terms on the same side of the equation.
4. Use the addition property to obtain an equation with the term containing the variable on one side of the equation and a constant on the other side. This will result in an equation of the form $ax = b$.
5. Use the multiplication property to isolate the variable.
6. Check the solution in the original equation.

When solving an equation you should always check your solution. To conserve space, we will not show all checks.

Consider the equation $2x + 4 = 10$, which contains no fractions, no parentheses, and no like terms on the same side of the equation. Therefore, we start with step 4.

$$2x + 4 = 10$$

$$2x + 4 - 4 = 10 - 4 \quad \text{Addition property (add } (-4) \text{ to both sides).}$$

$$2x = 6 \quad x\text{-term is now isolated on the left.}$$

Now we use the multiplication property (step 5) to solve for x.

$$2x = 6$$

$$\frac{2x}{2} = \frac{6}{2} \quad \text{Multiplication property (divide both sides by 2).}$$

$$1x = 3 \quad \text{We need not write the coefficient "1."}$$

$$x = 3 \quad \text{Final answer}$$

The solution to the equation $2x + 4 = 10$ is 3. Check this solution:

$$2x + 4 = 10$$

$$2(3) + 4 \stackrel{?}{=} 10$$

$$6 + 4 \stackrel{?}{=} 10$$

$$10 = 10 \quad \text{True}$$

EXAMPLE 1 Solve the equation $5x - 7 = 13$.

Solution Since the equation contains no fractions or parentheses, and since there are no like terms to be combined, we start with step 4.

Step 4

$$5x - 7 = 13$$

$$5x - 7 + 7 = 13 + 7 \quad \text{Add 7 to both sides.}$$

$$5x = 20$$

Step 5

$$\frac{5x}{5} = \frac{20}{5} \quad \text{Divide both sides by 5.}$$

$$x = 4$$

Step 6 Check

$$5x - 7 = 13$$

$$5(4) - 7 \stackrel{?}{=} 13$$

$$20 - 7 \stackrel{?}{=} 13$$

$$13 = 13 \quad \text{True}$$

Since the check is true, the solution is 4. Note that after completing step 4, we obtain $5x = 20$, which is an equation of the form $ax = b$. After completing step 5, we obtain the answer in the form $x =$ some real number.

Now Try Exercise 7

EXAMPLE 2 Solve the equation $-2r - 6 = -3$.

Solution

Step 4

$$-2r - 6 = -3$$

$$-2r - 6 + 6 = -3 + 6 \quad \text{Add 6 to both sides.}$$

$$-2r = 3$$

Step 5

$$\frac{-2r}{-2} = \frac{3}{-2} \quad \text{Divide both sides by } -2.$$

$$r = -\frac{3}{2}$$

Step 6 Check

$$-2r - 6 = -3$$
$$-2\left(-\frac{3}{2}\right) - 6 \stackrel{?}{=} -3$$
$$3 - 6 \stackrel{?}{=} -3$$
$$-3 = -3 \quad \text{True}$$

The solution is $-\frac{3}{2}$.

Now Try Exercise 15

Note that checks are always made with the *original* equation. In some of the following examples, the check will be omitted to save space. You should check all of your answers.

EXAMPLE 3 Solve the equation $16 = 4x + 6 - 2x$.

Solution Again we must isolate the variable, x. Since the right side of the equation has two like terms containing the variable, x, we will first combine these like terms.

Step 3 $\quad 16 = 4x + 6 - 2x$

$16 = 2x + 6$ — Like terms were combined.

Step 4 $\quad 16 - 6 = 2x + 6 - 6$ — Subtract 6 from both sides.

$10 = 2x$

Step 5 $\quad \frac{10}{2} = \frac{2x}{2}$ — Divide both sides by 2.

$5 = x$

Now Try Exercise 29

The preceding solution can be condensed as follows.

$16 = 4x + 6 - 2x$

$16 = 2x + 6$ — Like terms were combined.

$10 = 2x$ — 6 was subtracted from both sides.

$5 = x$ — Both sides were divided by 2.

EXAMPLE 4 Solve the equation $5x - 2(x + 4) = 3$.

Solution

Step 2 $\quad 5x - 2(x + 4) = 3$

$5x - 2x - 8 = 3$ — Distributive property was used.

Step 3 $\quad 3x - 8 = 3$ — Like terms were combined.

Step 4 $\quad 3x - 8 + 8 = 3 + 8$ — Add 8 to both sides.

$3x = 11$

Step 5 $\quad \frac{3x}{3} = \frac{11}{3}$ — Divide both sides by 3.

$x = \frac{11}{3}$

Now Try Exercise 57

The solution to Example 4 can be condensed as follows:

$$5x - 2(x + 4) = 3$$

$$5x - 2x - 8 = 3 \quad \text{Distributive property was used.}$$

$$3x - 8 = 3 \quad \text{Like terms were combined.}$$

$$3x = 11 \quad \text{8 was added to both sides.}$$

$$x = \frac{11}{3} \quad \text{Both sides were divided by 3.}$$

EXAMPLE 5 Solve the equation $3p - (2p + 5) = 7$.

Solution

$$3p - (2p + 5) = 7$$

$$3p - 2p - 5 = 7 \quad \text{Distributive property was used.}$$

$$p - 5 = 7 \quad \text{Like terms were combined.}$$

$$p = 12 \quad \text{5 was added to both sides.}$$

Now Try Exercise 61

2 Solve Equations Containing Decimal Numbers or Fractions

Example 6 illustrates two methods to solve an equation that contains decimal numbers.

EXAMPLE 6 Solve the equation $x + 1.24 - 0.07x = 4.96$.

Solution We will work this example using two methods. In method 1, we work with decimal numbers throughout the solving process. In method 2, we multiply both sides of the equation by a power of 10 to change the decimal numbers to integers.

Method 1

$$x + 1.24 - 0.07x = 4.96 \quad \text{Like terms were combined; } 1x - 0.07x = 0.93x.$$

$$0.93x + 1.24 = 4.96$$

$$0.93x + 1.24 - 1.24 = 4.96 - 1.24 \quad \text{Subtract 1.24 from both sides.}$$

$$0.93x = 3.72$$

$$\frac{0.93x}{0.93} = \frac{3.72}{0.93} \quad \text{Divide both sides by 0.93.}$$

$$x = 4$$

Method 2 We can eliminate the decimal numbers from the equation by multiplying both sides of the equation by 10 if the decimal numbers are given in tenths, by 100 if the decimal numbers are given in hundredths, and so on. In Example 6, since the decimal numbers are in hundredths, you can eliminate the decimals from the equation by multiplying both sides of the equation by 100. This alternate method would give the following.

Understanding Algebra

We can eliminate decimal numbers from an equation by multiplying *both* sides of the equation by 10 if the decimal numbers are given in tenths, by 100 if the decimal numbers are given in hundredths, and so on.

$$x + 1.24 - 0.07x = 4.96$$

$$100(x + 1.24 - 0.07x) = 100(4.96) \quad \text{Multiply both sides of equation by 100.}$$

$$100(x) + 100(1.24) - 100(0.07x) = 496 \quad \text{Distributive property}$$

$$100x + 124 - 7x = 496$$

$$93x + 124 = 496 \quad \text{Like terms were combined.}$$

$$93x = 372 \quad \text{124 was subtracted from both sides.}$$

$$x = 4 \quad \text{Both sides were divided by 93.}$$

Study both methods provided to see which method you prefer.

Now Try Exercise 33

Often, the first step in solving equations containing fractions is to multiply both sides of the equation by the LCD to eliminate the fractions from the equations. Examples 7–9 illustrate this procedure.

EXAMPLE 7 Solve $\frac{1}{5}(x + 1) = 1.$

Solution The LCD of the fraction is 5. We will begin by multiplying both sides of the equation by the LCD. This step will eliminate fractions from the equation.

> **Understanding Algebra**
>
> We can eliminate fractions from an equation by multiplying *both* sides of the equation by the least common denominator (LCD) of the fractions within the equation.

$$\frac{1}{5}(x + 1) = 1$$

Step 1 $$5\left[\frac{1}{5}(x + 1)\right] = 5 \cdot 1$$ Multiply both sides by the LCD, 5.

$$\cancel{5}\left(\frac{1}{\cancel{5}}\right)(x + 1) = 5$$

$$x + 1 = 5$$

Step 4 $$x = 4$$ 1 was subtracted from both sides.

Step 6 Check $$\frac{1}{5}(x + 1) = 1$$

$$\frac{1}{5}(4 + 1) \stackrel{?}{=} 1$$

$$\frac{1}{\cancel{5}}(\cancel{5}) \stackrel{?}{=} 1$$

$$1 = 1$$ True

The solution is 4.

Now Try Exercise 37

Example 7 could also be written as $\frac{x + 1}{5} = 1$. To solve this equation, we would begin by multiplying both sides of the equation by the LCD, 5, as follows.

$$\frac{x + 1}{5} = 1$$

$$5\left(\frac{x + 1}{5}\right) = 5 \cdot 1$$

$$x + 1 = 5$$

$$x = 4$$

EXAMPLE 8 Solve the equation $\frac{d}{2} + 3d = 14.$

Solution Multiply both sides of the equation by the LCD, 2. This step will eliminate fractions from the equation.

Step 1 $$2\left(\frac{d}{2} + 3d\right) = 2 \cdot 14$$ Multiply both sides by the LCD, 2.

Step 2 $$\cancel{2}\left(\frac{d}{\cancel{2}}\right) + 2 \cdot 3d = 2 \cdot 14$$ Distributive property

$$d + 6d = 28$$

Step 3 $$7d = 28$$ Like terms were combined.

Step 5 $d = 4$ Both sides were divided by 7.

Step 6 Check

$$\frac{d}{2} + 3d = 14$$

$$\frac{4}{2} + 3(4) \stackrel{?}{=} 14$$

$$2 + 12 \stackrel{?}{=} 14$$

$$14 = 14 \quad \text{True}$$

Now Try Exercise 81

EXAMPLE 9 Solve the equation $\frac{1}{5}x - \frac{3}{8}x = \frac{1}{10}$.

Solution The LCD of 5, 8, and 10 is 40. Multiply both sides of the equation by 40 to eliminate fractions from the equation.

$$\frac{1}{5}x - \frac{3}{8}x = \frac{1}{10}$$

Step 1 $40\left(\frac{1}{5}x - \frac{3}{8}x\right) = 40\left(\frac{1}{10}\right)$ Multiply both sides by the LCD, 40.

Step 2 $40\left(\frac{1}{5}x\right) - 40\left(\frac{3}{8}x\right) = 40\left(\frac{1}{10}\right)$ Distributive property

$$8x - 15x = 4$$

Step 3 $-7x = 4$ Like terms were combined.

Step 5 $x = -\frac{4}{7}$ Both sides were divided by -7.

Step 6 Check

$$\frac{1}{5}x - \frac{3}{8}x = \frac{1}{10}$$

$\frac{1}{5}\left(-\frac{4}{7}\right) - \frac{3}{8}\left(-\frac{4}{7}\right) \stackrel{?}{=} \frac{1}{10}$ Substitute $-\frac{4}{7}$ for each x.

$-\frac{4}{35} + \frac{3}{14} \stackrel{?}{=} \frac{1}{10}$ Divided out common factors, then multiplied fractions.

$-\frac{8}{70} + \frac{15}{70} \stackrel{?}{=} \frac{7}{70}$ Wrote each fraction with the LCD, 70.

$\frac{7}{70} = \frac{7}{70}$ True

Now Try Exercise 91

HELPFUL HINT

Some of the most commonly used terms in algebra are "evaluate," "simplify," "solve," and "check." Make sure you understand what each term means and when each term is used.

Evaluate: To *evaluate an expression* means to find its numerical value.

Evaluate

$$\begin{aligned} &16 \div 2^2 + 36 \div 4 \\ &= 16 \div 4 + 36 \div 4 \\ &= 4 + 36 \div 4 \\ &= 4 + 9 \\ &= 13 \end{aligned}$$

(Continued)

Evaluate $-x^2 + 3x - 2$ when $x = 4$

$$= -4^2 + 3(4) - 2$$
$$= -16 + 3(4) - 2$$
$$= -16 + 12 - 2$$
$$= -4 - 2$$
$$= -6$$

Simplify: To *simplify an expression* means to perform the operations and combine like terms.

Simplify $3(x - 2) - 4(2x + 3)$

$$3(x - 2) - 4(2x + 3) = 3x - 6 - 8x - 12$$
$$= -5x - 18$$

Note that when you simplify an expression containing variables you do not generally end up with just a numerical value unless all the variable terms happen to add to zero.

Solve: To *solve an equation* means to find the value or the values of the variable that make the equation a true statement.

Solve

$$2x + 3(x + 1) = 18$$
$$2x + 3x + 3 = 18$$
$$5x + 3 = 18$$
$$5x = 15$$
$$x = 3$$

Check: To *check the proposed solution to an equation,* substitute the value in the original equation. If this result is true, then the answer checks. For example, to check the solution to the equation just solved, we substitute 3 for x in the original equation.

Check

$$2x + 3(x + 1) = 18$$
$$2(3) + 3(3 + 1) \stackrel{?}{=} 18$$
$$2(3) + 3(4) \stackrel{?}{=} 18$$
$$6 + 12 \stackrel{?}{=} 18$$
$$18 = 18 \quad \text{True}$$

Since we obtained a true statement, the 3 checks.

It is important to realize that expressions may be evaluated or simplified (depending on the type of problem) and equations are solved and then checked.

2.4 Exercise Set MathXL® MyMathLab®

Warm-Up Exercises

Fill in the blanks with the appropriate word, phrase, or symbol(s) from the following list.

10	100	nonzero	dividing	multiplying
24	18	subtract	adding	LCD

1. The multiplication property of equality allows us to multiply or divide both sides of an equation by the same ________ number.

2. The addition property of equality allows us to add or ________ the same number from both sides of an equation.

3. In the equation $0.5x - 1.2 = 1.8$ we can eliminate the decimal numbers by multiplying both sides of the equation by ________.

4. In the equation $0.02x + 175 = 335.25$ we can eliminate the decimal numbers by multiplying both sides of the equation by ________.

5. We can eliminate fractions from an equation by ________ *both* sides of the equation by the least common denominator (LCD) of the fractions within the equation.

6. In the equation $\frac{2}{3}x - \frac{5}{8} = \frac{5}{6}$ we can eliminate the fractions by multiplying both sides of the equation by ________.

Practice the Skills

Solve each equation.

7. $5x - 6 = 19$
8. $2t - 4 = 8$
9. $-4w - 9 = 11$
10. $-5c - 7 = 8$
11. $3x + 6 = 12$
12. $2x + 9 = 21$
13. $5x - 2 = 10$
14. $-2t + 9 = 21$
15. $-5k - 9 = -19$
16. $-4x - 7 = -6$
17. $12 - x = 5$
18. $-13 - x = 8$
19. $8 + 3x = 19$
20. $5 + 8x = 18$
21. $16x + 5 = -14$
22. $-2x + 7 = -10$
23. $-9 + 5x = -9$
24. $-24 + 16x = -24$
25. $7r - 16 = -2$
26. $-2w + 4 = -8$
27. $60 = -5s + 9$
28. $45 = -6v - 8$
29. $14 = 5x + 8 - 3x$
30. $15 = 6x - 3 + 3x$
31. $2.3x - 9.34 = 6.3$
32. $1.5q - 1.05 = 3.9$
33. $0.91y + 2.25 - 0.01y = 5.85$
34. $0.15 = 0.05x - 1.35 - 0.20x$
35. $28.8 = x + 1.40x$
36. $x + 0.05x = 21$
37. $\frac{1}{7}(x + 6) = 4$
38. $\frac{1}{5}(x + 2) = -3$
39. $\frac{d + 3}{7} = 9$
40. $\frac{m - 6}{5} = 2$
41. $\frac{1}{3}(t - 7) = -7$
42. $\frac{2}{3}(n - 3) = 8$
43. $\frac{3}{4}(x - 5) = -12$
44. $-\frac{5}{8}(x + 3) = 10$
45. $\frac{x + 4}{7} = \frac{2}{7}$
46. $\frac{4x + 5}{6} = \frac{7}{2}$
47. $\frac{3}{4} = \frac{4m - 5}{6}$
48. $\frac{5}{6} = \frac{5t - 4}{2}$
49. $4(n + 3) = 12$
50. $3(x - 2) = 12$
51. $-2(x - 3) = 26$
52. $5(3 - x) = 15$
53. $-4 = -(x + 7)$
54. $-7 = -(x - 5)$
55. $12 = 4(x - 3)$
56. $9 = -2(a - 3)$
57. $2x - 3(x + 5) = 6$
58. $5(3x + 1) - 12x = -2$
59. $-3r - 4(r + 2) = 11$
60. $-2(x + 8) - 5 = 1$
61. $x - 3(2x + 3) = 36$
62. $3y - (y + 5) = 9$
63. $5x + 3x - 4x - 7 = 9$
64. $-7x + 4x - 2x + 11 = 26$
65. $0.7(x - 3) = 1.4$
66. $-0.5(x + 9) = -6.5$
67. $2.5(4q - 3) = 0.5$
68. $0.1(2.4x + 5) = 1.7$
69. $3 - 2(x + 3) + 2 = 1$
70. $7 - 3(x - 4) + 5 = 36$
71. $1 + (x + 3) + 6x = 6$
72. $6 - (x - 4) + 5x = 13$
73. $4.85 - 6.4x + 1.11 = 25.8$
74. $3.68 - 1.6x + 5.32 = 11.4$
75. $7 = 8 - 5(m + 3)$
76. $16 = -5 + 4(j + 7)$
77. $9 = \frac{2s + 9}{5}$
78. $12 = \frac{4d - 1}{3}$
79. $x + \frac{2}{3} = \frac{3}{5}$
80. $n - \frac{1}{4} = \frac{1}{2}$
81. $\frac{r}{3} + 2r = 6$
82. $\frac{x}{4} - 6x = 23$
83. $\frac{3}{7} = \frac{3t}{4} + 1$
84. $\frac{5}{8} = \frac{5t}{6} + 2$
85. $\frac{1}{2}r + \frac{1}{5}r = 7$
86. $\frac{x}{3} - \frac{3x}{4} = \frac{1}{12}$
87. $\frac{2}{8} + \frac{3}{4} = \frac{w}{5}$
88. $\frac{x}{4} - \frac{x}{6} = \frac{1}{4}$
89. $\frac{1}{2}x + 5 = \frac{1}{8}$
90. $\frac{4}{5} + n = \frac{1}{3}$
91. $\frac{4}{5}s - \frac{3}{4}s = \frac{1}{10}$
92. $\frac{1}{3}x - \frac{3}{4}x = \frac{1}{5}$
93. $\frac{4}{9} = \frac{1}{3}(n - 7)$
94. $-\frac{7}{16} = \frac{3}{8}(h + 5)$
95. $-\frac{3}{5} = -\frac{1}{6} - \frac{3}{4}q$
96. $-\frac{3}{5} = -\frac{1}{6} - \frac{5}{4}m$

Concept/Writing Exercises

97. a) Explain why it is easier to solve the equation $3x + 2 = 11$ by first subtracting 2 from both sides of the equation rather than by first dividing both sides of the equation by 3.

b) Solve the equation.

98. a) Explain why it is easier to solve the equation $5x - 3 = 12$ by first adding 3 to both sides of the equation rather than by first dividing both sides of the equation by 5.

b) Solve the equation.

Challenge Problems

For Exercises 99–101, solve the equation.

99. $3(x - 2) - (x + 5) - 2(3 - 2x) = 18$

100. $-6 = -(x - 5) - 3(5 + 2x) - 4(2x - 4)$

101. $4[3 - 2(x + 4)] - (x + 3) = 13$

102. Solve the equation □☺ − ∇ = @ for ☺.

Group Activity

In Chapter 3, we will discuss procedures for writing application problems as equations. Let's look at an application now.

Birthday Party John Logan purchased 2 large chocolate bars and a birthday card. The birthday card cost \$3. The total cost was \$9. What was the price of a single chocolate bar?

This problem can be represented by the equation $2x + 3 = 9$, which can be used to solve the problem. Solving the equation we find that x, the price of a single chocolate bar, is \$3.

For Exercises 103 and 104, each group member should do parts **a)** *and* **b)**. *Then do Part* **c)** *as a group.*

a) *Obtain an equation that can be used to solve the problem.*

b) *Solve the equation and answer the question.*

c) *Compare and check each other's work.*

103. Stationery Eduardo Verner purchased three boxes of stationery. He also purchased wrapping paper and thank-you cards. If the wrapping paper and thank-you cards together cost \$6, and the total he paid was \$42, find the cost of a box of stationery.

104. Candies Mahandi Ison purchased three rolls of peppermint candies and the local newspaper. The newspaper cost 50 cents. He paid \$2.75 in all. What did a roll of candies cost?

Cumulative Review Exercises

[1.4] **105.** True or false: Every real number is a rational number.

[1.9] **106.** Evaluate $[5(2 - 6) + 3(8 \div 4)^2]^2$.

[2.2] **107.** To solve an equation, what do you need to do to the variable?

[2.3] **108.** To solve the equation $7 = -4x$, would you add 4 to both sides of the equation or divide both sides of the equation by -4? Explain your answer.

MID-CHAPTER TEST: 2.1–2.4

To find out how well you understand the chapter material to this point, take this brief test. The answers, and the section where the material was initially discussed, are given in the back of the book. Review any questions that you answered incorrectly.

In Exercises 1 and 2, combine like terms.

1. $5x - 9y - 12 + 4y - 7x + 6$

2. $\frac{2}{5}x - 8 - \frac{3}{4}x + \frac{1}{2}$

In Exercises 3 and 4, use the distributive property to remove parentheses.

3. $-4(2a - 3b + 16)$

4. $1.6(2.1x - 3.4y - 5.2)$

5. Simplify $5(t - 3) - 3(t + 7) - 2$.

6. Is $x = 2$ a solution of $3(x - 4) = -2(x + 1)$?

7. Is $p = \frac{2}{5}$ a solution of $7p - 3 = 2p - 5$?

In Exercises 8–10, solve each equation and check your solution.

8. $x - 5 = -9$

9. $120 + x = -40$

10. $-16 = 7 + y$

11. When solving the equation $\frac{x}{4} = 5$, what would you do to isolate the variable? Explain.

In Exercises 12–15, solve each equation and check your solution.

12. $4 = 12y$

13. $\frac{x}{8} = 3$

14. $-\frac{x}{5} = -2$

15. $-x = \frac{3}{7}$

In Exercises 16–20, solve each equation.

16. $6x - 3 = 12$

17. $-4 = -2w - 7$

18. $\frac{3}{8} = \frac{4n - 1}{6}$

19. $-5(x + 4) - 7 = 3$

20. $8 - 9(y + 4) + 6 = -2$

2.5 Solving Linear Equations with the Variable on Both Sides of the Equation

1. Solve equations with the variable on both sides of the equation.
2. Solve equations containing decimal numbers or fractions.
3. Identify identities and contradictions.

1 Solve Equations with the Variable on Both Sides of the Equation

The equation $4x + 6 = 2x + 4$ contains the variable, x, on both sides of the equation. To solve equations of this type, we rewrite the equation so that all terms containing the variable are on only one side of the equation, and all terms not containing the variable are on the other side of the equation. The following steps in the procedure are only guidelines to use to solve these types of equations. For example, there may be times when you may choose to use the distributive property, step 2, before multiplying both sides of the equation by the LCD (step 1). We will illustrate this variation in Examples 9 and 10.

Understanding Algebra

As the equations get more complicated, our ultimate goal remains the same: *to isolate the variable on one side of the equation.*

To Solve Linear Equations with the Variable on Both Sides of the Equation

1. **Fractions:** If the equation contains fractions, multiply **both** sides of the equation by the least common denominator (LCD). This will eliminate fractions from the equation.
2. **Parentheses:** Use the distributive property to remove parentheses.
3. **Like terms:** Combine like terms on each side of the equation.
4. **Addition property:** Use the addition property to rewrite the equation with all terms containing the variable on one side of the equation and all terms not containing the variable on the other side of the equation. It may be necessary to use the addition property twice to accomplish this goal.
5. **Multiplication property:** Use the multiplication property to isolate the variable.
6. **Check:** Check the solution in the original equation.

Remember: *our goal in solving a linear equation is to isolate the variable, that is, to obtain an equation in which the variable is alone on one side of the equation.* For example, consider the equation $3x + 4 = x + 12$. It contains no fractions or parentheses nor are there any like terms on the same side of the equation. Therefore, we start with step 4 above and use the addition property twice. Notice in the steps outlined below that we subtract x from both sides and then subtract 4 from both sides:

$$3x + 4 = x + 12$$

$$3x - x + 4 = x - x + 12 \quad \text{Addition property (subtract } x \text{ from both sides).}$$

$$2x + 4 = 12 \quad \text{Variable appears only on left side of equation.}$$

Notice that the variable, x, now appears on only one side of the equation. However, $+4$ still appears on the same side.

$$2x + 4 = 12$$

$$2x + 4 - 4 = 12 - 4 \quad \text{Addition property (subtract 4 from both sides).}$$

$$2x = 8 \quad \text{Term with } x \text{ is now isolated (in the form } ax = b\text{).}$$

Finally, to solve for x, we use the multiplication property:

$$2x = 8$$

$$\frac{2x}{2} = \frac{\overset{4}{\cancel{8}}}{2} \quad \text{Multiplication property (divide both sides by 2).}$$

$$x = 4 \quad \text{Goal accomplished: } x \text{ is isolated.}$$

The solution to the equation is 4.

EXAMPLE 1 Solve the equation $4x + 6 = 2x + 4$.

Solution We start by getting all the terms with the variable on one side of the equation and all terms without the variable on the other side. The terms with the variable may be collected on either side of the equation. We will illustrate two methods of solving this equation.

Method 1 Isolate the variable term on the left.

$$4x + 6 = 2x + 4$$

Step 4 $\quad 4x - 2x + 6 = 2x - 2x + 4$ $\quad$ Subtract $2x$ from both sides.

$$2x + 6 = 4$$

Step 4 $\quad 2x + 6 - 6 = 4 - 6$ $\quad$ Subtract 6 from both sides.

$$2x = -2$$

Step 5 $\quad \dfrac{2x}{2} = \dfrac{-2}{2}$ $\quad$ Divide both sides by 2.

$$x = -1$$

Method 2 Isolate the variable term on the right.

$$4x + 6 = 2x + 4$$

Step 4 $\quad 4x - 4x + 6 = 2x - 4x + 4$ $\quad$ Subtract $4x$ from both sides.

$$6 = -2x + 4$$

Step 4 $\quad 6 - 4 = -2x + 4 - 4$ $\quad$ Subtract 4 from both sides.

$$2 = -2x$$

Step 5 $\quad \dfrac{2}{-2} = \dfrac{-2x}{-2}$ $\quad$ Divide both sides by -2.

$$-1 = x$$

The same answer is obtained whether we collect the terms with the variable on the left or right side.

Step 6 Check

$$4x + 6 = 2x + 4$$

$$4(-1) + 6 \stackrel{?}{=} 2(-1) + 4$$

$$-4 + 6 \stackrel{?}{=} -2 + 4$$

$$2 = 2 \quad \text{True}$$

Since the check is true, the solution is -1.

Now Try Exercise 19

EXAMPLE 2 Solve the equation $2x - 3 - 5x = 13 + 4x - 2$.

Solution We will choose to collect the terms containing the variable on the right side of the equation in order to create a positive coefficient of x. Since there are like terms *on the same side of the equation,* we will begin by combining these like terms.

Step 3 $\quad 2x - 3 - 5x = 13 + 4x - 2$

$-3x - 3 = 4x + 11$ — Combined like terms.

Step 4 $\quad -3x + 3x - 3 = 4x + 3x + 11$ — Add $3x$ to both sides.

$-3 = 7x + 11$

Step 4 $\quad -3 - 11 = 7x + 11 - 11$ — Subtract 11 from both sides.

$-14 = 7x$

Step 5 $\quad \dfrac{-14}{7} = \dfrac{7x}{7}$ — Divide both sides by 7.

$-2 = x$

Step 6 Check

$$2x - 3 - 5x = 13 + 4x - 2$$
$$2(-2) - 3 - 5(-2) \stackrel{?}{=} 13 + 4(-2) - 2$$
$$-4 - 3 + 10 \stackrel{?}{=} 13 - 8 - 2$$
$$-7 + 10 \stackrel{?}{=} 5 - 2$$
$$3 = 3 \quad \text{True}$$

Since the check is true, the solution is -2.

Now Try Exercise 25

The solution to Example 2 could be condensed as follows:

$2x - 3 - 5x = 13 + 4x - 2$

$-3x - 3 = 4x + 11$ — Combined like terms.

$-3 = 7x + 11$ — Added $3x$ to both sides.

$-14 = 7x$ — Subtracted 11 from both sides.

$-2 = x$ — Divided both sides by 7.

We solved Example 2 by moving the terms containing the variable to the right side of the equation. Now rework the problem by moving the terms containing the variable to the left side of the equation. You should obtain the same answer.

EXAMPLE 3 Solve the equation $2(p + 3) = -3p + 10$.

Solution $\quad 2(p + 3) = -3p + 10$

Step 2 $\quad 2p + 6 = -3p + 10$ — Distributive property was used.

Step 4 $\quad 2p + 3p + 6 = -3p + 3p + 10$ — Add $3p$ to both sides.

$5p + 6 = 10$

Step 4 $\quad 5p + 6 - 6 = 10 - 6$ — Subtract 6 from both sides.

$5p = 4$

Step 5 $\quad \dfrac{5p}{5} = \dfrac{4}{5}$ — Divide both sides by 5.

$p = \dfrac{4}{5}$

The solution is $\frac{4}{5}$.

Now Try Exercise 23

The solution to Example 3 could be condensed as follows:

$2(p + 3) = -3p + 10$	
$2p + 6 = -3p + 10$	Distributive property was used.
$5p + 6 = 10$	Added $3p$ to both sides.
$5p = 4$	Subtracted 6 from both sides.
$p = \frac{4}{5}$	Divided both sides by 5.

Understanding Algebra

In Step 3 of Example 4, we could also isolate the variable on the left side of the equation:

$$2x - 7 = 3x + 9$$
$$-x - 7 = 9$$
$$-x = 16$$
$$x = -16$$

EXAMPLE 4 Solve the equation $2(x - 5) + 3 = 3x + 9$.

Solution	$2(x - 5) + 3 = 3x + 9$	
Step 2	$2x - 10 + 3 = 3x + 9$	Distributive property was used.
Step 3	$2x - 7 = 3x + 9$	Combined like terms.
Step 4	$-7 = x + 9$	Subtracted $2x$ from both sides.
Step 4	$-16 = x$	Subtracted 9 from both sides.

The solution is -16.

Now Try Exercise 31

EXAMPLE 5 Solve the equation $2(x + 5) - 5(2x + 3) = -x + 9$

Solution	$2(x + 5) - 5(2x + 3) = -x + 9$	
Step 2	$2x + 10 - 10x - 15 = -x + 9$	Used the distributive property.
Step 3	$-8x - 5 = -x + 9$	Combined like terms.
Step 4	$-5 = 7x + 9$	Added $8x$ to both sides.
Step 4	$-14 = 7x$	Subtracted 9 from both sides.
Step 5	$-2 = x$	Divided both sides by 7.

The solution is -2.

Now Try Exercise 65

2 Solve Equations Containing Decimal Numbers or Fractions

Now we will solve an equation that contains decimal numbers. We will illustrate two procedures for solving Example 6.

EXAMPLE 6 Solve the equation $5.74x + 5.42 = 2.24x - 9.28$.

Solution

Method 1 Notice that there are no like terms on the same side of the equation that can be combined.

	$5.74x + 5.42 = 2.24x - 9.28$	
Step 4	$5.74x - 2.24x + 5.42 = 2.24x - 2.24x - 9.28$	Subtract $2.24x$ from both sides.
	$3.50x + 5.42 = -9.28$	
Step 4	$3.50x + 5.42 - 5.42 = -9.28 - 5.42$	Subtract 5.42 from both sides.
	$3.50x = -14.70$	
Step 5	$\frac{3.50x}{3.50} = \frac{-14.70}{3.50}$	Divide both sides by 3.50.
	$x = -4.20$	

The solution is -4.20.

Method 2 Since the given equation has numbers given in hundredths, we will multiply both sides of the equation by 100.

$$5.74x + 5.42 = 2.24x - 9.28$$

$$100(5.74x + 5.42) = 100(2.24x - 9.28) \quad \text{Multiply both sides by 100.}$$

$$100(5.74x) + 100(5.42) = 100(2.24x) - 100(9.28) \quad \text{Distributive property}$$

$$574x + 542 = 224x - 928$$

Step 4 $$574x + 542 - 542 = 224x - 928 - 542 \quad \text{Subtract 542 from both sides.}$$

$$574x = 224x - 1470$$

Step 4 $$574x - 224x = 224x - 224x - 1470 \quad \text{Subtract } 224x \text{ from both sides.}$$

$$350x = -1470$$

Step 5 $$\frac{350x}{350} = \frac{-1470}{350} \quad \text{Divide both sides by 350.}$$

$$x = -4.20$$

Notice we obtain the same answer using either method. You may use either method to solve equations of this type.

Now Try Exercise 35

Understanding Algebra

To eliminate decimal numbers from an equation, multiply by a power of 10:

If equation contains numbers to ...	... then multiply both sides by
tenths	10
hundredths	100
thousandths	1000

and so on.

Now let's solve some equations that contain fractions.

EXAMPLE 7 Solve the equation $\frac{1}{2}a = \frac{3}{4}a + \frac{1}{5}$.

Solution The least common denominator is 20. Begin by multiplying both sides of the equation by the LCD.

$$\frac{1}{2}a = \frac{3}{4}a + \frac{1}{5}$$

Step 1 $$20\left(\frac{1}{2}a\right) = 20\left(\frac{3}{4}a + \frac{1}{5}\right) \quad \text{Multiply both sides by the LCD, 20.}$$

Step 2 $$10a = \overset{5}{\cancel{20}}\left(\frac{3}{\underset{1}{\cancel{4}}}a\right) + \overset{4}{\cancel{20}}\left(\frac{1}{\underset{1}{\cancel{5}}}\right) \quad \text{Distributive property}$$

$$10a = 15a + 4$$

Step 4 $$-5a = 4 \quad \text{Subtracted } 15a \text{ from both sides.}$$

Step 5 $$a = -\frac{4}{5} \quad \text{Divided both sides by } -5.$$

Step 6 Check $$\frac{1}{2}a = \frac{3}{4}a + \frac{1}{5}$$

$$\frac{1}{2}\left(-\frac{4}{5}\right) \stackrel{?}{=} \frac{3}{4}\left(-\frac{4}{5}\right) + \frac{1}{5}$$

$$-\frac{2}{5} \stackrel{?}{=} -\frac{3}{5} + \frac{1}{5}$$

$$-\frac{2}{5} = -\frac{2}{5} \quad \text{True}$$

The solution is $-\frac{4}{5}$.

Now Try Exercise 43

Understanding Algebra

To eliminate fractions from equations, multiply both sides of the equation by the LCD of the fractions in the equation.

HELPFUL HINT

The equation in Example 7, $\frac{1}{2}a = \frac{3}{4}a + \frac{1}{5}$, could have been written as $\frac{a}{2} = \frac{3a}{4} + \frac{1}{5}$ because $\frac{1}{2}a$ is the same as $\frac{a}{2}$, and $\frac{3}{4}a$ is the same as $\frac{3a}{4}$. You would solve the equation $\frac{a}{2} = \frac{3a}{4} + \frac{1}{5}$ the same way you solved the equation in Example 7.

EXAMPLE 8 Solve the equation $\frac{x}{4} + 3 = 2(x - 2)$.

Solution We will begin by multiplying both sides of the equation by the LCD, 4.

$$\frac{x}{4} + 3 = 2(x - 2)$$

$$4\left(\frac{x}{4} + 3\right) = 4[2(x - 2)] \quad \text{Multiply both sides by the LCD, 4.}$$

$$4\left(\frac{x}{4}\right) + 4(3) = (4 \cdot 2)(x - 2) \quad \text{Used the distributive property (on left). Used the associative property (on right).}$$

$$x + 12 = 8(x - 2)$$

$$x + 12 = 8x - 16 \quad \text{Used the distributive property (on right).}$$

$$12 = 7x - 16 \quad \text{Subtracted } x \text{ from both sides.}$$

$$28 = 7x \quad \text{Added 16 to both sides.}$$

$$4 = x \quad \text{Divided both sides by 7.}$$

A check will show that 4 is the solution.

Now Try Exercise 61

In Example 8, we began the solution by multiplying both sides of the equation by the LCD. In Example 9, we will solve the same equation, but this time we will begin by using the distributive property.

EXAMPLE 9 Solve the equation in Example 8, $\frac{x}{4} + 3 = 2(x - 2)$, by first using the distributive property.

Solution Begin by using the distributive property.

$$\frac{x}{4} + 3 = 2(x - 2)$$

$$\frac{x}{4} + 3 = 2x - 4 \quad \text{Used the distributive property (on right).}$$

$$4\left(\frac{x}{4} + 3\right) = 4(2x - 4) \quad \text{Multiply both sides by the LCD, 4.}$$

$$4\left(\frac{x}{4}\right) + 4(3) = 4(2x) - 4(4) \quad \text{Distributive property (on left and right)}$$

$$x + 12 = 8x - 16$$

$$12 = 7x - 16 \quad x \text{ was subtracted from both sides.}$$

$$28 = 7x \quad \text{16 was added to both sides.}$$

$$4 = x \quad \text{Both sides were divided by 7.}$$

The solution is 4.

Now Try Exercise 71

Notice that we obtained the same answer in Examples 8 and 9.

EXAMPLE 10 Solve the equation $\frac{1}{2}(2x + 3) = \frac{2}{3}(x - 6) + 4$.

Notice that this equation contains one term on the left side of the equation and two terms on the right side of the equation.

Solution We will work this problem by first using the distributive property.

$$\frac{1}{2}(2x + 3) = \frac{2}{3}(x - 6) + 4$$

$$\frac{1}{2}(2x) + \frac{1}{2}(3) = \frac{2}{3}(x) - \frac{2}{3}(\overset{2}{\cancel{6}}) + 4 \quad \text{Distributive property (on left and right)}$$

$$x + \frac{3}{2} = \frac{2}{3}x - 4 + 4$$

$$x + \frac{3}{2} = \frac{2}{3}x \quad \text{Combined like terms.}$$

$$6\left(x + \frac{3}{2}\right) = 6\left(\frac{2}{3}x\right) \quad \text{Multiply both sides by the LCD, 6.}$$

$$6x + 6\left(\frac{3}{2}\right) = 6\left(\frac{2}{3}x\right) \quad \text{Used the distributive property.}$$

$$6x + \overset{3}{\cancel{6}}\left(\frac{3}{\cancel{2}}\right) = \overset{2}{\cancel{6}}\left(\frac{2}{\cancel{3}}x\right)$$

$$6x + 9 = 4x$$

$$2x + 9 = 0 \quad \text{Subtracted } 4x \text{ from both sides.}$$

$$2x = -9 \quad \text{Subtracted 9 from both sides.}$$

$$x = -\frac{9}{2} \quad \text{Divided both sides by 2.}$$

Check Substitute $-\frac{9}{2}$ for each x in the equation.

$$\frac{1}{2}(2x + 3) = \frac{2}{3}(x - 6) + 4$$

$$\frac{1}{2}\left[2\left(-\frac{9}{2}\right) + 3\right] \stackrel{?}{=} \frac{2}{3}\left(-\frac{9}{2} - 6\right) + 4$$

$$\frac{1}{2}[-9 + 3] \stackrel{?}{=} \frac{2}{3}\left(-\frac{9}{2} - \frac{12}{2}\right) + 4$$

$$\frac{1}{2}[-6] \stackrel{?}{=} \frac{2}{3}\left(-\frac{21}{2}\right) + 4$$

$$-3 \stackrel{?}{=} -7 + 4$$

$$-3 = -3 \quad \text{True}$$

The solution is $-\frac{9}{2}$.

Now Try Exercise 75

In Example 10, we began by using the distributive property. We could have also begun by multiplying both sides of the equation by the LCD, 6, before using the distributive property. Work Example 10 again now by first multiplying both sides of the equation by the LCD, 6.

Example 10 could have also been written as $\frac{2x+3}{2} = \frac{2(x-6)}{3} + 4$. If you were given the equation in this form, you could begin by using the distributive property on $2(x - 6)$ or you could begin by multiplying both sides of the equation by the LCD, 6. Because this is just another way of writing the equation in Example 10, the answer would be $-\frac{9}{2}$.

We will discuss solving equations containing fractions in more detail later in the book.

3 Identify Identities and Contradictions

Thus far each equation we have solved has had a single value for a solution. Such equations are called **conditional equations**. Additionally, there are two other types of equations based on the sets of solutions to the equations. All three types of equations are given below:

Types of Equations

A **conditional equation** is true for a specific value or values of the variable.

An **identity** is true for all values of the variable. When we have an identity, we will state the solution is *all real numbers*.

A **contradiction** is not true for any value of the variable. When we have a contradiction, we will state that there is *no solution*.

EXAMPLE 11 Solve the equation $5x - 5 - 2x = 3(x - 2) + 1$.

Solution

$$5x - 5 - 2x = 3(x - 2) + 1$$

$$5x - 5 - 2x = 3x - 6 + 1 \quad \text{Used the distributive property.}$$

$$3x - 5 = 3x - 5 \quad \text{Combined like terms.}$$

Since the same expression appears on both sides of the equation, the statement is true for infinitely many values of x. If we continue to solve this equation further, we might obtain

$$3x - 5 = 3x - 5$$

$$3x = 3x \quad \text{Added 5 to both sides.}$$

$$0 = 0 \quad \text{Subtracted } 3x \text{ from both sides.}$$

The answer is: all real numbers.

NOTE: *When solving an equation and we obtain an equation that is always true, like the equation in Example 11, we write the answer as "all real numbers."*

Now Try Exercise 47

Understanding Algebra

When we are solving an equation and we obtain an equation that is *always* true, such as $0 = 0$, then the original equation is an *identity* and we state that the solution is *all real numbers*.

EXAMPLE 12 Solve the equation $-2x + 5 + 3x = 5x - 4x + 7$.

Solution

$$-2x + 5 + 3x = 5x - 4x + 7$$

$$x + 5 = x + 7 \quad \text{Combined like terms.}$$

$$x - x + 5 = x - x + 7 \quad \text{Subtract } x \text{ from both sides.}$$

$$5 = 7 \quad \text{False}$$

The answer is: no solution.

NOTE: *When solving an equation and we obtain an equation that is never true, like the equation in Example 12, we write the answer as "no solution."*

Now Try Exercise 27

Understanding Algebra

When we are solving an equation and we obtain an equation that is *never* true, such as $5 = 7$, then the original equation is a *contradiction* and we state that there is *no solution*.

HELPFUL HINT

When solving equations, remember two important things: (1) your goal is to isolate the variable, and (2) whatever you do to one side of the equation you must also do to the other side. That is, you must treat both sides of the equation equally.

2.5 Exercise Set MathXL® MyMathLab®

Warm-Up Exercises

Fill in the blanks with the appropriate word, phrase, or symbol(s) from the following list.

denominator	numerator	real	always	contradiction	solution	associative
isolate	conditional	identity	power	never	distributive	

1. Our goal in solving a linear equation is to __________ the variable.
2. To eliminate decimal numbers from an equation we can multiply both sides of the equation by a __________ of ten.
3. To eliminate fractions from an equation we can multiply both sides of the equation by the least common __________ of the fractions in the equation.
4. A __________ equation is true for a specific value or values of the variable.
5. An __________ is an equation that is true for all values of the variable.
6. A __________ is an equation that is not true for any value of the variable.
7. When we have an identity, we will state the solution is all __________ numbers.
8. When we have a contradiction, we will state that there is no __________.
9. When solving an equation and we obtain an equation that is __________ true, we write the answer as *all real numbers*.
10. When solving an equation and we obtain an equation that is __________ true, we write the answer as *no solution*.

Practice the Skills

Solve each equation.

11. $5x + 3 = 6$
12. $9x - 7 = -5$
13. $-4x + 10 = 6x$
14. $-7a + 4 = -11a$
15. $3x = -2x + 10$
16. $-3x = -5x + 6$
17. $21 - 6p = 3p - 2p$
18. $8 - 3x = 4x + 50$
19. $2x - 8 = 3x - 6$
20. $5x + 7 = 3x + 5$
21. $6 - 2y = 9 - 8y + 6y$
22. $7 + 8y = 3y + 2 + 5y$
23. $5x + 3 = 2(x + 6)$
24. $x - 14 = 3(x + 2)$
25. $4y - 2 - 8y = 19 + 5y - 3$
26. $3x - 5 + 9x = 2 + 4x + 9$
27. $2(x - 2) = 4x - 6 - 2x$
28. $-9x + 5 + 2x = -7(x - 1) + 6$
29. $-(w + 2) = -6w + 11$
30. $7(-3m + 5) = 3(10 - 6m)$
31. $-3(2t - 5) + 5 = 3t + 13$
32. $4(x - 3) + 2 = 2x + 12$
33. $124.8 - 9.4x = 4.8x + 32.5$
34. $9 - 0.5x = 4.5x + 8.5$
35. $0.62x - 0.65 = 9.75 - 2.63x$
36. $8.71 - 2.44x = 11.02 - 5.74x$
37. $\frac{a}{5} = \frac{a - 3}{2}$
38. $\frac{b}{16} = \frac{b - 6}{4}$
39. $\frac{n}{10} = 9 - \frac{n}{5}$
40. $6 - \frac{x}{4} = \frac{x}{8}$
41. $\frac{7}{2} - \frac{x}{3} = 2x$
42. $\frac{x}{7} + \frac{5}{3} = -2x$
43. $\frac{5}{8} + \frac{1}{4}a = \frac{1}{2}a$
44. $\frac{5}{6}b + \frac{7}{3} = \frac{3}{2}b$
45. $0.1(x + 10) = 0.3x - 4$
46. $-0.3(5x - 1) = 4.3 - 2x$
47. $2(x + 4) = 4x + 3 - 2x + 5$
48. $3(y - 1) + 9 = 8y + 6 - 5y$
49. $5(3n + 3) = 2(5n - 4) + 6n$
50. $-4(-3z - 5) = -(10z + 8) - 2z$
51. $-(3 - p) = -(2p + 3)$
52. $-2(4 - p) = -8(p + 1)$
53. $-(x + 4) + 5 = 4x + 1 - 5x$
54. $-3y + 8(2y + 1) = 13y + 8$
55. $35(2x - 1) = 7(x + 4) + 3x$
56. $24(-2x + 1) = -7(5x + 2) - 3x$
57. $0.4(x + 0.7) = 0.6(x - 4.2)$
58. $0.5(6x - 8) = 1.4(x - 5) - 0.2$

59. $\frac{3}{5}x - 2 = x - \frac{1}{2}$

60. $-\frac{4}{7}k + 2 = \frac{1}{2}k - 1$

61. $\frac{y}{5} + 2 = 3(y - 4)$

62. $2(x - 4) = \frac{x}{5} + 10$

63. $12 - 3x + 7x = -2(-5x + 6)$

64. $-2x - 3 - x = -3(-2x + 7)$

65. $3(x - 6) - 4(3x + 1) = x - 22$

66. $-2(-3x + 5) + 6 = 4(x - 2)$

67. $5 + 2x = 6(x + 1) - 5(x - 3)$

68. $3 - 5x = 4(x - 3) - 7(x + 1)$

69. $7 - (-y - 5) = 2(y + 3) - 6(y + 3)$

70. $-3 - (4 - c) = 2(c - 7) - 5(2c + 1)$

71. $\frac{3}{5}(x - 6) = \frac{2}{3}(3x - 5)$

72. $\frac{1}{2}(2d + 4) = \frac{1}{3}(4d - 4)$

73. $\frac{3(2r - 5)}{5} = \frac{3r - 6}{3}$

74. $\frac{3(x - 4)}{4} = \frac{5(2x - 3)}{3}$

75. $\frac{2}{7}(5x + 4) = \frac{1}{2}(3x - 4) + 1$

76. $\frac{5}{12}(x + 2) = \frac{2}{3}\left(2x + 1\right) + \frac{1}{6}$

77. $\frac{a - 5}{2} = \frac{3a}{4} + \frac{a - 25}{6}$

78. $\frac{a - 7}{3} = \frac{a + 5}{2} - \frac{7a - 1}{6}$

Concept/Writing Exercises

79. a) Construct a *conditional equation* containing three terms on the left side of the equation and two terms on the right side of the equation.

b) Explain how you know your answer to part **a)** is a conditional equation.

c) Solve the equation.

80. a) Construct a *conditional equation* containing two terms on the left side of the equation and three terms on the right side of the equation.

b) Explain how you know your answer to part **a)** is a conditional equation.

c) Solve the equation.

81. a) Construct an *identity* containing three terms on the left side of the equation and two terms on the right side of the equation.

b) Explain how you know your answer to part **a)** is an identity.

c) What is the solution to the equation?

82. a) Construct an *identity* containing two terms on the left side of the equation and three terms on the right side of the equation.

b) Explain how you know your answer to part **a)** is an identity.

c) What is the solution to the equation?

83. a) Construct a *contradiction* containing three terms on the left side of the equation and two terms on the right side of the equation.

b) Explain how you know your answer to part **a)** is a contradiction.

c) What is the solution to the equation?

84. a) Construct a *contradiction* containing three terms on the left side of the equation and four terms on the right side of the equation.

b) Explain how you know your answer to part **a)** is a contradiction.

c) What is the solution to the equation?

Challenge Problems

85. Solve the equation 5✱ − 1 = 4✱ + 5✱ for ✱.

86. Solve the equation $2\triangle - 4 = 3\triangle + 5 - \triangle$ for $\triangle$.

87. Solve the equation 3☺ − 5 = 2☺ − 5 + ☺ for ☺.

88. Solve $-2(x + 3) + 5x = 3(4 - 2x) - (x + 2)$.

89. Solve $4 - [5 - 3(x + 2)] = x - 3$.

Group Activity

Discuss and answer Exercise 90 as a group. In the next chapter, we will be discussing procedures for writing application problems as equations. Let's get some practice now.

90. Chocolate Bars Consider the following word problem. Mary Kay purchased two large chocolate bars. The total cost of the two chocolate bars was equal to the cost of one chocolate bar plus \$6. Find the cost of one chocolate bar.

a) Each group member: Represent this problem as an equation with the variable x.

b) Each group member: Solve the equation you determined in part **a)**.

c) As a group, check your equation and your answer to make sure that it makes sense.

Cumulative Review Exercises

[1.5] **91.** Evaluate.

a) $|4|$ **b)** $|-7|$ **c)** $|0|$

[1.9] **92.** $\dfrac{x^a}{x^b}$

[2.1] **93.** Explain the difference between factors and terms.

94. Simplify $2(x - 3) + 4x - (4 - x)$.

[2.4] **95.** Solve $2(x - 3) + 4x - (4 - x) = 0$.

96. Solve $(x + 4) - (4x - 3) = 16$.

2.6 Formulas

1. Use the simple interest formula and the distance formula.
2. Use geometric formulas.
3. Solve for a variable in a formula.

Formula

A **formula** is an equation commonly used to express a specific relationship mathematically.

For example, the formula for the area of a rectangle is

$$\text{area} = \text{length} \cdot \text{width} \quad \text{or} \quad A = lw$$

To **evaluate a formula,** substitute the appropriate numerical values for the variables and perform the indicated operations.

1 Use the Simple Interest Formula and the Distance Formula

A formula commonly used in banking is the **simple interest formula.**

Understanding Algebra

Formulas are used in business, science, and many other disciplines. A formula is an equation that states a relationship involving several variables.

Simple Interest Formula

interest = principal · rate · time or $i = prt$

This formula is used to determine the simple interest, i, earned on some savings accounts, or the simple interest an individual must pay on certain loans. In the simple interest formula $i = prt$, p is the principal (the amount invested or borrowed), r is the interest rate in decimal form, and t is the amount of time of the investment or loan.

Understanding Algebra

$i = prt$

i interest received

p principal: amount invested

r rate: interest rate (as a decimal)

t time: time of investment

EXAMPLE 1 **Auto Loan** To buy a car, Mary Beth Orrange borrowed \$10,000 from a bank for 3 years. The bank charged 5% simple annual interest for the loan. How much interest will Mary Beth owe the bank?

Solution Understand and Translate Since the bank charged simple interest, we use the simple interest formula to solve the problem. We are given that the rate, r, is 5%, or 0.05 in decimal form. The principal, p, is \$10,000 and the time, t, is 3 years. We substitute these values in the simple interest formula and solve for the interest, i.

Carry Out

$$i = prt$$
$$i = 10{,}000(0.05)(3)$$
$$i = 1500$$

Check There are various ways to check this problem. First ask yourself "Is the answer realistic?" \$1500 is a realistic answer. The interest on \$10,000 for 1 year at 5% is \$500. Therefore for 3 years, an interest of \$1500 is correct.

Answer Mary Beth will pay \$1500 interest. After 3 years, when she repays the loan, she will pay the principal, \$10,000, plus the interest, \$1500, for a total of \$11,500.

Now Try Exercise 83

EXAMPLE 2 **Savings Account** John Kawai invests \$2500 in a savings account that earns simple interest for 4 years. If the interest earned from the account is \$200, determine the rate.

Solution Understand and Translate We will use the simple interest formula, $i = prt$. We are given the principal, p, the time, t, and the interest, i. We are asked to find the rate, r. We substitute the given values in the simple interest formula and solve the resulting equation for r.

$$i = prt$$
$$200 = 2500 \cdot r \cdot 4$$

Carry Out

$$200 = 10{,}000r$$
$$\frac{200}{10{,}000} = \frac{10{,}000r}{10{,}000}$$
$$0.02 = r$$

Check and Answer The simple interest rate of 0.02 or 2% per year is realistic. If we substitute $p = \$2500$, $r = 0.02$, and $t = 4$, we obtain the interest, $i = \$200$. Thus, the answer checks. The simple interest is 2%.

Now Try Exercise 85

Another important formula is the distance formula.

Distance Formula

distance = rate · time or $d = r \cdot t$

EXAMPLE 3 **Auto Race** At a NASCAR auto race, Brad Keselowski completed the race in 3.2 hours at an average speed of 156.25 miles per hour. Determine the distance of the race.

Solution Understand and Translate We are given the rate, 156.25 miles per hour, and the time is 3.2 hours. We are asked to find the distance.

$$\text{distance} = \text{rate} \cdot \text{time}$$

Carry Out

$$= (156.25)(3.2) = 500$$

Answer Thus, the distance of the race was 500 miles.

Now Try Exercise 89

Understanding Algebra

distance = rate × time

If you are traveling at a constant rate of 55 mph for 2 hours, you have traveled 55 × 2 = 110 miles.

Let's look at the units in Example 3. The rate is given in miles per hour and the time is given in hours. If we analyze the units (a process called *dimensional analysis*), we see that the answer is given in miles.

$$\text{distance} = \text{rate} \cdot \text{time}$$
$$= \frac{\text{miles}}{\cancel{\text{hour}}} \cdot \cancel{\text{hour}}$$
$$= \text{miles}$$

Now we will discuss geometric formulas that will be used throughout the book.

2 Use Geometric Formulas

The **perimeter,** P, is the sum of the lengths of the sides of a figure. Perimeters are measured in the same common unit as the sides. For example, perimeter may be measured in centimeters, inches, or feet. The **area,** A, is the measure of the amount of surface within the figure's boundaries. Areas are measured in square units. For example, area may be measured in square centimeters, square inches, or square feet. **Table 2.1** gives the formulas for finding the areas and perimeters of triangles and quadrilaterals. **Quadrilateral** is a general name for a four-sided figure.

In **Table 2.1,** the letter h is used to represent the *height* of the figure. In the figure of the trapezoid, the sides b and d are called the *bases* of the trapezoid. In the triangle, the side labeled b is called the *base* of the triangle.

Legend for Table 2.1

s = side
w = width
l = length
h = height
b = base
d = trapezoid's second base

TABLE 2.1 Formulas for Areas and Perimeters of Quadrilaterals and Triangles*

Figure	Sketch	Area	Perimeter
Square		$A = s^2$	$P = 4s$
Rectangle		$A = lw$	$P = 2l + 2w$
Parallelogram		$A = lh$	$P = 2l + 2w$
Trapezoid		$A = \frac{1}{2}h(b + d)$	$P = a + b + c + d$
Triangle		$A = \frac{1}{2}bh$	$P = a + b + c$

*See Appendix B for additional information on geometry and geometric figures.

EXAMPLE 4 **Building an Exercise Area** Dr. Alex Taurke, a veterinarian, decides to fence in a large rectangular area in the yard behind his office for exercising dogs that are boarded overnight. The part of the yard to be fenced in will be 40 feet long and 23 feet wide (see **Fig. 2.2**).

a) How much fencing is needed?

b) How large, in square feet, will the fenced-in area be?

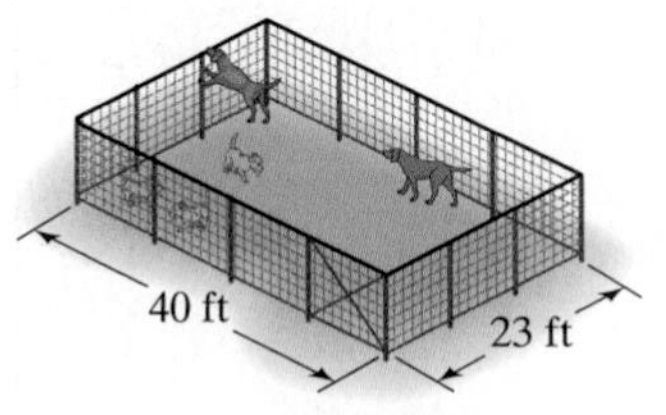

FIGURE 2.2

Solution

a) Understand To find the amount of fencing required, we need to find the perimeter of the rectangular area to be fenced in. To find the perimeter, P, substitute 40 for the length, l, and 23 for the width, w, in the perimeter formula, $P = 2l + 2w$.

$$P = 2l + 2w$$

Carry Out $$P = 2(40) + 2(23) = 80 + 46 = 126$$

Check and Answer By looking at **Figure 2.2,** we can see that a perimeter of 126 feet is a reasonable answer. Thus, 126 feet of fencing will be needed to fence in the area for the dogs to exercise.

b) To find the fenced-in area, substitute 40 for the length and 23 for the width in the formula for the area of a rectangle. Since we are multiplying an amount measured in feet by a second amount measured in feet, the answer will be in square feet (or ft^2).

$$\begin{aligned} A &= lw \\ &= 40(23) = 920 \text{ square feet (or } 920 \text{ ft}^2\text{)} \end{aligned}$$

Based upon the data given, an area of 920 ft^2 is reasonable. The area to be fenced in will be 920 square feet.

Now Try Exercise 93

EXAMPLE 5 **Panoramic Photo** Heather Hunter enlarges rectangular panoramic photos, like the one shown below. One of her enlarged panoramic photos has a perimeter of 116 inches and a length of 40 inches. Find the width of the photo.

Solution Understand and Translate The perimeter, P, is 116 inches and the length, l, is 40 inches. Substitute these values into the formula for the perimeter of a rectangle and solve for the width, w.

$$P = 2l + 2w$$
$$116 = 2(40) + 2w$$

Carry Out

$$116 = 80 + 2w$$
$$116 - 80 = 80 - 80 + 2w \quad \text{Subtract 80 from both sides.}$$
$$36 = 2w$$
$$\frac{36}{2} = \frac{2w}{2} \quad \text{Divide both sides by 2.}$$
$$18 = w$$

Check and Answer By comparing the length and width, you should realize that the dimensions of a length of 40 inches and a width of 18 inches is reasonable. The answer is, the width of the photo is 18 inches.

Now Try Exercise 25

EXAMPLE 6 **Sailboat** A small sailboat has a triangular sail that has an area of 30 square feet and a base of 5 feet (see **Fig. 2.3**). Determine the height of the sail.

Solution Understand and Translate We use the formula for the area of a triangle given in **Table 2.1** on page 139.

$$A = \frac{1}{2}bh$$
$$30 = \frac{1}{2}(5)h$$

Carry Out

$$2 \cdot 30 = 2 \cdot \frac{1}{2}(5)h \quad \text{Multiply both sides by 2.}$$
$$60 = 5h$$
$$\frac{60}{5} = \frac{5h}{5} \quad \text{Divide both sides by 5.}$$
$$12 = h$$

FIGURE 2.3

Check and Answer The height of the triangle is 12 feet. By looking at **Figure 2.3**, you may realize that a sail 12 feet tall and 5 feet wide at the base is reasonable. Thus, the height of the sail is 12 feet.

Now Try Exercise 97

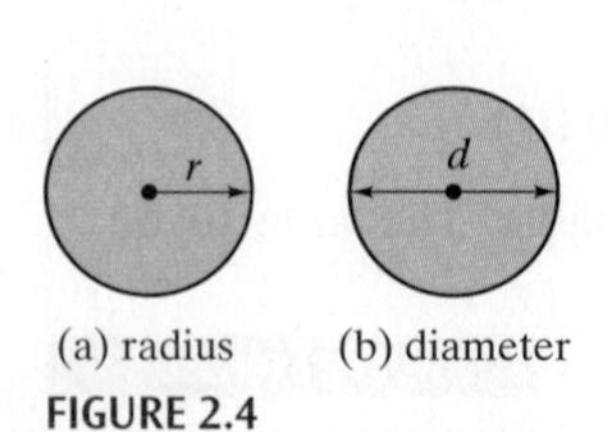

(a) radius (b) diameter

FIGURE 2.4

Another figure that we see and use daily is the circle. The **circumference**, C, is the length (or perimeter) of the curve that forms a circle. The **radius**, r, is the line segment from the center of the circle to any point on the circle (**Fig. 2.4a**). The **diameter** of a circle is a line segment through the center whose endpoints both lie on the circle (**Fig. 2.4b**). *Note that the length of the diameter is twice the length of the radius.*

Understanding Algebra

π has an infinite, non-repeating decimal expansion. It is sometimes approximated by $\frac{22}{7}$ or by 3.14.

Note: If you use 3.14 for π, your answers might vary slightly from the answers given in this book.

The formulas for both the area and the circumference of a circle are given in **Table 2.2**.

TABLE 2.2 Formulas for Circles

Circle	Area	Circumference
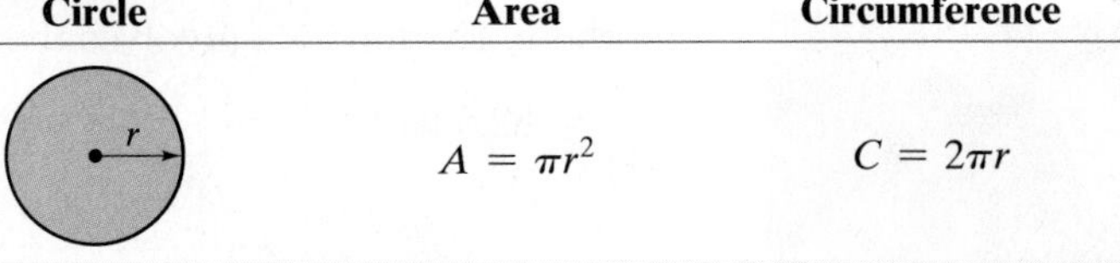	$A = \pi r^2$	$C = 2\pi r$

The value of **pi,** symbolized by the Greek lowercase letter π, is an irrational number that cannot be exactly expressed as a decimal number or a numerical fraction. Pi is *approximately* 3.14. Here it is approximated to 100 digits:

$\pi \approx$ 3.14159 26535 89793 23846 26433 83279 50288 41971 69399 37510 58209 74944 59230 78164 06286 20899 86280 34825 34211 7068...

EXAMPLE 7 **Pizza** A large pizza at Maria's Pizza House has a diameter of 14 inches. Determine the area and circumference of the pizza.

Solution The radius is half the diameter, so $r = \frac{14}{2} = 7$ inches.

$$A = \pi r^2 \qquad C = 2\pi r$$
$$A = \pi(7)^2 \qquad C = 2\pi(7)$$
$$A = \pi(49) \qquad C \approx 43.98 \text{ inches}$$
$$A \approx 153.94 \text{ square inches}$$

To obtain our answers 153.94 and 43.98, we used the $\boxed{\pi}$ key on a calculator and rounded our final answer to the nearest hundredth. If you do not have a calculator with a $\boxed{\pi}$ key and use 3.14 for π, your answer for the area would be 153.86.

Now Try Exercise 95

Table 2.3 below gives formulas for finding the volume of certain three-dimensional figures. **Volume** is the measure of the space occupied by a figure, and it is measured in cubic units, such as cubic centimeters or cubic feet.

Legend for Table 2.3

w = width
l = length
h = height
r = radius

TABLE 2.3 Formulas for Volumes of Three-Dimensional Figures

Figure	Sketch	Volume
Rectangular solid		$V = lwh$
Right circular cylinder		$V = \pi r^2 h$
Right circular cone		$V = \frac{1}{3}\pi r^2 h$
Sphere		$V = \frac{4}{3}\pi r^3$

EXAMPLE 8 **Spaceship Earth** The inside of Spaceship Earth at Epcot Center in Disney World, Florida, is a sphere with a diameter of 165 feet (see photo). Determine the volume of Spaceship Earth.

Spaceship Earth

Solution Understand and Translate **Table 2.3** gives the formula for the volume of a sphere. The formula involves the radius. Since the diameter is 165 feet, its radius is $\frac{165}{2} = 82.5$ feet.

$$V = \frac{4}{3}\pi r^3$$

Carry Out $V = \frac{4}{3}\pi(82.5)^3 = \frac{4}{3}\pi(561{,}515.625) \approx 2{,}352{,}071.15$

Check and Answer The volume inside the sphere is very large, so the answer of about 2,352,071.15 cubic feet is reasonable.

Now Try Exercise 107

3 Solve for a Variable in a Formula

Often in this course and in other mathematics and science courses, you will be given an equation or formula solved for one variable and have to solve it for a different variable.

To solve for a variable in a formula, treat each of the quantities, except the one for which you are solving, as if they were constants. Then solve for the desired variable by isolating it on one side of the equation.

EXAMPLE 9 **Rectangular Solid Formula** Solve the formula for the volume of a rectangular solid, $V = lwh$, for h.

Solution We must get h by itself on one side of the equation. Since h is multiplied by lw, we divide both sides of the equation by lw to isolate the h.

$$V = lwh$$

$$\frac{V}{lw} = \frac{lwh}{lw} \quad \text{Divide both sides by } lw.$$

$$\frac{V}{lw} = h$$

Therefore, $h = \frac{V}{lw}$.

Now Try Exercise 49

EXAMPLE 10 **Perimeter of Rectangle** The formula for the perimeter of a rectangle is $P = 2l + 2w$. Solve this formula for the length, l.

Solution We must get l all by itself on one side of the equation.

$$P = 2l + 2w$$

$$P - 2w = 2l + 2w - 2w \quad \text{Subtract } 2w \text{ from both sides.}$$

$$P - 2w = 2l$$

$$\frac{P - 2w}{2} = \frac{2l}{2} \quad \text{Divide both sides by 2.}$$

$$\frac{P - 2w}{2} = l \quad \left(\text{or } l = \frac{P}{2} - w\right)$$

Now Try Exercise 53

Some formulas contain fractions. When a formula contains a fraction, we can eliminate the fraction by multiplying both sides of the equation by the least common denominator.

EXAMPLE 11 The formula for the area of a triangle is $A = \frac{1}{2}bh$. Solve this formula for h.

Solution We begin by multiplying both sides of the equation by the LCD, 2, to eliminate the fraction.

$$A = \frac{1}{2}bh$$

$$2 \cdot A = 2 \cdot \frac{1}{2}bh \quad \text{Multiply both sides by 2.}$$

$$2A = bh$$

$$\frac{2A}{b} = \frac{bh}{b} \quad \text{Divide both sides by } b \text{ to isolate } h.$$

$$\frac{2A}{b} = h$$

Thus, $h = \frac{2A}{b}$.

Now Try Exercise 51

Write Equations in $y = mx + b$ Form

When discussing graphing later in this book, we will need to solve many equations for the variable y, and write the equation in the form $y = mx + b$, where m and b represent real numbers. Examples of equations in this form are $y = 2x + 4$, $y = -\frac{1}{2}x - 3$, and $y = \frac{4}{5}x + \frac{1}{3}$.

EXAMPLE 12 Solve the equation $6x + 3y = 12$ for y. Write the answer in $y = mx + b$ form.

Solution Begin by isolating the term containing the variable y.

$$6x + 3y = 12$$

$$6x - 6x + 3y = 12 - 6x \quad \text{Subtract } 6x \text{ from both sides.}$$

$$3y = 12 - 6x$$

$$\frac{3y}{3} = \frac{12 - 6x}{3} \quad \text{Divide both sides by 3.}$$

$$y = \frac{12 - 6x}{3}$$

$$y = \frac{12}{3} - \frac{6x}{3} \quad \text{Write as two fractions.}$$

$$y = 4 - 2x$$

$$y = -2x + 4$$

Now Try Exercise 69

HELPFUL HINT

Notice that in Example 12, when we obtained $y = \frac{12 - 6x}{3}$, we had solved the equation for y since the y was isolated on one side of the equation. When we wrote the answer as $y = -2x + 4$, we wrote the equation in $y = mx + b$ form.

EXAMPLE 13 Solve the equation $y - \frac{1}{3} = \frac{1}{4}(x - 6)$ for y. Write the answer in $y = mx + b$ form.

Solution Multiply both sides of the equation by the LCD, 12.

$$y - \frac{1}{3} = \frac{1}{4}(x - 6)$$

$$12\left(y - \frac{1}{3}\right) = 12 \cdot \frac{1}{4}(x - 6) \quad \text{Multiply both sides by 12.}$$

$$12y - 4 = 3(x - 6) \quad \text{Distributive property used on left.}$$

$$12y - 4 = 3x - 18 \quad \text{Distributive property used on right.}$$

$$12y = 3x - 14 \quad \text{Added 4 to both sides.}$$

$$y = \frac{3x - 14}{12} \quad \text{Divided both sides by 12.}$$

$$y = \frac{3x}{12} - \frac{14}{12} \quad \text{Write as two fractions.}$$

$$y = \frac{1}{4}x - \frac{7}{6}$$

Now Try Exercise 79

In Example 13, you may wish to use the distributive property on the right side of the equation before multiplying both sides of the equation by the LCD, 12. Try working the example using this method now to see which procedure you prefer.

2.6 Exercise Set MathXL® MyMathLab®

Warm-Up Exercises

Fill in the blanks with the appropriate word, phrase, or symbol(s) from the following list.

formula	quadrilateral	rate	radius	triangle	diameter
time	evaluating	cubic	square	decimal	equation

1. An equation used to express a specific relationship mathematically is called a ______________.
2. The process of substituting values and performing indicated operations on a formula is called ______________ the formula.
3. A four-sided figure is called a ______________.
4. Twice the radius of a circle is the ______________ of the circle.
5. Simple interest is found by multiplying the principal, the rate, and the ______________.
6. When calculating simple interest, the rate should be expressed as a ______________ number.
7. The circumference of a circle is found by multiplying 2π by the ______________.
8. Distance is found by multiplying the ______________ by the time.
9. A possible unit of measure for volume is ______________ feet.
10. A possible unit of measure for area is ______________ inches.

Practice the Skills

Use the formula to find the value of the variable indicated. Round decimal answers to the nearest hundredth.

11. $d = rt$ (distance formula); find d when $r = 60$ and $t = 8$.
12. $P = 4s$ (perimeter of a square); find P when $s = 6$.
13. $A = lw$ (area of a rectangle); find A when $l = 12$ and $w = 8$.
14. $A = s^2$ (area of a square); find A when $s = 9$.

15. $i = prt$ (simple interest formula); find i when $p = 2000$, $r = 0.06$, and $t = 3$.

16. $V = \frac{4}{3}\pi r^3$ (volume of a sphere); find V when $r = 8$.

17. $P = 2l + 2w$ (perimeter of a rectangle); find P when $l = 8$ and $w = 10$.

18. $V = \frac{1}{3}\pi r^2 h$ (volume of a cone); find V when $r = 6$ and $h = 7$.

19. $A = \pi r^2$ (area of a circle); find A when $r = 10$.

20. $A = \frac{1}{2}h(b + B)$ (area of a trapezoid); find A when $h = 6$, $b = 4$, and $B = 10$.

21. $A = \frac{m + n}{2}$ (mean of two values); find A when $m = 16$ and $n = 56$.

22. $A = \frac{a + b + c}{3}$ (mean of three values); find A when $a = 72$, $b = 81$, and $c = 93$.

23. $z = \frac{x - m}{s}$ (statistics formula for finding the z-score); find z when $x = 100$, $m = 80$, and $s = 10$.

24. $m = \frac{s - d}{d}$ (markup on cost); find m when $s = 135$ and $d = 100$.

25. $P = 2l + 2w$ (perimeter of a rectangle); find l when $P = 28$ and $w = 6$.

26. $V = lwh$ (volume of a rectangular solid); find w when $V = 27$, $l = \frac{1}{2}$, and $h = 6$.

27. $V = \pi r^2 h$ (volume of a cylinder); find h when $V = 678.24$ and $r = 6$.

28. $A = \frac{1}{2}h(b + d)$ (area of a trapezoid); find b when $A = 72$, $h = 12$, and $d = 7$.

In Exercises 29–42, use **Tables 2.1, 2.2,** *and* **2.3** *to find the formula for the area or volume of the figure. Then determine either the area or volume.*

29. 

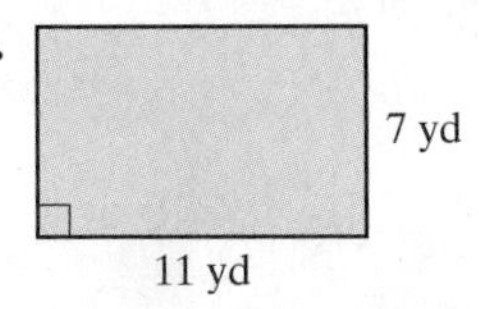

30.

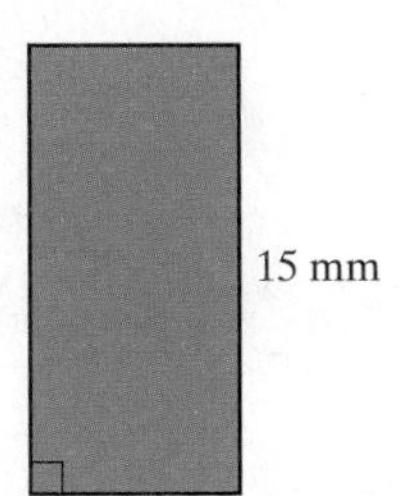

31.

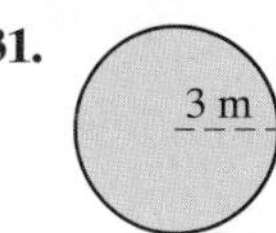

32.

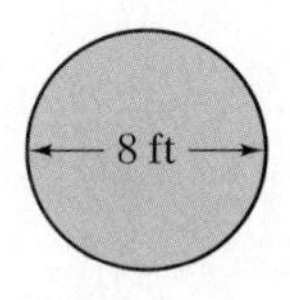

33.

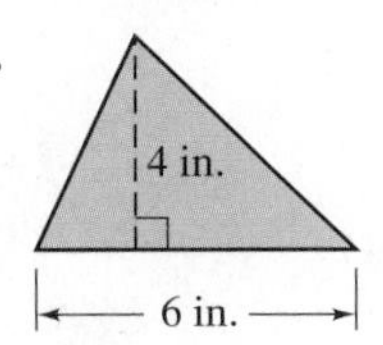

34.

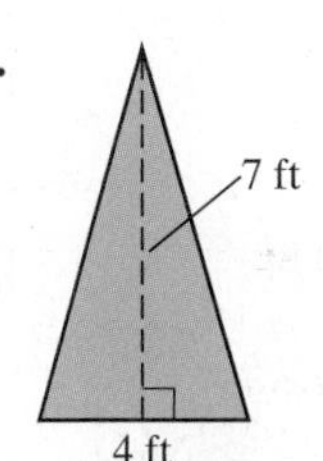

35.

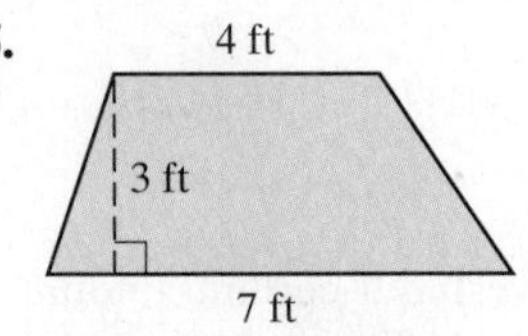

36.

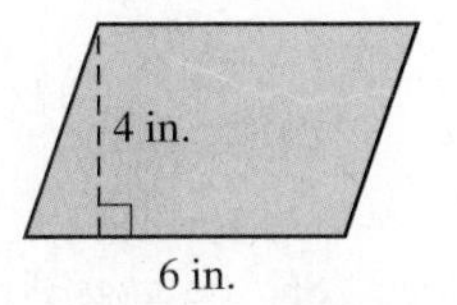

37.

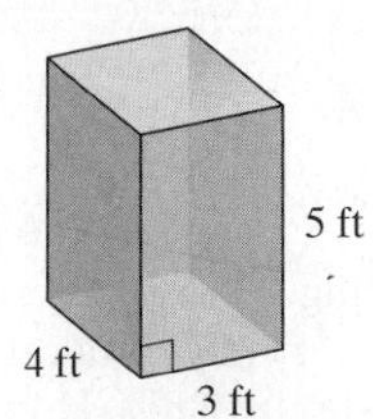

38.

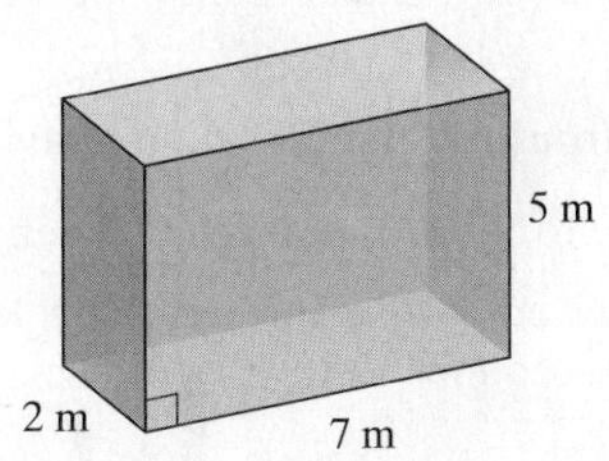

39.

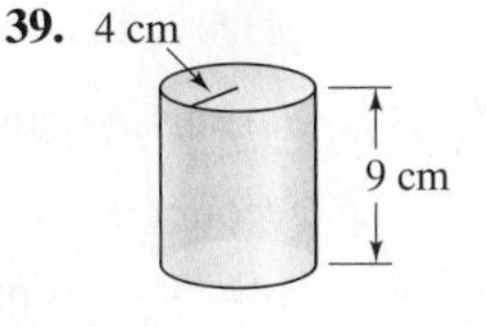

40.

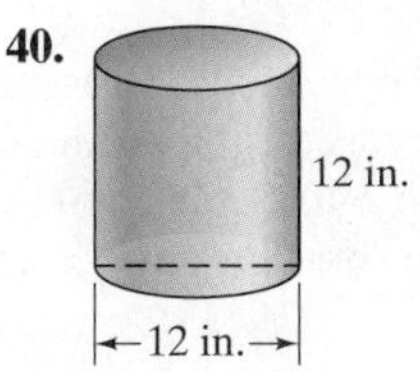

41.

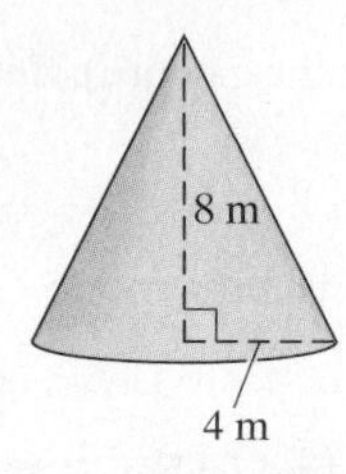

42.

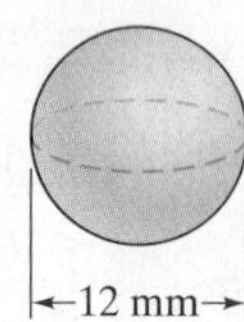

In Exercises 43 and 44, use the formula $C = \frac{5}{9}(F - 32)$ to find the Celsius temperature (C) equivalent to the given Fahrenheit temperature (F).

43. $F = 50°$

44. $F = 86°$

In Exercises 45 and 46, use the formula $F = \frac{9}{5}C + 32$, to find the Fahrenheit temperature (F) equivalent to the given Celsius temperature (C).

45. $C = 25°$

46. $C = -40°$

In Exercises 47–66, solve for the indicated variable.

47. $A = lw$, for l

48. $P = 4s$, for s

49. $i = prt$, for t

50. $C = \pi d$, for d

51. $A = \frac{1}{2}bh$, for b

52. $V = \frac{1}{3}Bh$, for h

53. $P = 2l + 2w$, for w

54. $4m + 5n = 25$, for n

55. $3 - 2r = n$, for r

56. $7 - 5p = q$, for p

57. $y = mx + b$, for b

58. $t = ps + w$, for w

59. $d = a + b + c$, for b

60. $r + s - t = w$, for s

61. $ax + by + c = 0$, for y

62. $ax + by = c$, for y

63. $V = \frac{1}{3}\pi r^2 h$, for h

64. $A = \frac{m + 2d}{3}$, for d

65. $A = \frac{m + d}{2}$, for m

66. $L = \frac{c + 2d}{4}$, for d

In Exercises 67–82, solve each equation for y. Write the answer in $y = mx + b$ form. See Examples 12 and 13.

67. $2x + y = 8$

68. $6x + 2y = -12$

69. $-3x + 3y = -18$

70. $-2y + 4x = -8$

71. $4x = 6y - 8$

72. $-3x = 9y - 6$

73. $5y = -10 + 3x$

74. $-7y = 21 - 9x$

75. $-6y = 15 - 3x$

76. $-2y = -3x - 18$

77. $5x - 3y = 11$

78. $4x + 3y = 20$

79. $y + 3 = -\frac{1}{3}(x - 4)$

80. $y - 3 = \frac{2}{3}(x + 4)$

81. $y - \frac{1}{5} = 2\left(x + \frac{1}{3}\right)$

82. $y + 5 = \frac{3}{4}\left(x + \frac{1}{2}\right)$

Problem Solving

In Exercises 83–86, use the simple interest formula.

83. Auto Loan Thang Tran decided to borrow \$6000 from Citibank to help pay for a car. His loan was for 3 years at a simple interest rate of 8%. How much interest will Thang pay?

84. Simple Interest Loan Holly Broesamle lent her brother \$4000 for a period of 2 years. At the end of the 2 years, her brother repaid the \$4000 plus \$640 interest. What simple interest rate did her brother pay?

85. Savings Account Mary Seitz invested a certain amount of money in a savings account paying 3% simple interest per year. When she withdrew her money at the end of 3 years, she received \$450 in interest. How much money did Mary place in the savings account?

86. Savings Account Peter Ostroushko put \$6000 in a savings account earning $3\frac{1}{2}\%$ simple interest per year. When he withdrew his money, he received \$840 in interest. How long had he left his money in the account?

In Exercises 87–90, use the distance formula.

87. Average Speed On her way from Omaha, Nebraska, to Kansas City, Kansas, Peg Hovde traveled 180 miles in 3 hours. What was her average speed?

88. Fastest Aircraft The world's fastest aircraft is the Falcon HTV-2. This aircraft can travel from Washington, DC, to London, England, a distance of about 3700 miles, in about $\frac{1}{3}$ hour. Determine the aircraft's average speed on this flight.

89. Fastest Car The fastest speed recorded on land was about 763.2 miles per hour by a jet-powered car called ThrustSSC. If, during the speed trial, the car traveled for 0.01 hour, how far had the car traveled?

90. Walk Lisa Feintech went for a walk where she walked at an average speed of 3.4 miles per hour for 2 hours. How far did she walk?

Use the formulas given in **Tables 2.1, 2.2,** *and* **2.3** *to work Exercises 91–110.*

91. DVD Player A portable DVD player has a screen with a length of 8 inches and a width (or height) of 6 inches. Determine the area of the screen.

92. Horse Pasture Ann Wolcott keeps her horses in a pasture that is in the shape of a rectangle with a length of 200 yards and a width of 75 yards. Determine the area of the horse pasture.

93. Television A plasma television has a rectangular screen with a length of 34.9 inches and a width (or height) of 19.6 inches. Determine the perimeter of the screen.

94. Fencing Milt McGowen has a rectangular lot that measures 100 feet by 60 feet. If Milt wants to fence in his lot, how much fencing will he need?

95. Swimming Pool A circular above-ground swimming pool has a diameter of 24 feet. Determine the circumference of the pool.

96. Banyan Tree The largest banyan tree in the continental United States is at the Edison House in Fort Myers, Florida. The circumference of the aerial roots of the tree is 390 feet. Find the *diameter* of the aerial roots to the nearest tenth of a foot.

See Exercise 96.

97. Kite Below we show a kite. Determine the area of the kite.

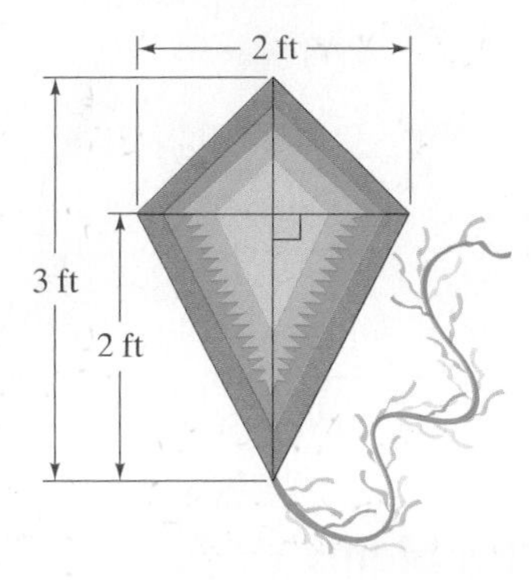

98. Yield Sign A yield traffic sign is triangular with a base of 36 inches and a height of 31 inches. Find the area of the sign.

99. Flower Garden Elizabeth Bonawitz has a circular flower garden with a radius of 7 feet. Determine the area of the flower garden.

100. Living Room Table A round living room table top has a diameter of 3 feet. Find the area of the table top.

101. Sand Volleyball Court Claire Medve has a backyard sand volleyball court. The court is in the shape of a rectangular solid with length 60 feet, width 30 feet, and depth of sand (or height) 1.5 feet. Determine the volume of the sand in the court.

102. Laptop The carrying case for a laptop computer measures 17 inches long by 14 inches wide by 6 inches deep. Determine the volume of the case.

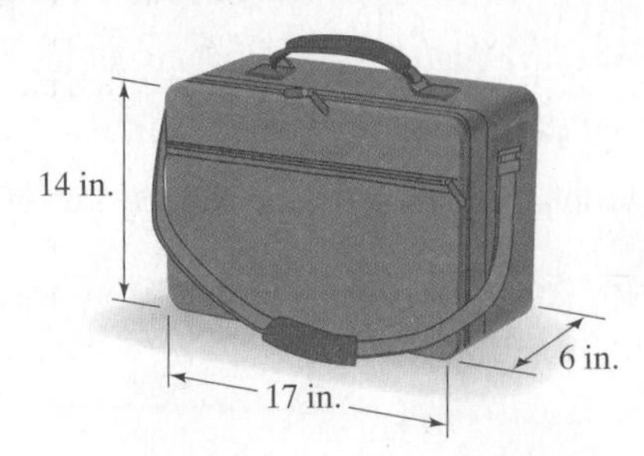

103. Hot Tub The inside of a circular hot tub is 8 feet in diameter. If the water inside the hot tub is 3 feet deep, determine, in cubic feet, the volume of water in the hot tub.

104. Oil Drum Roberto Sanchez has an empty oil drum that he uses for storage. The oil drum is 4 feet high and has a diameter of 24 inches. Find the volume of the drum in cubic feet.

105. Trapezoidal Sign Canter Martin made a sign to display at a baseball game. The sign was in the shape of a trapezoid. Its bases are 4 feet and 3 feet, and its height is 2 feet. Find the area of the sign.

106. Amphitheater The seats in an amphitheater are inside a trapezoidal area as shown in the figure. The bases of the trapezoidal area are 80 feet and 200 feet, and the height is 100 feet. Find the area of the floor occupied by seats.

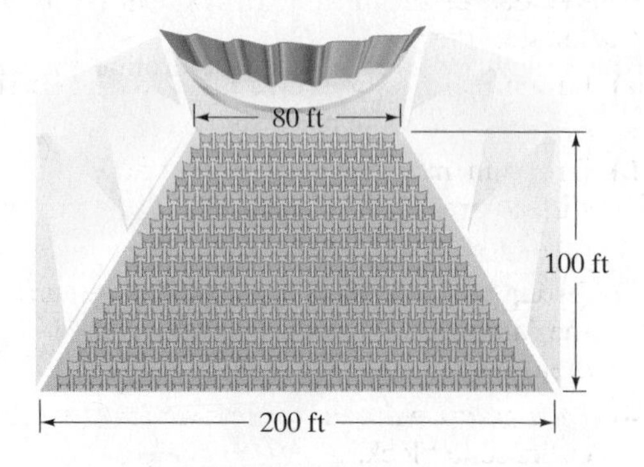

107. Basketball Find the volume of a basketball if its diameter is 9 inches.

108. Grain Bin Janet Evert has a grain bin on her farm that is in the shape of a right circular cone. The height of the cone is 15 feet and the radius is 8 feet. Determine the volume of the grain bin.

109. Body Mass Index A person's body mass index (BMI) is found by multiplying a person's weight, w, in pounds by 703, then dividing this product by the square of the person's height, h, in inches.

a) Write a formula to find the BMI.

b) Brandy Belmont is 5 feet 3 inches tall and weighs 135 pounds. Find her BMI.

110. Body Mass Index Refer to Exercise 109. Mario Guzza's weight is 162 pounds, and he is 5 feet 7 inches tall. Find his BMI.

Challenge Problems

111. Cereal Box A cereal box is to be made by folding the cardboard along the dashed lines as shown in the figure on the right.

a) Using the formula

$$\text{volume} = \text{length} \cdot \text{width} \cdot \text{height}$$

write an equation for the volume of the box.

b) Find the volume of the box when $x = 7$ cm.

c) Write an equation for the surface area of the box.

d) Find the surface area when $x = 7$ cm.

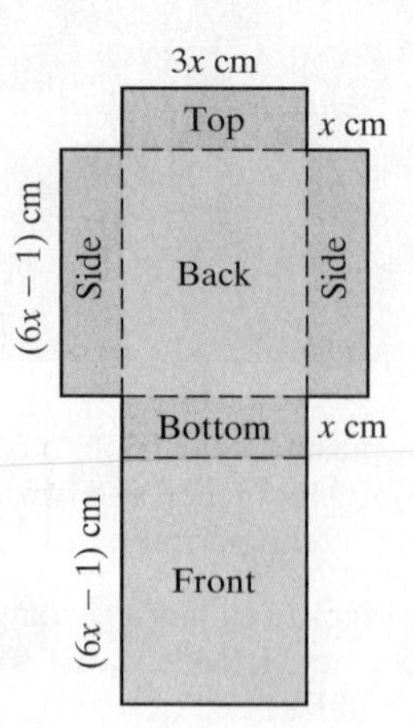

Concept/Writing Exercises

112. When using the distance formula, what happens to the distance if the rate is doubled and the time is halved? Explain.

113. When using the simple interest formula, what happens to the simple interest if both the principal and rate are doubled but the time is halved? Explain.

114. Consider the formula for the area of a square, $A = s^2$. If the length of the side of a square, s, is doubled, what is the change in its area? Explain.

115. Consider the formula for the volume of a cube, $V = s^3$. If the length of the side of a cube, s, is doubled, what is the change in its volume? Explain.

116. Which would have the greater area, a square whose side has a length of s inches, or a circle whose diameter has a length of s inches? Explain, using a sketch.

117. Which would have the greater area, a square whose diagonal has a length of s inches, or a circle whose diameter has a length of s inches? Explain, using a sketch.

Group Activity

118. Square Face on Cube Consider the following photo. The front of the figure is a square with a smaller black square

painted on the center of the larger square. Suppose the length of one side of the larger square is A, and the length of one side of the smaller (black) square is B. Also the thickness of the block is C.

a) Group member one: Determine an expression for the surface area of the black square.

b) Group member two: Determine an expression for the surface area of the larger square (which includes the smaller square).

c) Group member three: Determine the surface area of the larger square minus the black square (the purple area shown).

d) As a group, write an expression for the volume of the entire solid block.

e) As a group, determine the volume of the entire solid block if its length is 1.5 feet and its width is 0.8 feet.

Cumulative Review Exercises

[1.7] **119.** Evaluate $-\frac{4}{15} + \frac{3}{5}$.

120. Evaluate $-6 + 7 - 4 - 3$.

[1.9] **121.** Evaluate $[4(12 \div 2^2 - 3)^2]^2$.

[2.4] **122.** Solve the equation $\frac{r}{2} + 2r = 20$.

2.7 Ratios and Proportions

1. Understand ratios.
2. Solve proportions using cross-multiplication.
3. Solve applications.
4. Use proportions to change units.
5. Use proportions to solve problems involving similar figures.

1 Understand Ratios

Ratio

A **ratio** is a quotient of two quantities.

Ratios provide a way to compare two numbers or quantities. The ratio of the number a to the number b may be written

$$a \text{ to } b, \quad a{:}b, \quad \text{or} \quad \frac{a}{b}$$

where a and b are called the **terms of the ratio.** Notice that the symbol : can be used to indicate a ratio.

EXAMPLE 1 **Favorite Pets** Children at the Lakewood River Elementary School were asked to name their favorite pet. The results are indicated in **Figure 2.5.**

Favorite Pets

Pet	Number of children
Rabbits	12
Hamsters	28
Dogs	56
Cats	72

Number of children (axis: 0 10 20 30 40 50 60 70 80)

FIGURE 2.5

> **Understanding Algebra**
>
> A *ratio* is another word for a fraction. It can be denoted 3 different ways: $\frac{a}{b}$, $a : b$, or with the word "to" as "a to b."

a) Find the ratio of the number of children who selected dogs to those who selected hamsters. Then write the ratio in lowest terms.

b) Find the ratio of the number of children who selected cats to the total number in the survey.

Solution We will use the five-step problem-solving procedure.

a) Understand and Translate The ratio we are seeking is

Number who selected dogs : Number who selected hamsters

Carry Out We substitute the appropriate values from the bar graph into the ratio. This gives

$$56 : 28$$

To write the ratio in *lowest terms*, we simplify by dividing each number in the ratio by 28, the greatest number that divides both terms in the ratio. This gives

$$2 : 1$$

Check and Answer Our division is correct. Thus, the ratio of the number of children who selected dogs to the number of children who selected hamsters is 2 : 1.

b) We use the same procedure as in part **a**). Seventy-two children selected cats. There were $12 + 28 + 56 + 72$ or 168 students surveyed. The ratio is 72 : 168. This ratio can be simplified by dividing each of these numbers by 24. Thus the ratio simplifies to

$$3 : 7$$

Now Try Exercise 33

The answer in Example 1, part **a**) could have also been written $\frac{2}{1}$ or 2 to 1. The answer in part **b**) could have been written $\frac{3}{7}$ or 3 to 7.

EXAMPLE 2 **Cholesterol Level** There are two types of cholesterol: low-density lipoprotein, (LDL—considered the harmful type of cholesterol) and high-density lipoprotein (HDL—considered the healthful type of cholesterol). Some doctors recommend that the ratio of LDL to HDL be less than or equal to 4 : 1. Mr. Suarez's cholesterol test showed that his LDL was 167 milligrams per deciliter, and his HDL was 40 milligrams per deciliter. Is Mr. Suarez's ratio of LDL to HDL less than or equal to the recommended 4 : 1 ratio?

Solution Understand We need to determine if Mr. Suarez's LDL to HDL ratio is less than or equal to 4:1.

Translate Mr. Suarez's LDL to HDL ratio is 167:40. To make the second term equal to 1, we divide both terms in the ratio by the second term, 40.

Carry Out
$$\frac{167}{40} : \frac{40}{40}$$
$$\text{or } 4.175 : 1$$

Check and Answer Our division is correct. Therefore, Mr. Suarez's ratio is not less than or equal to the desired 4:1 ratio.

Now Try Exercise 87

EXAMPLE 3 **Gas-Oil Mixture** Some power equipment, such as chainsaws and blowers, use a gas–oil mixture to run the engine. The instructions on a particular chainsaw indicate that 5 gallons of gasoline should be mixed with 40 ounces of special oil to obtain the proper gas–oil mixture. Find the ratio of gasoline to oil in the proper mixture.

Solution Understand To express these quantities in a ratio, both quantities must be in the same units. We can either convert 5 gallons to ounces or 40 ounces to gallons.

Translate Let's change 5 gallons to ounces. Since there are 128 ounces in 1 gallon, 5 gallons of gas equals 5(128) or 640 ounces. The ratio we are seeking is

$$\text{ounces of gasoline} : \text{ounces of oil}$$

Carry Out
$$640 : 40$$
$$\text{or } 16 : 1 \quad \text{Divide both terms by 40 to simplify.}$$

Check and Answer Our simplification is correct. The correct ratio of gas to oil for this chainsaw is 16 : 1.

Now Try Exercise 23

EXAMPLE 4 **Gear Ratio** The *gear ratio* of two gears is defined as

$$\text{gear ratio} = \frac{\text{number of teeth on the driving gear}}{\text{number of teeth on the driven gear}}$$

Find the gear ratio of the gears shown in **Figure 2.6.**

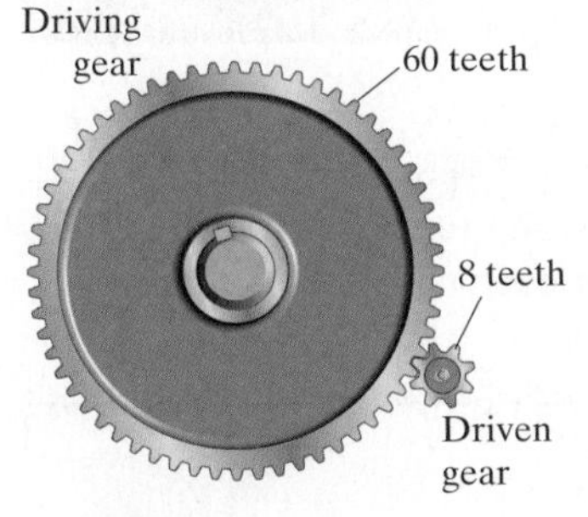

FIGURE 2.6

Solution Understand and Translate To find the gear ratio we need to substitute the appropriate values.

Carry Out $$\text{gear ratio} = \frac{\text{number of teeth on driving gear}}{\text{number of teeth on driven gear}} = \frac{60}{8} = \frac{15}{2}$$

Thus, the gear ratio is 15 : 2. Gear ratios are generally given as some quantity to 1. If we divide both terms of the ratio by the second term, we will obtain a ratio of some number to 1. Dividing both 15 and 2 by 2 gives a gear ratio of 7.5 : 1.

Check and Answer The gear ratio is 7.5 : 1. This means that as the driving gear goes around once the driven gear goes around 7.5 times. (A typical first gear ratio on a passenger car may be 3.545 : 1.)

Now Try Exercise 27

2 Solve Proportions Using Cross-Multiplication

Proportion

A **proportion** is a special type of equation. It is a statement of equality between two ratios.

One way of denoting a proportion is $a : b = c : d$, which is read "a is to b as c is to d." In this text we write proportions as

$$\frac{a}{b} = \frac{c}{d}$$

Understanding Algebra

When two ratios are set equal, we call it a *proportion*. For example, consider $\frac{3}{12} = \frac{1}{4}$. The numbers 3 and 4 are called the *extremes* and 12 and 1 are called the *means* of the proportion. *The product of the means* $(12 \cdot 1 = 12)$ *always equals the product of the extremes* $(3 \cdot 4 = 12)$.

The a and d are *referred* to as the **extremes,** and the b and c are referred to as the **means** of the proportion. In Sections 2.4 and 2.5, we solved equations containing fractions by multiplying both sides of the equation by the LCD to eliminate fractions. For example, for the proportion

$$\frac{x}{3} = \frac{35}{15}$$

$$15\left(\frac{x}{3}\right) = 15\left(\frac{35}{15}\right) \quad \text{Mulyiply both sides by the LCD, 15.}$$

$$5x = 35$$

$$x = 7$$

Another method that can be used to solve proportions is **cross-multiplication.**

Cross-Multiplication

If $\frac{a}{b} = \frac{c}{d}$, then $ad = bc$.

The product of the extremes, ad, is equal to the product of the means, bc.

If any three of the four quantities of a proportion are known, the fourth quantity can easily be found.

EXAMPLE 5 Solve $\frac{x}{3} = \frac{35}{15}$ for x by cross-multiplying.

Solution

$$\frac{x}{3} = \frac{35}{15}$$

$$x \cdot 15 = 3 \cdot 35$$

$$15x = 105$$

$$x = \frac{105}{15} = 7$$

Check

$$\frac{x}{3} = \frac{35}{15}$$

$$\frac{7}{3} \stackrel{?}{=} \frac{35}{15}$$

$$\frac{7}{3} = \frac{7}{3} \quad \text{True}$$

Now Try Exercise 37

Earlier in this chapter, we solved the proportions like $\frac{x}{3} = \frac{35}{15}$ by multiplying both sides of the equation by 15. In Example 5, we solved the same proportion using cross-multiplication. Notice we obtained the same solution, 7, in each case.

EXAMPLE 6 Solve $\frac{-8}{3} = \frac{64}{x}$ for x by cross-multiplying.

Solution

$$\frac{-8}{3} = \frac{64}{x}$$

$$-8 \cdot x = 3 \cdot 64$$

$$-8x = 192$$

$$\frac{-8x}{-8} = \frac{192}{-8}$$

$$x = -24$$

Check

$$\frac{-8}{3} = \frac{64}{x}$$

$$\frac{-8}{3} \stackrel{?}{=} \frac{64}{-24}$$

$$\frac{-8}{3} \stackrel{?}{=} \frac{8}{-3}$$

$$\frac{-8}{3} = \frac{-8}{3} \quad \text{True}$$

Now Try Exercise 41

3 Solve Applications

Often, practical problems can be solved using proportions. Below are specific directions for translating application problems into proportions.

To Solve Problems Using Proportions

1. Understand the problem.
2. Translate the problem into mathematical language.
   a) First, represent the unknown quantity by a variable.
   b) Second, set up the proportion by listing the given ratio on one side of the equation, and the variable and the other given quantity on the other side of the equation. When setting up the proportion, the same respective quantities should occupy the same respective positions on the left and the right. For example, an acceptable proportion might be

$$\text{Given ratio}\left\{\frac{\text{miles}}{\text{hour}}\right. = \frac{\text{miles}}{\text{hour}}$$

3. Carry out the mathematical calculations necessary to solve the problem.
   a) Once the proportion is correctly written, drop the units and cross-multiply.
   b) Solve the resulting equation and label with the appropriate units.
4. Check the answer obtained in step 3.
5. Make sure you have answered the question.

Understanding Algebra

When solving applied problems involving proportions, it is crucial that each ratio has the same units! For example, if one ratio is given in miles/hour and the second ratio is given in feet/hour, one of the ratio's units must be changed before setting up the proportion.

EXAMPLE 7 **Painting** A gallon of paint will cover a surface area of 575 square feet.

a) How many gallons of paint are needed to cover a house with a surface area of 6525 square feet?

b) If a gallon of paint costs \$24.99, what will it cost to paint the house?

Solution

a) Understand The given ratio is 1 gallon per 575 square feet. The unknown quantity is the number of gallons necessary to cover 6525 square feet.

Translate Let $x =$ number of gallons of paint needed.

$$\text{Given ratio}\left\{\frac{1\text{ gallon}}{575\text{ square feet}}\right. = \frac{x\text{ gallons}}{6525\text{ square feet}} \begin{array}{l}\longleftarrow \text{Unknown}\\ \longleftarrow \text{Given quantity}\end{array}$$

Note how the amount, in gallons, and the area, in square feet, are given in the same relative positions.

Carry Out

$$\frac{1}{575} = \frac{x}{6525}$$

$$1(6525) = 575x \quad \text{Cross-multiply.}$$

$$6525 = 575x \quad \text{Solve.}$$

$$\frac{6525}{575} = x$$

$$11.3 \approx x$$

Check Both ratios in the proportion, $\frac{1}{575}$ and $\frac{11.3}{6525}$, have approximately the same value of 0.00173. Thus, the answer of about 11.3 gallons checks.

Answer The amount of paint needed to cover an area of 6525 square feet is about 11.3 gallons.

b) Assuming the painter only buys full gallons of paint, he will need to buy 12 gallons in order to paint the house. Since each gallon costs \$24.99, the cost to paint the house is found by multiplication.

$$12 \times 24.99 = \$299.88$$

The cost to paint the house is \$299.88.

Now Try Exercise 61

EXAMPLE 8 At an amusement park, guests are waiting in a line to ride on a roller coaster. Maryann Justinger observes that 32 people get on the roller coaster every 5 minutes. If Maryann is the 272nd person in line, how long will she have to wait until she can ride on the roller coaster?

Solution The unknown quantity is the time Maryann has to wait until she can ride on the roller coaster. We are given that 32 people get on the roller coaster every 5 minutes. We will use this given ratio in setting up our proportion.

Translate We will let x represent the number of minutes Maryann has to wait.

$$\text{Given ratio } \left\{ \frac{32 \text{ people}}{5 \text{ minutes}} \right. = \frac{272 \text{ people}}{x \text{ minutes}}$$

Carry Out

$$\frac{32}{5} = \frac{272}{x}$$

$$32 \cdot x = 5 \cdot 272$$

$$32x = 1360$$

$$x = \frac{1360}{32} = 42.5$$

Check and Answer Both ratios in the proportion, $\frac{32}{5}$ and $\frac{272}{42.5}$, have the same value of 6.4. Thus, the answer checks, and Maryann would have to wait 42.5 minutes to ride on the roller coaster.

Now Try Exercise 55

EXAMPLE 9 **Drug Dosage** A nurse is to give a patient 250 milligrams of the drug simethicone. The drug is available only in a solution whose concentration is 40 milligrams of simethicone per 0.6 milliliter of solution. How many milliliters of solution should the nurse give the patient?

Solution **Understand and Translate** We can set up the proportion using the medication on hand as the given ratio and the number of milliliters needed to be given as the unknown.

$$\text{Given ratio } \left\{ \frac{40 \text{ milligrams}}{0.6 \text{ milliliter}} \right. = \frac{250 \text{ milligrams}}{x \text{ milliliters}} \begin{array}{l} \longleftarrow \text{Desired medication} \\ \longleftarrow \text{Unknown} \end{array}$$

Carry Out

$$\frac{40}{0.6} = \frac{250}{x}$$

$$40x = 0.6(250) \quad \text{Cross-multiply.}$$

$$40x = 150 \quad \text{Solve.}$$

$$x = \frac{150}{40} = 3.75$$

Check and Answer The nurse should administer 3.75 milliliters of the simethicone solution.

Now Try Exercise 69

AVOIDING COMMON ERRORS

When you are setting up a proportion, it does not matter which unit in the given ratio is in the numerator and which is in the denominator as long as the units in the other ratio are *in the same relative position*. For example,

$$\frac{60 \text{ miles}}{1.5 \text{ hours}} = \frac{x \text{ miles}}{4.2 \text{ hours}} \quad \text{and} \quad \frac{1.5 \text{ hours}}{60 \text{ miles}} = \frac{4.2 \text{ hours}}{x \text{ miles}}$$

will both give the same answer of 168 (try it and see). When setting up the proportion, set it up so that it makes the most sense to you. Notice that when setting up a proportion containing different units, the same units should not be multiplied by themselves during cross-multiplication.

CORRECT	INCORRECT
$\frac{\text{miles}}{\text{hour}} = \frac{\text{miles}}{\text{hour}}$	$\frac{\text{miles}}{\text{hour}} \times \frac{\text{hour}}{\text{miles}}$ (crossed out)

4 Use Proportions to Change Units

Proportions can also be used to convert from one quantity to another. For example, you can use a proportion to convert a measurement in feet to a measurement in meters, or to convert from pounds to kilograms.

EXAMPLE 10 **Converting Kilometers to Miles** There are approximately 1.6 kilometers in 1 mile. What is the distance, in miles, of 78 kilometers?

Solution Understand and Translate We know that 1 mile $\approx$ 1.6 kilometers. We use this known fact in one ratio of our proportion. In the second ratio, we set the quantities with the same units in the same respective positions. The unknown quantity is the number of miles, which we will call x.

$$\text{Known ratio} \left\{ \frac{1 \text{ mile}}{1.6 \text{ kilometers}} \right. = \frac{x \text{ miles}}{78 \text{ kilometers}}$$

Note that both numerators contain the same units, and both denominators contain the same units.

Carry Out Now solve for x by cross-multiplying.

$$\frac{1}{1.6} = \frac{x}{78}$$

$$1(78) = 1.6x \quad \text{Cross-multiply.}$$

$$78 = 1.6x \quad \text{Solve.}$$

$$\frac{78}{1.6} = \frac{1.6x}{1.6}$$

$$48.75 = x$$

Check and Answer Thus, 78 kilometers equals about 48.75 miles.

Now Try Exercise 75

EXAMPLE 11 **Exchanging Currency** When people travel to a foreign country they often need to exchange currency. Donna Boccio visited Cancun, Mexico. She stopped by a local bank and was told that $1 U.S. could be exchanged for 13.09 pesos.

a) How many pesos would she get if she exchanged $150 U.S.?

b) Later that same day, Donna went to the city market where she purchased a ceramic figurine. The price she negotiated for the figurine was 245 pesos. Using the exchange rate given, determine the cost of the figurine in U.S. dollars.

Solution

a) Understand We use the fact that \$1 U.S. can be exchanged for 13.09 Mexican pesos. We use this known fact for one ratio in our proportion. In the second ratio, we set the quantities with same units in the same respective positions.

Translate The unknown quantity is the number of pesos, which we shall call x.

$$\text{Given ratio}\ \left\{ \frac{\$1\text{ U.S.}}{13.09\text{ pesos}} \right. = \frac{\$150\text{ U.S.}}{x\text{ pesos}}$$

Note that both numerators contain U.S. dollars and both denominators contain pesos.

Carry Out

$$\frac{1}{13.09} = \frac{150}{x}$$
$$1x = 13.09(150)$$
$$x = 1963.5$$

Check and Answer Thus, \$150 U.S. could be exchanged for 1963.5 Mexican pesos.

b) Understand and Translate We use the same given ratio that we used in part **a)**. Now we must find the equivalent in U.S. dollars of 245 Mexican pesos. Let's call the equivalent U.S. dollars x.

$$\text{Given ratio}\ \left\{ \frac{\$1\text{ U.S.}}{13.09\text{ pesos}} \right. = \frac{\$x\text{ U.S.}}{245\text{ pesos}}$$

Carry Out

$$\frac{1}{13.09} = \frac{x}{245}$$
$$1(245) = 13.09x$$
$$245 = 13.09x$$
$$18.72 \approx x$$

Check and Answer The cost of the figurine in U.S. dollars is \$18.72.

Now Try Exercise 85

HELPFUL HINT

Some of the problems we have just worked using proportions could have been done without using proportions. However, when working problems of this type, students often have difficulty in deciding whether to multiply or divide to obtain the correct answer. By setting up a proportion, you may be better able to understand the problem and have more success in obtaining the correct answer.

5 Use Proportions to Solve Problems Involving Similar Figures

Understanding Algebra

Two figures are called *similar* if they have the same shape but (possibly) different size.

Similar Figures

Two figures are said to be **similar** when their corresponding angles are equal and their corresponding sides are in proportion.

Proportions can be used to solve problems involving similar figures.

EXAMPLE 12 The figures on the top of the next page are similar figures. Find the length of the side indicated by the x.

Solution We set up a proportion of corresponding sides to find the length of side x.

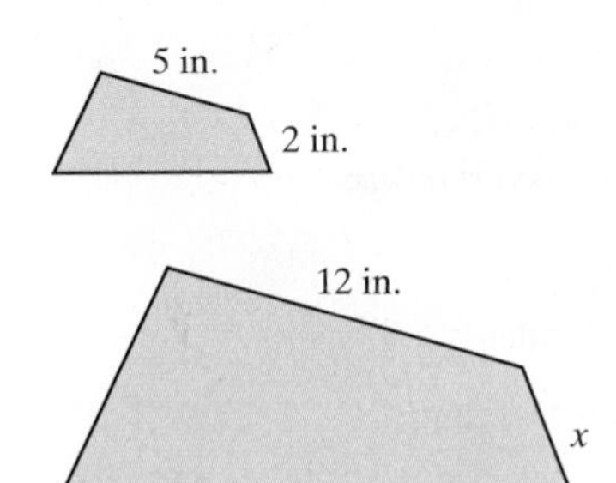

Lengths from smaller figure / Lengths from larger figure

5 inches and 12 inches are corresponding sides of similar figures. →
2 inches and x are corresponding sides of similar figures. →

$$\frac{5}{2} = \frac{12}{x}$$

$$5x = 24$$

$$x = \frac{24}{5} = 4.8$$

Thus, the side indicated by x is 4.8 inches in length.

Now Try Exercise 51

Note in Example 12 that the proportion could have also been set up as

$$\frac{5}{12} = \frac{2}{x}$$

because one pair of corresponding sides is in the numerators and another pair is in the denominators.

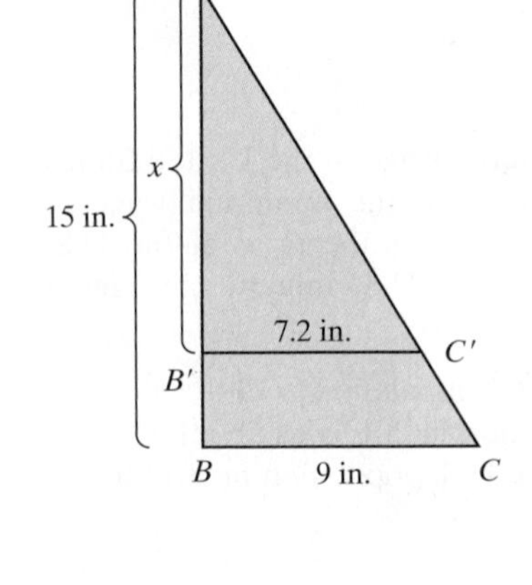

EXAMPLE 13 Triangles ABC and $AB'C'$ are similar triangles. Find the length of side AB'.

Solution We set up a proportion of corresponding sides to find the length of side AB'. We will let x represent the length of side AB'.

$$\frac{\text{length of } AB}{\text{length of } BC} = \frac{\text{length of } AB'}{\text{length of } B'C'}$$

Now we insert the proper values and solve for the variable, x.

$$\frac{15}{9} = \frac{x}{7.2}$$

$$(15)(7.2) = 9x$$

$$108 = 9x$$

$$12 = x$$

Thus, the length of side AB' is 12 inches.

Now Try Exercise 53

2.7 Exercise Set MathXL® MyMathLab®

Warm-Up Exercises

Fill in the blanks with the appropriate word, phrase, or symbol(s) from the following list.

ratio	cross	similar	$c:d$	$d:c$
units	proportion	means	extremes	

1. A ________ is a statement of equality between two ratios.
2. In the proportion $\frac{a}{b} = \frac{p}{q}$, a and q are called ________.
3. In the proportion $\frac{a}{b} = \frac{p}{q}$, b and p are called ________.
4. A method for solving proportions is ________ multiplying.
5. Two figures that have equal corresponding angles and have their corresponding sides in proportion are called ________ figures.
6. Another way of expressing the ratio $\frac{c}{d}$ is ________.
7. When solving an applied problem involving proportions, it is crucial that the two ratios have the same ________ in the same relative position.
8. The quotient of two quantities is called a ________.

In Exercises 9–12, is the proportion set up correctly?

9. $\frac{\text{gal}}{\text{min}} = \frac{\text{gal}}{\text{min}}$
10. $\frac{\text{mi}}{\text{hr}} = \frac{\text{mi}}{\text{hr}}$
11. $\frac{\text{ft}}{\text{sec}} = \frac{\text{sec}}{\text{ft}}$
12. $\frac{\text{tax}}{\text{cost}} = \frac{\text{cost}}{\text{tax}}$

Practice the Skills

The results of a mathematics examination are 6 A's, 4 B's, 9 C's, 3 D's, and 2 F's. Write the following ratios in lowest terms.

13. A's to C's

14. D's to A's

15. F's to total grades

16. Total grades to D's

17. Grades better than C to total grades

18. Grades better than C to grades less than C

Determine the following ratios. Write each ratio in lowest terms.

19. 7 gallons to 4 gallons

20. 3 ml to 47 ml

21. 5 ounces to 15 ounces

22. 18 liters to 24 liters

23. 3 hours to 30 minutes

24. 6 feet to 4 yards

25. 7 dimes to 12 quarters

26. 26 ounces to 4 pounds

In Exercises 27 and 28, find the gear ratio. Write the ratio as some quantity to 1. (See Example 4.)

27. Driving gear, 40 teeth; driven gear, 5 teeth

28. Driving gear, 30 teeth; driven gear, 8 teeth

In Exercises 29–32, **a)** *Determine the indicated ratio, and* **b)** *write the ratio as some quantity to 1.*

29. American Consumers Each year the average American consumer drinks approximately 50 gallons of soft drinks compared to 26 gallons of coffee, 23 gallons of milk, and less than 10 gallons of fruit juices. What is the ratio of the number of gallons of soft drinks consumed to the number of gallons of milk consumed?

30. Mail Letter In January 2013, the cost to mail a one-ounce letter was 46 cents and the cost to mail a two-ounce letter was 66 cents. What is the ratio of the cost to mail a one-ounce letter to the cost to mail a two-ounce letter?

31. Minimum Wage The minimum wage in the United States in 1996 was \$4.75 per hour and the minimum wage in 2012 was \$7.25 per hour. What is the ratio of the U.S. minimum wage in 2012 to the U.S. minimum wage in 1996?

32. Population The United States population in 1991 was about 250 million, and the population in 2013 was about 315 million. What is the ratio of the U.S. population in 2013 to the U.S. population in 1991?

Exercises 33–36 show graphs. For each exercise, find the indicated ratio.

33. Interstate Highways The following graph shows the five states with the greatest number of miles of interstate highway. The graph also shows the number of miles of interstate highway, to the nearest hundred miles, for each of the five states.

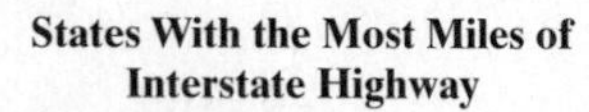

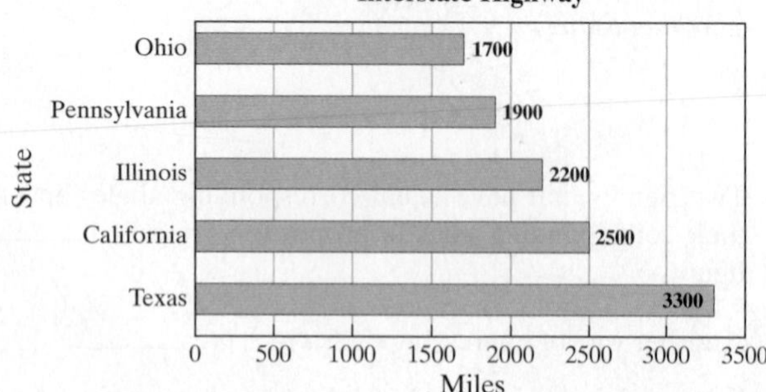

Source: United States Department of Transportation

a) Determine the ratio of miles of interstate highway in Texas to those in Ohio.

b) Determine the ratio of miles of interstate highway in California to those in Pennsylvania.

34. Passports Processed

a) Estimate the ratio of passports processed in the United States in 2012 to those processed in 2004.

b) Estimate the ratio of passports processed in the United States in 2007 to those processed in 2012.

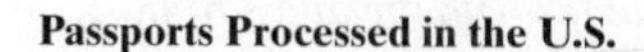

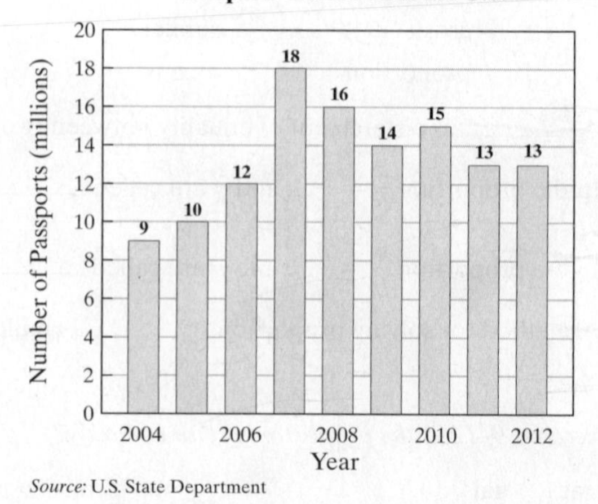

Source: U.S. State Department

35. Favorite Doughnut

a) Determine the ratio of people whose favorite doughnut is glazed to people whose favorite doughnut is filled.

b) Determine the ratio of people whose favorite doughnut is frosted to people whose favorite doughnut is plain.

Favorite Doughnut Flavors

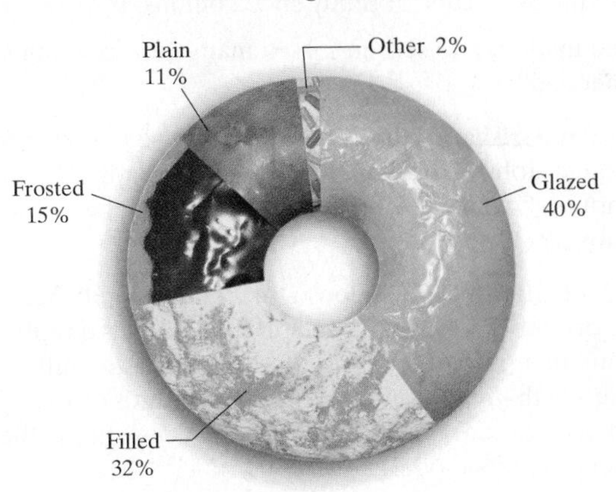

Source: The Heller Research Group

36. Commuting to Work

a) Determine the ratio of workers who drove alone to work to those who carpooled to work.

b) Determine the ratio of workers who walked to work to those who used public transportation.

Commuting to Work, 2012

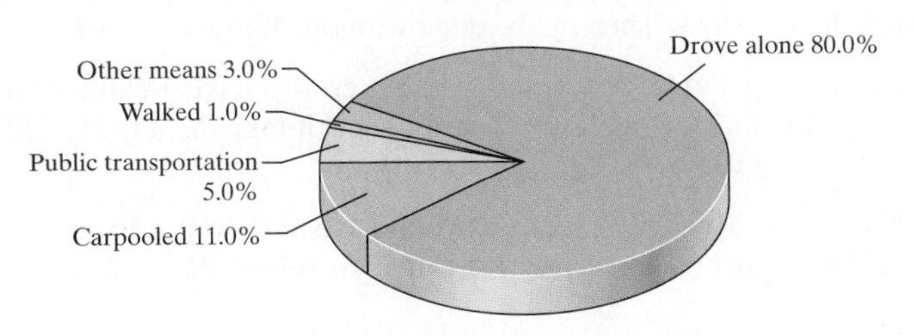

Source: United States Department of Transportation

Solve each proportion for the variable by cross-multiplying.

37. $\frac{x}{3} = \frac{20}{5}$ **38.** $\frac{x}{8} = \frac{24}{48}$ **39.** $\frac{5}{3} = \frac{75}{a}$ **40.** $\frac{x}{3} = \frac{90}{30}$

41. $\frac{-7}{3} = \frac{21}{p}$ **42.** $\frac{-12}{13} = \frac{36}{x}$ **43.** $\frac{15}{45} = \frac{x}{-6}$ **44.** $\frac{y}{6} = \frac{7}{42}$

45. $\frac{3}{z} = \frac{-1.5}{27}$ **46.** $\frac{3}{12} = \frac{-1.4}{z}$ **47.** $\frac{9}{12} = \frac{x}{8}$ **48.** $\frac{2}{20} = \frac{x}{200}$

The following figures are similar. For each pair, find the length of the side indicated by x.

49.

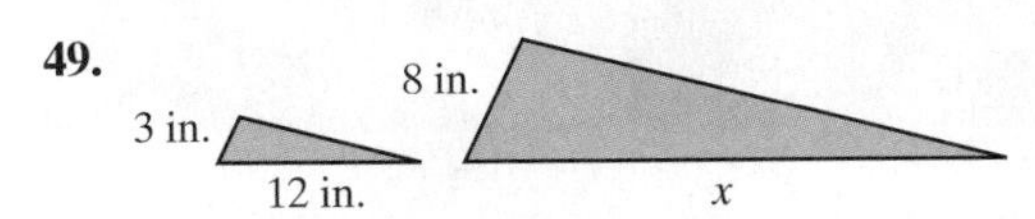

50.

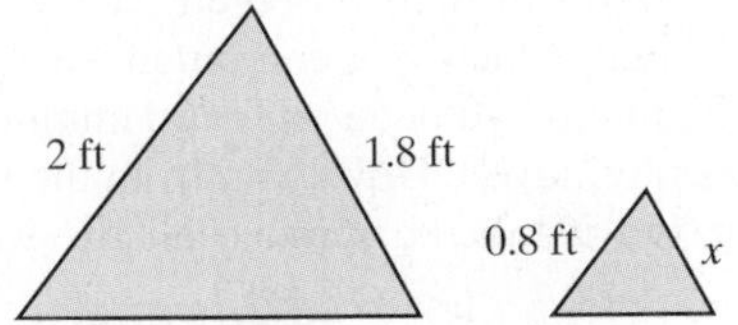

51.

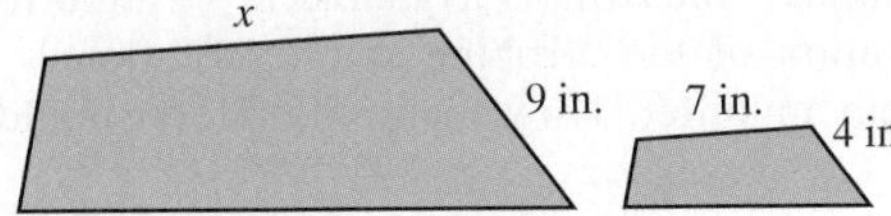

52.

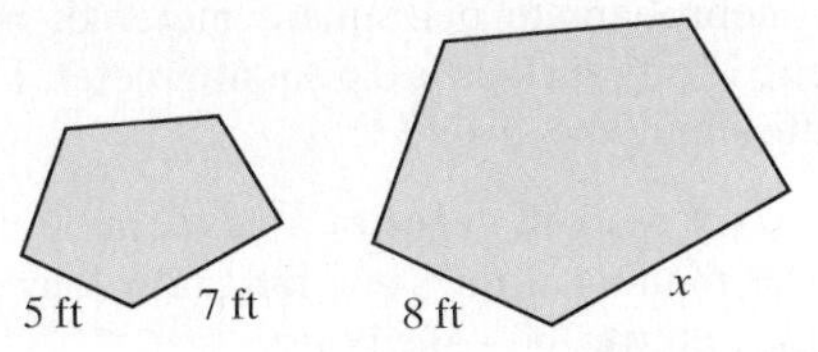

53.

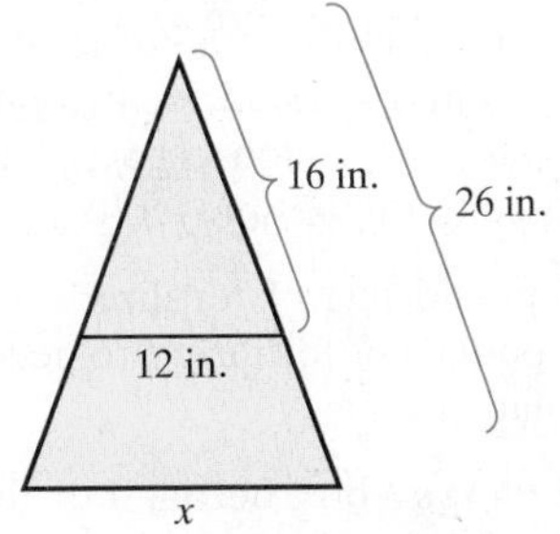

54.

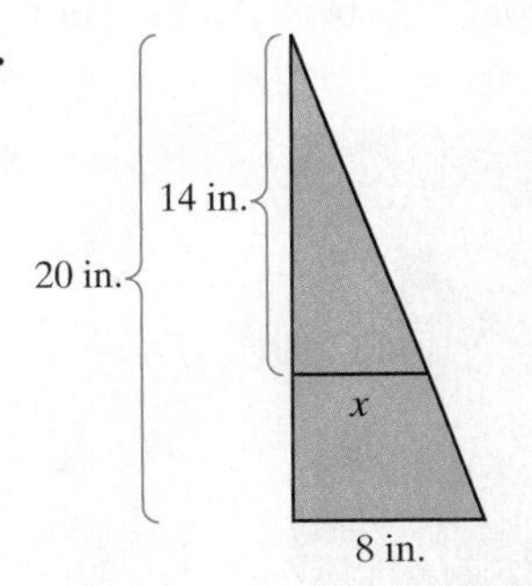

Problem Solving

In Exercises 55–74, write a proportion that can be used to solve the problem. Then solve the equation to obtain the answer.

55. **Washing Clothes** A bottle of liquid Tide contains 100 fluid ounces. If one wash load requires 4 ounces of the detergent, how many washes can be done with one bottle of Tide?

56. **Laying Cable** A telephone cable crew is laying cable at a rate of 42 feet an hour. How long will it take them to lay 252 feet of cable?

57. **Truck Mileage** A 2013 Toyota Tacoma is rated to get 25 miles per gallon. How far can it travel on 21.1 gallons of gas?

58. **Purchasing Stock** If 2 shares of stock can be purchased for \$38.25, how many shares can be purchased for \$344.25?

59. **Model Train** A model train set is in a ratio of 1 : 20. That is, one foot of the model represents 20 feet of the original train. If a caboose is 30 feet long, how long should the model be?

60. **Property Tax** The property tax for Manatee County, Florida, is \$15.9933 per \$1000 of assessed property value. If Jane Keller's property has an assessed value of \$333,716, determine Jane's property tax.

61. **Insecticide Application** The instructions on a bottle of liquid insecticide say "use 3 teaspoons of insecticide per gallon of water." If your sprayer has an 8-gallon capacity, how much insecticide should be used to fill the sprayer?

62. **Spreading Fertilizer** If a 40-pound bag of fertilizer covers 5000 square feet, how many pounds of fertilizer are needed to cover an area of 26,000 square feet?

63. **Blue Heron** The photograph shows a blue heron. If the blue heron that measures 3.5 inches in the photo is actually 3.75 feet tall, approximately how long is its beak if it measures 0.4 inch in the photo?

64. **Maps** On a map, 0.5 inch represents 22 miles. What will be the length on a map that corresponds to a distance of 55 miles?

65. **Onion Soup** A recipe for 6 servings of French onion soup requires $1\frac{1}{2}$ cups of thinly sliced onions. If the recipe were to be made for 15 servings, how many cups of onions would be needed?

66. **John Grisham Novel** Debbie Garrison is currently reading a John Grisham novel. If she reads 72 pages in 1.3 hours, how long will it take her to read the entire 656-page novel?

67. **Wall Street Bull** Suppose the famous bull near the New York Stock Exchange (see photo below) is a replica of a real bull in a ratio of 2.95 to 1. That is, the metal bull is 2.95 times larger than the regular bull. If the length of the Wall Street bull is 28 feet long, approximately how long is the bull that served as its model?

68. **Flood** When they returned home from vacation, the Duncans had a foot of water in their basement. They contacted their fire department, which sent equipment to pump out the water. After the pump had been on for 30 minutes, 3 inches of water had been removed. How long, from the time they started pumping, did it take to remove all the water from the basement?

69. **Drug Dosage** A nurse must administer 220 micrograms of atropine sulfate. The drug is available in solution form. The concentration of the atropine sulfate solution is 400 micrograms per milliliter. How many milliliters should be given?

70. **Dosage by Body Surface** A doctor asks a nurse to administer 0.7 gram of meprobamate per square meter of body surface. The patient's body surface is 0.6 square meter. How much meprobamate should be given?

71. **Reading a Novel** Mary read 40 pages of a novel in 30 minutes. If she continues reading at the same rate, how long will it take her to read the entire 760-page book?

72. **Swimming Laps** Terry Cheng swims 3 laps in 2.3 minutes. Approximately how long will it take him to swim 30 laps if he continues to swim at the same rate?

73. **Prader-Willi Syndrome** It is estimated that each year in the United States about 1 in every 12,000 $(1:12{,}000)$ people is born with a genetic disorder called Prader-Willi syndrome. If there were approximately 4,315,000 births in the United States in 2013, approximately how may children were born with Prader-Willi syndrome?

74. **Scrapbooking** Penelope Penna completed 4 scrapbook pages in 20.5 minutes. Approximately how long will it take her to complete 36 scrapbook pages if she continues to complete the scrapbook at the same rate?

In Exercises 75–86, use a proportion to make the conversion. Round your answers to two decimal places.

75. Convert 78 inches to feet.

76. Convert 22,704 feet to miles $(5280 \text{ feet} = 1 \text{ mile})$.

77. Convert 26.1 square feet to square yards $(9 \text{ square feet} = 1 \text{ square yard})$.

78. Convert 146.4 ounces to pounds.

79. **Newborn** One inch equals 2.54 centimeters. Find the length of a newborn baby, in inches, if the baby measures 50.8 centimeters.

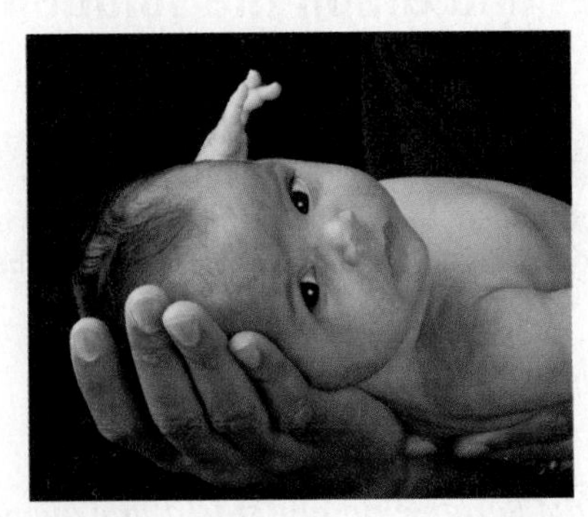

80. **Distance** One mile equals approximately 1.6 kilometers. Find the distance, in kilometers, from San Diego, California, to San Francisco, California—a distance of 520 miles.

81. **Baseball Hits Record** Ichiro Suzuki holds the major league record for the most hits, 262, in a single 162-game season. In the first 60 games of a season, how many hits would a player need to hit to be on pace to break Ichiro's record?

82. **Topsoil** A 40-pound bag of topsoil covers 12 square feet (one inch deep). How many pounds of the topsoil are needed to cover 350 square feet (one inch deep)?

83. **Interest on Savings** Jim Chao invests a certain amount of money in a savings account. If he earned \$110.52 in 180 days, how much interest would he earn in 500 days assuming the interest rate stays the same?

84. **Gold** If gold is selling for \$1728 per 480 grains (a troy ounce), what is the cost per grain?

85. **Mexican Pesos** Suppose that the exchange rate from U.S. dollars to Mexican pesos is \$1 per 13.09 pesos. How many pesos would Elizabeth Averbeck receive if she exchanged \$200 U.S.?

86. **Canadian Dollars** Suppose the exchange rate from U.S. dollars to Canadian dollars is 1 U.S. dollar per 1.0028 Canadian dollar. Barry Cogan exchanges \$350 Canadian dollars for U.S. dollars at this rate. How many U.S. dollars will Barry receive?

87. **Cholesterol** See Example 2. Ken Rauch's LDL is 127 milligrams per deciliter (mg/dL). His HDL is 60 mg/dL. Is Mr. Rauch's ratio of LDL to HDL less than or equal to the 4 : 1 recommended level?

88. **Cholesterol**
   a) Another ratio used by some doctors when measuring cholesterol level is the ratio of total cholesterol to HDL.* Is this ratio increased or decreased if the total cholesterol remains the same but the HDL is increased? Explain.
   b) Doctors recommend that the ratio of total cholesterol to HDL be less than or equal to 4.5 : 1. If Mike's total cholesterol is 220 mg/dL and his HDL is 50 mg/dL, is his ratio less than or equal to 4.5 : 1? Explain.

*Total cholesterol includes both LDL and HDL, plus other types of cholesterol.

Concept/Writing Exercises

89. For the proportion $\frac{a}{b} = \frac{c}{d}$, if a increases while b and d stay the same, what must happen to c? Explain.

90. For the proportion $\frac{a}{b} = \frac{c}{d}$, if a and c remain the same while d decreases, what must happen to b? Explain.

Challenge Problems

91. **Wear on Tires** A new Goodyear tire has a tread of about 0.34 inch. After 5000 miles the tread is about 0.31 inch. If the legal minimum amount of tread for a tire is 0.06 inch, how many more miles will the tires last?

92. **Apple Pie** The recipe for the filling for an apple pie calls for

12 cups sliced apples	$\frac{1}{4}$ teaspoon salt
$\frac{1}{2}$ cup flour	$1\frac{1}{2}$ cups sugar

1 teaspoon nutmeg 2 tablespoons butter or margarine
1 teaspoon cinnamon

Determine the amount of each of the other ingredients that should be used if only 8 cups of apples are available.

93. Insulin Insulin comes in 10-cubic-centimeter (cc) vials labeled in the number of units of insulin per cubic centimeter. Thus, a vial labeled U40 means there are 40 units of insulin per cubic centimeter of fluid. If a patient needs 25 units of insulin, how many cubic centimeters of fluid should be drawn up into a syringe from the U40 vial?

Group Activity

Discuss and answer Exercises 94 and 95 as a group.

94. a) Each group member: Find the ratio of your height to your arm span (finger tips to finger tips) when your arms are extended horizontally outward. You will need help from your group in getting these measurements.

b) If a box were to be drawn about your body with your arms extended, would the box be a square or a rectangle? If a rectangle, would the longer length be your arm span or your height measurement? Explain.

c) Compare these results with other members of your group.

d) What one ratio would you use to report the height to arm span for your group as a whole? Explain.

95. A special ratio in mathematics is called the *golden ratio*. Do research in a history of mathematics book or on the Internet, and as a group write a paper that explains what the golden ratio is and why it is important.

Cumulative Review Exercises

[1.10] *Name each illustrated property.*

96. $x + 3 = 3 + x$

97. $3(xy) = (3x)y$

98. $2(x - 3) = 2x - 6$

[2.5] **99.** Solve $3(4x - 3) = 6(2x + 1) - 15$

[2.6] **100.** Solve $y = mx + b$ for m.

2.8 Inequalities in One Variable

1 Solve linear inequalities and graph the solution on a number line.

2 Write inequality solutions in interval notation.

3 Solve linear inequalities that have all real numbers as their solution, or have no solution.

1 Solve Linear Inequalities and Graph the Solution on a Number Line

Inequalities

A mathematical statement containing one or more of the symbols $<$, $>$, $\leq$, or $\geq$ is called an **inequality.** The direction of the symbol is sometimes called the **sense** or **order of the inequality.**

Examples of Inequalities in One Variable

$x + 3 < 5 \qquad x + 4 \geq 2x - 6 \qquad 4 > -x + 3$

Understanding Algebra

An *inequality* expresses a relationship between two quantities.

Symbol	Read as
$<$	is less than
$>$	is greater than
$\leq$	is less than or equal to
$\geq$	is greater than or equal to

To solve an inequality, we must isolate the variable on one side of the inequality symbol. To do this, we make use of properties very similar to those used to solve equations. Here are four properties used to solve inequalities.

Properties Used to Solve Inequalities

For real numbers, a, b, and c:

1. If $a > b$, then $a + c > b + c$.
2. If $a > b$, then $a - c > b - c$.
3. If $a > b$ **and** $c > 0$, then $ac > bc$.
4. If $a > b$ **and** $c > 0$, then $\frac{a}{c} > \frac{b}{c}$.

When any of these four properties is used, *the direction of the inequality symbol does not change.*

Before we solve inequalities, we will explain how to graph the solution to an inequality on a number line. If an inequality contains the symbol $>$ or the symbol $<$, such as $x > 2$ or $x < -5$, then we will use an *open circle* on the number line to show that the endpoint *is not included* in the solution. If an inequality contains the symbol $\geq$ or the symbol $\leq$, such as $x \geq 2$ or $x \leq -5$, then we will use a *closed circle* on the number line to show that the endpoint *is included* in the solution. We will graph the solutions to the inequalities on number lines in the next three examples.

EXAMPLE 1 Solve the inequality $x - 5 > -2$, and graph the solution on a number line.

Solution To isolate the variable, x, add 5 to both sides of the inequality.

$$x - 5 > -2$$
$$x - 5 + 5 > -2 + 5 \quad \text{Add 5 to both sides.}$$
$$x > 3$$

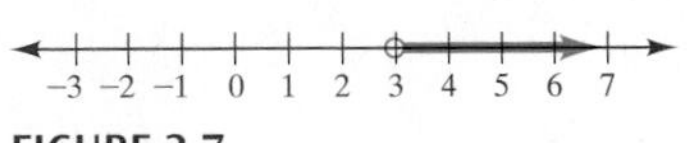

FIGURE 2.7

The solution is all real numbers greater than 3. We can illustrate the solution on a number line by placing an open circle at 3 on a number line and drawing an arrow to the right (**Fig. 2.7**).

The open circle at the 3 indicates that the 3 is *not* part of the solution. The arrow going to the right indicates that all the values greater than 3 are solutions to the inequality.

Now Try Exercise 13

EXAMPLE 2 Solve the inequality $2x + 6 \leq -2$, and graph the solution on a number line.

Solution To isolate the variable, subtract 6 from both sides of the inequality.

$$2x + 6 \leq -2$$
$$2x + 6 - 6 \leq -2 - 6 \quad \text{Subtract 6 from both sides.}$$
$$2x \leq -8$$
$$\frac{2x}{2} \leq \frac{-8}{2} \quad \text{Divide both sides by 2.}$$
$$x \leq -4$$

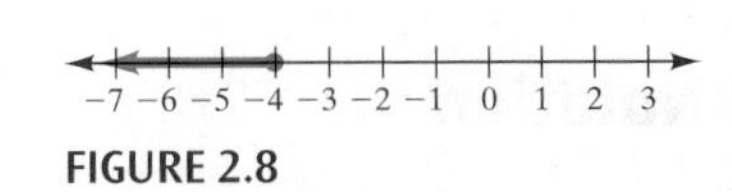

FIGURE 2.8

The solution is all real numbers less than or equal to -4. We can illustrate the solution on a number line by placing a closed circle at -4 and drawing an arrow to the left (**Fig. 2.8**.)

The closed circle at -4 indicates that -4 *is* a part of the solution. The arrow going to the left indicates that all the values less than -4 are also solutions to the inequality.

Now Try Exercise 21

Notice in properties 3 and 4 that we specified that $c > 0$. What happens when an inequality is multiplied or divided by a negative number? Example 3 will illustrate that *when an inequality is multiplied or divided by a negative number, the direction of the inequality symbol changes.*

EXAMPLE 3

a) Multiply both sides of the inequality $8 > -4$ by -2.

b) Divide both sides of the inequality $8 > -4$ by -2.

Understanding Algebra

Whenever we multiply or divide both sides of an inequality by a *negative* number, the direction of the inequality symbol changes.

Solution

a)
$$8 > -4$$
$$-2(8) < -2(-4) \quad \text{Change the direction of the inequality symbol.}$$
$$-16 < 8$$

b)
$$8 > -4$$
$$\frac{8}{-2} < \frac{-4}{-2} \quad \text{Change the direction of the inequality symbol.}$$
$$-4 < 2$$

Now Try Exercise 7

Now we state two additional properties, used when both sides of an inequality are multiplied or divided by a negative number.

Additional Properties Used to Solve Inequalities

5. If $a > b$ **and** $c < 0$, then $ac < bc$.
6. If $a > b$ **and** $c < 0$, then $\frac{a}{c} < \frac{b}{c}$.

EXAMPLE 4 Solve the inequality $-2x < 8$, and graph the solution on a number line.

Solution To isolate the variable, we can divide both sides of the inequality by -2. When we do this, however, we must remember to *change the direction* of the inequality symbol.

$$-2x < 8$$
$$\frac{-2x}{-2} > \frac{8}{-2} \quad \text{Divide both sides by } -2 \text{, and change the direction of the inequality symbol.}$$
$$x > -4$$

The solution is all real numbers greater than -4. The solution is graphed on a number line in **Figure 2.9**.

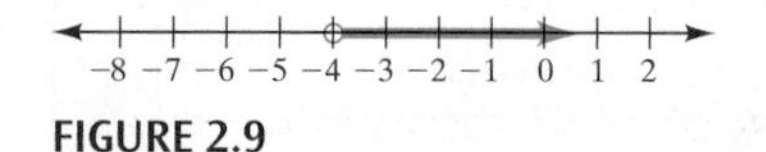

FIGURE 2.9

Now Try Exercise 9

2 Write Inequality Solutions in Interval Notation

We now have seen the solutions to several inequalities graphed on number lines. Another method used to represent the solutions to inequalities is by using *interval notation.* Interval notation uses either parentheses, ( ), or brackets, [ ], or a combination of the two, ( ] or [ ), to indicate the solution to an inequality. When an endpoint *is not included* in the solution a parenthesis is used. When an endpoint *is included* in the solution a bracket is used. **Figure 2.10** on the next page illustrates several examples of inequality solutions, the solutions graphed on the number line, and the solutions represented in interval notation. Note that whenever the infinity symbol, ∞, is used to show that the values continue indefinitely, a parenthesis must be used because there is no endpoint.

Inequality Symbol	Number Line Endpoints	Interval Notation Symbols	Is Endpoint Included in Solution?
$<$ or $>$	○	(or)	No
$\le$ or $\ge$	•	[or]	Yes

The symbol ∞ is read "infinity"; it indicates that the solution set continues indefinitely.

> **Understanding Algebra**
>
> Whenever ∞ is used in interval notation, a *parenthesis* must be used on the corresponding side of the interval notation.

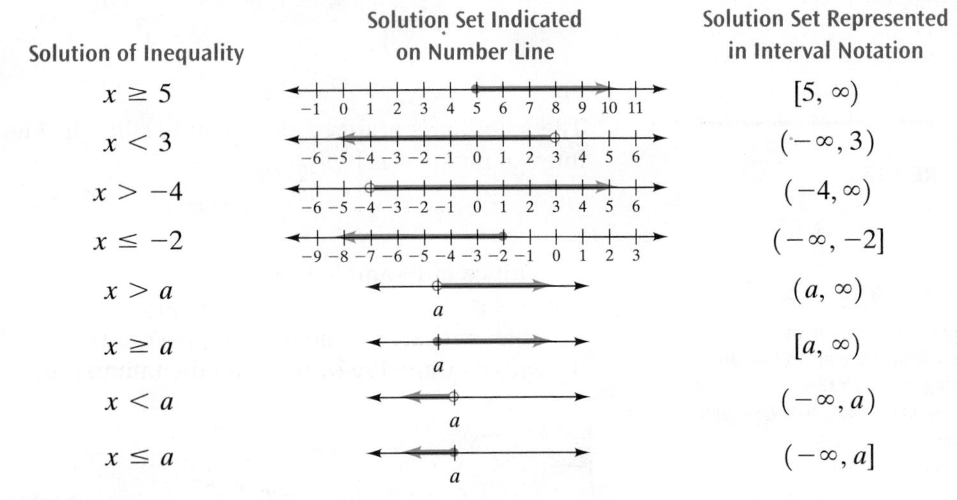

Solution of Inequality	Solution Set Indicated on Number Line	Solution Set Represented in Interval Notation
$x \geq 5$	−1 0 1 2 3 4 5 6 7 8 9 10 11	$[5, \infty)$
$x < 3$	−6 −5 −4 −3 −2 −1 0 1 2 3 4 5 6	$(-\infty, 3)$
$x > -4$	−6 −5 −4 −3 −2 −1 0 1 2 3 4 5 6	$(-4, \infty)$
$x \leq -2$	−9 −8 −7 −6 −5 −4 −3 −2 −1 0 1 2 3	$(-\infty, -2]$
$x > a$	a	(a, ∞)
$x \geq a$	a	$[a, \infty)$
$x < a$	a	$(-\infty, a)$
$x \leq a$	a	$(-\infty, a]$

FIGURE 2.10

EXAMPLE 5 Solve the inequality $-3x + 5 \leq -10$. Graph the solution on a number line and represent the solution in interval notation.

Solution

$$-3x + 5 \leq -10$$

$$-3x + 5 - 5 \leq -10 - 5 \quad \text{Subtract 5 from both sides.}$$

$$-3x \leq -15$$

$$\frac{-3x}{-3} \geq \frac{-15}{-3} \quad \text{Divide both sides by } -3\text{, and change the direction of the inequality symbol.}$$

$$x \geq 5$$

The solution is all real numbers greater than or equal to 5. To represent the solution on a number line we will place a closed circle, •, on the number line at 5 and then shade the number line to the right of 5 as in **Figure 2.11**. The closed circle indicates that we include the number 5 in the solution set.

To represent the solution in interval notation, first observe the number line in **Figure 2.11.** Note that the solution goes from 5, inclusive, to infinity, ∞. Since 5 is included in the solution, we will use a bracket next to the 5. Also, a parenthesis is always used with the symbol ∞. Therefore, the solution in interval notation is $[5, \infty)$.

−3 −2 −1 0 1 2 3 4 5 6 7 8 9

FIGURE 2.11

Now Try Exercise 19

EXAMPLE 6 Solve the inequality $4 \geq -5 - x$. Graph the solution on a number line and represent the solution in interval notation. We will illustrate two methods that can be used to solve this inequality.

Solution

Method 1

$$4 \geq -5 - x$$

$$4 + 5 \geq -5 + 5 - x \quad \text{Add 5 to both sides.}$$

$$9 \geq -x$$

$$-1(9) \leq -1(-x) \quad \text{Multiply both sides by } -1\text{, and change the direction of the inequality symbol.}$$

$$-9 \leq x$$

The inequality $-9 \leq x$ can also be written $x \geq -9$.

Method 2

$$4 \geq -5 - x$$
$$4 + x \geq -5 - x + x \quad \text{Add } x \text{ to both sides.}$$
$$4 + x \geq -5$$
$$4 - 4 + x \geq -5 - 4 \quad \text{Subtract 4 from both sides.}$$
$$x \geq -9$$

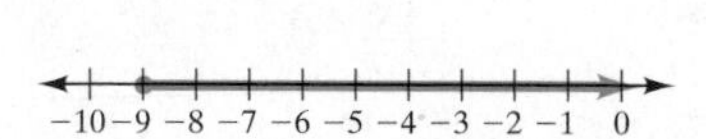

FIGURE 2.12

The solution is graphed on a number line in **Figure 2.12.** The solution written in interval notation is $[-9, \infty)$.

Now Try Exercise 17

Understanding Algebra

We can rewrite the inequality $3 > x$ as $x < 3$. They mean the same thing—in each case the inequality symbol points to the smaller quantity.

Notice in Example 6, Method 1, we wrote $-9 \leq x$ as $x \geq -9$. Although the solution $-9 \leq x$ is correct, it is customary to write the solution to an inequality with the variable on the left. One reason we write the variable on the left is that it often makes it easier to graph the solution on the number line.

HELPFUL HINT

$a > x$ means $x < a$ — Note that both inequality symbols point to x.

$a < x$ means $x > a$ — Note that both inequality symbols point to a.

Examples

$-3 > x$ means $x < -3$

$-5 \leq x$ means $x \geq -5$

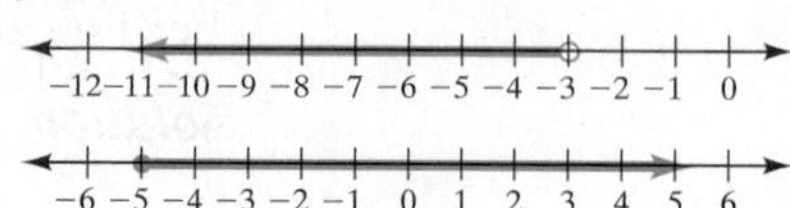

Now let's solve inequalities where the variable appears on both sides of the inequality symbol. We will use the same basic procedure that we used to solve equations. However, we must remember that whenever we multiply or divide both sides of an inequality by a negative number, we must change the direction of the inequality symbol.

EXAMPLE 7 Solve the inequality $-5p + 9 < -2p + 6$. Graph the solution on a number line and represent the solution in interval notation.

Solution

$$-5p + 9 < -2p + 6$$
$$-5p + 5p + 9 < -2p + 5p + 6 \quad \text{Add } 5p \text{ to both sides.}$$
$$9 < 3p + 6$$
$$9 - 6 < 3p + 6 - 6 \quad \text{Subtract 6 from both sides.}$$
$$3 < 3p$$
$$\frac{3}{3} < \frac{3p}{3} \quad \text{Divide both sides by 3.}$$
$$1 < p$$
$$\text{or} \quad p > 1$$

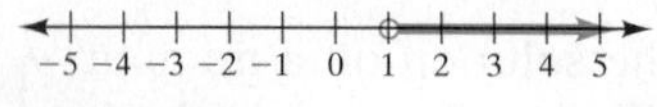

FIGURE 2.13

The solution is graphed in **Figure 2.13.** The solution written in interval notation is $(1, \infty)$

Now Try Exercise 33

EXAMPLE 8 Solve the inequality $\frac{1}{2}x + 3 \leq -\frac{1}{3}x + 7$. Graph the solution on a number line and represent the solution in interval notation.

Solution Since the inequality contains fractions, we begin by multiplying both sides of the inequality by the LCD, 6, to eliminate the fractions.

$$\frac{1}{2}x + 3 \le -\frac{1}{3}x + 7$$

$$6\left(\frac{1}{2}x + 3\right) \le 6\left(-\frac{1}{3}x + 7\right) \quad \text{Multiply both sides by the LCD, 6.}$$

$$3x + 18 \le -2x + 42 \quad \text{Distributive property}$$

$$5x + 18 \le 42 \quad 2x \text{ was added to both sides.}$$

$$5x \le 24 \quad \text{18 was subtracted from both sides.}$$

$$x \le \frac{24}{5} \quad \text{Both sides were divided by 5.}$$

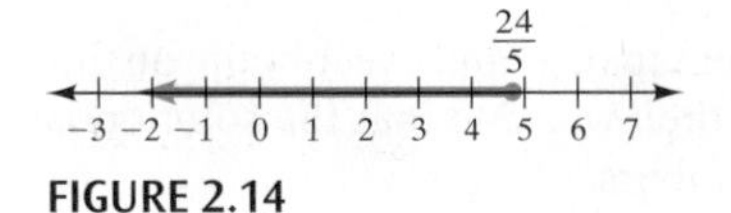

FIGURE 2.14

The solution is graphed in **Figure 2.14**. The solution written in interval notation is $\left(-\infty, \frac{24}{5}\right]$.

Now Try Exercise 53

3 Solve Linear Inequalities That Have All Real Numbers as Their Solution, or Have No Solution

In Examples 9 and 10, we illustrate two special types of inequalities. Example 9 is an inequality that is true for all real numbers, and Example 10 is an inequality that is never true for any real number.

EXAMPLE 9 Solve the inequality $2(x + 3) \le 5x - 3x + 8$. Graph the solution on a number line and represent the solution in interval notation.

Solution

$$2(x + 3) \le 5x - 3x + 8$$

$$2x + 6 \le 5x - 3x + 8 \quad \text{Distributive property was used.}$$

$$2x + 6 \le 2x + 8 \quad \text{Like terms were combined.}$$

$$2x - 2x + 6 \le 2x - 2x + 8 \quad \text{Subtract } 2x \text{ from both sides.}$$

$$6 \le 8$$

Understanding Algebra

When we are solving an inequality and we obtain an inequality that is ...

- *always true*, such as $6 \le 8$, then we state that the solution is *all real numbers*.
- *never* true, such as $4 > 5$, then we state that there is *no solution*.

Since 6 is always less than or equal to 8, the solution is *all real numbers* **(Fig. 2.15)**. The solution written in interval notation is $(-\infty, \infty)$.

FIGURE 2.15

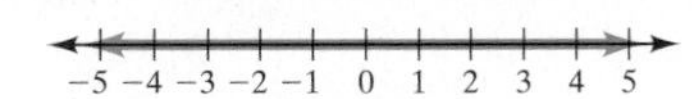

Now Try Exercise 37

EXAMPLE 10 Solve the inequality $4(x + 1) > x + 5 + 3x$, and graph the solution on a number line.

Solution

$$4(x + 1) > x + 5 + 3x$$

$$4x + 4 > x + 5 + 3x \quad \text{Distributive property was used.}$$

$$4x + 4 > 4x + 5 \quad \text{Like terms were combined.}$$

$$4x - 4x + 4 > 4x - 4x + 5 \quad \text{Subtract } 4x \text{ from both sides.}$$

$$4 > 5$$

Since 4 is never greater than 5, the answer is *no solution* **(Fig. 2.16).** There is no real number that makes the statement true. Also note that since there are no solutions, we cannot write the answer using interval notation.

FIGURE 2.16

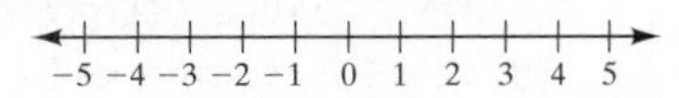

Now Try Exercise 43

2.8 Exercise Set MathXL® MyMathLab®

Warm-Up Exercises

Fill in the blanks with the appropriate word, phrase, or symbol(s) from the following list.

included	real	imaginary	inequality	direction
not	solution	equation	product	

1. A mathematical statement that includes the symbols $<$, $>$, $\leq$, or $\geq$ is called an ________________.

2. Whenever we multiply or divide both sides of an inequality by a negative number we must change the ________________ of the inequality symbol.

3. When representing a solution on a number line, an open circle, $\circ$, indicates that the endpoint is ________________ included in the solution.

4. When representing a solution on a number line, a closed circle, $\bullet$, indicates that the endpoint is ________________ in the solution.

5. When we are solving an inequality and we obtain an inequality that is *always* true then we state that the solution is all ________________ numbers.

6. When we are solving an inequality and we obtain an inequality that is *never* true then we state that there is no ________________.

Practice the Skills

7. a) Multiply both sides of $-7 < 3$ by -4.

b) Divide both sides of $-7 < 3$ by -4.

8. a) Multiply both sides of $12 > -5$ by -3.

b) Divide both sides of $12 > -5$ by -3.

Solve the inequality.

9. a) $-5x < 15$

b) $-7x \leq -28$

c) $-3x > 7$

d) $-9x \geq -15$

10. a) $-4x > 20$

b) $-8x \geq -56$

c) $-7x < 5$

d) $-8x \leq -12$

Solve each inequality. Graph the solution on a number line and represent the solution in interval notation when possible.

11. $x + 2 > 6$

12. $y + 9 \geq 6$

13. $x - 3 \geq -9$

14. $x - 5 > -1$

15. $-x + 3 < 8$

16. $-x - 4 \geq 2$

17. $8 \leq 2 - r$

18. $6 \leq -3 - x$

19. $-2x < 3$

20. $-5x > -3$

21. $2t + 3 \leq 5$

22. $3x - 4 < 5$

23. $-4x - 3 > 5$

24. $-7x + 9 \geq -12$

25. $4 - 6x > -5$

26. $-3 - 8x < 7$

27. $15 > -9x + 50$

28. $8 < 4 - 2q$

29. $7 > 2x + 10$

30. $-15 \leq 3x - 10$

31. $16s + 2 \leq 16s - 9$

32. $-5t + 3 < -5t - 7$

33. $x - 4 \leq 3x + 8$

34. $-4n - 6 > 4n - 20$

35. $-x + 4 < -3x + 6$

36. $-3x + 5 > -5x + 13$

37. $6(2m - 4) \geq 2(6m - 12)$

38. $-3(8x + 3) < 4(-6x + 1)$

39. $x + 3 < x + 4$

40. $y + 4 \geq y - 3$

41. $6(3 - x) < 2x + 15$

42. $2(3 - x) + 4x < -6$

43. $4x - 4 < 4(x - 5)$

44. $-2(-5 - x) > 3(x + 2) + 4 - x$

45. $5(2x + 3) \geq 6 + (x + 2) - 2x$

46. $-3(-2x + 12) < -4(x + 2) - 6$

47. $1.2x + 3.1 < 3.5x - 3.8$

48. $-5.3r - 6.7 \geq 2.3 - 6.5r$

49. $1.2(m - 3) \geq 4.6(2 - m) + 1.7$

50. $-4.6(4 - x) < 2.4(x - 3) - 0.2$

51. $\frac{x}{2} \geq \frac{x}{3} + 5$

52. $\frac{x}{7} - 1 \geq \frac{x}{8}$

53. $t + \frac{1}{6} > \frac{2}{3}t$

54. $\frac{3}{5}r - 9 < \frac{3}{8}r$

55. $\frac{1}{8}(4 - r) \leq \frac{1}{4}$

56. $\frac{1}{6}(5 - a) \geq \frac{1}{3}$

57. $\frac{2}{3}(t + 2) \leq \frac{1}{4}(2t - 6)$

58. $\frac{3}{4}(n - 4) \geq \frac{2}{3}(n - 4)$

Problem Solving

59. Chicago Temperatures The following chart shows the average high and low monthly temperatures in Chicago over a 126-year period. Notice that the months are not listed in order.

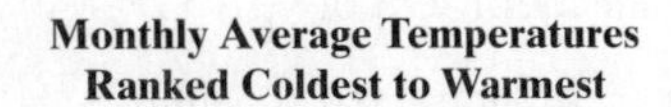

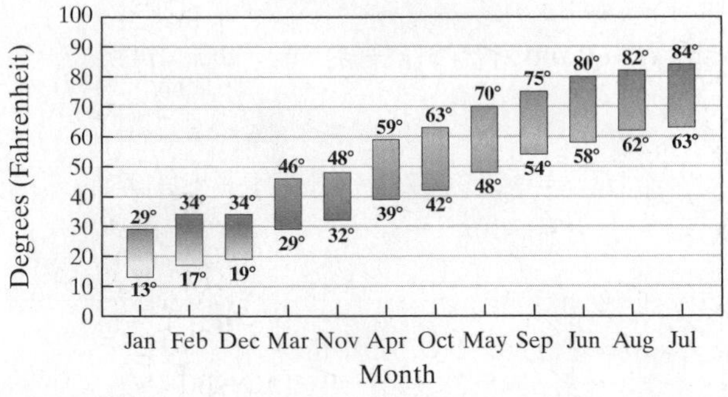

Note: Numbers indicate average monthly highs and lows.

Source: WGN-TV

a) In what months was the average high temperature >65°F?

b) In what months was the average high temperature ≤59°F?

c) In what months was the average low temperature <29°F?

d) In what months was the average low temperature ≥58°F?

60. Hours Worked The following graph indicates the average hours worked per person, per year.

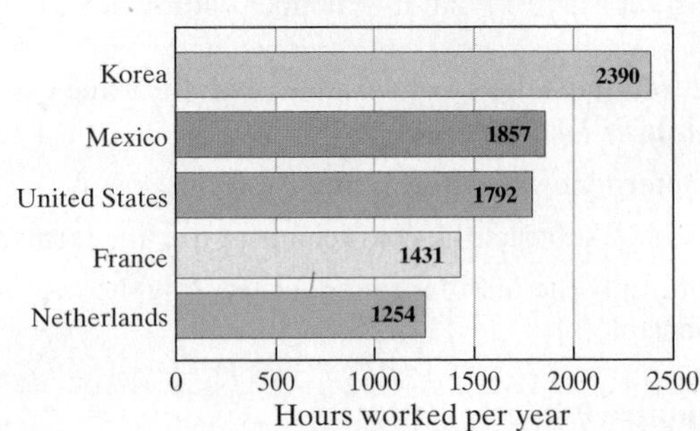

Source: Rochester Democrat & Chronicle

a) In which countries was the number of hours worked ≥1850?

b) In which countries was the number of hours worked ≥1500 but ≤1900?

c) In which countries was the number of hours worked >1300 but ≤1800?

d) In which countries was the number of hours worked ≥1792 and ≤1792?

61. The inequality symbols discussed so far are $<$, $\leq$, $>$, and $\geq$. Can you name an inequality symbol that we have not mentioned in this section?

62. Consider the inequality $-4x + 3 \leq 1$. Explain what is wrong with the following solution.

$$-4x + 3 \leq 1$$
$$-4x + 3 - 3 \leq 1 - 3$$
$$-4x \leq -2$$
$$\frac{-4x}{-4} \leq \frac{-2}{-4}$$
$$x \leq \frac{1}{2}$$

63. Consider the inequality $xy > 6$, where x and y represent real numbers. Explain why we *cannot* do the following step:

$$\frac{xy}{y} > \frac{6}{y} \quad \text{Divide both sides by } y.$$

Challenge Problems

64. Solve the following inequality.

$$3(2 - x) - 4(2x - 3) \leq 6 + 2x - 4x$$

65. Solve the following inequality.

$$6x - 6 > -4(x + 3) + 5(x + 6) - x$$

Cumulative Review Exercises

[1.9] **66.** Evaluate $-x^2$ for $x = 3$.

67. Evaluate $-x^2$ for $x = -5$.

[2.5] **68.** Solve $4 - 3(2x - 4) = 5 - (x + 3)$.

[2.7] **69. Electric Bill** The Milford Electric Company charges \$0.174 per kilowatt-hour of electricity. The Vega's monthly electric bill was \$87 for the month of July. How many kilowatt-hours of electricity did the Vegas use in July?

Chapter 2 Summary

IMPORTANT FACTS AND CONCEPTS	EXAMPLES
Section 2.1	
The **terms** of an expression are the parts that are added. The numerical part of a term is called its **numerical coefficient.** A **constant** is a term that is a number without a variable.	$2x^2 - 3xy + 5$ has 3 terms: $2x^2$, $-3xy$, and 5. The numerical coefficient of $\frac{3x}{4}$ is $\frac{3}{4}$. In $3x^2 + 2x - 7$, the -7 is a constant.
Like terms have the same variables with the same exponents. **To Combine Like Terms** **1.** Determine which terms are like terms. **2.** Add or subtract the coefficients of the like terms. **3.** Multiply the number found in step 2 by the common variable(s).	$7x$ and x; $6y^2$ and $2y^2$; $3(x + 4)$ and $-8(x + 4)$ $-3x^2 + 4y - 7x^2 + 6 - y - 9$ $= -3x^2 - 7x^2 + 4y - y + 6 - 9$ $= -10x^2 + 3y - 3$
Distributive Property For any real numbers a, b, and c, $a(b + c) = ab + ac$	$-5(3r - 6) = -15r + 30$
To Simplify an Expression **1.** Use the distributive property to remove any parentheses. **2.** Combine like terms.	$2(3c - 1) - 5(c + 4) - 6$ $= 6c - 2 - 5c - 20 - 6$ $= c - 28$
When two or more expressions are multiplied, each expression is a **factor** of the product.	Since $7 \cdot 8 = 56$, the 7 and the 8 are factors of 56.

IMPORTANT FACTS AND CONCEPTS	EXAMPLES
Section 2.2	
A **linear equation** in one variable is an equation that can be written in the form $ax + b = c$ where ***a, b,*** and ***c*** are real numbers and $a \neq 0$.	$9x - 2 = 16$
The **solution to an equation** is the number or numbers that when substituted for the variable or variables make the equation a true statement.	The solution to $2x + 3 = 9$ is 3.
The solution to an equation may be **checked** by substituting the value that is believed to be the solution for the variable in the original equation.	To check whether -2 is the solution to $-7x + 1 = 15$: $-7x + 1 = 15$; $-7(-2) + 1 \stackrel{?}{=} 15$; $14 + 1 \stackrel{?}{=} 15$; $15 = 15$ True. Thus, -2 is the solution.
Two or more equations with the same solution are called **equivalent equations.**	$-4x = 12$, $2x - 3 = -9$, and $x = -3$ are equivalent equations
Addition Property of Equality If $a = b$, then $a + c = b + c$ for any real numbers a, b, and c.	Solve the equation $x - 9 = -2$. $x - 9 = -2$; $x - 9 + 9 = -2 + 9$; $x = 7$
Section 2.3	
Two numbers are **reciprocals** of each other when their product is 1.	3 and $\frac{1}{3}$ are reciprocals since $3 \cdot \frac{1}{3} = 1$.
Multiplication Property of Equality If $a = b$, then $a \cdot c = b \cdot c$ for any real numbers a, b, and c.	Solve the equation $\frac{3}{7}x = 6$. $\frac{3}{7}x = 6$; $\frac{7}{3} \cdot \frac{3}{7}x = \frac{7}{3} \cdot 6$; $x = 14$
Section 2.4	
To Solve Linear Equations with a Variable on Only One Side of the Equation **1.** If the equation contains fractions, multiply *both* sides of the equation by the least common denominator (LCD). **2.** Use the distributive property to remove parentheses. **3.** Combine like terms on the same side of the equation. **4.** Use the addition property to obtain an equation with the term containing the variable on one side of the equation and a constant on the other side. **5.** Use the multiplication property to isolate the variable. **6.** Check the solution in the original equation.	Solve the equation $3(x - 5) - 6x = -2$. $3(x - 5) - 6x = -2$; $3x - 15 - 6x = -2$; $-3x - 15 = -2$; $-3x - 15 + 15 = -2 + 15$; $-3x = 13$; $\frac{-3x}{-3} = \frac{13}{-3}$; $x = -\frac{13}{3}$. A check will show that $-\frac{13}{3}$ is the solution.

IMPORTANT FACTS AND CONCEPTS	EXAMPLES
Section 2.5	
To Solve Linear Equations with the Variable on Both Sides of the Equation 1. If the equation contains fractions, multiply *both* sides of the equation by the LCD. 2. Use the distributive property to remove parentheses. 3. Combine like terms on the same side of the equation. 4. Use the addition property to rewrite the equation with all terms containing the variable on one side of the equation and all terms not containing the variable on the other side of the equation. 5. Use the multiplication property to isolate the variable. 6. Check the solution in the original equation.	Solve the equation $9 - 3x - 2(x+5) = 4x + 7 - x$. $9 - 3x - 2(x+5) = 4x + 7 - x$ $9 - 3x - 2x - 10 = 4x + 7 - x$ $-5x - 1 = 3x + 7$ $-5x + 5x - 1 = 3x + 5x + 7$ $-1 = 8x + 7$ $-1 - 7 = 8x + 7 - 7$ $-8 = 8x$ $\frac{-8}{8} = \frac{8x}{8}$ $-1 = x$ A check will show that -1 is the solution.
A **conditional equation** is an equation that has a single value for a solution.	$3x - 2 = 8$ is a conditional equation since its solution is $\frac{10}{3}$.
An **identity** is an equation that is true for infinitely many values of the variable.	$-4(x+3) = -5x - 12 + x$ is an identity because the equation is true for all real numbers.
A **contradiction** is an equation that has no solution.	$-9x + 7 + 6x = -5x + 1 + 2x$ is a contradiction because the equation is never true and has no solution.
Section 2.6	
Simple Interest Formula interest = principal · rate · time or $i = prt$	Determine the interest earned on a \$5000 investment at 3% simple interest for 2 years. $i = prt$ $i = 5000(0.03)(2)$ $i = \$300$
Distance Formula distance = rate · time or $d = r \cdot t$	Timothy John completed a snowmobile race in 2.4 hours at an average speed of 75 miles per hour. Determine the distance of the race. $d = rt$ $d = (75)(2.4)$ $d = 180$ miles
Area is the measure of the amount of surface within the figure's boundaries. Areas are measured in square units. **Perimeter** is the sum of the lengths of the sides of a figure. Perimeters are measured in the same common unit as the sides. Formulas for areas and perimeters of quadrilaterals and triangles can be found in **Table 2.1** on page 139.	Determine the perimeter and area of the following trapezoid. 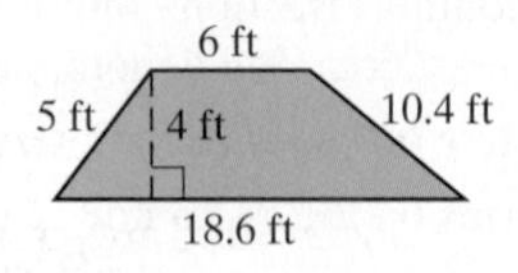 $P = a + b + c + d$ $P = 5 + 6 + 10.4 + 18.6$ $P = 40$ ft $A = \frac{1}{2}h(b+d)$ $A = \frac{1}{2}(4)(6 + 18.6)$ $A = 49.2 \text{ ft}^2$

IMPORTANT FACTS AND CONCEPTS	EXAMPLES
Section 2.6 (Continued)	
The **circumference** of a circle is the length (or perimeter) of the curve that forms a circle. Formulas for the area and circumference of a circle can be found in **Table 2.2** on page 141.	Determine the area and circumference of the following circle. 4 cm $A = \pi r^2$ $A = \pi(4)^2$ $A = \pi(16)$ $A \approx 50.27 \text{ cm}^2$ $C = 2\pi r$ $C = 2\pi(4)$ $C = 8\pi$ $C \approx 25.13 \text{ cm}$
Volume may be considered the space occupied by a figure. Volume is measured is cubic units. Volume formulas can be found in **Table 2.3** on page 141.	Determine the volume of the following figure. 7 m 3 m $V = \frac{1}{3}\pi r^2 h$ $V = \frac{1}{3}\pi(3)^2(7)$ $V = \frac{1}{3}\pi(9)(7)$ $V = 21\pi$ $V \approx 65.97 \text{ m}^3$
To **solve for a variable in a formula,** treat each of the quantities, except the one for which you are solving, as if they were constants. Then solve for the desired variable by isolating it on one side of the equation.	Solve $V = lwh$, for h. $V = lwh$ $\frac{V}{lw} = \frac{lwh}{lw}$ $\frac{V}{lw} = h$
Section 2.7	
A **ratio** is a quotient of two quantities.	3 to 5, $3:5$, $\frac{3}{5}$
A **proportion** is a statement of equality between two ratios. In the proportion $\frac{a}{b} = \frac{c}{d}$, the a and d are called the **extremes,** and the b and c are called the **means** of the proportion.	$\frac{7}{10} = \frac{21}{30}$ 7 and 30 are the extremes. 10 and 21 are the means.
Cross-Multiplication If $\frac{a}{b} = \frac{c}{d}$, then $ad = bc$.	Solve $\frac{-9}{2} = \frac{126}{x}$ for x by cross-multiplying. $\frac{-9}{2} = \frac{126}{x}$ $-9 \cdot x = 2 \cdot 126$ $-9x = 252$ $\frac{-9x}{-9} = \frac{252}{-9}$ $x = -28$

IMPORTANT FACTS AND CONCEPTS	EXAMPLES

Section 2.7 (Continued)

To Solve Problems Using Proportions

1. Understand the problem.
2. Translate the problem into mathematical language.
3. Carry out the mathematical calculations necessary to solve the problem.
4. Check the answer obtained in step 3.
5. Make sure you have answered the question.

See page 153 for more details on proportions.

Melanie Jo can type 40 words per minute. If she types for 20.5 minutes, how many words will she type?

$$\frac{40 \text{ words}}{1 \text{ minute}} = \frac{x \text{ words}}{20.5 \text{ minutes}}$$
$$40(20.5) = 1(x)$$
$$820 = x$$

Melanie Jo will type 820 words.

Similar figures are figures whose corresponding angles are equal and whose corresponding sides are in proportion.

These two figures are similar.

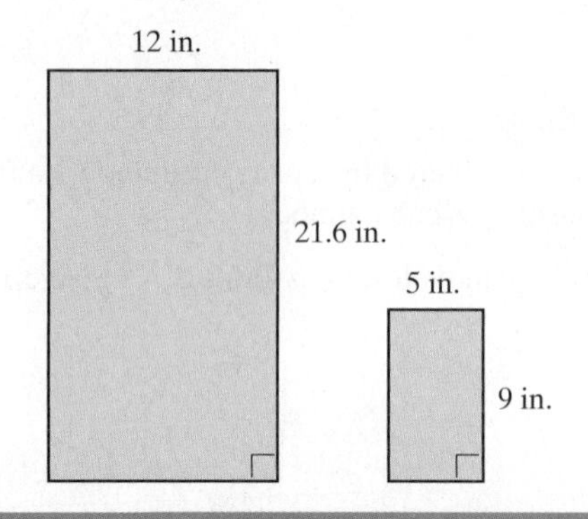

Section 2.8

An **inequality** is a mathematical statement containing one or more of the following symbols: $>, <, \geq, \leq$.

$$x + 7 \leq 3x - 5$$

Properties Used to Solve Inequalities

For real numbers, a, b, and c:

1. If $a > b$, then $a + c > b + c$.
2. If $a > b$, then $a - c > b - c$.
3. If $a > b$ **and** $c > 0$, then $ac > bc$.
4. If $a > b$ **and** $c > 0$, then $\frac{a}{c} > \frac{b}{c}$.
5. If $a > b$ **and** $c < 0$, then $ac < bc$.
6. If $a > b$ **and** $c < 0$, then $\frac{a}{c} < \frac{b}{c}$.

Examples:

1. If $x - 5 > 3$, then $x - 5 + 5 > 3 + 5$.
2. If $x + 4 \geq -9$, then $x + 4 - 4 \geq -9 - 4$.
3. If $\frac{1}{3}x > 2$, then $\left(\frac{1}{3}x\right)(3) > 2(3)$.
4. If $6x > 12$, then $\frac{6x}{6} > \frac{12}{6}$.
5. If $-\frac{1}{4}x > 8$, then $(-4)\left(-\frac{1}{4}x\right) < (-4)8$.
6. If $-7x \geq 35$, then $\frac{-7x}{-7} \leq \frac{35}{-7}$.

The solution to an inequality can be shown on a number line. The solution can also be written in interval notation.

Inequality	Number Line	Interval Notation
$x > -3$	−5 −4 −3 −2 −1 0 1 2 3	$(-3, \infty)$
$x \geq 2$	0 1 2 3 4 5 6 7 8	$[2, \infty)$
$x < 5$	−1 0 1 2 3 4 5 6 7	$(-\infty, 5)$
$x \leq -4$	−10 −9 −8 −7 −6 −5 −4 −3 −2	$(-\infty, -4]$

Chapter 2 Review Exercises

[2.1] *Use the distributive property to simplify.*

1. $3(x + 8)$

2. $5(x - 2)$

3. $-2(x + 4)$

4. $-(x + 2)$

5. $-(m + 8)$

6. $-4(4 - x)$

7. $5(5 - p)$

8. $6(4x - 5)$

9. $-5(5t - 5)$

10. $4(-x + 3)$

11. $\frac{1}{2}(2x + 4)$

12. $-\frac{1}{3}(3 + 6y)$

13. $-(x + 2y - z)$

14. $-3(2a - 5b + 7)$

[2.1] *Simplify.*

15. $10m - 6m$

16. $5 - 3y + 3$

17. $1 + 3x + 2x$

18. $-2x - x + 3y$

19. $4m + 2n + 4m + 6n$

20. $9x + 3y + 2$

21. $6x - 2x + 3y + 6$

22. $x + 8x - 9x + 3$

23. $-4x^2 - 8x^2 + 3$

24. $-2(3a^2 - 4) + 6a^2 - 8$

25. $2x + 3(x + 4) - 5$

26. $-4 + 2(3 - 2b) + b$

27. $6 - (-7x + 6) - 7x$

28. $2(2x + 5) - 10 - 4$

29. $-6(4 - 3x) - 18 + 4x$

30. $4.03x - 2(x + 1.6) + 6x^2 + 1.09$

31. $\frac{1}{4}d + 2 - \frac{3}{5}d + 5$

32. $3 - (a - b) + (a - b)$

33. $\frac{5}{6}x - \frac{1}{3}(2x - 6)$

34. $\frac{2}{3} - \frac{1}{4}n - \frac{1}{3}(n + 2)$

[2.2–2.5] *Solve.*

35. $-3x = -3$

36. $t + 6 = -7$

37. $x - 4 = 7$

38. $\frac{x}{3} = -9$

39. $-\frac{a}{5} = 3$

40. $14 = 3 + 2x$

41. $4c + 11 = -21$

42. $9 - 2a = 15$

43. $-x = -12$

44. $3(x - 2) = 6$

45. $-12 = 3(2x - 8)$

46. $4(6 + 2x) = 0$

47. $-6n + 2n + 6 = 0$

48. $-3 = 3w - (4w + 6)$

49. $6 - (2n + 3) - 4n = 6$

50. $6(2x - 3) - 4(3x - 2) = 13$

51. $5 + 3(x - 1) = 3(x + 1) - 1$

52. $8.4r - 6.3 = 6.3 + 2.1r$

53. $1.1x - 2.34 = 0.09x + 1.7$

54. $0.35(c - 5) = 0.45(c + 4)$

55. $0.2(x + 6) = -0.3(2x - 1)$

56. $-2.3(x - 8) = 3.7(x + 4)$

57. $\frac{p}{3} + 2 = \frac{1}{4}$

58. $\frac{d}{6} + \frac{1}{7} = 2$

59. $\frac{3}{5}(r - 6) = 3r$

60. $\frac{2}{3}w = \frac{1}{6}(w - 2)$

61. $8x - 5 = -4x + 19$

62. $-(w + 2) = 2(3w - 6)$

63. $2x + 6 = 3x + 9 - 3$

64. $-5a + 3 = 2a + 10$

65. $5p - 2 = -2(-3p + 6)$

66. $3x - 12x = 24 - 9x$

67. $4(2x - 3) + 4 = 8x - 8$

68. $4 - c - 2(4 - 3c) = 3(c - 4)$

69. $2(x + 7) = 6x + 9 - 4x$

70. $-5(3 - 4x) = -6 + 20x - 9$

71. $4(x - 3) - (x + 5) = 0$

72. $-2(4 - x) = 6(x + 2) + 3x$

73. $\frac{x + 3}{2} = \frac{x}{2}$

74. $\frac{y}{7} = \frac{y-5}{2}$

75. $\frac{1}{5}(3s+4) = \frac{1}{3}(2s-8)$

76. $\frac{2(2t-4)}{5} = \frac{3t+6}{4} - \frac{3}{2}$

77. $\frac{2}{5}(2-x) = \frac{1}{6}(-2x+2)$

78. $\frac{x}{4} + \frac{x}{6} = \frac{1}{2}(x+3)$

[2.6] *Use the formula to find the value of the variable indicated.*

79. $y = mx + b$ (slope-intercept form of a line); find m when $y = 7, x = 2$, and $b = 1$.

80. $A = \frac{1}{2}h(b+d)$ (area of a trapezoid); find A when $h = 12, b = 3$, and $d = 5$.

Determine the area or volume of the figure.

81.

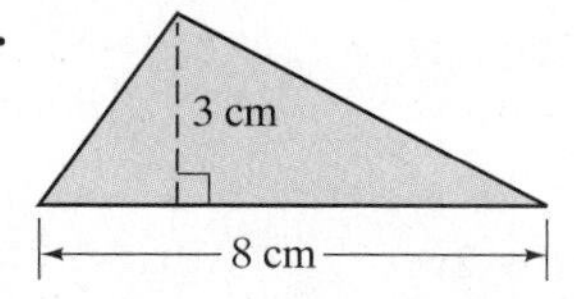

82.

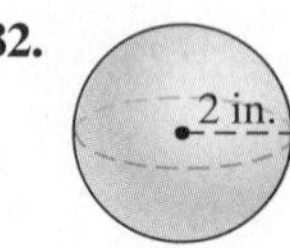

Solve for the indicated variable.

83. $P = 2l + 2w$, for l

84. $y - y_1 = m(x - x_1)$, for m

85. $-x + 3y = 2$, for y

86. Spring Break Yong Wolfer traveled to Florida for spring break at an average speed of 61.7 miles per hour for 5 hours. How far did he travel?

87. Flower Garden Chrishawn Miller has a rectangular flower garden that measures 20 feet by 12 feet. What is the area of Chrishawn's flower garden?

88. Tuna Fish Find the volume of a tuna fish can if its diameter is 4 inches and its height is 2 inches.

[2.7] *Determine the following ratios. Write each ratio in lowest terms.*

89. 15 inches to 20 inches

90. 80 ounces to 12 pounds

91. 4 minutes : 40 seconds

Solve each proportion.

92. $\frac{x}{4} = \frac{8}{16}$

93. $\frac{5}{20} = \frac{x}{80}$

94. $\frac{3}{t} = \frac{15}{45}$

95. $\frac{20}{45} = \frac{15}{q}$

96. $\frac{6}{5} = \frac{-12}{x}$

97. $\frac{b}{6} = \frac{8}{-3}$

98. $\frac{-7}{9} = \frac{-12}{y}$

99. $\frac{x}{-15} = \frac{30}{-5}$

The following pairs of figures are similar. For each pair, find the length of the side indicated by x.

100.

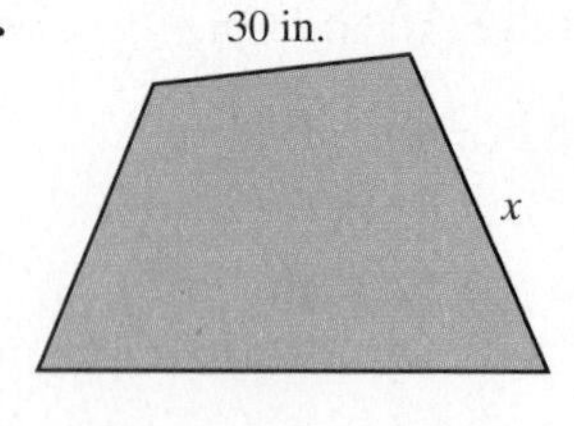

101.

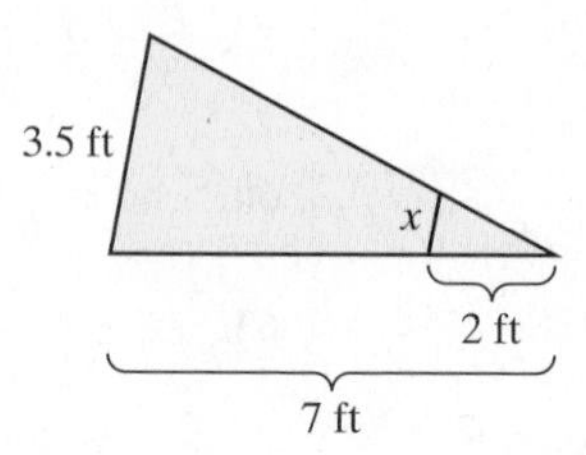

[2.8] *Solve each inequality. Graph the solution on a number line and represent the solution in interval notation.*

102. $3x + 4 \geq 10$

103. $-4a + 10 > 4a - 14$

104. $5 - 3r \leq 2r + 15$

105. $2(x+4) \leq 2x - 5$

106. $2(x+3) > 6x - 4x + 4$

107. $x + 6 > 9x + 30$

108. $x - 8 \le -3x + 11$

109. $-(y + 2) < -2(-2y + 5)$

110. $\frac{x}{2} < \frac{2}{3}(x + 3)$

111. $\frac{3}{10}(t - 2) \le \frac{3}{4}(4 + 2t)$

[2.7] *Set up a proportion and solve each problem.*

112. Boat Trip A boat travels 40 miles in 1.8 hours. If it travels at the same rate, how long will it take for it to travel 140 miles?

113. Washing Dishes If Adam Kloza can wash 12 dishes in 3.5 minutes, how many dishes can he wash in 21 minutes?

114. Copy Machine If a copy machine can copy 20 pages per minute, how many pages can be copied in 22 minutes?

115. Map Scale If the scale of a map is 1 inch to 60 miles, what distance on the map represents 380 miles?

116. Model Car Bryce Winston builds a model car to a scale of 1 inch to 1.5 feet. If the completed model is 10.5 inches, what is the size of the actual car?

117. Money Exchange Suppose that one U.S. dollar can be exchanged for 12.759 Mexican pesos. Find the value of 1 peso in terms of U.S. dollars.

118. Ketchup If a machine can fill and cap 80 bottles of ketchup in 50 seconds, how many bottles of ketchup can it fill and cap in 2 minutes?

Chapter 2 Practice Test

Chapter Test Prep Videos provide fully worked-out solutions to any of the exercises you want to review. Chapter Test Prep Videos are available via MyMathLab®, *or on* YouTube (www.youtube.com/user/AngelElementaryAlg).

Use the distributive property to simplify.

1. $-3(4 - 2x)$

2. $-(x + 3y - 4)$

Simplify.

3. $5x - 8x + 4$

4. $4 + 2x - 3x + 6$

5. $-y - x - 4x - 6$

6. $a - 2b + 6a - 6b - 3$

7. $2x^2 + 3 + 2(3x - 2)$

Solve Exercises 8–16.

8. $2.4x - 6.3 = 3.3$

9. $\frac{5}{6}(x - 2) = x - 3$

10. $2x - 3(-2x + 4) = -13 + x$

11. $3x - 4 - x = 2(x + 5)$

12. $-3(2x + 3) = -2(3x + 1) - 7$

13. $ax + by + c = 0$, for x

14. $-6x + 5y = -2$, for y

15. $\frac{1}{7}(2x - 5) = \frac{3}{8}x - \frac{5}{7}$

16. $\frac{9}{x} = \frac{3}{-15}$

17. What do we call an equation that has

a) exactly one solution,

b) no solution,

c) all real numbers as its solution?

Solve Exercises 18–21. Graph the solution on a number line and represent the solution in interval notation.

18. $2x - 4 < 4x + 10$

19. $3(x + 4) \ge 5x - 12$

20. $4(x + 3) + 2x < 6x - 3$

21. $-(x - 2) - 3x \le 4(1 - x) - 2$

22. The following figures are similar. Find the length of side x.

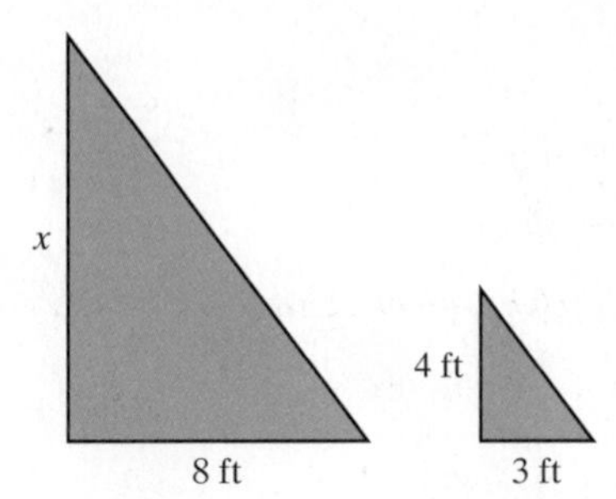

23. **Simple Interest Loan** Laura Hoye lent her sister \$2000 for a period of 1 year. At the end of 1 year, her sister repaid the \$2000 plus \$80 interest. What simple interest rate did her sister pay?

24. **Peanut Butter Pie** A peanut butter pie has a diameter of 9 inches. Determine the circumference of the pie.

25. **Travel Time** While traveling, you notice that you traveled 25 miles in 35 minutes. If your speed does not change, how long will it take you to travel 125 miles?

Cumulative Review Test

Take the following test and check your answers with those given in the back of the book. Review any questions that you answered incorrectly. The section where the material was covered is indicated after the answer.

1. Multiply $\frac{52}{15} \cdot \frac{10}{13}$.

2. Divide $\frac{5}{24} \div \frac{2}{9}$.

3. Insert $<$, $>$, or $=$ in the shaded area to make a true statement: $|-2| \; \square \; 1$.

4. Evaluate $-5 - (-4) + 12 - 8$.

5. Subtract -6 from -7.

6. Evaluate $20 - 6 \div 3 \cdot 2$.

7. Evaluate $3[6 - (4 - 8^2)] - 30$.

8. Evaluate $-2x^2 - 6x + 8$ when $x = -2$.

9. Name the illustrated property.

$$-5(x - 3y - 4z) = -5x + 15y + 20z$$

Simplify.

10. $8x + 2y + 4x - y$

11. $9 - \frac{2}{3}x + 16 + \frac{3}{4}x$

Solve.

12. $7t + 3 = -4$

13. $4(x - 2) = 5(x - 1) + 3x + 2$

14. $\frac{3}{4}n - \frac{1}{5} = \frac{2}{3}n$

15. $A = \frac{a + b + c}{3}$, for b

16. $\frac{40}{30} = \frac{3}{x}$

Solve. Graph the solution on a number line and represent the solution in interval notation.

17. $x - 3 > 7$

18. $2x - 7 \leq 3x + 5$

19. **Trampoline** A circular trampoline has a diameter of 22 feet. Determine the area of the trampoline.

20. **Earnings** If Samuel earns \$10.50 after working for 2 hours mowing a lawn, how much does he earn after 8 hours?

3 Applications of Algebra

The mixture of chemicals in the laboratory is an important process. In some cases, algebra can be used to determine the amounts of the chemicals needed for a specific process. In Example 6 on page 216 you will see how to determine the mixture proportions for a particular solution.

Goals of This Chapter

The emphasis of this chapter is to get you to express real-world problems mathematically and solve them. This process of representing real-life situations mathematically is called *modeling*. For many students, this chapter is the most important chapter in the entire text. The material presented in this chapter will not only help you succeed in this course, but will help you succeed throughout life!

3.1 Changing Application Problems into Equations

1. Translate phrases into mathematical expressions.
2. Express the relationship between two related quantities.
3. Write expressions involving multiplication.
4. Translate applications into equations.

1 Translate Phrases into Mathematical Expressions

HELPFUL HINT

Study Tip

It is important that you prepare for this chapter carefully. Make sure you read the book and work the examples carefully. *Attend class every day, and most of all, work all the exercises assigned to you.*

As you read through the examples in the rest of the chapter, think about how they can be expanded to other, similar problems. For example, in Example 1 **a)** we will state that the distance, d, increased by 10 miles, can be represented by $d + 10$. You can generalize this to other, similar problems. For example, a weight, w, increased by 15 pounds, can be represented as $w + 15$.

One practical advantage of knowing algebra is that you can use it to solve everyday problems by first *translating application problems into mathematical language*. One purpose of this section is to help you take an application problem, also referred to as a *word* or *verbal problem*, and write it as a mathematical equation.

Often the most difficult part of solving an application problem is translating it into an equation. Before you can translate a problem into an equation, you must understand the meaning of certain words and phrases and how they are expressed mathematically. **Table 3.1** is a list of selected words and phrases and the operations they imply. We used the variable x. However, any variable could have been used.

TABLE 3.1

Word or Phrase	Operation	Statement	Algebraic Form
Added to	Addition	A number *added to* 3	$3 + x$
More than		2 *more than* a number	$x + 2$
Increased by		A number *increased by* 7	$x + 7$
The sum of		The *sum of* 1 and a number	$1 + x$
Subtracted from	Subtraction	A number *subtracted from* 3	$3 - x$
Less than		5 *less than* a number	$x - 5$
Decreased by		A number *decreased by* 8	$x - 8$
The difference between		*The difference between* 4 and a number	$4 - x$
The difference in		*The difference in* a number and 6	$x - 6$
The difference of		*The difference of* x squared and y squared	$x^2 - y^2$
Multiplied by	Multiplication	8 *multiplied by* a number	$8x$
The product of		*The product of* 9 and a number	$9x$
Twice a number		*Twice a number*	$2x$
Three times a number		*Three times a number*	$3x$
Of (when used with a percent or fraction)		25% *of* a number	$0.25x$
		$\frac{1}{3}$ *of* a number	$\frac{1}{3}x$
Divided by	Division	13 *divided by* a number	$\frac{13}{x}$
The quotient of		*The quotient of* a number and 11	$\frac{x}{11}$
The ratio of		*The ratio of* a number and 17	$\frac{x}{17}$

Often a statement contains more than one operation. The following chart provides some examples of this.

Statement	Algebraic Form
Four more than twice a number	$\underbrace{2x}_{\text{Twice a number}} + 4$
Five less than 3 times a number	$\underbrace{3x}_{\text{Three times a number}} - 5$
Three times the sum of a number and 8	$3(\underbrace{x + 8}_{\text{The sum of a number and 8}})$
Twice the difference between a number and 4	$2(\underbrace{x - 4}_{\text{The difference between a number and 4}})$

Understanding Algebra

Subtraction is *not* commutative. For example, $6 - 9$ is not equal to $9 - 6$. *Thus, $x - 7$ is different from $7 - x$.*

AVOIDING COMMON ERRORS

Subtraction is not commutative. That is, $a - b \neq b - a$. Therefore, you must be very careful when writing expressions involving subtraction. Study the following examples.

5 less than 3 times a number

CORRECT	INCORRECT
$3x - 5$	~~$5 - 3x$~~

5 subtracted from 3 times a number

CORRECT	INCORRECT
$3x - 5$	~~$5 - 3x$~~

the difference between $3x$ and 5

CORRECT	INCORRECT
$3x - 5$	~~$5 - 3x$~~

Often an algebraic expression can be written in several different ways.

Algebraic Expression	Statements
$2x + 3$	Three more than twice a number The sum of twice a number and 3 Twice a number, increased by 3 Three added to twice a number
$3x - 4$	Four less than 3 times a number Three times a number, decreased by 4 The difference between 3 times a number and 4 Four subtracted from 3 times a number

EXAMPLE 1 Express each statement as an algebraic expression.

a) The distance, d, increased by 10 miles

b) Eight less than twice the area, a

c) Four pounds more than 5 times the weight, w

d) The difference in x miles and 3 miles

Solution

a) $d + 10$ **b)** $2a - 8$ **c)** $5w + 4$ **d)** $x - 3$

Now Try Exercise 13

2 Express the Relationship between Two Related Quantities

When two numbers are related to each other, we will often represent one number as x and the other number as an expression containing x.

Understanding Algebra

It is important to first determine what quantity we are representing with a variable. Then we should express it literally as "Let $x =$ "

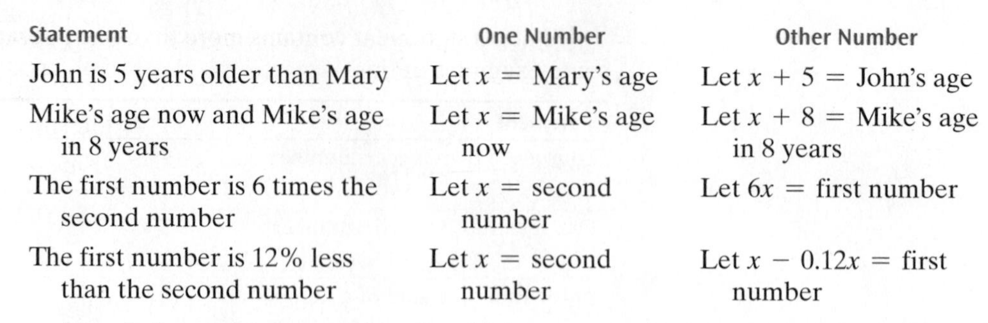

Statement	One Number	Other Number
John is 5 years older than Mary	Let x = Mary's age	Let $x + 5$ = John's age
Mike's age now and Mike's age in 8 years	Let x = Mike's age now	Let $x + 8$ = Mike's age in 8 years
The first number is 6 times the second number	Let x = second number	Let $6x$ = first number
The first number is 12% less than the second number	Let x = second number	Let $x - 0.12x$ = first number

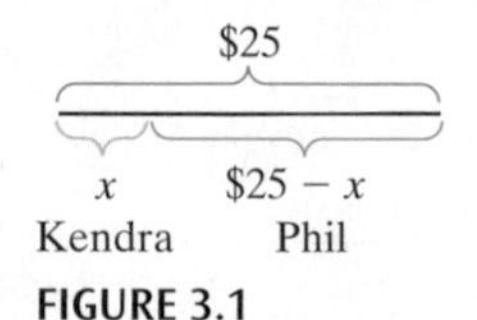

FIGURE 3.1

Now let's consider a problem in which one quantity is separated into two parts. For example, suppose \$25 is separated between Kendra and Phil.

If Kendra gets ...	Then Phil gets ...
\$20	\$25 − \$20 or \$5
\$15	\$25 − \$15 or \$10
\$8	\$25 − \$8 or \$17
\$2	\$25 − \$2 or \$23

Understanding Algebra

In general, if T represents the total to be separated into two parts, then if one part is called x, the other part will be $T - x$. See **Figure 3.2**.

In general, if we let x = the amount Kendra gets, then $25 - x$ = the amount Phil gets. Note that the sum of x and $25 - x$ is 25. (See **Figure 3.1**.)

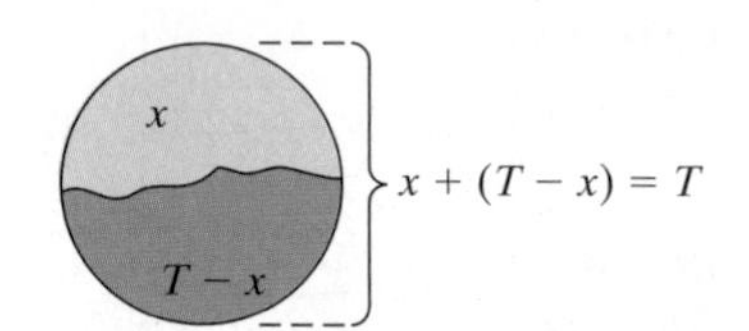

FIGURE 3.2

EXAMPLE 2 For each part, determine what to let x represent.

a) Mary weighs 15 pounds more than Sue.

b) The length of a rectangle is 4 inches more than twice its width.

c) Joe earns \$56.20 less than Larry.

d) The theater sold 525 tickets. Some were adults' tickets and some were children's tickets.

Solution In general, when writing expressions to represent word problems, if a quantity A is expressed in terms of a quantity B, then we let x = quantity B.

a) Since Mary's weight is expressed in terms of Sue's weight, we let x = Sue's weight.

b) Since the length of a rectangle is expressed in terms of its width, we let x = width of the rectangle.

c) Since the amount Joe earns is expressed in terms of what Larry earns, we let x = amount Larry makes.

d) Here, since neither quantity of adults' tickets nor children's tickets is expressed in terms of the other's quantity, we can let x = number of adults' tickets sold or let x = number of children's tickets sold. If we let x = number of adults' tickets sold, then $525 - x$ will equal the number of children's tickets sold. If we let x = number of children's tickets sold, then $525 - x$ will equal the number of adults' tickets sold.

Now Try Exercise 37

EXAMPLE 3 For each relationship, select a variable to represent one quantity and state what that variable represents. Then express the second quantity in terms of the variable selected.

a) The Bulldogs scored 12 points more than the Tigers.

b) An adult female gorilla is 50 times the weight of a baby gorilla.

c) Bill and Mary share \$75.

d) Kim has 7 more than 5 times the amount Sylvia has.

e) The length of a rectangle is 3 feet less than 4 times its width.

Solution To express the relationships, we must first decide which quantity we will let the variable represent. To give you practice with variables other than x, we will select different letters to represent the variable.

a) Since the number of points scored by the Bulldogs is expressed in terms of the number of points scored by the Tigers, we will select the variable t.

$$\text{Let } t = \text{number of points scored by the Tigers.}$$
$$\text{Then } t + 12 = \text{number of points scored by the Bulldogs.}$$

b) The weight of an adult female gorilla is given in terms of the weight of a baby gorilla.

$$\text{Let } w = \text{weight of a baby gorilla.}$$
$$\text{Then } 50w = \text{weight of an adult female gorilla.}$$

c) We are not told how much of the \$75 each person receives. In this case we can let the variable represent the amount either person receives. We will let a represent the amount Bill receives.

$$\text{Let } a = \text{amount Bill receives.}$$
$$\text{Then } 75 - a = \text{amount Mary receives.}$$

d) The amount Kim has is given in terms of the amount Sylvia has.

$$\text{Let } s = \text{amount Sylvia has.}$$
$$\text{Then } 5s + 7 = \text{amount Kim has.}$$

e) The length of the rectangle is given in terms of the width of the rectangle.

$$\text{Let } w = \text{width of the rectangle.}$$
$$\text{Then } 4w - 3 = \text{length of the rectangle.}$$

Now Try Exercise 47

3 Write Expressions Involving Multiplication

Consider the statement "the cost of 3 items at \$5 each." The cost would be 3 times \$5 and we could express the cost as $3 \cdot 5$ or $3(5)$. Now consider the statement "the cost of x items at \$5 each." The cost would be x times \$5 and we could express the cost as $x \cdot 5$ or $x(5)$. It is customary to write this product as $5x$. Thus, the cost of x items at \$5 each is represented as $5x$.

Finally, consider the statement "the cost of x items at y dollars each." We write the cost of x items at y dollars as $\boldsymbol{xy}$.

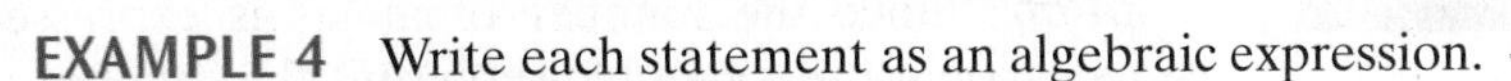

EXAMPLE 4 Write each statement as an algebraic expression.

a) The cost of purchasing x pens at \$2 each

b) A 5% commission on x dollars in sales

c) The dollar amount earned in h hours if a person earns \$6.50 per hour

d) The number of cents in q quarters

e) The number of ounces in x pounds

Solution

a) We can reason like this:

1 pen would cost	1(2) dollars	= \$2
2 pens would cost	2(2) dollars	= \$4
3 pens would cost	3(2) dollars	= \$6
⋮	⋮	⋮
x pens would cost	$x(2)$ dollars	or $\boldsymbol{2x}$ dollars

Thus the cost would be $2x$ dollars.

b) A 5% commission on \$1 sales would be 0.05(1), on \$2 sales 0.05(2), on \$3 sales 0.05(3), on \$4 sales 0.05(4), and so on. Therefore, the commission on sales of x dollars would be $0.05(x)$ or $0.05x$.

Note: If you need a review on changing a percent to a decimal number, review Appendix A.

c) In one hour the person would earn 1($6.50). In two hours the person would earn 2($6.50), and in h hours the person would earn h($6.50) or 6.50h$.

d) We know that each quarter is worth 25 cents. Thus, one quarter is 1(25) cents. Two quarters is 2(25) cents, and so on. Therefore, q quarters is $q(25)$ cents or $25q$ cents.

e) Each pound is equal to 16 ounces. Using the same reasoning as in part **d)**, we see that x pounds is $16x$ ounces.

Now Try Exercise 71

EXAMPLE 5 Truck Rental Maria Mears rented a truck for 1 day. She paid a daily fee of $88 and a mileage fee of 75 cents per mile. Write an expression that represents her total cost when she drives x miles.

Solution Maria's total cost consists of two parts, the daily fee and the mileage fee. Notice the daily fee is given in terms of dollars, and the mileage fee is given in cents. When writing an expression to represent the total cost, we want the units to be the same. Therefore, we will use a mileage fee of $0.75 per mile, which is equal to 75 cents per mile.

$$\text{Let } x = \text{number of miles driven.}$$
$$\text{Then } 0.75x = \text{cost of driving } x \text{ miles.}$$

$$\overbrace{\text{daily fee} + \text{mileage fee}}^{\text{total cost}}$$
$$88 + 0.75x$$

Thus, the expression that represents Maria's total cost is $88 + 0.75x$.

Now Try Exercise 75

Understanding Algebra

Caution! Be sure that the units are consistent throughout the problem! In Example 5, both fees are written in terms of dollars.

EXAMPLE 6 Write a Sum or Difference In a classroom the number of males was 3 more than twice the number of females. Write an expression for

a) the sum of the number of males and females

b) the difference between the number of males and females

c) the difference between the number of females and males

Solution Since the number of males is expressed in terms of the number of females, we let the variable represent the number of females. We will choose x to represent the variable.

$$\text{Let } x = \text{number of females.}$$
$$\text{Then } 2x + 3 = \text{number of males.}$$

a) The expression for the sum of the number of males and females is

$$\overbrace{(2x + 3)}^{\text{number of males}} + \overbrace{x}^{\text{number of females}}$$

b) The expression for the difference between the number of males and females is

$$\overbrace{(2x + 3)}^{\text{number of males}} - \overbrace{x}^{\text{number of females}}$$

c) The expression for the difference between the number of females and males is

$$\overbrace{x}^{\text{number of females}} - \overbrace{(2x + 3)}^{\text{number of males}}$$

Notice that parentheses are needed around the $2x + 3$ since both terms $2x$ and 3 are being subtracted.

Now Try Exercise 87

HELPFUL HINT

Using Parentheses When Writing Expressions

Sum: When writing the sum of two quantities, parentheses may be used to help in the understanding of the problem, but they are not necessary.

Examples	Answers
Find the sum of x and $2x - 3$.	$x + (2x - 3)$ or $x + 2x - 3$
Find the sum of $3c - 4$ and $c + 5$.	$(3c - 4) + (c + 5)$ or $3c - 4 + c + 5$

Difference: When writing the difference of two quantities, when only *a single term* is being subtracted, parentheses may be used in the understanding of the problem, but they are not necessary.

Examples	Answers
Subtract r from $3r - 2$.	$(3r - 2) - r$ or $3r - 2 - r$
Find the difference between $2s + 6$ and s.	$(2s + 6) - s$ or $2s + 6 - s$

When writing the difference of two quantities, *when two or more terms are being subtracted, parentheses **must** be placed around all the terms being subtracted,* since all the terms are being subtracted and not just the first term.

Examples	Answers
Subtract $x + 2$ from $3x$.	$3x - (x + 2)$
Subtract $3t - 4$ from $5t$.	$5t - (3t - 4)$
Subtract $r - 5$ from $2r + 3$.	$(2r + 3) - (r - 5)$ or $2r + 3 - (r - 5)$
Find the difference between 6 and $m + 3$.	$6 - (m + 3)$
Find the difference between $4n - 9$ and $2n - 3$.	$(4n - 9) - (2n - 3)$ or $4n - 9 - (2n - 3)$

Expressions Involving Percent

Example 7 involves percent. Whenever we perform a calculation involving percent, we change the percent to a decimal number or a fraction first.

When shopping we may see a "25% off" sign. We assume that this means 25% off of the *original cost,* even though this is not stated. If we let c represent the original cost, then 25% of the original cost would be represented as $0.25c$. Twenty-five percent off the original cost means the original cost, c, decreased by 25% of the original cost. Twenty-five percent off the original cost would be represented as $c - 0.25c$.

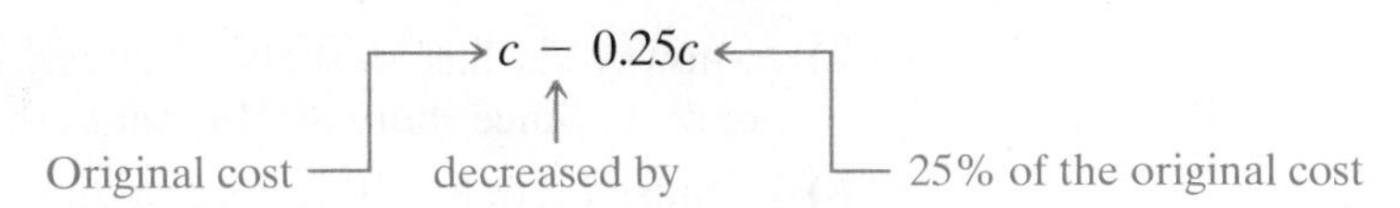

Now let's work an example involving percent.

EXAMPLE 7 Write each statement as an algebraic expression.

a) The cost of a pair of boots, c, increased by 6%

b) The population in the town of Brooksville, p, decreased by 12%

Solution

a) The question asks for the cost increased by 6%. We assume that this means the cost increased by 6% of the original cost. Therefore, the answer is $c + 0.06c$.

b) Using the same reasoning as in part **a)**, the answer is $p - 0.12p$.

Now Try Exercise 79

Understanding Algebra

If the sales tax rate is 6%, the *tax* on an item costing x dollars is $0.06x$.

The *total amount,* which is the cost of the item plus the tax on the item, is represented as $x + 0.06x$, which can also be written as $1.06x$.

AVOIDING COMMON ERRORS

In Example 7 **a)** we asked you to represent a cost, c, increased by 6%. Note, the answer is $c + 0.06c$. Often, students write the answer to this question as $c + 0.06$. It is important to realize that a percent of a quantity must always be a percent multiplied by some number or letter. Some phrases involving the word percent and the correct and incorrect interpretations follow.

PHRASE	CORRECT	INCORRECT
A $7\frac{1}{2}\%$ sales tax on c dollars	$0.075c$	~~0.075~~
The cost, c, increased by a $7\frac{1}{2}\%$ sales tax	$c + 0.075c$	~~$c + 0.075$~~
The cost, c, reduced by 25%	$c - 0.25c$	~~$c - 0.25$~~

4 Translate Applications into Equations

When writing application problems as equations, the word *is* often means *is equal to* and is represented by an equals sign. Some examples of statements written as equations follow.

Statement	Equation
Six times a number *is* 42.	$6x = 42$
Five more than twice a number *is* 4.	$2x + 5 = 4$
7 decreased by a number *is* 4 more than twice the number.	$7 - x = 2x + 4$
The sum of a number and the number increased by 4 *is* 60.	$x + (x + 4) = 60$
Twice the difference of a number and 3 *is* the sum of the number and 20.	$2(x - 3) = x + 20$
A number increased by 15% *is* 120.	$x + 0.15x = 120$
Six less than three times a number *is* one-fourth the number.	$3x - 6 = \frac{1}{4}x$

Now let's work some examples where we write equations.

EXAMPLE 8 Translate Words into Equations Write each problem as an equation.

a) A New York City subway car has 36 seats. The number of seats in s subway cars *is* 180.

b) The number of cents in d dimes *is* 120.

c) The cost of x gallons of gasoline at \$4.20 per gallon *is* \$35.20.

Solution

a) 1 subway car has 36 seats, 2 subway cars have 72 seats, and s subway cars have $36s$ seats. Since there are 180 seats, the equation is $36s = 180$.

b) Because there are 10 cents in a dime, the number of cents in d dimes is $d(10)$ or $10d$. Since the number of cents in d dime *is* 120, the equation is $10d = 120$.

c) Using similar reasoning as in parts **a)** and **b)**, the equation is $4.20x = 35.20$.

Now Try Exercise 111

HELPFUL HINT

In a written expression certain other words may be used in place of *is* to represent the equals sign. Some of these are *will be, was, yields,* and *gives*. For example,

"When 4 is added to a number, the sum *will be* 20" can be expressed as $x + 4 = 20$.

"Six subtracted from a number *was* $\frac{1}{2}$ the number" can be expressed as $x - 6 = \frac{1}{2}x$.

"A rental car cost \$75 per day. The cost for renting the car for x days *was* \$150" can be expressed as $75x = 150$.

EXAMPLE 9 Translate Words into an Equation Write the problem as an equation.

One number is 4 less than twice the other. Their sum is 14.

Solution

$$\text{Let } x = \text{one number.}$$
$$\text{Then } 2x - 4 = \text{second number.}$$

Now we write the equation using the information given.

$$\text{first number} + \text{second number} = 14$$
$$x + (2x - 4) = 14$$

Now Try Exercise 99

In Example 10 we will discuss *consecutive integers*. For example, the numbers 5 and 6 are consecutive integers. The numbers 103 and 104 are consecutive integers. Since the larger consecutive integer is always one more than the smaller consecutive integer, we let $x =$ the smaller consecutive integer and $x + 1 =$ the larger consecutive integer.

Understanding Algebra

For two *consecutive integers*:

Let $x =$ smaller consecutive integer

$x + 1 =$ larger consecutive integer

For two *consecutive* ***even*** *integers*:

Let $x =$ smaller consecutive even integer

$x + 2 =$ larger consecutive even integer

For two *consecutive* ***odd*** *integers*:

Let $x =$ smaller consecutive odd integer

$x + 2 =$ larger consecutive odd integer

EXAMPLE 10 Consecutive Integers Write the problem as an equation.

For two consecutive integers, the sum of the smaller and 3 times the larger is 23.

Solution First, we express the two consecutive integers in terms of the variable.

$$\text{Let } x = \text{smaller consecutive integer.}$$
$$\text{Then } x + 1 = \text{larger consecutive integer.}$$

Now we write the equation using the information given.

$$\text{smaller} + 3 \text{ times the larger} = 23$$
$$x + 3(x + 1) = 23$$

Now Try Exercise 107

In Example 10 we worked with consecutive integers. Other problems will require us to work with consecutive *even* integers. For example, 12 and 14 are consecutive even integers. Also, 128 and 130 are consecutive even integers. Since the larger consecutive even integer is always 2 more than the smaller consecutive even integer, we let $x =$ the smaller consecutive even integer and $x + 2 =$ the larger consecutive even integer. Still other problems will require us to work with consecutive *odd* integers. For example, 5 and 7 are consecutive odd integers. Also, 213 and 215 are consecutive odd integers. Since the larger consecutive odd integer is always 2 more than the smaller consecutive odd integer, we let $x =$ the smaller consecutive odd integer and $x + 2 =$ the larger consecutive odd integer.

EXAMPLE 11 Translate Words into an Equation Write the problem as an equation.

One train travels 3 miles more than twice the distance another train travels. The total distance traveled by both trains is 800 miles.

Solution First express the distance traveled by each train in terms of the variable.

$$\text{Let } x = \text{distance traveled by one train.}$$
$$\text{Then } 2x + 3 = \text{distance traveled by second train.}$$

Now write the equation using the information given.

$$\text{distance of train 1} + \text{distance of train 2} = \text{total distance}$$
$$x + (2x + 3) = 800$$

Now Try Exercise 123

EXAMPLE 12 **Translate Words into an Equation** Write the problem as an equation.

Lori Soushon is 4 years older than 3 times the age of her son Ron. The difference in their ages is 26 years.

Solution Since Lori's age is given in terms of Ron's age, we will let the variable represent Ron's age.

$$\text{Let } x = \text{Ron's age.}$$
$$\text{Then } 3x + 4 = \text{Lori's age.}$$

We are told that the difference in Lori's age and Ron's age is 26 years. The word *difference* indicates subtraction. Since Lori is older than Ron, we must subtract Ron's age from Lori's age to get a positive number.

$$\text{Lori's age} - \text{Ron's age} = 26$$
$$(3x + 4) - x = 26$$

Now Try Exercise 119

Example 13 will involve percent.

EXAMPLE 13 **Translate Words into an Equation** Write the problem as an equation.

The 2013 property tax for Danielle's house was 3.9% greater than her property tax in 2012. Her property tax in 2013 was $4008.

Solution In this example, we will choose to use the variable *t*, for tax. Since the 2013 property tax is based upon the 2012 property tax, we will let the variable represent the 2012 property tax.

$$\text{Let } t = 2012 \text{ property tax.}$$
$$\text{Then } t + \underbrace{0.039t}_{\text{this represents the 3.9\% increase}} = 2013 \text{ property tax.}$$

Since Danielle's 2013 property tax *was* $4008, the equation we write is $t + 0.039t = 4008$.

Now Try Exercise 125

HELPFUL HINT

It is important that you understand this section and work all your assigned homework problems. You will use the material learned in this section in the next three sections and throughout the book.

3.1 Exercise Set MathXL® MyMathLab®

Warm-Up Exercises

Fill in the blanks with the appropriate word, phrase, or symbol(s) from the following list.

$0.06y$	of	$x + 2$	$x + 1$	increased	quotient
$x - 8$	$8 - x$	product	sum	subtracted	$y + 0.06y$
even	odd	$y + 0.06$			

1. The algebraic expression $5 + a$ represents the phrase "5 ________ by *a*."

2. The algebraic expression $b - 7$ represents the phrase "7 ________ from *b*."

3. The algebraic expression $3 \cdot c$ represents the phrase "the ________ of 3 and *c*."

4. The algebraic expression $\frac{d}{4}$ represents the phrase "the ________ of *d* and 4."

5. An eight-foot rope is cut into two pieces. If x represents the length of one piece of rope then ________ represents the length of the other piece of rope.

6. The algebraic expression $\frac{2}{3}x$ represents the phrase "$\frac{2}{3}$ ________ a number."

7. If y represents the cost of a coffee mug and if the sales tax rate is 6%, then an expression for the total cost of the mug is ________.

8. If x represents the first consecutive integer then the expression for the next consecutive integer is ________.

9. If x represents the first consecutive even integer then the expression for the next consecutive even integer is ________.

10. If x represents the first consecutive odd integer then the expression for the next consecutive ________ integer is $x + 2$.

Practice the Skills

In Exercises 11–30, express the statement as an algebraic expression. See Example 1.

11. The height, h, increased by 4 inches
12. The weight, w, increased by 20 pounds
13. The age, a, decreased by 5 years
14. The time, t, decreased by 3 hours
15. Five times the height, h
16. Seven times the length, l
17. Twice the distance, d
18. Three times the rate, r
19. One-half the age, a
20. One-third the weight, w
21. Five subtracted from r
22. Nine subtracted from p
23. m subtracted from 12
24. n subtracted from 4
25. Eight pounds more than twice the weight, w
26. Six inches more than 3 times the height, h
27. Four years less than 5 times the age, a
28. One mile more than $\frac{1}{2}$ the distance, d
29. One-third the weight, w, decreased by 9 pounds
30. One-fifth the height, h, increased by 2 feet

In Exercises 31–44, determine what to let x represent. See Example 2.

31. Paul is 4 inches taller than Sonya.
32. The length of a rectangle is 5 inches greater than its width.
33. The length of Tortuga Beach is 60 feet shorter than the length of Jones Beach.

Tortuga Beach, Costa Rica

34. Wilma ran 4 miles per hour faster than Natasha.
35. The United States won 3 times the number of medals that Finland won.
36. The distance to Georgia is $\frac{1}{2}$ the distance to Tennessee.
37. The Cadillac costs $200 more than twice the cost of the Chevy.
38. Noah received 25 more votes than 3 times the number of votes that Tawnya received.
39. June's grade was 2 points less than twice Teri's grade.
40. Alberto's salary was $2000 greater than 4 times Nick's salary.
41. $60 is divided between Kristen and Yvonne.
42. Drawka has 25 marbles. They are either red or blue marbles.
43. Together Don and Angela weigh 270 pounds.
44. Together Oliver and Dalane have 1053 clients.

In Exercises 45–56, select a variable to represent one quantity and state what that variable represents. Express the second quantity in terms of the variable selected. See Example 3. Note that the variable you select may be different than the variable used in the answers in the back of the book.

45. The table costs 5 times as much as the chair.
46. Jim weighs twice as much as his daughter Aubrey.
47. The area of the living room is 20 square feet greater than twice the area of the kitchen.

48. The amount in Darla's savings account is \$250 less than 4 times the amount in Carmen's savings account.

49. The length of the rectangle is 2 inches less than 6 times the width.

50. The book *Golden Angels* sold 4 copies less than 5 times the amount the book *Flycatcher* sold.

51. A total of 20 medals were won by Sweden and Brazil.

52. A total of \$600 is to be divided between Evita and Brian.

53. Mike's age is 2 years more than $\frac{1}{2}$ George's age.

54. Our cat Charmin is 4 years older than $\frac{1}{2}$ the age of our dog Toby.

55. Jan and Edward used two different treadmills for exercise. The total distance walked between them was 6.4 miles.

56. On a 540-mile trip Cheng and Elsie shared the driving.

In Exercises 57–84, write the indicated expression. See Examples 3–7.

57. Age Dan Graber is n years old now. Write an expression that represents his age in 8 years.

58. Speed-Reading John Debruzzi was reading p words per minute. After taking a speed-reading course, his speed increased by 60 words per minute. Write an expression that represents his new reading speed.

59. Motorcycle Melissa Blum is selling her motorcycle. She was asking x dollars for the motorcycle but has cut the price in half. Write an expression that represents the new price.

60. Age Cathy Bennett's son is one-third as old as Cathy, c. Write an expression for her son's age.

61. Age Gayle Krzemien's age is one less than twice Mary Lou Baker's age, a. Write an expression for Gayle's age.

62. Calories The calories in a serving of mixed nuts is 280 calories less than twice the number of calories in a serving of cashew nuts, c. Write an expression for the number of calories in a serving of mixed nuts.

63. Temperature The average daily temperature in Jacksonville, Florida, in July is 30° less than twice its average daily temperature in January, t. Write an expression for the average daily temperature in July.

64. County Size Jefferson County is 250 square miles more than three times the size of Adams County. Write an expression for the size of Jefferson County.

65. Weight Anika Angel weighed p pounds at birth. At age 6 months her weight was 2.3 pounds less than twice her birth weight. Write an expression for Anika's weight at age 6 months.

66. Growth Pat Cook planted a walnut tree that had an original height of h feet. Now the tree has a height that is 5 feet less than 4 times the original height. Write an expression for the current height of the tree.

67. Profits Monica and Julia share in the profits of a toy store. If the total profit is \$80,000 and m is the amount Monica receives, write an expression for the amount Julia receives.

68. Charity Event A total of 83 men and women attended a charity event. If the number of men who attended is m, write an expression for the number of women who attended.

69. **Home Runs** Hank Aaron had 673 less than twice the number of home runs Babe Ruth had, r. Write an expression for the number of home runs Hank Aaron had.

70. **Life Expectancy** In 2011, the country with the highest life expectancy was Monaco and the country with the lowest life expectancy was Swaziland. The average life expectancy of Monaco was 25.9 years more than twice the average life expectancy of Swaziland, s. Write an expression for the life expectancy of Monaco. (*Source:* CIA World Factbook)

71. **Money** Carolyn Curley found that she had x dimes in her handbag. Write an expression that represents this quantity of money in cents.

72. **Weight** Jason Mahar's weight is w pounds. Write an expression that represents his weight in ounces.

73. **Soil** A total of six hundred pounds of soil is put onto two trucks. If a pounds of soil is placed onto one truck, write an expression for the amount of soil placed onto the other truck.

74. **Milk** Lisa DeLong has a pitcher that contains 26 ounces of milk. She gives m ounces to her daughter and the rest to her son. Write an expression for the amount given to Lisa's son.

75. **Truck Rental** Bob Melina rented a truck for a trip. He paid a daily fee of $45 and a mileage fee of 40 cents a mile. Write an expression that represents his total cost when he travels x miles in one day.

76. **Soil Delivery** Mary Vachon had topsoil delivered to her house. The total cost included a delivery charge of $48 plus $60 per cubic yard of soil. Write an expression for the total cost if Mary has x cubic yards of soil delivered.

77. **Sales Increase** Barry Cogan is a sales representative for a medical supply company. His 2013 sales increased by 20% over his 2012 sales, s. Write an expression for his 2013 sales.

78. **Salary Increase** Charles Idion, an engineer, had a salary increase of 15% over last year's salary, s. Write an expression for this year's salary.

79. **Electricity Use** Jean Olson's electricity use in 2013 decreased by 12% from her 2012 electricity use, e. Write an expression for her 2013 electricity use.

80. **Shirt Sale** At a 25% off everything sale, Bill Winchief purchased a new shirt. If c represents the original cost, write an expression for the sale price of the shirt.

81. **Car Cost** The cost of a new car purchased in Collier County included a 7% sales tax. If c represents the cost of the car before tax, write an expression for the total cost, including the sales tax.

82. **GPS Device** Deborah Doucette purchased a new global positioning system (GPS) device for z dollars. She also paid 7.5% sales tax on the GPS device. Write an expression for the total amount Deborah paid for the GPS device.

83. In 2010, the state with the greatest median age was Maine and the state with the lowest was Utah. The median age in Utah was 31.6% less than that of Maine. If m represents the median age in Maine, write an expression for the median age in Utah. (*Source:* U.S. Census Bureau)

84. **New Orleans** According to the U.S. Census Bureau, New Orleans, Louisiana, had the greatest population growth from 2010 to 2011 of any major U.S. city. Its population in 2011 was 4.9% greater than its population in 2010. If p represents the population of New Orleans in 2010, write an expression to represent the population of New Orleans in 2011.

New Orleans, LA

In Exercises 85–98, write the indicated expression. You will need to select the variable to use. Note that the variable you use may be different than the variable used in the answer section. See Example 6.

85. **Weight** Jennifer's weight is 15 pounds more than Frieda's weight. Write an expression for the sum of their weights.

86. **Length** The chestnut-mandibled toucan is 3 inches longer than the keel-billed toucan (shown right). Write an expression for the sum of their lengths.

87. **Height** Armando is taller than his son Luis. Armando is 1 inch less than twice Luis's height. Write an expression for the *difference* in Armando's and Luis's heights.

A Keel-billed Toucan, see Exercise 86.

88. **Profits** A company's profits in 2012 were \$100 less than the company's profits in 2013. Write an expression for the *difference* in the 2013 and 2012 profits.

89. **Numbers** A first number is 40 less than 3 times a second number. If the second number is represented by x, write an expression for the first number subtracted from the second number.

90. **Numbers** A first number is 16 less than twice a second number. If the second number is represented by n, write an expression for the first number subtracted from the second number.

91. **Weight** Jill's four-year-old child weighs 3 pounds less than twice the weight of her two-year-old child. Write an expression for the sum of their weights.

92. **Farm Size** William Echols owns two farms—one in Hempstead and the other in Rosenberg. The area of the farm in Hempstead is 125 acres less than twice the area of the farm in Rosenberg. Write an expression for the sum of the areas of both farms.

93. **Land Area** The largest state in area is Alaska and the smallest is Rhode Island. The area of Alaska is 462 square miles more than 479 times the area of Rhode Island. Write an expression for the sum of the areas of the two states.

94. **Assessed Value** The assessed value of Glen Sandifer's house is \$200 more than twice that of Olga Thompson's house. Write an expression for the sum of the assessed values.

95. **Stocks** The price of RCD Computer stock is 39.4% greater than the cost of Research in Motion stock. Write an expression for the sum of the prices of Research in Motion and RCD Computer stocks.

96. **Car Cost** The cost of a 2013 Chevrolet Corvette LT-1 coupe increased by 5.4% over the cost of a 2012 model. Write an expression for the sum of the costs of a 2012 and a 2013 Corvette.

97. **Bank Assets** The 2013 assets of Fifth Third Bancorp were 11.2% higher than their 2012 assets. Write an expression for the 2013 assets subtracted from the 2012 assets.

98. **Spam** The number of pieces of spam (or junk mail) that Laura Hoye received in 2013 was 12% less than the number she received in 2012. Write an expression for the difference between the numbers of pieces of spam she received in 2012 and 2013.

In Exercises 99–132, write an equation to represent the problem. See Examples 8–13.

99. **Two Numbers** One number is 4 times another. The sum of the two numbers is 20.

100. **Age** Marie is 6 years older than Denise. The sum of their ages is 48.

101. **Consecutive Integers** The sum of two consecutive integers is 41.

102. **Consecutive Integers** For two consecutive integers, the sum of the smaller and twice the larger is 29.

103. **Numbers** Twice a number, decreased by 8 is 12.

104. **Numbers** Five times a number increased by 7 is 42.

105. **Numbers** One-third of the sum of a number and 12 is 5.

106. **Numbers** One-fifth of the sum of a number and 10 is 150.

107. **Consecutive Even Integers** For two consecutive even integers, the sum of the smaller and twice the larger is 22.

108. **Consecutive Even Integers** For two consecutive even integers, the sum of three times the smaller and twice the larger is 24.

109. **Consecutive Odd Integers** For two consecutive odd integers, the sum of 3 times the smaller and the larger is 14.

110. **Consecutive Odd Integers** For two consecutive odd integers, the sum of twice the smaller and four times the larger is 38.

111. **Earnings** John Jones earns \$12.50 per hour. If he works h hours his earnings will be \$150.

112. **Employees** The T. W. Wilson company plans to increase its number of employees by 20 per year. The increase in the number of employees in t years will be 120.

113. **Top Soil** Abe Mantell purchased x bags of topsoil at a cost of \$2.99 a bag. The total he paid was \$17.94.

114. **Plants** Marjorie Moller purchased p plants at a nursery for \$5.99 each. The total he paid was \$65.89.

115. **Quarters** The number of cents in q quarters is 175.

116. **Seconds** The number of seconds in m minutes is 480.

117. **Age** Darta Aguilar is 1 year older than twice Julie Chesser's age. The sum of their ages is 52.

118. **Jogging** David Ostrow jogs 5 times as far as Jennifer Freer. The total distance traveled by both people is 8 miles.

119. **Baseball Cards** Jakob Meyer owns 300 more than twice the number of baseball cards that Saul Gonzales owns. The difference in the numbers of cards that Jakob and Saul own is 420.

120. **Horses** Marc Campbell owns more horses than Selina Jones owns. The number of horses that Marc owns is 3 less than five times the number that Selina owns. The difference in the numbers of horses owned by Marc and Selina is 5.

121. **Amtrak** An Amtrak train travels 4 miles less than twice the distance traveled by a Southern Pacific train. The total distance traveled by both trains is 890 miles.

122. **Wagon Ride** On a wagon ride the number of girls was 6 less than twice the number of boys. The total number of boys and girls on the wagon was 18.

123. **Distance Walked** Donna Douglas walked 2 miles less than 3 times as far as Malik Oamar walked. Together they walked a total of 12.6 miles.

124. **Distance** Lilia Orlova rollerbladed 4 miles less than twice the distance she ran. The total distance she traveled was 15 miles.

125. **Viper** The cost of a new Dodge Viper increased by 0.2% over last year's model. The new price is $89,560.

126. **Income** Dan Tadeo's 2013 income was 4.6% greater than his 2012 income. His income in 2013 was $56,900.

127. **Population** The population of the town of Tom's Valley decreased by 1.9%. The population after the decrease was 12,087.

128. **Blu Ray Player** At the Electronics Warehouse, Anne Long purchased a Blu Ray player that was reduced by 10% for $208.

129. **New Car** Carlotta Diaz bought a new car. The cost of the car plus a 7% sales tax was $32,600.

130. **Sport Coat** David Gillespie purchased a sport coat at a 25% off sale. He paid $195 for the sport coat.

131. **Cost of Meal** Beth Rechsteiner ate at a steakhouse. The cost of the meal plus a 15% tip was $42.50.

132. **Railroad** In a narrow gauge railway, the distance between the tracks is about 64% of the distance between the tracks in a standard gauge railroad. The difference in the distances between the tracks in the two types of railroads is about 1.67 feet.

Challenge Problem

133. **Time**

a) Write an algebraic expression for the number of seconds in d days, h hours, m minutes, and s seconds.

b) Use the expression found in part **a)** to determine the number of seconds in 4 days, 6 hours, 15 minutes, and 25 seconds.

Group Activity

Exercises 134 and 135 will help prepare you for the next section, where we set up and solve application problems. Discuss and work each exercise as a group. For each exercise, write down the quantity you are being asked to find and represent this quantity with a variable. Then write an equation containing your variable that can be used to solve the problem. Do not solve the equation.

134. **Water Usage** An average bath uses 30 gallons of water and an average shower uses 6 gallons of water per minute. How long a shower would result in the same water usage as a bath?

135. **Salary Plans** An employee has a choice of two salary plans. Plan A provides a weekly salary of $200 plus a 5% commission on the employee's sales. Plan B provides a weekly salary of $100 plus an 8% commission on the employee's sales. What must be the weekly sales for the two plans to give the same weekly salary?

Cumulative Review Exercises

[1.9] **136.** Evaluate $3[(4 - 16) \div 2] + 5^2 - 3$.

[2.6] **137.** $P = 2l + 2w$; find l when $P = 40$ and $w = 5$.

138. Solve $3x - 2y = 6$ for y.

[2.7] **139.** Solve the proportion $\frac{3.6}{x} = \frac{10}{7}$.

[2.8] **140.** Solve the inequality $2x - 4 > 3$. Graph the solution on a number line and write the solution in interval notation.

3.2 Solving Application Problems

1. Use the problem-solving procedure.
2. Set up and solve number application problems.
3. Set up and solve application problems involving money.
4. Set up and solve applications concerning percent.

1 Use the Problem-Solving Procedure

The general problem-solving procedure given in Section 1.2 can be used to solve all types of verbal problems. Below, we present the **five-step problem-solving procedure** again so you can easily refer to it. We have included some additional information under steps 1 and 2, since in this section we are going to emphasize translating application problems into equations.

Problem-Solving Procedure for Solving Applications

1. **Understand the problem.** Identify the quantity or quantities you are being asked to find.
2. **Translate the problem into mathematical language (express the problem as an equation).**
   a) Choose a variable to represent one quantity, *and write down exactly what it represents.* Represent any other quantity to be found in terms of this variable.
   b) Using the information from step a), write an equation that represents the application.
3. **Carry out the mathematical calculations (solve the equation).**
4. **Check the answer (using the *original* application).**
5. **Answer the question asked.**

Sometimes we will combine two steps in the problem-solving procedure when it helps to clarify the explanation. We may not show the check of a problem to save space. Even if we do not show a check, you should check the problem yourself and make sure your answer is reasonable and makes sense.

2 Set Up and Solve Number Application Problems

The examples presented here involve information and data but do not contain percents.

EXAMPLE 1 An Unknown Number Two subtracted from 4 times a number is 10. Find the number.

Solution Understand To solve this problem, we need to express the statement given as an equation. We are asked to find the unknown number.

Translate Let x = the unknown number. Now write the equation.

$$\underbrace{4x - 2}_{\text{2 subtracted from 4 times a number}} \underset{\text{is}}{=} \underset{10}{10}$$

Carry Out

$$4x = 12$$
$$x = 3$$

Check Substitute 3 for the number in the original problem, two subtracted from 4 times a number is 10.

$$4(3) - 2 \stackrel{?}{=} 10$$
$$10 = 10 \quad \text{True}$$

Answer Since the solution checks, the unknown number is 3.

Now Try Exercise 7

EXAMPLE 2 Number Problem The sum of two numbers is 26. Find the two numbers if the larger number is 2 less than three times the smaller number.

Solution Understand We are given that "the larger number is 2 less than three times the smaller number." Notice that the larger number is expressed in terms of the smaller number. Therefore, we will let the variable represent the smaller number.

Understanding Algebra

When solving application problems it is very important to note *how many* answers we are asked to find. Notice in Example 1 we are asked to "find the number," but in Example 2 we are asked to "find the two numbers." In Example 2, once we find the smaller number we must then find the larger number.

Translate

$$\text{Let } x = \text{smaller number.}$$
$$\text{Then } 3x - 2 = \text{larger number.}$$

The sum of the two numbers is 26. Therefore, we write the equation

$$\text{smaller number} + \text{larger number} = 26$$
$$x + (3x - 2) = 26$$

Carry Out Now we solve the equation.

$$4x - 2 = 26$$
$$4x = 28$$
$$x = 7$$

The smaller number is 7. Now we find the larger number.

$$\begin{aligned}\text{larger number} &= 3x - 2\\ &= 3(7) - 2 \quad \text{Substitute 7 for } x.\\ &= 19\end{aligned}$$

The larger number is 19.

Check The sum of the two numbers is 26.

$$7 + 19 \stackrel{?}{=} 26$$
$$26 = 26 \quad \text{True}$$

Answer The two numbers are 7 and 19.

Now Try Exercise 15

HELPFUL HINT

When reading a word problem, ask yourself, "How many answers are required?" In Example 2, the question asked for the two numbers. The answer is 7 and 19. It is important that you read the question and identify what you are being asked to find. If the question had asked "Find the *smaller* of the two numbers if the larger number is 2 less than three times the smaller number," then the answer would have been only the 7. If the question had asked to find the *larger* of the two numbers, then the answer would have been only 19. *Make sure you answer the question asked in the problem.*

EXAMPLE 3 2012 Summer Olympics In the 2012 Olympics in London, England, the United States won the most medals and China won the second greatest number of medals. The United States won 72 less than twice the number of medals won by China. If the difference between the numbers of medals won by the United States and China was 16, determine the number of medals won by the United States.

Solution Understand The word *difference* in the problem indicates that this problem will involve subtraction. We are asked to find the number of medals won by the United States. Since the number of medals won by the United States is given in terms of the number of medals won by China, we will let the variable represent medals won by China. We will use the variable c.

Translate

$$\text{Let } c = \text{number of medals won by China.}$$
$$\text{Then } 2c - 72 = \text{number of medals won by the United States.}$$

Since we are dealing with positive amounts, we must subtract the smaller quantity from the larger. Since the difference in medals between the United States and China is 16, we write the following equation.

$$\underbrace{\text{number of medals won by U.S.}}_{2c-72} - \underbrace{\text{number of medals won by China}}_{c} = 16$$

$$2c - 72 - c = 16$$

Carry Out

$$2c - 72 - c = 16$$
$$c - 72 = 16$$
$$c = 88$$

Check and Answer Remember c represents the number of medals won by China. We are asked to find the number of medals won by the United States, which we have represented as $2c - 72$. Now, we substitute the known value of 88 for c in the expression $2c - 72$: $2(88) - 72 = 104$. *The answer is that the United States won 104 medals*. Notice that the difference in the numbers of medals won by the United States and China is $104 - 88 = 16$, so the answer checks.

Now Try Exercise 27

EXAMPLE 4 Bicycles The Chain Wheel Drive Bicycle Company presently manufactures 800 bicycles a month. Each month after this month the company plans to increase production by 150 bicycles a month until its monthly production reaches 1700 bicycles. How long will it take the company to reach its production goal?

Solution Understand We are asked to find the *number of months* that it will take for the company's production to reach 1700 bicycles a month. Next month its production will increase by 150 bicycles. In two months, its production will increase by 2(150) over the present month's production. In n months, its production will increase by $n(150)$ or $150n$. We will use this information when we write the equation to solve the problem.

Translate

Let n = number of months.

Then $150n$ = increase in production over n months.

$$(\text{present production}) + \left(\begin{array}{c}\text{increased production}\\ \text{over } n \text{ months}\end{array}\right) = \text{future production}$$

$$800 + 150n = 1700$$

Carry Out

$$150n = 900$$
$$n = \frac{900}{150}$$
$$n = 6 \text{ months}$$

Check and Answer As a check, let's list the number of bicycles produced this month and for the next 6 months.

Presently	Next month	Month 2	Month 3	Month 4	Month 5	Month 6
↓	↓	↓	↓	↓	↓	↓
800	950	1100	1250	1400	1550	1700

Thus, in 6 months the company will produce 1700 bicycles per month.

Now Try Exercise 23

3 Set Up and Solve Application Problems Involving Money

When setting up an equation that involves money, you must make sure that all the monetary units entered into the equation are the same, either all dollars or all cents. When pieces of information are given in both dollars and cents, we generally convert the amount given in cents to an equivalent amount of dollars. For example, when renting a truck the cost may be $50 a day plus 90 cents a mile. When writing the equation, we would write the 90 cents a mile as \$0.90 a mile. The cost of traveling x miles at 90 cents a mile would be written $\$0.90x$.

EXAMPLE 5 **Replacing Sod** Part of Kim Martello's lawn was destroyed by grubs. She decided to purchase new sod to lay down. The cost of the sod is 45 cents per square foot plus a delivery charge of \$59. If the total cost of delivery plus the sod was \$284, how many square feet of sod was delivered?

Solution Understand The total cost consists of two parts, a cost of 45 cents per square foot of sod, plus a delivery charge of \$59. We need to determine the number of square feet of sod that will result in a total cost of \$284.

Translate

$$\text{Let } x = \text{number of square feet of sod.}$$
$$\text{Then } 0.45x = \text{cost of } x \text{ square feet of sod.}$$

$$\text{sod cost} + \text{delivery cost} = \text{total cost}$$
$$0.45x + 59 = 284 \quad \text{Subtract 59 from both sides.}$$

Carry Out

$$0.45x = 225$$
$$\frac{0.45x}{0.45} = \frac{225}{0.45}$$
$$x = 500$$

Check The cost of 500 square feet of sod at 45 cents a square foot is $500(0.45) = \$225$. Adding the \$225 to the delivery cost of \$59 gives \$284, so the answer checks.

Answer Five hundred square feet of sod was delivered.

Now Try Exercise 35

EXAMPLE 6 **Photo Printer** Elsie Newman is going to purchase a photo printer to print pictures from her digital camera. She is considering a Hewlett-Packard (HP) printer and a Lexmark printer. The HP printer costs \$419 and the cost for the ink and paper is 14 cents per photo printed. The Lexmark printer costs \$299 and the cost for the ink and paper is 18 cents per photo. How many photos would need to be printed for the total cost of the printers, ink, and paper to be the same?

Solution Understand The HP printer has a greater initial cost (\$419 versus \$299); however, its cost per photo printed is less (14 cents versus 18 cents). We are asked to find the number of photos printed so that the total cost of the two printers will be the same.

Translate

$$\text{Let } n = \text{number of photos.}$$
$$\text{Then } 0.14n = \text{cost for printing } n \text{ photos with the HP printer}$$
$$\text{and } 0.18n = \text{cost for printing } n \text{ photos with the Lexmark printer.}$$

$$\text{total cost of HP printer} = \text{total cost of Lexmark printer}$$
$$\begin{pmatrix}\text{initial}\\ \text{cost}\end{pmatrix} + \begin{pmatrix}\text{cost}\\ \text{for } n \text{ photos}\end{pmatrix} = \begin{pmatrix}\text{initial}\\ \text{cost}\end{pmatrix} + \begin{pmatrix}\text{cost}\\ \text{for } n \text{ photos}\end{pmatrix}$$
$$419 + 0.14n = 299 + 0.18n$$

Carry Out

$$120 + 0.14n = 0.18n \quad \text{299 was subtracted from both sides.}$$
$$120 = 0.04n \quad 0.14n \text{ was subtracted from both sides.}$$
$$\frac{120}{0.04} = \frac{0.04n}{0.04}$$
$$3000 = n$$

Check and Answer The total cost would be the same when 3000 photos were printed. We will leave the check of this answer for you.

Now Try Exercise 39

Understanding Algebra

Recall, the word *percent* means *per hundred*. So, 35% means $\frac{35}{100}$ or 0.35.

In application problems involving percent, we are always taking percents *of* quantities. Remember, "of" means multiply. Therefore, 35% of 40 means $0.35 \cdot 40$ or 14.

4 Set Up and Solve Applications Concerning Percent

Now we'll look at some application problems that involve percent. Remember that a percent is always a percent of something. Thus if the cost of an item, c, is increased by 8%, we would represent the new cost as $c + 0.08c$, and not $c + 0.08$. See the Avoiding Common Errors box on page 186.

EXAMPLE 7 Water Bike Rental At a beachfront hotel, the cost for a water bike rental is \$43 per half hour, which includes a $7\frac{1}{2}$% sales tax. Find the cost of the rental before tax.

Solution Understand We are asked to find the cost of the water bike rental before tax. The cost of the rental before tax plus the tax on the water bike must equal \$43.

Translate Let x = cost of the rental before tax.

Then $0.075x$ = tax on the rental.

$$(\text{cost of the water bike rental before tax}) + (\text{tax on the rental}) = 43$$

$$x + 0.075x = 43$$

Carry Out

$$1.075x = 43$$

$$x = \frac{43}{1.075}$$

$$x = 40$$

Check and Answer A check will show that if the cost of the rental is \$40, the cost of the rental including a $7\frac{1}{2}$% tax is \$43.

Now Try Exercise 47

EXAMPLE 8 Television Sale During a post-holiday sale, Stephan Delong pays \$380 for a new television. The price paid reflects a 24% discount off of the original price. Determine the original price of the television.

Solution Understand We are asked to find the original price of the television, before the discount. The amount of the discount is 24% *of* the original price. The price paid is \$380. The original price minus the amount of the discount is the price paid.

Translate Let x = the original price, before the discount.

Then $0.24x$ = the amount of the discount.

The original price minus the amount of the discount is the price paid.

$$x \quad - \quad 0.24x \quad = \quad 380$$

Carry Out

$$x - 0.24x = 380$$

$$(1 - 0.24)x = 380$$

$$0.76x = 380$$

$$x = \frac{380}{0.76}$$

$$x = 500$$

Check and Answer Since the original price, \$500, is more than the price paid, \$380, our answer is reasonable. The original price of the television is \$500.

Now Try Exercise 49

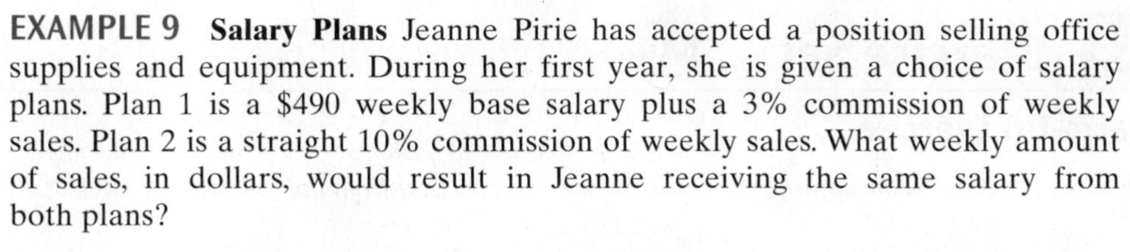

EXAMPLE 9 **Salary Plans** Jeanne Pirie has accepted a position selling office supplies and equipment. During her first year, she is given a choice of salary plans. Plan 1 is a \$490 weekly base salary plus a 3% commission of weekly sales. Plan 2 is a straight 10% commission of weekly sales. What weekly amount of sales, in dollars, would result in Jeanne receiving the same salary from both plans?

Solution Understand We are asked to find the *amount of sales*, in dollars, that will result in the same total salary from both plans. To solve this problem, we write expressions to represent the salary from each of the plans and set the salaries equal to one another.

Translate

$$\begin{aligned} \text{Let } x &= \text{amount of sales in dollars.} \\ \text{Then } 0.03x &= \text{commission from plan 1 sales.} \\ \text{and } 0.10x &= \text{commission from plan 2 sales.} \end{aligned}$$

$$\begin{aligned} \text{salary from plan 1} &= \text{salary from plan 2} \\ \text{base salary} + 3\% \text{ commission} &= 10\% \text{ commission} \\ 490 + 0.03x &= 0.10x \end{aligned}$$

Carry Out

$$\begin{aligned} 490 &= 0.07x \\ \text{or} \quad 0.07x &= 490 \\ \frac{0.07x}{0.07} &= \frac{490}{0.07} \\ x &= 7000 \end{aligned}$$

Check We will leave it up to you to show that sales of \$7000 result in Jeanne receiving the same weekly salary from both plans.

Answer Jeanne's weekly salary will be the same from both plans if she sells \$7000 worth of office supplies and equipment.

Now Try Exercise 59

HELPFUL HINT

Here are some suggestions if you find you are having some difficulty with application problems.

1. Instructor—Make an appointment to see your instructor. Make sure you have read the material in the book and attempted all the homework problems. Go with specific questions for your instructor.
2. Tutoring—If your college learning center offers free tutoring, you may wish to take advantage of tutoring.
3. Study Group—Form a study group with classmates. Exchange phone numbers and e-mail addresses. You may be able to help one another.
4. Student's Solutions Manual—If you get stuck on an exercise you may want to use the Student's Solutions Manual to help you understand a problem. Do not use the Solutions Manual in place of working the exercises. In general, the Solutions Manual should be used only to check your work.
5. MyMathLab—MyMathLab provides exercises correlated to the text. In addition, online tools such as video lectures, animations, and a multimedia textbook are available to help you understand the material.
6. Math XL®—MathXL is a powerful online homework, tutorial, and assessment system correlated specifically to this text. You can take chapter tests in MathXL and receive a personalized study plan based on your test results. The study plan links directly to tutorial exercises for the objectives you need to study or retest.

It is important that you keep trying! Remember, the more you practice, the better you will become at solving application problems.

3.2 Exercise Set MathXL® MyMathLab®

Warm-Up Exercises

Fill in the blanks with the appropriate word, phrase, or symbol(s) from the following list.

$2x + (2x + 2) = 20$	$x + (x + 2) = 20$	$x + (2x) = 20$	$(2x + 2) - x = 20$
$x + (2x + 2) = 20$	$(2x + 10) - x = 20$	$x + (10x - 2) = 20$	

1. The statement "a number (x) plus two more than twice that number is 20" can be represented by the equation __________.

2. The statement "a number (x) plus twice that number is 20" can be represented by the equation __________.

3. The statement "the sum of two consecutive odd numbers is 20" can be represented by the equation __________.

4. The statement "the sum of two numbers is 20 and the larger number is 2 less than ten times the smaller number (x)" can be represented by the equation __________.

5. The statement "the difference of two numbers is 20 and the larger number is 2 more than twice the smaller number (x)" can be represented by the equation __________.

6. The statement "the difference of two numbers is 20 and the larger number is 10 more than twice the smaller number (x)" can be represented by the equation __________.

Practice the Skills/Problem Solving

Exercises 7–32 involve finding a number or numbers. Review Examples 1–4, then set up an equation that can be used to solve the problem. Solve the equation and answer the question asked.

7. **Unknown Number** Three subtracted from 4 times a number is 17. Find the number.

8. **Unknown Number** Five subtracted from 6 times a number is 13. Find the number.

9. **Consecutive Integers** The sum of two consecutive integers is 87. Find the numbers.

10. **Consecutive Integers** The sum of two consecutive integers is 113. Find the numbers.

11. **Consecutive Odd Integers** The sum of two consecutive odd integers is 96. Find the numbers.

12. **Consecutive Odd Integers** For two consecutive odd integers, five times the smaller plus four times the larger is 89. Find the numbers.

13. **Consecutive Even Integers** For two consecutive even integers, three times the smaller plus eight times the larger is 104. Find the numbers.

14. **Consecutive Even Integers** The sum of two consecutive even integers is 146. Find the numbers.

15. **Sum of Numbers** One number is 3 more than twice a second number. Their sum is 27. Find the numbers.

16. **Sum of Numbers** One number is 5 less than 3 times a second number. Their sum is 43. Find the numbers.

17. **Difference of Numbers** The larger of two integers is 8 less than twice the smaller. When the smaller number is subtracted from the larger, the difference is 17. Find the two numbers.

18. **Difference of Numbers** The larger of two numbers is 4 less than five times the smaller. When the smaller number is subtracted from the larger, the difference is 4. Find the two numbers.

19. **Grandma's Gifts** Grandma gave some baseball cards to Richey and some to Erin. She gave 3 times the amount to Erin as she did to Richey. If the total amount she gave to both of them was 260 cards, how many cards did she give to Richey?

20. **Ski Shop** The Alpine Valley Ski Shop sold 6 times as many pairs of downhill skis as pairs of cross-country skis. Determine the number of pairs of cross-country skis sold if the total number of pairs of downhill and cross-country skis sold is 1806.

21. **Facing Pages** The sum of the two facing page numbers in an open book is 145. What are the page numbers?

22. **Facing Pages** The sum of two facing page numbers in a textbook is 1125. What are the page numbers?

23. **Collecting Frogs** Mary Shapiro collects ceramic and stuffed frogs. She presently has 422 frogs. She wishes to add 6 a week to her collection until her collection reaches a total of 500 frogs. How long will it take Mary's frog collection to reach 500 frogs?

24. **Population** The town of Dover currently has a population of 6500. If its population is increasing at a rate of 1200 people per year, how long will it take for the population to reach 20,600?

25. **Circuit Boards** The FGN Company produces circuit boards. It now has 4600 employees nationwide. It wishes to reduce the number of employees by 250 per year through retirements, until its total employment is 2200. How long will this take?

26. **Computers** The CTN Corporation has a supply of 3600 computers. It wishes to ship 128 computers each week until its supply drops to 2000. How long will this take?

27. **Tornados** According to The Weather Channel, the greatest number of tornados in the United States occurs in June and the fewest number occurs in December. The average number of tornados in June is 16 less than 11 times the average number of tornados in December. If the difference between the average numbers of tornados in June and December is 204, determine the average numbers of tornados in December and June.

28. **Lion Weights** The Racine Zoo has two lions, Leo and Nula. Leo's weight is 50 pounds less than twice Nula's weight. If Leo weighs more than Nula, and if the difference in their weights is 250 pounds, determine Leo's weight and Nula's weight.

29. **Bookstore Salaries** At the Book Nook bookstore the hourly wage for its managers is \$1.75 less than twice the hourly wage for its clerks. The managers' hourly wage is higher and the difference in their hourly wages is \$6.75. Determine the hourly wage for the managers.

30. **Best-Selling Albums** According to *Billboard Magazine* the top two best-selling albums of all time are Michael Jackson's *Thriller* and AC/DC's *Back in Black*, respectively. The number of *Thriller* albums sold is 12 million more than twice the number of *Back in Black* albums sold. If the difference in the number of *Thriller* albums and the number of *Back in Black* albums is 61 million albums, find the number of *Thriller* albums sold.

31. **Truck and Motorcycle** Dale Felkins owns a truck and a motorcycle. The truck weighs 400 pounds less than 10 times the weight of the motorcycle. If together the truck and the motorcycle have a total weight of 6090 pounds, what is the weight of the truck?

32. **Oil Consumption** According to the U.S. Energy Information Administration, oil and liquid fuel consumption was projected to be 4.8% higher in 2015 than in 2008. If the sum of consumption for these two years is 209.6 quadrillion Btu, determine the projected consumption for 2015.

Exercises 33–46 involve money. Read Examples 5–6, then set up an equation that can be used to solve the problem. Solve the equation and answer the question asked.

33. **Gasoline** Luvia Rivera has only \$48 to purchase gasoline. If gasoline costs \$3.84 per gallon, determine how many gallons of gasoline Luvia can purchase.

34. **Propane Gas** Reginald Fullwood has a propane gas tank in his back yard. Reginald pays \$2.40 per gallon to fill up his tank. If his bill comes to \$69, how many gallons of propane did Reginald purchase?

35. **Copy Machine** Yamil Bernz purchased a copy machine for \$2100 and a one-year maintenance protection plan that costs 2 cents per copy made. If he spends a total of \$2462 in a year, which includes the cost of the machine and the copies made, determine the number of copies he made.

36. **Gym Membership** At Goldies Gym there is a one-time membership fee of \$300 plus dues of \$40 per month. If Carlos Manieri has spent a total of \$700 for Goldies Gym, how long has he been a member?

37. **Television** Miles Potier's Time Warner cable bill includes a base charge of \$72.68 per month plus \$3.95 for each On Demand movie he watches that month. If his cable bill for December was \$96.38, determine the number of On Demand movies he watched in December.

38. **Hardwood Floors** Ruth Zasada is having hardwood floors installed in her living room. The cost for the material is \$2840 plus an installation charge of \$1.90 per square foot. If the total cost for the material plus installation is \$5120, determine the area of her living room.

39. **Truck Rental** Howard Sporn is considering two companies from which to rent a truck. American Truck Rental charges \$20 per day and 25 cents a mile. SavMor Truck Rental charges \$35 a day and 15 cents a mile. How far would Howard need to drive in one day for the both companies to have the same total cost?

40. **Washing Machines** Scott Montgomery is considering two washing machines, a Kenmore and a Neptune. The Neptune costs \$454 while the Kenmore costs \$362. The energy guides indicate that the Kenmore will cost an estimated \$84 per year to operate and the Neptune will cost an estimated \$38 per year to operate. How long will it be before the total cost is the same for both washing machines?

41. **Salaries** Brooke Mills is being recruited by a number of high-tech companies. Data Technology Corporation has offered her an annual salary of \$40,000 per year plus a \$2400 increase

per year. Nuteck has offered her an annual salary of $49,600 per year plus an $800 increase per year. In how many years will the salaries from the companies be the same?

42. Racquet Club The Coastline Racquet Club has two payment plans for its members. Plan 1 has a monthly fee of $20 plus $8 per hour for court time. Plan 2 has no monthly fee, but court time is $18 per hour. If court time is rented in 1-hour intervals, how many hours would you have to play per month so that plan 1 becomes a better buy?

43. Printers Hector Hanna will purchase one of two laser printers, a Hewlett-Packard (HP) or a Lexmark. The HP costs $499 and the Lexmark costs $419. The cost of printing a page on the HP is $0.06 per page and the cost of printing a page on the Lexmark is $0.08 per page. How many pages would need to be printed for the two printers to have the same total cost?

44. Satellite or Cable Sean Stewart is deciding whether to select a satellite receiver or cable for his television programming. The satellite receiver costs $298.90 and the monthly charge is $68.70. With cable there is no initial cost to purchase equipment, but the monthly charge for comparable channels is $74.80. After how many months will the total cost of the two systems be equal?

45. Newsletter Neil Simpson had a professional organization newsletter printed and sent out to all the members. The total cost included a $600 printing cost plus a 42.4 cents mailing cost for each envelope. If the total cost was $1448, determine how many newsletters were mailed.

46. Patio Pavers Elizabeth Chu is remodeling her patio and is having patio pavers installed. A & E Pavers charges $1500 for the pavers plus $40 per hour for labor. If the total cost of the job is $2190, how many hours of labor did Elizabeth pay for?

Exercises 47–68 involve percents. Read Examples 7–9, then set up an equation that can be used to solve the problem. Solve the equation and answer the question asked.

47. Airfare The airfare for a flight from Boston to Milwaukee cost $242.56, which includes a 22% surcharge (additional fee) to cover all taxes and airline fees. What is the price of the flight before the 22% surcharge?

48. New Car Yoliette Fournier purchased a new car. The cost of the car, including a 7.5% sales tax, was $24,600. What was the cost of the car before tax?

49. Salary Increase Zhen Tong just received a job offer that will pay him 30% more than his present job does. If the salary at his new job will be $30,200, determine his present salary.

50. New Headquarters Tarrach and Associates plans on increasing the size of its headquarters by 25%. If its new headquarters is to be 14,200 square feet, determine the size of its present headquarters.

51. Auto Exports According to the Federal Reserve Bank of Chicago, in 2011 exports of new light vehicles increased 52% over 2009 in the United States. If in 2011 about 1.6 million light vehicles were exported, determine the number of vehicles exported in 2009.

52. Teachers During the 2012 contract negotiations, the city school board approved a 5% pay increase for its teachers effective in 2013. If Dana Frick, a first-grade teacher, projects his 2013 annual salary to be $46,400, what was his 2012 salary?

53. Autographs A tennis star was hired to sign autographs at a convention. She was paid $3000 plus 3% of all admission fees collected at the door. The total amount she received for the day was $3750. Find the total amount collected at the door.

54. Sales Volume Mona Fabricant receives a weekly salary of $350. She also receives a 6% commission on the total sales she makes. What must her sales be in a week, if she is to make a total of $710?

55. Wage Cut A manufacturing plant is running at a deficit. To avoid layoffs, the workers agree on a temporary wage cut of 2%. If the average salary in the plant after the wage cut is $38,600, what was the average salary before the wage cut?

56. Retirement Income Ray and Mary Burnham have decided to retire. They estimate their annual income after retirement will be reduced by 15% from their pre-retirement income. If they estimate their retirement income to be $42,000, determine their pre-retirement income.

57. Presidents' Day Sale During a Presidents' Day Sale, Susan Grody paid $350 for an espresso machine. The price paid reflects a 30% discount off of the original price. Determine the original price of the espresso machine.

58. Sale At a 1-day 20% off sale, Jane Demsky purchased a hat for $25.99. What is the regular price of the hat?

59. Salary Plans Vince McAdams, a salesman, is given a choice of two salary plans. Plan 1 is a weekly salary of $600 plus 2% commission of sales. Plan 2 is a straight commission of 10% of sales. How much in sales must Vince make in a week for both plans to result in the same salary?

60. Financial Planning Belen Poltorade, a financial planner, is offering her customers two financial plans for managing their assets. With plan 1 she charges a planning fee of \$1000 plus 1% of the assets she will manage for the customers. With plan 2 she charges a planning fee of \$500 plus 2% of the assets she will manage. How much in customer assets would result in both plans having the same total fees?

61. Book The number of pages in the third edition of a book was 4% less than the number of pages in the second edition. If the number of pages in the third edition is 480, determine the number of pages in the second edition.

62. Reduced Fat Milk A single serving of reduced fat milk contains 4.5 grams of fat. This amount of fat is 43.75% less than the amount of fat in a single serving of whole milk. How many grams of fat are there in a single serving of whole milk?

63. Salary Plans Becky Schwartz, a saleswoman, is offered two salary plans. Plan 1 is \$400 per week salary plus a 2% commission of sales. Plan 2 is a \$250 per week salary plus a 16% commission of sales. How much would Becky need to make in sales for the salary to be the same from both plans?

64. Art Show Bill Rush wants to rent a building for a week to show his artwork and has been offered two rental plans. Plan 1 is a rental fee of \$500 plus 3% of the dollar sales he makes. Plan 2 is \$100 plus 15% of the dollar sales he makes. What dollar sales would result in both plans having the same total cost?

65. Sale and Coupon Abdollah Hajikandi purchases a calculator during a back-to-school sale, in which Office Station has discounted all calculators by 15%. Abdollah also uses a coupon to save an additional \$10 off the discounted price. If Abdollah pays \$92 for a calculator, what was the original price of the calculator?

66. Membership Fees The Holiday Health Club has reduced its annual membership fee by 10%. In addition, if you sign up on a Monday, the Club will take an additional \$20 off the already reduced price. If Jorge Sanchez purchases a year's membership on a Monday and pays \$250, what is the regular membership fee?

67. Estate Phil Dodge left an estate valued at \$140,000. In his will, he specified that his wife will get 25% more of his estate than his daughter. How much will his wife receive?

68. Charitable Giving Charles Ford made a \$200,000 cash contribution to two charities, the American Red Cross and the United Way. The amount received by the American Red Cross was 30% greater than the amount received by the United Way. How much did the United Way receive?

69. Oil Production In 2011, the top two oil producing countries were Russia and Saudi Arabia, respectively. Russia produced 0.27 million barrels of oil per day (bpd) more than Saudia Arabia and the total produced by both countries was 20.81 million bpd. Determine the amount of oil produced by Russia and by Saudi Arabia in 2011. (*Source:* International Energy Agency)

70. More Oil Production See Exercise 69. In 2011, the third and fourth highest oil producing countries were the United States and Iran, respectively. The United States produced 5.44 million bpd more than Iran and the total produced by both countries was 13.94 million bpd. Determine the amount of oil produced by the United States and by Iran in 2011.

Concept/Writing Exercises

71. Outline the five-step problem-solving procedure we use.

72. Explain the concept of *percent increase* to a friend. If you made \$8 per hour last month and got a 15% increase, what is your new hourly wage?

Challenge Problems

73. Average Value To find the *average* of a set of values, you find the sum of the values and divide the sum by the number of values.

a) If Paul Lavenski's first three test grades are 74, 88, and 76, write an equation that can be used to find the grade that Paul must get on his fourth exam to have an 80 average.

b) Solve the equation from part **a)** and determine the grade Paul must receive.

74. Driver Education A driver education course costs \$45 but saves those under age twenty-five 10% of their annual insurance premiums until they reach age twenty-five. Vicki Day has just turned 18, and her insurance costs, before the discount, \$600 per year.

a) How long will it take for the amount saved from insurance to equal the price of the course?

b) Including the cost of the course, when Vicki turns 25, how much will she have saved?

Cumulative Review Exercises

[1.9] **75.** Evaluate $4[(4 - 6) \div 2] + 3^2 - 1$.

[1.10] **76.** Name the following property: $3x + 4 = 4 + 3x$.

[2.6] **77.** Solve the formula $A = \frac{1}{2}bh$ for h.

[2.7] **78.** Solve the proportion $\frac{4.5}{6} = \frac{9}{x}$.

MID-CHAPTER TEST: 3.1–3.2

To find out how well you understand the chapter material to this point, take this brief test. The answers and the section where the material was initially discussed are given in the back of the book. Review any questions you answered incorrectly.

In Exercises 1–6, express each statement as an algebraic expression.

1. Six times the weight, w

2. Five inches more than 3 times the height, h

3. Represent the cost, c, increased by 20%, as a mathematical expression.

4. Dennis Donahue rents a truck for \$60 per day plus 95 cents per mile, m. Write an expression for the total cost of the rental for one day.

5. Write an expression for the number of cents in n half-dollars.

6. Twenty-five dollars is divided between Amy Keyser and Sherry Norris. If Amy gets x dollars, how much will Sherry get?

7. Explain why the cost, c, of an item at a 25% off sale is not $c - 25$. Write the correct algebraic expression for the cost of an item at a 25% off sale.

8. In the statement, determine what to let x represent.
A Gaudy Leaf Frog is 2 centimeters longer than 3 times the length of a Poison Dart Frog (see photos).

Gaudy Leaf Frog, see Exercise 8.

Poison Dart Frog, see Exercise 8.

9. Select a variable to represent one quantity and state what that variable represents. Express the second quantity in terms of the variable selected.
The distance Mary traveled is 6 miles more than 4 times the distance Pedro traveled.

10. Due to depreciation, the value of a car in 2013 was 18% less than its value in 2012. Write an expression for the difference in the values of the car from 2012 to 2013.

In Exercises 11 and 12, write the problem as an equation. Do not solve.

11. The population of Cedar Oaks increased by 12%. The population after the increase was 38,619.

12. For two consecutive odd integers, the sum of the smaller and 3 times the larger is 26.

In Exercises 13–20, write an equation that can be used to solve the problem. Solve the equation and answer the question asked.

13. Consecutive Integers The sum of two consecutive integers is 93. Find the numbers.

14. Numbers The larger of two integers is one less than 3 times the smaller. When the smaller number is subtracted from the larger, the difference is 7. Find the numbers.

15. Candy A candy manufacturer presently produces 240 boxes of candy a day and wants to increase production by 20 boxes per day until it produces 600 boxes of candy per day. How many days will it take for production to reach 600 boxes per day?

16. Tennis Kristina Schmid is considering joining one of two tennis clubs. At Dale's Tennis Club the monthly fee is \$90, and court time is \$4 per hour. At Abel's Tennis Club the monthly fee is \$30, but court time is \$8 per hour. How many hours in a month would Kristina need to play for the total cost to be the same with both clubs?

17. Television The cost of a television plus a 7% sales tax is \$749. Find the cost of the television before tax.

18. Clients Anita and Betty together have a total of 600 clients. If Anita has 12 more than twice the number of clients Betty has, determine the number of clients each person has.

19. Truck Rental A truck cost \$36 a day plus 18 cents a mile to rent. If the total cost for a one-day rental is \$45.36, how many miles were driven?

20. Salary Plans A salesman is offered two salary plans. Plan 1 is \$200 per week plus 8% commission of the dollar sales he makes. Plan 2 is \$300 per week plus 6% of the dollar sales he makes. How much in sales must the salesman make in a week for the two plans to have the same total salary?

3.3 Geometric Problems

1 Solve geometric problems.

1 Solve Geometric Problems

This section serves two purposes. One is to reinforce the geometric formulas introduced in Section 2.6. The second is to reinforce procedures for setting up and solving verbal problems discussed in Sections 3.1 and 3.2.

Understanding Algebra

Recall some useful geometry formulas:

Rectangle	Area: $A = l \cdot w$
	Perimeter: $P = 2l + 2w$
Triangle	Area: $A = \frac{1}{2} b \cdot h$
	Sum of the measures of the interior angles $= 180°$
Quadrilateral	Sum of the measures of the interior angles $= 360°$
Circle	Area $= \pi r^2$
	Circumference $= 2\pi r$

EXAMPLE 1 Sandbox Christine O'Connor is planning to build a sandbox for her daughter. She has 30 feet of lumber with which to build the perimeter. What should be the dimensions of the rectangular sandbox if the length is to be 3 feet longer than the width (**Fig. 3.3**)?

Solution Understand We are asked to find the dimensions of the sandbox that Christine plans to build. The perimeter of the sandbox will be 30 feet. Since the length is given in terms of the width, we will let w represent the width. Then we can express the length in terms of w. To solve this problem, we use the formula for the perimeter of a rectangle, $P = 2l + 2w$, where $P = 30$ feet.

Translate

Let w = width of the sandbox.

Then $w + 3$ = length of the sandbox

$$P = 2l + 2w$$

Carry Out

$$30 = 2(w + 3) + 2w$$
$$30 = 2w + 6 + 2w$$
$$30 = 4w + 6$$
$$24 = 4w$$
$$6 = w$$

FIGURE 3.3

Understanding Algebra

Additional geometry facts:

Equilateral triangle: has 3 equal sides and 3 equal angles (all 60°)

Isosceles triangles: has 2 equal sides

Congruent triangles: corresponding sides and corresponding angles are equal

Similar triangles: corresponding sides are in proportion and corresponding angles are equal

Complementary angles: angles whose measures add to 90°

Supplementary angles: angles whose measures add to 180°

The width is 6 feet. Since the length is 3 feet longer than the width, the length is $6 + 3 = 9$ feet.

Check We will check the solution by substituting the appropriate values in the perimeter formula.

$$P = 2l + 2w$$
$$30 \stackrel{?}{=} 2(9) + 2(6)$$
$$30 = 30 \quad \text{True}$$

Answer The width of the sandbox will be 6 feet and the length will be 9 feet.

Now Try Exercise 23

A triangle that contains two sides of equal length is called an **isosceles triangle.** In isosceles triangles, the angles opposite the two sides of equal length have equal measures.

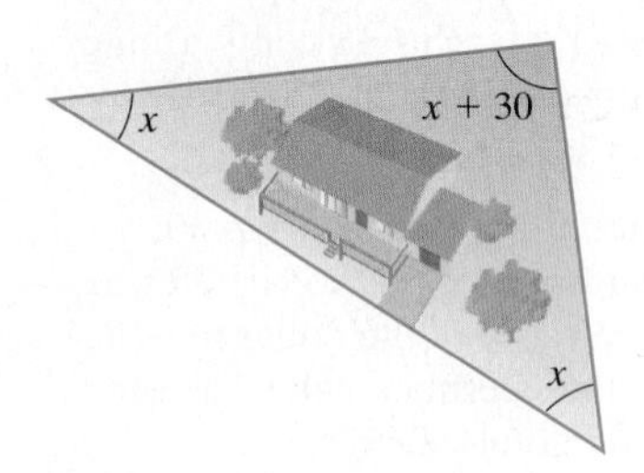

FIGURE 3.4

EXAMPLE 2 **Corner Lot** Mr. and Mrs. Harmon Katz have a corner lot that is in the shape of an isosceles triangle. Two angles of their triangular lot have the same measure and the measure of the third angle is 30° greater than the measure of the other two. Find the measure of all three angles (see **Fig. 3.4**).

Solution Understand To solve this problem, you must know that the sum of the angles of any triangle measures 180°. We are asked to find the measure of each of the three angles, where the two smaller angles have the same measure. We will let x represent the measure of the smaller angles, and then we will express the measure of the larger angle in terms of x.

Translate

Let x = the measure of each smaller angle.

Then $x + 30$ = the measure of the larger angle.

sum of the 3 angles = 180

Carry Out

$$x + x + (x + 30) = 180$$
$$3x + 30 = 180$$
$$3x = 150$$
$$x = \frac{150}{3} = 50$$

The two smaller angles each measure 50°. The measure of the larger angle is $x + 30°$ or $50° + 30° = 80°$.

Check and Answer Since $50° + 50° + 80° = 180°$, the answer checks. The two smaller angles each measure 50° and the larger angle measures 80°.

Now Try Exercise 9

Recall from Section 2.6 that a quadrilateral is a four-sided figure. Quadrilaterals include squares, rectangles, parallelograms, and trapezoids. The sum of the measures of the angles of any quadrilateral is 360°.

EXAMPLE 3 **Water Trough** Sarah Fuqua owns horses and uses a water trough whose ends are trapezoids. The measure of the two bottom angles of the trapezoid are the same, and the measure of the two top angles are the same. The bottom angles measure 15° less than twice the measure of the top angles. Find the measure of each angle.

FIGURE 3.5

Solution Understand To help visualize the problem, we draw a picture of the trapezoid, as in **Figure 3.5**. We use the fact that the sum of the measures of the four angles of a quadrilateral is 360°.

Translate Let x = the measure of each of the two smaller angles.
Then $2x - 15$ = the measure of each of the two larger angles.

$$\begin{pmatrix}\text{measure of the}\\ \text{two smaller angles}\end{pmatrix} + \begin{pmatrix}\text{measure of the}\\ \text{two larger angles}\end{pmatrix} = 360$$

$$x + x + (2x - 15) + (2x - 15) = 360$$

Carry Out

$$\begin{aligned} x + x + 2x - 15 + 2x - 15 &= 360 \\ 6x - 30 &= 360 \\ 6x &= 390 \\ x &= 65 \end{aligned}$$

Each smaller angle is 65°. Each larger angle is $2x - 15 = 2(65) - 15 = 115°$.

Check and Answer Since $65° + 65° + 115° + 115° = 360°$, the answer checks. Each smaller angle is 65° and each larger angle is 115°.

Now Try Exercise 27

EXAMPLE 4 **Fenced-In Area** Ronald Yates recently started an ostrich farm. He is separating the ostriches by fencing in three equal rectangular areas, as shown in **Figure 3.6**. The length of the fenced-in area, l, is to be 30 feet greater than the width and the total amount of fencing available is 660 feet. Find the length and width of the fenced-in area.

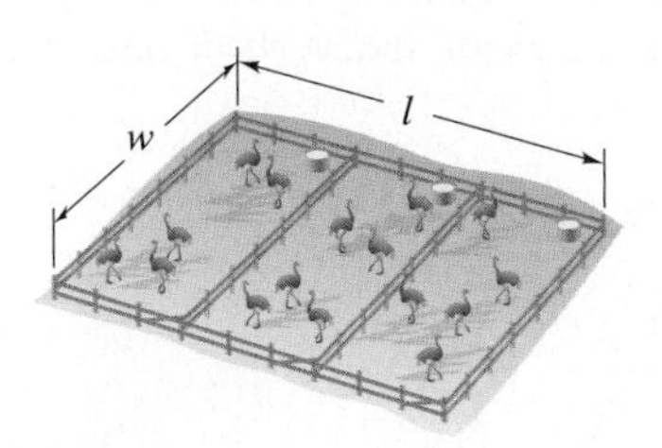

FIGURE 3.6

Solution Understand The fencing consists of four pieces of fence of length w, and two pieces of fence of length l.

Translate Let w = width of fenced-in area.
Then $w + 30$ = length of fenced-in area.

$$\begin{pmatrix}\text{4 pieces of fence}\\ \text{of length } w\end{pmatrix} + \begin{pmatrix}\text{2 pieces of fence}\\ \text{of length } w + 30\end{pmatrix} = 660$$

Carry Out

$$\begin{aligned} 4w + 2(w + 30) &= 660 \\ 4w + 2w + 60 &= 660 \\ 6w + 60 &= 660 \\ 6w &= 600 \\ w &= 100 \end{aligned}$$

Since the width is 100 feet, the length is $w + 30$ or $100 + 30$ or 130 feet.

Check and Answer Since $4(100) + 2(130) = 660$, the answer checks. The width of the fenced-in area is 100 feet and the length is 130 feet.

Now Try Exercise 37

3.3 Exercise Set MathXL® MyMathLab®

Warm-Up Exercises

Fill in the blanks with the appropriate word, phrase, or symbol(s) from the following list.

supplementary	complementary	270°	360°	180°
length	width	scalene	isosceles	equilateral

1. The formula for the area of a rectangle is Area = length · ______________.
2. The formula for the perimeter of a rectangle is Perimeter = 2 · ______________ + 2 · width.
3. If two angles have measures that add up to 90°, the angles are ______________ angles.
4. If two angles have measures that add up to 180°, the angle are ______________ angles.
5. The sum of the measures of the interior angles of a triangle is ______________.
6. The sum of the measures of the interior angles of a quadrilateral is ______________.
7. A triangle with two equal sides is called ______________.
8. A triangle with three equal sides is called ______________.

Practice the Skills/Problem Solving

*Solve the following geometric problems.**

9. **Isosceles Triangle** In an isosceles triangle, the measure of one angle is 42° greater than the measure of the other two equal angles. Find the measure of all three angles. See Example 2.

10. **Isosceles Triangle** In an isosceles triangle, the measure of one angle is 40° less than twice the measures of the other two equal angles. Find the measure of all three angles.

11. **Triangular Building** This building in New York City, referred to as the Flatiron Building, has a perimeter in the shape of an isosceles triangle. If the shortest side of the triangle is 50 feet shorter than the two longer sides, and the perimeter is 196 feet, determine the length of the three sides of the triangle.

12. **Transamerica Pyramid** One face of the Transamerica Pyramid in San Francisco has the shape of an isosceles triangle. The shortest side of the triangle is 682 feet shorter than the two longer sides and the perimeter of the triangle is 1889 feet. Determine the lengths of the three sides of the triangle.

13. **Equilateral Triangle** An **equilateral triangle** is a triangle that has three sides of the same length. The perimeter of an equilateral triangle is 34.5 inches. Find the length of each side.

14. **Equilateral Triangle** The perimeter of an equilateral triangle is 48.6 centimeters. Find the length of each side. See Exercise 13.

15. **Complementary Angles** Two angles are **complementary angles** if the sum of their measures is 90°. Angle A and angle B are complementary angles, and the measure of angle A is 21° more than twice the measure of angle B. Find the measures of angle A and angle B.

Complementary Angles

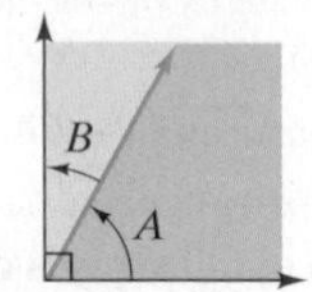

16. **Complementary Angles** Angles A and B are complementary angles, and the measure of angle B is 14° less than the measure of angle A. Find the measures of angle A and angle B. See Exercise 15.

17. **Supplementary Angles** Two angles are **supplementary angles** if the sum of their measures is 180°. Angle A and angle B are supplementary angles, and the measure of angle B is 8° less than three times the measure of angle A. Find the measures of angle A and angle B.

Supplementary Angles

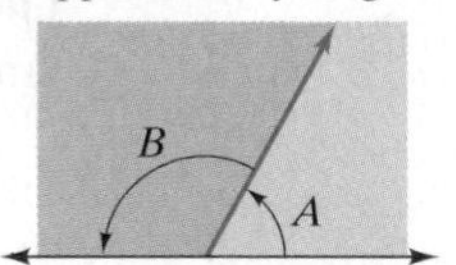

18. **Supplementary Angles** Angles A and B are supplementary angles and the measure of angle A is 2° more than 4 times the measure of angle B. Find the measures of angle A and angle B. See Exercise 17.

19. **Vertical Angles** When two lines intersect, the opposite angles are called **vertical angles.** Vertical angles have equal measures. Determine the measures of the vertical angles indicated in the following figure.

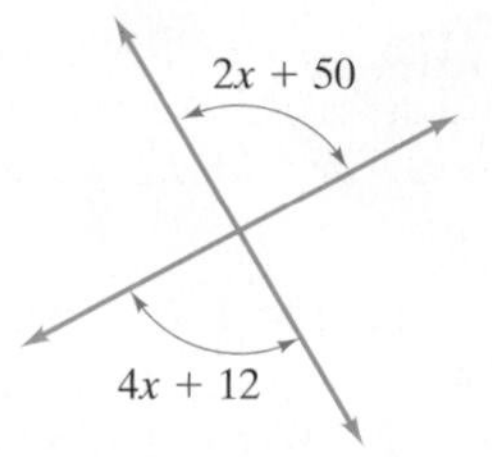

20. **Vertical Angles** A pair of vertical angles is indicated in the following figure. Determine the measure of the vertical angles indicated. See Exercise 19.

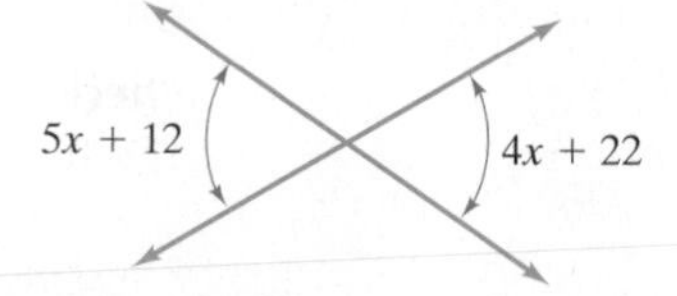

21. **Unknown Angles** The measure of one angle of a triangle is 10° greater than the measure of the smallest angle, and the measure of the third angle is 30° less than twice the measure of the smallest angle. Find the measures of the three angles.

22. **Unknown Angles** The measure of one angle of a triangle is 20° larger than the measure of the smallest angle, and the measure of the third angle is 6 times as large as the measure of the smallest angle. Find the measures of the three angles.

23. **Dimensions of Rectangle** The length of a rectangle is 6 feet more than its width. What are the dimensions of the rectangle if the perimeter is 44 feet?

24. **Dimensions of Rectangle** The perimeter of a rectangle is 120 feet. Find the length and width of the rectangle if the length is twice the width.

*See Appendix B for more information on geometry.

25. Tennis Court The length of a regulation tennis court is 6 feet greater than twice its width. The perimeter of the court is 228 feet. Find the length and width of the court.

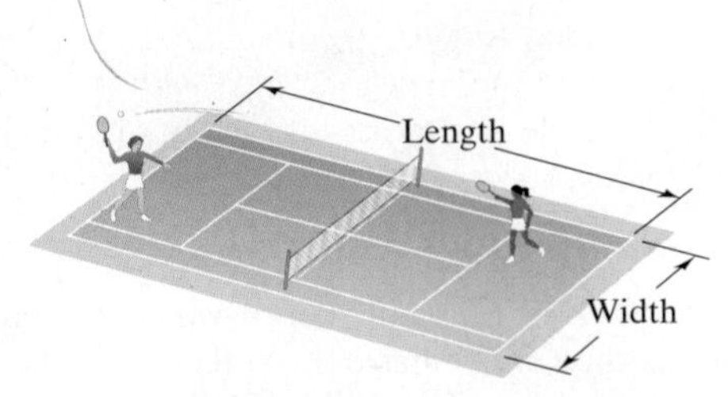

26. Patio Rikki Blair is building a rectangular patio. The perimeter of the patio is to be 96 feet. Determine the dimensions of the patio if the length is to be 6 feet less than twice the width.

27. Parallelogram In a parallelogram the opposite angles have the same measures. Each of the two larger angles in a parallelogram is 20° less than 3 times the smaller angles. Find the measure of each angle.

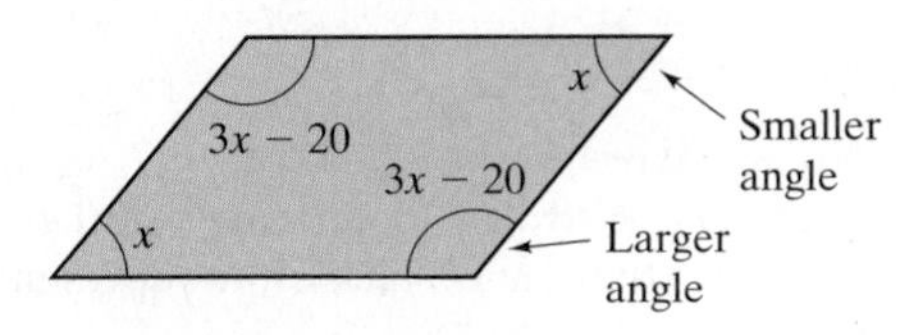

28. Parallelogram The two smaller angles of a parallelogram have equal measures, and the two larger angles each measure 27° less than twice each smaller angle. Find the measure of each angle.

29. Rhombus A rhombus is a parallelogram with four equal sides. Each of the two larger angles of a rhombus is 5 times as large as the two smaller angles. Find the measure of each of the four angles.

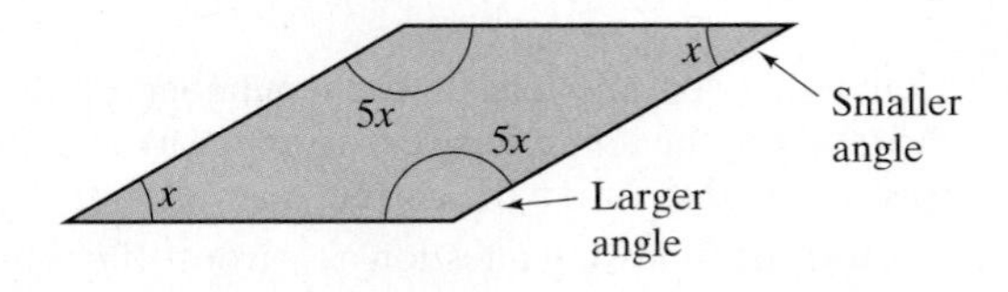

30. Rhombus Each of the two larger angles of a rhombus are 20° less than four times the two smaller angles. Find the measure of each of the four angles.

31. Quadrilateral The measure of one angle of a quadrilateral is 10° greater than the smallest angle; the third angle is 14° greater than twice the smallest angle; and the fourth angle is 21° greater than the smallest angle. Find the measures of the four angles of the quadrilateral.

32. Quadrilateral The measure of one angle of a quadrilateral is twice the smallest angle; the third angle is 20° greater than the smallest angle; and the fourth angle is 20° less than twice the smallest angle. Find the measures of the four angles of the quadrilateral.

33. Building a Bookcase A bookcase is to have four shelves, including the top, as shown. The height of the bookcase is to be 3 feet more than the width. Find the width and height of the bookcase if only 30 feet of lumber is available.

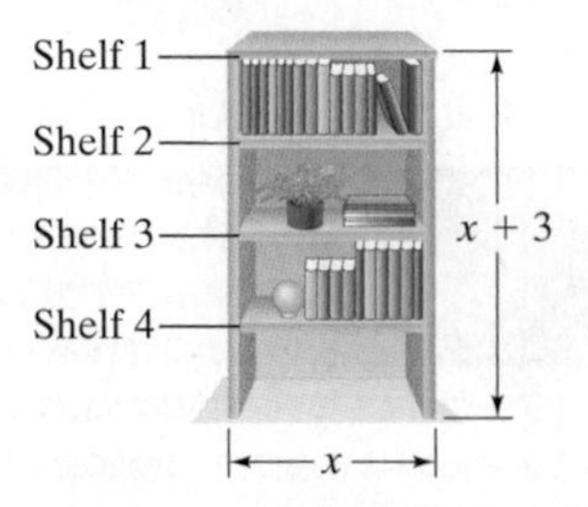

34. Bookcase A bookcase is to have four shelves as shown. The height of the bookcase is to be 2 feet more than the width, and only 22 feet of lumber is available. What should be the width and height of the bookcase?

35. Canadian Flag The dimensions of the Canadian flag appear in the figure below. The perimeter of this particular flag is 216 inches.

a) Determine its length and width.

b) The length (left-to-right) of each red rectangle is $\frac{1}{4}$ of the length of the entire flag. What is the length of each red rectangle?

c) The white area upon which the maple leaf rests is a square. What are the length and width of this square?

d) What is the area of this flag?

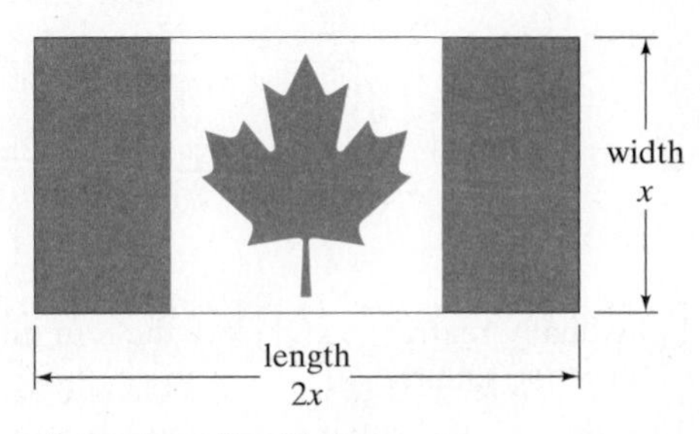

Source: Canadian Heritage

36. American Flag The dimensions of the American flag appear in the figure on the next page. The perimeter of this particular flag is 580 inches.

a) Determine its length and width.

b) What is the width of each of the stripes?

c) The length (left-to-right) of the blue rectangle is always 76% of the width of the flag. Determine the length of the blue rectangle.

d) The width (top-to-bottom) of the blue rectangle is always 53.85% of the width of the flag. Determine the width of the blue rectangle.

e) What is the area of this flag?

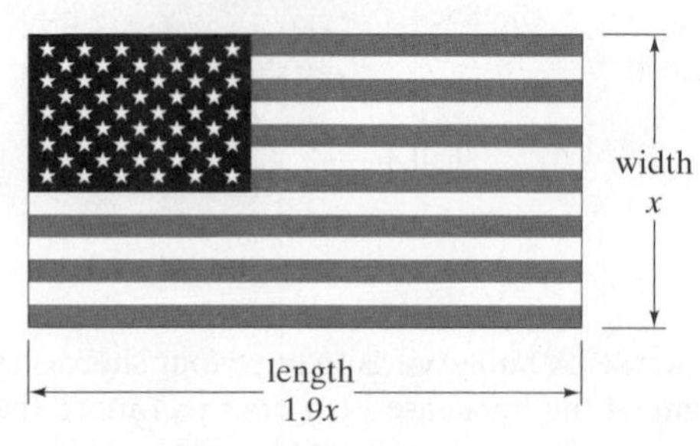

Source: www.usflag.org

37. **Fenced-In Area** A rectangular area is to be fenced in along a straight river bank as illustrated. The length of the fenced-in area is to be 5 feet greater than the width, and the total amount of fencing to be used is 71 feet. Find the width and length of the fenced-in area.

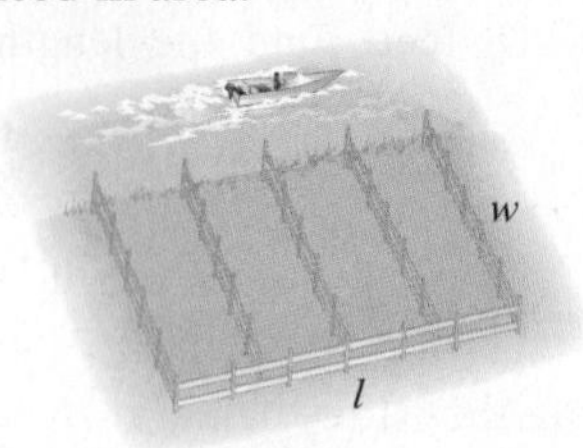

38. **Storage Shelves** Carlotta Perez plans to build storage shelves as shown. She has only 45 feet of lumber for the entire unit and wishes the width to be 3 times the height. Find the width and height of the unit.

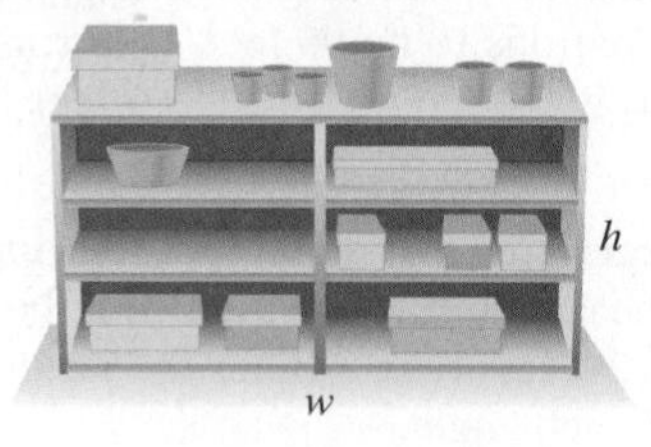

Challenge Problems

39. **Gardening** Trina Zimmerman is placing a border around and within a garden where she intends to plant flowers (see the figure). She has 60 feet of bordering, and the length of the garden is to be 2 feet greater than the width. Find the length and width of the garden. The red shows the location of all the bordering in the figure.

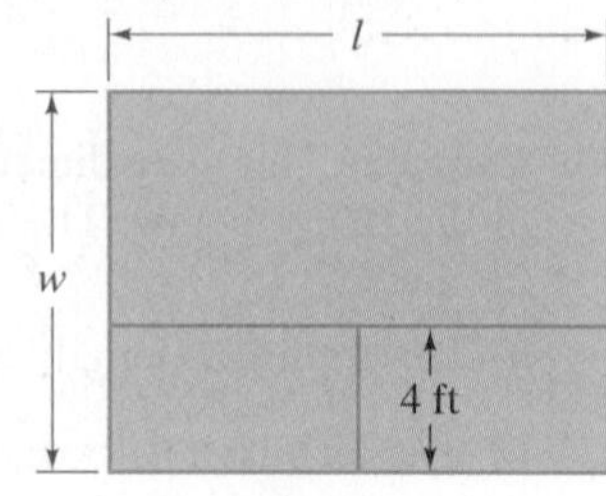

40. One way to express the area of the figure below is $(a + b)(c + d)$. Can you determine another expression, using the area of the four rectangles, to represent the area of the figure?

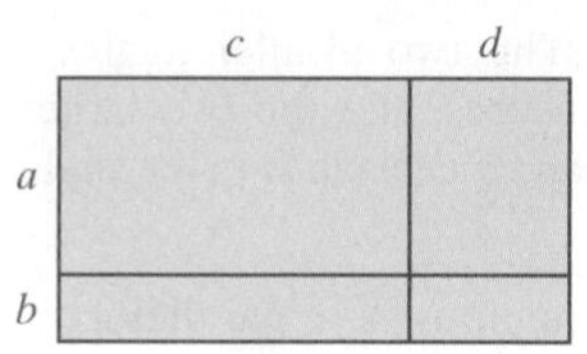

Group Activity

Discuss and answer Exercise 41 as a group.

41. Consider the four pieces shown. Two are squares and two are rectangles.

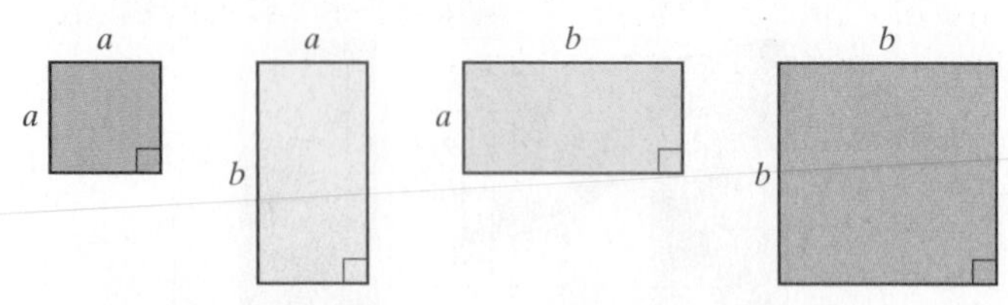

a) Individually, rearrange and place the four pieces together to form one square.

b) The area of the square you constructed is $(a + b)^2$. Write another expression for the area of the square by adding the four individual areas.

c) Compare your answers. If each member of the group did not get the same answers to parts **a)** and **b)**, work together to determine the correct answer.

d) Answer the following question as a group. If b is twice the length of a, and the perimeter of the square you created is 54 inches, find the length of a and b.

e) Use the values of a and b found in part **d)** to find the area of the square you created.

f) Use the values of a and b found in part **d)** to find the areas of the four individual pieces that make up the large square.

g) Does the sum of the areas of the four pieces found in part **f)** equal the area of the large square found in part **e)**? Is this what you expected? Explain.

Cumulative Review Exercises

Insert either $>$, $<$, or $=$ in each shaded area to make the statement true.

[1.5] 42. $-|-6| \quad \square \quad |-4|$

43. $|-3| \quad \square \quad -|3|$

[1.7] 44. Evaluate $-8 - (-2) + (-4)$.

[2.1] 45. Simplify $-7y + x - 3(x - 2) + 2y$.

[2.6] 46. Solve $6x + 3y = 9$ for y.

3.4 Motion, Money, and Mixture Problems

1. Solve motion problems involving two rates.
2. Solve money problems.
3. Solve mixture problems.

We now discuss three additional types of applications: motion, money, and mixture problems. These problems are grouped in the same section because, as you will learn shortly, you use the same general multiplication procedure to solve them. We begin by discussing motion problems.

Understanding Algebra

Motion problems involve *rates*. Units of rates include miles per hour (mph), feet per second (ft/s) and meters per second (m/s).

The formula we use in motion problems is:

distance = rate × time

or $d = r \cdot t$.

1 Solve Motion Problems Involving Two Rates

A **motion problem** is one in which an object is moving at a specific rate for a specific period of time. Examples of motion problems include a car traveling at a constant speed or a person walking at a constant speed or a boat being rowed at a constant speed. In this section, we will discuss motion problems that involve *two rates,* such as two trains traveling at different speeds. We will use the distance formula, distance = rate × time, and construct tables like the following one to organize the information. The formula at the top of the table shows how the distance in the last column is calculated.

	Rate	× Time	= Distance
Item	Rate	Time	Distance
Item 1			distance 1
Item 2			distance 2

Examples 1 and 2 illustrate the procedure used.

EXAMPLE 1 **Camping Trip** Maryanne and Paul Justinger and their son Danny are on a canoe trip on the Erie Canal. Danny is in one canoe and Paul and Maryanne are in a second canoe. Both canoes start at the same time from the same point and travel in the same direction. The parents paddle their canoe at 2 miles per hour and their son paddles his canoe at 4 miles per hour. In how many hours will the two canoes be 5 miles apart?

Solution Understand and Translate We are asked to find the time it takes for the canoes to become separated by 5 miles. We construct a table to aid us in setting up the problem.

Let t = time when canoes are 5 miles apart.

Draw a sketch to help visualize the problem (**Fig. 3.7**). When the two canoes are 5 miles apart, each has traveled for the same number of hours, t.

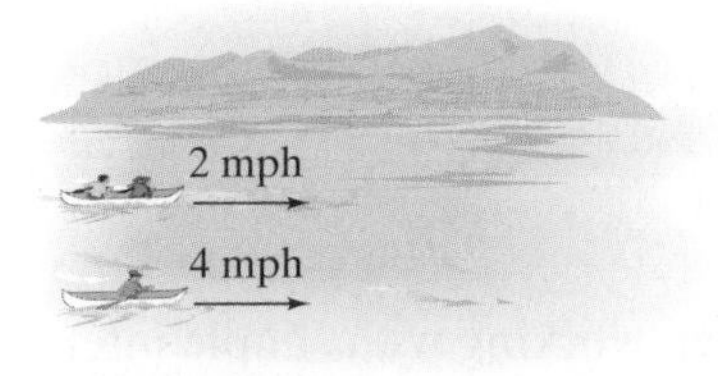

(a) Beginning of trip

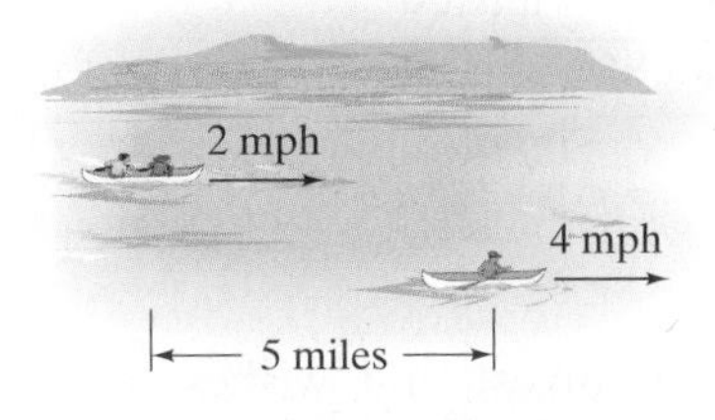

(b) After t hours

FIGURE 3.7

Canoe	Rate	Time	Distance
Parents	2	t	$2t$
Son	4	t	$4t$

Since the canoes are traveling in the same direction, the distance between them is found by subtracting the distance traveled by the slower canoe from the distance traveled by the faster canoe.

$$\begin{pmatrix}\text{distance traveled}\\ \text{by faster canoe}\end{pmatrix} - \begin{pmatrix}\text{distance traveled}\\ \text{by slower canoe}\end{pmatrix} = 5 \text{ miles}$$

$$4t - 2t = 5$$

Carry Out

$$2t = 5$$

$$t = 2.5$$

Answer After 2.5 hours the two canoes will be 5 miles apart.

Now Try Exercise 7

EXAMPLE 2 **Paving Roads** Two highway paving crews are 20 miles apart working toward each other. One crew paves 0.4 mile of road per day more than the other crew, and the two crews meet after 10 days. Find the rate at which each crew paves the road.

Solution Understand and Translate We are asked to find the two rates. We are told that both crews work for 10 days.

Let r = rate of slower crew.

Then $r + 0.4$ = rate of faster crew.

We make a sketch (**Fig. 3.8**) and set up a table of values.

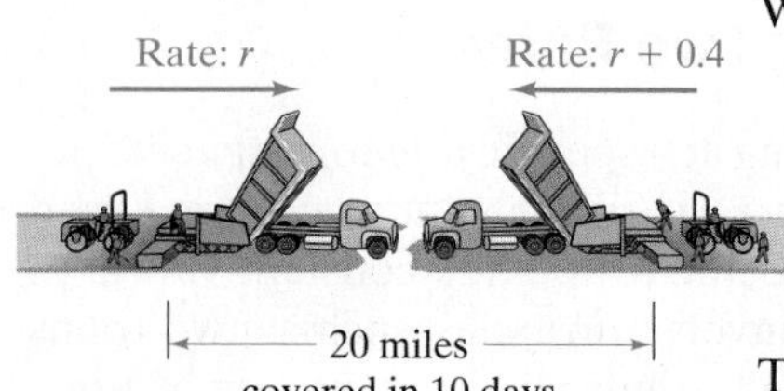

FIGURE 3.8

Crew	Rate	Time	Distance
Slower	r	10	$10r$
Faster	$r + 0.4$	10	$10(r + 0.4)$

The total distance covered by both crews is 20 miles. Since the crews are moving in opposite directions, the total miles of road paved is found by adding the two distances.

$$\begin{pmatrix}\text{distance covered}\\ \text{by slower crew}\end{pmatrix} + \begin{pmatrix}\text{distance covered}\\ \text{by faster crew}\end{pmatrix} = 20 \text{ miles}$$

$$10r + 10(r + 0.4) = 20$$

Carry Out

$$10r + 10r + 4 = 20$$
$$20r + 4 = 20$$
$$20r = 16$$
$$\frac{20r}{20} = \frac{16}{20}$$
$$r = 0.8$$

Answer The slower crew paves 0.8 mile of road per day and the faster crew paves $r + 0.4$ or $0.8 + 0.4 = 1.2$ miles of road per day.

Now Try Exercise 23

Understanding Algebra

Notice that in Example 2, the rate, r, is measured in *miles* (of road) *per day*.

HELPFUL HINT

When working with two different moving items, if the items are moving in the same direction, the solution will involve subtracting the smaller distance from the larger distance as in Example 1. If the items are moving in opposite directions, the solution will involve adding the distances together as in Example 2.

Understanding Algebra

The formula we use in simple interest problems is:

interest = principal × rate × time

2 Solve Money Problems

One type of money problem involves simple interest. When working with simple interest problems involving two amounts, we can use a table like the one below.

Understanding Algebra

In money problems, "*rate*" applies to rate of interest and is expressed as a decimal.

	Principal ×	Rate ×	Time =	Interest
Account	Principal	Rate	Time	Interest
Account 1				interest 1
Account 2				interest 2

EXAMPLE 3 **Investments** Olga Harrison has \$13,000 to invest and wishes to invest in two investments. The first investment is a loan to a small business that pays her 5% simple interest for one year. The second investment is a 1-year certificate of deposit (CD) that pays 2% simple interest. Olga wishes to earn a total of \$500 in interest in one year from the two investments. How much money should she put into each investment?

Solution Understand and Translate

$$\text{Let } x = \text{amount invested at } 5\%.$$
$$\text{Then } 13{,}000 - x = \text{amount invested at } 2\%.$$

We use the simple interest formula, interest = principal · rate · time, to solve this problem.

Investment	Principal	Rate	Time	Interest
Loan	x	0.05	1	$0.05x$
CD	$13{,}000 - x$	0.02	1	$0.02(13{,}000 - x)$

Since the sum of the interest from the two investments is \$500, we write the equation

$$\begin{pmatrix}\text{interest from}\\ 5\% \text{ loan}\end{pmatrix} + \begin{pmatrix}\text{interest from}\\ 2\% \text{ CD}\end{pmatrix} = \text{total interest}$$
$$0.05x + 0.02(13{,}000 - x) = 500$$

Carry Out

$$0.05x + 260 - 0.02x = 500$$
$$0.03x + 260 = 500$$
$$0.03x = 240$$
$$x = \frac{240}{0.03} = 8000$$

Check and Answer Thus, \$8000 should be invested in the loan at 5%. The amount to be invested in the CD at 2% is

$$13{,}000 - x = 13{,}000 - 8000 = 5000.$$

Therefore, \$5000 should be invested in the CD. The total amount invested is \$8000 + \$5000 = \$13,000, which checks with the information given.

Now Try Exercise 33

In Example 3, we let x represent the amount invested at 5%. If we had let x represent the amount invested at 2%, the answer would not have changed. Rework Example 3 now, letting x represent the amount invested at 2%.

In other types of problems involving two amounts of money, we generally set up similar tables, as illustrated in the next example.

EXAMPLE 4 **Rocking Chairs** Johnson's Patio Furniture Store sells two types of rocking chairs. The single-person rocking chair sells for \$130 each and the two-person rocking chair sells for \$240 each. On a given day 10 rocking chairs were sold for a total of \$1740. Determine the number of single-person and the number of two-person rocking chairs that were sold.

Solution Understand and Translate We are asked to find the number of each type of rocking chair sold.

$$\text{Let } x = \text{number of single–person rocking chairs sold.}$$
$$\text{Then } 10 - x = \text{number of two–person rocking chairs sold.}$$

The income received from the sale of the single-person rocking chairs is found by multiplying the number of single-person rocking chairs sold by the cost of a single-person rocking chair. The income received from the sale of the two-person rocking chairs is found by multiplying the number of two-person rocking chairs sold by the cost of a two-person rocking chair.

$$\begin{pmatrix}\text{Number of}\\ \text{rocking chairs}\end{pmatrix} \times \begin{pmatrix}\text{Cost of}\\ \text{rocking chairs}\end{pmatrix} = \begin{pmatrix}\text{Income from}\\ \text{rocking chairs}\end{pmatrix}$$

Rocking Chair	Number of Rocking Chairs	Cost	Income from Rocking Chairs
Single	x	130	$130x$
Double	$10 - x$	240	$240(10 - x)$

$$\begin{pmatrix}\text{income from}\\ \text{single-person}\\ \text{rocking chairs}\end{pmatrix} + \begin{pmatrix}\text{income from}\\ \text{two-person}\\ \text{rocking chairs}\end{pmatrix} = \text{total income}$$

$$130x + 240(10 - x) = 1740$$

Carry Out

$$\begin{aligned} 130x + 2400 - 240x &= 1740 \\ -110x + 2400 &= 1740 \\ -110x &= -660 \\ x &= \frac{-660}{-110} = 6 \end{aligned}$$

Check and Answer Six single-person rocking chairs and 10 − 6 or 4 two-person rocking chairs were sold.

Check

$$\begin{aligned} \text{income from 6 single–person rocking chairs} &= 130 \cdot 6 = 780 \\ \text{income from 4 two–person rocking chairs} &= 240 \cdot 4 = \underline{960} \\ \text{total} &= 1740 \quad \text{True} \end{aligned}$$

Now Try Exercise 43

3 Solve Mixture Problems

Any problem in which two or more quantities are combined to produce a single quantity or a single quantity is separated into two or more quantities may be considered a **mixture problem.**

Mixture problems in this section will generally be one of two types. In one type, we will mix two solids, as illustrated in **Figure 3.9a,** and be concerned about the value or cost of the mixture. In the second type, we will mix two liquids or solutions, as illustrated in **Figure 3.9b,** and be concerned about the content or strength of the mixture.

Type 1

Mixing solids together

Type 2

Mixing liquids, or solutions, together

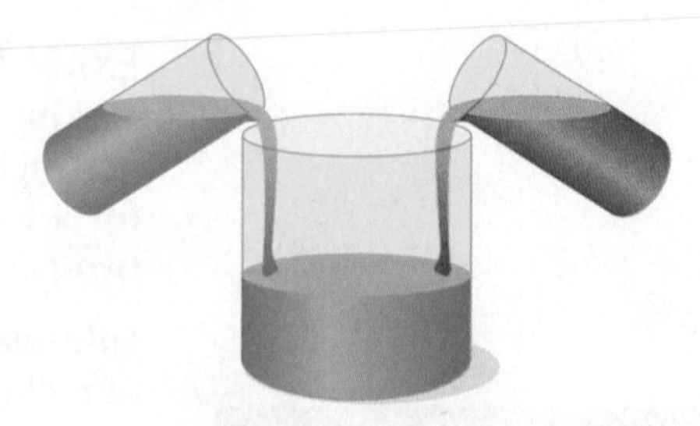

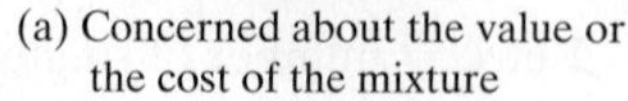

(a) Concerned about the value or the cost of the mixture

(b) Concerned about the content or the strength of the mixture

FIGURE 3.9

As we did with motion problems involving two rates and with money problems, we will use a table to help analyze mixture problems.

When we construct a table for mixture problems, our table will generally have three rows. One row will be for each of the two individual items being mixed, and the third row will be for the mixture of the two items.

Understanding Algebra

If we know the total weight of two items is 10 pounds and one item weighs x pounds, then the second item weighs $(10 - x)$ pounds.

Observe:

$$\begin{aligned} \text{item 1} + \text{item 2} &= \text{total} \\ x + (10 - x) &= 10 \end{aligned}$$

Type 1—Mixing Solids

When working with mixture problems involving solids, we generally use the fact that the value (or cost) of one part of the mixture plus the value (or cost) of the second part of the mixture is equal to the total value (or total cost) of the mixture.

When we are combining two solid items and are interested in the *value* of the mixture, the following table, or a variation of it, is often used.

Quantity × Price (per unit) = Value of item

Item	Quantity	Price	Value of Item
Item 1			value of item 1
Item 2			value of item 2
Mixture			value of mixture

When we use this table, we generally use the following formula to solve the problem.

value of item 1 + value of item 2 = value of mixture

Now let's look at a mixture problem where we discuss the value or cost of the mixture.

EXAMPLE 5 **Grass Seed** Troy's Family grass seed sells for \$2.75 per pound, and Troy's Spot Filler grass seed sells for \$2.35 per pound. How many pounds of each should be mixed to get a 10-pound mixture that sells for \$2.50 per pound?

Solution Understand and Translate We are asked to find the number of pounds of each type of grass seed.

$$\begin{aligned} \text{Let } x &= \text{number of pounds of Family grass seed.} \\ \text{Then } 10 - x &= \text{number of pounds of Spot Filler grass seed.} \end{aligned}$$

We make a sketch of the situation (**Fig. 3.10**), then construct a table.

FIGURE 3.10

The cost or value of the seeds is found by multiplying the number of pounds by the price per pound.

Type of Seed	Number of Pounds	Cost per Pound	Cost of Seed
Family	x	2.75	$2.75x$
Spot Filler	$10 - x$	2.35	$2.35(10 - x)$
Mixture	10	2.50	2.50(10)

$$\left(\begin{array}{c}\text{cost of}\\ \text{Family Seed}\end{array}\right) + \left(\begin{array}{c}\text{cost of Spot}\\ \text{Filler Seed}\end{array}\right) = \text{cost of mixture}$$

$$2.75x + 2.35(10 - x) = 2.50(10)$$

Carry Out

$$\begin{aligned} 2.75x + 23.5 - 2.35x &= 25.0 \\ 0.40x + 23.5 &= 25.0 \\ 0.40x &= 1.5 \\ x &= \frac{1.5}{0.40} = 3.75 \end{aligned}$$

Answer Thus, 3.75 pounds of the Family grass seed must be mixed with $10 - x$ or $10 - 3.75 = 6.25$ pounds of Spot Filler grass seed to make a mixture that sells for \$2.50 per pound.

Now Try Exercise 47

Type 2—Mixing Solutions

We generally solve mixture problems involving solutions by using the fact that the amount of one part of the mixture plus the amount of the second part of the mixture is equal to the total amount of the mixture.

When working with solutions, we use the following formula: *amount of substance in the solution* $=$ *quantity of solution* $\times$ *strength of solution (in percent written as a decimal).* When we are mixing two quantities and are interested in the *composition* of the mixture, we generally use the following table or a variation of the table.

Quantity $\times$ Strength $=$ Amount of substance

Solution	Quantity	Strength	Amount of Substance
Solution 1			amount of substance in solution 1
Solution 2			amount of substance in solution 2
Mixture			amount of substance in mixture

When using this table, we generally use the following formula to solve the problem.

$$\begin{pmatrix}\text{amount of substance}\\ \text{in solution 1}\end{pmatrix} + \begin{pmatrix}\text{amount of substance}\\ \text{in solution 2}\end{pmatrix} = \begin{pmatrix}\text{amount of substance}\\ \text{in mixture}\end{pmatrix}$$

Let us now look at an example of a mixture problem where two solutions are combined.

EXAMPLE 6 Mixing Acid Solutions Mr. Dave Lumsford needs a 10% acetic acid solution for a chemistry experiment. After checking the store room, he finds that there are only 5% and 20% acetic acid solutions available. Mr. Lumsford decides to make the 10% solution by combining the 5% and 20% solutions. How many liters of the 5% solution must he add to 8 liters of the 20% solution to get a solution that is 10% acetic acid?

Solution **Understand and Translate** We are asked to find the number of liters of the 5% acetic acid solution to mix with 8 liters of the 20% acetic acid solution.

Let $x =$ number of liters of 5% acetic acid solution.

Let's draw a sketch of the problem (**Fig. 3.11**).

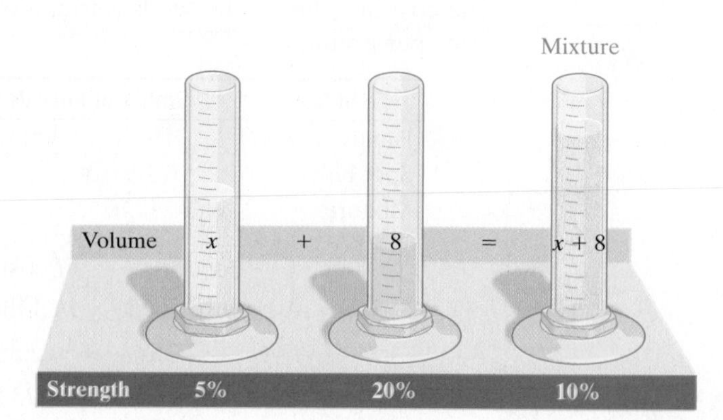

FIGURE 3.11

The amount of acid in a given solution is found by multiplying the number of liters by the percent strength.

Solution	Liters	Strength	Amount of Acetic Acid
5%	x	0.05	$0.05x$
20%	8	0.20	$0.20(8)$
Mixture	$x + 8$	0.10	$0.10(x + 8)$

$$\begin{pmatrix}\text{amount of acid} \\ \text{in 5\% solution}\end{pmatrix} + \begin{pmatrix}\text{amount of acid} \\ \text{in 20\% solution}\end{pmatrix} = \begin{pmatrix}\text{amount of acid} \\ \text{in 10\% mixture}\end{pmatrix}$$

$$0.05x \quad + \quad 0.20(8) \quad = 0.10(x + 8)$$

Carry Out

$$\begin{aligned} 0.05x + 1.6 &= 0.10x + 0.8 \\ 0.05x + 0.8 &= 0.10x \\ 0.8 &= 0.05x \\ \frac{0.8}{0.05} &= x \\ 16 &= x \end{aligned}$$

Answer Sixteen liters of 5% acetic acid solution must be added to the 8 liters of 20% acetic acid solution to get a 10% acetic acid solution. The total number of liters that will be obtained is 16 + 8 or 24.

Now Try Exercise 53

EXAMPLE 7 Nicole Pappas, a medical researcher, has 40% and 5% solutions of phenobarbital. How much of each solution must she mix to get 0.6 liter of a 20% phenobarbital solution?

Solution Understand and Translate We are asked to find how much of the 40% and 5% phenobarbital solutions must be mixed to get 0.6 liter of a 20% solution. We can choose to let x be the amount of either the 40% or the 5% solution. We will choose as follows:

Let x = number of liters of the 40% solution.

Then $0.6 - x$ = number of liters of the 5% solution.

Remember from Section 3.1 that if a total of 0.6 liter is divided in two, if one part is x, the other part is $0.6 - x$.

Let's draw a sketch of the problem (**Fig. 3.12**).

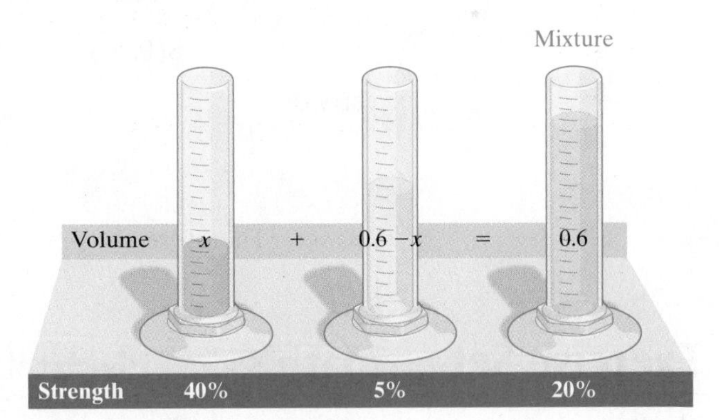

FIGURE 3.12

The amount of phenobarbital in a given solution is found by multiplying the number of liters by the percent strength.

Solution	Liters	Strength	Amount of Phenobarbital
40%	x	0.40	$0.40x$
5%	$0.6 - x$	0.05	$0.05(0.6 - x)$
Mixture	0.6	0.20	0.6(0.20)

$$\begin{pmatrix}\text{amount of phenobarbital} \\ \text{in 40\% solution}\end{pmatrix} + \begin{pmatrix}\text{amount of phenobarbital} \\ \text{in 5\% solution}\end{pmatrix} = \begin{pmatrix}\text{amount of phenobarbital} \\ \text{in mixture}\end{pmatrix}$$

$$0.40x \quad + \quad 0.05(0.6 - x) \quad = (0.6)(0.20)$$

Carry Out

$$0.40x + 0.03 - 0.05x = 0.12$$
$$0.35x + 0.03 = 0.12$$
$$0.35x = 0.09$$
$$x \approx 0.26$$

Answer Since the answer was less than 0.6 liter, the answer is reasonable. About 0.26 liter of the 40% solution must be mixed with about $0.6 - x = 0.60 - 0.26 = 0.34$ liter of the 5% solution to get 0.6 liter of the 20% mixture.

Now Try Exercise 57

In Example 7, we chose to let $x =$ number of liters of the 40% solution. We could have selected to let $x =$ number of liters of the 5% solution. Then $0.6 - x$ would be the number of liters of the 40% solution. Had you worked the problem out like this, you would have found that x was approximately 0.34 liter. Try reworking Example 7 now letting $x =$ number of liters of the 5% solution.

EXAMPLE 8 An orange punch contains 4% orange juice. If 5 ounces of water is added to 8 ounces of the punch, determine the percent of orange juice in the mixture.

Solution Understand and Translate We are asked to find the percent of orange juice in the mixture.

Let $x =$ percent of orange juice in the mixture.

We will again set up a table.

Solution	Ounces	Percent of Juice	Amount of Juice
Punch	8	0.04	8(0.04)
Water	5	0.00	5(0.00)
Mixture	13	x	$13x$

$$\begin{pmatrix}\text{amount of juice}\\ \text{in punch}\end{pmatrix} + \begin{pmatrix}\text{amount of juice}\\ \text{in water}\end{pmatrix} = \begin{pmatrix}\text{amount of juice}\\ \text{in mixture}\end{pmatrix}$$

$$8(0.04) + 5(0.00) = 13x$$

Carry Out

$$0.32 + 0.00 = 13x$$
$$0.32 = 13x$$
$$0.025 \approx x$$

Answer Therefore, the percent of juice in the mixture is about 2.5%.

Now Try Exercise 63

3.4 Exercise Set MathXL® MyMathLab®

Warm-Up Exercises

Fill in the blanks with the appropriate word, phrase, or symbol(s) from the following list.

$d = r \cdot t$	adding	subtracting	multiplying
$i = p \cdot r \cdot t$	$8 - x$	percent	

1. Solving a motion problem when the two items are traveling in the same direction usually involves ________ the distances.
2. In a mixture problem, if there is a total of 8 liters and the amount of one unknown is x, then the amount of the other unknown is ________.
3. A formula important in the solution of motion problems is ________.
4. Solving a money problem may include using the formula ________.
5. Solving a motion problem when the two items are traveling in different directions usually involves ________ the distances.
6. To find the amount of alcohol in an 8-liter solution, we multiply the quantity of solution times the ________ of alcohol in the solution.

Practice the Skills/Problem Solving

In Exercises 7–66, set up an equation that can be used to solve each problem. Solve the equation, and answer the question.

In Exercises 7–32, solve the motion problem. See Examples 1 and 2.

7. **Ferries** Two high-speed ferries leave at the same time from Ft. Myers, Florida, going to Key West, Florida. The first ferry, the *BigCat,* travels at 34 miles per hour. The second ferry, the *Atlantic Cat,* travels at 28 miles per hour. In how many hours will the two ferries be 6 miles apart?

8. **Trains** Two trains in New York City start at the same station going in the same direction on sets of parallel tracks. The local train stops often and averages 18.4 miles per hour. The express train stops less frequently and averages 30.2 miles per hour. In how many hours will the two trains be 5.9 miles apart?

9. **Horseback Riding** Two friends, Jodi Cotton and Abe Mantell, go horseback riding on the same trail in the same direction. Jodi's horse travels at 8 miles per hour while Abe's horse travels at a slower pace. After 2 hours they are 4 miles apart. Find the speed at which Abe's horse is traveling.

10. **Camel Riding** In the Outback in Australia, Betty Sue Adams and Carl Minieri go camel riding in the same direction along the same path. Betty Sue's camel travels at 6 miles per hour while Carl's camel travels at a slower pace. After 3 hours they are 2.4 miles apart. Find the speed of Carl's camel.

11. **Airplanes** A Jet Blue airplane leaves Chicago for New York at the same time a Southwest airplane leaves New York for Chicago. The distance from Chicago to New York is 798 miles. If the Jet Blue plane travels at 560 miles per hour and the Southwest plane travels at 580 miles per hour, how long into their flights will the two planes pass each other?

12. **Walking** Barb Dansky and Sandy Spears are at opposite ends of a shopping mall 5700 feet apart walking toward each other. If Barb walks 4.4 feet per second and Sandy walks 5.1 feet per second, how long will it be before they meet?

13. **Walkie-Talkies** Willie and Shanna Johnston have walkie-talkies that have a range of 16.8 miles. Willie and Shanna start at the same point and walk in opposite directions. If Willie walks 3 miles per hour and Shanna walks 4 miles per hour, how long will it take before they are out of range?

14. **Blue Angels** At a Navy Blue Angel air show two F/A-18 Hornet jets travel toward each other, both at a speed of 1000 miles per hour. After they pass each other, if they were to keep flying at the same speed in opposite directions, how long would it take for them to be 500 miles apart?

15. **Product Testing** The Goodyear Tire Company is testing new tires by placing them on a machine that can simulate the tires riding on a road. First, the machine runs the tires for 7.2 hours at 60 miles per hour. Then the tires are run for 6.8 hours at a different speed. After this 14-hour period, the machine indicates that the tires have traveled the equivalent of 908 miles. Find the second speed to which the machine was set.

16. **Ski Lifts** To get to the top of Whistler Mountain, people must use two different ski lifts. The first lift travels 4 miles per hour for 0.2 hours. The second lift travels for 0.3 hours to reach the top of the mountain. If the total distance traveled up the mountain is 1.4 miles, find the average speed of the second ski lift.

17. **Earthquakes** Earthquakes generate circular p-waves and s-waves, which travel outward (see the figure). Suppose the p-waves have a velocity of 3.6 miles per second and the s-waves have a velocity of 1.8 miles per second. How long after the earthquake will p-waves and s-waves be 81 miles apart?

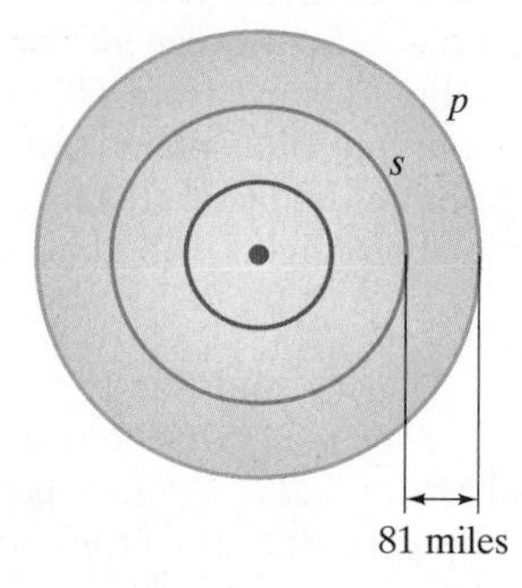

18. **Navigating O'Hare Airport** Sadie Bragg and Jose Cruz are in Chicago's O'Hare airport walking between terminals. Sadie walks on the moving walkway (like a flat escalator moving along the floor). Her speed (relative to the ground) is 220 feet per minute. Jose starts walking at the same time and walks alongside the walkway at a speed of 100 feet per minute. How long have they been walking when Sadie is 600 feet ahead of Jose?

19. **Disabled Boat** Two Coast Guard cutters are 225 miles apart traveling toward each other, one from the east and the other from the west, searching for a disabled boat. The eastbound cutter travels 5 miles per hour faster than the westbound cutter. If the two cutters pass each other after 3 hours, find the average speed of each cutter.

20. **Snowplowing** Two snowplowing crews are clearing snow from a 2.56-mile runway at an airport. Crew A starts at one end of the runway and crew B starts at the other end. They start clearing at the same time and they work toward each other. Crew A clears the runway at a speed of 0.4 miles per hour faster than crew B. If they meet 0.5 hours after they start, find the average speed of each snowplow.

21. **Round Trip** Samia Metwali walks for a time at 4 miles per hour, then slows down and walks at 3.2 miles per hour. The total distance she walked was 7 miles. If she walked for 0.5 hour more at 3.2 miles per hour than she did at 4 miles per hour, determine the time Samia walked at 4 miles per hour.

22. **Visit to Grandchild** Chuck Neumann drove from his house in Auburn Hills, Michigan, to visit his grandchild in Pasadena, Texas, a distance of 1343 miles. Part of the way he drove at 60 miles per hour and part of the way he drove at 70 miles per hour. If he drove for 0.5 hour more at 60 miles per hour than he did at 70 miles per hour, determine the time Chuck drove at 70 miles per hour.

23. **Paving Road** Two crews are laying blacktop on a road. They start at the same time at opposite ends of a 12-mile road and work toward one another. One crew lays blacktop at an average rate of 0.75 mile a day faster than the other crew. If the two crews meet after 3.2 days, find the rate of each crew.

24. **Beach Clean-Up** On Earth Day, two groups of people clean a 7-mile stretch of Myrtle Beach. Auturo's group and Jane's group start at the same time at opposite ends of the beach and walk toward each other. Auturo's group is traveling at a rate of 0.5 mile per hour faster than Jane's group, and they meet in 2 hours. Find the speed of each group.

25. **Sailing** Two sailboats are 9.8 miles apart and sailing toward each other. The larger boat, the *Pythagoras*, sails 4 miles per hour faster than the smaller boat, the *Apollo*. The two boats pass each other after 0.7 hour. Find the speed of each boat.

26. **Auto Travel** Two cars are 45 miles apart traveling toward one another. One is traveling at 70 mph and the other at 65 mph. How long will it take them to meet?

27. **Heavy Traffic** Laura Hoye drives along an interstate highway at 70 mph until she encounters heavy traffic and has to slow to 50 mph. The time she drove at 70 mph is 1 hour more than the time she drove at 50 mph and the distance she drove at 70 mph is greater than the distance she drove at 50 mph. If the difference in the distance she drove at 70 mph and the distance she drove at 50 mph is 100 miles, determine the time she spent driving at each speed.

28. **Treadmill Running** Barbara Hughes is running on a treadmill at her health club for 0.25 hour when her friend Mary Johnson begins running on another treadmill. Barbara is running at 6 mph and Mary is running at 5 mph. They finish at the same time and compare the distances they each ran. If the difference in their distances is 2.25 miles, determine the time Mary runs and the time Barbara runs.

29. **Bicycling** Dom Palmo leaves his house at noon and rides his bicycle south at a uniform rate. His wife, Sue, leaves at the same time heading due north. Dom's rate is 10 kilometers per hour faster than Sue's rate. At 5 PM they are 160 kilometers apart. Find the rate of travel for each cyclist.

30. **Salt Mine** The ore at a mine must travel on two different conveyer belts to be loaded onto a train. The second conveyer belt travels at a rate of 0.6 foot per second faster than the first conveyer belt. The ore travels 180 seconds on the

first belt and 160 seconds on the second belt. If the total distance traveled by the ore is 1116 feet, determine the speed of each belt.

31. Ironman Triathlon A triathlon consists of three parts: swimming, cycling, and running. Participants in the Ironman Triathlon must swim, cycle, and run certain distances. Jane Keller swam at an average of 2.64 miles per hour for 0.91 hours then cycled at an average of 22.96 miles per hour for 4.88 hours. Finally, she ran at an average of 7.62 miles per hour for 3.44 hours.

a) Estimate the distance that Jane swam.

b) Estimate the distance that Jane cycled.

c) Estimate the distance that Jane ran.

d) Estimate the total distance covered during the triathlon.

e) Find the total time of Jane's triathlon.

32. Ironman Triathlon Refer to Exercise 31, about the Ironman Triathlon. Milke Kerby swam at an average speed of 2.93 miles per hour for 0.82 hour, then cycled at an average speed of 23.85 miles per hour for 4.70 hours, then ran at an average speed of 8.94 miles per hour for 2.93 hours. Answer the questions in Exercise 31 **a)–e)** using Mike's data.

In Exercises 33–46, solve the money problem. See Examples 3 and 4.

33. Simple Interest Paul and Donna Petrie invested $12,000, part at 5% simple interest and the rest at 7% simple interest for a period of 1 year. How much did they invest at each rate if their total annual interest from both investments was $800? (Use interest = principal · rate · time.)

34. Simple Interest Jerry Correa invested $7000, part at 8% simple interest and the rest at 5% simple interest for a period of 1 year. If he received a total annual interest of $476 from both investments, how much did he invest at each rate?

35. Simple Interest Aleksandra Tomich invested $6000, part at 6% simple interest and part at 4% simple interest for a period of 1 year. How much did she invest at each rate if each account earned the same interest?

36. Simple Interest Mark Ernshausen invested $12,500, part at 7% simple interest and part at 6% simple interest for a period of 1 year. How much was invested at each rate if each account earned the same interest?

37. Simple Interest Míng Wang invested $10,000, part at 4% and part at 5% simple interest for a period of 1 year. How much was invested in each account if the interest earned in the 5% account was $320 greater than the interest earned in the 4% account?

38. Simple Interest Sharon Sledge invested $20,000, part at 5% and part at 7% simple interest for a period of 1 year. How much was invested in each account if the interest earned in the 7% account was $440 greater than the interest earned in the 5% account?

39. Cable TV Violet Kokola knows that her subscription rate for the basic tier of cable television increased from $18.20 to $19.50 at some point during the calendar year. She paid a total of $230.10 for the year to the cable company. Determine the month of the rate increase.

40. Rate Increase Patricia Burgess knows that at some point during the calendar year her basic monthly telephone rate increased from $17.10 to $18.40. If she paid a total of $207.80 for basic telephone service for the calendar year, in what month did the rate increase take effect?

41. Wages Mihály Sarett holds two part-time jobs. One job, at Home Depot, pays $8.50 an hour and the second job, at a veterinary clinic, pays $9.00 per hour. Last week Mihály worked a total of 18 hours and earned $158.00. How many hours did Mihály work at each job?

42. Ticket Sales At a movie theater an evening show cost $7.50 and a matinee cost $4.75. On one day there was one matinee and one evening showing. On that day a total of 310 tickets were sold, which resulted in ticket sales of $2022.50. How many people went to the matinee and how many went to the evening show?

43. Baseball Hall of Fame At the Baseball Hall of Fame in Cooperstown, New York, adult admission is $19.50 and child admission is $7.00. During one day a total of 2100 adult and child admissions were collected and $29,700 in admission fees was collected. How many adult admissions were collected?

Baseball Hall of Fame

44. Rock and Roll Hall of Fame At the Rock and Roll Hall of Fame in Cleveland, Ohio, adult admission is $22 and child admission is $13. During one day, a total of 1551 adult and child admissions were collected and $27,543 in admission fees was collected. How many child admissions were collected?

Rock and Roll Hall of Fame

45. Stock Purchase eBay stock is selling at \$52 a share and Facebook stock is selling at \$29 a share. Billy Jula has a maximum of \$10,000 to invest in these two stocks and he wishes to purchase five times as many shares of Facebook as shares of eBay. Only whole shares of stock can be purchased.

a) How many shares of each stock can Billy purchase?
b) How much money will Billy have left over?

46. Stock Purchase Gap stock is selling at \$36 a share and Best Buy stock is selling at \$14 a share. Sharon Griggs has a maximum of \$15,000 to invest in these two stocks and she wishes to purchase four times as many shares of Best Buy as shares of Gap. Only whole shares of stock can be purchased.

a) How many shares of each stock can Sharon purchase?
b) How much money will Sharon have left over?

In Exercises 47–66, solve the mixture problem. See Examples 5–8.

47. Nut Shop Jean Valjean owns a nut shop where walnuts cost \$6.80 per pound and almonds cost \$6.40 per pound. Jean gets an order that specifically requests a 30-pound mixture of walnuts and almonds that will cost \$6.65 per pound. How many pounds of each type of nut should Jean mix to get the desired mixture?

48. Grass Seed Buck's grass seed sells for \$2.45 per pound and Buck's Filler grass seed sells for \$2.10 per pound. How many pounds of each should be mixed to get a 10-pound mixture that sells for \$2.20 per pound?

49. Topsoil Strained topsoil sells for \$160 per cubic yard and unstrained topsoil sells for \$120 per cubic yard. How many cubic yards of each should be mixed to make 8 cubic yards of a mixture that sells for \$150 per cubic yard?

50. Coffee House Ruth Cordeff runs a coffee house where chocolate almond coffee beans sell for \$7.00 per pound and hazelnut coffee beans sell for \$6.10 per pound. A customer asks Ruth to make a 6-pound mixture using the two kinds of beans. How many pounds of each should be used if the mixture is to cost \$6.40 per pound?

51. Bulk Candies A grocery store sells certain candies in bulk. The Good and Plenty cost \$2.49 per pound and Sweet Treats cost \$2.89 per pound. If Jane Strange takes 3 scoops of Good and Plenty and mixes it with 5 scoops of Sweet Treats, how much per pound should the mixture sell for? Assume each scoop contained the same weight of candy.

52. Bird Food At Agway Gardens, bird food is sold in bulk. In one barrel are sunflower seeds that sell for \$1.80 per pound. In a second barrel is cracked corn that sells for \$1.40 per pound. If a mixture is made by taking 2.5 pounds of the sunflower seeds and 1 pound of the cracked corn, what should the mixture cost per pound?

53. Sulfuric Acid In chemistry class, Todd Corbin has 3 liters of a 20% sulfuric acid solution. How much of a 12% sulfuric acid solution must he mix with the 3 liters of 20% solution to make a 15% sulfuric acid solution?

54. Paint Nick Cooper has two cans of white paint, one with a 2% yellow pigment and the other with a 5% yellow pigment. Nick wants to mix the two paints to get paint with a 4% yellow pigment. How much of the 5% yellow pigment paint should be mixed with 0.4 gallon of the 2% yellow pigment paint to get the desired paint?

55. Alcohol Solution How many pints of a 12% isotrophic alcohol solution should Jenny Crawford mix with 15 pints of a 5% isotrophic alcohol solution to get a mixture that is 8% isotrophic alcohol?

56. Salt Concentration Suppose the dolphins at Sea World must be kept in salt water with an 0.8% salt content. After a week of warm weather, the salt content has increased to 0.9% due to water evaporation. How much water with 0% salt content must be added to 50,000 gallons of the 0.9% salt water to lower the salt concentration to 0.8%?

57. Bleach Clorox bleach is 5.25% sodium hypochlorite and swimming pool shock treatment is 10.5% sodium hypochlorite. How much of each item must be mixed to get 7 cups of a mixture that is 7.2% sodium hypochlorite?

58. Antifreeze Prestone antifreeze contains 12% ethylene glycose, and Zerex antifreeze contains 9% ethylene glycose. How much of each type antifreeze should be mixed to get 9 quarts of a mixture that is 10% ethylene glycose?

59. Tea Tree Oil Pola Sommers needs 4% tea tree oil solution to use as a topical treatment for mosquito bites. She has only 2% solution and 5% solution on hand. How many ounces of each of these solutions should she mix if she would like to have 6 ounces of 4% tea tree oil solution?

60. **Pharmacy** Susan Staples, a pharmacist, has a 60% solution of sodium iodide. She also has a 25% solution of the same drug. She gets a prescription calling for a 40% solution of the drug. How much of each solution should she mix to make 70 milliliters of the 40% solution?

61. **Mouthwash** The label on the Listerine Cool Mint Antiseptic mouthwash says that it is 21.6% alcohol by volume. The label on the Scope Original Mint mouthwash says that it is 15.0% alcohol by volume. If Hans mixes 6 ounces of the Listerine with 4 ounces of the Scope, what is the percent alcohol content of the mixture?

62. **Clorox** Clorox bleach is 5.25% sodium hypochlorite. The instructions on the Clorox bottle say to add 1 cup of Clorox to 4 cups (a quart) of water. Find the percent of sodium hypochlorite in the mixture.

63. **Orange Juice** Mary Ann Terwilliger has made 7 quarts of an orange juice punch for a party. The punch contains 12% orange juice. She feels that she may need more punch, but she has no more orange juice so she adds $\frac{1}{2}$ quart of water to the punch. Find the percent of orange juice in the new mixture.

See Exercise 63.

64. **Insecticide** The active ingredient in one type of Ortho insect spray is 50% malathion. The instructions on the bottle say to add 1 fluid ounce to a gallon (128 fluid ounces) of water. What percent of malathion will be in the mixture?

65. **Hawaiian Punch** The label on a 12-ounce can of frozen concentrate Hawaiian Punch indicates that when the can of concentrate is mixed with 3 cans (36 ounces) of cold water, the resulting mixture is 10% juice. Find the percent of pure juice in the concentrate.

66. **Plant Food** Miracle-Gro All Purpose liquid plant food has 12% nitrogen. Miracle-Gro Quick Start liquid plant food has 4% nitrogen. If 2 cups of the All Purpose plant food are mixed with 3 cups of the Quick Start plant food, determine the percent of nitrogen in the mixture.

Challenge Problems

67. **Fat Albert** The home base of the Navy's Blue Angels is in Pensacola, Florida. The Angels spend winters in El Centro, California. Assume they fly at about 900 miles per hour when they fly from Pensacola to El Centro in their F/A-18 Hornets. On every trip, their C-130 transport (affectionately called Fat Albert) leaves before them, carrying supplies and support personnel. The C-130 generally travels at about 370 miles per hour. On a trip from Pensacola to El Centro, how long before the Hornets leave should Fat Albert leave if it is to arrive 3 hours before the Hornets? The flying distance between Pensacola and El Centro is 1720 miles.

Group Activity

Discuss and answer Exercises 68 and 69 as a group.

68. **Race Horse** According to *Guinness World Records,* the fastest race horse speed recorded was by a horse called Winning Brew on May 14, 2008, in Grantville, Pennsylvania. Winning Brew ran $\frac{1}{4}$ mile in 20.57 seconds. Find Winning Brew's speed in miles per hour. Round your answer to the nearest hundredth.

69. **Garage Door Opener** An automatic garage door opener is designed to begin to open when a car is 100 feet from the garage. At what rate will the garage door have to open if it is to rise 6 feet by the time a car traveling at 4 miles per hour reaches it? (1 mile per hour $\approx$1.47 feet per second.)

Cumulative Review Exercises

[1.3] 70. **a)** Divide $2\frac{3}{4} \div 1\frac{5}{8}$.

b) Add $2\frac{3}{4} + 1\frac{5}{8}$.

[2.5] 71. Solve the equation $6(x - 3) = 4x - 18 + 2x$.

[2.7] 72. Solve the proportion $\frac{6}{x} = \frac{72}{9}$.

[2.8] 73. Solve the inequality $3x - 4 \leq -4x + 3(x - 1)$.

Chapter 3 Summary

IMPORTANT FACTS AND CONCEPTS	EXAMPLES
Section 3.1	
a subtracted from b means $b - a$. The difference between a and b means $a - b$.	3 subtracted from $5x$ is $5x - 3$. The difference between 3 and $5x$ is $3 - 5x$.
If T represents the total amount to be separated into two parts, and if x represents one of the parts, then $T - x$ represents the other part.	If \$800 is separated between Jay and Rose and if Rose gets x dollars, then Jay gets $800 - x$ dollars.
A percent is always a percent of some number.	An 8% sales tax on r dollars is $0.08r$. The cost of an item c increased by 5% is $c + 0.05c$.
When writing equations, the words *is, was, will be, yields, gives,* often mean =.	"Two less than 5 times a number *is* the number increased by 3" can be expressed as $5x - 2 = x + 3$.
Consecutive integers, such as 23 and 24, differ by 1 unit. **Consecutive even integers,** such as 24 and 26, differ by 2 units. **Consecutive odd integers,** such as 23 and 25, differ by 2 units.	x and $x + 1$ represent consecutive integers. x and $x + 2$ represent consecutive even integers. x and $x + 2$ represent consecutive odd integers.
Section 3.2	
Problem-Solving Procedure **1.** Understand the problem. **2.** Translate the problem into mathematical language. **3.** Carry out the mathematical calculation. **4.** Check the answer. **5.** Answer the question asked. (See page 194 for more detailed information.)	Three subtracted from 4 times a number is 17. Find the number. **Solution:** Understand We need to express the information given as an equation. We are asked to find the unknown number. Translate Let $x =$ the unknown number, then we can write the equation $4x - 3 = 17$ Carry out $4x = 20$ $x = 5$ Check Substitute 5 for the unknown number $4(5) - 3 \stackrel{?}{=} 17$ $17 = 17$ True Answer The unknown number is 5.
If you are having difficulty with this section, seek help.	See Helpful Hint on page 199 for possible sources for help.
Section 3.3	
An **isosceles triangle** has two sides of the same length. The angles opposite the sides of equal length have equal measures.	Two angles of an isosceles triangle are each 30° greater than the smallest angle. Find the measures of the three angles. **Solution:** Let $x =$ the smallest angle. Then $x + 30 =$ the measure of each larger angle. $x + (x + 30) + (x + 30) = 180$ $3x + 60 = 180$ $3x = 120$ $x = 40$ The smallest angle is 40°, and the two larger angles are 40° + 30° or 70°. Note that 40° + 70° + 70° = 180°, so the answer checks.
Section 3.4	
A **motion problem** can involve two rates.	Peter and Paul go walking. They start at the same point at the same time and walk in the same direction. Peter walks at 4 mph and Paul walks at 3.5 mph. In how many hours will they be 1 mile apart?

IMPORTANT FACTS AND CONCEPTS	EXAMPLES
Section 3.4 (Continued)	
	Solution: Let t = time when they are 1 mile apart. $4t - 3.5t = 1$ $0.5t = 1$ $t = 2$ In 2 hours they will be 1 mile apart.
A **money problem** can involve two rates of interest or two different costs.	John sold 12 paintings in one day. Some sold at \$80 and the others sold at \$125. If he collected a total of \$1275, how many of each type did he sell? Solution: Let x = number of \$80 paintings. Then $12 - x$ = number of \$125 paintings. $80x + 125(12 - x) = 1275$ $80x + 1500 - 125x = 1275$ $-45x + 1500 = 1275$ $-45x = -225$ $x = 5$ Five \$80 paintings and 12 − 5 or 7 \$125 paintings were sold.
A **mixture problem** may involve mixing different strengths or different types of solutions, or mixing solid items, such as nuts.	How many liters of an 8% acetic acid solution must be mixed with 6 liters of a 15% acetic acid solution to get a 10% acetic acid solution? Solution: Let x = number of liters of the 8% solution. $0.08x + 0.15(6) = 0.10(x + 6)$ $0.08x + 0.9 = 0.10x + 0.6$ $0.9 = 0.02x + 0.6$ $0.3 = 0.02x$ $15 = x$ Fifteen liters of the 8% solution must be mixed with the 6 liters of the 15% solution.

Chapter 3 Review Exercises

[3.1]

1. **Age** Charles's age is 7 more than 3 times Norman's age, n. Write an expression for Charles's age.

2. **Gasoline** The cost of a gallon of gasoline in California is 1.2 times the cost of a gallon of gasoline in Georgia, g. Write an expression for the cost of a gallon of gasoline in California.

3. **Dress** Write an expression for the cost of a dress, d, reduced by 25%.

4. **Pounds** Write an expression for the number of ounces in y pounds.

5. **Money** Two hundred dollars is divided between Jishing Wang and Norma Agras. If Jishing gets x dollars, write an expression for the amount Norma gets.

6. **Age** Mario is 6 years older than seven times Dino's age. Select a variable to represent one quantity and state what the variable represents. Express the second quantity in terms of the variable selected.

7. **Robberies** The number of robberies in 2013 was 12% less than the number of robberies in 2012. Write an expression for the difference in the numbers of robberies between 2012 and 2013.

8. **Numbers** The smaller of two numbers is 24 less than 3 times the larger. When the smaller is subtracted from the larger, the difference is 8. Write an equation to represent this information. Do not solve the equation.

[3.2] *In Exercises 9–18, set up an equation that can be used to solve the problem. Solve the equation and answer the question.*

9. **Numbers** One number is 8 more than the other. Find the two numbers if their sum is 74.

10. **Consecutive Integers** The sum of two consecutive integers is 237. Find the two integers.

11. **Numbers** The larger of two integers is 3 more than 5 times the smaller integer. Find the two numbers if the smaller subtracted from the larger is 31.

12. **New Car** Shaana recently purchased a new car. What was the cost of the car before tax if the total cost including a 7% tax was $23,260?

13. **Bagels** A bakery currently ships 520 bagels per month to various outlets. They wish to increase the shipment of bagels by 20 per month until reaching a shipment level of 900 bagels. How long will this take?

14. **Salary Comparison** Irene Doo, a salesperson, receives a salary of $600 per week plus a 3% commission on all sales she makes. Her company is planning on changing her salary to $500 per week, plus an 8% commission on all sales she makes. What weekly dollar sales would she have to make for the total salaries from each plan to be the same?

15. **Sale Price** During a going-out-of-business sale, all prices were reduced by 20%. If during the sale Kathy Golladay purchased an HDTV stand for $495, what was the original price of the HDTV stand?

16. **Landscaping** Two Brothers Nursery charges $400 for a tree and $45 per hour to plant the tree. ABC Nursery charges $200 for the same size tree and $65 per hour to plant the tree. How many hours for planting would result in the total price for both landscapers being the same?

17. **New Home** William Krant recently moved to a new house. The number of square feet in his new house is 27% greater than the number of square feet in his old house. If the new house has 3556 square feet, how many square feet did his old house have?

18. **Alaska Area** The area of Alaska is 125,000 square miles more than twice the area of Texas. The sum of the areas of these two states is 932,000 square miles. Determine the area of each state.

Alaska, see Exercise 18.

[3.3] *Solve each problem.*

19. **Unknown Angles** One angle of a triangle measures 10° greater than the smallest angle, and the third angle measures 10° less than twice the smallest angle. Find the measures of the three angles.

20. **Unknown Angles** One angle of a trapezoid measures 10° greater than the smallest angle; a third angle measures five times the smallest angle; and the fourth angle measures 20° greater than four times the smallest angle. Find the measure of the four angles.

21. **Garden** Steve Rodi has a rectangular garden whose length is 4 feet longer than its width. The perimeter of the garden is 70 feet. Find the width and length of the garden.

22. **Designing a House** The figure below shows plans for a rectangular basement of a house. The dots represents poles in the ground and the lines represent string that has been attached to the poles. The length of the basement is to be 30 feet greater than the width. If a total of 310 feet of string was used to mark off the rooms, find the width and length of the basement.

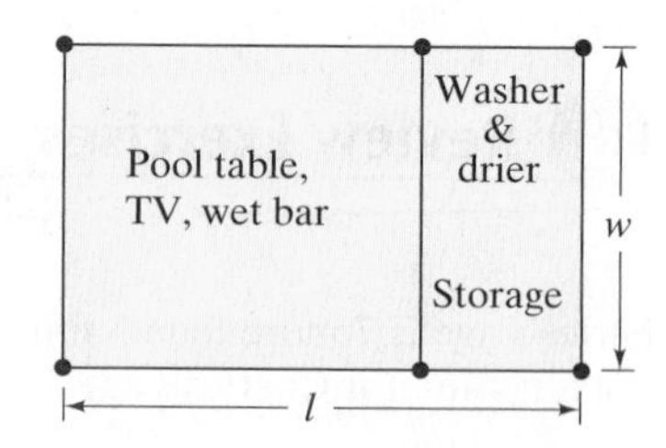

23. **Rhombus** The two larger angles of a rhombus are each 3 times the measure of the two smaller angles. Find the measure of each angle.

24. **Bookcase** Wade Ellis is building a bookcase with 4 shelves. The length of the bookcase is to be twice the height, as illustrated. If only 20 feet of lumber is available, what should be the length and height of the bookcase?

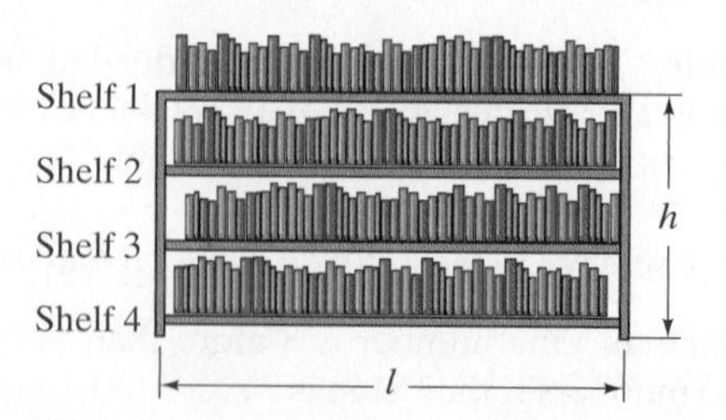

[3.4] *Solve each problem.*

25. Jogging Two joggers follow the same route. Harold Lowe jogs at 8 kilometers per hour and Susan Karney Fackert at 6 kilometers per hour. If they leave at the same time, how long will it take for them to be 4 kilometers apart?

26. Trains Leaving Two trains going in opposite directions leave from the same station on parallel tracks. One train travels at 50 miles per hour and the other at 60 miles per hour. How long will it take for the trains to be 440 miles apart?

27. Pittsburgh Incline The Duquesne Incline in Pittsburgh, Pennsylvania, is shown below. The two cars start at the same time at opposite ends of the incline and travel toward each other at the same speed. The length of the incline is 400 feet and the time it takes for the cars to be at the half-way point is about 22.73 seconds. Determine, in feet per second, the speed the cars travel.

Pittsburgh, PA

28. Savings Accounts Martha Goshaw wishes to place part of $12,000 into a savings account earning 8% simple interest and part into a savings account earning $7\frac{1}{4}$% simple interest. How much should she invest in each if she wishes to earn $900 in interest for the year?

29. Savings Accounts Aimee Tait invests $4000 into two savings accounts. One account pays 3% simple interest and the other account pays 3.5% simple interest. If the interest earned in the account paying 3.5% simple interest is $94.50 more than the interest earned in the 3% account, how much was invested in each account?

30. Party Punch Gayle Krzemine makes 2 gallons of a punch solution that contains 25% orange juice. How many gallons of pure orange juice should Gayle add so that the new solution contains 40% orange juice?

31. Wind Chimes Alan Carmell makes and then sells wind chimes. He makes two types, a smaller one that sells for $8 and a larger one that sells for $20. At an arts and crafts show he sells a total of 30 units, and his total receipts were $492. How many of each type of chime did he sell?

32. Acid Solution Bruce Kennan, a chemist, wishes to make 2 liters of an 8% acid solution by mixing a 10% acid solution and a 5% acid solution. How many liters of each should he use?

[3.2–3.4]. *Solve each problem.*

33. Numbers The sum of two consecutive odd integers is 208. Find the two integers.

34. Television What is the cost of a television before tax if the total cost, including a 6% tax, is $477?

35. Medical Supplies Mr. Chang sells medical supplies. He receives a weekly salary of $300 plus a 5% commission on the sales he makes. If Mr. Chang earned $900 last week, what were his sales in dollars?

36. Triangle One angle of a triangle is 8° greater than the smallest angle. The third angle is 4° greater than twice the smallest angle. Find the measure of the three angles of the triangle.

37. Increase Staff The Darchelle Leggett Company plans to increase its number of employees by 25 per year. If the company now has 427 employees, how long will it take before they have 627 employees?

38. Parallelogram The two larger angles of a parallelogram each measure 40° greater than the two smaller angles. Find the measure of the four angles.

39. Copy Centers Copy King charges a monthly fee of $20 plus 4 cents per copy. King Kopie charges a monthly fee of $25 plus 3 cents a copy. How many copies made in a month would result in both companies charging the same amount?

40. Swimming Rita Gonzales and Jim Ham are going swimming in Putnam Lake. They start swimming in the same direction at the same time. Rita swims at 1 mph and Jim swims at a slower pace. After 0.5 hour they are 0.2 mile apart. Find the speed that Jim is swimming.

41. **Butcher** A butcher combined ground beef that cost \$3.50 per pound with ground beef that cost \$4.10 per pound. How many pounds of each were used to make 80 pounds of a mixture that sells for \$3.65 per pound?

42. **Speed Traveled** Two brothers who are 230 miles apart start driving toward each other at the same time. The younger brother travels 5 miles per hour faster than the older brother, and the brothers meet after 2 hours. Find the speed traveled by each brother.

43. **Acid Solution** How many liters of a 30% acid solution must be mixed with 2 liters of a 12% acid solution to obtain a 15% acid solution?

44. **Fencing** Kathy Tomaino is partitioning a rectangular yard using fencing, as shown on the right. If the length is 1.5 times the width and if only 96 feet of fencing is available, find the length and width of the partitioned area.

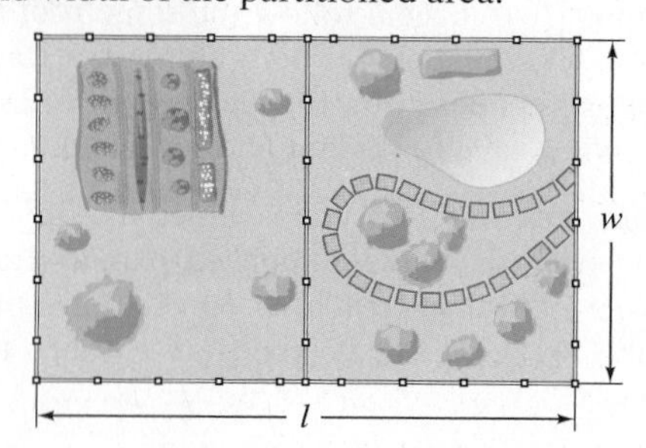

45. **Acid Solution** Six liters of a 3% sulfuric acid solution is mixed with an 8% sulfuric acid solution. If the mixture is a 4% sulfuric acid solution, determine how many liters of the 8% sulfuric acid solution were used in the mixture.

Chapter 3 Practice Test

Chapter Test Prep Videos provide fully worked-out solutions to any of the exercises you want to review. Chapter Test Prep Videos are available via MyMathLab®, *or on* YouTube™ (www.youtube.com/user/AngelElementaryAlg).

1. **Money** Five hundred dollars was divided between Boris and Monique. If Monique received n dollars, write an expression for the amount Boris received.

2. **Restaurants** The money Sally Sestini earned was \$6000 more than twice what William Rowley earned. Write an expression for what Sally earned.

3. Write an expression for the number of seconds in t minutes.

4. **Season Tickets** The cost of a season ticket for football games increased by 6% from the previous year's cost, c. Write an expression for the cost of a season ticket for this year.

See Exercise 4.

In Exercises 5–9, select a variable to represent one quantity and state what it represents. Express the second quantity in terms of the variable selected.

5. **Tic Tacs** The number of packages of peppermint flavored Tic Tacs® sold was 105 packages less than 7 times the number of packages of orange flavored Tic Tacs sold.

6. **Men and Women** At a play there were 600 men and women.

7. **Numbers** The larger of two numbers is 1 less than twice the smaller number. Write an expression for the smaller number subtracted from the larger number

8. **Liquid Tide** The number of fluid ounces in a large bottle of Tide® is 18 fluid ounces more than the amount in a smaller bottle. Write an expression for the sum of the amounts in a small and large bottle.

9. **Nuts** The cost of a can of Planters® Deluxe Nuts is 84% greater than the cost of a can of Planters Peanuts. Write an expression for the difference in costs between the Deluxe Nuts and the Peanuts.

In Exercises 10–25, set up an equation that can be used to solve the problem. Solve the problem and answer the question asked.

10. **Integers** The sum of two integers is 158. Find the two integers if the larger is 10 less than twice the smaller.

11. **Consecutive Odd Integers** For two consecutive odd integers, the sum of the smaller and 4 times the larger is 33. Find the integers.

12. **Numbers** One number is 12 less than 5 times the other. Find the two numbers if their sum is 42.

13. **Lawn Furniture** Dona Bishop purchased a set of lawn furniture. The cost of the furniture, including a 6% tax, was \$2650. Find the cost of the furniture before tax.

14. **Eating Out** Mark Sullivan has only $40. He wishes to leave a 15% tip. Find the price of the most expensive meal that he can order.

15. **Business Venture** Julie Burgmeier receives twice the profit in a business venture than Peter Ancona does. If the profit for the year was $120,000, how much will each receive?

16. **Snowplowing** William Echols is going to hire a snow plowing service. Elizabeth Suco charges an annual fee of $80, plus $5 each time she plows. Jon Wilkins charges an annual fee of $50, plus $10 each time he plows. How many times would the snow need to be plowed for the cost of both plans to be the same?

17. **Laser Printers** A Delta laser printer costs $499. The cost of printing each page of text is 1 cent. A TexMar laser printer costs $350. The cost of printing each page is 3 cents. How many pages would need to be printed for the total cost of both printers to be the same?

18. **Triangle** A triangle has a perimeter of 75 inches. Find the three sides if one side is 15 inches larger than the smallest side, and the third side is twice the smallest side.

19. **Flag** Paul Murphy's flag has a perimeter of 28 feet. Find the dimensions of the flag if the length is 4 feet less than twice its width.

20. **Trapezoid** The two larger angles of a trapezoid measure 3° more than twice the two smaller angles. Determine the four angles of the trapezoid.

21. **Laying Cable** Ellis and Harlene Matza are digging a shallow 67.2-foot-long trench to lay electrical cable to a new outdoor light fixture they just installed. They start digging at the same time at opposite ends of where the trench is to go, and dig toward each other. Ellis digs at a rate of 0.2 foot per minute faster than Harlene, and they meet after 84 minutes. Find the speed at which each digs.

22. **Running** Alice and Bonnie start running at the same time from the same point and run in the same direction. Alice runs at 8 mph while Bonnie runs at a slower pace. After 2 hours they are 4 miles apart. Determine the speed at which Bonnie is running.

23. **Bulk Candy** A candy shop sells candy in bulk. In one bin is Jelly Belly candy, which sells for $2.20 per pound, and in a second bin is Kits, which sells for $2.75 per pound. How much of each type should be mixed to obtain a 3-pound mixture that sells for $2.40 per pound?

24. **Salt Solution** How many liters of 20% salt solution must be added to 60 liters of 40% salt solution to get a solution that is 35% salt?

25. **Acid Solution** A chemist wishes to make 3 liters of a 6% acid solution by mixing an 8% acid solution and a 5% acid solution. How many liters of each should she mix?

Cumulative Review Test

Take the following test and check your answers with those given in the back of the book. Review any questions that you answered incorrectly. The section where the material was covered is indicated after the answer.

1. **Social Security** The following circle graph shows what the typical retiree receives in social security, as a percent of their total income.

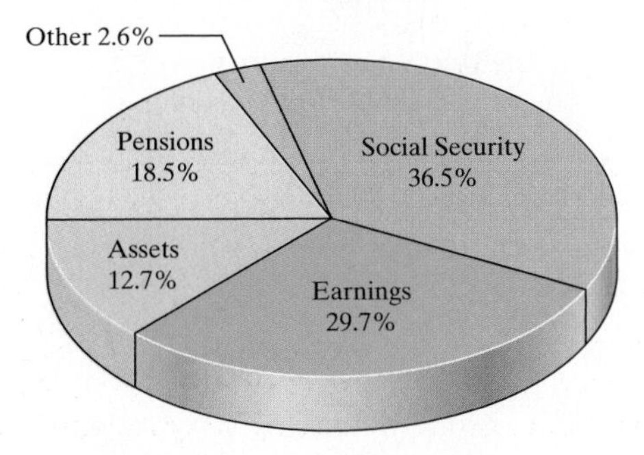

Source: ING Institutional Plan Services

If Emily receives $40,000 per year, and her income is typical of all social security recipients, how much is she receiving in social security?

2. **Olympic Gold Medals** The following graph illustrates the number of gold medals awarded in the 2012 London Olympics for the top 5 countries.

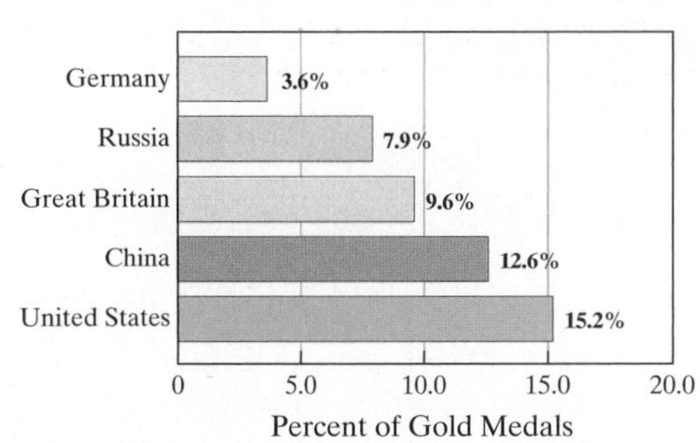

Source: United States Olympic Committee

a) What is the difference, in percent, between the number of gold medals awarded to the United States and Germany?

b) If 302 gold medals were awarded in the 2012 Olympics, how many were awarded to the United States?

3. **Carbon Dioxide Levels** David Warner, an environmentalist, was checking the level of carbon dioxide in the air. On five readings, he got the results indicated in the table below.

Test	Carbon Dioxide (parts per million)
1	5
2	6
3	8
4	12
5	5

   a) Find the mean level of carbon dioxide detected.

   b) Find the median level of carbon dioxide detected.

4. Evaluate $\frac{5}{12} \div \frac{3}{4}$.

5. How much larger is $\frac{2}{3}$ inch than $\frac{1}{8}$ inch?

6. **a)** List the set of natural numbers.

   b) List the set of whole numbers.

   c) What is a rational number?

7. **a)** Evaluate $|-4|$.

   b) Which is greater, $|-5|$ or $|-3|$? Explain.

8. Evaluate $2 - 6^2 \div 2 \cdot 2$.

9. Simplify $4(2x - 3) - 2(3x + 5) - 6$.

In Exercises 10–12, solve the equation.

10. $5x - 6 = x + 14$

11. $6r = 2(r + 3) - (r + 5)$

12. $2(x + 5) = 3(2x - 4) - 4x$

13. If $A = \pi r^2$, find A when $r = 6$.

14. Consider the equation $4x + 8y = 16$.

    a) Solve the equation for y and write the equation in $y = mx + b$ form.

    b) Find y when $x = -4$.

15. Solve the formula $P = 2l + 2w$ for w.

16. **Gas Needed** If Lisa Shough's car can travel 50 miles on 2 gallons of gasoline, how many gallons of gas will it need to travel 225 miles?

17. Solve the inequality $3x - 4 \leq -1$. Graph the solution on a number line and give the solution in interval notation.

18. **Calling Plan** Lori Sypher is considering two cellular telephone plans. Plan A has a monthly charge of $19.95 plus 35 cents per minute. Plan B has a monthly charge of $29.95 plus 10 cents per minute. How long would Lori need to talk in a month for the two plans to have the same total cost?

19. **Sum of Numbers** The sum of two numbers is 29. Find the two numbers if the larger is 11 greater than twice the smaller.

20. **Quadrilateral** One angle of a quadrilateral measures 5° larger than the smallest angle; the third angle measures 50° larger than the smallest angle; and the fourth angle measures 25° greater than 4 times the smallest angle. Find the measure of each angle of the quadrilateral.

4 Exponents and Polynomials

Many areas of science and technology deal with very small and very large numbers. A *googol* is a very large number, 1 followed by 100 zeros. The word googol was first coined by mathematician Edward Kasner (suggested by his 9-year old nephew). Kasner also proclaimed that the number of grains of sand on the beach at Coney Island was less than a googol. In Exercise 94 on page 260, we will examine the scientific notation that allows us to deal more readily with large numbers like a googol in this chapter.

Goals of This Chapter

The major emphasis of this chapter is to teach you how to work with exponents, polynomials, and scientific notation. You must understand the rules of exponents presented in the first two sections in order to be successful with the remaining material in the chapter.

You will use the rules learned in this chapter throughout the book, especially in Chapter 5, when factoring is introduced.

4.1 Exponents

1. Review exponents.
2. Learn the rules of exponents.
3. Simplify an expression before using the expanded power rule.

1 Review Exponents

Recall from Section 1.9, in the expression x^n, x is called the **base** and n is called the **exponent**. x^n is read "x to the nth power" or "x raised to the power n."

$$x^2 = \underbrace{x \cdot x}_{\text{2 factors of } x}$$

$$x^4 = \underbrace{x \cdot x \cdot x \cdot x}_{\text{4 factors of } x}$$

$$x^m = \underbrace{x \cdot x \cdot x \cdot \cdots \cdot x}_{m \text{ factors of } x}$$

Understanding Algebra

x^3 means $x \cdot x \cdot x$ so x is used as a factor 3 times. In x^3, the x is called the *base* and the 3 is called the *exponent*.

EXAMPLE 1 Write $xxxxyyy$ using exponents.

Solution

$$\underbrace{xxxx}_{\text{4 factors of } x} \quad \underbrace{yyy}_{\text{3 factors of } y} = x^4y^3$$

Now Try Exercise 9

Remember:

$$\left.\begin{aligned} x &= 1x \\ x^2y &= 1x^2y \end{aligned}\right\}$$ When a term containing a variable is given without a numerical coefficient, the numerical coefficient is assumed to be 1.

$$\left.\begin{aligned} x &= x^1, \\ xy &= x^1y^1, \\ x^2y &= x^2y^1, \\ 2xy^2 &= 2x^1y^2 \end{aligned}\right\}$$ When a variable or numerical value is given without an exponent, the exponent of that variable or numerical value is assumed to be 1.

2 Learn the Rules of Exponents

EXAMPLE 2 Multiply $x^4 \cdot x^3$.

Solution

$$\overbrace{x \cdot x \cdot x \cdot x}^{x^4} \cdot \overbrace{x \cdot x \cdot x}^{x^3} = x^7$$

Now Try Exercise 11

Example 2 illustrates the *product rule for exponents*.

Product Rule for Exponents

$$x^m \cdot x^n = x^{m+n}$$

When multiplying expressions with the same base, we keep the base and **add** the exponents.

Understanding Algebra

When multiplying expressions with the same base, we *add* the exponents:

$$y^2 \cdot y^4 = y^6$$

In Example 2, we showed that $x^4 \cdot x^3 = x^7$. This problem could also be done using the product rule: $x^4 \cdot x^3 = x^{4+3} = x^7$.

EXAMPLE 3 Multiply each expression using the product rule.

a) $3^2 \cdot 3$ **b)** $2^4 \cdot 2^2$ **c)** $x \cdot x^4$ **d)** $x^3 \cdot x^6$ **e)** $y^4 \cdot y^7$

Solution

a) $3^2 \cdot 3 = 3^2 \cdot 3^1 = 3^{2+1} = 3^3$ or 27

b) $2^4 \cdot 2^2 = 2^{4+2} = 2^6$ or 64

c) $x \cdot x^4 = x^1 \cdot x^4 = x^{1+4} = x^5$

d) $x^3 \cdot x^6 = x^{3+6} = x^9$

e) $y^4 \cdot y^7 = y^{4+7} = y^{11}$

Now Try Exercise 17

Understanding Algebra

Whenever we are working with an expression with a variable in the denominator, we will assume that the denominator does not equal zero. When we are introducing rules of exponents, we will remind you of this. For example, in the Quotient Rule, we will write $x \neq 0$.

AVOIDING COMMON ERRORS

Note in Example 3 **a)** that $3^2 \cdot 3^1$ is 3^3 and not 9^3. *When multiplying powers of the same base, do not multiply the bases.*

CORRECT	INCORRECT
$3^2 \cdot 3^1 = 3^3$	~~$3^2 \cdot 3^1 = 9^3$~~

Note also in Example 3 **b)** that $2^4 \cdot 2^2$ is 2^6 and not 2^8. *When multiplying powers of the same base, do not multiply the exponents.*

CORRECT	INCORRECT
$2^4 \cdot 2^2 = 2^6$	~~$2^4 \cdot 2^2 = 2^8$~~

Example 4 will help you understand the *quotient rule for exponents*.

EXAMPLE 4 Divide $x^5 \div x^3$.

Solution

$$\frac{x^5}{x^3} = \frac{\overset{1}{\cancel{x}} \cdot \overset{1}{\cancel{x}} \cdot \overset{1}{\cancel{x}} \cdot x \cdot x}{\underset{1}{\cancel{x}} \cdot \underset{1}{\cancel{x}} \cdot \underset{1}{\cancel{x}}} = \frac{1x^2}{1} = x^2$$

Now Try Exercise 23

When dividing expressions with the same base, keep the base and *subtract* the exponent in the denominator from the exponent in the numerator.

Understanding Algebra

When dividing expressions with the same base, we *subtract* the exponents:

$$t^7 \div t^4 = \frac{t^7}{t^4} = t^3, t \neq 0$$

Quotient Rule for Exponents

$$\frac{x^m}{x^n} = x^{m-n}, \qquad x \neq 0$$

When dividing expressions with the same base, we keep the base and *subtract* the exponent in the denominator from the exponent in the numerator.

In Example 4, we showed that $\frac{x^5}{x^3} = x^2$. This problem could also be done using the quotient rule: $\frac{x^5}{x^3} = x^{5-3} = x^2$.

EXAMPLE 5 Divide each expression using the quotient rule.

a) $\frac{3^5}{3^2}$ **b)** $\frac{z^8}{z^2}$ **c)** $\frac{x^{12}}{x^5}$ **d)** $\frac{y^{10}}{y^8}$ **e)** $\frac{6^4}{6}$

Solution

a) $\frac{3^5}{3^2} = 3^{5-2} = 3^3$ or 27

b) $\frac{z^8}{z^2} = z^{8-2} = z^6$

c) $\frac{x^{12}}{x^5} = x^{12-5} = x^7$

d) $\frac{y^{10}}{y^8} = y^{10-8} = y^2$

e) $\frac{6^4}{6} = \frac{6^4}{6^1} = 6^{4-1} = 6^3$ or 216

Now Try Exercise 25

AVOIDING COMMON ERRORS

Note in Example 5 **a)** that $\frac{3^5}{3^2}$ is 3^3 and not 1^3. *When dividing powers of the same base, do not divide out the bases.*

CORRECT

$$\frac{3^3}{3^1} = 3^2 \text{ or } 9$$

INCORRECT

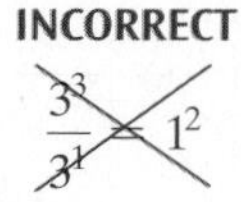

Also note in Example 5 **b)** that $\frac{z^8}{z^2} = z^6$ and not z^4. *When dividing powers of the same base, do not divide the exponents.*

The answer to Example 5 **c),** $\frac{x^{12}}{x^5}$, is x^7. We obtained this answer using the quotient rule. This answer could also be obtained by dividing out the common factors in both the numerator and denominator as follows.

$$\frac{x^{12}}{x^5} = \frac{(\cancel{x}\cdot\cancel{x}\cdot\cancel{x}\cdot\cancel{x}\cdot\cancel{x})\cdot x\cdot x\cdot x\cdot x\cdot x\cdot x\cdot x}{(\cancel{x}\cdot\cancel{x}\cdot\cancel{x}\cdot\cancel{x}\cdot\cancel{x})} = x^7$$

We divided out the product of five x's, which is x^5. We can indicate this process in shortened form as follows.

$$\frac{x^{12}}{x^5} = \frac{\overset{1}{\cancel{x^5}} \cdot x^7}{\underset{1}{\cancel{x^5}}} = x^7$$

To simplify an expression like $\frac{x^5}{x^{12}}$, where the exponent in the denominator is greater than the exponent in the numerator, we divide out common factors (in this case, the common factor is x^5).

Understanding Algebra

To divide expressions with the same base when the denominator's exponent is greater than the numerator's, divide out the common factors:

$$t^4 \div t^7 = \frac{t^4}{t^7} = \frac{\cancel{t^4}}{\cancel{t^4}\cdot t^3} = \frac{1}{t^3},\ t \neq 0$$

Note: Numerator and denominator must have the same base, x.

$$\frac{x^5}{x^{12}} = \frac{\overset{1}{\cancel{x^5}}}{\underset{1}{\cancel{x^5}} \cdot x^7} = \frac{1}{x^7}$$

We will now simplify some expressions by dividing out common factors.

EXAMPLE 6 Simplify each expression by dividing out a common factor in both the numerator and denominator.

a) $\frac{x^9}{x^{12}}$ **b)** $\frac{y^4}{y^9}$

Solution

a) Since the numerator is x^9, we write the denominator with a factor of x^9. Since $x^9 \cdot x^3 = x^{12}$, we rewrite x^{12} as $x^9 \cdot x^3$.

$$\frac{x^9}{x^{12}} = \frac{\overset{1}{\cancel{x^9}}}{\underset{1}{\cancel{x^9}} \cdot x^3} = \frac{1}{x^3}$$

b) $\frac{y^4}{y^9} = \frac{\overset{1}{\cancel{y^4}}}{\underset{1}{\cancel{y^4}} \cdot y^5} = \frac{1}{y^5}$

Now Try Exercise 27

In the next section, we will show another way to simplify expressions like $\frac{x^9}{x^{12}}$ by using the negative exponent rule.

Example 7 leads us to our next rule, the *zero exponent rule*.

EXAMPLE 7 Divide $\frac{x^3}{x^3}$.

Solution By the quotient rule,

$$\frac{x^3}{x^3} = x^{3-3} = x^0$$

However,

$$\frac{x^3}{x^3} = \frac{1x^3}{1x^3} = \frac{1 \cdot \cancel{x} \cdot \cancel{x} \cdot \cancel{x}}{1 \cdot \cancel{x} \cdot \cancel{x} \cdot \cancel{x}} = \frac{1}{1} = 1$$

Since $\frac{x^3}{x^3} = x^0$ and $\frac{x^3}{x^3} = 1$, then x^0 must equal 1.

Now Try Exercise 29

Zero Exponent Rule

$$x^0 = 1, \qquad x \neq 0$$

Any real number, except 0, raised to the zero power equals 1. Note that 0^0 is undefined.

EXAMPLE 8 Simplify each expression. Assume $x \neq 0$ and $z \neq 0$.

a) 3^0 **b)** x^0 **c)** $3x^0$ **d)** $(3x)^0$ **e)** $4x^2y^3z^0$

Solution

a) $3^0 = 1$

b) $x^0 = 1$

c) $3x^0 = 3(x^0)$
$= 3 \cdot 1 = 3$

Remember, the exponent refers only to the immediately preceding symbol unless parentheses are used.

d) $(3x)^0 = 1$

e) $4x^2y^3z^0 = 4x^2y^3 \cdot 1 = 4x^2y^3$

Now Try Exercise 37

AVOIDING COMMON ERRORS

An expression raised to the zero power is not equal to 0; it is equal to 1.

CORRECT	INCORRECT
$x^0 = 1$	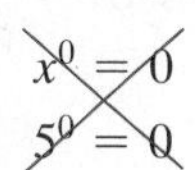
$5^0 = 1$	

The *power rule for exponents* will be explained with the aid of Example 9.

EXAMPLE 9 Simplify $(x^3)^2$.

Solution

$$(x^3)^2 = \underbrace{x^3 \cdot x^3}_{\text{2 factors of } x^3} = x^{3+3} = x^6$$

Now Try Exercise 45

Understanding Algebra

When raising an exponential expression to a power, *multiply* the exponents:

$$(z^2)^4 = z^8$$

Power Rule for Exponents

$$(x^m)^n = x^{m \cdot n}$$

When raising an exponential expression to a power, keep the base and *multiply* the exponents.

EXAMPLE 10 Simplify each expression.

a) $(x^3)^5$ **b)** $(3^4)^2$ **c)** $(y^5)^7$

Solution

a) $(x^3)^5 = x^{3\cdot 5} = x^{15}$ **b)** $(3^4)^2 = 3^{4\cdot 2} = 3^8$ **c)** $(y^5)^7 = y^{5\cdot 7} = y^{35}$

Now Try Exercise 49

HELPFUL HINT

Students often confuse the product and power rules. Note the difference carefully.

Product Rule	Power Rule
$x^m \cdot x^n = x^{m+n}$	$(x^m)^n = x^{m\cdot n}$
$2^3 \cdot 2^5 = 2^{3+5} = 2^8$	$(2^3)^5 = 2^{3\cdot 5} = 2^{15}$

The power rule can be extended to include powers of products and powers of quotients. Consider $(4x^2)^3$, the product $4x^2$ raised to the power 3:

$$\begin{aligned}(4x^2)^3 &= 4x^2 \cdot 4x^2 \cdot 4x^2\\ &= (4\cdot 4\cdot 4)(x^2\cdot x^2\cdot x^2)\\ &= 4^3(x^2)^3 \quad \text{Observe both 4 and } x^2 \text{ are raised to the power 3.}\\ &= 64x^6\end{aligned}$$

This leads us to the *power of a product rule:**

Power of a Product Rule

$$(xy)^n = x^n y^n$$

When a product is raised to a power, each factor in the product is raised to that power.

EXAMPLE 11 Simplify each expression.

a) $(a^2b^3)^4$ **b)** $\left(-\frac{2}{3}x^6y\right)^3$

Solution

a) $$\begin{aligned}(a^2b^3)^4 &= (a^2)^4\cdot(b^3)^4 \quad \text{Raise each factor to the power 4.}\\ &= a^8b^{12}\end{aligned}$$

b) $$\begin{aligned}\left(-\frac{2}{3}x^6y\right)^3 &= \left(-\frac{2}{3}\right)^3\cdot(x^6)^3\cdot(y)^3 \quad \text{Raise each factor to the power 3.}\\ &= -\frac{8}{27}x^{18}y^3 \quad \text{Because } \left(-\frac{2}{3}\right)^3 = \left(-\frac{2}{3}\right)\cdot\left(-\frac{2}{3}\right)\cdot\left(-\frac{2}{3}\right)\end{aligned}$$

Now Try Exercise 59

*When we write exponential expressions, it is always understood that the values of variables we use in definitions are always restricted to avoid 0^0. Thus, when we write $(xy)^n$, for example, n cannot be zero when either x or y is zero.

Just like we extended the power rule to a product, we can apply similar reasoning to extend the power rule to a quotient. Consider $\left(\frac{5}{x^3}\right)^2$, the quotient $\frac{5}{x^3}$ raised to the power 2:

$$\left(\frac{5}{x^3}\right)^2 = \frac{5}{x^3} \cdot \frac{5}{x^3}$$
$$= \frac{5 \cdot 5}{x^3 \cdot x^3}$$
$$= \frac{5^2}{(x^3)^2} \quad \text{Observe both the numerator and denominator are raised to the power 2.}$$
$$= \frac{25}{x^6}$$

This leads us to the *power of a quotient rule*:

Power of a Quotient Rule

$$\left(\frac{x}{y}\right)^n = \frac{x^n}{y^n}, \quad y \neq 0$$

When a quotient is raised to a power, both the numerator and the denominator are raised to that power.

EXAMPLE 12 Simplify each expression.

a) $\left(\frac{p^2}{q^3}\right)^4$ **b)** $\left(-\frac{x^2}{z^3}\right)^5$

Solution

a) $\left(\frac{p^2}{q^3}\right)^4 = \frac{(p^2)^4}{(q^3)^4}$ Raise numerator and denominator to the power 4.

$$= \frac{p^8}{q^{12}}$$

b) $\left(-\frac{x^2}{z^3}\right)^5 = \left(\frac{-x^2}{z^3}\right)^5$

$$= \frac{(-x^2)^5}{(z^3)^5} \quad \text{Raise numerator and denominator to the power 5.}$$
$$= \frac{-x^{10}}{z^{15}} \text{ or } -\frac{x^{10}}{z^{15}} \quad \text{Either answer is acceptable.}$$

Now Try Exercise 85

We now combine the previous two rules, the power rule for products and the power rule for quotients, into the *expanded power rule*.

Expanded Power Rule for Exponents

$$\left(\frac{ax}{by}\right)^m = \frac{a^m x^m}{b^m y^m}, \quad b \neq 0, y \neq 0$$

Every factor within parentheses is raised to the power outside the parentheses when the expression is simplified.

EXAMPLE 13 Simplify $\left(\frac{2r^2s}{t^4}\right)^3$.

Solution

$$\left(\frac{2r^2s}{t^4}\right)^3 = \frac{2^3(r^2)^3s^3}{(t^4)^3}$$ Expanded power rule; each factor is raised to the power 3.

$$= \frac{8r^6s^3}{t^{12}}$$

Now Try Exercise 87

3 Simplify an Expression Before Using the Expanded Power Rule

Whenever we have an expression raised to a power, it helps to simplify the expression in parentheses before using the expanded power rule.

EXAMPLE 14 Simplify $\left(\frac{9x^3y^2}{3xy^2}\right)^3$.

Solution We first simplify the expression within parentheses by dividing out common factors.

$$\left(\frac{9x^3y^2}{3xy^2}\right)^3 = \left(\frac{9}{3}\cdot\frac{x^3}{x}\cdot\frac{y^2}{y^2}\right)^3 = (3x^2)^3$$

Now we use the expanded power rule to simplify further.

$$(3x^2)^3 = 3^3(x^2)^3 = 27x^6$$

Thus, $\left(\frac{9x^3y^2}{3xy^2}\right)^3 = 27x^6$.

Now Try Exercise 93

HELPFUL HINT

Study Tip

Be very careful when writing exponents. Since exponents are generally smaller than regular text, take your time and write them clearly, and position them properly. If exponents are not written clearly it is very easy to confuse exponents such as 2 and 3, or 1 and 4, or 0 and 6. If you write down or carry an exponent from step to step incorrectly, you will obtain an incorrect answer.

EXAMPLE 15 Simplify $\left(-\frac{6c^8d^2}{9c^5d^6}\right)^4$.

Solution Begin by simplifying the expression within parentheses.

$$\left(-\frac{6c^8d^2}{9c^5d^6}\right)^4 = \left(\frac{-6}{9}\cdot\frac{c^8}{c^5}\cdot\frac{d^2}{d^6}\right)^4 = \left(\frac{-2}{3}\cdot\frac{c^3}{1}\cdot\frac{1}{d^4}\right)^4 = \left(\frac{-2c^3}{3d^4}\right)^4$$

Now use the expanded power rule to simplify further.

$$\left(\frac{-2c^3}{3d^4}\right)^4 = \frac{(-2)^4(c^3)^4}{3^4(d^4)^4} = \frac{16c^{12}}{81d^{16}}$$

Thus, $\left(-\frac{6c^8d^2}{9c^5d^6}\right)^4 = \frac{16c^{12}}{81d^{16}}$.

Now Try Exercise 95

AVOIDING COMMON ERRORS

Students sometimes make errors in simplifying expressions containing exponents. One of the most common errors follows.

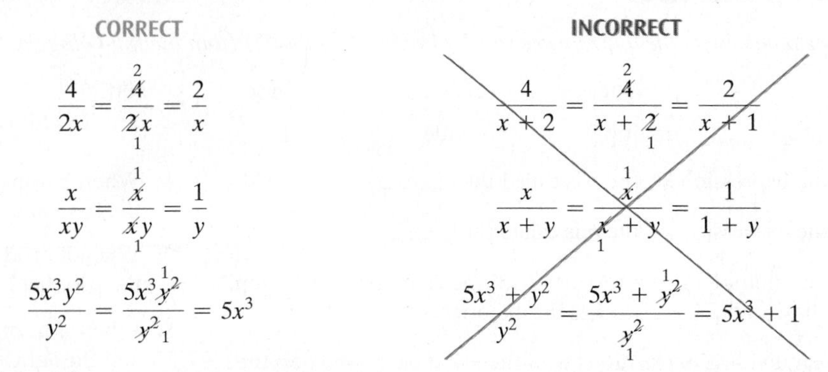

The simplifications on the right side are not correct because only common *factors* can be divided out (remember, factors are multiplied together). In the first denominator on the right, $x + 2$, the x and 2 are terms, not factors, since they are being added. Similarly, in the second denominator on the right, $x + y$, the x and the y are terms, not factors, since they are being added. Also, in the numerator $5x^3 + y^2$, the $5x^3$ and y^2 are terms, not factors. No common factors can be divided out in the fractions on the right.

EXAMPLE 16 Simplify $(2a^5b^3)^5(3a^2b)$.

Solution First simplify $(2a^5b^3)^5$ by using the expanded power rule.

$$(2a^5b^3)^5 = 2^5a^{5\cdot 5}b^{3\cdot 5} = 32a^{25}b^{15}$$

Now use the product rule to simplify further.

$$\begin{aligned}(2a^5b^3)^5(3a^2b) &= (32a^{25}b^{15})(3a^2b^1)\\ &= 32\cdot 3\cdot a^{25}\cdot a^2\cdot b^{15}\cdot b^1\\ &= 96a^{25+2}b^{15+1}\\ &= 96a^{27}b^{16}\end{aligned}$$

Thus, $(2a^5b^3)^5(3a^2b) = 96a^{27}b^{16}$.

Now Try Exercise 125

Summary of the Rules of Exponents Presented in This Section

1. Product Rule	$x^m \cdot x^n = x^{m+n}$
2. Quotient Rule	$\dfrac{x^m}{x^n} = x^{m-n},\quad x \neq 0$
3. Zero Exponent Rule	$x^0 = 1,\quad x \neq 0$
4. Power Rule	$(x^m)^n = x^{m\cdot n}$
5. Power of a Product Rule	$(xy)^n = x^ny^n$
6. Power of a Quotient Rule	$\left(\dfrac{x}{y}\right)^n = \dfrac{x^n}{y^n},\quad y \neq 0$
7. Expanded Power Rule	$\left(\dfrac{ax}{by}\right)^m = \dfrac{a^mx^m}{b^my^m},\quad b \neq 0,\ y \neq 0$

4.1 Exercise Set MathXL® MyMathLab®

Warm-Up Exercises

Fill in the blanks with the appropriate word, phrase, or symbol(s) from the following list.

factor	product	exponent	base	add	subtract	a^3b^5
denominator	multiply	divide	1	0	a^5b^3	a^5b^5

1. In the expression x^n, the x is called the ________.
2. In the expression x^n, the n is called the ________.
3. When multiplying expressions with the same base, we keep the base and ________ the exponents.
4. When dividing expressions with the same base, we keep the base and ________ the exponent in the denominator from the exponent in the numerator.
5. Any real number, except 0, raised to a power of ________ equals 1.
6. When raising an exponential expression to a power, keep the base and ________ the exponents.
7. When a product is raised to a power, each ________ in the product is raised to that power.
8. When a quotient is raised to a power, both the numerator and the ________ are raised to that power.
9. When *aaaaabbb* is rewritten using exponents, the expression becomes ________.
10. When *aaabbbbb* is rewritten using exponents, the expression becomes ________.

Practice the Skills

Multiply.

11. $x^6 \cdot x^4$
12. $x^6 \cdot x$
13. $-z^4 \cdot z$
14. $-x^3 \cdot x^4$
15. $y^3 \cdot y^2$
16. $t^7 \cdot t^2$
17. $3^2 \cdot 3^3$
18. $4^2 \cdot 4^3$
19. $z^3 \cdot z^5$
20. $2^4 \cdot 2^2$

Divide.

21. $\frac{6^2}{6}$
22. $\frac{2^6}{2}$
23. $\frac{x^{10}}{x^3}$
24. $\frac{y^9}{y}$
25. $\frac{3^6}{3^2}$
26. $\frac{4^5}{4^3}$
27. $\frac{y^4}{y^6}$
28. $\frac{a^7}{a^9}$
29. $\frac{c^4}{c^4}$
30. $\frac{5^4}{5^4}$
31. $\frac{q^3}{q^9}$
32. $\frac{x^9}{x^{13}}$

Simplify.

33. x^0
34. 5^0
35. $3x^0$
36. $-7x^0$
37. $4(5d)^0$
38. $-2(8t)^0$
39. $-9(-4y)^0$
40. $-(-x)^0$
41. $6x^3y^2z^0$
42. $-5xy^2z^0$
43. $-8r(st)^0$
44. $-3(a^2b^5c^3)^0$

Simplify.

45. $(x^4)^2$
46. $(a^5)^3$
47. $(x^5)^5$
48. $(y^5)^2$
49. $(x^4)^3$
50. $(x^5)^4$
51. $(3x)^2$
52. $(4y)^3$
53. $(a^2b)^3$
54. $(x^3y^4)^3$
55. $(-2w^2)^3$
56. $(-3t^3)^3$
57. $(-3x)^2$
58. $(-xy)^4$
59. $(4x^3y^2)^3$
60. $(3a^2b^4)^3$

Simplify.

61. $\left(\frac{x}{3}\right)^2$
62. $\left(\frac{2}{y}\right)^4$
63. $\left(\frac{y}{x}\right)^4$
64. $\left(\frac{p}{q}\right)^4$
65. $\left(\frac{-6}{x}\right)^3$
66. $\left(\frac{c}{-3}\right)^3$
67. $\left(\frac{2x}{y}\right)^3$
68. $\left(\frac{3s}{t^2}\right)^2$
69. $\left(\frac{4p}{5}\right)^2$
70. $\left(\frac{2y^3}{x}\right)^4$
71. $\left(\frac{-5z^3}{8}\right)^2$
72. $\left(\frac{-4x^2}{5}\right)^2$

Simplify.

73. $\frac{a^8 b}{ab^4}$
74. $\frac{x^3 y^5}{x^7 y}$
75. $\frac{5x^{12} y^2}{10xy^9}$
76. $\frac{10x^3 y^8}{2xy^{10}}$
77. $\frac{30y^5 z^3}{5yz^6}$
78. $\frac{54g^3 h^3}{45g^8 h}$
79. $\frac{35x^4 y^9}{15x^9 y^{12}}$
80. $\frac{6m^3 n^9}{9m^7 n^{12}}$
81. $-\frac{36xy^7 z}{12x^4 y^5 z}$
82. $-\frac{56a^5 b^2 c^7}{72a^2 b^8 c^7}$
83. $-\frac{6x^2 y^7 z}{3x^5 y^9 z^6}$
84. $-\frac{25x^4 y^{10}}{30x^3 y^7 z}$

Simplify.

85. $\left(\frac{10x^4}{5x^6}\right)^3$
86. $\left(\frac{4x^4}{8x^8}\right)^3$
87. $\left(\frac{6y^6}{2y^3}\right)^3$
88. $\left(\frac{4xy^5}{y}\right)^3$
89. $\left(\frac{6a^2 b^4}{3a^7 b^9}\right)^0$
90. $\left(\frac{34j^5 k^7}{51j^6 k^3}\right)^0$
91. $\left(\frac{x^4 y^3}{x^2 y^5}\right)^2$
92. $\left(\frac{2x^7 y^2}{4xy}\right)^3$
93. $\left(\frac{9y^2 z^7}{18y^9 z}\right)^4$
94. $\left(\frac{34j^5 k^7}{51j^6 k^3}\right)^3$
95. $\left(\frac{-25s^4 t}{5s^6 t^4}\right)^3$
96. $\left(\frac{-64xy^6}{32xy^9}\right)^4$

Simplify.

97. $(3xy^4)^2$
98. $(4ab^3)^3$
99. $(5x^2 y)(3xy^5)$
100. $(6xy^5)(3x^2 y^4)$
101. $(-2xy)(3xy)$
102. $(-3x^4 y^2)(5x^2 y)$
103. $(-8s^2 t^3)(-4st^4)$
104. $(-5xy)(-2xy^6)$
105. $(-3p^2 q)^2(-p^2 q)$
106. $(3x^2)^4(2xy^5)$
107. $(7r^3 s^2)^2(9r^3 s^4)^0$
108. $(2c^3 d^2)^2(3cd)^0$

Simplify.

109. $(-x)^2$
110. $(-y)^4$
111. $\left(\frac{x^5 y^5}{xy^5}\right)^3$
112. $\left(\frac{j^6 k^2}{j^6 k^7}\right)^3$
113. $(2.5x^3)^2$
114. $(0.4w^3)^2$
115. $\frac{x^9 y^3}{x^2 y^7}$
116. $\frac{x^2 y^6}{x^4 y}$
117. $\left(-\frac{m^4}{n^3}\right)^3$
118. $\left(\frac{-3x^3}{4}\right)^3$
119. $(-6x^3 y^2)^3$
120. $(-2s^5 t^3)^5$
121. $(-wx^2 y^3 z^4)^4$
122. $(-a^2 b^7 cd^3)^4$
123. $(9r^4 s^5)^3$
124. $(2xy^4)^3$
125. $(4x^2 y)(3xy^2)^3$
126. $(5x^4 z^{10})^2(2x^2 z^8)$
127. $(1.3x^2 y^4)^2$
128. $(2.1a^2 b^6)^2$
129. $(x^7 y^5)(xy^2)^4$
130. $(4c^3 d^2)(2c^5 d^3)^2$
131. $\left(\frac{-x^4 z^7}{x^2 z^5}\right)^4$
132. $\left(-\frac{12x}{16x^7 y^2}\right)^2$

Study the Avoiding Common Errors box on page 239. Simplify the following expressions by dividing out common factors. If the expression cannot be simplified by dividing out common factors, so state.

133. $\frac{a + b}{b}$
134. $\frac{xy}{x}$
135. $\frac{y^2 + 3}{y}$
136. $\frac{a + 9}{3}$
137. $\frac{6yz^4}{yz^2}$
138. $\frac{x}{x + 1}$
139. $\frac{a^2 + b^2}{a^2}$
140. $\frac{x^4}{x^2 y}$

Problem Solving

141. What is the value of $a^3 b$ if $a = 2$ and $b = 5$?

142. What is the value of xy^2 if $x = -3$ and $y = -4$?

143. What is the value of $(xy)^0$ if $x = -5$ and $y = 3$?

144. What is the value of $(xy)^0$ if $x = 2$ and $y = 4$?

145. Consider the expression $(-9x^4 y^6)^8$. When the power of a product rule is used to simplify the expression, will the *sign* of the simplified expression be positive or negative? Explain how you determined your answer.

146. Consider the expression $(-x^5 y^7)^9$. When the power of a product rule is used to simplify the expression, will the *sign* of the simplified expression be positive or negative? Explain how you determined your answer.

Write an expression for the total area of the figure or figures shown.

147.

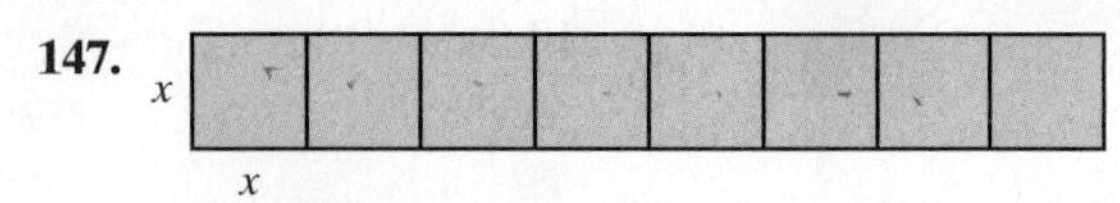

148.

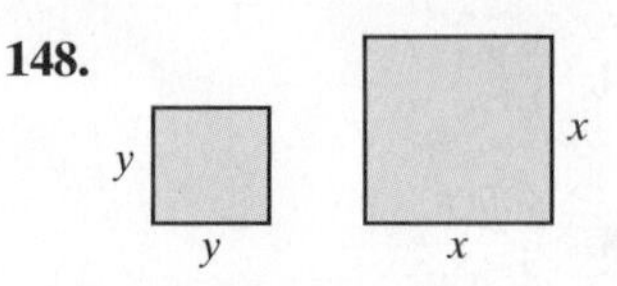

149.

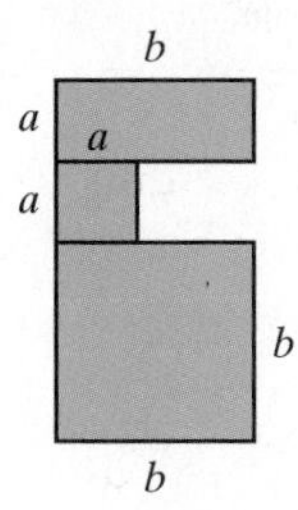

150.

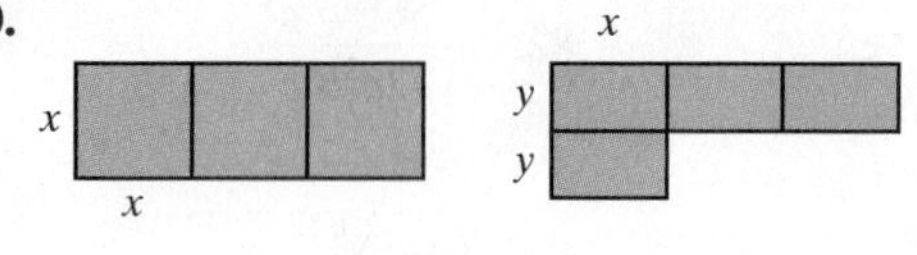

Concept/Writing Exercises

151. Explain the difference between the product rule and the power rule for exponents.

152. Explain the zero exponent rule. What restrictions are there on the base?

Challenge Problems

Simplify.

153. $(3yz^2)^2\left(\dfrac{2y^3z^5}{10y^6z^4}\right)^0(4y^2z^3)^3$

154. $\left(\dfrac{3x^4y^5}{6x^6y^8}\right)^3\left(\dfrac{9x^7y^8}{3x^3y^5}\right)^2$

Group Activity

Discuss and answer Exercise 155 as a group, according to the instructions.

155. In the next section we will be working with negative exponents. To prepare for that work, use the expression $\dfrac{3^2}{3^3}$ to work parts **a)** through **c)**. Work parts **a)** through **d)** individually, then part **e)** as a group.

a) Divide out common factors in the numerator and denominator and determine the value of the expression.

b) Use the quotient rule on the given expression and write down your results.

c) Write a statement of equality using the results of part **a)** and **b)** above.

d) Repeat parts **a)** through **c)** for the expression $\dfrac{2^3}{2^4}$.

e) As a group, compare your answers to parts **a)** through **d)**, then write an exponential expression for $\dfrac{1}{x^m}$.

Cumulative Review Exercises

[1.9] **156.** Evaluate $3^4 \div 3^3 - (5 - 8) + 7$.

[2.1] **157.** Simplify $-4(x - 3) + 5x - 2$.

[2.5] **158.** Solve the equation
$2(x + 4) - 3 = 5x + 4 - 3x + 1$.

[2.6] **159. a)** Use the formula $P = 2l + 2w$ to find the length of the sides of the rectangle shown if the perimeter of the rectangle is 26 inches.

b) Solve the formula $P = 2l + 2w$ for w.

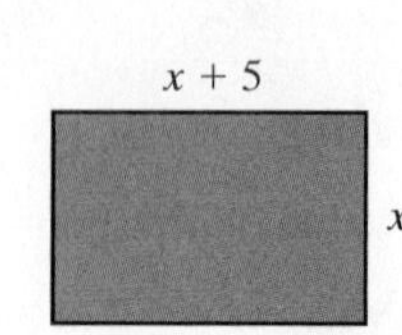

4.2 Negative Exponents

1. Understand the negative exponent rule.
2. Simplify expressions containing negative exponents.

1 Understand the Negative Exponent Rule

One additional rule that involves exponents is the negative exponent rule.

EXAMPLE 1 Simplify $\frac{x^3}{x^5}$ by **a)** using the quotient rule and **b)** dividing out common factors.

Solution

a) By the quotient rule,

$$\frac{x^3}{x^5} = x^{3-5} = x^{-2}$$

b) By dividing out common factors,

$$\frac{x^3}{x^5} = \frac{\cancel{x} \cdot \cancel{x} \cdot \cancel{x}}{\cancel{x} \cdot \cancel{x} \cdot \cancel{x} \cdot x \cdot x} = \frac{1}{x^2}$$

Now Try Exercise 35

In Example 1, we see that $\frac{x^3}{x^5}$ is equal to both x^{-2} and $\frac{1}{x^2}$. Therefore, x^{-2} must equal $\frac{1}{x^2}$. That is, $x^{-2} = \frac{1}{x^2}$. This is an example of the *negative exponent rule*.

Negative Exponent Rule

$$x^{-m} = \frac{1}{x^m}, \quad x \neq 0$$

When a variable or number is raised to a negative exponent, the expression may be rewritten as 1 divided by the variable or number raised to that positive exponent.

Examples

$$x^{-6} = \frac{1}{x^6} \qquad 4^{-2} = \frac{1}{4^2} = \frac{1}{16}$$

$$y^{-7} = \frac{1}{y^7} \qquad 5^{-3} = \frac{1}{5^3} = \frac{1}{125}$$

Understanding Algebra

Negative exponents have nothing to do with the sign of a quantity–they simply indicate "reciprocal":

$$(6)^{-2} = \frac{1}{6^2} = \frac{1}{36}$$

AVOIDING COMMON ERRORS

Students sometimes believe that a negative exponent automatically makes the value of the expression negative. This is not true.

EXPRESSION	CORRECT	INCORRECT	ALSO INCORRECT
3^{-2}	$\frac{1}{3^2} = \frac{1}{9}$	~~-3^2~~	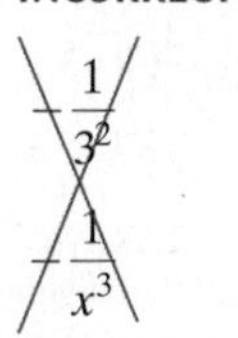
x^{-3}	$\frac{1}{x^3}$	~~$-x^3$~~	

To help you see that the negative exponent rule makes sense, consider the following sequence of exponential expressions and their corresponding values.

$$2^3 = 8, \quad 2^2 = 4, \quad 2^1 = 2, \quad 2^0 = 1, \quad 2^{-1} = \frac{1}{2^1} \text{ or } \frac{1}{2}, \quad 2^{-2} = \frac{1}{2^2} \text{ or } \frac{1}{4}, \quad 2^{-3} = \frac{1}{2^3} \text{ or } \frac{1}{8}$$

Note that each time the exponent decreases by 1, the value of the expression is halved. For example, when we go from 2^3 to 2^2, the value of the expression goes from 8 to 4. If we continue decreasing the exponents beyond $2^0 = 1$, the next exponent in the pattern is -1. If we take half of 1 we get $\frac{1}{2}$. This pattern illustrates that $x^{-m} = \frac{1}{x^m}$.

2 Simplify Expressions Containing Negative Exponents

Generally, *when you are asked to simplify an exponential expression your final answer should contain no negative exponents.* You may simplify exponential expressions using the negative exponent rule and the rules of exponents as shown in the next examples.

EXAMPLE 2 Use the negative exponent rule to write each expression with a positive exponent. Simplify the expressions further when possible.

a) y^{-5} **b)** x^{-4} **c)** 2^{-3} **d)** 6^{-1} **e)** -5^{-3} **f)** $(-5)^{-3}$

Solution

a) $y^{-5} = \frac{1}{y^5}$

b) $x^{-4} = \frac{1}{x^4}$

c) $2^{-3} = \frac{1}{2^3} = \frac{1}{8}$

d) $6^{-1} = \frac{1}{6}$

e) $-5^{-3} = -\frac{1}{5^3} = -\frac{1}{125}$

f) $(-5)^{-3} = \frac{1}{(-5)^3} = \frac{1}{-125} = -\frac{1}{125}$

Now Try Exercise 11

EXAMPLE 3 Use the negative exponent rule to write each expression with a positive exponent.

a) $\frac{1}{x^{-2}}$ **b)** $\frac{1}{4^{-1}}$

Solution First use the negative exponent rule on the denominator. Then simplify further.

a) $\frac{1}{x^{-2}} = \frac{1}{\frac{1}{x^2}} = 1 \div \frac{1}{x^2} = \frac{1}{1} \cdot \frac{x^2}{1} = x^2$

b) $\frac{1}{4^{-1}} = \frac{1}{\frac{1}{4}} = 1 \div \frac{1}{4} = \frac{1}{1} \cdot \frac{4}{1} = 4$

Now Try Exercise 15

Understanding Algebra

Negative exponent in denominator:

$$\frac{3}{x^{-4}} = 3 \div x^{-4} = 3 \div \frac{1}{x^4}$$
$$= 3 \cdot \frac{x^4}{1} = 3x^4$$

Recall that dividing by a quantity is the same as multiplying by its reciprocal.

HELPFUL HINT

From Examples 2 and 3, we can see that when a factor is moved from the denominator to the numerator or from the numerator to the denominator, the sign of the *exponent* changes.

$$x^{-4} = \frac{1}{x^4} \qquad \frac{1}{x^{-4}} = x^4$$

$$3^{-5} = \frac{1}{3^5} \qquad \frac{1}{3^{-5}} = 3^5$$

Now let's look at additional examples that combine two or more of the rules presented so far.

EXAMPLE 4 Simplify. **a)** $(z^{-5})^4$ **b)** $(4^2)^{-3}$

Solution

a) $(z^{-5})^4 = z^{(-5)(4)}$ Power rule

$= z^{-20}$

$= \dfrac{1}{z^{20}}$ Negative exponent rule

b) $(4^2)^{-3} = 4^{(2)(-3)}$ Power rule

$= 4^{-6}$

$= \dfrac{1}{4^6}$ Negative exponent rule

Now Try Exercise 25

EXAMPLE 5 Simplify. **a)** $x^3 \cdot x^{-5}$ **b)** $3^{-4} \cdot 3^{-7}$

Solution

a) $x^3 \cdot x^{-5} = x^{3+(-5)}$ Product rule

$= x^{-2}$

$= \dfrac{1}{x^2}$ Negative exponent rule

b) $3^{-4} \cdot 3^{-7} = 3^{-4+(-7)}$ Product rule

$= 3^{-11}$

$= \dfrac{1}{3^{11}}$ Negative exponent rule

Now Try Exercise 51

AVOIDING COMMON ERRORS

What is the sum of $3^2 + 3^{-2}$? Look carefully at the correct solution.

CORRECT

$3^2 + 3^{-2} = 9 + \dfrac{1}{9}$

$= 9\dfrac{1}{9}$

INCORRECT

~~$3^2 + 3^{-2} = 3^0 = 1$~~

Note that $3^2 \cdot 3^{-2} = 3^{2+(-2)} = 3^0 = 1$.

EXAMPLE 6 Simplify. **a)** $\dfrac{b^{-5}}{b^{13}}$ **b)** $\dfrac{6^{-8}}{6^{-5}}$

Solution

a) $\dfrac{b^{-5}}{b^{13}} = b^{-5-13}$ Quotient rule

$= b^{-18}$

$= \dfrac{1}{b^{18}}$ Negative exponent rule

b) $\dfrac{6^{-8}}{6^{-5}} = 6^{-8-(-5)}$ Quotient rule

$= 6^{-8+5}$

$= 6^{-3}$

$= \dfrac{1}{6^3}$ or $\dfrac{1}{216}$ Negative exponent rule

Now Try Exercise 81

Read the following Helpful Hint carefully.

HELPFUL HINT

An alternative way to simplify a fraction involving a variable with a negative exponent in the numerator is to move the variable with the negative exponent from the numerator to the denominator and change the sign of the exponent:

$$\frac{x^{-4}}{x^5} = \frac{1}{x^5 \cdot x^4} = \frac{1}{x^{5+4}} = \frac{1}{x^9}$$

Similarly, if we have a fraction involving a variable with a negative exponent in the denominator we can move the variable with the negative exponent from the denominator to the numerator and change the sign of the exponent:

$$\frac{y^3}{y^{-7}} = y^3 \cdot y^7 = y^{3+7} = y^{10}$$

Now consider a division problem where a number or variable has a negative exponent in both its numerator and its denominator, such as in Example 6 **b).** Another way to simplify such an expression is to move the variable with the more negative exponent from the numerator to the denominator, or from the denominator to the numerator, and change the sign of the exponent from negative to positive. For example,

$$\frac{x^{-8}}{x^{-3}} = \frac{1}{x^8 \cdot x^{-3}} = \frac{1}{x^{8-3}} = \frac{1}{x^5} \quad \text{Note that } -8 < -3.$$

$$\frac{y^{-4}}{y^{-7}} = y^7 \cdot y^{-4} = y^{7-4} = y^3 \quad \text{Note that } -7 < -4.$$

EXAMPLE 7 Simplify.

a) $7x^4(6x^{-9})$ **b)** $\dfrac{16r^3s^{-3}}{8rs^2}$ **c)** $\dfrac{2x^2y^5}{8x^7y^{-3}}$

Solution

a) $7x^4(6x^{-9}) = 7 \cdot 6 \cdot x^4 \cdot x^{-9} = 42x^{-5} = \dfrac{42}{x^5}$

b)
$$\frac{16r^3s^{-3}}{8rs^2} = \frac{16}{8} \cdot \frac{r^3}{r} \cdot \frac{s^{-3}}{s^2}$$
$$= 2 \cdot r^2 \cdot \frac{1}{s^5} = \frac{2r^2}{s^5} \quad s^{-3} \text{ was rewritten in the denominator as } s^3.$$

c)
$$\frac{2x^2y^5}{8x^7y^{-3}} = \frac{2}{8} \cdot \frac{x^2}{x^7} \cdot \frac{y^5}{y^{-3}}$$
$$= \frac{1}{4} \cdot \frac{1}{x^5} \cdot y^8 = \frac{y^8}{4x^5} \quad y^{-3} \text{ was rewritten in the numerator as } y^3.$$

Now Try Exercise 121

EXAMPLE 8 Simplify.

a) $(5x^{-3})^{-2}$ **b)** $(-5x^{-3})^{-2}$ **c)** $(-5x^{-3})^{-3}$

Solution Begin by using the power of a product rule.

a)
$$(5x^{-3})^{-2} = 5^{-2}x^{(-3)(-2)}$$
$$= 5^{-2}x^6$$
$$= \frac{1}{5^2}x^6$$
$$= \frac{x^6}{25}$$

b) $(-5x^{-3})^{-2} = (-5)^{-2}x^{(-3)(-2)}$

$$= \frac{1}{(-5)^2}x^6$$

$$= \frac{x^6}{25}$$

c) $(-5x^{-3})^{-3} = (-5)^{-3}x^{(-3)(-3)}$

$$= \frac{1}{(-5)^3}x^9$$

$$= \frac{1}{-125}x^9$$

$$= -\frac{x^9}{125}$$

Now Try Exercise 105

AVOIDING COMMON ERRORS

Can you explain why the simplification on the right is incorrect?

CORRECT

$$\frac{x^3y^{-2}}{w} = \frac{x^3}{wy^2}$$

INCORRECT

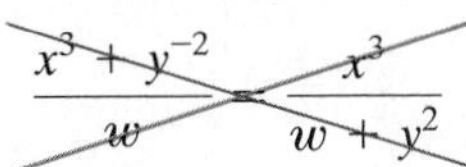

The simplification on the right is incorrect because in the numerator $x^3 + y^{-2}$, the y^{-2} *is not a factor,* it is a term.

EXAMPLE 9 Simplify $\left(\frac{2}{3}\right)^{-2}$.

Solution By the expanded power rule, we may write

$$\left(\frac{2}{3}\right)^{-2} = \frac{2^{-2}}{3^{-2}} = \frac{\frac{1}{2^2}}{\frac{1}{3^2}} = \frac{1}{2^2} \cdot \frac{3^2}{1} = \frac{3^2}{2^2} = \frac{9}{4}$$

Now Try Exercise 95

If we examine the results of Example 9, we see that

$$\left(\frac{2}{3}\right)^{-2} = \frac{3^2}{2^2} = \left(\frac{3}{2}\right)^2.$$

This example illustrates that $\left(\frac{a}{b}\right)^{-m} = \left(\frac{b}{a}\right)^m$ when $a \neq 0$ and $b \neq 0$. Thus, for example, $\left(\frac{3}{4}\right)^{-5} = \left(\frac{4}{3}\right)^5$ and $\left(\frac{5}{9}\right)^{-3} = \left(\frac{9}{5}\right)^3$. We can summarize this information as follows.

A Fraction Raised to a Negative Exponent Rule

For a fraction of the form $\frac{a}{b}$, $a \neq 0$ and $b \neq 0$, $\left(\frac{a}{b}\right)^{-m} = \left(\frac{b}{a}\right)^m$.

When a fraction is raised to a negative exponent, the expression may be rewritten as the reciprocal of the fraction raised to that positive exponent.

EXAMPLE 10 Simplify. **a)** $\left(\frac{4}{5}\right)^{-3}$ **b)** $\left(\frac{x^5}{y^7}\right)^{-4}$

Solution We use the above rule to simplify.

a) $\left(\frac{4}{5}\right)^{-3} = \left(\frac{5}{4}\right)^{3} = \frac{5^3}{4^3} = \frac{125}{64}$

b) $\left(\frac{x^5}{y^7}\right)^{-4} = \left(\frac{y^7}{x^5}\right)^{4} = \frac{(y^7)^4}{(x^5)^4} = \frac{y^{28}}{x^{20}}$

Now Try Exercise 97

EXAMPLE 11 Simplify. **a)** $\left(\frac{x^2y^{-3}}{z^4}\right)^{-5}$ **b)** $\left(\frac{2x^{-3}y^2z}{x^2}\right)^{2}$

Solution

a) We will work part **a)** using two different methods. In method 1, we begin by using the expanded power rule. In method 2, we use a fraction raised to a negative exponent rule before we use the expanded power rule. You may use either method.

Method 1

$$\left(\frac{x^2y^{-3}}{z^4}\right)^{-5} = \frac{x^{2(-5)}y^{(-3)(-5)}}{z^{4(-5)}} \quad \text{Expanded power rule}$$

$$= \frac{x^{-10}y^{15}}{z^{-20}} \quad \text{Multiplied exponents}$$

$$= \frac{y^{15}z^{20}}{x^{10}} \quad \text{Negative exponent rule}$$

Method 2

$$\left(\frac{x^2y^{-3}}{z^4}\right)^{-5} = \left(\frac{z^4}{x^2y^{-3}}\right)^{5} \quad \left(\frac{a}{b}\right)^{-m} = \left(\frac{b}{a}\right)^{m}$$

$$= \left(\frac{y^3z^4}{x^2}\right)^{5} \quad \text{Simplified expression within parentheses}$$

$$= \frac{y^{3\cdot5}z^{4\cdot5}}{x^{2\cdot5}} \quad \text{Expanded power rule}$$

$$= \frac{y^{15}z^{20}}{x^{10}} \quad \text{Multiplied exponents}$$

Understanding Algebra

Remember: a negative exponent indicates "reciprocal."

$$\left(\frac{2x^2}{3y^4}\right)^{-3} = \left(\frac{3y^4}{2x^2}\right)^{3} = \frac{27y^{12}}{8x^6}$$

b) First simplify the expression within parentheses, then square the results. To simplify, we note that $\frac{x^{-3}}{x^2}$ becomes $\frac{1}{x^5}$.

$$\left(\frac{2x^{-3}y^2z}{x^2}\right)^{2} = \left(\frac{2y^2z}{x^5}\right)^{2} = \frac{2^2y^{2\cdot2}z^{1\cdot2}}{x^{5\cdot2}} = \frac{4y^4z^2}{x^{10}}$$

Now Try Exercise 125

Summary of Rules of Exponents

1. Product Rule	$x^m \cdot x^n = x^{m+n}$
2. Quotient Rule	$\frac{x^m}{x^n} = x^{m-n}, \quad x \neq 0$
3. Zero Exponent Rule	$x^0 = 1, \quad x \neq 0$
4. Power Rule	$(x^m)^n = x^{m\cdot n}$
5. Power of a Product Rule	$(xy)^n = x^ny^n$

6. Power of a Quotient Rule	$\left(\frac{x}{y}\right)^n = \frac{x^n}{y^n}$,	$y \neq 0$
7. Expanded Power Rule	$\left(\frac{ax}{by}\right)^m = \frac{a^m x^m}{b^m y^m}$,	$b \neq 0, y \neq 0$
8. Negative Exponent Rule	$x^{-m} = \frac{1}{x^m}$,	$x \neq 0$
9. Fraction Raised to a Negative Exponent Rule	$\left(\frac{a}{b}\right)^{-m} = \left(\frac{b}{a}\right)^m$,	$a \neq 0, b \neq 0$

4.2 Exercise Set MathXL® MyMathLab®

Warm-Up Exercises

Fill in the blanks with the appropriate word, phrase, or symbol(s) from the following list.

positive	negative	multiplying	dividing	opposite	reciprocal	$\frac{1}{25}$
9	−9	$\frac{1}{9}$	$-\frac{1}{9}$	25	−25	$-\frac{1}{25}$

1. When an expression is raised to a negative exponent, the expression may be rewritten as 1 divided by the expression raised to that ________ exponent.
2. When a fraction is raised to a negative exponent, the expression may be rewritten as the ________ of the fraction raised to that positive exponent.
3. Dividing by a nonzero expression is the same as ________ by the reciprocal of the expression.
4. Generally, when you are asked to simplify an exponential expression your final answer should contain no ________ exponents.
5. The expression 3^{-2} in simplified form is ________.
6. The expression -3^2 in simplified form is ________.
7. The expression -3^{-2} in simplified form is ________.
8. The expression $\frac{1}{3^{-2}}$ in simplified form is ________.
9. The expression $\frac{1}{-5^{-2}}$ in simplified form is ________.
10. The expression -5^{-2} in simplified form is ________

Practice the Skills

Simplify.

11. x^{-6}
12. y^{-5}
13. 5^{-1}
14. 7^{-2}
15. $\frac{1}{t^{-3}}$
16. $\frac{1}{b^{-4}}$
17. $\frac{1}{a^{-1}}$
18. $\frac{1}{y^{-4}}$
19. $\frac{1}{6^{-2}}$
20. $\frac{1}{4^{-3}}$
21. $(x^{-2})^{10}$
22. $(x^4)^{-2}$
23. $(y^{-5})^4$
24. $(a^5)^{-4}$
25. $(m^{-5})^{-2}$
26. $(x^{-9})^{-2}$
27. $(3^{-2})^{-1}$
28. $(2^{-2})^{-2}$
29. $y^4 \cdot y^{-2}$
30. $x^7 \cdot x^{-5}$
31. $x^{-3} \cdot x^1$
32. $z^9 \cdot z^{-12}$
33. $3^{-2} \cdot 3^4$
34. $6^{-3} \cdot 6^6$
35. $\frac{r^5}{r^6}$
36. $\frac{x^6}{x^7}$
37. $\frac{p^0}{p^{-3}}$
38. $\frac{x^2}{x^{-1}}$
39. $\frac{x^{-7}}{x^{-3}}$
40. $\frac{n^{-9}}{n^{-3}}$
41. $\frac{3^2}{3^{-1}}$
42. $\frac{4^2}{4^{-1}}$
43. 5^{-3}
44. 3^{-4}
45. $\frac{1}{z^{-9}}$
46. $\frac{1}{b^{-4}}$

47. $(p^{-4})^{-6}$

48. $(x^{-3})^{-4}$

49. $(y^{-2})^{-3}$

50. $(a^{-4})^{-4}$

51. $x^3 \cdot x^{-7}$

52. $y^2 \cdot y^{-9}$

53. $x^{-8} \cdot x^{-7}$

54. $x^{-3} \cdot x^{-5}$

55. -4^{-2}

56. -2^{-3}

57. $(-4)^{-2}$

58. $(-2)^{-3}$

59. $-(-4)^{-2}$

60. $-(-2)^{-3}$

61. $\dfrac{1}{4^{-2}}$

62. $\dfrac{1}{2^{-3}}$

63. $\dfrac{x^{-5}}{x^5}$

64. $\dfrac{x^{-2}}{x^5}$

65. $\dfrac{n^{-5}}{n^{-7}}$

66. $\dfrac{z^{-11}}{z^{-12}}$

67. $\dfrac{9^{-3}}{9^{-3}}$

68. $\dfrac{7^{-9}}{7^{-9}}$

69. $(2^{-1} + 3^{-1})^0$

70. $(3^{-1} + 4^2)^0$

71. $\dfrac{2}{2^{-5}}$

72. $\dfrac{5}{5^{-3}}$

73. $(x^{-4})^{-2}$

74. $(z^{-5})^{-9}$

75. $(x^0)^{-2}$

76. $(x^{-7})^0$

77. $2^{-3} \cdot 2$

78. $8^{-3} \cdot 8^3$

79. $7^{-5} \cdot 7^3$

80. $5^7 \cdot 5^{-9}$

81. $\dfrac{x^{-1}}{x^{-4}}$

82. $\dfrac{z^{-3}}{z^{-7}}$

83. $(4^2)^{-1}$

84. $(2^{-3})^2$

85. $\dfrac{5}{5^{-2}}$

86. $\dfrac{2}{2^{-5}}$

87. $\dfrac{3^{-4}}{3^{-2}}$

88. $\dfrac{4^{-8}}{4^{-5}}$

89. $\dfrac{8^{-1}}{8^{-1}}$

90. $\dfrac{11^{-1}}{11^{-1}}$

91. $(-6x^2)^{-2}$

92. $(-3z^3)^{-2}$

93. $3x^{-2}y^2$

94. $-5x^4y^{-2}$

95. $\left(\dfrac{1}{2}\right)^{-2}$

96. $\left(\dfrac{1}{5}\right)^{-4}$

97. $\left(\dfrac{5}{4}\right)^{-3}$

98. $\left(\dfrac{3}{5}\right)^{-3}$

99. $\left(\dfrac{c^4}{d^2}\right)^{-2}$

100. $\left(\dfrac{x^2}{y}\right)^{-2}$

101. $-\left(\dfrac{r^4}{s}\right)^{-4}$

102. $-\left(\dfrac{m^3}{n^4}\right)^{-5}$

103. $-7a^{-3}b^{-4}$

104. $(3x^2y^3)^{-2}$

105. $(4x^5y^{-3})^{-3}$

106. $2w(3w^{-5})$

107. $(3z^{-4})(6z^{-5})$

108. $2x^5(3x^{-6})$

109. $4x^4(-2x^{-4})$

110. $(9x^5)(-3x^{-7})$

111. $(4x^2y)(3x^3y^{-1})$

112. $(7a^{-6}b^{-1})(a^9b^0)$

113. $(-5y^2)(4y^{-3}z^5)$

114. $(-3y^{-2})(5x^{-1}y^3)$

115. $\dfrac{24d^{12}}{3d^8}$

116. $\dfrac{8z^{-4}}{32z^{-2}}$

117. $\dfrac{36x^{-4}}{9x^{-2}}$

118. $\dfrac{18m^{-3}n^0}{6m^5n^9}$

119. $\dfrac{3x^4y^{-2}}{6y^3}$

120. $\dfrac{16x^{-7}y^{-2}}{4x^5y^2}$

121. $\dfrac{32x^4y^{-2}}{4x^{-2}y^0}$

122. $\dfrac{21x^{-3}z^2}{7xz^{-3}}$

123. $\left(\dfrac{5x^4y^{-7}}{z^3}\right)^{-2}$

124. $\left(\dfrac{b^4c^{-2}}{2d^{-3}}\right)^{-1}$

125. $\left(\dfrac{2r^{-5}s^9}{t^{12}}\right)^{-4}$

126. $\left(\dfrac{5m^{-1}n^{-3}}{p^2}\right)^{-3}$

127. $\left(\dfrac{x^3y^{-4}z}{y^{-2}}\right)^{-6}$

128. $\left(\dfrac{3p^{-1}q^{-2}r^3}{p^2}\right)^3$

129. $\left(\dfrac{p^6q^{-3}}{4p^8}\right)^2$

130. $\left(\dfrac{x^{12}y^5}{y^{-3}z}\right)^{-4}$

Problem Solving

131. a) Does $p^{-1}q^{-1} = \dfrac{1}{pq}$?

b) Does $p^{-1} + q^{-1} = \dfrac{1}{p+q}$?

132. a) Does $\dfrac{x^{-1}y^2}{z} = \dfrac{y^2}{xz}$?

b) Does $\dfrac{x^{-1} + y^2}{z} = \dfrac{y^2}{x+z}$?

Evaluate.

133. $4^2 + 4^{-2}$

134. $8^2 + 8^{-2}$

135. $5^3 + 5^{-3}$

136. $6^{-3} + 6^3$

Evaluate.

137. $5^0 - 3^{-1}$

138. $4^{-1} - 3^{-1}$

139. $2^{-3} - 2^3 \cdot 2^{-3}$

140. $2 \cdot 4^{-1} + 4 \cdot 3^{-1}$

141. $2 \cdot 4^{-1} - 4 \cdot 3^{-1}$

142. $2 \cdot 4^{-1} - 3^{-1}$

143. $3 \cdot 5^0 - 5 \cdot 3^{-2}$

144. $10^2 + 10^{-2}$

Determine the number that when placed in the shaded area makes the statement true.

145. $3^{\square} = \frac{1}{9}$

146. $\frac{1}{2^{\square}} = 64$

147. $\frac{1}{6^{\square}} = 216$

148. $4^{\square} = \frac{1}{256}$

Concept/Writing Exercises

149. Explain the difference between the product rule and the power rule. Give an example of each.

150. a) For what value of x is $x^0 \neq 1$? **b)** Write *pppqqqrrrrr* using exponents.

Challenge Problems

In Exercises 151–153, determine the number (or numbers) that when placed in the shaded area (or areas) make the statement true.

151. $(x^{\square}y^3)^{-2} = \frac{x^4}{y^6}$

152. $(x^4y^{-3})^{\square} = \frac{y^9}{x^{12}}$

153. $(\square x^{\square}y^{-2})^3 = \frac{8}{x^9y^6}$

154. For any nonzero real number a, if $a^{-1} = x$, describe the following in terms of x.

a) $-a^{-1}$ **b)** $\frac{1}{a^{-1}}$

155. Consider $(3^{-1} + 2^{-1})^0$. We know this is equal to 1 by the zero exponent rule. Determine the error in the following calculation. Explain your answer.

$$\begin{aligned}(3^{-1} + 2^{-1})^0 &= (3^{-1})^0 + (2^{-1})^0 \\ &= 3^{-1(0)} + 2^{-1(0)} \\ &= 3^0 + 2^0 \\ &= 1 + 1 = 2\end{aligned}$$

Group Activity

Discuss and answer Exercise 156 as a group.

156. Often problems involving exponents can be done in more than one way. Consider

$$\left(\frac{3x^2y^3}{x}\right)^{-2}$$

a) Group member 1: Simplify this expression by first simplifying the expression within parentheses.

b) Group member 2: Simplify this expression by first using the expanded power rule.

c) Group member 3: Simplify this expression by first using the negative exponent rule.

d) Compare your answers. If you did not all get the same answers, determine why.

e) As a group, decide which method–**a), b),** or **c)**–was the easiest way to simplify this expression.

Cumulative Review Exercises

[2.7] **157. Racing** If a race car travels 104 miles in 52 minutes, how far will it travel in 93 minutes (assuming all conditions stay the same)?

[3.2] **158. Even Integers** The sum of two consecutive even integers is 190. Find the numbers.

[3.3] **159. Shed** Michael Beattie is building a rectangular shed. The perimeter of the shed is to be 56 feet. Determine the dimensions of the shed if the length is to be 8 feet less than twice the width.

[3.4] **160. Simple Interest** Mia Kattee invested \$9000, part at 3% and part at 4% simple interest for a period of one year. How much was invested in each account if the interest earned in the 3% account was \$32 greater than the interest earned in the 4% account?

4.3 Scientific Notation

1 Convert numbers to and from scientific notation.

2 Recognize numbers in scientific notation with a coefficient of 1.

3 Do calculations using scientific notation.

1 Convert Numbers to and from Scientific Notation

Because it is difficult to work with and compute with really large or really small numbers, we often express such numbers using exponents. For example, the projected world population in 2050 of 8,909,000,000 people could be written as 8.909×10^9. An influenza virus with diameter 0.0000001 meter could be written as 1.0×10^{-7} meter. Numbers such as 8.909×10^9 and 1.0×10^{-7} are in a form called *scientific notation*.

Understanding Algebra

A number is in scientific notation if it is written in the form

$$a \times 10^b$$

where $1 \le a < 10$ and b is an integer.

Scientific Notation

A number written in **scientific notation** is written as a number greater than or equal to 1 and less than 10 $(1 \le a < 10)$ multiplied by some power of 10. The exponent on the 10 must be an integer.

Examples of Numbers in Scientific Notation

$$1.2 \times 10^6$$

$$3.762 \times 10^3$$

$$8.07 \times 10^{-2}$$

$$1.0 \times 10^{-5}$$

This is how we change the number 68,400 to scientific notation.

$68{,}400 = 6.84 \times 10{,}000$ — to go from 68,400 to 6.84 the decimal point was moved 4 places to the left

$= 6.84 \times 10^4$ — Note that $10{,}000 = 10 \cdot 10 \cdot 10 \cdot 10 = 10^4$.

Therefore, $68{,}400 = 6.84 \times 10^4$. Note that the exponent on the 10, the 4, is the same as the number of places the decimal point was moved to the left.

Following are general steps to write a number in scientific notation.

To Write a Number in Scientific Notation

1. Move the decimal point in the original number to the right of the first nonzero digit. This will give a number greater than or equal to 1 and less than 10.
2. Count the number of places you moved the decimal point to obtain the number in step 1. If the original number was 10 or greater, the count is considered positive. If the original number was less than 1, the count is considered negative.
3. Multiply the number obtained in step 1 by 10 raised to the count (power) found in step 2.

EXAMPLE 1 Write the following numbers in scientific notation.

a) 18,500 **b)** 0.0000416 **c)** 3,721,000 **d)** 0.0093

Solution

a) The original number is greater than 10; therefore, the exponent is positive. The decimal point in 18,500 belongs after the last zero.

$$18{,}500. = 1.85 \times 10^4$$

4 places

b) The original number is less than 1; therefore, the exponent is negative.

$$0.0000416 = 4.16 \times 10^{-5}$$

5 places

c) $3{,}721{,}000 = 3.721 \times 10^{6}$

6 places

d) $0.0093 = 9.3 \times 10^{-3}$

3 places

Now Try Exercise 13

When we write a number in scientific notation, we are allowed to leave our answer with a negative exponent, as in parts **b)** and **d)** of Example 1.

Now we explain how to write a number in scientific notation as a number without exponents, or in decimal form.

To Convert a Number from Scientific Notation to Decimal Form

1. Observe the exponent of the power of 10.
2. a) If the exponent is positive, move the decimal point in the number (greater than or equal to 1 and less than 10) to the right the same number of places as the exponent. It may be necessary to add zeros to the number. This will result in a number greater than or equal to 10.
   b) If the exponent is 0, do not move the decimal point. Drop the factor 10^0 since it equals 1. This will result in a number greater than or equal to 1 but less than 10.
   c) If the exponent is negative, move the decimal point in the number to the left the same number of places as the exponent. It may be necessary to add zeros to the number. This will result in a number less than 1.

EXAMPLE 2 Write each number without exponents.

a) 2.9×10^4 **b)** 6.28×10^{-3} **c)** 7.95×10^8

Solution

a) Move the decimal point four places to the right (because the exponent, 4, is positive).

$$2.9 \times 10^4 = 2.9 \times 10{,}000 = 29{,}000$$

b) Move the decimal point three places to the left (because the exponent, −3, is negative)

$$6.28 \times 10^{-3} = 0.00628$$

c) Move the decimal point eight places to the right.

$$7.95 \times 10^8 = 795{,}000{,}000$$

Now Try Exercise 29

Understanding Algebra

Some of the base units of the metric system include:

Unit	Abbreviation	Measure
meter	m	length
gram	g	mass
liter	L	volume
hertz	Hz	frequency
byte	B	computer memory

2 Recognize Numbers in Scientific Notation with a Coefficient of 1

The metric system is used throughout most of the world as the main system of measurement. Units of measurement in the metric system may include a *prefix* along with a *base unit*. For example, in the unit *kilometer*, the prefix is *kilo* and the base unit is *meter.*

The prefixes in the metric system indicate a power of 10. The prefix *kilo* indicates 10^3 or 1000. Thus 1 kilometer means 1×10^3 meters. An advantage of using metric units is that the units can be quickly converted into scientific notation. Some other prefixes* and their meanings are given in the following table.

Prefix	Meaning	Symbol	Meaning as a Decimal Number
nano	10^{-9}	n	$\frac{1}{1{,}000{,}000{,}000}$ or 0.000000001
micro	10^{-6}	μ	$\frac{1}{1{,}000{,}000}$ or 0.000001
milli	10^{-3}	m	$\frac{1}{1000}$ or 0.001
base unit**	10^0		1
kilo	10^3	k	1000
mega	10^6	M	1,000,000
giga	10^9	G	1,000,000,000

**The base unit is not a prefix. We included this row to include 10^0 in the chart.

The metric system is used in many applications in science, medicine, and technology. For example, a special kind of carbon molecule may have a diameter of 1.2 nanometers (nm). A cough medicine dosage might be 2.5 milliliters (mL). A computer processor's speed might be 3.2 gigahertz (GHz). These units are written in both scientific notation and decimal form in the table below.

Written Form	Scientific Notation	In Decimal Form
1.2 nm	1.2×10^{-9} meter	0.0000000012 meter
2.5 mL	2.5×10^{-3} liter	0.0025 liter
3.2 GHz	3.2×10^{9} hertz	3,200,000,000 hertz

Other metric measurements are converted into scientific notation in Example 3.

EXAMPLE 3 Write each quantity without the given metric prefix and then express the answer in scientific notation.

a) The diameter of a human hair may be 100 micrometers (μm)

b) Swazi, an elephant at the San Diego Zoo, has a mass of 3000 kilograms (kg).

c) A Dell Inspiron 660s computer has a 500 gigabyte (GB) hard drive.

Solution

a) $100\ \mu\text{m} = 100 \times 10^{-6}$ meters $= 0.0001$ meters $= 1 \times 10^{-4}$ meters

b) $3000\ \text{kg} = 3000 \times 10^{3}$ grams $= 3{,}000{,}000$ grams $= 3 \times 10^{6}$ grams

c) $500\ \text{GB} = 500 \times 10^{9}$ bytes $= 500{,}000{,}000{,}000$ bytes $= 5 \times 10^{11}$ bytes

Now Try Exercise 45

3 Do Calculations Using Scientific Notation

We can use the rules of exponents presented in Sections 4.1 and 4.2 when working with numbers written in scientific notation.

*There are other prefixes not listed. For example, centi is 10^{-2} or 0.01 times the base unit.

EXAMPLE 4 Multiply $(4.2 \times 10^6)(2.0 \times 10^{-4})$. Write the answer in decimal form.

Solution By the commutative and associative properties of multiplication we can rearrange the expression as follows.

$$\begin{aligned}(4.2 \times 10^6)(2.0 \times 10^{-4}) &= (4.2 \times 2.0)(10^6 \times 10^{-4}) \\ &= 8.4 \times 10^{6+(-4)} && \text{Product rule was used.} \\ &= 8.4 \times 10^2 && \text{Scientific notation} \\ &= 840 && \text{Decimal form}\end{aligned}$$

Now Try Exercise 55

EXAMPLE 5 Divide $\dfrac{3.2 \times 10^{-6}}{5.0 \times 10^{-3}}$. Write the answer in scientific notation.

Solution

$$\begin{aligned}\frac{3.2 \times 10^{-6}}{5.0 \times 10^{-3}} &= \left(\frac{3.2}{5.0}\right)\left(\frac{10^{-6}}{10^{-3}}\right) \\ &= 0.64 \times 10^{-6-(-3)} && \text{Quotient rule was used.} \\ &= 0.64 \times 10^{-6+3} \\ &= 0.64 \times 10^{-3} \\ &= 6.4 \times 10^{-4} && \text{Scientific notation}\end{aligned}$$

The answer to Example 5 in decimal form would be 0.00064.

Now Try Exercise 63

EXAMPLE 6 **Biggest Trees** The largest known tree (by volume) is a sequoia, called *General Sherman,* and has an estimated weight of 1.20×10^7 pounds. The largest coast redwood tree, called *Lost Monarch,* has an estimated weight of 1.11×10^6 pounds.

a) How much greater is the weight of the sequoia than the weight of the redwood?

b) How many times greater is the weight of the sequoia than the weight of the redwood?

Solution

a) Understand We need to subtract 1.11×10^6 from 1.20×10^7. To add or subtract numbers in scientific notation, we generally make the exponents on the 10's the same. This will allow us to add or subtract the numerical values preceding the base while maintaining the common base and exponent on the base.

Translate We can write 1.20×10^7 as 12.00×10^6. Now we subtract as follows.

Carry Out

$$\begin{array}{r} 12.00 \times 10^6 \\ -1.11 \times 10^6 \\ \hline 10.89 \times 10^6 \end{array} \quad \text{or } 1.089 \times 10^7$$

Notice that we did not subtract the 10^6's. Think of them as the unit "millions". In other words, 12.00 million minus 1.11 million is 10.89 million.

Check We can check by writing the numbers out in decimal form:

$$\begin{array}{r} 12{,}000{,}000 \\ -1{,}110{,}000 \\ \hline 10{,}890{,}000 \end{array} \quad \text{or } 1.089 \times 10^7$$

Answer The largest sequoia is 1.089×10^7 pounds heavier than the largest coast redwood.

b) Understand In part **b)** we are asked to find the *number of times* greater the weight of the sequoia is than the weight of the redwood. To find the number of times greater, we perform division.

Translate Divide the estimated weight of the sequoia by the estimated weight of the redwood.

Carry Out

$$\begin{aligned} \frac{1.20 \times 10^7}{1.11 \times 10^6} &= \frac{1.20}{1.11} \times \frac{10^7}{10^6} \\ &\approx 1.08 \times 10^{7-6} \quad \text{Quotient rule was used.} \\ &\approx 1.08 \times 10 \\ &\approx 10.8 \end{aligned}$$

Check We can check by writing the numbers out in decimal form:

$$\frac{12{,}000{,}000}{1{,}110{,}000} \approx 10.8$$

Answer The largest sequoia is 10.8 times heavier than the largest coast redwood.

Now Try Exercise 75

Cray Titan Computer

EXAMPLE 7 World's Fastest Computer As of December 4, 2012, the world's fastest computer is the Cray Titan located at the Oak Ridge National Laboratory in Tennessee. It can perform 1.76×10^{16} operations per second, which equates to taking about 0.0000000000000000568 second to perform one calculation. How long would it take this computer to perform 5 quadrillion (5,000,000,000,000,000) operations?

Solution **Understand** The computer can perform 1 calculation in $1 \cdot 0.0000000000000000568$ second, 2 calculations in $2 \cdot 0.0000000000000000568$ second, 3 calculations in $3 \cdot 0.0000000000000000568$ second, and 5 quadrillion calculations in $5{,}000{,}000{,}000{,}000{,}000 \cdot 0.0000000000000000568$ second.

Translate To compute $5{,}000{,}000{,}000{,}000{,}000 \cdot 0.0000000000000000568$, we convert each number to scientific notation.

Carry Out

$$\begin{aligned} 5{,}000{,}000{,}000{,}000{,}000 \cdot 0.0000000000000000568 &= (5 \times 10^{15}) \cdot (5.68 \times 10^{-17}) \\ &= (5 \cdot 5.68)(10^{15} \cdot 10^{-17}) \\ &= 28.4 \times 10^{-2} \\ &= 2.84 \times 10^{-1} \\ &= 0.284 \end{aligned}$$

Answer The Cray Titan would take about 0.284 of a second to perform 5 quadrillion calculations.

Now Try Exercise 79

4.3 Exercise Set MathXL® MyMathLab®

Warm-Up Exercises

Fill in the blanks with the appropriate word, phrase, or symbol(s) from the following list.

mega	milli	nonnegative	integer	greater	less	giga
nano	micro	fraction	negative			

1. Assume that $a \times 10^b$ represents a number written in scientific notation. Then a must be a number ________ than or equal to one and less than ten.

2. Assume that $a \times 10^b$ represents a number written in scientific notation. Then b must be an ________.

3. Assume that $a \times 10^b$ represents a number that is less than one written in scientific notation. Then b must be a ________ integer.

4. Assume that $a \times 10^b$ represents a number that is greater than one written in scientific notation. Then b must be a ________ integer.

5. In the metric system the prefix ________ indicates 10^{-9}.

6. In the metric system the prefix ________ indicates 10^{-6}.

7. In the metric system the prefix ________ indicates 10^{6}.

8. In the metric system the prefix ________ indicates 10^{9}.

Practice the Skills

Express each number in scientific notation.

9. 20,000 **10.** 4,000,000 **11.** 0.0005 **12.** 0.000007

13. 350,000 **14.** 3,610,000 **15.** 7950 **16.** 74,100

17. 0.053 **18.** 0.000089 **19.** 0.000726 **20.** 0.00000186

21. 5,260,000,000 **22.** 416,000 **23.** 0.00000914 **24.** 0.0125

25. 220,300 **26.** 4109 **27.** 0.005104 **28.** 0.00003007

Express each number in decimal form (without exponents).

29. 4.3×10^4 **30.** 9.0×10^6 **31.** 9.32×10^{-6} **32.** 1.63×10^{-4}

33. 2.13×10^{-5} **34.** 7.26×10^{-6} **35.** 6.25×10^5 **36.** 6.15×10^5

37. 2.891×10^1 **38.** 6.475×10^1 **39.** 5.35×10^2 **40.** 9.41×10^2

41. 7.73×10^{-7} **42.** 6.201×10^{-4} **43.** 1.0×10^4 **44.** 1.0×10^6

In Exercises 45–52, write each quantity without the given metric prefix and then express the answer in scientific notation. See Example 3.

45. A dietary supplement contains 180 milligrams (mg) of vitamin C.

46. A large locust may be 76 millimeters (mm) long.

47. A thunderstorm may generate 200 megawatts (MW) of electric power.

48. *WUSF* broadcasts its signal at 89.7 megahertz (MHz).

49. A dietary supplement contains 400 micrograms (μg) of folic acid.

50. A droplet of water from a vaporizer has a volume of about 90 nanoliters (nL).

51. The driving distance from Seattle to Miami is about 5400 kilometers (km).

52. Annually, the Hoover Dam produces about 4200 gigawatts (GW) of electric power.

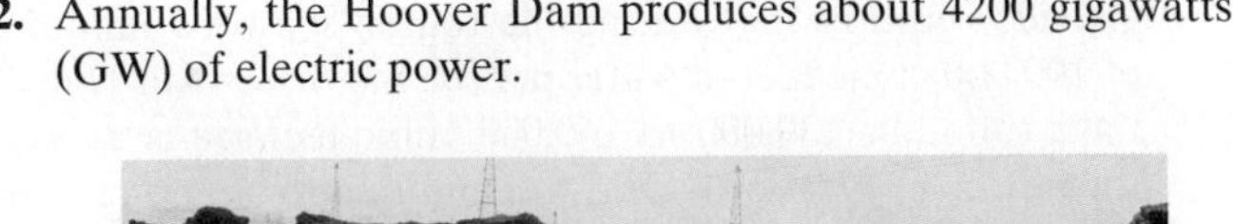

The Hoover Dam

Perform each indicated operation and express each number in decimal form (without exponents).

53. $(3.0 \times 10^2)(3.0 \times 10^5)$
54. $(2.8 \times 10^5)(3.0 \times 10^4)$
55. $(2.7 \times 10^{-6})(9.0 \times 10^4)$
56. $(1.3 \times 10^{-8})(1.74 \times 10^6)$
57. $(1.6 \times 10^{-2})(4.0 \times 10^{-3})$
58. $(4.4 \times 10^{-3})(2 \times 10^{-1})$
59. $\dfrac{7.5 \times 10^6}{3.0 \times 10^3}$
60. $\dfrac{1.4 \times 10^7}{4.0 \times 10^8}$
61. $\dfrac{2.0 \times 10^4}{8.0 \times 10^{-2}}$
62. $\dfrac{6.0 \times 10^{-3}}{3.0 \times 10^1}$
63. $\dfrac{3.9 \times 10^{-5}}{3.0 \times 10^{-2}}$
64. $\dfrac{5.6 \times 10^{-7}}{7.0 \times 10^{-10}}$

Perform each indicated operation by first converting each number to scientific notation. Write the answer in scientific notation.

65. (700,000)(8,000,000)
66. (67,000)(200,000)
67. (0.0004)(320)
68. (0.003)(0.00015)
69. $\dfrac{5{,}600{,}000}{8000}$
70. $\dfrac{0.00004}{200}$
71. $\dfrac{0.00035}{0.000002}$
72. $\dfrac{150{,}000}{0.0005}$

73. List the following numbers from smallest to largest: 7.3×10^2, 3.3×10^{-4}, 1.75×10^6, 5.3.

74. List the following numbers from smallest to largest: 4.8×10^5, 3.2×10^{-1}, 4.6, 8.3×10^{-4}.

Problem Solving

In Exercises 75–92, write the answer in scientific notation unless asked to do otherwise.

75. U.S. Population In 2013, the U.S. population was about 3.15×10^8 people and the world population was about 7.05×10^9 people.

a) How many people lived outside the United States in 2013?

b) How many times greater is the world population than the U.S. population? Write your answer in decimal form rounded to the nearest hundredth.

76. China Population See Exercise 75. In 2013, the China population was about 1.35×10^9.

a) How many people lived outside China in 2013?

b) How many times greater is the world population than the China population? Write your answer in decimal form rounded to the nearest hundredth.

77. Niagara Falls A treaty between the United States and Canada requires that during the tourist season a minimum of 100,000 cubic feet of water per second flows over Niagara Falls (another 130,000 to 160,000 cubic feet/sec is diverted for power generation). Find the minimum volume of water that will flow over the falls in a 24-hour period during the tourist season.

Niagara Falls

78. Hoover Dam To generate electricity, about 20,000 cubic feet of water pass through the Hoover Dam each second. At this rate, how many cubic feet of water will pass through the dam every day?

79. Computer Speed Mitchel Levy's computer can perform one calculation in 0.000000009 second. How long would it take Mitchel's computer to perform 10 trillion (10,000,000,000,000) calculations?

80. Computer Speed Mitzy Logan's computer can perform one calculation in 0.000000037 second. How long would it take Mitzy's computer to perform 10 billion (10,000,000,000) calculations?

81. Top Movies The approximate gross ticket sales of the top five movies in the United States as of December 3, 2012, are listed below.

Rank	Movie	Year Released	U.S. Gross Ticket Sales
1	*Avatar*	2009	$761,000,000
2	*Titanic*	1997	$659,000,000
3	*The Avengers*	2012	$623,000,000
4	*The Dark Knight*	2008	$533,000,000
5	*Star Wars*	1997	$461,000,000

Source: www.imdb.com

a) Write, in scientific notation, the gross ticket sales for *Avatar* and the gross ticket sales for *Star Wars*.

b) How much greater are the U.S. gross ticket sales for *Avatar* than for *Star Wars*?

c) How many times greater are the U.S. gross ticket sales for *Avatar* than for *Star Wars*? Write your answer in decimal form rounded to the nearest hundredth.

82. PGA Golfers The top five money winners from the 2012 Professional Golfers' Association Tour are listed below.

Rank	Golfer	Winnings
1	Rory McIlroy	\$8,000,000
2	Tiger Woods	\$6,100,000
3	Brandt Snedeker	\$5,000,000
4	Jason Dufner	\$4,900,000
5	Bubba Watson	\$4,600,000

a) Write, in scientific notation, the 2012 winnings for Rory McIlroy and the 2012 winnings for Bubba Watson.

b) How much greater are the winnings for Rory McIlroy than the winnings for Bubba Watson?

c) How many times greater are the winnings for Rory McIlroy than the winnings for Bubba Watson? Write your answer in decimal form rounded to the nearest hundredth.

83. Light from the Sun The sun is 9.3×10^7 miles from Earth. Light travels at a speed of 1.86×10^5 miles per second. How long in seconds, does it take light from the sun to reach Earth?

84. Proxima Centauri See Exercise 83. Other than our sun, the star closest to Earth is Proxima Centauri which is 2.48×10^{13} miles from Earth. How long in seconds does it take light from Proxima Centauri to reach Earth?

85. Earth and Moon The mass of Earth and the mass of the moon are given below.

Earth: 5,970,000,000,000,000,000,000 metric tons

Moon: 73,500,000,000,000,000,000 metric tons

Source: www.nasa.gov

a) Write, in scientific notation, the mass of Earth and the mass of the moon.

b) How many times greater is the mass of Earth than the mass of the moon? Write your answer in decimal form rounded to the nearest hundredth.

86. Jupiter and Mars See Exercise 85. The mass of the planets Jupiter and Mars are given below.

Jupiter: 1,900,000,000,000,000,000,000,000 metric tons

Mars: 642,000,000,000,000,000,000 metric tons

Source: www.nasa.gov

Jupiter

a) Write, in scientific notation, the mass of Jupiter and the mass of Mars.

b) How many times greater is the mass of Jupiter than the mass of Mars? Write your answer in decimal form rounded to the nearest hundredth.

87. Bill Gates In 2012 the wealthiest American was Bill Gates, with a net worth of \$$6.1 \times 10^{10}$. The 28th wealthiest American was Lauren Powell Jobs, with a net worth of \$$1.1 \times 10^{10}$.

Source: www.forbes.com

a) How much greater was the net worth of Gates than the net worth of Jobs?

b) How many times greater was the net worth of Gates than the net worth of Jobs? Write your answer in decimal form rounded to the nearest hundredth.

88. Warren Buffet In 2012 the second wealthiest American was Warren Buffet, with a net worth of \$$4.6 \times 10^{10}$. The 22nd wealthiest American was Michael Dell, with a net worth of \$$1.5 \times 10^{10}$.

Source: www.forbes.com

a) How much greater was the net worth of Buffet than the net worth of Dell?

b) How many times greater was the net worth of Buffet than the net worth of Dell? Write your answer in decimal form rounded to the nearest hundredth.

89. Population Projection The following bar graph shows the population projections for certain states for the year 2025.

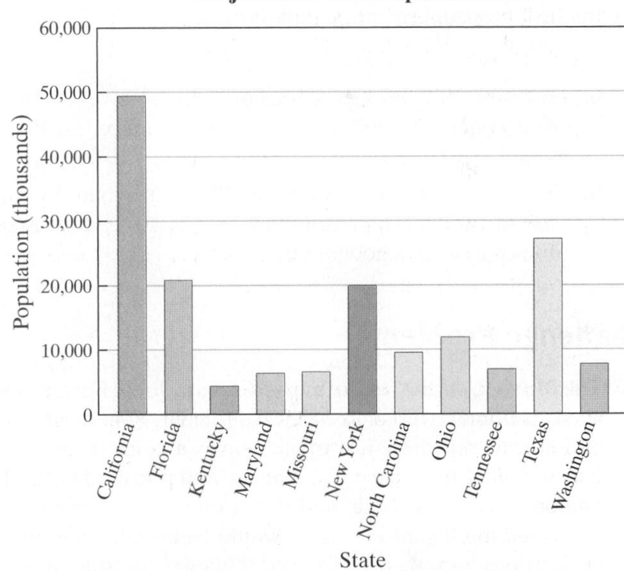

Source: www.census.gov

a) Write, in scientific notation, the approximate projected population of New York in 2025.

b) Determine, in scientific notation, the approximate sum of the projected populations of the two most populated states in 2025.

c) Determine, in scientific notation, the approximate difference between the projected population of the two most populated states in 2025.

90. Western Hemisphere The following bar graph shows the population projections for certain western hemisphere countries for the year 2050.

Projection of Total Population: 2050

Country	Population (millions)
Brazil	247
Mexico	147
Colombia	71
Argentina	55
Venezuela	42
Peru	42
Canada	40
Guatemala	27
Chile	23
Bolivia	17
Haiti	14

a) Write, in scientific notation, the projected population of Mexico and the projected population of Canada in 2050.

b) Determine, in scientific notation, the sum of the projected populations of Mexico and Canada in 2050.

c) Determine, in scientific notation, the difference between the projected populations of Mexico and Canada in 2050.

91. Structure of Matter An article in *Scientific American* states that physicists have created a *standard model* that describes the structure of matter down to 10^{-18} meters. If this number is written without exponents, how many zeroes would there be to the right of the decimal point?

92. A Large Number Avogadro's number, named after the nineteenth-century Italian chemist Amedeo Avogadro, is roughly 6.02×10^{23}. It represents the number of atoms in 12 grams of pure carbon.

a) If this number were written out in decimal form, how many digits would it contain?

b) What is the number of atoms in 1 gram of pure carbon? Write your answer in scientific notation.

Concept/Writing Exercises

93. Describe the form of a number given in scientific notation.

94. A **googol** is the number 1 followed by 100 zeros and is the largest number many calculators can handle. From what we learned in this section, a googol can be written as 10^{100} or, in scientific notation, as 1.0×10^{100}. What is a googol raised to the fifth power in scientific notation?

95. See Exercise 94.

a) Estimate the number of seconds that have occurred since January 1, 1900. Is that number more or less than a googol?

b) The total number of ways of filling in a nine-by-nine *Sudoku* grid is approximately 6.67×10^{21}. How many times greater is a googol than this?

96. The five planets closest to the sun are, in order, Mercury, Venus, Earth, Mars, and Jupiter. Their average distances from the sun are given below:

Planet	Average Distance to the Sun in Miles
Mercury	35,983,610
Venus	67,232,360
Earth	92,957,100
Mars	141,635,300
Jupiter	484,632,000

In scientific notation, give the ratio of Jupiter's average distance to the sun to Mercury's average distance to the sun.

Challenge Problems

97. The Movie *Contact* In the movie *Contact,* Jodie Foster plays an astronomer who makes the statement "There are 400 billion stars out there just in our universe alone. If only one out of a million of those had planets, and if just one out of a million of those had life, and if just one out of a million of those had intelligent life, there would be literally millions of civilizations out there." Do you believe this statement is correct? Explain your answer.

98. How many times smaller is 1 nanosecond than one millisecond?

99. How many times, either greater or smaller, is 10^{-12} meters than 10^{-18} meters?

100. Light Year Light travels at a speed of 1.86×10^5 miles per second. A **light year** is the distance light travels in one year. Determine the number of miles in a light year.

Group Activity

Discuss and answer Exercise 101 as a group.

101. A Million versus a Billion Do you have any idea of the difference in size between a million (1,000,000), a billion (1,000,000,000), and a trillion (1,000,000,000,000)?

a) Write a million, a billion, and a trillion in scientific notation.

b) Group member 1: Determine how long it would take to spend a million dollars if you spent \$1000 a day.

c) Group member 2: Repeat part **b)** for a billion dollars.

d) Group member 3: Repeat part **b)** for a trillion dollars.

e) As a group, determine how many times greater a billion dollars is than a million dollars.

Cumulative Review Exercises

[1.9] **102.** Evaluate $4x^2 + 3x + \frac{x}{2}$ when $x = 0$.

[2.3] **103. a)** If $-x = -\frac{3}{2}$, what is the value of x?

b) If $5x = 0$, what is the value of x?

[2.5] **104.** Solve the equation $2x - 3(x - 2) = x + 2$.

[4.1] **105.** Simplify $\left(\frac{-2x^5y^7}{8x^8y^3}\right)^3$.

MID-CHAPTER TEST: 4.1–4.3

To find out how well you understand the chapter material to this point, take this brief test. The answers, and the section where the material was initially discussed, are given in the back of the book. Review any questions that you answered incorrectly.

Simplify.

1. $y^{11} \cdot y^{20}$

2. $\frac{x^{13}}{x^{10}}$

3. $(-3x^5y^7)(-2xy^6)$

4. $\frac{6a^{12}b^8}{9a^7b^2}$

5. $(-4x^2y^4)^3$

6. $\left(\frac{-5s^4t^6}{10s^6t^3}\right)^2$

7. $(7x^9y^5)(-3xy^4)^2$

8. $\frac{p^{-3}}{p^5}$

9. $x^{-4} \cdot x^{-6}$

10. $(3^{-1} + 5^2)^0$

11. $\left(\frac{3}{7}\right)^{-2}$

12. $(8x^{-2}y^5)(4x^3y^{-6})$

13. $\frac{6m^{-4}n^{-7}}{2m^0n^{-1}}$

14. $\left(\frac{2x^{-3}y^{-4}}{x^3yz^{-2}}\right)^{-2}$

15. a) Describe how to write a number 10 or greater in scientific notation.

b) Describe how to write a number less than 1 in scientific notation.

16. Express 6,540,000,000 in scientific notation.

17. Express 3.27×10^{-5} in decimal form (without exponents).

18. Write 18.9 kilometers without metric prefixes.

19. Multiply $(3.4 \times 10^{-6})(7.0 \times 10^3)$ and express the answer in decimal form (without exponents).

20. Divide $\frac{0.00006}{200}$ by first converting to scientific notation. Write the answer in scientific notation.

4.4 Addition and Subtraction of Polynomials

1. Identify polynomials.
2. Add polynomials.
3. Subtract polynomials.
4. Subtract polynomials in columns.

1 Identify Polynomials

A polynomial in x is an expression that is the sum of a finite number of terms of the form ax^n, for any real number a and any *whole number n.*

Examples of Polynomials	Not Polynomials	
$8x$	$4x^{1/2}$	(Fractional exponent)
$\frac{1}{3}x - 4$	$3x^2 + 4x^{-1} + 5$	(Negative exponent)
$x^2 - 2x + 1$	$4 + \frac{1}{x}$	$\left(\frac{1}{x} = x^{-1}\text{, negative exponent}\right)$

Understanding Algebra

A polynomial is written in descending order when the *exponents* on the variable decrease from left to right. The polynomial

$$2x^3 - 5x^2 - 7x + 4$$

is written in descending order.

A polynomial is written in **descending order** when the exponents on the variable decrease from left to right.

Example of Polynomial in Descending Order

$$2x^4 + 4x^2 - 6x + 3$$

The constant term 3 is last because it can be written as $3x^0$. Remember that $x^0 = 1$.

A polynomial can be in more than one variable. For example, $3xy + 2$ is a polynomial in two variables, x and y.

Understanding Algebra

The prefix "poly" means many and "nomial" means "name" or "term."

The prefix "poly" means many and "nomial" means "name" or "term." A polynomial with one term is called a **monomial.** A **binomial** is a two-termed polynomial. A **trinomial** is a three-termed polynomial. Polynomials containing more than three terms usually are not given special names.

Type of Polynomial	Number of Terms	Examples
Monomial	One	$8,\ 4x,\ -6x^2$
Binomial	Two	$x + 5,\ x^2 - 6,\ 4y^2 - 5y$
Trinomial	Three	$x^2 - 2x + 3,\ 3z^2 - 6z + 7$

Degree

The **degree of a term in one variable** is the exponent on the variable in that term. The degree of $2x^3$ is 3.

The **degree of a term in two or more variables** is the sum of the exponents on the variables. The degree of $4x^2y^3$ is 5.

The **degree of a polynomial** is the same as that of its highest-degree term.

Term	Degree of Term	
$4x^2$	2	
$2x^2y^5$	7	$(2 + 5 = 7)$
$-23a^3b^4c$	8	$(3 + 4 + 1 = 8)$
$-5x$	1	$(-5x \text{ is } -5x^1)$
3	0	$(3 \text{ is } 3x^0)$

Polynomial	Degree of Polynomial	
$8x^3 + 2x^2 - 3x + 4$	3	($8x^3$ is highest-degree term.)
$x^2 - 4$	2	(x^2 is highest-degree term.)
$6x - 5$	1	($6x$ or $6x^1$ is highest-degree term.)
4	0	(4 or $4x^0$ is highest-degree term.)
$x^2y^4 + 2x + 3$	6	(x^2y^4 is highest-degree term.)

2 Add Polynomials

In Section 2.1, we stated that like terms are terms having the same variables and the same exponents. That is, like terms may differ only in their numerical coefficients.

Examples of Like Terms

$$3,\ -5$$
$$2x,\ x$$
$$-2x^2,\ 4x^2$$
$$3y^2,\ 5y^2$$
$$3xy^2,\ 5xy^2$$

To Add Polynomials

To add polynomials, combine the like terms of the polynomials.

EXAMPLE 1 Add $(4x^2 + 6x + 3) + (2x^2 + 5x - 1)$.

Solution Remember that $(4x^2 + 6x + 3) = 1(4x^2 + 6x + 3)$ and $(2x^2 + 5x - 1) = 1(2x^2 + 5x - 1)$.

$$\begin{aligned} &(4x^2 + 6x + 3) + (2x^2 + 5x - 1) \\ &= 1(4x^2 + 6x + 3) + 1(2x^2 + 5x - 1) \\ &= 4x^2 + 6x + 3 + 2x^2 + 5x - 1 && \text{Used the distributive property to remove parentheses.} \\ &= \underbrace{4x^2 + 2x^2}\ \underbrace{+ 6x + 5x}\ \underbrace{+ 3 - 1} && \text{Rearranged terms.} \\ &= 6x^2 + 11x + 2 && \text{Combined like terms.} \end{aligned}$$

Now Try Exercise 63

EXAMPLE 2 Add $(5a^2 + 3a + b) + (a^2 - 7a + 3)$.

Solution

$$\begin{aligned} &(5a^2 + 3a + b) + (a^2 - 7a + 3) \\ &= 5a^2 + 3a + b + a^2 - 7a + 3 && \text{Removed parentheses.} \\ &= \underbrace{5a^2 + a^2}\ \underbrace{+ 3a - 7a} + b + 3 && \text{Rearranged terms.} \\ &= 6a^2 - 4a + b + 3 && \text{Combined like terms.} \end{aligned}$$

Now Try Exercise 67

EXAMPLE 3 Add $(3x^2y - 4xy + y) + (x^2y + 2xy + 3y)$.

Solution

$$\begin{aligned} &(3x^2y - 4xy + y) + (x^2y + 2xy + 3y) \\ &= 3x^2y - 4xy + y + x^2y + 2xy + 3y && \text{Removed parentheses.} \\ &= \underbrace{3x^2y + x^2y}\ \underbrace{- 4xy + 2xy}\ \underbrace{+ y + 3y} && \text{Rearranged terms.} \\ &= 4x^2y - 2xy + 4y && \text{Combined like terms.} \end{aligned}$$

Now Try Exercise 69

Usually, when we add polynomials, we will do so horizontally as in Examples 1 through 3. Sometimes, it is easier to add polynomials in columns.

To Add Polynomials in Columns

1. Arrange polynomials in descending order, one under the other with like terms in the same columns.
2. Add the terms in each column.

EXAMPLE 4 Add $5x^2 - 9x - 3$ and $-3x^2 - 4x + 6$ using columns.

Solution

$$\begin{array}{r} 5x^2 - 9x - 3 \\ -3x^2 - 4x + 6 \\ \hline 2x^2 - 13x + 3 \end{array}$$

Now Try Exercise 83

EXAMPLE 5 Add $5w^3 + 2w - 4$ and $2w^2 - 6w - 3$ using columns.

Solution Since the polynomial $5w^3 + 2w - 4$ does not have a w^2 term, we will add the term $0w^2$ to the polynomial. This procedure helps in aligning like terms.

$$\begin{array}{r} 5w^3 + 0w^2 + 2w - 4 \\ 2w^2 - 6w - 3 \\ \hline 5w^3 + 2w^2 - 4w - 7 \end{array}$$

Now Try Exercise 85

3 Subtract Polynomials

To Subtract Polynomials

1. Use the distributive property to remove parentheses. This will have the effect of changing the sign of *every* term within the parentheses of the polynomial being subtracted.
2. Combine like terms.

EXAMPLE 6 Subtract $(3x^2 - 2x + 5) - (x^2 - 3x + 4)$.

Solution $(3x^2 - 2x + 5)$ means $1(3x^2 - 2x + 5)$ and $(x^2 - 3x + 4)$ means $1(x^2 - 3x + 4)$. We use this information in the solution, as shown below.

$$\begin{aligned}(3x^2 - 2x + 5) - (x^2 - 3x + 4) &= 1(3x^2 - 2x + 5) - 1(x^2 - 3x + 4)\\ &= 3x^2 - 2x + 5 - x^2 + 3x - 4 && \text{Removed parentheses.}\\ &= \underbrace{3x^2 - x^2}\ \underbrace{- 2x + 3x}\ \underbrace{+ 5 - 4} && \text{Rearranged terms.}\\ &= 2x^2 + x + 1 && \text{Combined like terms.}\end{aligned}$$

Now Try Exercise 95

EXAMPLE 7 Subtract $(-3x^2 - 5x + 3)$ from $(x^3 + 2x + 6)$.

Solution

$$\begin{aligned}&(x^3 + 2x + 6) - (-3x^2 - 5x + 3)\\ &= x^3 + 2x + 6 + 3x^2 + 5x - 3 && \text{Removed parentheses.}\\ &= x^3 + 3x^2 \underbrace{+ 2x + 5x}\ \underbrace{+ 6 - 3} && \text{Rearranged terms.}\\ &= x^3 + 3x^2 + 7x + 3 && \text{Combined like terms.}\end{aligned}$$

Now Try Exercise 107

Understanding Algebra

If a negative sign precedes parentheses, we must change the sign of each term of the polynomial when we remove the parentheses.

$$-(-2x^2 + 7x - 5)$$

becomes

$$2x^2 - 7x + 5$$

AVOIDING COMMON ERRORS

One of the most common mistakes occurs when subtracting polynomials. When subtracting one polynomial from another, the sign of each term in the polynomial being subtracted must be changed, not just the sign of the first term.

CORRECT

$$\begin{aligned}&6x^2 - 4x + 3 - (2x^2 - 3x + 4)\\ &= 6x^2 - 4x + 3 - 2x^2 + 3x - 4\\ &= 4x^2 - x - 1\end{aligned}$$

INCORRECT

$$\begin{aligned}&6x^2 - 4x + 3 - (2x^2 - 3x + 4)\\ &= 6x^2 - 4x + 3 - 2x^2 - 3x + 4\\ &= 4x^2 - 7x + 7\end{aligned}$$

Do not make this mistake!

4 Subtract Polynomials in Columns

To Subtract Polynomials in Columns

1. Write *the polynomial being subtracted* below the polynomial from which it is being subtracted. List like terms in the same column.
2. *Change the sign of each term* in the polynomial being subtracted. (This step can be done mentally, if you like.)
3. Add the terms in each column.

EXAMPLE 8 Subtract $(2x^2 - 4x + 6)$ from $(4x^2 + 5x + 8)$ using columns.

Solution Align like terms in columns (step 1).

$$\begin{array}{r}4x^2 + 5x + 8\\ -(2x^2 - 4x + 6)\\ \hline\end{array}$$ Align like terms.

Change *all* signs in the second row (step 2); then add (step 3).

$$\begin{array}{rl} 4x^2 + 5x + 8 & \\ \underline{-2x^2 + 4x - 6} & \text{Changed all signs.} \\ 2x^2 + 9x + 2 & \text{Added terms.} \end{array}$$

Now Try Exercise 113

EXAMPLE 9 Subtract $(2x^2 - 6)$ from $(-3x^3 + 4x - 3)$ using columns.

Solution To help align like terms, write each expression in descending order. If any power of x is missing, write that term with a numerical coefficient of 0.

$$-3x^3 + 4x - 3 = -3x^3 + 0x^2 + 4x - 3$$
$$2x^2 - 6 = 2x^2 + 0x - 6$$

Align like terms.

$$\begin{array}{r} -3x^3 + 0x^2 + 4x - 3 \\ \underline{-(2x^2 + 0x - 6)} \end{array}$$

Change all signs in the second row; then add the terms in each column.

$$\begin{array}{r} -3x^3 + 0x^2 + 4x - 3 \\ \underline{-2x^2 - 0x + 6} \\ -3x^3 - 2x^2 + 4x + 3 \end{array}$$

Now Try Exercise 115

NOTE: You may find that you can change the signs mentally and can therefore align and change the signs in one step.

4.4 Exercise Set MathXL® MyMathLab®

Warm-Up Exercises

Fill in the blanks with the appropriate word, phrase, or symbol(s) from the following list.

highest like monomial whole decrease combine
sum difference increase trinomial binomial exponent

1. A polynomial in x is an expression that is the sum of a finite number of terms of the form ax^n, for any real number a and any ________ number n.
2. A polynomial is written in descending order when the exponents on the variable ________ from left to right.
3. A polynomial with one term is called a ________.
4. A polynomial with two terms is called a ________.
5. A polynomial with three terms is called a ________.
6. The degree of a term in one variable is the ________ on the variable in that term.
7. The degree of a term in two or more variables is the __sum__ of the exponents on the variables.
8. The degree of a polynomial is the same as that of its term with the ________ degree.
9. Terms that have the same variables and the same exponents are called __like__ terms.
10. To add polynomials, ________ the like terms of the polynomials.

Practice the Skills

Indicate the degree of each term.

11. x^3 **12.** $6x^2$ **13.** $-2x$ **14.** y

15. 6 **16.** -7 **17.** $7s^4$ **18.** $4s^7$

19. x^5 **20.** z^{11} **21.** $-3b^8$ **22.** $5a^4$

23. x^2y **24.** a^4b^3 **25.** $-8x^3y^5z$ **26.** $-12p^4q^7r$

Indicate which expressions are polynomials. If the polynomial has a specific name–monomial, binomial, or trinomial–give that name.

27. $2x^2 - 6x + 7$

28. $-4x^2 - 7x + 21$

29. -6

30. 1

31. $9a^4 - 5$

32. $7x + 8$

33. $15x^{1/2} - 7$

34. $3x^{1/2} + 2x$

35. $a^{-1} + 4$

36. $5p^{-3}$

37. $6n^3 - 5n^2 + 4n - 3$

38. $5x^3 - 6x^2 - 7x + 2$

39. $\frac{2}{3}x^2 - \frac{1}{x}$

40. $\frac{3}{a^2} - \frac{2}{a} + 7$

41. $\frac{a^2}{3} - \frac{a}{2} + 7$

42. $0.6r^4 - \frac{1}{2}r^3 - 0.4r^2 - \frac{1}{3}$

Express each polynomial in descending order. If the polynomial is already in descending order, so state. Give the degree of each polynomial.

43. 18

44. 15

45. $4 + 5x$

46. $4 - 3p^3$

47. $-4 + x^2 - 2x$

48. $x + 3x^2 - 8$

49. $-a - 3$

50. $6x - 5$

51. $6w^2 - 5w + 9$

52. $2x^2 + 5x - 8$

53. $-4 + x - 3x^2 + 4x^3$

54. $1 - x^3 + 3x$

55. $5x + 3x^2 - 6 - 2x^4$

56. $-3r - 5r^2 + 2r^4 - 6$

Add.

57. $(9x - 2) + (4x - 7)$

58. $(5x - 6) + (2x - 3)$

59. $(-3x + 8) + (2x + 3)$

60. $(-7x - 9) + (-2x + 9)$

61. $(4m - 3) + (5m^2 - 4m + 7)$

62. $(-x^2 - 2x - 4) + (4x^2 + 3)$

63. $(2x^2 - 3x + 5) + (-x^2 + 6x - 8)$

64. $(x^2 - 6x + 7) + (-x^2 + 3x + 5)$

65. $(-7x^3 - 3x^2 + 4) + (4x + 5x^3 - 7)$

66. $(6x^3 - 4x^2 - 7) + (3x^2 + 3x - 3)$

67. $(8x^2 + 2x - y) + (3x^2 - 9x + 5)$

68. $(-7a^2 + 3a - b) + (4a^2 - 2a - 8)$

69. $(2x^2y + 2x - 3) + (3x^2y - 5x + 5)$

70. $(x^2y + x - y) + (2x^2y + 2x - 6y + 3)$

71. $(x^2 + 2.6x - 3) + (4x + 3.8)$

72. $(a^2 - 0.75a - 8.1) + (0.3a^2 + 0.05a + 8.02)$

73. $(5.2n^2 - 6n + 1.7) + (3n^2 + 1.2n - 2.3)$

74. $(8x^2 + 4) + (-2.6x^2 - 5x - 2.3)$

75. $\left(\frac{1}{8}x + \frac{5}{16}\right) + \left(\frac{3}{8}x - \frac{1}{16}\right)$

76. $\left(\frac{1}{12}x - \frac{5}{12}\right) + \left(\frac{7}{12}x + \frac{11}{12}\right)$

77. $(-x^2 - 4x + 8) + \left(5x - 2x^2 + \frac{1}{2}\right)$

78. $(8x^2 + 3x - 5) + \left(x^2 + \frac{1}{2}x + 2\right)$

Add using columns.

79. Add $8x - 7$ and $3x + 4$.

80. Add $-x + 5$ and $-4x - 5$.

81. Add $4y^2 - 2y + 4$ and $3y^2 + 1$.

82. Add $6m^2 - 2m + 1$ and $-10m^2 + 8$.

83. Add $-x^2 - 3x + 3$ and $5x^2 + 5x - 7$.

84. Add $-2s^2 - s + 5$ and $3s^2 - 6s$.

85. Add $2x^3 + 3x^2 + 6x - 9$ and $7 - 4x^2$.

86. Add $-3x^3 + 3x + 9$ and $2x^2 - 4$.

87. Add $4n^3 - 5n^2 + n - 6$ and $-n^3 - 6n^2 - 2n + 8$.

88. Add $7x^3 + 5x - 6$ and $3x^3 - 4x^2 - x + 8$.

Subtract.

89. $(4x - 4) - (2x + 2)$

90. $(6x - 5) - (2x - 3)$

91. $(-2x - 3) - (-5x - 7)$

92. $(10x - 3) - (-2x + 7)$

93. $(-r + 5) - (2r + 5)$

94. $(4x + 8) - (3x + 9)$

95. $(5x^2 - x - 1) - (-3x^2 - 2x - 5)$

96. $(-a^2 + 3a + 12) - (-4a^2 - 3)$

97. $(8x^3 - 2x^2 - 4x + 5) - (5x^2 + 8)$

98. $(-11x^3 + 2x^2 - 9x - 13) - (-4x^2 - 5x + 2)$

99. $(-y^2 + 4y - 5.2) - (5y^2 + 2.1y + 7.5)$

100. $(9x^2 + 7x - 5) - (3x^2 + 3.5)$

101. $(-4.1n^2 - 3n) - (2.3n^2 - 9n + 7.6)$

102. $(7x - 0.6) - (-2x^2 + 4x - 8)$

103. $(2x^3 - 4x^2 + 5x - 7) - \left(3x + \frac{3}{5}x^2 - 5\right)$

104. $(-3x^2 + 4x - 7) - \left(x^3 + 4x^2 - \frac{3}{4}x\right)$

105. Subtract $(7x + 4)$ from $(8x + 2)$.

106. Subtract $(-4x + 7)$ from $(-3x - 9)$.

107. Subtract $(5x - 6)$ from $(2x^2 - 4x + 8)$.

108. Subtract $(3x^2 - 5x - 3)$ from $(-x^2 + 3x + 10)$.

109. Subtract $(-2c^2 + 7c - 7)$ from $(-5c^3 - 6c^2 + 7)$.

110. Subtract $(4x^3 - 6x^2)$ from $(3x^3 + 5x^2 + 9x - 7)$.

Subtract using columns.

111. Subtract $(3x - 3)$ from $(6x + 5)$.

112. Subtract $(6x + 8)$ from $(2x - 5)$.

113. Subtract $(2a^2 + 3a - 9)$ from $(5a^2 - 13a + 19)$.

114. Subtract $(4x^2 - 7x + 3)$ from $(8x^2 + 5x - 9)$.

115. Subtract $(6x^2 - 1)$ from $(7x^2 - 3x - 4)$.

116. Subtract $(5n^3 + 7n - 9)$ from $(2n^3 - 6n + 3)$.

117. Subtract $(5x^2 + 4)$ from $(x^2 + 4)$.

118. Subtract $(-5m^2 + 6m)$ from $(m - 6)$.

119. Subtract $(x^2 + 6x - 7)$ from $(4x^3 - 6x^2 + 7x - 9)$.

120. Subtract $(2x^3 + 4x^2 - 9x)$ from $(-5x^3 + 4x - 12)$.

Problem Solving

121. Make up your own addition problem where the sum of two binomials is $-5x - 1$.

122. Make up your own addition problem where the sum of two trinomials is $3x^3 - 2x - 4$.

123. Make up your own subtraction problem where the difference of two trinomials is $7x - 3$.

124. Make up your own subtraction problem where the difference of two trinomials is $-x^2 + 4x - 5$.

125. When two binomials are added, will the sum always, sometimes, or never be a binomial? Explain your answer and give examples to support your answer.

126. When one binomial is subtracted from another, will the difference always, sometimes, or never be a binomial? Explain your answer and give examples to support your answer.

127. When two trinomials are added, will the sum always, sometimes, or never be a trinomial? Explain your answer and give examples to support your answer.

128. When one trinomial is subtracted from another, will the difference always, sometimes, or never be a trinomial? Explain your answer and give examples to support your answer.

129. Write a fourth-degree trinomial in the variable x that has neither a second-degree term nor a constant term.

130. Write a sixth-degree trinomial in the variable x that has no fifth-, fourth-, first-, or zero-degree terms.

Write a polynomial that represents the area of each figure shown.

131.

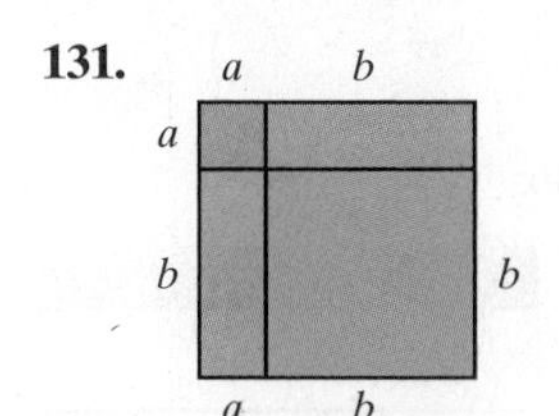

132.

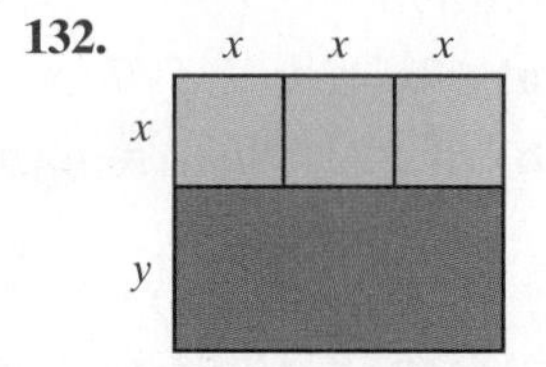

133.

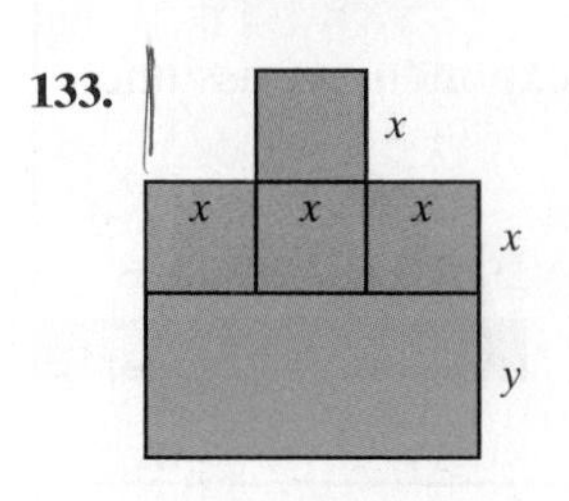

134.

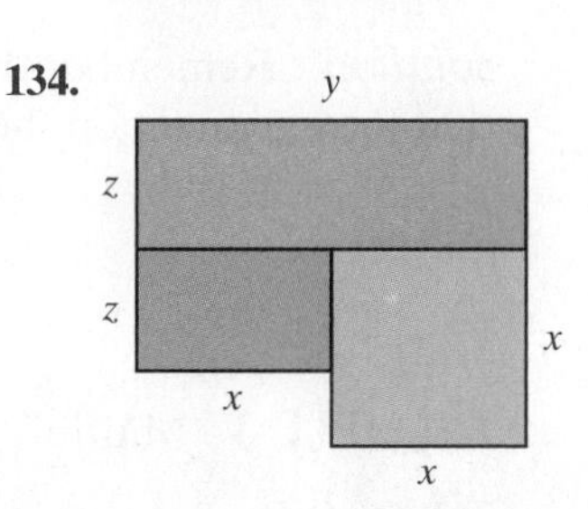

Concept/Writing Exercises

135. Is it possible to have a fourth-degree trinomial in x that has no third-, second-, or zero-degree terms and contains no like terms? Explain.

136. Is it possible to have a fifth-degree trinomial in x that has no fourth-, third-, second-, or first-degree terms and contains no like terms? Explain.

Challenge Problems

Simplify.

137. $(3x^2 - 6x + 3) - (2x^2 - x - 6) - (x^2 + 7x - 9)$

138. $3x^2y - 6xy - 2xy + 9xy^2 - 5xy + 3x$

139. $4(x^2 + 2x - 3) - 6(2 - 4x - x^2) - 2x(x + 2)$

Group Activity

Discuss and answer Exercise 140 as a group.

140. Make up a trinomial, a binomial, and a different trinomial such that (first trinomial) + (binomial) − (second trinomial) = 0.

Cumulative Review Exercises

[1.5] **141.** Insert either >, <, or = in the shaded area to make the statement true: $|-9| \;\square\; |-6|$.

[1.6–1.8] *Indicate whether each statement is true or false.*

142. The product of two negative numbers is always a positive number.

143. The sum of two negative numbers is always a negative number.

144. The difference of two negative numbers is always a negative number.

145. The quotient of two negative numbers is always a negative number.

[4.2] **146.** Simplify $\left(\dfrac{b^3c^{-4}}{2b^{-1}}\right)^{-2}$.

4.5 Multiplication of Polynomials

1. Multiply a monomial by a monomial.
2. Multiply a polynomial by a monomial.
3. Multiply binomials using the distributive property.
4. Multiply binomials using the FOIL method.
5. Multiply binomials using formulas for special products.
6. Multiply any two polynomials.

1 Multiply a Monomial by a Monomial

To multiply two monomials we use the associative and commutative properties of multiplication. Then we multiply the coefficients and use the product rule of exponents to determine the exponents on the variables.

EXAMPLE 1 Multiply.

a) $(7x^3)(6x^5)$ **b)** $(4b^2)(-9b^7)$

Solution

a) $(7x^3)(6x^5) = 7 \cdot x^3 \cdot 6 \cdot x^5 = 7 \cdot 6 \cdot x^3 \cdot x^5 = 42x^{3+5} = 42x^8$

b) $(4b^2)(-9b^7) = 4 \cdot b^2 \cdot (-9) \cdot b^7 = (4)(-9) \cdot b^2 \cdot b^7 = -36b^{2+7} = -36b^9$

Now Try Exercise 15

EXAMPLE 2 Multiply $5x^2y \cdot 8x^5y^4$.

Solution Remember that when a variable is given without an exponent we assume that the exponent on the variable is 1.

$$5x^2y \cdot 8x^5y^4 = 40x^{2+5}y^{1+4} = 40x^7y^5$$

Now Try Exercise 17

EXAMPLE 3 Multiply.

a) $6xy^2z^5(-3x^4y^7z)$ **b)** $(-4x^4z^9)(-3xy^7z^3)$

Solution

a) $6xy^2z^5(-3x^4y^7z) = -18x^5y^9z^6$

b) $(-4x^4z^9)(-3xy^7z^3) = 12x^5y^7z^{12}$

Now Try Exercise 19

2 Multiply a Polynomial by a Monomial

To multiply a polynomial by a monomial, we use the distributive property presented earlier.

$$a(b + c) = ab + ac$$

The distributive property can be expanded to

$$a(b + c + d + \cdots + n) = ab + ac + ad + \cdots + an$$

Understanding Algebra

When multiplying a monomial by a polynomial, the monomial is multiplied by each term of the polynomial:

$$3x^2(2x^3 - 5x + 7)$$
$$3x^2(2x^3) + 3x^2(-5x) + 3x^2(7)$$
$$6x^5 - 15x^3 + 21x^2$$

EXAMPLE 4 Multiply $3x(2x^2 + 4)$.

Solution

$$3x(2x^2 + 4) = (3x)(2x^2) + (3x)(4)$$
$$= 6x^3 + 12x$$

Now Try Exercise 29

EXAMPLE 5 Multiply $-3n(4n^2 - 2n - 1)$.

Solution

$$-3n(4n^2 - 2n - 1) = (-3n)(4n^2) + (-3n)(-2n) + (-3n)(-1)$$
$$= -12n^3 + 6n^2 + 3n$$

Now Try Exercise 33

EXAMPLE 6 Multiply $5x^2(4x^3 - 2x + 7)$.

Solution

$$5x^2(4x^3 - 2x + 7) = (5x^2)(4x^3) + (5x^2)(-2x) + (5x^2)(7)$$
$$= 20x^5 - 10x^3 + 35x^2$$

Now Try Exercise 37

EXAMPLE 7 Multiply $2x(3x^2y - 6xy + 5)$.

Solution

$$2x(3x^2y - 6xy + 5) = (2x)(3x^2y) + (2x)(-6xy) + (2x)(5)$$
$$= 6x^3y - 12x^2y + 10x$$

Now Try Exercise 39

In Example 8, we perform a multiplication where the monomial is placed to the right of the polynomial. Each term of the polynomial is multiplied by the monomial, as illustrated in the example.

EXAMPLE 8 Multiply $(3x^3 - 2xy + 3)4x$.

Solution

$$(3x^3 - 2xy + 3)4x = (3x^3)(4x) + (-2xy)(4x) + (3)(4x)$$
$$= 12x^4 - 8x^2y + 12x$$

Now Try Exercise 35

The problem in Example 8 could be written as $4x(3x^3 - 2xy + 3)$ by the commutative property of multiplication, and then simplified as in Examples 4 through 7.

3 Multiply Binomials Using the Distributive Property

Now we will discuss multiplying a binomial by a binomial. Before we explain how to do this, consider the multiplication problem $43 \cdot 12$.

$$\begin{array}{rcrcl} & & 43 & \longleftarrow & \text{Multiplicand} \\ & & \underline{12} & \longleftarrow & \text{Multiplier} \\ 2(4) & \longrightarrow & 86 & \longleftarrow & 2(3) \\ 1(4) & \longrightarrow & \underline{43} & \longleftarrow & 1(3) \\ & & 516 & \longleftarrow & \text{Product} \end{array}$$

Note how the 2 multiplies both the 3 and the 4, and the 1 also multiplies both the 3 and the 4. That is, every digit in the multiplier multiplies every digit in the multiplicand. We can also illustrate the multiplication process as follows.

$$\begin{aligned}(43)(12) &= (40 + 3)(10 + 2) \\ &= (40 + 3)(10) + (40 + 3)(2) \\ &= (40)(10) + (3)(10) + (40)(2) + (3)(2) \\ &= 400 + 30 + 80 + 6 \\ &= 516\end{aligned}$$

Understanding Algebra

When multiplying two polynomials, every term in the first polynomial multiplies every term in the second polynomial.

$$(2x + 5)(3x + 10)$$
$$(2x + 5)3x + (2x + 5)10$$
$$6x^2 + 15x + 20x + 50$$
$$6x^2 + 35x + 50$$

Whenever any two polynomials are multiplied, the same process must be followed. That is, *every term in one polynomial must multiply every term in the other polynomial.*

Consider multiplying $(a + b)(c + d)$. Treating $(a + b)$ as a single term and using the distributive property, we get

$$(a + b)(c + d) = (a + b)c + (a + b)d$$

Using the distributive property a second time gives

$$= ac + bc + ad + bd$$

Notice how each term of the first polynomial was multiplied by each term of the second polynomial, and all the products were added to obtain the answer.

EXAMPLE 9 Multiply $(x + 3)(x + 4)$.

Solution

$$\begin{aligned}(x + 3)(x + 4) &= (x + 3)\cdot x + (x + 3)\cdot 4 \\ &= x\cdot x + 3\cdot x + x\cdot 4 + 3\cdot 4 \\ &= x^2 + \underbrace{3x + 4x} + 12 \quad \text{Combine like terms} \\ &= x^2 + 7x + 12\end{aligned}$$

Now Try Exercise 43

Note that after performing the multiplication, like terms must be combined.

EXAMPLE 10 Multiply $(x - 4)(y + 3)$.

Solution

$$\begin{aligned}(x - 4)(y + 3) &= (x - 4)y + (x - 4)3 \\ &= xy - 4y + 3x - 12\end{aligned}$$

Now Try Exercise 65

Understanding Algebra

F	**O**	**I**	**L**
First	Outer	Inner	Last

When multiplying two binomials, multiply the **f**irst terms of each, the **o**uter terms of each, the **i**nner terms of each, and the **l**ast terms of each.

4 Multiply Binomials Using the FOIL Method

A commonly used method to multiply two binomials is the **FOIL method**. This procedure also results in each term of one binomial being multiplied by each term in the other binomial. The FOIL method is not actually a different method used to multiply binomials but rather an acronym to help students remember to correctly apply the distributive property.

The FOIL Method

Consider $(a + b)(c + d)$.

F stands for **first**—multiply the first terms of each binomial together:

$$(a + b)(c + d) \quad \text{product } ac$$

O stands for **outer**—multiply the two outer terms together:

$$(a + b)(c + d) \quad \text{product } ad$$

I stands for **inner**—multiply the two inner terms together:

$$(a + b)(c + d) \quad \text{product } bc$$

L stands for **last**–multiply the last terms together:

$$(a + b)(c + d) \quad \text{product } bd$$

The product of the two binomials is the sum of these four products.

$$(a + b)(c + d) = ac + ad + bc + bd$$

EXAMPLE 11 Using the FOIL method, multiply $(2x - 3)(x + 4)$.

Solution

$$(2x - 3)(x + 4)$$

$$\begin{aligned} &\quad\;\; \text{F} \qquad\quad \text{O} \qquad\quad \text{I} \qquad\qquad \text{L} \\ &= (2x)(x) + (2x)(4) + (-3)(x) + (-3)(4) \\ &= \;\; 2x^2 \;\;+\;\; 8x \;\;-\;\; 3x \;\;-\;\; 12 \\ &= 2x^2 + 5x - 12 \end{aligned}$$

Thus, $(2x - 3)(x + 4) = 2x^2 + 5x - 12$.

Now Try Exercise 45

EXAMPLE 12 Multiply $(4 - 2x)(6 - 5x)$.

Solution

$$(4 - 2x)(6 - 5x)$$

$$\begin{aligned} &\quad\; \text{F} \qquad\; \text{O} \qquad\qquad \text{I} \qquad\qquad \text{L} \\ &= 4(6) + 4(-5x) + (-2x)(6) + (-2x)(-5x) \\ &= \; 24 \;-\; 20x \;-\; 12x \;+\; 10x^2 \\ &= 10x^2 - 32x + 24 \end{aligned}$$

Thus, $(4 - 2x)(6 - 5x) = 10x^2 - 32x + 24$.

Now Try Exercise 63

EXAMPLE 13 Multiply $(4p + 5)(4p - 5)$.

Solution

$$\begin{aligned}
(4p + 5)(4p - 5) &= \overset{F}{(4p)(4p)} + \overset{O}{(4p)(-5)} + \overset{I}{(5)(4p)} + \overset{L}{(5)(-5)} \\
&= 16p^2 - 20p + 20p - 25 \\
&= 16p^2 - 25
\end{aligned}$$

Thus, $(4p + 5)(4p - 5) = 16p^2 - 25$.

Now Try Exercise 53

5 Multiply Binomials Using Formulas for Special Products

Example 13 illustrated a special product, the product of the sum and difference of the same two terms.

Special Product: Product of the Sum and Difference of the Same Two Terms

$$\underbrace{(a + b)}_{\text{sum of 2 terms}}\underbrace{(a - b)}_{\text{difference 2 terms}} = a^2 - b^2$$ where a and b represent terms.

The expression on the right side of the equals sign is called **the difference of two squares.** Since multiplication is commutative, $(a + b)(a - b) = a^2 - b^2$ can also be written $(a - b)(a + b) = a^2 - b^2$.

EXAMPLE 14 Use the rule for finding the product of the sum and difference of two quantities to multiply each expression.

a) $(x + 5)(x - 5)$ **b)** $(2x + 3)(2x - 3)$ **c)** $(3x - 2y)(3x + 2y)$

Solution

a) If we let $x = a$ and $5 = b$, then

$$\begin{aligned}
(a + b)(a - b) &= a^2 - b^2 \\
\downarrow \quad \downarrow \quad \downarrow \quad \downarrow \quad &\quad\ \downarrow \qquad \downarrow \\
(x + 5)(x - 5) &= (x)^2 - (5)^2 \\
&= x^2 - 25
\end{aligned}$$

b)

$$\begin{aligned}
(a + b)(a - b) &= a^2 - b^2 \\
\downarrow \quad \downarrow \quad \downarrow \quad \downarrow \quad &\quad\ \downarrow \qquad \downarrow \\
(2x + 3)(2x - 3) &= (2x)^2 - (3)^2 \\
&= 4x^2 - 9
\end{aligned}$$

c)

$$\begin{aligned}
(a - b)(a + b) &= a^2 - b^2 \\
\downarrow \quad \downarrow \quad \downarrow \quad \downarrow \quad &\quad\ \downarrow \qquad \downarrow \\
(3x - 2y)(3x + 2y) &= (3x)^2 - (2y)^2 \\
&= 9x^2 - 4y^2
\end{aligned}$$

Now Try Exercise 75

Example 14 could also have been done using the FOIL method.

EXAMPLE 15 Using the FOIL method, multiply $(x + 3)^2$.

Solution

$$(x + 3)^2 = (x + 3)(x + 3)$$

$$\begin{aligned} &\quad\ \ \text{F} \qquad\ \ \text{O} \qquad\quad \text{I} \qquad\quad \text{L} \\ &= x(x) + x(3) + 3(x) + (3)(3) \\ &= x^2 + 3x + 3x + 9 \\ &= x^2 + 6x + 9 \end{aligned}$$

Now Try Exercise 77

Example 15 illustrates the **square of a binomial,** another special product.

Special Product: Square of Binomial Formulas

$$(a + b)^2 = (a + b)(a + b) = a^2 + 2ab + b^2$$
$$(a - b)^2 = (a - b)(a - b) = a^2 - 2ab + b^2$$

To square a binomial, add the square of the first term, twice the product of the terms, and the square of the second term.

EXAMPLE 16 Use the square of a binomial formula to multiply each expression.

a) $(x + 5)^2$ **b)** $(2x - 3)^2$ **c)** $(3r + 2s)^2$ **d)** $(x - 3)(x - 3)$

Solution

a) If we let $x = a$ and $5 = b$, then

$$\begin{aligned} (a + b)(a + b) &= a^2 + 2ab + b^2 \\ (x + 5)^2 = (x + 5)(x + 5) &= (x)^2 + 2(x)(5) + (5)^2 \\ &= x^2 + 10x + 25 \end{aligned}$$

b)

$$\begin{aligned} (a - b)(a - b) &= a^2 - 2ab + b^2 \\ (2x - 3)^2 = (2x - 3)(2x - 3) &= (2x)^2 - 2(2x)(3) + (3)^2 \\ &= 4x^2 - 12x + 9 \end{aligned}$$

c)

$$\begin{aligned} (a + b)(a + b) &= a^2 + 2ab + b^2 \\ (3r + 2s)^2 = (3r + 2s)(3r + 2s) &= (3r)^2 + 2(3r)(2s) + (2s)^2 \\ &= 9r^2 + 12rs + 4s^2 \end{aligned}$$

d)

$$\begin{aligned} (a - b)(a - b) &= a^2 - 2ab + b^2 \\ (x - 3)(x - 3) = (x - 3)(x - 3) &= (x)^2 - 2(x)(3) + (3)^2 \\ &= x^2 - 6x + 9 \end{aligned}$$

Now Try Exercise 85

Example 16 could also have been done using the FOIL method.

AVOIDING COMMON ERRORS

CORRECT	INCORRECT
$(a + b)^2 = a^2 + 2ab + b^2$	~~$(a + b)^2 = a^2 + b^2$~~
$(a - b)^2 = a^2 - 2ab + b^2$	~~$(a - b)^2 = a^2 - b^2$~~

Do not forget the middle term when you square a binomial.

$$(x + 2)^2 \neq x^2 + 4$$
$$(x + 2)^2 = (x + 2)(x + 2)$$
$$= x^2 + 4x + 4$$

6 Multiply Any Two Polynomials

When multiplying any two polynomials, each term of one polynomial must be multiplied by each term of the other polynomial. For example.

$$(3x + 2)(4x^2 - 5x - 3)$$
$$= 3x(4x^2 - 5x - 3) + 2(4x^2 - 5x - 3) \quad \text{Use the distributive property.}$$
$$= 12x^3 - 15x^2 - 9x + 8x^2 - 10x - 6$$
$$= 12x^3 - 7x^2 - 19x - 6$$

Thus, $(3x + 2)(4x^2 - 5x - 3) = 12x^3 - 7x^2 - 19x - 6$.

Alternatively, you may prefer to multiply a polynomial by a polynomial using a vertical procedure. On page 269, we showed that when multiplying the number 43 by the number 12, we multiply each digit in the number 43 by each digit in the number 12. Review that example now. We can follow a similar procedure, being careful, however, to align like terms in the same column when performing the individual multiplications.

EXAMPLE 17 Multiply $(3x + 4)(2x + 5)$.

Solution First write the polynomials one beneath the other.

$$\begin{array}{r} 3x + 4 \\ \underline{2x + 5} \end{array}$$

Next, multiply each term in $(3x + 4)$ by 5.

$$\begin{array}{rr} & 3x + 4 \\ & \underline{2x + 5} \\ 5(3x + 4) \longrightarrow & 15x + 20 \end{array}$$

Next, multiply each term in $(3x + 4)$ by $2x$ and align like terms.

$$\begin{array}{rr} & 3x + 4 \\ & \underline{2x + 5} \\ & 15x + 20 \\ 2x(3x + 4) \longrightarrow & \underline{6x^2 + 8x \quad\quad} \\ & 6x^2 + 23x + 20 \end{array} \quad \text{Added like terms in columns.}$$

Now Try Exercise 51

The same answer for Example 17 would be obtained using the FOIL method.

EXAMPLE 18 Multiply $(4y + 3)(2y^2 - 7y - 5)$.

Solution For convenience, we place the shorter expression on the bottom, as illustrated.

$$\begin{array}{rrrrl} & 2y^2 & -\ 7y & -\ 5 & \\ & & 4y & +\ 3 & \\ \hline & 6y^2 & -\ 21y & -\ 15 & \text{Multiplied the top polynomial by 3.} \\ 8y^3 & -\ 28y^2 & -\ 20y & & \text{Multiplied the top polynomial by 4y; aligned like terms.} \\ \hline 8y^3 & -\ 22y^2 & -\ 41y & -\ 15 & \text{Added like terms in columns.} \end{array}$$

Now Try Exercise 95

EXAMPLE 19 Multiply $(x^2 - 3x + 2)(2x^2 - 3)$.

Solution

$$\begin{array}{rrrrrl} & & x^2 & -\ 3x & +\ 2 & \\ & & 2x^2 & & -\ 3 & \\ \hline & & -3x^2 & +\ 9x & -\ 6 & \text{Multiplied the top polynomial by } -3. \\ 2x^4 & -\ 6x^3 & +\ 4x^2 & & & \text{Multiplied the top polynomial by } 2x^2\text{; aligned like terms.} \\ \hline 2x^4 & -\ 6x^3 & +\ x^2 & +\ 9x & -\ 6 & \text{Added like terms in columns.} \end{array}$$

Now Try Exercise 103

EXAMPLE 20 Multiply $(3x^3 - 2x^2 + 4x + 6)(x^2 - 5x)$.

Solution

$$\begin{array}{rrrrrrl} & 3x^3 & -\ 2x^2 & +\ 4x & +\ 6 & & \\ & & & & x^2 & -\ 5x & \\ \hline & -15x^4 & +\ 10x^3 & -\ 20x^2 & -\ 30x & & \text{Multiplied the top polynomial by } -5x. \\ 3x^5 & -\ 2x^4 & +\ 4x^3 & +\ 6x^2 & & & \text{Multiplied the top polynomial by } x^2\text{; aligned like terms.} \\ \hline 3x^5 & -\ 17x^4 & +\ 14x^3 & -\ 14x^2 & -\ 30x & & \text{Added like terms in columns.} \end{array}$$

Now Try Exercise 105

4.5 Exercise Set MathXL® MyMathLab®

Warm-Up Exercises

Fill in the blanks with the appropriate word, phrase, or symbol(s) from the following list.

inverse	from	over	distributive	$a^2 + 2ab + b^2$	$a^2 - 2ab + b^2$
$a^2 + b^2$	$a^2 - b^2$	outer	last	first	inner

1. To multiply a polynomial by a monomial, we use the ______ property: $a(b + c) = ab + ac$
2. When multiplying two binomials using the FOIL method, the *F* stands for ______.
3. When multiplying two binomials using the FOIL method, the *O* stands for ______.
4. When multiplying two binomials using the FOIL method, the *I* stands for ______.
5. When multiplying two binomials using the FOIL method, the *L* stands for ______.
6. The special product $(a + b)(a - b) =$ ______.
7. The special product $(a + b)^2 =$ ______.
8. The special product $(a - b)^2 =$ ______.

Practice the Skills

Multiply.

9. $(3x^2)(5x^3)$
10. $(7x^5)(3x^4)$
11. $(-3x^2)(-10x^3)$
12. $(-7p^5)(-2p^3)$
13. $4x^3(-5x^3)$
14. $(-4z^{12})(5z^3)$
15. $-3y^7 \cdot 8y$
16. $-9z \cdot 4z^5$
17. $(5x^3y^5)(4x^2y)$

18. $(4a^3b^7)(6a^2b)$
19. $4xy^6(-7x^2y^9)$
20. $-5x^2y^4 \cdot 2x^3y^2$
21. $(6x^2y)\left(\frac{1}{2}x^4\right)$
22. $\frac{3}{4}x(8x^2y^3)$
23. $\frac{1}{3}x^2y^3 \cdot \frac{3}{5}x^6y$
24. $\frac{2}{7}x^8y \cdot \frac{7}{9}xy^5$
25. $(3.3x^4)(1.8x^4y^3)$
26. $(2.3x^5)(4.1x^2y^4)$

Multiply.

27. $9(x - 6)$
28. $4(x + 3)$
29. $-3x(2x - 2)$
30. $-4p(-3p + 6)$
31. $7z(2z^2 - z + 3)$
32. $2x(x^2 + 3x - 1)$
33. $-2x(x^2 - 2x + 5)$
34. $-6c(-3c^2 + 5c - 6)$
35. $(5z^2 - 4z + 2)z$
36. $(3x^2 + x - 6)x$
37. $0.5x^2(x^3 - 6x^2 - 1)$
38. $2.3b^2(2b^2 - b + 3)$
39. $-\frac{1}{3}m(12m^2 - 6m - 3)$
40. $-\frac{1}{2}x^3(2x^2 + 4x - 6y^2)$
41. $\frac{2}{3}n(3n^2 - 3n + 1)$
42. $\frac{1}{4}y^4(y^2 - 12y + 4x)$

Multiply.

43. $(x + 5)(x + 7)$
44. $(y + 4)(y + 9)$
45. $(5x - 2)(x + 4)$
46. $(2x - 3)(x + 5)$
47. $(2t + 5)(3t - 6)$
48. $(4a - 1)(a + 4)$
49. $(x - 2)(4x - 2)$
50. $(3a - 5)(a - 7)$
51. $(3k - 6)(4k - 2)$
52. $(3d - 5)(4d - 1)$
53. $(x - 2)(x + 2)$
54. $(2x + 1)(2x - 1)$
55. $(2x - 3)(2x - 3)$
56. $(3x - 1)(3x - 1)$
57. $(2x + 3)(2x + 3)$
58. $(3x + 4)(3x + 4)$
59. $(6x - 1)(-2x + 5)$
60. $(-x + 3)(2x + 5)$
61. $(6z - 4)(7 - z)$
62. $(6 - 2m)(5m - 3)$
63. $(9 - 2x)(7 - 4x)$
64. $(2 - 5x)(7 - 2x)$
65. $(x + 7)(y - 3)$
66. $(z + 2y)(4z - 3)$
67. $(2x - 3y)(3x + 2y)$
68. $(2x + 3)(2y - 5)$
69. $\left(\frac{3}{4}x + 1\right)(4x - 8)$
70. $(2x - 6)\left(\frac{3}{2}x + 5\right)$
71. $(x + 0.6)(x + 0.3)$
72. $(2x - 0.1)(x + 2.4)$
73. $(x + 4)\left(x - \frac{1}{2}\right)$
74. $\left(2x + \frac{1}{3}\right)(3x - 1)$

Multiply using a special product formula.

75. $(x + 3)(x - 3)$
76. $(x + 12)(x - 12)$
77. $(x + 4)^2$
78. $(x + 5)^2$
79. $(x - 6)^2$
80. $(x - 7)^2$
81. $(2x - 1)(2x + 1)$
82. $(3x - 5)(3x + 5)$
83. $(7x + 4)^2$
84. $(8x + 5)^2$
85. $(5x - 6)^2$
86. $(4x - 7)^2$
87. $(5x + y)(5x - y)$
88. $(7x + 3y)(7x - 3y)$
89. $\left(\frac{1}{2}x + 4y\right)^2$
90. $\left(3x + \frac{1}{2}y\right)^2$
91. $(0.3x - 7y)^2$
92. $(4x - 0.7y)^2$

Multiply.

93. $(x + 4)(3x^2 + 4x - 1)$

94. $(x - 1)(3x^2 + 3x + 2)$

95. $(4m + 3)(4m^2 - 5m + 6)$

96. $(3x + 2)(4x^2 - x + 5)$

97. $(-2x^2 - 4x + 1)(7x - 3)$

98. $(4x^2 + 9x - 2)(x - 2)$

99. $(a + b)(a^2 - ab + b^2)$

100. $(a - b)(a^2 + ab + b^2)$

101. $(3t^2 - 2t + 4)(2t^2 + 3t + 1)$

102. $(2d^2 - 3d - 4)(5d^2 - d + 2)$

103. $(x^2 - x + 3)(x^2 - 2x)$

104. $(x^2 - 2x + 3)(x^2 - 4)$

105. $(2x^3 - 6x^2 + x - 3)(x^2 + 4x)$

106. $(3y^3 + 4y^2 - y + 7)(y^2 - 5y)$

Determine the cube of each expression by writing the expression as the square of an expression multiplied by another expression. For example, $(x + 3y)^3 = (x + 3y)^2(x + 3y)$.

107. $(b - 1)^3$

108. $(x + 2)^3$

109. $(3a - 5)^3$

110. $(2z + 3)^3$

Concept/Writing Exercises

111. Will the product of a monomial and a monomial always be a monomial? Explain your answer.

112. Will the product of a monomial and a binomial ever be a trinomial? Explain your answer.

113. Will the product of two binomials after like terms are combined always be a trinomial? Explain your answer.

114. Will the product of any polynomial and a binomial always be a polynomial? Explain.

Problem Solving

Consider the multiplications in Exercises 115 and 116. Determine the exponents to be placed in the shaded areas.

115. $3x^2(2x^{\square} - 5x^{\square} + 3x^{\square}) = 6x^8 - 15x^5 + 9x^3$.

116. $4x^3(x^{\square} + 2x^{\square} - 5x^{\square}) = 4x^7 + 8x^5 - 20x^4$.

117. Suppose that one side of a rectangle is represented as $x + 2$ and a second side is represented as $2x + 1$.

a) Express the area of the rectangle in terms of x.

b) Find the area if $x = 4$ feet.

c) What value of x, in feet, would result in the rectangle being a square? Explain how you determined your answer.

118. Consider the figure below.

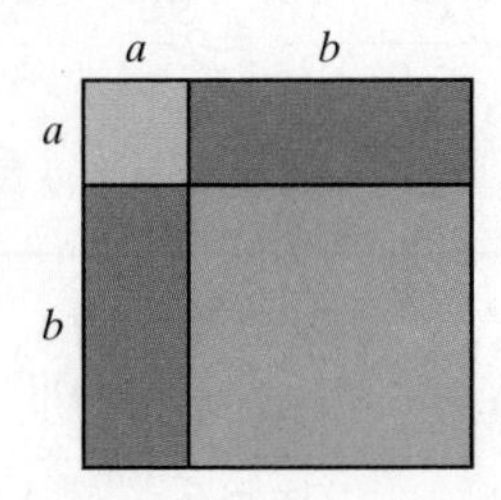

a) Write an expression for the length of the top.

b) Write an expression for the length of the left side.

c) Is this figure a square? Explain.

d) Express the area of this square as the square of a binomial.

e) Determine the area of the square by summing the areas of the four individual pieces.

f) Using the figure and your answer to part **e)**, complete the following.

$$(a + b)^2 = ?$$

119. Suppose that a rectangular solid has length $x + 5$, width $3x + 4$, and height $2x - 2$ (see the figure).

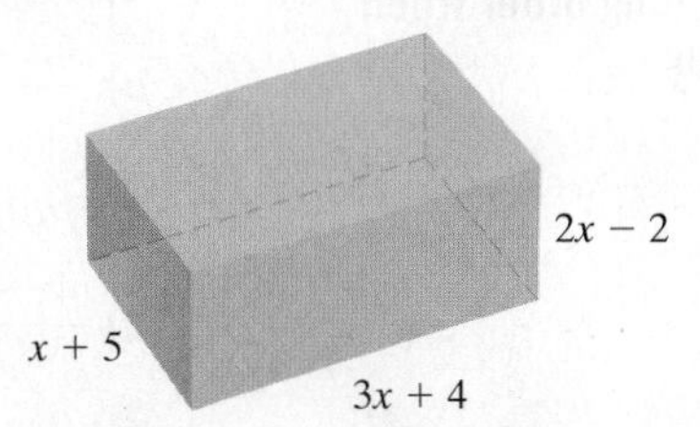

a) Write a polynomial that represents the area of the base by multiplying the length by the width.

b) The volume of the figure can be found by multiplying the area of the base by the height. Write a polynomial that represents the volume of the figure.

c) Using the polynomial in part **b)**, find the volume of the figure if x is 4 feet.

d) Using the binomials given for the length, width, and height, find the volume if x is 4 feet.

e) Are your answers to parts **c)** and **d)** the same? If not, explain why.

Challenge Problems

Multiply.

120. $(2x^3 - 6x^2 + 5x - 3)(3x^3 - 6x + 4)$

121. $\left(\frac{1}{2}x + \frac{2}{3}\right)\left(\frac{2}{3}x - \frac{2}{5}\right)$

122. $(x + 1)(x + 2)(x + 3)$

123. $(x - 1)(x - 2)(x - 3)$

Group Activity

124. Consider the trinomial $2x^2 + 7x + 3$.

a) As a group, determine whether there is a maximum number of pairs of binomials whose product is $2x^2 + 7x + 3$. That is, how many different pairs of binomials can go in the shaded areas?

$$2x^2 + 7x + 3 = (\quad)(\quad)$$

b) Individually, find a pair of binomials whose product is $2x^2 + 7x + 3$.

c) Compare your answer to part **b)** with the other members of your group. If you did not all arrive at the same answer, explain why.

Cumulative Review Exercises

[2.5] **125.** Solve the equation $3(x + 7) - 5 = 3x - 17$.

[3.3] **126. Complementary Angles** Angles C and D are complementary angles, and the measure of angle D is 16° less than the measure of angle C. Find the measures of angle C and angle D.

[4.1] **127.** Simplify $\left(\frac{3xy^4}{6y^6}\right)^4$.

[4.1–4.2] **128.** Evaluate the following.

a) -6^3

b) 6^{-3}

[4.4] **129.** Subtract $4x^2 - 4x - 9$ from $-x^2 - 6x + 5$.

4.6 Division of Polynomials

1. Divide a polynomial by a monomial.
2. Divide a polynomial by a binomial.
3. Check division of polynomial problems.
4. Write polynomials in descending order when dividing.

Now let's see how to divide polynomials.

1 Divide a Polynomial by a Monomial

To Divide a Polynomial by a Monomial

To divide a polynomial by a monomial, divide each term of the polynomial by the monomial.

EXAMPLE 1 Divide.

a) $\frac{4x + 20}{4}$ **b)** $\frac{9x^2 - 6x}{3x}$

Solution

a) $\frac{4x + 20}{4} = \frac{4x}{4} + \frac{20}{4} = x + 5$

b) $\frac{9x^2 - 6x}{3x} = \frac{9x^2}{3x} - \frac{6x}{3x} = 3x - 2$

Now Try Exercise 17

AVOIDING COMMON ERRORS

CORRECT

$$\frac{x + 2}{2} = \frac{x}{2} + \frac{2}{2} = \frac{x}{2} + 1$$

$$\frac{x + 2}{x} = \frac{x}{x} + \frac{2}{x} = 1 + \frac{2}{x}$$

INCORRECT

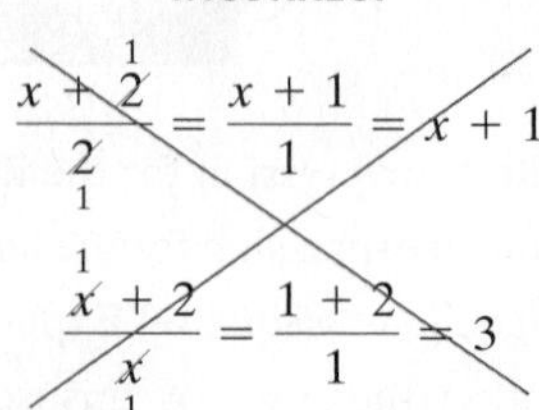

Can you explain why the procedures on the right are not correct?

EXAMPLE 2 Divide $\dfrac{4t^5 - 6t^4 + 8t - 3}{2t^2}$.

Solution

$$\frac{4t^5 - 6t^4 + 8t - 3}{2t^2} = \frac{4t^5}{2t^2} - \frac{6t^4}{2t^2} + \frac{8t}{2t^2} - \frac{3}{2t^2}$$

$$= 2t^3 - 3t^2 + \frac{4}{t} - \frac{3}{2t^2}$$

Now Try Exercise 37

EXAMPLE 3 Divide $\dfrac{3x^3 - 6x^2 + 4x - 1}{-3x}$.

Solution A negative sign appears in the denominator. Usually, it is easier to divide if the divisor is positive. We multiply both numerator and denominator by -1 to get a positive denominator.

$$\frac{(-1)(3x^3 - 6x^2 + 4x - 1)}{(-1)(-3x)} = \frac{-3x^3 + 6x^2 - 4x + 1}{3x}$$

$$= \frac{-3x^3}{3x} + \frac{6x^2}{3x} - \frac{4x}{3x} + \frac{1}{3x}$$

$$= -x^2 + 2x - \frac{4}{3} + \frac{1}{3x}$$

Now Try Exercise 41

2 Divide a Polynomial by a Binomial

We divide a polynomial by a binomial in much the same way as we perform long division.

EXAMPLE 4 Divide $\dfrac{x^2 + 6x + 8}{x + 2}$. ⟵ Dividend ⟵ Divisor

Solution Rewrite the division problem in the following form:

$$x + 2\overline{)x^2 + 6x + 8}$$

Divide x^2 (the first term in the dividend) by x (the first term in the divisor).

$$\frac{x^2}{x} = x$$

Place the quotient, x, above the like term containing x in the dividend.

$$\begin{array}{r} x \phantom{{}+8} \\ x + 2\overline{)x^2 + 6x + 8} \end{array}$$

Next, multiply the x by $x + 2$ as you would do in long division and place the terms of the product under their like terms.

Times

$$\begin{array}{r} x \phantom{{}+8} \\ x + 2\overline{)x^2 + 6x + 8} \\ \text{Equals} \quad \underline{x^2 + 2x} \phantom{{}+8} \end{array} \longleftarrow x(x + 2)$$

Now subtract $x^2 + 2x$ from $x^2 + 6x$. When subtracting, remember to change the sign of the terms being subtracted and then add the like terms.

$$\begin{array}{r} x \phantom{{}+8} \\ x + 2\overline{)x^2 + 6x + 8} \\ \underline{\overset{-}{x^2} \overset{-}{\not+} 2x} \phantom{{}+8} \\ 4x \phantom{{}+8} \end{array}$$

Next, bring down the 8, the next term in the dividend.

$$\begin{array}{r} x \\ x+2\overline{)x^2+6x+8} \\ \underline{x^2+2x} \\ 4x+8 \end{array}$$

Now divide $4x$, the first term at the bottom, by, x, the first term in the divisor.

$$\frac{4x}{x} = +4$$

Write the $+4$ in the quotient above the constant in the dividend.

$$\begin{array}{r} x+4 \\ x+2\overline{)x^2+6x+8} \\ \underline{x^2+2x} \\ 4x+8 \end{array}$$

Multiply the 4 by $x + 2$ and place the terms of the product under their like terms.

Times

$$\begin{array}{r} x+4 \\ x+2\overline{)x^2+6x+8} \\ \underline{x^2+2x} \\ 4x+8 \\ 4x+8 \end{array} \longleftarrow 4(x+2)$$

Equals

Now subtract.

$$\begin{array}{r} x+4 \\ x+2\overline{)x^2+6x+8} \\ \underline{x^2+2x} \\ 4x+8 \\ \underline{\overset{-}{4x} \not+ \overset{-}{8}} \\ 0 \end{array} \begin{array}{l} \longleftarrow \text{Quotient} \\ \\ \\ \\ \\ \longleftarrow \text{Remainder} \end{array}$$

Thus,

$$\frac{x^2 + 6x + 8}{x + 2} = x + 4$$

There is no remainder.

Now Try Exercise 43

> **Understanding Algebra**
>
> Dividing by binomials is similar to long division in arithmetic. There are five major steps:
>
> 1. Enter term in quotient
> 2. Multiply
> 3. Subtract
> 4. Bring down
> 5. Repeat

EXAMPLE 5 Divide $\dfrac{3x^2 + x - 12}{x + 2}$.

Solution First write the problem in the following form:

$$x+2\overline{)3x^2+x-12}$$

Since $3x^2$ divided by x is $3x$, place $3x$ above the x-term in the dividend.

$$\begin{array}{r} 3x \\ x+2\overline{)3x^2+x-12} \end{array}$$

Then multiply $3x(x + 2)$ and write the product $3x^2 + 6x$ as shown below. Then subtract to get a difference of $-5x$.

$$\begin{array}{r} 3x \\ x+2\overline{)3x^2+x-12} \\ \underline{\overset{-}{3x^2} \not+ \overset{-}{6x}} \\ -5x \end{array}$$

Next bring down the -12. Then divide $-5x$ by x, which gives -5. Place -5 over -12 in the dividend, as shown below. Then multiply $-5(x + 2)$. Write the product, $-5x - 10$ below the $-5x - 12$. Then subtract to get a remainder of -2.

$$\begin{array}{r} 3x - 5 \\ x + 2\overline{)3x^2 + x - 12} \\ \underline{3x^2 + 6x} \\ -5x - 12 \\ \underline{\overset{+}{\not-} 5x \overset{+}{\not-} 10} \\ -2 \end{array}$$

When there is a remainder, as in this example, list the quotient plus the remainder above the divisor. Thus,

$$\frac{3x^2 + x - 12}{x + 2} = 3x - 5 + \frac{-2}{x + 2} = 3x - 5 - \frac{2}{x + 2}$$

Now Try Exercise 45

EXAMPLE 6 Divide $\dfrac{6x^2 - 5x + 5}{2x + 3}$.

Solution

$$\begin{array}{r} \frac{6x^2}{2x} \quad \frac{-14x}{2x} \\ 3x - 7 \\ 2x + 3\overline{)\ 6x^2 - 5x + 5} \\ \underline{\overset{-}{}6x^2 \overset{-}{\not+} 9x} \quad \leftarrow 3x(2x + 3) \\ -14x + 5 \\ \underline{\overset{+}{\not-}14x \overset{+}{\not-}21} \quad \leftarrow -7(2x + 3) \\ 26 \quad \leftarrow \text{Remainder} \end{array}$$

Thus, $\dfrac{6x^2 - 5x + 5}{2x + 3} = 3x - 7 + \dfrac{26}{2x + 3}$.

Now Try Exercise 53

3 Check Division of Polynomial Problems

The answer to a division problem can be checked. Consider the division problem $13 \div 5$.

$$\begin{array}{r} 2 \\ 5\overline{)13} \\ \underline{10} \\ 3 \end{array}$$

Note that the divisor times the quotient, plus the remainder, equals the dividend:

$$(\text{divisor} \times \text{quotient}) + \text{remainder} = \text{dividend}$$

$$(5 \cdot 2) + 3 \stackrel{?}{=} 13$$

$$10 + 3 \stackrel{?}{=} 13$$

$$13 = 13 \quad \text{True}$$

This same procedure can be used to check all division problems.

To Check Division of Polynomials

(divisor × quotient) + remainder = dividend

Let's check the answer to Example 6. The divisor is $2x + 3$, the quotient is $3x - 7$, the remainder is 26, and the dividend is $6x^2 - 5x + 5$.

Check $\quad$ (divisor × quotient) + remainder = dividend

$$(2x + 3)(3x - 7) + 26 \stackrel{?}{=} 6x^2 - 5x + 5$$
$$(6x^2 - 5x - 21) + 26 \stackrel{?}{=} 6x^2 - 5x + 5$$
$$6x^2 - 5x + 5 = 6x^2 - 5x + 5 \quad \text{True}$$

4 Write Polynomials in Descending Order When Dividing

When dividing a polynomial by a binomial, both the polynomial and binomial should be listed in descending order. If there is no term for a certain power, we write that term with a numerical coefficient of 0 as a placeholder. This will help keep like terms aligned. For example, to divide $(6x^2 + x^3 - 4)/(x - 2)$, we begin by writing $(x^3 + 6x^2 + 0x - 4)/(x - 2)$.

EXAMPLE 7 Divide $(-x + 9x^3 - 28)$ by $(3x - 4)$.

Solution First we rewrite the dividend in descending order to get $(9x^3 - x - 28) \div (3x - 4)$. Since there is no x^2 term in the dividend, we will write $0x^2$ to help align like terms.

$$\frac{9x^3}{3x} \quad \frac{12x^2}{3x} \quad \frac{15x}{3x}$$

$$\begin{array}{r l}
 & 3x^2 + 4x + 5 \\
3x - 4 \overline{)} & 9x^3 + 0x^2 - x - 28 \\
 & \underline{9x^3 - 12x^2} \quad \leftarrow 3x^2(3x - 4) \\
 & 12x^2 - x \\
 & \underline{12x^2 - 16x} \quad \leftarrow 4x(3x - 4) \\
 & 15x - 28 \\
 & \underline{15x - 20} \quad \leftarrow 5(3x - 4) \\
 & -8 \quad \leftarrow \text{Remainder}
\end{array}$$

Thus, $\dfrac{-x + 9x^3 - 28}{3x - 4} = 3x^2 + 4x + 5 - \dfrac{8}{3x - 4}$. Check this division yourself using the procedure discussed on page 281.

Now Try Exercise 55

Understanding Algebra

When a term is "missing", we can represent the missing term with a zero coefficient as a placeholder.

For example, $9x^3 - x - 28$ has no x^2 term. But we can write the expression as $9x^3 + 0x^2 - x - 28$ so terms align in the division process.

4.6 Exercise Set MathXL® MyMathLab®

Warm-Up Exercises

Fill in the blanks with the appropriate word, phrase, or symbol(s) from the following list.

divisor	$6x^3 + 8x^2 + x + 16$	remainder	dividend
quotient	$6x^3 + 8x^2 + 0x + 16$	descending	ascending

1. In the division statement $(2x^2 + 11x + 12) \div (x + 5) = (2x + 1) + \dfrac{7}{x + 5}$, the quantity $2x^2 + 11x + 12$ is called the ________.

2. In the division statement $(2x^2 + 11x + 12) \div (x + 5) = (2x + 1) + \dfrac{7}{x + 5}$, the quantity $x + 5$ is called the ________.

3. In the division statement $(2x^2 + 11x + 12) \div (x + 5) = (2x + 1) + \dfrac{7}{x + 5}$, the quantity $2x + 1$ is called the ________.

4. In the division statement $(2x^2 + 11x + 12) \div (x + 5) = (2x + 1) + \dfrac{7}{x + 5}$, the quantity 7 is called the ________.

5. To prepare to do the division problem $\dfrac{6x^3 + 8x^2 + 16}{3x^2 - 2x + 4}$, the numerator $6x^3 + 8x^2 + 16$ is better written as ________.

6. When dividing a polynomial by a binomial, the terms of each should be written in ________ order.

In Exercises 7–10, decide if each division statement is true or false.

7. a) $\dfrac{4x + 8}{4} = x + 8$ b) $\dfrac{4x + 8}{4} = 4x + 2$ c) $\dfrac{4x + 8}{4} = x + 2$

8. a) $\dfrac{3x + 6}{3} = x + 6$ b) $\dfrac{3x + 6}{3} = 3x + 2$ c) $\dfrac{3x + 6}{3} = x + 2$

9. a) $\dfrac{5x + 6}{6} = 5x + 1$ b) $\dfrac{5x + 6}{6} = 5x$ c) $\dfrac{5x + 6}{6} = \dfrac{5}{6}x + 1$

10. a) $\dfrac{8x + 6}{6} = 8x + 1$ b) $\dfrac{8x + 6}{6} = 8x$ c) $\dfrac{8x + 6}{6} = \dfrac{4}{3}x + 1$

Rewrite each multiplication problem as a division problem. There is more than one correct answer.

11. $(x - 7)(x + 6) = x^2 - x - 42$
12. $(x + 3)(3x - 1) = 3x^2 + 8x - 3$
13. $(2x + 3)(x + 1) = 2x^2 + 5x + 3$
14. $(2x - 5)(x + 1) = 2x^2 - 3x - 5$
15. $(2x + 3)(2x - 3) = 4x^2 - 9$
16. $(3n + 4)(n - 5) = 3n^2 - 11n - 20$

Practice the Skills

Divide.

17. $\dfrac{3t + 6}{3}$
18. $\dfrac{5x + 15}{5}$
19. $\dfrac{3a + 7}{7}$
20. $\dfrac{2b + 5}{5}$
21. $\dfrac{7x + 6}{3}$
22. $\dfrac{5x + 14}{7}$
23. $\dfrac{-6x + 4}{2}$
24. $\dfrac{-24x - 18}{6}$
25. $\dfrac{-9x - 3}{-3}$
26. $\dfrac{8x - 3}{-8}$
27. $\dfrac{5y - 9}{5y}$
28. $\dfrac{2p - 3}{2p}$
29. $\dfrac{-10z + 8}{-8z}$
30. $\dfrac{15z - 9}{-9z}$
31. $(4x^2 + 8x - 12) \div 4x^2$
32. $(9p^2 + 21p - 15) \div 3p^2$
33. $\dfrac{-4x^5 + 6x + 8}{2x^2}$
34. $\dfrac{-20r^4 - 5r^2 + 15}{5r^2}$
35. $(x^5 + 3x^4 - 3) \div x^3$
36. $(6x^2 - 7x + 9) \div 3x$
37. $\dfrac{6x^5 - 4x^4 + 12x^3 - 5x^2}{2x^3}$
38. $\dfrac{-7s^3 + 15s^2 - 35s + 49}{7s^2}$
39. $\dfrac{8k^3 + 6k^2 - 8}{-4k}$
40. $\dfrac{-12x^4 + 6x^2 - 15x + 4}{-3x}$
41. $\dfrac{12x^5 + 3x^4 - 10x^2 - 9}{-3x^2}$
42. $\dfrac{-15m^3 - 6m^2 + 15}{-5m^3}$

Divide.

43. $\dfrac{x^2 + 4x + 3}{x + 1}$
44. $(2x^2 + 3x - 35) \div (x + 5)$
45. $\dfrac{5y^2 - 34y - 7}{y - 7}$
46. $\dfrac{2p^2 - 7p - 15}{p - 5}$
47. $\dfrac{6x^2 + 16x + 8}{3x + 2}$
48. $\dfrac{6t^2 - 7t - 20}{3t + 4}$
49. $(9x^2 - 4) \div (3x - 2)$
50. $(4a^2 - 25) \div (2a - 5)$
51. $(2x^2 + 7x - 18) \div (2x - 3)$
52. $(3x^2 - 4x - 9) \div (3x - 7)$
53. $(6x^2 + x - 10) \div (3x + 5)$
54. $(15x^2 + 11x - 7) \div (5x + 7)$
55. $\dfrac{-x + 9x^3 - 16}{3x - 4}$
56. $\dfrac{-x + 16x^3 - 25}{4x - 5}$
57. $\dfrac{2s^3 + 3s^2 + s + 1}{s + 2}$

58. $\dfrac{2x^3 - 3x^2 - 3x + 6}{x - 1}$

59. $\dfrac{7x^3 + 28x^2 - 5x - 20}{x + 4}$

60. $\dfrac{x^3 + 5x^2 + 2x - 8}{x + 2}$

61. $\dfrac{3s^3 + 9s - 5}{s + 1}$

62. $\dfrac{2x^3 + 6x - 4}{x + 4}$

63. $\dfrac{2x^3 - 17x^2 + 23x - 1}{2x - 3}$

64. $\dfrac{3x^3 - 17x^2 + 7x + 5}{3x - 2}$

65. $\dfrac{x^3 - 27}{x - 3}$

66. $\dfrac{x^3 + 64}{x + 4}$

67. $\dfrac{4x^3 - 5x}{2x - 1}$

68. $\dfrac{9x^3 - x + 3}{3x - 2}$

69. $\dfrac{-m^3 - 6m^2 + 2m - 3}{m - 1}$

70. $\dfrac{-x^3 + 3x^2 + 14x + 16}{x + 3}$

71. $\dfrac{4t^3 - t + 4}{t + 2}$

72. $\dfrac{4x^3 - 7x^2 - 5}{x - 2}$

Concept/Writing Exercises

73. When dividing a binomial by a monomial, must the quotient be a binomial? Explain and give an example to support your answer.

74. When dividing a trinomial by a monomial, must the quotient be a trinomial? Explain and give an example to support your answer.

Problem Solving

75. If the divisor is $x + 4$, the quotient is $2x + 3$, and the remainder is 4, find the dividend (or the polynomial being divided).

76. If the divisor is $2x - 3$, the quotient is $3x - 1$, and the remainder is -2, find the dividend.

77. If a polynomial of degree 4 in x is divided by a polynomial of degree 1 in x, what will be the degree of the quotient? Explain.

78. If a polynomial of degree 2 in x is divided by a polynomial of degree 1 in x, what will be the degree of the quotient? Explain.

Determine the monomial to be placed in the shaded area to make a true statement. Explain how you determined your answer.

79. $\dfrac{16x^4 + 20x^3 - 4x^2 + 12x}{\blacksquare} = 4x^3 + 5x^2 - x + 3$

80. $\dfrac{9x^5 - 6x^4 + 3x^2 + 12}{\blacksquare} = 3x^3 - 2x^2 + 1 + \dfrac{4}{x^2}$

Determine the exponents to be placed in the shaded areas to make a true statement. Explain how you determined your answer.

81. $\dfrac{8x^{\blacksquare} + 4x^{\blacksquare} - 20x^{\blacksquare} - 5x^{\blacksquare}}{2x^2} = 4x^3 + 2x - 10 - \dfrac{5}{2x}$

82. $\dfrac{15x^{\blacksquare} + 25x^{\blacksquare} + 5x^{\blacksquare} + 10x^{\blacksquare}}{5x^2} = 3x^5 + 5x^4 + x^2 + 2$

Challenge Problems

Divide. The quotients in Exercises 83 and 84 will contain fractions.

83. $\dfrac{3x^3 - 5}{3x - 2}$

84. $\dfrac{4x^3 - 4x + 6}{2x + 3}$

85. $\dfrac{3x^2 + 6x - 10}{-x - 3}$

Group Activity

Discuss and answer Exercises 86 and 87 as a group. Determine the polynomial that when substituted in the shaded area results in a true statement. Explain how you determined your answer.

86. $\dfrac{\blacksquare}{x + 4} = x + 2 + \dfrac{2}{x + 4}$

87. $\dfrac{\blacksquare}{x + 3} = x + 1 - \dfrac{1}{x + 3}$

Cumulative Review Exercises

[1.4] **88.** Consider the set of numbers

$$\left\{2, -5, 0, \sqrt{7}, \frac{2}{5}, -6.3, \sqrt{3}, -\frac{23}{34}\right\}.$$

List those that are

a) natural numbers;

b) whole numbers;

c) rational numbers;

d) irrational numbers;

e) real numbers.

[1.8] **89. a)** To what is $\frac{0}{1}$ equal?

b) How do we refer to an expression like $\frac{1}{0}$?

[1.9] **90.** Give the order of operations to be followed when evaluating a mathematical expression.

[2.5] **91.** Solve the equation $2(x + 3) + 2x = x + 4$.

[3.2] **92. Sale** At a 30% off sale Jennifer Lucking purchased a shirt for \$27.65. What was the original price of the shirt?

[4.2] **93.** Simplify $\frac{x^9}{x^{-4}}$.

See Exercise 92.

Chapter 4 Summary

IMPORTANT FACTS AND CONCEPTS	EXAMPLES
Section 4.1	
In the expression x^n, x is called the **base** and n is called the **exponent**.	base $\rightarrow 3^4 \nwarrow$ exponent
Rules of Exponents	Simplify.
1. Product Rule $x^m \cdot x^n = x^{m+n}$	**1.** $x^5 \cdot x^4 = x^{5+4} = x^9$
2. Quotient Rule $\frac{x^m}{x^n} = x^{m-n}, \quad x \neq 0$	**2.** $\frac{x^{12}}{x^7} = x^{12-7} = x^5$
3. Zero Exponent Rule $x^0 = 1, \quad x \neq 0$	**3.** $(-3ab^4)^0 = 1$
4. Power Rule $(x^m)^n = x^{m \cdot n}$	**4.** $(x^6)^3 = x^{6 \cdot 3} = x^{18}$
5. Power of a Product Rule $(xy)^n = x^n y^n$	**5.** $(5t)^2 = 5^2t^2 = 25t^2$
6. Power of a Quotient Rule $\left(\frac{x}{y}\right)^n = \frac{x^n}{y^n}, \quad y \neq 0$	**6.** $\left(\frac{x}{y}\right)^6 = \frac{x^6}{y^6}$
7. Expanded Power Rule $\left(\frac{ax}{by}\right)^m = \frac{a^m x^m}{b^m y^m}, \quad b \neq 0, \quad y \neq 0$	**7.** $\left(\frac{4x}{5y}\right)^2 = \frac{4^2x^2}{5^2y^2} = \frac{16x^2}{25y^2}$
Section 4.2	
Negative Exponent Rule $x^{-m} = \frac{1}{x^m}, \quad x \neq 0$	$x^{-2} = \frac{1}{x^2}$ $\frac{1}{y^{-6}} = y^6$
A Fraction Raised to a Negative Exponent Rule For a fraction of the form $\frac{a}{b}$, $a \neq 0$ and $b \neq 0$, $\left(\frac{a}{b}\right)^{-m} = \left(\frac{b}{a}\right)^m$	$\left(\frac{7}{8}\right)^{-2} = \left(\frac{8}{7}\right)^2 = \frac{8^2}{7^2} = \frac{64}{49}$
Section 4.3	
Each number written in **scientific notation** is written as a number greater than or equal to 1 and less than 10 multiplied by some power of 10.	1.3×10^7 4.76×10^{-2}
To Write a Number in Scientific Notation **1.** Move the decimal point in the original number to the right of the first nonzero digit. **2.** Count the number of places you moved the decimal point in step 1. If the original number was 10 or greater, the count is positive. If the original number was less than 1, the count is negative. **3.** Multiply the number obtained in step 1 by 10 raised to the count (power) found in step 2.	$25{,}700 = 2.57 \times 10^4$ $0.0000346 = 3.46 \times 10^{-5}$

IMPORTANT FACTS AND CONCEPTS	EXAMPLES
Section 4.3 (Continued)	
To Convert a Number from Scientific Notation to Decimal Form **1.** Observe the exponent of the power of 10. **2. a)** If the exponent is positive, move the decimal point in the number to the right the same number of places as the exponent. **b)** If the exponent is 0, do not move the decimal point. **c)** If the exponent is negative, move the decimal point in the number to the left the same number of places as the exponent.	$9.8 \times 10^6 = 9{,}800{,}000$ $5.17 \times 10^{-3} = 0.00517$
Section 4.4	
A **polynomial in** x is an expression containing the sum of a finite number of terms of the form ax^n, for any real number a and any whole number n.	$\frac{1}{5}x - 2$ and $x^2 - 4x + 7$ are both polynomials in x.
A polynomial is written in **descending order** when the exponents on the variable decrease from left to right.	$5x^4 - 3x^3 + 7x^2 - 6x + 9$ is written in descending order.
A **monomial** is a polynomial with one term.	$-7y^2$ is a monomial.
A **binomial** is a two-termed polynomial.	$x^2 - 8$ is a binomial.
A **trinomial** is a three-termed polynomial.	$4z^2 - 9z + 1$ is a trinomial.
The **degree of a term** of a polynomial in **one variable** is the exponent on the variable in that term.	$2y^6$ has degree six.
The **degree of a term** of a polynomial in **two or more variables** is the sum of the exponents on those variables.	$3x^2y^5$ has degree seven.
The **degree of a polynomial** is the same as that of its highest-degree term.	$9x^3 + 2x^2 - 5x + 4$ has degree three.
To Add Polynomials To add polynomials, combine the like terms of the polynomials.	$(3x^2 - 9x + 4) + (2x^2 - 3x - 5) = \underbrace{3x^2 + 2x^2} \underbrace{- 9x - 3x} \underbrace{+ 4 - 5}$ $= 5x^2 - 12x - 1$
To Subtract Polynomials **1.** Use the distributive property to remove parentheses. **2.** Combine like terms.	$(9a^2 - 6a + 1) - (a^2 - 5a - 3)$ $= 9a^2 - 6a + 1 - a^2 + 5a + 3$ $= \underbrace{9a^2 - a^2} \underbrace{- 6a + 5a} \underbrace{+ 1 + 3}$ $= 8a^2 - a + 4$
Section 4.5	
FOIL Method to Multiply Two Binomials (First, Outer, Inner, Last) 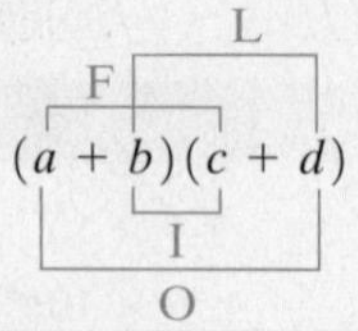	$(3x - 5)(x + 4) = \overset{F}{(3x)(x)} + \overset{O}{(3x)(4)} + \overset{I}{(-5)(x)} + \overset{L}{(-5)(4)}$ $= 3x^2 + 12x - 5x - 20$ $= 3x^2 + 7x - 20$
Product of Sum and Difference of the Same Two Terms (also called the difference of two squares): $(a + b)(a - b) = a^2 - b^2$	$(y + 6)(y - 6) = (y)^2 - (6)^2$ $= y^2 - 36$
Square of a Binomial $(a + b)^2 = a^2 + 2ab + b^2$ $(a - b)^2 = a^2 - 2ab + b^2$	$(x + 7)^2 = (x)^2 + 2(x)(7) + (7)^2$ $= x^2 + 14x + 49$ $(z - 3)^2 = (z)^2 - 2(z)(3) + (3)^2$ $= z^2 - 6z + 9$

IMPORTANT FACTS AND CONCEPTS	EXAMPLES
Section 4.5 (Continued)	
To Multiply Any Two Polynomials To multiply any two polynomials, each term of one polynomial must multiply each term of the second polynomial.	$(x^2 + 3x + 5)(x - 2)$ or $\begin{array}{r} x^2 + 3x + 5 \\ x - 2 \\ \hline -2x^2 - 6x - 10 \\ x^3 + 3x^2 + 5x \quad\quad \\ \hline x^3 + x^2 - x - 10 \end{array}$
Section 4.6	
To Divide a Polynomial by a Monomial To divide a polynomial by a monomial, divide each term of the polynomial by the monomial.	$\dfrac{6x + 24}{6} = \dfrac{6x}{6} + \dfrac{24}{6} = x + 4$
To Divide a Polynomial by a Binomial To divide a polynomial by a binomial we perform division in much the same way as we perform long division.	$\dfrac{x^2 - 4x + 3}{x + 2}$ $\begin{array}{r} x - 6 \\ x + 2\overline{)x^2 - 4x + 3} \\ \underline{x^2 + 2x} \quad\quad \\ -6x + 3 \\ \underline{-6x - 12} \\ 15 \end{array}$ $\dfrac{x^2 - 4x + 3}{x + 2} = x - 6 + \dfrac{15}{x + 2}$

Chapter 4 Review Exercises

[4.1] *Simplify.*

1. $x^5 \cdot x^3$
2. $x^2 \cdot x^4$
3. $3^2 \cdot 3^3$
4. $2^4 \cdot 2$
5. $\dfrac{x^4}{x}$
6. $\dfrac{a^5}{a^5}$
7. $\dfrac{5^5}{5^3}$
8. $\dfrac{4^4}{4}$
9. $\dfrac{x^6}{x^8}$
10. $\dfrac{y^4}{y}$
11. x^0
12. $7y^0$
13. $(-6z)^0$
14. 6^0
15. $(5x)^2$
16. $(3a)^3$
17. $(-3x)^3$
18. $(6s)^3$
19. $(2x^2)^4$
20. $(-t^4)^6$
21. $(-p^8)^4$
22. $\left(-\dfrac{2x^3}{y}\right)^2$
23. $\left(-\dfrac{5y^2}{2b}\right)^2$
24. $6x^2 \cdot 4x^3$
25. $\dfrac{16x^2y}{4xy^2}$
26. $2x(3xy^3)^3$
27. $\left(\dfrac{9x^2y}{3xy}\right)^2$
28. $(2x^2y)^3(3xy^4)$
29. $4x^2y^3(2x^3y^4)^2$
30. $3c^2(2c^4d^3)$
31. $\left(\dfrac{9a^3b^2}{3ab^7}\right)^3$
32. $\left(\dfrac{21x^4y^3}{7y^2}\right)^3$

[4.2] *Simplify.*

33. b^{-9}
34. 3^{-3}
35. 5^{-2}
36. $\dfrac{1}{z^{-2}}$
37. $\dfrac{1}{x^{-7}}$
38. $\dfrac{1}{4^{-2}}$
39. $y^5 \cdot y^{-8}$
40. $x^{-2} \cdot x^{-3}$
41. $p^{-6} \cdot p^4$
42. $a^{-2} \cdot a^{-3}$
43. $\dfrac{m^5}{m^{-5}}$
44. $\dfrac{x^5}{x^{-2}}$

45. $\dfrac{x^{-3}}{x^3}$ **46.** $(3x^4)^{-2}$ **47.** $(4x^{-3}y)^{-3}$ **48.** $(-2m^{-3}n)^2$

49. $6y^{-2} \cdot 2y^4$ **50.** $(-5y^{-3}z)^3$ **51.** $(-4x^{-2}y^3)^{-2}$ **52.** $2x(3x^{-2})$

53. $(5x^{-2}y)(2x^4y)$ **54.** $4y^{-2}(3x^2y)$ **55.** $4x^5(6x^{-7}y^2)$ **56.** $\dfrac{6xy^4}{2xy^{-1}}$

57. $\dfrac{12x^{-2}y^3}{3xy^2}$ **58.** $\dfrac{49x^2y^{-3}}{7x^{-3}y}$ **59.** $\dfrac{4x^8y^{-2}}{8x^7y^3}$ **60.** $\dfrac{36x^4y^7}{9x^5y^{-3}}$

[4.3] *Express each number in scientific notation.*

61. 1,720,000 **62.** 0.153 **63.** 0.00763

64. 47,000 **65.** 5760 **66.** 0.000314

Express each number without exponents.

67. 7.5×10^{-3} **68.** 6.52×10^{-4} **69.** 8.9×10^6

70. 5.12×10^4 **71.** 3.14×10^{-5} **72.** 1.103×10^7

Write each quantity as a base unit without metric prefixes and then write the quantity in scientific notation.

73. 92 milliliters **74.** 6 gigameters

75. 12.8 micrograms **76.** 19.2 kilograms

Perform each indicated operation and write your answer without exponents.

77. $(2.5 \times 10^2)(3.4 \times 10^{-4})$ **78.** $(4.2 \times 10^{-3})(3.0 \times 10^5)$ **79.** $(3.5 \times 10^{-2})(7.0 \times 10^3)$

80. $\dfrac{7.94 \times 10^6}{2.0 \times 10^{-2}}$ **81.** $\dfrac{1.5 \times 10^{-2}}{5.0 \times 10^2}$ **82.** $\dfrac{6.5 \times 10^4}{2.0 \times 10^6}$

Convert each number to scientific notation. Then calculate. Express your answer in scientific notation.

83. (14,000)(260,000) **84.** (0.00053)(40,000) **85.** (12,500)(400,000)

86. $\dfrac{250}{500{,}000}$ **87.** $\dfrac{0.000068}{0.02}$ **88.** $\dfrac{850{,}000}{0.025}$

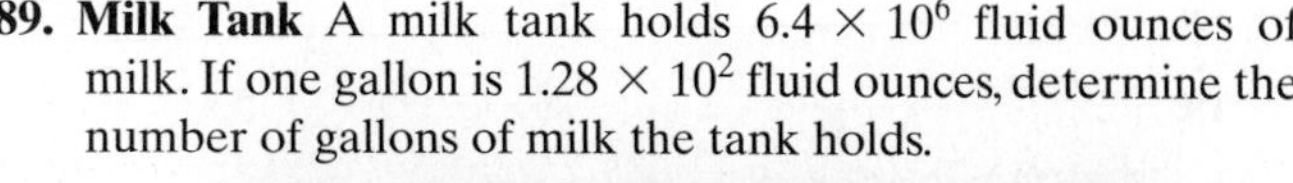

89. Milk Tank A milk tank holds 6.4×10^6 fluid ounces of milk. If one gallon is 1.28×10^2 fluid ounces, determine the number of gallons of milk the tank holds.

90. Indonesia Population In 2013, the population of Indonesia was about 2.38×10^8 people and the world population was about 7.05×10^9 people.

a) How many people lived outside of Indonesia in 2013?

b) How many times greater is the world population than the Indonesia population? Write your answer in decimal form rounded to the nearest hundredth.

[4.4] *Indicate whether each expression is a polynomial. If the polynomial is not written in descending order, rewrite it in descending order. If the polynomial has a specific name, give that name. State the degree of each polynomial.*

91. $x^{-4} - 8$ **92.** 7

93. $x^2 - 4 + 3x$ **94.** $-3 - x + 4x^2$

95. $4x^{1/2} - 6$

96. $13x^3 - 4$

97. $x - 4x^2$

98. $y^5 + y^{-3} - 9$

99. $2x^3 - 7 + 4x^2 - 3x$

[4.4–4.6] *Perform each indicated operation.*

100. $(x + 8) + (4x - 11)$

101. $(2d - 3) + (5d + 7)$

102. $(-x - 10) + (-2x + 5)$

103. $(-3x^2 + 9x + 5) + (-x^2 + 2x - 12)$

104. $(-m^2 + 5m - 8) + (6m^2 - 5m - 2)$

105. $(6.2p - 4.3) + (1.9p + 7.1)$

106. $(-6y - 7) - (-3y + 8)$

107. $(4x^2 - 9x) - (3x + 15)$

108. $(5a^2 - 6a - 9) - (2a^2 - a + 12)$

109. $(x^2 + 7x - 3) - (x^2 + 3x - 5)$

110. $(-2x^2 + 8x - 7) - (3x^2 + 12)$

111. $\frac{1}{7}x(21x + 21)$

112. $-3x(5x + 4)$

113. $3x(2x^2 - 4x + 7)$

114. $-c(2c^2 - 3c + 5)$

115. $-7b(-4b^2 - 3b - 5)$

116. $(x + 4)(x + 5)$

117. $(3x + 6)(-4x + 1)$

118. $(-5x + 3)^2$

119. $(6 - 2x)(2 + 3x)$

120. $(r + 5)(r - 5)$

121. $(x - 1)(3x^2 + 4x - 6)$

122. $(3x + 1)(x^2 + 2x + 4)$

123. $(-4x + 2)(3x^2 - x + 7)$

124. $\dfrac{2x + 4}{2}$

125. $\dfrac{12y + 18}{3}$

126. $\dfrac{8x^2 + 4x}{x}$

127. $\dfrac{6x^2 + 9x - 4}{3}$

128. $\dfrac{6w^2 - 5w + 3}{3w}$

129. $\dfrac{16x^6 - 8x^5 - 3x^3 + 1}{4x}$

130. $\dfrac{8m - 4}{-2}$

131. $\dfrac{5x^3 + 10x + 2}{2x^2}$

132. $\dfrac{5x^2 - 6x + 15}{3x}$

133. $\dfrac{x^2 + x - 12}{x - 3}$

134. $\dfrac{5x^2 + 28x - 10}{x + 6}$

135. $\dfrac{6n^2 + 19n + 3}{6n + 1}$

136. $\dfrac{4x^3 + 12x^2 + x - 12}{2x + 3}$

137. $\dfrac{4x^2 - 12x + 9}{2x - 3}$

Chapter 4 Practice Test

Chapter Test Prep Videos provide fully worked-out solutions to any of the exercises you want to review. Chapter Test Prep Videos are available via MyMathLab®, *or on* YouTube™ (www.youtube.com/user/AngelElementaryAlg).

Simplify each expression.

1. $5x^4 \cdot 3x^2$

2. $(3xy^2)^3$

3. $\dfrac{24p^7}{3p^2}$

4. $\left(\dfrac{3x^2y}{6xy^3}\right)^3$

5. $(2x^3y^{-2})^{-2}$

6. $(4x^0)(3x^2)^0$

7. $\dfrac{30x^6y^2}{45x^{-1}y}$

Convert each number to scientific notation and then determine the answer. Express your answer in scientific notation.

8. (285,000)(50,000)

9. $\dfrac{0.0008}{4000}$

Determine whether each expression is a polynomial. If the polynomial has a specific name, give that name.

10. $4x$

11. $-8c + 5$

12. $x^{-2} + 4$

13. Write the polynomial $-5 + 6x^3 - 2x^2 + 5x$ in descending order, and give its degree.

In Exercises 14–24, perform each indicated operation.

14. $(6x - 4) + (2x^2 - 5x - 3)$

15. $(y^2 - 7y + 3) - (4y^2 - 5y - 2)$

16. $(4x^2 - 5) - (x^2 + x - 8)$

17. $-5d(-3d + 8)$

18. $(5x + 8)(3x - 4)$

19. $(9 - 4c)(5 + 3c)$

20. $(3x - 5)(2x^2 + 4x - 5)$

21. $\dfrac{16x^2 + 8x - 4}{4}$

22. $\dfrac{-12x^2 - 6x + 5}{-3x}$

23. $\dfrac{8x^2 - 2x - 15}{2x - 3}$

24. $\dfrac{12x^2 + 7x - 12}{4x + 5}$

25. **Half-Life** The half-life of an element is the time it takes one half the amount of a radioactive element to decay. The half-life of carbon 14 (C^{14}) is 5730 years. The half-life of uranium 238 (U^{238}) is 4.47×10^9 years.

a) Write the half-life of C^{14} in scientific notation.

b) How many times longer is the half-life of U^{238} than C^{14}?

Cumulative Review Test

Take the following test and check your answers with those given in the back of the book. Review any questions that you answered incorrectly. The section where the material was covered is indicated after the answer.

1. Evaluate $12 + 8 \div 2^2 + 3$.

2. Simplify $7 - (2x - 3) + 2x - 8(1 - x)$.

3. Evaluate $-4x^2 + x - 7$ when $x = -2$.

4. Name each indicated property.

   a) $(5 + 2) + 7 = 5 + (2 + 7)$.

   b) $7 \cdot x = x \cdot 7$.

   c) $2(y + 9) = (y + 9)2$.

5. Solve the equation $5y + 7 = 2(y - 3)$.

6. Solve the equation $3(x + 2) + 3x - 5 = 4x + 1$.

7. Solve the inequality $3x - 11 < 5x - 2$. Graph the solution on a number line and give the answer in interval notation.

8. Solve the equation $3x - 2 = y - 7$ for y.

9. Solve $7x - 3y = 21$ for y, then find the value of y when $x = 6$.

10. Simplify $\left(\dfrac{5xy^{-3}}{x^{-2}y^5}\right)^2$.

11. Write the polynomial $-5x + 2 - 7x^2$ in descending order and give the degree.

Perform each indicated operation.

12. $(x^2 + 4x - 3) + (2x^2 + 5x + 1)$

13. $(6a^2 + 3a + 2) - (a^2 - 3a - 3)$

14. $(5t - 3)(2t - 1)$

15. $(2x - 1)(3x^2 - 5x + 2)$

16. $\dfrac{10d^2 + 12d - 8}{4d}$

17. $\dfrac{6x^2 + 11x - 10}{3x - 2}$

18. **Chicken Soup** At Art's Grocery Store, three cans of chicken soup sell for \$1.25. Find the cost of eight cans.

19. **Average Speed** Bob Dolan drives from Jackson, Mississippi, to Tallulah, Louisiana, a distance of 60 miles. At the same time, Nick Reide starts driving from Tallulah to Jackson along the same route. If Bob and Nick meet after 0.5 hour and Nick's average speed was 7 miles per hour greater than Bob's, find the average speed of each car.

20. **Rectangle** The length of a rectangle is 2 less than 3 times the width. Find the dimensions of the rectangle if its perimeter is 28 feet.

5 Factoring

What do Donald Duck, the Scarecrow from The Wizard of Oz, and a baseball diamond all have in common? The answer is the Pythagorean Theorem! Donald Duck is on a postage stamp talking about the Pythagorean Theorem, the Scarecrow talks about the Pythagorean Theorem in the movie *The Wizard of Oz* (see Exercise 53 on page 342), and the Pythagorean Theorem can be used to find the distance between home plate and second base in a baseball diamond.

Goals of This Chapter

The major emphasis of this chapter is to teach you how to factor polynomials. Factoring polynomials is the reverse process of multiplying polynomials.

In the first five sections of this chapter, you will learn how to factor a monomial from a polynomial, factor by grouping, factor trinomials of the form $ax^2 + bx + c$ when $a = 1$ and $a \neq 1$, and factor by using special factoring formulas. In the last two sections of this chapter, you will learn how to solve quadratic equations using factoring and how to solve applications of quadratic equations.

It is essential that you have a thorough understanding of factoring, especially Sections 5.3 through 5.5, to complete Chapter 6 successfully.

5.1 Factoring a Monomial from a Polynomial

1. Identify factors.
2. Determine the greatest common factor of two or more numbers.
3. Determine the greatest common factor of two or more terms.
4. Factor a monomial from a polynomial.

Understanding Algebra

Factoring is the reverse process of multiplication.

Multiplying:

$3x(2x^2+7) \overset{\text{becomes}}{\longrightarrow} 6x^3+21x$

Factoring:

$6x^3+21x \overset{\text{becomes}}{\longrightarrow} 3x(2x^2+7)$

1 Identify Factors

In Chapter 4, you learned how to multiply polynomials. In this chapter, we focus on factoring, the reverse process of multiplication.

Factor

To **factor an expression** means to write the expression as a product of its factors.
In general, if $a \cdot b = c$, then a and b are called factors of c.

For example, in Section 4.5 we showed that $3x(2x^2+4) = 6x^3+12x$. Thus, $3x$ and $2x^2+4$ are *factors* of $6x^3+12x$.

$3 \cdot 5 = 15$; so 3 and 5 are factors of 15.

$x^3 \cdot x^4 = x^7$; so x^3 and x^4 are factors of x^7.

$x(x+2) = x^2+2x$; so x and $x+2$ are factors of x^2+2x.

$(x-1)(x+3) = x^2+2x-3$; so $x-1$ and $x+3$ are factors of x^2+2x-3.

EXAMPLE 1 List the factors of $6x^3$.

Solution

Factors	Factors
$1 \cdot 6x^3 = 6x^3$	$x \cdot 6x^2 = 6x^3$
$2 \cdot 3x^3 = 6x^3$	$2x \cdot 3x^2 = 6x^3$
$3 \cdot 2x^3 = 6x^3$	$3x \cdot 2x^2 = 6x^3$
$6 \cdot x^3 = 6x^3$	$6x \cdot x^2 = 6x^3$

The factors of $6x^3$ are $1, 2, 3, 6, x, 2x, 3x, 6x, x^2, 2x^2, 3x^2, 6x^2, x^3, 2x^3, 3x^3$, and $6x^3$. The opposite (or negative) of each of these factors is also a factor, but these opposites are generally not listed unless specifically requested.

Now Try Exercise 7

Here are examples of multiplying and factoring. Notice again that factoring is the reverse process of multiplying.

Multiplying	Factoring
$3(2x+5) = 6x+15$	$6x+15 = 3(2x+5)$
$4y(y-7) = 4y^2-28y$	$4y^2-28y = 4y(y-7)$
$(x+1)(x+3) = x^2+4x+3$	$x^2+4x+3 = (x+1)(x+3)$

2 Determine the Greatest Common Factor of Two or More Numbers

To factor a monomial from a polynomial, we make use of the *greatest common factor (GCF)*. If you wish to see additional material on obtaining the GCF, you may wish to review Section 1.3, where we also discuss finding the GCF.

Before we examine how to find the GCF, a review of **prime numbers** is appropriate.

> **Understanding Algebra**
>
> When we write a number as a product of prime numbers, the product is called the *prime factorization* of the number. For example, $2 \cdot 2 \cdot 2 \cdot 2 \cdot 3$ or $2^4 \cdot 3$ is the prime factorization of 48.

Prime Numbers and Composite Numbers

A **prime number** is an integer greater than 1 that has exactly two factors, itself and 1.
The first 15 prime numbers are:

2, 3, 5, 7, 11, 13, 17, 19, 23, 29, 31, 37, 41, 43, 47

A **composite** number is an integer greater than 1 that is not prime.
The first 15 composite numbers are:

4, 6, 8, 9, 10, 12, 14, 15, 16, 18, 20, 21, 22, 24, 25

The number 1 is neither prime nor composite.

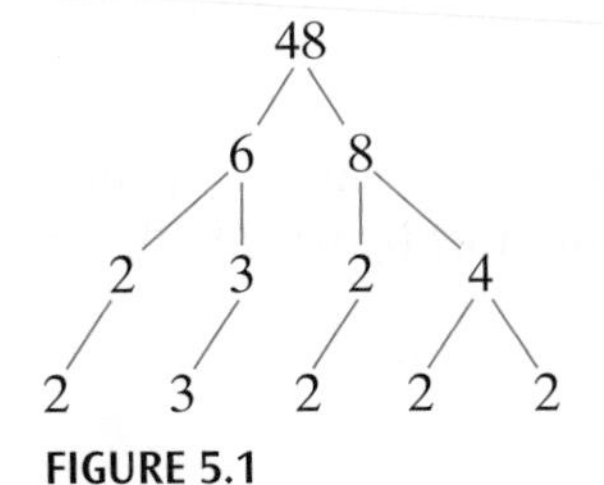

FIGURE 5.1

Examples 2 and 3 review the procedure for writing composite numbers as the product of prime numbers. Additional examples are given on page 21.

EXAMPLE 2 Write 48 as a product of prime numbers.

Solution Select any two numbers whose product is 48. Two possibilities are $6 \cdot 8$ and $4 \cdot 12$, but there are other choices. Continue breaking down the factors until all the factors are prime, as illustrated in **Figure 5.1**.

Note that no matter how you select your initial factors,

$$48 = 2 \cdot 2 \cdot 2 \cdot 2 \cdot 3 = 2^4 \cdot 3$$

Now Try Exercise 9

In Example 2, we found that $48 = 2 \cdot 2 \cdot 2 \cdot 2 \cdot 3 = 2^4 \cdot 3$. The $2 \cdot 2 \cdot 2 \cdot 2 \cdot 3$ or $2^4 \cdot 3$ may also be referred to as the **prime factorization** of 48.

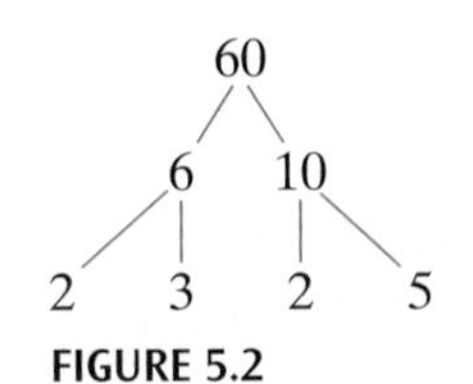

FIGURE 5.2

EXAMPLE 3 Write 60 as a product of its prime factors.

Solution One way to find the prime factors is shown in **Figure 5.2**. Therefore, $60 = 2 \cdot 2 \cdot 3 \cdot 5 = 2^2 \cdot 3 \cdot 5$.

Now Try Exercise 11

Recall from Section 1.3 that the **greatest common factor (GCF)** of two numbers is the greatest number that divides evenly into each of the numbers. The greatest common factor of 6 and 8 is 2 because 2 is the largest number that divides into both 6 and 8.

To find the GCF of two or more numbers, a method involving the numbers' prime factorizations is used.

To Determine the GCF of Two or More Numbers

1. Write each number as a product of prime factors.
2. Determine the prime factors common to all the numbers.
3. Multiply the common factors found in step 2. The product of these factors is the GCF.

EXAMPLE 4 Determine the greatest common factor of 48 and 60.

Solution From Examples 2 and 3, we know that

Step 1
$$48 = 2 \cdot 2 \cdot 2 \cdot 2 \cdot 3 = 2^4 \cdot 3$$
$$60 = 2 \cdot 2 \cdot 3 \cdot 5 = 2^2 \cdot 3 \cdot 5$$

Step 2 The common factors are circled. Two factors of 2 and one factor of 3 are common to both numbers. The product of these factors is the GCF of 48 and 60:

Step 3
$$\text{GCF} = 2 \cdot 2 \cdot 3 = 12$$

The GCF of 48 and 60 is 12. Twelve is the greatest number that divides evenly into both 48 and 60.

Now Try Exercise 15

> **Understanding Algebra**
>
> The greatest common factor (GCF) of 70 and 84 is 14 because 14 is the largest number that divides evenly into both 70 and 84.
>
> $70 = 2 \cdot 5 \cdot 7$
>
> $84 = 2 \cdot 2 \cdot 3 \cdot 7$

EXAMPLE 5 Determine the GCF of 18 and 24.

Solution

$$18 = 2 \cdot 3 \cdot 3 = 2 \cdot 3^2$$
$$24 = 2 \cdot 2 \cdot 2 \cdot 3 = 2^3 \cdot 3$$

One factor of 2 and one factor of 3 are common to both 18 and 24.

$$\text{GCF} = 2 \cdot 3 = 6$$

Now Try Exercise 19

3 Determine the Greatest Common Factor of Two or More Terms

Consider the terms x^3, x^4, x^5, and x^6. The GCF of these terms is x^3, since x^3 is the largest number of x's common to all four terms. We can illustrate this by writing the terms in factored form, with x^3 as one factor.

$$x^3 = x^3 \cdot 1$$
$$x^4 = x^3 \cdot x$$
$$x^5 = x^3 \cdot x^2$$
$$x^6 = x^3 \cdot x^3$$

GCF of all four terms is x^3.

Notice that x^3 evenly divides all four terms,

$$\frac{x^3}{x^3} = 1, \text{ and } \frac{x^4}{x^3} = x, \text{ and } \frac{x^5}{x^3} = x^2, \text{ and } \frac{x^6}{x^3} = x^3.$$

EXAMPLE 6 Determine the GCF of the terms m^9, m^5, m^7, and m^4.

Solution The GCF is m^4 because m^4 is the largest factor common to all the terms.

Now Try Exercise 21

EXAMPLE 7 Determine the GCF of the terms x^2y^3, x^3y^2 and xy^4.

Solution

$$x^2y^3 = x \cdot x \cdot y \cdot y \cdot y$$
$$x^3y^2 = x \cdot x \cdot x \cdot y \cdot y$$
$$xy^4 = x \cdot y \cdot y \cdot y \cdot y$$

One x is common to each term. Two y's are common to each term.

x^1 y^2

Thus, the GCF of the three terms is xy^2

Now Try Exercise 29

Greatest Common Factor of Two or More Terms

To determine the GCF of two or more terms

1. Find the GCF of the numerical coefficients of the terms.
2. Find the largest power of each variable that is common to all of the terms.
3. The GCF is the product of the number from step 1 and the variable expressions from step 2.

EXAMPLE 8 Determine the GCF of each group of terms.

a) $18y^2, 15y^3, 27y^5$ **b)** $-20x^2, 12x, 40x^3$ **c)** $5s^4, s^7, s^3$

Solution

a) The GCF of 18, 15, and 27 is 3. The GCF of y^2, y^3, and y^5 is y^2. Therefore, the GCF of the three terms is $3y^2$.

b) The GCF of -20, 12, and 40 is 4. The GCF of x^2, x, and x^3 is x. Therefore, the GCF of the three terms is $4x$.

c) The GCF of 5, 1, and 1 is 1. The GCF of s^4, s^7, and s^3 is s^3. Therefore, the GCF of the three terms is $1s^3$, which we write as s^3.

Now Try Exercise 27

EXAMPLE 9 Determine the GCF of $48x^2yz$ and $60x^3y^3$.

Solution

1. The numerical coefficients are 48 and 60. From Example 4, the GCF of 48 and 60 is 12.
2. The variables x and y are common to both terms. The variable z does not appear in the second term. The largest power of x common to both terms is x^2. The largest power of y common to both terms is y.
3. Thus, the GCF is the product of 12, x^2, and y, or $12x^2y$.

Now Try Exercise 35

EXAMPLE 10 Determine the GCF of each pair of terms.

a) $a(a - 6)$ and $3(a - 6)$
b) $t(t + 4)$ and $t + 4$
c) $3(p + q)$ and $4p(p + q)$

Solution

a) The GCF is $(a - 6)$.

b) $t + 4$ can be written as $1(t + 4)$. Therefore, the GCF of $t(t + 4)$ and $1(t + 4)$ is $(t + 4)$.

c) The GCF is $(p + q)$.

Now Try Exercise 39

4 Factor a Monomial from a Polynomial

Factoring is the reverse process of multiplying factors. As mentioned earlier, to *factor an expression* means to write the expression as a product of its factors.

To Factor a Monomial from a Polynomial

1. Determine the greatest common factor of all terms in the polynomial.
2. Write each term as the product of the GCF and its other factor.
3. Use the distributive property to factor out the GCF.

Understanding Algebra

With multiplication, we usually see the distributive property used as follows:

$$4(x+2) = 4 \cdot x + 4 \cdot 2 = 4x+8$$

But, with factoring, we reverse the process as follows:

$$4x+8 = 4 \cdot x + 4 \cdot 2 = 4(x+2)$$

In step 3 of the process, where we indicate that we use the distributive property, we factor the GCF out of the terms in the polynomial. For example, to factor $4 \cdot x + 4 \cdot 2$, we factor out the GCF, 4, to write $4(x + 2)$.

EXAMPLE 11 Factor $6x + 18$.

Solution The GCF is 6.

$$6x + 18 = 6 \cdot x + 6 \cdot 3 \quad \text{Write each term as a product of the GCF and its other factor.}$$

$$= 6(x + 3) \quad \text{Distributive property}$$

Now Try Exercise 49

HELPFUL HINT

Checking a factoring problem involves two steps. First, multiply the factored result—you should obtain the original expression. Second, be sure each factor in your factored result cannot be factored further. For example, if we factor $3x^2 + 6x$ as $x(3x + 6)$, the first part of the check—multiplying x by $(3x + 6)$—does in fact yield $3x^2 + 6x$. However, there is still a common factor, 3, in $(3x + 6)$, which needs to be factored out. The correct answer is that $3x^2 + 6x$ factors as $3x(x + 2)$. So, be sure the factors in your result do not have any remaining common factors.

EXAMPLE 12 Factor $15x - 20$.

Solution The GCF is 5.

$$\begin{aligned} 15x - 20 &= 5 \cdot 3x - 5 \cdot 4 \\ &= 5(3x - 4) \end{aligned}$$

The factored binomial no longer has any common factors and we can check that the factoring is correct by multiplying.

Now Try Exercise 55

EXAMPLE 13 Factor $6y^2 + 9y^5$.

Solution The GCF is $3y^2$.

$$\begin{aligned} 6y^2 + 9y^5 &= 3y^2 \cdot 2 + 3y^2 \cdot 3y^3 \\ &= 3y^2(2 + 3y^3) \end{aligned}$$

The factored binomial no longer has any common factors and we can check that the factoring is correct by multiplying.

Now Try Exercise 59

EXAMPLE 14 Factor $8q^3 - 20q^2 - 12q$.

Solution The GCF is $4q$.

$$\begin{aligned} 8q^3 - 20q^2 - 12q &= 4q \cdot 2q^2 - 4q \cdot 5q - 4q \cdot 3 \\ &= 4q(2q^2 - 5q - 3) \end{aligned}$$

Check $4q(2q^2 - 5q - 3) = 8q^3 - 20q^2 - 12q$

Now Try Exercise 81

EXAMPLE 15 Factor $35x^2 - 25x + 5$.

Solution The GCF is 5.

$$\begin{aligned} 35x^2 - 25x + 5 &= 5 \cdot 7x^2 - 5 \cdot 5x + 5 \cdot 1 \\ &= 5(7x^2 - 5x + 1) \end{aligned}$$

The factored trinomial no longer has any common factors and we can check that the factoring is correct by multiplying.

Now Try Exercise 83

Understanding Algebra

In Example 15, notice we rewrite the constant term 5 as $5 \cdot 1$ so that when 5 is factored out, a 1 remains.

EXAMPLE 16 Factor $4x^3 + x^2 + 8x^2y$.

Solution The GCF is x^2.

$$\begin{aligned} 4x^3 + x^2 + 8x^2y &= x^2 \cdot 4x + x^2 \cdot 1 + x^2 \cdot 8y \\ &= x^2(4x + 1 + 8y) \end{aligned}$$

The factored trinomial no longer has any common factors and we can check that the factoring is correct by multiplying.

Now Try Exercise 89

Notice in Examples 15 and 16 that when one of the terms is itself the GCF, we express it in factored form as the product of the term itself and 1.

EXAMPLE 17 Factor $x(5x - 2) + 7(5x - 2)$.

Solution The GCF of $x(5x - 2)$ and $7(5x - 2)$ is $(5x - 2)$. Factoring out the GCF gives

$$x(5x - 2) + 7(5x - 2) = (5x - 2)(x + 7)$$

This factored result no longer has any common factors and we can check that the factoring is correct by multiplying.

Now Try Exercise 95

EXAMPLE 18 Factor $4x(3x - 5) - 7(3x - 5)$.

Solution The GCF of $4x(3x - 5)$ and $-7(3x - 5)$ is $(3x - 5)$. Factoring out the GCF gives

$$4x(3x - 5) - 7(3x - 5) = (3x - 5)(4x - 7)$$

This factored result no longer has any common factors and we can check that the factoring is correct by multiplying.

Recall from Section 1.10 that the commutative property of multiplication states that the order in which any two real numbers are multiplied does not matter. Therefore, $(3x - 5)(4x - 7)$ can also be written $(4x - 7)(3x - 5)$. In the book, we will place the common factor on the left.

Now Try Exercise 97

EXAMPLE 19 Factor $2x(x + 3) - 5(x + 3)$.

Solution The GCF of $2x(x + 3)$ and $-5(x + 3)$ is $(x + 3)$. Factoring out the GCF gives

$$2x(x + 3) - 5(x + 3) = (x + 3)(2x - 5)$$

The factored result no longer has any common factors and we can check that the factoring is correct by multiplying.

Now Try Exercise 93

Understanding Algebra

When we are factoring *any* polynomial, the first step is to factor out the greatest common factor (if there is one).

Whenever you are factoring a polynomial by any of the methods presented in this chapter, the first step will always be to see if there is a common factor (other than 1) to all the terms in the polynomial. If so, factor the greatest common factor from each term using the distributive property.

5.1 Exercise Set MathXL® MyMathLab®

Warm-Up Exercises

Fill in the blanks with the appropriate word, phrase, or symbol(s) from the following list.

least factorization product quotient divides composite factors prime

1. To factor an expression means to write the expression as a __________ of its factors.
2. When we factor the expression $3x + 12$ as $3(x + 4)$, we say that 3 and $(x + 4)$ are __________ of $3x + 12$.
3. An integer that is greater than 1 that has exactly two factors, itself and 1, is called a __________ number.
4. An integer that is greater than 1 that is not a prime number is called a __________ number.
5. When we write a number as a product of prime numbers, the product is called the prime __________ of the number.
6. The greatest common factor of two numbers is the greatest number that __________ evenly into each of the numbers.

Practice the Skills

List the factors of the given term. Use only positive coefficients.

7. $4x^2$
8. $9y^2$
9. $6xy$
10. $10pq$

Write each number as a product of prime numbers.

11. 90
12. 120
13. 248
14. 540

Determine the greatest common factor for each pair of numbers.

15. 40, 56
16. 45, 27
17. 70, 98
18. 120, 96
19. 80, 126
20. 88, 160

Determine the greatest common factor for each group of terms.

21. x^5, x^3, x^2
22. y^3, y^5, y^2
23. $3x, 6x^2, 9x^3$
24. $6p, 4p^2, 8p^3$
25. a, ab, ab^2
26. c^2d, cd^2, d^3
27. $5q^3r, q^2r^2, qr^4$
28. $4x^2y^2, 3xy^4, 2xy^2$
29. $x^3y^7, x^7y^{12}, x^5y^5$
30. p^5q^2, p^4q^3, p^3q^4
31. $-3, 20x, 30x^2$
32. $-7, -35x^2, 54x^4$
33. $16x^9y^{12}, 8x^5y^3, 20x^4y^2$
34. $6p^4q^3, 9p^2q^5, 9p^4q^2$
35. $40x^3, 27x, 30x^4y^2$
36. $9x^3y^4, 8x^2y^4, 12x^4y^2$
37. $8(x - 4), 7(x - 4)$
38. $4(x - 5), 3x(x - 5)$
39. $x^2(2x - 3), 5(2x - 3)$
40. $3y(x + 2), 3(x + 2)$
41. $3w + 5, 6(3w + 5)$
42. $b(b + 3), b + 3$
43. $x - 4, y(x - 4)$
44. $x(9x - 3), 9x - 3$
45. $3(x - 1), 5(x - 1)^2$
46. $5(n + 2), 7(n + 2)^2$
47. $(x - 9)(x + 6), (x - 9)(x + 3)$
48. $3(b - 7)(b + 1), 2(b + 1)(b - 3)$

Factor the GCF from each term in the expression.

49. $4x + 12$
50. $5x + 30$
51. $9x + 3$
52. $10x + 5$
53. $5x - 40$
54. $6x - 54$
55. $8x - 2$
56. $56x - 7$
57. $9x^2 - 12x$
58. $16x^2 - 18x$
59. $3x^5 - 12x^2$
60. $26p^2 - 8p$
61. $36x^{12} + 24x^8$
62. $45y^{12} + 30y^{10}$
63. $27y^{15} - 9y^3$
64. $30w^5 + 25w^3$
65. $y + 6x^3y$
66. $w + 5w^3z$
67. $7a^4 + 3a^2$
68. $4x^2y - 6x$
69. $16xy^2z + 4x^3y$
70. $48m^4n^2 - 16mn^2$
71. $80x^5y^3z^4 - 36x^2yz^3$
72. $56xy^5z^{13} - 24y^4z^2$
73. $25x^2yz^3 + 25x^3yz$
74. $19x^4y^{12}z^{13} - 8x^5y^3z^9$
75. $13y^5z^3 - 11xy^2z^5$
76. $16r^4s^5t^3 - 20r^5s^4t$
77. $8c^2 - 4c - 32$
78. $x^3 - 4x^2 - 3x$
79. $9x^2 + 18x + 3$
80. $4x^2 + 8x + 24$
81. $4x^3 - 8x^2 + 12x$
82. $12a^3 - 16a^2 - 4a$
83. $40b^2 - 48c + 24$
84. $15p^2 - 6p + 9$
85. $5x^3 - xy^2 + x$
86. $45y^3 - 63y^2 + 27y$
87. $9a^4 - 6a^3 + 3ab$
88. $12s^4 - 9s^2 - 18st$
89. $8x^2y + 12xy^2 + 5xy$
90. $3n^5y^2 - 5n^4y^3 + 15n^2y^5$
91. $x(x - 7) + 6(x - 7)$
92. $9x(3x - 4) - 4(3x - 4)$
93. $3b(a - 2) - 4(a - 2)$
94. $3x(7x + 1) - 2(7x + 1)$
95. $4x(2x + 1) + 1(2x + 1)$
96. $4m(5m - 1) - 3(5m - 1)$
97. $5x(2x + 1) + 2x + 1$
98. $3x(4x - 5) + 4x - 5$
99. $3c(6c + 7) - 2(6c + 7)$
100. $5t(t - 2) - 3(t - 2)$

Problem Solving

Factor each expression, if possible. Treat the unknown symbol as if it were a variable.

101. $12\nabla - 6\nabla^2$

102. $3\bigstar + 6$

103. $12\square^3 - 4\square^2 + 4\square$

104. $© + 11\Delta$

Concept/Writing Exercises

105. Explain how to check a factoring problem.

106. What is the greatest common factor of two or more numbers?

Challenge Problems

107. Factor $6x^5(2x + 7) + 4x^3(2x + 7) - 2x^2(2x + 7)$.

108. Factor $4x^2(x - 3)^3 - 6x(x - 3)^2 + 4(x - 3)$.

109. Factor $x^2 + 2x + 3x + 6$. (*Hint:* Factor the first two terms, then factor the last two terms, then factor the resulting two terms. We will discuss factoring problems of this type in Section 5.2.)

Cumulative Review Exercises

[2.1] **110.** Simplify $2x - (x - 5) + 4(3 - x)$.

[2.5] **111.** Solve the equation $4 + 3(x - 8) = x - 4(x + 2)$.

[2.6] **112.** Solve the equation $4x - 5y = 20$ for y.

113. Find the volume of the cone shown below.

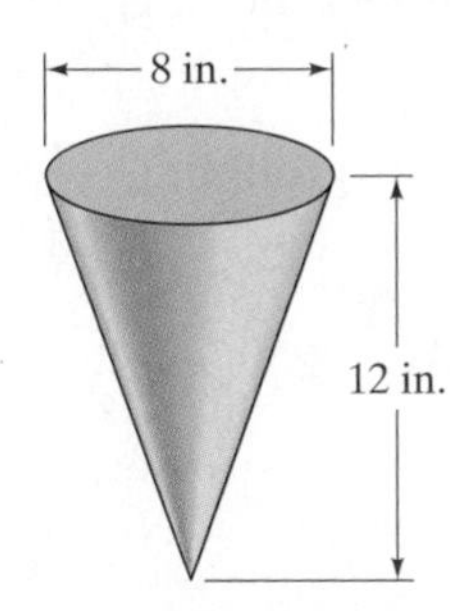

[3.2] **114.** The sum of two numbers is 41. Find the two numbers if the larger number is one less than twice the smaller number.

[4.1] **115.** Simplify $\left(\dfrac{3x^2y^3}{2x^5y^2}\right)^2$.

5.2 Factoring by Grouping

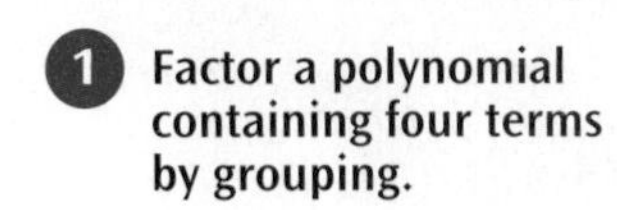

1 Factor a polynomial containing four terms by grouping.

1 Factor a Polynomial Containing Four Terms by Grouping

It may be possible to factor a polynomial containing four or more terms by removing common factors from groups of terms. This process is called **factoring by grouping,** which is illustrated in Example 1.

EXAMPLE 1 Factor $ax + ay + bx + by$ by grouping.

Solution There is no factor (other than 1) common to all four terms. However, a is common to the first two terms and b is common to the last two terms. Factor a from the first two terms and b from the last two terms.

$$ax + ay + bx + by = a(x + y) + b(x + y)$$

This factoring gives two terms, and $(x + y)$ is common to both terms. Proceed to factor $(x + y)$ from each term, as shown below.

$$a(x + y) + b(x + y) = (x + y)(a + b)$$

Notice that when $(x + y)$ is factored out we are left with $a + b$, which becomes the other factor. Thus, $ax + ay + bx + by = (x + y)(a + b)$.

Now Try Exercise 7

Understanding Algebra

$(ax + ay) \xrightarrow{\text{factor out } a} a(x + y)$

$(bx + by) \xrightarrow{\text{factor out } b} b(x + y)$

$a(x + y) + b(x + y) \xrightarrow{\text{factor out } (x+y)} (x + y)(a + b)$

To Factor a Four-Term Polynomial Using Grouping

1. Determine whether there are any factors common to all four terms. If so, factor the greatest common factor from each of the four terms.
2. If necessary, arrange the four terms so that the first two terms have a common factor and the last two have a common factor.
3. Use the distributive property to factor each group of two terms.
4. Factor the greatest common factor from the results of step 3.

EXAMPLE 2 Factor $x^2 + 3x + 4x + 12$ by grouping.

Solution No factor is common to all four terms. However, you can factor x from the first two terms and 4 from the last two terms.

$$x^2 + 3x + 4x + 12 = x(x + 3) + 4(x + 3)$$

$x + 3$ is common

$$x(x + 3) + 4(x + 3) = (x + 3)(x + 4)$$

Thus, $x^2 + 3x + 4x + 12 = (x + 3)(x + 4)$

Now Try Exercise 13

In Example 2, the $3x$ and $4x$ are like terms. However, if we were to combine them we would not be able to factor the four terms by grouping.

EXAMPLE 3 Factor $15x^2 + 10x + 12x + 8$ by grouping.

Solution

$$15x^2 + 10x + 12x + 8 = 5x(3x + 2) + 4(3x + 2)$$ Factor $5x$ from the first two terms and 4 from the last two terms.

$$= (3x + 2)(5x + 4)$$

Now Try Exercise 17

A factoring-by-grouping problem can be checked by multiplying the factors using the FOIL method. If you have not made a mistake, your result will be the polynomial you began with. Here is a check of Example 3.

Check

$$\begin{aligned}(3x + 2)(5x + 4) &= \overset{F}{(3x)(5x)} + \overset{O}{(3x)(4)} + \overset{I}{(2)(5x)} + \overset{L}{(2)(4)} \\ &= 15x^2 + 12x + 10x + 8 \\ &= 15x^2 + 10x + 12x + 8\end{aligned}$$

We can write $12x + 10x$ as $10x + 12x$ because of the commutative property of addition. Since this is the polynomial we started with, the factoring is correct.

EXAMPLE 4 Factor $15x^2 + 12x + 10x + 8$ by grouping.

Solution

$$\begin{aligned}15x^2 + 12x + 10x + 8 &= 3x(5x + 4) + 2(5x + 4) \\ &= (5x + 4)(3x + 2)\end{aligned}$$

Now Try Exercise 19

Notice that Example 4 is the same as Example 3 with the two middle terms interchanged. The answers to Examples 3 and 4 are equivalent since only the order of the factors are changed.

EXAMPLE 5 Factor $x^2 - 3x + x - 3$ by grouping.

Solution In the first two terms, x is the common factor. Is there a common factor in the last two terms? Yes; remember that 1 is a factor of every term. Factor 1 from the last two terms.

$$\begin{aligned} x^2 - 3x + x - 3 &= x^2 - 3x + 1 \cdot x - 1 \cdot 3 \\ &= x(x - 3) + 1(x - 3) \\ &= (x - 3)(x + 1) \end{aligned}$$

Note that $x - 3$ was expressed as $1 \cdot x - 1 \cdot 3 = 1(x - 3)$.

Now Try Exercise 21

Understanding Algebra

In Example 5, notice that $x - 3$ can be rewritten as $1 \cdot x - 1 \cdot 3$ and then as $1(x - 3)$.

EXAMPLE 6 Factor $6x^2 - 3x - 2x + 1$ by grouping.

Solution When $3x$ is factored from the first two terms, we get

$$6x^2 - 3x - 2x + 1 = 3x(2x - 1) - 2x + 1$$

What should we factor from the last two terms? We can factor -1 out of the last two terms so we can rewrite $-2x + 1$ as $-1(2x - 1)$.

Whenever we wish to change the sign of each term of an expression, we can factor out a negative number from each term. In this case, we factor out -1.

$$-2x + 1 = -1(2x - 1)$$

Next, we have

$$3x(2x - 1) - 2x + 1 = 3x(2x - 1) - 1(2x - 1)$$

Now we factor out the common factor $(2x - 1)$.

$$3x(2x - 1) - 1(2x - 1) = (2x - 1)(3x - 1)$$

Now Try Exercise 23

Understanding Algebra

It is important to recognize that $2x - 1$ is the opposite of $-2x + 1$. So, we can replace $-2x + 1$ with the equivalent $-1(2x - 1)$.

EXAMPLE 7 Factor $q^2 + 3q - q - 3$ by grouping.

Solution

$$\begin{aligned} q^2 + 3q - q - 3 &= q(q + 3) - q - 3 && \text{Factored out } q. \\ &= q(q + 3) - 1(q + 3) && \text{Factored out } -1. \\ &= (q + 3)(q - 1) && \text{Factored out } (q + 3). \end{aligned}$$

Now Try Exercise 25

EXAMPLE 8 Factor $3x^2 - 6x - 4x + 8$ by grouping.

Solution

$$\begin{aligned} 3x^2 - 6x - 4x + 8 &= 3x(x - 2) - 4(x - 2) \\ &= (x - 2)(3x - 4) \end{aligned}$$

Note: $-4x + 8 = -4(x - 2)$.

Now Try Exercise 27

HELPFUL HINT

When factoring four terms by grouping, if the coefficient of the third term is positive, as in Examples 2 through 5, you will generally factor out a positive coefficient from the last two terms. *If the coefficient of the third term is negative*, as in Examples 6 through 8, *you will generally factor out a negative coefficient from the last two terms.* The sign of the coefficient of the third term in the expression *must be included* so that the factoring results in two terms. For example,

$$2x^2 + 8x + 3x + 12 = 2x(x + 4) + 3(x + 4) = (x + 4)(2x + 3)$$

$$3x^2 - 15x - 2x + 10 = 3x(x - 5) - 2(x - 5) = (x - 5)(3x - 2)$$

When factoring four terms by grouping, the two middle terms need not be like terms. This is illustrated in Examples 9 and 10.

EXAMPLE 9 Factor $xy + 3x - 2y - 6$ by grouping.

Solution This problem contains two variables, x and y. Factor x from the first two terms and -2 from the last two terms.

$$\begin{aligned} xy + 3x - 2y - 6 &= x(y + 3) - 2(y + 3) \\ &= (y + 3)(x - 2) \quad \text{Factored out } (y + 3). \end{aligned}$$

Now Try Exercise 41

EXAMPLE 10 Factor $2xy + 4y + 3x + 6$.

Solution We will factor out $2y$ from the first two terms and 3 from the last two terms.

$$2xy + 4y + 3x + 6 = 2y(x + 2) + 3(x + 2)$$

Factor out the common factor $(x + 2)$ from each term on the right.

$$2y(x + 2) + 3(x + 2) = (x + 2)(2y + 3)$$

Check

$$\begin{aligned} (x + 2)(2y + 3) &= \overset{F}{(x)(2y)} + \overset{O}{(x)(3)} + \overset{I}{(2)(2y)} + \overset{L}{(2)(3)} \\ &= 2xy + 3x + 4y + 6 \\ &= 2xy + 4y + 3x + 6 \end{aligned}$$

Now Try Exercise 29

If Example 10 were given as $2xy + 3x + 4y + 6$, would the results be the same? Try it and see.

EXAMPLE 11 Factor $15a^2 - 10ab + 12ab - 8b^2$.

Solution Factor $5a$ from the first two terms and $4b$ from the last two terms.

$$\begin{aligned} 15a^2 - 10ab + 12ab - 8b^2 &= 5a(3a - 2b) + 4b(3a - 2b) \\ &= (3a - 2b)(5a + 4b) \end{aligned}$$

Now Try Exercise 31

EXAMPLE 12 Factor $2x^3 - 6x^2 + 4x - 12$.

Solution *The first step in any factoring problem is to determine whether all the terms have a common factor. If so, we factor out the greatest common factor (GCF).* In this polynomial, 2 is the GCF. Therefore, we begin by factoring out the 2.

$$2x^3 - 6x^2 + 4x - 12 = 2(x^3 - 3x^2 + 2x - 6)$$

Now we factor the polynomial in parentheses by grouping. We factor out x^2 from the first two terms and 2 from the last two terms.

$$\begin{aligned} 2(x^3 - 3x^2 + 2x - 6) &= 2[x^2(x - 3) + 2(x - 3)] \\ &= 2[(x - 3)(x^2 + 2)] \\ &= 2(x - 3)(x^2 + 2) \end{aligned}$$

Thus, $2x^3 - 6x^2 + 4x - 12 = 2(x - 3)(x^2 + 2)$.

Now Try Exercise 49

5.2 Exercise Set MathXL® MyMathLab®

Warm-Up Exercises

Fill in the blanks with the appropriate word, phrase, or symbol(s) from the following list.

division common negative positive FOIL grouping distributive associative

1. The factoring method we use to factor a polynomial such as $ax + ay + bx + by$ is called factoring by ________.
2. A factoring by grouping problem can be checked by multiplying the factors using the ________ method.
3. When factoring four terms by grouping, if the coefficient of the third term is positive, we will generally factor out a ________ coefficient from the last two terms.
4. When factoring four terms by grouping, if the coefficient of the third terms is negative, we will generally factor out a ________ coefficient from the last two terms.
5. The first step in any factoring problem is to determine whether all the terms have a ________ factor.
6. When factoring the polynomial $ax + ay + bx + by$ we first rewrite the polynomial as $a(x + y) + b(x + y)$ using the ________ property.

Practice the Skills

Factor by grouping.

7. $cw + cz + dw + dz$
8. $gj + gk + hj + hk$
9. $x^2 + 3x + 2x + 6$
10. $x^2 + 7x + 3x + 21$
11. $cw - cz + dw - dz$
12. $gj - gk + hj - hk$
13. $c^2 - 4c + 7c - 28$
14. $r^2 - 4r + 6r - 24$
15. $wy + wz - xy - xz$
16. $pr + ps - qr - qs$
17. $6x^2 + 8x + 15x + 20$
18. $8x^2 + 12x + 10x + 15$
19. $6x^2 + 8x - 15x - 20$
20. $8x^2 + 12x - 10x - 15$
21. $8x^2 + 32x + x + 4$
22. $3x^2 + 21x + x + 7$
23. $12t^2 - 8t - 3t + 2$
24. $16x^2 - 24x - 2x + 3$
25. $x^2 + 9x - x - 9$
26. $11x^2 + x - 11x - 1$
27. $6p^2 + 15p - 4p - 10$
28. $10c^2 + 25c - 6c - 15$
29. $x^2 + 2xy - 3xy - 6y^2$
30. $x^2 + 4xy - 3xy - 12y^2$
31. $3x^2 + 2xy - 9xy - 6y^2$
32. $3x^2 + 4xy - 18xy - 24y^2$
33. $10x^2 - 12xy - 25xy + 30y^2$
34. $21a^2 - 35ab - 24ab + 40b^2$
35. $x^2 - bx - ax + ab$
36. $x^2 + bx + ax + ab$
37. $xy + 9x - 5y - 45$
38. $x^2 + ax - 2x - 2a$
39. $a^2 + 3a + ab + 3b$
40. $6a^2 + 21a + 2ab + 7b$
41. $xy - x + 5y - 5$
42. $y^2 - yb + ya - ab$
43. $12 + 8y - 3x - 2xy$
44. $10 + 6y - 5z - 3yz$
45. $z^3 + 5z^2 + z + 5$
46. $x^3 - 3x^2 + 2x - 6$
47. $x^3 - 5x^2 + 8x - 40$
48. $x^3 - 3x^2 + 7x - 21$
49. $5x^3 - 15x^2 + 10x - 30$
50. $3x^3 - 6x^2 + 9x - 18$
51. $4x^2 + 8x + 8x + 16$
52. $16x^2 - 8x - 8x + 4$
53. $6x^3 + 9x^2 - 2x^2 - 3x$
54. $9x^3 + 6x^2 - 45x^2 - 30x$
55. $p^3 - 6p^2q + 2p^2q - 12pq^2$
56. $18x^2 + 27xy + 12xy + 18y^2$

Rearrange the terms so that the first two terms have a common factor and the last two terms have a common factor (other than 1). Then factor by grouping. There may be more than one way to arrange the factors. However, the answer should be equivalent regardless of the arrangement selected.

57. $6x + 5y + xy + 30$

58. $5m + 2w + mw + 10$

59. $ax - 21 - 3a + 7x$

60. $ax - 10 - 5x + 2a$

61. $5x + 5y + xy + 25$

62. $ax + by + ay + bx$

63. $rs - 42 + 6s - 7r$

64. $ca - 2b + 2a - cb$

65. $dc + 3c - ad - 3a$

66. $ac - bd - ad + bc$

Problem Solving

67. If you know that a polynomial with four terms is factorable by a specific arrangement of the terms, then will *any* arrangement of the terms be factorable by grouping? Explain, and support your answer with an example.

Factor each expression, if possible. Treat the unknown symbol as if it were a variable.

68. Ⓨ^{2} + 3Ⓨ + 4Ⓨ + 12

69. $\odot^2 + 3\odot - 5\odot - 15$

70. $\Delta^2 + 2\Delta - \Delta + 6$

Concept/Writing Exercises

71. A polynomial of four terms is factored by grouping and the result is $(x - 2y)(x - 3)$. Find the polynomial that was factored and explain how you determined the answer.

72. A polynomial of four terms is factored by grouping and the result is $(x - 2)(x + 4)$. Find the polynomial that was factored, and explain how you determined the answer.

Challenge Problems

In Section 5.4, we will factor trinomials of the form $ax^2 + bx + c, a \neq 1$, using grouping. To do this we rewrite the middle term of the trinomial, bx, as a sum or difference of two terms. Then we factor the resulting polynomial of four terms by grouping. For Exercises 73–78, **a)** *rewrite the trinomial as a polynomial of four terms by replacing the bx-term with the sum or difference given.* **b)** *Factor the polynomial of four terms. Note that the factors obtained are the factors of the trinomial.*

73. $2x^2 - 11x + 15, -11x = -5x - 6x$

74. $3x^2 + 10x + 8, 10x = 4x + 6x$

75. $2x^2 - 11x + 15, -11x = -6x - 5x$

76. $3x^2 + 10x + 8, 10x = 6x + 4x$

77. $4x^2 - 17x - 15, -17x = 3x - 20x$

78. $4x^2 - 17x - 15, -17x = -20x + 3x$

Factor each expression, if possible. Treat the unknown symbols as if they were variables.

79. $\bigstar\odot + 3\bigstar + 2\odot + 6$

80. $2\Delta^2 - 4\Delta\bigstar - 8\Delta\bigstar + 16\bigstar^2$

Cumulative Review Exercises

[2.5] **81.** Solve $5 - 3(2x - 7) = 4(x + 5) - 6$.

[3.4] **82. Mixture.** To celebrate the tenth anniversary of their candy store's opening, Harlan and Shelley Bricker decide to create a special candy mixture containing jelly beans and gumdrops. The jelly beans sell for \$6.25 per pound and the gumdrops sell for \$2.50 per pound. How many pounds of each type of candy will be needed to make a 50-pound mixture that will sell for \$4.75 per pound?

See Exercise 82.

[4.6] **83.** Divide $\dfrac{15x^3 - 6x^2 - 9x + 5}{3x}$.

84. Divide $\dfrac{a^2 - 16}{a + 4}$.

5.3 Factoring Trinomials of the Form $ax^2 + bx + c, a = 1$

1. Factor trinomials of the form $ax^2 + bx + c$, where $a = 1$.
2. Remove the greatest common factor from a trinomial.

1 Factor Trinomials of the Form $ax^2 + bx + c$, where $a = 1$

Now we discuss how to factor trinomials of the form $ax^2 + bx + c$, where a, the numerical coefficient of the squared term, is 1. Examples of such trinomials are

$$x^2 + 7x + 12 \qquad x^2 - 2x - 24$$
$$a = 1, b = 7, c = 12 \qquad a = 1, b = -2, c = -24$$

Recall that factoring is the reverse process of multiplication. We can show with the FOIL method of multiplying binomials that

$$(x + 3)(x + 4) = x^2 + 7x + 12 \quad \text{and} \quad (x - 6)(x + 4) = x^2 - 2x - 24$$

Therefore, $x^2 + 7x + 12$ and $x^2 - 2x - 24$ factor as follows:

$$x^2 + 7x + 12 = (x + 3)(x + 4) \quad \text{and} \quad x^2 - 2x - 24 = (x - 6)(x + 4)$$

In general, when we factor a trinomial of the form $x^2 + bx + c$ we will get a pair of binomial factors as follows:

$$x^2 + bx + c = (x + \square)(x + \square)$$

Numbers go here that add up to b and multiply to c.

To determine the numbers to place in the shaded areas, we list the different sets of factors of the constant c. We use those factors to create possible pairs of binomial factors and multiply each pair using the FOIL method, continuing until we find the pair whose sum of the products of the outer and inner terms is the same as the x-term in the trinomial. This method for factoring is called **trial and error** or the **reverse FOIL** method.

EXAMPLE 1 Factor $x^2 + 7x + 12$ by trial and error.

Solution Begin by listing the factors of 12. Then list the possible factors of the trinomial, and the products of these factors. Finally, determine which, if any, of these products gives the correct middle term, $7x$.

Factors of 12	Possible Factors of Trinomial	Products of Factors
$(1)(12)$	$(x + 1)(x + 12)$	$x^2 + 13x + 12$
$(2)(6)$	$(x + 2)(x + 6)$	$x^2 + 8x + 12$
$(3)(4)$	$(x + 3)(x + 4)$	$x^2 + 7x + 12$
$(-1)(-12)$	$(x - 1)(x - 12)$	$x^2 - 13x + 12$
$(-2)(-6)$	$(x - 2)(x - 6)$	$x^2 - 8x + 12$
$(-3)(-4)$	$(x - 3)(x - 4)$	$x^2 - 7x + 12$

In the last column, we find the trinomial we are seeking in the third line. Thus,

$$x^2 + 7x + 12 = (x + 3)(x + 4)$$

Now Try Exercise 11

Now let's consider an easier way to factor $x^2 + 7x + 12$. In Section 4.5, we illustrated how the FOIL method is used to multiply two binomials. Let's multiply $(x + 3)(x + 4)$ using the FOIL method.

F O I L

$$(x + 3)(x + 4) = x^2 + 4x + 3x + 12$$
$$= x^2 + 7x + 12$$

We see that $(x + 3)(x + 4) = x^2 + 7x + 12$.

Note that the *sum of the outer and inner terms is* $7x$ *and the product of the last terms is* 12. To factor $x^2 + 7x + 12$, we look for two numbers whose product is 12 and whose sum is 7. We list the factors of 12 first and then list the sum of the factors.

Factors of 12	Sum of Factors
$(1)(12) = 12$	$1 + 12 = 13$
$(2)(6) = 12$	$2 + 6 = 8$
$(3)(4) = 12$	$3 + 4 = 7$
$(-1)(-12) = 12$	$-1 + (-12) = -13$
$(-2)(-6) = 12$	$-2 + (-6) = -8$
$(-3)(-4) = 12$	$-3 + (-4) = -7$

The only factors of 12 whose sum is a positive 7 are 3 and 4. The factors of $x^2 + 7x + 12$ will therefore be $(x + 3)$ and $(x + 4)$.

$$x^2 + 7x + 12 = (x + 3)(x + 4)$$

In the previous illustration, all the possible factors of 12 were listed so that you could see them. However, once you find the specific factors you are seeking when working a problem, you need go no further.

To Factor Trinomials of the Form $ax^2 + bx + c$, where $a = 1$

1. Find two numbers whose product equals the constant, c, and whose sum equals the coefficient of the x-term, b.
2. Use the two numbers found in step 1, including their signs, to write the trinomial in factored form. The trinomial in factored form will be

$$(x + \text{first number})(x + \text{second number})$$

The sign of the constant, c, is the key in finding the sign of the two numbers to be placed in parentheses. See the Helpful Hint below.

HELPFUL HINT

To factor a trinomial of the form $x^2 + bx + c$, first observe the sign of the constant.

a) If the constant, c, is positive, both numbers in the factors will have the same sign, either both positive or both negative. If b is positive, both factors will contain positive numbers, and if b is negative, both factors will contain negative numbers.

Example:

$$x^2 + 7 + 12 = (x + 3)(x + 4)$$

The coefficient, b, is positive. The constant, c, is positive. Positive Positive

Both factors have positive numbers.

Example:

$$x^2 - 5x + 6 = (x - 2)(x - 3)$$

The coefficient, b, is negative. The constant, c, is positive. Negative Negative

Both factors have negative numbers.

b) If the constant is negative, the two numbers in the factors will have opposite signs. That is, one number will be positive and the other number will be negative.

Example:

$$x^2 + x - 6 = (x + 3)(x - 2)$$

The coefficient, b, is positive. The constant, c, is negative. Positive Negative

One factor has a positive number and the other factor has a negative number.

Example:

$$x^2 - 3x - 10 = (x + 2)(x - 5)$$

The coefficient, b, is negative. The constant, c, is negative. Positive Negative

One factor has a positive number and the other factor has a negative number.

We will use this information as a starting point when factoring trinomials.

EXAMPLE 2 Consider a trinomial of the form $x^2 + bx + c$. Use the signs of b and c given below to determine the signs of the numbers in the factors.

a) b is negative and c is positive

b) b is negative and c is negative

c) b is positive and c is negative

d) b is positive and c is positive

Solution In each case we look at the sign of the constant, c, first.

a) Since c is positive, both numbers must have the same sign. Since b is negative, both factors will contain negative numbers.

b) Since c is negative, one factor will contain a positive number and the other will contain a negative number.

c) Since c is negative, one factor will contain a positive number and the other will contain a negative number.

d) Since c is positive, both numbers must have the same sign. Since b is positive, both factors will contain positive numbers.

Now Try Exercise 5

Understanding Algebra

To factor $x^2 + bx + c$, first observe the sign of c, and then observe the sign of b:

$x^2 + 11x + 18 \rightarrow (x + 9)(x + 2)$

$x^2 - 11x + 18 \rightarrow (x - 9)(x - 2)$

If c is positive, numbers have same sign.

$x^2 - 7x - 18 \rightarrow (x - 9)(x + 2)$

$x^2 + 7x - 18 \rightarrow (x + 9)(x - 2)$

If c is negative, numbers have different signs.

EXAMPLE 3 Factor $x^2 + x - 6$.

Solution We must find two numbers whose product is the constant, -6, and whose sum is the coefficient of the x-term, 1. Remember that x means $1x$. Since the constant is negative, one number must be positive and the other negative. We now list the factors of -6 and look for the two factors whose sum is 1.

Factors of -6	Sum of Factors
$1(-6) = -6$	$1 + (-6) = -5$
$2(-3) = -6$	$2 + (-3) = -1$
$3(-2) = -6$	$3 + (-2) = 1$
$6(-1) = -6$	$6 + (-1) = 5$

The numbers 3 and -2 have a product of -6 and a sum of 1. Thus, the factors are $(x + 3)$ and $(x - 2)$.

$$x^2 + x - 6 = (x + 3)(x - 2)$$

The order of the factors is not crucial. Therefore, $x^2 + x - 6 = (x - 2)(x + 3)$ is also an acceptable answer.

Now Try Exercise 19

Understanding Algebra

Use FOIL to check the answer to Example 3.

$$(x + 3)(x - 2) = x^2 - 2x + 3x - 6$$
$$= x^2 + x - 6$$

Since the product of factors is identical to the original trinomial, the factoring is correct.

EXAMPLE 4 Factor $x^2 - x - 6$.

Solution The factors of -6 are illustrated in Example 3. The factors whose product is -6 and whose sum is -1 are 2 and -3.

Factors of -6	Sum of Factors
$2(-3) = -6$	$2 + (-3) = -1$

Therefore, $$x^2 - x - 6 = (x + 2)(x - 3)$$

Now Try Exercise 25

EXAMPLE 5 Factor $x^2 - 5x + 6$.

Solution We must find two numbers whose product is 6 and whose sum is -5. Since the constant, 6, is positive, both numbers must have the same sign. Since the coefficient of the x-term, -5, is negative, both numbers must be negative. We now list the negative factors of 6 and look for the pair whose sum is -5.

Understanding Algebra

To factor a trinomial, write the trinomial in descending order. To factor $-5x + x^2 + 6$, we must first rewrite it as $x^2 - 5x + 6$.

Factors of 6	Sum of Factors
$(-1)(-6)$	$-1 + (-6) = -7$
$(-2)(-3)$	$-2 + (-3) = -5$

The factors of 6 whose sum is -5 are -2 and -3.

$$x^2 - 5x + 6 = (x - 2)(x - 3)$$

Now Try Exercise 29

EXAMPLE 6 Factor $r^2 + 2r - 24$.

Solution We must find the two factors of -24 whose sum is 2. Since the constant is negative, one factor will be positive and the other factor will be negative.

Factors of -24	Sum of Factors
$(1)(-24)$	$1 + (-24) = -23$
$(2)(-12)$	$2 + (-12) = -10$
$(3)(-8)$	$3 + (-8) = -5$
$(4)(-6)$	$4 + (-6) = -2$
$(6)(-4)$	$6 + (-4) = 2$

Since we have found the two numbers, 6 and -4, whose product is -24 and whose sum is 2, we need go no further.

$$r^2 + 2r - 24 = (r + 6)(r - 4)$$

Now Try Exercise 33

EXAMPLE 7 Factor $x^2 - 8x + 16$.

Solution We must find the factors of 16 whose sum is -8. Both factors must be negative. The two factors whose product is 16 and whose sum is -8 are -4 and -4.

$$x^2 - 8x + 16 = (x - 4)(x - 4)$$
$$= (x - 4)^2$$

Now Try Exercise 41

Understanding Algebra

When factoring leads to a binomial factor multiplied by itself, we will write the answer as the binomial factor squared. For example, $x^2 + 6x + 9 = (x + 3)(x + 3)$, which we will write as $(x + 3)^2$.

Notice in Example 7 that the factoring led to a binomial factor multiplied by itself, $(x - 4)(x - 4)$. When this occurs we will write the answer as the binomial factor squared, $(x - 4)^2$.

EXAMPLE 8 Factor $x^2 - 11x - 60$.

Solution We must find two numbers whose product is -60 and whose sum is -11. Since the constant is negative, one number must be positive and the other negative. The desired numbers are -15 and 4 because $(-15)(4) = -60$ and $-15 + 4 = -11$.

$$x^2 - 11x - 60 = (x - 15)(x + 4)$$

Now Try Exercise 47

Not all trinomials are factorable using integer coefficients as Example 9 will show.

Prime Polynomials

A **prime polynomial** is a polynomial that cannot be factored using only integer coefficients.

For example, $x^2 + 3x + 8$ cannot be factored using integer coefficients because there are no two integers whose product is 8 and whose sum is 3. If asked to factor it, the correct answer is to say it is **prime.**

EXAMPLE 9 Factor $x^2 + 5x + 12$.

Solution Let's first find the two numbers whose product is 12 and whose sum is 5. Since both the constant and the coefficient of the x-term are positive, the two numbers must also be positive.

Factors of 12	Sum of Factors
$(1)(12)$	$1 + 12 = 13$
$(2)(6)$	$2 + 6 = 8$
$(3)(4)$	$3 + 4 = 7$

Note that there are no two integers whose product is 12 and whose sum is 5. Thus, $x^2 + 5x + 12$ is prime.

Now Try Exercise 31

Understanding Algebra

When factoring a trinomial of the form $x^2 + bx + c$, there is at most one pair of factors whose product is $x^2 + bx + c$. For example, when factoring $x^2 - 12x + 32$, the only factors are $(x - 4)$ and $(x - 8)$.

In Examples 10 and 11, we will factor trinomials of the form $x^2 + bxy + cy^2$, where b and c are real numbers. An example of a trinomial in this form is $x^2 - 2xy - 15y^2$. When factoring trinomials of this form, the factors must be of the form as follows.

$$x^2 + bxy + cy^2 = (x + \square y)(x + \square y)$$

Numbers go here that
- multiply to c
- add up to b

EXAMPLE 10 Factor $x^2 + 3xy + 2y^2$.

Solution In this problem, the second term contains two variables, x and y. The product of the first terms of the factors we are looking for must be x^2, and the product of the last terms of the factors must be $2y^2$.

We must find two numbers whose product is 2 (from $2y^2$) and whose sum is 3 (from $3xy$). The two numbers are 1 and 2. Thus,

$$x^2 + 3xy + 2y^2 = (x + 1y)(x + 2y) = (x + y)(x + 2y)$$

Now Try Exercise 65

EXAMPLE 11 Factor $x^2 - 2xy - 15y^2$.

Solution Find two numbers whose product is -15 and whose sum is -2. The numbers are -5 and 3. The last terms must be $-5y$ and $3y$ to obtain $-15y^2$.

$$x^2 - 2xy - 15y^2 = (x - 5y)(x + 3y)$$

Now Try Exercise 69

2 Remove the Greatest Common Factor from a Trinomial

Sometimes each term of a trinomial has a common factor. When this occurs, factor out the greatest common factor (GCF) first, as explained in Section 5.1. *The first step in any factoring problem is to factor out the GCF.* After factoring out the GCF, you should factor the remaining trinomial further, if possible, until it is completely factored.

EXAMPLE 12 Factor $2x^2 + 2x - 12$.

Solution Since 2 is the GCF, we factor it out.

$$2x^2 + 2x - 12 = 2(x^2 + x - 6)$$

Now we factor the remaining trinomial $x^2 + x - 6$ into $(x + 3)(x - 2)$. Thus,

$$2x^2 + 2x - 12 = 2(x + 3)(x - 2).$$

Note that the trinomial $2x^2 + 2x - 12$ is now completely factored into *three* factors: two binomial factors, $x + 3$ and $x - 2$, and a monomial factor, 2. After 2 has been factored out, it plays no part in the factoring of the remaining trinomial.

Understanding Algebra

Remember that the first step in factoring *any* polynomial is to factor out the greatest common factor (GCF).

Now Try Exercise 71

EXAMPLE 13 Factor $3n^3 + 24n^2 - 60n$.

Solution The GCF is $3n$. After factoring out the $3n$, we factor the remaining trinomial.

$$3n^3 + 24n^2 - 60n = 3n(n^2 + 8n - 20) \quad \text{Factored out the GCF.}$$
$$= 3n(n + 10)(n - 2) \quad \text{Factored the remaining trinomial.}$$

Now Try Exercise 79

5.3 Exercise Set MathXL® MyMathLab®

Warm-Up Exercises

positive	common	negative	whole	descending	reverse
integer	check	opposite	same	binomial	trinomial

1. The first step when factoring any polynomial is to look for a __________ factor.
2. In general, when we factor a trinomial of the form $x^2 + bx + c$ we will get a pair of __________ factors.
3. When factoring a trinomial of the form $x^2 + bx + c$, if c is positive, then both numbers in the factors will have the __________ sign.
4. When factoring a trinomial of the form $x^2 + bx + c$, if b and c are both positive, then both numbers in the factors will be __________.
5. When factoring a trinomial of the form $x^2 + bx + c$, if c is positive *and* b is negative, then both numbers in the factors will be __________.
6. When factoring a trinomial of the form $x^2 + bx + c$, if c is negative, then the numbers in the factors will have __________ signs.
7. To factor the trinomial $4x + x^2 - 21$ first rewrite the trinomial in __________ order.
8. A prime polynomial is a polynomial that cannot be factored using only __________ coefficients.
9. When factoring trinomials, we can multiply using FOIL to __________ our answers.
10. The trial-and-error method of factoring polynomials is also called the __________ FOIL method.

Practice the Skills

Factor each polynomial. If the polynomial is prime, so state.

11. $x^2 + 6x + 8$
12. $x^2 + 8x + 15$
13. $x^2 + 11x + 24$
14. $z^2 + 14z + 40$
15. $y^2 - 10y + 16$
16. $x^2 - 11x + 24$
17. $h^2 - 4h + 3$
18. $x^2 - 3x + 2$
19. $x^2 + 5x - 24$
20. $x^2 - x - 12$
21. $x^2 + 4x - 6$
22. $y^2 + 6y - 8$

23. $y^2 - 13y + 12$
24. $x^2 - 6x + 8$
25. $a^2 - 2a - 8$
26. $p^2 + 3p - 10$
27. $r^2 - 2r - 15$
28. $x^2 + 3x - 54$
29. $b^2 - 11b + 18$
30. $x^2 - 8x + 7$
31. $x^2 - 8x - 15$
32. $x^2 + 11x - 30$
33. $q^2 + 4q - 45$
34. $t^2 + 7t - 30$
35. $x^2 - 7x - 30$
36. $b^2 - 9b - 36$
37. $x^2 + 4x + 4$
38. $u^2 + 2u + 1$
39. $s^2 - 8s + 16$
40. $x^2 - 4x + 4$
41. $p^2 - 12p + 36$
42. $d^2 - 14d + 49$
43. $-18w + w^2 + 45$
44. $-11x + x^2 + 10$
45. $10x - 39 + x^2$
46. $15r - 34 + r^2$
47. $x^2 - x - 20$
48. $t^2 - 28t - 60$
49. $y^2 + 13y + 40$
50. $r^2 + 14r + 48$
51. $x^2 + 12x - 64$
52. $c^2 + 24c - 81$
53. $s^2 + 14s - 24$
54. $b^2 - 17b - 30$
55. $x^2 - 20x + 64$
56. $x^2 + 19x + 48$
57. $a^2 - 20a + 99$
58. $y^2 - 6y + 8$
59. $x^2 + 2 + 3x$
60. $m^2 - 11 - 10m$
61. $7w - 18 + w^2$
62. $30 + y^2 - 13y$
63. $x^2 - 8xy + 15y^2$
64. $b^2 - 2bc - 3c^2$
65. $m^2 - 6mn + 9n^2$
66. $x^2 - 2xy + y^2$
67. $x^2 + 8xy + 12y^2$
68. $x^2 + 16xy - 17y^2$
69. $m^2 - 5mn - 24n^2$
70. $c^2 + 2cd - 24d^2$

Factor completely.

71. $6x^2 - 30x + 24$
72. $2a^2 - 12a - 32$
73. $5x^2 + 20x + 15$
74. $7t^2 + 49t + 70$
75. $2x^2 - 18x + 40$
76. $3y^2 - 33y + 54$
77. $b^3 - 7b^2 + 10b$
78. $c^3 + 8c^2 - 48c$
79. $3z^3 - 21z^2 - 54z$
80. $3x^3 - 36x^2 + 33x$
81. $x^3 + 8x^2 + 16x$
82. $s^3 + 10s^2 + 25s$
83. $3x^3 + 3x^2y - 18xy^2$
84. $2x^3y - 12x^2y + 10xy$
85. $3r^3 + 6r^2t - 24rt^2$
86. $r^2s + 7rs^2 + 12s^3$
87. $x^4 - 4x^3 - 21x^2$
88. $2z^5 + 14z^4 + 12z^3$

Problem Solving

89. The first two columns in the following table describe the signs of the coefficient of the x-term and constant term of a trinomial of the form $x^2 + bx + c$. Determine whether the third column should contain "both positive," "both negative," or "one positive and one negative."

Sign of Coefficient of x-term	Sign of Constant Term	Sign of Constant Terms in the Binomial Factors
−	+	
−	−	
+	−	
+	+	

90. Assume that a trinomial of the form $x^2 + bx + c$ is factorable. Determine whether the constant terms in the factors are "both positive," "both negative," or "one positive and one negative" for the given signs of b and c.

a) $b < 0, c > 0$
b) $b > 0, c > 0$
c) $b > 0, c < 0$
d) $b < 0, c < 0$

91. Write a trinomial whose binomial factors contain constant terms whose sum is -12 and have a product of 32. Show the factoring of the trinomial.

92. Write a trinomial whose binomial factors contain constant terms whose sum is 5 and have a product of 4. Show the factoring of the trinomial.

93. Write a trinomial whose binomial factors contain constant terms whose sum is -2 and have a product of -35. Show the factoring of the trinomial.

94. Write a trinomial whose binomial factors contain constant terms whose sum is 5 and have a product of -14. Show the factoring of the trinomial.

Concept/Writing Exercises

95. On an exam, a student factored $2x^2 - 6x + 4$ as $(2x - 4)(x - 1)$. Even though $(2x - 4)(x - 1)$ does multiply out to $2x^2 - 6x + 4$, why did his or her professor deduct points?

96. Explain how to determine the factors when factoring a trinomial of the form $x^2 + bx + c$.

Challenge Problems

Factor.

97. $x^2 + 0.6x + 0.08$

98. $x^2 - 0.5x - 0.06$

99. $x^2 + \frac{2}{5}x + \frac{1}{25}$

100. $x^2 - \frac{2}{7}x + \frac{1}{49}$

101. $x^2 - 24x - 256$

102. $x^2 + 5x - 300$

Cumulative Review Exercises

[2.5] **103.** Solve the equation $4(2x - 4) = 5x + 11$.

[3.4] **104. Mixing Solutions** Karen Moreau, a chemist, mixes 4 liters of an 18% acid solution with 1 liter of a 26% acid solution. Find the strength of the mixture.

[4.5] **105.** Multiply $(2x^2 + 5x - 6)(x - 2)$.

[4.6] **106.** Divide $3x^2 - 10x - 10$ by $x - 4$.

[5.2] **107.** Factor $20x^2 + 8x - 15x - 6$ by grouping.

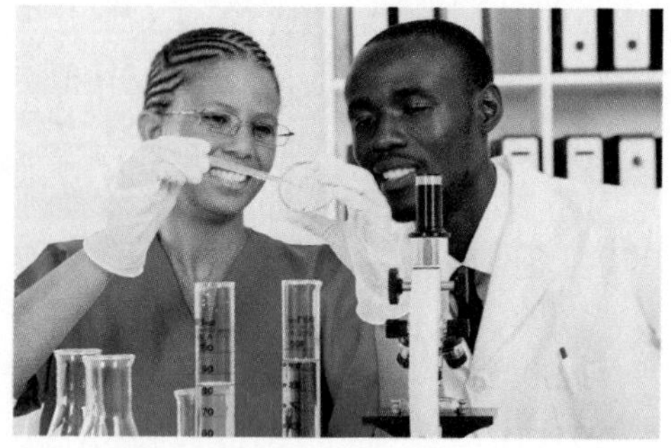

See Exercise 104.

5.4 Factoring Trinomials of the Form $ax^2 + bx + c, a \neq 1$

1. Factor trinomials of the form $ax^2 + bx + c, a \neq 1$, by trial and error.
2. Factor trinomials of the form $ax^2 + bx + c, a \neq 1$, by grouping.

An Important Note

In this section, we discuss two methods of factoring trinomials of the form $ax^2 + bx + c$, $a \neq 1$. That is, we will be factoring trinomials whose squared term has a numerical coefficient not equal to 1, after removing any common factors. Examples of trinomials with $a \neq 1$ are

$a \neq 1$

$2x^2 + 11x + 12$ (Notice $a = 2$) $4x^2 - 3x + 1$ (Notice $a = 4$)

The methods we discuss are (1) **factoring by trial and error** and (2) **factoring by grouping.** We present two different methods for factoring these trinomials because some students, and some instructors, prefer the first method, while others prefer the second method. You may use either method unless your instructor asks you to use a specific method. We will use the same examples to illustrate both methods so that you can make a comparison. Each method is treated independently of the other. Factoring by trial and error was introduced in Section 5.3 and factoring by grouping was introduced in Section 5.2.

Understanding Algebra

Use **FOIL** to multiply

$(2x + 3)(x + 5)$.

Product of ***First*** terms

$2x \cdot x = 2x^2$

Product of the ***Outer*** terms

$2x \cdot 5 = 10x$

Product of the ***Inner*** terms

$3 \cdot x = 3x$

Product of the ***Last*** terms

$3 \cdot 5 = 15$

$(2x + 3)(x + 5) =$
$2x^2 + 10x + 3x + 15 =$
$2x^2 + 13x + 15.$

Notice the sum of the products of the outer terms and the inner terms is $13x$.

1 Factor Trinomials of the Form $ax^2 + bx + c, a \neq 1$, by Trial and Error

Recall that factoring is the reverse of multiplying. Consider the product of the following two binomials:

$$\begin{aligned}(2x + 3)(x + 5) &= \overset{F}{2x(x)} + \overset{O}{(2x)(5)} + \overset{I}{3(x)} + \overset{L}{3(5)} \\ &= 2x^2 + 10x + 3x + 15 \\ &= 2x^2 + 13x + 15\end{aligned}$$

Note that $2x^2 + 13x + 15$ in factored form is $(2x + 3)(x + 5)$.

$$2x^2 + 13x + 15 = (2x + 3)(x + 5)$$

When factoring a trinomial of the form $ax^2 + bx + c$ by trial and error, the product of the x-terms in the binomial factors must equal the first term of the trinomial, ax^2. Also, the product of the constants in the binomial factors, including their signs, must equal the constant, c, of the trinomial.

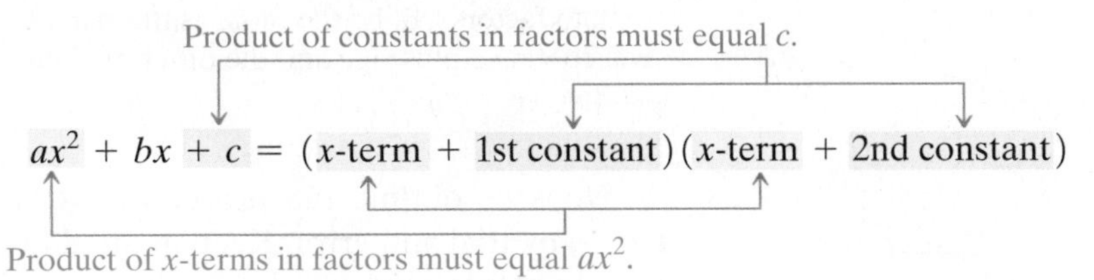

For example, when factoring the trinomial $2x^2 + 7x + 6$, each of the following pairs of factors has a product of the first terms equal to $2x^2$ and a product of the last terms equal to 6.

Trinomial	Possible Factors	Product of First Terms	Product of Last Terms
$2x^2 + 7x + 6$	$(2x + 1)(x + 6)$	$2x(x) = 2x^2$	$1(6) = 6$
	$(2x + 2)(x + 3)$	$2x(x) = 2x^2$	$2(3) = 6$
	$(2x + 3)(x + 2)$	$2x(x) = 2x^2$	$3(2) = 6$
	$(2x + 6)(x + 1)$	$2x(x) = 2x^2$	$6(1) = 6$

How do we determine which is the correct factoring of the trinomial $2x^2 + 7x + 6$? The key lies in the x-term. We need to find the pair of factors whose sum of the products of the outer and inner terms is equal to the x-term of the trinomial.

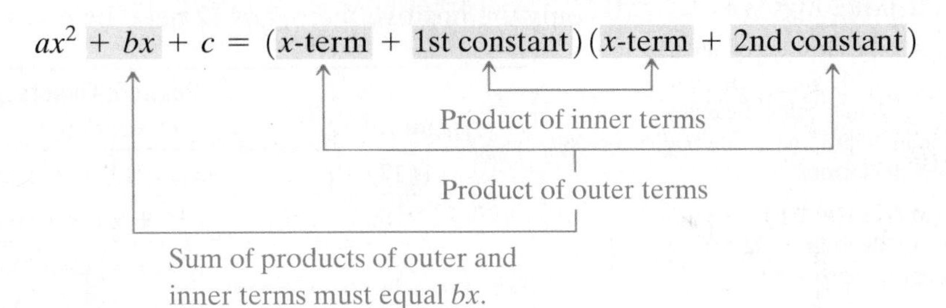

Now look at the possible pairs of factors we obtained for $2x^2 + 7x + 6$ to see if any yield the correct x-term, $7x$.

Trinomial	Possible Factors	Product of the First Terms	Product of the Last Terms	Sum of the Products of Outer and Inner Terms
$2x^2 + 7x + 6$	$(2x + 1)(x + 6)$	$2x^2$	6	$2x(6) + 1(x) = 13x$
	$(2x + 2)(x + 3)$	$2x^2$	6	$2x(3) + 2(x) = 8x$
	$(2x + 3)(x + 2)$	$2x^2$	6	$2x(2) + 3(x) = 7x$
	$(2x + 6)(x + 1)$	$2x^2$	6	$2x(1) + 6(x) = 8x$

Since $(2x + 3)(x + 2)$ yields the correct x-term, $7x$, the factors of the trinomial $2x^2 + 7x + 6$ are $(2x + 3)$ and $(x + 2)$.

$$2x^2 + 7x + 6 = (2x + 3)(x + 2)$$

We can check this factoring using the FOIL method.

Understanding Algebra

Notice that $2x^2 + 13x + 15$ factors as $(2x + 3)(x + 5)$. The product $(2x + 5)(x + 3)$ will yield a $2x^2$ term and a 15 term, but the middle term is $11x$ and not $13x$.

Check

$$\begin{aligned} & \quad\ \ \text{F} \qquad\quad \text{O} \qquad\quad \text{I} \qquad\quad \text{L} \\ (2x + 3)(x + 2) &= 2x(x) + 2x(2) + 3(x) + 3(2) \\ &= 2x^2 + 4x + 3x + 6 \\ &= 2x^2 + 7x + 6 \end{aligned}$$

Since we obtained the original trinomial, our factoring is correct.

HELPFUL HINT

When factoring a trinomial of the form $ax^2 + bx + c$, (with a positive) remember that the sign of the constant, c, and the sign of the x-term, bx, offer valuable information. When factoring a trinomial by trial and error, first check the sign of the constant. If it is positive, the signs in both factors will be the same as the sign of the x-term. If the constant is negative, one factor will contain a plus sign and the other a minus sign.

Now we outline the procedure to factor trinomials of the form $ax^2 + bx + c$, $a \neq 1$, by trial and error. Keep in mind that the more you practice, the better you will become at factoring.

To Factor Trinomials of the Form $ax^2 + bx + c, a \neq 1$, by Trial and Error

1. Factor out the greatest common factor (GCF), if any.
2. Write all pairs of factors of the coefficient of the squared term, a.
3. Write all pairs of factors of the constant term, c.
4. Try combinations of these factors until the correct middle term, bx, is found.

Understanding Algebra

We can check the factoring of any trinomial by multiplying the factors. This product should equal the original trinomial. In Example 1, use FOIL to multiply the factors.

$$(3x + 2)(x + 6) =$$
$$3x^2 + 18x + 2x + 12 =$$
$$3x^2 + 20x + 12.$$

Since we obtained the original trinomial, our factoring is correct.

EXAMPLE 1 Factor $3x^2 + 20x + 12$.

Solution We first determine that there is no GCF to factor out. Since the first term is $3x^2$, the factors will be of the form $(3x + \square)(x + \square)$. Now we must find the numbers to place in the shaded areas. The product of the last terms in the factors must be 12. Since the constant and the coefficient of the x-term are both positive, only the positive factors of 12 need be considered.

Factors of 12	Possible Factors of Trinomial	Sum of the Products of the Outer and Inner Terms
1(12)	$(3x + 1)(x + 12)$	$37x$
2(6)	$(3x + 2)(x + 6)$	$20x$
3(4)	$(3x + 3)(x + 4)$	$15x$
4(3)	$(3x + 4)(x + 3)$	$13x$
6(2)	$(3x + 6)(x + 2)$	$12x$
12(1)	$(3x + 12)(x + 1)$	$15x$

Since the product of $(3x + 2)$ and $(x + 6)$ yields the correct x-term, $20x$, they are the correct factors.

$$3x^2 + 20x + 12 = (3x + 2)(x + 6)$$

Now Try Exercise 5

EXAMPLE 2 Factor $5x^2 - 7x - 6$.

Solution There is no GCF to factor out. Since the first term is $5x^2$, one factor must contain a $5x$ and the other an x. List the factors of -6 and look for the pair of factors that yields $-7x$.

Factors of -6	Possible Factors	Sum of the Products of the Outer and Inner Terms
$-1(6)$	$(5x - 1)(x + 6)$	$29x$
$-2(3)$	$(5x - 2)(x + 3)$	$13x$
$-3(2)$	$(5x - 3)(x + 2)$	$7x$
$-6(1)$	$(5x - 6)(x + 1)$	$-x$

Understanding Algebra

When factoring a trinomial with a negative constant, if you obtain an x-term that is the opposite of the x-term you are seeking, *reverse the signs of the constants* in the factors. For example, suppose you are factoring $2x^2 + 3x - 5$ and you try the factors $(2x - 5)(x + 1)$. This product is $2x^2 - 3x - 5$. Since $+3x$ and $-3x$ are opposites, we know to reverse the signs of the constants in the factors to obtain the correct factoring: $(2x + 5)(x - 1)$.

Since we did not obtain the desired quantity, $-7x$, by writing the negative factor with the $5x$, we will now try listing the negative factor with the x.

Factors of -6	Possible Factors	Sum of the Products of the Outer and Inner Terms
$1(-6)$	$(5x + 1)(x - 6)$	$-29x$
$2(-3)$	$(5x + 2)(x - 3)$	$-13x$
$3(-2)$	$(5x + 3)(x - 2)$	$-7x$
$6(-1)$	$(5x + 6)(x - 1)$	x

We see that $(5x + 3)(x - 2)$ gives the $-7x$ we are looking for. Thus,

$$5x^2 - 7x - 6 = (5x + 3)(x - 2)$$

Again we listed all the possible combinations for you to study.

Now Try Exercise 9

HELPFUL HINT

In Example 2, we were asked to factor $5x^2 - 7x - 6$. When we considered $-3(2)$ in the first set of possible factors, we obtained

Factors of -6	Possible Factors	Sum of the Products of the Outer and Inner Terms
$-3(2)$	$(5x - 3)(x + 2)$	$7x$

Later in the solution we tried the factors $3(-2)$ and obtained the correct answer.

$3(-2)$	$(5x + 3)(x - 2)$	$-7x$

When factoring a trinomial with a *negative constant*, if you obtain an x-term that is the opposite of the x-term you are seeking, *reverse the signs of the constants* in the factors. This should give you the correct set of factors.

EXAMPLE 3 Factor $8x^2 + 33x + 4$.

Solution There is no GCF. Since the first term is $8x^2$, the possible factors may be of the form $(8x \quad)(x \quad)$ or $(4x \quad)(2x \quad)$. When this situation occurs, we will generally start with the middle-size pair of factors. Thus, we begin with $(4x \quad)(2x \quad)$. If this pair does not lead to the solution, we will then try $(8x \quad)(x \quad)$. We now list the factors of the constant, 4. Since all signs are positive, we list only the positive factors of 4.

Factors of 4	Possible Factors	Sum of the Products of the Outer and Inner Terms
$1(4)$	$(4x + 1)(2x + 4)$	$18x$
$2(2)$	$(4x + 2)(2x + 2)$	$12x$
$4(1)$	$(4x + 4)(2x + 1)$	$12x$

Since we did not obtain the correct factors with $(4x \quad)(2x \quad)$, we now try $(8x \quad)(x \quad)$.

Factors of 4	Possible Factors	Sum of the Products of the Outer and Inner Terms
$1(4)$	$(8x + 1)(x + 4)$	$33x$
$2(2)$	$(8x + 2)(x + 2)$	$18x$
$4(1)$	$(8x + 4)(x + 1)$	$12x$

Since the product of $(8x + 1)$ and $(x + 4)$ yields the correct x-term, $33x$, they are the correct factors.

$$8x^2 + 33x + 4 = (8x + 1)(x + 4)$$

Now Try Exercise 19

Understanding Algebra

In Example 3, another reason we can eliminate the factors

$(4x + 1)(2x + 4)$
$(4x + 2)(2x + 2)$
$(4x + 4)(2x + 1)$
$(8x + 2)(x + 2)$
$(8x + 4)(x + 1)$

is that each product has a binomial that contains a common factor.

Since the original trinomial did not have any common factors, there can be no common factors in the binomial factors.

EXAMPLE 4 Factor $25t^2 - 10t + 1$.

Solution The factors must be of the form $(25t \quad)(t \quad)$ or $(5t \quad)(5t \quad)$. We will start with the middle-size factors $(5t \quad)(5t \quad)$. Since the constant is positive and the coefficient of the x-term is negative, both factors must be negative.

Factors of 1	Possible Factors	Sum of the Products of the Outer and Inner Terms
$(-1)(-1)$	$(5t-1)(5t-1)$	$-10t$

Since we found the correct factors, we can stop.

$$25t^2 - 10t + 1 = (5t-1)(5t-1) = (5t-1)^2$$

Now Try Exercise 13

EXAMPLE 5 Factor $2x^2 + 3x + 7$.

Solution The factors will be of the form $(2x \quad)(x \quad)$. We need only consider the positive factors of 7.

Factors of 7	Possible Factors	Sum of the Products of the Outer and Inner Terms
$1(7)$	$(2x+1)(x+7)$	$15x$
$7(1)$	$(2x+7)(x+1)$	$9x$

Since we have tried all possible combinations and we have not obtained the x-term, $3x$, this trinomial *cannot be factored using integer coefficients*. Thus, the trinomial $2x^2 + 3x + 7$ is a *prime polynomial*.

Now Try Exercise 21

EXAMPLE 6 Factor $6x^2 + 19xy + 3y^2$.

Solution This trinomial is different from the other trinomials in that the last term is not a constant but contains y^2. The factoring process is the same, except that the second term of both factors will contain y. Consider factors of the form $(3x \quad)(2x \quad)$. If we cannot find the factors, then we try factors of the form $(6x \quad)(x \quad)$.

Factors of 3	Possible Factors	Sum of the Products of the Outer and Inner Terms
$1(3)$	$(3x+y)(2x+3y)$	$11xy$
$3(1)$	$(3x+3y)(2x+y)$	$9xy$
$1(3)$	$(6x+y)(x+3y)$	$19xy$
$3(1)$	$(6x+3y)(x+y)$	$9xy$

$$6x^2 + 19xy + 3y^2 = (6x+y)(x+3y)$$

Check $(6x+y)(x+3y) = 6x^2 + 18xy + xy + 3y^2 = 6x^2 + 19xy + 3y^2$

Now Try Exercise 55

EXAMPLE 7 Factor $6x^2 - 13xy - 8y^2$.

Solution We begin with factors of the form $(3x \quad)(2x \quad)$. If we cannot find the solution from these, we will try $(6x \quad)(x \quad)$. Since the last term, $-8y^2$, is negative, one factor will contain a plus sign and the other will contain a minus sign.

Factors of -8	Possible Factors	Sum of the Products of the Outer and Inner Terms
$1(-8)$	$(3x+y)(2x-8y)$	$-22xy$
$2(-4)$	$(3x+2y)(2x-4y)$	$-8xy$
$4(-2)$	$(3x+4y)(2x-2y)$	$2xy$
$8(-1)$	$(3x+8y)(2x-y)$	$13xy$

We are looking for $-13xy$. When we considered $8(-1)$, we obtained $13xy$. As explained in the Helpful Hint on page 315, if we reverse the signs of the second terms in the factors, we will obtain the factors we are seeking.

$$(3x + 8y)(2x - y) \quad \text{Product is } 6x^2 + 13xy - 8y^2.$$
$$(3x - 8y)(2x + y) \quad \text{Product is } 6x^2 - 13xy - 8y^2.$$

Therefore, $6x^2 - 13xy - 8y^2 = (3x - 8y)(2x + y)$.

Now Try Exercise 57

EXAMPLE 8 Factor $6x^3 + 15x^2 - 36x$.

Solution *The first step in any factoring problem is to factor out the GCF.* In this example, $3x$ is common to all three terms. We begin by factoring out the $3x$. Then we continue factoring by trial and error.

$$\begin{aligned} 6x^3 + 15x^2 - 36x &= 3x(2x^2 + 5x - 12) \\ &= 3x(2x - 3)(x + 4) \end{aligned}$$

Now Try Exercise 45

2 Factor Trinomials of the Form $ax^2 + bx + c, a \neq 1$, by Grouping

The steps in the box that follow give the procedure for factoring trinomials by grouping.

To Factor Trinomials of the Form $ax^2 + bx + c, a \neq 1$, by Grouping

1. Factor out the greatest common factor, if any.
2. Find two numbers whose product is equal to the product of a times c, and whose sum is equal to b.
3. Rewrite the middle term, bx, as the sum or difference of two terms using the numbers found in step 2.
4. Factor by grouping as explained in Section 5.2.

This process will be made clearer in Example 9. We will rework Examples 1 through 8 here using factoring by grouping. Example 9, which follows, is the same trinomial given in Example 1. After you study this method and try some exercises, you will gain a feel for which method you prefer using.

EXAMPLE 9 Factor $3x^2 + 20x + 12$.

Solution

1. There is no GCF to factor out.

$$a = 3 \quad b = 20 \quad c = 12$$

2. We must find two numbers whose product is $a \cdot c$ and whose sum is b. We must therefore find two numbers whose product equals $3 \cdot 12 = 36$ and whose sum equals 20. Only the positive factors of 36 need be considered since all signs of the trinomial are positive.

Factors of 36	Sum of Factors
$(1)(36)$	$1 + 36 = 37$
$(2)(18)$	$2 + 18 = 20$
$(3)(12)$	$3 + 12 = 15$
$(4)(9)$	$4 + 9 = 13$
$(6)(6)$	$6 + 6 = 12$

The desired factors are 2 and 18.

3. Rewrite $20x$ as the sum or difference of two terms using the values found in step 2. Therefore, we rewrite $20x$ as $2x + 18x$.

$$3x^2 + 20x + 12$$
$$= 3x^2 + 2x + 18x + 12$$

4. Now factor by grouping. Start by factoring out a common factor from the first two terms and a common factor from the last two terms. This procedure was discussed in Section 5.2.

x is common factor; 6 is common factor

$$3x^2 + 2x + 18x + 12$$
$$= x(3x + 2) + 6(3x + 2)$$
$$= (3x + 2)(x + 6)$$

Understanding Algebra

In Example 9, when listing the four terms, we could have written the middle terms as $18x + 2x$.

$$3x^2 + 18x + 2x + 12$$
$$3x(x + 6) + 2(x + 6)$$
$$(x + 6)(3x + 2)$$

This is an equivalent answer.

Now Try Exercise 7

EXAMPLE 10 Factor $5x^2 - 7x - 6$.

Solution There are no common factors other than 1.

$$a = 5, \quad b = -7, \quad c = -6$$

The product of a times c is $5(-6) = -30$. We must find two numbers whose product is -30 and whose sum is -7.

Factors of -30	Sum of Factors
$(-1)(30)$	$-1 + 30 = 29$
$(-2)(15)$	$-2 + 15 = 13$
$(-3)(10)$	$-3 + 10 = 7$
$(-5)(6)$	$-5 + 6 = 1$
$(-6)(5)$	$-6 + 5 = -1$
$(-10)(3)$	$-10 + 3 = -7$
$(-15)(2)$	$-15 + 2 = -13$
$(-30)(1)$	$-30 + 1 = -29$

Understanding Algebra

When factoring a trinomial with a negative constant, if you obtain an x-term that is the opposite of the x-term you are seeking, *reverse the signs of the constants* in the factors. For example, suppose you are factoring $5x^2 - 7x - 6$ and you mistakenly obtain the factors $(x + 2)(5x - 3)$. This product is $5x^2 + 7x - 6$. Since $-7x$ and $+7x$ are opposites, we know to reverse the signs of the constants in the factors to obtain the correct factoring: $(x - 2)(5x + 3)$.

Rewrite the middle term of the trinomial, $-7x$, as $-10x + 3x$.

$$5x^2 - 7x - 6$$
$$= 5x^2 - 10x + 3x - 6 \qquad \text{Now factor by grouping.}$$
$$= 5x(x - 2) + 3(x - 2)$$
$$= (x - 2)(5x + 3)$$

Now Try Exercise 31

In Example 10, we could have expressed the $-7x$ as $3x - 10x$ and obtained the same answer. Try working Example 10 by rewriting $-7x$ as $3x - 10x$.

HELPFUL HINT

In Example 10 we were looking for two factors of -30 whose sum was -7. When we considered the factors -3 and 10, we obtained a sum of 7, which is the opposite of -7. When trying pairs of factors to obtain the middle term, if you obtain the opposite of the coefficient you are seeking, reverse the signs in the factors. This should give you the coefficient you are seeking.

Understanding Algebra

In Example 11 we could also have written:

$$\begin{aligned}&8x^2 + 33x + 4\\ &= 8x^2 + x + 32x + 4\\ &= x(8x + 1) + 4(8x + 1)\\ &= (8x + 1)(x + 4)\end{aligned}$$

EXAMPLE 11 Factor $8x^2 + 33x + 4$.

Solution There is no GCF to factor out. We must find two numbers whose product is $8 \cdot 4$ or 32 and whose sum is 33. The numbers are 1 and 32.

Factors of 32	Sum of Factors
$(1)(32)$	$1 + 32 = 33$

Rewrite $33x$ as $32x + x$. Then factor by grouping.

$$\begin{aligned}&8x^2 + 33x + 4\\ &= 8x^2 + 32x + x + 4\\ &= 8x(x + 4) + 1(x + 4)\\ &= (x + 4)(8x + 1)\end{aligned}$$

Now Try Exercise 25

EXAMPLE 12 Factor $25t^2 - 10t + 1$.

Solution There is no GCF to factor out. We must find two numbers whose product is $25 \cdot 1$ or 25 and whose sum is -10. Since the product of a times c is positive and the coefficient of the t-term is negative, both numerical factors must be negative.

Factors of 25	Sum of Factors
$(-1)(-25)$	$-1 + (-25) = -26$
$(-5)(-5)$	$-5 + (-5) = -10$

The desired factors are -5 and -5.

$$\begin{aligned}&25t^2 - 10t + 1\\ &= 25t^2 - 5t - 5t + 1 && \text{Rewrote } -10t \text{ as } -5t - 5t.\\ &= 5t(5t - 1) - 5t + 1\\ &= 5t(5t - 1) - 1(5t - 1) && \text{Rewrote } -5t + 1 \text{ as } -1(5t - 1).\\ &= (5t - 1)(5t - 1) \text{ or } (5t - 1)^2\end{aligned}$$

Now Try Exercise 29

HELPFUL HINT

When attempting to factor a trinomial, if there are no two integers whose product equals $a \cdot c$ and whose sum equals b, the trinomial is prime.

EXAMPLE 13 Factor $2x^2 + 3x + 7$.

Solution There are no common factors other than 1. We must find two numbers whose product is 14 and whose sum is 3. We need consider only positive factors of 14. Why?

Factors of 14	Sum of Factors
$(1)(14)$	$1 + 14 = 15$
$(2)(7)$	$2 + 7 = 9$

Since there are no factors of 14 whose sum is 3, we conclude that this trinomial cannot be factored with integers. This is an example of a *prime polynomial*.

Now Try Exercise 17

EXAMPLE 14 Factor $6x^2 + 19xy + 3y^2$.

Solution There is no GCF to factor out. This trinomial contains two variables. It is factored in basically the same manner as the previous examples. Find two numbers whose product is $6 \cdot 3$ or 18 and whose sum is 19. The two numbers are 18 and 1.

$$
\begin{aligned}
&6x^2 + \overbrace{19xy}{} + 3y^2 \\
&= 6x^2 + 18xy + xy + 3y^2 \\
&= 6x(x + 3y) + y(x + 3y) \\
&= (x + 3y)(6x + y)
\end{aligned}
$$

Now Try Exercise 61

EXAMPLE 15 Factor $6x^2 - 13xy - 8y^2$.

Solution There is no GCF to factor out. Find two numbers whose product is $6(-8)$ or -48 and whose sum is -13. Since the product is negative, one factor must be positive and the other negative. Some factors are given below.

Product of Factors	Sum of Factors
$(1)(-48)$	$1 + (-48) = -47$
$(2)(-24)$	$2 + (-24) = -22$
$(3)(-16)$	$3 + (-16) = -13$

There are many other factors, but we have found the pair we were looking for. The two numbers whose product is -48 and whose sum is -13 are 3 and -16.

$$
\begin{aligned}
&6x^2 \overbrace{- 13xy}{} - 8y^2 \\
&= 6x^2 + 3xy - 16xy - 8y^2 \\
&= 3x(2x + y) - 8y(2x + y) \\
&= (2x + y)(3x - 8y)
\end{aligned}
$$

Check $(2x + y)(3x - 8y)$

$$
\begin{aligned}
&\quad\ \ \text{F} \qquad\qquad\quad \text{O} \qquad\qquad\quad \text{I} \qquad\qquad \text{L} \\
&= (2x)(3x) + (2x)(-8y) + (y)(3x) + (y)(-8y) \\
&= 6x^2 - 16xy + 3xy - 8y^2 \\
&= 6x^2 - 13xy - 8y^2
\end{aligned}
$$

Now Try Exercise 63

Remember that in any factoring problem our first step is to factor the GCF from each term. We then continue to factor the trinomial, if possible.

EXAMPLE 16 Factor $6x^3 + 15x^2 - 36x$.

Solution The factor $3x$ is common to all three terms. Factor the $3x$ from each term of the polynomial.

$$6x^3 + 15x^2 - 36x = 3x(2x^2 + 5x - 12)$$

Now continue by factoring $2x^2 + 5x - 12$. The two numbers whose product is $2(-12)$ or -24 and whose sum is 5 are 8 and -3.

$$
\begin{aligned}
&3x(2x^2 + \overbrace{5x}{} - 12) \\
&= 3x(2x^2 + 8x - 3x - 12) \\
&= 3x[2x(x + 4) - 3(x + 4)] \\
&= 3x(x + 4)(2x - 3)
\end{aligned}
$$

Now Try Exercise 49

HELPFUL HINT

Which Method Should You Use to Factor a Trinomial?

If your instructor asks you to use a specific method, you should use that method. If your instructor does not require a specific method, you should use the method you feel most comfortable with. You may wish to start with the trial-and-error method if there are only a few possible factors to try. If you cannot find the factors by trial and error or if there are many possible factors to consider, you may wish to use the grouping procedure. With time and practice you will learn which method you feel most comfortable with and which method gives you greater success.

5.4 Exercise Set MathXL® MyMathLab®

Warm-Up Exercises

Fill in the blanks with the appropriate word, phrase, or symbol(s) from the following list.

reverse opposite dividing multiplying opposite same

1. When factoring a trinomial of the form $ax^2 + bx + c$, if c is positive, then both numbers in the factors will have the __________ sign.

2. When factoring a trinomial of the form $ax^2 + bx + c$, if c is negative, then the numbers in the factors will have __________ signs.

3. We can check the factoring of any trinomial by __________ the factors.

4. When factoring a trinomial with a negative constant, if you obtain an x-term that is the opposite of the x-term you are seeking, __________ the signs of the constants in the factors.

Practice the Skills

Factor completely. If the polynomial is prime, so state.

5. $2x^2 + 11x + 5$
6. $3x^2 + 10x + 7$
7. $3x^2 + 14x + 8$
8. $7x^2 + 37x + 10$
9. $5x^2 - 9x - 2$
10. $3x^2 - 2x - 8$
11. $3r^2 + 13r - 10$
12. $2x^2 - x - 6$
13. $4z^2 - 12z + 9$
14. $9y^2 - 12y + 4$
15. $6z^2 + z - 12$
16. $4w^2 + 4w - 15$
17. $3x^2 + 11x + 4$
18. $5x^2 + 2x + 9$
19. $8x^2 + 19x + 6$
20. $6t^2 + 19t + 8$
21. $5a^2 - 12a + 6$
22. $7b^2 - 10b + 6$
23. $5y^2 - 16y + 3$
24. $7x^2 - 8x + 1$
25. $7x^2 + 43x + 6$
26. $5q^2 + 51q + 10$
27. $4x^2 + 4x - 15$
28. $9p^2 - 9p - 10$
29. $49t^2 - 14t + 1$
30. $16z^2 - 8z + 1$
31. $5z^2 - 6z - 8$
32. $2k^2 + 3k - 27$
33. $4y^2 + 5y - 6$
34. $6a^2 + 7a - 10$
35. $10x^2 - 27x + 5$
36. $4n^2 - 9n + 5$
37. $7s^2 + 10s - 3$
38. $11u^2 - 23u - 2$
39. $21m^2 - 13m - 20$
40. $12x^2 - 13x - 35$
41. $10t + 3 + 7t^2$
42. $16t + 5 + 11t^2$
43. $6x^2 + 16x + 10$
44. $12z^2 + 32z + 20$
45. $12x^3 + 28x^2 + 8x$
46. $15v^3 + 21v^2 + 6v$
47. $6x^3 - 5x^2 - 4x$
48. $8x^3 + 8x^2 - 6x$
49. $4x^3 - 2x^2 - 12x$
50. $18x^3 - 21x^2 - 9x$
51. $48c^2 + 8c - 16$
52. $300x^2 - 400x - 400$
53. $4p - 12 + 8p^2$
54. $5s - 10 + 15s^2$
55. $8c^2 + 41cd + 5d^2$
56. $10j^2 + 31jk + 3k^2$
57. $15x^2 - xy - 6y^2$
58. $10x^2 + 9xy - 9y^2$
59. $12x^2 + 10xy - 8y^2$
60. $9x^2 + 30xy - 24y^2$
61. $7p^2 + 13pq + 6q^2$
62. $5c^2 + 13cd + 8d^2$
63. $6m^2 - mn - 2n^2$
64. $6r^2 - 5rs - 6s^2$
65. $8x^3 + 10x^2y + 3xy^2$
66. $8a^2b + 10ab^2 + 3b^3$
67. $4x^4 + 8x^3y + 3x^2y^2$
68. $26u^2v + 6uv^2 + 24u^3$

Problem Solving

For Exercises 69–74, write a polynomial whose factors are listed.

69. $3x + 1, x - 7$

70. $6y - 5, 4y - 3$

71. $5, x + 3, 2x + 1$

72. $3, 2x + 3, x - 4$

73. $t^2, t + 4, 3t - 1$

74. $5x^2, 3x - 7, 2x + 3$

75. a) If you know one binomial factor of a trinomial, explain how you can use division to find the second binomial factor of the trinomial (see Section 4.6).

b) One factor of $18x^2 + 93x + 110$ is $3x + 10$. Use division to find the second factor.

76. One factor of $30x^2 - 17x - 247$ is $6x - 19$. Find the other factor.

Concept/Writing Exercises

77. Explain the relationship between the process of factoring trinomials and the process of multiplying binomials.

78. When factoring a trinomial of the form $ax^2 + bx + c$,

a) What must the product of the first terms of the binomial factors equal?

b) What must the product of the constants in the binomial factors equal?

Challenge Problems

In Exercises 79–84, factor each trinomial.

79. $18x^2 + 9x - 20$

80. $9p^2 - 104p + 55$

81. $15x^2 - 124x + 160$

82. $16x^2 - 62x - 45$

83. $105a^2 - 220a - 160$

84. $72x^2 + 417x - 420$

85. Two factors of $6x^3 + 235x^2 + 2250x$ are x and $3x + 50$. Determine the other factor. Explain how you determined your answer.

86. Two factors of the polynomial $2x^3 + 11x^2 + 3x - 36$ are $x + 3$ and $2x - 3$. Determine the third factor. Explain how you determined your answer.

Cumulative Review Exercises

[1.9] **87.** Evaluate $-x^2 - 4(y + 3) + 2y^2$ when $x = -3$ and $y = -5$.

[2.6] **88. Daytona 500** Matt Kenseth won the 2012 Daytona 500 in a time of about 3.6 hours. If the race covered 504.9 miles, find the average speed of Kenseth's car.

[5.1] **89.** Factor $36x^4y^3 - 12xy^2 + 24x^5y^6$.

[5.3] **90.** Factor $b^2 + 4b - 96$.

See Exercise 88.

MID-CHAPTER TEST: 5.1–5.4

To find out how well you understand the chapter material to this point, take this brief test. The answers, and the section where the material was initially discussed, are given in the back of the book. Review any questions that you answered incorrectly.

1. How may any factoring problem be checked?

2. Determine the greatest common factor of $18xy^2, 27x^3y^4$, and $12x^2y^3$.

In Exercises 3–5, factor the GCF from each term in the expression.

3. $4a^2b^3 - 24a^3b$

4. $5c(d - 6) - 3(d - 6)$

5. $7x(2x + 9) + 2x + 9$

In Exercises 6–10, factor by grouping.

6. $x^2 + 4x + 7x + 28$
7. $x^2 + 5x - 3x - 15$
8. $6a^2 + 15ab - 2ab - 5b^2$
9. $5x^2 - 2xy - 45x + 18y$
10. $8x^3 + 4x^2 - 48x^2 - 24x$

In Exercises 11–20, factor each polynomial completely. If the polynomial is prime, so state.

11. $x^2 - 10x + 21$
12. $t^2 + 9t + 20$
13. $p^2 - 3p - 8$
14. $x^2 + 16x + 64$
15. $m^2 - 4mn - 45n^2$
16. $3x^2 + 17x + 10$
17. $4z^2 - 11z + 6$
18. $3y^2 + 13y + 6$
19. $9x^2 - 6x + 1$
20. $6a^2 + 3ab - 3b^2$

5.5 Special Factoring Formulas and a General Review of Factoring

1. Factor the difference of two squares.
2. Factor the sum and difference of two cubes.
3. Learn the general procedure for factoring a polynomial.

The special formulas, for certain types of factoring problems, we focus on in this section are the *difference of two squares, the sum of two cubes, and the difference of two cubes. You will need to memorize the three highlighted formulas in this section* so that you can use them whenever you need them.

1 Factor the Difference of Two Squares

Consider the binomial $x^2 - 9$. Note that each term of the binomial can be expressed as the square of some expression.

$$x^2 - 9 = x^2 - 3^2$$

This is an example of a **difference of two squares**.

Understanding Algebra

$a^2 - b^2$ The difference of two squares is factorable.

$a^2 + b^2$ The sum of two squares is *not* factorable with real numbers.

Difference of Two Squares

$$a^2 - b^2 = (a + b)(a - b)$$

Notice that the above formula is used to factor binomials that are the *difference* of two squares, $a^2 - b^2$. Binomials that are the *sum* of two squares, $a^2 + b^2$, cannot be factored using real numbers.

EXAMPLE 1 Factor $x^2 - 9$.

Solution If we write $x^2 - 9$ as a difference of two squares, we have $(x)^2 - (3)^2$. Using the difference of two squares formula, where a is replaced by x and b is replaced by 3, we obtain the following:

$$\begin{array}{ccccc} & a^2 - b^2 & = & (a + b)(a - b) \\ & \downarrow \quad \downarrow & & \downarrow \;\; \downarrow \;\; \downarrow \;\; \downarrow \\ x^2 - 9 = & (x)^2 - (3)^2 & = & (x + 3)(x - 3) \end{array}$$

Thus, $x^2 - 9 = (x + 3)(x - 3)$.

Now Try Exercise 13

EXAMPLE 2 Factor using the difference of two squares formula.

a) $x^2 - 16$ **b)** $25x^2 - 4$ **c)** $36x^2 - 49y^2$

Solution

a) $x^2 - 16 = (x)^2 - (4)^2$
$= (x + 4)(x - 4)$

b) $25x^2 - 4 = (5x)^2 - (2)^2$
$= (5x + 2)(5x - 2)$

c) $36x^2 - 49y^2 = (6x)^2 - (7y)^2$
$= (6x + 7y)(6x - 7y)$

Now Try Exercise 21

EXAMPLE 3 Factor each difference of two squares.

a) $16x^4 - 9y^4$ **b)** $x^6 - y^4$

Solution

a) First, we rewrite $16x^4$ as $(4x^2)^2$ and $9y^4$ as $(3y^2)^2$, then use the difference of two squares formula.

$$16x^4 - 9y^4 = (4x^2)^2 - (3y^2)^2$$ Rewrote as a difference of two squares.

$$= (4x^2 + 3y^2)(4x^2 - 3y^2)$$ Difference of two squares formula was used.

b) Rewrite x^6 as $(x^3)^2$ and y^4 as $(y^2)^2$, then use the difference of two squares formula.

$$x^6 - y^4 = (x^3)^2 - (y^2)^2$$
$$= (x^3 + y^2)(x^3 - y^2)$$

Now Try Exercise 29

EXAMPLE 4 Factor $9x^2 - 36y^2$ using the difference of two squares formula.

Solution

$$9x^2 - 36y^2 = 9(x^2 - 4y^2)$$ Factored out the GCF, 9.

$$= 9[(x)^2 - (2y)^2]$$ Rewrote as difference of two squares.

$$= 9(x + 2y)(x - 2y)$$ Difference of two squares formula was used.

Now Try Exercise 31

Notice in Example 4 that $9x^2 - 36y^2$ is the difference of two squares, $(3x)^2 - (6y)^2$. If you factor this difference of squares without first factoring out the GCF, 9, the factoring may be more difficult. After you factor this difference of squares you will need to factor out the GCF, 3, from each binomial factor, as illustrated below.

$$9x^2 - 36y^2 = (3x)^2 - (6y)^2$$
$$= (3x + 6y)(3x - 6y)$$
$$= 3(x + 2y)3(x - 2y)$$
$$= 9(x + 2y)(x - 2y)$$

We obtain the same answer as we did in Example 4. However, since we did not factor out the common factor 9 first, we had to work a little harder to obtain the answer.

EXAMPLE 5 Factor $z^4 - 16$ using the difference of two squares formula.

Solution We rewrite z^4 as $(z^2)^2$ and 16 as $(4)^2$, then use the difference of two squares formula.

$$z^4 - 16 = (z^2)^2 - (4)^2$$
$$= (z^2 + 4)(z^2 - 4)$$

Notice that the second factor, $z^2 - 4$, is also the difference of two squares. To complete the factoring, we use the difference of two squares formula again to factor $z^2 - 4$.

$$= (z^2 + 4)(z^2 - 4)$$
$$= (z^2 + 4)(z + 2)(z - 2)$$

Now Try Exercise 35

AVOIDING COMMON ERRORS

The difference of two squares can be factored. However, a sum of two squares, where there is no common factor to the two terms, cannot be factored using real numbers.

CORRECT	INCORRECT
$a^2 - b^2 = (a + b)(a - b)$	~~$a^2 + b^2 = (a + b)(a + b)$~~

2 Factor the Sum and Difference of Two Cubes

Consider the product of $(a + b)(a^2 - ab + b^2)$.

$$
\begin{array}{rl}
a^2 - ab + b^2 & \\
a + b & \\
\hline
a^2b - ab^2 + b^3 & \leftarrow b(a^2 - ab + b^2) \\
a^3 - a^2b + ab^2 & \leftarrow a(a^2 - ab + b^2) \\
\hline
a^3 + b^3 & \leftarrow \text{Sum of terms}
\end{array}
$$

Thus, $(a + b)(a^2 - ab + b^2) = a^3 + b^3$. Since factoring is the opposite of multiplying, we may factor $a^3 + b^3$ as follows:

$$a^3 + b^3 = (a + b)(a^2 - ab + b^2)$$

We see, using the same procedure, that $a^3 - b^3 = (a - b)(a^2 + ab + b^2)$. The expression $a^3 + b^3$ is a sum of two cubes and the expression $a^3 - b^3$ is a difference of two cubes.

Sum of Two Cubes

$$a^3 + b^3 = (a + b)(a^2 - ab + b^2)$$

Difference of Two Cubes

$$a^3 - b^3 = (a - b)(a^2 + ab + b^2)$$

Note that the trinomials $a^2 - ab + b^2$ and $a^2 + ab + b^2$ cannot be factored further. Now let's solve some factoring problems using the sum and the difference of two cubes.

EXAMPLE 6 Factor $x^3 + 8$.

Solution We rewrite $x^3 + 8$ as a sum of two cubes: $x^3 + 8 = (x)^3 + (2)^3$. Using the sum of two cubes formula, where a is replaced by x and b is replaced by 2, we get

$$
\begin{aligned}
a^3 + b^3 &= (a + b)(a^2 - a \cdot b + b^2) \\
&\quad \downarrow \\
x^3 + 8 = (x)^3 + (2)^3 &= (x + 2)(x^2 - x \cdot 2 + 2^2) \\
&= (x + 2)(x^2 - 2x + 4)
\end{aligned}
$$

You can check the factoring by multiplying $(x+2)(x^2-2x+4)$. If factored correctly, the product of the factors will equal the original expression, $x^3 + 8$. Try it and see.

Now Try Exercise 41

HELPFUL HINT

To help remember the signs in the sum or difference of two cubes formulas, consider

$$a^3 + b^3 = (a + b)(a^2 - ab + b^2)$$

Same sign (the sign between a^3 and b^3 and the sign in $(a + b)$)

Opposite sign (the sign in $(a + b)$ and the sign before ab)

Always positive (the sign before b^2)

(Continued)

$$a^3 - b^3 = (a - b)(a^2 + ab + b^2)$$

Same sign

Opposite sign

Always positive

Understanding Algebra

When using the formulas

$$a^3 + b^3 = (a + b)(a^2 - ab + b^2)$$

and

$$a^3 - b^3 = (a - b)(a^2 + ab + b^2),$$

it is important to note that the trinomials $(a^2 - ab + b^2)$ and $(a^2 + ab + b^2)$ cannot be factored further. Therefore, in Example 7, the trinomial factor $(y^2 + 5y + 25)$ cannot be factored further.

EXAMPLE 7 Factor $y^3 - 125$.

Solution We rewrite $y^3 - 125$ as a difference of two cubes: $(y)^3 - (5)^3$. Using the difference of two cubes formula, where a is replaced by y and b is replaced by 5, we get

$$\begin{aligned} a^3 \quad - \quad b^3 &= (a - b)(a^2 + a \cdot b + b^2) \\ \downarrow \qquad\quad \downarrow \qquad & \quad\;\; \downarrow \quad\; \downarrow \quad \downarrow \qquad \downarrow \;\; \downarrow \;\; \downarrow \\ y^3 - 125 = (y)^3 - (5)^3 &= (y - 5)(y^2 + y \cdot 5 + 5^2) \\ &= (y - 5)(y^2 + 5y + 25) \end{aligned}$$

Now Try Exercise 43

EXAMPLE 8 Factor $64p^3 - q^3$.

Solution First, we rewrite $64p^3$ as $(4p)^3$ and q^3 as $(q)^3$, then use the difference of two cubes formula.

$$\begin{aligned} 64p^3 - q^3 &= (4p)^3 - (q)^3 \\ &= (4p - q)[(4p)^2 + (4p)(q) + (q)^2] \\ &= (4p - q)(16p^2 + 4pq + q^2) \end{aligned}$$

Now Try Exercise 49

EXAMPLE 9 Factor $8r^3 + 27s^3$.

Solution We rewrite $8r^3 + 27s^3$ as a sum of two cubes. Since $8r^3 = (2r)^3$ and $27s^3 = (3s)^3$, we write

$$\begin{aligned} 8r^3 + 27s^3 &= (2r)^3 + (3s)^3 \\ &= (2r + 3s)[(2r)^2 - (2r)(3s) + (3s)^2] \\ &= (2r + 3s)(4r^2 - 6rs + 9s^2) \end{aligned}$$

Now Try Exercise 53

AVOIDING COMMON ERRORS

Recall that $a^2 + b^2 \neq (a + b)^2$ and $a^2 - b^2 \neq (a - b)^2$. The same principle applies to the sum and difference of two cubes.

CORRECT	INCORRECT
$a^3 + b^3 = (a + b)(a^2 - ab + b^2)$	~~$a^3 + b^3 = (a + b)^3$~~
$a^3 - b^3 = (a - b)(a^2 + ab + b^2)$	~~$a^3 - b^3 = (a - b)^3$~~

Since $(a + b)^3 = (a + b)(a + b)(a + b)$, it cannot possibly equal $a^3 + b^3$. Also, since $(a - b)^3 = (a - b)(a - b)(a - b)$, it cannot possibly equal $a^3 - b^3$.

It may be easier to see that, for example, $a^3 + b^3 = (a + b)(a^2 - ab + b^2)$ and not $(a + b)^3$ by substituting numbers for a and b. Suppose $a = 3$ and $b = 4$, then

$$\begin{aligned} a^3 + b^3 &= (a + b)(a^2 - ab + b^2) \\ 3^3 + 4^3 &= (3 + 4)[3^2 - 3(4) + 4^2] \\ 27 + 64 &= 7(13) \\ 91 &= 91 \end{aligned}$$

$$\begin{aligned} \text{but} \quad a^3 + b^3 &\neq (a + b)^3 \\ 3^3 + 4^3 &\neq (3 + 4)^3 \\ 91 &\neq 343 \end{aligned}$$

3 Learn the General Procedure for Factoring a Polynomial

In this chapter, we have presented several methods of factoring. We now combine techniques from this and previous sections to give you an overview of a general factoring procedure.

Here is a general procedure for factoring any polynomial:

General Procedure for Factoring a Polynomial

1. If all the terms of the polynomial have a greatest common factor other than 1, factor it out.
2. If the polynomial has two terms (or is a binomial), determine whether it is a difference of two squares or a sum or a difference of two cubes. If so, factor using the appropriate formula.
3. If the polynomial has three terms, factor the trinomial using the methods discussed in Sections 5.3 and 5.4.
4. If the polynomial has more than three terms, try factoring by grouping.
5. As a final step, examine your factored polynomial to determine whether the terms in any factors have a common factor. If you find a common factor, factor it out at this point.

Understanding Algebra

Remember the first step when factoring *any* polynomial is to factor out the GCF. In many polynomials, doing so allows us to factor the remaining polynomial further as shown in Example 10.

EXAMPLE 10 Factor $3x^4 - 27x^2$.

Solution First determine whether the terms have a greatest common factor other than 1. Since $3x^2$ is common to both terms, factor it out.

$$3x^4 - 27x^2 = 3x^2(x^2 - 9)$$
$$= 3x^2(x + 3)(x - 3) \quad \text{Difference of two squares formula was used.}$$

Note that $x^2 - 9$ is a difference of two squares.

Now Try Exercise 69

EXAMPLE 11 Factor $2m^2n^2 + 6m^2n - 36m^2$.

Solution Begin by factoring the GCF, $2m^2$, from each term. Then factor the remaining trinomial.

$$2m^2n^2 + 6m^2n - 36m^2 = 2m^2(n^2 + 3n - 18)$$
$$= 2m^2(n + 6)(n - 3)$$

Now Try Exercise 81

EXAMPLE 12 Factor $15c^2d - 10cd + 20d$.

Solution

$$15c^2d - 10cd + 20d = 5d(3c^2 - 2c + 4)$$

Since $3c^2 - 2c + 4$ cannot be factored, we stop here.

Now Try Exercise 77

EXAMPLE 13 Factor $3xy + 6x + 3y + 6$.

Solution Always begin by determining whether all the terms in the polynomial have a common factor. In this example, 3 is the GCF. Factor 3 from each term.

$$3xy + 6x + 3y + 6 = 3(xy + 2x + y + 2)$$

Now factor by grouping.

$$= 3[x(y + 2) + 1(y + 2)]$$
$$= 3(y + 2)(x + 1)$$

Now Try Exercise 79

In Example 13, what would happen if we forgot to factor out the common factor 3? Let's rework the problem without first factoring out the 3, and see what happens. Factor $3x$ from the first two terms, and 3 from the last two terms.

$$\begin{aligned} 3xy + 6x + 3y + 6 &= 3x(y + 2) + 3(y + 2) \\ &= (y + 2)(3x + 3) \end{aligned}$$

In step 5 of the general factoring procedure on page 327, we are reminded to examine the factored polynomial to see whether the terms in any factor have a common factor. If we study the factors, we see that the factor $3x + 3$ has the common factor 3. If we factor out the 3 from $3x + 3$, we will obtain the same answer obtained in Example 13.

$$\begin{aligned} (y + 2)(3x + 3) &= (y + 2) \cdot 3 \cdot (x + 1) \\ &= 3(y + 2)(x + 1) \end{aligned}$$

EXAMPLE 14 Factor $12x^2 + 12x - 9$.

Solution First factor out the GCF, 3. Then factor the remaining trinomial by one of the methods discussed in Section 5.4 (either by grouping or trial and error).

$$\begin{aligned} 12x^2 + 12x - 9 &= 3(4x^2 + 4x - 3) \\ &= 3(2x + 3)(2x - 1) \end{aligned}$$

Now Try Exercise 59

EXAMPLE 15 Factor $2x^4y + 54xy$.

Solution First factor out the GCF, $2xy$.

$$\begin{aligned} 2x^4y + 54xy &= 2xy(x^3 + 27) \\ &= 2xy(x + 3)(x^2 - 3x + 9) \end{aligned}$$

Note that $x^3 + 27$ is a sum of two cubes.

Now Try Exercise 95

5.5 Exercise Set MathXL® MyMathLab®

Warm-Up Exercises

Fill in the blanks with the appropriate word, phrase, or symbol(s) from the following list.

distributive $\quad a^2 + ab + b^2 \quad a^2 - ab + b^2 \quad$ sum $\quad$ difference $\quad$ grouping $\quad$ common $\quad$ product

1. Binomials of the form $a^2 - b^2$ are the __________ of two squares and factor as $(a + b)(a - b)$.

2. Binomials of the form $a^2 + b^2$ are the __________ of two squares and cannot be factored using real numbers.

3. When using the formula $a^3 + b^3 = (a + b)(a^2 - ab + b^2)$ it is important to note that the trinomial __________ cannot be factored further.

4. When using the formula $a^3 - b^3 = (a - b)(a^2 + ab + b^2)$ it is important to note that the trinomial __________ cannot be factored further.

5. The first step when factoring *any* polynomial is to factor out the greatest __________ factor.

6. To factor a polynomial that has more than three terms, try factoring by __________.

In Exercises 7–12, the binomial is a sum of squares. There is no formula for factoring the sum of squares. However, sometimes a common factor can be factored out from a sum of squares. Factor those polynomials that are factorable. If the polynomial is not factorable, write the word prime.

7. $x^2 + 9$ **8.** $y^2 + 25$ **9.** $4z^2 + 16$

10. $16s^2 + 64t^2$ **11.** $49d^2 + 36$ **12.** $9y^2 + 16z^2$

Practice the Skills

Factor each difference of two squares.

13. $y^2 - 25$
14. $t^2 - 4$
15. $x^2 - 49$
16. $q^2 - 100$
17. $81 - z^2$
18. $64 - z^2$
19. $x^2 - y^2$
20. $c^2 - d^2$
21. $16x^2 - 9$
22. $64z^2 - 9$
23. $9y^2 - 25z^2$
24. $100x^2 - 81y^2$
25. $36 - 49x^2$
26. $121 - 25r^2$
27. $z^4 - 81x^2$
28. $g^4 - 49h^2$
29. $25x^4 - 49y^4$
30. $4x^4 - 25y^4$
31. $6x^2 - 150$
32. $10x^2 - 160$
33. $2x^4 - 50y^2$
34. $36x^4 - 4y^2$
35. $5x^4 - 405$
36. $6x^4 - 96$

Factor each sum or difference of two cubes.

37. $x^3 + y^3$
38. $a^3 + b^3$
39. $x^3 - y^3$
40. $a^3 - b^3$
41. $x^3 + 64$
42. $a^3 + 27$
43. $x^3 - 27$
44. $x^3 - 8$
45. $a^3 + 1$
46. $t^3 - 1$
47. $27x^3 - 1$
48. $125v^3 - 1$
49. $27k^3 - 125$
50. $64x^3 - 125y^3$
51. $27 - 8y^3$
52. $125 - 27z^3$
53. $64m^3 + 27n^3$
54. $27c^3 + 125d^3$
55. $16x^3 + 54y^3$
56. $81x^3 + 192y^3$

Factor completely.

57. $4t^2 - 24t + 36$
58. $2d^2 + 16d + 32$
59. $50x^2 - 10x - 12$
60. $3x^2 - 9x - 12$
61. $3x^2 + 9x + 12x + 36$
62. $3xy - 6x + 9y - 18$
63. $5x^2 - 10x - 15$
64. $12n^2 + 4n - 16$
65. $5x^2 - 20$
66. $3x^2 - 48$
67. $2x^2 - 50$
68. $4a^2y - 64y^3$
69. $2x^2y - 18y$
70. $3x^3 - 147x$
71. $3x^3y^2 + 3y^2$
72. $x^4 - 125x$
73. $2x^3 - 16$
74. $x^3 - 27y^3$
75. $18x^2 - 50$
76. $a^5b^2 - 4a^3b^4$
77. $6t^2r - 15tr + 21r$
78. $3c^6 + 12c^4d^2$
79. $6x^2 - 4x + 24x - 16$
80. $x^2y + 2xy - 6xy - 12y$
81. $2rs^2 - 10rs - 48r$
82. $4x^4 - 26x^3 + 30x^2$
83. $4x^2 + 5x - 6$
84. $12a^2 + 36a + 27$
85. $25b^2 - 100$
86. $3b^2 - 75c^2$
87. $5x^3 - 10x^2 + 15x - 30$
88. $3x^3 + 12x^2 - 15x - 60$
89. $5x^4 + 10x^3 + 5x^2$
90. $12x^2 + 36x - 3x - 9$
91. $x^3 + 25x$
92. $3x^3y + 147xy$
93. $y^4 - 16$
94. $m^4 - 81$
95. $16m^3 + 250$
96. $81n^3 + 24$
97. $ac + 2a + bc + 2b$
98. $2ab - 3b + 4a - 6$
99. $36x^4y^2 - 24x^3y^3 - 45x^2y^4$
100. $30x^5y^3 - 16x^4y^4 - 24x^3y^5$

Problem Solving

Factor each expression. Treat the unknown symbols as if they were variables.

101. ◆✱ + 2◆ + ☺✱ + 2☺
102. 2◆^{6} + 4◆^{4}✱^{2}
103. 4◆^{2}✱ − 6◆✱ − 20✱◆ + 30✱

Concept/Writing Exercises

104. Explain why the sum of two squares, $a^2 + b^2$, cannot be factored using real numbers.

105. Have you ever seen the proof that 1 is equal to 2? Here it is.

Let $a = b$, then square both sides of the equation:

$$a^2 = b^2$$
$$a^2 = b \cdot b$$
$$a^2 = ab \qquad \text{Substituted because } a = b.$$
$$a^2 - b^2 = ab - b^2 \qquad \text{Subtracted } b^2 \text{ from both sides of the equation.}$$
$$(a + b)(a - b) = b(a - b) \qquad \text{Factored both sides of the equation.}$$
$$\frac{(a + b)\cancel{(a - b)}}{\cancel{(a - b)}} = \frac{b\cancel{(a - b)}}{\cancel{(a - b)}} \qquad \text{Divided both sides of the equation by } (a - b) \text{ and divide out common factors.}$$
$$a + b = b$$
$$b + b = b \qquad \text{Substituted because } a = b.$$
$$2b = b$$
$$\frac{2\cancel{b}}{\cancel{b}} = \frac{\cancel{b}}{\cancel{b}} \qquad \text{Divided both sides of the equation by } b.$$
$$2 = 1$$

Obviously, $2 \neq 1$. Therefore, we must have made an error somewhere. Can you find it?

Challenge Problems

106. Factor $x^6 - 27y^9$.

107. Factor $x^6 + 1$.

108. Factor $x^2 - 6x + 9 - 4y^2$. (*Hint:* Write the first three terms as the square of a binomial.)

109. Factor $x^6 - y^6$. (*Hint:* Factor initially as the difference of two squares.)

110. Factor $x^2 + 10x + 25 - y^2 + 4y - 4$. (*Hint:* Group the first three terms and the last three terms.)

Cumulative Review Exercises

[2.8] **111.** Solve the inequality $3x - 2(x + 4) \geq 2x - 9$. Graph the solution on a number line and write the solution in interval notation.

[2.6] **112.** Use the formula $A = \frac{1}{2}h(b + d)$ to find h in the following trapezoid if the area of the trapezoid is 36 square inches.

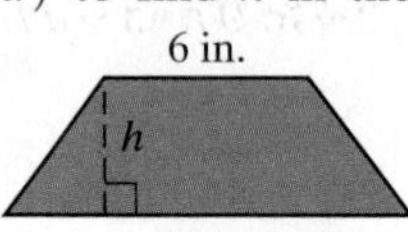

[4.1] **113.** Simplify $-9(a^3b^2c^6)^0$.

[4.1] **114.** Simplify $\left(\frac{4x^4y}{6xy^5}\right)^3$.

[4.2] **115.** Simplify $a^{-4}a^{-7}$.

5.6 Solving Quadratic Equations Using Factoring

1. Recognize quadratic equations.
2. Solve quadratic equations using factoring.

1 Recognize Quadratic Equations

In this section, we introduce **quadratic equations**, which are equations that contain a second-degree term and no term of a higher degree.

Quadratic Equation

Quadratic equations have the form

$$ax^2 + bx + c = 0$$

where a, b, and c are real numbers, $a \neq 0$.

Examples of Quadratic Equations

$$x^2 + 4x - 12 = 0$$
$$2x^2 - 5x = 0$$
$$3x^2 - 2 = 0$$

Quadratic equations like these, in which one side of the equation is written in descending order and the other side of the equation is 0, are said to be in **standard form**.

Some quadratic equations can be solved by factoring. To solve a quadratic equation by factoring, we use the *zero-factor property*.

Understanding Algebra

The zero-factor property states that if we multiply two expressions together and get a result of 0, then at least one of the expressions must be equal to 0.

Zero-Factor Property

If $m \cdot n = 0$, then $m = 0$ or $n = 0$.

The zero-factor property states that if the product of two expressions is 0, then at least one of the expression must equal 0. We now illustrate how the zero-factor property is used in solving equations.

EXAMPLE 1 Solve the equation $(x + 3)(x + 4) = 0$.

Solution Since the product of the factors equals 0, according to the zero-factor property, one or both factors must equal 0. Set each factor equal to 0, and solve each resulting equation.

$$\begin{aligned} x + 3 &= 0 & \text{or} \qquad x + 4 &= 0 \\ x + 3 - 3 &= 0 - 3 & x + 4 - 4 &= 0 - 4 \\ x &= -3 & x &= -4 \end{aligned}$$

Thus, if x is either -3 or -4, the product of the factors is 0. The solutions to the equation are -3 and -4.

Check $x = -3$ $\qquad$ $x = -4$

$$\begin{aligned} (x + 3)(x + 4) &= 0 & (x + 3)(x + 4) &= 0 \\ (-3 + 3)(-3 + 4) &\stackrel{?}{=} 0 & (-4 + 3)(-4 + 4) &\stackrel{?}{=} 0 \\ 0(1) &\stackrel{?}{=} 0 & -1(0) &\stackrel{?}{=} 0 \\ 0 &= 0 \quad \text{True} & 0 &= 0 \quad \text{True} \end{aligned}$$

Now Try Exercise 7

EXAMPLE 2 Solve the equation $(3x - 2)(4x + 1) = 0$.

Solution Set each factor equal to 0 and solve for x.

$$\begin{aligned} 3x - 2 &= 0 & \text{or} \qquad 4x + 1 &= 0 \\ 3x &= 2 & 4x &= -1 \\ x &= \frac{2}{3} & x &= -\frac{1}{4} \end{aligned}$$

The solutions to the equation are $\frac{2}{3}$ and $-\frac{1}{4}$.

Now Try Exercise 11

2 Solve Quadratic Equations Using Factoring

General Rules to Solve a Quadratic Equation Using Factoring

1. Write the equation in standard form with the squared term having a positive coefficient. This will result in one side of the equation being 0.
2. Factor the side of the equation that is not 0.
3. Set each factor *containing a variable* equal to 0 and solve each equation.
4. Check each solution found in step 3 in the *original* equation.

Now we will use factoring to solve quadratic equations.

EXAMPLE 3 Solve the equation $3x^2 = 12x$.

Solution

$$3x^2 = 12x$$

$$3x^2 - 12x = 12x - 12x \quad \text{Subtract } 12x \text{ from both sides to make one side 0.}$$

$$3x^2 - 12x = 0$$

$$3x(x - 4) = 0 \quad \text{Factored out the GCF, } 3x.$$

Understanding Algebra

To solve $x^2 = 8x$ it is *not* correct to divide both sides of the equation by x. Since it is a quadratic equation, we solve it by factoring:

$$x^2 = 8x$$

$$x^2 - 8x = 0$$

$$x(x - 8) = 0$$

$$x = 0 \quad \text{or} \quad x = 8$$

Now set each factor equal to 0.

$$3x = 0 \quad \text{or} \quad x - 4 = 0$$

$$x = \frac{0}{3} \qquad\qquad x = 4$$

$$x = 0$$

The solutions to the quadratic equation are 0 and 4. Check by substituting $x = 0$, then $x = 4$, into $3x^2 = 12x$.

Now Try Exercise 47

EXAMPLE 4 Solve the equation $x^2 + 10x + 28 = 4$.

Solution

$$x^2 + 10x + 28 = 4$$

$$x^2 + 10x + 24 = 0 \quad \text{Subtracted 4 from both sides.}$$

$$(x + 4)(x + 6) = 0 \quad \text{Factored.}$$

$$x + 4 = 0 \quad \text{or} \quad x + 6 = 0 \quad \text{Set each factor equal to 0.}$$

$$x = -4 \qquad\qquad x = -6$$

The solutions are -4 and -6. We will check these values in the original equation.

Check

$x = -4$

$$x^2 + 10x + 28 = 4$$

$$(-4)^2 + 10(-4) + 28 \stackrel{?}{=} 4$$

$$16 - 40 + 28 \stackrel{?}{=} 4$$

$$-24 + 28 \stackrel{?}{=} 4$$

$$4 = 4 \quad \text{True}$$

$x = -6$

$$x^2 + 10x + 28 = 4$$

$$(-6)^2 + 10(-6) + 28 \stackrel{?}{=} 4$$

$$36 - 60 + 28 \stackrel{?}{=} 4$$

$$-24 + 28 \stackrel{?}{=} 4$$

$$4 = 4 \quad \text{True}$$

Now Try Exercise 27

EXAMPLE 5 Solve the equation $4y^2 + 5y - 20 = -11y$.

Solution Since all terms are not on the same side of the equation, add $11y$ to both sides of the equation.

$$4y^2 + 5y - 20 = -11y$$

$$4y^2 + 16y - 20 = 0 \quad \text{Added } 11y \text{ to both sides.}$$

$$4(y^2 + 4y - 5) = 0 \quad \text{Factored out GCF, 4.}$$

$$4(y + 5)(y - 1) = 0 \quad \text{Factored.}$$

$$y + 5 = 0 \quad \text{or} \quad y - 1 = 0 \quad \text{Set each factor equal to 0.}$$

$$y = -5 \qquad\qquad y = 1$$

Understanding Algebra

Notice in Example 5 that the factored form of the equation $4y^2 + 16y - 20 = 0$ is

$$4(y + 5)(y - 1) = 0.$$

Since the 4 is a factor that does not contain a variable, we do not set it equal to 0. *Note that 4 is not a solution to this equation.* The solutions are -5 and 1.

Since 4 is a factor that does not contain a variable, we do not set it equal to 0. The solutions to the quadratic equation are -5 and 1.

Now Try Exercise 29

EXAMPLE 6 Solve the equation $-x^2 + 5x + 6 = 0$.

Solution When the squared term is negative, we generally make it positive by multiplying both sides of the equation by -1.

$$-1(-x^2 + 5x + 6) = -1 \cdot 0$$
$$x^2 - 5x - 6 = 0$$
$$(x - 6)(x + 1) = 0$$
$$x - 6 = 0 \quad \text{or} \quad x + 1 = 0 \qquad \text{Set each factor equal to 0.}$$
$$x = 6 \qquad\qquad x = -1$$

A check using the original equation will show that the solutions are 6 and -1.

Now Try Exercise 33

AVOIDING COMMON ERRORS

Be careful not to confuse factoring a polynomial with using factoring as a method to solve an equation.

CORRECT	INCORRECT
Factor: $x^2 + 3x + 2$	Factor: $x^2 + 3x + 2$
$(x + 2)(x + 1)$	$(x + 2)(x + 1)$
	~~$x + 2 = 0$ or $x + 1 = 0$~~
	~~$x = -2$ $\quad$ $x = -1$~~

The expression $x^2 + 3x + 2$ is a polynomial, not an equation. Since it is not an equation, it cannot be solved.

CORRECT

Solve:
$$x^2 + 3x + 2 = 0$$
$$(x + 2)(x + 1) = 0$$
$$x + 2 = 0 \quad \text{or} \quad x + 1 = 0$$
$$x = -2 \qquad\qquad x = -1$$

EXAMPLE 7 Solve the equation $x^2 = 49$.

Solution

$$x^2 = 49$$
$$x^2 - 49 = 0 \qquad \text{49 was subtracted from both sides.}$$
$$(x + 7)(x - 7) = 0 \qquad \text{Factored using the difference of two squares.}$$
$$x + 7 = 0 \quad \text{or} \quad x - 7 = 0 \qquad \text{Set each factor equal to 0.}$$
$$x = -7 \qquad\qquad x = 7$$

The solutions are -7 and 7.

Now Try Exercise 45

EXAMPLE 8 Solve the equation $(x - 3)(x + 1) = 5$.

Solution Notice this equation is not in standard form. Therefore, we must first multiply the binomials using the FOIL method. We can then subtract the 5 from both sides to put the equation into standard form.

$$(x - 3)(x + 1) = 5$$
$$x^2 - 2x - 3 = 5 \qquad \text{Factors were multiplied.}$$
$$x^2 - 2x - 8 = 0 \qquad \text{Wrote the equation in standard form.}$$
$$(x - 4)(x + 2) = 0 \qquad \text{Trinomial was factored.}$$
$$x - 4 = 0 \quad \text{or} \quad x + 2 = 0 \qquad \text{Zero-factor property}$$
$$x = 4 \qquad\qquad x = -2$$

The solutions are 4 and -2. We will check these values in the original equation.

Check

$x = 4$

$$(x - 3)(x + 1) = 5$$
$$(4 - 3)(4 + 1) \stackrel{?}{=} 5$$
$$1(5) \stackrel{?}{=} 5$$
$$5 = 5 \quad \text{True}$$

$x = -2$

$$(x - 3)(x + 1) = 5$$
$$(-2 - 3)(-2 + 1) \stackrel{?}{=} 5$$
$$(-5)(-1) \stackrel{?}{=} 5$$
$$5 = 5 \quad \text{True}$$

Now Try Exercise 51

HELPFUL HINT

In Example 8, you might have been tempted to start the problem by writing

$$x - 3 = 5 \quad \text{or} \quad x + 1 = 5.$$

This would lead to an incorrect solution. Remember, the zero-factor property only holds when one side of the equation is equal to 0. In Example 8, once we obtained $(x - 4)(x + 2) = 0$, we were able to use the zero-factor property.

5.6 Exercise Set MathXL® MyMathLab®

Warm-Up Exercises

Fill in the blanks with the appropriate word, phrase, or symbol(s) from the following list.

product quotient quadratic FOIL variable standard subtracting dividing

1. Equations that contain a second-degree term and no term of higher degree are called ________ equations.
2. A quadratic equation that has one side written in descending order and the other side equal to 0 is said to be in ________ form.
3. The zero-factor property states that if the ________ of two expressions is 0, then at least one of the expressions must equal 0.
4. When solving the equation $4(y + 5)(y - 1) = 0$, since 4 is a factor that does not contain a ________, we do not set it equal to 0.
5. When solving the equation $x^2 + 10x + 28 = 4$, we must first write the equation in standard form by ________ 4 from both sides of the equation.
6. In the equation $(x - 3)(x + 1) = 5$ we must first multiply the binomials using the ________ method.

Practice the Skills

Solve.

7. $(x + 8)(x - 7) = 0$
8. $(t + 3)(t + 5) = 0$
9. $7x(x - 8) = 0$
10. $-2x(x + 9) = 0$
11. $(3x + 7)(2x - 11) = 0$
12. $(3x - 2)(x - 5) = 0$
13. $x^2 - 12x = 0$
14. $x^2 + 7x = 0$
15. $7x^2 - 35x = 0$
16. $9x^2 + 27x = 0$
17. $d^2 - 5d - 24 = 0$
18. $z^2 - 4z - 12 = 0$
19. $3p^2 - 26p + 35 = 0$
20. $7p^2 + 33q - 10 = 0$
21. $x^2 - 8x + 16 = 0$
22. $x^2 + 12x + 36 = 0$
23. $x^2 - 16 = 0$
24. $y^2 - 9 = 0$
25. $x^2 + 12x = -20$
26. $k^2 + 3k = 40$
27. $x^2 + 12x + 22 = 2$
28. $p^2 + 12p - 2 = -29$
29. $2x^2 + 3x - 24 = 5x$
30. $3q^2 + 11q - 72 = 5q$
31. $-2x - 15 = -x^2$
32. $-9x + 20 = -x^2$
33. $-x^2 + 29x + 30 = 0$
34. $-j^2 - 27j + 28 = 0$
35. $-15 = 4m^2 + 17m$
36. $-7 = 9t^2 - 16t$
37. $9p^2 = -21p - 6$
38. $2x^2 - 5 = 3x$
39. $3r^2 + 13r = 10$
40. $3x^2 = 7x + 20$
41. $4x^2 + 4x - 48 = 0$
42. $6x^2 - 7x - 5 = 0$
43. $8x^2 + 2x = 3$
44. $2x^2 + 4x - 6 = 0$
45. $x^2 = 100$
46. $c^2 = 64$
47. $2x^2 = 50x$
48. $3h^2 = 108h$
49. $4x^2 - 25 = 0$
50. $9v^2 - 49 = 0$
51. $(x - 2)(x - 1) = 12$
52. $(x + 2)(x + 5) = -2$
53. $(3x + 2)(x + 1) = 4$
54. $(x - 1)(2x - 5) = 9$
55. $2(w^2 + 9) = 15w$
56. $x(x + 5) = 6$

Problem Solving

In Exercises 57–60, create a quadratic equation with the given solutions. Explain how you determined your answers.

57. 6, −4

58. −3, −5

59. 6, 0

60. 0, −9

61. The solutions to a quadratic equation are $\frac{1}{2}$ and $-\frac{1}{3}$.

a) Give two possible factors with integer coefficients that were set equal to 0 to obtain these solutions.

b) Write a quadratic equation whose solutions are $\frac{1}{2}$ and $-\frac{1}{3}$.

62. The solutions to a quadratic equation are $\frac{2}{3}$ and $-\frac{3}{4}$.

a) Give two possible factors with integer coefficients that were set equal to 0 to obtain these solutions.

b) Write a quadratic equation whose solutions are $\frac{2}{3}$ and $-\frac{3}{4}$.

Concept/Writing Exercises

63. Carefully observe the quadratic equations $x^2 - 7x + 12 = 0$, $2x^2 - 14x + 24 = 0$, and $3x^2 - 21x + 36 = 0$. Explain how you know that all three equations have the same solution.

64. Write the steps to the method you would use to solve $(x + 1)(x - 2) = 4$ as if you were writing an email to a friend taking an algebra course.

Challenge Problems

65. Solve the equation $(2x - 3)(x - 4) = (x - 5)(x + 3) + 7$.

66. Solve the equation $(x - 3)(x - 2) = (x + 5)(2x - 3) + 21$.

67. Solve the equation $x(x - 3)(x + 2) = 0$.

68. Solve the equation $x^3 - 10x^2 + 24x = 0$.

Cumulative Review Exercises

[1.7] **69.** Subtract $\frac{3}{5} - \frac{2}{9}$.

[2.5] **70. a)** What is the name given to an equation that has an infinite number of solutions?

b) What is the name given to an equation that has no solution?

[2.7] **71. Legoland** At Legoland, there is a long line of people waiting to go through the entrance. If 160 people are admitted in 13 minutes, how many people will be admitted in 60 minutes? Assume the rate stays the same.

[4.1] **72.** Simplify $\left(\frac{3p^5q^7}{p^9q^8}\right)^2$.

Legoland, see Exercise 71.

[4.4] *Identify the following as a monomial, binomial, trinomial or not a polynomial. If an expression is not a polynomial, explain why.*

73. $2x$

74. $x - 3$

75. $\frac{1}{x}$

76. $x^2 - 6x + 9$

5.7 Applications of Quadratic Equations

1 Solve applications by factoring quadratic equations.

2 Learn the Pythagorean Theorem.

1 Solve Applications by Factoring Quadratic Equations

In Section 5.6, we learned how to solve quadratic equations by factoring. In this section, we will discuss and solve application problems that require solving quadratic equations to obtain the answer. In Example 1, we will solve a problem involving a relationship between two numbers.

EXAMPLE 1 Number Problem The product of two numbers is 78. Find the two numbers if one number is 7 more than the other.

Solution Understand and Translate Our goal is to find the two numbers.

Let x = smaller number.

$x + 7$ = larger number.

Understanding Algebra

Remember when solving quadratic equations by factoring, you must first put the equation in standard form:

$$ax^2 + bx + c = 0.$$

	$x(x+7) = 78$	"Product" refers to multiplication.
Carry Out	$x^2 + 7x = 78$	Distributive property
	$x^2 + 7x - 78 = 0$	Standard quadratic form
	$(x-6)(x+13) = 0$	Factor.

$$x - 6 = 0 \quad \text{or} \quad x + 13 = 0$$
$$x = 6 \qquad\qquad x = -13$$

Remember that x represents the smaller of the two numbers. This problem has two possible solutions.

	Solution 1	Solution 2
Smaller number	6	-13
Larger number	$x + 7 = 6 + 7 = 13$	$x + 7 = -13 + 7 = -6$

Thus, the two possible solutions are 6 and 13, and -13 and -6.

Check	6 and 13	-13 and -6
Product of the two numbers is 78.	$6 \cdot 13 = 78$	$(-13)(-6) = 78$
One number is 7 more than the other number.	13 is 7 more than 6.	-6 is 7 more than -13.

Answer One solution is: smaller number 6, larger number 13. A second solution is: smaller number -13, larger number -6. You must give both solutions. If the question had stated "the product of two *positive* numbers is 78," the only solution would be 6 and 13.

Now Try Exercise 7

Now let us work an application problem involving geometry.

EXAMPLE 2 Advertising The marketing department of a large publishing company is planning to make a large rectangular sign to advertise a new book at a convention. They want the length of the sign to be 3 feet longer than the width (**Fig. 5.3**). Signs at the convention may have a maximum area of 54 square feet. Find the length and width of the sign if the area is to be 54 square feet.

Solution Understand and Translate We need to find the length and width of the sign. We will use the formula for the area of a rectangle.

FIGURE 5.3

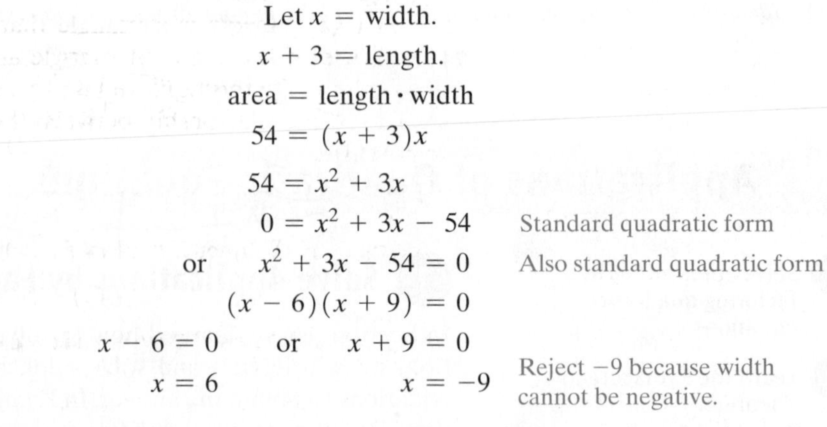

$$\text{Let } x = \text{width.}$$
$$x + 3 = \text{length.}$$
$$\text{area} = \text{length} \cdot \text{width}$$
$$54 = (x+3)x$$

Carry Out

$$54 = x^2 + 3x$$
$$0 = x^2 + 3x - 54 \quad \text{Standard quadratic form}$$
$$\text{or} \quad x^2 + 3x - 54 = 0 \quad \text{Also standard quadratic form}$$
$$(x-6)(x+9) = 0$$
$$x - 6 = 0 \quad \text{or} \quad x + 9 = 0$$
$$x = 6 \qquad\qquad x = -9$$

Reject -9 because width cannot be negative.

Check and Answer Since the width of the sign cannot be a negative number, the only solution is

$$\text{width} = x = 6 \text{ feet, length} = x + 3 = 6 + 3 = 9 \text{ feet}$$

The area, length · width, is 54 square feet, and the length is 3 feet more than the width, so the answer checks.

Now Try Exercise 17

EXAMPLE 3 Earth's Gravitational Field In Earth's gravitational field, the distance, d, in feet, that an object falls t seconds after it has been released is given by the formula $d = 16t^2$. While at the top of a roller coaster, a rider's eyeglasses slide off his head and fall out of the cart. How long does it take the eyeglasses to reach the ground 64 feet below?

Solution Understand and Translate Substitute 64 for d in the formula and then solve for t.

$$d = 16t^2$$
$$64 = 16t^2$$

Carry Out

$$\frac{64}{16} = t^2$$
$$4 = t^2$$

Pythagoras of Samos

Now subtract 4 from both sides of the equation and write the equation with 0 on the right side to put the quadratic equation in standard form.

$$4 - 4 = t^2 - 4$$
$$0 = t^2 - 4$$
$$\text{or} \quad t^2 - 4 = 0$$
$$(t + 2)(t - 2) = 0$$
$$t + 2 = 0 \quad \text{or} \quad t - 2 = 0$$
$$t = -2 \qquad t = 2$$

Reject -2 because time cannot be negative.

Check and Answer Since t represents the number of seconds, it must be a positive number. Thus, the only possible answer is 2 seconds. It takes 2 seconds for the eyeglasses (or any other object falling under the influence of gravity) to fall 64 feet.

Now Try Exercise 21

2 Learn The Pythagorean Theorem

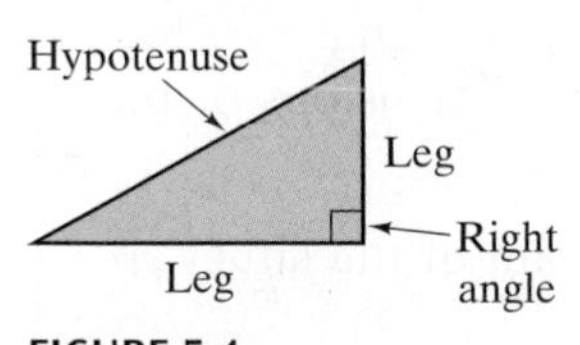

FIGURE 5.4

We now introduce the Pythagorean Theorem, which describes an important relationship between the length of the sides of a right triangle. The Pythagorean Theorem is named after Pythagoras of Samos ($\approx$569 B.C.–475 B.C.) who was born in Samos, Ionia. Pythagoras is often described as the first pure mathematician. Now let's discuss the Pythagorean Theorem.

A **right triangle** is a triangle that contains a right, or 90°, angle (**Fig. 5.4**). The two shorter sides of a right triangle are called the **legs** and the longest side, which is always opposite the right angle, is called the **hypotenuse**. The **Pythagorean Theorem** expresses the relationship between the lengths of the legs of a right triangle and its hypotenuse.

Understanding Algebra

The proof of the Pythagorean Theorem is often expressed visually:

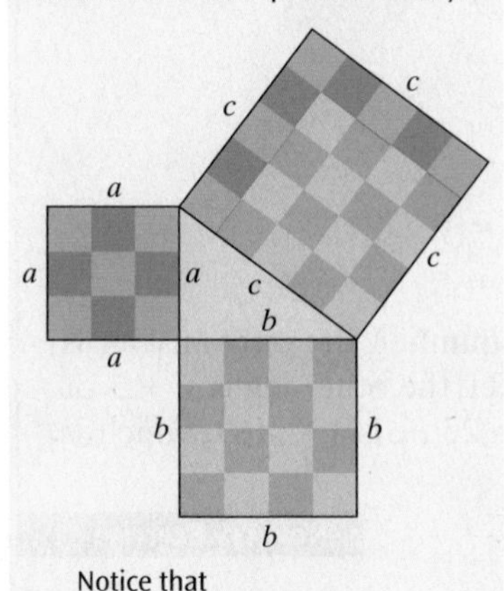

Notice that

$$a^2 + b^2 = c^2$$
$$3^2 + 4^2 = 5^2$$

Pythagorean Theorem

The square of the hypotenuse of a right triangle is equal to the sum of the squares of the two legs.

$$(\text{leg})^2 + (\text{leg})^2 = (\text{hypotenuse})^2$$

If a and b represent the legs, and c represents the hypotenuse, then

$$a^2 + b^2 = c^2$$

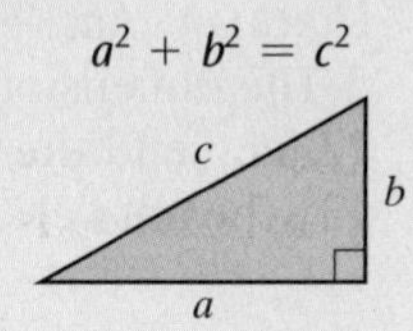

When you use the Pythagorean Theorem, it makes no difference which leg you designate as a and which leg you designate as b, but the hypotenuse, the longest side, is always designated as c.

EXAMPLE 4 Verifying Right Triangles Determine if a right triangle can have the following sides.

a) 3 inches, 4 inches, 5 inches **b)** 2 inches, 5 inches, 7 inches

Solution

a) Understand To determine if a right triangle can have the sides given, we will use the Pythagorean Theorem.

Translate We must always select the longest side to represent the hypotenuse, c. We will designate the length of leg a to be 3 inches and the length of leg b to be 4 inches. The length of the hypotenuse, c, will be 5 inches. See **Figure 5.5**.

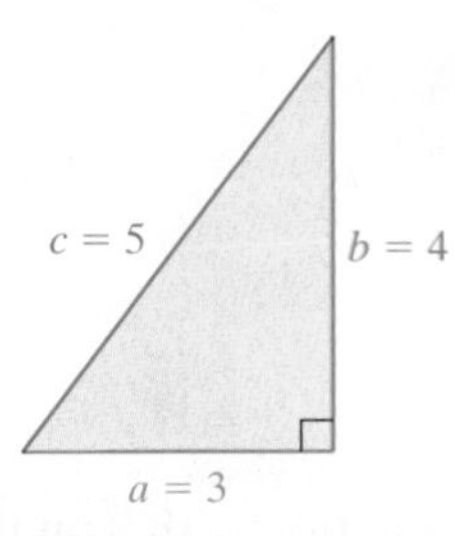

FIGURE 5.5

$$a^2 + b^2 = c^2$$
$$3^2 + 4^2 \stackrel{?}{=} 5^2$$

Carry Out
$$9 + 16 \stackrel{?}{=} 25$$
$$25 = 25 \quad \text{True}$$

Check and Answer Since using the Pythagorean Theorem results in a true statement, a right triangle can have the given sides.

b) We will let leg a have a length of 2 inches, leg b have a length of 5 inches, and the hypotenuse, c, have a length of 7 inches.

$$a^2 + b^2 = c^2$$
$$2^2 + 5^2 \stackrel{?}{=} 7^2$$
$$4 + 25 \stackrel{?}{=} 49$$
$$29 = 49 \quad \text{False}$$

Since 29 is not equal to 49, the Pythagorean Theorem does not hold for these lengths. Therefore, no right triangle can have sides with lengths of 2 inches, 5 inches, and 7 inches.

Now Try Exercise 23

EXAMPLE 5 Using the Pythagorean Theorem One leg of a right triangle is 7 feet longer than the other leg. The hypotenuse is 13 feet. Find the dimensions of the right triangle.

Solution Understand and Translate We will first draw a diagram of the situation. See **Figure 5.6**.

Now we will use the Pythagorean Theorem to determine the dimensions of the right triangle.

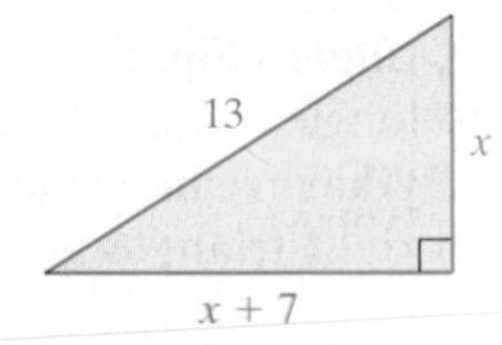

FIGURE 5.6

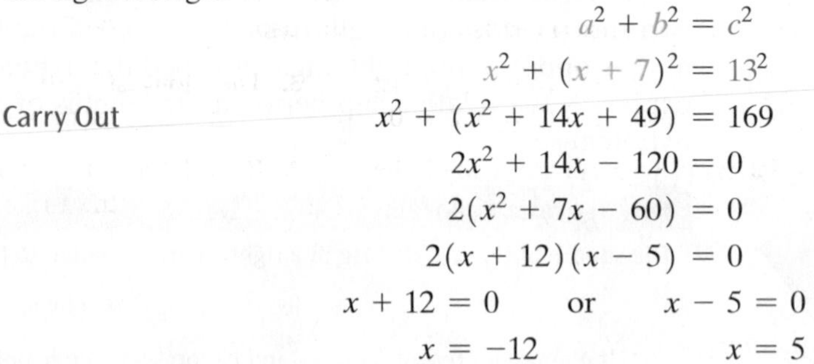

$$a^2 + b^2 = c^2$$
$$x^2 + (x + 7)^2 = 13^2$$

Carry Out
$$x^2 + (x^2 + 14x + 49) = 169$$
$$2x^2 + 14x - 120 = 0$$
$$2(x^2 + 7x - 60) = 0$$
$$2(x + 12)(x - 5) = 0$$
$$x + 12 = 0 \quad \text{or} \quad x - 5 = 0$$
$$x = -12 \qquad\qquad x = 5$$

Check and Answer Since a length cannot be a negative number, the only answer is 5. The dimensions of the right triangle are: one leg is 5 feet, the other leg is $x + 7$ or 12 feet, and the hypotenuse is 13 feet. Since $5^2 + 12^2 = 25 + 144 = 169$, which is 13^2, the answer checks.

Now Try Exercise 31

EXAMPLE 6 Sand and Water Table Clayton Jackson is building a rectangular table for his son to hold sand in one area and water in a different area (**Fig. 5.7** on the next page). The length of the table will be 2 feet less than twice the width. He is

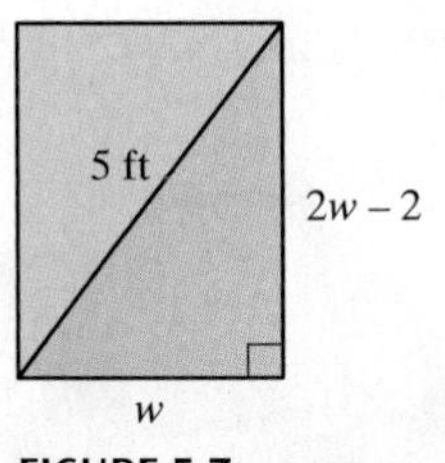

FIGURE 5.7

placing a divider that is 5 feet in length along the diagonal of the table to separate the sand from the water. Find the dimensions of the table.

Solution Understand and Translate Our goal is to find the dimensions of the table. **Figure 5.7** shows a right triangle. Therefore, we will use the Pythagorean Theorem to answer the question.

$$\text{Let } w = \text{width of the table.}$$
$$\text{Then } 2w - 2 = \text{length of the table.}$$

Now we use the Pythagorean Theorem. We will let w represent leg a, $2w - 2$ represent leg b, and 5 represent the hypotenuse, c.

$$a^2 + b^2 = c^2$$
$$w^2 + (2w - 2)^2 = 5^2$$

Carry Out

$$w^2 + 4w^2 - 8w + 4 = 25$$
$$5w^2 - 8w + 4 = 25$$
$$5w^2 - 8w - 21 = 0$$
$$(5w + 7)(w - 3) = 0$$
$$5w + 7 = 0 \quad \text{or} \quad w - 3 = 0$$
$$5w = -7 \qquad w = 3$$
$$w = -\frac{7}{5}$$

Check and Answer Since the width of the table cannot be a negative number, the only solution is 3. Therefore, the width of the table is 3 feet. The length of the table is $2w - 2 = 2(3) - 2 = 6 - 2 = 4$ feet.

Now Try Exercise 37

In the exercise set, we use the terms consecutive integers, consecutive even integers, and consecutive odd integers. Recall from Section 3.1 that *consecutive integers* may be represented as x and $x + 1$. *Consecutive even* or *consecutive odd integers* may be represented as x and $x + 2$.

5.7 Exercise Set MathXL® MyMathLab®

Warm-Up Exercises

Fill in the blanks with the appropriate word, phrase, or symbol(s) from the following list.

legs feet hypothesis hypotenuse standard $x + 1$ $x - 1$

1. When solving quadratic equations by factoring, you must first put the equation in ________ form: $ax^2 + bx + c = 0$
2. The two shorter sides of a right triangle are called the ________ of the triangle.
3. The longest side of a right triangle is called the ________ of the triangle.
4. If we let x = the smaller of two consecutive integers, then ________ = the larger of two consecutive integers.

Problem Solving

Express each problem as an equation, then solve.

5. **Positive Numbers** The product of two positive numbers is 363. Determine the two numbers if one number is triple the other.
6. **Positive Numbers** The product of two positive numbers is 245. Determine the two numbers if one number is 5 times the other.
7. **Product of Numbers** The product of two numbers is 98. Determine the two numbers if one is 7 less than the other.
8. **Product of Numbers** The product of two numbers is 221. Determine the two numbers if one is 4 more than the other.
9. **Positive Numbers** The product of two positive numbers is 84. Find the two numbers if one number is 2 more than twice the other.
10. **Positive Numbers** The product of two positive numbers is 68. Determine the two numbers if one number is 1 more than 4 times the other number.

11. **Consecutive Integers** The product of two consecutive positive integers is 56. Find the two integers.

12. **Consecutive Integers** The product of two consecutive positive integers is 132. Find the two integers.

13. **Consecutive Even Integers** The product of two consecutive positive even integers is 288. Find the two integers.

14. **Consecutive Odd Integers** The product of two consecutive positive odd integers is 143. Determine the two integers.

15. **Area of Rectangle** The area of a rectangle is 36 square feet. Determine the length and width if the length is 4 times the width.

16. **Rectangular Garden** Maureen Woolhouse has a rectangular garden whose width is two-thirds its length. If its area is 150 square feet, determine the length and width of the garden.

17. **Rectangular Scrapbook** A scrapbook page has an area of 180 square inches. Find the length and width if the width is 3 inches less than the length.

18. **Computer Screen** The area of a rectangular computer screen is 143 square inches. Find the length and width of the screen if the length is 2 inches more than the width.

See Exercise 18.

19. **Square** If each side of a square is increased by 4 meters, the area becomes 81 square meters. Determine the length of a side of the original square.

20. **Square** If the sides of a square is decreased by 7 inches, the area becomes 121 square inches. Determine the length of a side of the original square.

21. **Dropped Egg** How long would it take for an egg dropped from a helicopter to fall 256 feet to the ground? See Example 3.

22. **Falling Rock** How long would it take a rock that falls from a cliff 400 feet above the sea to hit the sea? See Example 3.

In Exercises 23–26, determine if a right triangle can have the following sides where a and b represent the legs and c represents the hypotenuse.

23. $a = 7, b = 24, c = 25$

24. $a = 9, b = 40, c = 41$

25. $a = 16, c = 20, b = 22$

26. $a = 13, b = 18, c = 28$

In Exercises 27–30, determine if the triangle shown is a right triangle.

27.

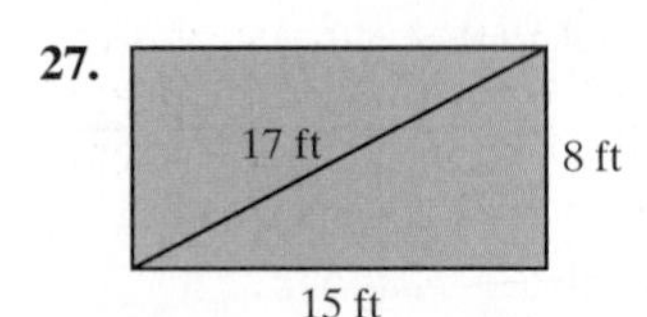

28.

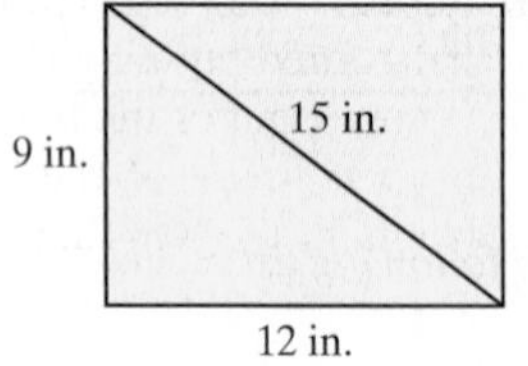

29.

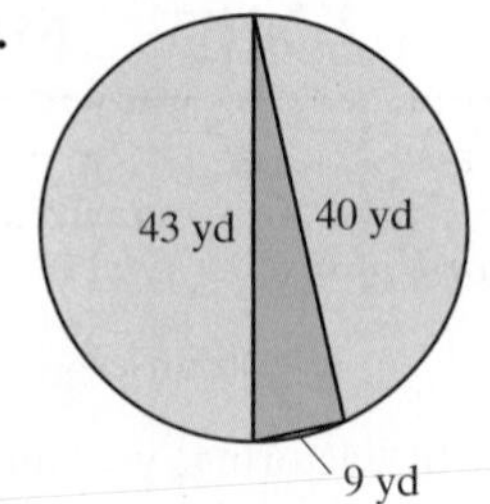

30.

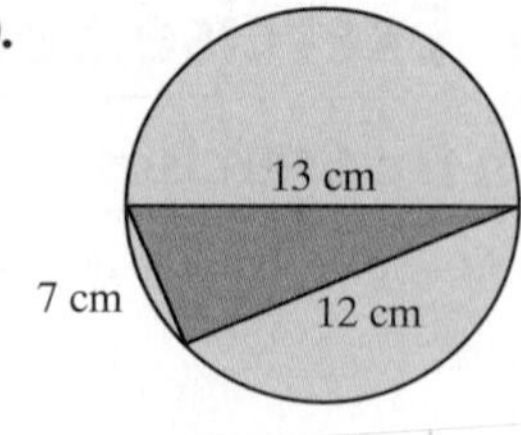

In Exercises 31–34, determine the value of x.

31.

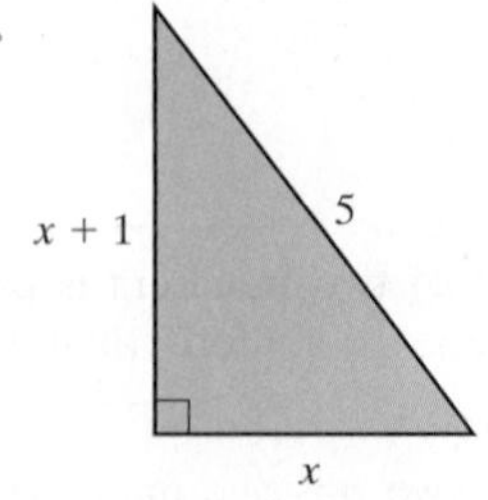

32.

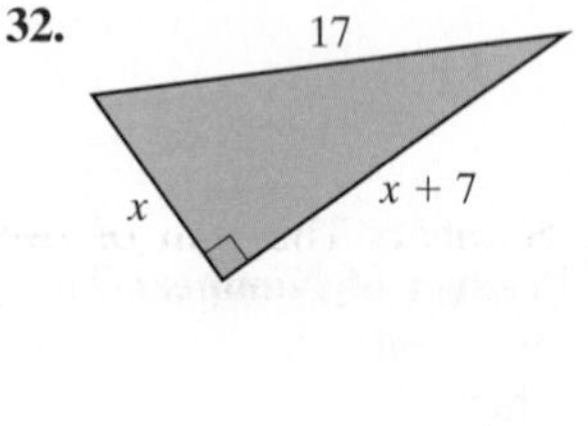

33.

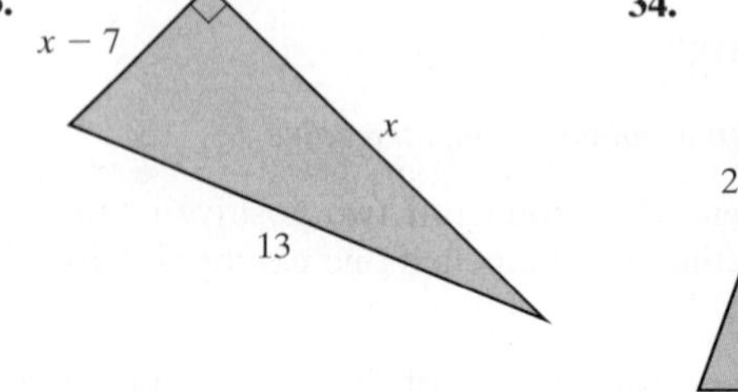

34.

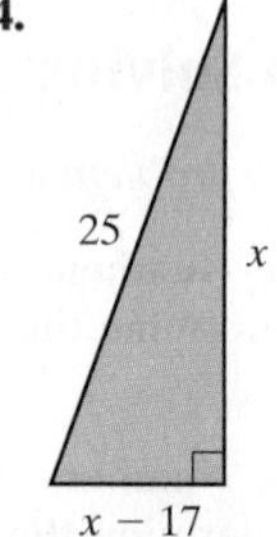

35. **Triangle** One leg of a right triangle is 2 feet longer than the other leg. The hypotenuse is 10 feet. Find the lengths of the three sides of the triangle.

36. **Triangle** One leg of a right triangle is two inches more than twice the other leg. The hypotenuse is 13 inches. Find the lengths of the three sides of the triangle.

37. **Artwork** Rachel bought a framed piece of artwork as a souvenir from her trip to Disney World. The diagonal of the frame is 15 inches. If the length of the frame is 3 inches greater than its width, find the dimensions of the frame.

38. **Laptop** The top of a new experimental rectangular laptop computer has a diagonal of 17 inches. If the length of the computer is 1 inch less than twice its width, find the dimensions of the computer.

39. **Sandbox** Jason Mahar is building a rectangular sandbox. The length is 2 feet more than the width. The diagonal is 4 feet more than the width. Find the length and width of the sandbox.

40. **Rectangular Garden** Mary Ann Tuerk has constructed a rectangular garden. The length of the garden is 3 feet more than three times its width. The diagonal of the garden is 4 feet more than three times the width. Find the length and width of the garden.

41. **Book Store** A book store owner finds that her daily profit, P, is approximated by the formula $P = x^2 - 15x - 50$, where x is the number of books she sells. How many books must she sell in a day for her profit to be $400?

42. **Water Sprinklers** The cost, C, for manufacturing x water sprinklers is given by the formula $C = x^2 - 27x - 20$. Determine the number of water sprinklers manufactured at a cost of $70.

43. **Sum of Numbers** The sum, s, of the first n even numbers is given by the formula $s = n^2 + n$. Determine n for the given sums:

a) $s = 20$ **b)** $s = 90$

44. **Telephone Lines** For a switchboard that handles n telephone lines, the maximum number of telephone connections, C, that it can make simultaneously is given by the formula

$$C = \frac{n(n-1)}{2}.$$

a) How many telephone connections can a switchboard make simultaneously if it handles 15 lines?

b) How many lines does a switchboard have if it can make 55 telephone connections simultaneously?

45. **Ladder** A 13-foot ladder is leaned against the very top of a flat-roofed building as shown in the diagram below. If the distance from the base of the ladder to the building is 7 feet less than the height of the building, determine the height of the building.

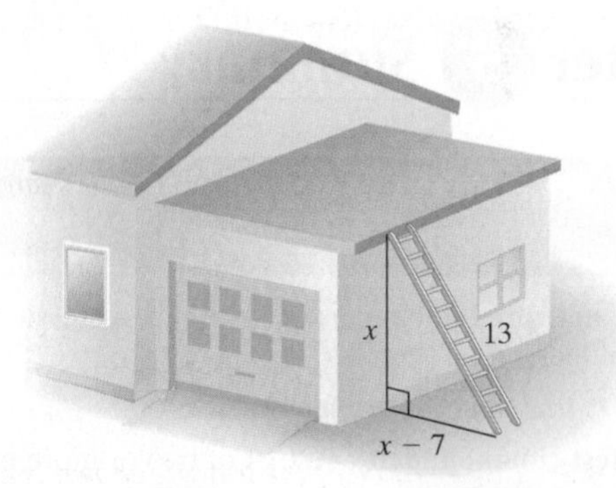

46. **Ladder** A 25-foot ladder is leaned against the very top of a flat-roofed building (see Exercise 45 for a similar diagram). If the height of the building is 17 feet more than the distance from the base of the ladder to the building, determine the height of the building.

Challenge Problems

47. **Area** Determine the value of x.

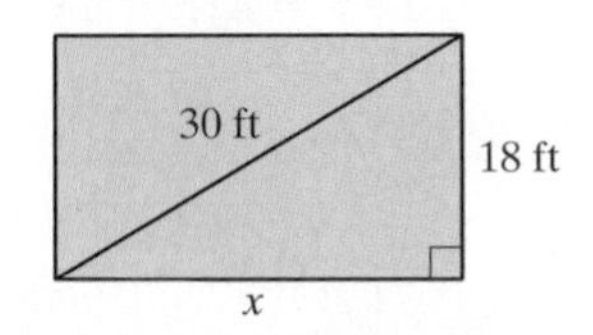

48. **Area** Determine the value of x.

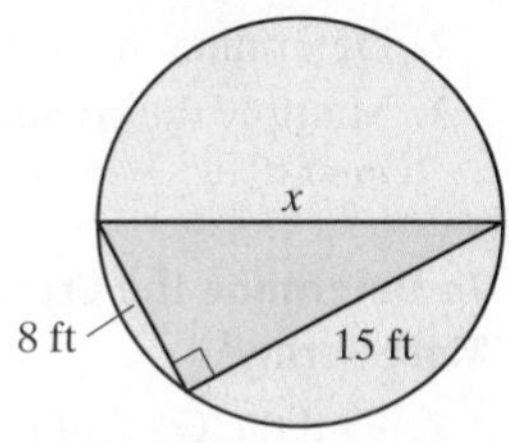

49. **Numbers** The product of two numbers is −63. Determine the numbers if their sum is −2.

50. **Numbers** The sum of two numbers is 9. The sum of the squares of the two numbers is 45. Determine the two numbers.

Group Activity

Discuss and solve Exercises 51–53 in groups.

51. **Cost and Revenue** The break-even point for a manufacturer occurs when its cost of production, C, is equal to its revenue, R. The cost equation for a company is $C = 2x^2 - 20x + 600$ and its revenue equation is $R = x^2 + 50x - 400$, where x is the number of units produced and sold. How many units must be produced and sold for the manufacturer to break even? There are two values.

52. Cannonball When a certain cannon is fired, the height, in feet, of the cannonball at time t can be found by using the formula $h = -16t^2 + 128t$.

a) Determine the height of the cannonball 3 seconds after being fired.

b) Determine the time it takes for the cannonball to hit the ground. (*Hint:* What is the value of h at impact?)

53. Wizard of Oz Near the end of the movie *The Wizard of Oz*, when the scarecrow receives his diploma, he starts talking rapidly and seems to impress us with his newfound intelligence. One of the things that the scarecrow *attempts* to say is the Pythagorean Theorem. The only problem is that he says it incorrectly! View the movie and write down exactly what the scarecrow says about the Pythagorean Theorem. Then see if you can rewrite the script to make it mathematically correct.

The scarecrow in *The Wizard of Oz*

Cumulative Review Exercises

[3.1] **54.** Express the statement "seven less than three times a number" as a mathematical expression.

[4.4] **55.** Subtract $x^2 - 4x + 6$ from $3x + 2$.

[4.5] **56.** Multiply $(3x^2 + 2x - 4)(2x - 1)$.

[4.6] **57.** Divide $\dfrac{6x^2 - 19x + 15}{3x - 5}$ by dividing the numerator by the denominator.

[5.4] **58.** Divide $\dfrac{6x^2 - 19x + 15}{3x - 5}$ by factoring the numerator and dividing out common factors.

Chapter 5 Summary

IMPORTANT FACTS AND CONCEPTS	EXAMPLES
Section 5.1	
To **factor an expression** means to write the expression as a product of its factors.	$x^2 + 2x - 35 = (x + 7)(x - 5)$
If $a \cdot b = c$, then a and b are **factors** of c.	6 and 4 are factors of 24.
The **greatest common factor (GCF)** of two or more numbers is the greatest number that divides into all the numbers.	The GCF of 36 and 48 is 12.
A **prime number** is an integer greater than 1 that has exactly two factors, itself and 1. A positive integer (other than 1) that is not prime is called **composite.** The number 1 is neither prime nor composite; it is called a **unit.**	23 is a prime number. 72 is a composite number.
To Determine the GCF of Two or More Numbers 1. Write each number as a product of prime factors. 2. Determine the prime factors common to all the numbers. 3. Multiply the common factors found in step 2. The product is the GCF.	$40 = 2^3 \cdot 5$ and $140 = 2^2 \cdot 5 \cdot 7$ The GCF of 40 and 140 is $2^2 \cdot 5 = 4 \cdot 5 = 20$.
To Determine the Greatest Common Factor of Two or More Terms 1. Find the GCF of the numerical coefficients of the terms. 2. Find the largest power of each variable that is common to all of the terms. 3. The GCF is the product of the number from step 1 and the variable expressions from step 2.	The GCF of $6xy^2$, $4x^3y^4$, and $8x^2y^3$ is $2xy^2$.
To Factor a Monomial from a Polynomial 1. Determine the greatest common factor of all terms in the polynomial. 2. Write each term as the product of the GCF and its other factor. 3. Use the distributive property to factor out the GCF.	$6a^4 + 27a^3 - 18a^2 = 3a^2(2a^2 + 9a - 6)$

IMPORTANT FACTS AND CONCEPTS	EXAMPLES
Section 5.2	
To Factor a Four-Term Polynomial Using Grouping 1. If all the terms have a GCF other than 1, factor it out. 2. If necessary, arrange the four terms so that the first two terms have a common factor and the last two have a common factor. 3. Use the distributive property to factor each group of two terms. 4. Factor the GCF from the results of step 3.	$xy + 5x - 3y - 15 = x(y + 5) - 3(y + 5)$ $= (y + 5)(x - 3)$
Section 5.3	
To Factor Trinomials of the Form $ax^2 + bx + c$, where $a = 1$ 1. Find two numbers whose product equals the constant, c, and whose sum equals the coefficient of the x-term, b. 2. Use the two numbers found in step 1, including their signs, to write the trinomial in factored form. The trinomial in factored form will be $(x + \text{first number})(x + \text{second number})$.	Factor $x^2 + 5x - 36$. The two numbers whose product is -36 and whose sum is 5 are 9 and -4. Therefore, $x^2 + 5x - 36 = (x + 9)(x - 4)$.
A **prime polynomial** is a polynomial that cannot be factored using only integer coefficients	$x^2 - 7x + 11$ is a prime polynomial.
Section 5.4	
To Factor Trinomials of the Form $ax^2 + bx + c$, $a \neq 1$, by Trial and Error 1. If all the terms have a GCF other than 1, factor it out. 2. Write all pairs of factors of the coefficient of the squared term, a. 3. Write all pairs of factors of the constant term, c. 4. Try various combinations of these factors until the correct middle term, bx, is found.	Factor $3x^2 - 2x - 8$. (see table below) Therefore, $3x^2 - 2x - 8 = (3x + 4)(x - 2)$.

Factors of -8	Possible Factors	Sum of the Products of the Outer and Inner Terms
$-1(8)$	$(3x - 1)(x + 8)$	$23x$
$-2(4)$	$(3x - 2)(x + 4)$	$10x$
$-4(2)$	$(3x - 4)(x + 2)$	$2x$
$-8(1)$	$(3x - 8)(x + 1)$	$-5x$
$4(-2)$	$(3x + 4)(x - 2)$	$-2x$

IMPORTANT FACTS AND CONCEPTS	EXAMPLES
To Factor Trinomials of the Form $ax^2 + bx + c$, $a \neq 1$, by Grouping 1. If all the terms have a GCF other than 1, factor it out. 2. Find two numbers whose product is equal to the product of a times c, and whose sum is equal to b. 3. Rewrite the middle term, bx, as the sum or difference of two terms using the numbers found in step 2. 4. Factor by grouping as explained in Section 5.2.	Factor $4x^2 + 19x - 30$. $ac = 4(-30) = -120$ Two numbers whose product is ac, or -120, and whose sum is b, or 19, are 24 and -5. $4x^2 + 19x - 30 = 4x^2 + \overbrace{24x - 5x}^{19x} - 30$ $= 4x(x + 6) - 5(x + 6)$ $= (x + 6)(4x - 5)$
Section 5.5	
Difference of Two Squares $a^2 - b^2 = (a + b)(a - b)$	$y^2 - 49 = (y + 7)(y - 7)$
Sum of Two Cubes $a^3 + b^3 = (a + b)(a^2 - ab + b^2)$	$8p^3 + q^3 = (2p)^3 + (q)^3$ $= (2p + q)[(2p)^2 - (2p)(q) + (q)^2]$ $= (2p + q)(4p^2 - 2pq + q^2)$

IMPORTANT FACTS AND CONCEPTS	EXAMPLES
Section 5.5 (Continued)	
Difference of Two Cubes $a^3 - b^3 = (a - b)(a^2 + ab + b^2)$	$8p^3 - q^3 = (2p)^3 - (q)^3$ $= (2p - q)[(2p)^2 + (2p)(q) + (q)^2]$ $= (2p - q)(4p^2 + 2pq + q^2)$
General Procedure for Factoring a Polynomial **1.** If all the terms have a GCF other than 1, factor it out. **2.** If the polynomial has two terms, determine whether it is a difference of two squares or a sum or a difference of two cubes. If so, factor using the appropriate formula from Section 5.5. **3.** If the polynomial has three terms, factor the trinomial using the methods discussed in Section 5.3 and 5.4. **4.** If the polynomial has more than three terms, try factoring by grouping as discussed in Section 5.2. **5.** Examine your factored polynomial to determine whether the terms in any factors have a common factor. If you find a common factor, factor it out.	Factor $3x^2 - 48$. $3x^2 - 48 = 3(x^2 - 16)$ $= 3(x + 4)(x - 4)$
Section 5.6	
Quadratic Equation Quadratic equations have the form $ax^2 + bx + c = 0$ where a, b, and c are real numbers, $a \neq 0$.	$6x^2 - 7x + 3 = 0$ is a quadratic equation in **standard form**.
Zero-Factor Property If $m \cdot n = 0$, then $m = 0$ or $n = 0$.	If $(x + 3)(x - 1) = 0$, then $x + 3 = 0$ or $x - 1 = 0$.
To Solve a Quadratic Equation Using Factoring **1.** Write the equation in standard form with the squared term having a positive coefficient. **2.** Factor the side of the equation that is not 0. **3.** Set each factor *containing a variable* equal to 0 and solve each equation. **4.** Check each solution found in step 3 in the *original* equation.	Solve the equation $x^2 - 3x - 52 = -12$. $x^2 - 3x - 52 = -12$ $x^2 - 3x - 40 = 0$ $(x - 8)(x + 5) = 0$ $x - 8 = 0$ or $x + 5 = 0$ $x = 8$ $\quad$ $x = -5$
Section 5.7	
A **right triangle** is a triangle that contains a right, or 90°, angle.	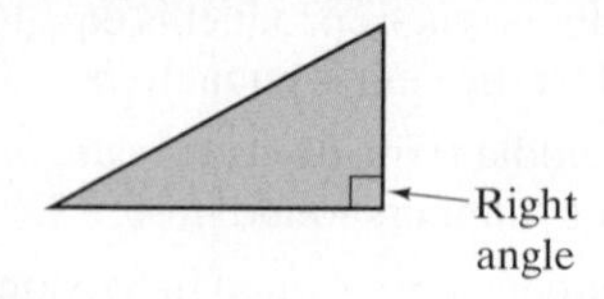
Pythagorean Theorem If a and b represent the legs of a right triangle, and c represents the hypotenuse, then $a^2 + b^2 = c^2$ 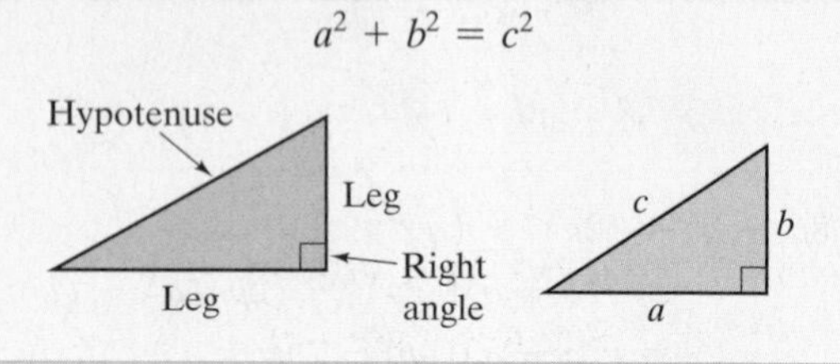 	Determine if the following triangle is a right triangle. 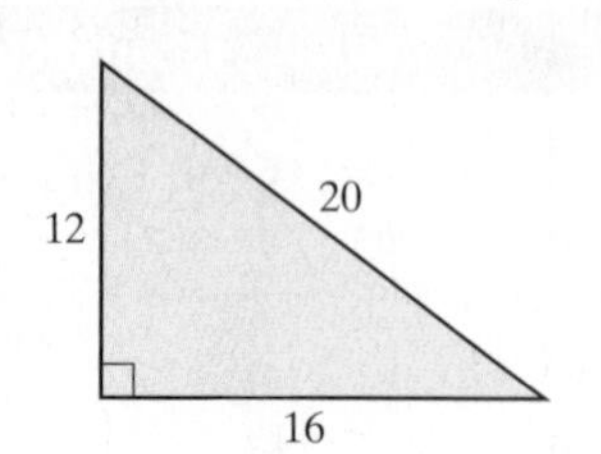 $(16)^2 + (12)^2 \stackrel{?}{=} (20)^2$ $256 + 144 \stackrel{?}{=} 400$ $400 = 400$ Yes, this is a right triangle.

Chapter 5 Review Exercises

[5.1] *Find the greatest common factor for each set of terms.*

1. $3y^5, y^4, 2y^3$
2. $3p, 6p^2, 9p^3$
3. $18c^4, 12c^2, 30c^5$
4. $20x^2y^3, 25x^3y^4, 10x^5y^2z$
5. $9xyz, 12xz, 36, x^2y$
6. $9st, 16s^2t, 24s, s^3t^3$
7. $8(x-3), x-3$
8. $x(x+5), x+5$

Factor each expression. If an expression is prime, so state.

9. $7x - 35$
10. $35x - 5$
11. $24y^2 - 4y$
12. $55p^3 - 20p^2$
13. $60a^2b - 36ab^2$
14. $9xy - 36x^3y^2$
15. $20x^3y^2 + 8x^9y^3 - 16x^5y^2$
16. $24x^2 - 13y^2 + 6xy$
17. $14a^2b - 7b - a^3$
18. $x(5x+3) - 2(5x+3)$
19. $3t(t-1) + 4(t-1)$
20. $2x(4x-3) + 4x - 3$

[5.2] *Factor by grouping.*

21. $x^2 + 6x + 2x + 12$
22. $x^2 - 5x + 4x - 20$
23. $y^2 - 6y - 6y + 36$
24. $3xy + 3x + 2y + 2$
25. $4a^2 - 4ab - a + b$
26. $2x^2 + 12x - x - 6$
27. $x^2 + 3x - 2xy - 6y$
28. $5x^2 - xy + 20xy - 4y^2$
29. $4x^2 + 12xy - 5xy - 15y^2$
30. $6a^2 - 10ab - 3ab + 5b^2$
31. $pq - 3q + 4p - 12$
32. $3x^2 - 9xy + 2xy - 6y^2$
33. $3x^3 - 4x^2 + 15x - 20$
34. $2x^3 - 6x^2 - 7x + 21$

[5.3] *Factor completely. If an expression is prime, so state.*

35. $x^2 + 5x + 6$
36. $x^2 + 4x - 15$
37. $x^2 + 11x + 18$
38. $n^2 + 3n - 40$
39. $b^2 + b - 20$
40. $x^2 - 15x + 56$
41. $c^2 - 10c - 20$
42. $y^2 - 10y - 22$
43. $x^3 - 17x^2 + 72x$
44. $t^3 - 5t^2 - 36t$
45. $x^2 - 2xy - 15y^2$
46. $4x^3 + 32x^2y + 60xy^2$

[5.4] *Factor completely. If an expression is prime, so state.*

47. $2x^2 - x - 15$
48. $6x^2 - 29x - 5$
49. $4x^2 - 9x + 5$
50. $5m^2 - 14m + 8$
51. $16y^2 + 8y - 3$
52. $5x^2 - 32x + 12$
53. $2t^2 + 14t + 9$
54. $5x^2 + 37x - 24$
55. $6s^2 + 13s + 5$
56. $6x^2 + 11x - 10$
57. $12x^2 + 2x - 4$
58. $25x^2 - 30x + 9$
59. $9x^3 - 12x^2 + 4x$
60. $18x^3 + 12x^2 - 16x$
61. $4a^2 - 16ab + 15b^2$
62. $16a^2 - 22ab - 3b^2$

[5.5] *Factor completely.*

63. $x^2 - 100$
64. $x^2 - 36$
65. $3x^2 - 48$
66. $81x^2 - 9y^2$
67. $81 - a^2$
68. $64 - x^2$
69. $16x^4 - 49y^2$
70. $64x^6 - 49y^6$
71. $a^3 + b^3$
72. $x^3 - y^3$
73. $x^3 - 1$
74. $x^3 + 8$
75. $a^3 + 27$
76. $b^3 - 64$
77. $125a^3 + b^3$
78. $27 - 8y^3$
79. $3x^3 - 192y^3$
80. $27x^4 - 75y^2$

[5.1–5.5] *Factor completely.*

81. $x^2 - 14x + 48$
82. $3x^2 - 18x + 27$
83. $5q^2 - 5$
84. $8x^2 + 16x - 24$
85. $4y^2 - 36$
86. $x^2 - 6x - 27$
87. $9x^2 - 6x + 1$
88. $7x^2 + 25x - 12$
89. $6b^3 - 6$
90. $x^3y - 27y$
91. $a^2b - 2ab - 15b$
92. $6x^3 + 30x^2 + 9x^2 + 45x$
93. $x^2 - 4xy + 3y^2$
94. $3m^2 + 2mn - 8n^2$
95. $4x^2 + 12xy + 9y^2$
96. $25a^2 - 49b^2$
97. $xy - 7x + 2y - 14$
98. $16y^5 - 25y^7$
99. $6x^2 + 5xy - 21y^2$
100. $4x^3 + 18x^2y + 20xy^2$
101. $16x^4 - 8x^3 - 3x^2$
102. $d^4 - 16$

[5.6] *Solve.*

103. $x(x+9) = 0$
104. $(a-2)(a+6) = 0$
105. $(x+5)(4x-3) = 0$
106. $x^2 + 7x = 0$
107. $6x^2 + 30x = 0$
108. $6x^2 + 18x = 0$

109. $r^2 + 9r + 18 = 0$

110. $x^2 - 3x = -2$

111. $x^2 - 12 = -x$

112. $15x + 12 = -3x^2$

113. $x^2 - 6x + 8 = 0$

114. $3p^2 + 6p = 45$

115. $8x^2 - 3 = -10x$

116. $3p^2 - 11p = 4$

117. $4x^2 - 16 = 0$

118. $49x^2 - 100 = 0$

119. $8x^2 - 14x + 3 = 0$

120. $-48x = -12x^2 - 45$

[5.7]

121. State the Pythagorean Theorem.

122. What is the longest side of a right triangle called?

In Exercises 123 and 124, determine the value of x.

123.

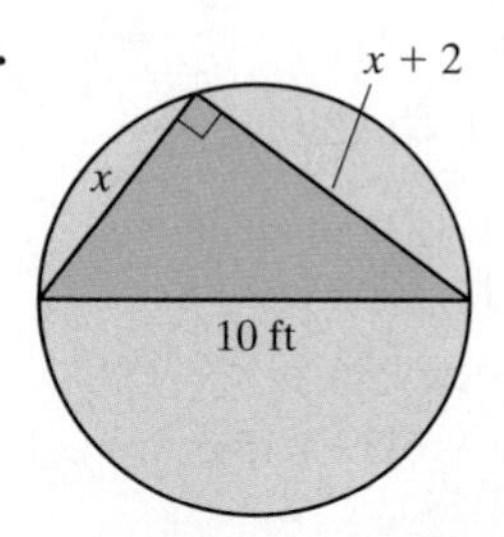

124.

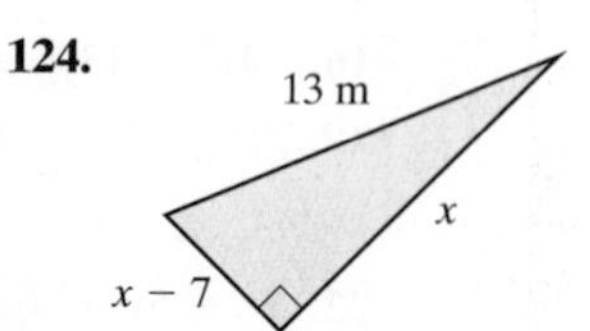

Express each problem as an equation, then solve.

125. Product of Integers The product of two consecutive positive odd integers is 99. Determine the two integers.

126. Product of Integers The product of two positive integers is 56. Determine the integers if the larger is 6 more than twice the smaller.

127. Area of Rectangle The area of a rectangle is 180 square feet. Determine the length and width of the rectangle if the length is 3 feet greater than the width.

128. Right Triangle One leg of a right triangle is 7 feet longer than the other leg. The hypotenuse is 9 feet longer than the shortest leg. Find the lengths of the three sides of the triangle.

129. Square The length of each side of a square is made smaller by 4 inches. If the area of the resulting square is 25 square inches, determine the length of a side of the original square.

130. Table Brian has a rectangular table. The length of the table is 2 feet greater than the width of the table. A diagonal across the table is 4 feet greater than the width of the table. Find the length of the diagonal across the table.

131. Falling Pear How long would it take a pear that falls off a 16-foot tree to hit the ground?

132. Baking Cookies The Pine Hills Neighborhood Association has determined that the cost, C, to make x dozen cookies can be estimated by the formula $C = x^2 - 79x + 20$. If they have $100 to be used to make the cookies, how many dozen cookies can the association make to sell at a fund-raiser?

Chapter 5 Practice Test

Chapter Test Prep Videos provide fully worked-out solutions to any of the exercises you want to review. Chapter Test Prep Videos are available via MyMathLab®, *or on* YouTube (www.youtube.com/user/AngelElementaryAlg).

1. Determine the greatest common factor of $9y^5$, $15y^3$, and $27y^4$.
2. Determine the greatest common factor of $8p^3q^2$, $32p^2q^5$, and $24p^4q^3$.

Factor completely.

3. $5x^2y^3 - 15x^5y^2$
4. $8a^3b - 12a^2b^2 + 28a^2b$
5. $4x^2 - 20x + x - 5$
6. $a^2 - 4ab - 5ab + 20b^2$
7. $r^2 + 5r - 24$
8. $25a^2 - 5ab - 6b^2$
9. $4x^2 - 16x - 48$
10. $2y^3 - y^2 - 3y$
11. $12x^2 - xy - 6y^2$
12. $x^2 - 9y^2$
13. $x^3 - 64$

Solve.

14. $(6x - 5)(x + 3) = 0$

15. $x^2 - 6x = 0$

16. $x^2 = 64$

17. $x^2 + 18x + 81 = 0$

18. $x^2 - 7x + 12 = 0$

19. $x^2 + 6 = -5x$

20. Right Triangle Determine the value of x.

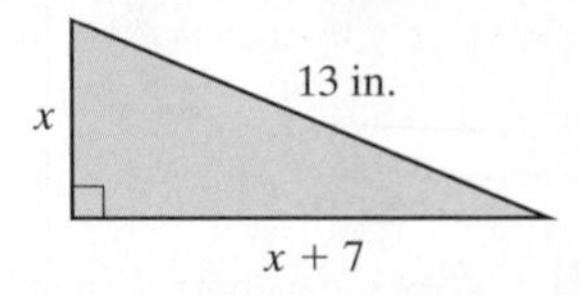

21. Right Triangle In a right triangle, one leg is 2 feet less than twice the length of the smaller leg. The hypotenuse is 2 feet more than twice the length of the smaller leg. Determine the hypotenuse of the triangle.

22. Product of Integers The product of two positive integers is 36. Determine the two integers if the larger is 1 more than twice the smaller.

23. Consecutive Even Integers The product of two positive consecutive even integers is 168. Determine the integers.

24. Rectangle The area of a rectangle is 24 square meters. Determine the length and width of the rectangle if its length is 2 meters greater than its width.

25. Fallen Object How long would it take for an object dropped from a hot air balloon to fall 1600 feet to the ground?

Cumulative Review Test

Take the following test and check your answers with those given in the back of the book. Review any questions that you answered incorrectly. The section where the material was covered is indicated after the answer.

1. Evaluate $4 - 5(2x + 4x^2 - 21)$ when $x = -4$.

2. Evaluate $5x^2 - 3y + 7(2 + y^2 - 4x)$ when $x = 3$ and $y = -2$.

3. **Motel Room** The cost of a motel room including a 12% state tax and 3% county tax is \$103.50. Determine the cost of the room before tax.

4. Consider the set of numbers

$$\left\{-6, -0.2, \frac{3}{5}, \sqrt{7}, -\sqrt{2}, 7, 0, -\frac{5}{9}, 1.34\right\}$$

List the elements that are

a) natural numbers.

b) rational numbers.

c) irrational numbers.

d) real numbers.

5. Which is greater, $|-8|$ or $-|8|$? Explain your answer.

6. Solve the equation $4x - 2 = 4(x - 7) + 2x$ for x.

7. Solve the proportion $\frac{5}{12} = \frac{8}{x}$ for x by cross-multiplying.

8. Solve the inequality $3x - 5 \geq 10(6 - x)$. Graph the solution on a number line and write the solution in interval notation.

9. Solve the equation $4x + 3y = 7$ for y. Write your answer in $y = mx + b$ form.

10. **Acid Solution** How many liters of a 10% acid solution must be mixed with three liters of a 4% acid solution to get an 8% acid solution?

11. **Consecutive Odd Integers** The sum of two consecutive odd integers is 96. Find the two integers.

12. **Cross-Country Skiing** Two cross-country skiers follow the same trail in a local park. Brooke Stoner skis at a rate of 8 kilometers per hour and Bob Thoresen skis at a rate of 4 kilometers per hour. How long will it take Brooke to catch Bob if she leaves 15 minutes after he does?

13. Simplify $\left(\frac{4x^3}{9y^4}\right)^2$.

14. Simplify $(2x^{-3})^{-2}(4x^{-3}y^2)^3$.

15. Subtract $(4x^3 - 3x^2 + 7)$ from $(x^3 - x^2 + 6x - 5)$.

Perform the operations indicated.

16. $(3x - 2)(x^2 + 5x - 6)$

17. $\dfrac{x^2 - 2x + 6}{x + 3}$

18. Factor $qr + 2q - 8r - 16$ by grouping.

19. Factor $5x^2 - 7x - 6$.

20. Factor $7y^3 - 63y$.

6 Rational Expressions and Equations

When two people paint a house, the job gets done faster than if each person painted the house separately. In Exercise 27 on page 399, we will see how algebra can be used to determine how much faster the job gets done.

Goals of This Chapter

You worked with rational numbers when you worked with fractions in arithmetic. Fractions that contain variables are often referred to as rational expressions. The same basic procedures that you used with arithmetic fractions will be used with rational expressions. You might wish to review Section 1.3 since the material presented in this chapter builds upon the procedures presented there.

Equations that contain rational expressions are called rational equations. We will solve rational equations in Section 6.6.

To be successful in this chapter, you need to have a complete understanding of factoring, which was presented in Chapter 5. Sections 5.3 and 5.4 are especially important.

6.1 Simplifying Rational Expressions

1. **Determine the values for which a rational expression is defined.**
2. **Understand the three signs of a fraction.**
3. **Simplify rational expressions.**
4. **Factor a negative 1 from a polynomial.**

1 Determine the Values for Which a Rational Expression Is Defined

We begin this chapter by defining a *rational expression.*

Rational Expression

A **rational expression** is an expression of the form $\frac{p}{q}$, where p and q are polynomials and $q \neq 0$.

Examples of Rational Expressions

$$\frac{4}{5}, \quad \frac{x+3}{x}, \quad \frac{x^2+8x}{x-3}, \quad \frac{x}{x^2-4}$$

The denominator of a rational expression cannot equal 0 since division by 0 is not defined.

Expression	Defined	Undefined
$\frac{x+3}{x}$	for all real numbers except 0	when $x = 0$
$\frac{x^2+8x}{x-3}$	for all real numbers except 3	when $x = 3$
$\frac{x}{x^2-4}$	for all real numbers except 2 and -2	when $x = 2$ or $x = -2$

Whenever a rational expression has a variable in the denominator, we always assume that the value or values of the variable that make the denominator 0 are excluded.

One method that can be used to determine the value or values of the variable that are excluded is to set the denominator equal to 0 and then solve the resulting equation for the variable.

EXAMPLE 1 Determine the value or values of the variable for which the rational expression is defined.

a) $\frac{7}{x-4}$ **b)** $\frac{3x+4}{2x-7}$ **c)** $\frac{x+3}{x^2+2x-3}$

Solution

a) To determine the conditions for which the denominator is zero, we set $x - 4$ equal to 0.

$$x - 4 = 0$$
$$x = 4$$

We say $\frac{7}{x-4}$ is defined for all real numbers except 4, or simply $x \neq 4$.

b) In $\frac{3x+4}{2x-7}$, we set the denominator $2x - 7$ equal to 0 to find values we need to exclude.

$$2x - 7 = 0$$
$$2x = 7$$
$$x = \frac{7}{2}$$

Thus when we consider the rational expression $\frac{3x+4}{2x-7}$, we say x cannot equal $\frac{7}{2}$, or $x \neq \frac{7}{2}$.

c) Again, we set the denominator equal to zero.

$$x^2 + 2x - 3 = 0$$
$$(x + 3)(x - 1) = 0$$
$$x + 3 = 0 \quad \text{or} \quad x - 1 = 0$$
$$x = -3 \quad \text{or} \quad x = 1$$

Thus, for $\dfrac{x + 3}{x^2 + 2x - 3}$, we say $x \neq -3$ and $x \neq 1$.

Now Try Exercise 19

In Example 1 **c)**, if we rewrite the original rational expression with the denominator in factored form we have $\dfrac{x + 3}{x^2 + 2x - 3} = \dfrac{x + 3}{(x + 3)(x - 1)}$. Notice that the factor $x + 3$ appears in both the numerator and the denominator. We will discuss how to simplify such rational expressions in Objective 3. For now, we need to realize this rational expression is not defined for any number that makes the denominator equal 0. Thus, this rational expression is not defined for $x = -3$ or $x = 1$.

Understanding Algebra

Since changing any two of a fraction's three signs does not change the fraction, we can write the following:

$$\frac{3}{-4} = \frac{-3}{4} = -\frac{3}{4}$$
$$\frac{-2x}{-3y^2} = -\frac{-2x}{3y^2} = \frac{2x}{3y^2}$$
$$-\frac{-4a}{-5b} = \frac{-4a}{5b} = -\frac{4a}{5b}$$

2 Understand the Three Signs of a Fraction

Three **signs** are associated with any fraction: the sign of the numerator, the sign of the denominator, and the sign of the fraction.

$$\text{Sign of fraction} \longrightarrow +\frac{-a}{+b}$$

(Sign of numerator: the sign of a; Sign of denominator: the sign of b)

Whenever any of the three signs is omitted, we assume it to be positive. For example,

$$\frac{a}{b} \quad \text{means} \quad +\frac{+a}{+b}$$
$$\frac{-a}{b} \quad \text{means} \quad +\frac{-a}{+b}$$
$$-\frac{a}{b} \quad \text{means} \quad -\frac{+a}{+b}$$

Negative Fractions

Changing any two of the three signs of a fraction does not change the value of a fraction.

$$\frac{-a}{b} = -\frac{a}{b} = \frac{a}{-b}$$

Generally, we do not write a fraction with a negative denominator. For example, the expression $\dfrac{2}{-5}$ would be written as either $\dfrac{-2}{5}$ or $-\dfrac{2}{5}$. The expression $\dfrac{x}{-(4 - x)}$ can be written $\dfrac{x}{x - 4}$ since $-(4 - x) = -4 + x$ or $x - 4$.

3 Simplify Rational Expressions

A rational expression is **simplified** or **reduced to its lowest terms** when the numerator and denominator have no common factors other than 1. The fraction $\dfrac{9}{12}$ is not simplified because 9 and 12 both contain the common factor 3. When the 3 is factored out, the simplified fraction is $\dfrac{3}{4}$.

$$\frac{9}{12} = \frac{\overset{1}{\cancel{3}} \cdot 3}{\underset{1}{\cancel{3}} \cdot 4} = \frac{3}{4}$$

Not simplified ($\frac{9}{12}$) — Simplified ($\frac{3}{4}$)

The rational expression $\frac{ab - b^2}{2b}$ is not simplified because both the numerator and denominator have a common factor, b. To simplify this expression, factor b from each term in the numerator, then divide it out.

$$\frac{ab - b^2}{2b} = \frac{\overset{1}{\cancel{b}}(a - b)}{2\underset{1}{\cancel{b}}} = \frac{a - b}{2}$$

Thus, $\frac{ab - b^2}{2b}$ becomes $\frac{a - b}{2}$ when simplified.

Understanding Algebra

We say that the simplified form of

$$\frac{ab - b^2}{2b} \text{ is } \frac{a - b}{2}.$$

We should keep in mind that in both original and simplified expressions, b cannot equal 0 because that would make the original denominator 0.

To Simplify Rational Expressions

1. Factor both the numerator and denominator as completely as possible.
2. Divide out any factors common to both the numerator and denominator.

EXAMPLE 2 Simplify $\frac{5x^3 + 10x^2 - 35x}{10x^2}$.

Solution

$$\frac{5x^3 + 10x^2 - 35x}{10x^2} \qquad \leftarrow \text{Observe that the GCF is } 5x.$$

$$= \frac{\overset{1}{\cancel{5x}}(x^2 + 2x - 7)}{\underset{1}{\cancel{5x}} \cdot 2x} \qquad \leftarrow \text{Factor } 5x \text{ from the numerator and the denominator and divide it out.}$$

$$= \frac{x^2 + 2x - 7}{2x}$$

Now Try Exercise 33

HELPFUL HINT

In Example 2 we *simplified* a polynomial divided by a monomial using factoring. In Section 4.6 we *divided* polynomials by monomials by writing each term in the numerator over the expression in the denominator. For example,

$$\frac{5x^3 + 10x^2 - 35x}{10x^2} = \frac{5x^3}{10x^2} + \frac{10x^2}{10x^2} - \frac{35x}{10x^2} = \frac{x}{2} + 1 - \frac{7}{2x}$$

The answer above, $\frac{x}{2} + 1 - \frac{7}{2x}$, is equivalent to the answer $\frac{x^2 + 2x - 7}{2x}$, which was obtained by factoring in Example 2. We show this below.

$$\frac{x}{2} + 1 - \frac{7}{2x}$$

$$= \frac{x}{x} \cdot \frac{x}{2} + \frac{2x}{2x} \cdot 1 - \frac{7}{2x} \qquad \text{Write each term with LCD } 2x.$$

$$= \frac{x^2}{2x} + \frac{2x}{2x} - \frac{7}{2x} = \frac{x^2 + 2x - 7}{2x}$$

When asked to *simplify* an expression we will factor numerators and denominators, when possible, then divide out common factors. This process was illustrated in Example 2 and will be further illustrated in Examples 3 through 5.

EXAMPLE 3 Simplify $\frac{x^2 - x - 12}{x + 3}$.

Solution Factor the numerator; then divide out the common factor.

$$\frac{x^2 - x - 12}{x + 3} = \frac{\overset{1}{\cancel{(x + 3)}}(x - 4)}{\underset{1}{\cancel{x + 3}}} = x - 4$$

Now Try Exercise 35

Notice in Example 3 that the value $x = -3$ produces a value of 0 in the denominator of the original rational expression. Therefore, both the original rational expression, $\frac{x^2 - x - 12}{x + 3}$, and the simplified expression, $x - 4$, are not defined at $x = -3$. *Whenever a rational expression is simplified, the simplified expression is not defined for the values for which the original rational expressions is not defined.*

EXAMPLE 4 Simplify $\frac{r^2 - 25}{r - 5}$.

Solution Factor the numerator; then divide out common factors.

$$\frac{r^2 - 25}{r - 5} = \frac{(r + 5)\cancel{(r - 5)}^{1}}{\cancel{r - 5}_{1}} = r + 5$$

Now Try Exercise 55

EXAMPLE 5 Simplify $\frac{3x^2 - 10x - 8}{x^2 + 3x - 28}$.

Solution Factor both the numerator and denominator, then divide out common factors.

$$\frac{3x^2 - 10x - 8}{x^2 + 3x - 28} = \frac{(3x + 2)\cancel{(x - 4)}^{1}}{(x + 7)\cancel{(x - 4)}_{1}} = \frac{3x + 2}{x + 7}.$$

Note that $\frac{3x + 2}{x + 7}$ cannot be simplified any further.

Now Try Exercise 41

AVOIDING COMMON ERRORS

Remember: Only common factors can be divided out from expressions.

CORRECT	INCORRECT
$\frac{\overset{5}{\cancel{20}}\ \overset{x}{\cancel{x^2}}}{\underset{1}{\cancel{4}}\ \underset{1}{\cancel{x}}} = 5x$	 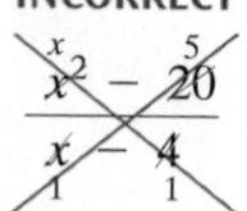

In the denominator of the example on the left, $4x$, the 4 and x are factors since they are *multiplied* together. The 4 and the x are also both factors of the numerator $20x^2$, since $20x^2$ can be written $4 \cdot x \cdot 5x$.

Some students incorrectly divide out *terms*. In the expression $\frac{x^2 - 20}{x - 4}$, the x and -4 are *terms* of the denominator, not factors, and therefore cannot be divided out.

4 Factor a Negative 1 from a Polynomial

Recall from Section 5.2 that when -1 is factored from a polynomial, the sign of each term in the polynomial changes.

Examples

$$-3x + 10 = -1(3x - 10) = -(3x - 10)$$
$$5 - 2x = -1(-5 + 2x) = -(2x - 5)$$
$$-2x^2 + 3x - 4 = -1(2x^2 - 3x + 4) = -(2x^2 - 3x + 4)$$

Whenever the terms in a numerator and denominator differ only in their signs, we can factor out -1 from either the numerator or denominator and then divide out the common factor.

EXAMPLE 6 Simplify $\frac{3x-8}{8-3x}$.

Solution Since each term in the numerator differs only in sign from its like term in the denominator, we will factor -1 from each term in the denominator.

Understanding Algebra

An expression divided by its *additive inverse* (or *opposite*) is -1:

$$\frac{a-b}{b-a} = -1 \text{ (provided } a \neq b)$$

$$\frac{3x-8}{8-3x} = \frac{3x-8}{-1(-8+3x)}$$

$$= \frac{\overset{1}{\cancel{3x-8}}}{-\underset{1}{\cancel{(3x-8)}}}$$

$$= -1$$

Now Try Exercise 57

EXAMPLE 7 Simplify $\frac{4n^2-23n-6}{6-n}$.

Understanding Algebra

In Example 7 we simplified $\frac{4n^2-23n-6}{6-n}$ to $-(4n+1)$. Notice $n \neq 6$ in the original expression. Therefore, it is assumed that $n \neq 6$ in the simplified expression, $-(4n+1)$, as well.

Solution $\frac{4n^2-23n-6}{6-n} = \frac{(4n+1)(n-6)}{6-n}$ The terms in $n-6$ differ only in sign from the terms in $6-n$.

$= (4n+1)(-1)$ Replaced $\frac{n-6}{6-n}$ with -1.

$= -(4n+1)$

Note that $-4n-1$ is also an acceptable answer.

Now Try Exercise 63

6.1 Exercise Set MathXL® MyMathLab®

Warm-Up Exercises

Fill in the blanks with the appropriate word, phrase, or symbol(s) from the following list.

$-\frac{2}{3}$	$\frac{2}{3}$	3	-3	-1	1
rational	reduced	excluded	included	common	radical

1. An expression of the form $\frac{p}{q}$, where p and q are polynomials and $q \neq 0$, is called a ________ expression.
2. Whenever a rational expression has a variable in the denominator, we always assume that the value(s) of the variable that make the denominator 0 are ________.
3. The fraction $\frac{-2}{-3}$ is equivalent to ________.
4. The fraction $\frac{2}{-3}$ is equivalent to ________.
5. A rational expression in which the numerator and denominator have no common factors is simplified or ________ to its lowest terms.
6. To simplify a rational expression, first factor both the numerator and denominator, then divide out any factors ________ to both the numerator and denominator.
7. The expression $\frac{x+4}{x+4}$ simplifies to ________.
8. The expression $\frac{x-4}{4-x}$ simplifies to ________.
9. The expression $\frac{5}{x-3}$ is defined for all real numbers except ________.
10. The expression $\frac{5}{x+3}$ is defined for all real numbers except ________.

Practice the Skills

Determine the value or values of the variable for which each expression is defined.

11. $\frac{5}{x-1}$

12. $\frac{6}{x-3}$

13. $\frac{x-2}{x}$

14. $\dfrac{x^2 + 3x - 4}{x}$

15. $\dfrac{7}{4n - 16}$

16. $\dfrac{7}{2x - 3}$

17. $\dfrac{x + 4}{x^2 - 4}$

18. $\dfrac{x^2 + 3x - 4}{x^2 + 4x - 5}$

19. $\dfrac{2x - 3}{2x^2 - 9x + 9}$

20. $\dfrac{x - 5}{2x^2 - 13x + 15}$

21. $\dfrac{x}{x^2 + 36}$

22. $\dfrac{x + 3}{x^2 + 9}$

23. $\dfrac{p + 8}{4p^2 - 25}$

24. $\dfrac{10}{9r^2 - 16}$

Simplify the following rational expressions that contain a monomial divided by a monomial. This material was covered in Sections 4.1 and 4.2, and it will help prepare you for the next section.

25. $\dfrac{8x^3y}{24x^2y^5}$

26. $\dfrac{18x^3y^2}{30x^4y^5}$

27. $\dfrac{(2a^4b^5)^3}{2a^{12}b^{20}}$

28. $\dfrac{(5r^2s^3)^2}{(3r^4s)^3}$

Simplify.

29. $\dfrac{5x}{5x + 15}$

30. $\dfrac{7x}{7x - 63}$

31. $\dfrac{5x + 15}{x + 3}$

32. $\dfrac{8x - 64}{x - 8}$

33. $\dfrac{2x^3 + 10x^2 - 22x}{8x^2}$

34. $\dfrac{5y^3 - 15y^2 + 35y}{30y^3}$

35. $\dfrac{r^2 - r - 2}{r - 2}$

36. $\dfrac{16x^2 + 24x + 9}{4x + 3}$

37. $\dfrac{x^2 + 2x}{x^2 + 4x + 4}$

38. $\dfrac{x^2 + 3x - 18}{4x - 12}$

39. $\dfrac{z^2 - 10z + 25}{z^2 - 25}$

40. $\dfrac{k^2 - 6k + 9}{k^2 - 9}$

41. $\dfrac{x^2 - 2x - 3}{x^2 - x - 6}$

42. $\dfrac{x^2 - x - 2}{x^2 + 3x + 2}$

43. $\dfrac{5x^3 + 30x^2}{5x^3 - 30x^2}$

44. $\dfrac{3x^2 + 6x}{3x^2 + 9x}$

45. $\dfrac{m - 2}{4m^2 - 13m + 10}$

46. $\dfrac{b - 6}{b^2 - 8b + 12}$

47. $\dfrac{x^2 - 25}{(x + 5)^2}$

48. $\dfrac{2x^2 - 13x + 21}{(x - 3)^2}$

49. $\dfrac{6x^2 - 13x + 6}{3x - 2}$

50. $\dfrac{6t^2 - 7t - 5}{3t - 5}$

51. $\dfrac{2x^2 - 8x + 3x - 12}{2x^2 + 8x + 3x + 12}$

52. $\dfrac{x^2 - 2x + 4x - 8}{2x^2 + 3x + 8x + 12}$

53. $\dfrac{a^3 - 8}{a - 2}$

54. $\dfrac{y^3 + 1}{y + 1}$

55. $\dfrac{9s^2 - 16t^2}{3s - 4t}$

56. $\dfrac{a + 6b}{a^2 - 36b^2}$

57. $\dfrac{4x - 8}{8 - 4x}$

58. $\dfrac{10a - 6}{3 - 5a}$

59. $\dfrac{x^2 - 2x - 8}{4 - x}$

60. $\dfrac{7 - s}{s^2 - 12s + 35}$

61. $\dfrac{x^2 + 3x - 18}{-2x^2 + 6x}$

62. $\dfrac{x^2 + 3x - 28}{-5x^2 - 35x}$

63. $\dfrac{2x^2 + 5x - 3}{1 - 2x}$

64. $\dfrac{3x^2 + 11x - 4}{1 - 3x}$

65. $\dfrac{p^3 - q^3}{q^2 - p^2}$

66. $\dfrac{r^3 - s^3}{s^2 - r^2}$

Problem Solving

Simplify the following expressions, if possible. Treat the unknown symbol as if it were a variable.

67. $\dfrac{3☺}{15}$

68. $\dfrac{☺}{☺ + 8☺^2}$

69. $\dfrac{7\Delta}{14\Delta + 63}$

70. $\dfrac{\Delta^2 + 2\Delta}{\Delta^2 + 4\Delta + 4}$

71. $\dfrac{3\Delta - 4}{4 - 3\Delta}$

72. $\dfrac{(\Delta - 3)^2}{\Delta^2 - 6\Delta + 9}$

Determine the denominator that will make each statement true. Explain how you obtained your answer.

73. $\dfrac{x^2 - x - 6}{\blacksquare} = x - 3$

74. $\dfrac{2x^2 + 11x + 12}{\blacksquare} = 2x + 3$

Determine the numerator that will make the statement true. Explain how you obtained your answer.

75. $\dfrac{\blacksquare}{x + 4} = x + 5$

Concept/Writing Exercises

76. Explain how to simplify a rational expression.

77. Explain how to determine the value or values of the variable that makes a rational expression undefined.

Challenge Problems

In Exercises 78 and 79, **a)** *determine the value or values that x cannot represent.* **b)** *Simplify the expression.*

78. $\dfrac{x - 4}{2x^2 - 5x - 8x + 20}$

79. $\dfrac{x + 5}{2x^3 + 7x^2 - 15x}$

Simplify. Explain how you determined your answer.

80. $\dfrac{\frac{1}{6}x^5 - \frac{2}{3}x^4}{x^4}$

81. $\dfrac{\frac{1}{5}x^5 - \frac{2}{3}x^4}{\frac{1}{5}x^5 - \frac{2}{3}x^4}$

82. $\dfrac{\frac{1}{5}x^5 - \frac{2}{3}x^4}{\frac{2}{3}x^4 - \frac{1}{5}x^5}$

Group Activity

Discuss and answer Exercise 83 as a group.

83. **a)** As a group, determine the values of the variable where the expression $\dfrac{x^2 - 25}{x^3 + 2x^2 - 15x}$ is undefined.

b) As a group, simplify the rational expression.

c) Group member 1: Substitute 6 in the *original expression* and evaluate.

d) Group member 2: Substitute 6 in the *simplified expression* from part **b)** and compare your result to that of Group member 1.

e) Group member 3: Substitute −2 in the original expression and in the simplified expression in part **b)**. Compare your answers.

f) As a group, discuss the results of your work in parts **c)–e)**.

g) Now, as a group, substitute −5 in the original expression and in the simplified expression. Discuss your results.

h) Is $\dfrac{x^2 - 25}{x^3 + 2x^2 - 15x}$ always equal to its simplified form for *any* value of x? Explain your answer.

Cumulative Review Exercises

[2.6] 84. Solve the formula $z = \dfrac{x - y}{4}$ for y.

[3.3] 85. **Triangle** Find the measures of the three angles of a triangle if one angle is 30° greater than the smallest angle, and the third angle is 10° greater than 3 times the smallest angle.

[4.1] 86. Simplify $\left(\dfrac{5x^2y^2}{9x^4y^3}\right)^2$.

[4.4] 87. Subtract $3x^2 - 4x - 8 - (-5x^2 + 6x + 11)$.

[5.3] 88. Factor $3a^2 - 6a - 72$ completely.

[5.7] 89. Find the length of the hypotenuse of the right triangle.

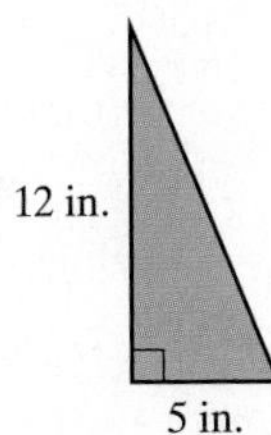

6.2 Multiplication and Division of Rational Expressions

1. Multiply rational expressions.
2. Divide rational expressions.

1 Multiply Rational Expressions

In Section 1.3 we reviewed multiplication of numerical fractions. Recall that to multiply two fractions we multiply their numerators together and multiply their denominators together.

To Multiply Two Fractions

$$\frac{a}{b}\cdot\frac{c}{d}=\frac{a\cdot c}{b\cdot d},\quad b\neq 0 \text{ and } d\neq 0$$

EXAMPLE 1 Multiply $\left(\frac{3}{5}\right)\left(\frac{-2}{9}\right)$.

Solution First divide out common factors; then multiply.

$$\frac{\overset{1}{\cancel{3}}}{5}\cdot\frac{-2}{\underset{3}{\cancel{9}}}=\frac{1\cdot(-2)}{5\cdot 3}=-\frac{2}{15}$$

Now Try Exercise 7

Understanding Algebra

Multiplying rational expressions in algebra is just like multiplying fractions in arithmetic.

Simplify by dividing out common factors first, then multiply numerators and multiply denominators.

Arithmetic:

$$\frac{2}{3}\times\frac{6}{7}=\frac{2\times\overset{2}{\cancel{6}}}{\underset{1}{\cancel{3}}\times 7}=\frac{4}{7}$$

Algebra:

$$\frac{x}{3}\cdot\frac{6}{7}=\frac{x\cdot\overset{2}{\cancel{6}}}{\underset{1}{\cancel{3}}\cdot 7}=\frac{2x}{7}$$

The same principles apply when multiplying rational expressions containing variables. Before multiplying, you should first divide out any factors common to both a numerator and a denominator.

To Multiply Rational Expressions

1. Factor all numerators and denominators completely.
2. Divide out common factors.
3. Multiply numerators together and multiply denominators together.

EXAMPLE 2 Multiply $\frac{3x^2}{2y}\cdot\frac{4y^3}{3x}$.

Solution This problem can be represented as

$$\frac{3x^2}{2y}\cdot\frac{4y^3}{3x}=\frac{3xx}{2y}\cdot\frac{4yyy}{3x}$$

$$=\frac{\overset{1}{\cancel{3}}\overset{1}{\cancel{x}}x}{2y}\cdot\frac{4yyy}{\underset{1}{\cancel{3}}\underset{1}{\cancel{x}}}\quad \text{Divide out the 3's and } x\text{'s.}$$

$$=\frac{\overset{1}{\cancel{3}}\overset{1}{\cancel{x}}x}{\underset{1}{\cancel{2}}\underset{1}{\cancel{y}}}\cdot\frac{\overset{2}{\cancel{4}}\overset{1}{\cancel{y}}yy}{\underset{1}{\cancel{3}}\underset{1}{\cancel{x}}}\quad \text{Divide both the 4 and the 2 by 2, and divide out the } y\text{'s.}$$

Now we multiply the remaining numerators together and the remaining denominators together.

$$=\frac{2xy^2}{1}\quad\text{or}\quad 2xy^2$$

Now Try Exercise 17

Rather than illustrating this entire process when multiplying rational expressions, we will often proceed as follows:

$$\frac{3x^2}{2y}\cdot\frac{4y^3}{3x}$$

$$=\frac{\overset{1}{\cancel{3}}\overset{x}{\cancel{x^2}}}{\underset{1}{\cancel{2}}\underset{1}{\cancel{y}}}\cdot\frac{\overset{2}{\cancel{4}}\overset{y^2}{\cancel{y^3}}}{\underset{1}{\cancel{3}}\underset{1}{\cancel{x}}}=2xy^2$$

EXAMPLE 3 Multiply $-\frac{5y^2}{2x^3} \cdot \frac{3x^2}{7y^2}$.

Solution

$$-\frac{\overset{1}{5\cancel{y^2}}}{\underset{x}{2\cancel{x^3}}} \cdot \frac{\overset{1}{3\cancel{x^2}}}{\underset{1}{7\cancel{y^2}}} = -\frac{15}{14x}$$

Now Try Exercise 19

Understanding Algebra

When we divide out common factors, it is not necessary to write the 1's as was done in Examples 2 and 3.

From this point on, instead of showing

$$-\frac{\overset{1}{5y^2}}{\underset{x}{2x^3}} \cdot \frac{\overset{1}{3x^2}}{\underset{1}{7y^2}} = -\frac{15}{14x}$$

we will generally show

$$-\frac{5y^2}{\underset{x}{2x^3}} \cdot \frac{3x^2}{7y^2} = -\frac{15}{14x}.$$

EXAMPLE 4 Multiply $(x - 6) \cdot \frac{4}{x^3 - 6x^2}$.

Solution

$$(x - 6) \cdot \frac{4}{x^3 - 6x^2} = \frac{\cancel{x - 6}}{1} \cdot \frac{4}{x^2\cancel{(x - 6)}} = \frac{4}{x^2}$$

Now Try Exercise 27

EXAMPLE 5 Multiply $\frac{(x + 2)^2}{6x^2} \cdot \frac{3x}{x^2 - 4}$.

Solution

$$\frac{(x + 2)^2}{6x^2} \cdot \frac{3x}{x^2 - 4} = \frac{(x + 2)(x + 2)}{6x^2} \cdot \frac{3x}{(x + 2)(x - 2)}$$

$$= \frac{\cancel{(x + 2)}(x + 2)}{\underset{2\ x}{\cancel{6}\cancel{x^2}}} \cdot \frac{\cancel{3x}}{\cancel{(x + 2)}(x - 2)} = \frac{x + 2}{2x(x - 2)}$$

Now Try Exercise 61

In the answer to Example 5 we could have multiplied the factors in the denominator to get $\frac{x + 2}{2x^2 - 4x}$. This is also a correct answer. In this section we will leave rational answers with the numerator as a polynomial (in unfactored form) and the denominators in factored form, as was given in Example 5.

EXAMPLE 6 Multiply $\frac{b - 4}{5b} \cdot \frac{10b}{4 - b}$.

Solution

$$\frac{b - 4}{\cancel{5}\cancel{b}} \cdot \frac{\overset{2}{\cancel{10}}\cancel{b}}{4 - b} = \frac{2(b - 4)}{4 - b}.$$

Next, we rewrite $4 - b$ as $-1(b - 4)$. Thus,

$$\frac{2(b - 4)}{4 - b} = \frac{2\cancel{(b - 4)}}{-1\cancel{(b - 4)}} = -2$$

Now Try Exercise 23

HELPFUL HINT

When only the signs differ in a numerator and denominator in a multiplication problem, factor out -1 *from either the numerator or denominator*; then divide out the common factor.

$$\frac{a - b}{x} \cdot \frac{y}{b - a} = \frac{\cancel{a - b}}{x} \cdot \frac{y}{-1\cancel{(a - b)}} = -\frac{y}{x}$$

EXAMPLE 7 Multiply $\frac{3x+2}{2x-1}\cdot\frac{4-8x}{3x+2}$.

Solution

$$\frac{3x+2}{2x-1}\cdot\frac{4-8x}{3x+2} = \frac{3x+2}{2x-1}\cdot\frac{4(1-2x)}{3x+2} \quad \text{Factor.}$$

$$= \frac{3x+2}{2x-1}\cdot\frac{4(1-2x)}{3x+2} \quad \text{Divide out common factors.}$$

Note that the factor $(1-2x)$ in the numerator of the second fraction differs only in sign from $2x-1$, the denominator of the first fraction. We will therefore factor -1 from each term of the $(1-2x)$ in the numerator of the second fraction.

$$= \frac{3x+2}{2x-1}\cdot\frac{4(-1)(2x-1)}{3x+2} \quad \text{Factored } -1 \text{ from the second numerator.}$$

$$= \frac{3x+2}{2x-1}\cdot\frac{-4(2x-1)}{3x+2} \quad \text{Divide out common factors.}$$

$$= \frac{-4}{1} = -4$$

Now Try Exercise 25

EXAMPLE 8 Multiply $\frac{2x^2+5x-12}{6x^2-11x+3}\cdot\frac{3x^2+2x-1}{x^2+5x+4}$.

Solution Factor all numerators and denominators, and then divide out common factors.

$$\frac{2x^2+5x-12}{6x^2-11x+3}\cdot\frac{3x^2+2x-1}{x^2+5x+4} = \frac{(2x-3)(x+4)}{(2x-3)(3x-1)}\cdot\frac{(3x-1)(x+1)}{(x+1)(x+4)}$$

$$= \frac{(2x-3)(x+4)}{(2x-3)(3x-1)}\cdot\frac{(3x-1)(x+1)}{(x+1)(x+4)} = 1$$

Now Try Exercise 67

EXAMPLE 9 Multiply $\frac{2x^3-18x^2+16x}{6y^2}\cdot\frac{-2y}{3x^2-3x}$.

Solution

$$\frac{2x^3-18x^2+16x}{6y^2}\cdot\frac{-2y}{3x^2-3x} = \frac{2x(x^2-9x+8)}{6y^2}\cdot\frac{-2y}{3x(x-1)}$$

$$= \frac{2x(x-8)(x-1)}{6y^2}\cdot\frac{-2y}{3x(x-1)}$$

$$= \frac{2x(x-8)(x-1)}{\underset{3\ \ y}{6y^2}}\cdot\frac{-2y}{3x(x-1)}$$

$$= \frac{-2(x-8)}{9y} = \frac{-2x+16}{9y}$$

Now Try Exercise 29

EXAMPLE 10 Multiply $\frac{x^2-y^2}{x+y}\cdot\frac{x+2y}{2x^2-3xy+y^2}$.

Solution

$$\frac{x^2-y^2}{x+y}\cdot\frac{x+2y}{2x^2-3xy+y^2} = \frac{(x+y)(x-y)}{x+y}\cdot\frac{x+2y}{(2x-y)(x-y)}$$

$$= \frac{(x+y)(x-y)}{x+y}\cdot\frac{x+2y}{(2x-y)(x-y)}$$

$$= \frac{x+2y}{2x-y}$$

Now Try Exercise 33

2 Divide Rational Expressions

Recall that to divide one fraction by a second fraction, we multiply the first fraction by the reciprocal of the second fraction (or by the reciprocal of the divisor).

To Divide Two Fractions

$$\frac{a}{b} \div \frac{c}{d} = \frac{a}{b} \cdot \frac{d}{c} = \frac{ad}{bc}, \quad b \neq 0, \quad d \neq 0, \quad \textit{and} \quad c \neq 0$$

EXAMPLE 11 Divide.

a) $\frac{2}{7} \div \frac{9}{7}$ **b)** $\frac{3}{4} \div \frac{10}{6}$

Solution

a) $\frac{2}{7} \div \frac{9}{7} = \frac{2}{\cancel{7}_1} \cdot \frac{\cancel{7}^1}{9} = \frac{2 \cdot 1}{1 \cdot 9} = \frac{2}{9}$

b) $\frac{3}{4} \div \frac{10}{6} = \frac{3}{\cancel{4}_2} \cdot \frac{\cancel{6}^3}{10} = \frac{3 \cdot 3}{2 \cdot 10} = \frac{9}{20}$

Now Try Exercise 13

Understanding Algebra

Dividing rational expressions in algebra is just like dividing fractions in arithmetic. Multiply the first rational expression by the reciprocal of the second rational expression, dividing out any common factors.

Arithmetic:

$$\frac{2}{3} \div \frac{6}{7} = \frac{2}{3} \cdot \frac{7}{6} = \frac{\cancel{2}^1 \cdot 7}{3 \cdot \cancel{6}_3} = \frac{7}{9}$$

Algebra:

$$\frac{x}{3} \div \frac{5}{6} = \frac{x}{\cancel{3}_1} \cdot \frac{\cancel{6}^2}{5} = \frac{2x}{5}$$

The same principles are used to *divide rational expressions*.

To Divide Rational Expressions

Multiply the first rational expression by the reciprocal of the second rational expression.

EXAMPLE 12 Divide $\frac{8x^3}{z} \div \frac{5z^3}{3}$.

Solution

$$\frac{8x^3}{z} \div \frac{5z^3}{3} = \frac{8x^3}{z} \cdot \frac{3}{5z^3}$$ Multiply $\frac{8x^3}{z}$ by $\frac{3}{5z^3}$, the reciprocal of $\frac{5z^3}{3}$.

$$= \frac{24x^3}{5z^4}$$ There are no common factors to divide out.

Now Try Exercise 35

EXAMPLE 13 Divide $\frac{x^2 - 9}{x + 4} \div \frac{x - 3}{x + 4}$.

Solution

$$\frac{x^2 - 9}{x + 4} \div \frac{x - 3}{x + 4} = \frac{x^2 - 9}{x + 4} \cdot \frac{x + 4}{x - 3}$$ Multiply the first fraction by the reciprocal of the second fraction.

$$= \frac{(x + 3)\cancel{(x - 3)}}{\cancel{x + 4}} \cdot \frac{\cancel{x + 4}}{\cancel{x - 3}}$$ Factor, and divide out common factors.

$$= x + 3$$

Now Try Exercise 41

EXAMPLE 14 Divide $\frac{-1}{2x-3} \div \frac{8}{3-2x}$.

Solution

$$\frac{-1}{2x-3} \div \frac{8}{3-2x} = \frac{-1}{2x-3} \cdot \frac{3-2x}{8}$$

Multiply the first fraction by the reciprocal of the second fraction.

$$= \frac{-1}{\cancel{2x-3}} \cdot \frac{-1\cancel{(2x-3)}}{8}$$

Factor out -1 then divide out common factors.

$$= \frac{(-1)(-1)}{(1)(8)} = \frac{1}{8}$$

Now Try Exercise 49

EXAMPLE 15 Divide $\frac{w^2 - 11w + 30}{w^2} \div (w-5)^2$.

Solution $(w-5)^2$ means $\frac{(w-5)^2}{1}$.

$$\frac{w^2 - 11w + 30}{w^2} \div (w-5)^2 = \frac{w^2 - 11w + 30}{w^2} \cdot \frac{1}{(w-5)^2}$$

$$= \frac{(w-6)\cancel{(w-5)}}{w^2} \cdot \frac{1}{\cancel{(w-5)}(w-5)}$$

$$= \frac{w-6}{w^2(w-5)}$$

Now Try Exercise 47

EXAMPLE 16 Divide $\frac{12x^2 - 22x + 8}{7x} \div \frac{3x^2 + 2x - 8}{2x^2 + 4x}$.

Solution

$$\frac{12x^2 - 22x + 8}{7x} \div \frac{3x^2 + 2x - 8}{2x^2 + 4x} = \frac{12x^2 - 22x + 8}{7x} \cdot \frac{2x^2 + 4x}{3x^2 + 2x - 8}$$

$$= \frac{2(6x^2 - 11x + 4)}{7x} \cdot \frac{2x(x+2)}{(3x-4)(x+2)}$$

$$= \frac{2\cancel{(3x-4)}(2x-1)}{7\cancel{x}} \cdot \frac{2\cancel{x}\cancel{(x+2)}}{\cancel{(3x-4)}\cancel{(x+2)}}$$

$$= \frac{4(2x-1)}{7} = \frac{8x-4}{7}$$

Now Try Exercise 45

6.2 Exercise Set MathXL® MyMathLab®

Warm-Up Exercises

Fill in the blanks with the appropriate word, phrase, or symbol(s) from the following list.

reciprocal opposite $\frac{ad}{bc}$ $\frac{ab}{cd}$ $\frac{ac}{bd}$ denominators

1. When multiplying rational expressions, first simplify by dividing out common factors, then multiply numerators together and multiply __________ together.

2. The product $\frac{a}{b} \cdot \frac{c}{d} =$ __________ for $b \neq 0$ and $d \neq 0$.

3. The quotient $\frac{a}{b} \div \frac{c}{d} = \frac{a}{b} \cdot \frac{d}{c} =$ __________ for $b \neq 0$, $d \neq 0$, and $c \neq 0$.

4. To divide rational expressions, multiply the first rational expression by the __________ of the second rational expression.

Practice the Skills

Multiply or divide as indicated.

5. $\left(\frac{1}{5}\right)\left(\frac{15}{19}\right)$

6. $\frac{21}{25}\cdot\frac{1}{7}$

7. $\left(-\frac{6}{35}\right)\left(\frac{10}{21}\right)$

8. $\left(\frac{22}{35}\right)\left(-\frac{21}{55}\right)$

9. $\left(-\frac{4}{11}\right)\left(-\frac{55}{64}\right)$

10. $\left(-\frac{12}{13}\right)\left(-\frac{65}{42}\right)$

11. $\frac{3}{7}\div\frac{5}{7}$

12. $\frac{3}{8}\div\frac{15}{44}$

13. $-\frac{2}{9}\div\frac{32}{39}$

14. $\frac{40}{27}\div\left(-\frac{32}{63}\right)$

15. $\left(-\frac{15}{14}\right)\div\left(-\frac{44}{35}\right)$

16. $\left(-\frac{3}{4}\right)\div\left(-\frac{15}{16}\right)$

Multiply.

17. $\frac{14ab^2}{3c}\cdot\frac{9c}{7ab^3}$

18. $\frac{15x^3y^2}{2z}\cdot\frac{2z}{5xy^3}$

19. $-\frac{10p^2}{q^3}\cdot\frac{5p}{3q^3}$

20. $\frac{14x^2}{y^4}\cdot\frac{5x^2}{y^2}$

21. $\frac{6x^5y^3}{5z^3}\cdot\frac{6x^4}{5yz^4}$

22. $\left(-\frac{6p^2q^3}{55r^4}\right)\left(-\frac{22p^2r^3}{15q^3}\right)$

23. $\frac{3x-2}{3x+2}\cdot\frac{x-1}{1-x}$

24. $\frac{3x-5}{5x-3}\cdot\frac{4-2x}{x-2}$

25. $\frac{x^2+7x+6}{x+6}\cdot\frac{1}{x+1}$

26. $\frac{b^2+7b+12}{6b}\cdot\frac{b^2-4b}{b^2-b-12}$

27. $\frac{a}{a^2-b^2}\cdot\frac{a+b}{a^2+ab}$

28. $\frac{t^2-36}{t^2+t-30}\cdot\frac{t-5}{4t}$

29. $\frac{6x^2-14x-12}{6x+4}\cdot\frac{2x+4}{2x^2-2x-12}$

30. $\frac{2x^2-9x+9}{8x-12}\cdot\frac{4x}{x^2-3x}$

31. $\frac{3x^2-13x-10}{x^2-2x-15}\cdot\frac{x^2+x-2}{3x^2-x-2}$

32. $\frac{2t^2-t-6}{2t^2-3t-2}\cdot\frac{2t^2-5t-3}{2t^2+11t+12}$

33. $\frac{x+9}{x-3}\cdot\frac{x^3-27}{x^2+3x+9}$

34. $\frac{x^3+8}{x^2-x-6}\cdot\frac{x+5}{x^2-2x+4}$

Divide.

35. $\frac{12x^3}{y^2}\div\frac{3x}{y^3}$

36. $\frac{35r^2}{8s^3}\div\frac{5r^5}{6s^2}$

37. $\frac{15xy^2}{4z}\div\frac{5x^2y^2}{12z^2}$

38. $\frac{36y}{5z^2}\div\frac{3xy}{2z}$

39. $-\frac{3xy^2}{7z^5}\div\frac{6x^2y}{35z^3}$

40. $\left(-\frac{6x^5y^3}{55z^5}\right)\div\left(-\frac{14yz^2}{33x^3}\right)$

41. $\frac{x^2-16}{3x^2-12x}\div\frac{5x^2+20x}{3x^2+6x}$

42. $\frac{z^2-4}{5z^2+10z}\div\frac{2z^2-6z}{3z^2-9z}$

43. $\frac{1}{x^2+7x-18}\div\frac{1}{x^2-17x+30}$

44. $\frac{1}{3x^2+13x-10}\div\frac{1}{3x^2+11x-20}$

45. $\frac{x^2-12x+32}{x^2-6x-16}\div\frac{x^2-x-12}{x^2-5x-24}$

46. $\frac{2x^2-13x+15}{x^2-2x-15}\div\frac{2x^2-3x-20}{x^2-x-12}$

47. $\frac{2x^2+9x+4}{x^2+7x+12}\div\frac{2x^2-x-1}{(x+3)^2}$

48. $\frac{(2x-3)^2}{2x^2+7x-15}\div\frac{2x^2-7x+6}{x^2+2x-15}$

49. $\frac{x^2-y^2}{x^2-2xy+y^2}\div\frac{x+y}{y-x}$

50. $\frac{x^2-4y^2}{(x+2y)^2}\div\frac{2y-x}{5x+10y}$

51. $\frac{5x^2-4x-1}{5x^2+6x+1}\div\frac{x^2-5x+4}{x^2+2x+1}$

52. $\frac{7n^2-15n+2}{n^2+n-6}\div\frac{n^2-3n-10}{n^2-2n-15}$

Perform each indicated operation.

53. $\frac{11z}{6y^2}\cdot\frac{24x^2y^4}{11z}$

54. $\frac{17ab^2}{c^5}\cdot\frac{5c^3}{34ab}$

55. $\frac{63a^2b^3}{20c^3}\cdot\frac{4c^4}{9a^3b^5}$

56. $\frac{56pq^2}{45r^3}\cdot\frac{15r}{32p^3q}$

57. $\frac{-xy}{a}\div\frac{-2ax}{6y}$

58. $\frac{-2xw}{y^5}\div\frac{8x^2}{y^6}$

59. $\left(\frac{39k^2m^2}{10n^2}\right)\left(-\frac{15mn^3}{13k}\right)$

60. $\frac{-18x^2y}{11z^2}\cdot\frac{22z^3}{x^2y^5}$

61. $\frac{(x+3)^2}{5x^2}\cdot\frac{10x}{x^2-9}$

62. $\frac{1}{4x-3}\cdot(24x-18)$

63. $\frac{1}{5x^2y^2}\div\frac{1}{35x^3y}$

64. $\frac{x^2y^5}{3z}\div\frac{7z}{2x}$

65. $\dfrac{(4m)^2}{8n^3} \div \dfrac{m^6n^8}{2}$

66. $\dfrac{(5a)^2}{3b^3} \div \dfrac{15a^3b^2}{8}$

67. $\dfrac{r^2 + 5r + 6}{r^2 + 9r + 18} \cdot \dfrac{r^2 + 4r - 12}{r^2 - 5r + 6}$

68. $\dfrac{z^2 - z - 20}{z^2 - 3z - 10} \cdot \dfrac{(z + 2)^2}{(z + 4)^2}$

69. $\dfrac{x^2 - 12x + 36}{x^2 - 8x + 12} \div \dfrac{x^2 - 7x + 12}{x^2 - 6x + 8}$

70. $\dfrac{p^2 - 5p + 6}{p^2 - 10p + 16} \div \dfrac{p^2 + 2p}{p^2 - 6p - 16}$

71. $\dfrac{2w^2 + 3w - 35}{w^2 - 7w - 8} \cdot \dfrac{w^2 - 5w - 24}{w^2 + 8w + 15}$

72. $\dfrac{3z^2 - 4z - 4}{z^2 - 4} \cdot \dfrac{2z^2 + 5z + 2}{2z^2 - 3z - 2}$

73. $\dfrac{q^2 - 11q + 30}{2q^2 - 7q - 15} \div \dfrac{q^2 - 2q - 24}{q^2 - q - 20}$

74. $\dfrac{2x^2 - 19x + 24}{x^2 - 12x + 32} \div \dfrac{2x^2 + x - 6}{x^2 + 7x + 10}$

75. $\dfrac{x^3 + y^3}{x^2 - y^2} \div \dfrac{x^2 - xy + y^2}{y - x}$

76. $\dfrac{x^2 - y^2}{(x - y)^2} \div \dfrac{x^3 + y^3}{-3x^2 + 3xy - 3y^2}$

Problem Solving

Perform each indicated operation. Treat Δ and ☺ as if they were variables.

77. $\dfrac{6\Delta^2}{13} \cdot \dfrac{13}{36\Delta^5}$

78. $\dfrac{\Delta - 7}{2\Delta + 5} \cdot \dfrac{3\Delta}{-\Delta + 7}$

79. $\dfrac{\Delta - \text{☺}}{9\Delta - 9\text{☺}} \div \dfrac{\Delta^2 - \text{☺}^2}{\Delta^2 + 2\Delta\text{☺} + \text{☺}^2}$

80. $\dfrac{\Delta^2 - \text{☺}^2}{\Delta^2 - 2\Delta\text{☺} + \text{☺}^2} \div \dfrac{\Delta + \text{☺}}{\text{☺} - \Delta}$

For each equation, fill in the shaded area with a binomial or trinomial to make the statement true. Explain how you determined your answer.

81. $\dfrac{\square}{x + 2} = x + 3$

82. $\dfrac{x + 5}{\square} = \dfrac{1}{x - 5}$

83. $\dfrac{\square}{x - 6} = x + 2$

84. $\dfrac{\square}{x^2 - 7x + 10} = \dfrac{1}{x - 2}$

85. $\dfrac{\square}{x^2 - 4} \cdot \dfrac{x + 2}{x - 1} = 1$

86. $\dfrac{x + 4}{x^2 + 9x + 20} \cdot \dfrac{\square}{x - 2} = 1$

Concept/Writing Exercises

87. Explain how to multiply two rational expressions.

88. Explain how to divide two rational expressions.

Challenge Problems

Simplify.

89. $\left(\dfrac{x^2 - x - 6}{2x^2 - 9x + 9} \div \dfrac{x^2 + x - 12}{x^2 + 3x - 4}\right) \cdot \dfrac{2x^2 - 5x + 3}{x^2 + x - 2}$

90. $\left(\dfrac{x^2 + 4x + 3}{x^2 - 6x - 16} \div \dfrac{x^2 + 5x + 6}{x^2 - 9x + 8}\right) \cdot \left(\dfrac{x^2 - 1}{x^2 + 4x + 4}\right)$

For Exercises 91 and 92, determine the polynomials that when placed in the shaded areas make the statement true. Explain how you determined your answer.

91. $\dfrac{\square}{\square} \cdot \dfrac{x^2 + 3x - 4}{x^2 - 4x + 3} = \dfrac{x - 2}{x - 5}$

92. $\dfrac{\square}{x^2 + x - 2} \cdot \dfrac{x^2 + 6x + 8}{\square} = \dfrac{x + 3}{x + 5}$

Group Activity

93. Consider the three problems that follow:

1. $\left(\dfrac{x + 2}{x - 3}\right) \div \left(\dfrac{x^2 - 5x + 6}{x - 2} \cdot \dfrac{x + 2}{x - 3}\right)$

2. $\left(\dfrac{x + 2}{x - 3} \div \dfrac{x^2 - 5x + 6}{x - 2}\right) \cdot \left(\dfrac{x + 2}{x - 3}\right)$

3. $\left(\dfrac{x + 2}{x - 3}\right) \div \left(\dfrac{x^2 - 5x + 6}{x - 2}\right) \cdot \left(\dfrac{x + 2}{x - 3}\right)$

a) Without working the problem, decide as a group which of the problems will have the same answer. Explain.

b) Individually, simplify each of the three problems.

c) Compare your answers to part **b)** with those of the other members of your group. If you did not get the same answers, determine why.

Cumulative Review Exercises

[3.4] **94. White Water Kayaking** Todd Smith leaves Idaho Falls paddling downstream at an average speed of 15 miles per hour toward Pocatello. On the return trip, paddling against the current, he averages 5 miles per hour. If the trip back to Idaho Falls took 2 hours longer than the trip out, find the time it took Todd to reach Pocatello.

[4.5] **95.** Multiply $(4x^3y^2z^4)(3xy^3z^7)$.

[4.6] **96.** Divide $\dfrac{4x^3 - 5x}{2x - 1}$.

[5.4] **97.** Factor $6x^2 - 18x - 60$.

[5.6] **98.** Solve $3x^2 - 9x - 30 = 0$.

6.3 Addition and Subtraction of Rational Expressions with a Common Denominator and Finding the Least Common Denominator

1. Add and subtract rational expressions with a common denominator.
2. Find the least common denominator.

1 Add and Subtract Rational Expressions with a Common Denominator

Recall that when adding or subtracting two fractions with a common denominator we add or subtract the numerators while keeping the common denominator.

To Add or Subtract Two Fractions with a Common Denominator

Adding: $\dfrac{a}{c} + \dfrac{b}{c} = \dfrac{a + b}{c}, c \neq 0$ Subtracting: $\dfrac{a}{c} - \dfrac{b}{c} = \dfrac{a - b}{c}, c \neq 0$

EXAMPLE 1 **a)** Add $\dfrac{7}{16} + \dfrac{8}{16}$. **b)** Subtract $\dfrac{4}{9} - \dfrac{1}{9}$.

Solution

a) $\dfrac{7}{16} + \dfrac{8}{16} = \dfrac{7 + 8}{16} = \dfrac{15}{16}$

b) $\dfrac{4}{9} - \dfrac{1}{9} = \dfrac{4 - 1}{9} = \dfrac{3}{9} = \dfrac{1}{3}$

Now Try Exercise 7

When adding or subtracting rational expressions, we use the same principles that we use when adding or subtracting fractions.

To Add or Subtract Rational Expressions with a Common Denominator

1. Add or subtract the numerators.
2. Place the sum or difference of the numerators found in step 1 over the common denominator.
3. Simplify the fraction if possible.

EXAMPLE 2 Add $\dfrac{3}{x-4}+\dfrac{x+8}{x-4}$.

Solution

$$\frac{3}{x-4}+\frac{x+8}{x-4}=\frac{3+(x+8)}{x-4}=\frac{x+11}{x-4}$$

Now Try Exercise 15

EXAMPLE 3 Add $\dfrac{2x^2+7}{x+3}+\dfrac{6x-7}{x+3}$.

Solution

$$\frac{2x^2+7}{x+3}+\frac{6x-7}{x+3}=\frac{(2x^2+7)+(6x-7)}{x+3}$$

$$=\frac{2x^2+7+6x-7}{x+3}$$

$$=\frac{2x^2+6x}{x+3}$$

Now factor $2x$ from each term in the numerator and simplify.

$$=\frac{2x\cancel{(x+3)}}{\cancel{x+3}}=2x$$

Now Try Exercise 19

EXAMPLE 4 Add $\dfrac{x^2+2x-2}{(x+5)(x-2)}+\dfrac{5x+12}{(x+5)(x-2)}$.

Solution

$$\frac{x^2+2x-2}{(x+5)(x-2)}+\frac{5x+12}{(x+5)(x-2)}=\frac{(x^2+2x-2)+(5x+12)}{(x+5)(x-2)}$$ Write as a single fraction.

$$=\frac{x^2+2x-2+5x+12}{(x+5)(x-2)}$$ Parentheses were removed in the numerator.

$$=\frac{x^2+7x+10}{(x+5)(x-2)}$$ Like terms were combined.

$$=\frac{\cancel{(x+5)}(x+2)}{\cancel{(x+5)}(x-2)}$$ Factor, divide out common factor.

$$=\frac{x+2}{x-2}$$

Now Try Exercise 27

When subtracting rational expressions, be sure to subtract the entire numerator of the fraction being subtracted.

AVOIDING COMMON ERRORS

Consider the subtraction

$$\frac{4x}{x-2}-\frac{2x+1}{x-2}$$

Many people begin problems of this type incorrectly. Here are the correct and incorrect ways of working this problem.

CORRECT

$$\frac{4x}{x-2}-\frac{2x+1}{x-2}=\frac{4x-(2x+1)}{x-2}$$

$$=\frac{4x-2x-1}{x-2}$$

$$=\frac{2x-1}{x-2}$$

INCORRECT

$$\frac{4x}{x-2}-\frac{2x+1}{x-2}=\frac{4x-2x+1}{x-2}$$ (crossed out)

Note that the entire numerator of the second fraction, not just the first term, must be subtracted. Also note that the sign of *each* term of the numerator being subtracted will change when the parentheses are removed.

EXAMPLE 5 Subtract $\frac{x^2 - 6x + 3}{x^2 + 7x + 12} - \frac{x^2 - 8x - 5}{x^2 + 7x + 12}$.

Solution

$$\frac{x^2 - 6x + 3}{x^2 + 7x + 12} - \frac{x^2 - 8x - 5}{x^2 + 7x + 12} = \frac{(x^2 - 6x + 3) - (x^2 - 8x - 5)}{x^2 + 7x + 12}$$ Write as a single fraction.

$$= \frac{x^2 - 6x + 3 - x^2 + 8x + 5}{x^2 + 7x + 12}$$ Removed parentheses.

$$= \frac{2x + 8}{x^2 + 7x + 12}$$ Combined like terms.

$$= \frac{2\cancel{(x + 4)}}{(x + 3)\cancel{(x + 4)}}$$ Factor, divide out common factor.

$$= \frac{2}{x + 3}$$

Now Try Exercise 41

Understanding Algebra

When subtracting rational expressions, we must be sure to subtract *each* term in the numerator of the second rational expression from the numerator of the first rational expression. In Example 5, the minus sign is applied to each term in the numerator:

$$-(x^2 - 8x - 5)$$
$$= -x^2 + 8x + 5.$$

EXAMPLE 6 Subtract $\frac{6r}{r - 5} - \frac{4r^2 - 17r + 15}{r - 5}$.

Solution

$$\frac{6r}{r - 5} - \frac{4r^2 - 17r + 15}{r - 5} = \frac{6r - (4r^2 - 17r + 15)}{r - 5}$$ Write as a single fraction.

$$= \frac{6r - 4r^2 + 17r - 15}{r - 5}$$ Parentheses were removed.

$$= \frac{-4r^2 + 23r - 15}{r - 5}$$ Like terms were combined.

$$= \frac{-(4r^2 - 23r + 15)}{r - 5}$$ Factored out -1.

$$= \frac{-(4r - 3)\cancel{(r - 5)}}{\cancel{r - 5}}$$ Factor, divide out common factor.

$$= -(4r - 3) \quad \text{or} \quad -4r + 3$$

Now Try Exercise 31

2 Find the Least Common Denominator

To add two fractions with unlike denominators, we first obtain the least common denominator. Now we explain how to find the *least common denominator* for rational expressions.

EXAMPLE 7 Add $\frac{4}{7} + \frac{2}{3}$.

Solution The least common denominator (LCD) of the fractions $\frac{4}{7}$ and $\frac{2}{3}$ is 21 because it is the smallest number that is divisible by both denominators, 7 and 3. Rewrite each fraction so that its denominator is 21.

$$\frac{4}{7} + \frac{2}{3} = \frac{3}{3} \cdot \frac{4}{7} + \frac{2}{3} \cdot \frac{7}{7}$$
$$= \frac{12}{21} + \frac{14}{21}$$
$$= \frac{26}{21}$$

Now Try Exercise 95

To Find the Least Common Denominator of Rational Expressions

1. Factor each denominator completely. Any factors that occur more than once should be expressed as powers. For example, $(x-3)(x-3)$ should be expressed as $(x-3)^2$.
2. List all different factors (other than 1) that appear in any of the denominators. When the same factor appears in more than one denominator, write that factor with the highest power that appears on it.
3. The least common denominator is the product of all the factors listed in step 2.

EXAMPLE 8 Find the least common denominator.

$$\frac{1}{7}+\frac{1}{y}$$

Solution The only factor (other than 1) of the first denominator is 7. The only factor (other than 1) of the second denominator is y. The LCD is therefore $7 \cdot y = 7y$.

Now Try Exercise 51

EXAMPLE 9 Find the LCD.

$$\frac{8}{x^2}-\frac{3}{5x}$$

Solution The factors that appear in the denominators are 5 and x. List each factor with its highest power. The LCD is the product of these factors.

Highest power of 5 ↓ ↓ Highest power of x

$$\text{LCD} = 5^1 \cdot x^2 = 5x^2$$

Now Try Exercise 59

EXAMPLE 10 Find the LCD.

$$\frac{11}{18x^3y}+\frac{5}{27x^2y^3}$$

Solution We first write the denominators in factored form. To determine the LCD, we multiply the highest power of each factor.

$18x^3y = 2 \cdot 3^3 \cdot x^3 \cdot y$ The prime factorization of 18 is $2 \cdot 3 \cdot 3$ or $2 \cdot 3^2$.

$27x^2y^3 = 3^3 \cdot x^2 \cdot y^3$ The prime factorization of 27 is $3 \cdot 3 \cdot 3$ or 3^3.

Highest power of 3 ↓ ↓ Highest power of x

$$\text{LCD} = 2 \cdot 3^3 \cdot x^3 \cdot y^3$$

Highest power of 2 ↑ ↑ Highest power of y

Thus, the LCD is $2 \cdot 3^3 \cdot x^3 \cdot y^3 = 54x^3y^3$

Now Try Exercise 63

Understanding Algebra

Numerical coefficients like those in Example 10 need to be factored also. To help you write numbers in their prime factored form, you may want to review page 21 or Section 5.1.

EXAMPLE 11 Find the LCD.

$$\frac{9}{x}-\frac{2z}{x+3}$$

Solution The factors in the denominators are x and $x+3$. *Note that the x in the second denominator, x + 3, is a term, not a factor.*

$$\text{LCD} = x(x+3)$$

Now Try Exercise 65

EXAMPLE 12 Find the LCD.

$$\frac{7}{3x^2 - 6x} + \frac{8x^2}{x^2 - 4x + 4}$$

Solution Factor both denominators.

$$\frac{7}{3x^2 - 6x} + \frac{8x^2}{x^2 - 4x + 4} = \frac{7}{3x(x - 2)} + \frac{8x^2}{(x - 2)(x - 2)}$$

$$= \frac{7}{3x(x - 2)} + \frac{8x^2}{(x - 2)^2}$$

The factors in the denominators are 3, x, and $x - 2$. List the highest power of each of these factors.

$$\text{LCD} = 3 \cdot x \cdot (x - 2)^2 = 3x(x - 2)^2.$$

Now Try Exercise 85

EXAMPLE 13 Find the LCD.

$$\frac{11x}{x^2 - x - 12} - \frac{6x^2}{x^2 - 7x + 12}$$

Solution Factor both denominators.

$$\frac{11x}{x^2 - x - 12} - \frac{6x^2}{x^2 - 7x + 12} = \frac{11x}{(x + 3)(x - 4)} - \frac{6x^2}{(x - 3)(x - 4)}$$

The factors in the denominators are $x + 3$, $x - 4$, and $x - 3$.

$$\text{LCD} = (x + 3)(x - 4)(x - 3)$$

Although $x - 4$ is a common factor of each denominator, the highest power of that factor that appears in each denominator is 1.

Now Try Exercise 81

EXAMPLE 14 Find the LCD.

$$\frac{6w}{w^2 - 14w + 45} + w + 8$$

Solution Factor the denominator of the first term.

$$\frac{6w}{w^2 - 14w + 45} + w + 8 = \frac{6w}{(w - 5)(w - 9)} + w + 8$$

Since the denominator of $w + 8$ is 1, the expression can be rewritten as

$$\frac{6w}{(w - 5)(w - 9)} + \frac{w + 8}{1}$$

The LCD is therefore $1(w - 5)(w - 9)$ or simply $(w - 5)(w - 9)$.

Now Try Exercise 89

6.3 Exercise Set MathXL® MyMathLab®

Warm-Up Exercises

Fill in the blanks with the appropriate word, phrase, or symbol(s) from the following list.

numerator denominator $\frac{a - b}{c}$ $\frac{a + b}{c}$ least $\frac{p - q - r}{s}$ $\frac{p - q + r}{s}$

1. When adding or subtracting two rational expressions with a common denominator, we add or subtract the numerators while keeping the common ___________.

2. $\frac{a}{c} + \frac{b}{c} =$ ________, $c \neq 0$.

3. $\frac{a}{c} - \frac{b}{c} =$ ________, $c \neq 0$.

4. To add two rational expressions with unlike denominators, we first obtain the ________ common denominator.

5. When subtracting rational expressions, we must be sure to subtract each term in the ________ of the second rational expression from the numerator of the first rational expression.

6. $\frac{p}{s} - \frac{q-r}{s} =$ ________.

Practice the Skills

Add or subtract.

7. $\frac{1}{7} + \frac{4}{7}$

8. $\frac{4}{15} + \frac{8}{15}$

9. $\frac{9}{16} - \frac{5}{16}$

10. $\frac{8}{5} - \frac{6}{5}$

11. $\frac{5r+2}{4} - \frac{3}{4}$

12. $\frac{3x+6}{2} - \frac{x}{2}$

13. $\frac{2}{x} + \frac{x+4}{x}$

14. $\frac{7}{z} + \frac{z+3}{z}$

15. $\frac{6}{n+1} + \frac{n+2}{n+1}$

16. $\frac{3x+1}{x+1} + \frac{6x+8}{x+1}$

17. $\frac{4a}{a+1} - \frac{2a+3}{a+1}$

18. $\frac{7}{t-2} - \frac{t+4}{t-2}$

19. $\frac{x}{x-3} + \frac{4x+9}{x-3}$

20. $\frac{b+7}{b-3} + \frac{b-10}{b-3}$

21. $\frac{3c+2}{c+4} - \frac{2c-5}{c+4}$

22. $\frac{4x-3}{x-7} - \frac{2x+8}{x-7}$

23. $\frac{4t+7}{5t^2} - \frac{3t+4}{5t^2}$

24. $\frac{x^2-6}{3x} - \frac{x^2+4x-11}{3x}$

25. $\frac{5x+4}{x^2-x-12} + \frac{-4x-1}{x^2-x-12}$

26. $\frac{3w+6}{w^2+2w+1} + \frac{-2w-5}{w^2+2w+1}$

27. $\frac{2m+5}{(m+4)(m-3)} - \frac{m+1}{(m+4)(m-3)}$

28. $\frac{x^2+3x}{(x+6)(x-3)} - \frac{x+15}{(x+6)(x-3)}$

29. $\frac{2p-6}{p-5} - \frac{p+6}{p-5}$

30. $\frac{5q-11}{11q-6} - \frac{3q-5}{11q-6}$

31. $\frac{x^2+4x+1}{x+2} - \frac{5x+7}{x+2}$

32. $\frac{2x^2+3x-21}{x-7} - \frac{x^2+6x+7}{x-7}$

33. $\frac{x^2-12}{x+5} - \frac{13}{x+5}$

34. $\frac{x^2}{x+4} - \frac{16}{x+4}$

35. $\frac{b^2-2b-2}{b^2-b-6} + \frac{b-4}{b^2-b-6}$

36. $\frac{c^2-5c+3}{c^2-2c-3} + \frac{c^2+c-9}{c^2-2c-3}$

37. $\frac{t-3}{t+3} - \frac{-3t-15}{t+3}$

38. $\frac{x+8}{3x+2} - \frac{x+8}{3x+2}$

39. $\frac{3x^2+15x}{x^3+2x^2-8x} + \frac{2x^2+5x}{x^3+2x^2-8x}$

40. $\frac{3x^2-9x}{4x^2-8x} + \frac{3x}{4x^2-8x}$

41. $\frac{x^2+3x-6}{x^2-5x+4} - \frac{-2x^2+4x-4}{x^2-5x+4}$

42. $\frac{5x^2+x+11}{x^3-7x^2-2x+14} - \frac{4x^2+11x-10}{x^3-7x^2-2x+14}$

43. $\frac{3x^2+4x-3}{x^3+4x^2+3x+12} - \frac{2x^2-5x-23}{x^3+4x^2+3x+12}$

44. $\frac{3x^2-4x+6}{3x^2+7x+2} - \frac{10x+11}{3x^2+7x+2}$

45. $\frac{x^2-2}{x^2+6x-7} - \frac{-4x+19}{x^2+6x-7}$

46. $\frac{4x^2+15}{9x^2-64} - \frac{x^2-x+39}{9x^2-64}$

47. $\frac{5x^2+30x+8}{x^2-64} + \frac{x^2+19x}{x^2-64}$

48. $\frac{2x^2+3x-20}{4x^2-64} + \frac{2x^2-7x-28}{4x^2-64}$

Find the least common denominator for each expression.

49. $\frac{x}{5} + \frac{x+4}{5}$

50. $\frac{2+r}{17} - \frac{12}{17}$

51. $\frac{3}{n} + \frac{1}{9n}$

52. $\frac{9}{z} - \frac{3}{7z}$

53. $\frac{2}{3} + \frac{5}{x+3}$

54. $\frac{6}{x+1} - \frac{4}{7}$

55. $\frac{6}{p} + \frac{9}{p^3}$

56. $\frac{k+7}{k^2} - \frac{k-3}{k^3}$

57. $\frac{m+3}{3m-4} + m$

58. $6x^2 + \frac{8x}{x-7}$

59. $\frac{t}{6t} + \frac{4}{t^2}$

60. $\frac{x}{5x^2} + \frac{9}{7x^3}$

61. $\frac{x-5}{x+8} - \frac{x+7}{x+4}$
62. $\frac{2x}{x+3} + \frac{6}{x-9}$
63. $\frac{4}{2r^4s^5} - \frac{5}{9r^3s^7}$
64. $\frac{5}{4w^5z^4} + \frac{4}{9wz^2}$
65. $\frac{3}{m} - \frac{17m}{m+2}$
66. $\frac{7}{h} - \frac{15h}{h-5}$
67. $\frac{5x-2}{x^2+x} - \frac{13}{x}$
68. $\frac{7}{5c^2-10c} + \frac{3}{5c}$
69. $\frac{n}{4n-1} + \frac{n-8}{1-4n}$
70. $\frac{3t}{t-5} + \frac{2}{5-t}$
71. $\frac{3}{4k-5r} - \frac{10}{-4k+5r}$
72. $\frac{3}{-2a+3b} - \frac{10}{2a-3b}$
73. $\frac{4}{2q^2+2q} - \frac{5}{9q}$
74. $\frac{p}{4p^2+2p} - \frac{7}{2p+1}$
75. $\frac{21}{24x^2y} + \frac{x+4}{15xy^3}$
76. $\frac{17x-7}{27x^3y^4} - \frac{3x+8}{45x^5y}$
77. $\frac{11}{3x+12} + \frac{3x+1}{2x+4}$
78. $\frac{7}{6x-18} + \frac{11}{15x-45}$
79. $\frac{9x+4}{x+1} - \frac{2x-6}{x+8}$
80. $\frac{3x+7}{5x-4} + \frac{2x-1}{15x+8}$
81. $\frac{x-2}{x^2-5x-24} + \frac{3}{x^2+11x+24}$
82. $\frac{x+3}{x^2+11x+18} - \frac{x^2-11}{x^2-3x-10}$
83. $\frac{5}{(a-4)^2} - \frac{a+2}{a^2-7a+12}$
84. $\frac{3x+5}{x^2-1} + \frac{x^2-18}{(x+1)^2}$
85. $\frac{9x}{x^2+6x+5} - \frac{5x^2}{x^2+4x+3}$
86. $\frac{6n}{n^2-4} - \frac{n-3}{n^2-5n-14}$
87. $\frac{3x-5}{x^2-6x+9} + \frac{3}{x-3}$
88. $\frac{6x+5}{x+2} + \frac{3x}{(x+2)^2}$
89. $\frac{8x^2}{x^2-7x+6} + x - 9$
90. $\frac{2x-1}{x^2-25} + x - 10$
91. $\frac{t-1}{3t^2+10t-8} - \frac{11}{3t^2+11t-4}$
92. $\frac{-4x+9}{2x^2+5x+2} + \frac{x^2}{3x^2+4x-4}$
93. $\frac{3x-1}{4x^2+4x+1} + \frac{x^2+x-9}{8x^2+10x+3}$
94. $\frac{3x+7}{6x^2+11x-10} + \frac{x^2-8}{9x^2-12x+4}$

Add or subtract using the technique from Example 7.

95. $\frac{1}{7} + \frac{2}{5}$
96. $\frac{3}{8} + \frac{1}{4}$
97. $\frac{2}{9} + \frac{3}{4}$
98. $\frac{5}{6} - \frac{1}{3}$
99. $\frac{5}{9} - \frac{1}{2}$
100. $\frac{6}{5} - \frac{3}{10}$

Problem Solving

List the polynomial to be placed in each shaded area to make a true statement. Explain how you determined your answer.

101. $\frac{x^2-6x+3}{x+3} + \frac{\blacksquare}{x+3} = \frac{2x^2-5x-6}{x+3}$
102. $\frac{4x^2-6x-7}{x^2-4} - \frac{\blacksquare}{x^2-4} = \frac{2x^2+x-3}{x^2-4}$
103. $\frac{-x^2-4x+3}{2x+5} + \frac{\blacksquare}{2x+5} = \frac{5x-7}{2x+5}$
104. $\frac{-3x^2-9}{(x+4)(x-2)} - \frac{\blacksquare}{(x+4)(x-2)} = \frac{x^2+3x}{(x+4)(x-2)}$

Find the least common denominator of each expression.

105. $\frac{3}{\text{☺}} + \frac{4}{5\text{☺}}$
106. $\frac{5}{8\Delta^2\text{☺}^2} + \frac{6}{5\Delta^4\text{☺}^5}$
107. $\frac{8}{\Delta^2-9} - \frac{2}{\Delta+3}$
108. $\frac{6}{\Delta+3} - \frac{\Delta+5}{\Delta^2-4\Delta+3}$

Concept/Writing Exercises

109. Explain how to add or subtract rational expressions with a common denominator.
110. Explain how to find the least common denominator of two rational expressions.

Challenge Problems

Perform each indicated operation.

111. $\frac{4x-1}{x^2-25} - \frac{3x^2-8}{x^2-25} + \frac{8x-7}{x^2-25}$
112. $\frac{x^2-8x+2}{x+7} + \frac{2x^2-5x}{x+7} - \frac{3x^2+7x+10}{x+7}$

Find the least common denominator for each expression.

113. $\dfrac{17}{6x^5y^9} - \dfrac{9}{2x^3y} + \dfrac{6}{5x^{12}y^2}$

114. $\dfrac{2x}{x-3} - \dfrac{3}{x^2-9} + \dfrac{5}{x+3}$

115. $\dfrac{3x}{x^2-x-12} + \dfrac{2}{x^2-6x+8} + \dfrac{3}{x^2+x-6}$

116. $\dfrac{9}{x^2-4} - \dfrac{8}{3x^2+5x-2} + \dfrac{7}{3x^2-7x+2}$

Cumulative Review Exercises

[1.3] **117.** Subtract $4\frac{3}{5} - 2\frac{5}{9}$.

[2.5] **118.** Solve $6x + 4 = -(x + 2) - 3x + 4$.

[2.7] **119. Hummingbird Food** The instructions on a bottle of concentrated hummingbird food indicate that 6 ounces of the concentrate should be mixed with 1 gallon (128 ounces) of water. If you wish to mix the concentrate with only 48 ounces of water, how much concentrate should you use?

[3.2] **120. Tennis Club** A Tennis Club has two payment plans. Plan 1 is a yearly membership fee of \$250 plus \$5.00 per hour for use of the tennis court. Plan 2 is an annual membership fee of \$600 with no charge for court time. How many hours would Malcolm Wu have to play in a year to make the cost of plan 1 equal to the cost of plan 2?

See Exercise 120.

[4.3] **121.** Use scientific notation to evaluate $\dfrac{840{,}000{,}000}{0.0021}$. Leave your answer in scientific notation.

[5.6] **122.** Solve $2x^2 - 3 = x$.

6.4 Addition and Subtraction of Rational Expressions

1 **Add and subtract rational expressions.**

In Section 6.3 we discussed how to add and subtract rational expressions with a common denominator. Now we discuss adding and subtracting rational expressions that are not given with a common denominator.

1 Add and Subtract Rational Expressions

The method used to add and subtract rational expressions with unlike denominators is outlined in Example 1.

EXAMPLE 1 Add $\dfrac{7}{x} + \dfrac{6}{y}$.

Solution First we determine the LCD as outlined in Section 6.3.

$$\text{LCD} = xy$$

We write each fraction with the LCD. We do this by multiplying *both* the numerator and denominator of each fraction by any factors needed to obtain the LCD.

$$\frac{7}{x} + \frac{6}{y} = \frac{7}{x} \cdot \frac{y}{y} + \frac{6}{y} \cdot \frac{x}{x}$$

Multiply $\frac{7}{x}$ by $\frac{y}{y}$ and multiply $\frac{6}{y}$ by $\frac{x}{x}$. Values do not change since $\frac{x}{x} = 1$ and $\frac{y}{y} = 1$.

$$= \frac{7y}{xy} + \frac{6x}{xy}$$

Understanding Algebra

The concept of adding rational expressions is the same as adding fractions in arithmetic. Rewrite the fractions as equivalent fractions with the least common denominator.

Arithmetic:

$$\frac{1}{8} + \frac{1}{2} = \frac{1}{8} + \frac{4}{8} = \frac{1+4}{8} = \frac{5}{8}$$

Algebra:

$$\frac{3}{x^2} + \frac{2}{x^3} = \frac{3x}{x^3} + \frac{2}{x^3} = \frac{3x+2}{x^3}$$

Now we add the numerators, and keep the LCD, xy.

$$\frac{7y}{xy} + \frac{6x}{xy} = \frac{7y + 6x}{xy} \quad \text{or} \quad \frac{6x + 7y}{xy}$$

Now Try Exercise 7

To Add or Subtract Two Rational Expressions with Unlike Denominators

1. Determine the LCD.
2. Rewrite each rational expression as an equivalent rational expression with the LCD. This is done by multiplying both the numerator and denominator of each rational expression by any factors needed to obtain the LCD.
3. Add or subtract the numerators while maintaining the LCD.
4. When possible, factor the remaining numerator and simplify the rational expression.

EXAMPLE 2 Add $\frac{1}{4x^2y} + \frac{3}{14xy^3}$.

Solution The LCD is $28x^2y^3$. We must write each rational expression with the denominator $28x^2y^3$.

$$\frac{1}{4x^2y} + \frac{3}{14xy^3} = \frac{1}{4x^2y} \cdot \frac{7y^2}{7y^2} + \frac{3}{14xy^3} \cdot \frac{2x}{2x}$$

Multiply $\frac{1}{4x^2y}$ by $\frac{7y^2}{7y^2}$ and multiply $\frac{3}{14xy^3}$ by $\frac{2x}{2x}$ to obtain the LCD of $28x^2y^3$.

$$= \frac{7y^2}{28x^2y^3} + \frac{6x}{28x^2y^3}$$

$$= \frac{7y^2 + 6x}{28x^2y^3} \quad \text{or} \quad \frac{6x + 7y^2}{28x^2y^3}$$

Now Try Exercise 15

EXAMPLE 3 Add $\frac{3}{x+2} + \frac{5}{x}$.

Solution The LCD is $x(x + 2)$.

$$\frac{3}{x+2} + \frac{5}{x} = \frac{3}{x+2} \cdot \frac{x}{x} + \frac{5}{x} \cdot \frac{x+2}{x+2}$$

Multiply $\frac{3}{x+2}$ by $\frac{x}{x}$ and multiply $\frac{5}{x}$ by $\frac{x+2}{x+2}$ to obtain the LCD of $x(x + 2)$.

$$= \frac{3x}{x(x+2)} + \frac{5(x+2)}{x(x+2)}$$

Each fraction has been written as an equivalent fraction with the LCD of $x(x + 2)$.

$$= \frac{3x}{x(x+2)} + \frac{5x+10}{x(x+2)}$$

Distributive property was used.

$$= \frac{3x + (5x + 10)}{x(x+2)}$$

Rewrote as a single fraction.

$$= \frac{3x + 5x + 10}{x(x+2)}$$

Parentheses were removed in the numerator.

$$= \frac{8x + 10}{x(x+2)}$$

Like terms were combined in the numerator.

Now Try Exercise 23

HELPFUL HINT

Look at the answer to Example 3, $\dfrac{8x + 10}{x(x + 2)}$. Notice that the numerator could have been factored to obtain $\dfrac{2(4x + 5)}{x(x + 2)}$. Also notice that the denominator could have been multiplied to get $\dfrac{8x + 10}{x^2 + 2x}$. All three of these answers are equivalent and each is correct. In this section, *when writing answers, unless there is a common factor in the numerator and denominator we will leave the numerator in unfactored form and the denominator in factored form.* If both the numerator and denominator have a common factor, we will factor the numerator and simplify the fraction.

EXAMPLE 4 Subtract $\dfrac{w}{w - 7} - \dfrac{6}{w - 4}$.

Solution The LCD is $(w - 7)(w - 4)$.

$$\frac{w}{w - 7} - \frac{6}{w - 4} = \frac{w - 4}{w - 4} \cdot \frac{w}{w - 7} - \frac{6}{w - 4} \cdot \frac{w - 7}{w - 7}$$

Multiply $\dfrac{w}{w - 7}$ by $\dfrac{w - 4}{w - 4}$ and multiply $\dfrac{6}{w - 4}$ by $\dfrac{w - 7}{w - 7}$.

$$= \frac{w(w - 4)}{(w - 4)(w - 7)} - \frac{6(w - 7)}{(w - 4)(w - 7)}$$

Rewrote each fraction as an equivalent fraction with the LCD.

$$= \frac{w^2 - 4w}{(w - 4)(w - 7)} - \frac{6w - 42}{(w - 4)(w - 7)}$$

Distributive property was used in the numerators.

$$= \frac{(w^2 - 4w) - (6w - 42)}{(w - 4)(w - 7)}$$

Wrote as a single fraction.

$$= \frac{w^2 - 4w - 6w + 42}{(w - 4)(w - 7)}$$

Parentheses were removed in the numerator.

$$= \frac{w^2 - 10w + 42}{(w - 4)(w - 7)}$$

Like terms were combined in the numerator.

Now Try Exercise 29

Understanding Algebra

Remember, when subtracting rational expressions, we must be sure to subtract *each* term in the numerator of the second rational expression from the numerator of the first rational expression. In Example 4, the minus sign is applied to each term in the numerator:

$$-(6w - 42) = -6w + 42.$$

AVOIDING COMMON ERRORS

Remember: When subtracting fractions, the subtraction symbol changes the sign of each term in the numerator following it.

CORRECT

$$\frac{5x}{x + 2} - \frac{x - 3}{x + 2} = \frac{5x - x + 3}{x + 2} = \frac{4x + 3}{x + 2}$$

INCORRECT

$$\frac{5x}{x + 2} - \frac{x - 3}{x + 2} = \frac{5x - x - 3}{x + 2} = \frac{4x - 3}{x + 2}$$

EXAMPLE 5 Subtract $\dfrac{x + 2}{x - 4} - \dfrac{x + 3}{x + 4}$.

Solution The LCD is $(x - 4)(x + 4)$.

$$\frac{x + 2}{x - 4} - \frac{x + 3}{x + 4} = \frac{x + 4}{x + 4} \cdot \frac{x + 2}{x - 4} - \frac{x + 3}{x + 4} \cdot \frac{x - 4}{x - 4}$$

$$= \frac{(x + 4)(x + 2)}{(x + 4)(x - 4)} - \frac{(x + 3)(x - 4)}{(x + 4)(x - 4)}$$

Rewrote each fraction as an equivalent fraction with the LCD.

Use the FOIL method to multiply each numerator.

$$= \frac{x^2 + 6x + 8}{(x + 4)(x - 4)} - \frac{x^2 - x - 12}{(x + 4)(x - 4)}$$

$$= \frac{(x^2 + 6x + 8) - (x^2 - x - 12)}{(x + 4)(x - 4)} \quad \text{Wrote as a single fraction.}$$

$$= \frac{x^2 + 6x + 8 - x^2 + x + 12}{(x + 4)(x - 4)} \quad \text{Parentheses were removed in the numerator.}$$

$$= \frac{7x + 20}{(x + 4)(x - 4)} \quad \text{Like terms were combined in the numerator.}$$

Now Try Exercise 37

Consider the problem

$$\frac{4}{x - 2} + \frac{x + 3}{2 - x}$$

Because the denominators are opposites, we can proceed by converting one of the denominators to the other by multiplying by -1. See the Helpful Hint below.

Understanding Algebra

The expressions $a - b$ and $b - a$ are opposites and, therefore, differ only by a factor of -1. Thus,

$$(b - a)(-1) = -b + a = a - b.$$

HELPFUL HINT

When adding or subtracting rational expressions whose denominators are opposites, multiply both the numerator *and* the denominator of *either* rational expression by -1. Then both rational expressions will have the same denominator.

$$\frac{x}{a - b} + \frac{y}{b - a} = \frac{x}{a - b} + \frac{y}{b - a} \cdot \frac{-1}{-1}$$

$$= \frac{x}{a - b} + \frac{-y}{a - b}$$

$$= \frac{x - y}{a - b}$$

EXAMPLE 6 Add $\frac{4}{x - 2} + \frac{x + 3}{2 - x}$.

Solution Since the denominators differ only in sign, we may multiply both the numerator and the denominator of either fraction by -1.

$$\frac{4}{x - 2} + \frac{x + 3}{2 - x} = \frac{4}{x - 2} + \frac{x + 3}{2 - x} \cdot \frac{-1}{-1} \quad \text{Multiply numerator and denominator of the second fraction by } -1.$$

$$= \frac{4}{x - 2} + \frac{(-x - 3)}{x - 2} \quad \text{Notice } (2 - x)(-1) = x - 2.$$

$$= \frac{4 + (-x - 3)}{x - 2} \quad \text{Wrote as a single fraction.}$$

$$= \frac{4 - x - 3}{x - 2} \quad \text{Parentheses were removed in the numerator.}$$

$$= \frac{-x + 1}{x - 2} \quad \text{Like terms were combined in the numerator.}$$

Now Try Exercise 31

Let's work another example where the denominators differ only in sign.

EXAMPLE 7 Subtract $\frac{a - 9}{3a - 4} - \frac{2a - 5}{4 - 3a}$.

Solution The denominators of the two rational expressions differ only in sign.

$$\frac{a - 9}{3a - 4} - \frac{2a - 5}{4 - 3a} = \frac{a - 9}{3a - 4} - \frac{2a - 5}{4 - 3a} \cdot \frac{-1}{-1} \quad \text{Multiply numerator and denominator of the second fraction by } -1.$$

$$= \frac{a-9}{3a-4} - \frac{(-2a+5)}{3a-4}$$ The common denominator is $3a - 4$.

$$= \frac{(a-9)-(-2a+5)}{3a-4}$$ Wrote as a single fraction.

$$= \frac{a-9+2a-5}{3a-4}$$ Parentheses were removed in the numerator.

$$= \frac{3a-14}{3a-4}$$ Like terms were combined in the numerator.

Now Try Exercise 33

EXAMPLE 8 Add $\frac{3}{x^2+5x+6} + \frac{1}{3x^2+8x-3}$.

Solution First, rewrite the denominators in factored form.

$$\frac{3}{x^2+5x+6} + \frac{1}{3x^2+8x-3} = \frac{3}{(x+2)(x+3)} + \frac{1}{(3x-1)(x+3)}$$

The LCD is $(x+2)(x+3)(3x-1)$.

$$= \frac{3x-1}{3x-1} \cdot \frac{3}{(x+2)(x+3)} + \frac{1}{(3x-1)(x+3)} \cdot \frac{x+2}{x+2}$$

$$= \frac{9x-3}{(3x-1)(x+2)(x+3)} + \frac{x+2}{(3x-1)(x+2)(x+3)}$$

$$= \frac{(9x-3)+(x+2)}{(3x-1)(x+2)(x+3)}$$

$$= \frac{9x-3+x+2}{(3x-1)(x+2)(x+3)}$$

$$= \frac{10x-1}{(3x-1)(x+2)(x+3)}$$

Now Try Exercise 55

EXAMPLE 9 Subtract $\frac{5}{x^2-5x} - \frac{x}{5x-25}$.

Solution First, rewrite the denominators in factored form.

$$\frac{5}{x^2-5x} - \frac{x}{5x-25} = \frac{5}{x(x-5)} - \frac{x}{5(x-5)}$$

The LCD is $5x(x-5)$.

$$= \frac{5}{5} \cdot \frac{5}{x(x-5)} - \frac{x}{5(x-5)} \cdot \frac{x}{x}$$

$$= \frac{25}{5x(x-5)} - \frac{x^2}{5x(x-5)}$$

$$= \frac{25-x^2}{5x(x-5)}$$ Factor the numerator.

$$= \frac{(5-x)(5+x)}{5x(x-5)}$$

$$= \frac{-1(x-5)(x+5)}{5x(x-5)}$$ $5 - x = -1(x - 5)$

$$= \frac{-1\cancel{(x-5)}(x+5)}{5x\cancel{(x-5)}}$$ Simplify.

$$= \frac{-1(x+5)}{5x} \quad \text{or} \quad -\frac{x+5}{5x}$$

Now Try Exercise 65

AVOIDING COMMON ERRORS

1. A common error in an addition or subtraction problem is to add or subtract the numerators and the denominators. Here is one such example.

CORRECT

$$\frac{1}{x} + \frac{x}{1} = \frac{1}{x} + \frac{x}{1} \cdot \frac{x}{x}$$

$$= \frac{1}{x} + \frac{x^2}{x}$$

$$= \frac{1 + x^2}{x} \quad \text{or} \quad \frac{x^2 + 1}{x}$$

INCORRECT

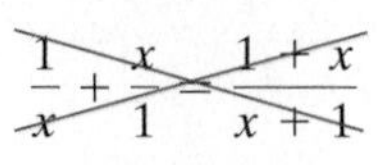

$$\frac{1}{x} - \frac{x}{1} = \frac{1 - x}{x - 1}$$

Remember that to add or subtract rational expressions you must first have a common denominator. Then you add or subtract the numerators while maintaining the common denominator.

2. Another common mistake is to treat an addition or subtraction problem as a multiplication problem. You can divide out common factors only when *multiplying* expressions, not when adding or subtracting them.

CORRECT

$$\frac{1}{x} \cdot \frac{x}{1} = \frac{1}{\cancel{x}_1} \cdot \frac{\cancel{x}^1}{1}$$

$$= 1 \cdot 1 = 1$$

INCORRECT

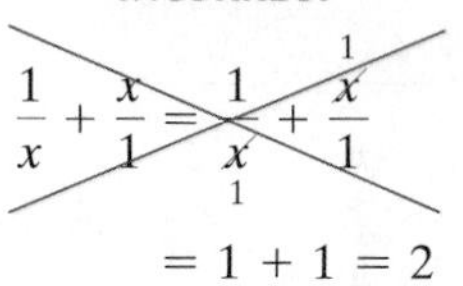

$$= 1 + 1 = 2$$

6.4 Exercise Set MathXL® MyMathLab®

Warm-Up Exercises

Fill in the blanks with the appropriate word, phrase, or symbol(s) from the following list.

simplify	maintaining	eliminating	numerator	denominator
-1	term	factor		

1. To add or subtract two rational expressions with unlike denominators, first determine the least common ________.
2. To rewrite each rational expression with the LCD, multiply both the ________ and the denominator of each rational expression by any factors needed to obtain the LCD.
3. To add or subtract two rational expressions with an LCD, add or subtract the numerators while ________ the LCD.
4. Once the numerators of two rational expressions with an LCD have been added or subtracted, when possible, factor the remaining numerator and ________ the fraction.
5. When adding or subtracting rational expressions whose denominators are opposites, multiply both the numerator and denominator of either rational expression by ________.
6. When subtracting rational expressions, the subtraction symbol changes the sign of each ________ in the numerator following it.

Practice the Skills

Add or subtract.

7. $\frac{2}{x} + \frac{5}{y}$
8. $\frac{7}{m} + \frac{3}{n}$
9. $\frac{3}{t^2} - \frac{4}{3t}$
10. $\frac{7}{s^3} - \frac{2}{5s^2}$
11. $3 + \frac{8}{x}$
12. $\frac{11}{n} + 5$
13. $\frac{2}{x^2} + \frac{3}{5x}$
14. $\frac{5}{6y} + \frac{3}{5y^2}$
15. $\frac{7}{6x^2y^4} - \frac{2}{15x^3y^3}$
16. $\frac{7}{12x^4y} - \frac{1}{5x^2y^3}$
17. $4y + \frac{x}{y}$
18. $x + \frac{2x}{y}$
19. $\frac{3a - 1}{2a} + \frac{5}{3a}$
20. $\frac{4d - 9}{4d} + \frac{5}{6d}$
21. $\frac{3}{y} - \frac{x - 4}{x}$
22. $\frac{x}{7} - \frac{x + 5}{y}$

23. $\frac{9}{p+3} + \frac{2}{p}$
24. $\frac{4}{x} + \frac{9}{x-3}$
25. $\frac{3}{2x} - \frac{x-3}{x^2}$
26. $\frac{5}{4x} - \frac{x+4}{2x^2}$
27. $\frac{a+b}{b} + \frac{b}{a-b}$
28. $\frac{2a}{a+b} + \frac{a-b}{a}$
29. $\frac{7}{x-7} - \frac{5}{x-4}$
30. $\frac{2}{x-3} - \frac{5}{x-1}$
31. $\frac{8}{p-3} + \frac{2}{3-p}$
32. $\frac{3}{p-5} + \frac{7}{5-p}$
33. $\frac{9}{x+7} - \frac{5}{-x-7}$
34. $\frac{6}{7x-1} - \frac{3}{1-7x}$
35. $\frac{8}{a-2} + \frac{a}{2a-4}$
36. $\frac{5}{b+7} + \frac{3b}{2b+14}$
37. $\frac{x+5}{x-5} - \frac{x-5}{x+5}$
38. $\frac{x+7}{x+3} - \frac{x+3}{x+7}$
39. $\frac{11}{2q+6} - \frac{3}{q}$
40. $\frac{5}{6n+3} - \frac{2}{n}$
41. $\frac{3}{2w+10} + \frac{6}{w+2}$
42. $\frac{5k}{4k-8} - \frac{k}{k+2}$
43. $\frac{2}{x-5} + \frac{3}{(x-5)^2}$
44. $\frac{5}{(x+4)^2} + \frac{3}{x+4}$
45. $\frac{x+2}{x^2-4} - \frac{2}{x+2}$
46. $\frac{x+3}{x^2+x-6} - \frac{4}{x+3}$
47. $\frac{3r+4}{r^2-10r+24} - \frac{2}{r-6}$
48. $\frac{x+9}{x^2-3x-10} - \frac{2}{x-5}$
49. $\frac{x-3}{x+4} - \frac{x+5}{x^2+8x+16}$
50. $\frac{x+8}{x^2-4x+4} - \frac{x+1}{x-2}$
51. $\frac{x-6}{x^2+10x+25} + \frac{x-3}{x+5}$
52. $\frac{x+2}{x^2+6x+9} + \frac{x-4}{x+3}$
53. $\frac{5}{a^2-9a+8} - \frac{6}{a^2-6a-16}$
54. $\frac{4}{a^2+2a-15} - \frac{1}{a^2-9}$
55. $\frac{2}{x^2+6x+9} + \frac{7}{x^2+x-6}$
56. $\frac{3}{x^2-4x+4} + \frac{4}{3x^2-2x-8}$
57. $\frac{x}{2x^2+7x+3} - \frac{5}{3x^2+7x-6}$
58. $\frac{x}{5x^2-9x-2} - \frac{1}{3x^2-7x+2}$
59. $\frac{x}{4x^2+11x+6} - \frac{2}{8x^2+2x-3}$
60. $\frac{x}{2x^2-5x-7} - \frac{3}{2x^2-x-21}$
61. $\frac{3w+12}{w^2+w-12} - \frac{2}{w-3}$
62. $\frac{5x+10}{x^2-5x-14} - \frac{4}{x-7}$
63. $\frac{4r}{2r^2-10r+12} + \frac{4}{r-2}$
64. $\frac{6m}{3m^2-24m+48} - \frac{2}{m-4}$
65. $\frac{4}{x^2-4x} - \frac{x}{4x-16}$
66. $\frac{6}{x^2-6x} - \frac{x}{6x-36}$

Problem Solving

For what value(s) of x is each expression defined?

67. $\frac{8}{x} + 6$
68. $\frac{6}{x-1} - \frac{5}{x}$
69. $\frac{3}{x-4} + \frac{7}{x+6}$
70. $\frac{4}{x^2-9} - \frac{9}{x+3}$

Add or subtract. Treat the unknown symbols as if they were variables.

71. $\frac{3}{\Delta-2} - \frac{4}{2-\Delta}$
72. $\frac{\Delta}{2\Delta^2+7\Delta-4} + \frac{2}{\Delta^2-\Delta-20}$

Challenge Problems

Under what conditions is each expression defined? Explain your answers.

73. $\frac{5}{a+b} + \frac{4}{a}$
74. $\frac{x+2}{x+5y} - \frac{y-2}{3x}$

Perform each indicated operation.

75. $\frac{x}{x^2-9} + \frac{2x}{x+3} + \frac{2x^2-5x}{9-x^2}$
76. $\frac{8x+9}{x^2+x-6} + \frac{x}{x+3} - \frac{5}{x-2}$

77. $\frac{x+6}{4-x^2} - \frac{x+3}{x+2} + \frac{x-3}{2-x}$

78. $\frac{3x-1}{x+2} + \frac{x}{x-3} - \frac{4}{2x+3}$

79. $\frac{2}{x^2-x-6} + \frac{3}{x^2-2x-3} + \frac{1}{x^2+3x+2}$

80. $\frac{3x}{x^2-4} + \frac{4}{x^3+8}$

Group Activity

Discuss and answer Exercise 81 as a group.

81. a) As a group, find the LCD of

$$\frac{x+3y}{x^2+3xy+2y^2} + \frac{y-x}{2x^2+3xy+y^2}$$

b) As a group, perform the indicated operation, but do not simplify your answer.

c) As a group, simplify your answer.

d) Group member 1: Substitute 2 for x and 1 for y in the fraction on the left in part **a)** and evaluate.

e) Group member 2: Substitute 2 for x and 1 for y in the fraction on the right in part **a)** and evaluate.

f) Group member 3: Add the numerical fractions found in parts **d)** and **e)**.

g) Individually, substitute 2 for x and 1 for y in the expression obtained in part **b)** and evaluate.

h) Individually, substitute 2 for x and 1 for y in the expression obtained in part **c)**, evaluate, and compare your answers.

i) As a group, discuss what you discovered from this activity.

j) Do you think your results would have been similar for any numbers substituted for x and y (for which the denominator is not 0)? Why?

Cumulative Review Exercises

[2.7] **82. White Pass Railroad** The White Pass Railroad is a narrow gauge railroad that travels slowly through the mountains of Alaska. If the train travels 22 miles in 0.8 hours, how long will it take to travel 42 miles? Assume the train travels at the same rate throughout the trip.

[2.8] **83.** Solve the inequality $3(x-2)+2 < 4(x+1)$. Graph the solution on a number line and write the solution using interval notation.

[4.6] **84.** Divide $(8x^2+6x-15) \div (2x+3)$.

[6.2] **85.** Multiply $\frac{x^2+xy-6y^2}{x^2-xy-2y^2} \cdot \frac{y^2-x^2}{x^2+2xy-3y^2}$.

MID-CHAPTER TEST: 6.1–6.4

To find out how well you understand the chapter material to this point, take this brief test. The answers, and the section where the material was initially discussed, are given in the back of the book. Review any questions that you answered incorrectly.

Determine the value or values of the variable where each expression is defined.

1. $\frac{9}{3x-2}$

2. $\frac{2x+1}{x^2-5x-14}$

Simplify each rational expression.

3. $\frac{9x+18}{x+2}$

4. $\frac{2x^2+13x+15}{3x^2+14x-5}$

5. $\frac{25r^2-36t^2}{5r-6t}$

Multiply or divide as indicated.

6. $\frac{15x^2}{2y} \cdot \frac{4y^4}{5x^5}$

7. $\frac{m-3}{m+4} \cdot \frac{m^2+8m+16}{3-m}$

8. $\frac{x^3+27}{x^2-2x-15} \cdot \frac{x^2-7x+10}{x^2-3x+9}$

9. $\frac{5x-1}{x^2+11x+10} \div \frac{10x-2}{x^2+17x+70}$

10. $\frac{5x^2+7x+2}{x^2+6x+5} \div \frac{7x^2-39x-18}{x^2-x-30}$

(Continued)

Add or subtract as indicated.

11. $\dfrac{x^2}{x+6} - \dfrac{36}{x+6}$

12. $\dfrac{2x^2 - 2x}{2x+5} + \dfrac{x-15}{2x+5}$

13. $\dfrac{3t^2 - t}{4t^2 - 9t + 2} - \dfrac{3t+4}{4t^2 - 9t + 2}$

Find the least common denominator.

14. $\dfrac{2m}{6m^2 + 3m} + \dfrac{m+7}{2m+1}$

15. $\dfrac{9x+8}{2x^2 - 5x - 12} + \dfrac{2x+3}{x^2 - 9x + 20}$

For Exercises 16–19, add or subtract as indicated.

16. $\dfrac{x+1}{2x} + \dfrac{4x-3}{5x}$

17. $\dfrac{2a+5}{a+3} - \dfrac{3a+1}{a-4}$

18. $\dfrac{x^2+5}{2x^2 + 13x + 6} + \dfrac{3x-1}{2x+1}$

19. $\dfrac{x}{x^2 + 3x + 2} - \dfrac{4}{x^2 - x - 6}$

20. To add the rational expressions $\dfrac{7}{x+1} + \dfrac{8}{x}$, Samuel Ditsi decided to add both numerators and then add both denominators to get $\dfrac{7+8}{(x+1)+x}$, which simplified to $\dfrac{15}{2x+1}$. This procedure is wrong. Why is it wrong? Explain your answer. Then add the rational expressions $\dfrac{7}{x+1} + \dfrac{8}{x}$ correctly.

6.5 Complex Fractions

1 Simplify complex fractions by simplifying numerator and denominator.

2 Simplify complex fractions using multiplication first to clear fractions.

In this section we discuss expressions called *complex fractions*. We will first define complex fractions and then we will discuss two methods to simplify complex fractions.

Complex Fraction

A **complex fraction** is a rational expression that has a rational expression in its numerator or its denominator or in both its numerator and denominator.

Examples of Complex Fractions

$$\frac{\frac{3}{5}}{7} \qquad \frac{\frac{x+9}{x}}{4x} \qquad \frac{\frac{x}{y}}{x+1} \qquad \frac{\frac{a+b}{a}}{\frac{a-b}{b}} \qquad \frac{\frac{2}{3}+\frac{4}{x}}{\frac{1}{x}-\frac{5}{6}}$$

The expression above the *main fraction bar* is the numerator of the complex fraction, and the expression below the main fraction bar is the denominator of the complex fraction.

Numerator of complex fraction $\left\{\dfrac{a+b}{a}\right.$

← Main fraction bar

Denominator of complex fraction $\left\{\dfrac{a-b}{b}\right.$

Understanding Algebra

The complex fraction

$$\frac{\frac{a}{b}}{\frac{c}{d}}$$

is another way to represent the division

$$\frac{a}{b} \div \frac{c}{d}.$$

We perform the division by multiplying the first fraction by the reciprocal of the second fraction:

$$\frac{a}{b} \cdot \frac{d}{c}$$

1 Simplify Complex Fractions by Simplifying Numerator and Denominator

There are two methods that may be used to simplify complex fractions. The first method we use to simplify complex fractions reinforces many of the concepts used in this chapter because we may need to add, subtract, multiply, and divide simpler fractions as we simplify the complex fraction.

Method 1—Simplify a Complex Fraction by Simplifying Numerator and Denominator

1. Add or subtract the rational expression in both the numerator and denominator of the complex fraction to obtain single rational expression in both the numerator and the denominator.
2. Multiply the rational expression in the numerator by the reciprocal of the rational expression in the denominator.
3. Simplify further if possible.

EXAMPLE 1 Simplify $\dfrac{\frac{ab^2}{c^3}}{\frac{a}{bc^2}}$.

Solution Since both numerator and denominator are already single rational expressions, we begin with step 2.

$$\frac{\frac{ab^2}{c^3}}{\frac{a}{bc^2}} = \frac{ab^2}{c^3} \div \frac{a}{bc^2} \quad \text{A fraction bar means "divided by."}$$

$$= \frac{ab^2}{c^3} \cdot \frac{bc^2}{a} \quad \text{Multiply by } \frac{bc^2}{a}\text{, the reciprocal of } \frac{a}{bc^2}.$$

$$= \frac{b^3}{c} \quad \text{Both } a \text{ and } c^2 \text{ were factored out of the numerator and denominator.}$$

Thus the expression simplifies to $\frac{b^3}{c}$.

Now Try Exercise 11

EXAMPLE 2 Simplify $\dfrac{a + \frac{1}{x}}{x + \frac{1}{a}}$.

Solution Step 1 is to express the numerator and denominator as single fractions. To obtain one fraction in the numerator, we notice that the LCD is x. Multiply a by $\frac{x}{x}$.

$$a + \frac{1}{x} = a \cdot \frac{x}{x} + \frac{1}{x} = \frac{ax + 1}{x} \quad \text{We use this as the numerator.}$$

To obtain one fraction in the denominator, we notice that the LCD is a. Multiply x by $\frac{a}{a}$.

$$x + \frac{1}{a} = x \cdot \frac{a}{a} + \frac{1}{a} = \frac{ax + 1}{a} \quad \text{We use this as the denominator.}$$

So

$$\frac{a + \frac{1}{x}}{x + \frac{1}{a}} = \frac{\frac{ax + 1}{x}}{\frac{ax + 1}{a}} = \frac{ax + 1}{x} \div \frac{ax + 1}{a}$$

$$= \frac{ax + 1}{x} \cdot \frac{a}{ax + 1} = \frac{a}{x}$$

Now Try Exercise 25

In Example 4 we will rework Example 2 using the second method.

2 Simplify Complex Fractions Using Multiplication First to Clear Fractions

The second method we use to simplify complex fractions involves multiplying both numerator and denominator by the least common denominator to clear the rational expressions. Many students prefer this method because the answer may be obtained more quickly.

Method 2—Simplify a Complex Fraction Using Multiplication First

1. Find the least common denominator of *all* the denominators appearing in the complex fraction.
2. Multiply both the numerator and denominator of the complex fraction by the LCD found in step 1.
3. Simplify when possible.

EXAMPLE 3 Simplify $\dfrac{\frac{2}{3}+\frac{1}{5}}{\frac{4}{5}-\frac{1}{3}}$.

Solution The denominators in the complex fraction are 3 and 5. Multiply both the numerator and denominator of the complex fraction by 15, the LCD of the complex fraction.

$$\frac{\frac{2}{3}+\frac{1}{5}}{\frac{4}{5}-\frac{1}{3}} = \frac{15}{15}\cdot\frac{\left(\frac{2}{3}+\frac{1}{5}\right)}{\left(\frac{4}{5}-\frac{1}{3}\right)} = \frac{15\left(\frac{2}{3}\right)+15\left(\frac{1}{5}\right)}{15\left(\frac{4}{5}\right)-15\left(\frac{1}{3}\right)}$$

Now simplify.

$$= \frac{10+3}{12-5} = \frac{13}{7}$$

Now Try Exercise 9

Understanding Algebra

The LCD of the complex fraction

$$\frac{\frac{a+b}{c}}{\frac{d+e}{f}}$$

is cf (c from the numerator and f from the denominator). To eliminate fractions within the complex fraction, we multiply both the numerator and the denominator by cf:

$$\frac{cf}{cf}\cdot\frac{\left(\frac{a+b}{c}\right)}{\left(\frac{d+e}{f}\right)} = \frac{f(a+b)}{c(d+e)}$$

Now we will rework Example 2 using method 2.

EXAMPLE 4 Simplify $\dfrac{a+\frac{1}{x}}{x+\frac{1}{a}}$.

Solution The denominators in the complex fraction are x and a. Multiply both the numerator and denominator of the complex fraction by ax, the LCD of the complex fraction.

$$\frac{a+\frac{1}{x}}{x+\frac{1}{a}} = \frac{ax}{ax}\cdot\frac{\left(a+\frac{1}{x}\right)}{\left(x+\frac{1}{a}\right)} = \frac{a^2x+a}{ax^2+x}$$

$$= \frac{a\cancel{(ax+1)}}{x\cancel{(ax+1)}} = \frac{a}{x}$$

Now Try Exercise 25

Note that the answers to Examples 2 and 4 are the same.

EXAMPLE 5 Simplify $\dfrac{y^2}{\frac{1}{x}+\frac{1}{y}}$.

Solution The denominators in the complex fraction are x and y. Multiply both the numerator and denominator of the complex fraction by xy, the LCD of the complex fraction.

$$\frac{y^2}{\frac{1}{x}+\frac{1}{y}} = \frac{xy}{xy}\cdot\frac{y^2}{\left(\frac{1}{x}+\frac{1}{y}\right)}$$

$$= \frac{xy^3}{xy\left(\frac{1}{x}\right)+xy\left(\frac{1}{y}\right)}$$

$$= \frac{xy^3}{y+x}$$

Now Try Exercise 33

> **HELPFUL HINT**
>
> We have presented two methods for simplifying complex fractions. Which method should you use? Although either method can be used to simplify complex fractions, most students prefer to use method 1 when both the numerator and denominator consist of a single term, as in Example 1. When the complex fraction has a sum or difference of expressions in either the numerator or denominator, as in Examples 2, 3, 4, or 5, most students prefer to use method 2.

6.5 Exercise Set MathXL® MyMathLab®

Warm-Up Exercises

Fill in the blanks with the appropriate word, phrase, or symbol(s) from the following list.

simple reciprocal fraction complex opposite denominator

1. A rational expression that has a rational expression in its numerator or its denominator or in both its numerator and denominator is called a __________ fraction.

2. The numerator of a complex fraction and the denominator of a complex fraction are separated by the main __________ bar.

3. To simplify a complex fraction using method 1, we multiply the numerator by the __________ of the denominator.

4. To simplify a complex fraction using method 2, we multiply both numerator and denominator by the least common __________.

Practice the Skills

Simplify.

5. $\dfrac{1 + \dfrac{2}{3}}{5 + \dfrac{1}{3}}$ **6.** $\dfrac{4 - \dfrac{3}{4}}{5 - \dfrac{1}{4}}$ **7.** $\dfrac{2 + \dfrac{3}{8}}{1 + \dfrac{1}{3}}$ **8.** $\dfrac{2 + \dfrac{4}{5}}{1 - \dfrac{9}{16}}$

9. $\dfrac{\dfrac{2}{3} + \dfrac{1}{4}}{\dfrac{5}{6} - \dfrac{1}{3}}$ **10.** $\dfrac{\dfrac{5}{8} + \dfrac{1}{3}}{\dfrac{11}{12} - \dfrac{1}{6}}$ **11.** $\dfrac{\dfrac{11a}{b^3}}{\dfrac{b^2}{4}}$ **12.** $\dfrac{\dfrac{xy^2}{7}}{\dfrac{3}{x^2}}$

13. $\dfrac{\dfrac{12p^3}{7r^2s}}{\dfrac{9p^4}{14r^3s^2}}$ **14.** $\dfrac{\dfrac{18x^4}{5y^4z^5}}{\dfrac{9xy^2}{15z^5}}$ **15.** $\dfrac{a - \dfrac{a}{b}}{\dfrac{3 + a}{b}}$ **16.** $\dfrac{\dfrac{j}{k} - 5}{\dfrac{j + 5}{k}}$

17. $\dfrac{\dfrac{9}{t} + \dfrac{3}{t^2}}{3 + \dfrac{1}{t}}$ **18.** $\dfrac{\dfrac{4}{a} + \dfrac{1}{2a}}{a + \dfrac{a}{2}}$ **19.** $\dfrac{5 - \dfrac{1}{x}}{4 - \dfrac{1}{x}}$ **20.** $\dfrac{7 + \dfrac{3}{n}}{9 - \dfrac{2}{n}}$

21. $\dfrac{\dfrac{m}{n} - \dfrac{n}{m}}{\dfrac{m + n}{n}}$ **22.** $\dfrac{\dfrac{g - h}{h}}{\dfrac{g}{h} - \dfrac{h}{g}}$ **23.** $\dfrac{\dfrac{a^2}{b} - b}{\dfrac{b^2}{a} - a}$ **24.** $\dfrac{b - \dfrac{b^2}{a}}{a - \dfrac{a^2}{b}}$

25. $\dfrac{2 - \dfrac{a}{b}}{\dfrac{a}{b} - 2}$ **26.** $\dfrac{\dfrac{x}{y} - 9}{\dfrac{-x}{y} + 9}$ **27.** $\dfrac{\dfrac{s - t}{t^2}}{\dfrac{s^2 - t^2}{t^3}}$ **28.** $\dfrac{\dfrac{a^2 - b^2}{a}}{\dfrac{a + b}{a^4}}$

29. $\dfrac{\dfrac{1}{a} - \dfrac{1}{b}}{\dfrac{1}{ab}}$ **30.** $\dfrac{\dfrac{2}{a} + \dfrac{3}{b}}{\dfrac{1}{a}}$ **31.** $\dfrac{\dfrac{a}{b} + \dfrac{1}{a}}{\dfrac{b}{a} + \dfrac{1}{a}}$ **32.** $\dfrac{\dfrac{1}{a} - \dfrac{1}{b}}{\dfrac{1}{a} + \dfrac{1}{b}}$

33. $\dfrac{x}{\frac{1}{x}-\frac{1}{y}}$

34. $\dfrac{\frac{1}{a}+\frac{1}{b}}{ab}$

35. $\dfrac{\frac{1}{b^2}-\frac{1}{a^2}}{\frac{1}{b^2}-\frac{1}{ab}}$

36. $\dfrac{\frac{a}{b^2}-\frac{b}{a^2}}{\frac{1}{b^2}-\frac{1}{a^2}}$

Problem Solving

For the complex fractions in Exercises 37–40,

a) *Determine which of the two methods discussed in this section you would use to simplify the fraction. Explain why.*

b) *Simplify by the method you selected in part* **a)**.

c) *Simplify by the method you did not select in part* **a)**. *If your answers to parts* **b)** *and* **c)** *are not the same, explain why.*

37. $\dfrac{5+\frac{3}{5}}{\frac{1}{8}-4}$

38. $\dfrac{\frac{x+y}{x^3}-\frac{1}{x}}{\frac{x-y}{x^5}+5}$

39. $\dfrac{\frac{x-y}{x+y}+\frac{6}{x+y}}{2-\frac{7}{x+y}}$

40. $\dfrac{\frac{25}{x-y}+\frac{2}{x+y}}{\frac{5}{x-y}-\frac{3}{x+y}}$

In Exercises 41 and 42, **a)** *write the complex fraction, and* **b)** *simplify the complex fraction.*

41. The numerator of the complex fraction consists of one term: 5 is divided by $12x$. The denominator of the complex fraction consists of two terms: 4 divided by $3x$ is subtracted from 8 divided by x^2.

42. The numerator of the complex fraction consists of two terms: 3 divided by $2x$ is subtracted from 6 divided by x. The denominator of the complex fraction consists of two terms: the sum of x and the quantity 1 divided by x.

Concept/Writing Exercises

43. Explain what a complex fraction is.

44. **a)** Select the method you prefer to simplify complex fractions and then write down a step-by-step procedure for simplifying complex fractions using that method.

b) Using your answer to part **a)**, simplify $\dfrac{\frac{4}{x}-\frac{3}{y}}{x+\frac{1}{y}}$.

Challenge Problems

Simplify. (Hint: Refer to Section 4.2, which discusses negative exponents.)

45. $\dfrac{x^{-1}+y^{-1}}{3}$

46. $\dfrac{x^{-1}+y^{-1}}{y^{-1}}$

47. $\dfrac{x^{-1}+y^{-1}}{x^{-1}y^{-1}}$

48. $\dfrac{x^{-2}-y^{-2}}{y^{-1}-x^{-1}}$

49. **Jack** The efficiency of a jack, E, is expressed by the formula $E=\dfrac{\frac{1}{2}h}{h+\frac{1}{2}}$, where h is determined by the pitch of the jack's thread. Determine the efficiency of a jack if h is

a) $\frac{2}{3}$ **b)** $\frac{4}{5}$

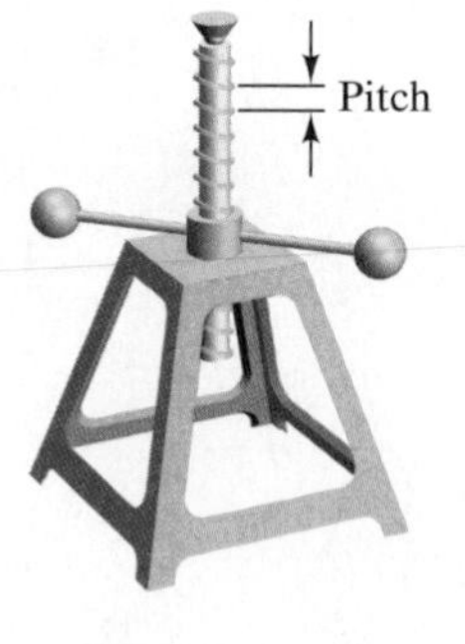

Simplify.

50. $\dfrac{\frac{x}{y}+\frac{y}{x}+\frac{2}{x}}{\frac{x}{y}+y}$

51. $\dfrac{\frac{a}{b}+b-\frac{1}{a}}{\frac{a}{b^2}-\frac{b}{a}+\frac{3}{a^2}}$

52. $\dfrac{x}{4+\frac{x}{1+x}}$

Cumulative Review Exercises

[2.5] 53. Solve the equation
$2x-8(5-x)=9x-3(x+2)$.

[4.4] 54. What is a polynomial?

[5.3] 55. Factor $x^2-13x+40$.

[6.4] 56. Subtract $\dfrac{x}{3x^2+17x-6}-\dfrac{2}{x^2+3x-18}$.

6.6 Solving Rational Equations

1. Solve rational equations with integer denominators.
2. Solve rational equations where a variable appears in a denominator.

1 Solve Rational Equations with Integer Denominators

A **rational equation** is one that contains one or more rational expressions. A rational equation may contain rational coefficients, such as $\frac{1}{2}x + \frac{3}{5}x = 8$ or $\frac{x}{2} + \frac{3x}{5} = 8$. A rational equation may also have a variable in a denominator, such as $\frac{4}{x-2} = 5$.

The emphasis of this section will be on solving rational equations where a variable appears in a denominator.

To Solve Rational Equations

1. Determine the least common denominator (LCD) of all fractions in the equation.
2. Multiply *both* sides of the equation by the LCD. *This will result in every term in the equation being multiplied by the LCD.*
3. Remove any parentheses and combine like terms on each side of the equation.
4. Solve the equation using the properties discussed in earlier chapters.
5. Check your solution in the *original* equation.

The purpose of multiplying both sides of the equation by the LCD (step 2) is to eliminate all rational expressions from the equation. After both sides of the equation are multiplied by the LCD, the resulting equation should contain no rational expressions.

EXAMPLE 1 Solve $\frac{t}{4} - \frac{t}{5} = 1$ for t.

Solution The LCD of 4 and 5 is 20. Multiply both sides of the equation by 20.

$$\frac{t}{4} - \frac{t}{5} = 1$$

$$20\left(\frac{t}{4} - \frac{t}{5}\right) = 20 \cdot 1 \qquad \text{Multiply both sides by the LCD, 20.}$$

$$20\left(\frac{t}{4}\right) - 20\left(\frac{t}{5}\right) = 20 \qquad \text{Distributive property}$$

$$5t - 4t = 20$$

$$t = 20$$

Check

$$\frac{t}{4} - \frac{t}{5} = 1$$

$$\frac{20}{4} - \frac{20}{5} \stackrel{?}{=} 1$$

$$5 - 4 \stackrel{?}{=} 1$$

$$1 = 1 \qquad \text{True}$$

The solution is 20.

Now Try Exercise 13

Understanding Algebra

The purpose of multiplying both sides of an equation by the least common denominator is to eliminate all rational expressions from the equation. In Example 1, after we multiply both sides of the equation by 20, the resulting equation, $5t - 4t = 20$, contains no rational expressions.

EXAMPLE 2 Solve $\frac{x-5}{30} = \frac{4}{5} - \frac{x-1}{10}$.

Solution Multiply both sides of the equation by the LCD, 30.

$$\frac{x-5}{30} = \frac{4}{5} - \frac{x-1}{10}$$

$$30\left(\frac{x-5}{30}\right) = 30\left(\frac{4}{5} - \frac{x-1}{10}\right) \quad \text{Multiply both sides by the LCD, 30.}$$

$$x - 5 = 30\left(\frac{4}{5}\right) - 30\left(\frac{x-1}{10}\right) \quad \text{Distributive property}$$

$$x - 5 = 24 - 3(x - 1)$$

$$x - 5 = 24 - 3x + 3 \quad \text{Distributive property}$$

$$x - 5 = -3x + 27 \quad \text{Combined like terms.}$$

$$4x - 5 = 27 \quad 3x \text{ was added to both sides.}$$

$$4x = 32 \quad \text{5 was added to both sides.}$$

$$x = 8 \quad \text{Both sides were divided by 4.}$$

A check will show that the answer is 8.

Now Try Exercise 27

2 Solve Rational Equations Where a Variable Appears in a Denominator

Recall from Section 6.1 that whenever a rational expression has a variable in the denominator, we always assume that the value or values of the variable that make the denominator 0 are excluded. Therefore, when solving a rational equation where a variable appears in any denominator, you *must* check your answer. *Whenever a variable appears in any denominator of a rational equation, it is necessary to check your answer in the original equation. If the answer obtained makes any denominator equal to zero, that value is not a solution to the equation and must be excluded.* Such values are called **extraneous roots** or **extraneous solutions**.

EXAMPLE 3 Solve $4 - \frac{5}{x} = \frac{3}{2}$.

Solution Multiply both sides of the equation by the LCD, $2x$.

$$2x\left(4 - \frac{5}{x}\right) = \left(\frac{3}{2}\right) \cdot 2x \quad \text{Multiply both sides by the LCD, } 2x.$$

$$2x(4) - 2x\left(\frac{5}{x}\right) = \left(\frac{3}{2}\right) \cdot 2x \quad \text{Distributive property}$$

$$8x - 10 = 3x$$

$$5x - 10 = 0 \quad 3x \text{ was subtracted from both sides.}$$

$$5x = 10 \quad \text{10 was added to both sides.}$$

$$x = 2$$

Check

$$4 - \frac{5}{x} = \frac{3}{2}$$

$$4 - \frac{5}{2} \stackrel{?}{=} \frac{3}{2}$$

$$\frac{8}{2} - \frac{5}{2} \stackrel{?}{=} \frac{3}{2}$$

$$\frac{3}{2} = \frac{3}{2} \quad \text{True}$$

Since 2 does check, it is the solution to the equation.

Now Try Exercise 17

Understanding Algebra

Whenever a variable appears in any denominator of a rational equation, it is necessary to check your answer in the *original* equation. If the answer obtained makes any denominator zero, that value is not a solution to the equation and must be excluded.

EXAMPLE 4 Solve $\dfrac{p-5}{p+3} = \dfrac{1}{5}$.

Solution The LCD is $5(p + 3)$. Multiply both sides of the equation by the LCD.

$$5(p+3) \cdot \frac{(p-5)}{p+3} = \frac{1}{5} \cdot 5(p+3)$$
$$5(p-5) = 1(p+3)$$
$$5p - 25 = p + 3$$
$$4p - 25 = 3$$
$$4p = 28$$
$$p = 7$$

A check will show that 7 is the solution.

Now Try Exercise 43

In Section 2.7 we stated that equations of the form

$$\frac{a}{b} = \frac{c}{d}$$

are called proportions, which can be solved by cross-multiplying to obtain $a \cdot d = b \cdot c$. Example 4 is a proportion and could have been solved by cross-multiplying. We will use cross-multiplication in Example 5.

EXAMPLE 5 Use cross-multiplication to solve $\dfrac{9}{x+1} = \dfrac{5}{x-3}$.

Solution

$$\frac{9}{x+1} = \frac{5}{x-3}$$
$$9(x-3) = 5(x+1) \quad \text{Cross-multiplied.}$$
$$9x - 27 = 5x + 5 \quad \text{Distributive property was used.}$$
$$4x - 27 = 5$$
$$4x = 32$$
$$x = 8$$

A check will show that 8 is the solution to the equation.

Now Try Exercise 41

Understanding Algebra

Equations of the form $\dfrac{a}{b} = \dfrac{c}{d}$ are called proportions and can be solved by cross-multiplying to obtain the equation $ad = bc$.

The following examples involve quadratic equations. Recall from Section 5.6 that quadratic equations have the form $ax^2 + bx + c = 0$, where $a \neq 0$.

EXAMPLE 6 Solve $x + \dfrac{12}{x} = -7$.

Solution

$$x + \frac{12}{x} = -7$$
$$x \cdot \left(x + \frac{12}{x}\right) = -7 \cdot x \quad \text{Multiply both sides by } x.$$
$$x(x) + x\left(\frac{12}{x}\right) = -7x \quad \text{Distributive property was used.}$$
$$x^2 + 12 = -7x$$
$$x^2 + 7x + 12 = 0 \quad 7x \text{ was added to both sides.}$$
$$(x+3)(x+4) = 0 \quad \text{Factored.}$$
$$x + 3 = 0 \quad \text{or} \quad x + 4 = 0 \quad \text{Zero-factor property}$$
$$x = -3 \qquad\qquad x = -4$$

Check

$x = -3$

$$x + \frac{12}{x} = -7$$
$$-3 + \frac{12}{-3} \stackrel{?}{=} -7$$
$$-3 + (-4) \stackrel{?}{=} -7$$
$$-7 = -7 \quad \text{True}$$

$x = -4$

$$x + \frac{12}{x} = -7$$
$$-4 + \frac{12}{-4} \stackrel{?}{=} -7$$
$$-4 + (-3) \stackrel{?}{=} -7$$
$$-7 = -7 \quad \text{True}$$

The solutions are -3 and -4.

Now Try Exercise 57

EXAMPLE 7 Solve $\dfrac{x^2 - 2x}{x - 6} = \dfrac{24}{x - 6}$.

Solution If we try to solve this equation using cross-multiplication we will get a cubic equation. We will solve this equation by multiplying both sides of the equation by the LCD, $x - 6$.

Understanding Algebra

When multiplying both sides of an equation by a variable expression, checking answers is crucial to determine if any of your answers are extraneous solutions that must be excluded from the final answer.

$$\frac{x^2 - 2x}{x - 6} = \frac{24}{x - 6}$$

$$\cancel{x-6} \cdot \frac{x^2 - 2x}{\cancel{x - 6}} = \frac{24}{\cancel{x - 6}} \cdot \cancel{x-6} \qquad \text{Multiply both sides by the LCD, } x - 6.$$

$$x^2 - 2x = 24$$

$$x^2 - 2x - 24 = 0 \qquad \text{24 was subtracted from both sides.}$$

$$(x + 4)(x - 6) = 0 \qquad \text{Factored.}$$

$$x + 4 = 0 \quad \text{or} \quad x - 6 = 0 \qquad \text{Zero-factor property}$$

$$x = -4 \qquad\qquad x = 6$$

Check

$x = -4$

$$\frac{x^2 - 2x}{x - 6} = \frac{24}{x - 6}$$
$$\frac{(-4)^2 - 2(-4)}{-4 - 6} \stackrel{?}{=} \frac{24}{-4 - 6}$$
$$\frac{16 + 8}{-10} \stackrel{?}{=} \frac{24}{-10}$$
$$\frac{24}{-10} = \frac{24}{-10} \quad \text{True}$$

$x = 6$

$$\frac{x^2 - 2x}{x - 6} = \frac{24}{x - 6}$$
$$\frac{6^2 - 2(6)}{6 - 6} \stackrel{?}{=} \frac{24}{6 - 6}$$
$$\frac{24}{0} = \frac{24}{0}$$

Since the denominator is 0, and we cannot divide by 0, 6 is not a solution.

Since $\dfrac{24}{0}$ is not a real number, 6 is an extraneous solution. Thus, this equation has only one solution, -4.

Now Try Exercise 47

HELPFUL HINT

Remember, when solving a rational equation in which a variable appears in a denominator, you must check *all* your answers to make sure that none is an extraneous root. If any of your answers make any denominator 0, that answer is an extraneous root and not a true solution.

EXAMPLE 8 Solve $\dfrac{5w}{w^2 - 4} + \dfrac{1}{w - 2} = \dfrac{4}{w + 2}$.

Solution First factor $w^2 - 4$.

$$\frac{5w}{(w + 2)(w - 2)} + \frac{1}{w - 2} = \frac{4}{w + 2}$$

Multiply both sides of the equation by the LCD, $(w + 2)(w - 2)$.

$$(w+2)(w-2)\left[\frac{5w}{(w+2)(w-2)} + \frac{1}{w-2}\right] = \frac{4}{w+2}\cdot(w+2)(w-2)$$

$$(w+2)(w-2)\cdot\frac{5w}{(w+2)(w-2)} + (w+2)(w-2)\cdot\frac{1}{w-2} = \frac{4}{w+2}\cdot(w+2)(w-2)$$

$$\cancel{(w+2)}\cancel{(w-2)}\cdot\frac{5w}{\cancel{(w+2)}\cancel{(w-2)}} + (w+2)\cancel{(w-2)}\cdot\frac{1}{\cancel{w-2}} = \frac{4}{\cancel{w+2}}\cdot\cancel{(w+2)}(w-2)$$

$$5w + (w + 2) = 4(w - 2)$$
$$6w + 2 = 4w - 8$$
$$2w + 2 = -8$$
$$2w = -10$$
$$w = -5$$

A check will show that -5 is the solution to the equation.

Now Try Exercise 61

Understanding Algebra

Remember, rational *expressions* ***do not*** contain an equals sign. Rational *equations* ***do*** contain an equals sign.

When adding and subtracting rational expressions, we usually end up with an algebraic expression.

When solving rational equations, the solution, if one exists, will be a numerical value or values.

HELPFUL HINT

Some students confuse adding and subtracting rational expressions with solving rational equations. Remember, rational *expressions* ***do not*** contain an equals sign. Rational *equations* ***do*** contain an equals sign. When adding or subtracting rational expressions, we must rewrite each expression with a common denominator. When solving a rational equation, we multiply both sides of the equation by the LCD to eliminate fractions from the equation. Consider the following two problems. Note that the one on the right is an equation because it contains an equals sign. We will work both problems. The LCD for both problems is $x(x + 4)$.

Adding Rational Expressions

$$\frac{x+2}{x+4} + \frac{3}{x}$$

We rewrite each fraction with the LCD, $x(x + 4)$.

$$= \frac{x}{x}\cdot\frac{x+2}{x+4} + \frac{3}{x}\cdot\frac{x+4}{x+4}$$

$$= \frac{x(x+2)}{x(x+4)} + \frac{3(x+4)}{x(x+4)}$$

$$= \frac{x^2+2x}{x(x+4)} + \frac{3x+12}{x(x+4)}$$

$$= \frac{x^2+2x+3x+12}{x(x+4)}$$

$$= \frac{x^2+5x+12}{x(x+4)}$$

Solving Rational Equations

$$\frac{x+2}{x+4} = \frac{3}{x}$$

We eliminate fractions by multiplying both sides of the equation by the LCD, $x(x + 4)$.

$$(x)\cancel{(x+4)}\left(\frac{x+2}{\cancel{x+4}}\right) = \frac{3}{\cancel{x}}\cancel{(x)}(x+4)$$

$$x(x + 2) = 3(x + 4)$$
$$x^2 + 2x = 3x + 12$$
$$x^2 - x - 12 = 0$$
$$(x - 4)(x + 3) = 0$$
$$x - 4 = 0 \quad \text{or} \quad x + 3 = 0$$
$$x = 4 \qquad\qquad x = -3$$

The numbers 4 and -3 on the right will both check and are thus solutions to the equation.

6.6 Exercise Set MathXL® MyMathLab®

Warm-Up Exercises

Fill in the blanks with the appropriate word, phrase, or symbol(s) from the following list.

cross extraneous quadratic check eliminate proportions equals rational linear

1. Equations that contain one or more rational expressions are called ________ equations.

2. The purpose of multiplying both sides of a rational equation by the LCD is to ________ all fractions from the equation.

3. When solving a rational equation where a variable appears in any denominator, you must __________ your answer for extraneous solutions.

4. If the answer to a rational equation makes any denominator equal to zero, that value is not a solution to the equation and is called an __________ root or solution.

5. Equations of the form $\frac{a}{b} = \frac{c}{d}$ are called __________.

6. Proportions can be solved by __________ multiplying.

7. Rational expressions do not contain an __________ sign.

8. Equations of the form $ax^2 + bx + c = 0$ are called __________ equations.

Practice the Skills

Solve each equation and check your solution. See Examples 1 and 2.

9. $\frac{x}{3} + \frac{x}{2} = 10$

10. $\frac{z}{5} + \frac{z}{2} = 7$

11. $\frac{y}{6} - \frac{y}{4} = \frac{1}{2}$

12. $\frac{x}{4} - \frac{x}{6} = \frac{1}{2}$

13. $\frac{x}{3} - \frac{x}{4} = 1$

14. $\frac{t}{5} - \frac{t}{6} = 2$

15. $\frac{r}{6} = \frac{r}{4} + \frac{1}{3}$

16. $\frac{n}{5} = \frac{n}{6} + \frac{2}{3}$

17. $\frac{z}{2} + 6 = \frac{z}{5}$

18. $\frac{w}{7} + 4 = \frac{w}{3}$

19. $\frac{z}{6} + \frac{2}{3} = \frac{z}{5} - \frac{1}{3}$

20. $\frac{p}{4} + \frac{1}{4} = \frac{p}{3} - \frac{1}{2}$

21. $\frac{p}{4} + \frac{p}{3} = \frac{7}{6}$

22. $\frac{q}{5} + \frac{q}{2} = \frac{21}{10}$

23. $\frac{k+3}{6} = \frac{5}{2} + \frac{k}{18}$

24. $\frac{m-2}{6} = \frac{2}{3} + \frac{m}{12}$

25. $\frac{a-3}{4} = \frac{a+1}{6}$

26. $\frac{b-4}{5} = \frac{b+2}{2}$

27. $\frac{x-5}{15} = \frac{3}{5} - \frac{x-4}{10}$

28. $\frac{y-4}{6} = \frac{5}{9} - \frac{y+2}{18}$

29. $\frac{-p+1}{4} + \frac{13}{20} = \frac{p}{5} - \frac{p-1}{2}$

30. $\frac{1}{10} - \frac{n+1}{6} = \frac{1}{5} - \frac{n+10}{15}$

31. $\frac{d-3}{4} + \frac{1}{15} = \frac{2d+1}{3} - \frac{34}{15}$

32. $\frac{t+4}{5} = \frac{5}{8} + \frac{t+7}{40}$

Solve each equation and check your solution. See Examples 3–8.

33. $2 + \frac{3}{x} = \frac{11}{4}$

34. $4 + \frac{3}{z} = \frac{9}{2}$

35. $7 - \frac{5}{x} = \frac{9}{2}$

36. $3 - \frac{1}{x} = \frac{14}{5}$

37. $\frac{4}{n} - \frac{3}{2n} = \frac{1}{2}$

38. $\frac{5}{3x} + \frac{2}{x} = 1$

39. $\frac{x-1}{x-5} = \frac{4}{x-5}$

40. $\frac{1}{y+7} = \frac{y+8}{y+7}$

41. $\frac{5}{a+3} = \frac{4}{a+1}$

42. $\frac{5}{x+2} = \frac{1}{x-4}$

43. $\frac{x}{x+6} = \frac{2}{5}$

44. $\frac{y+3}{y-3} = \frac{6}{4}$

45. $\frac{2x-3}{x-4} = \frac{5}{x-4}$

46. $\frac{7x-4}{3x+5} = \frac{4x-9}{3x+5}$

47. $\frac{x^2}{x-3} = \frac{9}{x-3}$

48. $\frac{x^2}{x+5} = \frac{25}{x+5}$

49. $\frac{a-3}{a+4} = \frac{a+2}{a-7}$

50. $\frac{x+5}{x+1} = \frac{x-6}{x-3}$

51. $\frac{1}{r} = \frac{3r}{8r+3}$

52. $\frac{1}{r} = \frac{2r}{r+15}$

53. $\frac{k}{k+2} = \frac{3}{k-2}$

54. $\frac{m}{m+2} = \frac{4}{m-3}$

55. $\frac{4}{r} + r = \frac{20}{r}$

56. $a + \frac{5}{a} = \frac{14}{a}$

57. $x + \frac{20}{x} = -9$

58. $x - \frac{32}{x} = 4$

59. $\frac{3a}{a-4} = 5 + \frac{12}{a-4}$

60. $\frac{5b}{b+2} = 3 - \frac{10}{b+2}$

61. $\frac{1}{x+3} + \frac{1}{x-3} = \frac{-5}{x^2-9}$

62. $\frac{3}{x-5} - \frac{4}{x+5} = \frac{11}{x^2-25}$

63. $\frac{c+1}{3c+9} + \frac{c}{2c+6} = \frac{3}{4c+12}$

64. $\frac{d-3}{6d-12} - \frac{d}{5d-10} = \frac{1}{2d-4}$

65. $\frac{2m^2+m-5}{m^2-8m+15} = \frac{m+3}{m-5} + \frac{m+4}{m-3}$

66. $\frac{2n^2-15}{n^2+n-6} = \frac{n+1}{n+3} + \frac{n-3}{n-2}$

67. $\frac{3x}{x^2-9} + \frac{1}{x-3} = \frac{3}{x+3}$

68. $\frac{3}{x+3} + \frac{5}{x+4} = \frac{12x+7}{x^2+7x+12}$

69. $\frac{1}{y-1} + \frac{1}{2} = \frac{2}{y^2-1}$

70. $\dfrac{1}{z+2}+\dfrac{4}{z^2-4}=\dfrac{1}{3}$

71. $\dfrac{2}{x-2}-\dfrac{1}{x+1}=\dfrac{2}{x^2-x-2}$

72. $\dfrac{3}{x+5}-\dfrac{5}{x-1}=\dfrac{2}{x^2+4x-5}$

Problem Solving

In Exercises 73–78, determine the solution by observation. Explain how you determined your answer.

73. $\dfrac{3}{x-2}=\dfrac{x-2}{x-2}$

74. $\dfrac{1}{2}+\dfrac{x}{2}=\dfrac{5}{2}$

75. $\dfrac{x}{x-6}+\dfrac{x}{x-6}=0$

76. $\dfrac{x}{4}+\dfrac{3x}{4}=x$

77. $\dfrac{x-2}{3}+\dfrac{x-2}{3}=\dfrac{2x-4}{3}$

78. $\dfrac{3}{x}-\dfrac{1}{x}=\dfrac{2}{x}$

79. Optics A formula frequently used in optics is

$$\frac{1}{p}+\frac{1}{q}=\frac{1}{f}$$

where p represents the distance of the object from a mirror (or lens), q represents the distance of the image from the mirror (or lens), and f represents the focal length of the mirror (or lens). If a mirror has a focal length of 10 centimeters, how far from the mirror will the image appear when the object is 30 centimeters from the mirror?

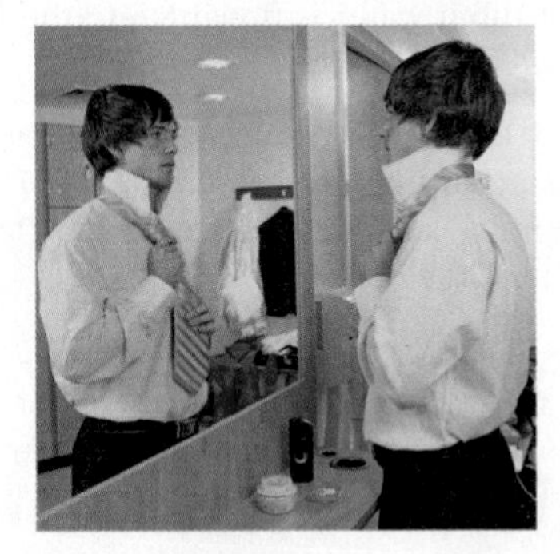

See Exercise 79.

Challenge Problems

80. a) Explain why the equation $\dfrac{x^2}{x-3}=\dfrac{9}{x-3}$ cannot be solved by cross-multiplying using the material presented in the book.

b) Solve the equation given in part **a**).

81. Solve the equation $\dfrac{x-4}{x^2-2x}=\dfrac{-4}{x^2-4}$

82. Electrical Resistance In electronics the total resistance, R_T, of resistors wired in a parallel circuit is determined by the formula

$$\frac{1}{R_T}=\frac{1}{R_1}+\frac{1}{R_2}+\frac{1}{R_3}+\cdots+\frac{1}{R_n}$$

where $R_1, R_2, R_3, \ldots, R_n$ are the resistances of the individual resistors (measured in ohms) in the circuit.

a) Find the total resistance if two resistors, one of 200 ohms and the other of 300 ohms, are wired in a parallel circuit.

b) If three identical resistors are to be wired in parallel, what should be the resistance of each resistor if the total resistance of the circuit is to be 300 ohms?

83. Can an equation of the form $\dfrac{a}{x}+1=\dfrac{a}{x}$ have a real number solution for any real number a? Explain your answer.

Group Activity

Discuss and answer Exercise 84 as a group.

84. a) As a group, discuss two different methods you can use to solve the equation $\dfrac{x+3}{5}=\dfrac{x}{4}$.

b) Group member 1: Solve the equation by obtaining a common denominator.

Group member 2: Solve the equation by cross-multiplying.

Group member 3: Check the results of group member 1 and group member 2.

c) Individually, create another equation by taking the reciprocal of each term in the equation in part **a**). Compare your results. Do you think that the reciprocal of the answer you found in part **b)** will be the solution to this equation? Explain.

d) Individually, solve the equation you found in part **c)** and check your answer. Compare your work with the other group members. Was the conclusion you came to in part **c)** correct? Explain.

e) As a group, solve the equation $\dfrac{1}{x}+\dfrac{1}{3}=\dfrac{2}{x}$. Check your result.

f) As a group, create another equation by taking the reciprocal of each term of the equation in part **e**). Do you think that the reciprocal of the answer you found in part **e)** will be the solution to this equation? Explain.

g) Individually, solve the equation you found in part **f)** and check your answer. Compare your work with the other group members. Did your group make the correct conclusion in part **f)**? Explain.

h) As a group, discuss the relationship between the solution to the equation $\dfrac{7}{x-9}=\dfrac{3}{x}$ and the solution to the equation $\dfrac{x-9}{7}=\dfrac{x}{3}$. Explain your answer.

Cumulative Review Exercises

[3.2] **85. Internet Plans** An Internet service offers two plans for its customers. One plan includes 5 hours of use and costs \$7.95 per month. Each additional minute after the 5 hours costs \$0.15. The second plan costs \$19.95 per month and provides unlimited Internet access. How many hours would Jake LaRue have to use the Internet monthly to make the second plan less expensive?

86. Filling a Hot Tub How long will it take to fill a 600-gallon hot tub if water is flowing into the hot tub at a rate of 4 gallons a minute?

See Exercise 86.

[3.3] **87. Supplementary Angles** Two angles are supplementary angles if the sum of their measures is 180°. Find the two supplementary angles if the smaller angle is 30° less than half the larger angle.

[4.6] **88.** Multiply $(3.4 \times 10^{-5})(2 \times 10^{13})$.

6.7 Rational Equations: Applications and Problem Solving

1. Set up and solve applications containing rational expressions.
2. Set up and solve motion problems.
3. Set up and solve work problems.

1 Set Up and Solve Applications Containing Rational Expressions

Many applications of algebra involve rational equations. After we represent the application as an equation, we solve the rational equation as we did in Section 6.6.

The first type of application we will consider is a *geometry problem*.

EXAMPLE 1 A New Rug Mary and Larry Armstrong are interested in purchasing a carpet whose area is 60 square feet. Determine the length and width if the width is 5 feet less than $\frac{3}{5}$ of the length. See **Figure 6.1.**

FIGURE 6.1

Solution

Understand and Translate

$$\text{Let } x = \text{length.}$$
$$\text{Then } \frac{3}{5}x - 5 = \text{width.}$$
$$\text{area} = \text{length} \cdot \text{width}$$
$$60 = x\left(\frac{3}{5}x - 5\right)$$

Carry Out

$$60 = \frac{3}{5}x^2 - 5x$$
$$5(60) = 5\left(\frac{3}{5}x^2 - 5x\right) \quad \text{Multiply both sides by 5.}$$
$$300 = 3x^2 - 25x \quad \text{Distributive property was used.}$$
$$0 = 3x^2 - 25x - 300 \quad \text{Subtracted 300 from both sides.}$$
$$\text{or } 3x^2 - 25x - 300 = 0$$
$$(3x + 20)(x - 15) = 0 \quad \text{Factored.}$$
$$3x + 20 = 0 \quad \text{or} \quad x - 15 = 0 \quad \text{Zero-factor property}$$
$$3x = -20 \qquad x = 15$$
$$x = -\frac{20}{3}$$

Check and Answer Since the length of a rectangle cannot be negative, we can eliminate $-\frac{20}{3}$ as an answer to our problem.

$$\text{length} = x = 15 \text{ feet}$$

$$\text{width} = \frac{3}{5}(15) - 5 = 4 \text{ feet}$$

Check

$$a = lw$$
$$60 \stackrel{?}{=} 15(4)$$
$$60 = 60 \quad \text{True}$$

Therefore, the length is 15 feet and the width is 4 feet.

Now Try Exercise 5

Now we will work with a problem that expresses the relationship between two numbers. Problems like this are sometimes referred to as *number problems*.

EXAMPLE 2 **Reciprocals** One number is 4 times another number. The sum of their reciprocals is $\frac{5}{2}$. Determine the numbers.

Solution Understand and Translate

$$\text{Let } x = \text{first number.}$$
$$\text{Then } 4x = \text{second number.}$$

The reciprocal of the first number is $\frac{1}{x}$ and the reciprocal of the second number is $\frac{1}{4x}$. The sum of their reciprocals is $\frac{5}{2}$, thus, $\frac{1}{x} + \frac{1}{4x} = \frac{5}{2}$.

Carry Out

$$\frac{1}{x} + \frac{1}{4x} = \frac{5}{2}$$

$$4x\left(\frac{1}{x} + \frac{1}{4x}\right) = 4x\left(\frac{5}{2}\right) \quad \text{Multiply both sides by the LCD, } 4x.$$

$$4x\left(\frac{1}{x}\right) + 4x\left(\frac{1}{4x}\right) = 10x \quad \text{Distributive property}$$

$$4 + 1 = 10x$$
$$5 = 10x$$
$$\frac{5}{10} = x$$
$$\frac{1}{2} = x$$

Check The first number is $\frac{1}{2}$. The second number is therefore $4x = 4\left(\frac{1}{2}\right) = 2$.

Let's now check if the sum of the reciprocals is $\frac{5}{2}$. The reciprocal of $\frac{1}{2}$ is 2. The reciprocal of 2 is $\frac{1}{2}$. The sum of the reciprocals is

$$2 + \frac{1}{2} = \frac{4}{2} + \frac{1}{2} = \frac{5}{2}$$

Answer Since the sum of the reciprocals is $\frac{5}{2}$, the two numbers are 2 and $\frac{1}{2}$.

Now Try Exercise 11

Understanding Algebra

The distance formula is usually written as:

$$\text{distance} = \text{rate} \cdot \text{time}$$

However, it is sometimes convenient to solve the formula for *time*:

$$\frac{\text{distance}}{\text{rate}} = \frac{\text{rate} \cdot \text{time}}{\text{rate}}$$

$$\frac{\text{distance}}{\text{rate}} = \text{time}$$

$$\text{or} \quad \text{time} = \frac{\text{distance}}{\text{rate}}$$

2 Set Up and Solve Motion Problems

In Chapter 3 we discussed *motion problems.* Recall that

$$\text{distance} = \text{rate} \cdot \text{time}$$

If we solve this equation for time, we obtain

$$\text{time} = \frac{\text{distance}}{\text{rate}} \quad \text{or} \quad t = \frac{d}{r}$$

This equation is useful in solving motion problems when the total time of travel for two objects or the time of travel between two points is known.

EXAMPLE 3 **Canoeing** Cindy Kilborn went canoeing in the Colorado River. The current in the river was 2 miles per hour. If it took Cindy the same amount of time to travel 10 miles downstream as 2 miles upstream, determine the speed at which Cindy's canoe would travel in still water.

Solution Understand and Translate

$$\begin{aligned} \text{Let } r &= \text{the canoe's speed in still water.} \\ \text{Then } r + 2 &= \text{the canoe's speed traveling downstream (with current)} \\ \text{and } r - 2 &= \text{the canoe's speed traveling upstream (against current.)} \end{aligned}$$

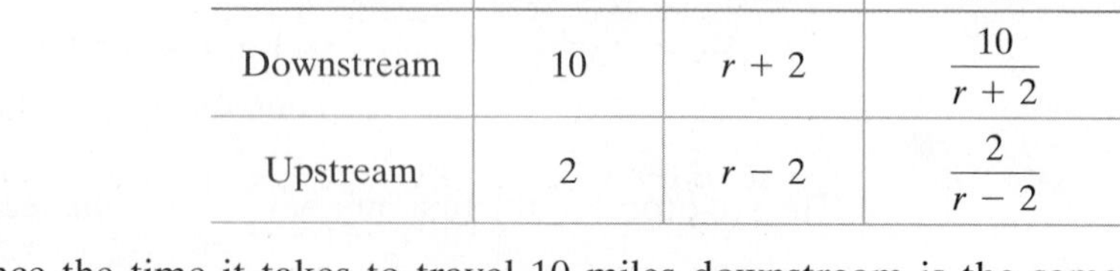

Direction	Distance	Rate	Time
Downstream	10	$r + 2$	$\frac{10}{r+2}$
Upstream	2	$r - 2$	$\frac{2}{r-2}$

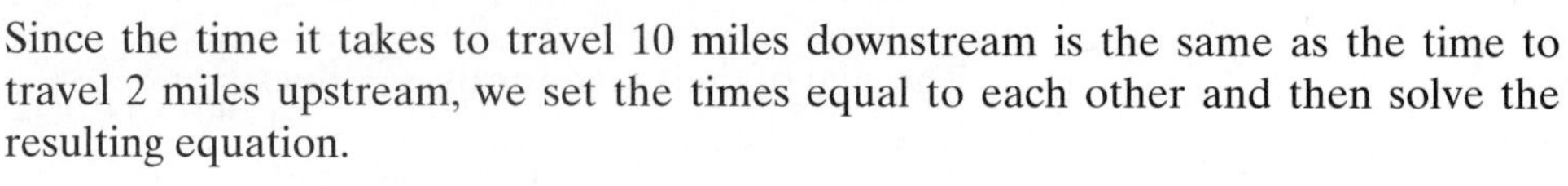

Since the time it takes to travel 10 miles downstream is the same as the time to travel 2 miles upstream, we set the times equal to each other and then solve the resulting equation.

$$\text{time downstream} = \text{time upstream}$$

$$\frac{10}{r+2} = \frac{2}{r-2}$$

Carry Out

$$\begin{aligned} 10(r-2) &= 2(r+2) \quad \text{Cross-multiplied.} \\ 10r - 20 &= 2r + 4 \\ 8r &= 24 \\ r &= 3 \end{aligned}$$

Check and Answer A check will show that 3 satisfies the equation. Thus, the canoe would travel at 3 miles per hour in still water.

Now Try Exercise 15

EXAMPLE 4 **Rollerblading** While rollerblading, Jason Mahar warms up at a slower speed for 1 mile and then increases his speed and travels for an additional 12 miles. Jason's faster speed is 3 times his warm-up speed. If the total time Jason spends rollerblading is 1 hour, determine his warm-up speed and his faster speed.

Solution Understand and Translate

$$\begin{aligned} \text{Let } r &= \text{Jason's warm-up speed.} \\ \text{Then } 3r &= \text{Jason's faster speed.} \end{aligned}$$

	Distance	Rate	Time
Warming up	1	r	$\frac{1}{r}$
Rollerblading faster	12	$3r$	$\frac{12}{3r}$

Since the total time Jason spends rollerblading is 1 hour, we write

$$\text{Time warming up} + \text{Time rollerblading faster} = 1$$

$$\frac{1}{r} + \frac{12}{3r} = 1$$

Carry Out $$3r\left(\frac{1}{r} + \frac{12}{3r}\right) = 3r(1)$$ Multiplied both sides by the LCD, $3r$.

$$3r\left(\frac{1}{r}\right) + 3r\left(\frac{12}{3r}\right) = 3r(1)$$ Distributive property

$$3 + 12 = 3r$$

$$15 = 3r$$

$$5 = r$$

Answer Therefore, Jason warms up at a speed of 5 miles per hour and then rollerblades at a faster speed of $3 \cdot 5$ or 15 miles per hour.

Now Try Exercise 19

EXAMPLE 5 Distance of a Race At a fund-raising race participants can either bike, walk, or run. Kim Clark, who rode a bike, completed the entire distance of the race with an average speed of 16 kilometers per hour (kph). Steve Schwartz, who jogged, completed the entire distance with an average speed of 5 kph. If Kim completed the race in 2.75 hours less time than Steve did, determine the distance the race covered.

Solution Understand and Translate Let d = the distance from the start to the finish of the race. Then we can construct the following table. To determine the time, we divide the distance by the rate.

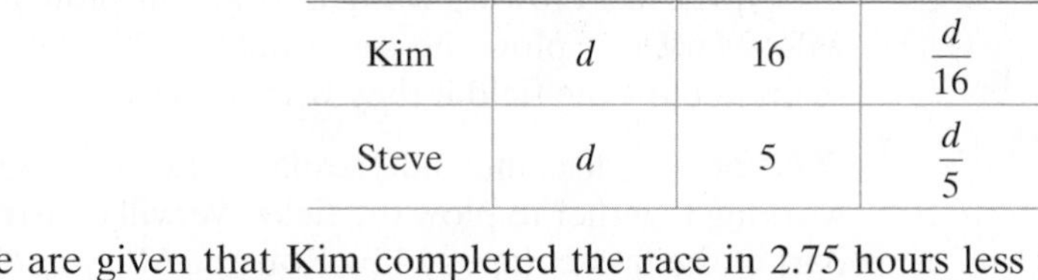

Person	Distance	Rate	Time
Kim	d	16	$\frac{d}{16}$
Steve	d	5	$\frac{d}{5}$

We are given that Kim completed the race in 2.75 hours less time than Steve did. Therefore, to make Kim's and Steve's times equal, we need to subtract 2.75 hours from Steve's time (or add 2.75 hours to Kim's time).

$$\text{Time for Kim} = \text{Time for Steve} - 2.75 \text{ hours}$$

$$\frac{d}{16} = \frac{d}{5} - 2.75$$

Carry Out $$80\left(\frac{d}{16}\right) = 80\left(\frac{d}{5} - 2.75\right)$$ Multiply both sides by the LCD, 80.

$$5d = 80\left(\frac{d}{5}\right) - 80(2.75)$$ Distributive property was used.

$$5d = 16d - 220$$

$$-11d = -220$$

$$d = 20$$

Check and Answer To check this answer we will determine the times it took Kim and Steve to complete the race and see if the difference between the times is 2.75 hours. To determine the times, divide the distance, 20 kilometers, by the rate.

$$\text{Kim's time} = \frac{d}{16} = \frac{20}{16} = 1.25 \text{ hours}$$

$$\text{Steve's time} = \frac{d}{5} = \frac{20}{5} = 4 \text{ hours}$$

Since $4 - 1.25 = 2.75$ hours, the answer checks. Therefore the distance the race covered is 20 kilometers.

Now Try Exercise 25

3 Set Up and Solve Work Problems

When two machines or two people work together to get a job done, the situation leads to solving a *work problem*. To solve work problems, we use the fact summarized in the following diagram:

$$\begin{pmatrix}\text{part of task done}\\ \text{by one person}\\ \text{or machine}\end{pmatrix} + \begin{pmatrix}\text{part of task done}\\ \text{by second person}\\ \text{or machine}\end{pmatrix} = \begin{pmatrix}1\\ \text{(one completed)}\\ \text{task}\end{pmatrix}$$

To determine the part of the task completed by each person or machine, we use the following formula:

rate of work · time worked = part of task completed

An important step in solving these problems is determining the *rate* of work.

Consider the following examples:

- If Joe can do a task by himself in 5 hours, his *rate* is $\frac{1}{5}$ of the task per hour.
- If Yoko can do a task by herself in 4 hours, her *rate* is $\frac{1}{4}$ of the task per hour.
- If a pump can empty a tank in 10 hours, its *rate* is $\frac{1}{10}$ of the tank per hour.

In general, if a task can get done in x hours, then the *rate* for completing that task is $\frac{1}{x}$ of the task per hour.

Understanding Algebra

If JoAnn can perform a task in 6 hours, then her *rate* of work is $\frac{1}{6}$ task per hour.

If she then works t hours, the *amount* of work she has completed is represented as $\frac{1}{6} \cdot t$ or $\frac{t}{6}$ of the task.

EXAMPLE 6 Plowing a Field Bob can plow a field by himself in 20 hours. His wife, Mary, can plow the same field by herself in 30 hours. How long will it take them to plow the field if they work together?

Solution Understand and Translate Let $t =$ the time, in hours, for Bob and Mary working together to plow the field. We will construct a table to help us in finding the part of the task completed by Bob and Mary in t hours.

Person	Rate of Work (part of the task completed per hour)	Time Worked	Part of Task
Bob	$\frac{1}{20}$	t	$\frac{t}{20}$
Mary	$\frac{1}{30}$	t	$\frac{t}{30}$

$$\begin{pmatrix}\text{part of the field plowed}\\ \text{by Bob in } t \text{ hours}\end{pmatrix} + \begin{pmatrix}\text{part of the field plowed}\\ \text{by Mary in } t \text{ hours}\end{pmatrix} = 1(\text{entire field plowed})$$

$$\frac{t}{20} + \frac{t}{30} = 1$$

Carry Out Now multiply both sides of the equation by the LCD, 60.

$$60\left(\frac{t}{20} + \frac{t}{30}\right) = 60 \cdot 1$$

$$\overset{3}{\cancel{60}}\left(\frac{t}{\cancel{20}}\right) + \overset{2}{\cancel{60}}\left(\frac{t}{\cancel{30}}\right) = 60 \qquad \text{Distributive property}$$

$$3t + 2t = 60$$

$$5t = 60$$

$$t = 12$$

Answer Thus, Bob and Mary working together can plow the field in 12 hours. We leave the check for you.

Now Try Exercise 27

HELPFUL HINT

In Example 6, Bob could plow the field by himself in 20 hours, and Mary could plow the field by herself in 30 hours. We determined that together they could plow the field in 12 hours. Does this answer make sense? Since you would expect the time to plow the field together to be less than the time either of them could plow it alone, the answer makes sense.

EXAMPLE 7 **Storing Wine** At a winery in Napa Valley, California, one pipe can fill a tank with wine in 3 hours and another pipe can empty the tank in 5 hours. If the valves to both pipes are open, how long will it take to fill the empty tank?

Solution Understand and Translate Let t = amount of time to fill the tank with the valves to both pipes open.

Pipe	Rate of Work	Time	Part of Task
Pipe filling tank	$\frac{1}{3}$	t	$\frac{t}{3}$
Pipe emptying tank	$\frac{1}{5}$	t	$\frac{t}{5}$

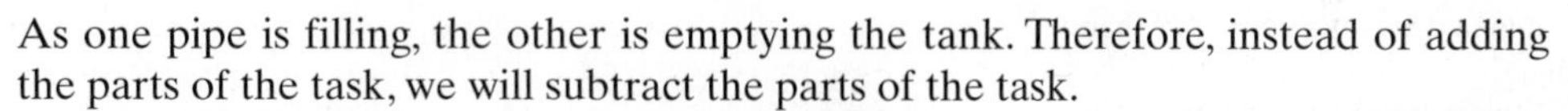

As one pipe is filling, the other is emptying the tank. Therefore, instead of adding the parts of the task, we will subtract the parts of the task.

$$\left(\begin{array}{c}\text{part of tank}\\ \text{filled in } t \text{ hours}\end{array}\right) - \left(\begin{array}{c}\text{part of tank}\\ \text{emptied in } t \text{ hours}\end{array}\right) = 1(\text{total tank filled})$$

$$\frac{t}{3} - \frac{t}{5} = 1$$

Carry Out

$$15\left(\frac{t}{3} - \frac{t}{5}\right) = 15 \cdot 1 \quad \text{Multiply both sides by the LCD, 15.}$$

$$\overset{5}{\cancel{15}}\left(\frac{t}{\cancel{3}}\right) - \overset{3}{\cancel{15}}\left(\frac{t}{\cancel{5}}\right) = 15 \quad \text{Distributive property was used.}$$

$$5t - 3t = 15$$

$$2t = 15$$

$$t = 7\frac{1}{2}$$

Check and Answer The tank will be filled in $7\frac{1}{2}$ hours. This answer is reasonable because we expect it to take longer than 3 hours when the tank is being drained at the same time.

Now Try Exercise 35

EXAMPLE 8 **Cleaning Service** Linda and John Franco own a house cleaning service. When Linda cleans Damon's house by herself, it takes 7 hours. When Linda and John work together, they can clean the house in 4 hours. How long will it take John to clean the house by himself?

Solution Let t = time for John to clean the house by himself. Then John's rate is $\frac{1}{t}$. Since Linda can clean the house by herself in 7 hours, her rate is $\frac{1}{7}$ of the job per hour. In the table below, we use the fact that together they can clean the house in 4 hours.

Worker	Rate of Work	Time	Part of Task
Linda	$\frac{1}{7}$	4	$\frac{4}{7}$
John	$\frac{1}{t}$	4	$\frac{4}{t}$

$$\left(\begin{array}{c}\text{part of house}\\ \text{cleaned by Linda}\end{array}\right)+\left(\begin{array}{c}\text{part of house}\\ \text{cleaned by John}\end{array}\right)=1$$

$$\frac{4}{7}+\frac{4}{t}=1$$

Carry Out

$$7t\left(\frac{4}{7}+\frac{4}{t}\right)=7t\cdot 1 \quad \text{Multiply both sides by the LCD, } 7t.$$

$$7t\left(\frac{4}{7}\right)+7t\left(\frac{4}{t}\right)=7t \quad \text{Distributive property was used.}$$

$$4t+28=7t$$

$$28=3t$$

$$\frac{28}{3}=t$$

$$9\frac{1}{3}=t$$

Understanding Algebra

In Example 8, notice that if two people were to work at Linda's rate, they would clean the house in $\frac{1}{2}\times 7=\frac{7}{2}$, or $3\frac{1}{2}$, hours.

Since Linda and John together clean the house in 4 hours, we know that John cleans at a slower rate than Linda.

This is one way to determine that our answer makes sense.

Check and Answer Thus, it takes John $9\frac{1}{3}$ hours, or 9 hours 20 minutes, to clean the house by himself. This answer is reasonable because we expect it to take longer for John to clean the house by himself than it would for Linda and John working together.

Now Try Exercise 37

EXAMPLE 9 **Thank-You Notes** Peter and Kaitlyn Kewin are handwriting thank-you notes to guests who attended their 20th wedding anniversary party. Kaitlyn by herself could write all the notes in 6 hours and Peter could write all the notes by himself in 10 hours. After Kaitlyn has been writing thank-you notes for 4 hours by herself, she must leave town on business. Peter then continues the task of writing the thank-you notes. How long will it take Peter to finish writing the remaining notes?

Solution Understand and Translate Let t = time it will take Peter to finish writing the notes.

Person	Rate of Work	Time	Part of Task
Kaitlyn	$\frac{1}{6}$	4	$\frac{4}{6}=\frac{2}{3}$
Peter	$\frac{1}{10}$	t	$\frac{t}{10}$

$$\left(\begin{array}{c}\text{part of notes written}\\ \text{by Kaitlyn}\end{array}\right)+\left(\begin{array}{c}\text{part of notes written}\\ \text{by Peter}\end{array}\right)=1$$

$$\frac{2}{3}+\frac{t}{10}=1$$

Carry Out

$$30\left(\frac{2}{3}+\frac{t}{10}\right) = 30 \cdot 1 \qquad \text{Multiply both sides by the LCD, 30.}$$

$$\overset{10}{\cancel{30}}\left(\frac{2}{\underset{1}{\cancel{3}}}\right) + \overset{3}{\cancel{30}}\left(\frac{t}{\underset{1}{\cancel{10}}}\right) = 30 \qquad \text{Distributive property}$$

$$20 + 3t = 30$$

$$3t = 10$$

$$t = \frac{10}{3} \quad \text{or} \quad 3\frac{1}{3}$$

Answer Thus, it will take Peter $3\frac{1}{3}$ hours to complete the notes.

Now Try Exercise 41

6.7 Exercise Set MathXL® MyMathLab®

Warm-Up Exercises

Fill in the blanks with the appropriate word, phrase, or symbol(s) from the following list.

time $\quad \frac{1}{5} \quad \frac{1}{9}$ $\quad$ rate $\quad \frac{\text{distance}}{\text{rate}} \quad \frac{\text{rate}}{\text{distance}}$

1. If we solve the distance formula, distance = rate · time, for time, we obtain the equation time = ________.

2. To determine the part of the task completed by each person or machine, we use the formula ________ of work · time worked = part of task completed.

3. If Lynette Meslinski can complete a task in 5 hours, then her rate of work is ________ of the task per hour.

4. If a pump can empty a tank in 9 hours, then the rate of work for the pump is ________ of the tank per hour.

Practice the Skills/Problem Solving

In Exercises 5–42, solve the problem and answer the question.

Geometry Problems; see Example 1.

5. **Vegetable Garden** Constance Meade is planting a rectangular vegetable garden in her backyard. The garden is to have an area of 42 square meters. The width of the garden is 4 meters less than $\frac{1}{2}$ the length. Determine the length and the width of the garden. Use $A = lw$.

6. **Nature Preserve** The new River Run housing development includes a rectangular piece of land for a nature preserve. The preserve is to have an area of 15 square miles. The width of the preserve is 1 mile less than $\frac{1}{4}$ the length. Determine the length and width of the preserve. Use $A = lw$.

7. **Triangles of Dough** Pillsbury Crescent Rolls are packaged in tubes that contain perforated triangles of dough. The base of the triangular piece of dough is about 5 centimeters more than its height. Determine the base and height of a piece of dough if the area is about 42 square centimeters. Use $A = \frac{1}{2}bh$.

8. **Dinghy Sail** The area of a triangular sail on a dinghy sail boat is 20 square feet. The base of the sail is 1 foot more than $\frac{1}{2}$ the height. Determine the base and height of the sail. Use $A = \frac{1}{2}bh$.

9. Park Sign A sign for Wyalusing State Park is in the shape of a trapezoid. The area of the sign is 24 square feet. Find the height of the sign if one base (b) is 1 foot more than the height and the other base (d) is 3 feet more than the height. Use $A = \frac{1}{2}h(b + d)$.

10. Dog Pen A dog recreation area in Elizabeth Morrison's backyard is in the shape of a trapezoid. The area of the trapezoid is 1800 ft^2. Find the height of the trapezoid if one base (b) is 4 times the height and the other base (d) is 5 times the height. Use $A = \frac{1}{2}h(b + d)$.

Number Problems; see Example 2.

11. Difference of Numbers One number is 9 times larger than another. The difference of their reciprocals is 1. Determine the two numbers.

12. Sum of Numbers One number is 3 times larger than another. The sum of their reciprocals is $\frac{4}{3}$. Determine the two numbers.

13. Increased Numerator The numerator of the fraction $\frac{3}{4}$ is increased by an amount so that the value of the resulting fraction is $\frac{5}{2}$. Determine the amount by which the numerator was increased.

14. Decreased Denominator The denominator of the fraction $\frac{8}{21}$ is decreased by an amount so that the value of the resulting fraction is $\frac{1}{2}$. Determine the amount by which the denominator was decreased.

Motion Problems; see Examples 3–5.

15. Paddleboat Ride In the Mississippi River near New Orleans, the Creole Queen paddleboat travels 6 miles upstream (against the current) in the same amount of time it travels 12 miles downstream (with the current). If the current of the river is 3 miles per hour, determine the speed of the Creole Queen in still water.

16. Kayak Ride Kathy Boothby-Sestak can paddle her kayak 6 miles per hour in still water. It takes her as long to paddle 5 miles upstream as 10 miles downstream in the Wabash River near Lafayette, Indiana. Determine the river's current.

17. Ultralight Aircraft Charlotte Newsome's ultralight aircraft has a speed in still air of 60 miles per hour. One day Charlotte flies 25 miles with the wind in the same amount of time that she flies 15 miles against the wind. Determine the speed of the wind.

18. Flying Elsie Newman is flying her airplane when the speed of the wind is 30 miles per hour. Elsie can travel 320 miles when flying with the wind in the same amount of time that she can travel 200 miles when flying against the wind. Determine the speed of Elsie's airplane in still air.

19. Flat Tire Jean Olson rode her bicycle 24 miles, then her bike got a flat tire. She then walked her bike 2 miles to a bicycle shop to repair the tire. Jean's riding speed was 4 times her walking speed. If the total time Jean spent riding and walking was two hours, determine her walking speed and her riding speed.

20. Traffic Charles Odion is driving to visit his mother. At first he encounters heavy traffic for 5 miles before the traffic clears and he is able to drive for 135 miles at a faster speed. His speed after the traffic clears was 3 times his speed while driving in traffic. If the total time Charles was driving was 2 hours, determine his speed while in traffic and his speed after the traffic cleared.

21. Jet Flight Elenore Morales traveled 1600 miles by commercial jet from Kansas City, Missouri, to Spokane, Washington. She then traveled an additional 500 miles on a private propeller plane from Spokane to Billings, Montana. If the speed of the jet was 4 times the speed of the propeller plane and the total time in the air was 6 hours, determine the speed of each plane.

22. Exercise Regimen Chris Barker walks a distance of 2 miles on an indoor track and then jogs at twice his walking speed for another 2 miles. If the total time spent on the track was one hour, determine the speeds at which he walks and jogs.

23. The Tail of the Dragon One stretch of US 129 in Tennessee is a very popular road for motorcyclists because the road has many curves and is called The Tail of the Dragon. Larry Gilligan rode his motorcycle in one direction and averaged 22 miles per hour. On the return trip it was raining and he only averaged 11 miles per hour. If the round trip took 1.5 hours, what is the length of The Tail of the Dragon?

24. **Thalys Train** The Thalys train in Europe has been known to travel an average 240 kilometers per hour (kph). Prior to using the Thalys trains in Europe, trains traveled an average speed of 120 kph. If a Thalys train traveling from Brussels to Amsterdam can complete its trip in 0.88 hour less time than an older train, determine the distance from Brussels to Amsterdam.

Thalys train

25. **Water Skiers** At a water show a boat pulls a water skier at a speed of 30 feet per second. When it reaches the end of the lake, more skiers are added to be pulled by the boat, so the boat's speed drops to 25 feet per second. If the boat traveled the same distance on both trips and the trip back with the additional skiers took 8 seconds longer than the trip with the single skier, how far, in feet, in one direction, had the boat traveled?

26. **Cross-Country Skiing** Alana Bradley and her father Tim begin skiing the same cross-country ski trail in Elmwood Park in Sioux Falls, South Dakota, at the same time. If Alana, who averages 9 miles per hour, finishes the trail 0.25 hour sooner than her father, who averages 6 miles per hour, determine the length of the trail.

Work Problems; see Examples 6–9.

27. **Painting a Room** Eric Kweeder can paint a room in 60 minutes. His brother, Jessup, can paint the same room in 40 minutes. How long will it take them working together to paint the room?

28. **Conveyor Belt** At a salt mine, one conveyor belt requires 20 minutes to fill a large truck with ore. A second conveyor belt requires 30 minutes to fill the same truck with ore. How long would it take if both conveyor belts were working together to fill the truck with ore?

29. **Wallpaper** Reynaldo and Felicia Fernandez decide to wallpaper their family room. Felicia, who has wallpapering experience, can wallpaper the room in 6 hours. Reynaldo can wallpaper the same room in 8 hours. How long will it take them to wallpaper the family room if they work together?

30. **Watering Plants** In a small nursery, Becky Hailey can water all the plants in 30 minutes. Her co-worker, Karen Grizzaffi, can water all the plants in 20 minutes. How long will it take them working together to water the plants?

31. **Hot Tub** Pam and Loren Fornieri know that their hot tub can be filled in 40 minutes and drained completely in 60 minutes. If the water is turned on and the drain is left open, how long would it take the tub to fill completely?

32. **Filling a Tank** During a rainstorm, the rain is flowing into a large holding tank. At the rate the rain is falling, the empty tank would fill in 8 hours. At the bottom of the tank is a spigot to dispense water. Typically, it takes about 12 hours with the spigot wide open to empty the water in a full tank. If the tank is empty and the spigot has been accidentally left open, and the rain falls at the constant rate, how long would it take for the tank to fill completely?

33. **Milk Tank** At a milk processing plant, one pipe can fill a milk tank in 15 hours and another pipe can drain the tank in 20 hours. If the valves to both pipes are open, how long will it take to fill the empty tank?

34. **Sanitation Plant** At a city sanitation plant, a septic tank can be filled in 5 hours. The same septic tank can be drained in 30 hours. If the tank is being filled and drained at the same time, how long will it take to fill the empty tank?

35. **Payroll Checks** At the Community Savings Bank, it takes a computer 40 minutes to process and print payroll checks. When a second computer is used and the two computers work together, the checks can be processed and printed in 24 minutes. How long would it take the second computer by itself to process and print the payroll checks?

36. **Flowing Water** When the water is turned on and passes through a small hose, a pool can be filled in 6 hours. When the water is turned on at two spigots and passes through both the small hose and a large hose, the pool can be filled in 2 hours. How long would it take to fill the pool using only the large hose?

37. **Hotel Attendants** At the Ritz-Carlton Hotel, Pat Hutton can clean and prepare a room in 25 minutes. If Chris Armstrong helps Pat, together they can clean and prepare a room in 20 minutes. How long would it take Chris, working alone, to clean and prepare a room?

38. **Restaurant Vacuuming** At Hungry Howie's restaurant, Glenn Sandifer can vacuum the dining room in 16 minutes. If Brian Sanders helps Glenn, together they can vacuum the dining room in 12 minutes. How long would it take Brian, working alone, to vacuum the dining room?

39. **Digging a Trench** A construction company with two backhoes has contracted to dig a long trench for drainage pipes. The larger backhoe can dig the entire trench by itself in 12 days. The smaller backhoe can dig the entire trench by itself in

15 days. The large backhoe begins working on the trench by itself, but after 5 days it is transferred to a different job and the smaller backhoe begins working on the trench. How long will it take for the smaller backhoe to complete the job?

40. Delivery of Food Ian and Nicole Murphy deliver food to various restaurants. If Ian drove the entire trip, the trip would take about 10 hours. If Nicole drove the entire trip, the trip would take about 8 hours. After Nicole had been driving for 4 hours, Ian takes over the driving. About how much longer will Ian drive before they reach their final destination?

41. Snowstorm Following a snowstorm, Ken and Bettina Reeves must clear their driveway and sidewalk. Ken can clear the snow by himself in 4 hours, and Bettina can clear the snow by herself in 6 hours. After Bettina has been working for 3 hours, Ken is able to join her. How much longer will it take them working together to remove the rest of the snow?

See Exercise 41.

42. Lawn Mowing Jearme Pirie and Behnaz Rouhani are responsible for mowing the lawn at Sunset Park Place retirement home. Jearme can mow the lawn, working alone, in 3 hours and Behnaz can mow the lawn, working alone, in 5 hours. Behnaz is working alone for 2 hours when Jearme arrives and begins working. How long will it take them, working together, to finish mowing the lawn?

Challenge Problems

43. Reciprocal of a Number If 2 times a number is added to 3 times the reciprocal of the number, the answer is 7. Determine the number(s).

44. Determine a Number The reciprocal of the difference of a certain number and 5 is twice the reciprocal of the difference of twice the number and 10. Determine the number(s).

45. Picking Blueberries Ed and Samantha Weisman, whose parents own a fruit farm, must each pick the same number of pints of blueberries each day during the season. Ed picks an average of 8 pints per hour, while Samantha picks an average of 4 pints per hour. If Ed and Samantha begin picking blueberries at the same time, and Samantha finishes 1 hour after Ed, how many pints of blueberries must each pick?

46. Sorting Mail A mail processing machine can sort a large bin of mail in 1 hour. A newer model can sort the same quantity of mail in 30 minutes. If they operate together, how long will it take them to sort the bin of mail?

Cumulative Review Exercises

[2.1] **47.** Simplify $\frac{1}{2}(x + 3) - (2x + 15)$.

[5.2] **48.** Factor $y^2 + 6y - y - 6$ by grouping.

[6.2] **49.** Divide $\frac{x^2 - 14x + 48}{x^2 - 5x - 24} \div \frac{2x^2 - 13x + 6}{2x^2 + 5x - 3}$.

[6.4] **50.** Subtract $\frac{x}{6x^2 - x - 15} - \frac{5}{9x^2 - 12x - 5}$.

6.8 Variation

1 Set up and solve direct variation problems.

2 Set up and solve inverse variation problems.

Variation equations show how one quantity changes in relation to another quantity or quantities. In this section we will discuss two types of variation: *direct* and *inverse*. In the exercises, we will address two additional types of variation: *joint* and *combined*.

1 Set Up and Solve Direct Variation Problems

Direct variation *involves two variables that increase together or decrease together.* For example, consider a car traveling 80 miles per hour on an interstate highway. The car travels

- 80 miles in 1 hour,
- 160 miles in 2 hours,
- 240 miles in 3 hours, and so on.

As the *time* increases, the *distance* also increases.

The formula used to calculate distance traveled is

$$\text{distance} = \text{rate} \cdot \text{time}$$
$$d = rt$$

Understanding Algebra

Direct variation involves two variables that increase together or decrease together.

The phrases

- "y varies directly as x" and
- "y is directly proportional to x"

are both represented by the direct variation equation

$$y = kx.$$

Since the rate in the example above is constant, the formula can be written

$$d = 80t$$

We say distance *varies directly* as time or that distance is *directly proportional* to time.

Direct Variation

If a variable y varies directly as a variable x, then

$$y = kx$$

where k is the **constant of proportionality** or the **variation constant.**

EXAMPLE 1 **Heating Up a Hot Tub** When the heater is turned on to warm the water in a hot tub, the temperature of the water, w, increases directly with the length of time in minutes, t, the heater is on.

a) Write the variation equation.

b) If the constant of proportionality, k, is 0.8, find the increase in temperature of the water after 40 minutes.

Solution

a) We are told that the water temperature varies directly with the time. Thus we set up the direct variation equation as follows.

$$w = kt$$

b) To find the increase in water temperature, we will substitute 0.8 for k and 40 for t.

$$w = kt$$
$$w = 0.8(40) = 32$$

Thus, after 40 minutes the water temperature has increased by 32°.

Now Try Exercise 31

In many variation problems you will first have to solve for the constant of proportionality before you can solve for the variable you are asked to find. To determine the constant of proportionality, substitute the values given for the variables, and solve for k.

EXAMPLE 2 **Direct Variation Problem** s varies directly as the square of m. If $s = 125$ when $m = 5$, find s when $m = 12$.

Solution Understand and Translate We begin by setting up the variation equation. Notice that we are told that s varies directly as the square of m. The square of m is written m^2. Therefore, the equation is $s = km^2$. Since we are not given the constant of proportionality, we find it by substituting the values we are given for the variables.

$$s = km^2$$
$$125 = k(5^2) \quad \text{Substituted values.}$$
$$125 = k(25)$$
$$125 = 25k$$
$$5 = k$$

Now that we have determined k, we can answer the question by substituting 5 for k, and 12 for m.

Carry Out
$$s = km^2$$
$$s = 5(12)^2$$
$$s = 5(144)$$
$$s = 720$$

Answer Thus, when $m = 12, s = 720$.

Now Try Exercise 27

EXAMPLE 3 **Drug Dosage** The amount of a drug, d, given to a person is directly proportional to the person's weight, w. If an adult who weighs 75 kilograms is given 300 milligrams (mg) of the drug, determine how many milligrams of the drug are given to an adult who weighs 96 kg.

Solution Understand and Translate We are told that the amount of the drug is directly proportional to the person's weight. Thus we set up the equation

$$d = kw$$

Now we determine k by substituting the values given for d and w.

Carry Out
$$d = kw$$
$$300 = k(75)$$
$$\frac{300}{75} = k$$
$$4 = k$$

Now we proceed to find the number of milligrams of the drug to be given by substituting 4 for k and 96 for w.

$$d = kw$$
$$d = 4(96) = 384$$

Check and Answer Since we expect the amount of the drug to be greater than 300 milligrams, our answer is reasonable. A 96-kg adult should be given 384 milligrams of the drug.

Now Try Exercise 35

2 Set Up and Solve Inverse Variation Problems

Inverse variation *involves two variables in which one variable increases as the other decreases and vice versa.* For example, consider traveling 120 miles in a car. If the car is traveling

- 30 miles per hour, the trip takes 4 hours,
- 40 miles per hour, the trip takes 3 hours,
- 60 miles per hour, the trip takes 2 hours, and so on.

As the *rate* increases, the time to travel 120 miles decreases.

The formula used to calculate time, given the distance and the rate is

$$\text{time} = \frac{\text{distance}}{\text{rate}}$$

Since the distance in the example above is constant, the formula can be rewritten

$$\text{time} = \frac{120}{\text{rate}}$$

We say time *varies inversely* as rate or that time is *inversely proportional* to rate.

Inverse Variation

If a variable y varies inversely as x, then

$$y = \frac{k}{x}$$

where k is the **constant of proportionality** or the **variation constant.**

Understanding Algebra

Inverse variation involves two variables in which one variable increases as the other decreases and vice versa.

The phrases

- "y varies inversely as x" and
- "y is inversely proportional to x"

are both represented by the inverse variation equation

$$y = \frac{k}{x}.$$

EXAMPLE 4 Chartering a Sailboat The cost per person for chartering a sailboat, c, is inversely proportional to the number of people chartering the boat, n. If 8 friends decide to charter the boat, the cost per person is \$60. Determine the cost per person if 15 friends decide to charter the boat.

Solution Understand and Translate We are told this is an example of inverse variation. Therefore, we will set up an equation to represent the inverse proportion.

$$c = \frac{k}{n}$$

Since we are not given the constant of proportionality, we determine k by substituting the values given for c and n.

$$60 = \frac{k}{8}$$

$$480 = k$$

Now we can determine the answer to the question by using $k = 480$ and $n = 15$.

Carry Out

$$c = \frac{k}{n}$$

$$c = \frac{480}{15}$$

$$c = 32$$

Check and Answer The cost to each person would be \$32 if 15 friends decided to charter the sailboat.

Now Try Exercise 37

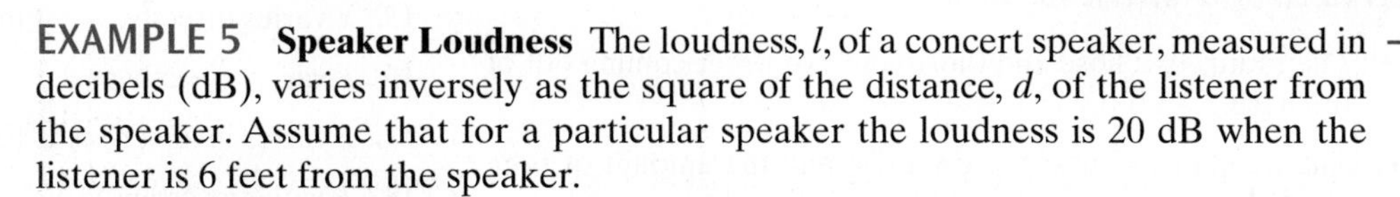

EXAMPLE 5 Speaker Loudness The loudness, l, of a concert speaker, measured in decibels (dB), varies inversely as the square of the distance, d, of the listener from the speaker. Assume that for a particular speaker the loudness is 20 dB when the listener is 6 feet from the speaker.

a) Determine an equation that expresses the relationship between the loudness and the distance.

b) Using the equation obtained in part **a)**, determine the loudness when a person is 3 feet from the speaker.

Solution This problem is broken down into two parts. The first part asks us to find a general formula, while the second part asks us to use the formula.

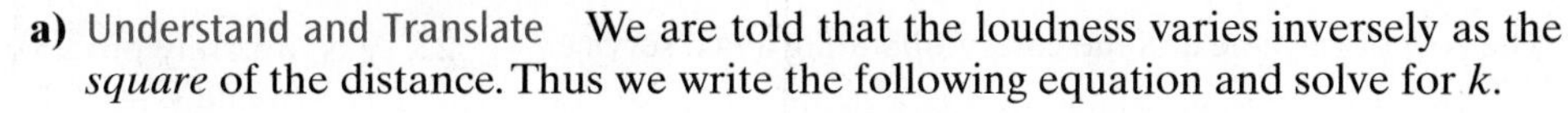

a) Understand and Translate We are told that the loudness varies inversely as the *square* of the distance. Thus we write the following equation and solve for k.

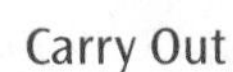

Carry Out

$$l = \frac{k}{d^2}$$

$$20 = \frac{k}{6^2}$$

$$20 = \frac{k}{36}$$

$$720 = k$$

Check and Answer The constant of proportionality, k, is 720. Since for this speaker $k = 720$, the equation we are seeking is

$$l = \frac{720}{d^2}$$

b) Understand and Translate In part **a)** we determined the equation used to find the loudness. We substitute 3 for d in the formula and solve for l.

$$l = \frac{720}{d^2}$$

$$l = \frac{720}{3^2}$$

Understanding Algebra

Direct variation:

As x increases so does y and as x decreases, so does y; we write

$$y = kx.$$

Inverse variation:

As x increases, y decreases and as x decreases, y increases; we write

$$y = \frac{k}{x}.$$

k is called the *constant of proportionality* in each case.

Carry Out $$l = \frac{720}{9}$$

$$l = 80$$

Check and Answer Thus at 3 feet the loudness is 80 decibels. This is reasonable because at a shorter distance (3 feet versus 6 feet) the sound will be louder.

Now Try Exercise 49

6.8 Exercise Set MathXL® MyMathLab®

Warm-Up Exercises

Fill in the blanks with the appropriate word, phrase, or symbol(s) from the following list.

inversely directly increases decrease constant

1. Direct variation involves two variables that increase together or ________ together.
2. Inverse variation involves two variables in which one variable ________ as the other decreases, and vice versa.
3. If a variable y varies directly as x, then $y = kx$ where k is the ________ of proportionality.
4. If a variable y varies ________ as x, then $y = \frac{k}{x}$ where k is the constant of proportionality.

Practice the Skills

In Exercises 5–14, determine if the following are examples of direct variation or inverse variation.

5. The radius of a hose and the amount of water coming out of the hose.
6. The number of questions on a test and the amount of time needed to complete the test.
7. The speed of a turtle and the length of time it takes the turtle to cross a road.
8. The amount of rain received in a month and the amount of watering needed to keep your lawn green.
9. The temperature of water and the time it takes for an ice cube placed in the water to melt.
10. The cost of gasoline and the cost of operating a taxi service.

11. The length of a roll of Scotch tape and the number of 2-inch strips that can be obtained from the roll.
12. The cubic-inch displacement, in liters, and the horsepower of the engine.
13. A person's reading speed and the time it takes to read a novel.
14. The speed of a riding lawn mower and the time it takes to cut the lawn.

In Exercises 15–22, find the quantity indicated.

15. x varies directly as z. Find x when $z = 11$ and $k = 40$.
16. x varies directly as y. Find x when $y = 9$ and $k = 6$.
17. x varies inversely as y. Find x when $y = 25$ and $k = 5$.
18. R varies inversely as W. Find R when $W = 80$ and $k = 120$.
19. C is directly proportional to the square of Z. Find C when $Z = 5$ and $k = 3$.
20. L is directly proportional to the square of R. Find L when $R = 9$ and $k = 2$.
21. y is inversely proportional to the square of x. Find y when $x = 10$ and $k = 250$.
22. y is inversely proportional to the square of w. Find y when $w = 8$ and $k = 288$.

For Exercises 23–30, find the quantity indicated.

23. y varies directly as x. If $y = 18$ when $x = 2$, find y when $x = 6$.
24. Z varies directly as W. If $Z = 7$ when $W = 21$, find Z when $W = 51$.
25. C varies inversely as J. If $C = 7$ when $J = 1$, find C when $J = 2$.
26. H varies inversely as L. If $H = 15$ when $L = 60$, find H when $L = 10$.
27. y varies directly as the square of R. If $y = 4$ when $R = 4$, find y when $R = 12$.
28. A varies directly as the square of B. If $A = 245$ when $B = 7$, find A when $B = 9$.
29. L varies inversely as the square of P. If L is 320 when $P = 20$, find L when $P = 40$.
30. x varies inversely as the square of P. If $x = 10$ when $P = 6$, find x when $P = 20$.

Problem Solving

In Exercises 31–50, determine the quantity you are asked to find.

31. **Distance and Speed** The distance, d, a car travels is directly proportional to the speed, s, the car is traveling. Determine the distance traveled if the constant of proportionality, k, is 2 and the speed is 55 miles per hour.

32. **Swimming Pool** The time, t, it takes to fill an inground pool is inversely proportional to the amount of water coming out of the hose, w. Determine the time it takes a hose to fill the pool if the constant of proportionality is 12,000 and the amount of water coming out of the hose is 200 gallons per hour.

33. **Commute Time** The time for Rebecca Shantz to drive to work, t, is inversely proportional to the average speed driven, r. Determine the time required to drive to work if the average speed driven is 70 miles per hour and the constant of proportionality is 35.

34. **Light through Water** The percent of light that filters through water, l, is inversely proportional to the depth of the water, d. Determine the percent of light that filters down to a depth of 10 feet if the constant of proportionality is 200.

35. **Kiddie Train** The income, I, for a kiddie train at an amusement park is directly proportional to the number of tickets sold, n. If the income, I, is \$33 when 22 tickets are sold, determine the income when 38 tickets are sold.

36. **Lawn Mowing** The time it takes Sue to mow her lawn, t, is directly proportional to the area of the lawn, A. If it takes Sue 1 hour to cut an area of 2400 square feet, how long will it take her to cut an area of 1800 square feet?

37. **Roofing** The time, t, it takes to nail in shingles on a roof is inversely proportional to the number of people nailing in the shingles, n. When three people are nailing in the shingles it takes 7 hours to complete the job. How long will it take to complete the job if five people are nailing in the shingles?

38. **Baking a Turkey** The time, t, it takes to bake a turkey is inversely proportional to the oven temperature, T. If it takes 3 hours to bake a turkey at 300°F, how long will it take to bake the turkey at 250°F?

39. **Baseball Gate Receipts** The receipts r, at an International League baseball park are directly proportional to the number of people attending the game, n. If the receipts for a game are \$37,200 when 1200 people attend, determine how many people attend if the receipts for a game are \$31,000.

40. **Hooke's Law** Hooke's law states that the length a spring will stretch, S, varies directly with the force (or weight), F, attached to the spring. If a spring stretches 1.4 inches when 20 pounds is attached, how far will it stretch when 10 pounds is attached?

41. **Volume of Gas** The volume of a gas, V, varies inversely as its pressure, P. If the volume, V, is 800 cubic centimeters when the pressure is 200 millimeters (mm) of mercury, find the volume when the pressure is 25 mm of mercury.

42. **Cleaning Windows** The time, t, it takes to clean all the windows in a large office building is inversely proportional to the number of teams, n, of window workers used. If 6 teams can clean all the windows in 20 days, how many teams are used if the windows are cleaned in 12 days?

43. **Area of a Circle** The area of a circle, A, is directly proportional to the square of the radius of a circle, r. If the area of a circle is about 78.5 square inches when the radius is 5 inches, determine the area when the radius is 12 inches.

44. **Falling Distance** The distance an object falls from rest, d, is directly proportional to the square of the time the object has been falling, t. If an object that has been falling for 2 seconds has fallen 64 feet, how far will an object fall in 5 seconds?

45. **Volume of a Cylinder** For a cylinder of a specific volume, the height, h, of the cylinder is inversely proportional to the square of the radius of the cylinder, r. When the radius is 6 inches, the height is 10 inches. Determine the height when the radius is 5 inches.

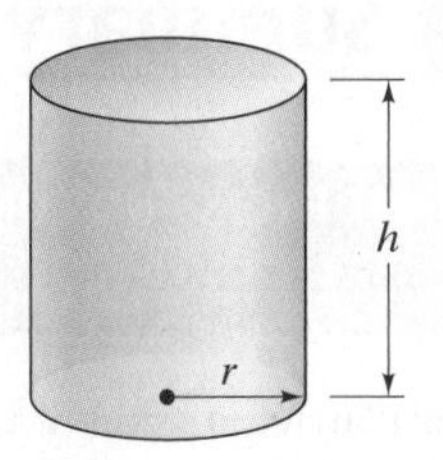

46. **Electrical Circuit** In an electrical circuit (with constant power) the resistance, r, is inversely proportional to the square of the current, c. If the resistance in a circuit is 5 ohms, the current is 4 amps. What is the resistance if the current is 5 amps?

47. **Powerboats** The horsepower it takes to propel a speedboat, P, is directly proportional to the square of the velocity, v, of the boat. If it takes 900 horsepower for the boat to travel at 45 mph, what horsepower is needed to propel the speedboat at 54 mph?

48. **Equilateral Triangle** The area A of an equilateral triangle (a triangle with all three sides of equal length) is directly proportional to the square of the length of one side of the triangle, s. If an equilateral triangle with a side length of 2 centimeters has an area of about 1.73 square centimeters, determine the area of an equilateral triangle with side length of 4 centimeters.

49. **Light Illumination** The intensity of illumination from a light source, I, is inversely proportional to the square of the distance, d, from the light source. If a car headlight produces an intensity of 3.75 footcandles at a distance of 40 feet, what is the intensity of the same headlight at 100 feet?

50. **Sound Intensity** The sound intensity, I, varies inversely as the square of the distance, d, from the sound source. If the sound intensity is 30 watts per square inch at a distance of 8 meters from a stereo speaker, what is the sound intensity at 4 meters?

Concept/Writing Exercises

51. Assume a varies directly as b. If b is doubled, how will it affect a? Explain.

52. Assume a varies directly as b^2. If b is doubled, how will it affect a? Explain.

53. Assume y varies inversely as x. If x is doubled, how will it affect y? Explain.

54. Assume y varies inversely as a^2. If a is doubled, how will it affect y? Explain.

55. In variation problems, the constant of proportionality often has a unit attached to it. In Hooke's Law (Exercise 40) for example, where $F = kS$, F is measured in pounds, and S is measured in inches, what is the unit of k?

56. Consider the relationship between the resistance, r, and current, c, in Exercise 46, $r = \frac{k}{c^2}$. If r is measured in ohms and c is measured in amperes, what is the unit of k?

Challenge Problems

In addition to direct variation and inverse variation, there is also joint variation and combined variation. In ***joint variation****, one quantity may vary directly (as the product of) two or more quantities. In* ***combined variation****, one quantity may vary directly with some variables and inversely with other variables. Exercise 57 is a joint variation problem, and Exercise 58 is a combined variation problem. For Exercises 57 and 58,* **a)** *write the variation equation, and* **b)** *find the quantity indicated.*

57. x varies jointly as y and z. If x is 72 when $y = 18$ and $z = 2$, find x when $y = 36$ and $z = 3$.

58. T varies directly as the square of D and inversely as F. If $T = 18$ when $D = 6$ and $F = 4$, find T when $D = 8$ and $F = 8$.

Cumulative Review Exercises

[4.6] 59. Divide $\frac{8x^2 + 6x - 21}{4x + 9}$.

[5.1] 60. Factor $y(z - 2) + 8(z - 2)$.

[5.6] 61. Solve $3x^2 - 24 = -6x$.

[6.2] 62. Multiply $\frac{x + 8}{x - 3} \cdot \frac{x^3 - 27}{x^2 + 3x + 9}$.

Chapter 6 Summary

IMPORTANT FACTS AND CONCEPTS	EXAMPLES
Section 6.1	
A **rational expression** is an expression of the form $\frac{p}{q}$, where p and q are polynomials and $q \neq 0$. Whenever we have a rational expression containing a variable in the denominator, we always assume that the value or values of the variable that make the denominator 0 are excluded.	$\frac{x + 2}{x}$ and $\frac{x^2 + x}{x - 1}$ are rational expressions. The rational expression $\frac{x - 5}{x - 3}$ is defined for all real numbers except 3.

IMPORTANT FACTS AND CONCEPTS	EXAMPLES
Section 6.1 (Continued)	
A rational expression is simplified or reduced to its lowest terms when the numerator and denominator have no common factors other than 1.	$\frac{1}{8}$ and $\frac{x^2+x+1}{2x^2+3x+7}$ are expressions reduced to lowest terms.
To Simplify Rational Expressions **1.** Factor both the numerator and denominator completely. **2.** Divide out common factors.	$\frac{12x^2-11x-5}{9x^2+6x+1}=\frac{(3x+1)(4x-5)}{(3x+1)(3x+1)}=\frac{4x-5}{3x+1}$
Section 6.2	
To Multiply Two Fractions $\frac{a}{b}\cdot\frac{c}{d}=\frac{a\cdot c}{b\cdot d},\quad b\neq 0 \text{ and } d\neq 0$	$\frac{1}{3}\cdot\frac{4}{5}=\frac{1\cdot 4}{3\cdot 5}=\frac{4}{15}$
To Multiply Rational Expressions **1.** Factor all numerators and denominators completely. **2.** Divide out common factors. **3.** Multiply numerators together and multiply denominators together.	$\frac{4x^3}{7y^2}\cdot\frac{14y^3}{6x}=\frac{\overset{2}{\cancel{4}}\overset{x^2}{\cancel{x^3}}\cdot\overset{2}{\cancel{14}}\overset{y}{\cancel{y^3}}}{\cancel{7}\cancel{y^2}\cdot\underset{3}{\cancel{6}}\cancel{x}}=\frac{4x^2y}{3}$
To Divide Two Fractions $\frac{a}{b}\div\frac{c}{d}=\frac{a}{b}\cdot\frac{d}{c}=\frac{ad}{bc},\quad b\neq 0,\ d\neq 0,\ \text{and}\ c\neq 0$	$\frac{3}{5}\div\frac{6}{7}=\frac{\overset{1}{\cancel{3}}}{5}\cdot\frac{7}{\underset{2}{\cancel{6}}}=\frac{7}{10}$
To Divide Rational Expressions Multiply the first rational expression by the reciprocal of the second rational expression.	$\frac{x+4}{x+3}\div\frac{3x+12}{x+3}=\frac{\cancel{x+4}}{\cancel{x+3}}\cdot\frac{\cancel{x+3}}{3\cancel{(x+4)}}=\frac{1}{3}$
Section 6.3	
To Add or Subtract Two Fractions $\frac{a}{c}+\frac{b}{c}=\frac{a+b}{c}, c\neq 0 \qquad \frac{a}{c}-\frac{b}{c}=\frac{a-b}{c}, c\neq 0$	$\frac{3}{11}+\frac{4}{11}=\frac{7}{11},\quad \frac{18}{19}-\frac{5}{19}=\frac{13}{19}$
To Add or Subtract Rational Expressions with a Common Denominator **1.** Add or subtract the numerators. **2.** Place the sum or difference of the numerators over the common denominator. **3.** Simplify the rational expression if possible.	$\frac{2}{x-3}+\frac{x+5}{x-3}=\frac{2+x+5}{x-3}=\frac{x+7}{x-3}$
To Find the Least Common Denominator of Rational Expressions **1.** Factor each denominator completely. **2.** List all different factors of each denominator. When the same factor appears in more than one denominator, write that factor with the highest power that appears on it. **3.** The least common denominator is the product of all the factors listed in step 2.	Find the least common denominator. $\frac{1}{9x^3y^4}+\frac{1}{6x^5y^3}$ $9x^3y^4=3\cdot 3x^3y^4$ $6x^5y^3=3\cdot 2x^5y^3$ LCD is $2\cdot 3^2x^5y^4=18x^5y^4$.
Section 6.4	
To Add or Subtract Two Rational Expressions with Unlike Denominators **1.** Determine the LCD. **2.** Rewrite each rational expression as an equivalent rational expression with the LCD. **3.** Add or subtract the numerators while maintaining the LCD. **4.** When possible, factor the remaining numerator and simplify the rational expression.	$\frac{9}{m}+\frac{5}{m-1}=\frac{m-1}{m-1}\cdot\frac{9}{m}+\frac{5}{m-1}\cdot\frac{m}{m}$ $=\frac{9(m-1)}{m(m-1)}+\frac{5m}{m(m-1)}$ $=\frac{9m-9+5m}{m(m-1)}$ $=\frac{14m-9}{m(m-1)}$

IMPORTANT FACTS AND CONCEPTS	EXAMPLES
Section 6.5	
A **complex fraction** is one that has a rational expression in its numerator or its denominator or in both its numerator and denominator.	$\dfrac{\frac{2}{3}}{\frac{4}{7}}$, $\dfrac{\frac{1}{x}+\frac{1}{y}}{\frac{1}{a}+\frac{1}{b}}$
Method 1—Simplify a Complex Fraction by Simplifying Numerator and Denominator **1.** Add or subtract the fractions in both the numerator and denominator of the complex fraction to obtain single fractions in both. **2.** Multiply the fraction in the numerator by the reciprocal of the fraction in the denominator. **3.** Simplify further if possible.	$\dfrac{1+\frac{1}{x}}{x}=\dfrac{\frac{x}{x}+\frac{1}{x}}{x}=\dfrac{\frac{x+1}{x}}{x}$ $=\dfrac{x+1}{x}\cdot\dfrac{1}{x}=\dfrac{x+1}{x^2}$
Method 2—Simplify a Complex Fraction Using Multiplication First **1.** Find the LCD of *all* the denominators appearing in the complex fraction. **2.** Multiply both the numerator and denominator of the complex fraction by the LCD found in step 1. **3.** Simplify when possible.	$\dfrac{1+\frac{1}{x}}{x}=\dfrac{x}{x}\cdot\dfrac{1+\frac{1}{x}}{x}=\dfrac{x(1)+x\left(\frac{1}{x}\right)}{x(x)}=\dfrac{x+1}{x^2}$ Note: LCD $= x$.
Section 6.6	
A **rational equation** is an equation that contains one or more rational expressions.	$\frac{1}{3}x-\frac{1}{7}x=10$, $\quad x+\frac{9}{x}=\frac{1}{3}$
To Solve Rational Equations **1.** Determine the LCD of all fractions in the equation. **2.** Multiply *both* sides of the equation by the LCD. **3.** Remove any parentheses and combine like terms on each side of the equation. **4.** Solve the equation using the properties discussed in earlier chapters. **5.** Check your solution in the *original* equation.	$\frac{x}{5}-\frac{x}{8}=1$ $40\left(\frac{x}{5}-\frac{x}{8}\right)=40\,(1)$ $8x-5x=40$ $3x=40$ $x=\frac{40}{3}$ A check shows that $\frac{40}{3}$ is the solution.
Section 6.7	
Applications A **geometry problem** involves geometric figures and formulas.	A rectangle has an area of 70 square meters. Find the dimensions if the width is 3 meters shorter than the length. The answer is 7 meters by 10 meters.
A **motion problem** involves distance, rate, and time and uses the formula distance = rate · time or $\quad \text{time}=\dfrac{\text{distance}}{\text{rate}}$ or $\quad \text{rate}=\dfrac{\text{distance}}{\text{time}}$	A cyclist can travel 20 miles with the wind to his back in the same time he can travel 12 miles going into the wind. If the wind is blowing at 2 miles per hour, find the speed of the cyclist without any wind. The answer is 8 miles per hour.
A **work problem** involves two or more machines or people working together to complete a specific task.	Tom can paint a room in 6 hours and Bill can paint the same room in 4 hours. How long will it take them working together to paint this room? The answer is 2.4 hours.
Section 6.8	
A **variation equation** is an equation that relates one variable to one or more other variables using the operations of multiplication or division.	

IMPORTANT FACTS AND CONCEPTS	EXAMPLES
Section 6.8 (Continued)	
Direct Variation In direct variation, as one variable increases, so does the other, and as one variable decreases, so does the other. If a variable y varies directly as a variable x, then $y = kx$ where k is the *constant of proportionality* (or the variation constant).	m varies directly as the square of n. Find m when $n = 5$ and $k = 4$. $m = kn^2$ $= 4(5)^2$ $= 4(25)$ $= 100$
Inverse Variation In inverse variation, as one variable increases, the other quantity decreases, and vice versa. If a variable y varies inversely as a variable x, then $y = \frac{k}{x}$ (or $xy = k$) where k is the constant of proportionality.	y varies inversely as the square root of x. Find y when $x = 4$ and $k = 30$. $y = \frac{k}{\sqrt{x}}$ $= \frac{30}{\sqrt{4}}$ $= \frac{30}{2}$ $= 15$

Chapter 6 Review Exercises

[6.1] *Determine the values of the variable for which the following expressions are defined.*

1. $\frac{5}{2x - 38}$ **2.** $\frac{2x + 1}{x^2 - 8x + 15}$ **3.** $\frac{7x - 1}{5x^2 + 4x - 1}$

Simplify.

4. $\frac{y}{xy - 8y}$ **5.** $\frac{x^3 + 5x^2 + 12x}{x}$ **6.** $\frac{9x^2 + 3xy}{3x}$

7. $\frac{x^2 + 2x - 8}{x - 2}$ **8.** $\frac{a^2 - 81}{a - 9}$ **9.** $\frac{-2x^2 + 7x + 4}{x - 4}$

10. $\frac{b^2 - 7b + 10}{b^2 - 3b - 10}$ **11.** $\frac{4x^2 - 11x - 3}{4x^2 - 7x - 2}$ **12.** $\frac{2x^2 - 21x + 40}{4x^2 - 4x - 15}$

[6.2] *Multiply.*

13. $\frac{5a^2}{6b} \cdot \frac{2}{4a^2b}$ **14.** $\frac{30x^2y^3}{3z} \cdot \frac{6z^3}{5xy^3}$ **15.** $\frac{20a^3b^4}{7c^3} \cdot \frac{14c^7}{5a^5b}$

16. $\frac{2}{x - 4} \cdot \frac{4 - x}{9}$ **17.** $\frac{-m + 4}{15m} \cdot \frac{10m}{m - 4}$ **18.** $\frac{a - 2}{a + 3} \cdot \frac{a^2 + 4a + 3}{a^2 - a - 2}$

Divide.

19. $\frac{9x^6}{y^2} \div \frac{x^4}{4y}$ **20.** $\frac{5xy^2}{z} \div \frac{x^4y^2}{4z^2}$ **21.** $\frac{6a + 6b}{a^2} \div \frac{a^2 - b^2}{a^2}$

22. $\frac{1}{a^2 + 8a + 15} \div \frac{8}{a + 5}$ **23.** $(t + 8) \div \frac{t^2 + 5t - 24}{t - 3}$ **24.** $\frac{x^2 + xy - 2y^2}{2y} \div \frac{x + 2y}{12y^2}$

[6.3] *Add or subtract.*

25. $\frac{n}{n + 5} - \frac{2}{n + 5}$ **26.** $\frac{4x}{x + 7} + \frac{28}{x + 7}$ **27.** $\frac{5x - 4}{x + 8} + \frac{44}{x + 8}$

28. $\frac{7x - 3}{x^2 + 7x - 30} - \frac{3x + 9}{x^2 + 7x - 30}$ **29.** $\frac{5h^2 + 12h - 1}{h + 5} - \frac{h^2 - 5h + 14}{h + 5}$ **30.** $\frac{6x^2 - 4x}{2x - 3} - \frac{-3x + 12}{2x - 3}$

Find the least common denominator for each expression.

31. $\frac{a}{8} + \frac{5a}{3}$

32. $\frac{10}{x+3} + \frac{2x}{x+3}$

33. $\frac{10}{4xy^3} - \frac{11}{10x^2y}$

34. $\frac{6}{x-3} - \frac{2}{x}$

35. $\frac{8}{n+5} + \frac{2n-3}{n-4}$

36. $\frac{5x-12}{x^2+2x} - \frac{4}{x+2}$

37. $\frac{2r+1}{r-s} - \frac{6}{r^2-s^2}$

38. $\frac{3x^2}{x-9} + 10x^3$

39. $\frac{19x-5}{x^2+2x-35} + \frac{-10x+1}{x^2+9x+14}$

[6.4] *Add or subtract.*

40. $\frac{5}{3y^2} + \frac{y}{2y}$

41. $\frac{3x}{xy} + \frac{1}{4x}$

42. $\frac{5x}{3xy} - \frac{6}{x^2}$

43. $7 - \frac{2}{x+2}$

44. $\frac{x-y}{y} - \frac{x+y}{x}$

45. $\frac{7}{x+4} + \frac{2}{x}$

46. $\frac{2}{3x} - \frac{3}{3x-6}$

47. $\frac{1}{(z+5)} + \frac{9}{(z+5)^2}$

48. $\frac{x+2}{x^2-x-6} + \frac{x-3}{x^2-8x+15}$

[6.2–6.4] *Perform each indicated operation.*

49. $\frac{x+4}{x+6} - \frac{x-5}{x+2}$

50. $2 + \frac{x}{x-4}$

51. $\frac{a+2}{b} \div \frac{a-2}{5b^2}$

52. $\frac{x+5}{x^2-9} + \frac{2}{x+3}$

53. $\frac{6p+12q}{p^2q} \cdot \frac{p^5}{p+2q}$

54. $\frac{8}{(x+2)(x-3)} - \frac{6}{(x-2)(x+2)}$

55. $\frac{x+7}{x^2+9x+14} - \frac{x-10}{x^2-49}$

56. $\frac{x-y}{x+y} \cdot \frac{xy+x^2}{x^2-y^2}$

57. $\frac{3x^2-27y^2}{30} \div \frac{(x-3y)^2}{6}$

58. $\frac{a^2-11a+30}{a-6} \cdot \frac{a^2-8a+15}{a^2-10a+25}$

59. $\frac{a}{a^2-1} - \frac{3}{3a^2-2a-5}$

60. $\frac{2x^2+6x-20}{x^2-2x} \div \frac{x^2+7x+10}{2x^2-8}$

[6.5] *Simplify each complex fraction.*

61. $\dfrac{5+\frac{1}{3}}{\frac{3}{4}}$

62. $\dfrac{1+\frac{5}{8}}{3-\frac{9}{16}}$

63. $\dfrac{\frac{12ab}{9c}}{\frac{4a}{c^2}}$

64. $\dfrac{\frac{18x^4y^2}{9xy^5}}{4z^2}$

65. $\dfrac{a-\frac{a}{b}}{\frac{1+a}{b}}$

66. $\dfrac{r^2+\frac{7}{s}}{s^2}$

67. $\dfrac{\frac{3}{x}+\frac{2}{x^2}}{5-\frac{1}{x}}$

68. $\dfrac{\frac{x}{x+y}}{\frac{x^2}{4x+4y}}$

69. $\dfrac{\frac{9}{x}}{\frac{9}{x^2}}$

70. $\dfrac{\frac{1}{a}+3}{\frac{1}{a}+\frac{3}{a}}$

71. $\dfrac{\frac{1}{x^2}-\frac{1}{x}}{\frac{1}{x^2}+\frac{1}{x}}$

72. $\dfrac{\frac{8x}{y}-x}{\frac{y}{x}-1}$

[6.6] *Solve.*

73. $\frac{5}{8} = \frac{10}{x+3}$

74. $\frac{x}{4} = \frac{x-3}{2}$

75. $\frac{12}{n} + 2 = \frac{n}{4}$

76. $\frac{10}{m} + \frac{3}{2} = \frac{m}{10}$

77. $\frac{-4}{d} = \frac{3}{2} + \frac{4-d}{d}$

78. $\frac{1}{x-7} + \frac{1}{x+7} = \frac{1}{x^2-49}$

79. $\frac{x-3}{x-2} + \frac{x+1}{x+3} = \frac{2x^2+x+1}{x^2+x-6}$

80. $\frac{a}{a^2-64} + \frac{4}{a+8} = \frac{3}{a-8}$

81. $\frac{d}{d-4} - 4 = \frac{4}{d-4}$

[6.7] *Solve.*

82. Sandcastles It takes John and Amy Brogan 6 hours to build a sandcastle. It takes Paul and Cindy Carter 4 hours to make the same sandcastle. How long will it take all four people together to build the sandcastle?

83. Filling a Pool One hose can fill a swimming pool in 7 hours. A second hose can siphon all the water out of a full pool in 12 hours. How long will it take to fill the pool if, while one hose is filling the pool, the other hose is siphoning water from the pool?

84. Sum of Numbers One number is six times as large as another. The sum of their reciprocals is 7. Determine the numbers.

85. Rollerblading and Bicycling Robert Johnston can travel 3 miles on his rollerblades in the same time Tran Lee can travel 8 miles on his mountain bike. If Tran's speed on his bike is 3.5 miles per hour faster than that of Robert on his rollerblades, determine Robert's and Tran's speeds.

[6.8]

86. Drug Dosage The recommended dosage, d, of the antibiotic drug vancomycin is directly proportional to a person's weight, w. If Carmen Brown, who is 132 pounds, is given 182 milligrams, find the recommended dosage for Bill Glenn, who is 198 pounds.

87. Boyle's Law When a gas is kept at a constant temperature, its volume, V, is inversely proportional to the pressure, p, on the gas. If the pressure on 18 cubic inches of argon gas is 4 pounds per square inch, determine the volume of the gas when the pressure is 6 pounds per square inch.

Chapter 6 Practice Test

Chapter Test Prep Videos provide fully worked-out solutions to any of the exercises you want to review. Chapter Test Prep Videos are available via MyMathLab®, *or on* YouTube (www.youtube.com/user/AngelElementaryAlg).

Simplify.

1. $\dfrac{-8 + x}{x - 8}$

2. $\dfrac{x^3 - 1}{x^2 - 1}$

Perform each indicated operation.

3. $\dfrac{20x^2y^3}{4z^2} \cdot \dfrac{8xz^3}{5xy^4}$

4. $\dfrac{a^2 - 9a + 14}{a - 2} \cdot \dfrac{a^2 - 4a - 21}{(a - 7)^2}$

5. $\dfrac{x^2 - x - 6}{x^2 - 9} \cdot \dfrac{x^2 - 6x + 9}{x^2 + 4x + 4}$

6. $\dfrac{x^2 - 1}{x + 2} \cdot \dfrac{x + 2}{1 - x^2}$

7. $\dfrac{x^2 - 4y^2}{5x + 20y} \div \dfrac{x + 2y}{x + 4y}$

8. $\dfrac{15}{y^2 + 2y - 15} \div \dfrac{5}{y - 3}$

9. $\dfrac{m^2 + 3m - 18}{m - 3} \div \dfrac{m^2 - 8m + 15}{3 - m}$

10. $\dfrac{4x + 3}{8y} + \dfrac{2x - 5}{8y}$

11. $\dfrac{7x^2 - 4}{x + 3} - \dfrac{6x + 9}{x + 3}$

12. $\dfrac{2}{xy} - \dfrac{8}{xy^3}$

13. $3 - \dfrac{5z}{z - 5}$

14. $\dfrac{x - 5}{x^2 - 16} - \dfrac{x - 2}{x^2 + 2x - 8}$

Simplify.

15. $\dfrac{2 + \frac{1}{2}}{3 - \frac{1}{5}}$

16. $\dfrac{x + \frac{x}{y}}{\frac{7}{x}}$

17. $\dfrac{4 + \frac{3}{x}}{\frac{9}{x} - 5}$

Solve.

18. $2 + \dfrac{8}{x} = 6$

19. $\dfrac{2x}{3} - \dfrac{x}{4} = x + 1$

20. $\dfrac{x}{x - 8} + \dfrac{6}{x - 2} = \dfrac{x^2}{x^2 - 10x + 16}$

Solve.

21. **Working Together** Mr. Jackson, on his tractor, can clear a 1-acre field in 10 hours. Mr. Hackett, on his tractor, can clear a 1-acre field in 15 hours. If they work together, how long will it take them to clear a 1-acre field?

22. **Determine a Number** The sum of a positive number and its reciprocal is 2. Determine the number.

23. **Area of Triangle** The area of a triangle is 30 square inches. If the height is 2 inches less than 2 times the base, determine the height and base of the triangle.

24. **Exercising** LaConya Bertrell exercises for $1\frac{1}{2}$ hours each day. During the first part of her routine, she rides a bicycle and averages 10 miles per hour. For the remainder of the time, she rollerblades and averages 4 miles per hour. If the total distance she travels is 12 miles, how far did she travel on the rollerblades?

25. **Making Music** The wavelength of sound waves, w, is inversely proportional to the frequency, f (or pitch). If a frequency of 263 cycles per second (middle C on a piano) produces a wavelength of about 4.3 feet, determine the length of a wavelength of a frequency of 1000 cycles per second.

Cumulative Review Test

Take the following test and check your answers with those given in the back of the book. Review any questions that you answered incorrectly. The section where the material was covered is indicated after the answer.

1. Evaluate $3x^2 - 5xy^2 - 7$, when $x = -4$ and $y = -2$.
2. Solve $5z + 4 = -3(z - 7)$.
3. Simplify $\left(\dfrac{10x^6y^3}{2x^5y^5}\right)^3$.
4. The cost of a 2013 Ford Mustang increased by 1.8% over the cost of the 2012 Mustang. Write an expression for the sum of the costs of a 2013 and a 2012 Mustang.
5. Simplify $(6x^2 - 3x - 5) - (-2x^2 - 8x - 19)$.
6. Multiply $(3n^2 - 4n + 3)(2n - 5)$.
7. Factor $8a^2 - 8a - 5a + 5$.
8. Factor $13x^2 + 26x - 39$.
9. Evaluate $\{6 - [3(8 \div 4)]^2 + 9 \cdot 4\}^2$
10. Solve $2(x + 4) \le -(x + 3) - 1$. Graph the solution on a real number line and write the answer in interval notation.
11. Divide $\dfrac{4x - 38}{8}$.
12. Solve $2x^2 = 11x - 12$.
13. Multiply $\dfrac{x^2 + x - 12}{x^2 - x - 6} \cdot \dfrac{x^2 - 2x - 8}{2x^2 - 7x - 4}$.
14. Subtract $\dfrac{r}{r + 2} - \dfrac{3}{r - 5}$.
15. Add $\dfrac{4}{x^2 - 3x - 10} + \dfrac{6}{x^2 + 5x + 6}$.
16. Solve the equation $\dfrac{x}{9} - \dfrac{x}{6} = \dfrac{1}{12}$.
17. Solve the equation $\dfrac{7}{x + 3} + \dfrac{5}{x + 2} = \dfrac{5}{x^2 + 5x + 6}$.
18. **Medical Plans** A school district allows its employees to choose from two medical plans. With plan 1, the employee pays 10% of all medical bills (the school district pays the balance). With plan 2, the employee pays the school district a one-time payment of $150, then the employee pays 5% of all medical bills. What total medical bills would result in the employee paying the same amount with the two plans?
19. **Bird Seed** A feed store owner wishes to make his own store-brand mixture of bird seed by mixing sunflower seed that costs $0.50 per pound with a premixed assorted seed that costs $0.20 per pound. How many pounds of each will he have to use to make a 50-pound mixture that will cost $16.00?
20. **Sailing** During the first leg of a race, the sailboat *Thumper* sailed at an average speed of 6.5 miles per hour. During the second leg of the race, the winds increased and *Thumper* sailed at an average speed of 9.5 miles per hour. If the total distance sailed by *Thumper* was 12.75 miles, and the total time spent racing was 1.5 hours, determine the distance traveled by *Thumper* on each leg of the race.

7 Graphing Linear Equations

We see graphs daily. They are very important in both mathematics and in everyday living. Graphs are used to display information. For example, in Exercise 75 on page 429, we use a graph to illustrate the total monthly cost for telephone calls.

Goals of This Chapter

In this chapter you will learn how to graph linear equations. The graphs of linear equations are straight lines. Graphing is one of the most important topics in mathematics.

The two most important concepts we will discuss in this chapter are slopes and functions. Functions are a unifying concept in mathematics.

This chapter contains concepts that are central to mathematics. If you plan to take more mathematics courses, graphs and functions will probably be a significant part of those courses.

The Cartesian Coordinate System and Linear Equations in Two Variables

1 Plot points in the Cartesian coordinate system.

2 Determine whether an ordered pair is a solution to a linear equation.

René Descartes

1 Plot Points in the Cartesian Coordinate System

Many algebraic relationships are easier to understand if we can see a picture of them. A **graph** shows the relationship between two variables in an equation. In this chapter we discuss several procedures that can be used to draw graphs using the **Cartesian (or rectangular) coordinate system**. The Cartesian coordinate system is named for its developer, the French mathematician and philosopher René Descartes (1596–1650).

The Cartesian coordinate system provides a means of locating and identifying points just as the coordinates on a map help us find cities and other locations. Consider the map of the Great Smoky Mountains (see **Fig. 7.1**). Can you find Cades Cove on the map? If we tell you that it is in grid A3, you can probably find it much more quickly and easily.

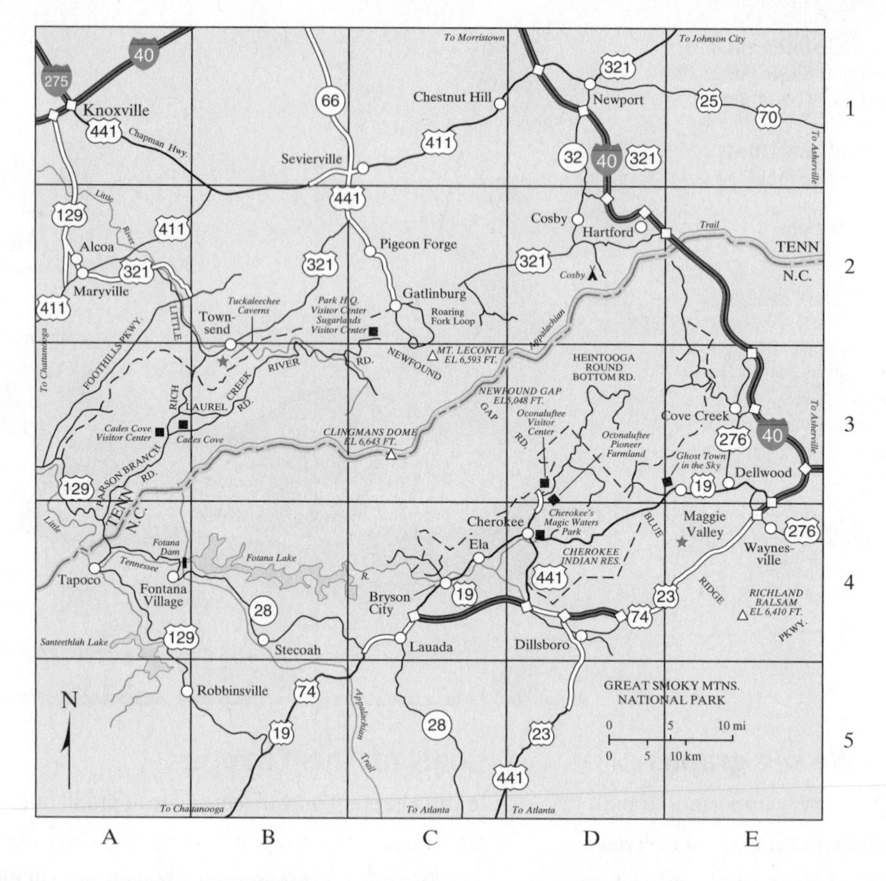

FIGURE 7.1

The Cartesian coordinate system is a grid system, like that of a map, except that it is formed by two axes (or number lines) drawn perpendicular to each other. The two intersecting axes form four **quadrants**, numbered I through IV in **Figure 7.2.**

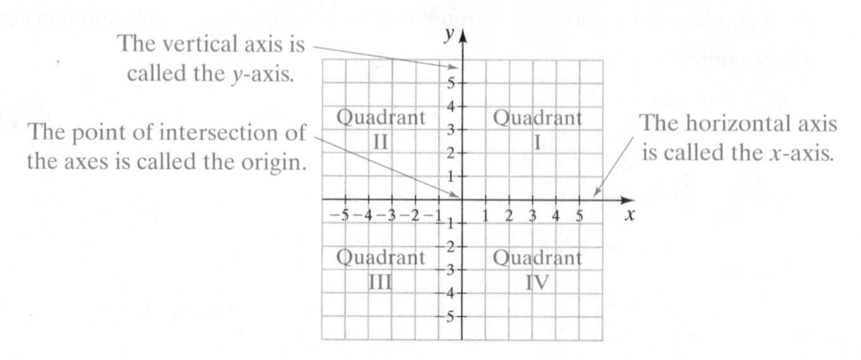

FIGURE 7.2

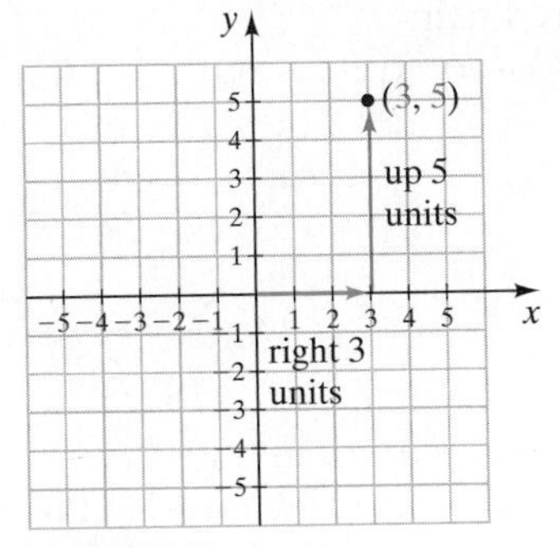

FIGURE 7.3

To locate a point in the Cartesian coordinate system, we will use an **ordered pair** of the form (x, y). The first number, x, is called the ***x*-coordinate** and the second number, y, is called the ***y*-coordinate.**

To plot the point $(3, 5)$, start at the origin (see **Fig. 7.3**),

$(3, 5)$

x-coordinate is 3 means go *right* 3 units; y-coordinate is 5 means then go *up* 5 units

To plot the point, $(-4, -2)$, start at the origin (see **Fig. 7.4**),

$(-4, -2)$

go *left* 4 units; *then go down* 2 units

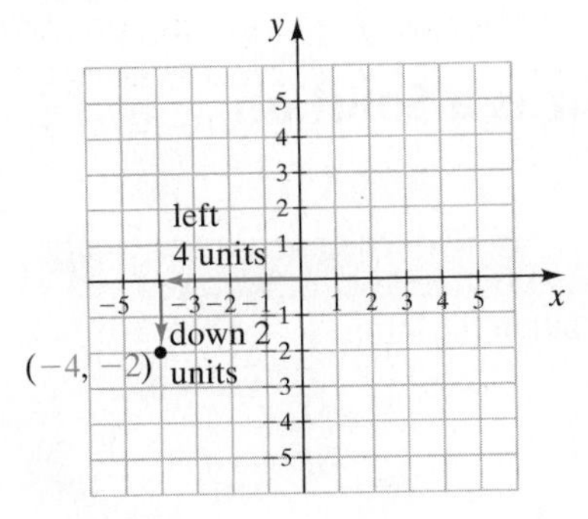

FIGURE 7.4

Understanding Algebra

Remember that the order of the numbers in an *ordered pair* is very important. For example, $(3, 5)$ is a different point than $(5, 3)$. The x-coordinate is always listed first in an ordered pair.

EXAMPLE 1 Plot each point on the same axes.

a) $A(5, 3)$ **b)** $B(2, 4)$ **c)** $C(-3, 1)$

d) $D(4, 0)$ **e)** $E(-2, -5)$ **f)** $F(0, -3)$

g) $G(0, 2)$ **h)** $H\left(6, -\frac{9}{2}\right)$ **i)** $I\left(-\frac{3}{2}, -\frac{5}{2}\right)$

Solution The first number in each ordered pair is the x-coordinate and the second number is the y-coordinate. The points are plotted in **Figure 7.5.**

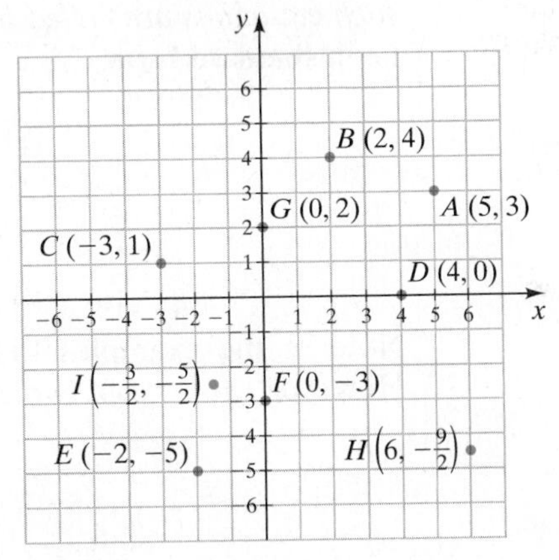

FIGURE 7.5

Note that when the x-coordinate is 0, as in Example 1 **f)** and 1 **g)**, the point is on the y-axis. When the y-coordinate is 0, as in Example 1 **d)**, the point is on the x-axis.

Now Try Exercise 23

EXAMPLE 2 List the ordered pairs for each point shown in **Figure 7.6.**

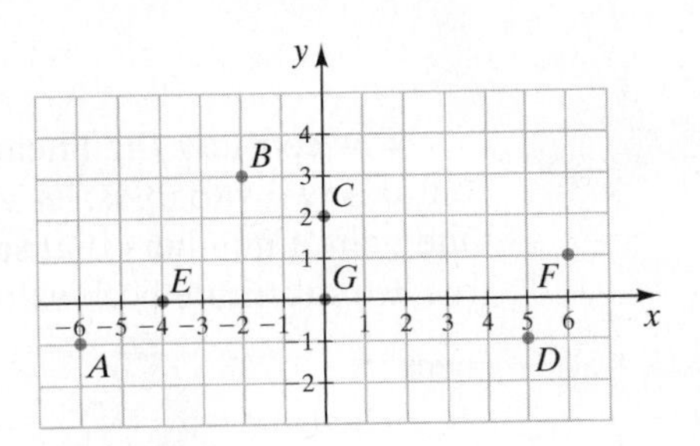

FIGURE 7.6

Solution Remember to give the x-value first in the ordered pair.

Point	Ordered Pair
A	$(-6, -1)$
B	$(-2, 3)$
C	$(0, 2)$
D	$(5, -1)$
E	$(-4, 0)$
F	$(6, 1)$
G	$(0, 0)$

Now Try Exercise 21

2 Determine Whether an Ordered Pair Is a Solution to a Linear Equation

Linear Equations in Two Variables

A **linear equation in two variables** is an equation that can be put in the form

$$ax + by = c$$

where a, b, and c are real numbers.

The graphs of equations of the form $ax + by = c$ are straight lines. For this reason such equations are called linear. A linear equation in the form $ax + by = c$ is said to be in **standard form**.

Examples of Linear Equations

$$4x - 3y = 12$$
$$y = 5x + 3$$
$$x - 3y + 4 = 0$$

Note in the examples that only the equation $4x - 3y = 12$ is in standard form. However, the bottom two equations can be written in standard form, as follows:

$$y = 5x + 3 \qquad\qquad x - 3y + 4 = 0$$
$$-5x + y = 3 \qquad\qquad x - 3y = -4$$

Most of the equations we have discussed thus far have contained only one variable. Consider the linear equation in *one* variable, $2x + 3 = 5$. What is its solution?

$$2x + 3 = 5$$
$$2x = 2$$
$$x = 1$$

This equation has only one solution, 1.

Check

$$2x + 3 = 5$$
$$2(1) + 3 \stackrel{?}{=} 5$$
$$5 = 5 \quad \text{True}$$

Now consider the linear equation in *two* variables, $y = x + 1$. Since the equation contains two variables, its solutions must contain two numbers, one for each variable. One pair of numbers that satisfies this equation is $x = 1$ and $y = 2$. To see that this is true, we substitute both values into the equation.

Check

$$y = x + 1$$
$$2 \stackrel{?}{=} 1 + 1$$
$$2 = 2 \quad \text{True}$$

We write this answer as an ordered pair by writing the x- and y-values within parentheses separated by a comma. Therefore, one solution to this equation is the ordered pair $(1, 2)$. The equation $y = x + 1$ has other solutions.

	Solution	Solution	Solution
	$x = 2, y = 3$	$x = -3, y = -2$	$x = -\frac{1}{3}, y = \frac{2}{3}$
Check	$y = x + 1$	$y = x + 1$	$y = x + 1$
	$3 \stackrel{?}{=} 2 + 1$	$-2 \stackrel{?}{=} -3 + 1$	$\frac{2}{3} \stackrel{?}{=} -\frac{1}{3} + 1$
	$3 = 3$ True	$-2 = -2$ True	$\frac{2}{3} = \frac{2}{3}$ True

Solution Written as an Ordered Pair

$(2, 3)$	$(-3, -2)$	$\left(-\frac{1}{3}, \frac{2}{3}\right)$

How many possible solutions does the equation $y = x + 1$ have? The equation $y = x + 1$ has an unlimited or *infinite number* of possible solutions. Since it is not possible to list all the specific solutions, the solutions are illustrated with a graph.

Graph of an Equation

A **graph** of an equation in two variables is an illustration of the set of points whose coordinates satisfy the equation.

Figure 7.7a shows the points $(2, 3)$, $(-3, -2)$, and $\left(-\frac{1}{3}, \frac{2}{3}\right)$ plotted in the Cartesian coordinate system. **Figure 7.7b** shows a straight line drawn through the three points. Arrowheads are placed at the ends of the line to show that the line continues in both directions. Every point on this line will satisfy the equation $y = x + 1$, so this graph illustrates all the solutions of $y = x + 1$. The ordered pair $(1, 2)$, which is on the line, also satisfies the equation.

In **Figure 7.7b**, what do you notice about the points $(2, 3)$, $(1, 2)$, $\left(-\frac{1}{3}, \frac{2}{3}\right)$, and $(-3, -2)$? You probably noticed that they are in a straight line. A set of points that are on a line are said to be **collinear**.

Understanding Algebra

The equation $y = x + 1$ has an infinite number of solutions. Each solution is represented by a point on the line in **Figure 7.7b.** Also, each point on the line represents a solution to the equation $y = x + 1$.

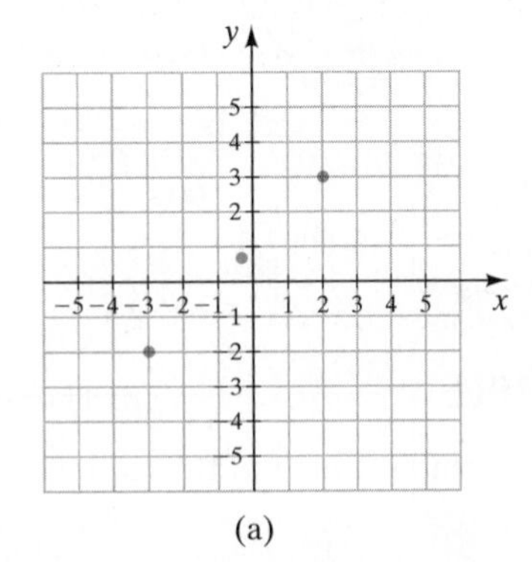

(a)

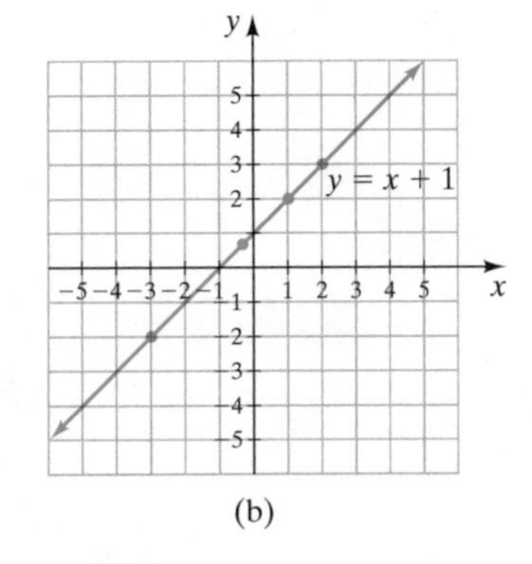

(b)

FIGURE 7.7

EXAMPLE 3 Determine whether the three points appear to be collinear.

a) $(2, 7)$, $(0, 3)$, and $(-2, -1)$

b) $(0, 5)$, $\left(\frac{5}{2}, 0\right)$, and $(5, -5)$

c) $(-2, -5)$, $(0, 1)$, and $(6, 8)$

Solution We plot the points to determine whether they appear to be collinear. The solution is shown in **Figure 7.8.**

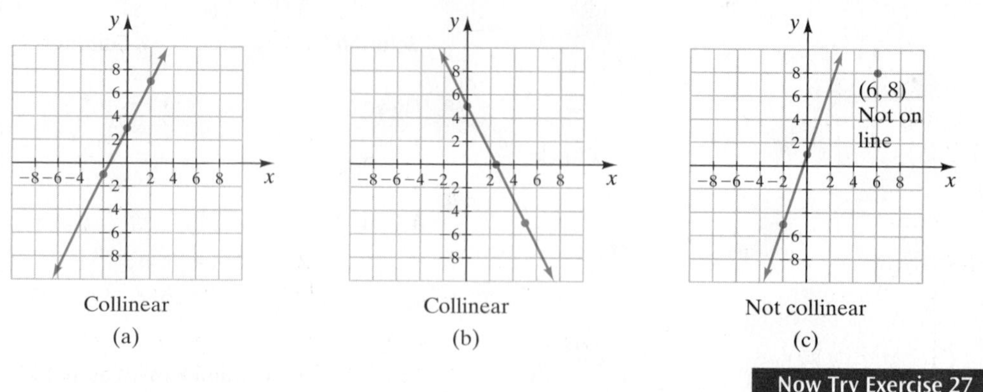

FIGURE 7.8 Collinear (a) Collinear (b) Not collinear (c)

Now Try Exercise 27

To graph an equation, you will need to determine ordered pairs that satisfy the equation and then plot the points.

Understanding Algebra

Although only two points are needed to graph a linear equation, it is always a good idea to use at least three points. As shown in Figure 7.9, if you only use two points and one of the points is incorrect, the graph will also be incorrect. However, if you plot three points and the points appear to be collinear, you probably have not made a mistake.

HELPFUL HINT

Only two points are needed to graph a linear equation because the graph of every linear equation is a straight line. However, if you graph a linear equation using only two points and you have made an error in determining or plotting one of those points, your graph will be wrong and you will not know it. In **Figures 7.9a** and **b** we plot only two points to show that if only one of the two points plotted is incorrect, the graph will be wrong. In both **Figures 7.9a** and **b** we use the ordered pair $(-2, -2)$. However, in **Figure 7.9a** the second point is $(1, 2)$, while in **Figure 7.9b** the second point is $(2, 1)$. Notice how the two graphs differ.

If you use at least three points to plot your graph, as in Figure 7.7b on page 417, and they appear to be collinear, you probably have not made a mistake.

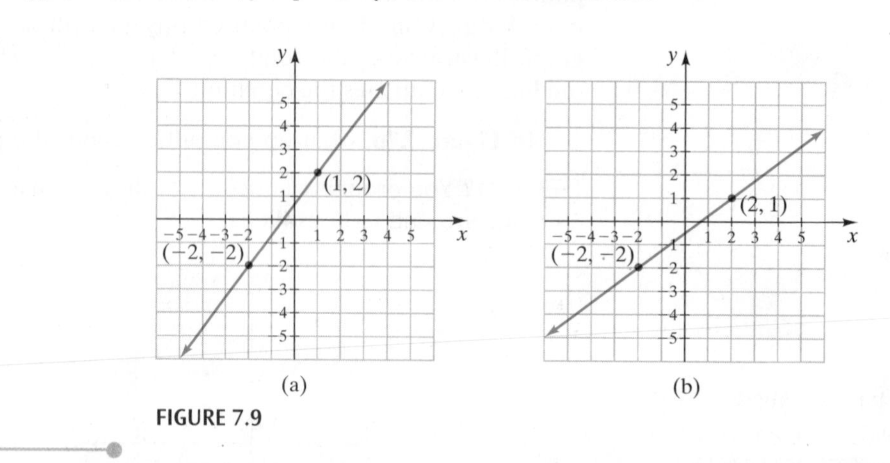

(a) (b)

FIGURE 7.9

EXAMPLE 4

a) Determine which of the following ordered pairs satisfy the equation $2x + y = 4$.

$$(2, 0),\ (0, 4),\ (3, 1),\ (-1, 6)$$

b) Plot all the points that satisfy the equation on the same axes and draw a straight line through the points.

c) What does this line represent?

Solution

a) We substitute values for x and y into the equation $2x + y = 4$ and determine whether they check.

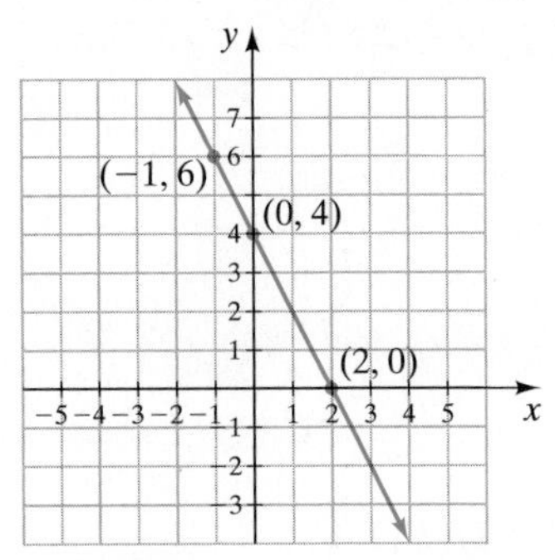

FIGURE 7.10

Check

(2, 0)		(0, 4)	
$2x + y = 4$		$2x + y = 4$	
$2(2) + 0 \stackrel{?}{=} 4$		$2(0) + 4 \stackrel{?}{=} 4$	
$4 = 4$	True	$4 = 4$	True
(3, 1)		(−1, 6)	
$2x + y = 4$		$2x + y = 4$	
$2(3) + 1 \stackrel{?}{=} 4$		$2(-1) + 6 \stackrel{?}{=} 4$	
$7 = 4$	False	$4 = 4$	True

The ordered pairs $(2, 0)$, $(0, 4)$, and $(-1, 6)$ satisfy the equation. The ordered pair $(3, 1)$ does not satisfy the equation.

b) **Figure 7.10** shows the three points that satisfy the equation. A straight line drawn through the three points shows that they appear to be collinear. Note that the ordered pair $(3, 1)$, which is not a solution to the equation, is not a point on this line.

c) The line represents all solutions of $2x + y = 4$. The coordinates of every point on this line satisfy the equation $2x + y = 4$.

Now Try Exercise 31

7.1 Exercise Set MathXL® MyMathLab®

Warm-Up Exercises

Fill in the blanks with the appropriate word, phrase, or symbol(s) from the following list.

line	solution	quadratic	*y*-axis	linear
collinear	*x*-axis	Cartesian	*x*-coordinate	*y*-coordinate

1. In an ordered pair the __________ is listed first.

2. In an ordered pair the __________ is listed second.

3. The __________ is the horizontal axis.

4. The __________ is the vertical axis.

5. A __________ equation can be put into the form $ax + by = c$.

6. The graph of a linear equation is a straight __________.

7. A __________ to an equation with two variables is an ordered pair that makes the equation true.

8. Three or more points that are on the same line are called __________.

Practice the Skills

Indicate the quadrant in which each of the points belongs.

9. $(-2, 5)$ **10.** $(-4, 4)$ **11.** $(5, -6)$ **12.** $(2, -3)$

13. $(3, 6)$ **14.** $(4, 30)$ **15.** $(-17, -87)$ **16.** $(-124, -132)$

17. $(63, 47)$ **18.** $(75, 200)$ **19.** $(-8, 42)$ **20.** $(76, -92)$

21. List the ordered pairs corresponding to each point.

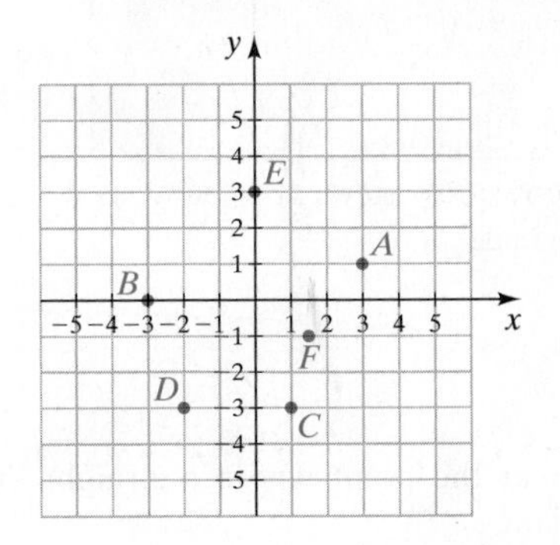

22. List the ordered pairs corresponding to each point.

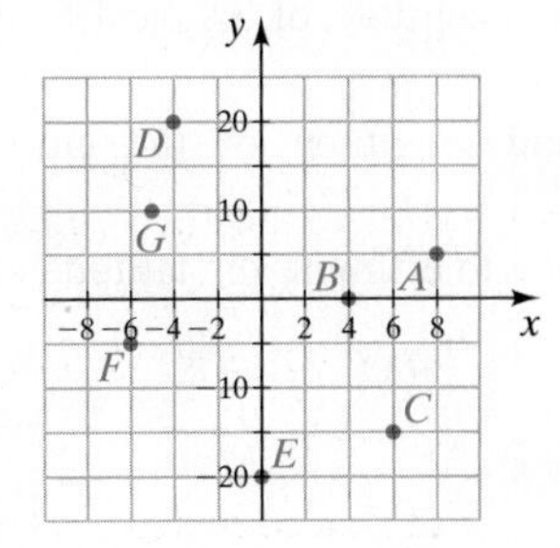

Plot each point on the same axes.

23. $A(3, 2), B(-4, 1), C(0, -3), D(-2, 0), E(-3, -4), F\left(-4, -\frac{5}{2}\right)$

24. $A(-3, -1), B(2, 0), C(3, 2), D\left(\frac{1}{2}, -4\right), E(-4, 2), F(0, 5)$

25. $A(4, 0), B(-1, 3), C(2, 4), D(0, -2), E(-3, -3), F(2, -3)$

26. $A(-3, 4), B(2, 3), C(0, 3), D(-1, 0), E(-2, -2), F(2, -4)$

Plot the following points. Then determine whether they appear to be collinear.

27. $A(1, -1), B(5, 3), C(-3, -5), D(0, -2), E(2, 0)$

28. $A(1, -1), B(3, 5), C(0, -3), D(-2, -7), E(2, 1)$

29. $A(1, 5), B\left(-\frac{1}{2}, \frac{1}{2}\right), C(0, 2), D(-5, -3), E(-2, -4)$

30. $A(1, -2), B(0, -5), C(4, 1), D(-1, -8), E\left(\frac{1}{2}, -\frac{7}{2}\right)$

In Exercises 31–36, **a)** *determine which of the four ordered pairs does not satisfy the given equation.* **b)** *Plot all the points that satisfy the equation on the same axes and draw a straight line through the points.*

31. $y = x + 2$, **a)** $(2, 4)$ **b)** $(-2, 0)$ **c)** $(-1, 5)$ **d)** $(0, 2)$

32. $y = \frac{1}{2}x + 2$, **a)** $(0, 2)$ **b)** $(-4, 3)$ **c)** $(-2, 1)$ **d)** $(4, 4)$

33. $3x - 2y = 6$, **a)** $(4, 0)$ **b)** $(2, 0)$ **c)** $\left(\frac{2}{3}, -2\right)$ **d)** $\left(\frac{4}{3}, -1\right)$

34. $2x + y = -4$, **a)** $(-2, 0)$ **b)** $(2, 3)$ **c)** $(0, -4)$ **d)** $(-1, -2)$

35. $\frac{1}{2}x + 4y = 4$, **a)** $(-2, 3)$ **b)** $\left(2, \frac{3}{4}\right)$ **c)** $(0, 1)$ **d)** $\left(-4, \frac{3}{2}\right)$

36. $4x - 3y = 0$, **a)** $(3, 4)$ **b)** $(-3, -4)$ **c)** $(0, 0)$ **d)** $(2, 5)$

Problem Solving

Consider the linear equation $y = 3x - 4$. *In Exercises 37–40, find the value of y that makes the given ordered pair a solution to the equation.*

37. $(2, y)$

38. $(-1, y)$

39. $(0, y)$

40. $(3, y)$

Consider the linear equation $2x + 3y = 12$. *In Exercises 41–44, find the value of x that makes the given ordered pair a solution to the equation.*

41. $(x, 0)$

42. $(x, -2)$

43. $\left(x, \frac{11}{3}\right)$

44. $\left(x, -\frac{2}{3}\right)$

45. **Longitude and Latitude** Another type of coordinate system that is used to identify a location or position on Earth's surface involves *latitude* and *longitude*. On a globe, the longitudinal lines are lines that go from top to bottom; on a world map they go up and down. The latitudinal lines go around the globe, or left to right on a world map. The locations of Hurricane Georges and Tropical Storm Hermine are indicated on the map on the right.

    a) Estimate the latitude and longitude of Hurricane Georges.

    b) Estimate the latitude and longitude of Tropical Storm Hermine.

    c) Estimate the latitude and longitude of the city of Miami.

    d) Use either a map or a globe to estimate the latitude and longitude of your college.

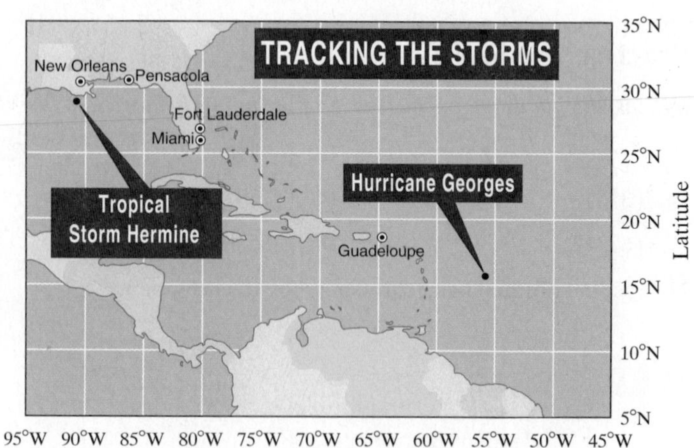

Source: National Weather Service

Concept/Writing Exercises

46. What is the value of y at the point where a straight line crosses the x-axis? Explain.

47. What is the value of x at the point where a straight line crosses the y-axis? Explain.

48. What does the graph of a linear equation illustrate?

49. Why are arrowheads added to the ends of graphs of linear equations?

50. How many solutions does a linear equation in two variables have?

Group Activity

In Section 7.2 we discuss how to find ordered pairs to plot when graphing linear equations. Let's see if you can draw some graphs now. Individually work parts **a)** *through* **c)** *in Exercises 51–54.*

a) *Select any three values for x and find the corresponding values of y.*

b) *Plot the points (they should appear to be collinear).*

c) *Draw the graph.*

d) *As a group, compare your answers. You should all have the same lines.*

51. $y = 2x$

52. $y = x$

53. $y = 2x + 1$

54. $y = -2x$

Cumulative Review Exercises

[2.5] **55.** Solve the equation $\frac{1}{2}(x - 3) = \frac{1}{3}x + 2$.

[2.6] **56.** Solve the equation $3x - 2y = 4$ for y.

[4.1] **57.** Simplify $(2x^3)^4$

[5.3] **58.** Factor $x^2 - 6x - 27$.

[5.6] **59.** Solve $y(y - 7) = 0$.

[6.4] **60.** Add $\frac{6}{x^2} + \frac{5}{3x}$.

7.2 Graphing Linear Equations

1. Graph linear equations by plotting points.
2. Graph linear equations of the form $ax + by = 0$.
3. Graph linear equations using the x- and y-intercepts.
4. Graph horizontal and vertical lines.
5. Study applications of graphs.

Now we are ready to graph linear equations.

1 Graph Linear Equations by Plotting Points

Graphing by plotting points is the most versatile method of graphing because we can also use it to graph other types of equations.

To Graph Linear Equations by Plotting Points

1. **Solve for y:** Solve the equation for the variable y. That is, get y by itself on the left side of the equation.
2. **Substitute a number in for x:** Select a number and substitute it into the equation for x and find the corresponding value for y. Record the ordered pair (x, y).
3. **Repeat step 2:** Select two different values for x. This will give you two additional ordered pairs.
4. **Plot the points:** Plot the three ordered pairs. The three points should be collinear. If they are not, recheck your work.
5. **Draw the line:** With a straightedge, draw a straight line through the three points. Draw an arrowhead on each end of the line.

HELPFUL HINTS

- If you have forgotten how to solve an equation for y, review Section 2.6.
- When selecting numbers for x, you should select numbers that result in integer values for y.
- Select numbers for x that are small enough so that the ordered pairs obtained can be plotted easily on the axes.
- A good number to select for x is always 0.

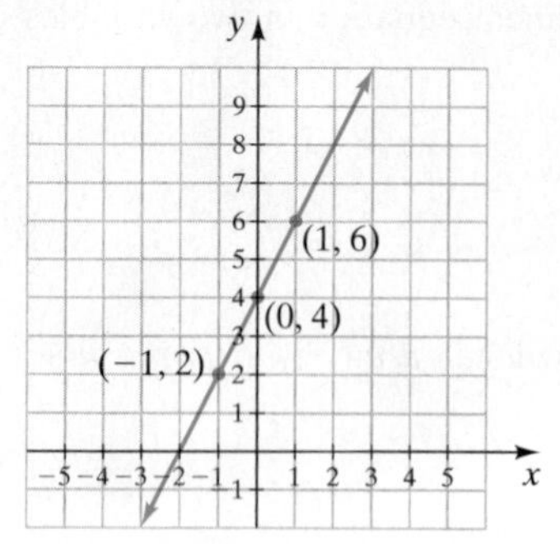

FIGURE 7.11

EXAMPLE 1 Graph $y = 2x + 4$.

Solution The equation is already solved for y. We will arbitrarily select the numbers -1, 0, and 1 and substitute them into the equation for x and find the corresponding values for y. The calculations that follow show that when $x = -1$, $y = 2$, when $x = 0$, $y = 4$, and when $x = 1$, $y = 6$.

x	$y = 2x + 4$	Ordered Pair
-1	$y = 2(-1) + 4 = 2$	$(-1, 2)$
0	$y = 2(0) + 4 = 4$	$(0, 4)$
1	$y = 2(1) + 4 = 6$	$(1, 6)$

x	y
-1	2
0	4
1	6

We then plot the three ordered pairs on the same axes **(Fig. 7.11).** Notice that the points are collinear. Connect the points with a straight line and place the arrowheads at the ends of the line.

Now Try Exercise 25

If we select any point on this line, the ordered pair represented by that point will be a solution to the equation $y = 2x + 4$. Similarly, any solution to the equation will be represented by a point on the line. Notice that $(-2, 0)$ and $(2, 8)$ are points on the line (see **Figure 7.12**).

Understanding Algebra

Remember, a graph of an equation is an illustration of the set of points whose coordinates satisfy the equation.

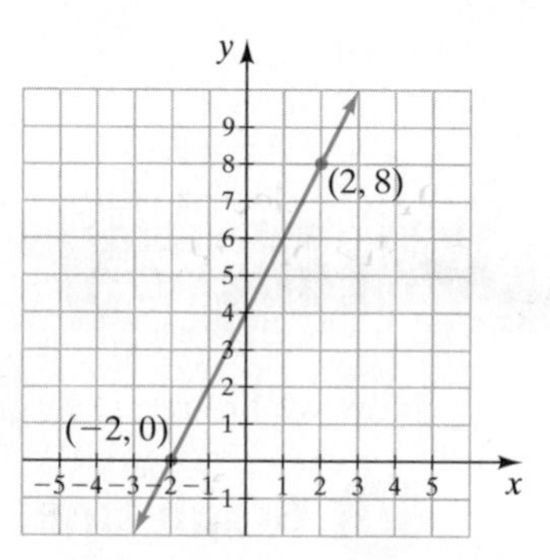

FIGURE 7.12

Let's verify that the ordered pairs $(-2, 0)$ and $(2, 8)$ are solutions to the equation

Check $(-2, 0)$	Check $(2, 8)$
$y = 2x + 4$	$y = 2x + 4$
$0 = 2(-2) + 4$	$8 = 2(2) + 4$
$0 = -4 + 4$	$8 = 4 + 4$
$0 = 0$ True	$8 = 8$ True

EXAMPLE 2 Graph $3y = 5x - 6$.

Solution We begin by solving the equation for y.

$$3y = 5x - 6$$

$$y = \frac{5x - 6}{3} \quad \text{Divided both sides by 3.}$$

$$y = \frac{5x}{3} - \frac{6}{3} \quad \text{Wrote as two fractions.}$$

$$y = \frac{5}{3}x - 2$$

Now we can see that if we select values for x that are multiples of the denominator, 3, the values we obtain for y will be integers. Let's select the values -3, 0, and 3 for x.

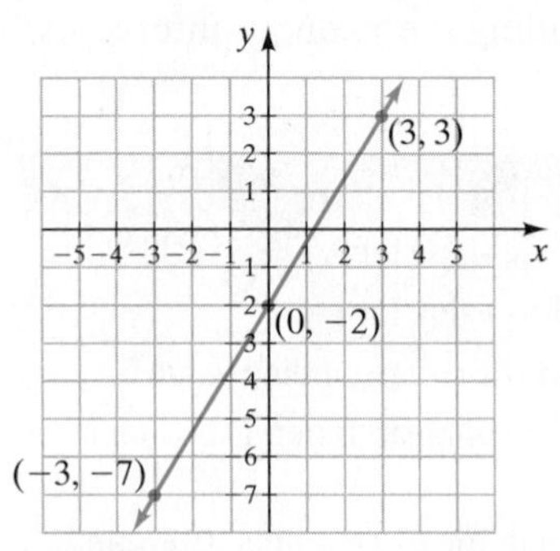

FIGURE 7.13

$$y = \frac{5}{3}x - 2$$

Let $x = -3$. $\quad y = \frac{5}{3}(-3) - 2 = -5 - 2 = -7$

Let $x = 0$. $\quad y = \frac{5}{3}(0) - 2 = 0 - 2 = -2$

Let $x = 3$. $\quad y = \frac{5}{3}(3) - 2 = 5 - 2 = 3$

x	y
−3	−7
0	−2
3	3

Finally, we plot the points and draw the straight line (**Figure 7.13**).

Now Try Exercise 31

2 Graph Linear Equations of the Form $ax + by = 0$

EXAMPLE 3 Graph $2x + 5y = 0$.

Solution We begin by solving the equation for y.

$$2x + 5y = 0$$
$$5y = -2x$$
$$y = -\frac{2x}{5} \quad \text{or} \quad y = -\frac{2}{5}x$$

If we select values for x that are multiples of 5, we will get integer values for y. We will arbitrarily select the values $x = -5$, $x = 0$, and $x = 5$.

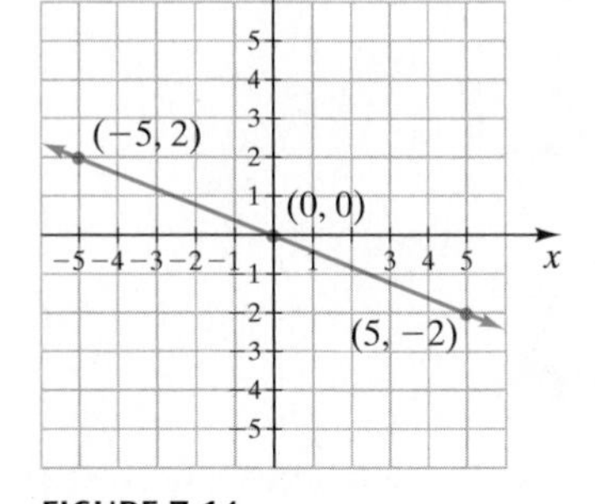

FIGURE 7.14

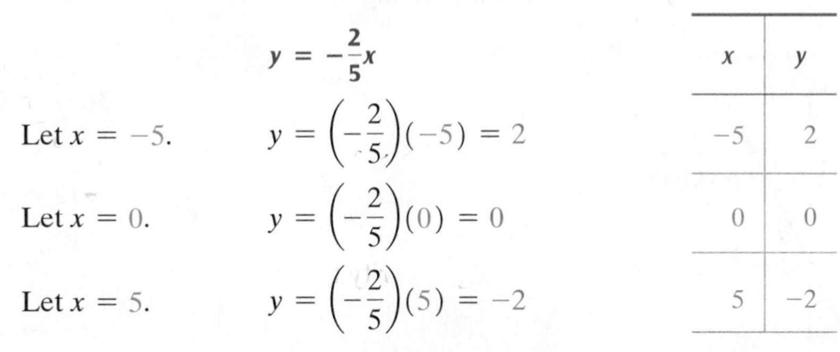

$$y = -\frac{2}{5}x$$

Let $x = -5$. $\quad y = \left(-\frac{2}{5}\right)(-5) = 2$

Let $x = 0$. $\quad y = \left(-\frac{2}{5}\right)(0) = 0$

Let $x = 5$. $\quad y = \left(-\frac{2}{5}\right)(5) = -2$

x	y
−5	2
0	0
5	−2

Now we plot the points and draw the graph **(Fig. 7.14)**.

Now Try Exercise 37

Understanding Algebra

The graph of every linear equation with a constant of 0 (equations of the form, $ax + by = 0$) will pass through the origin.

Note that the graph in Example 3 passes through the origin.

3 Graph Linear Equations Using the x- and y-Intercepts

We now discuss the intercepts of a graph.

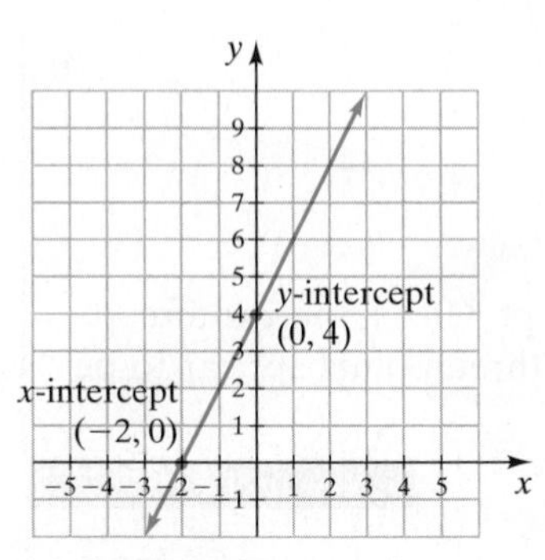

FIGURE 7.15

x- and y-intercepts

- The **x-intercept** is the point at which a graph crosses the x-axis.
- The **y-intercept** is the point at which a graph crosses the y-axis.

Consider the graph in **Figure 7.15**, which is the graph we drew in Example 1. Note that the line crosses the x-axis at −2. Therefore, the x-intercept is $(-2, 0)$. Also note that the line crosses the y-axis at 4. Therefore, the y-intercept is $(0, 4)$.

> **Understanding Algebra**
>
> In general terms, the x-intercept is $(x, 0)$ and the y-intercept is $(0, y)$.

It is often convenient to graph linear equations by finding the x- and y-intercepts.

To Graph Linear Equations Using the x- and y-intercepts

1. **Find the y-intercept.** Set x equal to 0 and find the corresponding value for y.
2. **Find the x-intercept.** Set y equal to 0 and find the corresponding value for x.
3. **Determine a checkpoint.** Select a nonzero value for x and find the corresponding value for y.
4. **Plot the intercepts and checkpoint.** The three points should be collinear. If not, check your work.
5. **Draw the line.** Using a straightedge, draw a straight line through the three points. Draw an arrowhead on each end of the line.

EXAMPLE 4 Graph $3y = 6x + 12$ using the x- and y-intercepts.

Solution To find the y-intercept, set $x = 0$ and find the corresponding value of y.

$$
\begin{aligned}
3y &= 6x + 12 \\
3y &= 6(0) + 12 \\
3y &= 0 + 12 \\
3y &= 12 \\
y &= \frac{12}{3} = 4
\end{aligned}
$$

Therefore, the y-intercept is $(0, 4)$. To find the x-intercept, set $y = 0$ and find the corresponding value of x.

$$
\begin{aligned}
3y &= 6x + 12 \\
3(0) &= 6x + 12 \\
0 &= 6x + 12 \\
-12 &= 6x \\
\frac{-12}{6} &= x \\
-2 &= x
\end{aligned}
$$

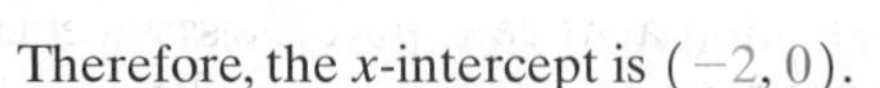

Therefore, the x-intercept is $(-2, 0)$.

Next, to find a checkpoint, we will select a nonzero value for x and find the corresponding value for y. We will select $x = 2$.

$$
\begin{aligned}
x &= 2 \\
3y &= 6x + 12 \\
3y &= 6(2) + 12 \\
3y &= 12 + 12 \\
3y &= 24 \\
y &= \frac{24}{3} = 8
\end{aligned}
$$

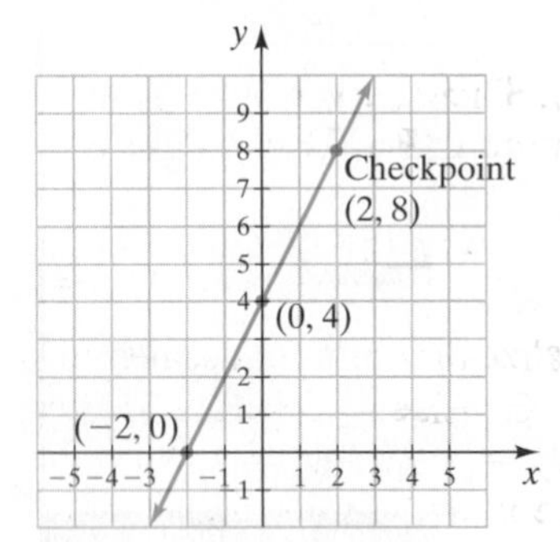

FIGURE 7.16

Therefore, the checkpoint is $(2, 8)$. Now plot the y-intercept, $(0, 4)$, the x-intercept, $(-2, 0)$, and the checkpoint $(2, 8)$ (**Fig. 7.16**). Since the three points appear to be collinear, draw the straight line through all three points.

Now try Exercise 55

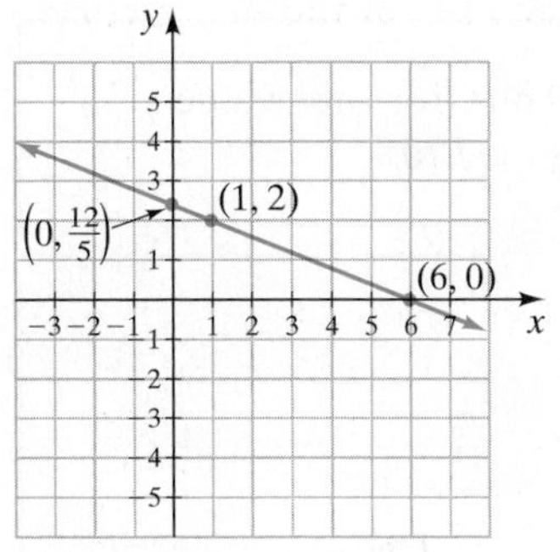

FIGURE 7.17

EXAMPLE 5 Graph $2x + 5y = 12$ using the x- and y-intercepts.

Solution

Find y-intercept	Find x-intercept	Checkpoint
Let $x = 0$.	Let $y = 0$.	Let $x = 1$.
$2x + 5y = 12$	$2x + 5y = 12$	$2x + 5y = 12$
$2(0) + 5y = 12$	$2x + 5(0) = 12$	$2(1) + 5y = 12$
$0 + 5y = 12$	$2x + 0 = 12$	$2 + 5y = 12$
$5y = 12$	$2x = 12$	$5y = 10$
$y = \frac{12}{5}$	$x = 6$	$y = 2$

The three ordered pairs are $\left(0, \frac{12}{5}\right)$, $(6, 0)$, and $(1, 2)$.

The three points appear to be collinear. Draw a straight line through all three points (**Fig. 7.17**).

Now Try Exercise 45

EXAMPLE 6 Graph $y = 20x + 60$ using the x- and y-intercepts.

Solution

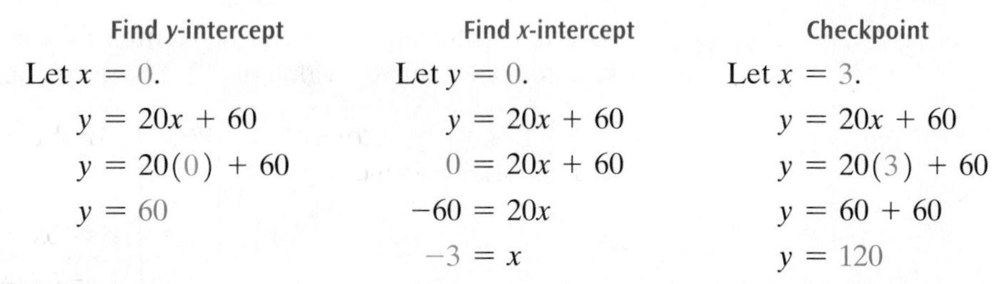

Find y-intercept	Find x-intercept	Checkpoint
Let $x = 0$.	Let $y = 0$.	Let $x = 3$.
$y = 20x + 60$	$y = 20x + 60$	$y = 20x + 60$
$y = 20(0) + 60$	$0 = 20x + 60$	$y = 20(3) + 60$
$y = 60$	$-60 = 20x$	$y = 60 + 60$
	$-3 = x$	$y = 120$

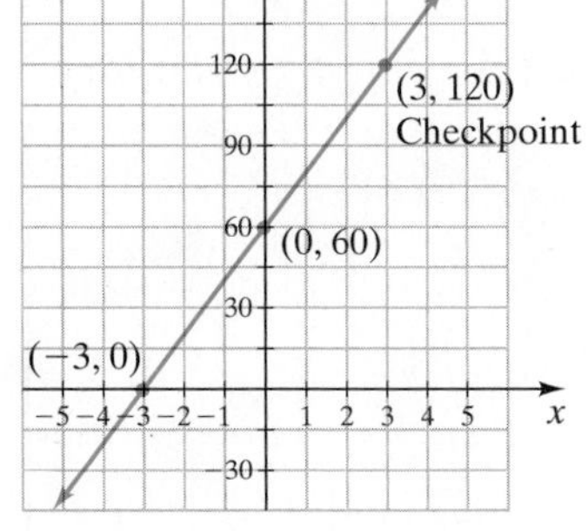

FIGURE 7.18

The three ordered pairs are $(0, 60)$, $(-3, 0)$, and $(3, 120)$. Since the values of y are large, we let each interval on the y-axis be 15 units rather than 1 (**Fig. 7.18**). Now we plot the points and draw the graph.

Now Try Exercise 59

When selecting the scales for your axes, you should realize that different scales will result in the same equation having a different appearance. Consider the graphs shown in **Figure 7.19**. Both graphs represent the same equation, $y = x$. In **Figure 7.19a** both the x- and y-axes have the same scale. In **Figure 7.19b**, the x- and y-axes do not have the same scale. Both graphs are correct in that each represents the graph of $y = x$. The difference in appearance is due to the difference in scales on the x-axis.

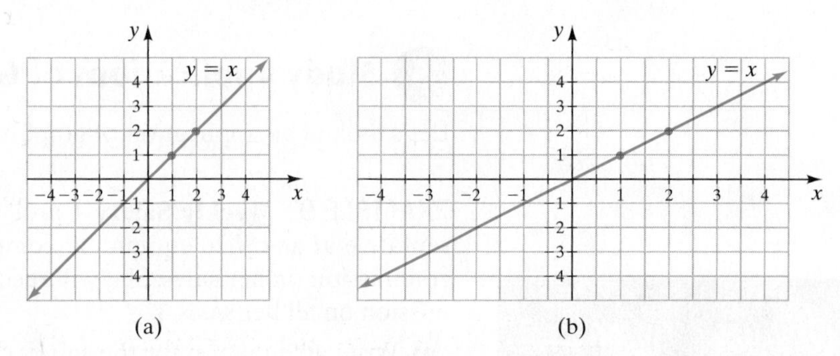

FIGURE 7.19

4 Graph Horizontal and Vertical Lines

When a linear equation contains only one variable, its graph will be either a horizontal or a vertical line, as is explained in Examples 7 and 8.

EXAMPLE 7 Graph $y = 3$.

Solution This equation can be written as $y = 3 + 0x$. Thus, for any value of x selected, y will be 3. The graph of $y = 3$ is illustrated in **Figure 7.20**.

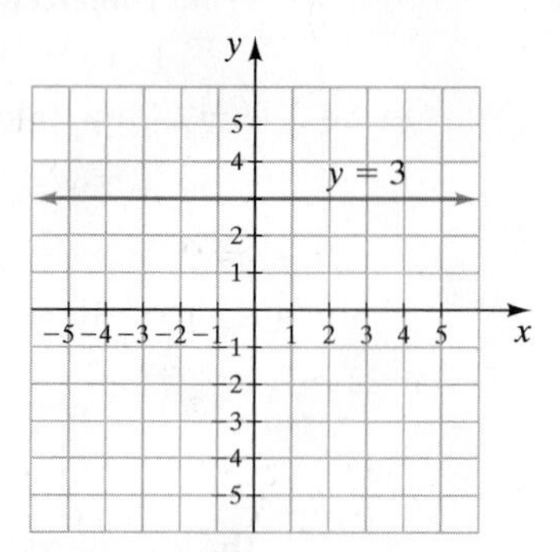

FIGURE 7.20

Now Try Exercise 67

Horizontal Line

The graph of an equation of the form $y = b$ is a **horizontal line** whose y-intercept is $(0, b)$.

EXAMPLE 8 Graph $x = -2$.

Solution This equation can be written as $x = -2 + 0y$. Thus, for any value of y selected, x will have a value of -2. The graph of $x = -2$ is illustrated in **Figure 7.21**.

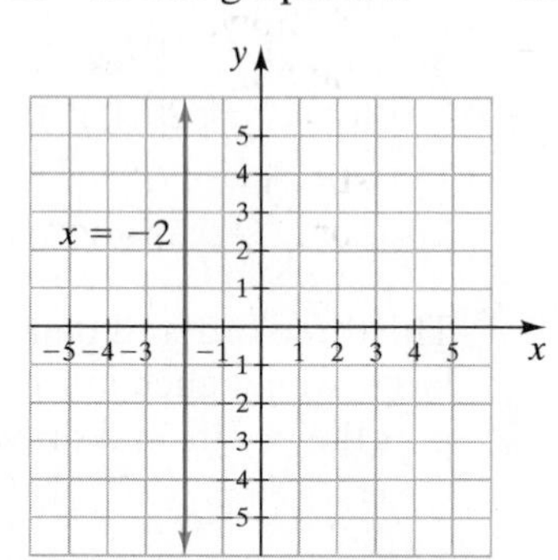

FIGURE 7.21

Now Try Exercise 65

Vertical Line

The graph of an equation of the form $x = a$ is a **vertical line** whose x-intercept is $(a, 0)$.

5 Study Applications of Graphs

Let's look at an application of graphing a linear equation.

EXAMPLE 9 **Weekly Salary** Carol Waters accepted a position as a sales representative at an office equipment company where she is paid a weekly salary plus a commission on her sales. She will receive a salary of \$300 per week plus a 7% commission on all her sales, s.

a) Write an equation for the salary Carol will receive, R, in terms of the sales, s.

b) Graph the salary for sales of \$0 up to and including \$20,000.

c) From the graph, estimate Carol's salary if her weekly sales are \$15,000.

d) From the graph, estimate the sales needed for Carol to earn a weekly salary of \$900.

Solution

a) Since s is the amount of sales, a 7% commission on s dollars in sales is $0.07s$.

$$\text{salary received} = \$300 + \text{commission}$$
$$R = 300 + 0.07s$$

b) We select three values for s and find the corresponding values of R.

$$R = 300 + 0.07s$$

Let $s = 0$. $\quad R = 300 + 0.07(0) = 300$

Let $s = 10{,}000$. $\quad R = 300 + 0.07(10{,}000) = 1000$

Let $s = 20{,}000$. $\quad R = 300 + 0.07(20{,}000) = 1700$

s	R
0	300
10,000	1000
20,000	1700

The graph is illustrated in **Figure 7.22**. Notice that since we only graph the equation for values of s from \$0 to \$20,000, we do not place arrowheads on the ends of the graph.

c) To determine Carol's weekly salary on sales of \$15,000, locate \$15,000 on the sales axis. Then draw a vertical line up to where it intersects the graph, the *purple* line in **Figure 7.22**. Now draw a horizontal line across to the salary axis. Since the horizontal line crosses the salary axis at about \$1350, weekly sales of \$15,000 would result in a weekly salary of about \$1350. We can find the exact salary by substituting 15,000 for s in the equation $R = 300 + 0.07s$ and finding the value of R. Do this now.

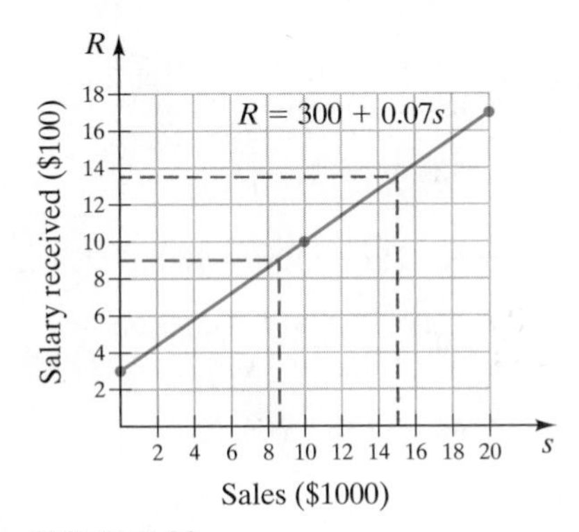

FIGURE 7.22

d) To find the sales needed for Carol to earn a weekly salary of \$900, we find \$900 on the salary axis. We then draw a horizontal line from that point to the graph, as shown with the *green* line in **Figure 7.22**. We then draw a vertical line from the point of intersection of the graph to the sales axis. Thus, sales of about \$8600 per week would result in a salary of \$900. We can find an exact answer by substituting 900 for R in the equation $R = 300 + 0.07s$ and solving the equation for s. Do this now.

Now Try Exercise 75

7.2 Exercise Set MathXL® MyMathLab®

Warm-Up Exercises

Fill in the blanks with the appropriate word, phrase, or symbol(s) from the following list.

$(x, 0)$ horizontal y-intercept x-axis diagonal $(0, y)$ x-intercept vertical y-axis

1. The point at which a graph crosses the ________ is called the x-intercept.

2. The point at which a graph crosses the ________ is called the y-intercept.

3. To find the ________ of a graph of a linear equation, set x equal to 0 and find the corresponding value for y.

4. To find the ________ of a graph of a linear equation, set y equal to 0 and find the corresponding value for x.

5. The x-intercept of a graph will have coordinates ________.

6. The y-intercept of a graph will have coordinates ________.

7. The graph of an equation of the form $y = b$ is a ________ line.

8. The graph of an equation of the form $x = a$ is a ________ line.

Practice the Skills

Find the missing coordinate if the ordered pair is to be a solution to the equation $3x + y = 9$.

9. $(2, ?)$

10. $(-2, ?)$

11. $(?, -6)$

12. $(?, -9)$

13. $\left(-\frac{1}{2}, ?\right)$

14. $\left(\frac{3}{2}, ?\right)$

Find the missing coordinate in the given solutions for $3x - 2y = 8$.

15. $(4, ?)$ **16.** $(0, ?)$ **17.** $(?, 0)$

18. $(?, -3)$ **19.** $\left(?, \frac{1}{2}\right)$ **20.** $\left(?, -\frac{5}{2}\right)$

Graph by plotting points. Plot at least three points for each graph.

21. $y = x$ **22.** $y = -x$ **23.** $y = -\frac{1}{2}x$ **24.** $y = \frac{1}{2}x$

25. $y = 3x - 1$ **26.** $y = -x + 3$ **27.** $y = 4x - 2$ **28.** $y = x - 4$

29. $x + 2y = 6$ **30.** $-3x + 3y = 6$ **31.** $3x - 2y = 4$ **32.** $3x - 2y = 6$

33. $4x + 3y = -9$ **34.** $6y - 12x = 18$ **35.** $6x + 5y = 30$ **36.** $2x + 3y = -6$

37. $-4x + 5y = 0$ **38.** $3x + 2y = 0$ **39.** $y = -20x + 60$ **40.** $2y - 100x = 50$

41. $y = \frac{4}{3}x$ **42.** $y = -\frac{3}{5}x$ **43.** $y = \frac{1}{2}x + 4$ **44.** $y = -\frac{2}{5}x + 2$

Graph using the x- and y-intercepts.

45. $y = 3x + 3$ **46.** $y = -3x + 6$ **47.** $y = -4x + 2$ **48.** $y = -2x + 5$

49. $y = 4x + 16$ **50.** $y = -5x + 4$ **51.** $4y + 6x = 24$ **52.** $4x = 3y - 9$

53. $\frac{1}{2}x + 2y = 4$ **54.** $x + \frac{1}{2}y = 2$ **55.** $12x - 24y = 48$ **56.** $25x + 50y = 100$

57. $8y = 6x - 12$ **58.** $6y = -4x + 12$ **59.** $y = 15x + 45$ **60.** $y = -10x + 30$

61. $\frac{1}{3}x + \frac{1}{4}y = 12$ **62.** $\frac{1}{4}x - \frac{2}{3}y = 60$ **63.** $\frac{1}{2}x = \frac{2}{5}y - 80$ **64.** $\frac{2}{3}y = \frac{5}{4}x + 120$

Graph each equation.

65. $x = -3$ **66.** $x = \frac{3}{2}$ **67.** $y = 4$ **68.** $y = -\frac{5}{3}$

Write the equation represented by the given graph.

69.

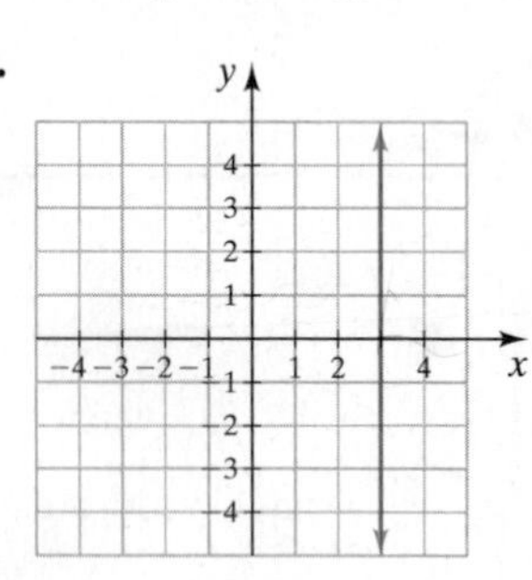

70.

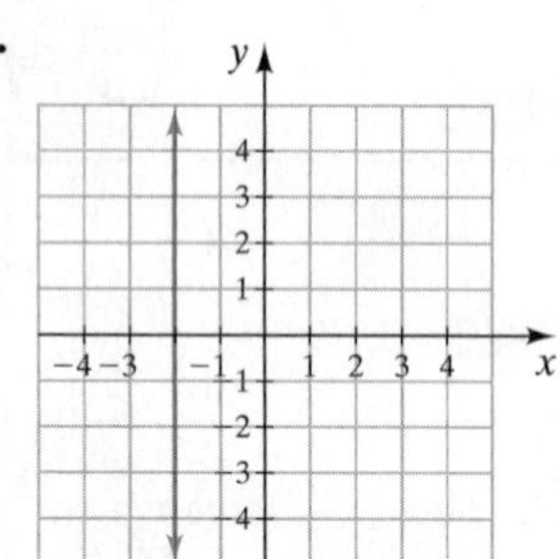

71.

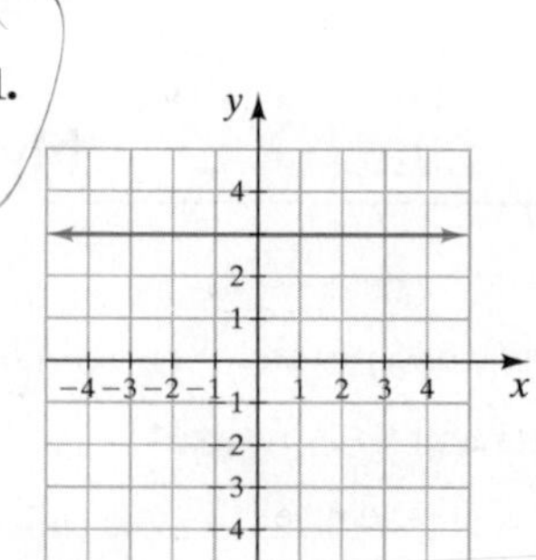

72.

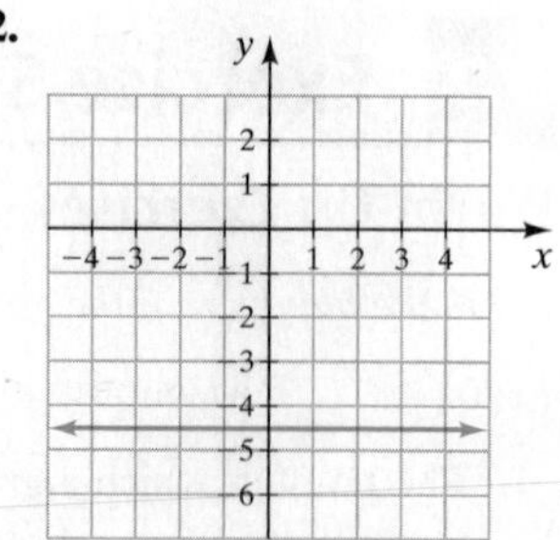

Problem Solving

The bar graphs in Exercises 73 and 74 display information. State whether the graph displays a linear relationship. Explain your answer.

73.

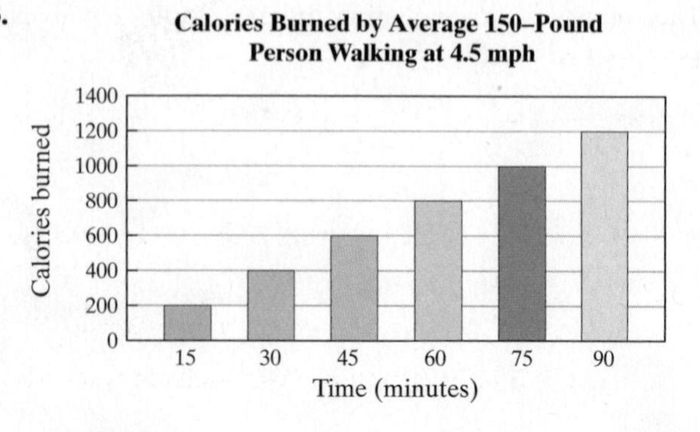

74.

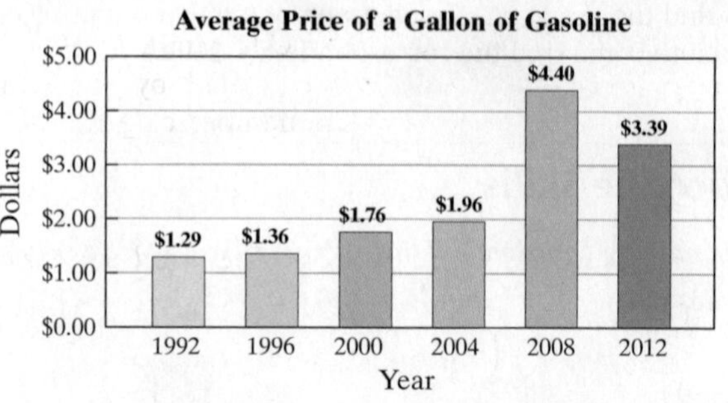

Source: American Petroleum Institute

Review Example 9 before working Exercises 75–80.

75. **Telephone Calls** Alexus Judd's telephone plan consists of a monthly fee of \$15 plus 10 cents per minute for long-distance calls made.
   a) Write an equation for the total monthly cost, C, when n minutes are used for long-distance calls.
   b) Graph the equation for up to and including 100 minutes of long-distance calls made.
   c) Estimate the total monthly cost if 40 minutes of long-distance calls are made.
   d) If the total monthly bill is \$25, estimate the number of minutes used for long-distance calls.

76. **Distance Traveled** Distance traveled is calculated using the formula,

$$\text{distance} = \text{rate} \cdot \text{time or } d = rt.$$

Assume the rate of a car is a constant 30 miles per hour.
   a) Write an equation for the distance, d, in terms of time, t.
   b) Graph the equation for times of 0 to 20 hours inclusive.
   c) Estimate the distance traveled in 12 hours.
   d) If the distance traveled is 150 miles, estimate the time traveled.

77. **Truck Rental** Lynn Brown needs a large truck to move some furniture. She found that the cost, C, of renting a truck is \$40 per day plus \$1 per mile, m.
   a) Write an equation for the cost in terms of the miles driven.
   b) Graph the equation for values up to and including 100 miles.
   c) Estimate the cost of driving 60 miles in one day.
   d) Estimate the miles driven if the cost for one day is \$70.

78. **Simple Interest** Simple interest is calculated by the simple interest formula,

$$\text{interest} = \text{principal} \cdot \text{rate} \cdot \text{time or } I = prt.$$

Suppose the principal is \$10,000 and the rate is 5%.
   a) Write an equation for simple interest in terms of time.
   b) Graph the equation for times of 0 to 20 years inclusive.
   c) What is the simple interest for 10 years?
   d) If the simple interest is \$500, find the length of time.

79. **Video Game Store Profit** The weekly profit, P, of a video game rental store can be approximated by the formula $P = 1.5n - 200$, where n is the number of games rented weekly.
   a) Draw a graph of profit in terms of game rentals for up to and including 1000 games.
   b) Estimate the weekly profit if 500 games are rented.
   c) Estimate the number of games rented if the week's profit is \$1000.

80. **Playing Tennis** The cost, C, of playing tennis in the Downtown Tennis Club includes an annual \$200 membership fee plus \$10 per hour, h, of court time.
   a) Write an equation for the annual cost of playing tennis at the Downtown Tennis Club in terms of hours played.
   b) Graph the equation for up to and including 300 hours.
   c) Estimate the cost for playing 50 hours in a year.
   d) If the annual cost for playing tennis was \$1700, estimate how many hours of tennis were played.

81. What is the value of a if the graph of $ax + 2y = 15$ is to have an x-intercept of $(3, 0)$?

82. What is the value of a if the graph of $ax - 4y = 8$ is to have an x-intercept of $(4, 0)$?

83. What is the value of b if the graph of $3x + by = 14$ is to have a y-intercept of $(0, 7)$?

84. What is the value of b if the graph of $4x + by = 15$ is to have a y-intercept of $(0, -3)$?

Determine the coefficients to be placed in the shaded areas so that the graph of the equation will be a line with the x- and y-intercepts specified. Explain how you determined your answer.

85. $\square x + \square y = 6$; x-intercept at 2; y-intercept at 3

86. $\square x + \square y = 18$; x-intercept at -3, y-intercept at 6

87. $\square x - \square y = -12$; x-intercept at -2, y-intercept at 3

88. $\square x - \square y = 30$; x-intercept at -10, y-intercept at -15

Challenge Problems

89. Consider the following equations: $y = 2x - 1$, $y = -x + 5$.

a) Carefully graph both equations on the same axes.

b) Determine the point of intersection of the two graphs.

c) Substitute the values for x and y at the point of intersection into each of the two equations and determine whether the point of intersection satisfies each equation.

d) Do you believe there are any other ordered pairs that satisfy both equations? Explain your answer. (We will study equations like these, called systems of equations, in Chapter 8.)

90. In Chapter 10 we will be graphing quadratic equations. The graphs of quadratic equations are *not* straight lines. Graph the quadratic equation $y = x^2 - 4$ by selecting values for x and find the corresponding values of y, then plot the points. Make sure you plot a sufficient number of points to get an accurate graph.

Group Activity

Discuss and answer Exercise 91 as a group.

91. Let's study the graphs of the equations $y = 2x + 4$, $y = 2x + 2$, and $y = 2x - 2$ to see how they are similar and how they differ. Each group member should start with the same axes.

a) Group member 1: Graph $y = 2x + 4$.

b) Group member 2: Graph $y = 2x + 2$.

c) Group member 3: Graph $y = 2x - 2$.

d) Now transfer all three graphs onto the same axes. (You can use one of the group members' graphs or you can construct new axes.)

e) Explain what you notice about the three graphs.

f) Explain what you notice about the y-intercepts.

Cumulative Review Exercises

[1.9] **92.** Evaluate $2[6 - (4 - 5)] \div 2 - 8^2$.

[2.7] **93. House Cleaning** According to the instructions on a bottle of concentrated household cleaner, 8 ounces of the cleaner should be mixed with 3 gallons of water. If your bucket holds only 2.5 gallons of water, how much cleaner should you use?

[3.2] **94. Refrigerator Purchase** Kristin Runde purchased a new refrigerator. The cost of the refrigerator, including a 6.5% sales tax, was \$1491. What was the price of the refrigerator before tax?

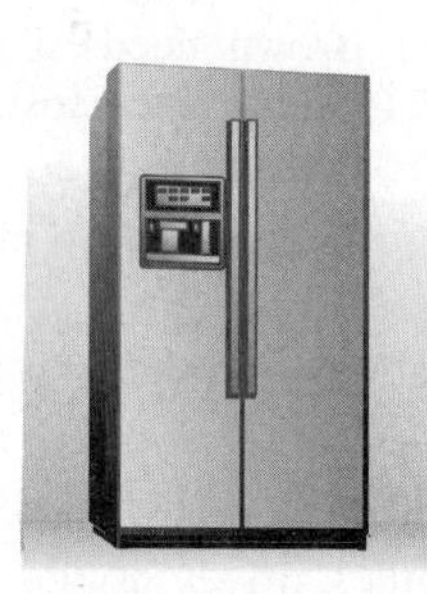

See Exercise 94.

[6.2] **95.** Divide $\dfrac{3xy^3}{z} \div \dfrac{x^2y^2}{5z^3}$.

[6.4] **96.** Add $\dfrac{3}{x-2} + \dfrac{4}{x-3} + 2$

[6.6] **97.** Solve $\dfrac{3}{x-2} + \dfrac{4}{x-3} = 3$

7.3 Slope of a Line

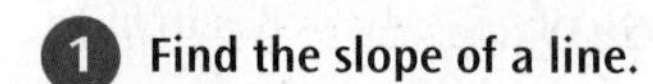

1 Find the slope of a line.

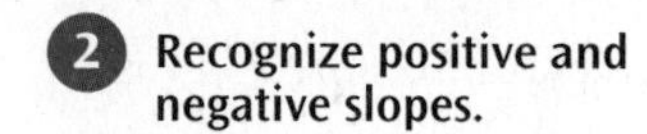

2 Recognize positive and negative slopes.

3 Examine the slopes of horizontal and vertical lines.

4 Examine the slopes of parallel and perpendicular lines.

1 Find the Slope of a Line

The *slope of a line* is a measure of the *steepness* of the line. The slope of a line is an important concept in many areas of mathematics. A knowledge of slope is helpful in understanding linear equations.

Slope of a Line

- The **slope of a line, m,** is the ratio of the vertical change, or *rise,* to the horizontal change, or *run,* between any two selected points on the line.
- $m = \textbf{slope} = \dfrac{\text{vertical change}}{\text{horizontal change}} = \dfrac{\text{rise}}{\text{run}}$

Understanding Algebra

Slope is often described with the phrase "*rise over run.*"

Slope can also be thought of as the **rate of change** of one variable, y, with respect to another variable, x. For example, in the Helpful Hint below, we describe how slope is used when we describe the *pitch* of a roof or the *grade* of a hill. In each case, slope describes the rate of vertical change, y, with respect to the horizontal change, x.

HELPFUL HINT

Slope

We often come across slope in everyday life. A road may have a slope, called the *grade*, of 8%. A roof may have a slope, called the *pitch*, of $\frac{6}{15}$. Suppose a road has an 8% grade. Since $8\% = \frac{8}{100}$, this means the road rises 8 feet for each 100 feet of horizontal length. A roof pitch of $\frac{6}{15}$ means the roof rises 6 feet for each 15 feet of horizontal length.

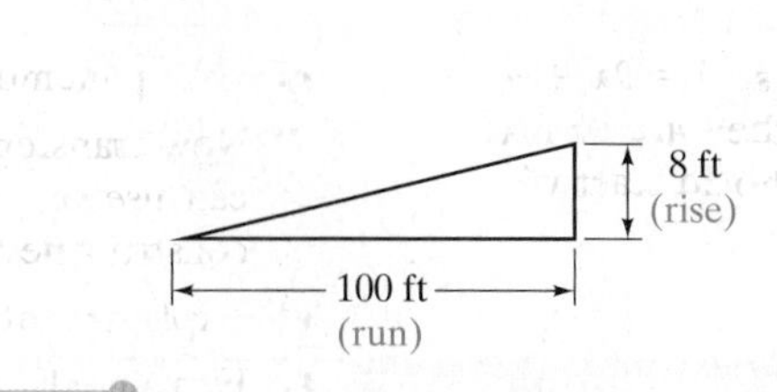

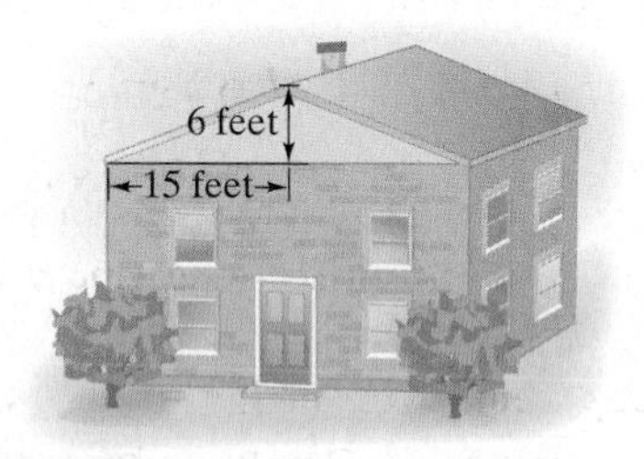

As an example of slope, consider the line that goes through the two points $(1, 2)$ and $(3, 6)$ (see **Fig. 7.23a**). From **Figure 7.23b,** we can see that the vertical change is $6 - 2$, or 4 units. The horizontal change is $3 - 1$, or 2 units.

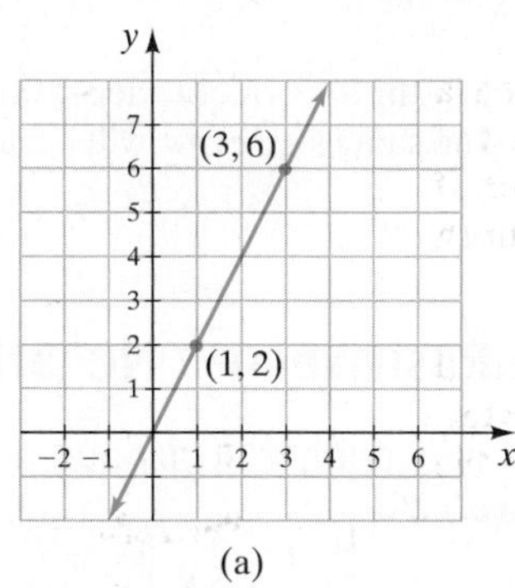

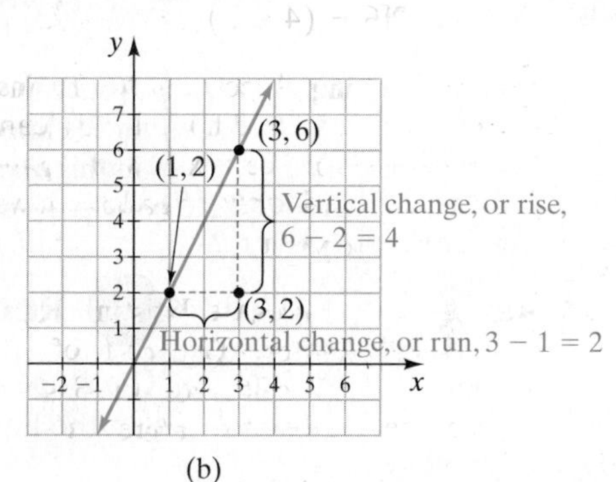

FIGURE 7.23

$$m = \text{slope} = \frac{\text{vertical change}}{\text{horizontal change}} = \frac{\text{rise}}{\text{run}} = \frac{4}{2} = 2$$

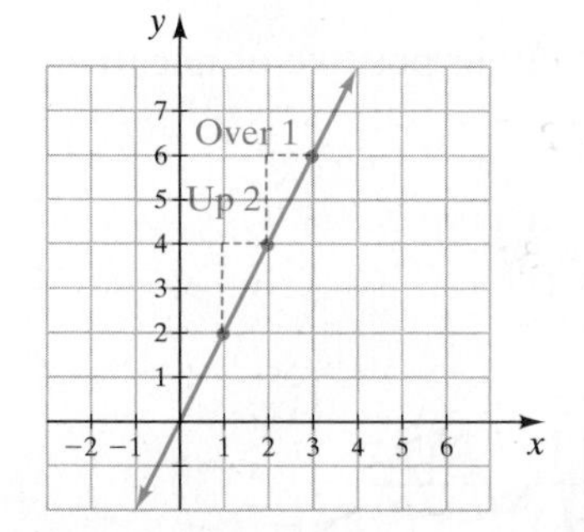

FIGURE 7.24

Thus, the slope of the line through these two points is 2. By examining the line connecting these two points, we can see that as the graph moves up 2 units it moves to the right 1 unit (**Fig. 7.24**).

Now we present the formula to find the slope of a line given any two points (x_1, y_1) and (x_2, y_2) on the line. Look at **Figure 7.25.** The *rise* is the difference between y_2 and y_1 and the *run* is the difference between x_2 and x_1.

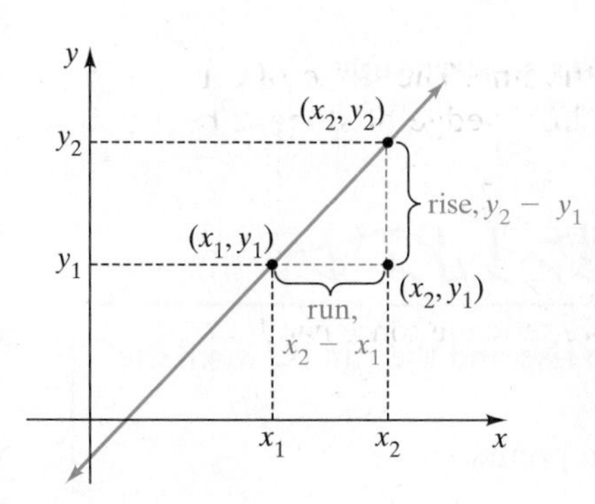

FIGURE 7.25

Slope of a Line Through the Points (x_1, y_1) and (x_2, y_2)

$$m = \text{slope} = \frac{\text{vertical change}}{\text{horizontal change}} = \frac{\text{rise}}{\text{run}} = \frac{y_2 - y_1}{x_2 - x_1}$$

HELPFUL HINT

It makes no difference which two points on a line are selected when finding the slope of the line. It also makes no difference which point you choose as (x_1, y_1) or (x_2, y_2).

Understanding Algebra

The Greek letter delta, Δ, is often used to represent the phrase "the change in." So, Δy represents "the change in y" and Δx represents "the change in x" and the formula for slope can be given as

$$m = \frac{\Delta y}{\Delta x} = \frac{y_2 - y_1}{x_2 - x_1}$$

EXAMPLE 1 Find the slope of the line through the points $(-6, 1)$ and $(3, 5)$.

Solution We will designate $(-6, 1)$ as (x_1, y_1) and $(3, 5)$ as (x_2, y_2).

$$\begin{aligned} m &= \frac{y_2 - y_1}{x_2 - x_1} \\ &= \frac{5 - 1}{3 - (-6)} \\ &= \frac{5 - 1}{3 + 6} = \frac{4}{9} \end{aligned}$$

Thus, the slope is $\frac{4}{9}$.

If we had designated $(3, 5)$ as (x_1, y_1) and $(-6, 1)$ as (x_2, y_2), we would have obtained the same results.

$$\begin{aligned} m &= \frac{y_2 - y_1}{x_2 - x_1} \\ &= \frac{1 - 5}{-6 - 3} = \frac{-4}{-9} = \frac{4}{9} \end{aligned}$$

Now Try Exercise 13

AVOIDING COMMON ERRORS

Students sometimes subtract the x's and y's in the slope formula in the wrong order. For instance, using the problem in Example 1:

$$m = \frac{\cancel{y_2 - y_1}}{\cancel{x_1 - x_2}} = \frac{5 - 1}{-6 - 3} = \frac{4}{-9} = -\frac{4}{9}$$

Notice that subtracting in this incorrect order results in a negative slope, when the actual slope of the line is positive. The same sign error will occur each time subtraction is done incorrectly in this manner.

2 Recognize Positive and Negative Slopes

The slope of a line that is neither horizontal nor vertical is either positive or negative. See the lines in **Figure 7.26**.

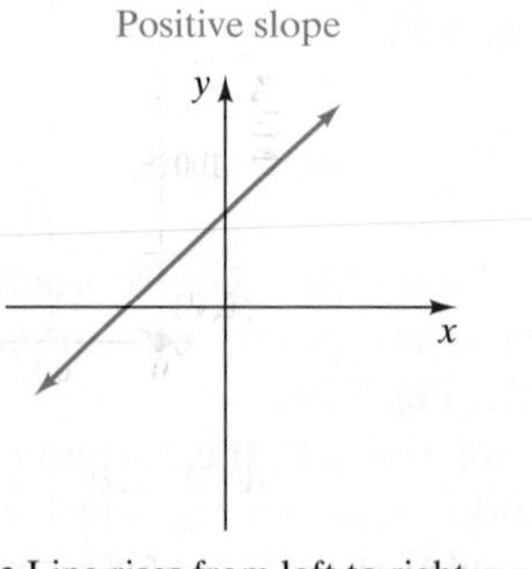

- Line rises from left to right
- As x increases, y increases

(a)

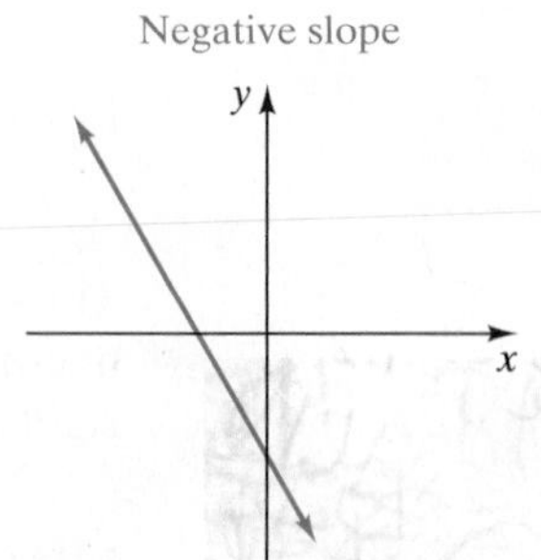

- Line falls from left to right
- As x increases, y decreases

(b)

FIGURE 7.26

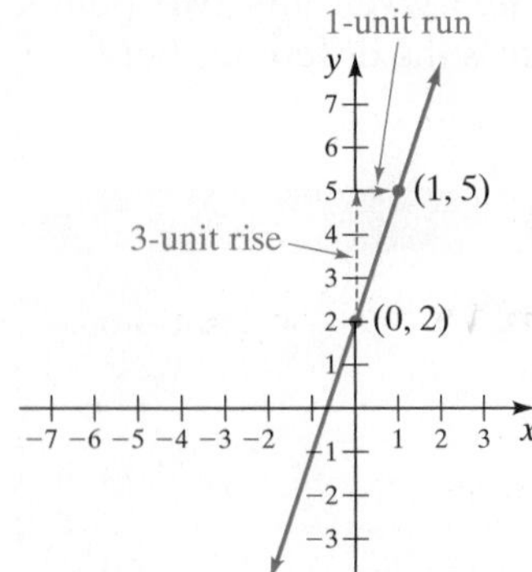

FIGURE 7.27

EXAMPLE 2 Consider the line in **Figure 7.27**.

a) Determine the slope of the line by observing the rise and the run between the points $(1, 5)$ and $(0, 2)$ on the graph.

b) Calculate the slope of the line using the two given points.

Solution

a) Notice that the slope is positive since the line rises from left to right. From **Figure 7.27** we see that the rise is +3 units and the run is +1 unit. Thus, the slope of the line is $\frac{3}{1}$ or 3.

b) We will use the ordered pairs $(1, 5)$ and $(0, 2)$.

Let (x_2, y_2) be $(1, 5)$. Let (x_1, y_1) be $(0, 2)$.

$$m = \frac{y_2 - y_1}{x_2 - x_1} = \frac{5 - 2}{1 - 0} = \frac{3}{1} = 3$$

Note that the slope obtained in part **b)** agrees with the slope obtained in part **a)**. If we had designated $(1, 5)$ as (x_1, y_1) and $(0, 2)$ as (x_2, y_2), the slope would not have changed. Try it and see that you will still obtain a slope of 3.

Now Try Exercise 25

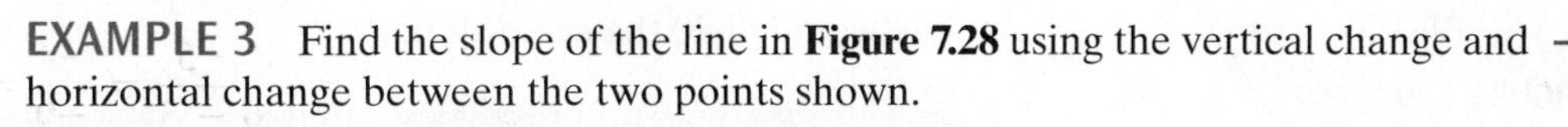

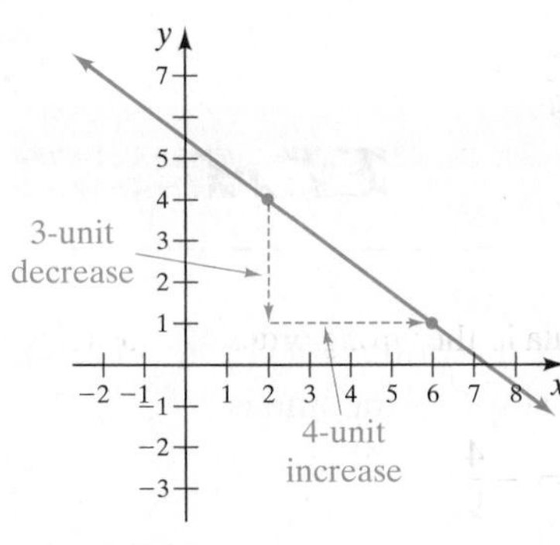

FIGURE 7.28

EXAMPLE 3 Find the slope of the line in **Figure 7.28** using the vertical change and horizontal change between the two points shown.

Solution Since the graph falls from left to right, the line has a negative slope. From **Figure 7.28** we see that the vertical change between the two given points is −3 units and the horizontal change between the two given points is 4 units. Thus, the slope is $\frac{-3}{4}$ or $-\frac{3}{4}$.

Now Try Exercise 29

Using the two points shown in **Figure 7.28** and the definition of slope, calculate the slope of the line in Example 3. You should obtain the same answer.

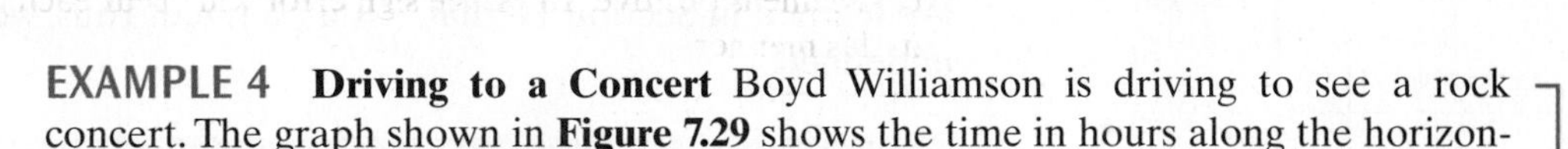

EXAMPLE 4 **Driving to a Concert** Boyd Williamson is driving to see a rock concert. The graph shown in **Figure 7.29** shows the time in hours along the horizontal axis and the distance in miles along the vertical axis.

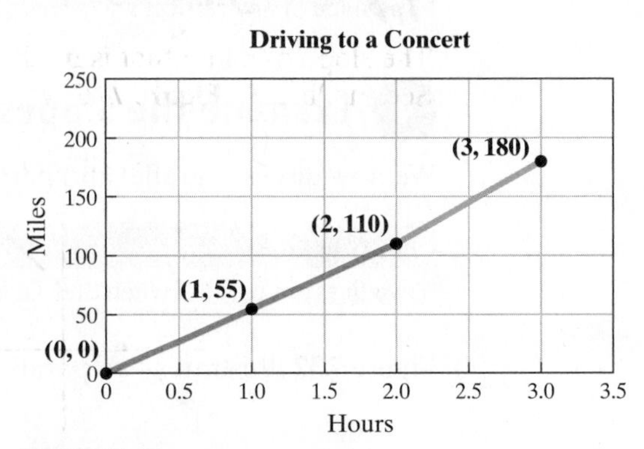

FIGURE 7.29

a) Determine the slope of the red line segment.

b) Determine the slope of the blue line segment.

Solution

a) We can choose any two points on the red line segment. We will use the points $(0, 0)$ as (x_1, y_1) and $(1, 55)$ as (x_2, y_2).

$$m = \frac{y_2 - y_1}{x_2 - x_1} = \frac{55 - 0}{1 - 0} = \frac{55}{1} = 55$$

Thus, the slope of the red line segment is 55.

b) We will use the points $(2, 110)$ as (x_1, y_1) and $(3, 180)$ as (x_2, y_2).

$$m = \frac{y_2 - y_1}{x_2 - x_1} = \frac{180 - 110}{3 - 2} = \frac{70}{1} = 70$$

Thus, the slope of the blue line segment is 70.

Now Try Exercise 71

3 Examine the Slopes of Horizontal and Vertical Lines

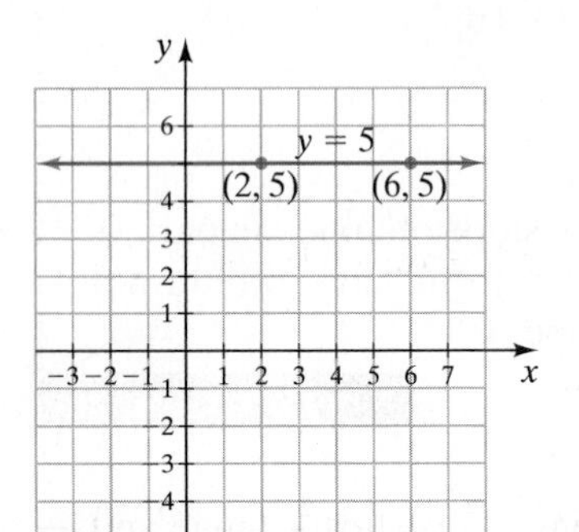

FIGURE 7.30

Consider the graph of $y = 5$ (**Fig. 7.30**). What is its slope?

The graph is parallel to the x-axis and goes through the points $(2, 5)$ and $(6, 5)$. Select $(6, 5)$ as (x_2, y_2) and $(2, 5)$ as (x_1, y_1). Then the slope of the line is

$$m = \frac{y_2 - y_1}{x_2 - x_1} = \frac{5 - 5}{6 - 2} = \frac{0}{4} = 0$$

Since there is no change in y, this line has a slope of 0. Note that *any* two points on the line would yield the same slope, 0.

Slope of a Horizontal Line

Every horizontal line has a slope of 0.

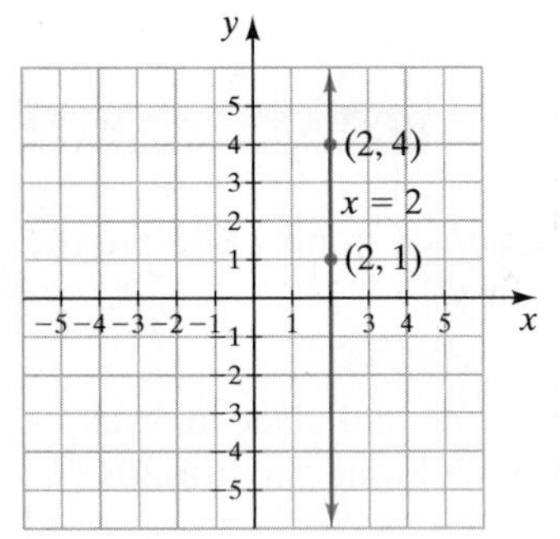

FIGURE 7.31

Consider the graph of $x = 2$ (**Fig. 7.31**). What is its slope?

The graph is parallel to the y-axis and goes through the points $(2, 1)$ and $(2, 4)$. Select $(2, 4)$ as, (x_2, y_2) and $(2, 1)$ as (x_1, y_1). Then the slope of the line is

$$m = \frac{y_2 - y_1}{x_2 - x_1} = \frac{4 - 1}{2 - 2} = \frac{3}{0}$$

We learned in Section 1.8 that $\frac{3}{0}$ is undefined. Thus, we say that the slope of this line is undefined.

Slope of a Vertical Line

The slope of any vertical line is undefined.

4 Examine the Slopes of Parallel and Perpendicular Lines

We now discuss parallel and perpendicular lines.

Parallel Lines

Two lines are **parallel** when they lie in the same plane but do not intersect.

Figure 7.32 illustrates two parallel lines.

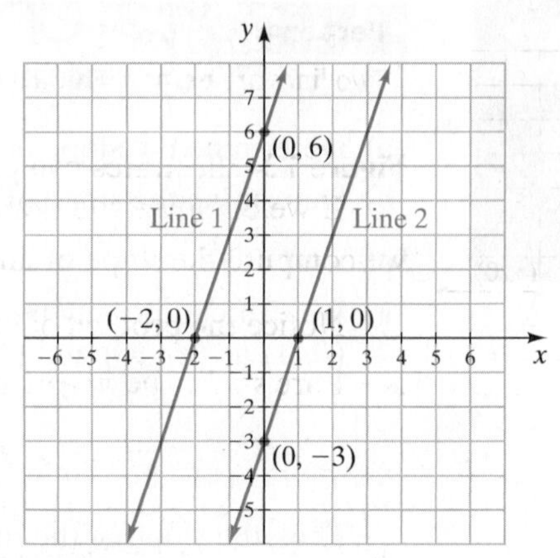

FIGURE 7.32

If we compute the slope of line 1 in **Figure 7.32** using the given points, we obtain a slope of 3. If we compute the slope of line 2, we obtain a slope of 3. Notice both lines have the same slope. Any two nonvertical lines that have the same slope are parallel lines.

Slopes of Parallel Lines

Two lines are parallel when they have the same slope and different y-intercepts. Any two vertical lines are parallel to each other.

EXAMPLE 5

a) Draw a line with a slope of $\frac{1}{2}$ through the point $(2, 3)$.

b) On the same set of axes, draw a line with a slope of $\frac{1}{2}$ through the point $(-1, -3)$.

c) Are the two lines in parts **a)** and **b)** parallel? Explain.

Solution

a) Place a point at $(2, 3)$. Because the slope is a positive $\frac{1}{2}$, from the point $(2, 3)$ move *up* 1 unit and to the *right* 2 units to get a second point. Draw a line through the two points; see the blue line in **Figure 7.33**.

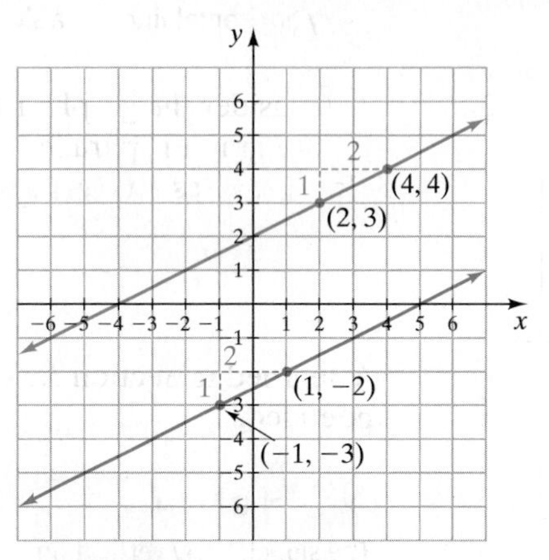

FIGURE 7.33

b) Place a point at $(-1, -3)$. From the point $(-1, -3)$ move up 1 unit and to the right 2 units to get a second point. Draw a line through the two points; see the red line in **Figure 7.33**.

c) The lines appear to be parallel on the graph. Since both lines have the same slope, $\frac{1}{2}$, they are parallel lines.

Now Try Exercise 73

Perpendicular Lines

Two lines are **perpendicular** when they intersect and form right (90°) angles.

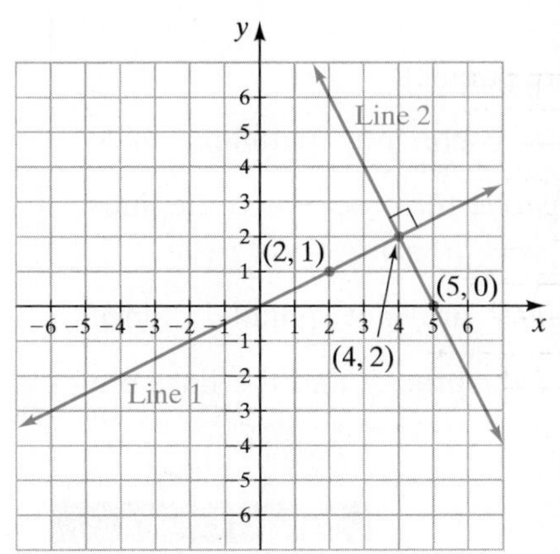

FIGURE 7.34

Figure 7.34 illustrates two perpendicular lines.

If we compute the slope of line 1 using the given points, we obtain a slope of $\frac{1}{2}$. If we compute the slope of line 2 using the given points, we obtain a slope of -2.

Notice the product of their slopes, $\frac{1}{2}(-2)$, is -1. Any two numbers whose product is -1 are said to be **negative reciprocals** of each other.

Slopes of Perpendicular Lines

Two lines are perpendicular when their slopes are negative reciprocals. Any vertical line is perpendicular to any horizontal line.

EXAMPLE 6

a) Draw a line with a slope of -3 through the point $(2, 3)$.

b) On the same set of axes, draw a line with a slope of $\frac{1}{3}$ through the point $(-1, -3)$.

c) Are the two lines in parts **a)** and **b)** perpendicular? Explain.

Solution

a) Place a point at $(2, 3)$. A slope of -3 means $\frac{-3}{1}$. So, from the point $(2, 3)$ move *down* 3 units and to the *right* 1 unit to get a second point. Draw a line through the two points; see the blue line in **Figure 7.35**.

b) Place a point at $(-1, -3)$. The slope is $\frac{1}{3}$, so from this point move *up* 1 unit and to the *right* 3 units. Draw a line through the two points; see the red line in **Figure 7.35**.

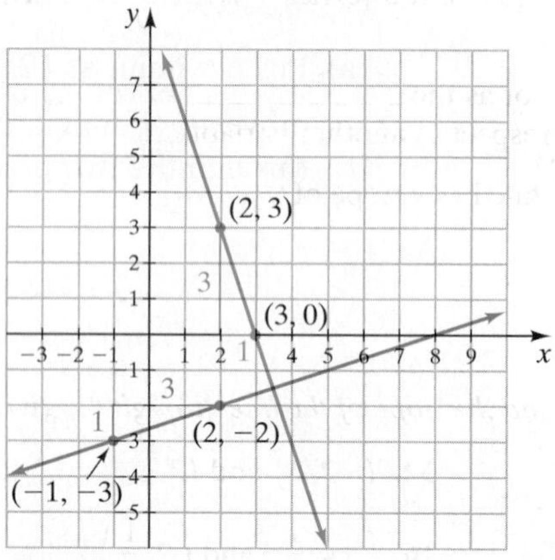

FIGURE 7.35

c) The lines appear to be perpendicular on the graph. To determine if they are perpendicular, multiply the slopes of the two lines together.

$$(-3)\left(\frac{1}{3}\right) = -1$$

Since the slopes are negative reciprocals, the two lines are perpendicular.

Now Try Exercise 75

Understanding Algebra

In general, if m represents a number, its negative reciprocal will be $-\frac{1}{m}$ because $m\left(-\frac{1}{m}\right) = -1$.

EXAMPLE 7

If m_1 represents the slope of line 1 and m_2 represents the slope of line 2, determine if line 1 and line 2 are parallel, perpendicular or neither.

a) $m_1 = \frac{5}{6}, m_2 = \frac{5}{6}$ **b)** $m_1 = \frac{2}{5}, m_2 = 4$ **c)** $m_1 = \frac{3}{5}, m_2 = -\frac{5}{3}$

Solution

a) Since the slopes are the same, both $\frac{5}{6}$, the lines are parallel.

b) Since the slopes are not the same, the lines are not parallel. Since $m_1 \cdot m_2 = \left(\frac{2}{5}\right)(4) \neq -1$, the slopes are not negative reciprocals and the lines are not perpendicular. Thus the answer is neither.

c) Since the slopes are not the same, the lines are not parallel. Since $m_1 \cdot m_2 = \frac{3}{5}\left(-\frac{5}{3}\right) = -1$, the slopes are negative reciprocals and the lines are perpendicular.

Now Try Exercise 53

7.3 Exercise Set MathXL® MyMathLab®

Warm-Up Exercises

Fill in the blanks with the appropriate word, phrase, or symbol(s) from the following list.

horizontal	rate	vertical	equal
perpendicular	run	opposites	ratio
rise	parallel	product	negative

1. The vertical change between two points on a line is known as the ________.
2. The horizontal change between two points on a line is known as the ________.
3. Slope is the ________ of the vertical change to the horizontal change.
4. Slope can be thought of as the ________ of change of one variable, y, with respect to another variable, x.
5. Every ________ line has a slope of 0.
6. Every ________ line has an undefined slope.
7. Two lines are ________ when they lie in same plane but do not intersect.
8. Two line are ________ when they intersect and form right angles.
9. The slopes of parallel lines are ________ to each other.
10. The slopes of perpendicular lines are ________ reciprocals of each other.

Practice the Skills

Using the slope formula, find the slope of the line through the given points.

11. $(2, 1)$ and $(5, 7)$
12. $(-2, 3)$ and $(3, 8)$
13. $(8, 0)$ and $(4, -2)$
14. $(-3, 2)$ and $(-6, 1)$
15. $(3, 5)$ and $(-1, 5)$
16. $\left(-4, \frac{1}{2}\right)$ and $\left(2, \frac{1}{2}\right)$
17. $(5, -6)$ and $(8, -3)$
18. $(9, 3)$ and $(5, -6)$
19. $(6, 4)$ and $(6, 2)$
20. $(-2, 3)$ and $(-2, -5)$
21. $(6, 0)$ and $(-2, 3)$
22. $(-7, 8)$ and $(3, -1)$
23. $\left(0, \frac{5}{2}\right)$ and $\left(-\frac{3}{4}, 2\right)$
24. $(-1, 8)$ and $\left(\frac{1}{3}, -1\right)$

By observing the vertical and horizontal change of the line between the two points indicated, determine the slope of each line.

25.

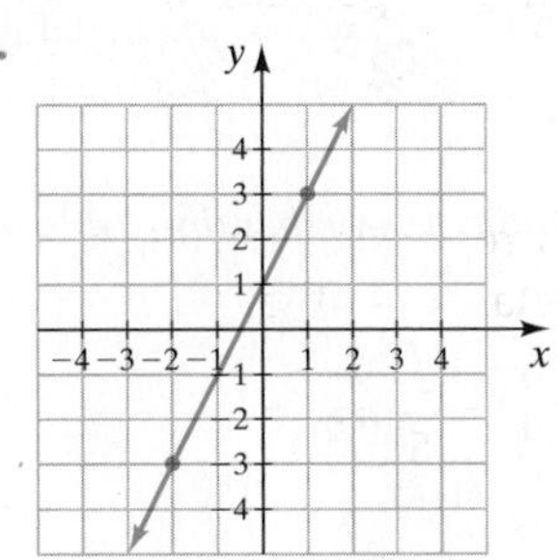

26.

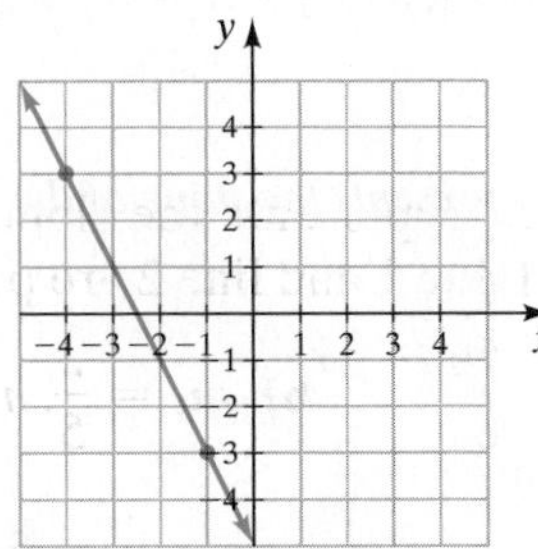

27.

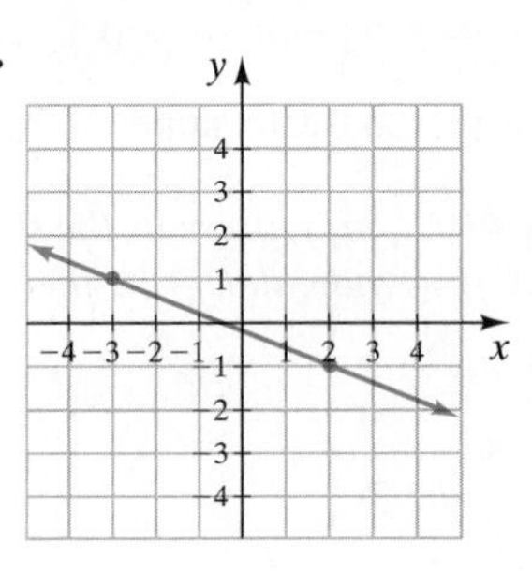

28.

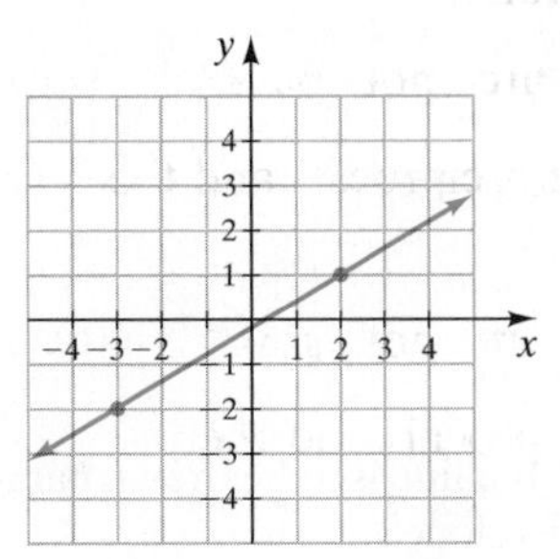

29.

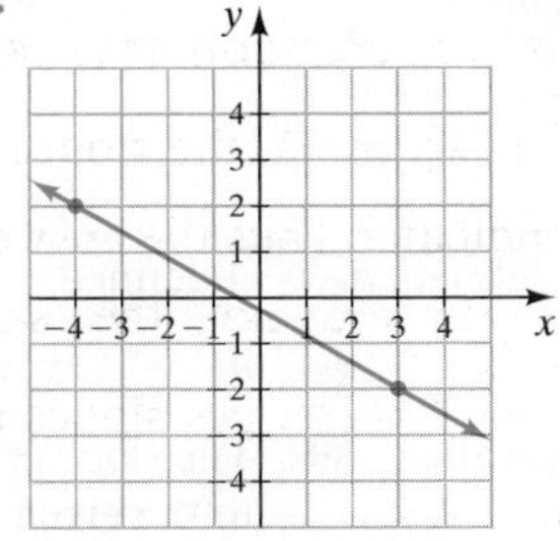

30.

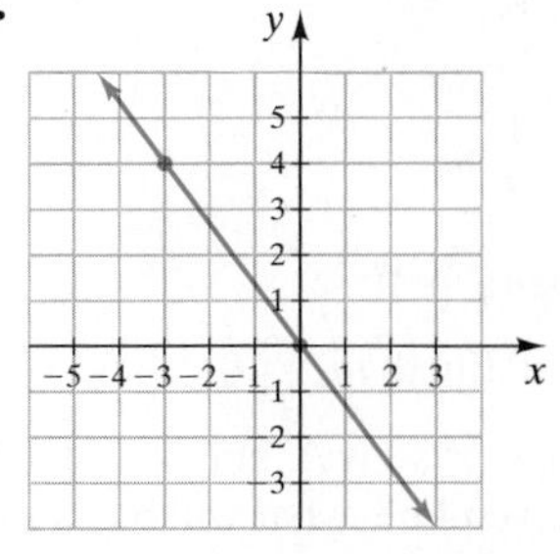

31.

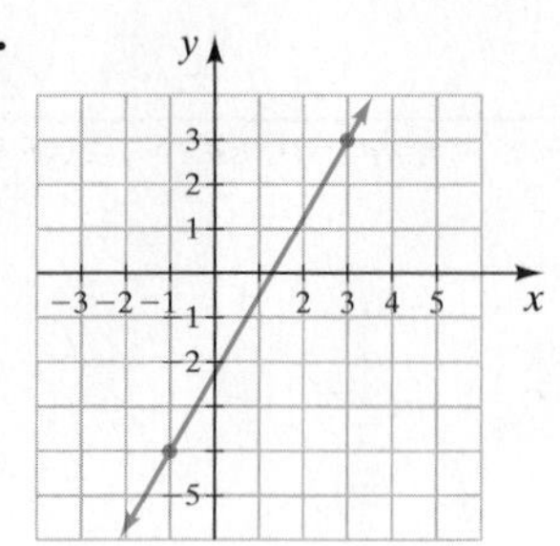

32.

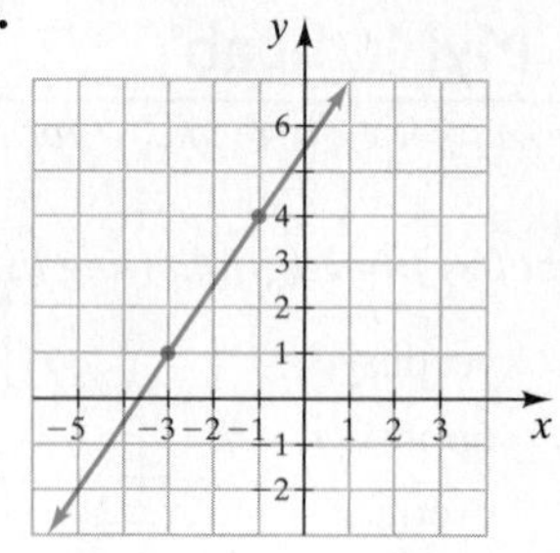

33.

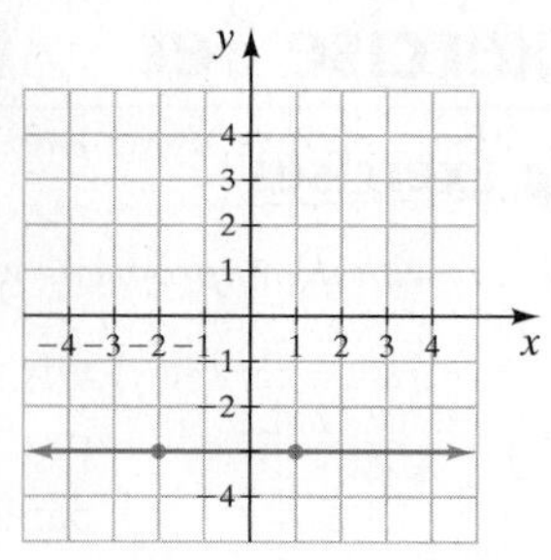

34.

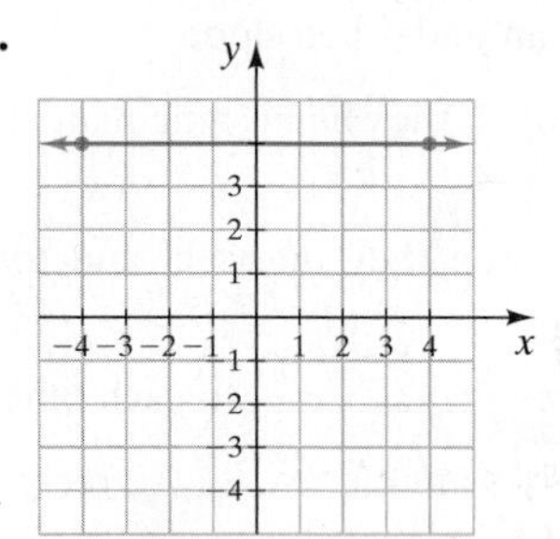

35.

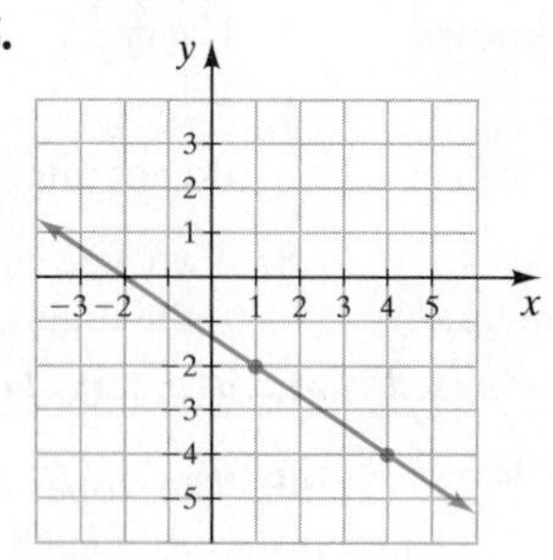

36.

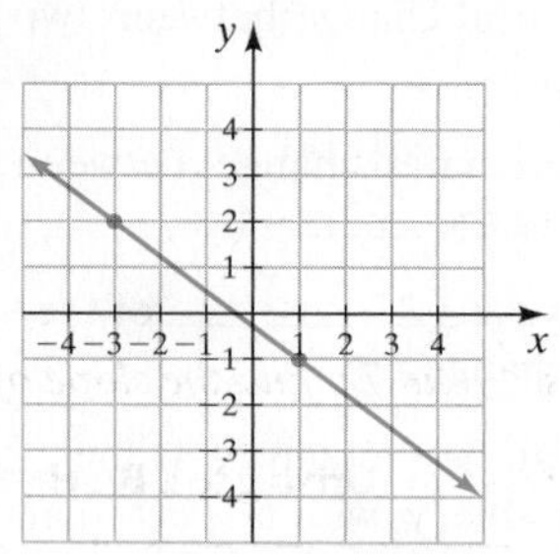

37.

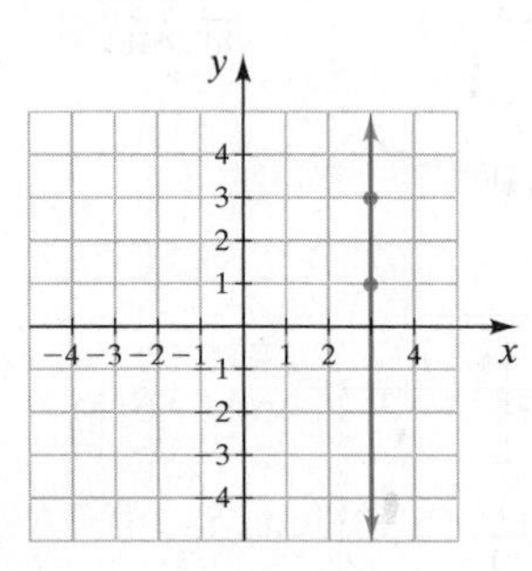

38.

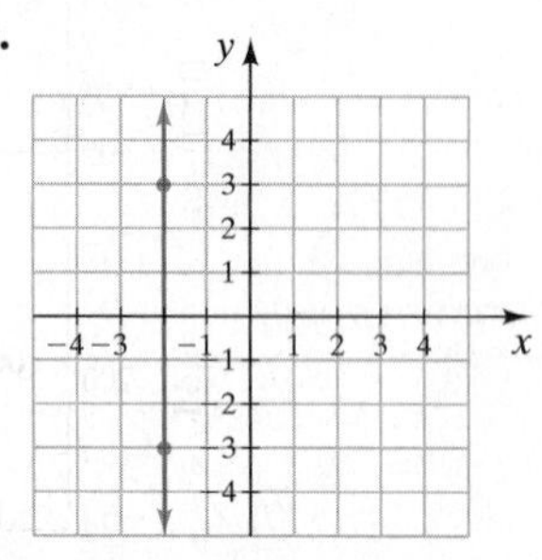

In Exercises 39–48, graph the line with the given slope that goes through the given point.

39. Through $(3, -1)$ with $m = 2$.

40. Through $(-1, -2)$ with $m = -2$

41. Through $(0, -2)$ with $m = \frac{1}{2}$

42. Through $(0, 2)$ with $m = -\frac{1}{2}$

43. Through $(0, 0)$ with $m = -\frac{1}{3}$

44. Through $(-3, 4)$ with $m = -\frac{2}{3}$

45. Through $(-3, 2)$ with $m = 0$

46. Through $(-1, 3)$ with $m = 0$

47. Through $(2, -2)$ with slope undefined

48. Through $(-1, 5)$ with slope undefined

In Exercises 49–64, m_1 represents the slope of line 1, and m_2 represents the slope of the distinct line, line 2. Indicate whether line 1 and line 2 are parallel, perpendicular, or neither.

49. $m_1 = 5, m_2 = 5$

50. $m_1 = -1, m_2 = -1$

51. $m_1 = 7, m_2 = -7$

52. $m_1 = -\frac{5}{6}, m_2 = \frac{5}{6}$

53. $m_1 = \frac{2}{3}, m_2 = -\frac{3}{2}$

54. $m_1 = -\frac{7}{8}, m_2 = \frac{8}{7}$

55. $m_1 = 6, m_2 = \frac{2}{3}$

56. $m_1 = 0, m_2 = -\frac{2}{5}$

57. $m_1 = \frac{1}{4}, m_2 = 4$

58. $m_1 = -\frac{1}{3}, m_2 = -3$

59. $m_1 = \frac{1}{4}, m_2 = -4$

60. $m_1 = 6, m_2 = -\frac{1}{6}$

61. $m_1 = 0, m_2 = 0$

62. m_1 is undefined, m_2 is undefined

63. m_1 is undefined, $m_2 = 0$

64. $m_1 = 0$, m_2 is undefined

65. The slope of a given line is 3. If a line is to be drawn parallel to the given line, what will be its slope?

66. The slope of a given line is −2. If a line is to be drawn parallel to the given line, what will be its slope?

67. The slope of a given line is −4. If a line is to be drawn perpendicular to the given line, what will be its slope?

68. The slope of a given line is 5. If a line is to be drawn perpendicular to the given line, what will be its slope?

Problem Solving

In Exercises 69 and 70, determine which line (the first or second) has the greater slope. Explain your answer. Notice that the scales on the x- and y-axes are different.

69.

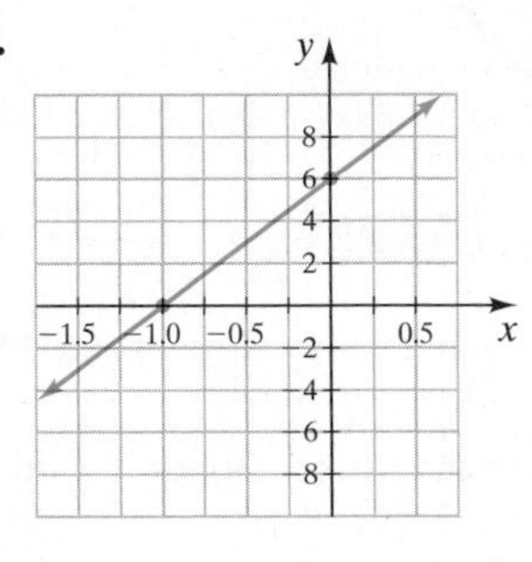

70.

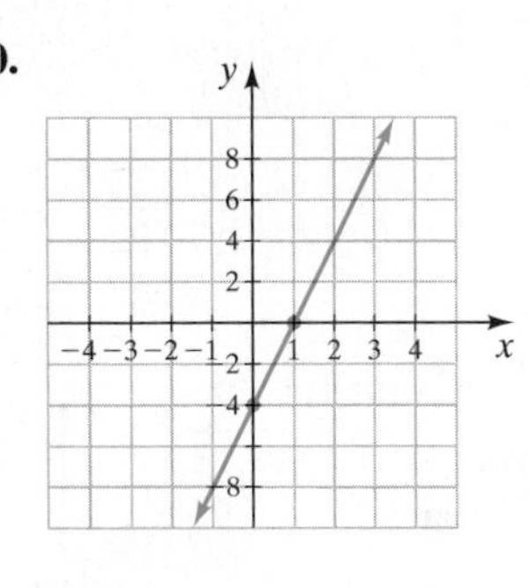

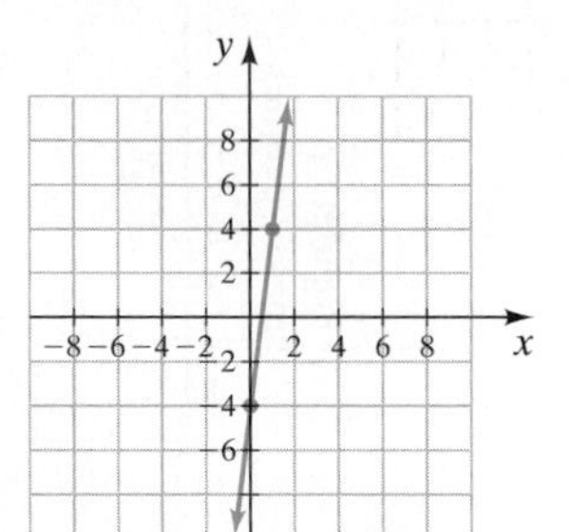

In Exercises 71 and 72, find the slope of the line segments indicated in **a)** *red and* **b)** *blue.*

71.

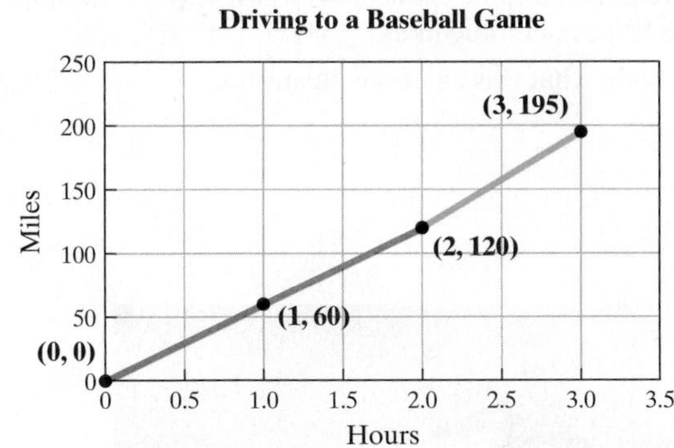

72.

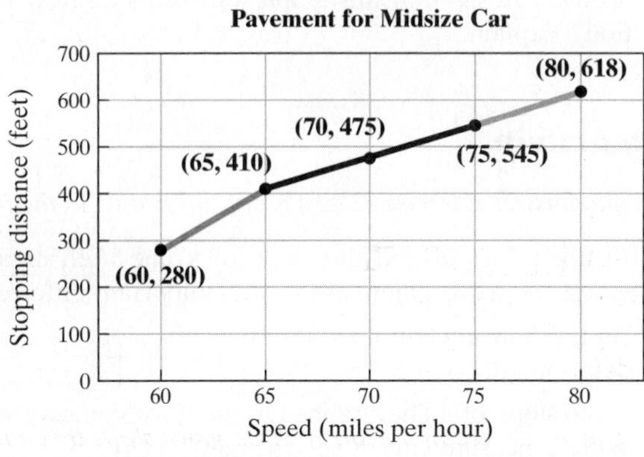

73. A given line goes through the points $(2, 6)$ and $(4, -2)$. If a line is to be drawn parallel to the given line, what will be its slope?

74. A given line goes through the points $(-2, 5)$ and $(4, 7)$. If a line is to be drawn parallel to the given line, what will be its slope?

75. A given line goes through the points $(1, -7)$ and $(2, 1)$. If a line is to be drawn perpendicular to the given line, what will be its slope?

76. A given line goes through the points $(-3, 0)$ and $(-2, 3)$. If a line is to be drawn perpendicular to the given line, what will be its slope?

Challenge Problems

77. A line contains the points $(1, b)$ and $(4, 9)$. If the line has slope $m = 2$, what is the value of b?

78. A line contains the points $(1, 2)$ and $(a, -4)$. If the line has slope $m = 3$, what is the value of a?

79. A quadrilateral (a four-sided figure) has four vertices (the points where the sides meet). Vertex A is at $(0, 1)$, vertex B is at $(6, 2)$, vertex C is at $(5, 4)$, and vertex D is at $(1, -1)$.

a) Graph the quadrilateral in the Cartesian coordinate system.

b) Find the slopes of sides AC, CB, DB, and AD.

c) Do you think this figure is a parallelogram? Explain.

80. Population The following graph shows the world's population estimated to the year 2016.

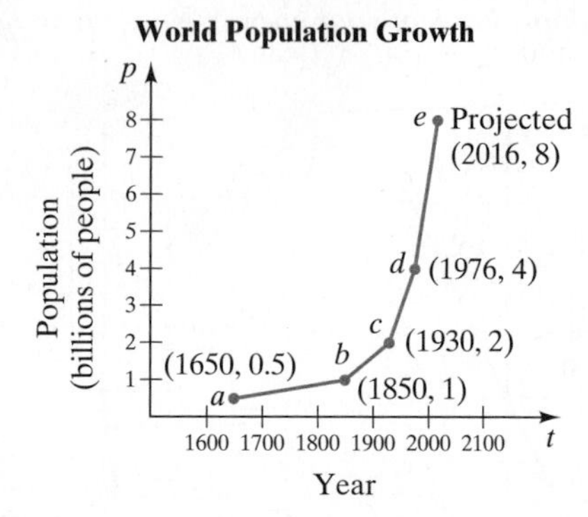

a) Find the slope of the line segment between each pair of points, that is, *ab*, *bc*, and so on. Remember, the second coordinate is in billions. Thus, for example, 0.5 billion is actually 500,000,000.

b) Would you say that this graph represents a linear equation? Explain.

81. Consider the graph below.

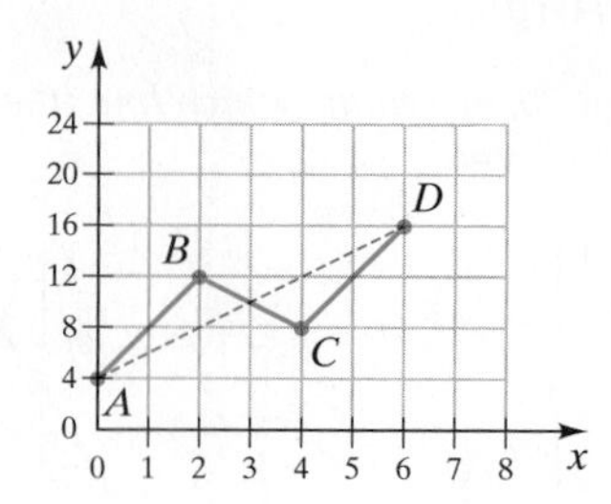

a) Determine the slope of each of the three solid blue lines.

b) Determine the average of the three slopes found in part **a)**.

c) Determine the slope of the red dashed line from A to D.

d) Determine whether the slope of the red dashed line from A to D is the same as the (mean) average of the slopes of the three solid blue lines.

e) Explain what this example illustrates.

Group Activity

Discuss and answer Exercise 82 as a group, according to the instructions.

82. The slope of a hill and the slope of a line both measure steepness. However, there are several important differences.

a) As a group, explain how you think the slope of a hill is determined.

b) Is the slope of a line, graphed in the Cartesian coordinate system, measured in any specific unit?

c) Is the slope of a hill measured in any specific unit?

Cumulative Review Exercises

[1.9] **83.** Evaluate $-5x^2 + 3x + 4$ when $x = 1$.

[2.3] **84. a)** If $-x = 3$, what is the value of x?

b) If $-4x = 0$, what is the value of x?

[4.4] **85.** Subtract $(2x - 9)$ from $(4x + 7)$.

[6.6] **86.** Solve the equation $\frac{2x}{x - 3} = 2 + \frac{3}{x}$

[7.2] **87.** Find the x- and y-intercepts for the line whose equation is $5x - 3y = 30$.

MID-CHAPTER TEST: 7.1–7.3

To find out how well you understand the chapter material to this point, take this brief test. The answers, and the section where the material was initially discussed, are given in the back of the book. Review any questions that you answered incorrectly.

1. In which quadrant does the point $(3, -4)$ belong?

2. Plot the points $A(2, 6), B(-3, 1), C(-5, -2), D(0, -4), E(4, -7)$ on the same axes.

3. Determine which of the three ordered pairs does not satisfy the equation $\frac{1}{3}x + y = -2$.

a) $(3, -3)$ **b)** $(0, 2)$ **c)** $(-6, 0)$

4. Find the value of y that makes the ordered pair $(-1, y)$ a solution to the linear equation $y = 5x + 1$.

5. Find the value of x that makes the ordered pair $(x, 2)$ a solution to the equation $3x - 4y = 1$.

6. What does the graph of an equation illustrate?

Graph each equation.

7. $x = \frac{5}{2}$

8. $y = -2$

Graph by plotting points.

9. $y = 3x + 1$

10. $y = -\frac{1}{2}x + 4$

Graph using the x- and y-intercepts.

11. $3x - 4y = 12$

12. $\frac{1}{2}x + \frac{1}{5}y = 10$

Find the slope of the line through each pair of points.

13. $(-1, 5)$ and $(6, 3)$ **14.** $(4, 2)$ and $(7, 2)$

15. $(-3, 0)$ and $(-3, 5)$

Graph the line with the given slope that goes through the given point.

16. Through $(-2, 3)$ with $m = -\frac{1}{2}$

17. Through $(4, 1)$ with $m = \frac{3}{5}$.

Indicate whether the lines with the following slopes are parallel, perpendicular, or neither.

18. $m_1 = 5$ and $m_2 = \frac{1}{5}$

19. $m_1 = \frac{6}{7}$ and $m_2 = -\frac{7}{6}$

20. Interest The following graph illustrates the interest obtained when $1000 is invested for 1 year at various interest rates from 0% to 10%. Determine the slope of the line in the graph.

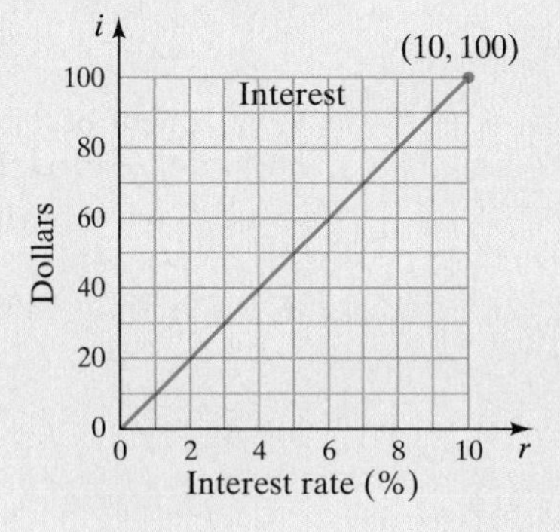

7.4 Slope-Intercept and Point-Slope Forms of a Linear Equation

1. Write a linear equation in slope-intercept form.
2. Graph a linear equation using the slope and y-intercept.
3. Use the slope-intercept form to determine the equation of a line.
4. Use the point-slope form to determine the equation of a line.
5. Compare the three methods of graphing linear equations.

In Section 7.1 we introduced the *standard form* of a linear equation, $ax + by = c$. In this section we introduce two more forms, the slope-intercept form and the point-slope form.

1 Write a Linear Equation in Slope-Intercept Form

A very important form of a linear equation is the **slope-intercept form, $y = mx + b$.** The graph of an equation of the form $y = mx + b$ will always be a straight line with a **slope of m** and a **y-intercept $(0, b)$.**

Slope-Intercept Form of a Linear Equation

$$y = mx + b$$

where m is the slope, and $(0, b)$ is the y-intercept of the line.

Slope → m; y-intercept is $(0, b)$ → b

$$y = mx + b$$

Equations in Slope-Intercept Form	Slope, m	y-Intercept $(0, b)$
$y = 4x - 6$	4	$(0, -6)$
$y = \frac{1}{2}x + \frac{3}{2}$	$\frac{1}{2}$	$\left(0, \frac{3}{2}\right)$
$y = -5x + 3$	-5	$(0, 3)$

Writing an Equation in Slope-Intercept Form

To write a linear equation in slope-intercept form, solve the equation for y.

Once the equation is solved for y, the numerical coefficient of the x-term will be the slope, and the constant term will give the y-intercept.

EXAMPLE 1 Write the equation $-3x + 4y = 8$ in slope-intercept form. State the slope and y-intercept.

Solution To write this equation in slope-intercept form, we solve the equation for y.

$$-3x + 4y = 8$$

$$4y = 3x + 8 \quad \text{Added } 3x \text{ to both sides.}$$

$$y = \frac{3x + 8}{4} \quad \text{Divided both sides by 4.}$$

$$y = \frac{3}{4}x + \frac{8}{4} \quad \text{Wrote as two fractions.}$$

$$y = \frac{3}{4}x + 2 \quad \text{Equation in slope-intercept form}$$

$$y = \frac{3}{4}x + 2$$

$$m = \frac{3}{4} \qquad b = 2$$

Thus, the slope is $\frac{3}{4}$, and the y-intercept is $(0, 2)$.

Now Try Exercise 11

EXAMPLE 2 Determine whether the two equations represent lines that are parallel, perpendicular, or neither.

a) $2x + y = 9$
$2y = -4x + 5$

b) $3x - 2y = 7$
$6y + 4x = -6$

Solution Two lines that have the same slope but different y-intercepts are parallel lines, and two lines whose slopes are negative reciprocals are perpendicular lines. We can determine the slope of each line by solving each equation for y. The coefficient of the x term will be the slope.

a) $2x + y = 9$

$$y = -2x + 9$$

$$m = -2$$

$$2y = -4x + 5$$

$$y = \frac{-4x + 5}{2}$$

$$y = -2x + \frac{5}{2}$$

$$m = -2$$

Since both equations have the same slope, -2, the equations represent lines that are parallel. Notice the equations represent two different lines because their y-intercepts are different.

b) $3x - 2y = 7$

$$-2y = -3x + 7$$

$$y = \frac{-3x + 7}{-2}$$

$$y = \frac{3}{2}x - \frac{7}{2}$$

$$m = \frac{3}{2}$$

$$6y + 4x = -6$$

$$6y = -4x - 6$$

$$y = \frac{-4x - 6}{6}$$

$$y = -\frac{2}{3}x - 1$$

$$m = -\frac{2}{3}$$

The slope of one line is $\frac{3}{2}$ and the slope of the other line is $-\frac{2}{3}$. Since the slopes are negative reciprocals, the lines are perpendicular.

Now Try Exercise 39

Understanding Algebra

When graphing using the slope-intercept form, our starting point is always the *y*-intercept.

2 Graph a Linear Equation Using the Slope and *y*-Intercept

In Section 7.2 we discussed graphing a linear equation by (1) plotting points and by (2) using the *x*- and *y*-intercepts. Now we present a third method. This method makes use of the slope and the *y*-intercept. We graph equations using the slope and *y*-intercept in a manner very similar to the way we worked Examples 5 and 6 in Section 7.3.

To Graph Linear Equations Using the Slope and *y*-intercept

1. If necessary, solve the equation for *y*.
2. Plot the *y*-intercept.
3. Use the slope to find two more points on the line.
4. Using a straightedge, draw a straight line through the three points. Draw an arrowhead on each end of the line.

EXAMPLE 3 Graph $-3x + 4y = 8$ by using the slope and *y*-intercept.

Solution In Example 1 we solved $-3x + 4y = 8$ for *y*. We found that

$$y = \frac{3}{4}x + 2$$

The slope of the line is $\frac{3}{4}$ and the *y*-intercept is $(0, 2)$. We plot the *y*-intercept at 2 on the *y*-axis **(Fig. 7.36)**. Now we use the slope, $\frac{3}{4}$, to find a second point. Since the slope is positive, we move 3 units up and 4 units to the right to find a second point at $(4, 5)$. We continue this process to obtain a third point at $(8, 8)$. Now we draw a straight line through the three points.

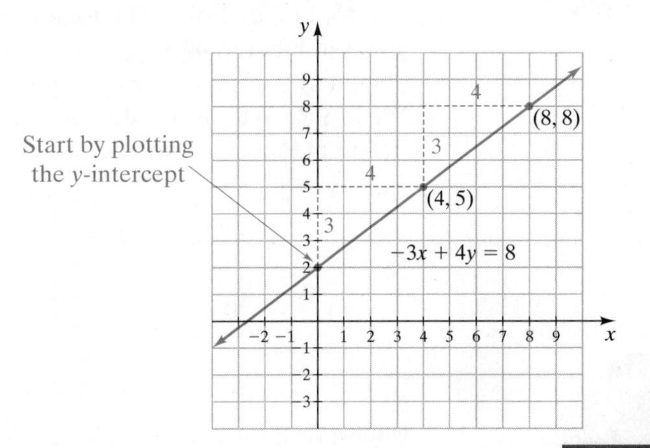

FIGURE 7.36

Now Try Exercise 19

EXAMPLE 4 Graph $5x + 3y = 12$ by using the slope and *y*-intercept.

Solution First, solve the equation for *y*.

$$\begin{aligned} 5x + 3y &= 12 \\ 3y &= -5x + 12 \\ y &= \frac{-5x + 12}{3} \\ &= -\frac{5}{3}x + 4 \end{aligned}$$

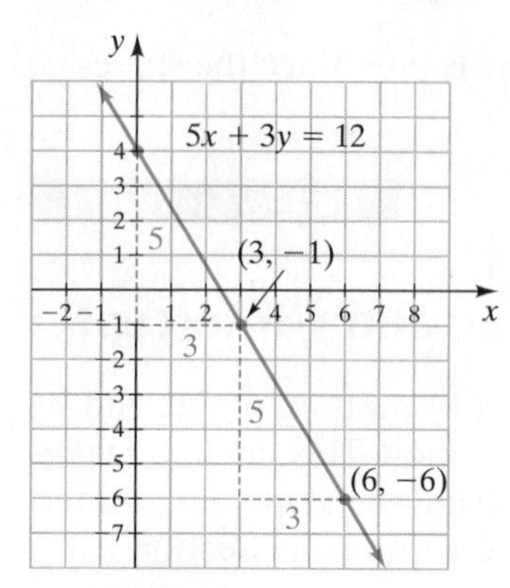

FIGURE 7.37

Thus, the slope is $-\frac{5}{3}$ and the y-intercept is $(0, 4)$. Plot the y-intercept at 4 on the y-axis **(Fig. 7.37)**. Then, since $-\frac{5}{3} = \frac{-5}{3}$, move 5 units down and 3 units to the right to determine the next point. You can follow this procedure again to obtain a third point. Finally, draw the straight line between the plotted points.

Now Try Exercise 21

3 Use the Slope-Intercept Form to Determine the Equation of a Line

We can also use the slope-intercept form to write the equation of a given line. To do so, we need to determine the slope, m, and the y-intercept, $(0, b)$. We then can substitute m and b into $y = mx + b$ to get the equation of the given line.

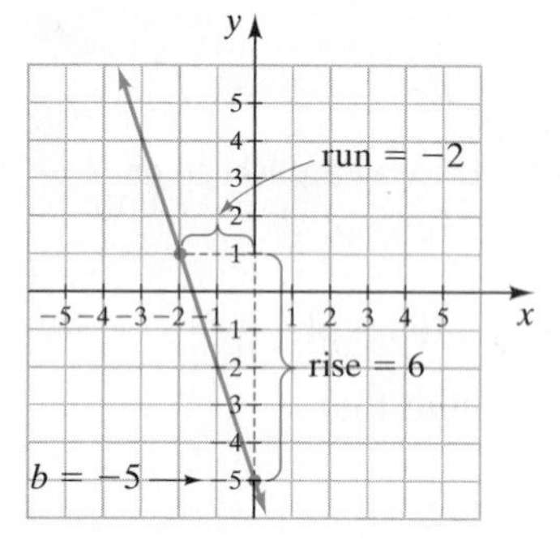

FIGURE 7.38

EXAMPLE 5 Determine the equation of the line shown in **Figure 7.38**

Solution The graph shows that the y-intercept is at -5 so $b = -5$. Since the line falls from left to right it has a negative slope. Looking at the point $(0, -5)$ and $(-2, 1)$, we can see that the rise is 6 and the run is -2. So,

$$m = \frac{\text{rise}}{\text{run}} = \frac{6}{-2} = -3$$

Substituting -3 for m and -5 for b into the slope-intercept form gives us the equation $y = -3x - 5$, which is the equation of the line shown.

Now Try Exercise 29

EXAMPLE 6 **Artistic Vases** Kris, a pottery artist, makes ceramic vases that he sells at art shows. His business has a fixed monthly cost (booth rental, advertising, cell phone, etc.) and a variable cost per vase made (cost of materials, cost of labor, cost for kiln use, etc.). The total monthly cost for making x vases is illustrated in the graph in **Figure 7.39**.

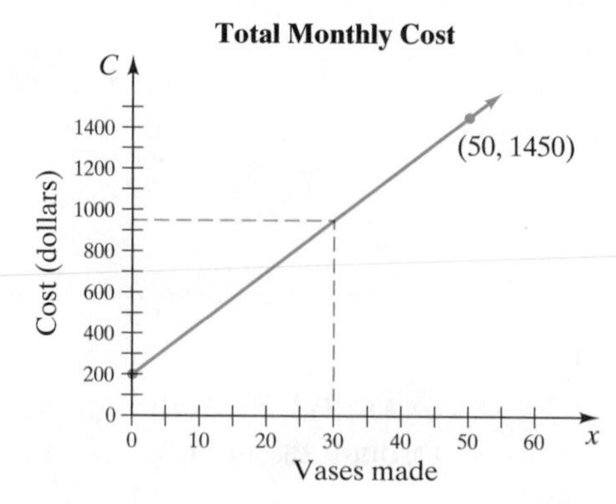

FIGURE 7.39

a) Find the equation of the total monthly cost when x vases are made.

b) Use the equation found in part **a)** to find the total monthly cost if 30 vases are made.

c) Use the graph in **Figure 7.39** to see whether your answer in part **b)** appears correct.

Solution

a) Understand and Translate Notice that the vertical axis is cost, C, and not y. Since y is replaced by C, we will use $C = mx + b$, in which b is where the graph crosses the vertical or C-axis. Note that the graph crosses the vertical axis at 200. Thus, b is 200. Now use the points $(0, 200)$ and $(50, 1450)$ to find the slope.

Carry Out
$$m = \frac{y_2 - y_1}{x_2 - x_1}$$
$$= \frac{1450 - 200}{50 - 0} = \frac{1250}{50} = 25$$

Answer The slope is 25. So $m = 25$ and $b = 200$ and the equation in slope-intercept form is

$$C = mx + b$$
$$= 25x + 200$$

b) To find the monthly cost when 30 vases are made, we substitute 30 for x.

$$C = 25x + 200$$
$$= 25(30) + 200$$
$$= 750 + 200 = 950$$

The monthly cost when 30 vases are made is $950.

c) If we draw a vertical line up from 30 on the x-axis (the red line), we see that the corresponding cost is about $950. Thus, our answer in part **b)** appears correct.

Now Try Exercise 67

4 Use the Point-Slope Form to Determine the Equation of a Line

Understanding Algebra

Forms of Linear Equations

1. Standard Form

   $ax + by = c$

2. Slope-Intercept Form

   $y = mx + b$

3. Point-Slope Form

   $y - y_1 = m(x - x_1)$

When the slope of a line and a point on the line are known, we can use the point-slope form to determine the equation of the line. The **point-slope form** can be obtained by beginning with the slope between any selected point (x, y) and a fixed point (x_1, y_1) on a line.

$$m = \frac{y - y_1}{x - x_1} \quad \text{or} \quad \frac{m}{1} = \frac{y - y_1}{x - x_1}$$

Now cross-multiply to obtain

$$m(x - x_1) = y - y_1 \quad \text{or} \quad y - y_1 = m(x - x_1)$$

Point-Slope Form of a Linear Equation

$$y - y_1 = m(x - x_1)$$

where m is the slope of the line and (x_1, y_1) is a point on the line.

EXAMPLE 7 Write an equation, in slope-intercept form, of the line that goes through the point $(1, 3)$ and has a slope of 2.

Solution Since we are given a *point* on the line and the *slope* of the line, we begin by writing the equation in *point-slope* form. Since the slope is 2 and the point on the line is $(1, 3)$, we have $m = 2$, $x_1 = 1$, and $y_1 = 3$. We substitute into the point-slope form and solve for y as follows.

$$y - y_1 = m(x - x_1)$$
$$y - 3 = 2(x - 1) \quad \text{Equation in point-slope form}$$
$$y - 3 = 2x - 2 \quad \text{Distributive property}$$
$$y = 2x + 1 \quad \text{Equation in slope-intercept form}$$

Now Try Exercise 51

EXAMPLE 8 Write an equation, in slope-intercept form, of the line that goes through the point $(1, -2)$ and has a slope of $\frac{3}{4}$.

Solution We will begin with the point-slope form of a line, where $m = \frac{3}{4}$, $x_1 = 1$, and $y_1 = -2$.

$$y - y_1 = m(x - x_1)$$

$$y - (-2) = \frac{3}{4}(x - 1) \quad \text{Equation in point-slope form}$$

$$4(y + 2) = 4 \cdot \frac{3}{4}(x - 1) \quad \text{Multiply both sides by 4.}$$

$$4y + 8 = 3(x - 1) \quad \text{Distributive property (left side of equation)}$$

$$4y + 8 = 3x - 3 \quad \text{Distributive property (right side of equation)}$$

$$4y = 3x - 11 \quad \text{Subtracted 8 from both sides.}$$

$$y = \frac{3x - 11}{4} \quad \text{Divided both sides by 4.}$$

$$y = \frac{3}{4}x - \frac{11}{4} \quad \text{Equation in slope-intercept form}$$

Now Try Exercise 55

Understanding Algebra

To simplify an equation that contains a fraction, multiply both sides of the equation by the denominator of the fraction.

HELPFUL HINTS

We have discussed three forms of a linear equation. We summarize the three forms below. It is important that you memorize these forms.

Standard Form

$ax + by = c$

Examples

$2x - 3y = 8$

$-5x + y = -2$

Slope-Intercept Form

$y = mx + b$

m is the slope, $(0, b)$ is the y-intercept

Examples

$y = 2x - 5$

$y = -\frac{3}{2}x + 2$

Point-Slope Form

$y - y_1 = m(x - x_1)$

m is the slope, (x_1, y_1) is a point on the line

Examples

$y - 3 = 2(x + 4)$

$y + 5 = -4(x - 1)$

We now discuss how to use the point-slope form to determine the equation of a line when two points on the line are known.

EXAMPLE 9 Find an equation of the line through the points $(-1, 3)$ and $(-5, 1)$. Write the equation in slope-intercept form.

Solution To use the point-slope form, we must first find the slope of the line through the two points.

$$m = \frac{y_2 - y_1}{x_2 - x_1} = \frac{1 - 3}{-5 - (-1)} = \frac{1 - 3}{-5 + 1} = \frac{-2}{-4} = \frac{1}{2}$$

Thus $m = \frac{1}{2}$. We can use either point in determining the equation of the line. This example will be worked out using each of the points to show that the solutions obtained are identical.

Using the point $(-1, 3)$ as (x_1, y_1),

$$y - y_1 = m(x - x_1)$$
$$y - 3 = \frac{1}{2}[x - (-1)]$$
$$y - 3 = \frac{1}{2}(x + 1)$$
$$2 \cdot (y - 3) = 2 \cdot \frac{1}{2}(x + 1) \quad \text{Multiply both sides by the LCD, 2.}$$
$$2y - 6 = x + 1$$
$$2y = x + 7$$
$$y = \frac{x + 7}{2} \quad \text{or} \quad y = \frac{1}{2}x + \frac{7}{2}$$

Using the point $(-5, 1)$ as (x_1, y_1),

$$y - y_1 = m(x - x_1)$$
$$y - 1 = \frac{1}{2}[x - (-5)]$$
$$y - 1 = \frac{1}{2}(x + 5)$$
$$2 \cdot (y - 1) = 2 \cdot \frac{1}{2}(x + 5) \quad \text{Multiply both sides by the LCD, 2.}$$
$$2y - 2 = x + 5$$
$$2y = x + 7$$
$$y = \frac{x + 7}{2} \quad \text{or} \quad y = \frac{1}{2}x + \frac{7}{2}$$

Note that the equations are identical.

Now Try Exercise 61

HELPFUL HINTS

In the Exercises, you will be asked to write a linear equation in slope-intercept form. Even though you will eventually write the equation in slope-intercept form, you may need to start your work with the point-slope form. Below we indicate the initial form to use to solve the problem.

1. Begin with the **slope-intercept form** if you know
   - The slope of the line and the y-intercept
2. Begin with the **point-slope form** if you know
   a) The slope of the line and a point on the line other than the y-intercept or
   b) Two points on the line (first find the slope, then use the point-slope form)

5 Compare the Three Methods of Graphing Linear Equations

We have discussed three methods to graph a linear equation: (1) plotting points, (2) using the x- and y-intercepts, and (3) using the slope and y-intercept. No single method is always the easiest to use. If the equation is given in slope-intercept form, $y = mx + b$, then graphing by plotting points or by using the slope and y-intercept might be easier. If the equation is given in standard form, $ax + by = c$, then graphing using the intercepts might be easier.

EXAMPLE 10 Graph $3x - 2y = 8$

a) by plotting points;
b) using the x- and y-intercepts;
c) using the slope and y-intercept.

Solution For parts **a)** and **c)** we will write the equation in slope-intercept form by solving for y.

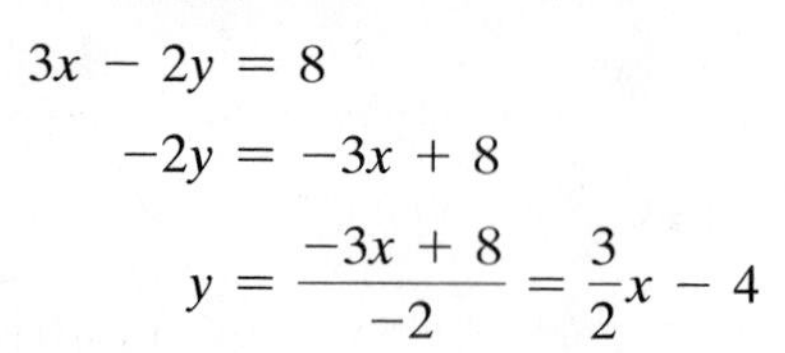

$$3x - 2y = 8$$
$$-2y = -3x + 8$$
$$y = \frac{-3x + 8}{-2} = \frac{3}{2}x - 4$$

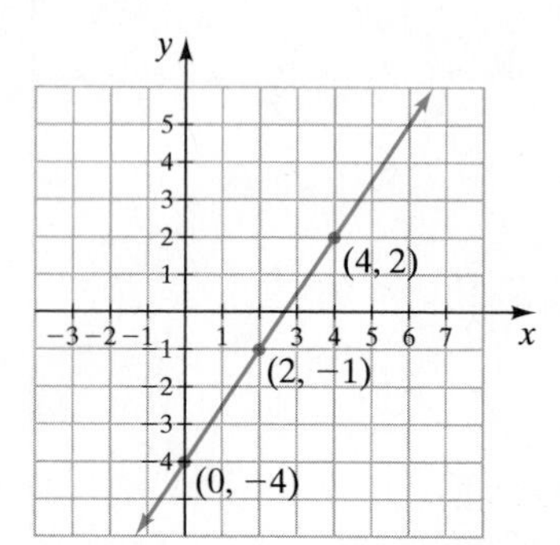

FIGURE 7.40

a) Plotting Points We substitute values for x and find the corresponding values of y. Three ordered pairs are indicated in the following table. Next we plot the ordered pairs and draw the graph **(Fig. 7.40)**.

$$y = \frac{3}{2}x - 4$$

x	y
0	−4
2	−1
4	2

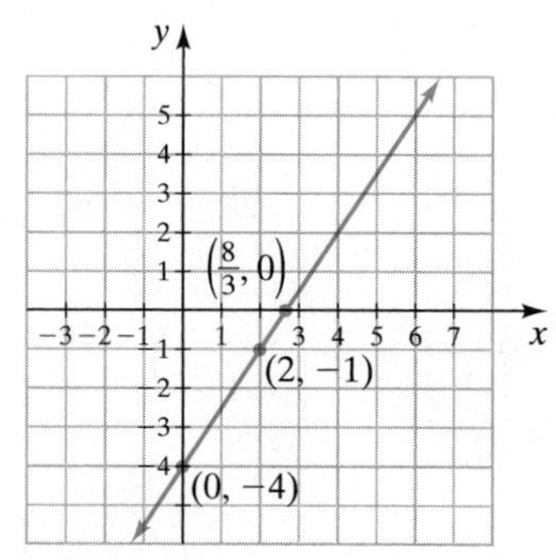

FIGURE 7.41

b) Intercepts We find the x- and y-intercepts and a checkpoint. Then we plot the points and draw the graph **(Fig. 7.41)**.

$$3x - 2y = 8$$

***x*-Intercept**

Let $y = 0$.
$$3x - 2y = 8$$
$$3x - 2(0) = 8$$
$$3x = 8$$
$$x = \frac{8}{3}$$

***y*-Intercept**

Let $x = 0$.
$$3x - 2y = 8$$
$$3(0) - 2y = 8$$
$$-2y = 8$$
$$y = -4$$

Checkpoint

Let $x = 2$.
$$3x - 2y = 8$$
$$3(2) - 2y = 8$$
$$6 - 2y = 8$$
$$-2y = 2$$
$$y = -1$$

The three ordered pairs are $\left(\frac{8}{3}, 0\right)$, $(0, -4)$, and $(2, -1)$. The graph is shown in **Figure 7.41**.

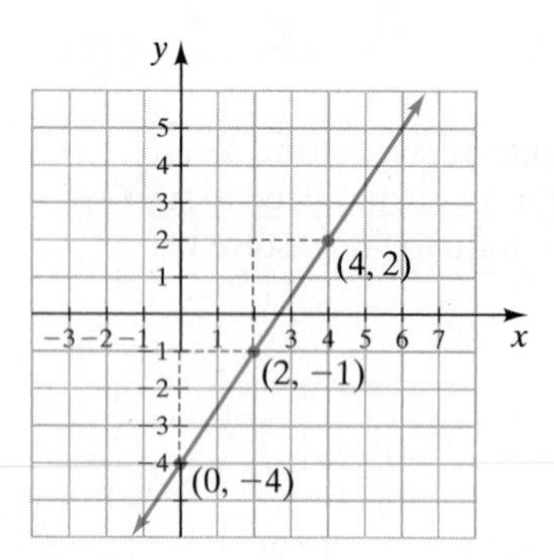

FIGURE 7.42

c) Slope and *y*-intercept The y-intercept is $(0, -4)$; therefore, we place a point at -4 on the y-axis. Since the slope is $\frac{3}{2}$, we obtain a second point by moving 3 units up and 2 units to the right. Thus, we place a point at $(2, -1)$. We can obtain a third point by again moving 3 units up and 2 units to the right. We place a third point at $(4, 2)$. The graph is illustrated in **Figure 7.42**. Notice that we get the same line by all three methods.

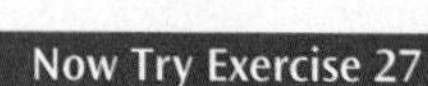

7.4 Exercise Set MathXL® MyMathLab®

Warm-Up Exercises

Fill in the blanks with the appropriate word, phrase, or symbol(s) from the following list.

numerator	y-intercept	standard	point-slope	slope
slope-intercept	denominator	point	y	

1. The __________ form of a linear equation is $y - y_1 = m(x - x_1)$.

2. The __________ form of a linear equation is $ax + by = c$.

3. The __________ form of a linear equation is $y = mx + b$.

4. To write a linear equation in slope-intercept form, solve the equation for __________.

5. When graphing a linear equation using the slope and y-intercept, you first plot the ________.

6. To simplify an equation that contains a fraction, multiply both sides of the equation by the ________ of the fraction.

7. If you are asked to find the equation of a line and if you know the ________ and the y-intercept of the line, you should use the slope-intercept form.

8. If you are asked to find the equation of a line and if you know a ________ on the line and the slope of the line, you should use the point-slope form.

Practice the Skills

Determine the slope and y-intercept of the line represented by the given equation.

9. $y = 3x + 1$

10. $y = -3x + 17$

11. $4x - 3y = 21$

12. $7x = 5y + 25$

Determine the slope and y-intercept of the line represented by each equation. Graph the line using the slope and y-intercept.

13. $y = x - 3$

14. $y = -x + 5$

15. $y = 3x + 2$

16. $3x + y = 4$

17. $y = -4x$

18. $y = 2x$

19. $-2x + y = -3$

20. $3x + 3y = 9$

21. $5x - 2y = 10$

22. $-x + 2y = 8$

23. $6x + 12y = 18$

24. $4x = 6y + 9$

25. $-6x + 2y - 8 = 0$

26. $-9x + 3y + 6 = 0$

27. $3x = 2y - 4$

28. $16y = 8x + 32$

Determine the equation of each line.

29.

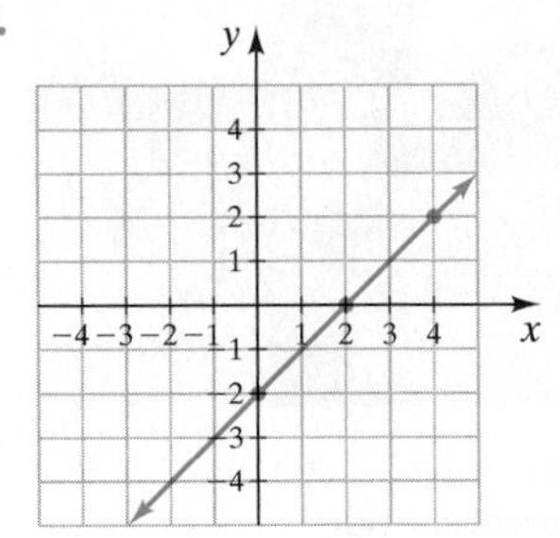

30.

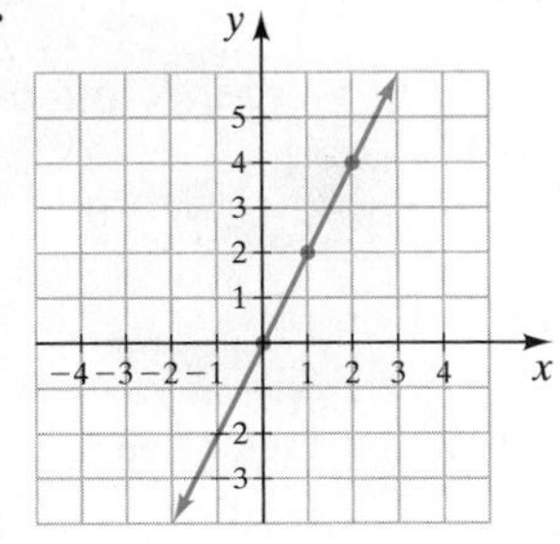

31.

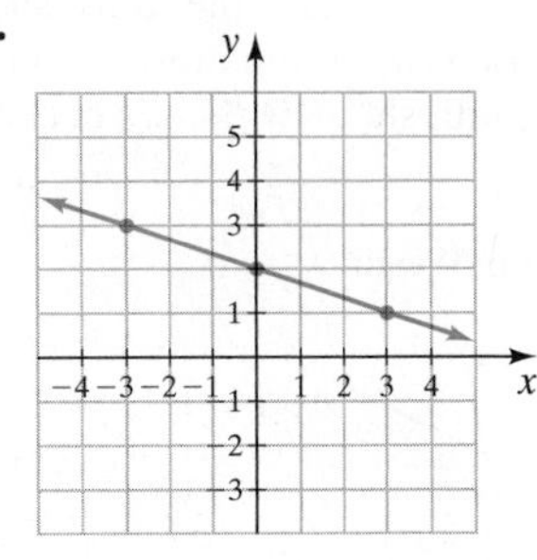

32.

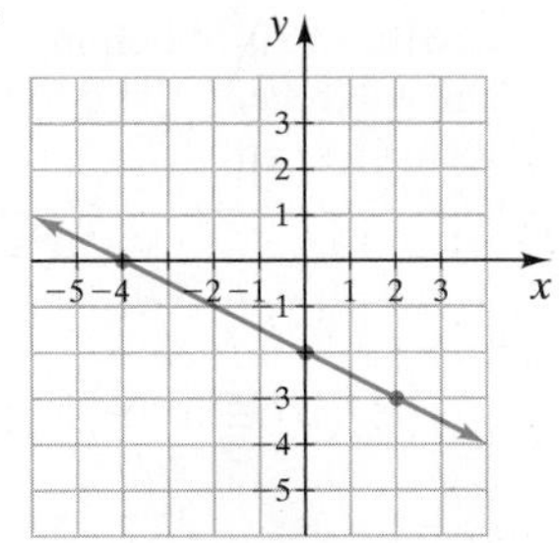

33.

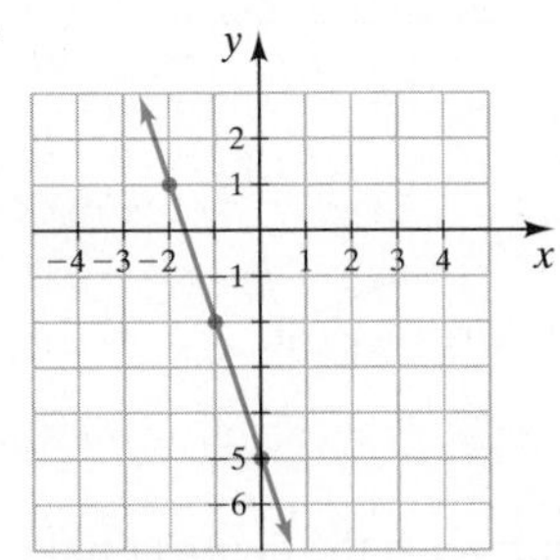

34.

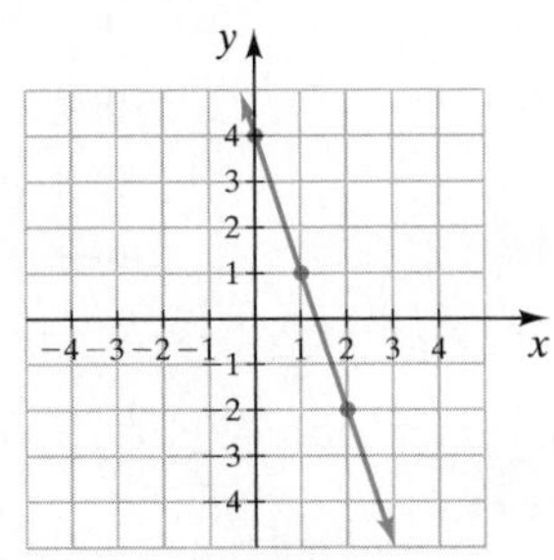

35.

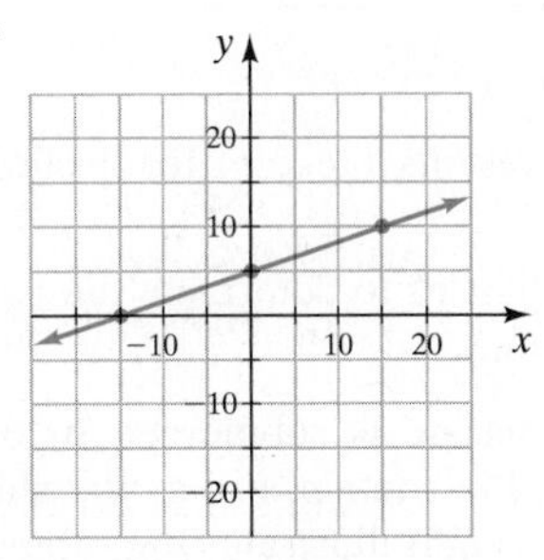

36.

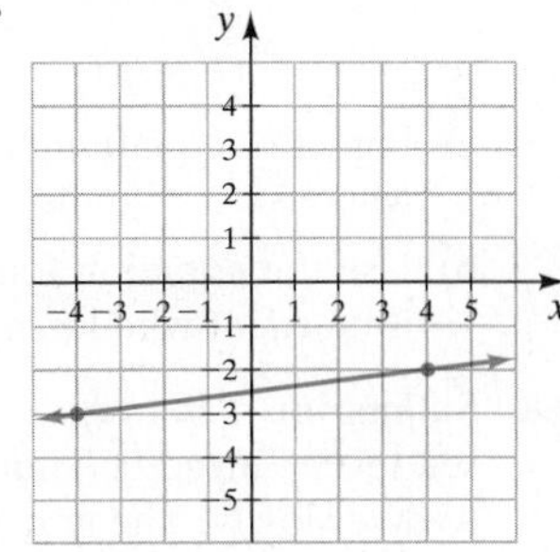

Determine whether each pair of lines are parallel, perpendicular, or neither.

37. $y = 3x - 2$
$y = 3x + 2$

38. $y = -\frac{5}{7}x + \frac{4}{3}$
$y = -\frac{5}{7}x - \frac{3}{4}$

39. $4x + 2y = 7$
$4x = 8y + 12$

40. $8x + 2y = 10$
$x - 7 = 4y$

41. $3x + 5y = 9$
$6x = -10y + 9$

42. $3y - 4 = -5x$
$y = -\frac{5}{3}x - 3$

43. $y = \frac{1}{2}x - 2$
$2y = 6x + 9$

44. $3x - 5y = 7$
$5y + 3x = 4$

45. $5y = 2x + 9$
$-10x = 4y + 11$

46. $2x + y = 7$
$x - 2y = 5$

Problem Solving

Write the equation of each line, with the given properties, in slope-intercept form.

47. Slope $= 3$, through $(0, 2)$

48. Slope $= -2$, through $(0, 5)$

49. Slope $= \frac{2}{3}$, y-intercept is $(0, 6)$

50. Slope $= \frac{1}{9}$, y-intercept is $\left(0, -\frac{2}{5}\right)$

51. Slope $= -3$, through $(-4, 5)$

52. Slope $= 2$, through $(4, 3)$

53. Slope $= -\frac{5}{6}$, through $(12, 1)$

54. Slope $= \frac{7}{2}$, through $(-4, -5)$

55. Slope $= \frac{1}{2}$, through $(-1, -3)$

56. Slope $= -\frac{2}{3}$, through $(4, -5)$

57. Through $(10, 3)$ and $(0, -2)$

58. Through $(5, -4)$ and $(0, 3)$

59. Through $(2, 3)$ and $(4, 7)$

60. Through $(7, 4)$ and $(6, 3)$

61. Through $(-4, -2)$ and $(-2, 4)$

62. Through $(-7, 13)$ and $(8, -17)$

63. Through $(-6, 9)$ and $(8, -12)$

64. Through $(4, 3)$ and $(-8, -6)$

65. Through $(3, 2)$ and $(10, -4)$

66. Through $(-6, -2)$ and $(5, -3)$

67. Weight Loss Clinic Stacy Best owns a weight loss clinic. She charges her clients a one-time membership fee. She also charges per pound of weight lost. Therefore, the more successful she is at helping clients lose weight, the more income she will receive. The following graph shows a client's cost of losing weight.

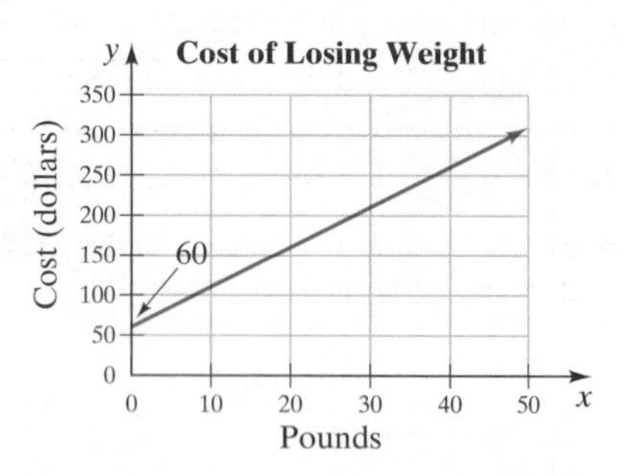

a) Find the equation that represents the cost for a client who loses x pounds.

b) Use the equation found in part **a)** to determine the cost for a client who loses 30 pounds.

68. Submarine Submerges A submarine is submerged below sea level. Burnette Thompson, the captain, orders the ship to dive slowly. The graph on the right illustrates the submarine's depth at a time t minutes after the submarine begins to dive.

See Exercise 68.

Submarine's Depth

Time (minutes)

0 10 20 30 40 50 t

0, −200, −300, −400, −600, −800, −1000

Depth (feet) d

a) Find the equation that represents the depth at time t.

b) Use the equation found in part **a)** to find the submarine's depth after 30 minutes.

Concept/Writing Exercises

69. Assume the slope of a line is 2 and two points on the line are $(-5, -4)$ and $(3, 12)$.

a) If you use $(-5, -4)$ as (x_1, y_1) and then $(3, 12)$ as (x_1, y_1) will the appearance of the two equations be the same in point-slope form? Explain.

b) Find the equation, in point-slope form, using $(-5, -4)$ as (x_1, y_1).

c) Find the equation, in point-slope form, using $(3, 12)$ as (x_1, y_1).

d) Write the equation obtained in part **b)** in slope-intercept form.

e) Write the equation obtained in part **c)** in slope-intercept form.

f) Are the equations obtained in parts **d)** and **e)** the same? If not, explain why.

70. Assume the slope of a line is -3 and two points on the line are $(-1, 8)$ and $(2, -1)$.

a) If you use $(-1, 8)$ as (x_1, y_1) and then $(2, -1)$ as (x_1, y_1) will the appearance of the two equations be the same in point-slope form? Explain.

b) Find the equation, in point-slope form, using $(-1, 8)$ as (x_1, y_1).

c) Find the equation, in point-slope form, using $(2, -1)$ as (x_1, y_1).

d) Write the equation obtained in part **b)** in slope-intercept form.

e) Write the equation obtained in part **c)** in slope-intercept form.

f) Are the equations obtained in parts **d)** and **e)** the same? If not, explain why.

Challenge Problems

71. Unit Conversions The following graph shows the approximate relationship between speed in miles per hour and feet per second.

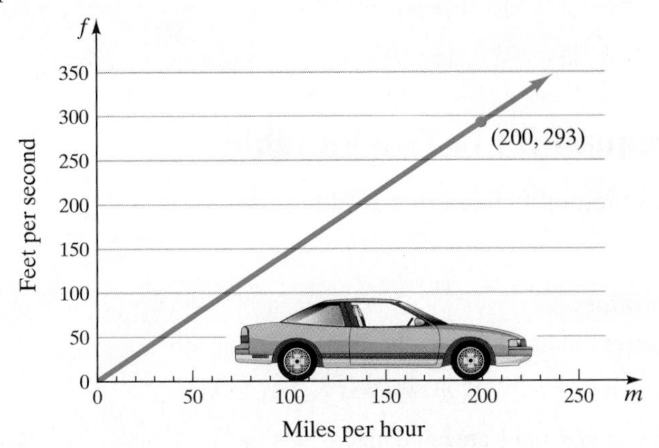

a) Determine the slope of the line.

b) Determine the equation of the line.

c) At the 2012 Daytona 500 race, winner Matt Kenseth had an average speed of 140.3 miles per hour. Use the equation you obtained in part **b)** to determine his average speed in feet per second. Round your answer to one decimal place.

d) Use the graph to estimate a speed of 100 miles per hour in feet per second.

e) Use the graph to estimate a speed of 80 feet per second in miles per hour.

72. Temperature The following graph shows the relationship between Fahrenheit temperature and Celsius temperature.

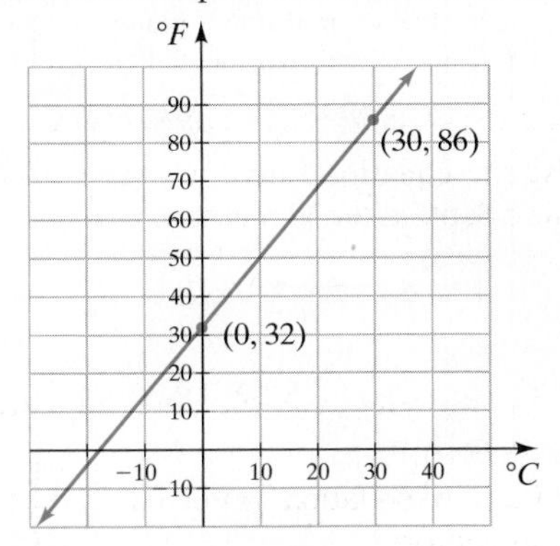

a) Determine the slope of the line.

b) Determine the equation of the line in slope-intercept form.

c) Use the equation (or formula) you obtained in part **b)** to find the Fahrenheit temperature when the Celsius temperature is 20°.

d) Use the graph to estimate the Celsius temperature when the Fahrenheit temperature is 100°.

e) Estimate the Celsius temperature that corresponds to a Fahrenheit temperature of 0°.

73. Determine the equation of the line with y-intercept at 5 that is parallel to the line whose equation is $2x + y = 6$. Explain how you determined your answer.

74. Will a line through the points $(60, 30)$ and $(20, 90)$ be parallel to the line with x-intercept at 2 and y-intercept at 3? Explain how you determined your answer.

75. Write an equation of the line parallel to the graph of $3x - 4y = 6$ that passes through the point $(-8, -1)$.

76. Determine the equation of the straight line that intersects the greatest number of shaded points on the following graph.

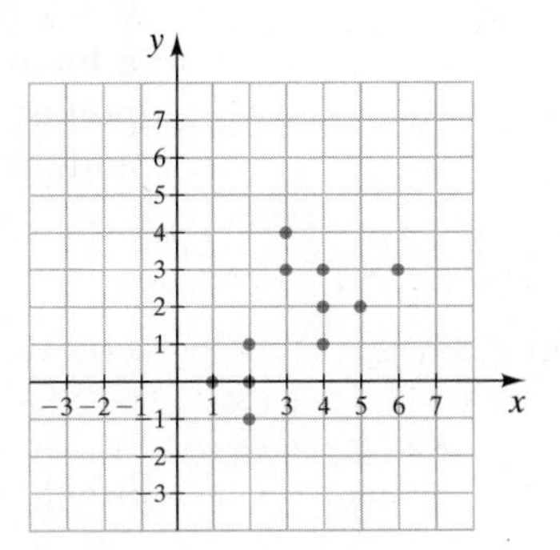

Group Activity

Discuss and answer Exercise 77 as a group, according to the instructions.

77. Consider the equation $-3x + 2y = 4$.

a) Group member 1: Explain how to graph this equation by plotting points. Then graph the equation by plotting points.

b) Group member 2: Explain how to graph this equation using the intercepts. Then graph the equation using the intercepts.

c) Group member 3: Explain how to graph this equation using the slope and y-intercept. Then graph the equation using the slope and y-intercept.

d) As a group, compare your graphs. Did you all obtain the same graph? If not, determine why.

Cumulative Review Exercises

[1.5] **78.** Insert either $>$, $<$, or $=$ in the shaded area to make the statement true: $|-4| \; \square \; |-9|$.

[2.6] **79.** Solve $i = prt$ for r.

[2.8] **80.** Solve $2(x - 3) \geq 5x + 6$. Graph the solution on a number line and write the solution using interval notation.

[5.2] **81.** Factor $x^2 - 4xy + 3xy - 12y^2$ by grouping.

[6.6] **82.** Solve $\frac{x}{3} - \frac{3x + 2}{6} = \frac{1}{2}$.

7.5 Graphing Linear Inequalities

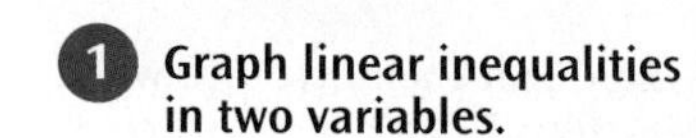

1 Graph linear inequalities in two variables.

1 Graph Linear Inequalities in Two Variables

In Section 2.8, we introduced inequalities in *one* variable. Now we introduce inequalities in *two* variables.

Understanding Algebra

The following symbols are used in linear inequalities:

- $<$ is less than
- $>$ is greater than
- $\leq$ is less than or equal to
- $\geq$ is greater than or equal to

Linear Inequality in Two Variables

A **linear inequality in two variables** can be written in one of the following forms:

$$ax + by < c, \quad ax + by > c, \quad ax + by \leq c, \quad ax + by \geq c$$

where a, b, and c are real numbers and a and b are not both 0.

Examples of Linear Inequalities in Two Variables

$2x + 3y > 2$	$3y < 4x - 9$	$y > 2x - 5$
$-x - 2y \leq 3$	$5x \geq 2y - 7$	$y \leq -\frac{2}{3}x + 1$

Consider the graph of the equation $x + y = 3$ shown in **Figure 7.43**. The line acts as a **boundary** between two *half-planes* and divides the plane into three distinct sets of points: the points on the line itself, the points in the half-plane on one side of the line, and the points in the half-plane on the other side of the line.

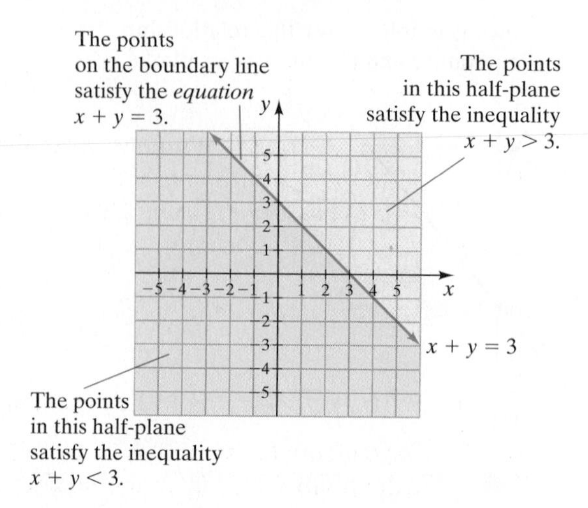

FIGURE 7.43

When we graph linear inequalities, we generally will shade only one of the two half-planes established by the boundary line. If the inequality contains $<$ or $>$, we draw a dashed line to indicate that the boundary line is not part of the solution. If the inequality contains $\leq$ or $\geq$, we will draw a solid line to indicate that the boundary line is part of the solution.

Understanding Algebra

If a linear inequality contains

- $<$ or $>$, draw a *dashed* boundary line.
- $\le$ or $\ge$, draw a *solid* boundary line.

To Graph a Linear Inequality in Two Variables

1. To obtain the equation of the boundary line, replace the inequality symbol with an equals sign.
2. Draw the graph of the equation in step 1. If the original inequality contains a $\ge$ or $\le$ symbol, draw the boundary line using a solid line. If the original inequality contains a $>$ or $<$ symbol, draw the boundary line using a dashed line.
3. Select any point not on the boundary line and determine if this point is a solution to the original inequality. If the point selected is a solution, shade the half-plane on the side of the line containing this point. If the selected point does not satisfy the inequality, shade the half-plane on the side of the line not containing this point.

EXAMPLE 1 Graph the inequality $y < 2x - 4$.

Solution First we graph the boundary line by graphing the equation $y = 2x - 4$ (**Fig. 7.44**). Since the original inequality contains the symbol $<$, we use a dashed line when drawing the boundary line. The dashed line indicates that the points on this line are not solutions to the inequality $y < 2x - 4$.

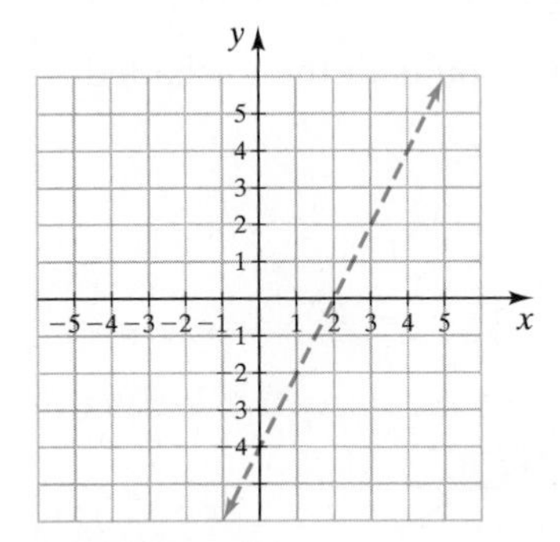

FIGURE 7.44

Next we select a point not on the line and determine whether this point satisfies the inequality. Often the easiest point to use is the origin, $(0, 0)$. In the check we will use the symbol $\overset{?}{<}$ until we determine whether the statement is true or false.

$$y < 2x - 4$$

Checkpoint

$$0 \overset{?}{<} 2(0) - 4$$

$$0 \overset{?}{<} 0 - 4$$

$$0 < -4 \quad \text{False}$$

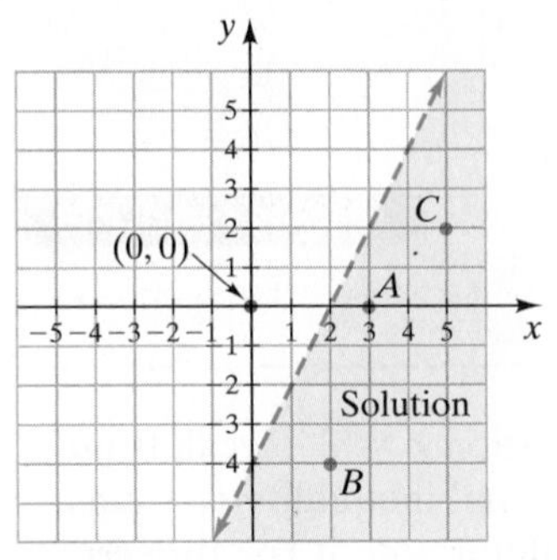

FIGURE 7.45

Since 0 is not less than -4, the ordered pair $(0, 0)$ does not satisfy the inequality. The solution will therefore be all the points in the half-plane on the opposite side of the line from the point $(0, 0)$. We shade this region (**Fig. 7.45**).

Every point in the shaded region satisfies the given inequality. Let's check a few selected points A, B, and C.

Point A	Point B	Point C
$(3, 0)$	$(2, -4)$	$(5, 2)$
$y < 2x - 4$	$y < 2x - 4$	$y < 2x - 4$
$0 \overset{?}{<} 2(3) - 4$	$-4 \overset{?}{<} 2(2) - 4$	$2 \overset{?}{<} 2(5) - 4$
$0 < 2$ True	$-4 < 0$ True	$2 < 6$ True

All points in the shaded region in **Figure 7.45** satisfy the inequality $y < 2x - 4$. The points in the unshaded region as well as the points on the boundary line itself do not satisfy the inequality $y < 2x - 4$.

Now Try Exercise 5

EXAMPLE 2 Graph the inequality $y \ge -\frac{1}{2}x$.

Solution We graph the boundary line by graphing the equation $y = -\frac{1}{2}x$. Since the inequality symbol is $\ge$, we will use a solid line (**Fig. 7.46** on the next page). Since the point $(0, 0)$ is on the line, we cannot select it as our test point. Let's select the point $(3, 1)$.

$$y \geq -\frac{1}{2}x$$

Checkpoint $$1 \overset{?}{\geq} -\frac{1}{2}(3)$$

$$1 \geq -\frac{3}{2} \quad \text{True}$$

Since the ordered pair $(3, 1)$ satisfies the inequality, every point in the half-plane on the same side of the line as $(3, 1)$ will also satisfy the inequality $y \geq -\frac{1}{2}x$. We shade this region (**Fig. 7.47**). Every point in the shaded region as well as every point on the boundary line satisfies the inequality $y \geq -\frac{1}{2}x$. The points in the unshaded region do not satisfy the inequality.

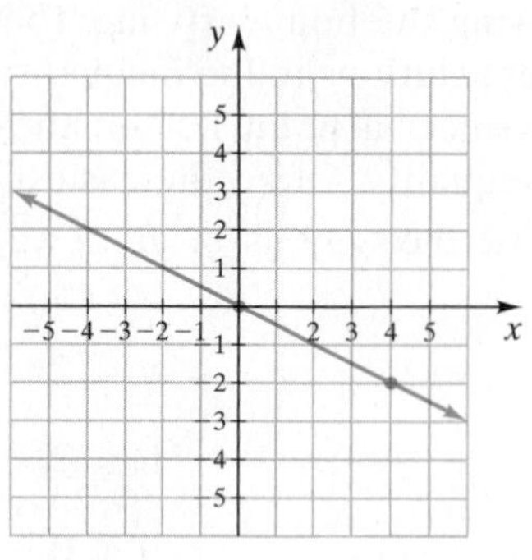

FIGURE 7.46

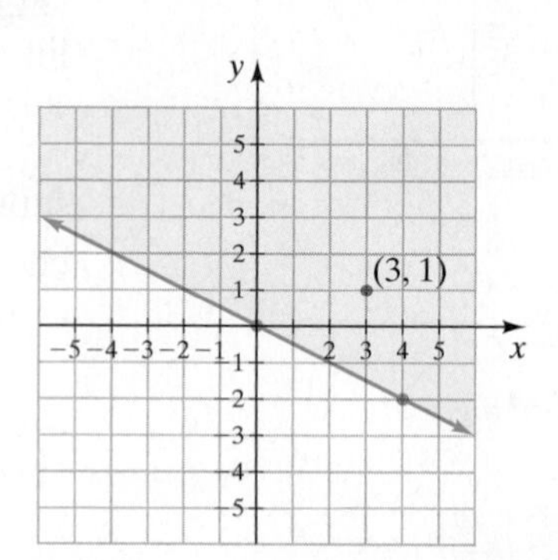

FIGURE 7.47

Now Try Exercise 13

EXAMPLE 3 Graph the inequality $x < 3$.

Solution We graph the boundary line by graphing the equation $x = 3$. Recall from Section 7.2 that $x = 3$ is a vertical line. Also, since the original inequality contains the symbol $<$, we use a dashed line (**Fig. 7.48**). Next, we determine if the ordered pair $(0, 0)$ satisfies the inequality $x < 3$. Since the original inequality does not contain the variable y, we only replace the x with the number 0.

$$x < 3$$

Checkpoint $$0 < 3 \quad \text{True}$$

Since the ordered pair $(0, 0)$ satisfies the inequality, every point in the half-plane on the same side of the boundary line as $(0, 0)$ will also satisfy the inequality $x < 3$. We shade this region (**Fig. 7.49**). Every point in the shaded region satisfies the inequality $x < 3$. The points in the unshaded region and the points on the boundary line itself do not satisfy the inequality $x < 3$.

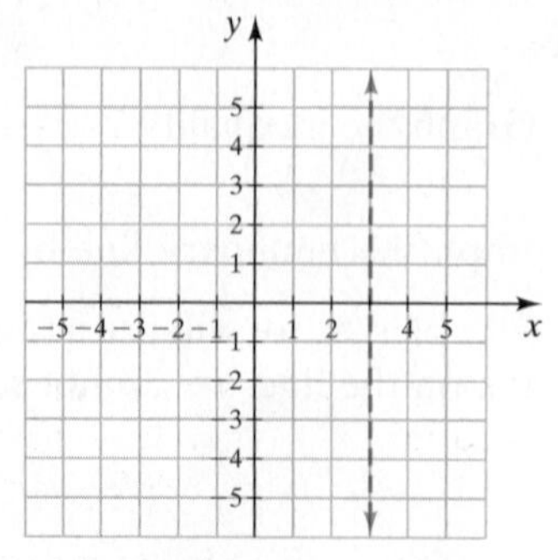

FIGURE 7.48

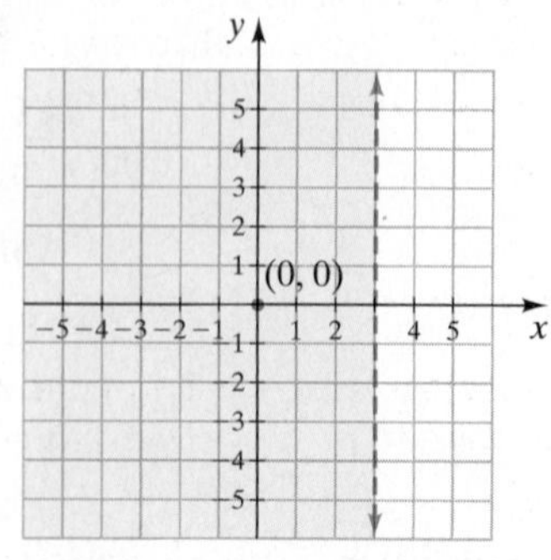

FIGURE 7.49

Now Try Exercise 23

HELPFUL HINT

Earlier in this chapter we discussed three methods of graphing linear equations: (1) plotting points, (2) using the x- and y-intercepts, and (3) using the slope and y-intercept. You may use any of these methods to graph the corresponding equations when graphing inequalities. Example 10 on page 448 provides a review of each of the methods.

7.5 Exercise Set MathXL® MyMathLab®

Warm-Up Exercises

Fill in the blanks with the appropriate word, phrase, or symbol(s) from the following list.

contains does not contain dashed solid solutions

1. When graphing an inequality with the symbol $\leq$ or $\geq$, use a ________ boundary line.

2. When graphing an inequality with the symbol $<$ or $>$, use a ________ boundary line.

3. The points in the shaded region of the graph of an inequality represent ________ to the inequality.

4. If the checkpoint satisfies the inequality, shade the region that ________ the point.

Practice the Skills

Graph each inequality.

5. $y < x - 4$
6. $y < 2x + 1$
7. $y \geq \frac{1}{2}x - 4$
8. $y \geq 2x - 3$
9. $y < -3x + 4$
10. $y > -\frac{x}{2} + 2$
11. $y > \frac{1}{2}x - 2$
12. $y > \frac{1}{3}x + 1$
13. $y \leq 3x$
14. $y > -2x$
15. $y < -\frac{2}{3}x$
16. $y \geq \frac{3}{4}x$
17. $2x + y \leq 3$
18. $x + y > -2$
19. $2x - 3y > 6$
20. $3x + 4y \leq -12$
21. $y > -3$
22. $y \leq \frac{7}{2}$
23. $x \geq \frac{3}{2}$
24. $x < -1$

Problem Solving

25. Determine whether $(4, 2)$ is a solution to each inequality.

a) $2x + 4y < 16$
b) $2x + 4y > 16$
c) $2x + 4y \geq 16$
d) $2x + 4y \leq 16$

26. Determine whether $(-3, 7)$ is a solution to each inequality.

a) $-2x + 3y < 9$
b) $-2x + 3y > 9$
c) $-2x + 3y \geq 9$
d) $-2x + 3y \leq 9$

27. Determine whether the given phrase means: less than, less than or equal to, greater than, or greater than or equal to.

a) no more than
b) no less than
c) at most
d) at least

28. Consider the two inequalities $2x + 1 > 5$ and $2x + y > 5$.

a) How many variables does the inequality $2x + 1 > 5$ contain?

b) How many variables does the inequality $2x + y > 5$ contain?

c) What is the solution to $2x + 1 > 5$? Indicate the solution on a number line.

d) Graph $2x + y > 5$.

Concept/Writing Exercises

29. When graphing inequalities that contain either $\leq$ or $\geq$, explain why the points on the line will be solutions to the inequality.

30. When graphing inequalities that contain either $<$ or $>$, explain why the points on the line will not be solutions to the inequality.

31. If an ordered pair is not a solution to the inequality $ax + by < c$, must the ordered pair be a solution to $ax + by > c$? Explain.

32. If an ordered pair is not a solution to the inequality $ax + by \leq c$, must the ordered pair be a solution to $ax + by > c$? Explain.

33. If an ordered pair is a solution to $ax + by > c$, is it possible for the ordered pair to be a solution to $ax + by \leq c$? Explain.

34. Is it possible for an ordered pair to be a solution to both $ax + by < c$ and $ax + by > c$? Explain.

35. How do the graphs of $2x + 3y > 6$ and $2x + 3y < 6$ differ?

36. Which of the following inequalities have the same graphs? Explain how you determined your answer.

a) $2x - y > 4$ **b)** $-2x + y < -4$

c) $y < 2x - 4$ **d)** $-2y + 4x < -8$

Cumulative Review Exercises

[1.4] **37.** Consider the set of numbers

$$\left\{2, -5, 0, \sqrt{7}, \frac{2}{5}, -6.3, \sqrt{3}, -\frac{23}{34}\right\}$$

List those that are

a) natural numbers;

b) whole numbers;

c) rational numbers;

d) irrational numbers;

e) real numbers.

[2.5] **38.** Solve $3(x - 2) + 4x = 5x - 2$.

[4.6] **39.** Divide $\dfrac{10x^2 - 15x + 25}{5x}$.

[6.1] **40.** Simplify $\dfrac{3x}{3x^2 + 6xy}$.

7.6 Functions

1 Find the domain and range of a relation.

2 Recognize functions.

3 Evaluate functions.

4 Graph linear functions.

In this section we introduce relations and functions. A function is a special type of relation. Functions are a common thread in mathematics courses from algebra through calculus. In this section we give an informal introduction to relations and functions.

1 Find the Domain and Range of a Relation

First we will discuss **relations.**

Relation

A **relation** is any set of ordered pairs.

Since a relation is *any* set of points, *every graph will represent a relation.*

Examples of Relations

$\{(2, 5), (4, 6), (5, 9), (7, 12)\}$

$\{(1, 2), (2, 2), (3, 2), (4, 2)\}$

$\{(3, 2), (3, 3), (3, 4), (3, 5), (3, 6)\}$

In the ordered pair (x, y), the x and y are called the **components of the ordered pair**. The **domain** of a relation is the set of *first components* in the set of ordered pairs. For example,

Relation	Domain
$\{(2, 5), (4, 6), (5, 9), (7, 12)\}$	$\{2, 4, 5, 7\}$
$\{(1, 2), (2, 2), (3, 2), (4, 2)\}$	$\{1, 2, 3, 4\}$
$\{(3, 2), (3, 3), (3, 4), (3, 5), (3, 6)\}$	$\{3\}$

The **range** of a relation is the set of *second components* in the set of ordered pairs. For example,

Relation	Range
$\{(2, 5), (4, 6), (5, 9), (7, 12)\}$	$\{5, 6, 9, 12\}$
$\{(1, 2), (2, 2), (3, 2), (4, 2)\}$	$\{2\}$
$\{(3, 2), (3, 3), (3, 4), (3, 5), (3, 6)\}$	$\{2, 3, 4, 5, 6\}$

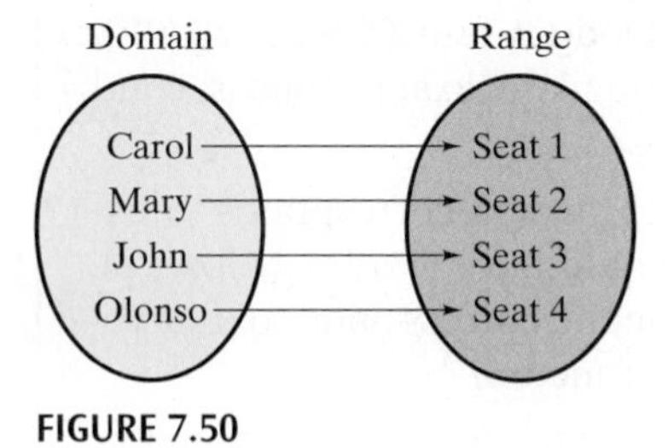

FIGURE 7.50

In relations, the sets can contain elements other than numbers. For example,

Relation

$$\{(\text{Carol, Seat } 1), (\text{Mary, Seat } 2), (\text{John, Seat } 3), (\text{Olonso, Seat } 4)\}$$

Domain

$$\{\text{Carol, Mary, John, Olonso}\}$$

Range

$$\{\text{Seat } 1, \text{Seat } 2, \text{Seat } 3, \text{Seat } 4\}$$

Figure 7.50 illustrates the relation between the person and the seat number.

2 Recognize Functions

Consider the relation shown in **Figure 7.50**. Notice that each member in the domain corresponds with exactly one member of the range. That is, each person is assigned to exactly one seat. This is an example of a **function**.

Function

A **function** is a relation in which each first component corresponds to exactly one second component.

Since a function is a special type of relation, our discussion of domain and range also applies to functions. In the definition of a function, *the set of first components represents the domain of the function and the set of second components represents the range of the function.*

EXAMPLE 1 Consider the relations in **Figures 7.51a** through **d**. Which relations are functions?

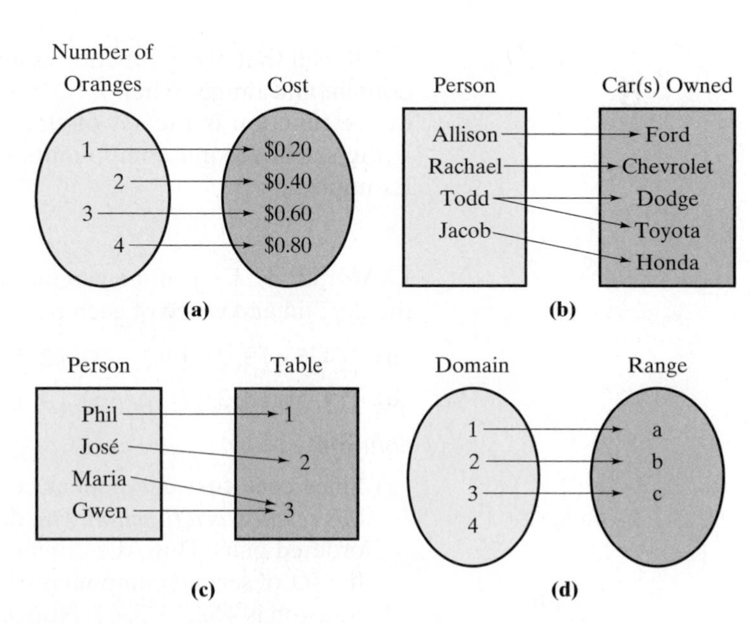

FIGURE 7.51

Solution

a) If we wished, we could represent the information given in the figure as the following set of ordered pairs: $\{(1, \$0.20), (2, \$0.40), (3, \$0.60), (4, \$0.80)\}$. Notice that each first component corresponds to exactly one second component. Therefore, this relation is a function.

b) If we look at **Figure 7.51b**, we can see that Todd does *not* correspond to exactly one car. If we were to list the set of ordered pairs to represent this relation,

Understanding Algebra

It is important to realize that *any* set of ordered pairs is a relation. Therefore, *every function is also a relation*. However, *not every relation is a function.*

the set would contain the ordered pairs (Todd, Dodge) and (Todd, Toyota). Therefore, each first component does not correspond to exactly one second component and this relation is not a function.

c) Although both Maria and Gwen share a table, each person corresponds to exactly one table. If we listed the ordered pairs we would have (Phil, 1), (José, 2), (Maria, 3), (Gwen, 3). Note that each *first* component corresponds to exactly one second component. Therefore, this relation is a function.

d) Since the number 4 in the domain does not correspond to any component in the range, this is not a function. *Every* component in the domain must correspond to exactly one component in the range for it to be a function.

Now Try Exercise 19

The functions given in Example 1 **a)** and 1 **c)** were determined by looking at correspondences in figures. Most functions have an infinite number of ordered pairs and are usually defined with an equation (or rule) that tells how to obtain the second component when you are given the first component. In Example 2, we determine a function from the information provided.

EXAMPLE 2 **Cost of Matchbox Cars** At a toy store, matchbox cars cost \$0.99 each. Write a function to determine the cost, c, when n matchbox cars are purchased.

Solution When one car is purchased, the cost is \$0.99. When two cars are purchased, the cost is $2(\$0.99)$, and when n cars are purchased, the cost is $n(\$0.99)$ or $\$0.99n$. The function $c = 0.99n$ will give the cost, c, in dollars, when n cars are purchased. Note that for any value of n, there is exactly one value of c.

Now Try Exercise 51

Recall that every function is also a relation. Therefore, every function also has a domain and range. When a function is written as a set of ordered pairs the domain of the function is the set of first components in the ordered pairs, and the range is the set of second components in the ordered pairs. We will demonstrate this in Example 3.

EXAMPLE 3 Determine whether the following relations are functions. Then state the domain and range of each relation or function.

a) $\{(4, 5), (3, 2), (-2, -3), (2, 5), (1, 6)\}$

b) $\{(4, 5), (3, 2), (-2, -3), (4, 1), (5, -2)\}$

Solution

a) Since each first component corresponds with exactly one second component, *this relation is a function.* The domain is the set of first components in the set of ordered pairs. Thus, the domain of the function is $\{4, 3, -2, 2, 1\}$. The range is the set of second components in the set of ordered pairs. Thus, the range of the function is $\{5, 2, -3, 6\}$. Notice that when listing the range, we only include the number 5 once, even though it appears in both $(4, 5)$ and $(2, 5)$.

b) The ordered pairs $(4, 5)$ and $(4, 1)$ contain the same first component, 4, but different second components, 5 and 1, respectively. Therefore, each first component does not correspond to exactly one second component and *this relation is not a function.* The domain of the relation is $\{4, 3, -2, 5\}$. The range of the relation is $\{5, 2, -3, 1, -2\}$.

Now Try Exercise 13

In **Figures 7.52a** and **7.52b** we plot the ordered pairs from Example 3 **a)** and 3 **b)**, respectively.

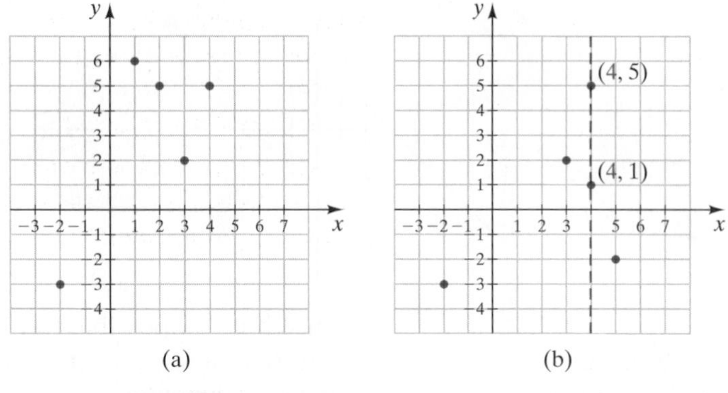

FIGURE 7.52 (a) Function (b) Not a function

The graphs in **Figure 7.52** are sets of points. *Every* graph consists of a set of points, and every point on a graph corresponds to an ordered pair. Thus, we can think of any graph as a set of ordered pairs and therefore as a relation. However, only certain graphs will represent functions.

Consider **Figure 7.52a**. If a vertical line is drawn through each point, no vertical line intersects more than one point. This indicates that each value of x in the domain corresponds to exactly one value of y in the range. Therefore, this set of points represents a function.

Now look at **Figure 7.52b**. The dashed red vertical line intersects both $(4, 5)$ and $(4, 1)$. Each element in the domain *does not* correspond to exactly one element in the range. Therefore, this set of ordered pairs does *not* represent a function.

To determine whether a graph represents a function, we can use the **vertical line test.**

Vertical Line Test

If a vertical line can be drawn so that it intersects a graph at more than one point, then the graph does not represent a function.

When a relation or a function is represented by a graph, the domain is the set of x-values that correspond to points on the graph. The range is the set of y-values that correspond to points on the graph. Recall that every graph represents a relation, but only certain graphs represent functions. We will use the vertical line test in Example 4 to determine whether a graph also represents a function. In Example 4 we will also use interval notation to describe the domain and range of graphs. We discussed interval notation in Section 2.8. You may wish to review this topic now.

EXAMPLE 4 The following graphs represent relations. Use the vertical line test to determine which graphs also represent functions. Then state the domain and range of each relation or function.

a)

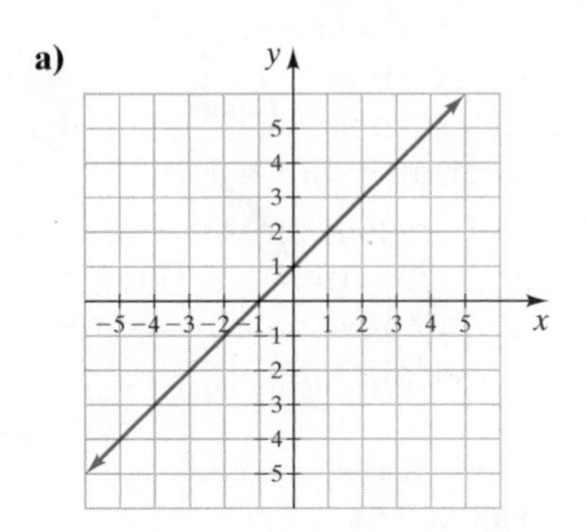

b)

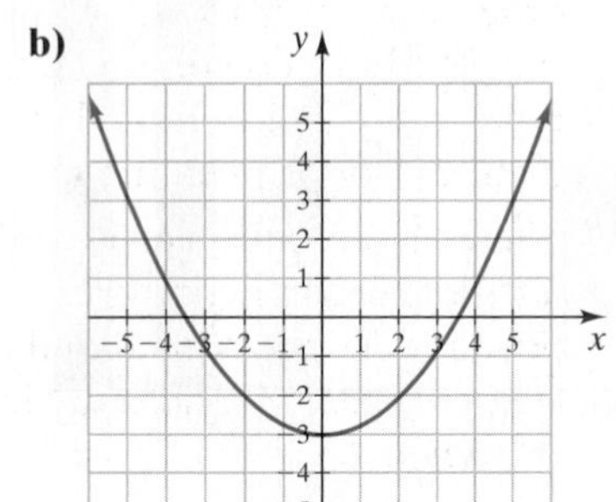

c)

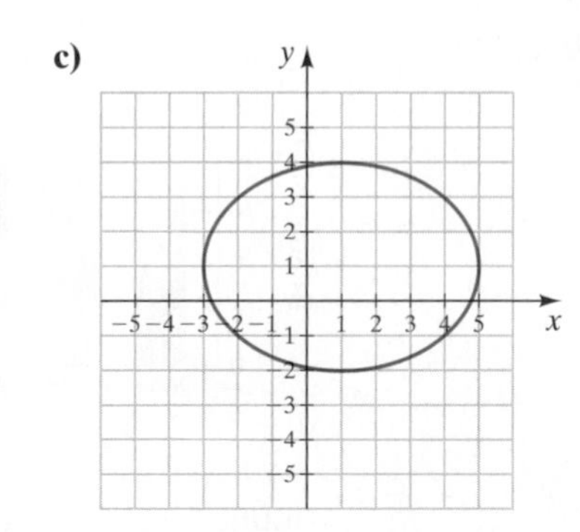

Solution

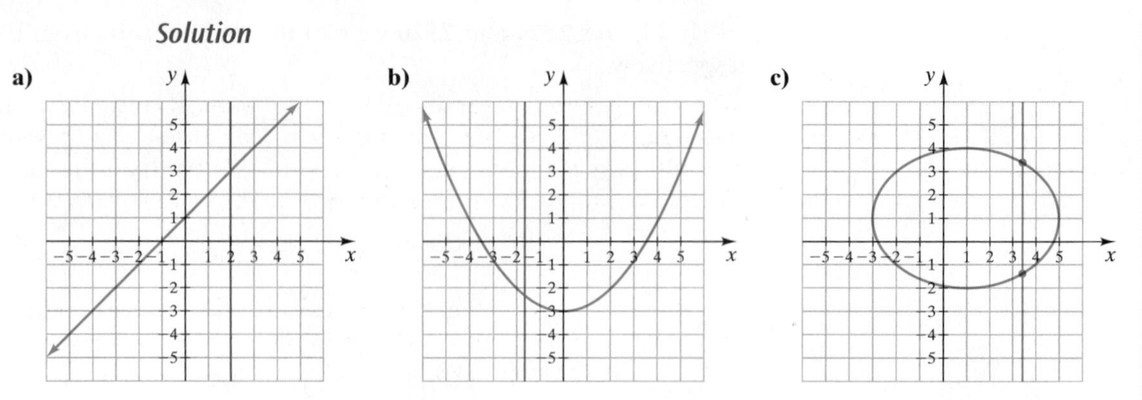

a) Since it is not possible to draw a vertical line that intersects the graph at more than one point, this graph represents a function. The domain is the set of the first components, or x-coordinates, of the ordered pairs that correspond to points on the graph. Since the graph continues indefinitely to the left and to the right, the domain is all real numbers. Using interval notation, the domain is $(-\infty, \infty)$. The range is the set of all of the second components, or y-coordinates, of the ordered pairs that correspond to points on the graph. Since the graph continues indefinitely downward and upward, the range is all real numbers. Using interval notation, the range is $(-\infty, \infty)$.

b) Since it is not possible to draw a vertical line that intersects the graph at more than one point, this graph represents a function. Since the graph continues indefinitely to the left and to the right, the domain is $(-\infty, \infty)$. The set of y-coordinates for this graph go from -3 and continue indefinitely upward. Therefore, the range is $[-3, \infty)$.

c) Since a vertical line can be drawn that intersects the graph at more than one point, this graph does not represent a function. The domain is the set of the x-coordinates of the ordered pairs that correspond to points on the graph. Therefore, the domain is $[-3, 5]$. The range is the set of y-coordinates of the ordered pairs that correspond to points on the graph. Therefore, the range is $[-2, 4]$.

Now Try Exercise 31

Understanding Algebra

Every graph consists of a set of points. Each of these points corresponds to an ordered pair. Therefore, every graph represents a set of ordered pairs and every graph represents a relation. However, only the graphs that pass the vertical line test represent functions.

TABLE 7.1 Monthly Income

Sales (dollars)	Income (dollars)
0	$1500
$5000	$1800
$10,000	$2100
$15,000	$2400
$20,000	$2700
$25,000	$3000

Consider the information provided in **Table 7.1**. This table of values is a function because each amount of sales corresponds to exactly one income.

Consider the graph shown in **Figure 7.53**. This graph represents a function. Notice that each year corresponds to exactly one value of exports to China. Both graphs shown in **Figure 7.54**, represent functions. Each individual graph passes the vertical line test.

FIGURE 7.53

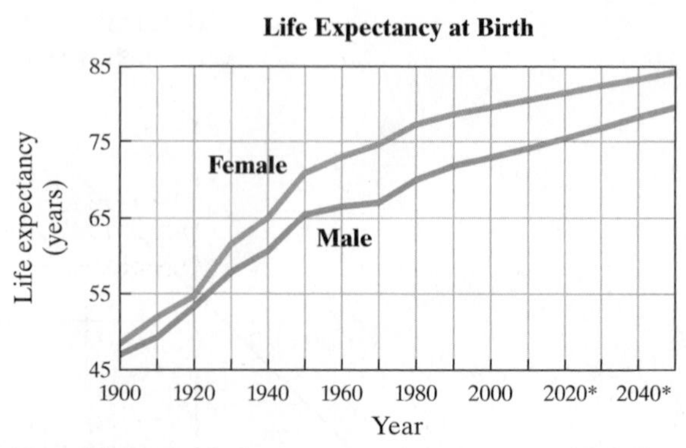

FIGURE 7.54

Understanding Algebra

The notation $y = f(x)$, pronounced "f of x," is used to show that y is a function of the variable x. It does *not* mean f times x.

3 Evaluate Functions

When a function is represented by an equation, it is often convenient to use **function notation**, $f(x)$. If we were to graph $y = x + 2$, we would see that it is a function because its graph passes the vertical line test. The value of y *depends on* the value of x. Therefore, we say that *y is a function of x*, and we can write $y = f(x)$. We can write

$$y = f(x) = x + 2 \quad \text{or simply} \quad f(x) = x + 2$$

To evaluate a function for a specific value of x, we substitute that value for x everywhere the x appears in the function. For example, to evaluate the function $f(x) = x + 2$ at $x = 1$, we do the following:

$$f(x) = x + 2$$
$$f(1) = 1 + 2 = 3$$

Thus, when x is 1, $f(x)$ or y is 3.

When $x = 4$, $f(x)$ or $y = 6$, as illustrated below.

$$y = f(x) = x + 2$$
$$y = f(4) = 4 + 2 = 6$$

The notation $f(1)$ is read "f of 1" and $f(4)$ is read "f of 4."

EXAMPLE 5 For the function $f(x) = x^2 + 4x - 9$, find **a)** $f(3)$ and **b)** $f(-6)$. **c)** If $x = -1$, determine the value of y.

Solution

a) Substitute 3 for each x in the function, and then evaluate.

$$f(x) = x^2 + 4x - 9$$
$$f(3) = (3)^2 + 4(3) - 9$$
$$= 9 + 12 - 9 = 12$$

b)

$$f(x) = x^2 + 4x - 9$$
$$f(-6) = (-6)^2 + 4(-6) - 9$$
$$= 36 - 24 - 9 = 3$$

c) Since $y = f(x)$, we evaluate $f(x)$ at -1.

$$f(x) = x^2 + 4x - 9$$
$$f(-1) = (-1)^2 + 4(-1) - 9$$
$$= 1 - 4 - 9 = -12$$

Thus, when $x = -1$, $y = -12$.

Now Try Exercise 39

4 Graph Linear Functions

The graphs of all equations of the form $y = ax + b$ will be straight lines that are functions. Therefore, we may refer to equations of the form $y = f(x) = ax + b$ as **linear functions.**

EXAMPLE 6 Graph $f(x) = 2x + 4$.

Solution Since $f(x)$ is the same as y, write $y = f(x) = 2x + 4$. Select values for x and find the corresponding values for y or $f(x)$.

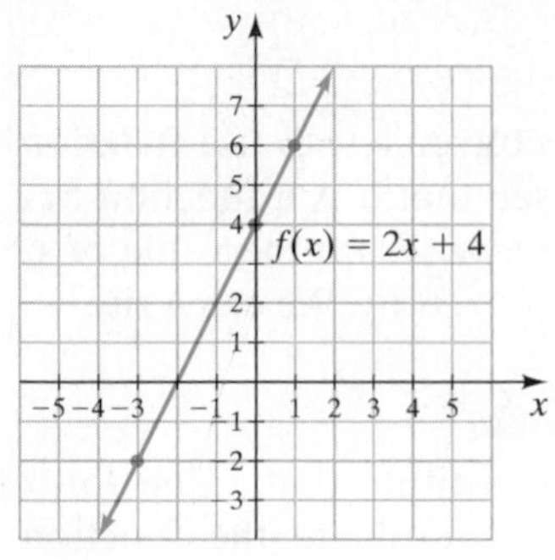

FIGURE 7.55

$$y = f(x) = 2x + 4$$

Let $x = -3$. $\quad y = f(-3) = 2(-3) + 4 = -2$

Let $x = 0$. $\quad y = f(0) = 2(0) + 4 = 4$

Let $x = 1$. $\quad y = f(1) = 2(1) + 4 = 6$

x	y
-3	-2
0	4
1	6

Now plot the points and draw the graph of the function (**Fig. 7.55**).

Now Try Exercise 45

EXAMPLE 7 Ice Skating The weekly profit, p, of an ice skate rental company is a function of the number of pairs of skates rented per week, n. The function approximating the profit is $p = f(n) = 8n - 600$, where $0 \le n \le 400$.

a) Construct a graph showing the relationship between the number of pairs of skates rented and the weekly profit.

b) Estimate the profit if there are 200 pairs of skates rented in a given week.

Rockefeller Plaza in New York City

Solution

a) Select values for n, and find the corresponding values for p. Then draw the graph (**Fig. 7.56**). Notice there are no arrowheads on the line because the function is defined only for values of n between 0 and 400 inclusive.

$$p = f(n) = 8n - 600$$

Let $n = 0$. $\quad p = f(0) = 8(0) - 600 = -600$

Let $n = 100$. $\quad p = f(100) = 8(100) - 600 = 200$

Let $n = 400$. $\quad p = f(400) = 8(400) - 600 = 2600$

n	p
0	-600
100	200
400	2600

FIGURE 7.56

b) Using the red dashed line on the graph, we can see that if there are 200 pairs of skates rented the weekly profit is \$1000.

Now Try Exercise 57

7.6 Exercise Set MathXL® MyMathLab®

Warm-Up Exercises

Fill in the blanks with the appropriate word, phrase, or symbol(s) from the following list.

relation function x-component domain horizontal range $f(x)$ vertical y-components

1. The ____________ consists of the x-components of a set of ordered pairs.
2. The range consists of the ____________ of a set of ordered pairs.
3. A ____________ is any set of ordered pairs.
4. A ____________ is a relation in which each first component corresponds to exactly one second component.

5. If a ____________ line intersects a graph in more than one point, then this graph does not represent a function.

6. The notation ____________ is pronounced "f of x."

7. In a function, every ____________ corresponds to exactly one y-component.

8. In a function, every element in the domain must correspond to exactly one element in the ____________.

Practice the Skills

Determine which of the relations are also functions. Then state the domain and range of each relation or function.

9. $\{(5,4),(2,2),(3,5),(1,3),(4,1)\}$

10. $\{(2,3),(4,0),(9,1),(3,2),(5,8)\}$

11. $\{(5,5),(3,0),(3,2),(1,4),(2,4),(7,-2)\}$

12. $\{(-2,1),(1,-3),(3,4),(4,5),(-2,0)\}$

13. $\{(5,0),(4,-4),(0,-1),(3,2),(1,1)\}$

14. $\{(-6,3),(-3,4),(0,3),(5,2),(3,5),(2,3)\}$

15. $\{(0,3),(1,3),(2,3),(3,3),(4,3)\}$

16. $\{(4,2),(3,2),(2,2),(1,2),(0,2)\}$

17. $\{(3,0),(3,1),(3,2),(3,3),(3,4)\}$

18. $\{(2,4),(2,3),(2,2),(2,1),(2,0)\}$

In the figures in Exercises 19–22, the domain and range of a relation are illustrated. **a)** *Construct a set of ordered pairs that represent the relation.* **b)** *Determine whether the relation is a function.*

19. Domain Range

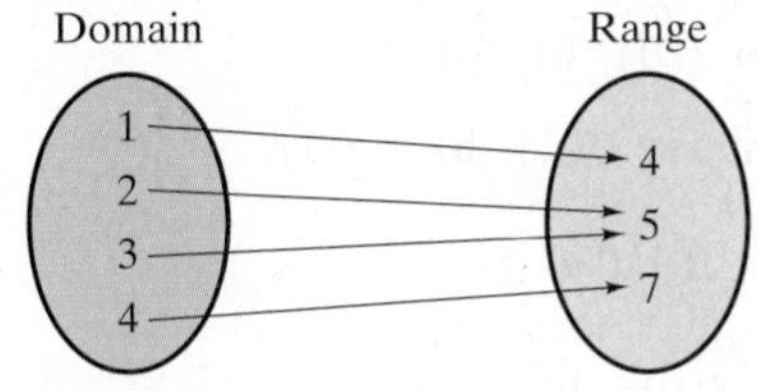

20. Domain Range

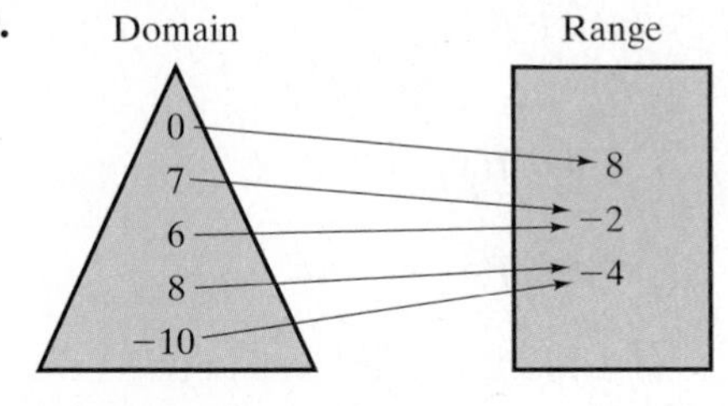

21. Domain Range

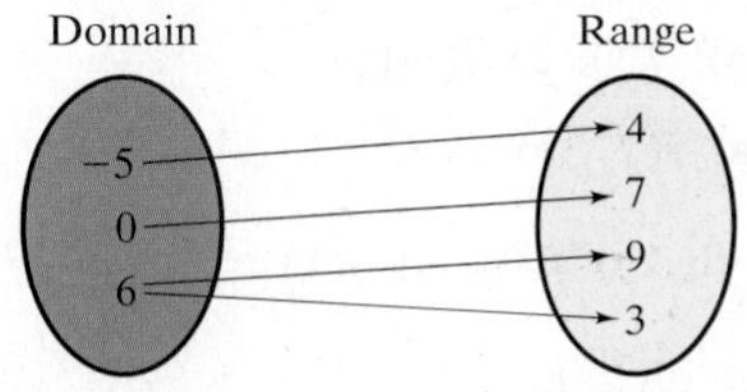

22. Domain Range

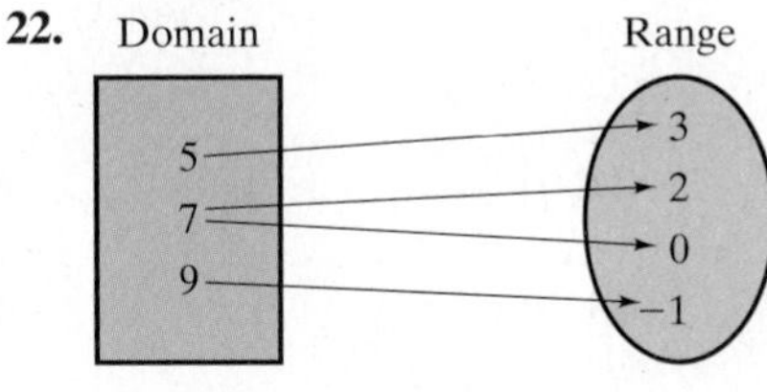

Use the vertical line test to determine whether each relation is also a function. Then state the domain and range of each relation or function.

23.

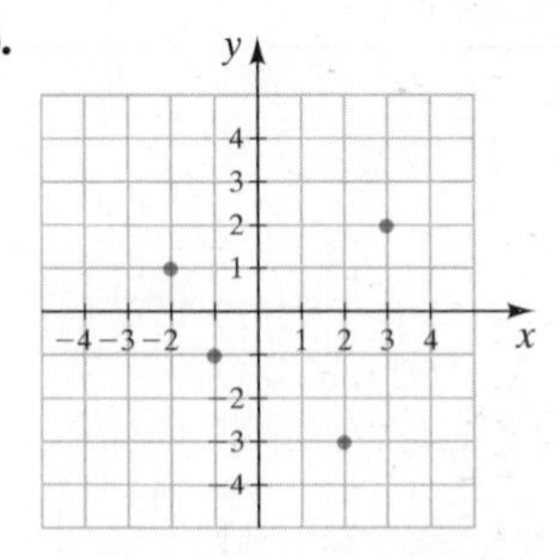

24.

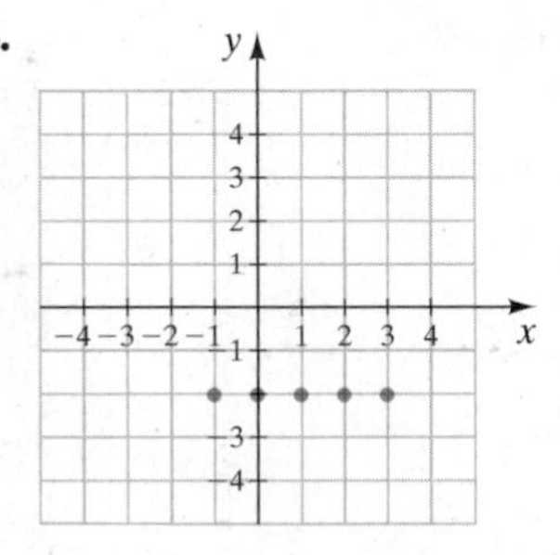

25.

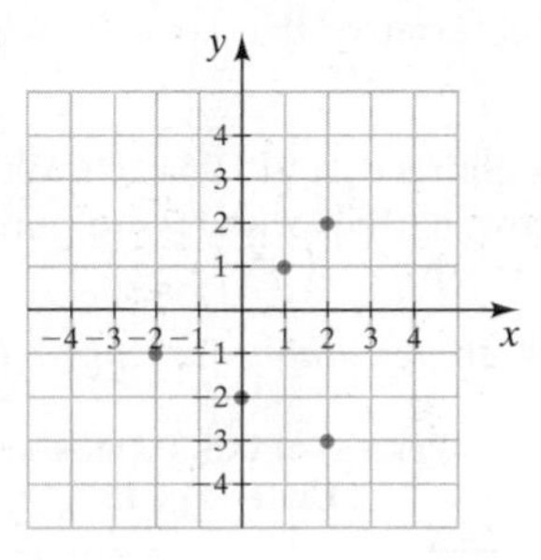

26.

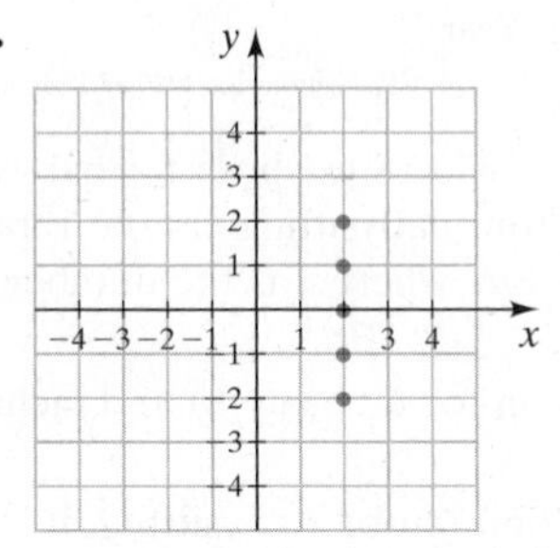

27.

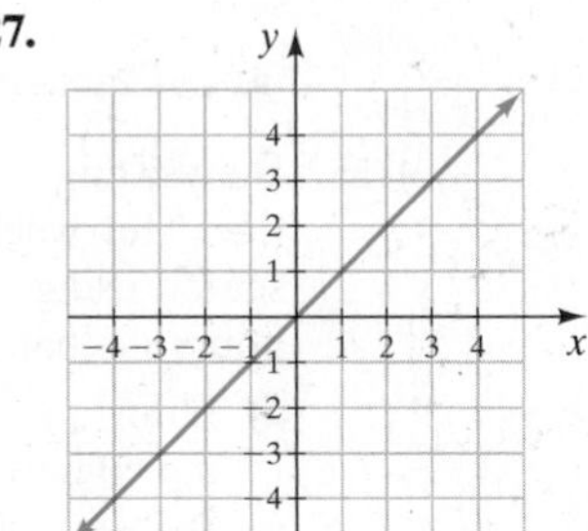

28.

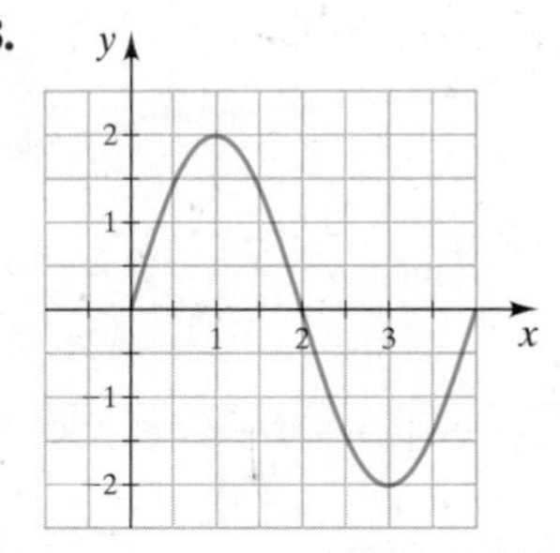

29.

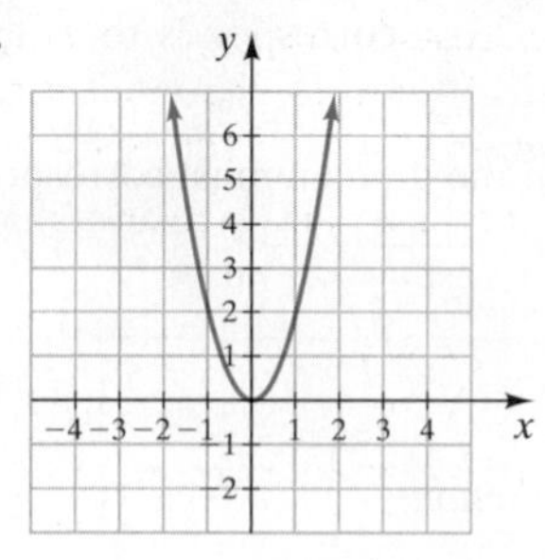

30.

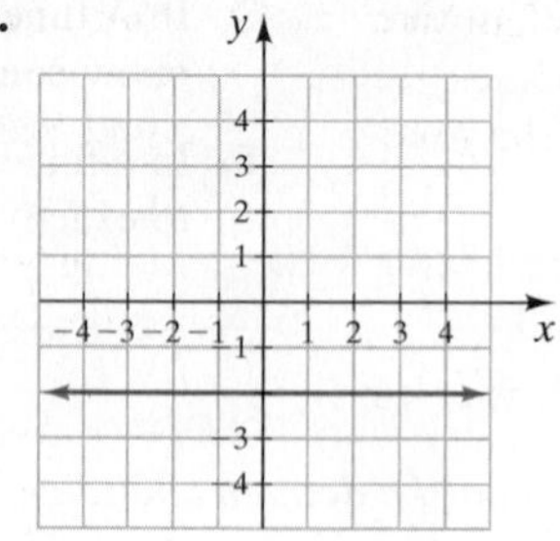

31.

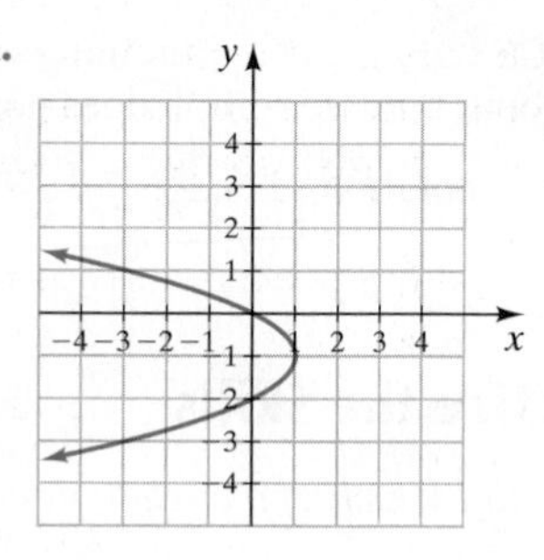

32.

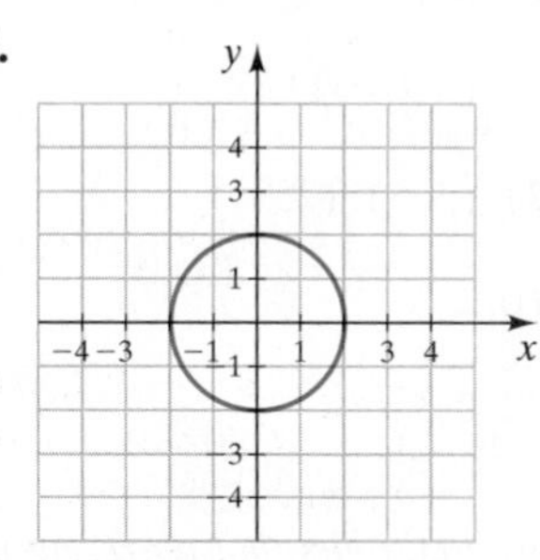

33.

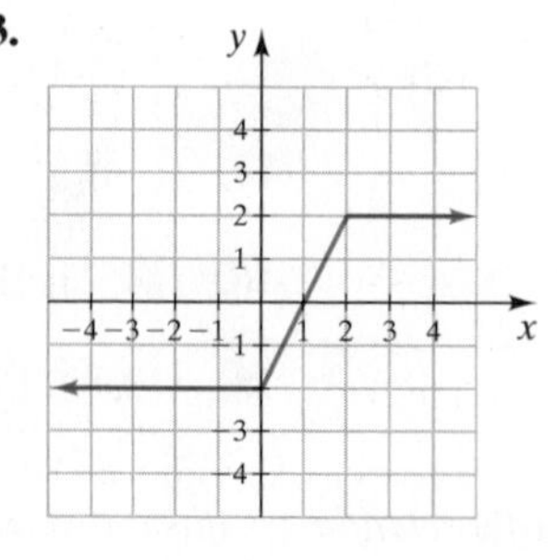

34.

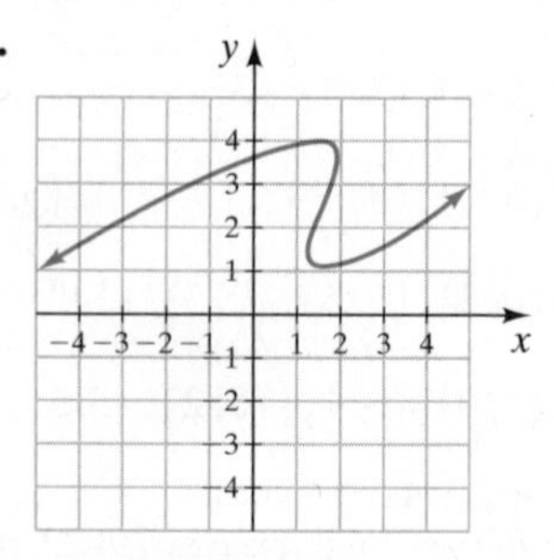

Evaluate each function at the indicated values.

35. $f(x) = 4x + 2$; find **a)** $f(3)$, **b)** $f(-1)$

36. $f(x) = -4x + 7$; find **a)** $f(0)$, **b)** $f(4)$,

37. $f(x) = x^2 + 4$; find **a)** $f(2)$, **b)** $f(-2)$

38. $f(x) = 2x^2 + 3x - 4$; find **a)** $f(2)$, **b)** $f(-3)$

39. $f(x) = 3x^2 - x + 4$; find **a)** $f(0)$, **b)** $f(1)$

40. $f(x) = \frac{1}{2}x^2 + 6$; find **a)** $f(4)$, **b)** $f(-6)$

41. $f(x) = \frac{x + 4}{2}$; find **a)** $f(2)$, **b)** $f(12)$

42. $f(x) = \frac{1}{2}x - 4$; find **a)** $f(10)$, **b)** $f(-8)$

Graph each function.

43. $f(x) = x + 3$

44. $f(x) = -x + 4$

45. $f(x) = -2x + 4$

46. $f(x) = 2x - 4$

47. $f(x) = -\frac{1}{2}x + 2$

48. $f(x) = \frac{2}{3}x + 1$

49. $f(x) = 2x$

50. $f(x) = -4x$

Problem Solving

51. Cost of Oranges Oranges cost \$0.35 each. Write a function to determine the cost, c, when n oranges are purchased.

52. Cost of Shirts Shirts cost \$19.95 each. Write a function to determine the cost, c, when n shirts are purchased.

In Exercises 53–54, are the graphs functions? Explain.

53.

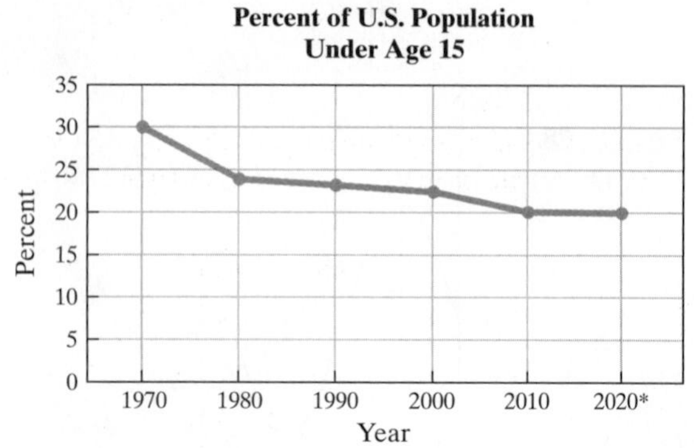

54.

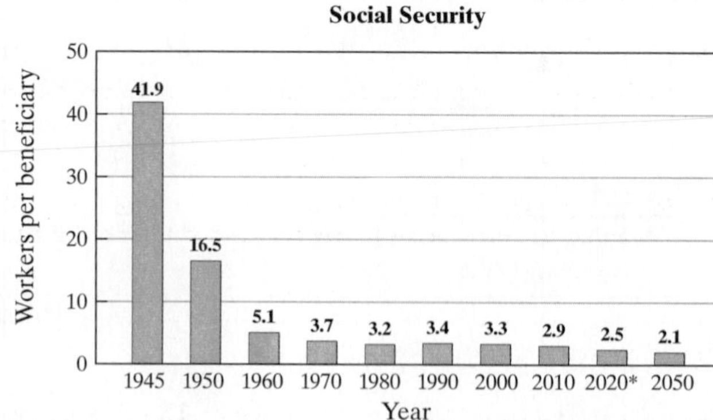

55. Babysitting Mary Vachon gets paid \$8 per hour for babysitting. Her weekly income, I, from babysitting can be represented using the function $I = 8h$, where h is the number of hours she babysits.

a) Draw a graph of the function for times up to and including 60 hours.

b) Estimate her weekly income from babysitting if she babysits for 20 hours per week.

56. Bike Riding Ronald Yates and Andrea Vorwalk go bike riding together. If they ride at a constant rate of 8 miles per hour, the distance they travel, d, can be found by the function $d = 8t$, where t is time in hours.

a) Draw a graph of the function for times up to and including 6 hours.

b) Estimate the distance they will travel if they ride for 4 hours.

57. Selling a House The Ferreras are selling their house. The cost they will pay to the realtor is 6% of the selling price of the house, p. They expect to have \$2000 in other selling costs. Therefore, the total cost, C, to the Ferreras for selling their house can be represented by the function

$$C = 2000 + 0.06p$$

a) Draw a graph of the function for selling prices from \$0 up to and including \$200,000.

b) Estimate the total cost if the selling price is \$150,000.

58. Vacationing Daniel Fahringer plans to take a vacation in the San Diego/Los Angeles area. He determines that the cost, C, of the vacation can be estimated using the function $C = 350n + 400$, where n is the number of days spent in the area.

a) Draw a graph of the function for up to and including 10 days.

b) Estimate the cost of a 5-day vacation in the area.

Los Angeles, CA

59. Photo Printer A photo printer costs \$120. The cost per picture printed, n, is 40 cents each. Therefore, the total cost, C, of the printer plus the photos printed is $C = 120 + 0.40n$.

a) Draw a graph showing the total cost for up to and including 500 photos.

b) If 300 photos are printed, estimate the total cost.

60. Auto Registration A state's auto registration fee, f, is \$20 plus \$15 per 1000 pounds of the vehicle's gross weight. The registration fee is a function of the vehicle's weight, $f = 20 + 0.015w$, where w is the weight of the vehicle in pounds.

a) Draw a graph of the function for vehicle weights up to and including 10,000 pounds.

b) Estimate the registration fee of a vehicle whose gross weight is 6000 pounds.

61. Singing Sensation A new singing group, Three Forks and a Spoon, signs a recording contract with the Smash Record label. The contract provides a signing bonus of \$10,000, plus an 8% royalty on the sales, s, of the group's new record, *There's Mud in Your Eye!* The group's income, i, is a function of its sales, $i = 10{,}000 + 0.08s$.

a) Draw a graph of the function for sales of up to and including \$100,000.

b) Estimate the group's income if its sales are \$20,000.

62. Electric Bill A monthly electric bill, m, in dollars, consists of a \$20 monthly fee plus \$0.07 per kilowatt-hour, k, of electricity used. The amount of the bill is a function of the kilowatt-hours used, $m = 20 + 0.07k$.

a) Draw a graph for up to and including 3000 kilowatt-hours of electricity used in a month.

b) Estimate the bill if 2100 kilowatt-hours of electricity are used.

Concept/Writing Exercises

63. a) If two distinct ordered pairs in a relation have the same first coordinate, can the relation be a function? Explain.

b) In a function is it necessary for each value of y in the range to correspond to exactly one value of x in the domain? Explain.

64. a) If a relation consists of six ordered pairs and the domain of the relation consists of five values of x, can the relation be a function? Explain.

b) If a relation consists of six ordered pairs and the range of the relation consists of five values of y, can the relation be a function? Explain.

Challenge Problems

Consider the following graphs. Recall from Section 2.8 that an open circle at the end of a line segment means that the endpoint is not included in the answer. A solid circle at the end of a line segment indicates that the endpoint is included in the answer. Determine whether the following graphs are functions. Explain your answer.

65.

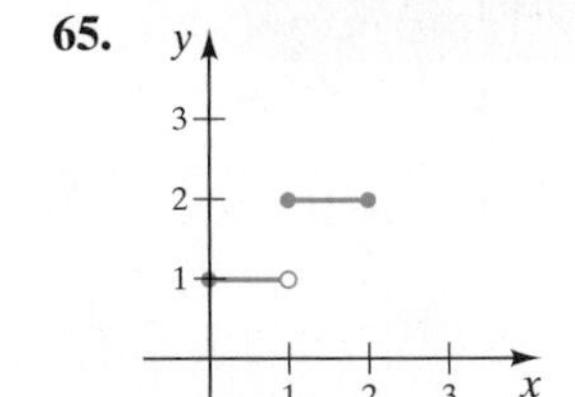

66.

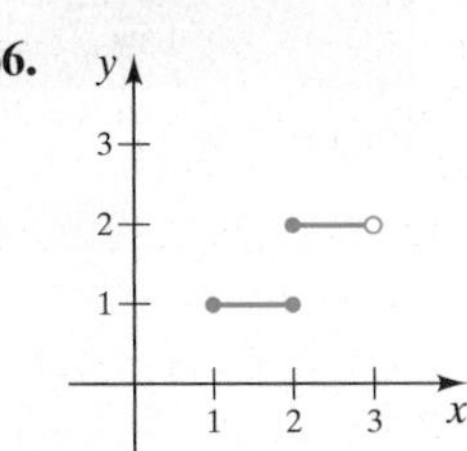

67.

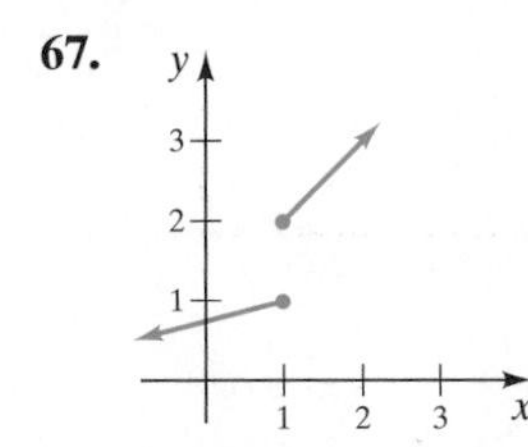

68.

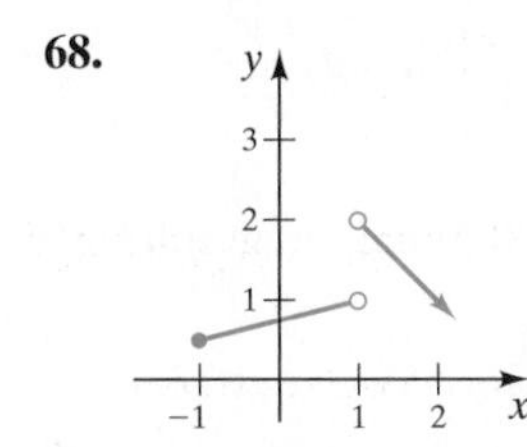

69. $f(x) = \frac{1}{2}x^2 - 3x + 5$; find

a) $f\left(\frac{1}{2}\right)$,

b) $f\left(\frac{2}{3}\right)$,

c) $f(0.4)$.

70. $f(x) = x^2 + 2x - 4$; find

a) $f(1)$,

b) $f(2)$,

c) $f(a)$.

Explain how you determined your answer to part **c)**.

Group Activity

Discuss and answer Exercises 71 and 72 as a group.

71. Submit three real-life examples (different from those already given) of a quantity that is a function of another. Write each as a function, and indicate what each variable represents.

72. Postage The graph on the right shows the cost of mailing a first-class letter on January 1, 2013. The x-axis shows the weight of the letter and the y-axis shows the cost of mailing the letter.

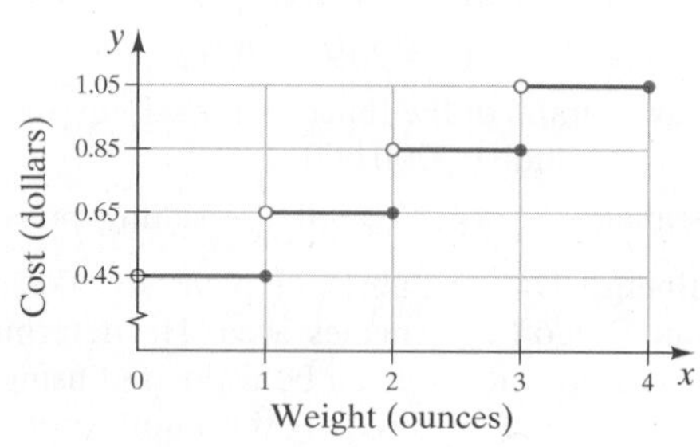

Source: United States Postal Service

a) Does the graph represent a function? Explain your answer.

b) Use the graph to determine the cost of mailing a letter that weighs 0.5 ounces.

c) Use the graph to determine the cost of mailing a letter that weighs 1.5 ounces.

d) Use the graph to determine the cost of mailing a letter that weighs 2.5 ounces.

e) Use the graph to determine the cost of mailing a letter that weighs 3.25 ounces.

Cumulative Review Exercises

[1.3] **73.** Evaluate $\frac{4}{9} - \frac{3}{7}$.

[2.4] **74.** Solve $2x - 3(x + 2) = 8$.

[3.2] **75. Taxi Ride** The cost of a taxi ride is $2.00 for the first mile and $1.50 for each additional mile or part thereof. Find the maximum distance Andrew Collins can ride in the taxi if he has only $20.

[5.5] **76.** Factor $25x^2 - 49y^2$.

[6.5] **77.** Simplify $\dfrac{\frac{28x}{y^2}}{\frac{7}{xy}}$.

[7.1] **78.** What is a graph?

See Exercise 75.

Chapter 7 Summary

IMPORTANT FACTS AND CONCEPTS	EXAMPLES
Section 7.1	
The **Cartesian coordinate system** is formed by two axes drawn perpendicular to each other. The point of intersection is called the **origin.** The horizontal axis is called the **x-axis.** The vertical axis is called the **y-axis. Ordered pairs** are of the form (x, y).	**Cartesian coordinate system** 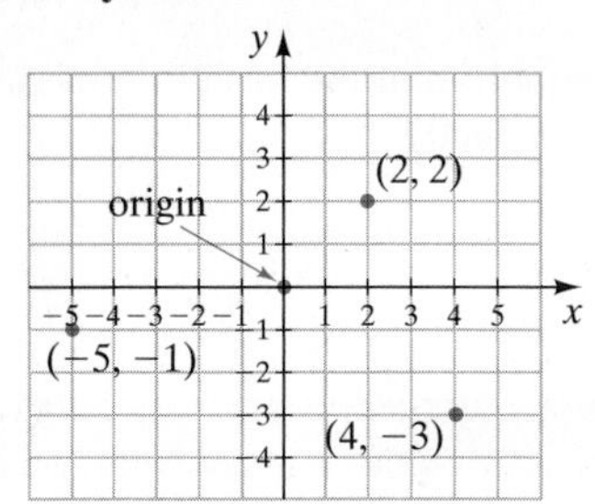
A **linear equation in two variables** is an equation that can be put in the form $ax + by = c$ where *a, b,* and *c* are real numbers. This form is called the **standard form** for a linear equation.	$3x + 7y = 2, \quad -2x - y = 9$
A **graph** of an equation in two variables is an illustration of the set of points whose coordinates satisfy the equation. Points that lie in a straight line are **collinear.**	Every point on the graph satifies the equation $y = 2x - 1$. The points on the graph are collinear and the graph is a straight line. 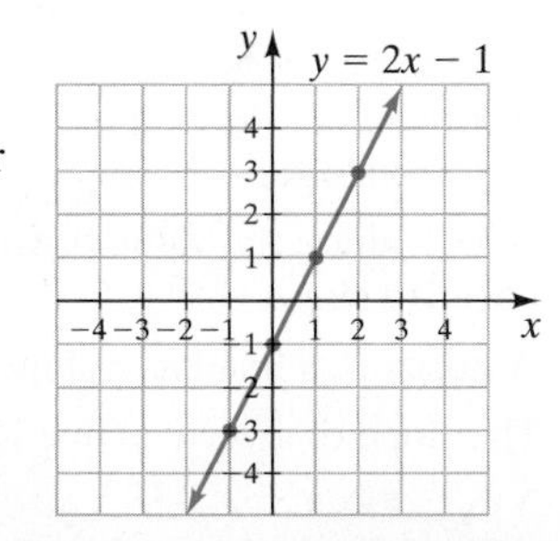
Section 7.2	
To Graph Linear Equations by Plotting Points 1. Solve the linear equation for the variable y. 2. Select a value for the variable x. Substitute this value in the equation for x and find the corresponding value of y. Record the ordered pair (x, y). 3. Repeat step 2 with two different values of x. 4. Plot the three ordered pairs. 5. Draw a straight line through the three points. Draw an arrowhead on each end of the line.	**Table** x: −1, 0, 2; y: 3, 2, 0 Graph $y = -x + 2$. 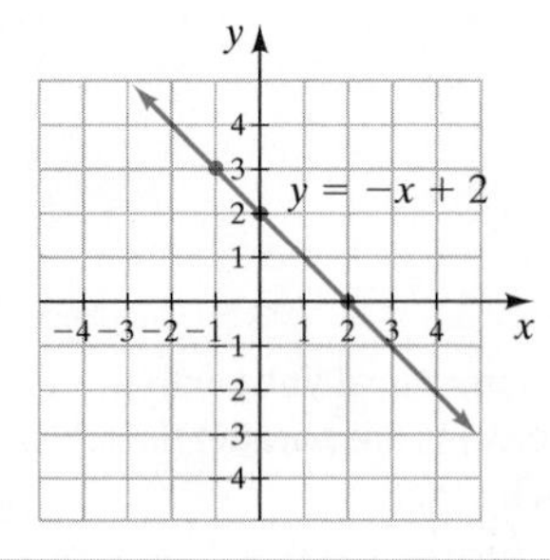
The ***x*-intercept** is the point where the graph crosses the x-axis. The ***y*-intercept** is the point where the graph crosses the y-axis.	On the graph directly above, the x-intercept is $(2, 0)$ and the y- intercept is $(0, 2)$.
To Graph Linear Equations using the *x*- and *y*-Intercepts 1. Find the y-intercept by setting x in the given equation equal to 0 and finding the corresponding value of y. 2. Find the x-intercept by setting y in the given equation equal to 0 and finding the corresponding value of x. 3. Determine a checkpoint by selecting a nonzero value for x and finding the corresponding value of y. 4. Plot the y-intercept, the x-intercept, and the checkpoint. 5. Draw a straight line through the three points. Draw an arrowhead on each end of the line.	Graph $4x + 2y = 8$ using the x- and y-intercepts. Let $x = 0$: $4(0) + 2y = 8$ $2y = 8$ $y = 4.$ y-intercept: $(0, 4)$ Let $y = 0$: $4x + 2(0) = 8$ $4x = 8$ $x = 2$ x-intercept: $(2, 0)$ 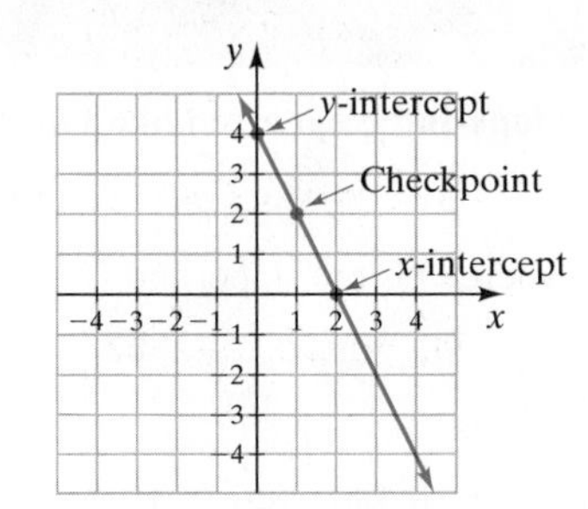 Checkpoint: $(1, 2)$

IMPORTANT FACTS AND CONCEPTS	EXAMPLES

Section 7.2 (Continued)

Horizontal Line

The graph of an equation of the form $y = b$ is a horizontal line whose y-intercept is $(0, b)$.

Vertical Line

The graph of an equation of the form $x = a$ is a vertical line whose x-intercept is $(a, 0)$.

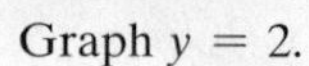

Graph $y = 2$.

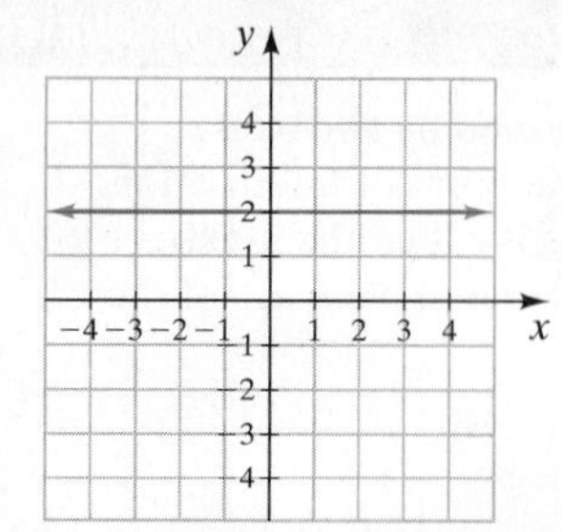

Graph $x = -4$.

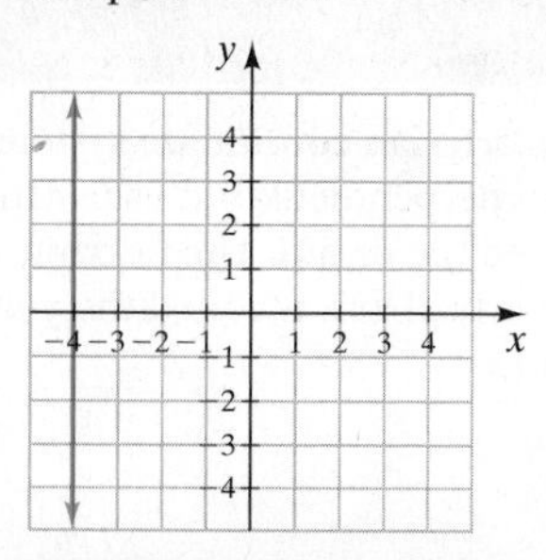

Section 7.3

The **slope of a line** is the ratio of the vertical change to the horizontal change between any two selected points on the line. The slope of a line through the points (x_1, y_1) and (x_2, y_2) is

$$m = \text{slope} = \frac{\text{change in } y \text{ (vertical change)}}{\text{change in } x \text{ (horizontal change)}} = \frac{\text{rise}}{\text{run}} = \frac{y_2 - y_1}{x_2 - x_1}$$

The slope of the line through $(-1, 3)$ and $(5, 7)$ is

$$m = \frac{7 - 3}{5 - (-1)} = \frac{4}{6} = \frac{2}{3}$$

A line where the value of y increases as x increases has a **positive slope.**

A line where the value of y decreases as x increases has a **negative slope.**

A horizontal line has a **slope of 0.**

The slope of a vertical line is **undefined.**

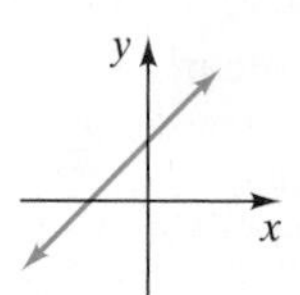

Positive slope (rises to right)

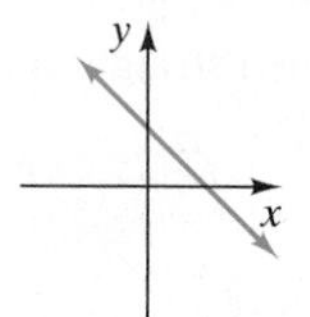

Negative slope (falls to right)

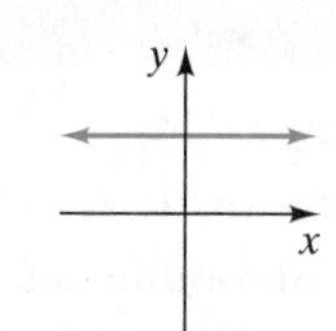

Slope is 0. (horizontal line)

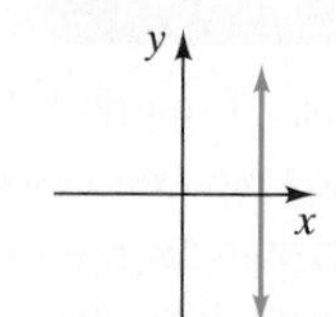

Slope is undefined. (vertical line)

Two nonvertical lines with the same slope and different y-intercepts are **parallel lines.** Any two vertical lines are parallel to each other.

The graphs of the equations $y = 2x + 3$ and $y = 2x + 4$ are parallel lines since the graphs have the same slope, 2, and different y-intercepts.

Two lines whose slopes are negative reciprocals of each other are **perpendicular lines.** Any vertical line is perpendicular to any horizontal line.

The graphs of the equations $y = 2x + 4$ and $y = -\frac{1}{2}x + 3$ are perpendicular lines since the slopes of the graphs are negative reciprocals of each other.

Section 7.4

Slope-Intercept Form of a Linear Equation

$$y = mx + b$$

where m is the slope, and $(0, b)$ is the y-intercept of the line.

The graph of $y = 3x - 4$ has a slope of 3 and a y-intercept of $(0, -4)$.

The equation of a line with a slope of $-\frac{1}{2}$ and a y-intercept of $(0, 6)$ is $y = -\frac{1}{2}x + 6$.

IMPORTANT FACTS AND CONCEPTS	EXAMPLES
Section 7.4 (Continued)	
To Graph Linear Equations Using the Slope and y-intercept 1. If necessary, solve the equation for y. 2. Plot the y-intercept. 3. Use the slope to find two more points on the line. 4. Using a straightedge, draw a straight line through the three points. Draw an arrowhead on each end of the line.	$2x + 4y = 8$ written in slope-intercept form is $y = -\frac{1}{2}x + 2$. The slope is $-\frac{1}{2}$ and the y-intercept is $(0, 2)$. 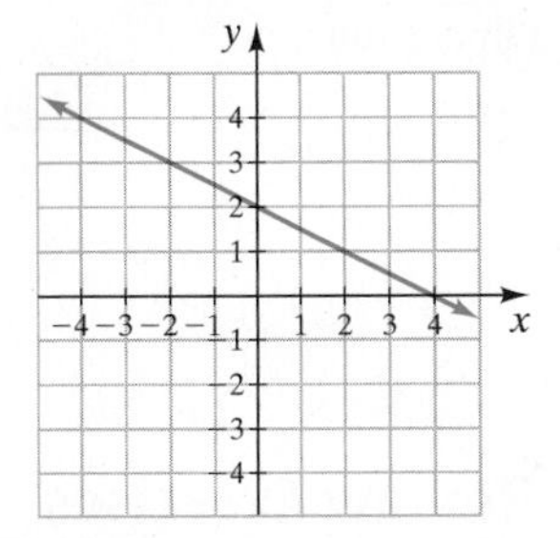
Point-Slope Form of a Linear Equation $y - y_1 = m(x - x_1)$ where m is the slope of the line and (x_1, y_1) is a point on the line.	The graph of $y - 7 = \frac{5}{6}(x - 3)$ has a slope of $\frac{5}{6}$ and contains the point $(3, 7)$.
Section 7.5	
A **linear inequality in two variables** can be written in one of the following forms: $ax + by < c,\quad ax + by > c,\quad ax + by \leq c,\quad ax + by \geq c$ Where a, b, and c are real numbers and a and b are not both 0.	$4x - 3y > 1, \qquad -x + 2y \leq 7$
To Graph a Linear Inequality in Two Variables 1. To get the equation of the boundary line, replace the inequality symbol with an equals sign. 2. Draw the graph of the equation in step 1. If the original inequality contains a $\geq$ or $\leq$ symbol, draw the boundary line using a solid line. If the original inequality contains a $>$ or $<$ symbol, draw the boundary line using a dashed line. 3. Select any point not on the boundary line and determine if this point is a solution to the original inequality. If the point selected is a solution, shade the half-plane on the side of the line containing this point. If the selected point does not satisfy the inequality, shade the half-plane on the side of the line not containing this point.	Graph $y < x + 3$. 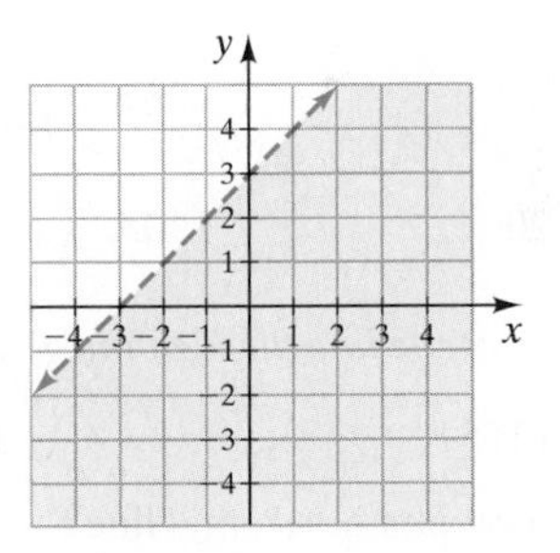
Section 7.6	
A **relation** is any set of ordered pairs. A **function** is a relation in which each first component corresponds to exactly one second component. The **domain** of a relation or function is the set of first components in the set of ordered pairs. The **range** of a relation or function is the set of second components in the set of ordered pairs.	**Relation** $\{(-1, 1), (3, 2), (3, 5), (0, 4)\}$ **Function** $\{(1, 2), (2, -1), (6, 5), (0, 3)\}$ Domain: $\{0, 1, 2, 6\}$, Range: $\{-1, 2, 3, 5\}$
Vertical Line Test If a vertical line intersects a graph at more than one point, then the graph does not represent a function.	**Function** 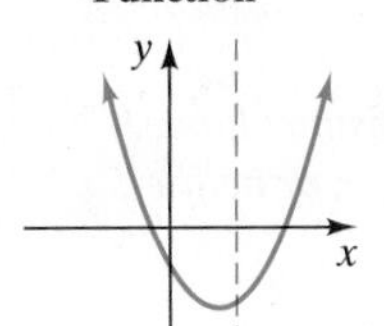 **Not a Function** 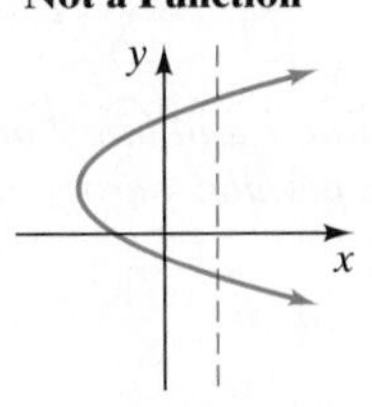

IMPORTANT FACTS AND CONCEPTS	EXAMPLES
Section 7.6 (Continued)	
When a function is represented by an equation, we can use function notation, $f(x)$. Given the function $f(x)$, to compute $f(a)$, replace x by a. $y = ax + b$ is a linear equation. $f(x) = ax + b$ is a linear function.	The function $y = 3x^2 - x + 7$ can be written as $f(x) = 3x^2 - x + 7$. If $f(x) = 3x^2 - x + 7$, then $f(-1) = 3(-1)^2 - (-1) + 7 = 11$. On the previous page we graphed the equation $y = -\frac{1}{2}x + 2$. Both $y = -\frac{1}{2}x + 2$ and $f(x) = -\frac{1}{2}x + 2$ have the same graph since y can be replaced by $f(x)$.

Chapter 7 Review Exercises

[7.1]

1. Plot each ordered pair on the same axes.

a) $A(5, 3)$ **b)** $B(0, 6)$ **c)** $C\left(5, \frac{1}{2}\right)$

d) $D(-4, 3)$ **e)** $E(-6, -1)$ **f)** $F(-2, 0)$

2. Determine whether the following points are collinear.

$$(7, 1), (6, 8), (-2, 0), (4, 5)$$

3. Which of the following ordered pairs satisfy the equation $2x + 3y = 9$?

a) $(3, 1)$ **b)** $\left(5, -\frac{1}{3}\right)$ **c)** $(-2, 4)$ **d)** $\left(2, \frac{5}{3}\right)$

[7.2]

4. Find the missing coordinate in the following solutions to $3x - 2y = 8$.

a) $(-2, ?)$ **b)** $(0, ?)$ **c)** $(?, 5)$ **d)** $(?, 0)$

Graph each equation using the method of your choice.

5. $y = 4$

6. $x = 2$

7. $y = 3x$

8. $y = 2x - 1$

9. $y = -2x + 5$

10. $2y + x = 8$

11. $-2x + 3y = 6$

12. $5x + 2y + 10 = 0$

13. $5x + 10y = 20$

14. $\frac{2}{3}x = \frac{1}{4}y + 20$

[7.3] *Find the slope of the line through the given points.*

15. $(6, -4)$ and $(1, 5)$

16. $(-4, -6)$ and $(8, -7)$

17. $(-2, -3)$ and $(-4, 1)$

18. What is the slope of a horizontal line?

19. What is the slope of a vertical line?

20. Define the slope of a straight line.

Find the slope of each line.

21.

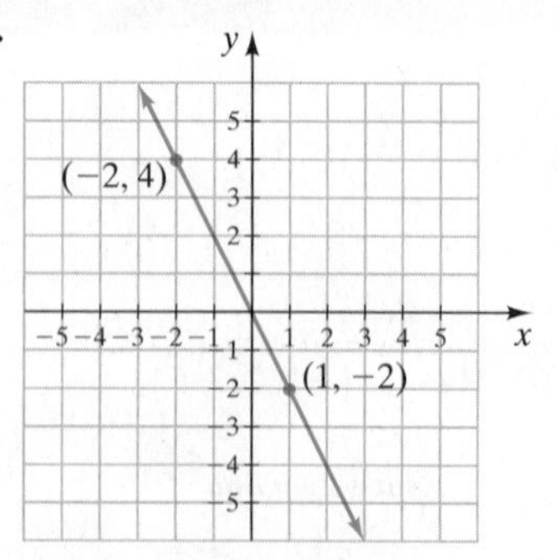

22.

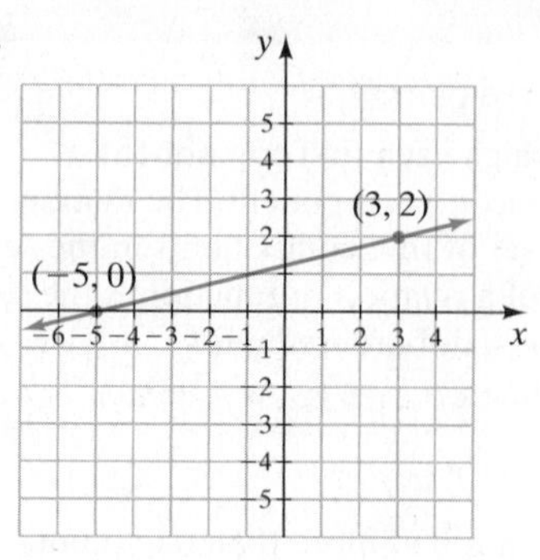

Assume that line 1 and line 2 are distinct lines. If m_1 represents the slope of line 1 and m_2 represents the slope of line 2, determine if line 1 and line 2 are parallel, perpendicular, or neither.

23. $m_1 = \frac{7}{8}, m_2 = -\frac{7}{8}$

24. $m_1 = -3, m_2 = \frac{1}{3}$

25. David Nezelek is riding his bicycle in a charity fund raiser. The graph on the right shows his riding time, in hours, along the horizontal axis, and the distance traveled, in miles, along the vertical axis.
Find the slope of the line segment in

a) red

b) blue

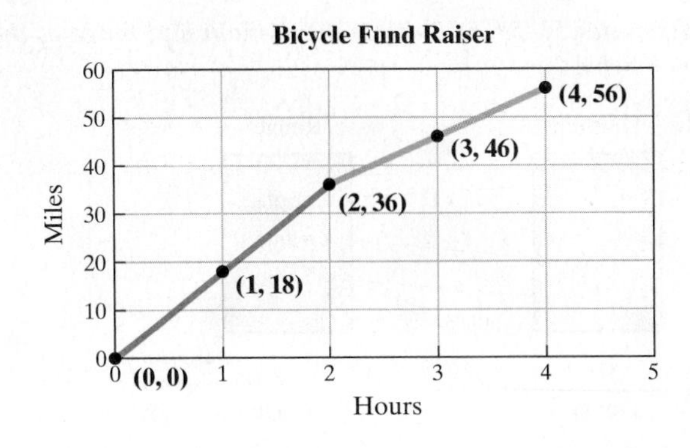

[7.4] *Determine the slope and y-intercept of the graph of each equation.*

26. $6x + 7y = 21$

27. $2x + 7 = 0$

28. $4y + 12 = 0$

Write the equation of each line.

29.

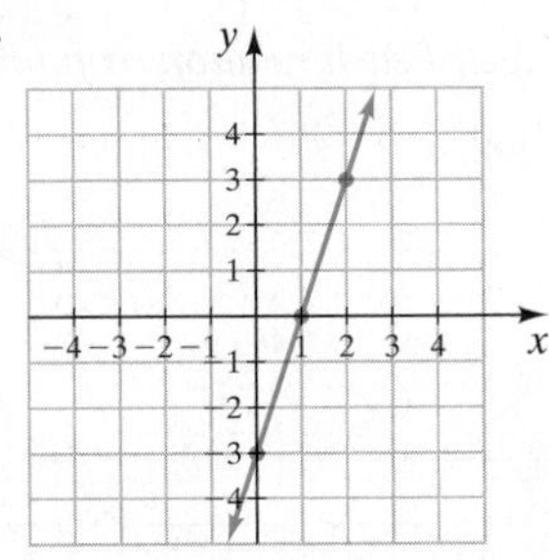

30.

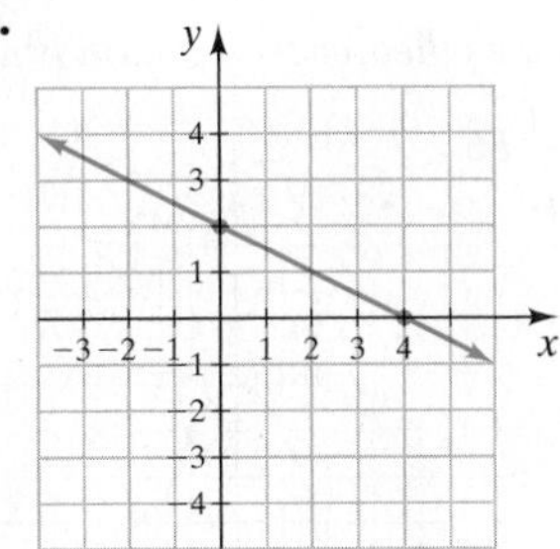

Determine whether each pair of lines is parallel, perpendicular, or neither.

31. $y = 2x - 7$
$3y = 6x + 18$

32. $2x - 3y = 15$
$3x + 2y = 12$

Find the equation of each line with the given properties.

33. Slope $= 3$ through $(2, 7)$

34. Slope $= -\frac{2}{3}$, through $(3, 2)$

35. Slope $= 0$, through $(6, 2)$

36. Slope is undefined, through $(4, 1)$

37. Through $(-2, 4)$ and $(0, -3)$

38. Through $(-5, -2)$ and $(-5, 3)$

[7.5] *Graph each inequality.*

39. $y \geq 1$

40. $x < 4$

41. $y < 3x$

42. $y > 2x + 1$

43. $-6x + y \geq 5$

44. $3y + 6 \leq x$

[7.6] *Determine which of the following relations are also functions. Then state the domain and range of each.*

45. $\{(3, 2), (4, -3), (1, 5), (2, -1), (6, 4)\}$

46. $\{(3, 1), (4, 2), (4, 5), (6, 1), (7, 0)\}$

47. $\{(3, 1), (4, 1), (5, 1), (6, 2), (3, -3)\}$

48. $\{(5, -2), (3, -2), (4, -2), (9, -2), (-2, -2)\}$

In Exercises 49 and 50, the domain and range of a relation are illustrated. **a)** *Construct a set of ordered pairs that represent the relation.* **b)** *Determine whether the relation is a function. Explain your answer.*

49. Domain Range

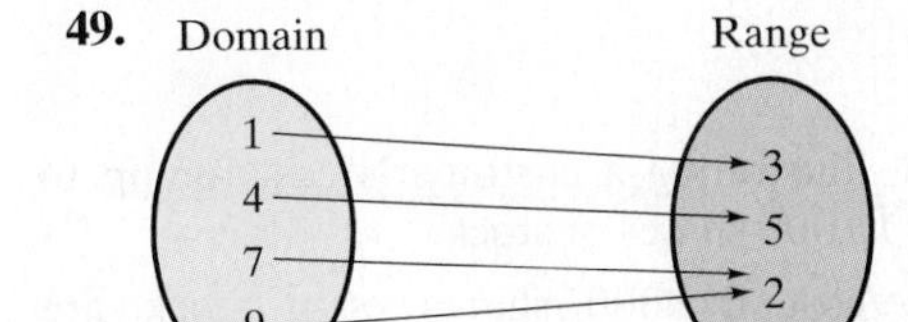

50. Domain Range

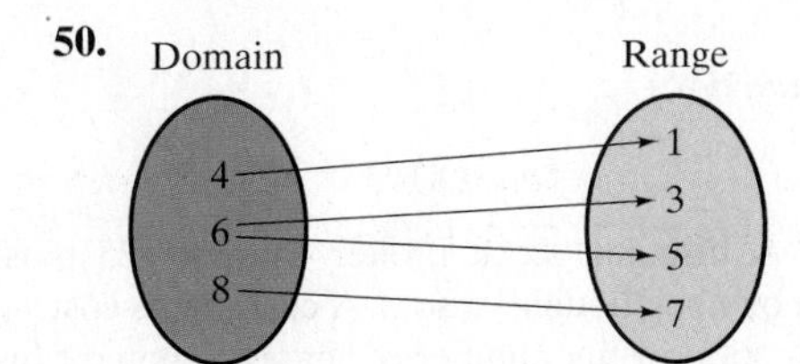

In Exercises 51–54, **a)** *indicate the domain and range of the relation, and* **b)** *indicate if the relation is a function. If it is not a function, explain why.*

51.

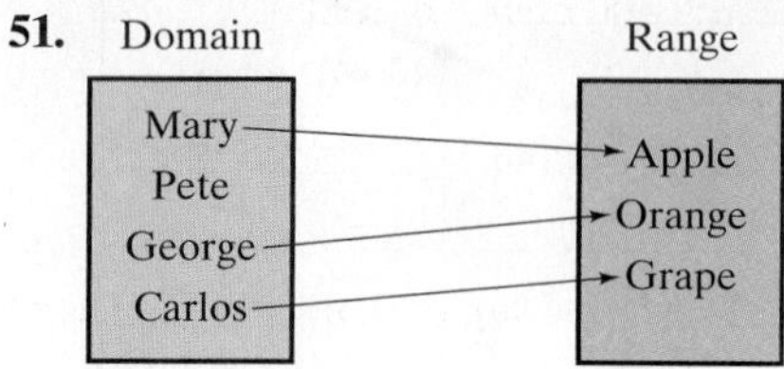

52.

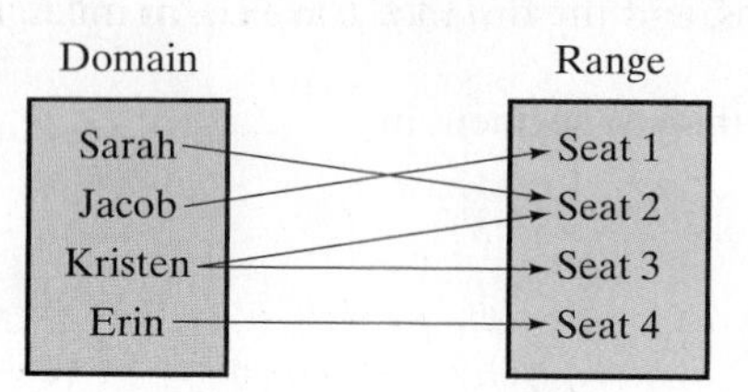

53.

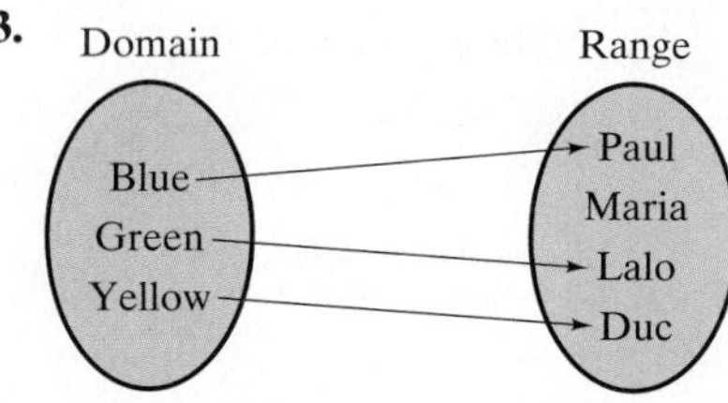

54.

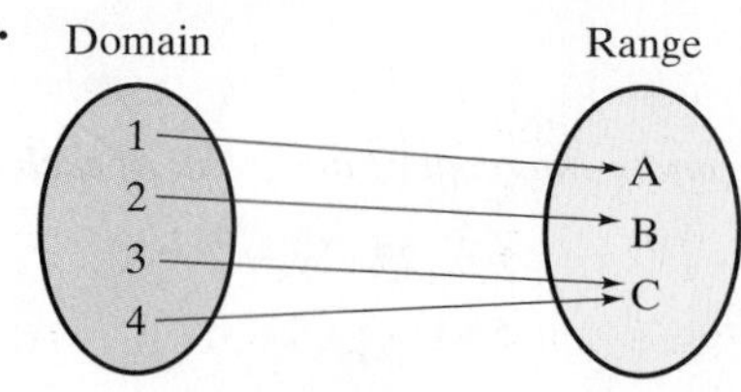

Use the vertical line test to determine whether the relation is also a function. Then state the domain and range of each relation or function.

55.

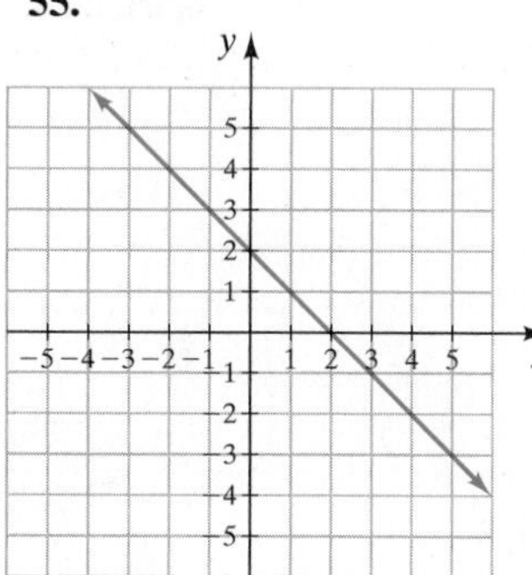

56.

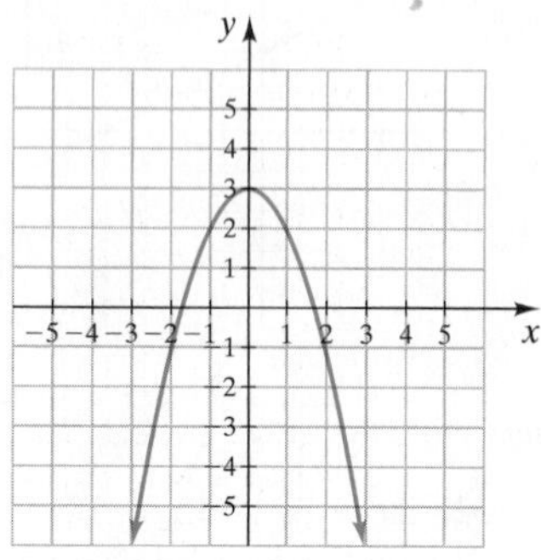

57.

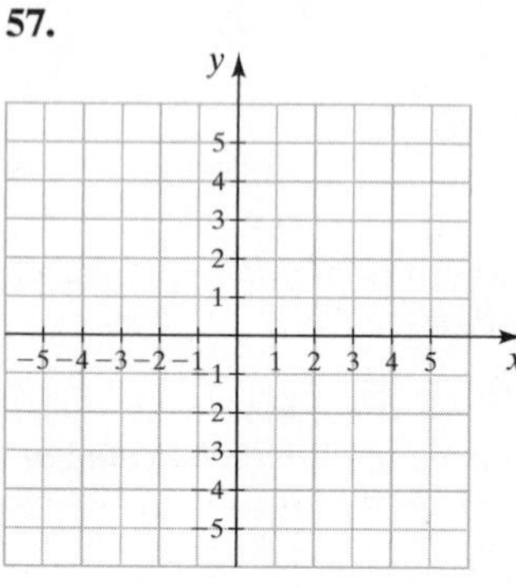

58.

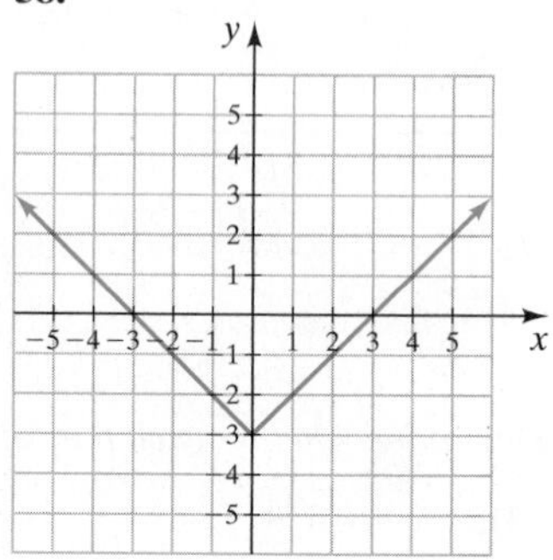

Evaluate each function at the indicated values.

59. $f(x) = 6x - 1$; find
a) $f(1)$,
b) $f(-5)$

60. $f(x) = -4x - 7$; find
a) $f(-4)$,
b) $f(8)$

61. $f(x) = \frac{1}{3}x - 5$; find
a) $f(3)$,
b) $f(-9)$

62. $f(x) = 2x^2 - 4x + 6$; find
a) $f(3)$,
b) $f(-5)$

Determine whether the following graphs are functions. Explain your answer.

63.

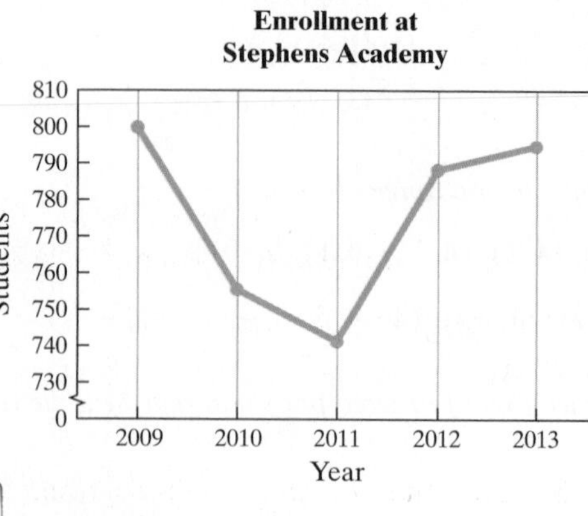

64.

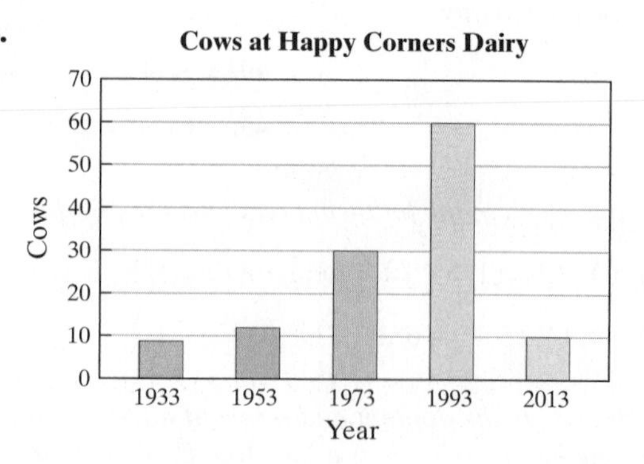

Graph the following functions.

65. $f(x) = 3x - 5$

66. $f(x) = -2x + 3$

67. Stock Purchase A discount stock broker charges \$25 plus 3 cents per share of stock bought or sold. A customer's cost, c, in dollars, is a function of the number of shares, n, bought or sold, $c = 25 + 0.03n$.

a) Draw a graph illustrating a customer's cost for up to and including 10,000 shares of stock.

b) Estimate the cost if 4000 shares of a stock are purchased.

68. Dollar Store The monthly profit, p, of an Everything for a Dollar store can be estimated by the function $p = 4x - 1600$, where x represents the number of items sold.

a) Draw a graph of the function for up to and including 1000 items sold.

b) Estimate the profit if 500 items are sold.

Chapter 7 Practice Test

Chapter Test Prep Videos provide fully worked-out solutions to any of the exercises you want to review. Chapter Test Prep Videos are available via MyMathLab®, *or on* YouTube (www.youtube.com/user/AngelElementaryAlg)

1. What is a graph?

2. In which quadrants do the following points lie?

   a) $(3, -5)$ **b)** $\left(-2, \frac{1}{2}\right)$

3. **a)** What is the standard form of a linear equation?

   b) What is the slope-intercept form of a linear equation?

   c) What is the point-slope form of a linear equation?

4. Which of the following ordered pairs satisfy the equation $3y = 5x - 9$?

   a) $(4, 2)$ **b)** $\left(\frac{9}{5}, 0\right)$

   c) $(-1, -10)$ **d)** $(0, -3)$

5. Find the slope of the line through the points $(-2, 5)$ and $(4, -3)$.

6. Find the slope and y-intercept of $4x - 9y = 15$

7. Write an equation of the graph in the accompanying figure.

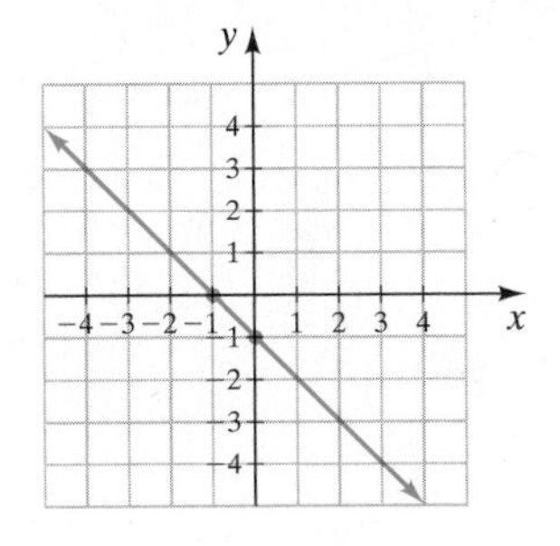

8. Graph $x = -4$.

9. Graph $y = 2$.

10. Graph $y = 3x - 2$ by plotting points.

11. **a)** Solve the equation $3x - 6y = 12$ for y.

    b) Graph the equation by plotting points.

12. Graph $3x + 5y = 15$ using the intercepts.

13. Write, in slope-intercept form, an equation of the line with a slope of 4 passing through the point $(2, -5)$

14. Write, in slope-intercept form, an equation of the line passing through the points $(3, -1)$ and $(-4, 2)$.

15. Determine whether the following equations represent parallel lines. Explain how you determined your answer.

$$2y = 3x - 6 \quad \text{and} \quad y - \frac{3}{2}x = -5$$

16. Graph $y = 3x - 4$ using the slope and y-intercept.

17. Graph $4x - 2y = 6$ using the slope and y-intercept.

18. Define a function.

19. **a)** Determine whether the following relation is a function. Explain your answer.

$$\{(1, 2), (3, -4), (5, 3), (3, 0), (6, 5)\}$$

    b) Give the domain and range of the relation or function.

20. Use the vertical line test to determine whether the relation is also a function. Then state the domain and range of each relation or function.

    a)

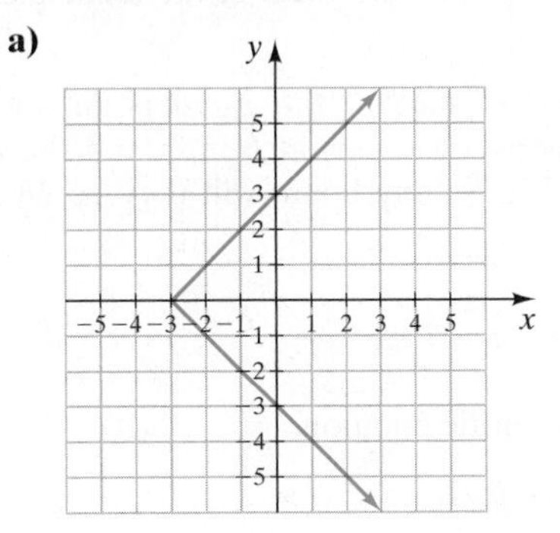

    b)

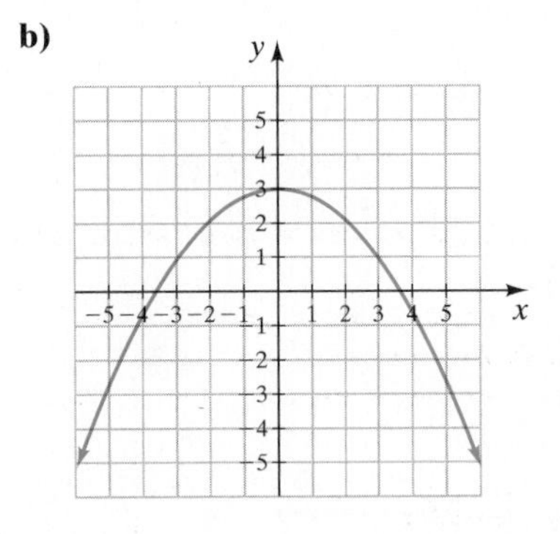

21. If $f(x) = 2x^2 + 3x + 1$ find **a)** $f(2)$ and **b)** $f(-3)$.

22. Graph the function $f(x) = 2x - 4$.

23. Graph $y \geq -3x + 5$.

24. Graph $y < 4x - 2$.

25. **Weekly Income** Kate Moore, a salesperson, has a weekly income, i, that can be determined by the function $i = 200 + 0.05s$, where s is her weekly sales.

    a) Draw a graph of her weekly income for sales from \$0 to \$10,000.

    b) Estimate her weekly income if her sales are \$5000.

Cumulative Review Test

Take the following test and check your answers with those given in the back of the book. Review any questions that you answered incorrectly. The section where the material was covered is indicated after the answer.

1. Write the set of
   a) natural numbers.
   b) whole numbers.
2. Name each indicated property.
   a) $3(x + 2) = 3x + 3 \cdot 2$
   b) $a + b = b + a$
3. Solve $2x + 5 = 3(x - 5)$.
4. Solve $3(x - 1) - (x + 4) = 2x - 7$
5. Solve the inequality $2x - 14 > 5x + 1$. Graph the solution on a number line and write the solution using interval notation.
6. **Chicken Soup** At Tsong Hsu's Grocery Store, 3 cans of chicken soup sell for $1.50. Find the cost of 8 cans.
7. **Rectangle** The length of a rectangle is 3 more than twice the width. Find the length and width of the rectangle if its perimeter is 36 feet.
8. **Running** Two runners start at the same point and run in opposite directions. One runs at 6 mph and the other runs at 8 mph. In how many hours will they be 28 miles apart?
9. Simplify $\dfrac{x^{-4}}{x^{11}}$.
10. Express 652.3 in scientific notation.
11. Factor $2x^2 - 12x + 10$.
12. Factor $4a^2 + 4a - 35$
13. Solve $3x^2 = 21x$.
14. Simplify $\dfrac{2r - 7}{14 - 4r}$.
15. Multiply $\dfrac{x - 2}{3x + 7} \cdot \dfrac{8x}{x - 2}$.
16. Solve $\dfrac{y^2}{y - 6} = \dfrac{36}{y - 6}$.
17. Graph $6x - 3y = -12$ using the x- and y-intercepts.
18. Graph $y = \dfrac{2}{3}x - 3$ using the slope and y-intercept.
19. Write the equation, in point-slope form, of the line with a slope of 3 passing through the point $(5, 2)$.
20. Determine whether the following relations are functions. Then state the domain and range of each relation or function.
    a) $\{(-4, 1), (5, 3), (7, 3), (-2, 0)\}$
    b)

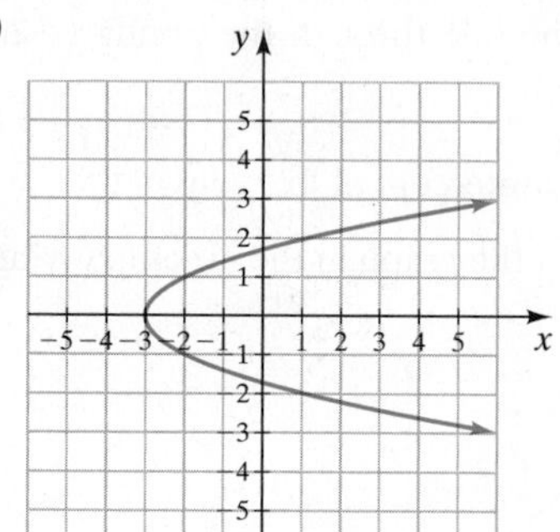

Systems of Linear Equations

Almost daily we make financial decisions. Often an understanding of systems of equations is helpful in making such decisions. For instance, in Exercise 39 on page 489, we use a system of equations to decide whether it is beneficial for a homeowner to refinance her home mortgage. As you read through this chapter, you will see many real-life applications of systems of equations.

Goals of This Chapter

In this chapter we learn how to express real-life application problems as systems of linear equations and how to solve systems of linear equations. The solution to a system of equations is the value or values that satisfy all equations in the system.

We explain three procedures for solving systems of equations: using graphs, using substitution, and using the addition (or elimination) method. We also solve systems of linear *inequalities* graphically.

8.1 Solving Systems of Equations Graphically

1. Determine if an ordered pair is a solution to a system of equations.
2. Determine if a system of equations is consistent, inconsistent, or dependent.
3. Solve a system of equations graphically.

1 Determine If an Ordered Pair Is a Solution to a System of Equations

System of Linear Equations

When two or more linear equations are considered simultaneously, the equations are called a **system of linear equations**.

For example,

System of linear equations $\begin{cases} y = x + 5 & \text{Equation 1} \\ y = 2x + 4 & \text{Equation 2} \end{cases}$

Understanding Algebra

A solution to a system of two linear equations with variables x and y is an ordered pair (x, y) that satisfies *both* equations.

Solution

A **solution to a system of linear equations in two variables** is an ordered pair that satisfies each equation in the system.

The solution to the system above is the ordered pair $(1, 6)$.

Check

In Equation 1	In Equation 2
$(1, 6)$	$(1, 6)$
$y = x + 5$	$y = 2x + 4$
$6 \stackrel{?}{=} 1 + 5$	$6 \stackrel{?}{=} 2(1) + 4$
$6 = 6$ True	$6 = 6$ True

Because the ordered pair $(1, 6)$ satisfies *both* equations, it is a solution to the system of equations. Notice that the ordered pair $(3, 8)$ satisfies the first equation but does not satisfy the second equation.

Check

In Equation 1	In Equation 2
$(3, 8)$	$(3, 8)$
$y = x + 5$	$y = 2x + 4$
$8 \stackrel{?}{=} 3 + 5$	$8 \stackrel{?}{=} 2(3) + 4$
$8 = 8$ True	$8 = 10$ False

Since the ordered pair $(3, 8)$ does not satisfy *both* equations, it is *not* a solution to the system of equations.

EXAMPLE 1 Determine whether each of the following ordered pairs satisfies the system of equations.

$$y = 2x - 8$$
$$2x + y = 4$$

a) $(1, -6)$ **b)** $(3, -2)$

Solution

a) Substitute 1 for x and -6 for y in each equation.

$y = 2x - 8$	$2x + y = 4$
$-6 \stackrel{?}{=} 2(1) - 8$	$2(1) + (-6) \stackrel{?}{=} 4$
$-6 \stackrel{?}{=} 2 - 8$	$2 - 6 \stackrel{?}{=} 4$
$-6 \stackrel{?}{=} -6$ True	$-4 = 4$ False

Since $(1, -6)$ does not satisfy both equations, it is not a solution to the system of equations.

b) Substitute 3 for x and -2 for y in each equation.

$$\begin{aligned} y &= 2x - 8 \\ -2 &\stackrel{?}{=} 2(3) - 8 \\ -2 &\stackrel{?}{=} 6 - 8 \\ -2 &= -2 \quad \text{True} \end{aligned} \qquad \begin{aligned} 2x + y &= 4 \\ 2(3) + (-2) &\stackrel{?}{=} 4 \\ 6 - 2 &\stackrel{?}{=} 4 \\ 4 &= 4 \quad \text{True} \end{aligned}$$

Since $(3, -2)$ satisfies both equations, it is a solution to the system of linear equations.

Now Try Exercise 9

We next discuss the three possible outcomes when graphing systems of linear equations in two variables.

2 Determine If a System of Equations Is Consistent, Inconsistent, or Dependent

Understanding Algebra

The solution to a linear system of equations is the ordered pair or ordered pairs represented by the point or points at the intersection of the graphs of the equations in the system.

Recall from Chapter 7 that the graph of a linear equation is the set of points whose ordered pairs are solutions to that equation. Thus, the solution to a system of linear equations is represented by the point or points that are on the graphs of *each* of the equations in the system. When two linear equations are graphed, three situations are possible, as illustrated in **Figure 8.1**. The system of equations may be *consistent*, *inconsistent*, or *dependent*.

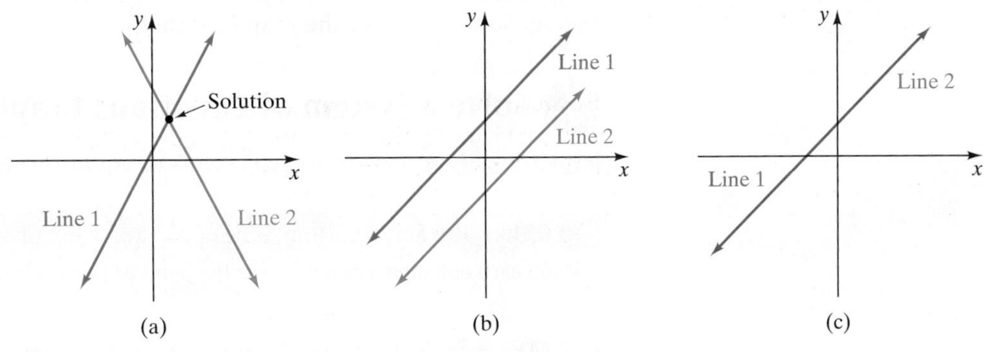

FIGURE 8.1

Line 1 *Intersects* Line 2
- Exactly one solution
- The solution is at the point of intersection.
- **Consistent system**
- Lines have different slopes

Line 1 is *Parallel* to Line 2
- No solution
- Since parallel lines do not intersect, there is no solution.
- **Inconsistent system**
- Lines have same slopes and different y-intercepts

Line 1 is the *Same Line* as Line 2
- Infinite number of solutions
- Every point on the common line is a solution.
- **Dependent system**
- Lines have same slope and the same y-intercept

Understanding Algebra

- A **consistent system of equations** has at least one solution.
- An **inconsistent system of equations** has no solution.
- A **dependent system of equations** has an infinite number of solutions.

We can determine if a system of linear equations is consistent, inconsistent, or dependent by writing each equation in slope-intercept form and comparing the slopes and y-intercepts.

EXAMPLE 2 Determine whether the following system of equations has exactly one solution, no solution, or an infinite number of solutions.

$$y = \frac{1}{2}x + 4$$
$$x - 2y = 6$$

Solution The first equation is given in slope-intercept form:

$$y = \frac{1}{2}x + 4$$

the slope is $\frac{1}{2}$ — the y-intercept is $(0, 4)$

Next, write the second equation in slope-intercept form by solving for y.

$$x - 2y = 6$$
$$-2y = -x + 6$$
$$y = \frac{1}{2}x - 3$$

the slope is $\frac{1}{2}$ — the y-intercept is $(0, -3)$

The lines that represent the two equations in the system have the same slope and different y-intercepts. Thus, the two lines are parallel. The system is inconsistent and has no solution.

Now Try Exercise 29

In this chapter we discuss three methods for finding the solution to a system of equations: the *graphical method*, the *substitution method*, and the *addition method*. In this section we discuss the graphical method.

3 Solve a System of Equations Graphically

Now we will see how to solve systems of equations graphically.

To Obtain the Solution to a System of Equations Graphically

Graph each equation and determine the point or points of intersection.

EXAMPLE 3 Solve the following system of equations graphically.

$$x + 2y = 4$$
$$y = 2x - 3$$

Solution From Chapter 7 we learned we can graph linear equations by using the x- and y-intercepts (see Section 7.2) or by using the slope and y-intercept (see Section 7.4). We will graph the first equation by using x- and y-intercepts.

$x + 2y = 4$	**Ordered Pair**
To find the x-intercept, let $y = 0$; then $x = 4$.	$(4, 0)$
To find the y-intercept, let $x = 0$; then $y = 2$.	$(0, 2)$

We now plot these points and draw a straight line through these points. The graph of the equation $x + 2y = 4$ is shown in blue in **Figure 8.2** on the next page.

Next we will graph the second equation by using the slope and y-intercept.

$$y = 2x - 3$$

The graph of this equation has a slope of 2 or $\frac{2}{1}$ and a y-intercept of $(0, -3)$. We first plot the point $(0, -3)$. Then we use the slope to move 2 units up and 1 unit to the right to find a second point at $(1, -1)$. We continue this process to obtain a third point at $(2, 1)$. Next we draw a straight line through these three points. The graph of the equation $y = 2x - 3$ is shown in red in **Figure 8.2**.

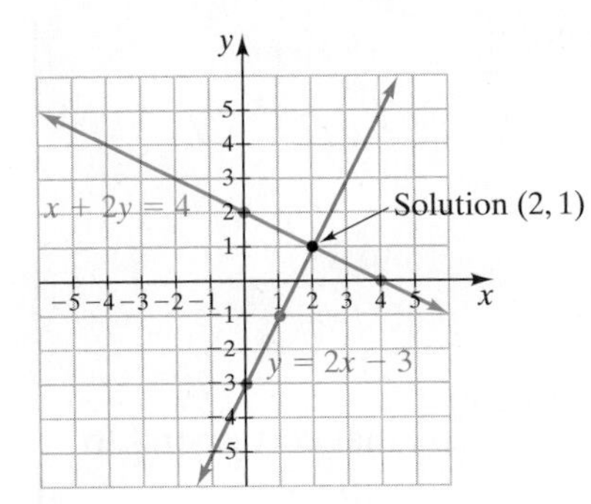

FIGURE 8.2

The two lines in **Figure 8.2** appear to intersect at the point $(2, 1)$. To be sure that this is the solution, we must check to see that $(2, 1)$ satisfies *both* equations.

Check

$$x + 2y = 4 \qquad\qquad y = 2x - 3$$
$$2 + 2(1) \stackrel{?}{=} 4 \qquad\qquad 1 \stackrel{?}{=} 2(2) - 3$$
$$4 = 4 \quad \text{True} \qquad\qquad 1 = 1 \quad \text{True}$$

Since the ordered pair $(2, 1)$ satisfies both equations, it is the solution to the system of equations. This system of equations is consistent.

Now Try Exercise 39

EXAMPLE 4 Solve the following system of equations graphically.

$$2x + y = 3$$
$$4x + 2y = 12$$

Solution Find the x- and y-intercepts of each graph; then draw the graphs.

$2x + y = 3$	**Ordered Pair**
Let $y = 0$; then $x = \frac{3}{2}$.	$\left(\frac{3}{2}, 0\right)$
Let $x = 0$; then $y = 3$.	$(0, 3)$

$4x + 2y = 12$	**Ordered Pair**
Let $y = 0$; then $x = 3$.	$(3, 0)$
Let $x = 0$; then $y = 6$.	$(0, 6)$

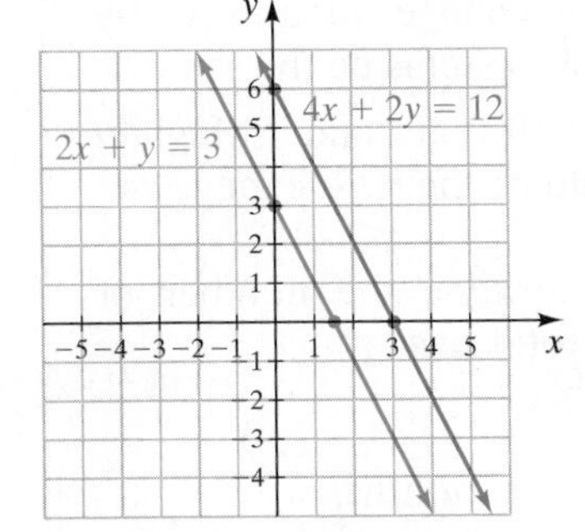

FIGURE 8.3

The two lines (**Fig. 8.3**) appear to be parallel.

To show that the two lines are indeed parallel, write each equation in slope-intercept form.

$$2x + y = 3 \qquad\qquad 4x + 2y = 12$$
$$y = -2x + 3 \qquad\qquad 2y = -4x + 12$$
$$y = -2x + 6$$

Both equations have the same slope, -2, and different y-intercepts. Thus the lines are parallel. The system is inconsistent, and has no solution.

Now Try Exercise 55

EXAMPLE 5 Solve the following system of equations graphically.

$$x - \frac{1}{2}y = 2$$
$$y = 2x - 4$$

Solution Find the x- and y-intercepts of each graph; then draw the graphs.

$x - \frac{1}{2}y = 2$	**Ordered Pair**
Let $x = 0$; then $y = -4$.	$(0, -4)$
Let $y = 0$; then $x = 2$.	$(2, 0)$

$y = 2x - 4$	**Ordered Pair**
Let $x = 0$; then $y = -4$.	$(0, -4)$
Let $y = 0$; then $x = 2$.	$(2, 0)$

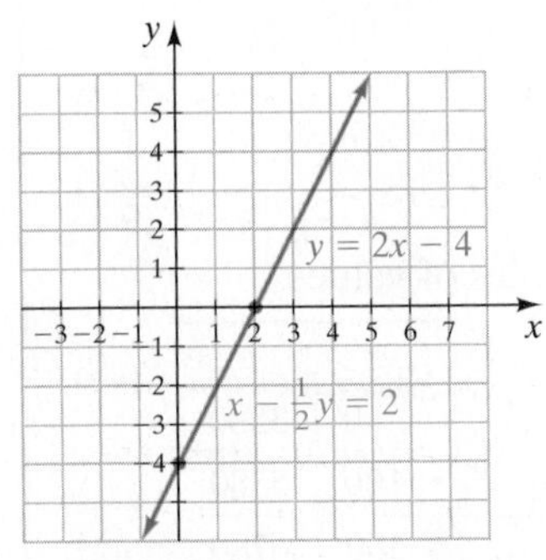

FIGURE 8.4

Because the lines have the same x- and y-intercepts, both equations represent the same line (**Fig. 8.4**). When the equations are written in slope-intercept form, it becomes clear that the equations are represented by the same line. The system is dependent, and there are an infinite number of solutions.

$$x - \frac{1}{2}y = 2 \qquad\qquad y = 2x - 4$$
$$2\left(x - \frac{1}{2}y\right) = 2(2)$$
$$2x - y = 4$$
$$-y = -2x + 4$$
$$y = 2x - 4$$

The solution to this system of equations is all the ordered pairs that correspond to the points on the line.

Now Try Exercise 53

Can you determine the solution to the system of equations shown in **Figure 8.5?** You may estimate the solution to be $\left(\frac{7}{10}, \frac{3}{2}\right)$ when it may actually be $\left(\frac{4}{5}, \frac{8}{5}\right)$. The accuracy of your answer will depend on how carefully you draw the graphs and on the scale of the graph paper used. In Section 8.2 we present an algebraic method that gives exact solutions to systems of equations.

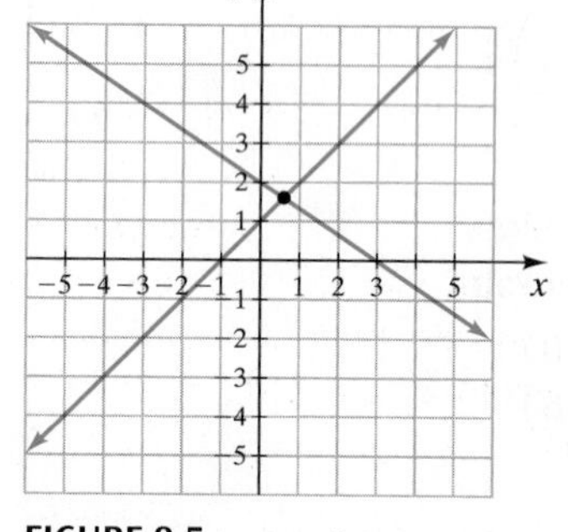

FIGURE 8.5

Next, we work an application problem using two variables and illustrate the solution in the form of a graph. A graph may help you better visualize the total picture in application problems.

EXAMPLE 6 Security Systems Meghan O'Donnell plans to install a security system in her house. She has narrowed her choices to two security dealers: Moneywell and Doile. Moneywell's system costs \$3580 to install and the monitoring fee is \$20 per month. Doile's equivalent system costs only \$2620 to install, but the monitoring fee is \$32 per month.

a) Assuming that the monthly monitoring fees do not change, in how many months would the total cost of Moneywell's and Doile's systems be the same?

b) If both dealers guarantee not to raise monthly fees for 10 years, and if Meghan plans to use the system for 10 years, which system would be the least expensive?

Solution **a)** Understand and Translate We need to determine the number of months at which both systems will have reached the same total cost.

Let n = number of months.

Let c = total cost of the security system over n months.

Now we can write an equation to represent the cost of each system using the two variables c and n.

Moneywell

$$\text{Total cost} = \begin{pmatrix}\text{initial}\\ \text{cost}\end{pmatrix} + \begin{pmatrix}\text{fees over}\\ n \text{ months}\end{pmatrix}$$
$$c = 3580 + 20n$$

Doile

$$\text{Total cost} = \begin{pmatrix}\text{initial}\\ \text{cost}\end{pmatrix} + \begin{pmatrix}\text{fees over}\\ n \text{ months}\end{pmatrix}$$
$$c = 2620 + 32n$$

Thus, our system of equations is

$$c = 3580 + 20n$$
$$c = 2620 + 32n$$

Carry Out Now let's graph each equation. Following are tables of values.

$c = 3580 + 20n$

Let $n = 0$. $\quad c = 3580 + 20(0) = 3580$

Let $n = 100$. $\quad c = 3580 + 20(100) = 5580$

Let $n = 160$. $\quad c = 3580 + 20(160) = 6780$

n	c
0	3580
100	5580
160	6780

	$c = 2620 + 32n$
Let $n = 0$.	$c = 2620 + 32(0) = 2620$
Let $n = 100$.	$c = 2620 + 32(100) = 5820$
Let $n = 160$.	$c = 2620 + 32(160) = 7740$

n	c
0	2620
100	5820
160	7740

Figure 8.6 shows the graphs of the equations.

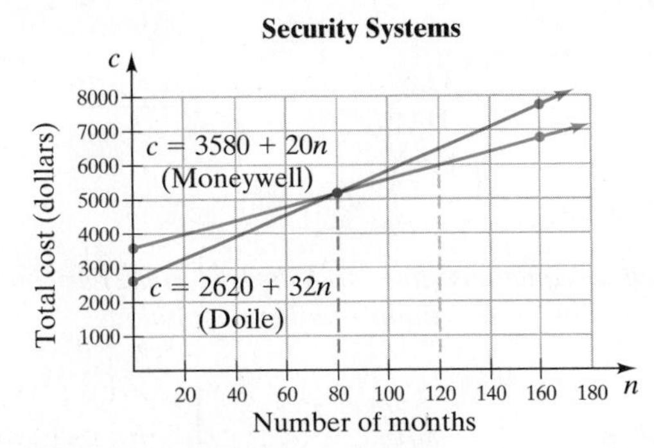

FIGURE 8.6

Check and Answer The graph (**Fig. 8.6**) shows that the total cost of the two security systems would be the same in 80 months.

b) Since 10 years is 120 months, we draw a dashed vertical line at $n = 120$ months and see where it intersects the two lines. Since at 120 months the Doile line is higher than the Moneywell line, the cost for the Doile system for 120 months is more than the cost of the Moneywell system. Therefore, the cost of the Moneywell system would be less expensive for 10 years.

Now Try Exercise 63

8.1 Exercise Set MathXL® MyMathLab®

Warm-Up Exercises

Fill in the blanks with the appropriate word, phrase, or symbol(s) from the following list.

no system infinite one pair intersection intercept

1. When two or more linear equations are considered simultaneously, the equations are called a ________ of linear equations.
2. A solution to a system of linear equations in two variables is an ordered ________ that satisfies each equation in the system.
3. A consistent system of linear equations has at least ________ solution.
4. An inconsistent system of linear equations has ________ solution.
5. A dependent system of linear equations has an ________ number of solutions.
6. To obtain the solution to a system of linear equations graphically, graph each equation and determine the point or points of ________.

Practice the Skills

Determine which, if any, of the following ordered pairs satisfy each system of linear equations.

7. $y = 5x - 8$
$y = -3x$
a) $(2, 2)$ **b)** $(0, 0)$ **c)** $(1, -3)$

8. $y = -4x$
$y = -2x + 8$
a) $(0, 8)$ **b)** $(-4, 16)$ **c)** $(3, -12)$

9. $3x - 2y = -2$
$y = -x + 6$
a) $(2, 4)$ **b)** $(0, 1)$ **c)** $(3, 3)$

10. $x + 2y = 4$
$y = 3x - 5$
a) $(0, 2)$ **b)** $(2, 1)$ **c)** $(4, 0)$

11. $4x + y = 15$
$5x + y = 10$
a) $(3, 3)$ **b)** $(2, 0)$ **c)** $(-1, 19)$

12. $x + 3y = -15$
$2x + 6y = 6$
a) $(0, -5)$ **b)** $(0, 1)$ **c)** $(9, -8)$

13. $4x - 6y = 12$
$y = \frac{2}{3}x - 2$
a) $(3, 0)$ **b)** $(9, 4)$ **c)** $(-3, -4)$

14. $y = -x + 5$
$2y = -2x + 10$
a) $(6, -1)$ **b)** $(0, 5)$ **c)** $(-2, 7)$

15. $2x + 3y = 2$
$y = -\frac{4}{3}x + 1$
a) $(1, 0)$ **b)** $\left(0, \frac{2}{3}\right)$ **c)** $\left(\frac{1}{2}, \frac{1}{3}\right)$

16. $-3x + 5y = 0$
$y = -\frac{12}{5}x + 1$
a) $\left(0, \frac{4}{5}\right)$ **b)** $\left(\frac{5}{12}, 0\right)$ **c)** $\left(\frac{1}{3}, \frac{1}{5}\right)$

Identify each system of linear equations (lines are labeled 1 and 2) as consistent, inconsistent, or dependent. State whether the system has exactly one solution, no solution, or an infinite number of solutions.

17.
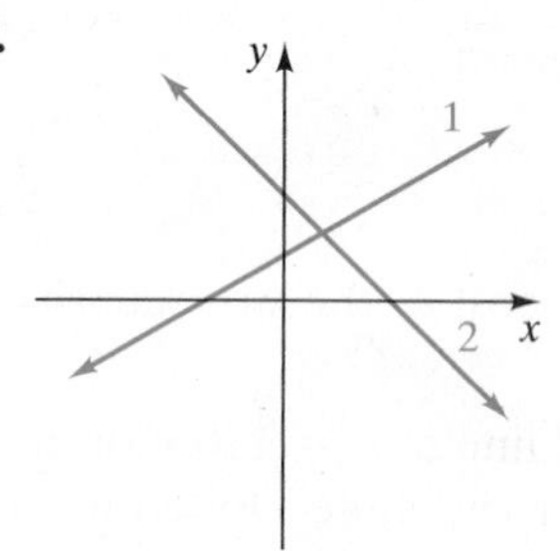

18.
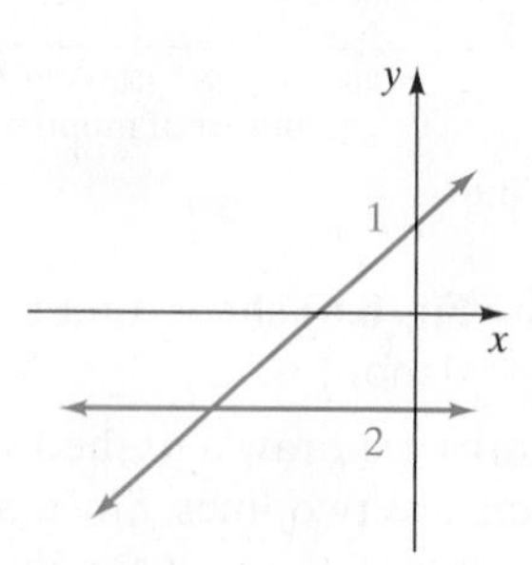

19.
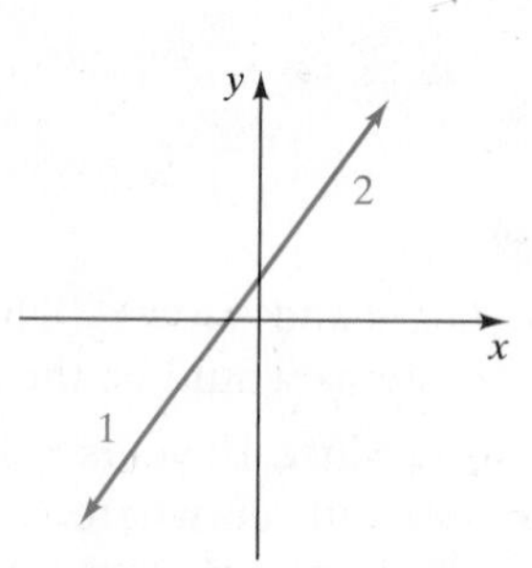

20.
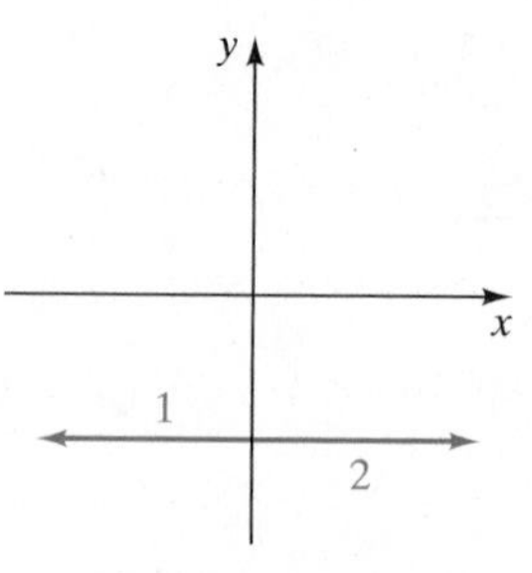

21.
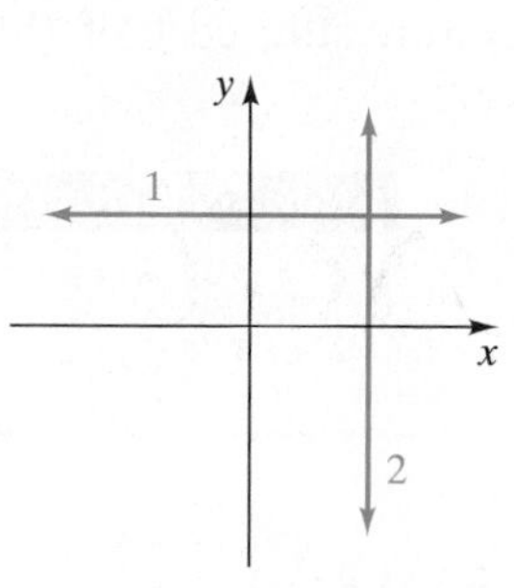

22.
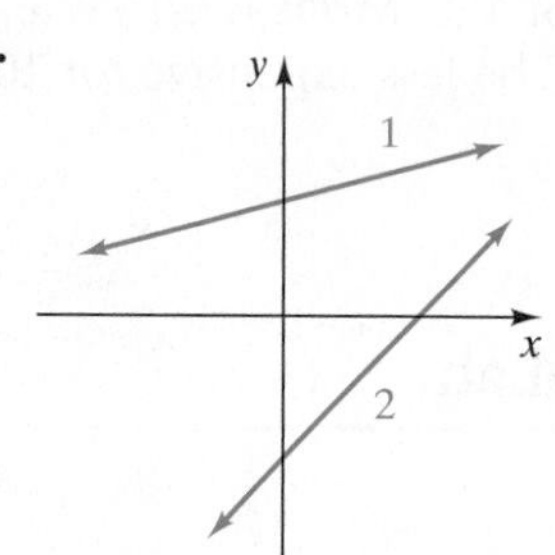

23.
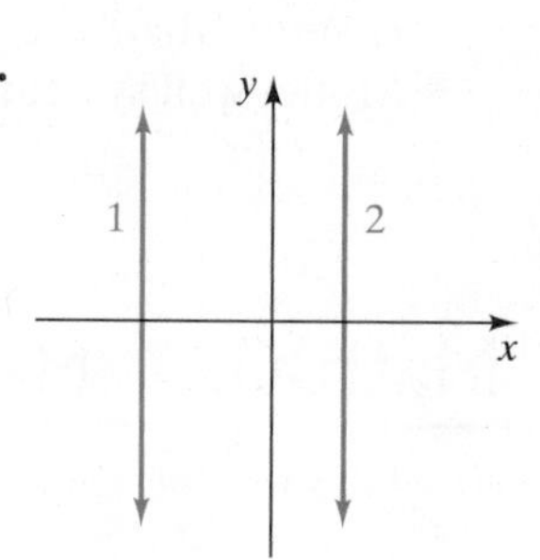

24.
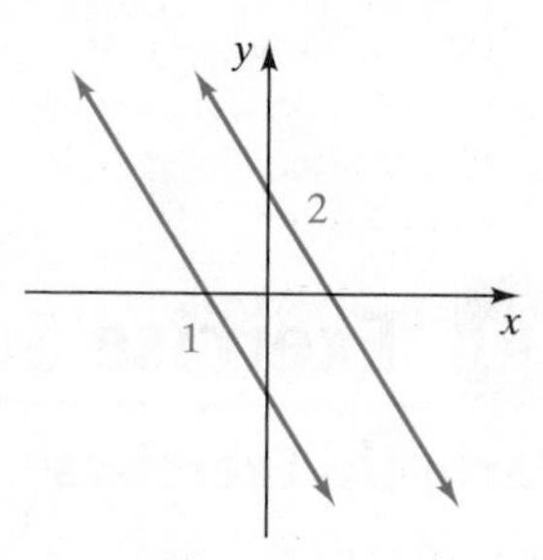

Express each equation in slope-intercept form. Without graphing the equations, state whether the system of equations has exactly one solution, no solution, or an infinite number of solutions.

25. $y = \frac{1}{2}x - 5$
$3x - 4y = 8$

26. $x + y = 8$
$x - y = 8$

27. $2y = 3x + 3$
$y = \frac{3}{2}x - 2$

28. $y = \frac{3}{2}x + \frac{1}{2}$
$3x - 2y = \frac{5}{2}$

29. $2x = y - 6$
$3x = 3y + 5$

30. $x + 3y = 6$
$4x + y = 4$

31. $3x + 5y = -7$
$-3x - 5y = -10$

32. $x - y = 2$
$2x - 2y = -6$

33. $x = 3y + 5$
$2x - 6y = 10$

34. $y = \frac{1}{2}x + 3$
$2y = x + 6$

35. $x - y = 3$
$\frac{1}{2}x - 2y = -8$

36. $2y = \frac{7}{3}x - 9$
$4y = 8x + 9$

Determine the solution to each system of equations graphically. If the system is dependent or inconsistent, so state.

37. $y = x + 3$
$y = -x + 3$

38. $y = -3x + 9$
$y = 2x - 6$

39. $y = 3x - 6$
$x + y = 6$

40. $3x - y = 4$
$y = -x$

41. $3x - y = -5$
$2y = 2x + 6$

42. $2x - y = 7$
$2y = 2x - 6$

43. $x + y = 5$
$-x + y = 1$

44. $-x + 2y = 7$
$2x - y = -2$

45. $y = -\frac{1}{2}x + 4$
$x + 2y = 6$

46. $y = \frac{3}{4}x - 3$
$3x - 4y = 4$

47. $x + 2y = 8$
$5x + 2y = 0$

48. $-x + 2y = 0$
$2x - y = -3$

49. $2x + 3y = 6$
$4x = -6y + 12$

50. $2x + 6y = 12$
$y = -\frac{1}{3}x + 2$

51. $y = 3$
$y = 2x - 3$

52. $x = 5$
$y = 2x - 8$

53. $x - 2y = 4$
$2x - 4y = 8$

54. $4x - y = 6$
$2y = 8x - 12$

55. $2x + y = -2$
$6x + 3y = 6$

56. $y = 2x - 1$
$2y = 4x + 5$

57. $4x = 8$
$y = -3$

58. $-3x = 12$
$4y + 3 = -5$

59. $2x - 3y = 0$
$x + 2y = 0$

60. $2x - 5y = 0$
$2x + 3y = 0$

Problem Solving

In Exercises 61–64, find each solution by graphing the system of equations.

61. **Furnace Repair** Edith Hall's furnace is 10 years old and has a problem. The furnace repair company indicates that it will cost Edith \$600 to repair her furnace. She can purchase a new, more efficient furnace for \$1800. Her present furnace averages about \$650 per year for energy cost and the new furnace would average about \$450 per year.

We can represent the total cost, c, of repair or replacement, plus energy cost over n years by the following system of equations.

(repair) $c = 600 + 650n$
(replacement) $c = 1800 + 450n$

In how many years would the total cost of repair equal the total cost of replacement?

62. **Security Systems** Juan Varges is considering the two security systems discussed in Example 6. If Moneywell's system costs \$4400 plus \$15 per month and Doile's system costs \$3400 plus \$25 per month, after how many months would the total cost of the two systems be the same?

63. **Boat Ride** Rudy has visitors at his home and wants to take them out on a pontoon boat for a day. There are two pontoon boat rental agencies on the lake. Bob's Boat Rental charges \$25 per hour for the boat rental, which includes all the gasoline used. Hopper's Rental charges \$21 per hour plus a flat charge of \$28 for the gasoline used. The equations that represent the total cost, c, follow. In the equations h represents the number of hours the boats are rented.

$$c = 25h$$
$$c = 21h + 28$$

Determine the number of hours the boats must be rented for the total cost to be the same.

64. **Landscaping** The Evergreen Landscape Service charges a consultation fee of \$200 plus \$50 per hour for labor. The Out of Sight Landscape Service charges a consultation fee of \$300 plus \$40 per hour for labor. We can represent this situation with the system of equations

$$c = 200 + 50h$$
$$c = 300 + 40h$$

where c is the total cost and h is the number of hours of labor. Find the number of hours of labor for the two services to have the same total cost.

Concept/Writing Exercises

65. Given the system of equations $6x - 4y = 12$ and $12y = 18x - 24$, determine without graphing whether the graphs of the two equations will be parallel lines. Explain how you determined your answer.

66. Given the system of equations $4x - 8y = 12$ and $2x - 8 = 4y$, determine without graphing whether the graphs of the two equations will be parallel lines. Explain how you determined your answer.

67. If a system of linear equations has solutions $(4, 3)$ and $(6, 5)$, how many solutions does the system have? Explain.

68. If the slope of one line in a system of linear equations is 2 and the slope of the second line in the system is 3, how many solutions does the system have? Explain.

69. If two distinct lines are parallel, how many solutions does the system have? Explain.

70. If two different lines in a linear system of equations pass through the origin, must the solution to the system be $(0, 0)$? Explain.

71. Consider the system $x = 5$ and $y = 3$. How many solutions does the system have? What is the solution?

72. A system of linear equations has $(3, -1)$ as its solution. If one line in the system is vertical and the other line is horizontal, determine the equations in the system.

Group Activity

Discuss and answer Exercises 73–78 as a group. Suppose that a system of three linear equations in two variables is graphed on the same axes. Find the maximum number of points where two or more of the lines can intersect if

73. the three lines have the same slope but different y-intercepts.

74. the three lines have the same slope and the same y-intercept.

75. two lines have the same slope but different y-intercepts and the third line has a different slope.

76. the three lines have different slopes but the same y-intercept.

77. the three lines have different slopes but two have the same y-intercept.

78. the three lines have different slopes and different y-intercepts.

Cumulative Review Exercises

[2.1] **79.** Simplify $7x - (x - 6) + 4(3 - x)$.

[2.5] **80.** Solve the equation $2(x + 3) - x = 5x + 2$.

[6.1] **81.** Simplify $\dfrac{x^2 - 9x + 14}{2 - x}$.

[6.6] **82.** Solve the equation $\dfrac{2}{b} + b = \dfrac{19}{3}$.

[7.2] **83. a)** Find the x- and y-intercepts of $2x + 3y = 12$.

b) Use the intercepts to graph the equation.

8.2 Solving Systems of Equations by Substitution

1 Solve systems of equations by substitution.

Although solving systems of equations graphically helps us visualize a system of equations and its solution, an exact solution is sometimes difficult to determine from the graph. For this reason it may be necessary to use algebraic methods to find the solution to a system of linear equations in two variables. The first of these methods is **substitution.**

1 Solve Systems of Equations by Substitution

To Solve a System of Equations by Substitution

1. If necessary, solve for a variable in either equation. If possible, solve for a variable with a numerical coefficient of 1 or -1 to avoid working with fractions.
2. Substitute the expression found for the variable in step 1 into the other equation.
3. Solve the equation determined in step 2 to find the value of one variable.
4. Substitute the value found in step 3 into the equation obtained in step 1 to find the value of the other variable.
5. Check by substituting both values in both original equations.

EXAMPLE 1 Solve the following system of equations by substitution.

$$x + 2y = 4$$
$$y = 2x - 3$$

Solution

Step 1 We can choose to solve for either variable in either equation. Since the second equation is already solved for y, we choose the second equation.

$y = 2x - 3$ Second equation already solved for y.

Step 2 Next, since y is equal to the expression $2x - 3$, we will substitute the expression $2x - 3$ in for y in the *other equation*, $x + 2y = 4$.

$$x + 2y = 4 \quad \text{First Equation}$$

$$x + 2\overbrace{(2x - 3)} = 4 \quad \text{Substitute } 2x - 3 \text{ in for } y.$$

Step 3 This equation now has only one variable, x. We will now solve for x.

$$\begin{aligned} x + 2(2x - 3) &= 4 \\ x + 4x - 6 &= 4 \\ 5x - 6 &= 4 \\ 5x &= 10 \\ x &= 2 \end{aligned}$$

Step 4 The x-coordinate of the solution is 2. To find the value of the y-coordinate, substitute $x = 2$ in the second equation.

$$\begin{aligned} y &= 2x - 3 \\ y &= 2(2) - 3 \\ y &= 4 - 3 \\ y &= 1 \end{aligned}$$

Step 5 Check the solution $(2,1)$ by substituting $x = 2$ and $y = 1$ into both equations.

$$\begin{aligned} x + 2y &= 4 \\ 2 + 2(1) &= 4 \\ 2 + 2 &= 4 \\ 4 &= 4 \quad \text{True} \end{aligned} \qquad \begin{aligned} y &= 2x - 3 \\ 1 &= 2(2) - 3 \\ 1 &= 4 - 3 \\ 1 &= 1 \quad \text{True} \end{aligned}$$

Since $(2,1)$ satisfies both equations, it is the solution to the system of equations.

Now Try Exercise 5

The system from Example 1 was solved graphically in Example 3 of Section 8.1. **Figure 8.2** on page 479 shows that the solution $(2, 1)$ is the point of intersection of the graphs of the two lines.

EXAMPLE 2 Solve the following system of equations by substitution.

$$\begin{aligned} 2x + y &= 3 \\ 4x + 2y &= 12 \end{aligned}$$

Solution First solve for y in $2x + y = 3$.

Step 1

$$\begin{aligned} 2x + y &= 3 \\ y &= -2x + 3 \end{aligned}$$

Now substitute the expression $-2x + 3$ for y in the *other equation*, $4x + 2y = 12$, and solve for x.

Step 2

$$\begin{aligned} 4x + 2y &= 12 \\ 4x + 2\overbrace{(-2x + 3)} &= 12 \quad \text{Substitution; this is now an equation in only one variable, } x. \\ 4x - 4x + 6 &= 12 \\ 6 &= 12 \quad \text{False} \end{aligned}$$

Since the statement $6 = 12$ is false, the system has no solution. Therefore, the graphs of the equations will be parallel lines and the system is inconsistent because it has no solution.

Now Try Exercise 13

Understanding Algebra

If when solving a system of equations you obtain an equation that is always

- *false*, such as $6 = 12$ or $0 = 6$, the system is inconsistent and has no solution.
- *true*, such as $2 = 2$ or $0 = 0$, the system is dependent and has an infinite number of solutions.

Note that the solution in Example 2 is identical to the graphical solution obtained in Example 4 of Section 8.1. **Figure 8.3** on page 479 shows the parallel lines.

EXAMPLE 3 Solve the following system of equations by substitution.

$$x - \frac{1}{2}y = 2$$
$$y = 2x - 4$$

Solution The equation $y = 2x - 4$ is already solved for y. Substitute $2x - 4$ for y in the other equation, $x - \frac{1}{2}y = 2$, and solve for x.

$$x - \frac{1}{2}y = 2$$

Step 2

$$x - \frac{1}{2}(2x - 4) = 2$$
$$x - x + 2 = 2$$
$$2 = 2 \quad \text{True}$$

Since the statement $2 = 2$ is true, this system has an infinite number of solutions. Therefore, the graphs of the equations represent the same line and the system is dependent.

Now Try Exercise 15

Note that the solution in Example 3 is identical to the solution obtained graphically in Example 5 of Section 8.1. **Figure 8.4** on page 480 shows that the graphs of both equations are the same line.

EXAMPLE 4 Solve the following system of equations by substitution.

$$3x + 6y = 9$$
$$2x - 3y = 6$$

Solution None of the variables in either equation has a numerical coefficient of 1. However, since the numbers 3, 6, and 9 are all divisible by 3, if you solve the first equation for x, you will avoid having to work with fractions.

$$3x + 6y = 9$$
$$3x = -6y + 9$$
$$\frac{3x}{3} = \frac{-6y + 9}{3}$$
$$x = -\frac{6}{3}y + \frac{9}{3}$$
$$x = -2y + 3$$

Now substitute $-2y + 3$ for x in the other equation, $2x - 3y = 6$, and solve for the remaining variable, y.

$$2x - 3y = 6$$
$$2(-2y + 3) - 3y = 6$$
$$-4y + 6 - 3y = 6$$
$$-7y + 6 = 6$$
$$-7y = 0$$
$$y = 0$$

Finally, solve for x by substituting $y = 0$ in the equation previously solved for x.

$$x = -2y + 3$$
$$x = -2(0) + 3 = 0 + 3 = 3$$

The solution is $(3, 0)$.

Now Try Exercise 11

HELPFUL HINT

Remember that a solution to a system of linear equations must contain both an x- and a y-value. Write the solution as an ordered pair.

EXAMPLE 5 Solve the following system of equations by substitution.

$$6x + 8y = 3$$
$$3x = 3y + 5$$

Solution We will solve for x in the second equation.

$$3x = 3y + 5$$
$$x = \frac{3y + 5}{3}$$
$$x = y + \frac{5}{3}$$

Now substitute $y + \frac{5}{3}$ for x in the other equation.

$$6x + 8y = 3$$
$$6\left(y + \frac{5}{3}\right) + 8y = 3$$
$$6y + 10 + 8y = 3$$
$$14y + 10 = 3$$
$$14y = -7$$
$$y = \frac{-7}{14} = -\frac{1}{2}$$

Finally, find the value of x.

$$x = y + \frac{5}{3}$$
$$x = -\frac{1}{2} + \frac{5}{3} = -\frac{3}{6} + \frac{10}{6} = \frac{7}{6}$$

The solution is the ordered pair $\left(\frac{7}{6}, -\frac{1}{2}\right)$.

Now Try Exercise 29

EXAMPLE 6 **Wisconsin Towns** The total population of the towns of Dickeyville and Muscoda, Wisconsin, is 2462. If the population of Muscoda is 372 more than the population of Dickeyville, find the population of each town.

Solution Understand and Translate We will use two variables to solve this problem.

Let x = population of Dickeyville.
Let y = population of Muscoda.

Since we use two variables we must write two equations. The first equation comes from the phrase

the total population of the towns . . . is 2462

$$x + y = 2462$$

The second equation comes from the phrase

the population of Muscoda is 372 more than the population of Dickeyville

$$y = x + 372$$

Thus we have the system of equations

$$x + y = 2462$$
$$y = x + 372$$

Carry Out Because the second equation, $y = x + 372$, is already solved for y, we will substitute $x + 372$ for y in the first equation.

$$\begin{aligned} x + y &= 2462 \\ x + \overbrace{x + 372} &= 2462 \\ 2x + 372 &= 2462 \\ 2x &= 2090 \\ x &= 1045 \end{aligned}$$

Check and Answer The population of Dickeyville is 1045. The population of Muscoda is therefore $x + 372 = 1045 + 372 = 1417$. We check the solution by noting the total population of the two towns is $1045 + 1417 = 2462$.

Now Try Exercise 37

8.2 Exercise Set

Warm-Up Exercises

Fill in the blanks with the appropriate word, phrase, or symbol(s) from the following list.

factor	x-intercept	unique	infinite
intersection	solution	coefficient	constant

1. When solving a system of linear equations by substitution, if possible, solve for a variable with a ________________ of 1 or −1 to avoid working with fractions.

2. The solution to a system of linear equations occurs at the point of ____________ of the graphs of the equations in the system.

3. If when solving a system of linear equations you obtain an equation that is always false, the system is inconsistent and has no ________________.

4. If when solving a system of linear equations you obtain an equation that is always true, the system is dependent and has an ________________ number of solutions.

Practice the Skills

Find the solution to each system of equations by substitution.

5. $x = y + 1$, $4x + 2y = -14$

6. $y = -2x + 7$, $x + 4y = 0$

7. $x - 3y = -1$, $2x + 4y = 8$

8. $5x - 2y = -7$, $8 = y - 4x$

9. $3x + 4y = 8$, $y = \frac{1}{2}x - 3$

10. $x = \frac{1}{3}y + 1$, $9x - 5y = -3$

11. $2x + 5y = 9$, $6x - 2y = 10$

12. $3x - 5y = -4$, $2x - 8y = 2$

13. $3x + y = 3$, $3x + y + 5 = 0$

14. $3x - y = 8$, $6x - 2y = 10$

15. $y = \frac{1}{3}x - 2$, $x - 3y = 6$

16. $x - \frac{1}{2}y = 7$, $y = 2x - 14$

17. $x = 4y - 1$, $x = -3y + 13$

18. $y = 2x + 7$, $y = -x - 5$

19. $x = 3$, $x + y + 5 = 0$

20. $y = 2x + 4$, $y = -2$

21. $2x - y = 3$, $2y = 4x - 6$

22. $x + 2y = 6$, $4y = 12 - 2x$

23. $2x + 3y = 3$
$2x + 3y = -3$

24. $3x - 4y = 15$
$-6x + 8y = -14$

25. $\frac{1}{4}x - \frac{1}{3}y = \frac{1}{2}$
$\frac{1}{2}x - \frac{1}{4}y = -\frac{1}{4}$

26. $\frac{1}{4}x - \frac{1}{10}y = \frac{1}{5}$
$\frac{1}{21}x - \frac{1}{7}y = -\frac{1}{3}$

27. $3x - 2y = 8$
$x + 4y = 5$

28. $2x + 3y = 7$
$6x - y = 1$

29. $4x + 5y = -6$
$2x - \frac{10}{3}y = -4$

30. $4x - y = 1$
$10x + \frac{1}{2}y = 1$

Problem Solving

31. **Positive Integers** The sum of two positive integers is 63. The first number is twice the second number. Find the two numbers.

32. **Positive Integers** The difference of two positive integers is 44. Find the integers if the larger number is twice the smaller number.

33. **Rectangle** The perimeter of a rectangle is 50 feet. Find the dimensions of the rectangle if the length is 9 feet greater than the width.

34. **Rectangle** The perimeter of a rectangle is 60 feet. Find the dimensions of the rectangle if its length is 5 times its width.

35. **Soccer Game** A total of 550 adults and students attended a soccer game at Bayshore High School. If there were 150 more students than adults, how many adults and how many students attended the game?

36. **Wooden Horse** To buy a statue of a wooden horse Billy and Jean combined their money. Together they had \$530. If Jean had \$130 more than Billy, how much money did each have?

37. **Legal Settlement** After a legal settlement, the client's portion of the award was three times as much money as the attorney's portion. If the total award was \$40,000, how much did the client get?

38. **Jelly Beans** A candy store mixed green and red jelly beans in a barrel. The barrel contains 42 pounds of the mixture. If there are 3 times as many pounds of the green jelly beans as red jelly beans, find the number of pounds of green jelly beans and the number of pounds of red jelly beans in the barrel.

39. **Refinancing** Dona Lee is considering refinancing her house. The cost of refinancing is a one-time charge of \$1280. With her reduced mortgage rate, her monthly interest and principal payments would be \$794 per month. Her total cost, c, for n months could be represented by $c = 1280 + 794n$. At her current rate her mortgage payments are \$874 per month and the total cost for n months can be represented by $c = 874n$.

a) In how many months would both mortgage plans have the same total cost?

b) If Dona plans to remain in her house for 12 years, should she refinance?

40. **Temperatures** In Seattle the temperature is 86°F, but it is decreasing by 2 degrees per hour. The temperature, T, at time, t, in hours, is represented by $T = 86 - 2t$. In Spokane the temperature is 59°F, but it is increasing by 2.5 degrees per hour. The temperature, T, can be represented by $T = 59 + 2.5t$.

a) If the temperature continues decreasing and increasing at the same rate in these cities, how long will it be before both cities have the same temperature?

b) When both cities have the same temperature, what will that temperature be?

Seattle, Washington

41. **Traveling by Car** Jean Woody's car is at the 80 mile marker on a highway. Roberta Kieronski's car is 15 miles behind Jean's car. Jean's car is traveling at 60 miles per hour. The mile marker that Jean's car will be at in t hours can be found by the equation $m = 80 + 60t$. Roberta's car is traveling at 72 miles per hour. The mile marker that Roberta's car will be at in t hours can be found by the equation $m = 65 + 72t$.

a) Determine the time it will take for Roberta's car to catch up with Jean's car.

b) At which mile marker will they be when they meet?

42. Computer Store Will Worthy's present salary consists of a fixed weekly salary of \$300 plus a \$20 bonus for each computer system he sells. His weekly salary can be represented by $s = 300 + 20n$, where n is the number of computer systems he sells. He is considering another position where his weekly salary would be \$400 plus a \$10 bonus for each computer system he sells. The other position's weekly salary can be represented by $s = 400 + 10n$. How many computer systems would Will need to sell in a week for his salary to be the same with both employers?

Challenge Problems

Answer parts **a)** *through* **d)** *on your own.*

43. Heat Transfer In a laboratory during an experiment on heat transfer, a large metal ball is heated to a temperature of 180°F. This metal ball is then placed in a gallon of oil at a temperature of 20°F. Assume that when the ball is placed in the oil it loses temperature at the rate of 10 degrees per minute while the oil's temperature rises at a rate of 6 degrees per minute.

a) Write an equation that can be used to determine the ball's temperature t minutes after being placed in the oil.

b) Write an equation that can be used to determine the oil's temperature t minutes after the ball is placed in it.

c) Determine how long it will take for the ball and oil to reach the same temperature.

d) When the ball and oil reach the same temperature, what will the temperature be?

Group Activity

Discuss and work Exercise 44 as a group.

44. In intermediate algebra you may solve systems containing three equations with three variables. As a group, solve the system of equations on the right. Your answer will be in the form of an **ordered triple** (x, y, z).

$$x = 4$$
$$2x - y = 6$$
$$-x + y + z = -3$$

Cumulative Review Exercises

[2.7] **45. Willow Tree** The diameter of a willow tree grows about 1.2 inches per year. What is the approximate age of a willow tree whose diameter is 27.6 inches?

[4.5] **46.** Multiply $(6x + 7)(3x - 2)$.

[7.2] **47.** Graph $4x - 8y = 16$ using the intercepts.

[7.4] **48.** Find the slope and y-intercept of the graph of the equation $3x - 5y = 25$.

49. Determine the equation of the line shown at right.

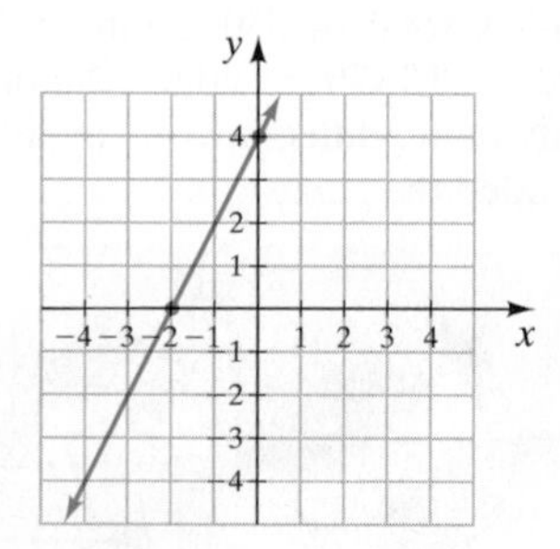

8.3 Solving Systems of Equations by the Addition Method

1 Solve systems of equations by the addition method.

1 Solve Systems of Equations by the Addition Method

A third method of solving a system of linear equations, and the second algebraic method, is the **addition (or elimination) method.**

In the addition method, we use the fact that if $a = b$ and $c = d$, then $a + c = b + d$. Suppose we have the following two equations.

$$x - 2y = 9$$
$$3x + 2y = 11$$

Notice the like terms of the two equations are aligned. Also notice that $-2y$ and $+2y$ are opposites of each other. If we *add* the like terms of the two equations together we are able to *eliminate* the y variable.

Understanding Algebra

The goal of both the substitution method and the addition method is to obtain a single equation with only one variable. With the addition method, we add two equations together so that one of the variables is eliminated and we are able to solve for the other variable.

$$\begin{aligned} x - 2y &= 9 \\ 3x + 2y &= 11 \\ \hline 4x \quad &= 20 \end{aligned}$$ Add the like terms of the two equations.

Now we are able to solve this single equation for x.

$$4x = 20$$
$$x = 5$$

We can now substitute 5 for x in either of the original equations to find y. We will substitute 5 for x in $x - 2y = 9$.

$$x - 2y = 9$$
$$5 - 2y = 9 \quad \text{Substitute } x = 5 \text{ into first equation.}$$
$$-2y = 4 \quad \text{Solve for } y.$$
$$y = -2$$

The solution to the system is $(5, -2)$. This ordered pair will check in both equations. Check this solution now.

To Solve a System of Equations Using the Addition (or Elimination) Method

1. If necessary, rewrite each equation so that the terms containing variables appear on the left side of the equation and any constants appear on the right side of the equation.
2. If necessary, multiply one or both equations by a constant(s) so that when the equations are added the resulting sum will contain only one variable.
3. Add the equations. This will result in a single equation containing only one variable.
4. Solve for the variable in the equation from step 3.
5. **a)** Substitute the value found in step 4 into either of the original equations. Solve that equation to find the value of the remaining variable.
   or
   b) Repeat steps 2–4 to eliminate the other variable.
6. Check the values obtained in all original equations.

Understanding Algebra

In step 5, we generally use

- method **a)** if the value obtained for the first variable is an integer.
- method **b)** if the value obtained for the first variable is a fraction.

EXAMPLE 1 Solve the following system of equations using the addition method.

$$x + y = 6$$
$$2x - y = 3$$

Solution Note that the first equation contains $+y$ and the second equation contains $-y$. By *adding* the equations, we can *eliminate* the variable y and obtain one equation containing only one variable, x.

$$\begin{aligned} x + y &= 6 \\ 2x - y &= 3 \\ \hline 3x \quad &= 9 \end{aligned}$$ Add the like terms of the two equations.

Now we solve for the remaining variable, x.

$$\frac{3x}{3} = \frac{9}{3} \quad \text{Solve for } x.$$
$$x = 3$$

Finally, we solve for y by substituting $x = 3$ in either of the original equations.

$$x + y = 6$$
$$3 + y = 6$$
$$y = 3$$

The solution is $(3,3)$.

Check We check the ordered pair $(3, 3)$ in *both* equations.

$$\begin{aligned} x + y &= 6 \\ 3 + 3 &\stackrel{?}{=} 6 \\ 6 &= 6 \quad \text{True} \end{aligned} \qquad \begin{aligned} 2x - y &= 3 \\ 2(3) - 3 &\stackrel{?}{=} 3 \\ 6 - 3 &\stackrel{?}{=} 3 \\ 3 &= 3 \quad \text{True} \end{aligned}$$

Now Try Exercise 7

Understanding Algebra

Adding the two equations in the system

$$3x + 2y = 6$$
$$x - y = 8$$

will not eliminate either variable. However, if we first multiply the second equation by -3, the system becomes

$$3x + 2y = 6$$
$$-3x + 3y = -24$$

Now, adding the equations will eliminate the variable x.

When solving systems of equations with the addition method, it is often necessary to multiply one or both equations in the system by a constant. This is done to get opposite terms that will be eliminated when the equations are added.

For example, consider the system of equations below.

$$\begin{aligned} 3x + 2y &= 6 \qquad eq.\ 1 \\ x - y &= 8 \qquad eq.\ 2 \end{aligned}$$

To get the x-terms to be opposites of each other, we can multiply *eq.* 2 by -3.

$$3x + 2y = 6 \xrightarrow{eq.\ 1 \text{ stays the same}} 3x + 2y = 6$$
$$-3(x - y) = -3(8) \xrightarrow{eq.\ 2 \text{ multiplied by } -3} -3x + 3y = -24$$

After this step, we can continue to solve the system in a manner similar to what we did in Example 1. We could have also multiplied *eq.* 2 by 2 so that the y-terms would be opposite and would be eliminated when the equations were added.

EXAMPLE 2 Solve the following system of equations using the addition method.

$$\begin{aligned} x + 3y &= 13 \qquad eq.\ 1 \\ x + 4y &= 18 \qquad eq.\ 2 \end{aligned}$$

Solution Our goal is to rewrite the system so that we have opposite terms that are eliminated when the two equations are added together. We will multiply *eq.* 1 by -1.

$$-1(x + 3y) = -1 \cdot 13 \xrightarrow{eq.\ 1 \text{ multiplied by } -1} -x - 3y = -13$$
$$x + 4y = 18 \xrightarrow{eq.\ 2 \text{ stays the same}} x + 4y = 18$$

Understanding Algebra

When using the addition method, if you need to multiply an equation by a constant, remember that *both sides of the equation must be multiplied* by the constant. This will produce an equivalent equation and will not change the solution to the system of equations.

Now add the two equations.

$$\begin{aligned} -x - 3y &= -13 \\ x + 4y &= 18 \\ \hline y &= 5 \end{aligned}$$

Next, we solve for x in either of the original equations. We will substitute $y = 5$ into *eq.* 1.

$$\begin{aligned} x + 3y &= 13 \\ x + 3(5) &= 13 \\ x + 15 &= 13 \\ x &= -2 \end{aligned}$$

A check will show that the solution is $(-2, 5)$.

Now Try Exercise 9

EXAMPLE 3 Solve the following system of equations using the addition method.

$$\begin{aligned} x + 2y &= 4 \\ y &= 2x - 3 \end{aligned}$$

Solution Our first step is to rewrite the second equation so the variable terms are on the left side of the equation and the constant is on the right side of the equation.

$$x + 2y = 4 \xrightarrow{\text{First equation stays the same}} x + 2y = 4 \qquad eq.\,1$$

$$y = 2x - 3 \xrightarrow{\text{Subtracted } 2x \text{ from both sides}} -2x + y = -3 \qquad eq.\,2$$

Next, to eliminate the variable x we will multiply *eq.* 1 by 2 and then add the two equations.

$$2(x + 2y) = 2(4) \xrightarrow{eq.\ 1 \text{ multiplied by } 2} 2x + 4y = 8$$

$$-2x + y = -3 \xrightarrow{eq.\ 2 \text{ stays the same}} -2x + y = -3$$

Now add the two equations:

$$\begin{aligned} 2x + 4y &= 8 \\ -2x + y &= -3 \\ \hline 5y &= 5 \\ y &= 1 \end{aligned}$$

Next, solve for x by substituting $y = 1$ into the *original* first equation.

$$\begin{aligned} x + 2y &= 4 \\ x + 2(1) &= 4 \\ x + 2 &= 4 \\ x &= 2 \end{aligned}$$

Understanding Algebra

In Example 3, we could have solved the system by multiplying *eq.* 2 by -2 to eliminate the y-terms. Try this now. You should get the same answer.

The solution is $(2,1)$.

Now Try Exercise 15

Note that the solution in Example 3 is the same as the solution obtained graphically in Example 3 of Section 8.1 (see **Fig. 8.2** *on page 479) and by substitution in Example 1 of Section 8.2 on page 484.*

EXAMPLE 4 Solve the system of equations using the addition method.

$$3x = 2y + 5$$
$$6y = -5x + 27$$

Solution Our first step is to rewrite each equation so the variable terms are on the left side and the constants are on the right side of the equation.

$$3x = 2y + 5 \xrightarrow{\text{Subtracted } 2y \text{ from the both sides}} 3x - 2y = 5 \qquad eq.\,1$$

$$6y = -5x + 27 \xrightarrow{\text{Add } 5x \text{ to both sides}} 5x + 6y = 27 \qquad eq.\,2$$

Now, continue to solve the system. To eliminate the y-terms, multiply *eq.* 1 by 3.

$$3(3x - 2y) = 3 \cdot 5 \xrightarrow{eq.\ 1 \text{ multiplied by } 3} 9x - 6y = 15$$

$$5x + 6y = 27 \xrightarrow{eq.\ 2 \text{ stays the same}} 5x + 6y = 27$$

Next, add the two equations and solve for x.

$$\begin{aligned} 9x - 6y &= 15 \\ 5x + 6y &= 27 \\ \hline 14x \quad &= 42 \\ x &= 3 \end{aligned}$$

Solve for y by substituting $x = 3$ into *eq.* 2.

$$\begin{aligned} 6y &= -5x + 27 \\ 6y &= -5(3) + 27 \\ 6y &= -15 + 27 \\ 6y &= 12 \\ y &= 2 \end{aligned}$$

The solution is $(3, 2)$. Check the solution by substituting $x = 3$ and $y = 2$ into both of the *original* equations.

Now Try Exercise 31

HELPFUL HINT

When solving a system of equations, if a mistake is made midway through the solution process, you will discover the error by substituting the values for the variables back into the *original* equations during the *check* phase.

EXAMPLE 5 Solve the following system of equations using the addition method.

$$\begin{aligned} 2x + 3y &= 6 \qquad \textit{eq. } 1 \\ 5x - 4y &= -8 \qquad \textit{eq. } 2 \end{aligned}$$

Solution Observe the coefficients on the x-terms and the y-terms in the system. We will have to multiply each equation in the system by a different constant to get opposite terms. The variable x can be eliminated by multiplying *eq.* 1 by -5 and *eq.* 2 by 2 and then adding the equations.

$$-5(2x + 3y) = -5 \cdot 6 \xrightarrow{\textit{eq. } 1 \text{ multiplied by } -5} -10x - 15y = -30$$

$$2(5x - 4y) = 2 \cdot (-8) \xrightarrow{\textit{eq. } 2 \text{ multiplied by } 2} 10x - 8y = -16$$

$$\begin{aligned} -10x - 15y &= -30 \\ 10x - 8y &= -16 \\ \hline -23y &= -46 \\ y &= 2 \end{aligned}$$

Solve for x by substituting $y = 2$ into *eq.* 1.

$$\begin{aligned} 2x + 3y &= 6 \\ 2x + 3(2) &= 6 \\ 2x + 6 &= 6 \\ 2x &= 0 \\ x &= 0 \end{aligned}$$

The solution is $(0, 2)$.

Understanding Algebra

In Example 5, we also could have solved the system by first multiplying *eq.* 1 by 4 and *eq.* 2 by 3 to eliminate the y-terms. Try this now. You should get the same answer.

Now Try Exercise 23

EXAMPLE 6 Solve the following system of equations using the addition method.

$$\begin{aligned} 2x + y &= 3 \qquad \textit{eq. } 1 \\ 4x + 2y &= 12 \qquad \textit{eq. } 2 \end{aligned}$$

Solution The variable y can be eliminated by multiplying *eq.* 1 by -2 and then adding the two equations.

$$-2(2x + y) = -2 \cdot 3 \xrightarrow{\textit{eq. } 1 \text{ multiplied by } -2} -4x - 2y = -6$$

$$4x + 2y = 12 \xrightarrow{\textit{eq. } 2 \text{ stays the same}} 4x + 2y = 12$$

$$\begin{aligned} -4x - 2y &= -6 \\ 4x + 2y &= 12 \\ \hline 0 &= 6 \quad \text{False} \end{aligned}$$

Since $0 = 6$ is a false statement, this system has no solution. The system is inconsistent. The graphs of the equations will be parallel lines.

Now Try Exercise 17

Note that the solution in Example 6 is identical to the solutions obtained by graphing in Example 4 of Section 8.1 (see **Fig. 8.3** *on page 479) and by substitution in Example 2 of Section 8.2 on page 485.*

EXAMPLE 7 Solve the following system of equations using the addition method.

$$x - \frac{1}{2}y = 2$$
$$y = 2x - 4$$

Solution First align the x- and y-terms on the left side of the equation by subtracting $2x$ from both sides of the second equation.

$$x - \frac{1}{2}y = 2 \qquad eq.\,1$$
$$-2x + y = -4 \qquad eq.\,2$$

Now proceed as in the previous examples. Begin by multiplying *eq.* 1 by 2.

$$2\left(x - \frac{1}{2}y\right) = 2 \cdot 2 \xrightarrow{eq.\,1 \text{ multiplied by } 2} 2x - y = 4$$
$$-2x + y = -4 \xrightarrow{eq.\,2 \text{ stays the same}} -2x + y = -4$$

$$\begin{aligned} 2x - y &= 4 \\ -2x + y &= -4 \\ \hline 0 &= 0 \quad \text{True} \end{aligned}$$

Understanding Algebra

If when solving a system of equations you obtain an equation that is always

- *false*, such as $0 = 6$, the system is inconsistent and has no solution.
- *true*, such as $0 = 0$, the system is dependent and has an infinite number of solutions.

Since $0 = 0$ is a true statement, the system is dependent and has an infinite number of solutions. When graphed, both equations will be the same line.

Now Try Exercise 21

The solution in Example 7 is the same as the solutions obtained by graphing in Example 5 of Section 8.1 (see **Fig. 8.4** *on page 480) and by substitution in Example 3 of Section 8.2 on page 486.*

EXAMPLE 8 Solve the following system of equations using the addition method.

$$2x + 3y = 7 \qquad eq.\,1$$
$$5x - 7y = -3 \qquad eq.\,2$$

Solution We can eliminate the variable x by multiplying *eq.* 1 by -5 and *eq.* 2 by 2.

$$-5(2x + 3y) = -5 \cdot 7 \xrightarrow{eq.\,1 \text{ multiplied by } -5} -10x - 15y = -35$$
$$2(5x - 7y) = 2(-3) \xrightarrow{eq.\,2 \text{ multiplied by } 2} 10x - 14y = -6$$

$$\begin{aligned} -10x - 15y &= -35 \\ 10x - 14y &= -6 \\ \hline -29y &= -41 \end{aligned}$$

$$y = \frac{41}{29}$$

We can now find x by substituting $y = \frac{41}{29}$ into one of the original equations and solving for x. An easier method of solving for x is to go back to the original equations and eliminate the variable y. We can do this by multiplying *eq.* 1 by 7 and *eq.* 2 by 3.

$$7(2x + 3y) = 7 \cdot 7 \quad \text{or} \quad 14x + 21y = 49$$
$$3(5x - 7y) = 3(-3) \quad \text{or} \quad 15x - 21y = -9$$

$$\begin{aligned} 14x + 21y &= 49 \\ 15x - 21y &= -9 \\ \hline 29x \quad &= 40 \\ x &= \frac{40}{29} \end{aligned}$$

The solution is $\left(\frac{40}{29}, \frac{41}{29}\right)$.

Now Try Exercise 37

HELPFUL HINT

In Sections 8.1, 8.2, and 8.3 we have introduced three different methods for solving linear systems of equations in two variables. Which method should you use? The table below can help you decide.

Method	Comments
Graphing	• Graphing is used when you wish to see a visual representation of the system of equations and the solution. • Graphing may not be the best method to use when an exact answer is needed–especially if the solution includes fractions.
Substitution	• Substitution will give you an exact answer. • Substitution works well if at least one of the variable terms in at least one of the equations has a coefficient of 1 or −1.
Addition (or Elimination)	• Addition will give you an exact answer. • Addition works well when neither of the variable terms in either equation has a coefficient of 1 or −1.

8.3 Exercise Set MathXL® MyMathLab®

Warm-Up Exercises

Fill in the blanks with the appropriate word, phrase, or symbol(s) from the following list.

factor eliminate unique infinite intersection solution constant

1. When solving a system of linear equations using the addition method, our goal is to __________ one of the variables when we add the two equations together.
2. The solution to a system of linear equations occurs at the point of __________ of the graphs of the equations in the system.
3. If, when solving a system of linear equations by the addition method, you obtain an equation that is always true, the system is dependent and has an __________ number of solutions.
4. If, when solving a system of linear equations by the addition method, you obtain an equation that is always false, the system is inconsistent and has no __________.

Practice the Skills

Solve each system of equations using the addition method.

5. $x + y = 6$
   $x - y = 4$
6. $x - y = 3$
   $x + y = 5$
7. $x - 3y = 1$
   $x + 3y = -5$
8. $x + 2y = 21$
   $2x - 2y = -6$
9. $-5x + y = 14$
   $-3x + y = -2$
10. $6x + 3y = 30$
    $2x + 3y = 18$

11. $5x + 3y = 12$
$3x - 6y = 15$

12. $-4x + 3y = 0$
$7x - 6y = 3$

13. $2x + 5y = 11$
$6x + 7y = 9$

14. $3x - 4y = 15$
$5x - 12y = 41$

15. $3x + 2y = 5$
$y = -5x - 1$

16. $x = -5y + 1$
$2x + 7y = -1$

17. $4x + y = 6$
$-8x - 2y = 20$

18. $x - y = 2$
$3x - 3y = 1$

19. $2x + y = -6$
$2x - 2y = 3$

20. $8x - 4y = 12$
$2x - 8y = 3$

21. $-2y = -4x + 12$
$y = 2x - 6$

22. $6x - 3y = 18$
$4y = 8x - 24$

23. $4x + 9y = -3$
$5x - 2y = -17$

24. $2x + 3y = -3$
$-3x - 5y = 7$

25. $5x - 4y = -3$
$7y = 2x + 12$

26. $2x - 3y = 11$
$5y = 3x - 17$

27. $\frac{2}{9}x + \frac{1}{3}y = \frac{1}{9}$
$\frac{1}{3}x - \frac{1}{5}y = \frac{13}{15}$

28. $-\frac{x}{4} + \frac{y}{3} = \frac{17}{12}$
$\frac{3}{10}x - \frac{y}{2} = -\frac{19}{10}$

29. $3x - y = 4$
$2x - \frac{2}{3}y = 8$

30. $\frac{3}{7}y - x = \frac{12}{7}$
$y - \frac{7}{3}x = \frac{1}{3}$

31. $6x = 4y + 12$
$3y - 5x = -6$

32. $3x + 1 = -5y$
$16 - 8y = 7x$

33. $\frac{x}{5} - \frac{y}{2} = \frac{1}{10}$
$2x - 5y = 1$

34. $-5x + 6y = -12$
$\frac{5}{3}x - 4 = 2y$

35. $-5x + 4y = -20$
$3x - 2y = 15$

36. $4x - 3y = -4$
$3x - 5y = 10$

37. $4x + 5y = 0$
$3x = 6y + 4$

38. $5x = 2y - 4$
$3x - 5y = 6$

39. $x - \frac{1}{2}y = 4$
$3x + y = 6$

40. $4x - 3y = 8$
$-3x + 4y = 9$

Problem Solving

41. Sum of Numbers The sum of two numbers is 60. When the second number is subtracted from the first number, the difference is 38. Find the two numbers.

42. Sum of Numbers The sum of two numbers is 46. When the first number is subtracted from the second number, the difference is 6. Find the two numbers.

43. Sum of Numbers The sum of a number and twice a second number is 14. When the second number is subtracted from the first number, the difference is 2. Find the two numbers.

44. Sum of Numbers The sum of two numbers is 9. Twice the first number subtracted from three times the second number is 7. Find the two numbers.

45. Rectangles When the length of a rectangle is x inches and the width is y inches, the perimeter is 18 inches. If the length is doubled and the width is tripled, the perimeter becomes 42 inches. Find the length and width of the original rectangle.

46. Perimeter of a Rectangle When the length of a rectangle is x inches and the width is y inches, the perimeter is 28 inches. If the length is doubled and the width is tripled, the perimeter becomes 66 inches. Find the length and width of the original rectangle.

47. Photograph A photograph has a perimeter of 36 inches. The difference between the photograph's length and width is 2 inches. Find the length and width of the photograph.

48. Rectangular Garden Montreal Botanical Garden has a number of rectangular flower gardens, all with a perimeter of 82 feet. The difference between the garden's length and width is 11 feet. Determine the length and width of the gardens.

Montreal Botanical Garden

Concept/Writing Exercises

49. Construct a system of two equations that has no solution. Explain how you know the system has no solution.

50. Construct a system of two equations that has an infinite number of solutions. Explain how you know the system has an infinite number of solutions.

51. a) Solve the system of equations

$$4x + 2y = 1000$$
$$2x + 4y = 800$$

b) If we divide all the terms in the top equation by 2, we get the system shown at right:

$$2x + y = 500$$
$$2x + 4y = 800$$

How will the solution to this system compare to the solution in part **a)**? Explain and then check your explanation by solving this system.

52. Suppose we divided all the terms in both equations given in Exercise 51 **a)** by 2, and then solved the system. How will the solution to this system compare to the solution in part **a)**? Explain and then check your explanation by solving each system.

Challenge Problems

In Exercises 53 and 54, solve each system of equations using the addition method. (Hint: First remove all fractions by multiplying both sides of the equation by the LCD.)

53. $\dfrac{x+2}{2} - \dfrac{y+4}{3} = 4$

$\dfrac{x+y}{2} = \dfrac{1}{2} + \dfrac{x-y}{3}$

54. $\dfrac{5}{2}x + 3y = \dfrac{9}{2} + y$

$\dfrac{1}{4}x - \dfrac{1}{2}y = 6x + 12$

In intermediate algebra you may solve systems of three equations with three unknowns. Solve the following system.

55.

$$x + 2y - z = 2$$
$$2x - y + z = 3$$
$$3x + y + z = 8$$

Hint: Work with *one pair* of equations to get one equation in two unknowns. Then work with *a different pair* of the original equations to get another equation in the same two unknowns. Then solve the system of two equations in two unknowns. List your answer as an *ordered triple* of the form (x, y, z).

Group Activity

Work parts **a)** *and* **b)** *of Exercise 56 on your own. Then discuss and work parts* **c)** *and* **d)** *as a group.*

56. How difficult is it to construct a system of linear equations that has a specific solution? It is really not too difficult to do. Consider:

$$2(3) + 4(5) = 26$$
$$4(3) - 7(5) = -23$$

The system of equations

$$2x + 4y = 26$$
$$4x - 7y = -23$$

has solution $(3, 5)$.

a) Using the information provided, determine another system of equations that has $(3, 5)$ as a solution.

b) Determine a system of linear equations that has $(2, 3)$ as a solution.

c) Compare your answer with the answers of the other members of your group.

d) As a group, determine the number of systems of equations that have $(2, 3)$ as a solution.

Cumulative Review Exercises

[1.9] **57.** Evaluate 5^3.

[2.5] **58.** Solve the equation $2(2x - 5) = 2x + 4$.

[4.4] **59.** Simplify $(4x^2y - 3xy + y) - (2x^2y + 6xy - 7y)$.

[4.5] **60.** Multiply $(9a^4b^2c)(4a^2b^7c^4)$.

[5.2] **61.** Factor $xy + xc - ay - ac$ by grouping.

[7.6] **62.** If $f(x) = 2x^2 - 13$, find $f(-3)$.

MID-CHAPTER TEST: 8.1–8.3

To find out how well you understand the chapter material to this point, take this brief test. The answers, and the section where the material was initially discussed, are given in the back of the book. Review any questions that you answered incorrectly.

Determine which of the following ordered pairs satisfy each system of equations.

1. $4x + 3y = -1$
$x - 2y = 8$

a) $(-1, 1)$ **b)** $(2, -3)$

2. $6x - y = -2$
$7x + \frac{1}{2}y = 6$

a) $\left(\frac{1}{2}, 5\right)$ **b)** $\left(\frac{1}{3}, 4\right)$

Without graphing, state whether the system of equations has exactly one solution, no solution, or an infinite number of solutions.

3. $2x + y = 8$
$3x - 4y = 1$

4. $\frac{1}{2}x - 3y = 5$
$-2x + 12y = -20$

5. $y = \frac{3}{2}x + \frac{5}{2}$
$3x - 2y = 7$

Determine the solution to each system of equations graphically. If the system is dependent or inconsistent, so state.

6. $y = 2x + 1$
$y = -x + 4$

7. $x = 5$
$y = -3$

Solve each system of equations by substitution.

8. $3x + y = -2$
$2x - 3y = -16$

9. $x - 3y = 2$
$4x + 9y = 1$

10. $3x - y = 5$
$x - \frac{1}{3}y = 2$

11. Rectangle The perimeter of a rectangle is 44 feet. Find the length and the width if the length is 8 feet greater than the width.

Solve each system of equations using the addition method.

12. $x + 3y = 1$
$2x - 3y = 11$

13. $4x + 3y = 4$
$-8x + 5y = 14$

14. $5x - 2y = 1$
$-10x + 4y = -2$

15. In solving the system of equations

$$3x - 5y = -16$$
$$2x + 3y = 21,$$

Hugo Platt stated that the solution was $x = 3$. This is incorrect. Why? Explain your answer. Give the correct solution to the system.

8.4 Systems of Equations: Applications and Problem Solving

1 Use systems of equations to solve application problems.

1 Use Systems of Equations to Solve Application Problems

Many of the applications solved in earlier chapters using only one variable can now be solved using two variables. *Whenever we use two variables to solve an application problem, we must write a system of two equations.*

EXAMPLE 1 **Building a Patio** Phil Mahler is building a rectangular patio out of concrete (**Fig. 8.7**). When he measures the two angles formed by a diagonal, he finds that angle x is 22° greater than angle y. Find the two angles.

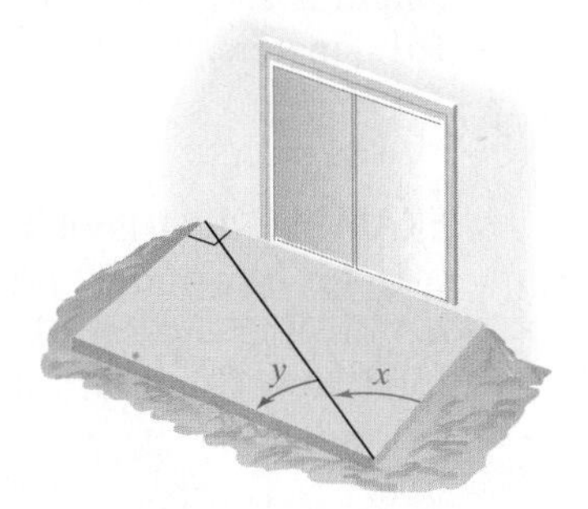

FIGURE 8.7

Understanding Algebra

- **Complementary angles** are angles whose sum measures 90°. **Figure 8.8a** illustrates complementary angles.
- **Supplementary angles** are angles whose sum measures 180°. **Figure 8.8b** illustrates supplementary angles.

Complementary Angles

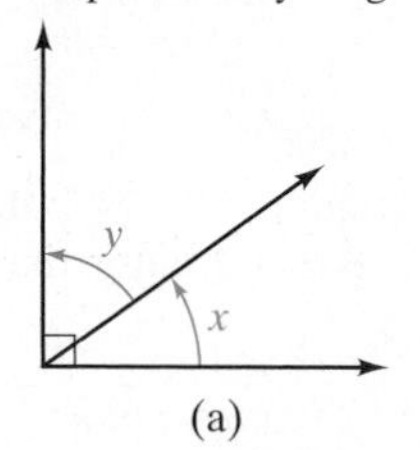

(a)

Supplementary Angles

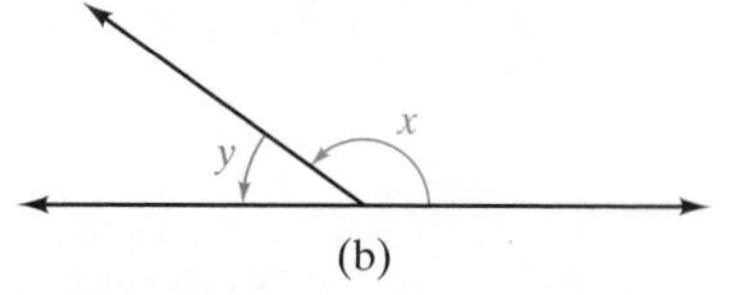

(b)

FIGURE 8.8

Solution Understand and Translate A rectangle has 4 right (or 90°) angles. Angles x and y are therefore complementary, and the sum of angles x and y is 90°. Thus one equation in the system of equations is $x + y = 90$. Since angle x is 22° greater than angle y, the second equation is $x = y + 22$.

$$\text{System of equations} \quad \begin{cases} x + y = 90 \\ x = y + 22 \end{cases}$$

Carry Out Subtract y from each side of the second equation. Then use the addition method to solve.

$$\begin{aligned} x + y &= 90 \\ x - y &= 22 \\ \hline 2x \quad &= 112 \\ x &= 56 \end{aligned}$$

Now substitute 56 for x in the first equation and solve for y.

$$\begin{aligned} x + y &= 90 \\ 56 + y &= 90 \\ y &= 34 \end{aligned}$$

Check and Answer Angle x is 56° and angle y is 34°. Note that their sum is 90° and angle x is 22° greater than angle y.

Now Try Exercise 5

Example 1 could also have been solved using the substitution method. Try solving the system now using substitution.

Temple of Hera in Paestum, Italy

EXAMPLE 2 **Greek Ruins** The perimeter of the rectangular floor of the Temple of Hera is 161.2 meters. The length is 2.1 times the width. Determine the length and width of the floor of the Temple of Hera.

Solution Understand and Translate We will let $l =$ the length and $w =$ the width of the floor of the Temple. The formula for the perimeter of a rectangle is $P = 2l + 2w$. Since the perimeter of the floor is 161.2 meters, the first equation is $161.2 = 2l + 2w$. Since the length is 2.1 times the width, the second equation is $l = 2.1w$.

$$\text{System of equations} \quad \begin{cases} 161.2 = 2l + 2w \\ l = 2.1w \end{cases}$$

Carry Out We will solve this system by substitution. Since $l = 2.1w$, substitute $2.1w$ for l in the equation $161.2 = 2l + 2w$ to obtain

$$\begin{aligned} 161.2 &= 2l + 2w \\ 161.2 &= 2(2.1w) + 2w \\ 161.2 &= 4.2w + 2w \\ 161.2 &= 6.2w \\ 26 &= w \end{aligned}$$

Check and Answer The width is 26 meters. Since the length is 2.1 times the width, the length is $2.1(26)$ or 54.6 meters. Notice that the perimeter is $2(26) + 2(54.6) =$ 161.2 meters.

Now Try Exercise 9

EXAMPLE 3 **Island Toll Bridge** The toll for the bridge to Sanibel Island, Florida, is \$6 for cars and \$2 for motorcycles. On a Sunday morning a total of 100 cars and motorcycles crossed the bridge. The total amount of tolls from these vehicles was \$540. How many cars and how many motorcycles crossed the bridge on that Sunday morning?

Understanding Algebra

Whenever we use *two* variables to solve an application problem, we must determine *two* equations to solve the problem.

Solution Understand and Translate We are asked to find the number of cars and the number of motorcycles.

$$\text{Let } x = \text{the number of cars.}$$
$$\text{Let } y = \text{the number of motorcycles.}$$

Since a total of 100 cars and motorcycles crossed the bridge, one equation is $x + y = 100$. The second equation comes from the tolls collected.

$$\text{Tolls from cars} + \text{Tolls from motorcycles} = 540$$
$$6x + 2y = 540$$

$$\text{System of equations} \begin{cases} x + y = 100 \\ 6x + 2y = 540 \end{cases}$$

Carry Out Since the first equation can be easily solved for y, we will solve this system by substitution. Solving for y in the equation $x + y = 100$ gives $y = 100 - x$. Substitute the expression $(100 - x)$ for y in the second equation and solve for x.

$$\begin{aligned} 6x + 2y &= 540 \\ 6x + 2(100 - x) &= 540 \\ 6x + 200 - 2x &= 540 \\ 4x + 200 &= 540 \\ 4x &= 340 \\ x &= 85 \end{aligned}$$

Understanding Algebra

The system of equations in Example 3 could also have been solved using the elimination method. Try this now. You should get the same answer.

Answer Therefore, 85 cars crossed the bridge. The total number of cars and motorcycles that crossed the bridge was 100. Therefore, $100 - 85$ or 15 motorcycles crossed the bridge that Sunday morning. A check will show that the total collected for these vehicles is \$540.

Now Try Exercise 11

EXAMPLE 4 **Swans** Kim Enterprise is sponsoring the construction of a swan sculpture and is considering two artists for the job. One artist, Pam, has indicated the materials needed will cost \$2200, and she will charge \$80 per hour to paint the swan. The other artist, Mike, indicated the materials will cost \$1825, and he will charge \$95 per hour to paint the swan.

a) How many hours would the painting need to take for the total cost for both artists to be the same?

b) If both artists estimate the time required for painting the swan to be 20 hours, which artist would be less expensive?

Solution **a)** Understand and Translate We will write equations for the total cost for both Pam and Mike.

$$\text{Let } n = \text{number of hours for painting.}$$
$$\text{Let } c = \text{total cost of swan.}$$

Pam	Mike
total cost = materials + labor	total cost = materials + labor
$c = 2200 + 80n$	$c = 1825 + 95n$

Now we have our system of equations.

$$\text{System of equations} \begin{cases} c = 2200 + 80n \\ c = 1825 + 95n \end{cases}$$

Carry Out

$$\text{total cost for Pam} = \text{total cost for Mike}$$
$$2200 + 80n = 1825 + 95n$$
$$375 + 80n = 95n$$
$$375 = 15n$$
$$25 = n$$

Answer If 25 hours of painting is required, the cost would be equal.

b) If the painting takes 20 hours, Mike would be less expensive, as shown below.

Pam	Mike
$c = 2200 + 80n$	$c = 1825 + 95n$
$c = 2200 + 80(20)$	$c = 1825 + 95(20)$
$c = 3800$	$c = 3725$

Now Try Exercise 17

Motion Problems with Two Rates

We introduced the distance formula, distance = rate · time or $d = rt$, in Section 2.6. Now we introduce a method using two variables and a system of equations to solve motion problems that involve two rates.

EXAMPLE 5 Meeting for Lunch Bob and Jim live 420 miles apart. Sometimes they meet for lunch at a restaurant somewhere between the towns where they live. On one occasion, they left their houses at the same time. They both arrived at the restaurant 4 hours after they began. Determine Bob's and Jim's speed if Bob drove at an average speed of 5 miles per hour faster than Jim.

Solution Understand and Translate We are asked to find Bob's and Jim's speed. We are given sufficient information in the example to obtain two equations for our system of equations.

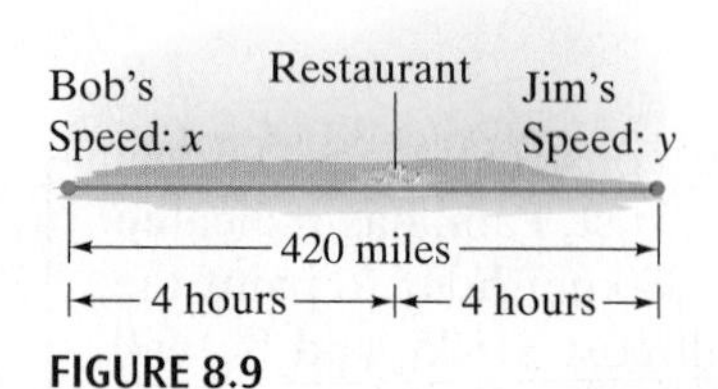

FIGURE 8.9

$$\text{Let } x = \text{Bob's speed.}$$
$$\text{Let } y = \text{Jim's speed.}$$

We will draw a sketch to help understand the problem. See **Figure 8.9**. We will use the formula, distance = rate · time. They both traveled for 4 hours.

Traveler	Rate	Time	Distance
Bob	x	4	$4x$
Jim	y	4	$4y$

Since the total distance covered is 420 miles, our first equation is

$$4x + 4y = 420$$

The second equation comes from the fact that Bob's speed was 5 miles per hour greater than Jim's speed. Therefore, we can add 5 miles per hour to Jim's speed to get Bob's speed.

$$x = y + 5$$

Our system of equations is

$$4x + 4y = 420$$
$$x = y + 5$$

Carry Out The equation $x = y + 5$ is already solved for x. Substituting $y + 5$ for x in the first equation gives

$$\begin{aligned} 4x + 4y &= 420 \\ 4(y + 5) + 4y &= 420 \\ 4y + 20 + 4y &= 420 \\ 8y + 20 &= 420 \\ 8y &= 400 \\ y &= 50 \end{aligned}$$

Jim's speed is 50 miles per hour. Bob's speed is

$$\begin{aligned} x &= y + 5 \\ x &= 50 + 5 \\ x &= 55 \end{aligned}$$

Check and Answer The answer seems reasonable. We can check to see if the equation $4x + 4y = 420$ holds true for $x = 55$ and $y = 50$.

$$\begin{aligned} 4x + 4y &= 420 \\ 4(55) + 4(50) &\stackrel{?}{=} 420 \\ 220 + 200 &\stackrel{?}{=} 420 \\ 420 &= 420 \quad \text{True} \end{aligned}$$

Thus Bob's average speed is 55 miles per hour and Jim's average speed is 50 miles per hour.

Now Try Exercise 23

Mixture Problems

Mixture problems were solved with one variable in Section 3.4. Now we will solve mixture problems using two variables and systems of equations.

Understanding Algebra

Any problem in which two or more quantities are combined to produce a different quantity, or a single quantity is separated into two or more quantities, may be considered a *mixture* problem.

EXAMPLE 6 Candy Shop A customer at a candy shop explains that he needs a mixture of chocolate-covered cherries and chocolate-covered pecans. The chocolate-covered cherries sell for \$7.50 per pound and the chocolate-covered pecans sell for \$6.00 per pound.

a) How many pounds of the chocolate-covered pecans must be mixed with 2 pounds of chocolate-covered cherries to obtain a mixture that sells for \$6.50 per pound?

b) How many pounds of the mixture will there be?

Solution **a)** Understand and Translate We are asked to find the number of pounds of chocolate-covered pecans to be mixed with 2 pounds of chocolate-covered cherries. We are given sufficient information in the example to obtain two equations for our system of equations.

Let x = number of pounds of chocolate-covered pecans.

Let y = number of pounds of mixture.

Often it is helpful to make a sketch of the situation. After we draw a sketch, we will construct a table. In our sketch we will use bins to mix the candy (**Fig. 8.10** on page 504).

FIGURE 8.10

The value of the candy is found by multiplying the number of pounds by the price per pound.

Candy	Price	Number of Pounds	Value of Candy
Pecans	6	x	$6x$
Cherries	7.50	2	7.50(2)
Mixture	6.50	y	$6.50y$

Our two equations come from the following information:

$$\begin{pmatrix}\text{number of pounds}\\ \text{of pecans}\end{pmatrix} + \begin{pmatrix}\text{number of pounds}\\ \text{of cherries}\end{pmatrix} = \begin{pmatrix}\text{number of pounds}\\ \text{of mixture}\end{pmatrix}$$

$$x \quad + \quad 2 \quad = \quad y$$

$$\text{value of pecans} + \text{value of cherries} = \text{value of mixture}$$

$$6x \quad + \quad 7.50(2) \quad = \quad 6.50y$$

$$\text{System of equations} \quad \begin{cases} x + 2 = y \\ 6x + 7.50(2) = 6.50y \end{cases}$$

Carry Out Since $y = x + 2$, we substitute $x + 2$ for y in the second equation and solve for x.

$$\begin{aligned} 6x + 7.50(2) &= 6.50y \\ 6x + 7.50(2) &= 6.50(x + 2) \\ 6x + 15 &= 6.50x + 13 \\ 15 &= 0.50x + 13 \\ 2 &= 0.50x \\ 4 &= x \end{aligned}$$

Answer Thus, 4 pounds of the chocolate-covered pecans must be mixed with 2 pounds of the chocolate-covered cherries.

b) The total mixture will weigh 4 + 2 or 6 pounds.

Now Try Exercise 33

Now we will work an example similar to Example 7 in Section 3.4, but this time we will use a system of equations to solve the problem.

EXAMPLE 7 **Chemistry Lab** Chemistry professor Linda Barton needs 60 liters of 15% sulfuric acid solution but only has 5% and 30% sulfuric acid solutions on hand. To obtain her desired mixture, how many liters of 5% sulfuric acid solution and how many liters of 30% sulfuric acid solution should Linda mix?

Solution Understand and Translate Linda will combine the 5% and 30% solutions to get 60 liters of a 15% solution of sulfuric acid. We need to determine how much of each should be mixed.

Let x = number of liters of 5% solution.

Let y = number of liters of 30% solution.

The problem is displayed in **Figure 8.11** and the information is summarized in the following table.

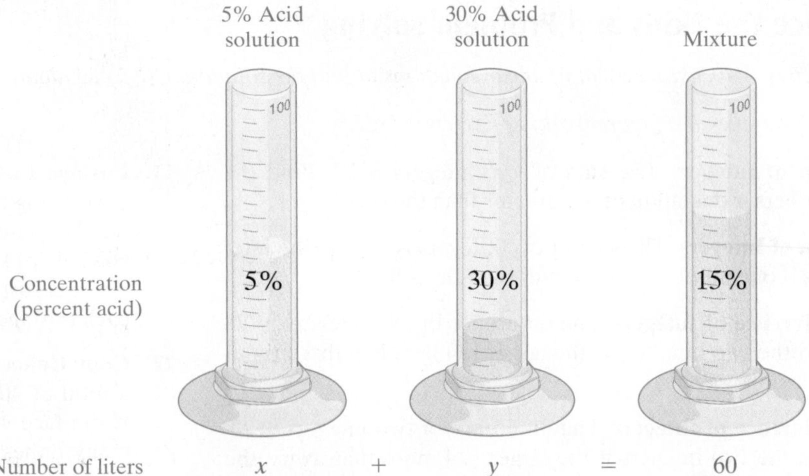

FIGURE 8.11

Solution	Number of Liters	Concentration	Acid Content
5% Solution	x	0.05	$0.05x$
30% Solution	y	0.30	$0.30y$
Mixture	60	0.15	0.15(60)

Because the total volume of the combination is 60 liters, we have

$$x + y = 60$$

From the table we see that

$$\begin{pmatrix}\text{acid content of}\\ \text{5\% solution}\end{pmatrix} + \begin{pmatrix}\text{acid content of}\\ \text{30\% solution}\end{pmatrix} = \begin{pmatrix}\text{acid content}\\ \text{of mixture}\end{pmatrix}$$

$$0.05x \quad + \quad 0.30y \quad = \quad 0.15(60)$$

System of equations $\begin{cases} x + y = 60 \\ 0.05x + 0.30y = 0.15(60) \end{cases}$

Carry Out We will solve this system by substitution. First we solve for y in the first equation.

$$\begin{aligned} x + y &= 60 \\ y &= 60 - x \end{aligned}$$

Then we substitute $60 - x$ for y in the second equation.

$$\begin{aligned} 0.05x + 0.30y &= 0.15(60) \\ 0.05x + 0.30(60 - x) &= 9 \\ 0.05x + 18 - 0.30x &= 9 \\ -0.25x + 18 &= 9 \\ -0.25x &= -9 \\ x &= \frac{-9}{-0.25} = 36 \end{aligned}$$

Now we solve for y.

$$\begin{aligned} y &= 60 - x \\ y &= 60 - 36 \\ y &= 24 \end{aligned}$$

Answer Thus, 36 liters of the 5% acid solution should be mixed with 24 liters of the 30% acid solution to obtain 60 liters of a 15% acid solution.

Now Try Exercise 31

8.4 Exercise Set MathXL® MyMathLab®

Practice the Skills and Problem Solving

In Exercises 1–40, use a system of linear equations to find the solution. Use a calculator where appropriate.

Review Examples 1–4 before working Exercises 1–22.

1. **Sum of Integers** The sum of two integers is 77. Find the numbers if one number is 7 greater than the other.

2. **Sum of Integers** The sum of two integers is 4. Find the integers if one number is 20 greater than the other.

3. **Difference of Integers** The difference of two integers is 20. Find the two numbers if the larger is 4 less than three times the smaller.

4. **Difference of Integers** The difference of two integers is 23. Find the two integers if the larger is 4 more than twice the smaller.

5. **Complementary Angles** Angles A and B are complementary angles. If angle B is 18° greater than angle A, find the measure of each angle.

6. **Complementary Angles** The larger of two complementary angles is 6° greater than 6 times the smaller. Find the two angles.

7. **Supplementary Angles** If angles A and B are supplementary angles and angle A is 44° greater than angle B, find the measure of each angle.

8. **Supplementary Angles** If angles A and B are supplementary angles, and angle A is five times as large as angle B, find the measure of each angle.

9. **American Flag** The satellite photo shown is of a floral flag that covers 6.65 acres in Lompoc, California. The flag maintains the proper flag dimensions. The perimeter of the flag is 2260 feet. Determine the flag's length and width if the length is 350 feet greater than the width.

10. **Vegetable Garden** Fred MacDonald bought 60 feet of fence to make a rectangular vegetable garden. What dimensions will the vegetable garden have if the length is 12 feet greater than the width?

11. **Postage Cost** Peter Collins, a financial planner, is mailing advertising information to potential clients. Some of the letters he mails require 46 cents of postage and some require 66 cents of postage. If Peter mails 70 letters and if the total postage cost is $40.20, find how many letters were mailed at each postage rate.

12. **Coin Collection** Darren collects coins. His collection has a total of 40 old coins containing only dimes and quarters. If the face value of the coins in his collection is $7.30, how many dimes and how many quarters are in the collection?

13. **Kayaking** Shane Stagg is kayaking in the St. Lawrence River. He can paddle 4.7 miles per hour with the current and 3.4 miles per hour against the current. Find the speed of the kayak in still water and the current.

14. **Airline Flights** Two commercial airplanes are flying in the same vicinity but in opposite directions. The American Airlines jet, flying with the wind, is flying at 570 miles per hour. The US Airways jet, flying against the wind, is flying at 500 miles per hour. If it were not for the wind, the two planes would be flying at identical speeds. Find the speed of the planes in still air and the speed of the wind.

15. **Park Tickets** The Meyer family had a family reunion in Texas. They all purchased tickets to visit an amusement park. The adults' tickets cost $40 and the children's cost $30. If 27 tickets were purchased, and the total cost for all the adults' and all the children's tickets was $930, how many adults' and how many children's tickets were purchased?

16. **Farming** Celeste Nossiter plants corn and wheat on her 100-acre farm near Albuquerque, New Mexico. She estimates that her income after deducting expenses is $500 per acre of corn and $475 per acre of wheat. Find the number of acres of corn and wheat planted if her total income after expenses is $49,000.

17. **Copier Contract** Carol Juncker just purchased a high speed copier for her office and wants a service contract on the copier. She is considering two sources for the contract. The Kate Spence Copier Sales and Service Company charges $18 a month plus 2 cents a copy. Office Copier Depot charges $25 a month but only 1.5 cents a copy.

    a) Assuming the prices do not change, how many copies would Carol need to make for the monthly cost of both plans to be the same?

    b) If Carol plans to make 2500 copies a month, which plan would be the least expensive?

18. **Sales Position** Susan Summerlin, a salesperson, is considering two job offers. At the Medtec Company, Susan's salary would be $300 per week plus a 5% commission of sales. At the Genzone Company, her salary would be $210 per week plus an 8% commission of sales.

    a) What weekly dollar volume of sales would Susan need to make for the total income from both companies to be the same?

    b) If she expects to make sales of $4000, which company would give the greater salary?

19. **Savings Accounts** Mr. and Mrs. Vinny McAdams invest a total of $8000 in two savings accounts. One account yields 5% simple interest and the other 4% simple interest. Find the amount placed in each account if they receive a total of $375 in interest after 1 year. Use interest = principal · rate · time.

20. **Investments** Carol Horton invested a total of $10,000. Part of the money was placed in a savings account paying 5.5% simple interest. The rest was placed in a fixed annuity paying 6% simple interest. If the total interest received for the year was $570, how much had been invested in each account?

21. **Custom Publishing** Many bookstores now offer print-on-demand service to produce paper copies of out-of-print books. University Bookstore offers this service for a $25 set-up fee and then 9 cents per page printed. The Frugal Muse Bookstore offers this service for a $40 set-up fee and then 5 cents per page printed.

    a) Determine how many pages a book must have so the total cost for printing the book is the same at both bookstores.

    b) Karen Egedy wants to reproduce a copy of *The American Farmhouse*, a 265-page out-of-print book. Which bookstore would print the book for the lowest cost?

22. **Job Opportunity** Jorge Perez is a recent college graduate and is considering job offers as a sales representative from two different companies. American Office Equipment offers Jorge an annual salary of $21,000 plus 15% commission on all equipment he sells. Business Solutions offers Jorge an annual salary of $27,000 plus 12% commission on all equipment he sells.

    a) How much in sales would Jorge need to have in one year for the two companies to pay him the same total annual compensation?

    b) If Jorge projects that he will have about $175,000 in sales per year, which company would provide him with the higher annual compensation?

Review Example 5 before working Exercises 23–30.

23. **Road Trip** Dave Visser started driving from Columbus, Ohio, toward Lincoln, Nebraska—a distance of 903 miles. At the same time Alice Harra started driving to Columbus, Ohio, from Lincoln, Nebraska. If the two meet after 7 hours and Alice's speed averages 15 miles per hour greater than Dave's speed, find the speed of each car.

24. **Trains** Two trains are 560 miles apart on a parallel set of tracks traveling toward each other. One train is traveling 10 miles per hour faster than the other. The trains will pass each other in 4 hours. Find the speed of the trains.

25. **Speed Boat** During a race, Elizabeth Kell's speed boat travels 4 miles per hour faster than Melissa Suarez's boat. If Elizabeth's boat finishes the race in 3 hours and Melissa's finishes the race in 3.4 hours, find the speed of each boat.

26. **Bicycle Race** During a bicycle race, Sharon Hamsa rides at 3 miles per hour slower than Cynthia Harrison. If Sharon finished the race in 5 hours and Cynthia finished the race in 4 hours, determine the speed of each cyclist.

27. **Jogging** Amanda Ginter and Delores Melendez go jogging along the River Walk Trail in San Antonio, Texas. They start at the same point, but Amanda starts 0.25 hours before Delores does. If Amanda jogs at a rate of 4 miles per hour and Delores jogs at a rate of 5 miles per hour, how long after Delores starts will Delores catch up to Amanda?

San Antonio, Texas

28. **Dude Ranch** Kate and Ernie Danforth are at a dude ranch. Kate has been horseback riding for 0.5 hours on the trail at 6 miles per hour when Ernie starts riding his horse on the trail. If Ernie rides at 10 miles per hour, how long will it be before he catches up to Kate?

29. **Rollerblading** Christopher Yarish rollerblades north at a rate of 8 miles per hour. One-quarter hour later, Judy Kasabian starts at the same point and rollerblades south at a rate of 10 miles per hour. How long after Judy leaves will Christopher and Judy be 15.5 miles apart?

30. **Horseback Riding** Bill Leonard trots his horse Trixie east at 8 miles per hour. One half-hour later, Mary Mullaley starts at the same point and canters her horse Pegarno west at 12 miles per hour. How long after Mary starts riding will Mary and Bill be separated by 14 miles?

Review Examples 6 and 7 before working Exercises 31–40.

31. **Chemist** Karl Schmid, a chemist, has a 15% hydrochloric acid solution and a 40% hydrochloric acid solution. How many liters of each should he mix to get 20 liters of a hydrochloric acid solution with a 30% acid concentration?

32. **Pharmacist** Jacque Williams, a pharmacist, needs 500 milliliters of a 10% phenobarbital solution. She has only 5% and 25% phenobarbital solutions available. How many milliliters of each solution should she mix to obtain the desired solution?

33. **Juice** The All Natural Juice Company sells apple juice for 16 cents an ounce and apple drink for 6 cents an ounce. They wish to market and sell for 10 cents an ounce cans of juice drink that are part juice and part drink. How many ounces of each will be used if the juice drink is to be sold in 8-ounce cans?

34. **Laying Tile** Julie Hildebrand is selecting tile for her foyer and living room. She wants to make a pattern using two different colors and types of tile. One type costs $3 per tile (per square foot) and the other type costs $5 per tile (per square foot). She needs a total of 380 tiles and wants to spend $1500 on the tile. How many of each type of tile should she purchase?

35. **Dairy Farm** Wayne Froelich has milk that is 5% butterfat and skim milk without butterfat. How much 5% milk and how much skim milk should he mix to make 200 gallons of milk that is 3.5% butterfat?

36. **Melted Ice** A fruit drink is 8% juice by volume. Ice is added to the drink. After the ice melts the juice contents of the 12 ounce mixture is 6%. Find the volume of fruit drink and the volume of ice that was added.

37. **Soybean Mix** Lynn Hicks wishes to mix soybean meal that is 18% protein and cornmeal that is 6% protein to get a 150-pound mixture that is 10% protein. How much of each should be used?

38. **Quiche** Pierre LaRue's recipe for quiche lorraine calls for 510 ml of light cream, which is 20% butterfat. It is often difficult to find light cream with 20% butterfat at the supermarket. What is commonly found is heavy cream, which is 36% butterfat, and half-and-half, which is 10.5% butterfat. How many milliliters of the heavy cream and how much of the half-and-half should be mixed to obtain 510 milliliters of light cream that is 20% butterfat?

39. **Concession Stand Purchase** At Braden River Little League, Kevin Harris purchased 3 hot dogs and 2 bags of sunflower seeds for $6. Paul Thomas purchased 5 hot dogs and 3 bags of sunflower seeds for $9.75. Determine the price of a hot dog and the price of a bag sunflower seeds.

40. **Play Ticket Prices** Michael Schieb purchased 5 adult tickets and 8 student tickets to see the play *Chicago* at State College of Florida for $90. Gemma Hynds purchased 3 adult tickets and 7 student tickets for $65. Determine the price of an adult ticket and the price of a student ticket.

Challenge Problems

41. **Jogging** A brother and sister, Sean and Moira O'Donnell, jog to school daily. Sean, who is older, jogs at 9 miles per hour. Moira jogs at 5 miles per hour. When Sean reaches the school, Moira is $\frac{1}{2}$ mile away. How far is the school from their house?

42. **Alloys** By weight, an alloy of brass is 70% copper and 30% zinc. Another alloy of brass is 40% copper and 60% zinc. How many grams of each of these alloys must be melted and combined to obtain 300 grams of a brass alloy that is 60% copper and 40% zinc?

43. Pressurized Tanks Two pressurized tanks are connected by a controlled pressure valve, as shown in the figure. Initially, the internal pressure in tank 1 is 200 pounds per square inch, and the internal pressure in tank 2 is 20 pounds per square inch. The pressure valve is opened slightly to reduce the pressure in tank 1 by 2 pounds per square inch per minute. This increases the pressure in tank 2 by 2 pounds per square inch per minute. At this rate, how long will it take for the pressure to be equal in both tanks?

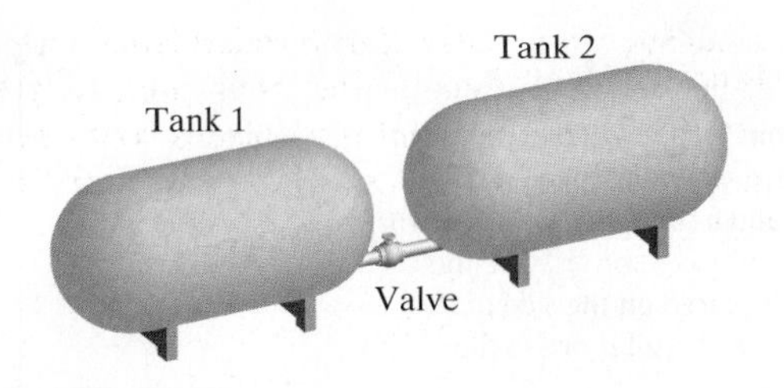

Group Activity

Discuss and work Exercise 44 as a group.

44. Car Purchase Debby Patterson is considering 2 cars for purchase. Car A has a list price of $23,000 and gets an average of 40 miles per gallon. Car B has a list price of $21,500 and gets an average of 20 miles per gallon. Being a conservationist, Debby wishes to purchase car A but is concerned about its greater initial cost. She plans to keep the car for many years. If she purchases car A, how many miles would she need to drive for the total cost of car A to equal the total cost of car B? Assume gasoline costs $3.00 per gallon.

Cumulative Review Exercises

[1.10] **45.** Name the properties illustrated.

a) $x + 4 = 4 + x$

b) $(3x)y = 3(xy)$

c) $4(x + 2) = 4x + 8$

[3.3] **46. Perimeter** The perimeter of a rectangle is 22 feet. Find the length and width of the rectangle if the length is two more than twice the width.

[6.6] **47.** Solve the equation $x + \frac{2}{x} = \frac{6}{x}$.

[7.1] **48.** What is a graph?

8.5 Solving Systems of Linear Inequalities

1 Solve systems of linear inequalities graphically.

1 Solve Systems of Linear Inequalities Graphically

In Section 7.5, we learned how to graph linear inequalities in two variables. In Section 8.1, we learned how to solve systems of equations graphically. In this section, we discuss how to solve systems of linear inequalities

To Solve a System of Linear Inequalities Graphically

Graph each inequality on the same axes. The **solution to a system of linear inequalities** is the set of points whose coordinates satisfy all the inequalities in the system.

Although a system of linear inequalities may contain more than two inequalities, in this book we will focus on systems with only two inequalities.

EXAMPLE 1 Determine the solution to the following system of inequalities.

$$x + 2y \leq 6$$
$$y > 2x - 4$$

Solution First graph the inequality $x + 2y \leq 6$ (**Fig. 8.12** on the next page) as explained in Section 7.5. Note the boundary line is solid. Now, on the same axes, graph the inequality $y > 2x - 4$ (**Fig. 8.13** on the next page). Note that the boundary line is dashed.

The solution is the set of points common to both inequalities—the part of the graph that contains both shadings (the purple color). The dashed line is not part of the solution. However, the part of the solid line that satisfies both inequalities is part of the solution.

> **Understanding Algebra**
>
> Recall from Section 7.5 that a dashed line is used when the inequality is $<$ or $>$ and a solid line is used when the inequality is $\le$ or $\ge$. Also, the shading is placed on the side of the line that has the solutions to the inequality.

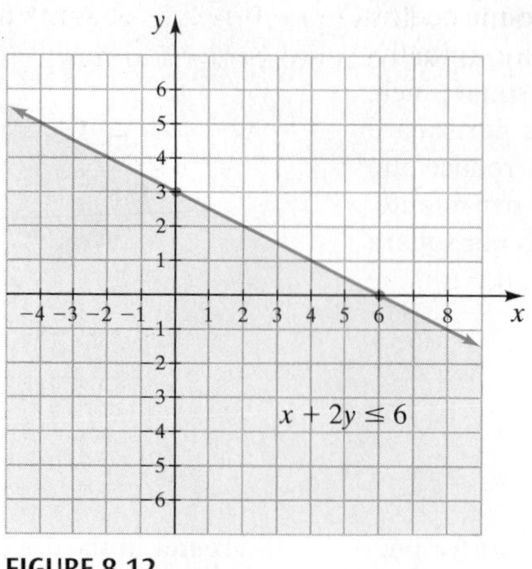

FIGURE 8.12

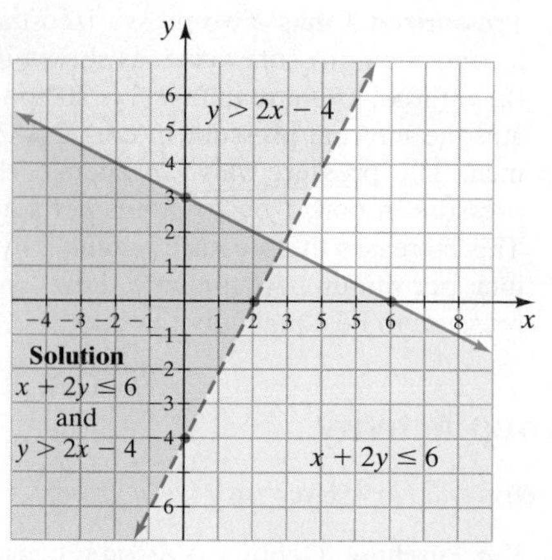

FIGURE 8.13

Now Try Exercise 5

In Example 1, the ordered pairs of each of the points in the solution (the purple colored area) satisfies both inequalities in the system. We can choose any point in the shaded region, substitute its x and y values into *both* inequalities, and obtain true statements. Let us choose the origin, which is in the purple shaded area. Let (x, y) be $(0, 0)$.

$x + 2y \le 6$	$y > 2x - 4$
$0 + 2(0) \overset{?}{\le} 6$	$0 \overset{?}{>} 2(0) - 4$
$0 \le 6$ True	$0 > -4$ True

As we expected, the ordered pair $(0, 0)$ satisfies both inequalities.

EXAMPLE 2 Determine the solution to the following system of inequalities.

$$2x + 3y > 4$$
$$2x - y \ge -6$$

Solution Graph $2x + 3y > 4$ (**Fig. 8.14**). Graph $2x - y \ge -6$ on the same axes (**Fig. 8.15**). The solution is the part of the graph with both shadings and the part of the solid line that satisfies both inequalities.

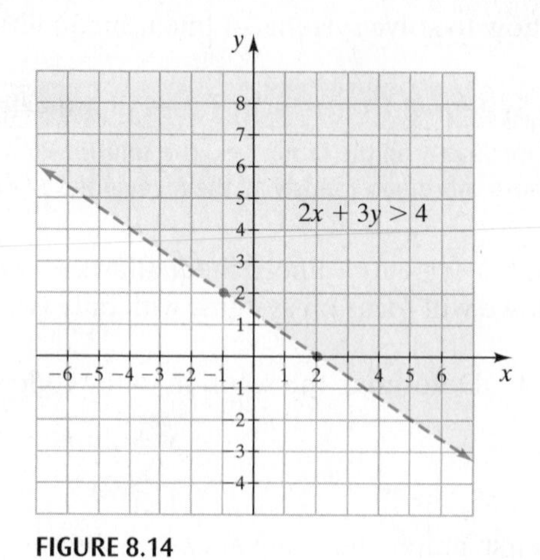

FIGURE 8.14

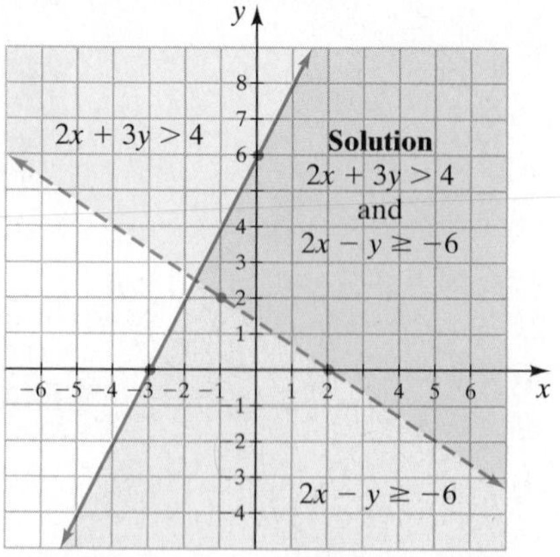

FIGURE 8.15

Now Try Exercise 15

EXAMPLE 3 Determine the solution to the following system of inequalities.

$$y < 2$$
$$x > -3$$

Solution Graph both inequalities on the same axes (**Fig. 8.16**). The solution is the part of the graph that contains both shadings.

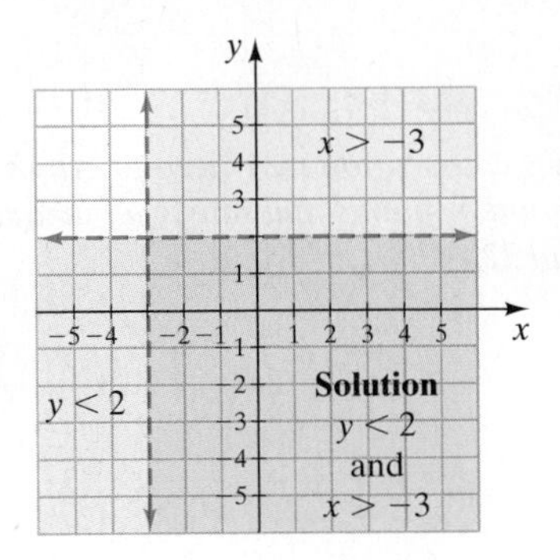

FIGURE 8.16

Now Try Exercise 11

It is possible for a system of linear inequalities to have no solution. This can occur if the graphs of the inequalities are lines that have the same slope and are parallel to each other. For example, the system of inequalities $x + y \le 2$ and $x + y \ge 4$ has no solution.

8.5 Exercise Set MathXL® MyMathLab®

Warm-Up Exercises

Fill in the blanks with the appropriate word, phrase, or symbol(s) from the following list.

solutions intercepts solid dashed satisfy vertical horizontal equal

1. When graphing a linear inequality, if the inequality is $<$ or $>$, use a ________ line.
2. When graphing a linear inequality, if the inequality is $\le$ or $\ge$, use a ________ line.
3. The points in the shaded region of the graph of a linear inequality represent the ________ to the inequality.
4. The solution to a system of linear inequalities is the set of points whose coordinates ________ all the inequalities in the system.

Practice the Skills

Determine the solution to each system of inequalities.

5. $x - 2y < 6$, $y \le -x + 4$
6. $3x - 4y \le 12$, $y > -x + 4$
7. $x + y > 2$, $x - y < 2$
8. $x - 2y \le 4$, $x + 2y > -2$
9. $y \le x$, $y < -2x + 2$
10. $y \le 3x - 2$, $y > -4x$
11. $x \le 4$, $y \ge -2$
12. $x \le 0$, $y \le 0$
13. $4x + 5y < 20$, $x \ge -3$
14. $y \le 3x + 4$, $y < 2$
15. $2x - y \le 3$, $4x + 2y > 8$
16. $-2x + 3y \ge 6$, $x + 4y \ge 4$
17. $2x + 4y > 6$, $4x + 8y < 4$
18. $x - 3y > 3$, $2x - 6y < 6$
19. $x > -3$, $y > 1$
20. $y \le 4$, $x > -1$
21. $2x - y < 4$, $4x - 2y \ge -10$
22. $x - 3y \le 6$, $x + 3y \le 6$
23. $y < -2x - 3$, $y \ge 3x + 2$
24. $y \le -\frac{2}{3}x + 4$, $y > \frac{1}{3}x + 1$

Concept/Writing Exercises

25. If an ordered pair satisfies both inequalities in a system of linear inequalities, must that ordered pair be in the solution to the system? Explain.
26. If an ordered pair satisfies only one inequality in a system of linear inequalities, is it possible for that ordered pair to be in the solution to the system? Explain.
27. Can a system of linear inequalities have no solution? Explain your answer with the use of your own example.
28. Is it possible to construct a system of two nonparallel linear inequalities that has no solution? Explain.

29. Is it possible for a system of two linear inequalities to have only one solution? Explain.

30. Construct a system of two linear inequalities that has no solution. Explain how you determined your answer.

Group Activity

In more advanced mathematics courses you may need to graph more than two linear inequalities. When a system has more than two inequalities, the solution is the point or points that satisfy all inequalities in the system. As a group, determine the solutions to the systems of inequalities in Exercises 31 and 32.

31. $x + 2y \leq 6$
$2x - y < 2$
$y > 2$

32. $x \geq 0$
$y \geq 0$
$y \leq 2x + 4$
$y \leq -x + 6$

Cumulative Review Exercises

[2.8] **33.** Solve the inequality. Graph the solution on a number line and write the solution using interval notation: $6(x - 2) < 4x - 3 + 2x$.

[2.6] **34.** Solve the equation $3x - 5y = 6$ for y.

[5.6] **35.** Solve the equation $4x^2 - 11x - 3 = 0$.

[4.2] **36.** Simplify $\dfrac{x^{-6}y^2}{x^3y^{-5}}$.

Chapter 8 Summary

IMPORTANT FACTS AND CONCEPTS	EXAMPLES
Section 8.1	
A **system of linear equations** is a system having two or more linear equations. The **solution to a system of equations** is the ordered pair or pairs that satisfy all the equations in the system.	$y = 2x - 5$ $y = -4x + 7$ The solution to the above system is $(2, -1)$.
A linear system of equations may have exactly one solution (intersecting lines), no solution (parallel lines), or an infinite number of solutions (same line).	Consistent, exactly 1 solution (a) Inconsistent, no solution (b) Dependent, infinite number of solutions (c)

IMPORTANT FACTS AND CONCEPTS	EXAMPLES

Section 8.1 (Continued)

To **solve a system of equations graphically**, graph each equation and determine the point or points of intersection.

Solve the system of equations graphically.

$$y = -x + 5$$
$$y = x + 3$$

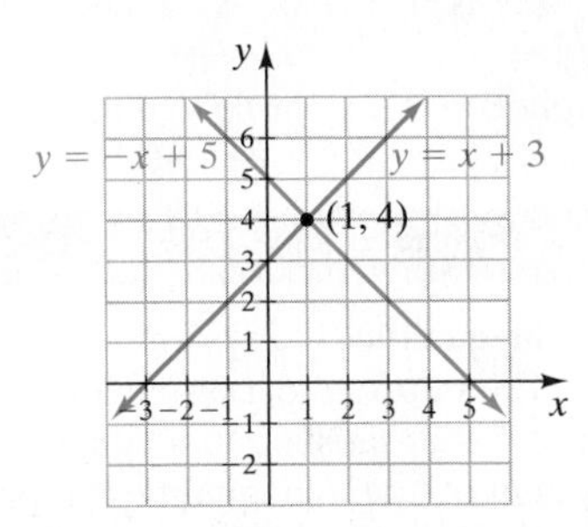

The solution is $(1, 4)$.

Section 8.2

To Solve a System of Equations by Substitution

1. If necessary, solve for a variable in either equation.
2. Substitute the expression found for the variable in step 1 into the other equation.
3. Solve the equation determined in step 2 to find the value of one variable.
4. Substitute the value found in step 3 into the equation obtained in step 1 to find the value of the other variable.
5. Check by substituting both values in both original equations.

Solve the system of equations by substitution

$$y = -x + 5$$
$$y = x + 3$$

Substitute $x + 3$ for y in the first equation.

$$x + 3 = -x + 5$$
$$2x + 3 = 5$$
$$2x = 2$$
$$x = 1$$

Now, solve for y by substituting $x = 1$ into the first equation.

$$y = -x + 5$$
$$y = -1 + 5$$
$$= 4$$

The solution is $(1, 4)$.

Section 8.3

To Solve a System of Equations by the Addition (or Elimination) Method

1. If necessary, rewrite each equation so that the terms containing variables appear on the left side of the equation and any constants appear on the right side of the equation.
2. If necessary, multiply one or both equations by a constant(s) so that when the equations are added the resulting sum will contain only one variable.
3. Add the equations.
4. Solve for the variable in the equation from step 3.
5. Substitute the value found in step 4 into either of the original equations. Solve that equation to find the value of the remaining variable.
6. Check the values obtained in all original equations.

Solve the system of equations by the addition method.

$$2x + 3y = -4 \quad \textit{eq. 1}$$
$$3x - y = -17 \quad \textit{eq. 2}$$

Multiply *eq.* 2 by 3, then add:

$$2x + 3y = -4 \quad \textit{eq. 1 stays the same}$$
$$\underline{9x - 3y = -51} \quad \textit{eq. 2 multiplied by 3}$$
$$11x = -55$$
$$x = -5$$

Substitute -5 for x into *eq.* 2.

$$3x - y = -17$$
$$3(-5) - y = -17$$
$$-15 - y = -17$$
$$-y = -2$$
$$y = 2$$

The solution is $(-5, 2)$.

IMPORTANT FACTS AND CONCEPTS	EXAMPLES
Section 8.4	
Complementary Angles and Supplementary Angles Two angles are complementary angles if the sum of their measures is 90°. Two angles are supplementary angles if the sum of their measures is 180°.	If angle A is 64° and angle B is 26°, then angles A and B are complementary angles. If angle A is 135° and angle B is 45°, then angles A and B are supplementary angles.
Many real-life applications may be solved using systems of equations	See Examples 1–7 in Section 8.4.
Section 8.5	
A **system of linear inequalities** is a system having two or more linear inequalities. The **solution** to a system of linear inequalities is the set of points that satisfies all inequalities in the system. To solve a system of linear inequalities graphically, graph each inequality on the same axes. The solution is the set of points that satisfies all the inequalities in the system.	Determine the solution to the system of inequalities: $x + y < 4$ $y \geq 2x - 3$ 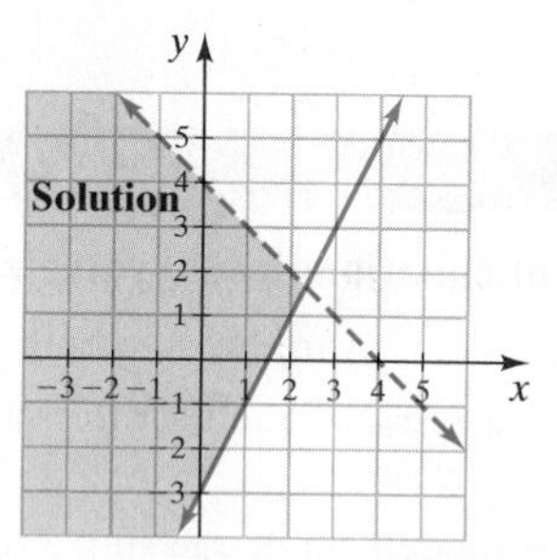

Chapter 8 Review Exercises

[8.1] *Determine which, if any, of the ordered pairs satisfy each system of equations.*

1. $y = 4x - 2$
$2x + 3y = 8$
a) $(0, -2)$ **b)** $(4, 0)$ **c)** $(1, 2)$

2. $y = -x + 4$
$3x + 5y = 15$
a) $\left(\frac{5}{2}, \frac{3}{2}\right)$ **b)** $(-1, 5)$ **c)** $\left(\frac{1}{2}, \frac{3}{5}\right)$

Identify each system of linear equations as consistent, inconsistent, or dependent. State whether the system has exactly one solution, no solution, or an infinite number of solutions.

3.
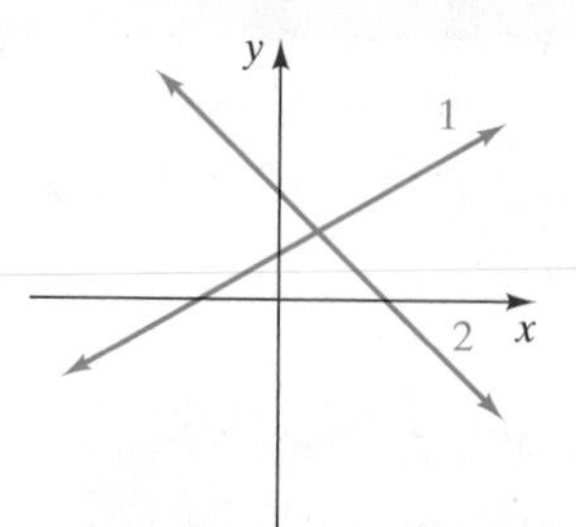

4.
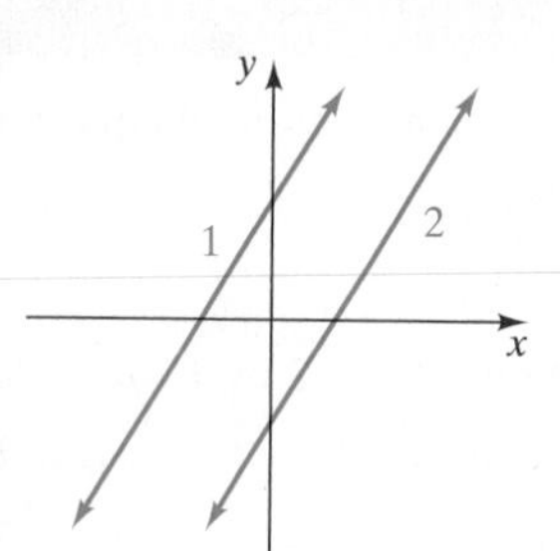

5.
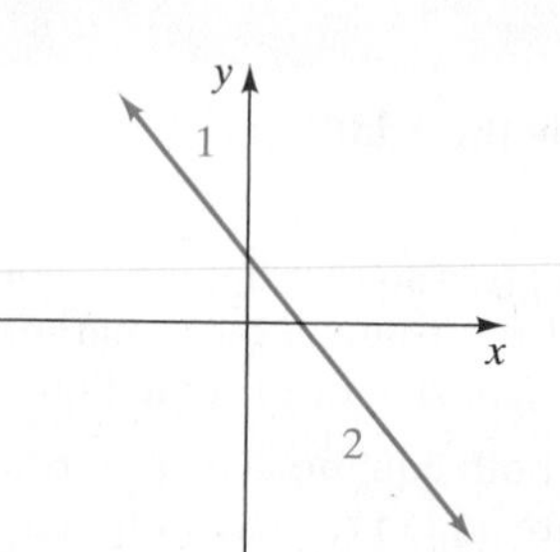

6.
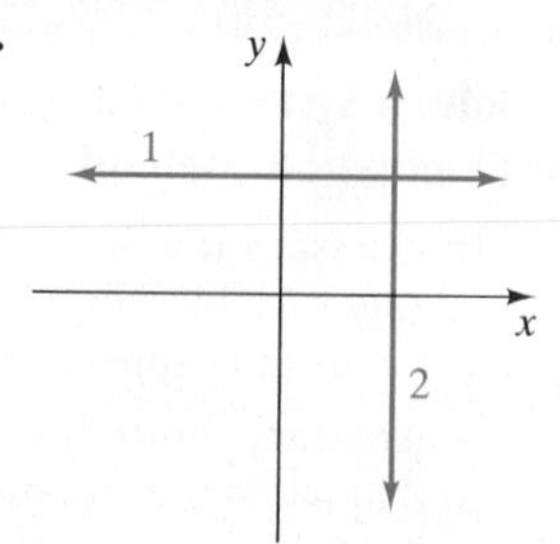

Write each equation in slope-intercept form. Without graphing or solving the system of equations, state whether the system of linear equations has exactly one solution, no solution, or an infinite number of solutions.

7. $x + 2y = 10$
$4x = -8y + 16$

8. $y = 4x - 3$
$3x - y = 7$

9. $y = \frac{1}{3}x + \frac{2}{3}$
$6y - 2x = 4$

10. $6x = 4y - 20$
$4x = 6y + 20$

Determine the solution to each system of equations graphically.

11. $y = x - 4$
$y = 2x - 7$

12. $x = -2$
$y = 5$

13. $x + 2y = 8$
$2x - y = -4$

14. $x + 4y = 8$
$y = 2$

15. $y = 3$
$y = -2x + 5$

16. $y = x - 3$
$3x - 3y = 9$

17. $3x + y = 0$
$3x - 3y = 12$

18. $x + 5y = 10$
$\frac{1}{5}x + y = -1$

[8.2] *Find the solution to each system of equations by substitution.*

19. $y = 2x - 3$
$3x - 5y = 1$

20. $x = 3y - 9$
$x + 2y = 1$

21. $2x - y = 7$
$x + 2y = 6$

22. $x = -3y$
$x + 4y = 5$

23. $4x - 2y = 7$
$y = 2x + 3$

24. $2x - 4y = 7$
$-4x + 8y = -14$

25. $2x - 3y = 8$
$6x - 5y = 20$

26. $3x - y = -5$
$x + 2y = 8$

[8.3] *Find the solution to each system of equations using the addition method.*

27. $x - y = -4$
$-x + 6y = -6$

28. $x + 2y = -3$
$5x - 2y = 9$

29. $x + y = 12$
$2x + y = 5$

30. $4x - 3y = 2$
$2x + 5y = 14$

31. $-2x + 3y = 15$
$7x + 3y = 6$

32. $2x + y = 3$
$-4x - 2y = 5$

33. $3x = -4y + 15$
$8y = -6x + 30$

34. $2x - 5y = 12$
$3x - 4y = -6$

[8.4] *Use a system of linear equations to find the solution.*

35. Sum of Integers The sum of two integers is 40. Find the two numbers if the larger is 8 less than twice the smaller.

36. Plane Flight A Boeing 757 flies 580 miles per hour with the wind and 520 miles per hour against the wind. Find the speed of the wind and the speed of the plane in still air.

37. Truck Rental Katz's Truck Rental charges \$35 per day plus 50 cents per mile, while Willie's Truck Rental charges \$40 per day plus 40 cents per mile. How far would you have to travel in one day for the total cost from both rental companies to be the same?

38. Savings Account Moura Hakala invested a total of \$16,000. Part of the money was placed in a savings account paying 4% simple interest. The rest was placed in a savings account paying 6% simple interest. If the total interest received for the year was \$760, how much had she invested in each account?

39. Road Trip Liz Wood drives from Charleston, South Carolina, to Louisville, Kentucky—a distance of 600 miles. At the same time, Mary Mayer starts driving from Louisville to Charleston along the same route. If the two meet after driving 5 hours and Mary's average speed was 6 miles per hour greater than Liz's, find the average speed of each car.

40. Grass Seed Green Turf's grass seed costs 60 cents a pound and Agway's grass seed costs 45 cents a pound. How many pounds of each were used to make a 50-pound mixture that cost \$25.50?

41. Chemist A chemist has a 25% acid solution and a 55% acid solution. How much of each must be mixed to get 10 liters of a 40% acid solution?

[8.5] *Determine the solution to each system of inequalities.*

42. $2x - 6y > 6$
$x > -2$

43. $x < 2$
$y \geq -3$

44. $2x + y > 2$
$2x - y \leq 4$

45. $2x - 3y \leq 6$
$2x + 8y > 8$

Chapter 8 Practice Test

Chapter Test Prep Videos provide fully worked-out solutions to any of the exercises you want to review. Chapter Test Prep Videos are available via MyMathLab®, *or on* YouTube (www.youtube.com/user/AngelElementaryAlg).

1. Determine which, if any, of the ordered pairs satisfy the system of equations.

$x + 2y = -6$
$3x + 2y = -12$

a) $(-6, 0)$ **b)** $\left(-3, -\frac{3}{2}\right)$ **c)** $(2, -4)$

Identify each system as consistent, inconsistent, or dependent. State whether the system has exactly one solution, no solution, or an infinite number of solutions.

2.

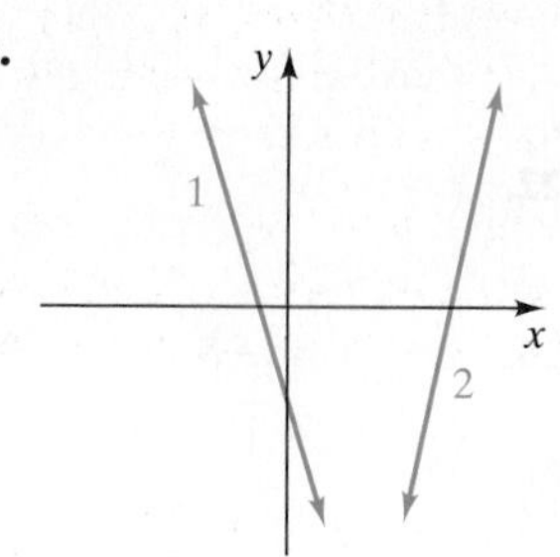

3.

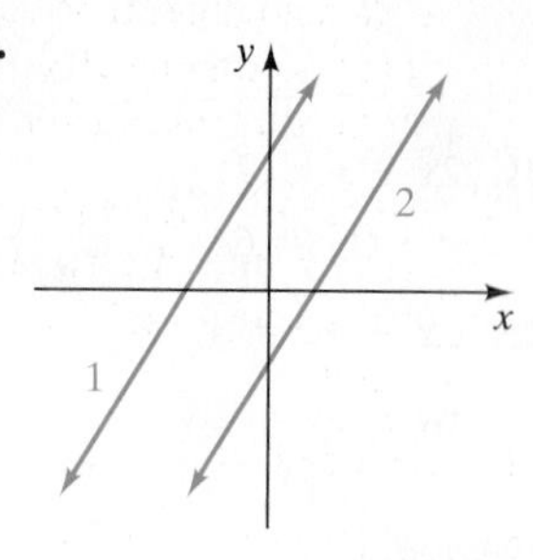

4.

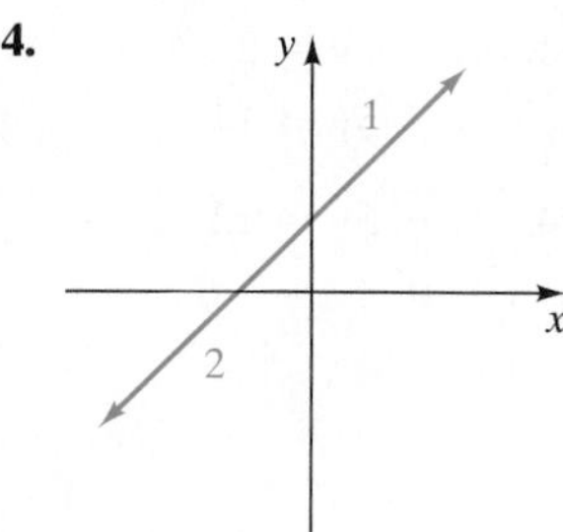

Write each equation in slope-intercept form. Then determine, without graphing or solving the system, whether the system of equations has exactly one solution, no solution, or an infinite number of solutions.

5. $-3y = 6x - 10$
$2x + y = 5$

6. $5x + 2y = 12$
$5x - 2y = 16$

7. $4x = 6y - 12$
$2x - 3y = -6$

8. When solving a system of linear equations by the substitution or the addition methods, how will you know if the system is

a) inconsistent,

b) dependent?

Solve each system of equations graphically.

9. $y = 2x - 4$
$y = -2x + 8$

10. $3x - 2y = -3$
$3x + y = 6$

11. $y = 2x + 4$
$4x - 2y = 6$

Solve each system of equations by substitution.

12. $y = 5x - 9$
$y = 3x + 3$

13. $3x + y = 8$
$x - y = 6$

14. $3x - 2y = 2$
$-5x + 4y = 0$

Solve each system of equations using the addition method.

15. $4x + y = -6$
$x + 3y = 4$

16. $4x - y = 8$
$10x + 3y = -13$

17. $5x - 10y = 20$
$2x = 4y + 8$

Solve each system of equations using the method of your choice.

18. $y = 5x - 13$
$y = -2x + 8$

19. $3x - 4y = 1$
$-2x + 5y = -10$

20. $3x + 5y = 20$
$6x + 3y = -12$

Use a system of linear equations to find the solution.

21. Truck Rental Charley's Rent a Truck Agency charges $47 per day plus 8 cents per mile to rent a certain truck. Hugh's Rent a Truck charges $40 per day plus 15 cents per mile to rent the same truck. How many miles will have to be driven in one day for the cost of Charley's truck to equal the cost of Hugh's truck?

22. Candies Albert's Grocery sells individually wrapped lemon candies for $6.00 a pound and individually wrapped butterscotch candies for $4.50 a pound. How much of each must Albert mix to get 20 pounds of a mixture that he can sell for $5.00 per pound?

23. Boat Race During a race, Dante Hull's speed boat travels 4 miles per hour faster than Deja Rocket's speed boat. If Dante's boat finishes the race in 2.5 hours and Deja's finishes the race in 3 hours, find the speed of each boat.

Determine the solution to each system of inequalities.

24. $x + 3y \geq 6$
$y < 3$

25. $2x + 4y < 8$
$x - 3y \geq 6$

Cumulative Review Test

Take the following test and check your answers with those given in the back of the book. Review any questions that you answered incorrectly. The section where the material was covered is indicated after the answer.

1. Women in the Military The number of women serving in the military has grown dramatically since 1967 when the military abolished the 2% cap on women serving in the armed forces. The graphs on the next page provide information about women and the military.

a) Which branches of the military are composed of at least 15% women?

b) Which branches of the military are composed of at least 10% women *and* have at least 70% of the military positions open to women?

Percent of Military Positions Filled by Women

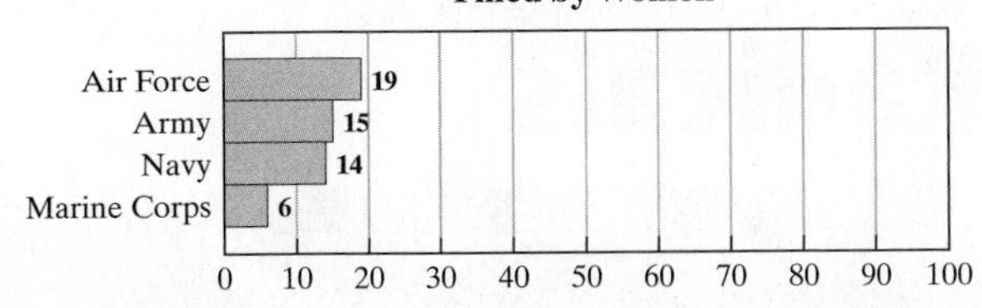

Percent of Military Positions Open to Women

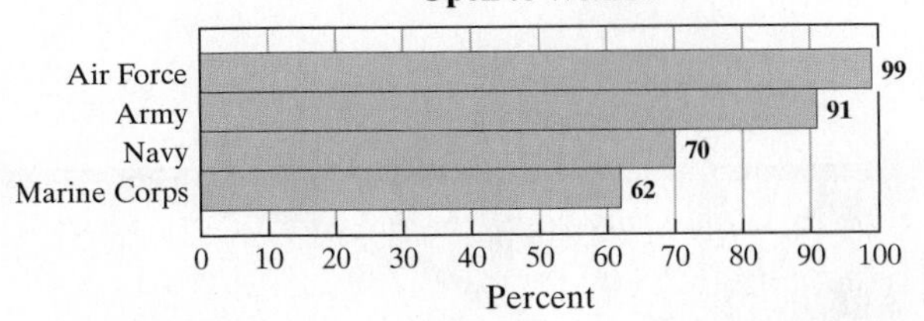

Source: Defense Department

2. Consider the set of numbers

$$\left\{-5, -0.6, \frac{3}{5}, \sqrt{10}, -\sqrt{2}, 7, 0, -\frac{5}{9}, 1.34\right\}$$

List the elements that are

a) natural numbers.

b) rational numbers.

c) irrational numbers.

d) real numbers.

3. Simplify $-115 + 63 + (-192)$.

4. Simplify $10 - (3a - 2) + 4(a + 3)$.

5. Solve the equation $2(x - 4) + 2 = 3x - 4$.

6. Solve the inequality $3x - 2 \leq x + 8$. Graph the solution on a number line and write the solution using internal notation.

7. Cutting a Board A 39-foot length of board is cut into two pieces. If one of the pieces has length y, what is the length of the other piece?

8. Select a Salary Plan Maria Gentile recently graduated from college and has accepted a position selling software. She is given a choice of two salary plans. Plan A is a straight 12% commission on sales. Plan B is \$348 per week plus 6% commission on sales. How much must Maria sell in a week for both plans to have the same weekly salary?

9. Angles of a Triangle One angle of a triangle measures 30° greater than the smallest angle. The third angle measures 8 times the smallest angle. Find the measures of all three angles.

10. Simplify $(4x^3y^2)^3(x^2y)$.

11. Multiply $(x - 2y)^2$.

12. Factor $2n^2 - 5n - 12$.

13. Solve $x^2 - 3x - 54 = 0$.

14. Multiply $\dfrac{x^2 - 4x - 12}{7x} \cdot \dfrac{x^2 - x}{x^2 - 7x + 6}$.

15. Solve $\dfrac{2x + 3}{5} = \dfrac{x - 2}{2}$.

16. Graph $2x - 4y = 8$.

17. Graph $\frac{1}{3}x + \frac{1}{2}y = 12$.

18. Determine if the system of equations

$$3x - y = 6$$
$$\frac{3}{2}x - 3 = \frac{1}{2}y$$

has one solution, no solution, or an infinite number of solutions. Explain how you determined your answer.

19. Solve the following system of equations graphically.

$$2x + y = 5$$
$$x - 2y = 0$$

20. Solve the following system of equations using the addition method.

$$3x + 5y = 6$$
$$-6x + 7y = -29$$

9 Roots and Radicals

When we determine a distance or other items that are measured, the answer obtained often involves a radical (for example, a square root), which we often round off to a decimal number. Many scientific and mathematical formulas involve square roots. For example, when finding the radius of a circular helicopter pad in Exercise 31 on page 558, the answer contains a square root.

Goals of This Chapter

In this chapter we study roots, radical expressions, and radical equations, with an emphasis on square roots. Square roots are one type of radical expressions. We also present some real-life applications of square roots. Radical expressions and equations play an important role in mathematics and the sciences. If allowed by your instructor, we suggest that you use a scientific calculator or graphing calculator for this chapter and the next chapter.

9.1 Evaluating Square Roots

1. Evaluate square roots of real numbers.
2. Recognize that not all square roots represent real numbers.
3. Determine whether the square root of a real number is rational or irrational.
4. Write square roots as exponential expressions.

1 Evaluate Square Roots of Real Numbers

Earlier we discussed determining the square of a number. Recall that when we square a number we multiply the number by itself. For example,

$$\text{If } a = 5, \text{ then } a^2 = 5 \cdot 5 = 25.$$
$$\text{If } a = -5, \text{ then } a^2 = (-5)(-5) = 25.$$

In this chapter we consider the opposite problem. That is, if we know the square of a number, what is the number? For example,

$$\text{If } a^2 = 25, \text{ then what are the values of } a?$$

To find a in the above statement, we must find the values that when multiplied by themselves result in a product of 25. Since $5 \cdot 5 = 25$ and $(-5)(-5) = 25$, a has two values, -5 and 5. Every real number greater than 0 has two square roots, a positive square root and a negative square root. For example, the number 25 has two square roots, -5 and 5. We use the radical symbol $\sqrt{\ }$ to indicate the *positive* or *principal square root* of a number. For example,

$$\sqrt{25} = 5 \quad \text{(The principal square root of 25 is 5.)}$$

The symbol $-\sqrt{\ }$ is used to denote the negative square root of a number. For example,

$$-\sqrt{25} = -5 \quad \text{(The negative square root of 25 is } -5.)$$

The negative square root is the additive inverse or opposite of the principal square root.

Understanding Algebra

Finding the *square root* of a given number is the reverse process of squaring a number. When we find the square root of a given number, we are determining what numbers, when multiplied by themselves, result in the given number.

Square Root

The **positive** or **principal square root** of a positive number a is written as $\sqrt{a}$. The **negative square root** is written as $-\sqrt{a}$.

$$\sqrt{a} = b \quad \text{if} \quad b^2 = a$$

Also, the square root of 0 is 0, written $\sqrt{0} = 0$.

Note that the principal square root of a positive number a is the positive number whose square equals a. *Whenever we use the term* square root *in this book, we mean the positive or principal square root.*

The $\sqrt{\ }$ is called the **radical sign**. The number or expression inside the radical sign is called the **radicand**.

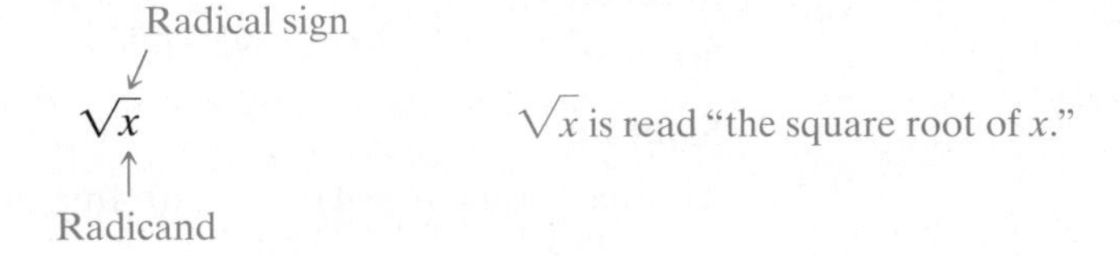

Understanding Algebra

$\sqrt{25}$ is read as "The square root of 25" or as "radical 25."

The entire expression, including the radical sign and radicand, is called the **radical expression**.

Another part of a radical expression is its *index*. The **index** tells the "root" of the expression. Square roots have an index of 2. The index of a square root is generally not written.

Index

$$\sqrt{x} \quad \text{means} \quad \sqrt[2]{x}$$

Radical expressions with different indices will be discussed later in this chapter.

Examples of Square Roots

$\sqrt{25} = 5$ since $5^2 = 5 \cdot 5 = 25$ (The square root of 25 is 5.)

$\sqrt{64} = 8$ since $8^2 = 8 \cdot 8 = 64$ (The square root of 64 is 8.)

$\sqrt{\frac{1}{4}} = \frac{1}{2}$ since $\left(\frac{1}{2}\right)^2 = \left(\frac{1}{2}\right)\left(\frac{1}{2}\right) = \frac{1}{4}$ $\left(\text{The square root of } \frac{1}{4} \text{ is } \frac{1}{2}.\right)$

$\sqrt{\frac{4}{9}} = \frac{2}{3}$ since $\left(\frac{2}{3}\right)^2 = \left(\frac{2}{3}\right)\left(\frac{2}{3}\right) = \frac{4}{9}$ $\left(\text{The square root of } \frac{4}{9} \text{ is } \frac{2}{3}.\right)$

EXAMPLE 1 Evaluate. **a)** $\sqrt{81}$ **b)** $\sqrt{121}$ **c)** $\sqrt{\frac{9}{100}}$

Solution

a) $\sqrt{81} = 9$ since $9^2 = (9)(9) = 81$

b) $\sqrt{121} = 11$ since $(11)^2 = (11)(11) = 121$

c) $\sqrt{\frac{9}{100}} = \frac{3}{10}$ since $\left(\frac{3}{10}\right)^2 = \left(\frac{3}{10}\right)\left(\frac{3}{10}\right) = \frac{9}{100}$

Now Try Exercise 23

EXAMPLE 2 Evaluate. **a)** $-\sqrt{81}$ **b)** $-\sqrt{100}$

Solution

a) $\sqrt{81} = 9$. Now we take the opposite of both sides to get

$$-\sqrt{81} = -9.$$

b) Similarly, $-\sqrt{100} = -10$.

Now Try Exercise 21

Understanding Algebra

Remember that $-\sqrt{9}$ and $\sqrt{-9}$ are different numbers. $-\sqrt{9}$ is -3, but $\sqrt{-9}$ is an imaginary number. The location of the minus sign is very important.

2 Recognize That Not All Square Roots Represent Real Numbers

You must understand that *square roots of negative numbers are not real numbers.* Consider $\sqrt{-9}$. To evaluate $\sqrt{-9}$, we must find some number whose square equals -9. But we know that the square of any nonzero real number must be a positive number. Therefore, no real number squared equals -9, so $\sqrt{-9}$ has no real value. Numbers like $\sqrt{-9}$, or square roots of any negative numbers, are called **imaginary numbers**. Imaginary numbers are discussed further in Section 10.5.

EXAMPLE 3 Indicate whether the radical expression is a real or an imaginary number.

a) $-\sqrt{16}$ **b)** $\sqrt{-16}$ **c)** $\sqrt{-43}$ **d)** $-\sqrt{43}$

Solution

a) Real (equal to -4) **b)** Imaginary **c)** Imaginary **d)** Real

Now Try Exercise 43

HELPFUL HINT

The square root of any nonnegative number will be a real number. The square root of any negative number will be an imaginary number.

Examples of *Real* Numbers

$-\sqrt{9}$, $-\sqrt{\frac{1}{2}}$, $-\sqrt{6.74}$, $-\sqrt{16}$

↑ ↑ ↑ ↑

Radicands are positive numbers

Examples of *Imaginary* Numbers

$\sqrt{-9}$, $\sqrt{-\frac{1}{2}}$, $\sqrt{-6.74}$, $\sqrt{-16}$

↑ ↑ ↑ ↑

Radicands are negative numbers

Suppose we have an expression like $\sqrt{x}$ where x represents some real number. For the radical to be a real number, and not imaginary, we must assume that x is a nonnegative real number.

In this chapter, unless stated otherwise, we will assume that all expressions that are radicands represent nonnegative real numbers.

3 Determine Whether the Square Root of a Real Number Is Rational or Irrational

To help in our discussion of rational and irrational numbers, we will define perfect squares. The numbers 1, 4, 9, 16, 25, 36, 49, . . . are called **perfect squares** because each number is *the square of a natural number*. When a perfect square is a factor of a radicand, we may refer to it as a **perfect square factor**.

$$\begin{array}{llllllll} 1, & 2, & 3, & 4, & 5, & 6, & 7, & \ldots \quad \text{Natural numbers} \\ 1^2, & 2^2, & 3^2, & 4^2, & 5^2, & 6^2, & 7^2, & \ldots \quad \text{The squares of the natural numbers} \\ 1, & 4, & 9, & 16, & 25, & 36, & 49, & \ldots \quad \text{Perfect squares} \end{array}$$

Note that the square root of a perfect square is an integer. That is, $\sqrt{1} = 1$, $\sqrt{4} = 2$, $\sqrt{9} = 3$, $\sqrt{16} = 4$, and so on.

Table 9.1 illustrates the first 20 perfect squares. You may wish to refer to this table when simplifying radical expressions.

TABLE 9.1

Perfect Square	Square Root of Perfect Square		Value	Perfect Square	Square Root of Perfect Square		Value
1	$\sqrt{1}$	=	1	121	$\sqrt{121}$	=	11
4	$\sqrt{4}$	=	2	144	$\sqrt{144}$	=	12
9	$\sqrt{9}$	=	3	169	$\sqrt{169}$	=	13
16	$\sqrt{16}$	=	4	196	$\sqrt{196}$	=	14
25	$\sqrt{25}$	=	5	225	$\sqrt{225}$	=	15
36	$\sqrt{36}$	=	6	256	$\sqrt{256}$	=	16
49	$\sqrt{49}$	=	7	289	$\sqrt{289}$	=	17
64	$\sqrt{64}$	=	8	324	$\sqrt{324}$	=	18
81	$\sqrt{81}$	=	9	361	$\sqrt{361}$	=	19
100	$\sqrt{100}$	=	10	400	$\sqrt{400}$	=	20

In Section 1.4 we discussed real, rational, and irrational numbers. We now continue this discussion.

Rational Number

A **rational number** is a number that can be written in the form $\frac{a}{b}$, where a and b are integers, and $b \neq 0$.

Examples of rational numbers are $\frac{1}{2}, \frac{3}{5}, -\frac{9}{2}$, 4, and 0. All integers are rational numbers since they can be expressed with a denominator of 1. For example, $4 = \frac{4}{1}$ and $0 = \frac{0}{1}$. The square roots of perfect squares are also rational numbers since each is an integer. For example, the twenty square roots shown in **Table 9.1** are all rational numbers.

Irrational Numbers

Real numbers that are not rational numbers are called **irrational numbers**.

Understanding Algebra

Rational vs. Irrational

Rational Numbers

- Terminating decimal numbers
- Repeating decimal numbers
- Square roots of perfect squares

Examples: $\frac{1}{4}, \frac{1}{3}, \sqrt{25}$

Irrational Numbers

- Real numbers that are not rational numbers
- Nonterminating, nonrepeating decimal numbers.
- Square roots of nonperfect squares

Examples: $\sqrt{2}, \sqrt{3}, \sqrt{50}$

Irrational numbers when written as decimals are nonterminating, nonrepeating decimals. The square root of every positive integer that is not a perfect square is an irrational number. For example, $\sqrt{2}$ and $\sqrt{3}$ are irrational numbers.

Next, consider $\sqrt{51}$. Since 51 is not a perfect square, $\sqrt{51}$ is an irrational number. An estimation of the value of $\sqrt{51}$ can be determined as follows:

$$49 < 51 < 64, \quad \text{therefore}$$
$$\sqrt{49} < \sqrt{51} < \sqrt{64} \quad \text{and}$$
$$7 < \sqrt{51} < 8$$

So $\sqrt{51}$ is between 7 and 8. If you evaluate $\sqrt{51}$ on a calculator, you will see that $\sqrt{51}$ is indeed approximately equal to 7.14.

When evaluating radicals, we may use the *is approximately equal to* symbol, $\approx$. For example, we may write $\sqrt{2} \approx 1.414$. This is read "the square root of 2 is approximately equal to 1.414." Recall that $\sqrt{2}$ is not a perfect square, so its square root cannot be evaluated exactly.

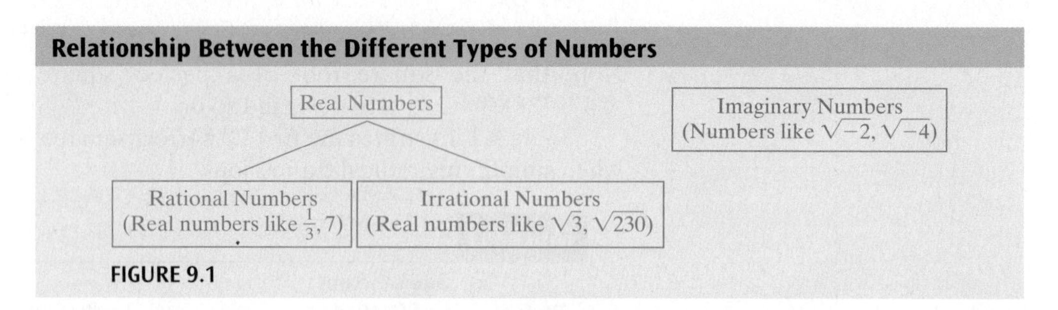

FIGURE 9.1

From **Figure 9.1**, we see that the rational numbers plus the irrational numbers form the real numbers, and imaginary numbers are not real numbers, and vice versa.

EXAMPLE 4 Use **Table 9.1**, on page 521, to determine whether the following square roots are rational or irrational numbers.

a) $\sqrt{73}$ **b)** $\sqrt{196}$ **c)** $\sqrt{238}$ **d)** $\sqrt{324}$

Solution

a) Irrational **b)** Rational, equal to 14

c) Irrational **d)** Rational, equal to 18

Now Try Exercise 47

4 Write Square Roots as Exponential Expressions

Radical expressions can be written in exponential form. Since we are discussing square roots, we will show how to write square roots in exponential form.

Recall that the index of square roots is 2. For example,

$$\sqrt{x} \quad \text{means} \quad \sqrt[2]{x}$$

We use the index, 2, when writing square roots in exponential form. To change from an expression in square root form to an expression in exponential form, simply write the radicand of the square root to the $1/2$ power, as follows:

Writing a Square Root in Exponential Form

$$\sqrt{\blacksquare} = \blacksquare^{1/2} \leftarrow \text{Index of square root}$$

(↑ Radicand ↑)

For example, $\sqrt{8}$ in exponential form is $8^{1/2}$, and $\sqrt{5ab} = (5ab)^{1/2}$. Other examples are

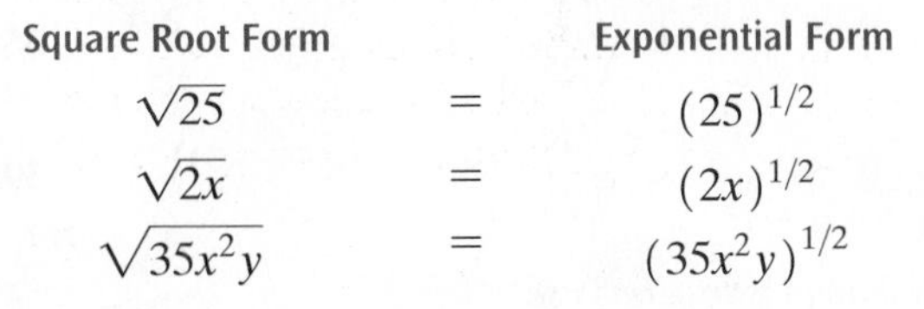

Square Root Form		Exponential Form
$\sqrt{25}$	=	$(25)^{1/2}$
$\sqrt{2x}$	=	$(2x)^{1/2}$
$\sqrt{35x^2y}$	=	$(35x^2y)^{1/2}$

EXAMPLE 5 Write each radical expression in exponential form.

a) $\sqrt{3}$ **b)** $\sqrt{26x}$

Solution

a) $3^{1/2}$ **b)** $(26x)^{1/2}$

Now Try Exercise 57

We can also convert an expression from exponential form to radical form. For example, $(8x)^{1/2}$ can be written $\sqrt{8x}$ and $(20x^4)^{1/2}$ can be written $\sqrt{20x^4}$.

The rules of exponents presented in Sections 4.1 and 4.2 (see summary on page 285) apply to rational (or fractional) exponents. For example,

$$(x^2)^{1/2} = x^{2(1/2)} = x^1 = x$$
$$(xy)^{1/2} = x^{1/2}y^{1/2}$$
and $$x^{1/2} \cdot x^{3/2} = x^{(1/2)+(3/2)} = x^{4/2} = x^2$$

9.1 Exercise Set MathXL® MyMathLab®

Warm-Up Exercises

Fill in the blanks with the appropriate word, phrase, or symbol(s) from the following list.

imaginary	real	sign	rational	exponential	index	approximately	negative
radicand	radical	principal	irrational	subscript	perfect	integer	

1. The expression $\sqrt{a}$ is called a ____________ expression.
2. In the expression $\sqrt{a}$ the symbol $\sqrt{\ }$ is called the radical ____________.
3. In the expression $\sqrt{a}$ the a is called the ____________.
4. The expression $\sqrt{a}$ means $\sqrt[2]{a}$ where the 2 is called the ____________ of the radical.
5. Every real number greater than 0 has two square roots, a positive square root and a ____________ square root.
6. We use the radical symbol $\sqrt{\ }$ to indicate the positive or ____________ square root of a number.
7. The square root of a negative number is an ____________ number.
8. The numbers 1, 4, 9, 16, 25, 36, 49, . . . are called ____________ squares.
9. Numbers that can be written as the ratio of two integers with the denominator not equal to 0 are called ____________ numbers.
10. Real numbers that are not rational numbers are called ____________ numbers.
11. When we write the radical expression $\sqrt{a}$ as $a^{1/2}$ we are using ____________ form.
12. The symbol $\approx$ means is ____________ equal to.

Practice the Skills

Evaluate each square root.

13. $\sqrt{0}$ **14.** $\sqrt{1}$ **15.** $\sqrt{4}$ **16.** $\sqrt{36}$

17. $-\sqrt{49}$ **18.** $-\sqrt{81}$ **19.** $\sqrt{400}$ **20.** $\sqrt{121}$

21. $-\sqrt{16}$ **22.** $-\sqrt{36}$ **23.** $\sqrt{144}$ **24.** $\sqrt{49}$

25. $\sqrt{169}$ **26.** $\sqrt{225}$ **27.** $-\sqrt{1}$ **28.** $-\sqrt{100}$

29. $-\sqrt{144}$ **30.** $-\sqrt{196}$ **31.** $-\sqrt{121}$ **32.** $-\sqrt{400}$

33. $\sqrt{\frac{1}{4}}$ **34.** $\sqrt{\frac{1}{25}}$ **35.** $-\sqrt{\frac{25}{36}}$ **36.** $-\sqrt{\frac{49}{144}}$

37. $\sqrt{\frac{25}{16}}$ **38.** $\sqrt{\frac{49}{25}}$ **39.** $-\sqrt{\frac{100}{9}}$ **40.** $-\sqrt{\frac{196}{169}}$

Indicate whether each statement is true or false.

41. $\sqrt{25}$ is a rational number.

42. $\sqrt{36}$ is an irrational number.

43. $\sqrt{6}$ is an irrational number.

44. $\sqrt{28}$ is a rational number.

45. $\sqrt{-25}$ is a real number.

46. $\sqrt{-4}$ is an imaginary number.

47. $\sqrt{\frac{1}{4}}$ is a rational number.

48. $\sqrt{\frac{49}{4}}$ is an irrational number.

49. $\sqrt{7}$ is an imaginary number.

50. $\sqrt{-17}$ is an irrational number.

51. $\sqrt{(12)^2}$ is an integer.

52. $\sqrt{(13)^2}$ is an integer.

Write in exponential form.

53. $\sqrt{7}$ **54.** $\sqrt{13}$ **55.** $\sqrt{29}$ **56.** $\sqrt{73}$

57. $\sqrt{8x}$ **58.** $\sqrt{5x}$ **59.** $\sqrt{12x^2}$ **60.** $\sqrt{25x^2y}$

61. $\sqrt{21ab^2}$ **62.** $\sqrt{93x^3y}$ **63.** $\sqrt{38n^3}$ **64.** $\sqrt{59x^3y^3}$

Problem Solving

65. Classify each number as rational, irrational, or imaginary.

$$9.83, \sqrt{-4}, \frac{3}{5}, 0.333\ldots, 5, \sqrt{\frac{4}{49}}, \frac{3}{7}, \sqrt{\frac{5}{16}}, -\sqrt{9}, -\sqrt{-16}$$

66. Classify each number as rational, irrational, or imaginary.

$$\sqrt{5}, 8.23, \sqrt{-7}, 10, \frac{1}{3}, 0.33, \sqrt{\frac{25}{64}}, -\sqrt{90}$$

67. Between what two integers is the square root of 47? Do not use **Table 9.1**. Explain how you determined your answer.

68. Between what two integers is the square root of 88? Do not use **Table 9.1**. Explain how you determined your answer.

69. a) Explain how you can determine without using a calculator whether 4.6 or $\sqrt{20}$ is greater.

b) Without using a calculator, determine which is greater.

70. a) Explain how you can determine without using a calculator whether 7.2 or $\sqrt{58}$ is greater.

b) Without using a calculator, determine which is greater.

71. Arrange the following list from smallest to largest. Do not use **Table 9.1**.

$$5, -\sqrt{9}, -\sqrt{7}, 12, 2.5, -\frac{1}{2}, 4.01, \sqrt{16}$$

72. Arrange the following list from smallest to largest. Do not use **Table 9.1**.

$$-\frac{1}{3}, -\sqrt{9}, 5, 0, \sqrt{9}, 8, -2, 3.25$$

73. Match each number in the column on the left with the corresponding answer in the column on the right.

$\sqrt{4}$	imaginary number
$(36)^{1/2}$	2
$-\sqrt{9}$	6
$-(25)^{1/2}$	-5
$(81)^{1/2}$	-3
$(-4)^{1/2}$	9

74. Match each number in the column on the left with the corresponding answer in the column on the right.

$-\sqrt{36}$	10
$(49)^{1/2}$	-7
$\sqrt{100}$	imaginary number
$(9)^{1/2}$	7
$-(49^{1/2})$	3
$(-16)^{1/2}$	-6

75. Is $\sqrt{0}$

a) a real number?

b) a positive number?

c) a negative number?

d) a rational number?

e) an irrational number? Explain your answer.

Concept/Writing Exercises

76. Explain the difference between a rational number and an irrational number.

77. Whenever we see an expression in a square root, what assumption do we make about the expression? Why do we make this assumption?

Challenge Problems

We discuss the following concepts in Sections 9.2 and 9.3.

78. a) Is $\sqrt{4}\cdot\sqrt{9}$ equal to $\sqrt{4\cdot 9}$?

b) Is $\sqrt{9}\cdot\sqrt{25}$ equal to $\sqrt{9\cdot 25}$?

c) Using these two examples, can you guess what $\sqrt{a}\cdot\sqrt{b}$ is equal to (provided that $a \geq 0, b \geq 0$)?

d) Create your own problem like those given in parts **a)** and **b)** and see if the answer you gave in part **c)** works with your numbers.

79. a) Is $\sqrt{3^2}$ equal to 3?

b) Is $\sqrt{8^2}$ equal to 8?

c) Using these two examples, can you guess what $\sqrt{a^2}, a \geq 0$, is equal to?

d) Create your own problem like those given in parts **a)** and **b)** and see if the answer you gave in part **c)** works with your numbers.

80. a) Is $\frac{\sqrt{16}}{\sqrt{4}}$ equal to $\sqrt{\frac{16}{4}}$?

b) Is $\frac{\sqrt{36}}{\sqrt{49}}$ equal to $\sqrt{\frac{36}{49}}$?

c) Using these two examples, can you guess what $\frac{\sqrt{a}}{\sqrt{b}}$ is equal to (provided that $a \geq 0, b > 0$)?

d) Create your own problem like those given in parts **a)** and **b)** and see if the answer you gave in part **c)** works with your numbers.

The rules of exponents we discussed in Chapter 4 also apply with rational exponents. Use the rules of exponents to simplify the following expressions. We will discuss problems like this in Section 9.7.

81. $(x^3)^{1/2}$

82. $(x^6)^{1/2}$

83. $x^{1/2}\cdot x^{5/2}$

84. $x^{7/2}\cdot x^{1/2}$

Cumulative Review Exercises

[3.2] **85. Jumping on a Trampoline** Allison jumped on the trampoline for a minute, and then her mother Elizabeth jumped on the trampoline for a minute. If Allison made twice as many jumps as her mother did, and the total number of jumps was 78, determine the number of jumps Allison made.

[6.6] *Solve.*

86. $\frac{2x}{x^2 - 4} + \frac{1}{x - 2} = \frac{2}{x + 2}$

87. $\frac{4x}{x^2 + 6x + 9} - \frac{2x}{x + 3} = \frac{x + 1}{x + 3}$

[7.3] **88.** Determine the slope of the line through the points $(-5, 2)$ and $(6, 7)$.

[7.6] **89.** If $f(x) = x^2 - 4x - 15$, find $f(-3)$.

9.2 Simplifying Square Roots

1. Use the product rule to simplify square roots containing constants.
2. Use the product rule to simplify square roots containing variables.

1 Use the Product Rule to Simplify Square Roots Containing Constants

To simplify square roots we will make use of the **product rule for square roots.**

Product Rule for Square Roots

$$\sqrt{ab} = \sqrt{a}\cdot\sqrt{b} \text{ and } \sqrt{a}\cdot\sqrt{b} = \sqrt{ab}, \text{ provided } a \geq 0, b \geq 0 \quad \textbf{Rule 1}$$

Notice the square root of a product is equal to the product of the square roots of the factors. The product rule applies only when both a and b are nonnegative, since the square root of negative numbers are not real numbers. We will use this rule to discuss the number $\sqrt{60}$.

Examples of the Product Rule

$$\sqrt{60} = \begin{cases} \sqrt{1 \cdot 60} = \sqrt{1} \cdot \sqrt{60} \\ \sqrt{2 \cdot 30} = \sqrt{2} \cdot \sqrt{30} \\ \sqrt{3 \cdot 20} = \sqrt{3} \cdot \sqrt{20} \\ \sqrt{4 \cdot 15} = \sqrt{4} \cdot \sqrt{15} \\ \sqrt{5 \cdot 12} = \sqrt{5} \cdot \sqrt{12} \\ \sqrt{6 \cdot 10} = \sqrt{6} \cdot \sqrt{10} \end{cases}$$

Note that $\sqrt{60}$ can be factored into any of these forms. We will use this list in Example 1.

To Simplify the Square Root of a Constant

1. Write the radicand as a product of the largest perfect square factor and another factor.
2. Use the product rule to write the expression as a product of square roots, with each square root containing one of the factors.
3. Find the square root of the perfect square factor.

EXAMPLE 1 Simplify $\sqrt{60}$.

Solution Examine the list of factors for $\sqrt{60}$, above. Notice that among all of the factors of 60, the only perfect square is 4. We will use the fact that $\sqrt{60} = \sqrt{4 \cdot 15}$ to simplify $\sqrt{60}$.

$$\begin{aligned} \sqrt{60} &= \sqrt{4 \cdot 15} && \text{Write 60 as } 4 \cdot 15. \\ &= \sqrt{4} \cdot \sqrt{15} && \text{Product Rule} \\ &= 2\sqrt{15} && \text{Simplify: } \sqrt{4} = 2 \end{aligned}$$

Since 15 is not a perfect square and has no perfect square factors, this expression cannot be simplified further. The expression $2\sqrt{15}$ is read "two times the square root of fifteen."

Now Try Exercise 15

EXAMPLE 2 Simplify $\sqrt{12}$.

Solution First, list the positive factors of 12: 1, 2, 3, 4, 6, and 12. Note that the only perfect square factor of 12 is 4. Also note that $12 = 4 \cdot 3$. Therefore, to simplify $\sqrt{12}$ we will begin with $\sqrt{12} = \sqrt{4 \cdot 3}$.

$$\begin{aligned} \sqrt{12} = \sqrt{4 \cdot 3} &= \sqrt{4} \cdot \sqrt{3} \\ &= 2\sqrt{3} \end{aligned}$$

Now Try Exercise 17

EXAMPLE 3 Simplify $\sqrt{80}$.

Solution First, list the positive factors of 80: 1, 2, 4, 5, 8, 10, 16, 20, 40, and 80. Note that both 4 and 16 are perfect square factors of 80. However, when simplifying square roots we will choose the largest perfect square factor (see the Helpful Hint on the next page). Therefore, we will use the fact that $80 = 16 \cdot 5$ to simplify $\sqrt{80}$.

$$\begin{aligned} \sqrt{80} = \sqrt{16 \cdot 5} &= \sqrt{16} \cdot \sqrt{5} \\ &= 4\sqrt{5} \end{aligned}$$

Now Try Exercise 19

HELPFUL HINT

When simplifying a square root, it is not uncommon to use a perfect square factor that is not the *largest* perfect square factor of the radicand. Let's consider Example 3 again. Four is also a perfect square factor of 80.

$$\sqrt{80} = \sqrt{4 \cdot 20} = \sqrt{4} \cdot \sqrt{20} = 2\sqrt{20}$$

Since 20 itself contains a perfect square factor of 4, the solution is not complete. Rather than starting the entire solution again, you can continue the simplification process as follows.

$$\sqrt{80} = 2\sqrt{20} = 2\sqrt{4 \cdot 5} = 2\sqrt{4} \cdot \sqrt{5} = 2 \cdot 2 \cdot \sqrt{5} = 4\sqrt{5}$$

Now the result checks with the answer in Example 3.

EXAMPLE 4 Simplify $\sqrt{245}$.

Solution

$$\begin{aligned}\sqrt{245} = \sqrt{49 \cdot 5} &= \sqrt{49} \cdot \sqrt{5} \\ &= 7\sqrt{5}\end{aligned}$$

Now Try Exercise 21

EXAMPLE 5 Simplify $\sqrt{135}$.

Solution

$$\begin{aligned}\sqrt{135} = \sqrt{9 \cdot 15} &= \sqrt{9} \cdot \sqrt{15} \\ &= 3\sqrt{15}\end{aligned}$$

Although 15 can be factored into $5 \cdot 3$, neither of these factors is a perfect square. Thus, the answer cannot be simplified any further.

Now Try Exercise 29

2 Use the Product Rule to Simplify Square Roots Containing Variables

In Section 9.1 we noted that certain numbers were *perfect squares*. We will also refer to certain expressions that contain a variable as perfect squares. When a radical contains a variable (or number) raised to an *even exponent*, that variable (or number) and exponent together also form a perfect square. For example, in the expression $\sqrt{x^6}$, the x^6 is a perfect square since the exponent, 6, is even.

$$\sqrt{x^6}$$

↑

x^6 is a perfect square because the exponent, 6, is even.

In the expression $\sqrt{x^7}$, the x^7 is not a perfect square since the exponent is odd. However, x^6 is a *perfect square factor* of x^7 because x^6 is a perfect square and x^6 is a factor of x^7. Note that $x^7 = x^6 \cdot x$.

To evaluate square roots when the radicand is a perfect square, we use the following rule.

Square Root of a Perfect Square

$$\sqrt{a^{2 \cdot n}} = a^n, \qquad a \geq 0 \qquad \textbf{Rule 2}$$

The square root of a variable raised to an even power equals the variable raised to one-half that power. To explain this rule, we can write the square root expression $\sqrt{a^{2n}}$ in exponential form, and then simplify as follows.

$$\sqrt{a^{2n}} = (a^{2n})^{1/2} = a^{2n(1/2)} = a^n$$

Examples of Rule 2 follow. Remember that we are assuming the variables represent nonnegative real numbers.

Examples

$\sqrt{x^2} = x$

$\sqrt{a^4} = a^2$

$\sqrt{y^{20}} = y^{10}$

$\sqrt{z^{28}} = z^{14}$

EXAMPLE 6 Simplify. **a)** $\sqrt{x^{32}}$ **b)** $\sqrt{x^4y^8}$ **c)** $\sqrt{a^{10}b^2}$ **d)** $\sqrt{y^{12}z^{16}}$

Solution

a) $\sqrt{x^{32}} = x^{16}$

b) $\sqrt{x^4y^8} = \sqrt{x^4}\sqrt{y^8} = x^2y^4$

c) $\sqrt{a^{10}b^2} = \sqrt{a^{10}}\sqrt{b^2} = a^5b$

d) $\sqrt{y^{12}z^{16}} = \sqrt{y^{12}}\sqrt{z^{16}} = y^6z^8$

Now Try Exercise 35

Understanding Algebra

$\sqrt{x^7} = \sqrt{x^6 \cdot x^1} = x^3\sqrt{x}$

↑ Perfect square factor of x^7.

To Simplify the Square Root of a Radicand Containing a Variable Raised to an Odd Power

1. Express the radicand as the product of two factors, one of which has an exponent of 1 (the other will therefore have an even exponent).
2. Use the product rule to simplify.

The radicand of your simplified answer should not contain any perfect square factors or any variables with an exponent greater than 1.

Examples 7 and 8 illustrate the above procedure.

EXAMPLE 7 Simplify. **a)** $\sqrt{x^3}$ **b)** $\sqrt{y^7}$ **c)** $\sqrt{a^{49}}$

Solution

a) $\sqrt{x^3} = \sqrt{x^2 \cdot x} = \sqrt{x^2} \cdot \sqrt{x}$

$= x \cdot \sqrt{x} = x\sqrt{x}$ Remember that x means x^1.

b) $\sqrt{y^7} = \sqrt{y^6 \cdot y^1} = \sqrt{y^6} \cdot \sqrt{y}$

$= y^3\sqrt{y}$

c) $\sqrt{a^{49}} = \sqrt{a^{48} \cdot a} = \sqrt{a^{48}} \cdot \sqrt{a}$

$= a^{24}\sqrt{a}$

Now Try Exercise 33

Understanding Algebra

Compare $\sqrt{49}$ to $\sqrt{a^{49}}$. Be careful when simplifying square roots of expressions that have perfect square exponents. $\sqrt{49} = 7$, but $\sqrt{a^{49}} \neq a^7$. To simplify $\sqrt{a^{49}}$ study Example 7 c).

Radicals that contain both numbers and variables can be simplified using the product rule for radicals and the rules discussed in this section.

EXAMPLE 8 Simplify. **a)** $\sqrt{36x^3}$ **b)** $\sqrt{50x^2}$ **c)** $\sqrt{50x^3}$

Solution Write each expression as the product of square roots, one of which has a radicand that is a perfect square. Then simplify.

a) $\sqrt{36x^3} = \sqrt{36x^2} \cdot \sqrt{x} = 6x\sqrt{x}$

b) $\sqrt{50x^2} = \sqrt{25x^2} \cdot \sqrt{2} = 5x\sqrt{2}$

c) $\sqrt{50x^3} = \sqrt{25x^2} \cdot \sqrt{2x} = 5x\sqrt{2x}$

Now Try Exercise 41

EXAMPLE 9 Simplify. **a)** $\sqrt{72x^2y}$ **b)** $\sqrt{45x^3y^4}$ **c)** $\sqrt{98a^9b^7}$

Solution

a) $\sqrt{72x^2y} = \sqrt{36x^2} \cdot \sqrt{2y}$ — Write $\sqrt{72x^2y}$ as a product of a perfect square factor and another factor.

$= 6x\sqrt{2y}$ — Simplify the perfect square factor.

b) $\sqrt{45x^3y^4} = \sqrt{9x^2y^4} \cdot \sqrt{5x}$ — Write $\sqrt{45x^3y^4}$ as a product of a perfect square factor and another factor.

$= 3xy^2\sqrt{5x}$ — Simplify the perfect square factor.

c) $\sqrt{98a^9b^7} = \sqrt{49a^8b^6} \cdot \sqrt{2ab}$ — Write $\sqrt{98a^9b^7}$ as a product of a perfect square factor and another factor.

$= 7a^4b^3\sqrt{2ab}$ — Simplify the perfect square factor.

Now Try Exercise 49

Now let's look at an example where we use the product rule to multiply two radicals before simplifying.

EXAMPLE 10 Multiply and then simplify.

a) $\sqrt{2} \cdot \sqrt{8}$ **b)** $\sqrt{2x} \cdot \sqrt{8}$ **c)** $(\sqrt{5x})^2$

Solution

a) $\sqrt{2} \cdot \sqrt{8} = \sqrt{2 \cdot 8} = \sqrt{16} = 4$

b) $\sqrt{2x} \cdot \sqrt{8} = \sqrt{16x} = \sqrt{16} \cdot \sqrt{x} = 4\sqrt{x}$

c) $(\sqrt{5x})^2 = \sqrt{5x} \cdot \sqrt{5x} = \sqrt{25x^2} = 5x$

Now Try Exercise 55

EXAMPLE 11 Multiply and then simplify.

a) $\sqrt{8x^3y}\sqrt{4xy^5}$ **b)** $\sqrt{7ab^8}\sqrt{6a^5b}$

Solution

a) $\sqrt{8x^3y}\sqrt{4xy^5} = \sqrt{32x^4y^6} = \sqrt{16x^4y^6} \cdot \sqrt{2}$
$= 4x^2y^3\sqrt{2}$

b) $\sqrt{7ab^8}\sqrt{6a^5b} = \sqrt{42a^6b^9} = \sqrt{a^6b^8} \cdot \sqrt{42b}$
$= a^3b^4\sqrt{42b}$

In part **b)**, 42 can be factored in many ways. However, none of the factors are perfect squares, so we leave the answer as given.

Now Try Exercise 63

9.2 Exercise Set MathXL® MyMathLab®

Warm-Up Exercises

Fill in the blanks with the appropriate word, phrase, or symbol(s) from the following list.

squares real one-half greater factors even perfect nonnegative radicand

1. The product rule for square roots states that the square root of a product is equal to the product of the square roots of the ________.

2. The product rule for square roots applies only when both a and b are ________ numbers.

3. The square roots of negative numbers are not ________ numbers.

4. To simplify the square root of a constant, we first write the radicand as a product of the largest ________ square factor and another factor.

5. The numbers 1, 4, 9, 16, 25, 36, . . . are called perfect ____________.

6. When a radical contains a variable raised to an ____________ exponent, that variable and exponent together form a perfect square.

7. The square root of a variable raised to an even power equals the variable raised to ____________ that power.

8. When simplifying a square root, the radicand of the simplified answer should not contain any perfect square factors or any variables with an exponent ____________ than one.

Practice the Skills

Simplify each square root. Assume all variables represent positive numbers. If the square root is already simplified, so state.

9. $\sqrt{19}$ **10.** $\sqrt{23}$ **11.** $\sqrt{8}$ **12.** $\sqrt{18}$

13. $\sqrt{35}$ **14.** $\sqrt{77}$ **15.** $\sqrt{28}$ **16.** $\sqrt{50}$

17. $\sqrt{32}$ **18.** $\sqrt{48}$ **19.** $\sqrt{20}$ **20.** $\sqrt{24}$

21. $\sqrt{160}$ **22.** $\sqrt{90}$ **23.** $\sqrt{44}$ **24.** $\sqrt{99}$

25. $\sqrt{72}$ **26.** $\sqrt{147}$ **27.** $\sqrt{84}$ **28.** $\sqrt{140}$

29. $\sqrt{125}$ **30.** $\sqrt{192}$ **31.** $\sqrt{x^6}$ **32.** $\sqrt{x^4}$

33. $\sqrt{z^7}$ **34.** $\sqrt{y^{11}}$ **35.** $\sqrt{x^2y^8}$ **36.** $\sqrt{x^2y^4}$

37. $\sqrt{a^{12}b^9}$ **38.** $\sqrt{x^{16}y^5}$ **39.** $\sqrt{a^2b^4c}$ **40.** $\sqrt{x^4y^5z^{10}}$

41. $\sqrt{6n^3}$ **42.** $\sqrt{24p}$ **43.** $\sqrt{108a^3b^2}$ **44.** $\sqrt{75m^4n^3}$

45. $\sqrt{243x^3y^4}$ **46.** $\sqrt{300a^5b^{11}}$ **47.** $\sqrt{64xyz^7}$ **48.** $\sqrt{600ab^4c^3}$

49. $\sqrt{192a^2b^9c}$ **50.** $\sqrt{405x^4y^7z}$ **51.** $\sqrt{180r^3s^{10}t^5}$ **52.** $\sqrt{98x^6y^4z}$

Multiply and then simplify. Assume all variables represent positive numbers.

53. $\sqrt{7}\cdot\sqrt{7}$ **54.** $\sqrt{15}\cdot\sqrt{15}$ **55.** $\sqrt{6}\cdot\sqrt{2}$

56. $\sqrt{21}\cdot\sqrt{7}$ **57.** $\sqrt{15}\cdot\sqrt{6}$ **58.** $\sqrt{35}\cdot\sqrt{10}$

59. $\sqrt{2x}\,\sqrt{7x}$ **60.** $\sqrt{6x^3}\,\sqrt{6x}$ **61.** $\sqrt{4a^2}\,\sqrt{12ab^2}$

62. $\sqrt{30b^2}\,\sqrt{6b^5}$ **63.** $\sqrt{3xy^3}\sqrt{24x^2y}$ **64.** $\sqrt{20xy^4}\,\sqrt{8x^5}$

65. $\sqrt{3r^4s^7}\,\sqrt{33r^6s^5}$ **66.** $\sqrt{27x^3y}\,\sqrt{5x^3y^5}$ **67.** $\sqrt{15xy^6}\,\sqrt{6xyz}$

68. $\sqrt{14xyz^5}\,\sqrt{5xy^2z^6}$ **69.** $\sqrt{6a^2b^4}\,\sqrt{9a^4b^6}$ **70.** $\sqrt{6a^4b^5c^6}\,\sqrt{3a^3bc^6}$

71. $(\sqrt{5x})^2$ **72.** $(\sqrt{7x^2})^2$ **73.** $(\sqrt{13x^4y^6})^2$

74. $(\sqrt{17p^3q^2})^2$ **75.** $(\sqrt{6a})^2(\sqrt{7a})^2$ **76.** $(\sqrt{5ab})^2(\sqrt{9ab})^2$

Problem Solving

In Exercises 77–82, which coefficients and exponents should be placed in the shaded areas to make a true statement? Explain how you obtained your answer.

77. $\sqrt{25x^{\blacksquare}y^6} = 5x^4y^3$

78. $\sqrt{\blacksquare x^4y^{\blacksquare}} = 6x^2y^4$

79. $\sqrt{4x^{\blacksquare}y^{\blacksquare}} = 2x^6y^2\sqrt{y}$

80. $\sqrt{3x^4y^{\blacksquare}}\cdot\sqrt{3x^{\blacksquare}y^5} = 3x^5y^7\sqrt{xy}$

81. $\sqrt{2x^{\blacksquare}y^5}\cdot\sqrt{\blacksquare x^3y^{\blacksquare}} = 4x^7y^6\sqrt{x}$

82. $\sqrt{32x^4z^{\blacksquare}}\cdot\sqrt{\blacksquare x^{\blacksquare}z^{12}} = 8x^5z^9\sqrt{z}$

83. **a)** Showing all steps, simplify $(\sqrt{7x^4})^2$.

b) Showing all steps, simplify $\sqrt{(7x^4)^2}$.

c) Compare your results in part **a)** and part **b)**. Are they the same?

84. **a)** Showing all steps, simplify $(\sqrt{23x^3})^2$.

b) Showing all steps, simplify $\sqrt{(23x^3)^2}$.

c) Compare your results in part **a)** and part **b)**. Are they the same?

Concept/Writing Exercises

85. **a)** Explain how to simplify a square root containing only a constant.

b) Simplify $\sqrt{20}$ using the procedure you gave in part **a)**.

86. Explain why the product rule cannot be used to simplify the problem $\sqrt{-4}\cdot\sqrt{-9}$.

87. We learned that for $a \geq 0$, $\sqrt{a^{2\cdot n}} = a^n$. Explain what this means.

88. **a)** Explain how to simplify the square root of a radical containing a variable raised to an odd power.

b) Simplify $\sqrt{x^{13}}$ using the procedure you gave in part **a)**.

Challenge Problems

Below we show two simplifications involving square roots.

$$\sqrt{x^4} = (x^4)^{1/2} = x^{4(1/2)} = x^2$$
$$\sqrt{x^{2/4}} = (x^{2/4})^{1/2} = x^{(2/4)(1/2)} = x^{1/4}$$

In Section 9.1 we indicated that the square root of an expression may be written as the expression to the 1/2 power. The rules for exponents that were discussed in Section 4.1 also apply when the exponents are rational numbers. Use the two examples illustrated and the rules for exponents to simplify Exercises 89–92.

89. $\sqrt{x^{2/6}}$ **90.** $\sqrt{y^{10/14}}$ **91.** $\sqrt{4x^{4/5}}$ **92.** $\sqrt{25y^{8/3}}$

93. Is $\sqrt{6.25}$ a rational or an irrational number? Explain how you determined your answer.

94. a) In Section 9.4 we will be multiplying expressions like $(\sqrt{a} + \sqrt{b})(\sqrt{a} - \sqrt{b})$ using the FOIL method. Can you find this product now?

b) Multiply $(\sqrt{7} + \sqrt{3})(\sqrt{7} - \sqrt{3})$.

95. The area of a square is found by the formula $A = s^2$. We will learn later that we can rewrite this formula as $s = \sqrt{A}$.

a) If the area is 16 square feet, what is the length of a side?

b) If the area is doubled, is the length of a side doubled? Explain.

c) To double the length of a side of a square, how much must the area be increased? Explain.

96. We know that $\sqrt{a} \cdot \sqrt{b} = \sqrt{a \cdot b}$ if $a \ge 0$ and $b \ge 0$. Does $\sqrt{\frac{a}{b}} = \frac{\sqrt{a}}{\sqrt{b}}$ if $a \ge 0$ and $b > 0$? Try several pairs of values for a and b and see.

97. a) Will the product of two rational numbers always be a rational number? Explain and give an example to support your answer.

b) Will the product of two irrational numbers always be an irrational number? Explain and give an example to support your answer.

Group Activity

We learned earlier that $\sqrt{\square} = \square^{1/2}$. *For example,* $\sqrt{x^6} = (x^6)^{1/2} = x^3$. *We can simplify* $\sqrt{x^{2n}}$ *by writing the expression in exponential form,* $\sqrt{x^{2n}} = (x^{2n})^{1/2} = x^{(2n)(1/2)} = x^n$. *As a group, simplify the following square roots by writing the expression in exponential form. Show all the steps in the simplification process.*

98. $\sqrt{x^{10a}}$ **99.** $\sqrt{x^{8b}}$ **100.** $\sqrt{x^{4a}y^{12b}}$ **101.** $\sqrt{x^{10b}y^{6c}}$

Cumulative Review Exercises

[2.6] **102. Area** Find the area of the trapezoid.

Use $A = \frac{1}{2}h(b + d)$.

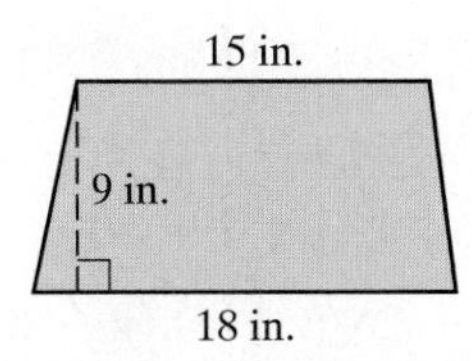

103. Volume Find the volume of the cone.

Use $V = \frac{1}{3}\pi r^2 h$.

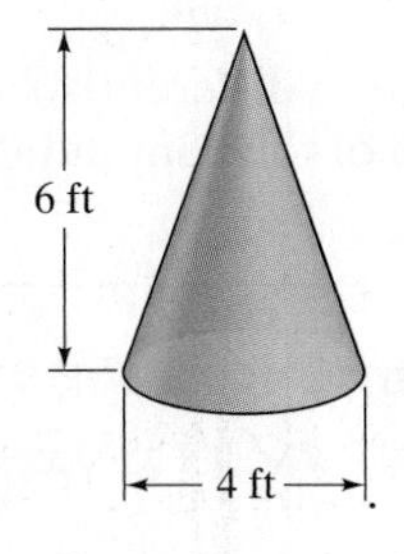

[6.2] **104.** Divide $\frac{3x^2 - 16x - 12}{3x^2 - 10x - 8} \div \frac{x^2 - 7x + 6}{3x^2 - 11x - 4}$.

[7.4] **105.** Write the equation $5x + 10y = 15$ in slope-intercept form and indicate the slope and the y-intercept.

[7.5] **106.** Graph $6x - 5y \ge 30$.

[8.3] **107.** Solve the system of equations.

$$2x + 3y = -8$$
$$6x - y = 6$$

9.3 Adding, Subtracting, and Multiplying Square Roots

1. Add and subtract square roots.
2. Multiply square roots.

1 Add and Subtract Square Roots

Like Radicals and Like Square Roots

Like radicals are radicals that have the same index and the same radicand. **Like square roots** are square roots that have the same radicand.

Like square roots are added in much the same manner that like terms are added, as illustrated below.

Examples of Adding Like Terms

$$2x + 3x = (2 + 3)x = 5x$$
$$4x + x = 4x + 1x = (4 + 1)x = 5x$$

Examples of Adding Like Square Roots

$$2\sqrt{7} + 3\sqrt{7} = (2 + 3)\sqrt{7} = 5\sqrt{7}$$
$$8\sqrt{x} + \sqrt{x} = 8\sqrt{x} + 1\sqrt{x} = (8 + 1)\sqrt{x} = 9\sqrt{x}$$

Note that adding like square roots is an application of the distributive property.

$$2\sqrt{7} + 3\sqrt{7} = (2 + 3)\sqrt{7}$$
$$= 5\sqrt{7}$$

Understanding Algebra

$5\sqrt{x}$ and $7\sqrt{x}$ are *like square roots* just as $3x$ and $4x$ are *like terms.*

Other Examples of Adding and Subtracting Like Square Roots

$$2\sqrt{5} - 8\sqrt{5} = (2 - 8)\sqrt{5} = -6\sqrt{5}$$
$$\sqrt{x} + \sqrt{x} = 1\sqrt{x} + 1\sqrt{x} = (1 + 1)\sqrt{x} = 2\sqrt{x}$$
$$9\sqrt{2} + 3\sqrt{2} - \sqrt{2} = (9 + 3 - 1)\sqrt{2} = 11\sqrt{2}$$
$$\frac{2\sqrt{3}}{5} + \frac{1\sqrt{3}}{5} = \left(\frac{2}{5} + \frac{1}{5}\right)\sqrt{3} = \frac{3}{5}\sqrt{3} \quad \text{or} \quad \frac{3\sqrt{3}}{5}$$

EXAMPLE 1 Simplify if possible.

a) $4\sqrt{5} + 3\sqrt{5} - 8$

b) $\sqrt{6} - 4\sqrt{6} + 10$

c) $5\sqrt{3} + 4\sqrt{y} - 2\sqrt{3} - 6\sqrt{y}$

d) $2\sqrt{3} + 9\sqrt{2}$

Solution

a) $4\sqrt{5} + 3\sqrt{5} - 8 = (4 + 3)\sqrt{5} - 8$ Only $4\sqrt{5}$ and $3\sqrt{5}$ can be combined.

$= 7\sqrt{5} - 8$ Simplify.

b) $\sqrt{6} - 4\sqrt{6} + 10 = (1 - 4)\sqrt{6} + 10 = -3\sqrt{6} + 10$

c) $5\sqrt{3} + 4\sqrt{y} - 2\sqrt{3} - 6\sqrt{y} = 5\sqrt{3} - 2\sqrt{3} + 4\sqrt{y} - 6\sqrt{y}$ Place like radicals together.

$= 3\sqrt{3} - 2\sqrt{y}$ Simplify.

d) Cannot be simplified since the radicands are different.

Understanding Algebra

When adding or subtracting radicals, *like radicals* are combined.

Now Try Exercise 13

The answers in Example 1 could be written differently. For example, the answer to part **a)** could be written $-8 + 7\sqrt{5}$ because of the commutative property of addition.

EXAMPLE 2 Simplify.

a) $3\sqrt{x} - 4\sqrt{x} + 6\sqrt{x}$

b) $2\sqrt{a} + a + 9\sqrt{a}$

c) $x + \sqrt{x} + 7\sqrt{x} + 3$

d) $x\sqrt{x} + 5\sqrt{x} + x$

e) $\sqrt{xy} + 8\sqrt{xy} - \sqrt{x}$

Solution

a) $3\sqrt{x} - 4\sqrt{x} + 6\sqrt{x} = (3 - 4 + 6)\sqrt{x}$ All three are like radicals.
$= 5\sqrt{x}$

b) $2\sqrt{a} + a + 9\sqrt{a} = a + 2\sqrt{a} + 9\sqrt{a}$ Place like radicals together.
$= a + 11\sqrt{a}$ Simplify.

c) $x + \sqrt{x} + 7\sqrt{x} + 3 = x + 1\sqrt{x} + 7\sqrt{x} + 3$ $\sqrt{x}$ means $1\sqrt{x}$.
$= x + 8\sqrt{x} + 3$ Add $1\sqrt{x} + 7\sqrt{x}$ to get $8\sqrt{x}$.

d) $x\sqrt{x} + 5\sqrt{x} + x = (x + 5)\sqrt{x} + x$ Only $x\sqrt{x}$ and $5\sqrt{x}$ can be combined.

e) $\sqrt{xy} + 8\sqrt{xy} - \sqrt{x} = 9\sqrt{xy} - \sqrt{x}$ Only $\sqrt{xy}$ and $8\sqrt{xy}$ can be combined.

Now Try Exercise 21

Unlike Radicals and Unlike Square Roots

Unlike radicals are radicals that have different indices or different radicands. **Unlike square roots** are square roots having different radicands.

It is sometimes possible to change unlike square roots into like square roots by simplifying the radicals in an expression. After simplifying, square roots that contain the same radicand can be combined.

EXAMPLE 3 Simplify $\sqrt{3} + \sqrt{75}$.

Solution Since 75 has a perfect square factor, 25, we write 75 as a product of the perfect square factor and another factor.

$$\sqrt{3} + \sqrt{75} = \sqrt{3} + \sqrt{25 \cdot 3}$$
$$= \sqrt{3} + \sqrt{25} \cdot \sqrt{3} \quad \text{Product rule}$$
$$= \sqrt{3} + 5\sqrt{3} \quad \text{Now the radicals are like radicals.}$$
$$= 6\sqrt{3} \quad \text{Add like square roots.}$$

Now Try Example 25

EXAMPLE 4 Simplify $\sqrt{63} - \sqrt{28}$.

Solution Write each radicand as a product of a perfect square factor and another factor.

$$\sqrt{63} - \sqrt{28} = \sqrt{9 \cdot 7} - \sqrt{4 \cdot 7}$$
$$= \sqrt{9} \cdot \sqrt{7} - \sqrt{4} \cdot \sqrt{7} \quad \text{Product rule}$$
$$= 3\sqrt{7} - 2\sqrt{7} \quad \text{Now the radicals are like radicals.}$$
$$= \sqrt{7} \quad \text{Subtract like square roots.}$$

Now Try Exercise 27

EXAMPLE 5 Simplify.

a) $2\sqrt{8} - \sqrt{32}$ **b)** $3\sqrt{12} + 5\sqrt{27} + 6$ **c)** $\sqrt{120} - \sqrt{125}$

Solution In each part we begin by writing each square root as a product of its largest perfect square factor and another factor. Then we simplify the perfect square factor.

a) $$2\sqrt{8} - \sqrt{32} = 2\sqrt{4 \cdot 2} - \sqrt{16 \cdot 2}$$
$$= 2\sqrt{4}\,\sqrt{2} - \sqrt{16}\,\sqrt{2} \quad \text{Product rule}$$
$$= 2 \cdot 2\sqrt{2} - 4\sqrt{2} \quad \text{Simplify perfect square factors.}$$
$$= 4\sqrt{2} - 4\sqrt{2} \quad \text{Simplify.}$$
$$= 0$$

b) $3\sqrt{12} + 5\sqrt{27} + 6 = 3\sqrt{4\cdot 3} + 5\sqrt{9\cdot 3} + 6$

$= 3\sqrt{4}\,\sqrt{3} + 5\sqrt{9}\,\sqrt{3} + 6$ Product rule

$= 3\cdot 2\sqrt{3} + 5\cdot 3\sqrt{3} + 6$ Simplify perfect square factors.

$= 6\sqrt{3} + 15\sqrt{3} + 6$ Simplify.

$= 21\sqrt{3} + 6$

c) $\sqrt{120} - \sqrt{125} = \sqrt{4\cdot 30} - \sqrt{25\cdot 5}$

$= \sqrt{4}\,\sqrt{30} - \sqrt{25}\,\sqrt{5}$ Product rule

$= 2\sqrt{30} - 5\sqrt{5}$ Simplify perfect square factors.

Since 30 has no perfect square factors and since the radicands are different, the expression $2\sqrt{30} - 5\sqrt{5}$ cannot be simplified any further.

Now Try Exercise 33

AVOIDING COMMON ERRORS

The product rule presented in Section 9.2 was $\sqrt{a}\cdot\sqrt{b} = \sqrt{a\cdot b}$. *The same principle does not apply to addition.*

INCORRECT

~~$\sqrt{a} + \sqrt{b} = \sqrt{a+b}$~~

For example, to evaluate $\sqrt{9} + \sqrt{16}$,

CORRECT	INCORRECT
$\sqrt{9} + \sqrt{16} = 3 + 4$	~~$\sqrt{9} + \sqrt{16} = \sqrt{9+16}$~~
$= 7$	~~$= \sqrt{25}$~~
	~~$= 5$~~

2 Multiply Square Roots

We introduced the product rule for radicals earlier. Now we will expand on multiplying radical expressions.

The distributive property can be used to multiply radical expressions. When the distributive property is used, each quantity within the parentheses is multiplied by the number preceding the parentheses. Some examples follow.

$$\sqrt{6}(\sqrt{6} + 3) = (\sqrt{6})(\sqrt{6}) + (\sqrt{6})(3) = 6 + 3\sqrt{6}$$

$$\sqrt{5}(\sqrt{x} + y) = (\sqrt{5})(\sqrt{x}) + (\sqrt{5})(y) = \sqrt{5x} + y\sqrt{5}$$

Now let's work an example.

EXAMPLE 6 Multiply.

a) $\sqrt{3}(\sqrt{6} - 8)$ **b)** $\sqrt{2x}(\sqrt{x} + \sqrt{2y})$

Solution

a) Use the distributive property. This gives

$\sqrt{3}(\sqrt{6} - 8) = (\sqrt{3})(\sqrt{6}) - (\sqrt{3})(8)$

$= \sqrt{18} - 8\sqrt{3}$ Product rule

$= \sqrt{9\cdot 2} - 8\sqrt{3}$

$= \sqrt{9}\cdot\sqrt{2} - 8\sqrt{3}$ Product rule

$= 3\sqrt{2} - 8\sqrt{3}$

b)
$$\begin{aligned}\sqrt{2x}\left(\sqrt{x}+\sqrt{2y}\right) &= \left(\sqrt{2x}\right)\left(\sqrt{x}\right)+\left(\sqrt{2x}\right)\left(\sqrt{2y}\right)\\ &= \sqrt{2x^2}+\sqrt{4xy} && \text{Product rule}\\ &= \sqrt{x^2}\cdot\sqrt{2}+\sqrt{4}\cdot\sqrt{xy} && \text{Product rule}\\ &= x\sqrt{2}+2\sqrt{xy}\end{aligned}$$

Now Try Exercise 41

In Section 4.5 we discussed using the FOIL method for multiplying two binomials. In the following examples, we will use the FOIL method to multiply radical expressions.

EXAMPLE 7 Multiply $\left(3+\sqrt{5}\right)\left(4-2\sqrt{5}\right)$ using the FOIL method.

Solution

$$\begin{aligned}\left(3+\sqrt{5}\right)\left(4-2\sqrt{5}\right) &= \overset{F}{3(4)}+\overset{O}{3\left(-2\sqrt{5}\right)}+\overset{I}{4\sqrt{5}}+\overset{L}{\sqrt{5}\left(-2\sqrt{5}\right)}\\ &= 12-6\sqrt{5}+4\sqrt{5}-2\sqrt{25}\\ &= 12-2\sqrt{5}-2(5)\\ &= 12-2\sqrt{5}-10\\ &= 2-2\sqrt{5}\end{aligned}$$

Understanding Algebra

Remember that FOIL means

F: First

O: Outer

I: Inner

L: Last

Now Try Exercise 55

EXAMPLE 8 Multiply using the FOIL method.

a) $\left(\sqrt{2}-\sqrt{7}\right)\left(\sqrt{2}+\sqrt{7}\right)$ **b)** $\left(x-\sqrt{y}\right)\left(x+\sqrt{y}\right)$

Solution

a)
$$\begin{aligned}\left(\sqrt{2}-\sqrt{7}\right)\left(\sqrt{2}+\sqrt{7}\right) &= \overset{F}{\left(\sqrt{2}\right)\left(\sqrt{2}\right)}+\overset{O}{\left(\sqrt{2}\right)\left(\sqrt{7}\right)}+\overset{I}{\left(-\sqrt{7}\right)\left(\sqrt{2}\right)}+\overset{L}{\left(-\sqrt{7}\right)\left(\sqrt{7}\right)}\\ &= \sqrt{4}+\sqrt{14}-\sqrt{14}-\sqrt{49}\\ &= \sqrt{4}-\sqrt{49}\\ &= 2-7=-5\end{aligned}$$

b)
$$\begin{aligned}\left(x-\sqrt{y}\right)\left(x+\sqrt{y}\right) &= (x)(x)+x\left(\sqrt{y}\right)+\left(-\sqrt{y}\right)(x)+\left(-\sqrt{y}\right)\left(\sqrt{y}\right)\\ &= x^2+x\sqrt{y}-x\sqrt{y}-\sqrt{y^2}\\ &= x^2-\sqrt{y^2}\\ &= x^2-y\end{aligned}$$

Now Try Exercise 73

Understanding Algebra

Notice that the factors in Example 8 were of the form $(a+b)(a-b)$. From Section 4.5 we know that $(a+b)(a-b)=a^2-b^2$.

In Section 4.5 we learned that $(a+b)(a-b)=a^2-b^2$. In part **a)** of Example 8 if we let $a=\sqrt{2}$ and $b=\sqrt{7}$, then

$$a^2-b^2=\left(\sqrt{2}\right)^2-\left(\sqrt{7}\right)^2=2-7=-5$$

In part **b)** if we let $a=x$ and $b=\sqrt{y}$, then

$$a^2-b^2=(x)^2-\left(\sqrt{y}\right)^2=x^2-y$$

Both answers agree with the answers obtained in Example 8. In Section 9.4 we will use products such as those in Example 8 to simplify other radical expressions.

9.3 Exercise Set MathXL® MyMathLab®

Warm-Up Exercises

Fill in the blanks with the appropriate word, phrase, or symbol(s) from the following list.

product FOIL index radicand associative distributive

1. Like radicals are radicals that have the same ________ and the same radicand.

2. Like square roots are square roots that have the same ________.

3. To multiply the radical expressions $\sqrt{15}(\sqrt{5}+\sqrt{10})$ we use the ________ property.

4. To multiply the radical expressions $(2+\sqrt{7})(3-\sqrt{11})$ we use the ________ method.

Practice the Skills

Simplify each expression.

5. $9\sqrt{2}+6\sqrt{2}$
6. $7\sqrt{5}-4\sqrt{5}$
7. $\sqrt{3}+5\sqrt{3}-8\sqrt{3}$
8. $10\sqrt{13}-6\sqrt{13}+16\sqrt{13}$
9. $\sqrt{7}+4\sqrt{7}-3\sqrt{7}+5$
10. $3\sqrt{7}-4\sqrt{7}-\sqrt{7}+9$
11. $15\sqrt{x}-9\sqrt{x}$
12. $6\sqrt{x}+\sqrt{x}$
13. $-\sqrt{y}+3\sqrt{y}-6\sqrt{y}$
14. $2\sqrt{p}-7\sqrt{p}+5\sqrt{p}$
15. $-6\sqrt{t}+2\sqrt{t}-19$
16. $3\sqrt{y}-8\sqrt{y}+12$
17. $\sqrt{x}+\sqrt{y}+x+7\sqrt{y}$
18. $\sqrt{j}-\sqrt{k}+k-5\sqrt{k}$
19. $-3\sqrt{7}+\sqrt{7}-2\sqrt{x}-7\sqrt{x}$
20. $4\sqrt{5}-2\sqrt{5}+3\sqrt{y}-9\sqrt{y}$
21. $13+4\sqrt{m}-6\sqrt{m}+5m-2$
22. $3\sqrt{a}-5+3a+2\sqrt{a}+2$

Simplify each expression.

23. $\sqrt{8}-\sqrt{27}$
24. $\sqrt{40}+\sqrt{18}$
25. $\sqrt{300}-\sqrt{12}$
26. $\sqrt{125}+\sqrt{45}$
27. $\sqrt{75}+\sqrt{108}$
28. $\sqrt{15}-\sqrt{60}$
29. $4\sqrt{50}-\sqrt{72}+\sqrt{8}$
30. $-4\sqrt{99}+3\sqrt{44}+7\sqrt{11}$
31. $-3\sqrt{125}+4\sqrt{75}$
32. $7\sqrt{40}-2\sqrt{50}$
33. $2\sqrt{360}+3\sqrt{160}$
34. $3\sqrt{250}+5\sqrt{160}$
35. $8\sqrt{16}-\sqrt{48}$
36. $5\sqrt{180}-2\sqrt{108}$

Multiply. Assume all variables represent positive numbers.

37. $\sqrt{3}(1+\sqrt{3})$
38. $\sqrt{2}(7+\sqrt{2})$
39. $4(\sqrt{x}-\sqrt{5})$
40. $9(\sqrt{7}-\sqrt{x})$
41. $y(\sqrt{y}+y)$
42. $z(z^2-\sqrt{z})$
43. $\sqrt{3}(\sqrt{6}+2)$
44. $\sqrt{15}(\sqrt{5}+\sqrt{3})$
45. $\sqrt{x}(\sqrt{x}+\sqrt{3})$
46. $\sqrt{y}(\sqrt{y}-\sqrt{2})$
47. $\sqrt{a}(9-\sqrt{2a})$
48. $\sqrt{d}(8+\sqrt{2d})$
49. $x(x+3\sqrt{y})$
50. $x(11x-3\sqrt{y})$
51. $3x(4x-3\sqrt{x})$
52. $6y(y+\sqrt{y})$

Multiply. Assume all variables represent positive numbers.

53. $(6+\sqrt{5})(1-\sqrt{2})$
54. $(7+\sqrt{3})(6-\sqrt{2})$
55. $(\sqrt{5}-4)(\sqrt{6}+3)$
56. $(\sqrt{10}+6)(\sqrt{10}-2)$
57. $(6-2\sqrt{7})(8-2\sqrt{7})$
58. $(5+3\sqrt{6})(5+3\sqrt{6})$
59. $(8+3\sqrt{k})(8+3\sqrt{k})$
60. $(5-\sqrt{x})(5-\sqrt{x})$
61. $(\sqrt{3z}-4)(\sqrt{5z}+1)$
62. $(\sqrt{2a}+3)(\sqrt{7a}-6)$
63. $(r+6\sqrt{s})(2r-3\sqrt{s})$
64. $(g-5\sqrt{h})(3g+2\sqrt{h})$
65. $(3x-\sqrt{2y})(2x-2\sqrt{2y})$
66. $(n-6\sqrt{w})(2n+4\sqrt{w})$
67. $(4p-2\sqrt{3q})(p+2\sqrt{3q})$
68. $(8n-\sqrt{7x})(2n-5\sqrt{7x})$

Multiply. Assume all variables represent positive numbers.

69. $(1 + \sqrt{3})(1 - \sqrt{3})$

70. $(7 - \sqrt{2})(7 + \sqrt{2})$

71. $(\sqrt{13} - 3)(\sqrt{13} + 3)$

72. $(\sqrt{10} - 4)(\sqrt{10} + 4)$

73. $(\sqrt{x} + 2)(\sqrt{x} - 2)$

74. $(\sqrt{a} + 9)(\sqrt{a} - 9)$

75. $(\sqrt{5} + m)(\sqrt{5} - m)$

76. $(\sqrt{6} - y)(\sqrt{6} + y)$

77. $(\sqrt{5x} + \sqrt{y})(\sqrt{5x} - \sqrt{y})$

78. $(\sqrt{3x} - \sqrt{2y})(\sqrt{3x} + \sqrt{2y})$

79. $(2\sqrt{3w} + 4\sqrt{5z})(2\sqrt{3w} - 4\sqrt{5z})$

80. $(4\sqrt{2c} + \sqrt{3d})(4\sqrt{2c} - \sqrt{3d})$

Problem Solving

81. a) Is $\sqrt{10}$ twice as large as $\sqrt{5}$?

b) What number is twice as large as $\sqrt{5}$? Explain how you determined your answer.

82. a) Is $\sqrt{21}$ three times as large as $\sqrt{7}$?

b) What number is three times as large as $\sqrt{7}$? Explain how you determined your answer.

83. If $x > -3$, what is $\sqrt{x+3} \cdot \sqrt{x+3}$ equal to?

84. If $x > -6$, what is $\sqrt{x+6} \cdot \sqrt{x+6}$ equal to?

85. Is the sum or difference of two rational expressions always a rational expression? Explain and give an example.

86. Is the sum or difference of two irrational expressions always an irrational expression? Explain and give an example.

Fill in the shaded area to make each statement true. Explain how you determined your answer.

87. $\sqrt{\blacksquare} - \sqrt{63} = 4\sqrt{7}$

88. $\sqrt{180} + \sqrt{\blacksquare} = 9\sqrt{5}$

Find the perimeter and area of the following figures.

89.

$\sqrt{2} + \sqrt{3}$

$\sqrt{2} + \sqrt{3}$

90.

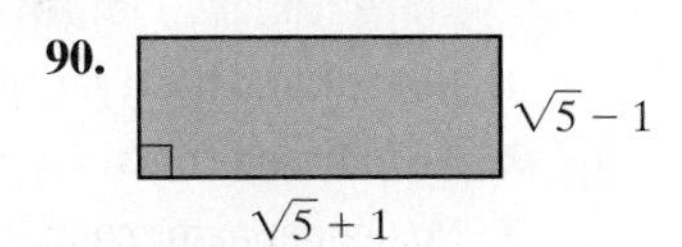

91.

$\sqrt{7}$

$\sqrt{22}$

$\sqrt{15}$

92.

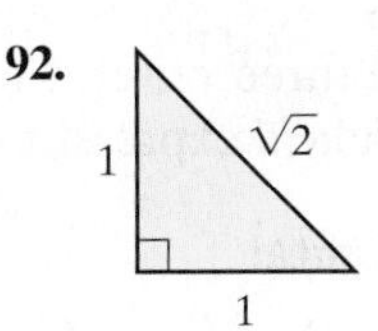

In Section 5.6, we solved quadratic equations of the form $ax^2 + bx + c = 0$, by factoring. Not all quadratic equations can be solved by factoring. In Chapter 10, we will be introducing another way to solve quadratic equations, called the quadratic formula. The quadratic formula contains the expression $\sqrt{b^2 - 4ac}$. Evaluate this expression for the following quadratic equations. Note, for example, in the quadratic equation $2x^2 - 4x - 5 = 0$ that $a = 2$, $b = -4$, and $c = -5$.

93. $x^2 + 3x + 2 = 0$

94. $x^2 + 7x + 12 = 0$

95. $x^2 - 14x - 5 = 0$

96. $x^2 + 4x - 3 = 0$

97. $-2x^2 + 4x + 7 = 0$

98. $-3x^2 + 6x + 2 = 0$

Concept/Writing Exercises

99. What are like square roots? Give an example.

100. What are unlike square roots? Give an example.

101. Under what conditions can two square roots be added or subtracted?

102. Explain how to add like square roots.

Challenge Problems

In Exercises 103 and 104, fill in the shaded area to make each statement true. Explain how you determined your answer.

103. $-5\sqrt{\blacksquare} + \sqrt{3} + 3\sqrt{27} = -10\sqrt{3}$

104. $20\sqrt{2} - 4\sqrt{\blacksquare} + 5\sqrt{18} = 15\sqrt{2}$

Cumulative Review Exercises

[3.4] **105. Sledding** Riding a sled down a hill, Jason averages 10 miles per hour. Walking up the same hill to where he started, he averages 2 miles per hour. If it took Jason 0.2 hours longer to come up the hill than to go down the hill, how long was Jason's sled ride?

[5.4] **106.** Factor $3x^2 - 12x - 96$.

[6.1] **107.** Simplify $\frac{x^2 - 1}{x^3 - 1}$.

[6.6] **108.** Solve the equation $x - \frac{24}{x} = 2$.

[8.1] **109.** Solve the system of equations graphically.

$$y = 2x - 2$$
$$2x + 3y = 10$$

9.4 Dividing Square Roots

1. Understand what it means for a square root to be simplified.
2. Use the quotient rule to simplify square roots.
3. Rationalize denominators.
4. Rationalize denominators using conjugates.

1 Understand What It Means for a Square Root to Be Simplified

In this section we will use a new rule, the quotient rule, to simplify square roots containing fractions. Before we do that, let's discuss what it means for a square root to be simplified.

A Square Root Is Simplified When

1. No radicand has a factor that is a perfect square.
2. No radicand contains a fraction.
3. No denominator contains a square root.

All three criteria must be met for an expression to be simplified. Let's look at some radical expressions that *are not simplified*.

Understanding Algebra

$\frac{2}{\sqrt{5}}$ has an irrational denominator since $\sqrt{5}$ is an irrational number. $\frac{2\sqrt{5}}{5}$ has a rational denominator since 5 is a rational number.

Radical	Reason Not Simplified	
$\sqrt{8}$	Contains perfect square factor, 4. $(\sqrt{8} = \sqrt{4} \cdot \sqrt{2} = 2\sqrt{2})$	(Criterion 1)
$\sqrt{x^3}$	Contains perfect square factor, x^2. $(\sqrt{x^3} = \sqrt{x^2} \cdot \sqrt{x} = x\sqrt{x})$	(Criterion 1)
$\sqrt{\frac{1}{3}}$	Radicand contains a fraction.	(Criterion 2)
$\frac{2}{\sqrt{5}}$	Square root in the denominator.	(Criterion 3)

Before we discuss the procedure to divide and simplify square roots containing fractions, we need to understand when common factors can and cannot be divided out.

Common factors in the numerator and denominator of a fraction within a radical sign *can* be divided out. For example, the following are correct.

$$\sqrt{\frac{6}{3}} = \sqrt{\frac{3 \cdot 2}{3}} = \sqrt{2} \quad \text{and} \quad \sqrt{\frac{x^3}{x^2}} = \sqrt{\frac{x^2 \cdot x}{x^2}} = \sqrt{x}$$

3 is the common factor; x^2 is the common factor

However, we cannot divide out common factors directly if one expression containing the common factor is within a radical sign and the other expression is not.

AVOIDING COMMON ERRORS

An expression under a radical sign *cannot* be divided by an expression not under a radical sign.

CORRECT

$\dfrac{\sqrt{6}}{3}$ cannot be simplified any further.

$\dfrac{\sqrt{x^3}}{x} = \dfrac{\sqrt{x^2}\sqrt{x}}{x} = \dfrac{\cancel{x}\sqrt{x}}{\cancel{x}} = \sqrt{x}$

INCORRECT

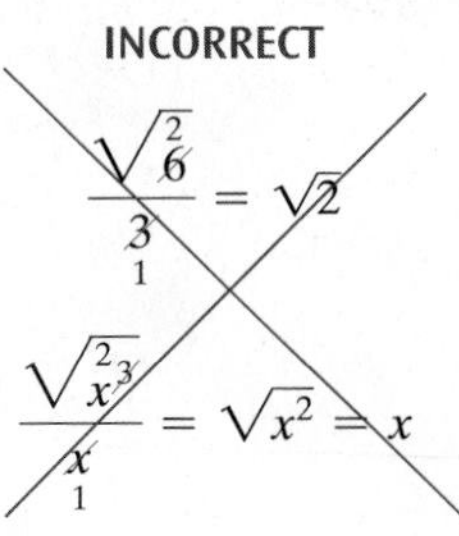

Each of the following simplifications is correct because the constant or variable divided out is not under a radical sign.

CORRECT

$\dfrac{\overset{2}{\cancel{6}}\sqrt{2}}{\underset{1}{\cancel{3}}} = 2\sqrt{2}$

$\dfrac{\overset{1}{\cancel{4}}\sqrt{3}}{\underset{2}{\cancel{8}}} = \dfrac{\sqrt{3}}{2}$

CORRECT

$\dfrac{\cancel{x}\sqrt{2}}{\cancel{x}} = \sqrt{2}$

$\dfrac{3\overset{x}{\cancel{x^2}}\sqrt{5}}{\cancel{x}} = 3x\sqrt{5}$

Now we will discuss the quotient rule.

2 Use the Quotient Rule to Simplify Square Roots

The square root of a quotient is equal to the quotient of two square roots.

Quotient Rule for Square Roots

$$\sqrt{\frac{a}{b}} = \frac{\sqrt{a}}{\sqrt{b}} \text{ and } \frac{\sqrt{a}}{\sqrt{b}} = \sqrt{\frac{a}{b}} \quad \text{provided} \quad a \geq 0, b > 0 \qquad \textbf{Rule 3}$$

Examples 1 through 4 illustrate how the quotient rule can be used to simplify square roots.

EXAMPLE 1 Simplify. **a)** $\sqrt{\dfrac{30}{5}}$ **b)** $\sqrt{\dfrac{64}{4}}$ **c)** $\sqrt{\dfrac{4}{49}}$

Solution When the square root contains a fraction, divide out any factor common to both the numerator and denominator. If the square root still contains a fraction, use the quotient rule to simplify.

a) $\sqrt{\dfrac{30}{5}} = \sqrt{6}$ **b)** $\sqrt{\dfrac{64}{4}} = \sqrt{16} = 4$ **c)** $\sqrt{\dfrac{4}{49}} = \dfrac{\sqrt{4}}{\sqrt{49}} = \dfrac{2}{7}$

Now Try Exercise 21

EXAMPLE 2 Simplify. **a)** $\sqrt{\dfrac{10x^2}{5}}$ **b)** $\sqrt{\dfrac{80a^2b}{2b}}$ **c)** $\sqrt{\dfrac{3x^2y^4}{48x^4}}$ **d)** $\sqrt{\dfrac{35ab^5c^2}{7a^5bc}}$

Solution First divide out any factors common to both the numerator and denominator; then simplify.

a) $\sqrt{\dfrac{10x^2}{5}} = \sqrt{2x^2}$ Simplify the radicand.

$= \sqrt{x^2}\sqrt{2}$ Product rule

$= x\sqrt{2}$ Simplify.

b) $\sqrt{\dfrac{80a^2b}{2b}} = \sqrt{40a^2}$ Simplify the radicand.

$= \sqrt{4a^2}\sqrt{10}$ Product rule

$= 2a\sqrt{10}$ Simplify.

c) $\sqrt{\dfrac{3x^2y^4}{48x^4}} = \sqrt{\dfrac{y^4}{16x^2}}$ Simplify the radicand.

$= \dfrac{\sqrt{y^4}}{\sqrt{16x^2}}$ Quotient rule

$= \dfrac{y^2}{4x}$ Simplify.

d) $\sqrt{\dfrac{35ab^5c^2}{7a^5bc}} = \sqrt{\dfrac{5b^4c}{a^4}} = \dfrac{\sqrt{5b^4c}}{\sqrt{a^4}} = \dfrac{\sqrt{b^4}\sqrt{5c}}{\sqrt{a^4}} = \dfrac{b^2\sqrt{5c}}{a^2}$

Now Try Exercise 27

When you are given a fraction containing a radical expression in both the numerator and the denominator, use the quotient rule to simplify, as in Examples 3 and 4.

EXAMPLE 3 Simplify. **a)** $\dfrac{\sqrt{2}}{\sqrt{8}}$ **b)** $\dfrac{\sqrt{147}}{\sqrt{3}}$

Solution

a) $\dfrac{\sqrt{2}}{\sqrt{8}} = \sqrt{\dfrac{2}{8}}$ Quotient rule

$= \sqrt{\dfrac{1}{4}}$ Simplify the radicand.

$= \dfrac{\sqrt{1}}{\sqrt{4}}$ Quotient rule

$= \dfrac{1}{2}$ Simplify.

b) $\dfrac{\sqrt{147}}{\sqrt{3}} = \sqrt{\dfrac{147}{3}} = \sqrt{49} = 7$

Now Try Exercise 13

EXAMPLE 4 Simplify. **a)** $\dfrac{\sqrt{32x^4y^3}}{\sqrt{8xy}}$ **b)** $\dfrac{\sqrt{125x^8y^4}}{\sqrt{5x^5y^8}}$

Solution

a) $\dfrac{\sqrt{32x^4y^3}}{\sqrt{8xy}} = \sqrt{\dfrac{32x^4y^3}{8xy}}$ Quotient rule

$= \sqrt{4x^3y^2}$ Simplify radicand.

$= \sqrt{4x^2y^2}\sqrt{x}$ Product rule

$= 2xy\sqrt{x}$ Simplify.

b) $\dfrac{\sqrt{125x^8y^4}}{\sqrt{5x^5y^8}} = \sqrt{\dfrac{125x^8y^4}{5x^5y^8}} = \sqrt{\dfrac{25x^3}{y^4}} = \dfrac{\sqrt{25x^3}}{\sqrt{y^4}} = \dfrac{\sqrt{25x^2}\sqrt{x}}{\sqrt{y^4}} = \dfrac{5x\sqrt{x}}{y^2}$

Now Try Exercise 33

If you observe all the answers in Examples 1 through 4 you will notice they all meet the three criteria needed for a square root to be simplified.

3 Rationalize Denominators

Suppose we are asked to simplify $\frac{1}{\sqrt{3}}$. To simplify this expression means to rewrite it as an equivalent expression that does not contain a radical in the denominator. This process is called **rationalizing the denominator.** We next give the process to rationalize a denominator that contains a square root of a number that is not a perfect square.

Rationalizing a Denominator with a Square Root

To rationalize a denominator with a square root, multiply *both* the numerator and the denominator of the fraction by the square root of a number so that the radicand in the denominator becomes a perfect square.

To simplify the expression $\frac{1}{\sqrt{3}}$, we must rationalize the denominator. We will multiply both the numerator and denominator by $\sqrt{3}$. Note that when we multiply, the denominator becomes $\sqrt{3} \cdot \sqrt{3}$ or $\sqrt{9}$, and 9 is a perfect square. This allows us to write the expression without a radical in the denominator.

$$\frac{1}{\sqrt{3}} = \frac{1}{\sqrt{3}} \cdot \frac{\sqrt{3}}{\sqrt{3}} = \frac{\sqrt{3}}{\sqrt{9}} = \frac{\sqrt{3}}{3}$$

Notice we multiplied both the numerator and the denominator by $\sqrt{3}$. This is equivalent to multiplying the original expression by 1, which does not change the value of the original expression.

EXAMPLE 5 Simplify $\frac{1}{\sqrt{5}}$.

Solution Since $\sqrt{5} \cdot \sqrt{5} = \sqrt{25} = 5$, we multiply both the numerator and denominator by $\sqrt{5}$.

$$\frac{1}{\sqrt{5}} = \frac{1}{\sqrt{5}} \cdot \frac{\sqrt{5}}{\sqrt{5}} = \frac{\sqrt{5}}{\sqrt{25}} = \frac{\sqrt{5}}{5}$$

The answer $\frac{\sqrt{5}}{5}$ is simplified because it satisfies the three requirements stated earlier.

Now Try Exercise 41

Understanding Algebra

Historically, denominators were rationalized so that obtaining a decimal approximation would be easier. Example 5 showed that $\frac{1}{\sqrt{5}} = \frac{\sqrt{5}}{5}$.

Approximating $\frac{1}{\sqrt{5}}$ using long division requires you to divide 1 by an approximation of $\sqrt{5}$ such as 2.236.

However, approximating $\frac{\sqrt{5}}{5}$ requires you to divide 2.236 by 5, a much easier task!

In Example 5, multiplying both the numerator and denominator by $\sqrt{5}$ is equivalent to multiplying the fraction by 1, which does not change its value.

EXAMPLE 6 Simplify. **a)** $\sqrt{\frac{2}{3}}$ **b)** $\sqrt{\frac{x}{18}}$

Solution

a)

$$\sqrt{\frac{2}{3}} = \frac{\sqrt{2}}{\sqrt{3}}$$ Quotient rule

$$= \frac{\sqrt{2}}{\sqrt{3}} \cdot \frac{\sqrt{3}}{\sqrt{3}}$$ Multiply numerator and denominator by $\sqrt{3}$ to rationalize the denominator.

$$= \frac{\sqrt{6}}{\sqrt{9}}$$ Product rule

$$= \frac{\sqrt{6}}{3}$$ Simplify.

b) Our first step is to apply the quotient rule: $\sqrt{\frac{x}{18}} = \frac{\sqrt{x}}{\sqrt{18}}$. Next, we need to multiply both the numerator and the denominator by a square root so that the radicand in the denominator becomes a perfect square. We have several choices; however $\sqrt{2}$ is the smallest such number. Note that $18 \cdot 2 = 36$ and 36 is a perfect square. We therefore will multiply both numerator and denominator by $\sqrt{2}$.

$$\sqrt{\frac{x}{18}} = \frac{\sqrt{x}}{\sqrt{18}} = \frac{\sqrt{x}}{\sqrt{18}} \cdot \frac{\sqrt{2}}{\sqrt{2}} = \frac{\sqrt{2x}}{\sqrt{36}} = \frac{\sqrt{2x}}{6}$$

Now Try Exercise 59

In Example 6 **b)** we can also first simplify the square root in the denominator.

$$\sqrt{\frac{x}{18}} = \frac{\sqrt{x}}{\sqrt{18}} = \frac{\sqrt{x}}{\sqrt{9}\sqrt{2}} = \frac{\sqrt{x}}{3\sqrt{2}}$$

Then we can rationalize the denominator by multiplying both numerator and denominator by $\sqrt{2}$.

$$\frac{\sqrt{x}}{3\sqrt{2}} \cdot \frac{\sqrt{2}}{\sqrt{2}} = \frac{\sqrt{2x}}{3\sqrt{4}} = \frac{\sqrt{2x}}{3 \cdot 2} = \frac{\sqrt{2x}}{6}$$

Notice this is the same result we obtained in Example 6 **b).**

4 Rationalize Denominators Using Conjugates

When the denominator contains a sum or difference involving radicals, such as the denominator in $\frac{8}{2 + \sqrt{3}}$, we rationalize the denominator by multiplying both numerator and denominator by the conjugate of the denominator. The radical expression $2 + \sqrt{3}$ and the radical expression $2 - \sqrt{3}$ are **conjugates** of each other. In general terms, $a + b$ and $a - b$ are conjugates of each other. Other examples of conjugates are shown below.

Radical Expression	Conjugate
$8 + \sqrt{2}$	$8 - \sqrt{2}$
$\sqrt{3} - 4$	$\sqrt{3} + 4$
$9\sqrt{3} - \sqrt{5}$	$9\sqrt{3} + \sqrt{5}$
$-x + \sqrt{3}$	$-x - \sqrt{3}$

Understanding Algebra

When an expression $a + b$ is multiplied by its conjugate $a - b$, we have the following result

$$(a + b)(a - b) = a^2 - ab + ab - b^2 = a^2 - b^2.$$

When a radical expression is multiplied by its conjugate using the FOIL method, the product simplifies to an expression that has no square roots. For example,

$$(2 + \sqrt{3})(2 - \sqrt{3}) = \underset{\text{F}}{4} - \underset{\text{O}}{2\sqrt{3}} + \underset{\text{I}}{2\sqrt{3}} - \underset{\text{L}}{3} = 4 - 3 = 1$$

The product we just evaluated will be used in the next example to rationalize the denominator.

EXAMPLE 7 Simplify $\frac{8}{2 + \sqrt{3}}$.

Solution To rationalize the denominator, multiply both the numerator and the denominator of the fraction by $2 - \sqrt{3}$, which is the conjugate of $2 + \sqrt{3}$.

$$\frac{8}{2 + \sqrt{3}} = \frac{8}{2 + \sqrt{3}} \cdot \frac{2 - \sqrt{3}}{2 - \sqrt{3}}$$ Multiply numerator and denominator by $2 - \sqrt{3}$.

$$= \frac{8(2 - \sqrt{3})}{(2 + \sqrt{3})(2 - \sqrt{3})}$$

$$= \frac{8(2-\sqrt{3})}{4-3}$$ Multiply factors in denominator, as discussed previously.

$$= \frac{8(2-\sqrt{3})}{1}$$ Simplify.

$$= 8(2-\sqrt{3}) = 16 - 8\sqrt{3}$$

Note that $-8\sqrt{3} + 16$ is also an acceptable answer.

Now Try Exercise 83

EXAMPLE 8 Simplify $\frac{9}{\sqrt{2}-\sqrt{5}}$.

Solution

$$\frac{9}{\sqrt{2}-\sqrt{5}} = \frac{9}{\sqrt{2}-\sqrt{5}} \cdot \frac{\sqrt{2}+\sqrt{5}}{\sqrt{2}+\sqrt{5}}$$ Multiply numerator and denominator by $\sqrt{2} + \sqrt{5}$.

$$= \frac{9(\sqrt{2}+\sqrt{5})}{(\sqrt{2}-\sqrt{5})(\sqrt{2}+\sqrt{5})}$$

$$= \frac{9(\sqrt{2}+\sqrt{5})}{2-5}$$ Multiply factors in denominator.

$$= \frac{9(\sqrt{2}+\sqrt{5})}{-3}$$ Simplify.

$$= \frac{-\overset{3}{\cancel{9}}(\sqrt{2}+\sqrt{5})}{\cancel{3}}$$

$$= -3(\sqrt{2}+\sqrt{5}) = -3\sqrt{2} - 3\sqrt{5}$$

Now Try Exercise 87

EXAMPLE 9 Simplify $\frac{\sqrt{3}}{2-\sqrt{6}}$.

Solution

$$\frac{\sqrt{3}}{2-\sqrt{6}} = \frac{\sqrt{3}}{2-\sqrt{6}} \cdot \frac{2+\sqrt{6}}{2+\sqrt{6}}$$ Multiply numerator and denominator by $2 + \sqrt{6}$.

$$= \frac{\sqrt{3}(2+\sqrt{6})}{4-6}$$ Multiply factors in denominator.

$$= \frac{2\sqrt{3}+\sqrt{3}\sqrt{6}}{-2}$$ Distributive property

$$= \frac{2\sqrt{3}+\sqrt{18}}{-2}$$ Product rule

$$= \frac{2\sqrt{3}+\sqrt{9}\sqrt{2}}{-2}$$ Product rule

$$= \frac{2\sqrt{3}+3\sqrt{2}}{-2}$$ Simplify.

$$= \frac{-2\sqrt{3}-3\sqrt{2}}{2}$$ Multiply numerator and denominator by -1.

Now Try Exercise 85

EXAMPLE 10 Simplify $\dfrac{x}{x - \sqrt{y}}$.

Solution Multiply both the numerator and the denominator of the fraction by $x + \sqrt{y}$, the conjugate of the denominator.

$$\frac{x}{x-\sqrt{y}} = \frac{x}{x-\sqrt{y}} \cdot \frac{x+\sqrt{y}}{x+\sqrt{y}} = \frac{x(x+\sqrt{y})}{x^2-y} = \frac{x^2+x\sqrt{y}}{x^2-y}$$

Remember, you cannot divide out the x^2 terms because they are not factors.

Now Try Exercise 93

9.4 Exercise Set MathXL® MyMathLab®

Warm-Up Exercises

conjugates	irrational	multiply	square	radicand
numerator	denominator	rational	index	fraction

1. For a square root to be simplified, the radicand cannot have a factor that is a perfect ________.
2. For a square root to be simplified, the radicand cannot contain a ________.
3. For a square root to be simplified, the ________ cannot contain a square root.
4. The number $\dfrac{2}{\sqrt{5}}$ has a denominator that is an ________ number.
5. The number $\dfrac{2\sqrt{5}}{5}$ has a denominator that is a ________ number.
6. To rationalize a fraction that has a square root in the denominator, multiply both the numerator and the denominator of the fraction by a square root so that the ________ in the denominator becomes a perfect square.
7. The two numbers $5 + \sqrt{3}$ and $5 - \sqrt{3}$ are called ________ of each other.
8. To rationalize the denominator of the expression $\dfrac{3}{2 + \sqrt{7}}$, we ________ both numerator and denominator by the expression $2 - \sqrt{7}$.

Practice the Skills

Simplify each expression. Assume all variables represent positive numbers.

9. $\sqrt{\dfrac{27}{3}}$
10. $\sqrt{\dfrac{32}{8}}$
11. $\sqrt{\dfrac{125}{5}}$
12. $\sqrt{\dfrac{300}{3}}$
13. $\dfrac{\sqrt{18}}{\sqrt{2}}$
14. $\dfrac{\sqrt{28}}{\sqrt{7}}$
15. $\dfrac{\sqrt{72}}{\sqrt{2}}$
16. $\dfrac{\sqrt{75}}{\sqrt{3}}$
17. $\sqrt{\dfrac{9}{121}}$
18. $\sqrt{\dfrac{81}{100}}$
19. $\dfrac{\sqrt{20}}{\sqrt{2000}}$
20. $\dfrac{\sqrt{160}}{\sqrt{640}}$
21. $\sqrt{\dfrac{24}{8}}$
22. $\sqrt{\dfrac{35}{5}}$
23. $\sqrt{\dfrac{13}{25}}$
24. $\sqrt{\dfrac{7}{144}}$
25. $\sqrt{\dfrac{75x^6y^2}{3x^4y^2}}$
26. $\sqrt{\dfrac{200p^7q^5}{8p^3q^3}}$
27. $\sqrt{\dfrac{16x^5y^3}{81x^7y}}$
28. $\sqrt{\dfrac{18a^9b^2}{32a^3b^{10}}}$
29. $\sqrt{\dfrac{48x^3}{2x}}$
30. $\sqrt{\dfrac{135c^5d^7}{3cd^5}}$
31. $\sqrt{\dfrac{45x^2}{16x^2y^4}}$
32. $\sqrt{\dfrac{48s^2t^{11}}{81s^{10}t}}$
33. $\dfrac{\sqrt{32n^5}}{\sqrt{8n}}$
34. $\dfrac{\sqrt{52x^2y^2}}{\sqrt{13x^2y^4}}$
35. $\dfrac{\sqrt{50ab^6}}{\sqrt{10ab^4c^2}}$
36. $\dfrac{\sqrt{81}}{\sqrt{27m^4n^6}}$
37. $\dfrac{\sqrt{125a^6b^8}}{\sqrt{5a^2b^2}}$
38. $\dfrac{\sqrt{84x^{60}y^{32}}}{\sqrt{7x^{40}y^{18}}}$
39. $\dfrac{\sqrt{54x^7y^5z}}{\sqrt{3x^4y^7z^9}}$
40. $\dfrac{\sqrt{72xy^9z^3}}{\sqrt{6xy^6z^9}}$

Simplify each expression. Assume all variables represent positive numbers.

41. $\frac{1}{\sqrt{7}}$ **42.** $\frac{1}{\sqrt{6}}$ **43.** $\frac{12}{\sqrt{3}}$ **44.** $\frac{10}{\sqrt{2}}$

45. $\frac{6}{\sqrt{12}}$ **46.** $\frac{30}{\sqrt{50}}$ **47.** $\sqrt{\frac{3}{5}}$ **48.** $\sqrt{\frac{2}{7}}$

49. $\sqrt{\frac{5}{15}}$ **50.** $\sqrt{\frac{6}{30}}$ **51.** $\sqrt{\frac{5}{8}}$ **52.** $\sqrt{\frac{7}{27}}$

53. $\sqrt{\frac{6}{x}}$ **54.** $\sqrt{\frac{5}{a}}$ **55.** $\frac{\sqrt{a}}{\sqrt{b}}$ **56.** $\frac{\sqrt{m}}{\sqrt{n}}$

57. $\sqrt{\frac{c}{d}}$ **58.** $\sqrt{\frac{r}{7}}$ **59.** $\sqrt{\frac{5p}{12}}$ **60.** $\sqrt{\frac{7w}{18}}$

61. $\sqrt{\frac{x^2}{5}}$ **62.** $\sqrt{\frac{x^2}{3}}$ **63.** $\sqrt{\frac{z^3}{8}}$ **64.** $\sqrt{\frac{a^3}{27}}$

65. $\sqrt{\frac{t^5}{6}}$ **66.** $\sqrt{\frac{x^3}{10}}$ **67.** $\sqrt{\frac{a^8}{14b}}$ **68.** $\sqrt{\frac{a^7b}{24b^4}}$

69. $\sqrt{\frac{7c^2d^4}{35c^2d^7}}$ **70.** $\sqrt{\frac{27xz^4}{6y^5}}$ **71.** $\sqrt{\frac{25yz}{12x^4y^5z^9}}$ **72.** $\frac{\sqrt{21x^5}}{\sqrt{84xy^7}}$

73. $\frac{\sqrt{90x^4y}}{\sqrt{2x^5y^5}}$ **74.** $\frac{\sqrt{40xyz^2}}{\sqrt{3xy^4}}$

Find the product of the given binomial and its conjugate. Assume all variables represent positive numbers.

75. $6 + \sqrt{3}$ **76.** $\sqrt{6} - 2$ **77.** $\sqrt{3} - \sqrt{11}$ **78.** $\sqrt{7} + \sqrt{5}$

79. $\sqrt{x} - y$ **80.** $x + \sqrt{y}$ **81.** $\sqrt{a} + \sqrt{b}$ **82.** $\sqrt{x} - \sqrt{y}$

Simplify each expression. Assume all variables represent positive numbers.

83. $\frac{3}{\sqrt{5} + 2}$ **84.** $\frac{7}{\sqrt{10} - 3}$ **85.** $\frac{4}{\sqrt{6} - 1}$

86. $\frac{6}{4 - \sqrt{3}}$ **87.** $\frac{12}{\sqrt{3} + \sqrt{5}}$ **88.** $\frac{15}{\sqrt{2} + \sqrt{5}}$

89. $\frac{8}{\sqrt{5} - \sqrt{8}}$ **90.** $\frac{11}{\sqrt{2} - \sqrt{6}}$ **91.** $\frac{4}{\sqrt{y} + 3}$

92. $\frac{6}{6 - \sqrt{x}}$ **93.** $\frac{7}{4 - \sqrt{y}}$ **94.** $\frac{11}{3 + \sqrt{x}}$

95. $\frac{16}{\sqrt{y} + x}$ **96.** $\frac{10}{r + \sqrt{t}}$ **97.** $\frac{5}{\sqrt{x} + \sqrt{y}}$

98. $\frac{\sqrt{6}}{\sqrt{x} - \sqrt{6}}$ **99.** $\frac{\sqrt{5}}{\sqrt{5} - \sqrt{n}}$ **100.** $\frac{2}{\sqrt{x} - y}$

101. $\frac{\sqrt{x}}{6 - \sqrt{x}}$ **102.** $\frac{8\sqrt{r}}{2 + \sqrt{r}}$

Problem Solving

103. Will the quotient of two rational numbers (denominator not equal to 0) always be a rational number? Explain and give an example to support your answer.

104. Will the quotient of two irrational numbers (denominator not equal to 0) always be an irrational number? Explain and give an example to support your answer.

In Exercises 105–108, if $\sqrt{5} \approx 2.236$ *and* $\sqrt{10} \approx 3.162$, *find to the nearest hundredth:*

105. $\sqrt{5} + \sqrt{10}$ **106.** $\sqrt{5} \cdot \sqrt{10}$ **107.** $\frac{\sqrt{5}}{\sqrt{10}}$ **108.** $\frac{\sqrt{10}}{\sqrt{5}}$

In Exercises 109–112, if $\sqrt{7} \approx 2.646$ *and* $\sqrt{21} \approx 4.583$, *find to the nearest hundredth:*

109. $\sqrt{7} + \sqrt{21}$

110. $\sqrt{7} \cdot \sqrt{21}$

111. $\dfrac{\sqrt{7}}{\sqrt{21}}$

112. $\dfrac{\sqrt{21}}{\sqrt{7}}$

In Exercises 113 and 114, find the width of each rectangle whose area and length are given. Give the answer in simplified form with integer denominator.

113. $A = 24$, $l = 4 + \sqrt{3}$

114. $A = \sqrt{70}$, $l = 7 - \sqrt{2}$

Concept/Writing Exercises

In Exercises 115–120, indicate if a common factor can be divided out from both the numerator and denominator of the fraction. If a common factor can be divided out, do so and show the simplified answer. See the Avoiding Common Errors box on page 539.

115. $\dfrac{\sqrt{7}}{7}$

116. $\dfrac{4\sqrt{5}}{2}$

117. $\dfrac{x^2\sqrt{3}}{x}$

118. $\dfrac{\sqrt{6}}{2}$

119. $\dfrac{\sqrt{x}}{x}$

120. $\dfrac{\sqrt{15}}{5}$

121. What are the three requirements for a square root to be considered simplified?

122. Explain why each expression is not simplified.

a) $\sqrt{27}$ **b)** $\sqrt{\dfrac{1}{3}}$ **c)** $\dfrac{8}{\sqrt{5}}$

Challenge Problems

Fill in the shaded area to make the expression true. Explain how you determined your answer.

123. $\dfrac{1}{\sqrt{\blacksquare}} = \dfrac{\sqrt{2}}{2}$

124. $\dfrac{9x}{\sqrt{\blacksquare}} = \dfrac{9\sqrt{2x}}{2}$

125. $\sqrt{\dfrac{\blacksquare}{4x^2}} = 4x^4$

126. $\dfrac{\sqrt{32x^5}}{\sqrt{\blacksquare}} = 2x^2$

127. Simplify $\dfrac{\sqrt{x}}{1 - \sqrt{3}}$ by rationalizing the denominator.

128. Simplify $\dfrac{\sqrt{x}}{2 + \sqrt{2}}$ by rationalizing the denominator.

Group Activity

129. Consider $\dfrac{\sqrt{x}}{x + \sqrt{x}}$, where $x > 0$.

a) Individually, simplify $\dfrac{\sqrt{x}}{x + \sqrt{x}}$ by rationalizing the denominator. Compare and correct your answer if necessary.

b) Group Member 1: Substitute $x = 4$ in the original expression and in the results found in part **a)**. Evaluate each expression.

c) Group Member 2: Repeat part **b)** for $x = 9$.

d) Group Member 3: Repeat part **b)** for $x = 25$.

e) As a group, determine what relationship exists between the given expression and the rationalized expression you obtained in part **a)**.

Cumulative Review Exercises

[4.6] **130.** Divide $\dfrac{3x^2 + 4x - 26}{x + 4}$.

[5.6] **131.** Solve the equation $2x^2 - x - 36 = 0$.

[6.4] **132.** Subtract $\dfrac{1}{x^2 - 4} - \dfrac{3}{x - 2}$.

[6.7] **133. Stacking Wood** Mark DeGroat can stack a cord of wood in 20 minutes. With his wife's help, they can stack the wood in 12 minutes. How long would it take his wife, Terry, to stack the wood by herself?

See Exercise 133.

MID-CHAPTER TEST: 9.1–9.4

To find out how well you understand the chapter material to this point, take this brief test. The answers, and the section where the material was initially discussed, are given in the back of the book. Review any questions that you answered incorrectly.

Evaluate each square root.

1. $\sqrt{49}$ **2.** $-\sqrt{121}$ **3.** $\sqrt{\dfrac{169}{81}}$

4. Determine whether the following numbers are rational, irrational, or imaginary.

a) $\sqrt{17}$ **b)** $\sqrt{36}$

c) $\sqrt{-36}$ **d)** $-\sqrt{36}$

5. Write $\sqrt{59xy^2}$ in exponential form.

Simplify.

6. $\sqrt{40}$ **7.** $\sqrt{63}$

8. $\sqrt{a^2b^5c^8}$ **9.** $\sqrt{128x^9y^{14}}$

10. $\sqrt{6x^2y^3}\sqrt{9x^3y^6}$

11. Add $3\sqrt{x} + 2\sqrt{5} + 9\sqrt{x} + 4\sqrt{5} + 17$.

12. Subtract $4\sqrt{75} - 2\sqrt{108}$.

13. Multiply $(5 - \sqrt{2})(7 + \sqrt{3})$.

14. Multiply $(\sqrt{x} - 4z)(\sqrt{x} + 5z)$.

15. Multiply $(\sqrt{3a} + \sqrt{2b})(\sqrt{3a} - \sqrt{2b})$.

16. When asked to add $\sqrt{8} + \sqrt{32}$, Robert Jones decided to add the two radicands together to obtain $\sqrt{8 + 32}$ or $\sqrt{40}$. This is incorrect. Why is this incorrect? Explain your answer. Then, add $\sqrt{8} + \sqrt{32}$ correctly.

Divide.

17. $\dfrac{\sqrt{17}}{\sqrt{68}}$ **18.** $\dfrac{\sqrt{54x^5y^6}}{\sqrt{6xy^8}}$

Rationalize the denominator.

19. $\dfrac{7}{\sqrt{6} - 2}$ **20.** $\dfrac{4\sqrt{x}}{\sqrt{x} + 5}$

9.5 Solving Radical Equations

1. Solve radical equations containing only one square root.
2. Solve radical equations containing two square roots.

1 Solve Radical Equations Containing Only One Square Root

Radical Equation

A **radical equation** is an equation that contains a variable in a radicand.

In this section, we will only solve radical equations that contain square roots. Some examples of radical equations are

$$\sqrt{x} = 5 \qquad \sqrt{x + 4} = 6 \qquad \sqrt{x - 2} = x - 8$$

Notice in each equation there is a variable in the radicand. In this section, we assume that all radicands represent nonnegative numbers. This allows us to solve equations by squaring both sides of the equation. To do this, we make use of the following fact. *If x is any nonnegative real number, then* $(\sqrt{x})^2 = x$.

Consider the radical equation $\sqrt{x} = 3$. To solve this equation, we need to eliminate the square root. *We can eliminate the square root by squaring both sides of the equation as demonstrated below.*

$$\sqrt{x} = 3$$
$$(\sqrt{x})^2 = (3)^2 \quad \text{Square both sides of the equation.}$$
$$x = 9$$

Now we check this solution.

$$\sqrt{9} = 3$$
$$3 = 3 \quad \text{True}$$

The following provides a general procedure for solving equations that contain one square root.

Understanding Algebra

When squaring both sides of an equation, we sometimes introduce a false, or extraneous, solution. Therefore, we *must* check the solutions when we square both sides of an equation.

To Solve a Radical Equation Containing Only One Square Root

1. Rewrite the equation with the square root by itself on one side of the equation. We call this **isolating the square root**.
2. Combine like terms.
3. Square both sides of the equation to eliminate the square root.
4. Solve the equation for the variable.
5. Check the solution in the *original* equation for extraneous roots.

EXAMPLE 1 Solve the equation $\sqrt{x} = 7$.

Solution The square root containing the variable is already by itself on one side of the equation. Square both sides of the equation.

$$\sqrt{x} = 7 \quad \text{Square root is isolated.}$$
$$(\sqrt{x})^2 = 7^2 \quad \text{Square both sides.}$$
$$x = 49$$

Check

$$\sqrt{x} = 7$$
$$\sqrt{49} \stackrel{?}{=} 7$$
$$7 = 7 \quad \text{True}$$

Now Try Exercise 17

EXAMPLE 2 Solve the equation $\sqrt{x-3} = 4$.

Solution The square root containing the variable is already by itself on one side of the equation. Square both sides of the equation.

$$\sqrt{x-3} = 4$$
$$(\sqrt{x-3})^2 = 4^2 \quad \text{Square both sides.}$$
$$x - 3 = 16 \quad \text{Simplify.}$$
$$x = 19 \quad \text{3 was added to both sides.}$$

Check

$$\sqrt{x-3} = 4$$
$$\sqrt{19-3} \stackrel{?}{=} 4$$
$$\sqrt{16} \stackrel{?}{=} 4$$
$$4 = 4 \quad \text{True}$$

Now Try Exercise 21

In the next example, the square root term containing the variable is not by itself on one side of the equation.

EXAMPLE 3 Solve the equation $\sqrt{a} + 7 = 10$.

Solution Begin by isolating the square root.

$$\sqrt{a} + 7 = 10$$
$$\sqrt{a} + 7 - 7 = 10 - 7 \quad \text{Subtract 7 from both sides.}$$
$$\sqrt{a} = 3 \quad \text{Square root is isolated.}$$
$$(\sqrt{a})^2 = 3^2 \quad \text{Square both sides.}$$
$$a = 9 \quad \text{Simplify.}$$

A check will show that 9 is the solution.

Now Try Exercise 25

When we square both sides of an equation, we may introduce *extraneous roots*. An **extraneous root** is a number obtained when solving an equation that is not a solution to the original equation. Equations where both sides are squared in the process of finding their solutions should always be checked for extraneous roots by substituting the numbers found back into the *original* equation.

Consider the equation

$$x = 6$$

Now square both sides.

$$x^2 = 36$$

Note that the original equation $x = 6$ is true only when x is 6. However, the equation $x^2 = 36$ is true for both 6 and -6. When we squared $x = 6$, we introduced the extraneous root -6.

Our next example involves an extraneous root.

EXAMPLE 4 Solve the equation $\sqrt{x} = -5$.

Solution Begin by squaring both sides of the equation

$$\sqrt{x} = -5$$
$$(\sqrt{x})^2 = (-5)^2$$
$$x = 25$$

Check

$$\sqrt{x} = -5$$
$$\sqrt{25} \stackrel{?}{=} -5$$
$$5 = -5 \quad \text{False. The number 25 is an extraneous root.}$$

Since the check results in a false statement, the number 25 is an extraneous root and is not a solution.

The equation $\sqrt{x} = -5$ has *no real solution*.

Now Try Exercise 23

In Example 4, you may have realized before trying to solve it that there is no solution to the equation. The left side of the equation, $\sqrt{x}$, cannot be a negative number. However, the right side of the equation, -5, is a negative number. Therefore, there is no solution to this equation.

$$\underbrace{\underset{\sqrt{x}\text{ cannot be negative}}{\underset{\uparrow}{\sqrt{x}}} \quad = \quad \underset{\text{negative number}}{\underset{\uparrow}{-5}}}_{\text{The equation cannot have a solution}}$$

EXAMPLE 5 Solve the equation $\sqrt{2x - 3} = x - 3$.

Solution

$$\sqrt{2x - 3} = x - 3$$
$$(\sqrt{2x - 3})^2 = (x - 3)^2 \quad \text{Square both sides.}$$
$$2x - 3 = x^2 - 6x + 9 \quad \text{Simplify on left. Write } (x - 3)^2 \text{ as } x^2 - 6x + 9.$$

Now solve the quadratic equation as explained in Section 5.6. Make one side of the equation equal to zero by subtracting $2x$ and adding 3 to both sides of the equation. This gives the following quadratic equation.

$$0 = x^2 - 8x + 12 \quad \text{or} \quad x^2 - 8x + 12 = 0$$

Solve for x by factoring.

$$x^2 - 8x + 12 = 0$$

$$(x-6)(x-2) = 0 \quad \text{Factor.}$$

$$x - 6 = 0 \quad \text{or} \quad x - 2 = 0 \quad \text{Zero-factor property}$$

$$x = 6 \qquad x = 2 \quad \text{Solve for } x.$$

Check $x = 6$

$$\sqrt{2x-3} = x - 3$$

$$\sqrt{2(6)-3} \stackrel{?}{=} 6 - 3$$

$$\sqrt{9} \stackrel{?}{=} 3$$

$$3 = 3 \quad \text{True}$$

$x = 2$

$$\sqrt{2x-3} = x - 3$$

$$\sqrt{2(2)-3} \stackrel{?}{=} 2 - 3$$

$$\sqrt{1} \stackrel{?}{=} -1$$

$$1 = -1 \quad \text{False}$$

The solution is 6, but 2 is not a solution to the equation.

Now Try Exercise 37

Remember, when solving a radical equation, begin by isolating the radical, as is shown in Example 6.

EXAMPLE 6 Solve the equation $2x - 5\sqrt{x} - 3 = 0$.

Solution First rewrite the equation so that the square root containing the variable is by itself on one side of the equation.

$$2x - 5\sqrt{x} - 3 = 0$$

$$-5\sqrt{x} = -2x + 3 \quad \text{Isolate square root.}$$

$$\text{or} \quad 5\sqrt{x} = 2x - 3 \quad \text{Multiplied both sides by } -1.$$

Now square both sides of the equation.

$$(5\sqrt{x})^2 = (2x-3)^2 \quad \text{Square both sides.}$$

$$5^2(\sqrt{x})^2 = (2x-3)^2$$

$$25x = 4x^2 - 12x + 9 \quad \text{Simplify on left. Write } (2x-3)^2 \text{ as } 4x^2 - 12x + 9.$$

$$0 = 4x^2 - 37x + 9 \quad \text{Set one side of equation equal to zero.}$$

$$\text{or} \quad 4x^2 - 37x + 9 = 0$$

$$(4x-1)(x-9) = 0 \quad \text{Factor.}$$

$$4x - 1 = 0 \quad \text{or} \quad x - 9 = 0 \quad \text{Zero-factor property}$$

$$4x = 1 \qquad x = 9 \quad \text{Solve for } x.$$

$$x = \frac{1}{4}$$

Check $x = \frac{1}{4}$

$$2x - 5\sqrt{x} - 3 = 0$$

$$2\left(\frac{1}{4}\right) - 5\sqrt{\frac{1}{4}} - 3 \stackrel{?}{=} 0$$

$$\frac{1}{2} - 5\left(\frac{1}{2}\right) - 3 \stackrel{?}{=} 0$$

$$\frac{1}{2} - \frac{5}{2} - 3 \stackrel{?}{=} 0$$

$$-\frac{4}{2} - 3 \stackrel{?}{=} 0$$

$$-2 - 3 \stackrel{?}{=} 0$$

$$-5 = 0 \quad \text{False}$$

$x = 9$

$$2x - 5\sqrt{x} - 3 = 0$$

$$2(9) - 5\sqrt{9} - 3 \stackrel{?}{=} 0$$

$$18 - 5(3) - 3 \stackrel{?}{=} 0$$

$$18 - 15 - 3 \stackrel{?}{=} 0$$

$$0 = 0 \quad \text{True}$$

The solution is 9, but $\frac{1}{4}$ is not a solution.

Now Try Exercise 43

2 Solve Radical Equations Containing Two Square Roots

Consider the radical equations

$$\sqrt{x-4} = \sqrt{x+9} \quad \text{and} \quad \sqrt{2x+5} - \sqrt{3x+8} = 0$$

These equations are different from those previously discussed because they have two square roots containing the variable x. To solve equations of this type, rewrite the equation, when necessary, so that there is only one square root on each side of the equation.

EXAMPLE 7 Solve the equation $\sqrt{2x+4} = \sqrt{3x-2}$.

Solution Since each side of the equation already contains only one square root, it is not necessary to rewrite the equation.

$$\left(\sqrt{2x+4}\right)^2 = \left(\sqrt{3x-2}\right)^2 \quad \text{Square both sides.}$$
$$2x + 4 = 3x - 2 \quad \text{Solve for } x.$$
$$4 = x - 2$$
$$6 = x$$

Check

$$\sqrt{2x+4} = \sqrt{3x-2}$$
$$\sqrt{2(6)+4} \stackrel{?}{=} \sqrt{3(6)-2}$$
$$\sqrt{16} \stackrel{?}{=} \sqrt{16}$$
$$4 = 4 \quad \text{True}$$

The solution is 6.

Now Try Exercise 29

EXAMPLE 8 Solve the equation $4\sqrt{r-4} - \sqrt{10r+14} = 0$.

Solution Add $\sqrt{10r+14}$ to both sides of the equation to get one square root on each side of the equation. Then square both sides of the equation.

$$4\sqrt{r-4} - \sqrt{10r+14} + \sqrt{10r+14} = 0 + \sqrt{10r+14}$$
$$4\sqrt{r-4} = \sqrt{10r+14}$$
$$\left(4\sqrt{r-4}\right)^2 = \left(\sqrt{10r+14}\right)^2 \quad \text{Square both sides.}$$
$$16(r-4) = 10r + 14 \quad \text{Simplify.}$$
$$16r - 64 = 10r + 14 \quad \text{Distributive property}$$
$$6r - 64 = 14 \quad 10r \text{ was subtracted from both sides.}$$
$$6r = 78 \quad 64 \text{ was added to both sides.}$$
$$r = 13$$

Check

$$4\sqrt{r-4} - \sqrt{10r+14} = 0$$
$$4\sqrt{13-4} - \sqrt{10(13)+14} \stackrel{?}{=} 0$$
$$4\sqrt{9} - \sqrt{130+14} \stackrel{?}{=} 0$$
$$4(3) - \sqrt{144} \stackrel{?}{=} 0$$
$$12 - 12 = 0 \quad \text{True}$$

The solution is 13.

Now Try Exercise 51

HELPFUL HINT

In Example 6 when we simplified $(5\sqrt{x})^2$ we obtained $25x$ and in Example 8 when we simplified $(4\sqrt{r-4})^2$ we obtained $16(r-4)$. Remember, by the power rule for exponents (discussed in Section 4.1) that when a product of factors is raised to a power, each of the factors is raised to that power. Thus, we see that

$$(5\sqrt{x})^2 = 5^2(\sqrt{x})^2 = 25x \quad \text{and} \quad (4\sqrt{r-4})^2 = 4^2(\sqrt{r-4})^2 = 16(r-4)$$

9.5 Exercise Set MathXL® MyMathLab®

Warm-Up Exercises

Fill in the blanks with the appropriate word, phrase, or symbol(s) from the following list.

isolate variable extraneous real rational original eliminate constant

1. A radical equation is an equation that contains a ________ in the radicand.
2. When solving an equation with only one square root, after we isolate the square root we can ________ the square root by squaring both sides of the equation.
3. When squaring both sides of an equation, we sometimes introduce a false, or ________, solution.
4. Whenever we square both sides of an equation, we must check the solutions in the ________ equation.
5. When solving a radical equation with one square root, we first should ________ the radical.
6. The equation $\sqrt{x} = -5$ has no ________ solution.

Practice the Skills

Determine whether the given value is the solution to the equation. If it is not a solution, state why.

7. $\sqrt{x} = 7, x = 49$
8. $\sqrt{b} = 9, b = 81$
9. $\sqrt{x} = -7, x = 49$
10. $\sqrt{b} = -9, b = 81$
11. $\sqrt{3x-5} = x-3, x = 7$
12. $\sqrt{2x+7} = x-4, x = 9$
13. $\sqrt{3x-5} = x-3, x = 2$
14. $\sqrt{2x+7} = x-4, x = 1$
15. $\sqrt{2x+3} = \sqrt{5x+6}, x = -1$
16. $\sqrt{3x-7} = \sqrt{7x-19}, x = 3$

Solve each equation. If the equation has no real solution, so state.

17. $\sqrt{x} = 5$
18. $\sqrt{x} = 10$
19. $\sqrt{x} = -8$
20. $\sqrt{m} = -3$
21. $\sqrt{x} + 5 = 3$
22. $\sqrt{k} - 7 = 4$
23. $\sqrt{x} + 5 = -3$
24. $8 + \sqrt{n} = 4$
25. $11 = 6 + \sqrt{x}$
26. $10 + \sqrt{w} = 13$
27. $4 - \sqrt{x} = -2$
28. $5 = 7 - \sqrt{x}$
29. $\sqrt{2r-3} = \sqrt{r+3}$
30. $\sqrt{2x+4} = \sqrt{x+10}$
31. $\sqrt{4x+4} = \sqrt{6x-2}$
32. $\sqrt{3x-5} = \sqrt{x+9}$
33. $\sqrt{x^2+5} = x+1$
34. $\sqrt{x^2-36} = x+6$
35. $\sqrt{3x+1} = x-1$
36. $\sqrt{3x+4} = x-2$
37. $\sqrt{2x-5} = x-4$
38. $\sqrt{x+8} = x+2$
39. $6\sqrt{x-4} = 24$
40. $2\sqrt{3x-2} = 8$
41. $3\sqrt{x} = \sqrt{x+8}$
42. $\sqrt{3k+5} = 2\sqrt{k}$
43. $4\sqrt{x} = x+3$
44. $6\sqrt{x} = x+8$
45. $7 + \sqrt{x-7} = x$
46. $\sqrt{p+3} - 3 = p$
47. $3 + \sqrt{3x-5} = x$
48. $\sqrt{4x+5} + 5 = 2x$
49. $2 + \sqrt{4x-11} = x$
50. $\sqrt{7x-3} - 1 = x$
51. $2\sqrt{3b-5} - \sqrt{2b+10} = 0$
52. $2\sqrt{m+4} - \sqrt{7m+1} = 0$
53. $4\sqrt{2w+3} = 4w$
54. $9\sqrt{z+6} = 9z$
55. $2\sqrt{5x-1} = x+4$
56. $2\sqrt{4c-7} = c+2$

Problem Solving

Exercises 57–60 show rectangles with areas. Find the value of the indicated variable.

57.

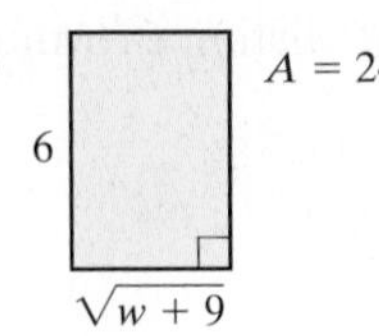

58.

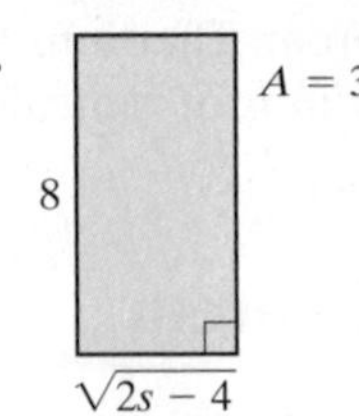

59. $A = 37.2$; sides $\sqrt{3n+3}$ and 6.2

60. $A = 29.2$; sides $\sqrt{3p+1}$ and 7.3

In Section 9.1, we indicated that we could write a square root expression in exponential form. For example, $\sqrt{25}$ can be written $(25)^{1/2}$. Similarly, we can write certain exponential expressions as square roots. For example $(25)^{1/2}$ can be written $\sqrt{25}$, and $(x-5)^{1/2}$ can be written $\sqrt{x-5}$. Rewrite the following equations as equations containing square roots. Then solve each equation.

61. $(x+4)^{1/2} = 7$

62. $(x-8)^{1/2} = 5$

63. $(x-2)^{1/2} = (2x-9)^{1/2}$

64. $3(x-9)^{1/2} = 21$

In Exercises 65–68, **a)** *Use the FOIL method to multiply the factors on the left-hand side of the equation.* **b)** *Solve the equation. Remember to check your answer in the original equation.*

65. $(\sqrt{x}-3)(\sqrt{x}+3) = 40$

66. $(\sqrt{x}+5)(\sqrt{x}-5) = 11$

67. $(7-\sqrt{x})(5+\sqrt{x}) = 35$

68. $(6-\sqrt{x})(2+\sqrt{x}) = 15$

69. The sum of a natural number and its square root is 2. Find the number.

70. The product of 3 and the square root of a number is 15. Find the number.

Challenge Problems

Solve each equation. (Hint: You will need to square both sides of the equation, then isolate the square root containing the variable, then square both sides of the equation again.)

71. $\sqrt{x}+1 = \sqrt{x+11}$

72. $\sqrt{x+1} = 2-\sqrt{x}$

73. $\sqrt{x+7} = 3+\sqrt{x-8}$

Group Activity

Discuss and answer Exercises 74–76 as a group.

74. Radical equations in two variables can be graphed by selecting values for x and finding the corresponding values of y, as was done in Section 7.2 when we graphed linear equations.

a) As a group, discuss what values would be appropriate to select for x if you wanted to graph $y = \sqrt{x}$. Explain your answer.

b) Select four values for x that will result in y being a whole number. List your values for x and y in a table.

c) Individually, plot the points and sketch the graph. Compare your graphs.

d) Is the graph linear? Explain.

e) Is the graph a function? (See Section 7.6.) Explain.

f) Can you determine whether $y = \sqrt{x}$ has an x-intercept and a y-intercept from the graph? If so, give the intercepts.

As a group, repeat parts **a)–f)** *of Exercise 74 for each the following equations.*

75. $y = \sqrt{x-2}$

76. $y = \sqrt{x+4}$

Cumulative Review Exercises

[8.1] **77.** Solve the following system of equations graphically.

$$3x - 2y = 6$$
$$y = 2x - 4$$

[8.2] **78.** Solve the following system of equations by substitution.

$$3x - 2y = 6$$
$$y = 2x - 4$$

[8.3] **79.** Solve the following system of equations using the addition method.

$$3x - 2y = 6$$
$$y = 2x - 4$$

[8.4] **80. Boat Ride** A ferryboat on the Hudson River can travel at a speed of 18 miles per hour with the current and 14 miles per hour against the current. Find the speed of the ferryboat in still water and the speed of the current.

9.6 Radicals: Applications and Problem Solving

1. Use the Pythagorean Theorem.
2. Use the distance formula.
3. Use radicals to solve application problems.

In this section we will use the Pythagorean Theorem to introduce the distance formula, and then give a few additional applications of radicals.

1 Use the Pythagorean Theorem

In Section 5.7 we introduced the Pythagorean Theorem.

Pythagorean Theorem

The square of the hypotenuse of a right triangle is equal to the sum of the squares of the two legs.

$$(\text{leg})^2 + (\text{leg})^2 = (\text{hypotenuse})^2$$

If a and b represent the legs, and c represents the hypotenuse, then

$$a^2 + b^2 = c^2$$

hypotenuse c, b leg, a leg

Understanding Algebra

When using the Pythagorean Theorem, be sure to identify the *longest* side of the triangle as the hypotenuse, or side c. The Pythagorean Theorem can be thought of as follows:

$$\begin{pmatrix}\text{first}\\ \text{shorter}\\ \text{side}\end{pmatrix}^2 + \begin{pmatrix}\text{second}\\ \text{shorter}\\ \text{side}\end{pmatrix}^2 = \begin{pmatrix}\text{the}\\ \text{longest}\\ \text{side}\end{pmatrix}^2$$

When we solve problems using the Pythagorean Theorem, we will take the square root of both sides of the equation. Since lengths are positive, the values of a, b, and c in the Pythagorean Theorem must represent positive values.

EXAMPLE 1 **Right Triangle Problem** Find the hypotenuse of a right triangle whose legs are $\sqrt{3}$ feet and 4 feet.

Solution Draw a picture of the problem (**Fig. 9.2**). It makes no difference which leg is called a and which leg is called b.

$$a^2 + b^2 = c^2$$

$$4^2 + \left(\sqrt{3}\right)^2 = c^2 \qquad \text{Substitute values for } a \text{ and } b.$$

$$16 + 3 = c^2$$

$$19 = c^2$$

$$\sqrt{19} = \sqrt{c^2} \qquad \text{Take the square root of both sides.}$$

$$\sqrt{19} = c \qquad \text{Since } c > 0, \sqrt{c^2} = c.$$

$$c = \sqrt{19} \approx 4.36$$

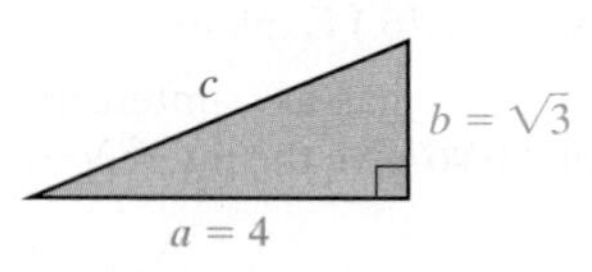

FIGURE 9.2

Thus the hypotenuse is $\sqrt{19}$ feet or approximately 4.36 feet.

Check

$$a^2 + b^2 = c^2$$

$$4^2 + \left(\sqrt{3}\right)^2 \stackrel{?}{=} \left(\sqrt{19}\right)^2$$

$$16 + 3 \stackrel{?}{=} 19$$

$$19 = 19 \qquad \text{True}$$

Now Try Exercise 13

EXAMPLE 2 **Basketball Court** A basketball court is a rectangle whose overall dimensions are 94 feet by 50 feet. Find the length of the diagonal of the court.

Solution Understand and Translate First, we draw the court (**Fig. 9.3** on the next page). We are asked to find the length of the diagonal. This length is the hypotenuse, c, of the triangle shown in **Figure 9.4** on the next page.

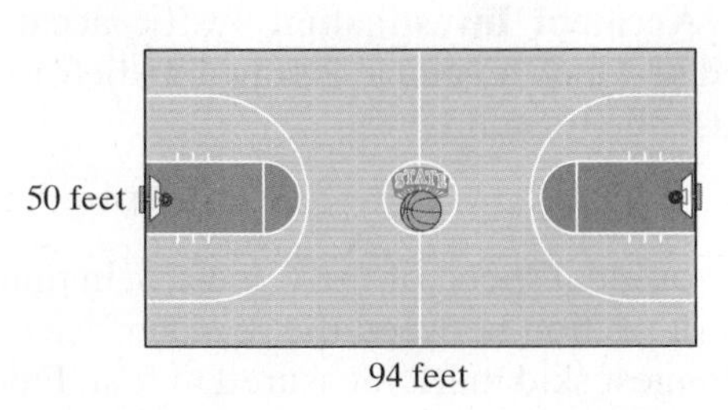

FIGURE 9.3

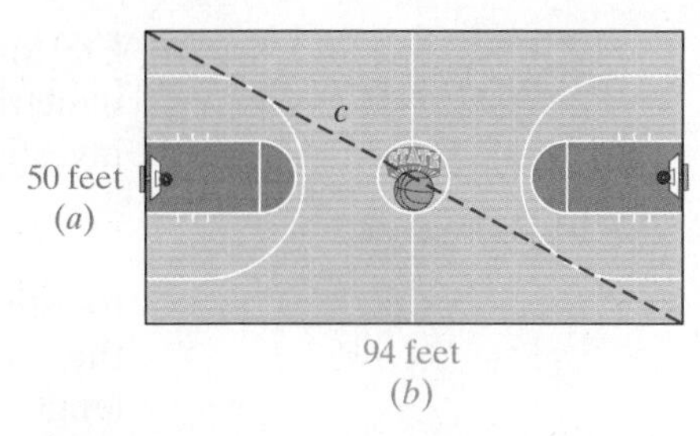

FIGURE 9.4

$$a^2 + b^2 = c^2$$

Carry Out

$$(50)^2 + (94)^2 = c^2$$
$$2500 + 8836 = c^2$$
$$11{,}336 = c^2$$
$$c = \sqrt{11{,}336} \approx 106.47$$

Answer The length of the diagonal of the court is approximately 106.47 feet.

Now Try Exercise 25

2 Use the Distance Formula

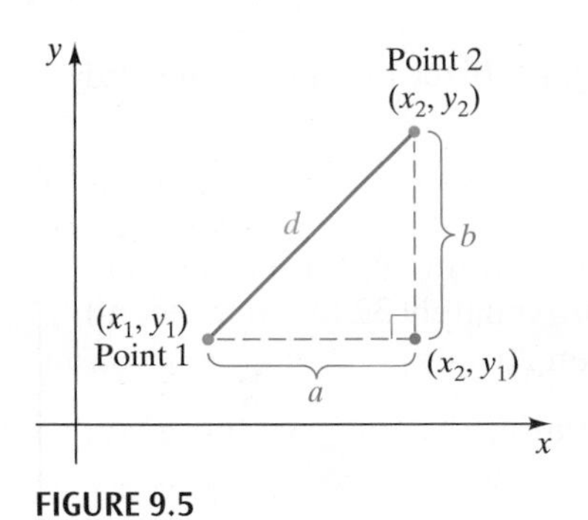

FIGURE 9.5

The **distance formula** can be used to find the distance between two points, (x_1, y_1) and (x_2, y_2), in the Cartesian coordinate system.

It can be derived using the Pythagorean Theorem. In **Figure 9.5**, three points are indicated. The length of side a is $x_2 - x_1$ and the length of side b is $y_2 - y_1$. The distance between points 1 and 2 can be determined using the Pythagorean Theorem. In this case we let the hypotenuse be represented by the letter d, for distance.

$$d^2 = a^2 + b^2$$
$$d^2 = (x_2 - x_1)^2 + (y_2 - y_1)^2$$
$$d = \sqrt{(x_2 - x_1)^2 + (y_2 - y_1)^2}$$

Distance Formula

$$d = \sqrt{(x_2 - x_1)^2 + (y_2 - y_1)^2}$$

EXAMPLE 3 Find the length of the line segment between the points $(-1, -4)$ and $(5, -2)$.

Solution The two points are illustrated in **Figure 9.6**. It makes no difference which point is labeled (x_1, y_1), and which is labeled (x_2, y_2). Let $(5, -2)$ be (x_2, y_2) and $(-1, -4)$ be (x_1, y_1). Thus $x_2 = 5, y_2 = -2$ and $x_1 = -1, y_1 = -4$.

$$\begin{aligned} d &= \sqrt{(x_2 - x_1)^2 + (y_2 - y_1)^2} \\ &= \sqrt{[5 - (-1)]^2 + [-2 - (-4)]^2} \\ &= \sqrt{(5 + 1)^2 + (-2 + 4)^2} \\ &= \sqrt{6^2 + 2^2} \\ &= \sqrt{36 + 4} = \sqrt{40}, \quad \text{or approximately } 6.32 \end{aligned}$$

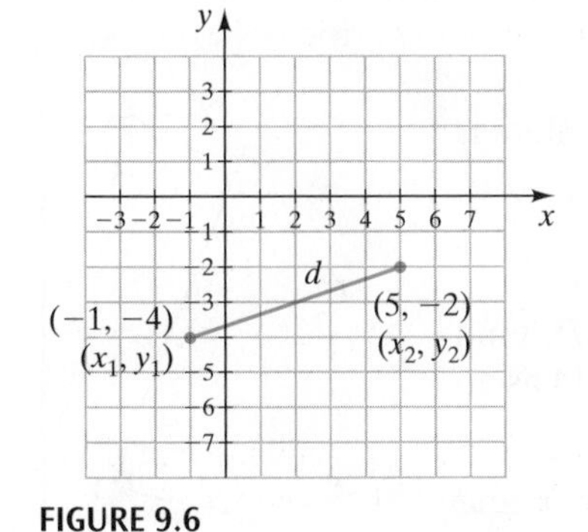

FIGURE 9.6

Thus, the distance between $(-1, -4)$ and $(5, -2)$ is approximately 6.32 units.

Now Try Exercise 21

3 Use Radicals to Solve Application Problems

Examples 4 through 6 illustrate some scientific applications that use radicals.

EXAMPLE 4 Accident Investigation Traffic accident investigators try to determine the speed at which a vehicle is traveling before hitting an object. The accident investigators use the formula

$$S = 5.5\sqrt{cl}$$

to estimate the original speed of the vehicle, S, in miles per hour, where c represents the coefficient of friction between the road surface and the tire, and l represents the length of the longest skid mark measured in feet. Find the speed of a car that leaves a 40-foot-long skid mark before stopping. Use $c = 0.72$.

Solution Understand and Translate Begin by substituting 0.72 for c and 40 for the length in the given formula.

Carry Out

$$\begin{aligned} S &= 5.5\sqrt{cl} \\ &= 5.5\sqrt{0.72 \times 40} \\ &= 5.5\sqrt{28.8} \\ &\approx 29.5 \end{aligned}$$

Answer A car leaving a 40-foot-long skid mark on a dry surface before stopping is originally traveling at approximately 29.5 miles per hour.

Now Try Exercise 45

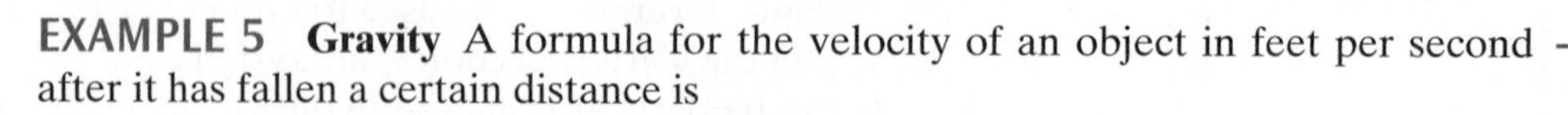

EXAMPLE 5 Gravity A formula for the velocity of an object in feet per second after it has fallen a certain distance is

$$v = \sqrt{2gh}$$

where g is the acceleration due to gravity and h is the height the object has fallen in feet. On Earth the acceleration due to gravity, g, is approximately 32 feet per second squared. Find the velocity of a coconut after it has fallen 20 feet.

Solution Understand and Translate Begin by substituting 32 for g in the given equation.

$$v = \sqrt{2gh} = \sqrt{2(32)(h)} = \sqrt{64h}$$

Carry Out At $h = 20$ feet,

$$v = \sqrt{64(20)} = \sqrt{1280} \approx 35.78$$

Answer After a coconut has fallen 20 feet, its velocity is approximately 35.78 feet per second.

Now Try Exercise 49

EXAMPLE 6 A Pendulum The *period of a pendulum* is the time required for the pendulum to make one complete swing both back and forth. The period depends on the length of the pendulum.

The formula for the period, T (in seconds), of a pendulum is

$$T = 2\pi\sqrt{\frac{L}{32}}$$

where L is the length of the pendulum in feet (**Fig. 9.7**). Find the period of the pendulum if its length is 8 feet. Use 3.14 as an approximation for π.

Solution Understand and Translate Substitute 3.14 for π and 8 for L in the formula.

$$T \approx 2(3.14)\sqrt{\frac{8}{32}}$$

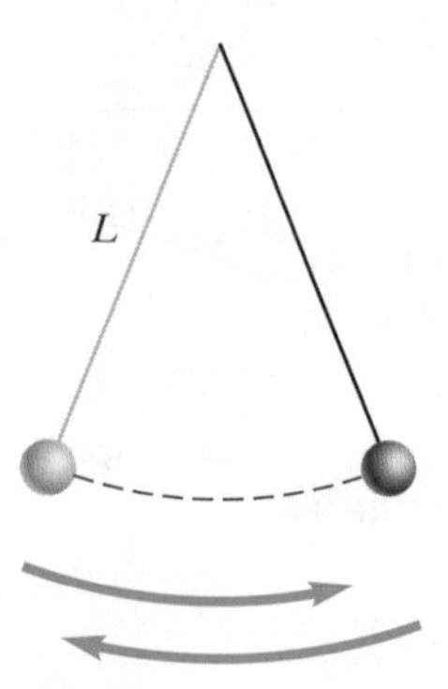

FIGURE 9.7

Carry Out $\approx 6.28\sqrt{\dfrac{1}{4}}$

$\approx 6.28\left(\dfrac{1}{2}\right) = 3.14$ seconds

Answer A pendulum 8 feet long takes about 3.14 seconds to make one complete swing.

Now Try Exercise 39

9.6 Exercise Set MathXL® MyMathLab®

Warm-Up Exercises

Fill in the blanks with the appropriate word, phrase, or symbol(s) from the following list.

distance leg period Pythagorean diagonal hypothesis hypotenuse

1. The formula describing the relationship of the sides of a right triangle as $(\text{leg})^2 + (\text{leg})^2 = (\text{hypotenuse})^2$ is called the ___________ Theorem.

2. The longest side of a right triangle is called the ___________.

3. Each of the shorter sides of a right triangle is called a ___________.

4. In a rectangle, the line that connects one corner to the opposite corner is called the ___________.

5. The formula $d = \sqrt{(x_2 - x_1)^2 + (y_2 - y_1)^2}$ is known as the ___________ formula.

6. The time required for a pendulum to make one complete swing back and forth is called the ___________ of the pendulum.

Practice the Skills

Use the Pythagorean Theorem to find each indicated quantity. Give the exact value, then round your answer to the nearest hundredth.

7.

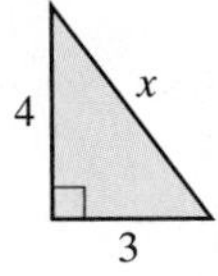

8.

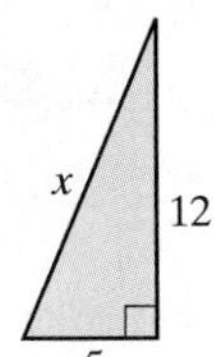

9.

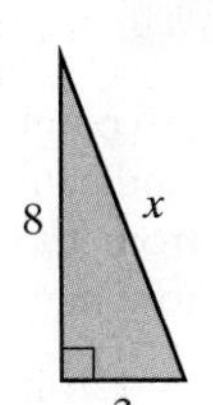

10.

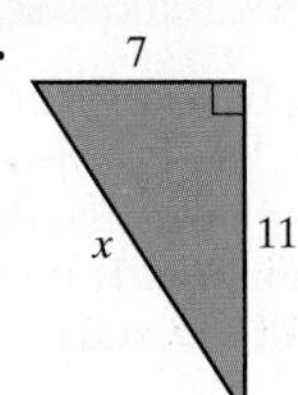

11.

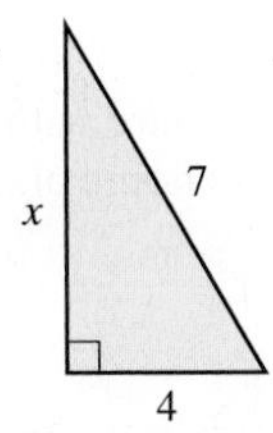

12.

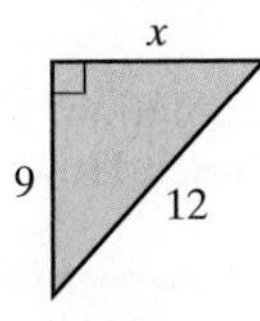

13.

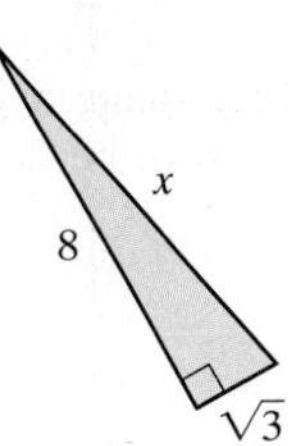

14.

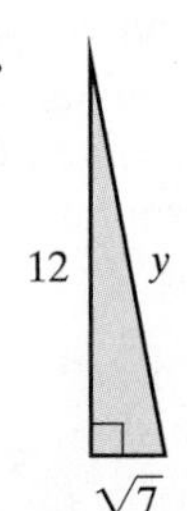

15.

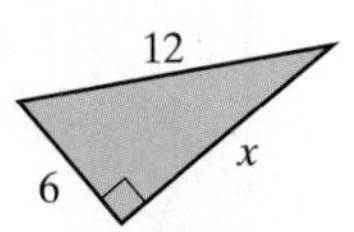

16.

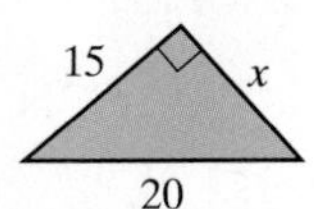

17.

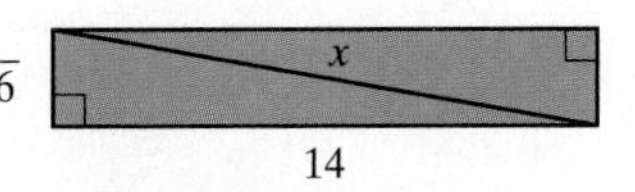

18.

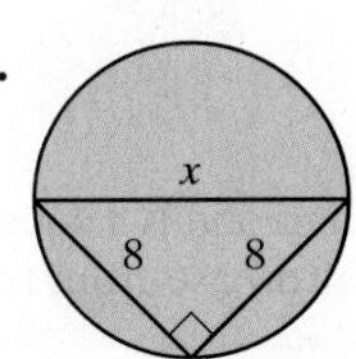

19.

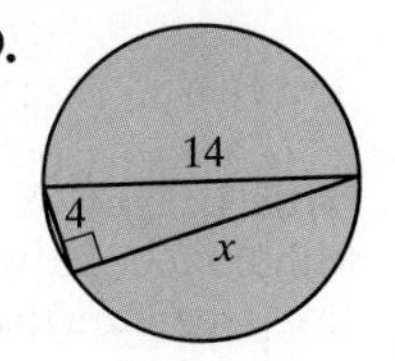

20.

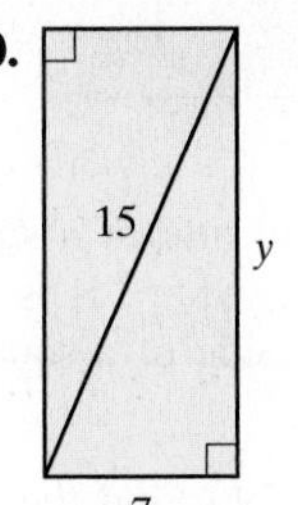

Use the distance formula to find the length of the line segments between each pair of points. Give the exact value, then round your answer to the nearest hundreath.

21. $(8, 7)$ and $(3, 10)$

22. $(6, -1)$ and $(-3, 8)$

23. $(-8, 4)$ and $(4, 11)$

24. $(2, 7)$ and $(-4, -2)$

Problem Solving

In Exercises 25–41, give the exact value, then, when appropriate, round to the nearest hundreath.

25. **Football Field Dimensions** A football field is 120 yards long from end zone to end zone. Find the length of the diagonal, to the nearest hundredth, from one end zone to the opposite corner of the other if the width of the field is 53.3 yards.

26. **Rugby Field** The dimensions of a rugby field are 69 meters by 100 meters. What is the length of the diagonal, to the nearest hundredth, from one corner of the field to the opposite corner of the field?

27. **Ladder on a House** Pedro and Vanessa are planning to paint their house. The instructions on an 8-meter ladder indicate that the base of the ladder should be two meters from the house when in use. If they follow the instructions, how high will the top of the ladder reach on the house?

28. **Wind Turbine Guy Wire** A guy wire on the pole supporting a wind turbine is 30 meters long and is attached to the pole at a height of 25 meters above the ground (see diagram). How far from the base of the pole will the guy wire be anchored into the ground?

29. **Boxing Ring** The length of the side of a square, s, can be found by the formula $s = \sqrt{A}$, where A is the area of the square. Find the length of the side of a square boxing ring whose area is 256 square feet.

30. **Square Window** (See Exercise 29.) Find the length of a side of a square window having an area 1600 square centimeters.

31. **Helicopter Pad** The formula for the area of a circle is $A = \pi r^2$, where π is approximately 3.14 and r is the radius of the circle. Find the radius of a circular helicopter pad if its area is 1965 square feet.

32. **Radius of a Circle** Find the radius of a circle whose area is 80 square feet.

33. **Birdbath** A birdbath can hold water 2 inches deep, as shown in the figure. If the volume of water that can be held in the birdbath is 402 cubic inches, find the radius of the birdbath. Water volume is calculated by finding the area of the circular bottom of the birdbath and then multiplying the area by the depth of the water. Use the formula $V = \pi r^2 h$.

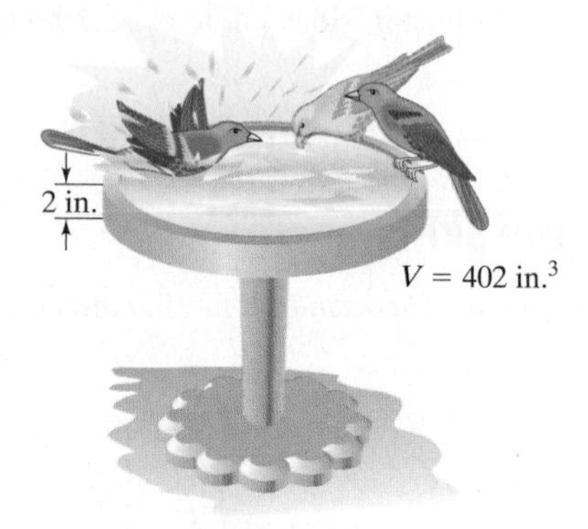

34. **Cake Pan** A round cake pan can hold batter 4 centimeters deep. If the volume of the batter that can be held in the pan is 1018 cubic centimeters, find the radius of the cake pan. (See Exercise 33.)

35. **Baseball** A regulation baseball diamond is a square with 90 feet between bases. How far is second base from home plate?

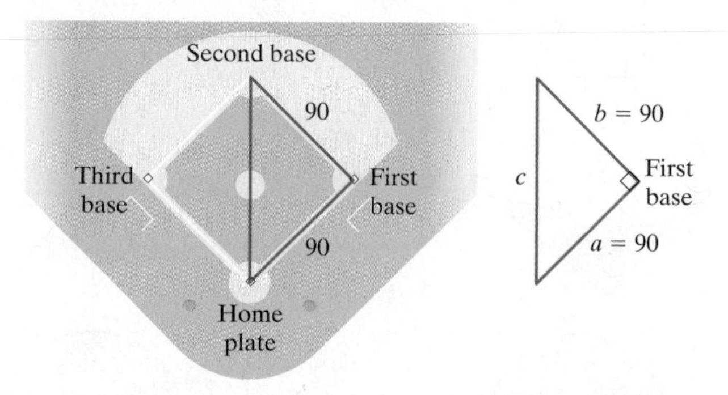

36. **Little League** A Little League baseball diamond is a square with 60 feet between bases. How far is third base from first base (see Exercise 35).

37. **Diagonal of Rectangular Solid** The length of the diagonal of a rectangular solid (see the figure on the next page) is given by

$$d = \sqrt{a^2 + b^2 + c^2}$$

Find the length of the diagonal of the suitcase shown below.

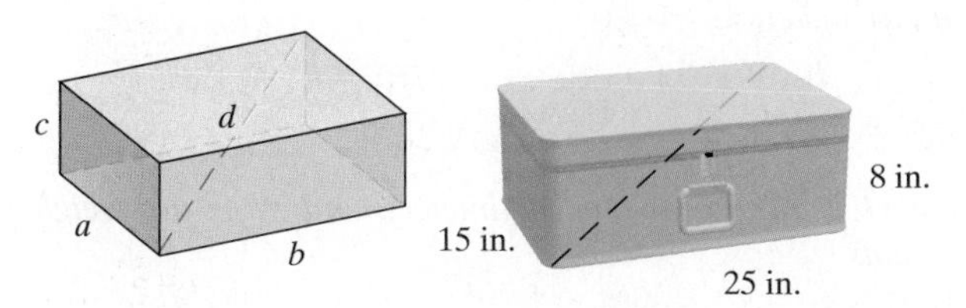

38. Carry-On Luggage The maximum dimensions of a carry-on suitcase on most airlines is length: 24 inches, height: 16 inches, and depth: 10 inches. If the suitcase is lying flat, determine the length of the diagonal of the suitcase. (See Exercise 37.)

39. **Pendulum** The period of a simple pendulum is determined by the length of the string, not the material composing the bob (or ball). Mikel Finn made a pendulum with a string measuring 43 inches. What is the period for this pendulum? (Use $T = 2\pi\sqrt{L/32}$; refer to Example 6.)

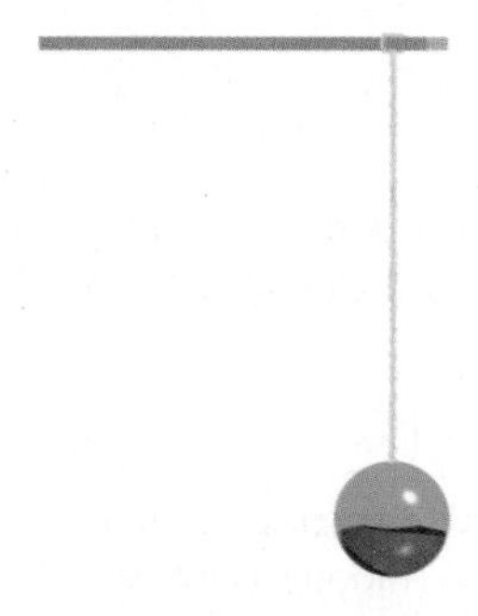

40. Foucault Pendulum An example of a Foucault pendulum is on display in the Smithsonian Institute in Washington, D.C. It is used to demonstrate Earth's rotation. It has a cable 52 feet long and a symmetrical, hollow brass bob weighing about 240 pounds. Find the period of the pendulum. (See Exercise 39.)

41. Another Pendulum The Houston Museum of Natural Science also has a Foucault pendulum on display. It has a cable 61.6 feet long and a bob weighing 180 pounds. Find the period of the pendulum. (See Exercise 39.)

Earth Days *For any planet, its "year" is the time it takes for the planet to revolve once around the sun. The number of "Earth days," N, in a given planet's year, is approximated by the formula*

$$N = 0.2(\sqrt{R})^3$$

where R is the mean distance from the sun in millions of kilometers. The mean distance from the sun, in millions of kilometers, for three planets is illustrated below.

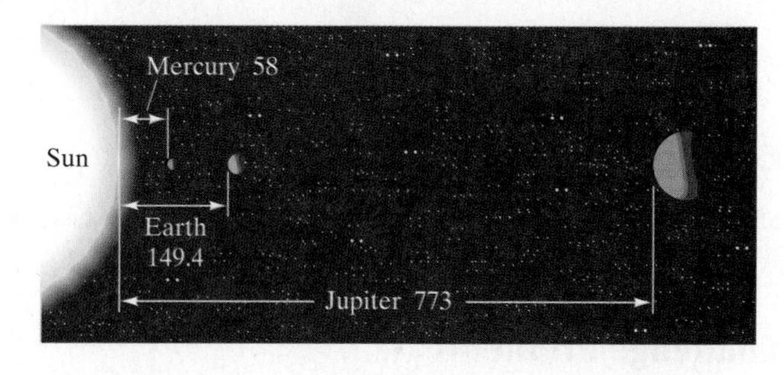

42. Find the number of Earth days in the year of Mercury.

43. Find the number of Earth days in the year of Earth.

44. Find the number of Earth days in the year of Jupiter.

Speed *In Exercises 45–48, use the formula $S = 5.5\sqrt{cl}$ to find the speed of a vehicle, S, in miles per hour given the coefficient of friction, c, between the road surface and the tires, and the length, l, of the skid mark, measured in feet; see Example 4.*

45. $c = 0.72$
$l = 50$ feet

46. $c = 0.70$
$l = 70$ feet

47. $c = 0.75$
$l = 120$ feet

48. $c = 0.65$
$l = 180$ feet

Escape velocity, v_e, is the velocity needed for a spacecraft to escape a planet's gravitational field. This is found by the formula $v_e = \sqrt{2gR}$, where g is the acceleration due to gravity of the planet and R is the radius of the planet in meters.

49. **Earth's Escape Velocity** Find the escape velocity for Earth where $g = 9.81$ meters per second squared and $R =$ 6,370,000 meters.

50. Mars's Escape Velocity What is the approximate velocity needed for a spacecraft to escape the gravitational field of Mars? The acceleration due to gravity on Mars is approximately 3.6835 meters per second squared. The radius of Mars is approximately 3,393,500 meters.

51. Freefall If a pair of glasses was dropped from Willis Tower at a height of 1431 feet, determine the velocity of the glasses when they strike the ground. (Use $v = \sqrt{2gh}$; refer to Example 5.)

Willis Tower in Chicago, Illinois, see Exercise 51.

52. **Resultant Force** When two forces, F_1 and F_2, pull at right angles to each other as illustrated, the resultant, or the effective force, R, can be found by the formula $R = \sqrt{F_1^2 + F_2^2}$.

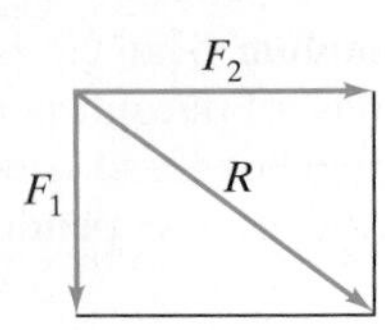

Two cars at a 90° angle to each other are trying to pull a third car out of the mud, as shown. If car A exerts 600 pounds of force and car B exerts 800 pounds of force, find the resulting force on the car stuck in the mud.

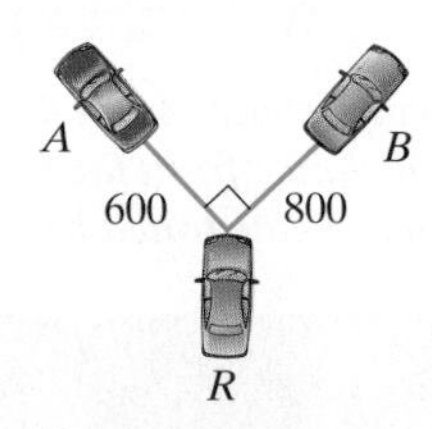

Body Surface Area *The amount of skin that covers our bodies is called* **body surface area** *(BSA). The formula to calculate BSA, in square meters, is*

$$\text{BSA} = \sqrt{\frac{HW}{3600}}$$

where H is the height, in centimeters, and W is the weight, in kilograms.

In Exercises 53 and 54, calculate the BSA for a person with the given height and weight.

53. $H = 180$ centimeters
$W = 85$ kilograms

54. $H = 150$ centimeters
$W = 70$ kilograms

Concept/Writing Exercises

55. Can the Pythagorean Theorem be used with any triangle? Explain.

56. What do the following ordered pairs represent in the distance formula: (x_1, y_1) and (x_2, y_2)?

Challenge Problems

57. **Rectangle** The length of a rectangle is 3 inches more than its width. If the length of the diagonal is 15 inches, find the dimensions of the rectangle.

58. **Gravity on the Moon** The acceleration due to gravity on the moon is $\frac{1}{6}$ of that on Earth. If a camera falls from a rocket 100 feet above the surface of the moon, with what velocity will it strike the moon? (Use $v = \sqrt{2gh}$; see Example 5.)

59. **Length of a Pendulum** Find the length of a pendulum if the period is 2 seconds. (Use $T = 2\pi\sqrt{L/32}$; refer to Example 6.)

Cumulative Review Exercises

[2.8] 60. Solve the inequality $2(x + 3) < 4x - 6$.

[4.2] 61. Simplify $(x^{-4}y^3)^{-1}$.

[6.6] 62. Solve $5 + \frac{6}{x} = \frac{2}{3x}$.

[8.3] 63. Solve the following system of equations using the addition method.

$$3x + 4y = 12$$
$$\frac{1}{2}x - 2y = 8$$

9.7 Higher Roots and Rational Exponents

1 Evaluate cube and fourth roots.

2 Simplify cube and fourth roots.

3 Write radical expressions in exponential form.

1 Evaluate Cube and Fourth Roots

In this section we will use the same basic concepts used in Sections 9.1 through 9.4 to work with radicals with indices of 3 and 4. Now we introduce **cube** and **fourth roots**.

Cube Root and Fourth Root

Cube root: $\sqrt[3]{a}$ is read "the cube root of a."

$$\sqrt[3]{a} = b \quad \text{if} \quad b^3 = a$$

Fourth root: $\sqrt[4]{a}$ is read "the fourth root of a."

$$\sqrt[4]{a} = b \quad \text{if} \quad b^4 = a, \quad a \geq 0$$

Examples

$\sqrt[3]{8} = 2$ since $2^3 = 2 \cdot 2 \cdot 2 = 8$

$\sqrt[3]{-8} = -2$ since $(-2)^3 = (-2)(-2)(-2) = -8$

$\sqrt[3]{27} = 3$ since $3^3 = 3 \cdot 3 \cdot 3 = 27$

$\sqrt[4]{16} = 2$ since $2^4 = 2 \cdot 2 \cdot 2 \cdot 2 = 16$

$\sqrt[4]{81} = 3$ since $3^4 = 3 \cdot 3 \cdot 3 \cdot 3 = 81$

$\sqrt[3]{\frac{1}{27}} = \frac{1}{3}$ since $\left(\frac{1}{3}\right)^3 = \frac{1}{3} \cdot \frac{1}{3} \cdot \frac{1}{3} = \frac{1}{27}$

EXAMPLE 1 Evaluate.

a) $\sqrt[3]{-64}$ **b)** $\sqrt[3]{216}$

Solution

a) To find $\sqrt[3]{-64}$, we must find the number that when cubed is -64.

$$\sqrt[3]{-64} = -4 \qquad \text{since } (-4)^3 = -64$$

b) To find $\sqrt[3]{216}$, we must find the number that when cubed is 216.

$$\sqrt[3]{216} = 6 \qquad \text{since } 6^3 = 216$$

Now Try Exercise 11

Note that the cube root of a positive number is a positive number and the cube root of a negative number is a negative number. The radicand of a fourth root (or any even root) must be a nonnegative number for the expression to be a real number. For example, $\sqrt[4]{-16}$ is not a real number because no real number raised to the fourth power can be a negative number.

It will be helpful in the explanations that follow if we define perfect cubes. A **perfect cube** is a number that is the cube of a natural number.

Natural Numbers	1,	2,	3,	4,	5,	6,	7,	8,	9,	10,	...
Cubes of Natural Numbers	1^3,	2^3,	3^3,	4^3,	5^3,	6^3,	7^3,	8^3,	9^3,	10^3,	...
Perfect Cubes	1,	8,	27,	64,	125,	216,	343,	512,	729,	1000,	...

Note that $\sqrt[3]{1} = 1$, $\sqrt[3]{8} = 2$, $\sqrt[3]{27} = 3$, $\sqrt[3]{64} = 4$, and so on.

Perfect fourth powers can be expressed in a similar manner.

Natural Numbers	1,	2,	3,	4,	5,	6,	...
Fourth Powers of Natural Numbers	1^4,	2^4,	3^4,	4^4,	5^4,	6^4,	...
Perfect Fourth Powers	1,	16,	81,	256,	625,	1296,	...

Note that $\sqrt[4]{1} = 1$, $\sqrt[4]{16} = 2$, $\sqrt[4]{81} = 3$, $\sqrt[4]{256} = 4$, and so on.

Table 9.2 on the next page illustrates some perfect cube and perfect fourth power numbers. This table may prove helpful when simplifying cube and fourth power roots.

You may wish to refer to **Table 9.2** when evaluating cube and fourth roots.

TABLE 9.2

Perfect Cubes	Cube Root		Value	Perfect Fourth Powers	Fourth Root		Value
1	$\sqrt[3]{1}$	=	1	1	$\sqrt[4]{1}$	=	1
8	$\sqrt[3]{8}$	=	2	16	$\sqrt[4]{16}$	=	2
27	$\sqrt[3]{27}$	=	3	81	$\sqrt[4]{81}$	=	3
64	$\sqrt[3]{64}$	=	4	256	$\sqrt[4]{256}$	=	4
125	$\sqrt[3]{125}$	=	5	625	$\sqrt[4]{625}$	=	5
216	$\sqrt[3]{216}$	=	6	1296	$\sqrt[4]{1296}$	=	6
343	$\sqrt[3]{343}$	=	7				
512	$\sqrt[3]{512}$	=	8				

2 Simplify Cube and Fourth Roots

The product rule used in simplifying square roots can be expanded to indices greater than 2.

Product Rule for Radicals

When n is even: $\sqrt[n]{ab} = \sqrt[n]{a} \cdot \sqrt[n]{b}$ and $\sqrt[n]{a} \cdot \sqrt[n]{b} = \sqrt[n]{ab}$, provided $a \geq 0, b \geq 0$

When n is odd: $\sqrt[n]{ab} = \sqrt[n]{a} \cdot \sqrt[n]{b}$ and $\sqrt[n]{a} \cdot \sqrt[n]{b} = \sqrt[n]{ab}$

To simplify a cube root whose radicand is a constant, write the radicand as the product of a perfect cube and another number. Then simplify, using the product rule.

EXAMPLE 2 Simplify. **a)** $\sqrt[3]{40}$ **b)** $\sqrt[3]{54}$ **c)** $\sqrt[4]{48}$

Solution

a) Eight is a perfect cube that is a factor of the radicand, 40. Therefore, we simplify as follows:

$$\begin{aligned} \sqrt[3]{40} &= \sqrt[3]{8 \cdot 5} \\ &= \sqrt[3]{8}\sqrt[3]{5} && \text{Product rule} \\ &= 2\sqrt[3]{5} && \text{Simplify.} \end{aligned}$$

b) Twenty-seven is a perfect cube factor of 54. Therefore we simplify as follows.

$$\sqrt[3]{54} = \sqrt[3]{27 \cdot 2} = \sqrt[3]{27}\sqrt[3]{2} = 3\sqrt[3]{2}$$

c) Write $\sqrt[4]{48}$ as a product of a perfect fourth power and another number, then simplify. From **Table 9.2**, we see that 16 is a perfect fourth power. Since 16 is a factor of 48, we simplify as follows:

$$\begin{aligned} \sqrt[4]{48} &= \sqrt[4]{16 \cdot 3} \\ &= \sqrt[4]{16}\sqrt[4]{3} && \text{Product rule} \\ &= 2\sqrt[4]{3} && \text{Simplify.} \end{aligned}$$

Now Try Exercise 25

3 Write Radical Expressions in Exponential Form

In Section 9.1 we showed that expressions involving square roots can be rewritten in exponential form using the fact that $\sqrt{a} = a^{1/2}$. We next can generalize this fact to write any radical expression in **exponential form** using the following rule.

Exponential Form of a Radical Expression

$\sqrt[n]{a} = a^{1/n}$, $a \geq 0$ and n is even **Rule 4**

$\sqrt[n]{a} = a^{1/n}$, when n is odd

Examples

$\sqrt{8} = 8^{1/2}$ $\sqrt{x} = x^{1/2}$

$\sqrt[3]{7} = 7^{1/3}$ $\sqrt[4]{9} = 9^{1/4}$

$\sqrt[3]{x} = x^{1/3}$ $\sqrt[4]{y} = y^{1/4}$

$\sqrt[3]{5z^2} = (5z^2)^{1/3}$ $\sqrt[4]{3y^2} = (3y^2)^{1/4}$

Notice $\sqrt{8} = 8^{1/2}$ and $\sqrt{x} = x^{1/2}$, which is consistent with what we learned in Section 9.1. This concept can be expanded as follows.

Understanding Algebra

Rule 5 shows that any radical expression can be rewritten using exponents. Thus, the rules of exponents we studied in Chapter 4 apply to radicals as well. See page 285 for a summary of the rules of exponents.

Exponential Form of a Radical Expression

Power ↘ ↙ Index

$\sqrt[n]{a^m} = (\sqrt[n]{a})^m = a^{m/n}$, for $a \geq 0$ and m and n integers **Rule 5**

As long as the radicand is nonnegative, we can change from one form to another.

Examples

$\sqrt[3]{27^4} = (\sqrt[3]{27})^4 = 3^4 = 81$ $\sqrt[3]{x^3} = x^{3/3} = x^1 = x$

$8^{2/3} = (\sqrt[3]{8})^2 = 2^2 = 4$ $\sqrt[4]{y^8} = y^{8/4} = y^2$

EXAMPLE 3 Write each radical in exponential form.

a) $\sqrt[3]{a^5}$ **b)** $\sqrt[4]{y^7}$ **c)** $\sqrt[4]{x^{13}}$

Solution

a) $\sqrt[3]{a^5} = a^{5/3}$ **b)** $\sqrt[4]{y^7} = y^{7/4}$ **c)** $\sqrt[4]{x^{13}} = x^{13/4}$

Now Try Exercise 31

EXAMPLE 4 Simplify. **a)** $\sqrt[4]{x^{12}}$ **b)** $\sqrt[3]{y^{21}}$

Solution Write each radical expression in exponential form, then simplify.

a) $\sqrt[4]{x^{12}} = x^{12/4} = x^3$ **b)** $\sqrt[3]{y^{21}} = y^{21/3} = y^7$

Now Try Exercise 41

EXAMPLE 5 Evaluate. **a)** $8^{4/3}$ **b)** $16^{5/4}$ **c)** $8^{-2/3}$

Solution To evaluate, we write each exponential expression in radical form.

a) $8^{4/3} = (\sqrt[3]{8})^4 = 2^4 = 16$ **b)** $16^{5/4} = (\sqrt[4]{16})^5 = 2^5 = 32$

c) Recall from Section 4.2 that $x^{-m} = \dfrac{1}{x^m}$. Thus,

$$8^{-2/3} = \frac{1}{8^{2/3}} = \frac{1}{(\sqrt[3]{8})^2} = \frac{1}{2^2} = \frac{1}{4}$$

Now Try Exercise 71

AVOIDING COMMON ERRORS

Students may make mistakes simplifying expressions that contain negative exponents. Be careful when working such problems. The following is a common error.

CORRECT	INCORRECT
$27^{-2/3} = \dfrac{1}{27^{2/3}}$	~~$27^{-2/3} = -27^{2/3}$~~

The expression $27^{-2/3}$ simplifies to $\frac{1}{9}$.

EXAMPLE 6 Simplify. **a)** $\sqrt[4]{36^2}$ **b)** $\sqrt[6]{27^2}$

Solution Write each expression in exponential form, then simplify.

a) $\sqrt[4]{36^2} = 36^{2/4} = 36^{1/2} = \sqrt{36} = 6$

b) $\sqrt[6]{27^2} = 27^{2/6} = 27^{1/3} = \sqrt[3]{27} = 3$

Now Try Exercise 79

EXAMPLE 7 Simplify and write the answer in radical form.

a) $\sqrt[8]{y^4}$ **b)** $\sqrt[9]{z^3}$

Solution Write the expression in exponential form, then simplify.

a) $\sqrt[8]{y^4} = y^{4/8} = y^{1/2} = \sqrt{y}$ **b)** $\sqrt[9]{z^3} = z^{3/9} = z^{1/3} = \sqrt[3]{z}$

Now Try Exercise 53

EXAMPLE 8 Simplify. **a)** $\sqrt{x} \cdot \sqrt[4]{x}$ **b)** $\left(\sqrt[4]{a^2}\right)^6$

Solution To simplify, we change each radical expression to exponential form, then apply the rules of exponents.

a)
$$\begin{aligned}\sqrt{x} \cdot \sqrt[4]{x} &= x^{1/2} \cdot x^{1/4}\\ &= x^{(1/2)+(1/4)}\\ &= x^{(2/4)+(1/4)}\\ &= x^{3/4} \quad \left(\text{or } \sqrt[4]{x^3} \text{ in radical form}\right)\end{aligned}$$

b)
$$\begin{aligned}\left(\sqrt[4]{a^2}\right)^6 &= \left(a^{2/4}\right)^6\\ &= \left(a^{1/2}\right)^6\\ &= a^3\end{aligned}$$

Now Try Exercise 91

Summary of Rules for Radicals

Rule 1: Product Rule for Square Roots

$$\sqrt{ab} = \sqrt{a} \cdot \sqrt{b} \text{ and } \sqrt{a} \cdot \sqrt{b} = \sqrt{ab}, \text{ provided } a \geq 0, b \geq 0$$

Rule 2: Square Root of a Perfect Square

$$\sqrt{a^{2 \cdot n}} = a^n, a \geq 0$$

Rule 3: Quotient Rule for Square Roots

$$\sqrt{\frac{a}{b}} = \frac{\sqrt{a}}{\sqrt{b}} \text{ and } \frac{\sqrt{a}}{\sqrt{b}} = \sqrt{\frac{a}{b}}, \text{ provided } a \geq 0, b > 0$$

Rule 4: Exponential Form of a Radical Expression

$$\sqrt[n]{a} = a^{1/n}, \quad a \geq 0 \text{ and } n \text{ is even}$$
$$\sqrt[n]{a} = a^{1/n}, \quad \text{when } n \text{ is odd}$$

Rule 5: Exponential Form of a Radical Expression

Power ↘ ↙ Index

$$\sqrt[n]{a^m} = \left(\sqrt[n]{a}\right)^m = a^{m/n}, \text{ for } a \geq 0 \text{ and } m \text{ and } n \text{ integers}$$

9.7 Exercise Set MathXL® MyMathLab®

In this exercise set, assume all variables represent nonnegative real numbers.

Warm-Up Exercises

Fill in the blanks with the appropriate word, phrase, or symbol(s) from the following list.

real	fourth	perfect	cube	positive
negative	index	power	imaginary	real

1. The radical expression $\sqrt[3]{a}$ is read "the ________ root of a."
2. The radical expression $\sqrt[4]{a}$ is read "the ________ root of a.".
3. The cube root of a positive number is a ________ number.
4. The cube root of a negative number is a ________ number.
5. The radicand of any even root must be a nonnegative number for the expression to be a ________ number.
6. In the exponential expression $a^{m/n}$, the m represents the ________.
7. In the exponential expression $a^{m/n}$, the n represents the ________.
8. The numbers 1, 8, 27, 64, 125, 216, ... are called ________ cubes.

Practice the Skills

Evaluate.

9. $\sqrt[3]{8}$
10. $\sqrt[3]{27}$
11. $\sqrt[3]{-27}$
12. $\sqrt[3]{-8}$
13. $-\sqrt[3]{-8}$
14. $-\sqrt[3]{-27}$
15. $\sqrt[3]{216}$
16. $\sqrt[3]{-1000}$
17. $\sqrt[4]{81}$
18. $\sqrt[4]{16}$
19. $\sqrt[4]{256}$
20. $\sqrt[4]{625}$

Simplify.

21. $\sqrt[3]{40}$
22. $\sqrt[3]{54}$
23. $\sqrt[3]{16}$
24. $\sqrt[3]{81}$
25. $\sqrt[3]{128}$
26. $\sqrt[3]{192}$
27. $\sqrt[4]{48}$
28. $\sqrt[4]{80}$
29. $\sqrt[5]{96}$
30. $\sqrt[5]{160}$

Write each radical in exponential form.

31. $\sqrt[3]{x^7}$
32. $\sqrt[3]{x^4}$
33. $\sqrt[5]{a^3}$
34. $\sqrt[5]{x^{11}}$
35. $\sqrt[4]{y^{15}}$
36. $\sqrt[4]{w^{13}}$
37. $\sqrt[3]{y^8}$
38. $\sqrt[4]{x^7}$

Write each radical in exponential form and then simplify.

39. $\sqrt[3]{x^3}$
40. $\sqrt[3]{y^9}$
41. $\sqrt[3]{x^{15}}$
42. $\sqrt[3]{m^{21}}$
43. $\sqrt[4]{x^4}$
44. $\sqrt[4]{m^8}$
45. $\sqrt[4]{a^{24}}$
46. $\sqrt[4]{a^{60}}$
47. $\sqrt[5]{k^5}$
48. $\sqrt[5]{p^{20}}$

Write each radical in exponential form and then simplify. Write the answer in simplified radical form.

49. $\sqrt[4]{m^2}$
50. $\sqrt[8]{y^4}$
51. $\sqrt[10]{c^5}$
52. $\sqrt[18]{z^6}$
53. $\sqrt[9]{x^3}$
54. $\sqrt[12]{r^3}$
55. $\sqrt[14]{k^2}$
56. $\sqrt[16]{z^4}$
57. $\sqrt[6]{x^4}$
58. $\sqrt[12]{p^9}$
59. $\sqrt[8]{z^6}$
60. $\sqrt[15]{k^{10}}$

Write each expression in radical form and then evaluate.

61. $4^{5/2}$
62. $25^{3/2}$
63. $9^{3/2}$
64. $64^{3/2}$
65. $8^{4/3}$
66. $27^{2/3}$
67. $64^{2/3}$
68. $125^{4/3}$
69. $8^{-1/3}$
70. $16^{-3/4}$
71. $27^{-4/3}$
72. $64^{-2/3}$
73. $125^{-2/3}$
74. $216^{-2/3}$
75. $256^{-3/4}$
76. $625^{-3/4}$

Write each radical in exponential form and then simplify.

77. $\sqrt[4]{9^2}$
78. $\sqrt[4]{49^2}$
79. $\sqrt[6]{4^3}$
80. $\sqrt[6]{25^3}$

81. $\sqrt[8]{64^4}$ **82.** $\sqrt[8]{81^4}$ **83.** $\sqrt[4]{7^8}$ **84.** $\sqrt[4]{8^8}$

85. $\sqrt[4]{3^{12}}$ **86.** $\sqrt[6]{2^{18}}$ **87.** $\sqrt[5]{6^{10}}$ **88.** $\sqrt[5]{5^{15}}$

Write each radical in exponential form and then simplify. Write the answer in exponential form.

89. $\sqrt[4]{x}\cdot\sqrt[4]{x^3}$ **90.** $\sqrt[3]{x^4}\cdot\sqrt[3]{x^5}$ **91.** $\sqrt[3]{x}\cdot\sqrt[4]{x}$ **92.** $\sqrt[4]{x}\sqrt[5]{x}$

93. $\left(\sqrt[3]{r^4}\right)^6$ **94.** $\left(\sqrt[4]{x^3}\right)^{12}$ **95.** $\left(\sqrt[4]{x^8}\right)^3$ **96.** $\left(\sqrt[3]{x^6}\right)^2$

97. Show that $\left(\sqrt[3]{x}\right)^2 = \sqrt[3]{x^2}$ for $x = 8$. **98.** Show that $\left(\sqrt[4]{x}\right)^3 = \sqrt[4]{x^3}$ for $x = 16$.

Problem Solving

The product $\sqrt[3]{2}\cdot\sqrt[3]{2}$ is not an integer, but $\sqrt[3]{2}\cdot\sqrt[3]{2^2} = \sqrt[3]{2^3} = 2^{3/3} = 2$. Use this information to determine by what radical expression you can multiply each radical so that the result is an integer.

99. $\sqrt[3]{3}$ **100.** $\sqrt[3]{5^2}$ **101.** $\sqrt[3]{6}$

102. $\sqrt[4]{11^3}$ **103.** $\sqrt[4]{7}$ **104.** $\sqrt[4]{3^2}$

Fill in the shaded area(s) in each equation to make a true statement.

105. $\sqrt[\square]{6^2}\cdot\sqrt[\square]{6} = 6$ **106.** $\sqrt[\square]{7^2}\cdot\sqrt[\square]{7^2} = 7$ **107.** $\sqrt[3]{5^{\square}}\cdot\sqrt[3]{5^2} = 5$ **108.** $\sqrt[5]{2^3}\cdot\sqrt[5]{2^{\square}} = 2$

Challenge Problems

Simplify.

109. $\sqrt[3]{xy}\cdot\sqrt[3]{x^2y^2}$ **110.** $\sqrt[4]{3x^2y}\cdot\sqrt[4]{27x^6y^3}$ **111.** $\sqrt[4]{32} - \sqrt[4]{2}$ **112.** $\sqrt[3]{3x^3y} + \sqrt[3]{24x^3y}$

113. a) Explain why multiplying both the numerator and denominator of $\dfrac{1}{\sqrt[3]{2}}$ by $\sqrt[3]{2^2}$ will give an integer in the denominator.

b) Rationalize the denominator of $\dfrac{1}{\sqrt[3]{2}}$ by multiplying both the numerator and denominator by $\sqrt[3]{2^2}$.

Cumulative Review Exercises

[1.9] **114.** Evaluate $-x^2 + 4xy - 8$ when $x = 2$ and $y = -4$.

[5.4] **115.** Factor $3x^2 - 28x + 32$.

[7.4] **116.** Determine the slope and the y-intercept of the graph of the equation $2x - 3y = 4$.

[9.4] **117.** Simplify $\sqrt{\dfrac{36x^3y^7}{2x^4}}$.

Chapter 9 Summary

IMPORTANT FACTS AND CONCEPTS	EXAMPLES
Section 9.1	
The **positive** or **principal square root** of a positive number a is written as $\sqrt{a}$. The negative square root is written as $-\sqrt{a}$. $\sqrt{a} = b$ if $b^2 = a$ The $\sqrt{\ }$ is called the **radical sign**. The number or expression inside the radical sign is called the **radicand**.	$\sqrt{36} = 6$ since $6\cdot 6 = 36$ $\sqrt{\dfrac{25}{49}} = \dfrac{5}{7}$ since $\dfrac{5}{7}\cdot\dfrac{5}{7} = \dfrac{25}{49}$ In the radical expression $\sqrt{50}$, the 50 is the radicand.
If $a \geq 0$, then $\sqrt{a}$ is a real number. If $a < 0$, then $\sqrt{a}$ is an imaginary number.	$\sqrt{31}$, $\sqrt{19.42}$ are real numbers. $\sqrt{-101}$, $\sqrt{-1.63}$ are imaginary numbers.
The square of a natural number is a perfect square. The square root of a perfect square is an integer.	$1, 4, 9, 16, 25, 36, \ldots$ are perfect squares $\sqrt{16} = 4$, $\sqrt{81} = 9$

IMPORTANT FACTS AND CONCEPTS	EXAMPLES
Section 9.1 (Continued)	
A **rational number** is a number that can be written in the form $\frac{a}{b}$, where a and b are integers and $b \neq 0$. When written as a decimal, a rational number is either a **terminating** or **repeating decimal**.	$\frac{1}{4} = 0.25$ and $\frac{21}{99} = 0.212121\ldots = 0.\overline{21}$ are rational numbers.
Irrational numbers are real numbers that are not rational.	$\sqrt{15}$, $\sqrt{23}$ are irrational numbers.
When writing a square root in exponential form, $\sqrt{a} = a^{1/2}$	$\sqrt{36} = 36^{1/2}$, $\sqrt{3x} = (3x)^{1/2}$
Section 9.2	
Rule 1: Product Rule for Square Roots $\sqrt{ab} = \sqrt{a}\cdot\sqrt{b}$ and $\sqrt{a}\cdot\sqrt{b} = \sqrt{ab}$, provided $a \geq 0, b \geq 0$	$\sqrt{60} = \sqrt{4}\cdot\sqrt{15}$ $\sqrt{5}\cdot\sqrt{7} = \sqrt{35}$
To Simplify the Square Root of a Constant 1. Write the radicand as a product of the largest perfect square factor and another factor. 2. Use the product rule to write the expression as a product of square roots, with each square root containing one of the factors. 3. Find the square root of the perfect square factor.	Simplify $\sqrt{54}$. $\sqrt{54} = \sqrt{9\cdot 6} = \sqrt{9}\cdot\sqrt{6} = 3\sqrt{6}$
Rule 2: Square Root of a Perfect Square $\sqrt{a^{2n}} = a^n, a \geq 0$	$\sqrt{x^6} = x^3$
To Simplify the Square Root of a Radicand Containing a Variable Raised to an Odd Power 1. Express the radicand as the product of two factors, one of which has an exponent of 1 (the other will therefore have an even exponent). 2. Use the product rule to simplify.	$\sqrt{75y^7} = \sqrt{25y^6\cdot 3y} = \sqrt{25y^6}\cdot\sqrt{3y}$ $= 5y^3\sqrt{3y}$
Section 9.3	
Like square roots are square roots having the same radicands. To add, combine like square roots.	$3\sqrt{5} + 8\sqrt{5} + \sqrt{5} = 12\sqrt{5}$ $2\sqrt{x} + y + 7\sqrt{x} = 9\sqrt{x} + y$
Unlike square roots are square roots having different radicands. Sometimes unlike square roots can be changed to like square roots by simplifying the radicals in the expression.	$3\sqrt{2} + \sqrt{32} = 3\sqrt{2} + \sqrt{16}\cdot\sqrt{2}$ $= 3\sqrt{2} + 4\sqrt{2}$ $= 7\sqrt{2}$
To multiply radical expressions, use the distributive property.	$\sqrt{5x}\left(\sqrt{x} - \sqrt{3y}\right) = \left(\sqrt{5x}\right)\left(\sqrt{x}\right) - \left(\sqrt{5x}\right)\left(\sqrt{3y}\right)$ $= \sqrt{5x^2} - \sqrt{15xy}$ $= x\sqrt{5} - \sqrt{15xy}$
The FOIL method can be used to multiply radical expressions that contain sums or differences.	$\left(2 + \sqrt{3}\right)\left(5 - 4\sqrt{3}\right) = 2\cdot 5 + 2\left(-4\sqrt{3}\right) + \sqrt{3}(5)$ $+ \left(\sqrt{3}\right)\left(-4\sqrt{3}\right)$ $= 10 - 8\sqrt{3} + 5\sqrt{3} - 12$ $= -2 - 3\sqrt{3}$

IMPORTANT FACTS AND CONCEPTS	EXAMPLES
Section 9.4	
A Square Root Is Simplified When **1.** No radicand has a factor that is a perfect square. **2.** No radicand contains a fraction. **3.** No denominator contains a square root.	**1.** $\sqrt{32}$ is not simplified. **2.** $\sqrt{\frac{1}{3}}$ is not simplified. **3.** $\frac{1}{\sqrt{11}}$ is not simplified.
Rule 3: Quotient Rule for Square Roots $\sqrt{\frac{a}{b}} = \frac{\sqrt{a}}{\sqrt{b}}$ and $\frac{\sqrt{a}}{\sqrt{b}} = \sqrt{\frac{a}{b}}$, provided $a \geq 0, b > 0$	$\sqrt{\frac{7}{4}} = \frac{\sqrt{7}}{\sqrt{4}} = \frac{\sqrt{7}}{2}$ $\frac{\sqrt{35}}{\sqrt{5}} = \sqrt{\frac{35}{5}} = \sqrt{7}$
To **rationalize a denominator** means to remove all radicals from the denominator. To rationalize a denominator with a square root, multiply *both* the numerator and the denominator of the fraction by the square root that appears in the denominator or by the square root of a number that makes the denominator a perfect square.	$\frac{1}{\sqrt{7}} = \frac{1}{\sqrt{7}} \cdot \frac{\sqrt{7}}{\sqrt{7}} = \frac{\sqrt{7}}{7}$ $\frac{9}{\sqrt{c}} = \frac{9}{\sqrt{c}} \cdot \frac{\sqrt{c}}{\sqrt{c}} = \frac{9\sqrt{c}}{c}$
The **conjugate** of a radical expression of the form $a + b$ is $a - b$. To rationalize a denominator that has a radical expression with a sum or difference, multiply the numerator and denominator by the conjugate of the denominator.	$13 - \sqrt{2}$ is the conjugate of $13 + \sqrt{2}$ $-\sqrt{5} + \sqrt{3}$ is the conjugate of $-\sqrt{5} - \sqrt{3}$ $\frac{9}{5+\sqrt{2}} = \frac{9}{5+\sqrt{2}} \cdot \frac{5-\sqrt{2}}{5-\sqrt{2}} = \frac{9(5-\sqrt{2})}{25-2} = \frac{45-9\sqrt{2}}{23}$ $\frac{y}{\sqrt{x}+y} = \frac{y}{\sqrt{x}+y} \cdot \frac{\sqrt{x}-y}{\sqrt{x}-y} = \frac{y(\sqrt{x}-y)}{x-y^2} = \frac{y\sqrt{x}-y^2}{x-y^2}$
Section 9.5	
A **radical equation** is an equation that contains a variable in a radicand.	$\sqrt{x} = 5, \quad \sqrt{y - 10} = 8$
To Solve a Radical Equation Containing Only One Square Root Terms **1.** Rewrite the equation with the square root term by itself on one side of the equation. **2.** Combine like terms. **3.** Square both sides of the equation to eliminate the square root. **4.** Solve the equation for the variable. **5.** Check the solution in the *original* equation for extraneous roots.	Solve $\sqrt{x} + 4 = 6$. $\sqrt{x} + 4 = 6$ $\sqrt{x} = 2$ $(\sqrt{x})^2 = 2^2$ $x = 4$ A check shows that 4 is the solution.
Section 9.6	
Pythagorean Theorem In a right triangle, if a and b represent the legs, and c represents the hypotenuse, then $a^2 + b^2 = c^2$.	Find the hypotenuse of a right triangle whose legs are $\sqrt{15}$ and 3. $a^2 + b^2 = c^2$ $(\sqrt{15})^2 + 3^2 = c^2$ $15 + 9 = c^2$ $24 = c^2$ $\sqrt{24} = c$ or $c = 2\sqrt{6}$
Distance Formula The distance between two points (x_1, y_1) and (x_2, y_2) is $d = \sqrt{(x_2 - x_1)^2 + (y_2 - y_1)^2}$	The length of the line segment joining the points $(-3, 4)$ and $(1, 7)$ is $d = \sqrt{[1 - (-3)]^2 + (7 - 4)^2}$ $= \sqrt{4^2 + 3^2} = \sqrt{16 + 9} = \sqrt{25} = 5$

IMPORTANT FACTS AND CONCEPTS	EXAMPLES
Section 9.7	
The **cube root of a** is $\sqrt[3]{a}$. $\sqrt[3]{a} = b$ if $b^3 = a$ The **fourth root of a** is $\sqrt[4]{a}$. $\sqrt[4]{a} = b$ if $b^4 = a, a \geq 0$ $\sqrt[4]{a}$ is not a real number if $a < 0$.	$\sqrt[3]{27} = 3, \sqrt[3]{-27} = -3$ $\sqrt[4]{16} = 2,$ $\sqrt[4]{-16}$ is not a real number
Product Rule for Radicals When n is even: $\sqrt[n]{ab} = \sqrt[n]{a} \cdot \sqrt[n]{b}$ and $\sqrt[n]{a} \cdot \sqrt[n]{b} = \sqrt[n]{ab}$, provided $a \geq 0, b \geq 0$ When n is odd: $\sqrt[n]{ab} = \sqrt[n]{a} \cdot \sqrt[n]{b}$ and $\sqrt[n]{a} \cdot \sqrt[n]{b} = \sqrt[n]{ab}$	$\sqrt[4]{96} = \sqrt[4]{16 \cdot 6} = \sqrt[4]{16} \cdot \sqrt[4]{6} = 2\sqrt[4]{6}$ $\sqrt[3]{4} \cdot \sqrt[3]{2} = \sqrt[3]{8} = 2$
Rule 4: Exponential Form of a Radical Expression $\sqrt[n]{a} = a^{1/n}$, for $a \geq 0$ and n is even $\sqrt[n]{a} = a^{1/n}$, when n is odd **Rule 5: Exponential Form of a Radical Expression** $\sqrt[n]{a^m} = \left(\sqrt[n]{a}\right)^m = a^{m/n}$, for $a \geq 0$ and m and n integers	$\sqrt{z} = z^{1/2}, \sqrt[3]{7b^2} = (7b^2)^{1/3}$ **Rule 4** $\sqrt[3]{27^2} = \left(\sqrt[3]{27}\right)^2 = 3^2 = 9,$ $8^{4/3} = \left(\sqrt[3]{8}\right)^4 = 2^4 = 16$ **Rule 5** $\sqrt[4]{x^{20}} = x^{20/4} = x^5$
Negative Exponent Rule $x^{-m} = \dfrac{1}{x^m}$	$3^{-4} = \dfrac{1}{3^4} = \dfrac{1}{81}$

Chapter 9 Review Exercises

[9.1] *Evaluate.*

1. $\sqrt{81}$
2. $\sqrt{144}$
3. $-\sqrt{100}$

Write in exponential form.

4. $\sqrt{21}$
5. $\sqrt{43z}$
6. $\sqrt{58x^2y}$

[9.2] *Simplify.*

7. $\sqrt{98}$
8. $\sqrt{56}$
9. $\sqrt{48x^7y^5}$
10. $\sqrt{125x^4y^8}$
11. $\sqrt{52ab^5c^4}$
12. $\sqrt{72a^2b^2c^7}$

Simplify.

13. $\sqrt{64}\sqrt{10}$
14. $\sqrt{7a}\sqrt{7a}$
15. $\sqrt{32x}\sqrt{2xy}$
16. $\sqrt{25x^2y}\sqrt{3y}$
17. $\sqrt{12a^3b^4}\sqrt{3b^4}$
18. $\sqrt{2ab^3}\sqrt{50ab^4}$

[9.3] *Simplify.*

19. $9\sqrt{3} - 5\sqrt{3}$
20. $7\sqrt{5} - 6\sqrt{5} - 3\sqrt{5}$
21. $21\sqrt{x} - 5\sqrt{x}$
22. $\sqrt{k} + 4\sqrt{k} - 8\sqrt{k}$
23. $3\sqrt{18} - \sqrt{27}$
24. $8\sqrt{40} - 3\sqrt{10}$
25. $2\sqrt{98} - 4\sqrt{72}$
26. $7\sqrt{50} + 2\sqrt{18} - 4\sqrt{32}$

Multiply.

27. $\sqrt{7}(2 + \sqrt{7})$
28. $\sqrt{3}(\sqrt{3} + 11)$
29. $\sqrt{y}(x - 9\sqrt{y})$
30. $4a(3a + \sqrt{2a})$
31. $(\sqrt{7} - 8)(\sqrt{7} + 8)$
32. $(10 - \sqrt{5})(10 + \sqrt{5})$
33. $(x - 5\sqrt{y})(x + 5\sqrt{y})$
34. $(\sqrt{c} - 4\sqrt{d})(\sqrt{c} + 4\sqrt{d})$
35. $(m + 2\sqrt{r})(m - 5\sqrt{r})$
36. $(\sqrt{t} + 2s)(3\sqrt{t} - s)$
37. $(\sqrt{5m} + 6\sqrt{n})(3\sqrt{5m} - \sqrt{n})$
38. $(\sqrt{7} - 3\sqrt{p})(2\sqrt{7} - 3\sqrt{p})$

[9.4] *Simplify.*

39. $\frac{\sqrt{80}}{\sqrt{5}}$
40. $\sqrt{\frac{5}{245}}$
41. $\sqrt{\frac{3}{75}}$
42. $\frac{2}{\sqrt{3}}$
43. $\sqrt{\frac{n}{11}}$
44. $\sqrt{\frac{10a}{24}}$
45. $\sqrt{\frac{x^2}{7}}$
46. $\sqrt{\frac{z^6}{8}}$
47. $\sqrt{\frac{42x^3y^7}{6x^3y^3}}$
48. $\sqrt{\frac{34x^4y}{17x^2y^4}}$
49. $\frac{\sqrt{60}}{\sqrt{27a^3b^2}}$
50. $\frac{\sqrt{2a^4bc^4}}{\sqrt{7a^5bc^2}}$

Simplify.

51. $\frac{2}{1 - \sqrt{6}}$
52. $\frac{4}{4 - \sqrt{5}}$
53. $\frac{\sqrt{3}}{2 + \sqrt{y}}$
54. $\frac{3}{\sqrt{x} - 8}$
55. $\frac{\sqrt{10}}{\sqrt{x} + \sqrt{3}}$
56. $\frac{\sqrt{11}}{\sqrt{5} - x}$

[9.5] *Solve.*

57. $\sqrt{x} = 9$
58. $\sqrt{g} = -2$
59. $\sqrt{h - 8} = 3$
60. $\sqrt{3x + 4} = 5$
61. $\sqrt{5x + 7} = \sqrt{3x + 11}$
62. $4\sqrt{x} - x = 4$
63. $\sqrt{x^2 + 7} = x + 1$
64. $\sqrt{4x + 8} - \sqrt{7x - 13} = 0$
65. $\sqrt{4p + 1} = 2p - 1$

[9.6] *Find each length indicated by x. Round answers to nearest hundredth.*

66.

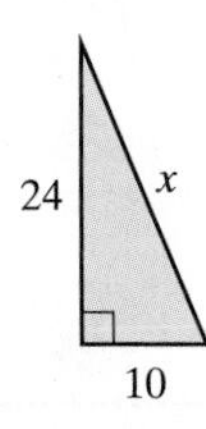

67.

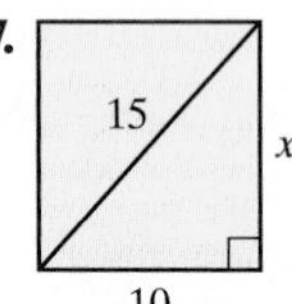

68.

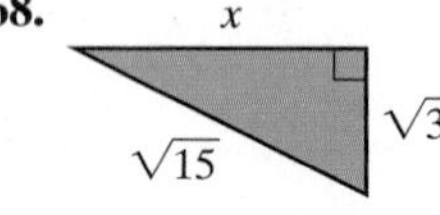

69.

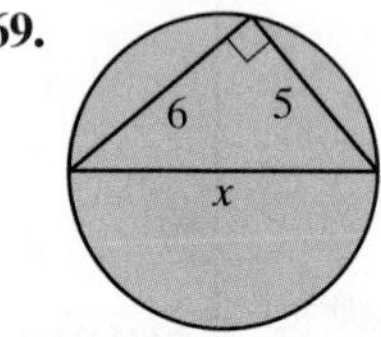

70. **Ladder on a Cottage** Adam Kurtz leans a 12-foot ladder against his grandfather's cottage. If the base of the ladder is 6 feet from the cottage, how high is the ladder on the cottage?

71. **Diagonal of a Rectangle** Find the diagonal of a rectangle of length 15 inches and width 10 inches.

72. **Distance Between Points** Find the distance between the points $(4, -3)$ and $(1, 7)$.

73. **Length of a Line Segment** Find the length of the line segment between the points $(6, 5)$ and $(-6, 8)$.

74. **Yield Sign** An **equilateral triangle** is a triangle with three sides of the same length. A yield sign is an equilateral triangle with each side, s, measuring 36 inches. Use the formula for the area of an equilateral triangle

$$\text{Area} = \frac{s^2\sqrt{3}}{4}$$

to find an approximation of the area of a yield sign. Round the area to the nearest hundredth.

75. **Distance Seen** The farthest distance, in miles, that a person looking at the horizon can see is approximated by the formula *distance* $= \sqrt{(3/2)h}$, where h represents the height of the vantage point measured in feet. Approximately how far can a person see, correct to the nearest hundredth of a mile, if the person is at a vantage point 40 feet high?

76. **Luxor Hotel** The Luxor Hotel in Las Vegas, Nevada, is the world's second largest hotel. It is a pyramid with a square base and a height of 350 feet. The length, s, of each side of the base of any pyramid with a square base can be found by the formula $s = \sqrt{(3V)/h}$, where V represents the volume, in cubic feet, of the pyramid and h represents the height, in feet. The volume of the Luxor is approximately 48,686,866.67 cubic feet. Find the length of each side of the square base of the Luxor Hotel.

Luxor Hotel in Las Vegas, Nevada

[9.7] *Evaluate.*

77. $\sqrt[3]{8}$ **78.** $\sqrt[3]{-64}$ **79.** $\sqrt[4]{16}$ **80.** $\sqrt[4]{81}$

Simplify.

81. $\sqrt[3]{16}$ **82.** $\sqrt[3]{32}$ **83.** $\sqrt[3]{56}$ **84.** $\sqrt[4]{32}$

85. $\sqrt[3]{54}$ **86.** $\sqrt[4]{80}$ **87.** $\sqrt[3]{x^{21}}$ **88.** $\sqrt[4]{s^{48}}$

Evaluate.

89. $27^{2/3}$ **90.** $4^{7/2}$ **91.** $27^{-2/3}$ **92.** $64^{4/3}$ **93.** $125^{-4/3}$ **94.** $9^{5/2}$

Write in exponential form.

95. $\sqrt{x^5}$ **96.** $\sqrt{a^7}$ **97.** $\sqrt[3]{z^{11}}$ **98.** $\sqrt[3]{a^{13}}$ **99.** $\sqrt[4]{b^{17}}$ **100.** $\sqrt[4]{m^6}$

Simplify.

101. $\sqrt[3]{x} \cdot \sqrt[3]{x^2}$ **102.** $\sqrt[4]{x} \cdot \sqrt[4]{x}$ **103.** $\sqrt[3]{a^4} \cdot \sqrt[3]{a^8}$ **104.** $\sqrt[4]{x^2} \cdot \sqrt[4]{x^6}$

105. $(\sqrt[3]{q^3})^3$ **106.** $(\sqrt[4]{b^2})^4$ **107.** $(\sqrt[4]{x^8})^3$ **108.** $(\sqrt[4]{x^5})^8$

Chapter 9 Practice Test

Chapter Test Prep Videos provide fully worked-out solutions to any of the exercises you want to review. Chapter Test Prep Videos are available via MyMathLab®, *or on* YouTube (www.youtube.com/user/AngelElementaryAlg)

1. Write $\sqrt{5x}$ in exponential form.

2. Write $x^{3/4}$ in radical form.

Simplify.

3. $\sqrt{169}$ **4.** $\sqrt{90}$

5. $\sqrt{12x^2}$ **6.** $\sqrt{75x^7y^3}$

7. $\sqrt{4x^2y}\sqrt{20xy}$ **8.** $\sqrt{10xy^2}\sqrt{5x^3y^3}$

9. $\sqrt{\dfrac{5}{125}}$ **10.** $\dfrac{\sqrt{3c^4d}}{\sqrt{3d^3}}$

11. $\dfrac{1}{\sqrt{5}}$ **12.** $\sqrt{\dfrac{9r}{5}}$

13. $\sqrt{\dfrac{40x^2y^5}{3x^3y^7}}$ **14.** $\dfrac{6}{2 - \sqrt{7}}$

15. $\dfrac{7}{\sqrt{x} - 3}$ **16.** $\sqrt{48} + 5\sqrt{12} + 2\sqrt{3}$

17. $9\sqrt{y} - 3\sqrt{y} - \sqrt{y}$

Solve.

18. $\sqrt{x - 10} = 4$ **19.** $2\sqrt{x - 4} + 4 = x$

Solve.

20. Find the value of x in the right triangle.

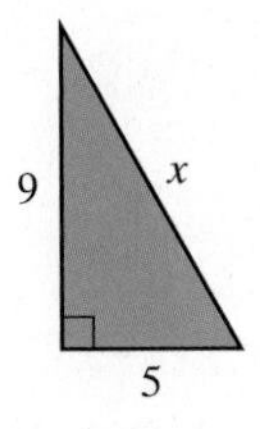

21. Find the length of the line segment between the points $(3, 2)$ and $(-4, -1)$.

22. Evaluate $27^{-4/3}$.

23. Simplify $\sqrt[4]{x^5} \cdot \sqrt[4]{x^7}$.

24. Side of a Square Find the side of a square whose area is 169 square meters.

25. Dropping Eggs Mary Ellen Baker and her brother, Michael, were in the tree house in their backyard. They were dropping raw eggs (much to their mother's dismay) from a ledge in the tree house 10 feet in the air. At what velocity did the eggs hit the ground? Use the formula $v = \sqrt{2gh}$ with $g = 32$ feet per second squared. The velocity will be in feet per second.

Cumulative Review Test

Take the following test and check your answers with those given in the back of the book. Review any questions that you answered incorrectly. The section where the material was covered is indicated after the answer.

1. Consider the set of numbers

$$\left\{-5, \sqrt{12}, 735, 0.5, 4, \frac{1}{2}\right\}.$$

List those that are

a) integers;

b) whole numbers;

c) rational numbers;

d) irrational numbers;

e) real numbers.

2. Evaluate $7a^2 - 4b^2 + 3ab$ when $a = -3$ and $b = 2$.

3. Solve $-7(3 - x) = 4(x + 2) - 3x$.

4. Solve the inequality $3(x + 2) > 5 - 4(2x - 7)$. Graph the solution on a number line and write the solution using interval notation.

5. Factor $4x^3 + x^2 + 12x + 3$.

6. Factor $2x^2 - 21x + 27$.

7. Use factoring to solve $r^2 - 8r = 0$.

8. Simplify $\dfrac{4a^3b^{-5}}{20a^8b}$.

9. Graph $4x - 6y = 24$.

10. Write the equation of the graph below.

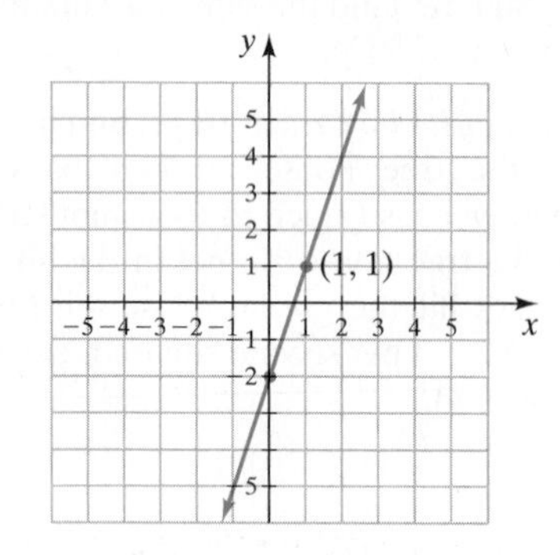

11. Find the equation of the line, in slope–intercept form, that has a slope of $\frac{2}{5}$ and goes through the point $(5,6)$.

12. Solve the following system of equations.

$$-2x + 3y = 6$$
$$4x - 2y = -4$$

13. Simplify $\dfrac{y + 9}{8} + \dfrac{2y - 14}{8}$.

14. Solve $\dfrac{3}{5} + \dfrac{1}{z} = 2$.

15. Simplify $12\sqrt{11} - 5\sqrt{11}$.

16. Simplify $\sqrt{\dfrac{3z}{28y^5}}$.

17. Solve $\sqrt{x + 10} = 6$.

18. Making a Fruitcake Susan Effing is planning her first holiday meal. She has decided to make her grandmother's famous fruitcake. The recipe calls for 10 cups of flour and will make an 11-pound fruitcake. She would like to make a 3-pound fruitcake. How many cups of flour should Susan use?

19. Special Promotion During a special promotion, the rooms in the Hyatt Hotel in Oklahoma City cost \$75. If this is a 40% discount off the hotel's regular room rate, determine the Hyatt's regular room rate.

20. Distance between Cities Alan Heard is in the process of getting his pilot's license. As part of his training, he flew his instructor from a small airport near Pasadena, California, to another small airport near San Diego at an average speed of 100 miles per hour. On the return trip, his average speed was 125 miles per hour. How far apart were the two airports if Alan's total flying time was 2 hours?

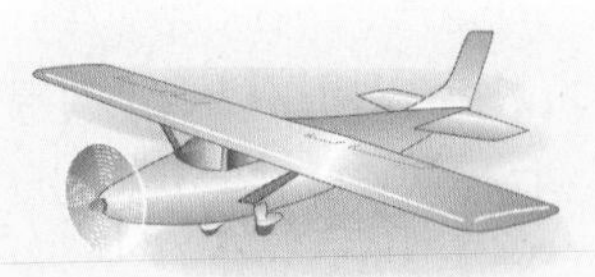

10 Quadratic Equations

We see objects projected upward often but rarely think of the mathematics that describes the projected motion. For example, when a ball hits a bat, or a child kicks a football or a soccer ball, the ball is projected upward. In Exercise 62 on page 603, we will determine the height of a ball projected upward.

Goals of This Chapter

Quadratic equations were introduced and solved by factoring in Section 5.6. Recall that quadratic equations are of the form $ax^2 + bx + c = 0$, where $a \neq 0$. Not every quadratic equation can be solved by factoring. In this chapter we present two additional methods to solve quadratic equations: completing the square and the quadratic formula. We also present additional applications of quadratic equations.

10.1 The Square Root Property

1. Know that every positive real number has two square roots.
2. Solve quadratic equations using the square root property.

In Section 5.6 we solved quadratic equations by factoring. Recall that **quadratic equations** are equations of the form

$$ax^2 + bx + c = 0$$

where a, b, and c are real numbers, $a \neq 0$. A quadratic equation in this form is said to be in **standard form**. Solving quadratic equations by factoring is the preferred technique when the factors can be found quickly.

However, not every quadratic equation can be factored easily, and many cannot be factored at all. In this chapter we give two techniques, completing the square and the quadratic formula, for solving quadratic equations that cannot be solved by factoring.

1 Know That Every Positive Real Number Has Two Square Roots

In Section 9.1 we stated that every positive number has two square roots. For example, the positive or principal square root of 49 is 7.

$$\sqrt{49} = 7$$

The negative square root of 49 is -7.

$$-\sqrt{49} = -7$$

Understanding Algebra

When we refer to both square roots of a number we will use the symbol $\pm$. For example, the *square roots* of 25 can be written as $\pm\sqrt{25}$ or ± 5. The symbol ± 5 is read "plus or minus 5" and refers to both of the numbers 5 *and* -5.

The two square roots of 49 are $+7$ and -7. A convenient way to indicate the two square roots of a number is to use the plus or minus symbol, $\pm$. For example, the square roots of 49 can be indicated ± 7, read "plus or minus 7."

Number	Both Square Roots
64	± 8
100	± 10
3	$\pm\sqrt{3}$
7	$\pm\sqrt{7}$

An approximation of a number like $-\sqrt{5}$ can be found by evaluating $\sqrt{5}$ on your calculator and then taking its opposite or negative value.

$$\sqrt{5} \approx 2.24 \quad \text{(rounded to the nearest hundredth)}$$
$$-\sqrt{5} \approx -2.24$$

Now consider the equation

$$x^2 = 49$$

We can see by substitution that this equation has two solutions, 7 and -7.

Check

$x = 7$	$x = -7$
$x^2 = 49$	$x^2 = 49$
$7^2 \stackrel{?}{=} 49$	$(-7)^2 \stackrel{?}{=} 49$
$49 = 49$ True	$49 = 49$ True

Therefore, the solutions to the equation $x^2 = 49$ are 7 and -7 (or ± 7).

2 Solve Quadratic Equations Using the Square Root Property

In general, for any quadratic equation of the form $x^2 = a$, we can use the **square root property** to obtain the solution.

Square Root Property

If $x^2 = a$, then $x = \sqrt{a}$ or $x = -\sqrt{a}$ (abbreviated $x = \pm\sqrt{a}$).

For example, if $x^2 = 7$, then by the square root property, $x = \sqrt{7}$ or $x = -\sqrt{7}$. We may also write $x = \pm\sqrt{7}$.

EXAMPLE 1 Solve the equation $x^2 - 25 = 0$.

Solution Before we use the square root property we must isolate the squared variable. Add 25 to both sides of the equation to get the variable by itself on one side of the equation.

$$x^2 - 25 = 0$$
$$x^2 = 25 \quad \text{Added 25 to both sides.}$$
$$x = \pm\sqrt{25} \quad \text{Square root property}$$
$$x = \pm 5$$

Check in the original equation.

Check

$x = 5$	$x = -5$
$x^2 - 25 = 0$	$x^2 - 25 = 0$
$5^2 - 25 \stackrel{?}{=} 0$	$(-5)^2 - 25 \stackrel{?}{=} 0$
$25 - 25 \stackrel{?}{=} 0$	$25 - 25 \stackrel{?}{=} 0$
$0 = 0$ True	$0 = 0$ True

Now Try Exercise 11

EXAMPLE 2 Solve the equation $x^2 + 10 = 74$.

Solution

$$x^2 + 10 = 74$$
$$x^2 = 64 \quad \text{Subtracted 10 from both sides.}$$
$$x = \pm\sqrt{64} \quad \text{Square root property}$$
$$x = \pm 8$$

Now Try Exercise 13

EXAMPLE 3 Solve the equation $a^2 - 13 = 0$.

Solution

$$a^2 - 13 = 0$$
$$a^2 = 13 \quad \text{Added 13 to both sides.}$$
$$a = \pm\sqrt{13} \quad \text{Square root property}$$

Now Try Exercise 15

Understanding Algebra

When the solutions to an equation are not rational numbers, leave the solutions in radical form unless you are instructed to approximate the solutions. In Example 3, we leave the solutions as $a = \pm\sqrt{13}$.

EXAMPLE 4 Solve the equation $(x - 3)^2 = 4$.

Solution Begin by using the square root property.

$$(x - 3)^2 = 4$$
$$x - 3 = \pm\sqrt{4} \quad \text{Square root property}$$
$$x - 3 = \pm 2$$
$$x - 3 + 3 = 3 \pm 2 \quad \text{Add 3 to both sides.}$$
$$x = 3 \pm 2$$
$$x = 3 + 2 \quad \text{or} \quad x = 3 - 2$$
$$x = 5 \qquad\qquad x = 1$$

The solutions are 1 and 5.

Now Try Exercise 27

EXAMPLE 5 Solve the equation $(5x + 4)^2 - 2 = 16$.

Solution We must first isolate the squared term.

$$(5x + 4)^2 - 2 = 16$$

$$(5x + 4)^2 = 18 \qquad \text{Added 2 to both sides to isolate the squared term.}$$

$$5x + 4 = \pm\sqrt{18} \qquad \text{Square root property}$$

$$5x + 4 = \pm\sqrt{9}\sqrt{2} \qquad \text{Simplify } \sqrt{18}.$$

$$5x + 4 = \pm 3\sqrt{2}$$

$$5x + 4 - 4 = -4 \pm 3\sqrt{2} \qquad \text{Subtract 4 from both sides.}$$

$$5x = -4 \pm 3\sqrt{2}$$

$$x = \frac{-4 \pm 3\sqrt{2}}{5} \qquad \text{Divide both sides by 5.}$$

The solutions are $\frac{-4 + 3\sqrt{2}}{5}$ and $\frac{-4 - 3\sqrt{2}}{5}$.

Now Try Exercise 45

Now let's look at one of many applications of quadratic equations.

EXAMPLE 6 **Creating Advertisements** Antoinette LeMans designed a magazine advertisement for her company in the shape of a rectangle whose length is 1.62 times its width. Find the dimensions of the advertisement if it is to have an area of 20 square inches. See **Figure 10.1**.

Solution Understand and Translate

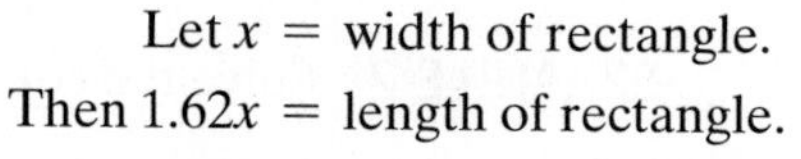

$$\text{Let } x = \text{width of rectangle.}$$

$$\text{Then } 1.62x = \text{length of rectangle.}$$

$$\text{area} = \text{length} \cdot \text{width}$$

$$20 = (1.62x)x$$

FIGURE 10.1

Carry Out

$$20 = 1.62x^2$$

$$\text{or} \quad 1.62x^2 = 20$$

$$x^2 = \frac{20}{1.62} \approx 12.3$$

$$x \approx \pm\sqrt{12.3} \approx \pm 3.51 \text{ inches}$$

Check and Answer Since the width cannot be negative, the width, x, is approximately 3.51 inches. The length is about $1.62(3.51) = 5.69$ inches.

Check

$$\text{area} = \text{length} \cdot \text{width}$$

$$20 \stackrel{?}{=} (5.69)(3.51)$$

$$20 \approx 19.97 \qquad \text{True}$$

(There is a slight round-off error due to rounding off decimal answers.)

Now Try Exercise 51

10.1 Exercise Set MathXL® MyMathLab®

Warm-Up Exercises

Fill in the blanks with the appropriate word, phrase, or symbol(s) from the following list.

root rational plus quadratic square cube

1. Equations that can be put into the form $ax^2 + bx + c = 0$ are called ________ equations.
2. Every positive number has two ________ roots.
3. The symbol $\pm$ is read ________ or minus.
4. If $x^2 = a$, then $x = \pm\sqrt{a}$ is known as the square ________ property.

Practice the Skills

Solve.

5. $x^2 = 81$

6. $x^2 = 4$

7. $y^2 = 169$

8. $z^2 = 225$

9. $x^2 = 75$

10. $x^2 = 98$

11. $x^2 - 64 = 0$

12. $x^2 - 49 = 0$

13. $x^2 + 3 = 103$

14. $x^2 - 8 = 56$

15. $x^2 + 10 = 30$

16. $w^2 + 20 = 44$

17. $7x^2 = 28$

18. $3x^2 = 48$

19. $3w^2 = 51$

20. $5k^2 = 75$

21. $3z^2 + 2 = 29$

22. $3x^2 - 4 = 8$

23. $9w^2 + 5 = 20$

24. $16x^2 - 7 = 66$

25. $3y^2 + 13 = 97$

26. $2x^2 + 15 = 63$

Solve.

27. $(x - 4)^2 = 1$

28. $(y - 2)^2 = 81$

29. $(a + 3)^2 = 36$

30. $(x + 5)^2 = 49$

31. $(x + 2)^2 = 25$

32. $(x + 4)^2 = 100$

33. $(r + 8)^2 = 32$

34. $(x + 5)^2 = 18$

35. $(d + 1)^2 = 45$

36. $(k - 11)^2 = 40$

37. $(n - 8)^2 = 64$

38. $(p - 7)^2 = 49$

39. $(3n + 5)^2 = 81$

40. $(7n - 2)^2 = 36$

41. $(3x - 8)^2 - 5 = 11$

42. $(5s - 6)^2 - 30 = 70$

43. $(2x + 3)^2 = 18$

44. $(3a - 2)^2 = 29$

45. $(2p - 7)^2 - 8 = 10$

46. $(5x + 9)^2 + 20 = 60$

Problem Solving

If necessary, round answers to two decimal places.

47. Product of Numbers The product of two positive numbers is 68. Determine the numbers if the larger number is 4.25 times the smaller number.

48. Product of Numbers The product of two positive numbers is 130. Determine the numbers if the larger number is 5.2 times the smaller number.

49. Newspaper Page The area of an opened newspaper page is 661.25 square inches. Determine the width and length of the page if the length is 1.25 times the width.

50. Table Top The area of a rectangular kitchen table top is 2400 square inches. Determine the width and length of the table top if the length is 1.5 times the width.

51. **Planting a Garden** Paul Martin decided to plant a rectangular garden so that the length is 1.6 times the width. Determine the length and width of the garden if it is to have an area of 2000 square feet.

52. Mailing Envelope The length of a priority mailing envelope is about 1.33 times its width. Determine the length and width of the envelope if its area is approximately 112 square inches.

53. Write an equation that has the solutions 7 and -7.

54. Write an equation that has the solutions $\sqrt{3}$ and $-\sqrt{3}$.

55. Fill in the shaded area to make a true statement. The equation $x^2 - \square = 17$ has the solutions 6 and -6.

56. Fill in the shaded area to make a true statement. The equation $x^2 + \square = 40$ has the solutions 8 and -8.

57. a) Rewrite $-3x^2 + 9x - 6 = 0$ so that the coefficient of the x^2-term is positive.

b) Rewrite $-3x^2 + 9x - 6 = 0$ so that the coefficient of the x^2-term is 1.

58. a) Rewrite $-\frac{1}{4}x^2 + 3x - 4 = 0$ so that the coefficient of the x^2-term is positive.

b) Rewrite $-\frac{1}{4}x^2 + 3x - 4 = 0$ so that the coefficient of the x^2-term is 1.

59. Comparing Areas Consider the two squares with sides x and $x + 3$ shown below.

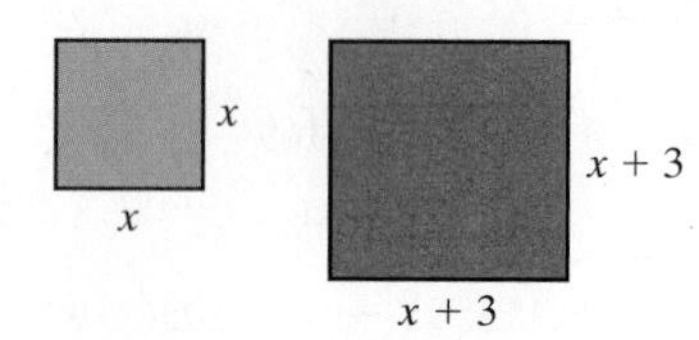

a) Write a quadratic expression for the area of each square.

b) If the area of the blue square is 36 square inches, what is the length of each side of the square?

c) If the area of the blue square is 50 square inches, what is the length of each side of the square?

d) If the area of the red square is 81 square inches, what is the length of each side of the square?

e) If the area of the red square is 92 square inches, what is the length of each side of the square?

60. Area Consider the figure below. If the area shaded in pink is approximately 153.94 square inches, find x (to the nearest hundredth).

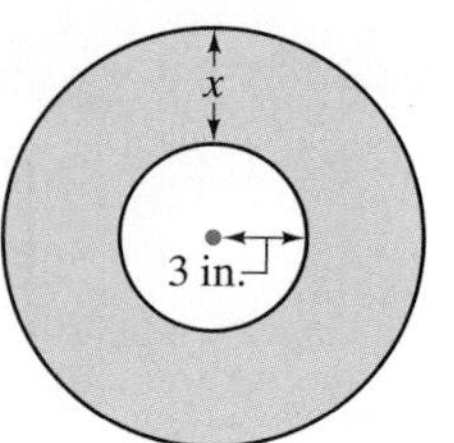

Challenge Problems

Use the square root property to solve for the indicated variable. Assume that all variables represent positive numbers. You may wish to review Section 2.6 before working these problems. List only the positive square root.

61. $A = s^2$, for s

62. $I = p^2r$, for p

63. $A = \pi r^2$, for r

64. $a^2 + b^2 = c^2$, for b

65. $I = \dfrac{k}{d^2}$, for d

66. $A = p(1 + r)^2$, for r

Cumulative Review Exercises

[5.4] **67.** Factor $6x^2 - 15x - 36$.

[6.5] **68.** Simplify $\dfrac{5 - \frac{1}{y}}{6 - \frac{1}{y}}$.

[7.4] **69.** Determine the equation of the line illustrated in the graph on the right.

[9.4] **70.** Simplify $\dfrac{\sqrt{135a^4b}}{\sqrt{3a^5b^7}}$.

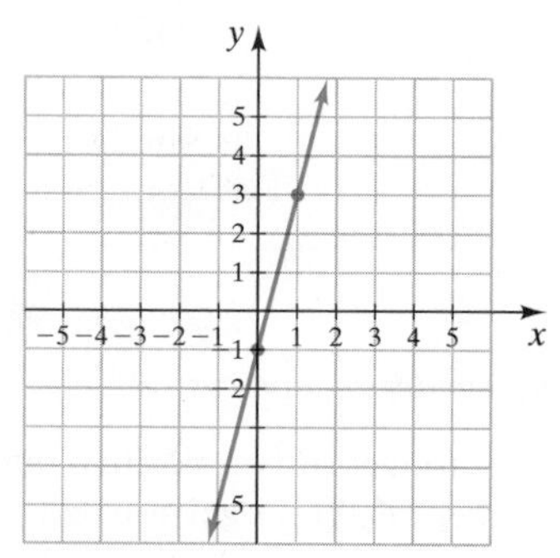

See Exercise 69.

10.2 Solving Quadratic Equations by Completing the Square

1 Write perfect square trinomials.

2 Solve quadratic equations by completing the square.

Quadratic equations that cannot be solved by factoring can be solved by completing the square or by the quadratic formula. In this section we focus on completing the square.

1 Write Perfect Square Trinomials

Understanding Algebra

A trinomial that can be written as a square of a binomial is called a *perfect square trinomial.*

Perfect Square Trinomial

A **perfect square trinomial** is a trinomial that can be expressed as the square of a binomial.

Some examples follow.

Perfect Square Trinomials		Factors		Square of a Binomial
$x^2 + 6x + 9$	=	$(x + 3)(x + 3)$	=	$(x + 3)^2$
$x^2 - 6x + 9$	=	$(x - 3)(x - 3)$	=	$(x - 3)^2$
$x^2 + 10x + 25$	=	$(x + 5)(x + 5)$	=	$(x + 5)^2$
$x^2 - 10x + 25$	=	$(x - 5)(x - 5)$	=	$(x - 5)^2$

In a perfect square trinomial, *when the coefficient of the squared term is 1, the constant term is the square of one-half the coefficient of the x-term.*

Consider the perfect square trinomial $x^2 + 6x + 9$.

$$x^2 + 6x + 9$$

6 is the coefficient of $6x$. 9 is the constant.

$$\left[\frac{1}{2}(6)\right]^2 = 3^2 = 9 \quad \text{Take } \frac{1}{2} \text{ of 6, and then square.}$$

Consider the perfect square trinomial $x^2 - 10x + 25$.

$$x^2 - 10x + 25$$

-10 is the coefficient of $-10x$. 25 is the constant.

$$\left[\frac{1}{2}(-10)\right]^2 = (-5)^2 = 25 \quad \text{Take } \frac{1}{2} \text{ of } -10\text{, and then square.}$$

Consider the expression $x^2 + 8x + \square$. Can you determine what number must be placed in the colored box to make the trinomial a perfect square trinomial? If you answered 16, you answered correctly.

$$x^2 + 8x + \square$$

$$\left[\frac{1}{2}(8)\right]^2 = 4^2 = 16 \quad \text{The constant is 16.}$$

The perfect square trinomial is $x^2 + 8x + 16$. Note that $x^2 + 8x + 16 = (x + 4)^2$.

Let's examine perfect square trinomials a little further.

Perfect Square Trinomial		Square of a Binomial
$x^2 + 6x + 9$	$=$	$(x + 3)^2$
	$\frac{1}{2}(6) = 3$	
$x^2 - 10x + 25$	$=$	$(x - 5)^2$
	$\frac{1}{2}(-10) = -5$	

Note that when a perfect square trinomial is written as the square of a binomial *the constant in the binomial is one-half the value of the coefficient of the x-term in the perfect square trinomial.*

2 Solve Quadratic Equations by Completing the Square

In the following example, we solve a quadratic equation by **completing the square**. Several of the examples in this section *could* be solved by factoring, but we will solve them by completing the square to illustrate the procedure before solving more difficult problems.

EXAMPLE 1 Solve the equation $x^2 + 6x - 7 = 0$ by completing the square.

Solution First note that the squared term has a coefficient of 1. Next, to get the terms containing a variable by themselves on the left side of the equation, we add 7 to both sides of the equation.

$$x^2 + 6x - 7 = 0$$
$$x^2 + 6x = 7$$

Now determine one-half the numerical coefficient of the x-term. In this example, the x-term is $6x$.

$$\frac{1}{2}(6) = 3$$

Square this number.

$$(3)^2 = (3)(3) = 9$$

Then add this product to both sides of the equation.

$$x^2 + 6x + 9 = 7 + 9$$

or

$$x^2 + 6x + 9 = 16$$

By following this procedure, we produce a perfect square trinomial on the left side of the equation. The expression $x^2 + 6x + 9$ is a perfect square trinomial that can be expressed as $(x + 3)^2$. Therefore,

$$x^2 + 6x + 9 = 16$$

can be written

$$(x + 3)^2 = 16$$

Now we use the square root property,

$$x + 3 = \pm\sqrt{16}$$
$$x + 3 = \pm 4$$

Finally, we solve for x by subtracting 3 from both sides of the equation.

$$x + 3 - 3 = -3 \pm 4$$
$$x = -3 \pm 4$$
$$x = -3 + 4 \quad \text{or} \quad x = -3 - 4$$
$$x = 1 \qquad\qquad x = -7$$

Thus, the solutions are 1 and -7. We check both solutions in the original equation.

Check $\quad x = 1 \qquad\qquad x = -7$

$$x^2 + 6x - 7 = 0 \qquad\qquad x^2 + 6x - 7 = 0$$
$$(1)^2 + 6(1) - 7 \stackrel{?}{=} 0 \qquad\qquad (-7)^2 + 6(-7) - 7 \stackrel{?}{=} 0$$
$$1 + 6 - 7 \stackrel{?}{=} 0 \qquad\qquad 49 - 42 - 7 \stackrel{?}{=} 0$$
$$0 = 0 \quad \text{True} \qquad\qquad 0 = 0 \quad \text{True}$$

Now Try Exercise 7

Understanding Algebra

In Example 1, the coefficient of the x-term is 6. We take $\frac{1}{2}$ of 6; this gives us 3. We then square 3 to get 9. We add 9 to both sides of the equation. This procedure will produce a perfect square trinomial on the left side of the equation.

Now let's summarize the procedure.

To Solve a Quadratic Equation by Completing the Square

1. Use the multiplication (or division) property of equality if necessary to make the numerical coefficient of the squared term equal to 1.
2. Rewrite the equation with the constant by itself on the right side of the equation.
3. Take one-half the numerical coefficient of the first-degree term, square it, and add this quantity to both sides of the equation.
4. Replace the trinomial with its equivalent squared binomial.
5. Use the square root property.
6. Solve for the variable.
7. Check your answers in the *original* equation.

EXAMPLE 2 Solve the equation $x^2 - 10x + 21 = 0$ by completing the square.

Solution

$$x^2 - 10x + 21 = 0$$
$$x^2 - 10x = -21 \quad \text{Step 2}$$

Take half the numerical coefficient of the x-term and then square it. You will add this product to both sides of the equation.

$$\frac{1}{2}(-10) = -5, \quad (-5)^2 = 25$$

Now add 25 to both sides of the equation.

$$x^2 - 10x + 25 = -21 + 25 \quad \text{Step 3}$$
$$x^2 - 10x + 25 = 4$$
$$(x - 5)^2 = 4 \quad \text{Step 4}$$
$$x - 5 = \pm\sqrt{4} \quad \text{Step 5}$$
$$x - 5 = \pm 2$$
$$x = 5 \pm 2 \quad \text{Step 6}$$
$$x = 5 + 2 \quad \text{or} \quad x = 5 - 2$$
$$x = 7 \qquad\qquad x = 3$$

A check will show that the solutions are 7 and 3.

Understanding Algebra

An important step in solving a quadratic equation by completing the square is to get the variable terms on the left side and the constant term on the right side of the equation. After that, complete the square with the terms on the left side.

Now Try Exercise 9

EXAMPLE 3 Solve the equation $x^2 = 3x + 18$ by completing the square.

Solution Begin by subtracting $3x$ from both sides of the equation.

$$x^2 = 3x + 18$$
$$x^2 - 3x = 18 \quad \text{Step 2}$$

Take half the numerical coefficient of the x-term, square it, and add this product to both sides of the equation.

$$\frac{1}{2}(-3) = -\frac{3}{2}, \left(-\frac{3}{2}\right)^2 = \frac{9}{4}$$

$$x^2 - 3x + \frac{9}{4} = 18 + \frac{9}{4} \quad \text{Step 3}$$
$$\left(x - \frac{3}{2}\right)^2 = 18 + \frac{9}{4} \quad \text{Step 4}$$
$$\left(x - \frac{3}{2}\right)^2 = \frac{72}{4} + \frac{9}{4}$$
$$\left(x - \frac{3}{2}\right)^2 = \frac{81}{4}$$
$$x - \frac{3}{2} = \pm\sqrt{\frac{81}{4}} \quad \text{Step 5}$$
$$x - \frac{3}{2} = \pm\frac{9}{2}$$
$$x = \frac{3}{2} \pm \frac{9}{2} \quad \text{Step 6}$$
$$x = \frac{3}{2} + \frac{9}{2} \quad \text{or} \quad x = \frac{3}{2} - \frac{9}{2}$$
$$x = \frac{12}{2} = 6 \qquad x = -\frac{6}{2} = -3$$

The solutions are 6 and -3.

Understanding Algebra

Remember the square of a negative number is a positive number. Thus,

$$\left(-\frac{3}{2}\right)^2 = \frac{9}{4} \textit{ not } -\frac{9}{4}$$

Now Try Exercise 19

In the following examples we will not show some of the intermediate steps.

EXAMPLE 4 Solve the equation $x^2 - 16x + 12 = 0$ by completing the square.

Solution

$$x^2 - 16x + 12 = 0$$

$$x^2 - 16x = -12 \qquad \text{Step 2}$$

$$x^2 - 16x + 64 = -12 + 64 \qquad \text{Step 3}$$

$$(x - 8)^2 = 52 \qquad \text{Step 4}$$

$$x - 8 = \pm\sqrt{52} \qquad \text{Step 5}$$

$$x - 8 = \pm\sqrt{4}\sqrt{13}$$

$$x - 8 = \pm 2\sqrt{13}$$

$$x = 8 \pm 2\sqrt{13} \qquad \text{Step 6}$$

The solutions are $8 + 2\sqrt{13}$ and $8 - 2\sqrt{13}$.

Now Try Exercise 29

EXAMPLE 5 Solve the equation $5z^2 - 25z + 10 = 0$ by completing the square.

Solution Since the coefficient of the squared term is 5, we multiply both sides of the equation by $\frac{1}{5}$ (or divide every term by 5) to make the coefficient equal to 1.

$$5z^2 - 25z + 10 = 0$$

$$\frac{1}{5}(5z^2 - 25z + 10) = \frac{1}{5}(0) \qquad \text{Step 1}$$

$$z^2 - 5z + 2 = 0$$

Now we proceed as in earlier examples.

$$z^2 - 5z = -2 \qquad \text{Step 2}$$

$$z^2 - 5z + \frac{25}{4} = -2 + \frac{25}{4} \qquad \text{Step 3}$$

$$\left(z - \frac{5}{2}\right)^2 = -\frac{8}{4} + \frac{25}{4} \qquad \text{Step 4}$$

$$\left(z - \frac{5}{2}\right)^2 = \frac{17}{4}$$

$$z - \frac{5}{2} = \pm\sqrt{\frac{17}{4}} \qquad \text{Step 5}$$

$$z - \frac{5}{2} = \pm\frac{\sqrt{17}}{2}$$

$$z = \frac{5}{2} \pm \frac{\sqrt{17}}{2} \qquad \text{Step 6}$$

$$z = \frac{5}{2} + \frac{\sqrt{17}}{2} \quad \text{or} \quad z = \frac{5}{2} - \frac{\sqrt{17}}{2}$$

$$z = \frac{5 + \sqrt{17}}{2} \qquad z = \frac{5 - \sqrt{17}}{2}$$

The solutions are $\frac{5 + \sqrt{17}}{2}$ and $\frac{5 - \sqrt{17}}{2}$.

Now Try Exercise 33

10.2 Exercise Set MathXL® MyMathLab®

Warm-Up Exercises

Fill in the blanks with the appropriate word, phrase, or symbol(s) from the following list.

original constant one-half quadratic twice perfect coefficient root

1. A trinomial that can be expressed as the square of a binomial is called a ________ square trinomial.
2. In a perfect square trinomial, when the coefficient of the squared term is 1, the constant term is the square of ________ the coefficient of the x-term.
3. To solve a quadratic equation by completing the square, when necessary, we first use the multiplication property of equality to make the ________ of the squared term equal to 1.
4. When solving a quadratic equation by completing the square, once the coefficient of the squared term is equal to 1, we rewrite the equation with the ________ by itself on the right side of the equation.
5. When solving a quadratic equation by completing the square, once we rewrite the perfect square trinomial as a binomial squared, we solve the equation by using the square ________ property.
6. When solving a quadratic equation by completing the square, once we have solved for the variable, we can check our answers in the ________ equation.

Practice the Skills

Solve by completing the square.

7. $x^2 + 10x + 24 = 0$
8. $x^2 + 10x + 16 = 0$
9. $x^2 - 8x + 7 = 0$
10. $r^2 - 2r - 35 = 0$
11. $x^2 = 2x + 15$
12. $x^2 = 2x + 35$
13. $3x^2 + 6x - 9 = 0$
14. $5x^2 + 20x - 25 = 0$
15. $x^2 + 7x + 10 = 0$
16. $x^2 + 9x + 18 = 0$
17. $x^2 - 3x - 4 = 0$
18. $x^2 - 7x + 12 = 0$
19. $x^2 = -5x - 6$
20. $-40 = n^2 + 13n$
21. $3h^2 - 15h = 18$
22. $2x^2 + 2x - 24 = 0$
23. $n^2 = -6n - 9$
24. $k^2 = 14k - 49$
25. $z^2 - 4z = -2$
26. $z^2 + 2z = 10$
27. $w^2 + 6w = -3$
28. $g^2 - 2g = 9$
29. $3x^2 + 12x - 27 = 0$
30. $5x^2 + 60x + 15 = 0$
31. $m^2 + 7m + 2 = 0$
32. $x^2 + 3x - 3 = 0$
33. $2x^2 + 18x + 4 = 0$
34. $3x^2 + 21x - 12 = 0$
35. $3x^2 - 11x - 4 = 0$
36. $3x^2 - 8x + 4 = 0$
37. $6x^2 + 7x - 3 = 0$
38. $6x^2 + x - 12 = 0$
39. $9t^2 + 6t = 6$
40. $6x^2 - 3x = 15$
41. $2x^2 - 16x = 0$
42. $6x^2 - 42x = 0$
43. $3x^2 = 27x$
44. $7x^2 = 28x$

Problem Solving

45. **Numbers** When 3 times a number is added to the square of a number, the sum is 4. Find the number(s).
46. **Numbers** When 5 times a number is subtracted from 2 times the square of a number, the difference is 12. Find the number(s).
47. **Product** The product of two positive numbers is 21. Find the two numbers if the larger is 4 greater than the smaller.
48. **Product** The product of two positive numbers is 84. Find the two numbers if the smaller is 8 less than the larger.
49. **Ladder** A 26-foot ladder is leaning against a house as shown on the right. Determine the vertical distance from the ground to where the ladder rests on the house.

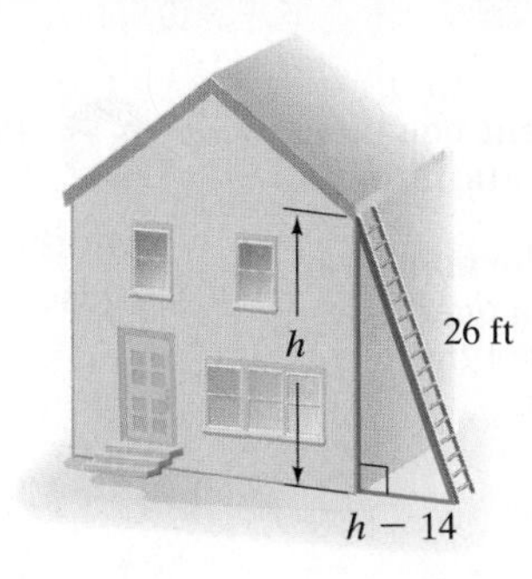

50. Supporting a Pole A guy wire 20 feet long is supporting a pole as shown in the figure. Determine the height of the pole.

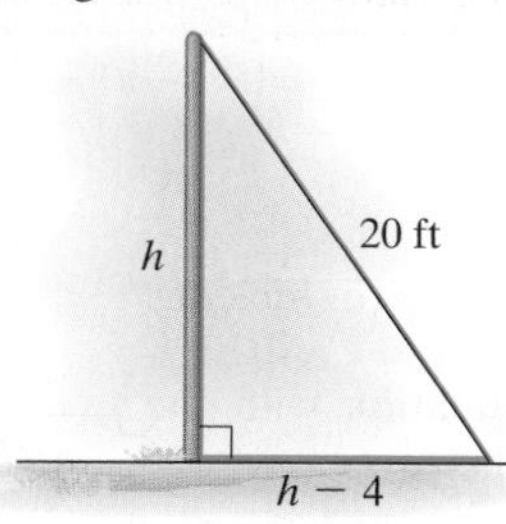

51. Sum of Even Natural Numbers The sum of the first n even natural numbers, s, can be found by the formula $s = n^2 + n$. Find the value of n if the sum is 110.

52. Sum of Natural Numbers The sum of the first n natural numbers, s, can be found by the formula $s = \dfrac{n^2 + n}{2}$. Find the value of n if the sum is 28.

53. Height of an Object When an object is thrown straight up from Earth with an initial velocity of 128 feet per second, its height above the ground, s, in feet, in t seconds is given by the formula $s = -16t^2 + 128t$. How long will it take the object to reach a height of 240 feet? (Therefore, $s = 240$.)

54. Height of an Object Repeat Exercise 53 for a height of 112 feet.

55. a) Write a perfect square trinomial that has a term of $10x$.

b) Explain how you constructed your perfect square trinomial.

56. a) Write a perfect square trinomial that has a term of $-16x$.

b) Explain how you constructed your perfect square trinomial.

Challenge Problems

57. Fill in the shaded area to make a perfect square trinomial.

$$x^2 \quad\square\quad + 81$$

58. Fill in the shaded area to make a perfect square trinomial.

$$x^2 \quad\square\quad + \frac{9}{100}$$

59. a) Solve the equation $x^2 - 14x - 1 = 0$ by completing the square.

b) Check your solution (it will not be a rational number) by substituting the value(s) you obtained in part **a)** for each x in the equation in part **a)**.

60. a) Solve the equation $x^2 + 3x - 7 = 0$ by completing the square.

b) Check your solution (it will not be a rational number) by substituting the value(s) you obtained in part **a)** for each x in the equation in part **a)**.

Solve by completing the square.

61. $x^2 + \frac{3}{5}x - \frac{1}{2} = 0$

62. $x^2 - \frac{2}{3}x - \frac{1}{5} = 0$

63. $3x^2 + \frac{1}{2}x = 4$

64. $0.1x^2 + 0.2x - 0.54 = 0$

65. $-5.26x^2 + 7.89x + 15.78 = 0$

Cumulative Review Exercises

[6.4] **66.** Simplify $\dfrac{x^2}{x^2 - x - 6} - \dfrac{x - 2}{x - 3}$.

[7.4] **67.** Explain how you can determine whether two equations represent parallel lines without graphing the equations.

[8.2, 8.3] **68.** Solve the following system of equations.

$$2x + 3y = 6$$
$$-x + 4y = 19$$

[9.5] **69.** Solve the equation $\sqrt{2x + 3} = 2x - 3$.

10.3 Solving Quadratic Equations by the Quadratic Formula

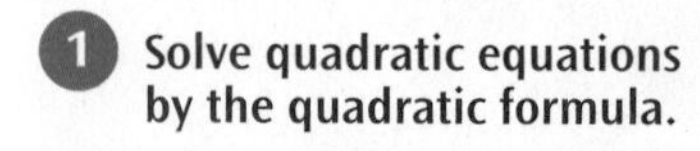

1. Solve quadratic equations by the quadratic formula.
2. Determine the number of solutions to a quadratic equation using the discriminant.

1 Solve Quadratic Equations by the Quadratic Formula

Another method that can be used to solve any quadratic equation is the **quadratic formula**. It is the most versatile method of solving quadratic equations.

Quadratic Equation in Standard Form	Values of a, b, and c		
$x^2 - 5x + 6 = 0$	$a = 1,$	$b = -5,$	$c = 6$
$5x^2 + 3x = 0$	$a = 5,$	$b = 3,$	$c = 0$
$-\frac{1}{2}x^2 + 5 = 0$	$a = -\frac{1}{2},$	$b = 0,$	$c = 5$

Understanding Algebra

The standard form of a quadratic equation is $ax^2 + bx + c = 0$.

If we complete the square we develop the quadratic formula:

$$x = \frac{-b \pm \sqrt{b^2 - 4ac}}{2a}$$

where a, b, and c are the numerical coefficients from the equation in standard form.

We can develop the quadratic formula by starting with a quadratic equation in standard form and completing the square, as discussed in the preceding section.

$$ax^2 + bx + c = 0 \qquad \text{Standard form of quadratic equation}$$

$$\frac{ax^2}{a} + \frac{b}{a}x + \frac{c}{a} = 0 \qquad \text{Divide both sides by a.}$$

$$x^2 + \frac{b}{a}x = -\frac{c}{a} \qquad \frac{c}{a} \text{ was subtracted from both sides.}$$

$$x^2 + \frac{b}{a}x + \frac{b^2}{4a^2} = -\frac{c}{a} + \frac{b^2}{4a^2} \qquad \text{Take } \frac{1}{2} \text{ of } \frac{b}{a}\text{; and square it to get } \frac{b^2}{4a^2}. \text{ Then add this expression to both sides.}$$

$$\left(x + \frac{b}{2a}\right)^2 = \frac{b^2}{4a^2} - \frac{c}{a} \qquad \text{Rewrite the left side of the equation as the square of a binomial.}$$

$$\left(x + \frac{b}{2a}\right)^2 = \frac{b^2 - 4ac}{4a^2} \qquad \text{Write the right side with a common denominator.}$$

$$x + \frac{b}{2a} = \pm\sqrt{\frac{b^2 - 4ac}{4a^2}} \qquad \text{Square root property}$$

$$x + \frac{b}{2a} = \pm\frac{\sqrt{b^2 - 4ac}}{2a} \qquad \text{Quotient rule for radicals, } \sqrt{4a^2} = 2a$$

$$x = \frac{-b}{2a} \pm \frac{\sqrt{b^2 - 4ac}}{2a} \qquad \frac{b}{2a} \text{ was subtracted from both sides.}$$

$$x = \frac{-b \pm \sqrt{b^2 - 4ac}}{2a} \qquad \text{Write with a common denominator to get the quadratic formula.}$$

To Solve a Quadratic Equation by the Quadratic Formula

1. Write the equation in standard form, $ax^2 + bx + c = 0$, and determine the numerical values for a, b, and c.
2. Substitute the values for a, b, and c from step 1 into the quadratic formula below and then evaluate to obtain the solution.

THE QUADRATIC FORMULA

$$x = \frac{-b \pm \sqrt{b^2 - 4ac}}{2a}$$

EXAMPLE 1 Use the quadratic formula to solve the equation $x^2 + 4x + 3 = 0$.

Solution In this equation $a = 1, b = 4$, and $c = 3$.

$$x = \frac{-b \pm \sqrt{b^2 - 4ac}}{2a}$$

$$= \frac{-(4) \pm \sqrt{(4)^2 - 4(1)(3)}}{2(1)} \qquad \text{Substitute values for } a, b, \text{ and } c.$$

$$= \frac{-4 \pm \sqrt{16 - 12}}{2} \qquad \text{Evaluate.}$$

$$= \frac{-4 \pm \sqrt{4}}{2}$$

$$= \frac{-4 \pm 2}{2}$$

$$x = \frac{-4 + 2}{2} \quad \text{or} \quad x = \frac{-4 - 2}{2}$$

$$= \frac{-2}{2} = -1 \qquad = \frac{-6}{2} = -3$$

Understanding Algebra

Always identify the values of a, b, and c before using them in the quadratic formula.

Check

$x = -1$

$$\begin{aligned} x^2 + 4x + 3 &= 0 \\ (-1)^2 + 4(-1) + 3 &\stackrel{?}{=} 0 \\ 1 - 4 + 3 &\stackrel{?}{=} 0 \\ 0 &= 0 \quad \text{True} \end{aligned}$$

$x = -3$

$$\begin{aligned} x^2 + 4x + 3 &= 0 \\ (-3)^2 + 4(-3) + 3 &\stackrel{?}{=} 0 \\ 9 - 12 + 3 &\stackrel{?}{=} 0 \\ 0 &= 0 \quad \text{True} \end{aligned}$$

Now Try Exercise 27

AVOIDING COMMON ERRORS

The *entire numerator* of the quadratic formula must be divided by $2a$.

CORRECT

$$x = \frac{-b \pm \sqrt{b^2 - 4ac}}{2a}$$

INCORRECT

$$x = -b \pm \frac{\sqrt{b^2 - 4ac}}{2a}$$

$$x = \frac{-b}{2a} \pm \sqrt{b^2 - 4ac}$$

EXAMPLE 2 Use the quadratic formula to solve the equation $8x^2 + 2x - 1 = 0$.

Solution

$$8x^2 + 2x - 1 = 0$$

$$a = 8, \quad b = 2, \quad c = -1$$

$$\begin{aligned} x &= \frac{-b \pm \sqrt{b^2 - 4ac}}{2a} \\ &= \frac{-(2) \pm \sqrt{(2)^2 - 4(8)(-1)}}{2(8)} \\ &= \frac{-2 \pm \sqrt{4 + 32}}{16} \\ &= \frac{-2 \pm \sqrt{36}}{16} \\ &= \frac{-2 \pm 6}{16} \end{aligned}$$

$$x = \frac{-2 + 6}{16} \quad \text{or} \quad x = \frac{-2 - 6}{16}$$

$$= \frac{4}{16} = \frac{1}{4} \qquad = \frac{-8}{16} = -\frac{1}{2}$$

Check

$x = \frac{1}{4}$

$$\begin{aligned} 8x^2 + 2x - 1 &= 0 \\ 8\left(\frac{1}{4}\right)^2 + 2\left(\frac{1}{4}\right) - 1 &\stackrel{?}{=} 0 \\ 8\left(\frac{1}{16}\right) + \left(\frac{1}{2}\right) - 1 &\stackrel{?}{=} 0 \\ \frac{1}{2} + \frac{1}{2} - 1 &\stackrel{?}{=} 0 \\ 0 &= 0 \quad \text{True} \end{aligned}$$

$x = -\frac{1}{2}$

$$\begin{aligned} 8x^2 + 2x - 1 &= 0 \\ 8\left(-\frac{1}{2}\right)^2 + 2\left(-\frac{1}{2}\right) - 1 &\stackrel{?}{=} 0 \\ 8\left(\frac{1}{4}\right) - 1 - 1 &\stackrel{?}{=} 0 \\ 2 - 1 - 1 &\stackrel{?}{=} 0 \\ 0 &= 0 \quad \text{True} \end{aligned}$$

Now Try Exercise 35

HELPFUL HINT

- Be sure you learn the quadratic formula, as it will be used to solve many problems and applications in algebra and in math courses beyond algebra.
- Always identify the values of a, b, and c before using them in the quadratic formula.

EXAMPLE 3 Use the quadratic formula to solve the equation $2w^2 + 6w - 3 = 0$.

Solution The variable in this equation is w. The procedure to solve the equation is the same.

$$a = 2, \quad b = 6, \quad c = -3$$

$$\begin{aligned} w &= \frac{-b \pm \sqrt{b^2 - 4ac}}{2a} \\ &= \frac{-(6) \pm \sqrt{(6)^2 - 4(2)(-3)}}{2(2)} \\ &= \frac{-6 \pm \sqrt{36 + 24}}{4} \\ &= \frac{-6 \pm \sqrt{60}}{4} \\ &= \frac{-6 \pm \sqrt{4}\sqrt{15}}{4} \\ &= \frac{-6 \pm 2\sqrt{15}}{4} \end{aligned}$$

Now factor out 2 from both terms in the numerator; then divide out common factors as explained in Section 9.4.

$$w = \frac{\overset{1}{\cancel{2}}(-3 \pm \sqrt{15})}{\underset{2}{\cancel{4}}} \qquad \text{Factor. Divide out common factors.}$$

$$w = \frac{-3 \pm \sqrt{15}}{2}$$

Thus, the solutions are $w = \dfrac{-3 + \sqrt{15}}{2}$ and $w = \dfrac{-3 - \sqrt{15}}{2}$.

Now Try Exercise 53

Now let's try two examples where the equation is not in standard form.

EXAMPLE 4 Use the quadratic formula to solve the equation $x^2 = 6x - 4$.

Solution First write the equation in standard form.

$$x^2 - 6x + 4 = 0 \qquad \text{Set one side of the equation equal to zero.}$$

$$a = 1, \quad b = -6, \quad c = 4$$

$$\begin{aligned} x &= \frac{-b \pm \sqrt{b^2 - 4ac}}{2a} \\ &= \frac{-(-6) \pm \sqrt{(-6)^2 - 4(1)(4)}}{2(1)} \qquad \text{Substitute.} \end{aligned}$$

$$= \frac{6 \pm \sqrt{36 - 16}}{2} \quad \text{Simplify.}$$

$$= \frac{6 \pm \sqrt{20}}{2}$$

$$= \frac{6 \pm \sqrt{4}\sqrt{5}}{2} \quad \text{Product rule}$$

$$= \frac{6 \pm 2\sqrt{5}}{2}$$

$$= \frac{\overset{1}{\cancel{2}}(3 \pm \sqrt{5})}{\underset{1}{\cancel{2}}} \quad \text{Factor out 2.}$$

$$= 3 \pm \sqrt{5}$$

The solutions are $x = 3 + \sqrt{5}$ and $x = 3 - \sqrt{5}$.

Now Try Exercise 49

AVOIDING COMMON ERRORS

Many students solve quadratic equations correctly until the last step, when they make an error. Do not make the mistake of trying to simplify an answer that cannot be simplified any further. The following are answers that cannot be simplified, along with some common errors.

ANSWERS THAT CANNOT BE SIMPLIFIED

$$\frac{3 + 2\sqrt{5}}{2}$$

$$\frac{4 + 3\sqrt{5}}{2}$$

INCORRECT

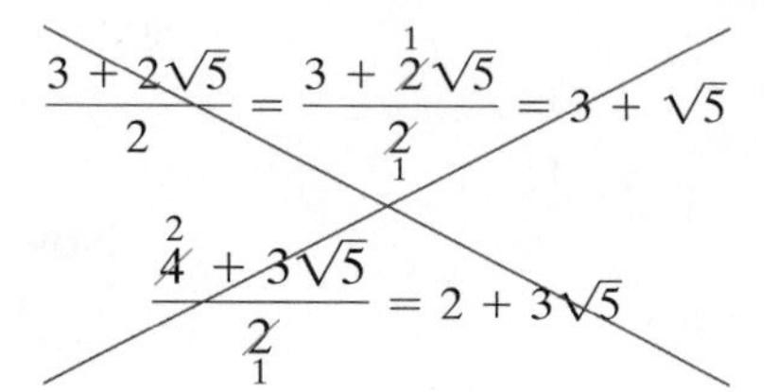

EXAMPLE 5 Use the quadratic formula to solve the equation $t^2 = 36$.

Solution First write the equation in standard form.

$$t^2 - 36 = 0 \quad \text{Set one side of the equation equal to 0.}$$

$$a = 1, \quad b = 0, \quad c = -36$$

$$t = \frac{-b \pm \sqrt{b^2 - 4ac}}{2a}$$

$$= \frac{-0 \pm \sqrt{0^2 - 4(1)(-36)}}{2(1)} \quad \text{Substitute.}$$

$$= \frac{\pm\sqrt{144}}{2} = \frac{\pm 12}{2} = \pm 6 \quad \text{Simplify and solve for t.}$$

Thus, the solutions are 6 and -6.

Now Try Exercise 57

The solution to Example 5 could have been solved more quickly by factoring or by using the square root property. We worked Example 5 using the quadratic formula to give you more practice using the formula.

The next example illustrates a quadratic equation that has no real number solution.

EXAMPLE 6 Use the quadratic formula to solve the equation $3x^2 = x - 1$.

Solution

$$3x^2 - x + 1 = 0$$

$$a = 3, \quad b = -1, \quad c = 1$$

$$x = \frac{-b \pm \sqrt{b^2 - 4ac}}{2a}$$

$$= \frac{-(-1) \pm \sqrt{(-1)^2 - 4(3)(1)}}{2(3)}$$

$$= \frac{1 \pm \sqrt{1 - 12}}{6}$$

$$= \frac{1 \pm \sqrt{-11}}{6}$$

Since $\sqrt{-11}$ is not a real number, we stop here. This equation has no real number solution. *When given a problem of this type, your answer should be "no real number solution."*

Now Try Exercise 51

Equations with no real solutions, like the equation in Example 6, will be discussed again in Section 10.5 when we introduce complex numbers.

2 Determine the Number of Solutions to a Quadratic Equation Using the Discriminant

Discriminant

The expression under the square root sign in the quadratic formula is called the **discriminant**.

$$\underbrace{b^2 - 4ac}_{\text{Discriminant}}$$

The discriminant can be used to determine the number of real solutions to a quadratic equation, as shown below.

Quadratic Equation Solutions

For a quadratic equation in standard form, $ax^2 + bx + c = 0$, when the discriminant is:

1. **Greater than zero,** $b^2 - 4ac > 0$, the quadratic equation has *two distinct real number solutions.*
2. **Equal to zero,** $b^2 - 4ac = 0$, the quadratic equation has *one real number solution.*
3. **Less than zero,** $b^2 - 4ac < 0$, the quadratic equation has *no real number solution.*

We indicate this information in a shortened form in the chart below.

If $b^2 - 4ac$ is	Then the number of solutions is
Positive	Two distinct real number solutions
0	One real number solution
Negative	No real number solution

EXAMPLE 7

a) Find the discriminant of the equation $x^2 - 12x + 36 = 0$.

b) Use the discriminant to determine the number of solutions to the equation.

c) Use the quadratic formula to find the solutions, if any exist.

Solution

a) $a = 1, \quad b = -12, \quad c = 36$

$$b^2 - 4ac = (-12)^2 - 4(1)(36) = 144 - 144 = 0$$

b) Since the discriminant is equal to zero, there is one real number solution.

c)

$$x = \frac{-b \pm \sqrt{b^2 - 4ac}}{2a}$$

$$= \frac{-(-12) \pm \sqrt{0}}{2(1)}$$

$$= \frac{12 \pm 0}{2} = \frac{12}{2} = 6$$

The only solution is 6.

Now Try Exercise 13

EXAMPLE 8 Without actually finding the solutions, determine whether the following equations have two distinct real number solutions, one real number solution, or no real number solution.

a) $4x^2 - 4x + 1 = 0$ **b)** $2x^2 + 13x = -10$ **c)** $6p^2 = -5p - 3$

Solution We use the discriminant of the quadratic formula to answer these questions.

a) $b^2 - 4ac = (-4)^2 - 4(4)(1) = 16 - 16 = 0$

Since the discriminant is equal to zero, this equation has one real number solution.

b) First, rewrite $2x^2 + 13x = -10$ in standard form as $2x^2 + 13x + 10 = 0$.

$$b^2 - 4ac = (13)^2 - 4(2)(10) = 169 - 80 = 89$$

Since the discriminant is positive, this equation has two distinct real number solutions.

c) First rewrite $6p^2 = -5p - 3$ in standard form as $6p^2 + 5p + 3 = 0$

$$b^2 - 4ac = (5)^2 - 4(6)(3) = 25 - 72 = -47$$

Since the discriminant is negative, this equation has no real number solution.

Now Try Exercise 17

Many applications may be solved using the quadratic formula.

EXAMPLE 9 **Building a Border** The Johnsons have a rectangular swimming pool that measures 30 feet by 16 feet. They want to add a concrete border of uniform width around all sides of the pool. How wide can they make the border if they want the area of the border to be 200 square feet?

Solution Let's make a diagram of the pool; see **Figure 10.2.** Let x = uniform width of the border. Then the total length of the pool and border is $2x + 30$. The total width of the pool and border is $2x + 16$. The area of the border can be found by subtracting the area of the pool (the smaller rectangle area) from the area of the pool and border (the larger rectangle area).

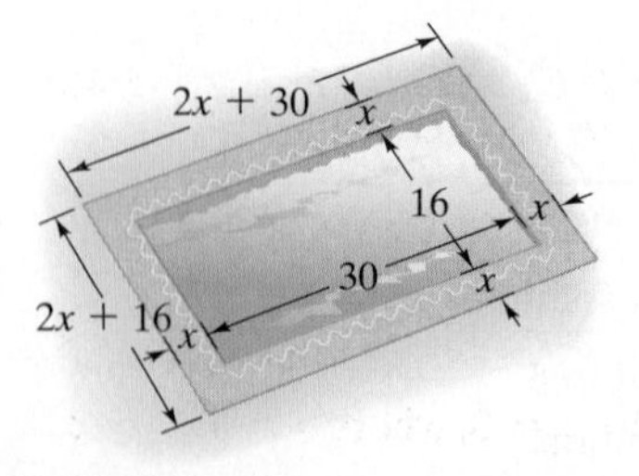

FIGURE 10.2

$$\text{area of pool} = l \cdot w = (30)(16) = 480$$

$$\text{area of pool and border} = l \cdot w = (2x + 30)(2x + 16)$$

$$= 4x^2 + 92x + 480$$

$$\begin{aligned} \text{area of border} &= \text{area of pool and border} - \text{area of pool} \\ &= (4x^2 + 92x + 480) - 480 \\ &= 4x^2 + 92x \end{aligned}$$

The total area of the border is 200 square feet. Therefore,

$$\begin{aligned} \text{area of border} &= 4x^2 + 92x \\ 200 &= 4x^2 + 92x \\ \text{or} \quad 4x^2 + 92x - 200 &= 0 && \text{Write equation in standard form.} \\ 4(x^2 + 23x - 50) &= 0 && \text{Factor out 4.} \\ \frac{1}{\cancel{4}} \cdot \cancel{4}(x^2 + 23x - 50) &= \frac{1}{4} \cdot 0 && \text{Multiply both sides by } \frac{1}{4} \text{ to eliminate 4.} \\ x^2 + 23x - 50 &= 0 \end{aligned}$$

Now use the quadratic formula.

$$a = 1 \quad b = 23 \quad c = -50$$

$$\begin{aligned} x &= \frac{-b \pm \sqrt{b^2 - 4ac}}{2a} \\ &= \frac{-23 \pm \sqrt{(23)^2 - 4(1)(-50)}}{2(1)} \\ &= \frac{-23 \pm \sqrt{529 + 200}}{2} \\ &= \frac{-23 \pm \sqrt{729}}{2} \\ &= \frac{-23 \pm 27}{2} \end{aligned}$$

$$x = \frac{-23 - 27}{2} = \frac{-50}{2} = -25 \quad \text{or} \quad x = \frac{-23 + 27}{2} = \frac{4}{2} = 2$$

Since lengths are positive, the only possible answer is $x = 2$. The uniform concrete border will be 2 feet wide all around the pool.

Now Try Exercise 61

Many times when working with quadratic application problems the answer is an irrational number. When this occurs in the exercise set we will round answers to two decimal places.

HELPFUL HINT

If all the terms in a quadratic equation have a common factor, it is easier to factor it out first so that you will have smaller numbers when you use the quadratic formula. Consider the quadratic equation $4x^2 + 8x - 12 = 0$.

In this equation $a = 4$, $b = 8$, and $c = -12$. If you solve this equation with the quadratic formula, after simplification you will get the solutions -3 and 1. Try this and see. If you factor out 4 to get

$$\begin{aligned} 4x^2 + 8x - 12 &= 0 \\ 4(x^2 + 2x - 3) &= 0 \end{aligned}$$

and then use the quadratic formula with the equation $x^2 + 2x - 3 = 0$, where $a = 1$, $b = 2$, and $c = -3$, you get the same solution. Try this and see.

10.3 Exercise Set MathXL® MyMathLab®

Warm-Up Exercises

Fill in the blanks with the appropriate word, phrase, or symbol(s) from the following list.

discriminant	coefficients	one	standard	-3	3	two
5	7	-7	no	quadratic	three	

1. When a quadratic equation is written in the form $ax^2 + bx + c = 0$ it is said to be in ________ form.

2. In the equation $ax^2 + bx + c = 0$, the numbers a, b, and c are called the numerical ________.

3. In the equation $5x^2 - 3x - 7 = 0$,
   a) The numerical coefficient $a =$ ________.
   b) The numerical coefficient $b =$ ________.
   c) The numerical coefficient $c =$ ________.

4. The formula $x = \dfrac{-b \pm \sqrt{b^2 - 4ac}}{2a}$ is called the ________ formula.

5. The expression $b^2 - 4ac$ is called the ________.

6. If the discriminant is greater than zero, then the quadratic equation has ________ distinct real number solutions.

7. If the discriminant is equal to zero, then the quadratic equation has ________ real number solution.

8. If the discriminant is less than zero, then the quadratic equation has ________ real number solution.

Practice the Skills

Determine whether each equation has two distinct real number solutions, one real number solution, or no real number solution.

9. $x^2 + 5x - 9 = 0$
10. $x^2 + 2x - 7 = 0$
11. $2x^2 + x + 1 = 0$
12. $r^2 + 3r + 5 = 0$
13. $x^2 + 6x + 9 = 0$
14. $x^2 + 10x + 25 = 0$
15. $z^2 = 5z + 11$
16. $5x^2 - 4x = 3$
17. $4x = 8 + x^2$
18. $5x - 8 = 3x^2$
19. $14x = x^2 + 49$
20. $-16 = w^2 + 8w$
21. $2x^2 - 7x + 10 = 0$
22. $2x^2 - 6x + 5 = 0$
23. $2.1x^2 - 0.5 = 0$
24. $0.6x^2 - 1.3x = 0$
25. $18 = -2t^2 + 12t$
26. $3x^2 + 66x + 363 = 0$

Use the quadratic formula to solve each equation. If the equation has no real number solution, so state.

27. $x^2 + 9x + 18 = 0$
28. $x^2 + 3x + 2 = 0$
29. $x^2 + 2x - 8 = 0$
30. $x^2 + 5x - 24 = 0$
31. $x^2 - 10x + 24 = 0$
32. $x^2 - 10x + 9 = 0$
33. $x^2 - 7x - 8 = 0$
34. $x^2 - 3x - 10 = 0$
35. $2x^2 + 7x + 3 = 0$
36. $3x^2 + 7x + 2 = 0$
37. $6x^2 = -x + 1$
38. $8p^2 + 10p - 3 = 0$
39. $2s^2 - 4s + 5 = 0$
40. $3w^2 + 2 = 4w$
41. $2x^2 = 5x + 7$
42. $4r^2 - 5 = r$
43. $x^2 + 7x + 3 = 0$
44. $x^2 + 9x + 7 = 0$
45. $x^2 + x - 10 = 0$
46. $x^2 + 3x - 1 = 0$
47. $x^2 - 4x + 2 = 0$
48. $x^2 - 6x + 7 = 0$
49. $x^2 = 6x + 3$
50. $x^2 = 8x + 4$
51. $6y^2 + 9 = -5y$
52. $15 = -5a^2 - 5a$
53. $2y^2 - 7y + 4 = 0$
54. $3x^2 + 5x + 1 = 0$
55. $2t^2 - 6t - 56 = 0$
56. $2r^2 - 18r + 36 = 0$
57. $7x^2 - 3 = 0$
58. $11x^2 - 7 = 0$

Problem Solving

59. **Dimensions of Rectangle** The length of a rectangle is 3 feet smaller than twice its width. Find the length and width of the rectangle if its area is 20 square feet.

60. **Dimensions of Rectangle** The length of a rectangle is 6 feet longer than its width. Find the dimensions of the rectangle if its area is 55 square feet.

61. **Swimming Pool** Harold Goldstein and his wife Elaine recently installed a built-in rectangular swimming pool measuring 25 feet by 35 feet. They want to add a decorative tile border of uniform width around all sides of the pool. How wide can they make the tile border if they purchased enough tile to cover 256 square feet?

62. **Rectangular Garden** Sean McDonald's rectangular garden measures 20 feet by 30 feet. He wishes to build a uniform-width brick walkway around his garden that covers an area of 336 square feet. What will be the width of the walkway?

63. **Playground** Judy Kasabian has a rectangular-shaped playground in her backyard that measures 17 feet by 23 feet.

Judy wishes to plant a grass border of uniform width around the playground. If Judy wishes the area of the border to be 400 square feet, what will the width be?

64. Pottery Studio Julie Bonds is planning to plant a grass lawn of uniform width around her rectangular pottery studio, which measures 48 feet by 36 feet. How far will the lawn extend from the studio if Julie has only enough seed to plant 4000 square feet of grass?

65. Diagonals in a Polygon The number of diagonals, d, in a polygon with n sides is given by the formula $d = \dfrac{n^2 - 3n}{2}$. For example, a pentagon, a polygon with 5 sides, has $d = \dfrac{5^2 - 15}{2} = 5$ diagonals; see the figure.

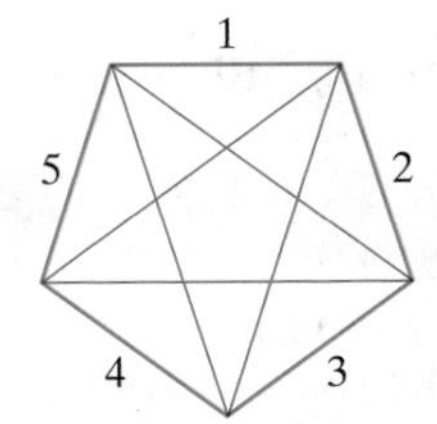

If a polygon has 14 diagonals, how many sides does it have?

66. Diagonals in a Polygon If a polygon has 20 diagonals, how many sides does it have? See Exercise 65.

67. Flags The cost, c, for manufacturing x American flags is given by $c = x^2 - 16x + 40$. Find the number of flags manufactured if the cost is \$120.

68. Manufacturing Cost Repeat Exercise 67 for a cost of \$2680.

69. Model Rocket Phil Chefetz launches a model rocket from the ground. The height, s, of the rocket above the ground at time t seconds after it is launched can be found by the formula $s = -16t^2 + 90t$. Find how long it will take for the rocket to reach a height of 80 feet.

70. Model Rocket Repeat Exercise 69 for a height of 100 feet.

Challenge Problems

Find all the values of c that will result in each equation having **a)** *two real number solutions,* **b)** *one real number solution, and* **c)** *no real number solution.*

71. $x^2 + 8x + c = 0$

72. $2x^2 + 3x + c = 0$

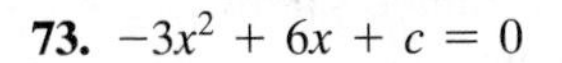

73. $-3x^2 + 6x + c = 0$

74. Fenced in Area Farmer Justina Wells wishes to form a rectangular region along a river bank by constructing fencing on three sides, as illustrated in the diagram on the right. If she has only 400 feet of fencing and wishes to enclose an area of 15,000 square feet, find the dimensions of the rectangular region.

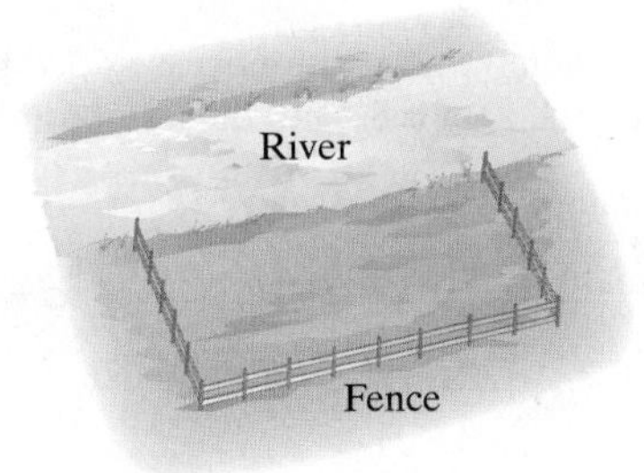

Group Activity

75. In Section 10.4 we will graph quadratic equations. We will learn that the graphs of quadratic equations are *parabolas*. The graph of the quadratic equation $y = x^2 - 2x - 8$ is illustrated on the right.

a) Each member of the group, copy the graph in your notebook.

b) Group Member 1: List the ordered pairs corresponding to points A and B. Verify that each ordered pair is a solution to the equation $y = x^2 - 2x - 8$.

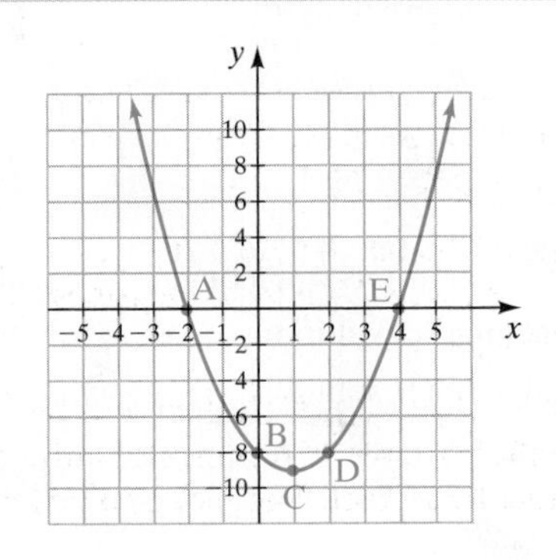

c) Group Member 2: List the ordered pairs corresponding to points C and D. Verify that each ordered pair is a solution to the equation $y = x^2 - 2x - 8$.

d) Group Member 3: List the ordered pair corresponding to point E. Verify that the ordered pair is a solution to the equation $y = x^2 - 2x - 8$.

e) Individually, graph the equation $y = 2x - 3$ on the same axes that you used in part **a)**. Compare your graphs with the other members of your group.

f) The two graphs represent the system of equations

$$y = x^2 - 2x - 8$$
$$y = 2x - 3$$

As a group, estimate the points of intersection of the graphs.

g) If we set the two equations equal to each other, we obtain the following quadratic equation in only the variable x.

$$x^2 - 2x - 8 = 2x - 3$$

As a group, solve this quadratic equation. Does your answer agree with the x-coordinates of the points of intersection from part **f)**?

h) As a group, use the values of x found in part **g)** to find the values of y in $y = x^2 - 2x - 8$ and $y = 2x - 3$. Does your answer agree with the y-coordinates of the points of intersection from part **f)**?

Cumulative Review Exercises

[5.6, 10.2, 10.3] *Solve the following quadratic equations by* **a)** *factoring,* **b)** *completing the square, and* **c)** *the quadratic formula. If the equation cannot be solved by factoring, so state.*

76. $x^2 - 14x + 40 = 0$ **77.** $6x^2 + 11x - 35 = 0$ **78.** $2x^2 + 3x - 4 = 0$ **79.** $3x^2 = 48$

[6.4] **80.** Subtract $\dfrac{x}{2x^2 + 7x - 4} - \dfrac{2}{x^2 - x - 20}$

MID-CHAPTER TEST: 10.1–10.3

To find out how well you understand the chapter material to this point, take this brief test. The answers, and the section where the material was initially discussed, are given in the back of the book. Review any questions that you answered incorrectly.

Solve.

1. $x^2 = 49$ **2.** $a^2 = 21$

3. $16m^2 + 10 = 25$ **4.** $(y - 3)^2 = 4$

5. $(z + 6)^2 = 81$ **6.** $(b - 7)^2 = 24$

7. Numbers The product of two positive numbers is 40. Determine the numbers if the larger number is 2.5 times the smaller number.

8. Numbers When 2 times a number is added to the square of the number, the sum is 8. Find the number(s).

Solve by completing the square.

9. $x^2 + 2x - 15 = 0$ **10.** $x^2 + 11x + 18 = 0$

11. $p^2 - 8p = 0$ **12.** $h^2 + 2h - 6 = 0$

13. $x^2 - 9x + 1 = 0$

Solve by using the quadratic formula.

14. $x^2 - 3x - 40 = 0$

15. $x^2 + 13x + 42 = 0$

16. $m^2 - 5m - 2 = 0$

17. Under what conditions will a quadratic equation have

a) two real number solutions

b) one real number solution,

c) no real number solution?

Determine whether each equation has two distinct real number solutions, one real number solution, or no real number solution.

18. $3x^2 - x - 2 = 0$

19. $\dfrac{1}{2}x^2 + 4x + 11 = 0$

20. Rectangle The length of the rectangular screen on a portable DVD player is 3 inches longer than the width. Find the length and width if the screen's area is 88 square inches.

10.4 Graphing Quadratic Equations

1. Graph quadratic equations in two variables.
2. Find the coordinates of the vertex of a parabola.
3. Use symmetry to graph quadratic equations.
4. Find the intercepts of the graph of a quadratic equation.

1 Graph Quadratic Equations in Two Variables

HELPFUL HINT

Study Tip

In this section we will graph quadratic equations. Quadratic equations can be graphed by plotting points, as we did in Chapter 7 when we graphed linear equations. However, there are certain things we can do to make graphing quadratic equations easier. These include finding the vertex of the graph, using symmetry to draw the graph, and finding the intercepts of the graph.

In Section 7.2 we learned how to graph linear equations. In this section we graph quadratic equations of the form

$$y = ax^2 + bx + c, \quad a \neq 0$$

The graph of every quadratic equation is a **parabola**. The graph of $y = ax^2 + bx + c$ will have one of the shapes indicated in **Figure 10.3**.

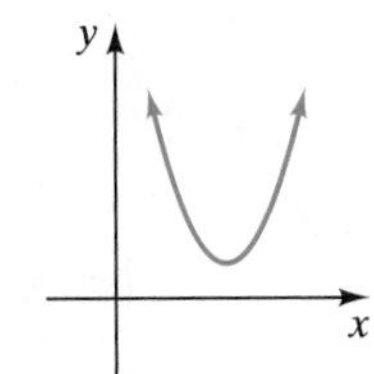

When a is positive, the parabola opens upward

(a)

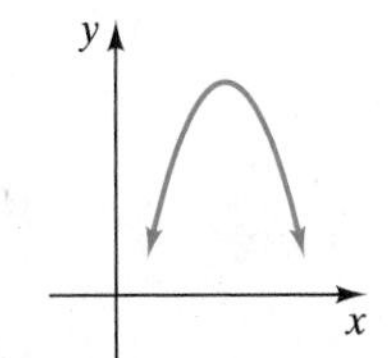

When a is negative, the parabola opens downward

(b)

FIGURE 10.3

The **vertex** is the lowest point on a parabola that opens upward or the highest point on a parabola that opens downward (**Fig. 10.4**). Graphs of quadratic equations of the form $y = ax^2 + bx + c$ have **symmetry** about a vertical line through the vertex. This means that if we fold the paper along this imaginary line, called the **axis of symmetry**, the right and left sides of the graph match up.

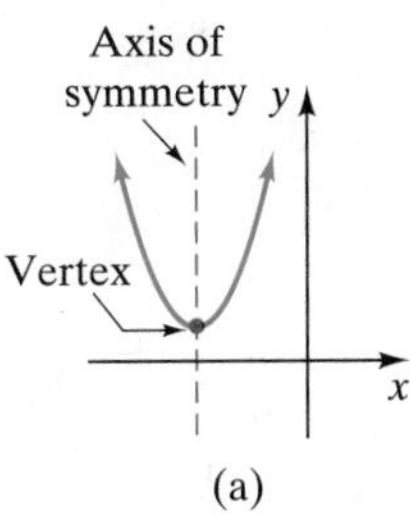

(a)

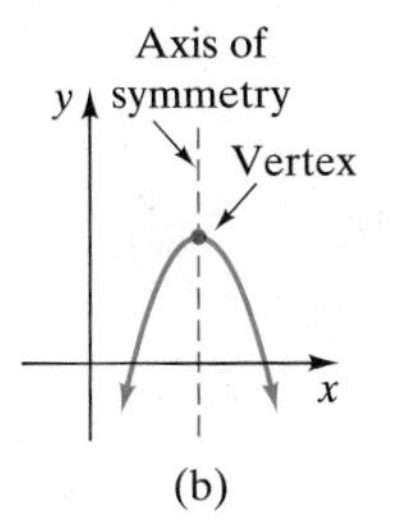

(b)

FIGURE 10.4

One method to graph a quadratic equation is to plot it point by point. Select values for x and determine the corresponding values for y. Then plot the ordered pairs.

EXAMPLE 1 Graph $y = x^2$.

Solution Since $a = 1$, which is positive, this parabola opens upward.

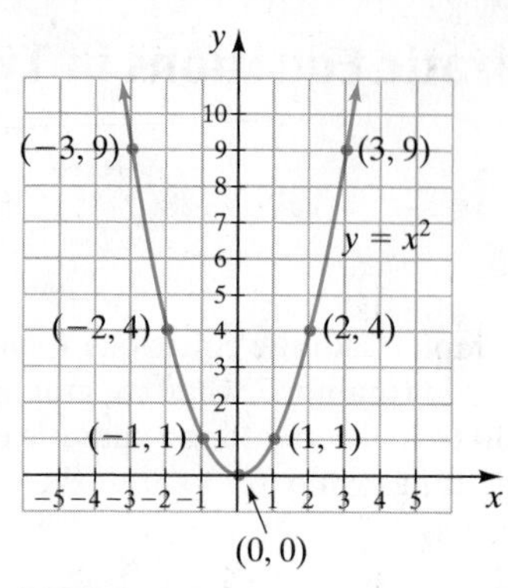

FIGURE 10.5

$$y = x^2$$

Let $x = 3$, $y = (3)^2 = 9$
Let $x = 2$, $y = (2)^2 = 4$
Let $x = 1$, $y = (1)^2 = 1$
Let $x = 0$, $y = (0)^2 = 0$
Let $x = -1$, $y = (-1)^2 = 1$
Let $x = -2$, $y = (-2)^2 = 4$
Let $x = -3$, $y = (-3)^2 = 9$

x	y
3	9
2	4
1	1
0	0
−1	1
−2	4
−3	9

Connect the points with a smooth curve (**Fig. 10.5**). Note how the graph is symmetric about the line $x = 0$ (or the y-axis). Thus the equation of the axis of symmetry is $x = 0$.

Now Try Exercise 21

EXAMPLE 2 Graph $y = -2x^2 + 4x + 6$.

Solution Since $a = -2$, which is negative, this parabola opens downward.

$$y = -2x^2 + 4x + 6$$

Let $x = 5$, $y = -2(5)^2 + 4(5) + 6 = -24$
Let $x = 4$, $y = -2(4)^2 + 4(4) + 6 = -10$
Let $x = 3$, $y = -2(3)^2 + 4(3) + 6 = 0$
Let $x = 2$, $y = -2(2)^2 + 4(2) + 6 = 6$
Let $x = 1$, $y = -2(1)^2 + 4(1) + 6 = 8$
Let $x = 0$, $y = -2(0)^2 + 4(0) + 6 = 6$
Let $x = -1$, $y = -2(-1)^2 + 4(-1) + 6 = 0$
Let $x = -2$, $y = -2(-2)^2 + 4(-2) + 6 = -10$
Let $x = -3$, $y = -2(-3)^2 + 4(-3) + 6 = -24$

x	y
5	−24
4	−10
3	0
2	6
1	8
0	6
−1	0
−2	−10
−3	−24

Understanding Algebra

Recall from Section 7.2 that the equations of vertical lines are of the form $x = a$, where a is the x-coordinate of all of the points on the vertical line. Therefore, the equation of the axis of symmetry in Example 1 is $x = 0$ and the equation of the axis of symmetry in Example 2 is $x = 1$.

The graph of $y = -2x^2 + 4x + 6$ is shown in **Figure 10.6**. The graph is symmetric about the line $x = 1$ (which is dashed since it is not part of the graph).

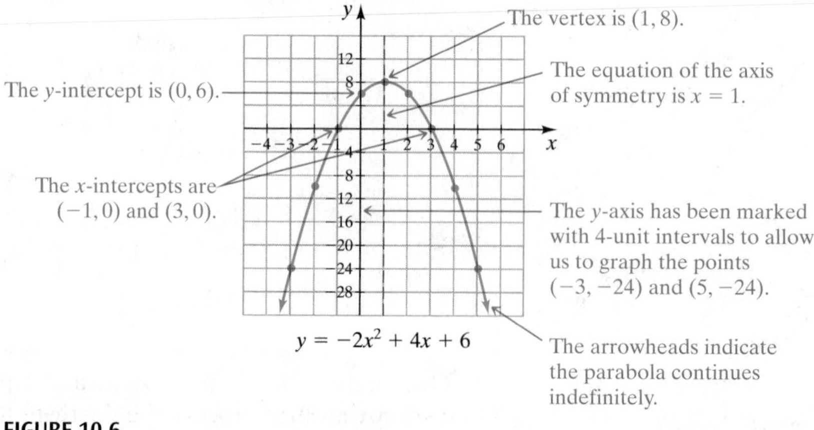

FIGURE 10.6

Now Try Exercise 29

2 Find the Coordinates of the Vertex of a Parabola

When the location of the axis of symmetry and the vertex are known, it is easier to decide which values to use for x when plotting points.

Axis of Symmetry and x-Coordinate of the Vertex

For a quadratic equation of the form $y = ax^2 + bx + c$, both the equation of the axis of symmetry and the x-coordinate of the vertex can be found using the following formula:

$$x = -\frac{b}{2a}$$

Understanding Algebra

The vertex of a parabola is always on the axis of symmetry.

In the quadratic equation in Example 2, $a = -2$, $b = 4$, and $c = 6$. Substituting these values in the formula for the axis of symmetry gives

$$x = -\frac{b}{2a} = -\frac{4}{2(-2)} = -\frac{4}{-4} = 1$$

Thus, the equation of the axis of symmetry is $x = 1$. The graph is symmetric about the vertical line $x = 1$. Also, the x-coordinate of the vertex is 1.

The y-coordinate of the vertex can be found by substituting the value of the x-coordinate of the vertex into the quadratic equation and solving for y.

$$\begin{aligned} y &= -2x^2 + 4x + 6 \\ &= -2(1)^2 + 4(1) + 6 \\ &= -2(1) + 4 + 6 \\ &= -2 + 4 + 6 \\ &= 8 \end{aligned}$$

The vertex is at the point $(1, 8)$.

The y-coordinate of the vertex can also be found using the following formula.

y-Coordinate of the Vertex

For a quadratic equation of the form $y = ax^2 + bx + c$, the y-coordinate of the vertex can be found using the following formula:

$$y = \frac{4ac - b^2}{4a}$$

In Example 2,

$$\begin{aligned} y &= \frac{4ac - b^2}{4a} \\ &= \frac{4(-2)(6) - 4^2}{4(-2)} \\ &= \frac{-48 - 16}{-8} = \frac{-64}{-8} = 8 \end{aligned}$$

You may use the method of your choice to find the y-coordinate of the vertex. Both methods result in the same value of y.

3 Use Symmetry to Graph Quadratic Equations

When graphing parabolas, first determine the axis of symmetry and the vertex of the graph. Then select nearby values of x on either side of the axis of symmetry. When plotting points, make use of the symmetry of the graph.

EXAMPLE 3

a) Find the equation of the axis of symmetry of the graph of the equation $y = x^2 + 6x + 5$.

b) Find the vertex of the graph.

c) Graph the equation.

Solution

a) $a = 1, \quad b = 6, \quad c = 5.$

$$x = -\frac{b}{2a} = -\frac{6}{2(1)} = -3$$

The equation of the axis of symmetry is $x = -3$. The x-coordinate of the vertex is -3.

b) Now find the y-coordinate of the vertex. Substitute -3 for x in the quadratic equation.

$$y = x^2 + 6x + 5$$
$$y = (-3)^2 + 6(-3) + 5 = 9 - 18 + 5 = -4$$

The vertex is at the point $(-3, -4)$.

c) Since the equation of the axis of symmetry is $x = -3$, we will select values for x that are greater than -3. It is often helpful to plot each point as it is determined. If a point does not appear to lie on the parabola, check it.

$$y = x^2 + 6x + 5$$

Let $x = -2$, $\quad y = (-2)^2 + 6(-2) + 5 = -3$

Let $x = -1$, $\quad y = (-1)^2 + 6(-1) + 5 = 0$

Let $x = 0$, $\quad y = (0)^2 + 6(0) + 5 = 5$

x	y
-2	-3
-1	0
0	5

These points are plotted in **Figure 10.7a**. Note how we use symmetry to complete the graph in **Figure 10.7b**. The points $(-2, -3)$ and $(-4, -3)$ are each 1 horizontal unit from the axis of symmetry, $x = -3$. The points $(-1, 0)$ and $(-5, 0)$ are each 2 horizontal units from the axis of symmetry, and the points $(0, 5)$ and $(-6, 5)$ are each 3 horizontal units from the axis of symmetry.

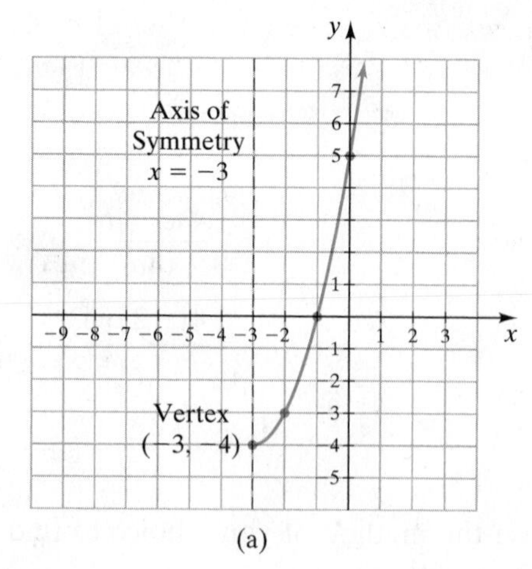

(a)

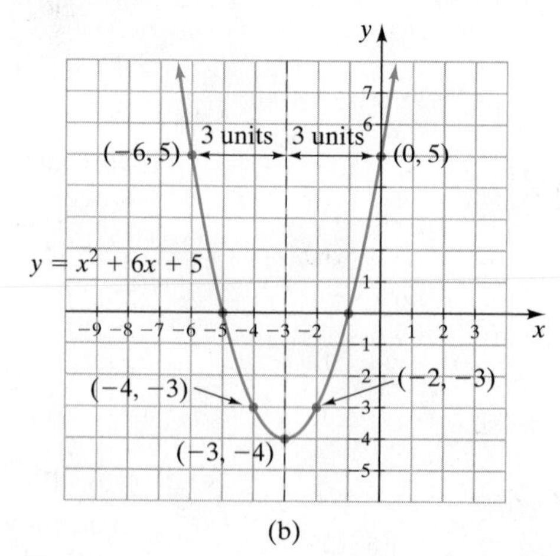

(b)

FIGURE 10.7

Now Try Exercise 25

EXAMPLE 4 Graph $y = -2x^2 + 5x - 4$.

Solution $a = -2, \quad b = 5, \quad c = -4.$

Since $a < 0$, this parabola will open downward.

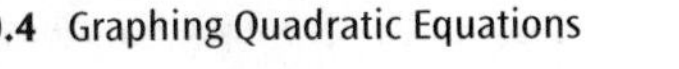

$$\text{Axis of symmetry: } x = -\frac{b}{2a}$$

$$= -\frac{5}{2(-2)} = -\frac{5}{-4} = \frac{5}{4} \quad \left(\text{or } 1\frac{1}{4}\right)$$

Since the x-value of the vertex is a fraction, we would need to add fractions if we wished to find y by substituting $\frac{5}{4}$ for x in the given equation. Therefore, we will use the formula to find the y-coordinate of the vertex.

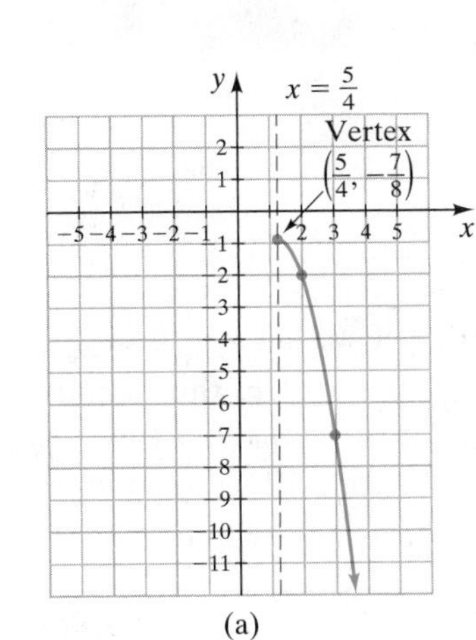

(a)

$$y = \frac{4ac - b^2}{4a}$$

$$= \frac{4(-2)(-4) - 5^2}{4(-2)} = \frac{32 - 25}{-8} = \frac{7}{-8} = -\frac{7}{8}$$

The vertex of this graph is at the point $\left(\frac{5}{4}, -\frac{7}{8}\right)$. Since the axis of symmetry is $x = \frac{5}{4}$, we will begin by selecting values of x that are greater than $\frac{5}{4}$, or $1\frac{1}{4}$.

	$y = -2x^2 + 5x - 4$
Let $x = 2$,	$y = -2(2)^2 + 5(2) - 4 = -2$
Let $x = 3$,	$y = -2(3)^2 + 5(3) - 4 = -7$
Let $x = 4$,	$y = -2(4)^2 + 5(4) - 4 = -16$

x	y
2	−2
3	−7
4	−16

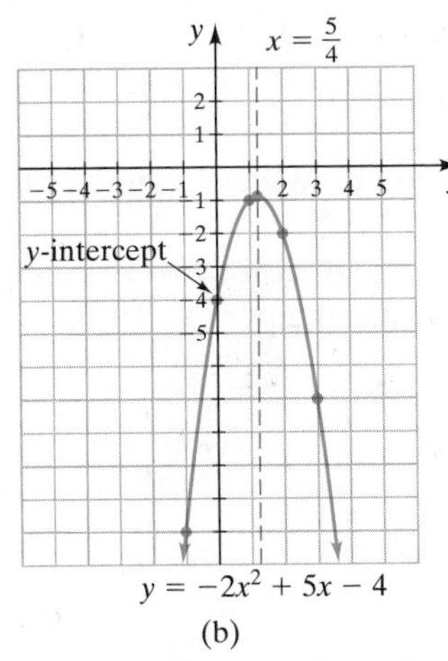

(b)

FIGURE 10.8

When the axis of symmetry is a fractional value, be very careful when constructing the graph. You should plot as many additional points as needed. Following we determine some values of y when x is less than $\frac{5}{4}$.

Let $x = 1$	$y = -2(1)^2 + 5(1) - 4 = -1$
Let $x = 0$	$y = -2(0)^2 + 5(0) - 4 = -4$
Let $x = -1$	$y = -2(-1)^2 + 5(-1) - 4 = -11$

x	y
1	−1
0	−4
−1	−11

Figure 10.8a shows the points plotted on the right side of the axis of symmetry. **Figure 10.8b** shows the completed graph. The point $(4, -16)$ is not shown on the graphs.

Now Try Exercise 45

(a)

(b)

FIGURE 10.9 Not every shape that resembles a parabola is a parabola. For example, the St. Louis Arch, **Figure 10.9a** resembles a parabola, but it is not a parabola. However, the bridge over the Mississippi near Jefferson Barracks, Missouri, which connects Missouri and Illinois, **Figure 10.9b**, is a parabola.

4 Find the Intercepts of the Graph of a Quadratic Equation

When graphing parabolas, knowing the location of the intercepts is very helpful. We include finding the intercepts as a part of a general strategy for graphing quadratic equations.

x-intercepts, y-intercepts

An **x-intercept** is a point where a graph crosses the x-axis. An x-intercept will always have the form $(x, 0)$.

A **y-intercept** is a point where a graph crosses the y-axis. A y-intercept will always have the form $(0, y)$.

Understanding Algebra

- To find the x-intercept(s), if they exist, set $y = 0$ and solve for x.
- To find the y-intercept, set $x = 0$ and solve for y.

To find the y-intercept, set $x = 0$ and solve for y. Notice that if we set $x = 0$ in the equation $y = ax^2 + bx + c$, we get $y = a(0)^2 + b(0) + c = 0 + 0 + c$. Thus, *a quadratic equation in the form $y = ax^2 + bx + c$ will always have y-intercept $(0, c)$.*

To find the x-intercept(s), if they exist, set $y = 0$ and solve for x. If we set $y = 0$ in the equation $y = ax^2 + bx + c$ we get $0 = ax^2 + bx + c$ or $ax^2 + bx + c = 0$. To solve this equation we can use one of three methods:

Method 1: Factoring, as explained in Section 5.6

Method 2: Completing the square, as explained in Section 10.2

Method 3: The quadratic formula, as explained in Section 10.3

The number of x-intercepts that a parabola has can be determined by the discriminant, $b^2 - 4ac$.

A quadratic equation of the form $y = ax^2 + bx + c$ will have either two distinct x-intercepts (**Fig. 10.10a**), one x-intercept (**Fig. 10.10b**), or no x-intercept (**Fig. 10.10c**).

Understanding Algebra

For the equation $y = ax^2 + bx + c = 0$, when the discriminant, $b^2 - 4ac$, is

- *positive:* there are two distinct x-intercepts.
- *equal to zero:* there is one x-intercept.
- *negative:* there is no x-intercept.

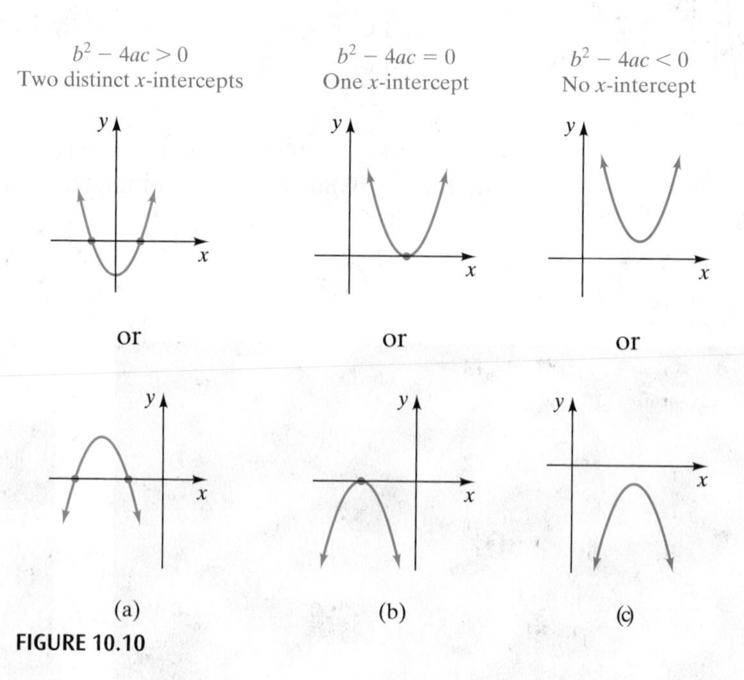

FIGURE 10.10

The x-intercepts can be found algebraically by setting y equal to 0 and solving the resulting equation for x, as we will show in Example 5.

EXAMPLE 5

a) Find the y-intercept of the graph of the equation $y = x^2 - 6x - 7$.

b) Find the x-intercepts of the graph of the equation $y = x^2 - 6x - 7$ by factoring, by completing the square, and by the quadratic formula.

c) Graph the equation.

Solution

a) To find the y-intercept we set $x = 0$ to get

$$y = (0)^2 - 6(0) - 7 = 0 - 0 - 7 = -7.$$

Thus the y-intercept is $(0, -7)$. We could also have noted that the constant term c was -7, therefore the y-intercept is $(0, -7)$.

b) To find the x-intercepts we set y equal to 0 and solve the resulting equation, $x^2 - 6x - 7 = 0$. We will solve this equation by all three algebraic methods.

Method 1: Factoring.

$$\begin{aligned} x^2 - 6x - 7 &= 0 \\ (x - 7)(x + 1) &= 0 \\ x - 7 = 0 \quad &\text{or} \quad x + 1 = 0 \\ x = 7 \quad & \quad\quad\quad x = -1 \end{aligned}$$

Method 2: Completing the square.

$$\begin{aligned} x^2 - 6x - 7 &= 0 \\ x^2 - 6x &= 7 \\ x^2 - 6x + 9 &= 7 + 9 \\ (x - 3)^2 &= 16 \\ x - 3 &= \pm 4 \\ x &= 3 \pm 4 \\ x = 3 + 4 \quad &\text{or} \quad x = 3 - 4 \\ x = 7 \quad & \quad\quad\quad x = -1 \end{aligned}$$

Method 3: Quadratic formula.

$$x^2 - 6x - 7 = 0$$

$$a = 1, \qquad b = -6, \qquad c = -7$$

$$\begin{aligned} x &= \frac{-b \pm \sqrt{b^2 - 4ac}}{2a} \\ &= \frac{-(-6) \pm \sqrt{(-6)^2 - 4(1)(-7)}}{2(1)} \\ &= \frac{6 \pm \sqrt{36 + 28}}{2} \\ &= \frac{6 \pm \sqrt{64}}{2} \\ &= \frac{6 \pm 8}{2} \end{aligned}$$

$$x = \frac{6 + 8}{2} \quad \text{or} \quad x = \frac{6 - 8}{2}$$

$$= \frac{14}{2} = 7 \qquad\qquad = \frac{-2}{2} = -1$$

Note that the same solutions, 7 and -1, were obtained by all three methods. The graph of the equation $y = x^2 - 6x - 7$ will cross the x-axis at 7 and -1. The x-intercepts are $(7, 0)$ and $(-1, 0)$.

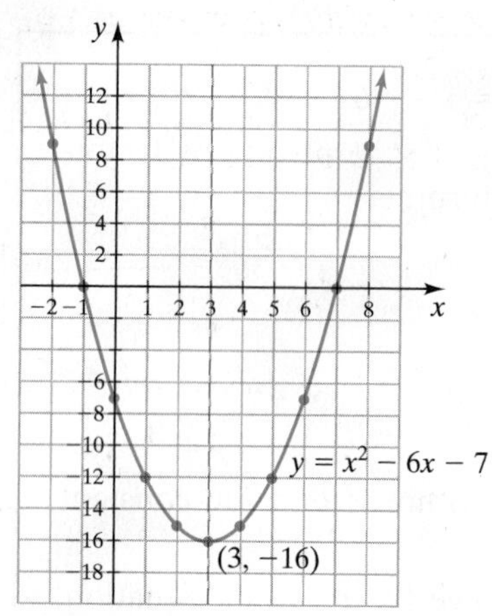

FIGURE 10.11

c) Since $a > 0$, this parabola opens upward.

$$\text{axis of symmetry:} \quad x = -\frac{b}{2a} = -\frac{-6}{2(1)} = \frac{6}{2} = 3$$

$$y = x^2 - 6x - 7$$

Let $x = 3$, $\quad y = 3^2 - 6(3) - 7 = -16$

Let $x = 4$, $\quad y = 4^2 - 6(4) - 7 = -15$

Let $x = 5$, $\quad y = 5^2 - 6(5) - 7 = -12$

Let $x = 6$, $\quad y = 6^2 - 6(6) - 7 = -7$

Let $x = 7$, $\quad y = 7^2 - 6(7) - 7 = 0$

Let $x = 8$, $\quad y = 8^2 - 6(8) - 7 = 9$

x	y
3	−16
4	−15
5	−12
6	−7
7	0
8	9

The vertex is at $(3, -16)$. Again we use symmetry to complete the graph (**Fig. 10.11**). The y-intercept is $(0, -7)$ and the x-intercepts are $(7, 0)$ and $(-1, 0)$. This agrees with the answer obtained in parts **a)** and **b)**.

Now Try Exercise 31

Our discussion of graphing quadratic equations can be summarized with the following guidelines.

Graphing Quadratic Equations of the Form $y = ax^2 + bx + c, a \neq 0$

1. If necessary, rewrite the equation in the form $y = ax^2 + bx + c$.
2. Examine the numerical coefficient a. If a is
   - Positive, then the parabola will open upward.
   - Negative, then the parabola will open downward.
3. Determine the equation of the axis of symmetry using $x = -\frac{b}{2a}$.
4. Determine the vertex of the parabola.
   - The x-coordinate of the vertex is the value determined in step 3.
   - The y-coordinate can be determined by either substituting the x-coordinate into the original equation, or by using the formula $y = \frac{4ac - b^2}{4a}$.
5. Determine the intercepts of the graph.
   - The y-intercept is $(0, c)$ where c is the constant term.
   - The x-intercepts are determined by setting $y = 0$ and solving for x by factoring, completing the square, or using the quadratic formula.
6. Begin your graph by graphing the axis of symmetry, vertex, and intercepts.
7. Select values of x and substitute them into the original equation to find the corresponding y-coordinates. Plot these points on your graph.
8. Use symmetry to plot more points on your graph.
9. Connect the points with a smooth curve.

10.4 Exercise Set MathXL® MyMathLab®

Warm-Up Exercises

Fill in the blanks with the appropriate word, phrase, or symbol(s) from the following list.

two	downward	one	symmetry	parabola
no	y-coordinate	upward	vertex	x-coordinate

1. The graph of $y = ax^2 + bx + c$ is a ________.

2. The graph of $y = 3x^2 - \frac{1}{2}x + 7$ opens ________.

3. The graph of $y = -\frac{1}{4}x^2 + x - 4$ opens ________.

4. The lowest point on a parabola that opens upward is the ________ of the parabola.

5. The formula $x = -\dfrac{b}{2a}$ is used to find the ________ of the vertex.

6. The formula $y = \dfrac{4ac - b^2}{4a}$ is used to find the ________ of the vertex.

7. The graph of a parabola is symmetric about the axis of ________.

8. If $b^2 - 4ac = 0$, the graph of the quadratic equation $y = ax^2 + bx + c$ has ________ x-intercept(s).

Practice the Skills

Indicate the equation of the axis of symmetry, the coordinates of the vertex, and whether the parabola opens upward or downward.

9. $y = x^2 + 2x - 7$

10. $y = x^2 + 4x - 3$

11. $y = 4x^2 + 8x + 3$

12. $y = -x^2 + 8x - 11$

13. $y = -3x^2 + 2x + 1$

14. $y = x^2 + 3x - 5$

15. $y = -x^2 + 3x - 4$

16. $y = 3x^2 - 2x + 6$

17. $y = 2x^2 + 3x + 2$

18. $y = -2x^2 - 6x - 1$

19. $y = -x^2 + x + 8$

20. $y = -5x^2 + 6x - 3$

Graph each quadratic equation and determine the x-intercepts, if they exist.

21. $y = x^2 + 3$

22. $y = x^2 - 1$

23. $y = -x^2 + 5$

24. $y = -x^2 + 9$

25. $y = x^2 + 4x + 3$

26. $y = 2x^2 + 1$

27. $y = x^2 + 4x + 4$

28. $y = x^2 - 4x + 4$

29. $y = -x^2 - 5x - 4$

30. $y = -x^2 - 5x + 6$

31. $y = x^2 + 5x - 6$

32. $y = x^2 - 4x - 5$

33. $y = x^2 + 5x - 14$

34. $y = x^2 - 5x + 4$

35. $y = x^2 - 6x + 9$

36. $y = x^2 + 8x + 16$

37. $y = x^2 - 6x$

38. $y = -x^2 + 5x$

39. $y = x^2 - 2x + 1$

40. $y = 2x^2 + 4x + 2$

41. $y = -x^2 + 7x - 10$

42. $y = -x^2 + 5x - 6$

43. $y = 4x^2 + 12x + 9$

44. $y = 2x^2 + 3x - 2$

45. $y = -2x^2 + 3x - 2$

46. $y = -4x^2 - 6x + 4$

47. $y = 2x^2 - x - 15$

48. $y = x^2 - x + 1$

Using the discriminant, determine the number of x-intercepts the graph of each equation will have. Do not graph the equation.

49. $y = 4x^2 - 2x - 7$

50. $y = 4x^2 - 6x - 5$

51. $y = x^2 - 22x + 121$

52. $y = x^2 - 6x + 9$

53. $y = -x^2 - x - 3$

54. $y = -x^2 - 3$

55. $y = -8x^2 - 5x + 1$

56. $y = -3x^2 + 7x - 2$

Problem Solving

The graph of a quadratic equation of the form $y = ax^2 + bx + c$ is a parabola. The value of a (the coefficient of the squared term in the equation) and the vertex of the parabola are given. In Exercises 57–60, determine the number of x-intercepts the parabola will have. Explain how you determined your answer.

57. $a = -2$, vertex at $(0, -4)$

58. $a = 2$, vertex at $(1, -3)$

59. $a = -3$, vertex at $(-5, 0)$

60. $a = -1$, vertex at $(2, -7)$

61. Height above Ground An object is projected upward from the ground. The height of the object above the ground, in feet, at time t, in seconds, is illustrated in the following graph.

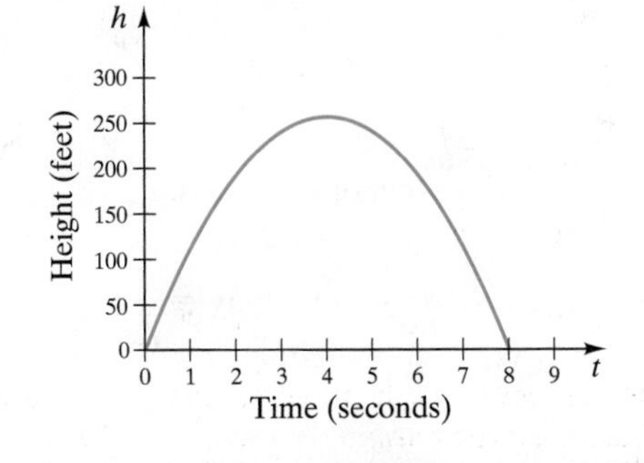

a) Estimate the maximum height the object will obtain.

b) How long will it take for the object to reach its maximum height?

c) Estimate how long it will take for the object to strike the ground.

d) Estimate the object's height at 2 seconds and at 5 seconds.

62. Height above Ground A ball is projected upward from the top of a building that is 100 feet above the ground. The height of the ball above the ground, in feet, at time t, in seconds, is illustrated in the following graph.

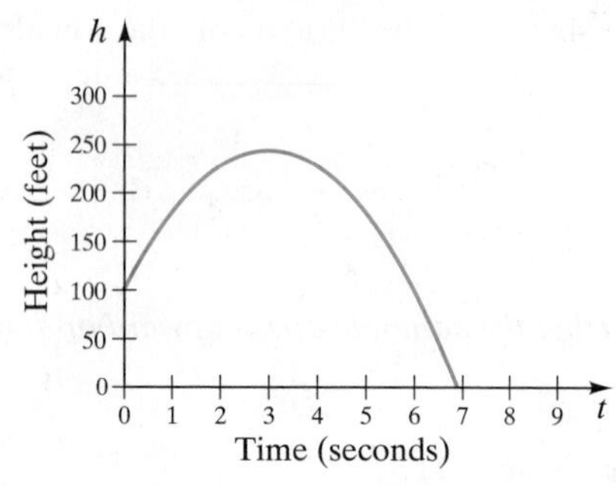

a) Estimate the maximum height the ball will obtain.

b) How long will it take for the ball to reach its maximum height?

c) Approximately how long will it take for the ball to strike the ground?

d) Estimate the ball's height above the ground at 2 seconds and at 5 seconds.

63. Maximum Area An area is to be fenced in along a river as shown in the figure on the right. Only 180 feet of fencing is available. The fenced-in area is $A = \text{length} \cdot \text{width}$ or $A = x(180 - 2x) = -2x^2 + 180x$.

a) Graph $A = -2x^2 + 180x$.

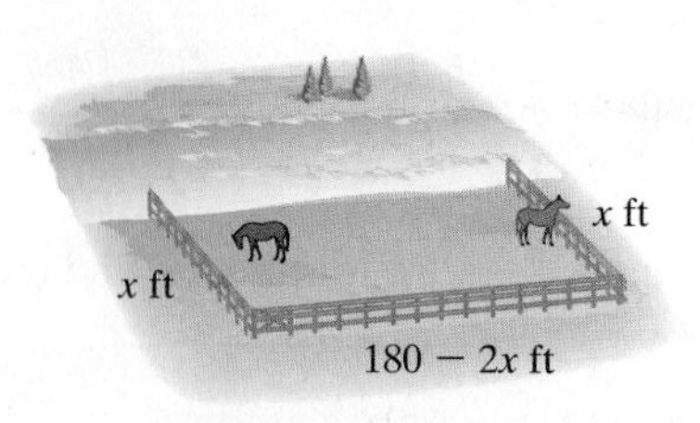

b) Using the graph, estimate the value of x that will yield the maximum area.

c) Estimate the maximum area.

64. Maximum Area See Exercise 63. If 260 feet of fencing is available, the fenced-in area is $A = x(260 - 2x) = -2x^2 + 260x$.

a) Graph $A = -2x^2 + 260x$.

b) Using the graph, estimate the value of x that yields the maximum area.

c) Estimate the maximum area.

65. Will the equations below have the same x-intercepts when graphed? Explain how you determined your answer.

$$y = x^2 - 2x - 15 \quad \text{and} \quad y = -x^2 + 2x + 15$$

66. a) How will the graphs of the following equations compare? Explain how you determined your answer.

$$y = x^2 - 2x - 8 \quad \text{and} \quad y = -x^2 + 2x + 8$$

b) Graph $y = x^2 - 2x - 8$ and $y = -x^2 + 2x + 8$ on the same axes.

Concept/Writing Exercises

67. Explain how to find the coordinates of the vertex of a parabola.

68. What determines whether the graph of a quadratic equation of the form $y = ax^2 + bx + c, a \neq 0$, is a parabola that opens upward or downward? Explain your answer.

69. a) What are the x-intercepts of a graph?

b) How can you find the x-intercepts of a graph algebraically?

70. What does it mean when we say that graphs of quadratic equations of the form $y = ax^2 + bx + c$ have symmetry about the axis of symmetry?

71. When graphing a quadratic equation of the form $y = ax^2 + bx + c$, what is the equation of the axis of symmetry?

72. How many x-intercepts will the graph of a quadratic equation have if the discriminant has a value of

a) 19

b) -5

c) 0

Challenge Problems

73. a) Graph the quadratic equation $y = -x^2 + 6x$.

b) On the same axes, graph the quadratic equation $y = x^2 - 2x$.

c) Estimate the points of intersection of the graphs. The points represent the solution to the system of equations.

74. a) Graph the quadratic equation $y = x^2 + 2x - 3$.

b) On the same axes, graph the quadratic equation $y = -x^2 + 1$.

c) Estimate the points of intersection of the graphs. The points represent the solution to the system of equations.

Cumulative Review Exercises

[6.4] **75.** Subtract $\frac{5}{x+3} - \frac{x-2}{x-4}$.

[6.6] **76.** Solve $\frac{1}{3}(x + 6) = 3 - \frac{1}{4}(x - 5)$.

[8.3] **77.** Solve the system of equations.

$$2x + 3y = -3$$
$$3x + 5y = -7$$

[9.5] **78.** Solve $\sqrt{x + 9} - x = -3$.

10.5 Complex Numbers

1. Write complex numbers using i.
2. Add and subtract complex numbers.
3. Solve quadratic equations with complex number solutions.

1 Write Complex Numbers Using i

Square roots of negative numbers, such as $\sqrt{-11}$, are called **imaginary numbers**. Such numbers are called imaginary (or nonreal) because when they were introduced many mathematicians refused to believe that they existed. Although they do not belong to the set of real numbers, the imaginary numbers do exist and are very useful in mathematics and science.

Every imaginary number has the number $\sqrt{-1}$ as a factor. The number $\sqrt{-1}$, called the **imaginary unit**, is denoted by the letter i.

Imaginary Unit

$$i = \sqrt{-1}$$

We can therefore write

$$\sqrt{-4} = \sqrt{4}\sqrt{-1} = 2\sqrt{-1} = 2i$$
$$\sqrt{-9} = \sqrt{9}\sqrt{-1} = 3\sqrt{-1} = 3i$$
$$\sqrt{-11} = \sqrt{11}\sqrt{-1} = \sqrt{11}i \quad \text{or} \quad i\sqrt{11}$$

We will generally write $i\sqrt{11}$ rather than $\sqrt{11}i$ to avoid confusion with $\sqrt{11i}$.

To help in writing square roots of negative numbers using i, we give the following rule.

Square Root of a Negative Number

For any positive real number n,

$$\sqrt{-n} = i\sqrt{n}$$

Examples

$$\sqrt{-16} = i\sqrt{16} = i \cdot 4 = 4i \qquad \sqrt{-6} = i\sqrt{6}$$
$$\sqrt{-100} = i\sqrt{100} = i \cdot 10 = 10i \qquad \sqrt{-10} = i\sqrt{10}$$

The real number system is a part of a larger number system, called the *complex number system*. Now we will discuss **complex numbers**.

Complex Number

Every number of the form

$$a + bi$$

where a and b are real numbers, and $i = \sqrt{-1}$, is a **complex number**.

A complex number has two parts: a real part, a, and an imaginary part, b.

Real part → a ; Imaginary part → bi

$$a + bi$$

Every real number and every imaginary number are also complex numbers. If $b = 0$, the complex number is a real number. If $a = 0$, the complex number is a **pure imaginary number**.

Understanding Algebra

The complex number $a + bi$ has two parts: The real part, a, and the imaginary part b.

Example of Complex Numbers

$3 + 8i$	$a = 3,$	$b = 8$	
$5 - i\sqrt{3}$	$a = 5,$	$b = -\sqrt{3}$	
2	$a = 2,$	$b = 0$	(real number, $b = 0$)
$9i$	$a = 0,$	$b = 9$	(pure imaginary number, $a = 0$)
$-i\sqrt{7}$	$a = 0,$	$b = -\sqrt{7}$	(pure imaginary number, $a = 0$)

We stated that all real numbers and imaginary numbers are also complex numbers. The relationship between the various sets of numbers is illustrated in **Figure 10.12**. As the figure illustrates, the real numbers include both the rational numbers and the irrational numbers, and the complex numbers include both the real numbers and the nonreal numbers.

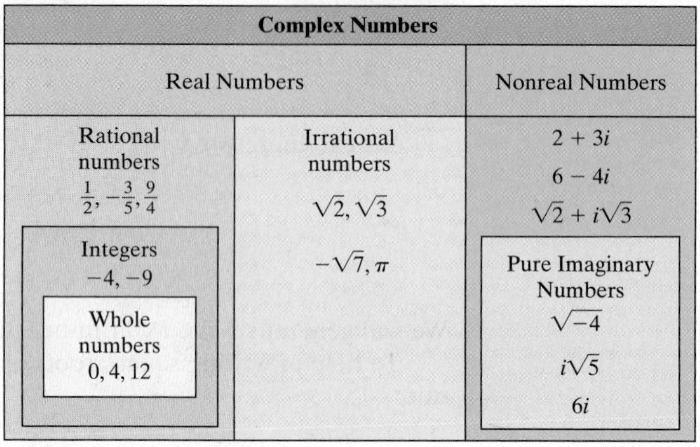

Complex Numbers

Real Numbers

Rational numbers $\frac{1}{2}, -\frac{3}{5}, \frac{9}{4}$

Integers $-4, -9$

Whole numbers $0, 4, 12$

Irrational numbers $\sqrt{2}, \sqrt{3}$, $-\sqrt{7}, \pi$

Nonreal Numbers

$2 + 3i$, $6 - 4i$, $\sqrt{2} + i\sqrt{3}$

Pure Imaginary Numbers $\sqrt{-4}$, $i\sqrt{5}$, $6i$

FIGURE 10.12

EXAMPLE 1 Write each of the following complex numbers in the form $a + bi$.

a) $5 + \sqrt{-4}$ **b)** $9 - \sqrt{-12}$

Solution

a) $5 + \sqrt{-4} = 5 + \sqrt{4}\sqrt{-1}$.
$= 5 + 2i$

b) $9 - \sqrt{-12} = 9 - \sqrt{12}\,\sqrt{-1}$
$= 9 - \sqrt{4}\,\sqrt{3}\,\sqrt{-1}$
$= 9 - 2\sqrt{3}i$ or $9 - 2i\sqrt{3}$

Now Try Exercise 21

In Section 10.3, Example 6, we obtained the answers $\frac{1 + \sqrt{-11}}{6}$ and $\frac{1 - \sqrt{-11}}{6}$. We indicated that since $\sqrt{-11}$ is not a real number we stopped there. Now we can rewrite these answers as $\frac{1 + i\sqrt{11}}{6}$ and $\frac{1 - i\sqrt{11}}{6}$, or $\frac{1 \pm i\sqrt{11}}{6}$.

2 Add and Subtract Complex Numbers

Complex numbers can be added, subtracted, multiplied, and divided. To add (or subtract) complex numbers, add (or subtract) their real parts and add (or subtract) their imaginary parts.

EXAMPLE 2 Add or subtract the following complex numbers.

a) $(4 + 3i) + (5 - 8i)$ **b)** $(7 - 2i) - (5 - 3i)$

Solution In each case we will remove the parentheses and combine the real parts and combine the imaginary parts. This is similar to combining like terms when adding and subtracting polynomials.

a) $$\begin{aligned}(4 + 3i) + (5 - 8i) &= 4 + 3i + 5 - 8i \\ &= 4 + 5 + 3i - 8i = 9 - 5i\end{aligned}$$

b) $$\begin{aligned}(7 - 2i) - (5 - 3i) &= 7 - 2i - 5 + 3i \\ &= 7 - 5 - 2i + 3i = 2 + i\end{aligned}$$

Now Try Exercise 37

Complex numbers can also be multiplied and divided, but we will not multiply and divide complex numbers in this text.

3 Solve Quadratic Equations with Complex Number Solutions

Before we leave this section, let's solve a quadratic equation that does not have real numbers as solutions. We will write the answers as complex numbers.

EXAMPLE 3 Solve $2x^2 - 4x + 7 = 0$.

Solution We will use the quadratic formula.

$$a = 2, \quad b = -4, \quad c = 7$$

$$x = \frac{-b \pm \sqrt{b^2 - 4ac}}{2a} \qquad \text{Quadratic formula}$$

$$= \frac{-(-4) \pm \sqrt{(-4)^2 - 4(2)(7)}}{2(2)}$$

$$= \frac{4 \pm \sqrt{16 - 56}}{4}$$

$$= \frac{4 \pm \sqrt{-40}}{4} \qquad \text{These are complex numbers.}$$

$$= \frac{4 \pm \sqrt{4}\sqrt{10}\sqrt{-1}}{4}$$

$$= \frac{4 \pm 2i\sqrt{10}}{4} \qquad \text{Write complex numbers using } i.$$

$$= \frac{\overset{1}{\cancel{2}}(2 \pm i\sqrt{10})}{\underset{2}{\cancel{4}}} \qquad \text{Factor numerator, divide out common factors.}$$

$$= \frac{2 \pm i\sqrt{10}}{2}$$

The solutions, $\frac{2 + i\sqrt{10}}{2}$ and $\frac{2 - i\sqrt{10}}{2}$, are complex numbers.

Now Try Exercise 51

10.5 Exercise Set MathXL® MyMathLab®

Warm-Up Exercises

Fill in the blanks with the appropriate word, phrase, or symbol(s) from the following list.

real negative unit pure imaginary positive complex irrational

1. Square roots of ________ numbers, such as $\sqrt{-11}$, are called imaginary numbers.
2. The number $\sqrt{-1}$, is called the imaginary ________.
3. Every number of the form $a + bi$, where a and b are real numbers and $i = \sqrt{-1}$, is a ________ number.
4. In the complex number $5 + 3i$, the number 5 is the ________ part.
5. In the complex number $5 + 3i$, the number 3 is the ________ part.
6. In the complex number $a + bi$, if $a = 0$, the complex number is a ________ imaginary number.

Practice the Skills

Write each imaginary number in terms of i.

7. $\sqrt{-16}$
8. $\sqrt{-81}$
9. $\sqrt{-100}$
10. $\sqrt{-144}$
11. $\sqrt{-23}$
12. $\sqrt{-13}$
13. $\sqrt{-10}$
14. $\sqrt{-37}$
15. $\sqrt{-32}$
16. $\sqrt{-18}$
17. $\sqrt{-45}$
18. $\sqrt{-300}$

Write each complex number in the form of a + bi.

19. $8 + \sqrt{-4}$
20. $9 + \sqrt{-49}$
21. $10 + \sqrt{-16}$
22. $11 + \sqrt{-64}$
23. $-3 - \sqrt{-9}$
24. $-4 - \sqrt{-81}$
25. $-9 - \sqrt{-15}$
26. $22 - \sqrt{-6}$
27. $5.2 + \sqrt{-48}$
28. $21.9 - \sqrt{-32}$
29. $\frac{1}{6} + \sqrt{-75}$
30. $-\frac{2}{5} + \sqrt{-80}$

Add or subtract the following complex numbers.

31. $(2 + 3i) + 8$
32. $(3 - 5i) + 4$
33. $(13 - 2i) - 9i$
34. $(11 - 5i) - 7i$
35. $(2 + 3i) + (9 + 2i)$
36. $(4 + i) + (7 + 9i)$
37. $(4 - 3i) - (6 + 4i)$
38. $(-2 + 5i) - (-6 - 3i)$
39. $(-9 + 2i) + (-4 - i)$
40. $(7 - 5i) + (8 - 2i)$
41. $(27 - 3i) - (36 + i)$
42. $(6 - 5i) - (11 + 8i)$

Solve the quadratic equation and write the solutions in terms of i.

43. $x^2 = -16$
44. $x^2 = -100$
45. $4x^2 = -120$
46. $5a^2 = -60$
47. $2r^2 + 3r + 5 = 0$
48. $3w^2 - 5w + 4 = 0$
49. $x^2 - 4x + 5 = 0$
50. $x^2 - 2x + 5 = 0$
51. $2p^2 + 4p + 7 = 0$
52. $4m^2 - 6m + 3 = 0$
53. $-5x^2 + 2x - 3 = 0$
54. $-3z^2 - 2z - 10 = 0$

Concept/Writing Exercises

55. Under what conditions will an equation of the form $x^2 = c$ have imaginary number solutions?
56. Can an equation of the form $x^2 = c$ ever have only one imaginary solution? Explain.
57. Under what conditions will an equation of the form $ax^2 + bx + c = 0$ have nonreal number solutions?
58. Is it possible for a quadratic equation to have only one nonreal number solution? Explain.

Cumulative Review Exercises

[1.9] 59. Evaluate $-\{[3(x - 4)^2 - 5] - 2x\}$ when $x = 6$.

[2.5] 60. Solve the equation $\frac{1}{2}x + \frac{3}{5}x = \frac{1}{2}(x - 2)$.

[6.6] 61. Solve the equation $\frac{w - 2}{w - 5} = \frac{3}{w - 6} + \frac{3}{4}$.

[9.5] 62. Solve the equation $2\sqrt{r - 4} + 5 = 11$.

Chapter 10 Summary

IMPORTANT FACTS AND CONCEPTS	EXAMPLES
Section 10.1	
A quadratic equation in standard form is one of the form $ax^2 + bx + c = 0, a \neq 0$.	$3x^2 - x + 10 = 0$
Square Root Property If $x^2 = a$, then $x = \sqrt{a}$ or $x = -\sqrt{a}$ (abbreviated $x = \pm\sqrt{a}$).	If $x^2 = 5$, then $x = \sqrt{5}$ or $x = -\sqrt{5}$ (or $x = \pm\sqrt{5}$).
Section 10.2	
A perfect square trinomial is a trinomial that can be expressed as the square of a binomial.	$x^2 - 10x + 25 = (x - 5)^2$
To Solve a Quadratic Equation by Completing the Square **1.** Use the multiplication (or division) property of equality if necessary to make the numerical coefficient of the squared term equal to 1. **2.** Rewrite the equation with the constant by itself on the right side of the equation. **3.** Take one-half the numerical coefficient of the first-degree term, square it, and add this quantity to both sides of the equation. **4.** Replace the trinomial with its equivalent squared binomial. **5.** Use the square root property. **6.** Solve for the variable. **7.** Check your answers in the *original* equation.	Solve $x^2 + 2x - 35 = 0$ by completing the square. $x^2 + 2x - 35 = 0$ $x^2 + 2x = 35$ $x^2 + 2x + 1 = 35 + 1$ $(x + 1)^2 = 36$ $x + 1 = \pm\sqrt{36}$ $x + 1 = \pm 6$ $x = -1 \pm 6$ $x = -1 + 6 = 5$ or $x = -1 - 6 = -7$. A check shows that 5 and -7 are the solutions.
Section 10.3	
To Solve a Quadratic Equation by the Quadratic Formula **1.** Write the equation in standard form, $ax^2 + bx + c = 0$, and determine the numerical values for a, b, and c. **2.** Substitute the values for a, b, and c from step 1 into the quadratic formula below and then evaluate to obtain the solution. **The Quadratic Formula** $x = \frac{-b \pm \sqrt{b^2 - 4ac}}{2a}$	Solve $x^2 + 4x - 21 = 0$ by the quadratic formula. $x^2 + 4x - 21 = 0$ $a = 1, \quad b = 4, \quad c = -21$ $x = \frac{-b \pm \sqrt{b^2 - 4ac}}{2a}$ $= \frac{-4 \pm \sqrt{4^2 - 4(1)(-21)}}{2(1)}$ $= \frac{-4 \pm \sqrt{100}}{2} = \frac{-4 \pm 10}{2}$ $x = \frac{-4 + 10}{2} = \frac{6}{2} = 3$ or $x = \frac{-4 - 10}{2} = \frac{-14}{2} = -7$. A check shows that 3 and -7 are the solutions.
The expression under the square root sign in the quadratic formula, $b^2 - 4ac$, is called the **discriminant**. **When the Discriminant Is** **1. Greater than zero,** $b^2 - 4ac > 0$, the quadratic equation has *two distinct real number solutions*. **2. Equal to zero,** $b^2 - 4ac = 0$, the quadratic equation has *one real number solution*. **3. Less than zero,** $b^2 - 4ac < 0$, the quadratic equation has *no real number solution*.	$5x^2 - 3x + 2 = 0$: $b^2 - 4ac = (-3)^2 - 4(5)(2) = -31$ The discriminant is -31. Therefore, the equation has no real number solution. $t^2 - 18t + 81 = 0$: $b^2 - 4ac = (-18)^2 - 4(1)(81) = 0$ The discriminant is 0. Therefore, the equation has one real number solution. $2d^2 + 7d - 3 = 0$: $b^2 - 4ac = (7)^2 - 4(2)(-3) = 73$ The discriminant is 73. Therefore, the equation has two real number solutions.

IMPORTANT FACTS AND CONCEPTS	EXAMPLES
Section 10.4	
The graph of an equation of the form $y = ax^2 + bx + c, a \neq 0$, is called a **parabola**. The parabola **opens upward** when $a > 0$ and **downward** when $a < 0$. The **vertex** is the lowest point on a parabola that opens upward or the highest point on a parabola that opens downward. The **axis of symmetry** is a vertical line through the vertex where the left and right sides have symmetry.	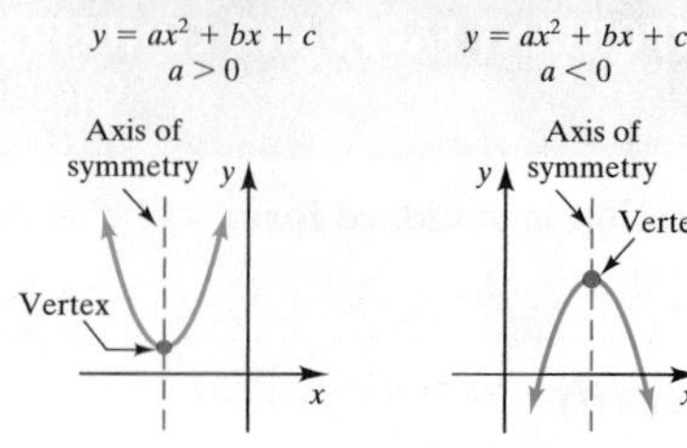
Equation of the Axis of Symmetry and *x*-Coordinate of the Vertex $x = -\frac{b}{2a}$ **y-Coordinate of the Vertex** $y = \frac{4ac - b^2}{4a}$	Find the axis of symmetry and the vertex of the graph of $y = x^2 - 4x + 9$. $x = -\frac{b}{2a} = -\frac{-4}{2(1)} = 2$ axis of symmetry: $x = 2$ $y = \frac{4ac - b^2}{4a} = \frac{4(1)(9) - (-4)^2}{4(1)} = \frac{36 - 16}{4} = \frac{20}{4} = 5$ vertex: $(2,5)$
Section 10.5	
The **imaginary unit, *i*,** is defined as $i = \sqrt{-1}$. For any positive number n, $\sqrt{-n} = i\sqrt{n}$.	$\sqrt{-4} = 2i$, $\sqrt{-10} = i\sqrt{10}$
A **complex number** is of the form $a + bi$, where a and b are real numbers and $i = \sqrt{-1}$.	$2 + 6i$, $9 - i\sqrt{13}$
To add (or subtract) complex numbers, add (or subtract) their real parts and add (or subtract) their imaginary parts.	$(2 + 3i) + (8 - 7i) = (2 + 8) + (3 - 7)i = 10 - 4i$

Chapter 10 Review Exercises

[10.1] *Solve using the square root property.*

1. $x^2 = 25$
2. $x^2 = 144$
3. $x^2 + 7 = 12$
4. $4x^2 = 24$
5. $2x^2 - 3 = 11$
6. $x^2 + 10 = 30$
7. $4a^2 - 31 = 1$
8. $(r - 5)^2 = 24$
9. $(4t - 9)^2 = 50$
10. $(3x + 4)^2 = 30$

[10.2] *Solve by completing the square.*

11. $x^2 - 7x + 10 = 0$
12. $x^2 - 11x + 28 = 0$
13. $x^2 + 15x + 14 = 0$
14. $x^2 + 18x + 80 = 0$
15. $t^2 - 4t - 45 = 0$
16. $x^2 = -5x + 6$
17. $2x^2 - 8x = 64$
18. $30 = 2n^2 - 4n$
19. $x^2 + 2x - 9 = 0$
20. $y^2 - 5y - 2 = 0$
21. $3b^2 + b = 2$
22. $12a^2 + 19a = -4$

[10.3] *Determine whether each equation has two distinct real number solutions, one real number solution, or no real number solution.*

23. $-4x^2 + 4x - 5 = 0$
24. $-3x^2 + 4x = 10$
25. $x^2 - 18x + 81 = 0$
26. $y^2 + 5y - 8 = 0$
27. $4z^2 - 3z = 2$
28. $3x^2 - 2x + 4 = 0$
29. $-3x^2 - 4x + 1 = 0$
30. $x^2 - 9x + 6 = 0$

Use the quadratic formula to solve. If an equation has no real number solution, so state.

31. $x^2 - 10x + 16 = 0$
32. $x^2 - 12x = -27$
33. $x^2 - 7x - 44 = 0$
34. $4x^2 - 9x = -5$
35. $35 = r^2 + 2r$
36. $x^2 - x + 3 = 0$
37. $10x^2 - 7x = 12$
38. $-2x^2 + 3x + 6 = 0$
39. $-2x^2 - 4x + 3 = 0$
40. $y^2 - 6y + 3 = 0$
41. $3x^2 - 4x + 8 = 0$
42. $3x^2 - 6x - 8 = 0$
43. $7x^2 - 5x = 0$
44. $12z^2 = 10z$

[10.1–10.3] *Solve each quadratic equation using the method of your choice.*

45. $x^2 - 11x + 28 = 0$
46. $x^2 - 14x + 24 = 0$
47. $r^2 - 4r - 60 = 0$
48. $x^2 + 6x = 27$
49. $x^2 + 15x + 44 = 0$
50. $x^2 + 13x + 40 = 0$
51. $y^2 + 9y - 36 = 0$
52. $t^2 = 8t$
53. $x^2 = 81$
54. $x^2 = 196$
55. $3x^2 = 11x - 10$
56. $6x^2 + 5x = 6$
57. $5x^2 - 11x = 0$
58. $3x^2 + 13x = 0$
59. $-5w = 3w^2 - 8$
60. $3x^2 - 11x + 10 = 0$
61. $6n = 2n^2 - 9$
62. $x^2 + 3x = 6$

[10.4] *Indicate the equation of the axis of symmetry, the coordinates of the vertex, and whether the parabola opens upward or downward.*

63. $y = x^2 - 4x - 5$
64. $y = x^2 - 12x + 22$
65. $y = x^2 - 3x + 7$
66. $y = -x^2 - 2x + 9$
67. $y = 3x^2 + 7x + 3$
68. $y = -x^2 - 7x$
69. $y = -x^2 - 12$
70. $y = -2x^2 - x + 15$
71. $y = -4x^2 + 8x + 9$
72. $y = 3x^2 + 5x - 6$

Graph each quadratic equation and determine the x-intercepts, if they exist. If they do not exist, so state.

73. $y = x^2 - 2x$
74. $y = -x^2 - 6x$
75. $y = -x^2 - 1$
76. $y = -3x^2 + 6$
77. $y = x^2 + x + 1$
78. $y = x^2 + 5x + 4$
79. $y = -x^2 + 2x - 3$
80. $y = x^2 - 2x - 16$
81. $y = x^2 + 4x + 3$
82. $y = 3x^2 - 4x - 8$
83. $y = -2x^2 + 7x - 3$
84. $y = 6x^2 + 10x - 4$

[10.1–10.4] *Solve.*

85. Product of Integers The product of two positive consecutive odd integers is 99. Find the integers.

86. Product of Integers The product of two positive integers is 78. Find the two integers if the larger one is 7 greater than the smaller.

87. Rectangular Table Samuel Jones is making a table with a rectangular top for the kitchen. He determines that the length of the table top should be 6 inches more than twice its width. What will be the dimensions of the table top if its area is to be 920 square inches?

88. Wood Desk Jordan and Patricia Wells recently purchased an oak desk for their son's room. To protect the rectangular writing surface of the desk, they decided to cover the top with a piece of glass. The length of the desktop is 16 inches greater than its width. Find the dimensions of the glass piece Jordan and Patricia should order if the area of the desktop is 960 square inches.

[10.5] *Write each imaginary number or complex number in terms of i.*

89. $\sqrt{-9}$

90. $\sqrt{-26}$

91. $8 - \sqrt{-25}$

92. $9 - \sqrt{-60}$

Add or subtract the complex numbers.

93. $(4 - 6i) + (8 - 3i)$

94. $(9 + 5i) - (6 - 3i)$

Solve the quadratic equations and write the complex solutions in terms of i.

95. $5x^2 = -80$

96. $7a^2 = -42$

97. $2r^2 - 5r + 6 = 0$

98. $4w^2 - 8w + 9 = 0$

Chapter 10 Practice Test

Chapter Test Prep Videos provide fully worked-out solutions to any of the exercises you want to review. Chapter Test Prep Videos are available via MyMathLab®, *or on* YouTube™ (www.youtube.com/user/AngelElementaryAlg).

1. Solve $x^2 - 2 = 30$ using the square root property.

2. Solve $(2p - 3)^2 = 13$ using the square root property.

3. Solve $x^2 - 5x = 50$ by completing the square.

4. Solve $r^2 + 8r = 33$ by completing the square.

5. Solve $k^2 = 13k - 42$ by the quadratic formula.

6. Solve $2x^2 + 5 = -8x$ by the quadratic formula.

7. Solve $16x^2 = 25$ by the method of your choice.

8. Write the quadratic formula.

9. Give an example of a perfect square trinomial.

10. Determine whether $-2x^2 - 4x + 7 = 0$ has two distinct real solutions, one real solution, or no real solution. Explain your answer.

11. Determine whether $x^2 + 10x + 25 = 0$ has two distinct real solutions, one real solution, or no real solution. Explain your answer.

12. Find the equation of the axis of symmetry of the graph of $y = -x^2 - 8x + 11$.

13. Find the equation of the axis of symmetry of the graph of $y = 4x^2 - 16x + 9$.

14. Determine whether the graph of $y = -x^2 - 10x + 3$ opens upward or downward. Explain your answer.

15. Determine whether the graph of $y = 5x^2 - 2x + 9$ opens upward or downward. Explain your answer.

16. What is the vertex of the graph of a parabola?

17. Find the vertex of the graph of $y = -x^2 - 4x - 12$.

18. Find the vertex of the graph of $y = 3x^2 - 8x + 9$.

19. Graph the equation $y = x^2 + 2x - 8$ and determine the x-intercepts, if they exist.

20. Graph the equation $y = 2x^2 - 6x$ and determine the x-intercepts, if they exist.

21. Graph the equation $y = -x^2 + 6x - 9$ and determine the x-intercepts, if they exist.

22. Wall Mural The length of a rectangular wall mural is 2 feet greater than 3 times its width. Find the length and width of the mural if its area is 33 square feet.

23. Consecutive Odd Integers The product of two positive consecutive odd integers is 35. Find the larger of the two integers.

24. Shawn's Age Shawn Goodwin is 6 years older than his cousin, Aaron. The product of their ages is 55. How old is Shawn?

25. Solve the quadratic equation $3p^2 - 2p + 6 = 0$ and write the answer in terms of i.

Cumulative Review Test

Take the following test and check your answers with those given in the back of the book. Review any questions that you answered incorrectly. The section where the material was covered is indicated after the answer.

1. Evaluate $-5x^2y + 3y^2 + xy$ when $x = 4$ and $y = -3$.

2. Solve $\frac{1}{2}z - \frac{2}{7}z = \frac{1}{5}(3z - 1)$.

3. Find the length of side x.

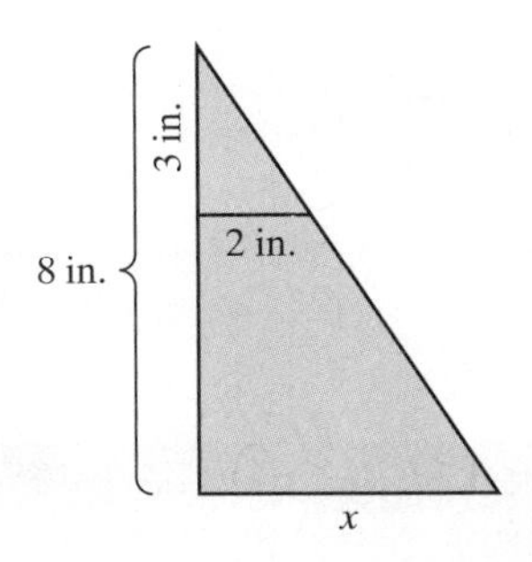

4. Solve the inequality $2(x - 3) \le 6x - 5$. Graph the solution on a number line and write the solution using interval notation.

5. Solve the formula $A = \frac{m + n + P}{2}$ for P.

6. Simplify $(2a^4b^5)^3(3a^2b^5)^2$.

7. Divide $\frac{x^2 + 6x - 7}{x + 2}$.

8. Factor by grouping $2x^2 - 3xy - 4xy + 6y^2$.

9. Factor $6x^2 - 27x - 54$.

10. Add $\frac{3}{a^2 - 16} + \frac{2}{(a - 4)^2}$.

11. Solve the equation $x + \frac{24}{x} = 11$.

12. Graph the equation $4x - y = 8$.

13. Solve the system of equations by the addition method.

$$5x - 3y = 12$$
$$4x - 2y = 6$$

14. Simplify $\sqrt{\frac{4x^2y^3}{24x}}$.

15. Add $2\sqrt{28} - 3\sqrt{7} + \sqrt{63}$.

16. Solve the equation $x + 1 = \sqrt{x^2 + 9}$.

17. Solve the equation $2x^2 + 3x - 8 = 0$ using the quadratic formula.

18. Fertilizer If 4 pounds of fertilizer can fertilize 500 square feet of lawn, how many pounds of fertilizer are needed to fertilize 3200 square feet of lawn?

19. Vegetable Garden The length of a rectangular vegetable garden is 3 feet less than four times its width. Find the width and length of the garden if its perimeter is 64 feet.

20. Jogging Robert McCloud jogs 3 miles per hour faster than he walks. He jogs for 2 miles and then walks for 2 miles. If the total time of his outing is 1 hour, find the rate at which he walks and jogs.

Appendices

A Review of Decimals and Percent

B Geometry

Appendix A Review of Decimals and Percent

Decimals

To Add or Subtract Numbers Containing Decimal Points

1. Align the numbers by the decimal points.
2. Add or subtract the numbers as if they were whole numbers.
3. Place the decimal point in the sum or difference directly below the decimal points in the numbers being added or subtracted.

EXAMPLE 1 Add 4.6 + 13.813 + 9.02.

Solution

$$\begin{array}{r} 4.600 \\ 13.813 \\ +\ 9.020 \\ \hline 27.433 \end{array}$$

EXAMPLE 2 Subtract 3.062 from 34.9.

Solution

$$\begin{array}{r} 34.900 \\ -\ 3.062 \\ \hline 31.838 \end{array}$$

To Multiply Numbers Containing Decimal Points

1. Multiply as if the factors were whole numbers.
2. Determine the total number of digits to the right of the decimal points in the factors.
3. Place the decimal point in the product so that the product contains the same number of digits to the right of the decimal as the total found in step 2. For example, if there are a total of three digits to the right of the decimal points in the factors, there must be three digits to the right of the decimal point in the product.

EXAMPLE 3 Multiply 2.34 × 1.9.

Solution

$$\begin{array}{rl} 2.34 & \longleftarrow \text{two digits to the right of the decimal point} \\ \times\ \ 1.9 & \longleftarrow \text{one digit to the right of the decimal point} \\ \hline 2106 & \\ 234 & \\ \hline 4.446 & \longleftarrow \text{three digits to the right of the decimal point in the product} \end{array}$$

EXAMPLE 4 Multiply 2.13 × 0.02.

Solution

$$\begin{array}{r l} 2.13 & \longleftarrow \text{ two digits to the right of the decimal point} \\ \times\ 0.02 & \longleftarrow \text{ two digits to the right of the decimal point} \\ \hline 0.0426 & \longleftarrow \text{ four digits to the right of the decimal point in the product} \end{array}$$

Note that it was necessary to add a zero preceding the digit 4 in the answer in order to have four digits to the right of the decimal point.

To Divide Numbers Containing Decimal Points

1. Multiply both the dividend and divisor by a power of 10 that will make the divisor a whole number.
2. Divide as if working with whole numbers.
3. Place the decimal point in the quotient directly above the decimal point in the dividend.

To make the divisor a whole number, multiply *both* the dividend and divisor by 10 if the divisor is given in tenths, by 100 if the divisor is given in hundredths, by 1000 if the divisor is given in thousandths, and so on. Multiplying both the numerator and denominator by the same nonzero number is the same as multiplying the fraction by 1. Therefore, the value of the fraction is unchanged.

EXAMPLE 5 Divide $\frac{1.956}{0.12}$.

Solution Since the divisor, 0.12, is twelve-hundredths, we multiply both the divisor and dividend by 100.

$$\frac{1.956}{0.12} \times \frac{100}{100} = \frac{195.6}{12.}$$

Now we divide.

$$\begin{array}{r} 16.3 \\ 12\overline{)195.6} \\ \underline{12} \\ 75 \\ \underline{72} \\ 3\,6 \\ \underline{3\,6} \\ 0 \end{array}$$

The decimal point in the answer is placed directly above the decimal point in the dividend. Thus, $\frac{1.956}{0.12} = 16.3$.

EXAMPLE 6 Divide 0.26 by 10.4.

Solution First, multiply both the dividend and divisor by 10.

$$\frac{0.26}{10.4} \times \frac{10}{10} = \frac{2.6}{104.}$$

Now divide.

$$\begin{array}{r} 0.025 \\ 104\overline{)2.600} \\ \underline{2\,08} \\ 520 \\ \underline{520} \\ 0 \end{array}$$

Note that a zero had to be placed before the digit 2 in the quotient.

$$\frac{0.26}{10.4} = 0.025$$

Rounding Decimal Numbers

Now we will explain how to round decimal numbers. The explanation of the procedure will refer to the positional values to the right of the decimal point, as illustrated here:

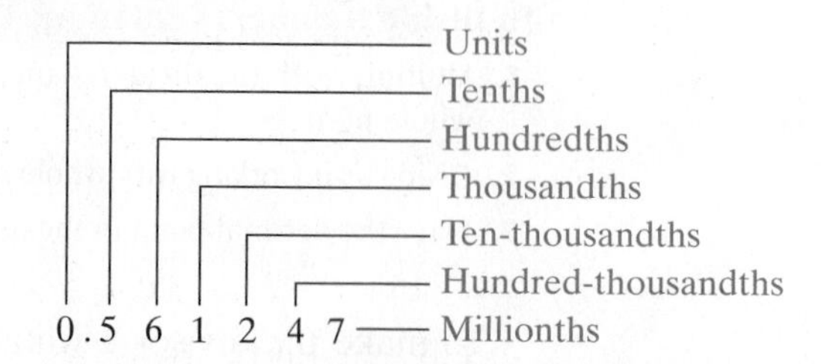

To Round a Decimal Number

1. Identify the number in the problem to the right of the positional value to which the number is to be rounded. For example, if you wish to round to tenths, you identify the number in the hundredths position. If you wish to round to hundredths, you identify the number in the thousandths position, and so on.
2. **a)** If the number you identified in step 1 is greater than or equal to 5, the number in the preceding position is increased by 1 unit, and all numbers to the right of the number that has been increased by 1 unit are eliminated.
   b) If the number you identified in step 1 is less than 5, the number you identified in step 1 and all numbers to its right are eliminated.

EXAMPLE 7 Round 4.863 to tenths.

Solution Since we are rounding to tenths, we identify the number in the hundredths position as 6. Because the number 6 is greater than or equal to 5, we increase the number in the preceding position by 1 unit and drop the remaining digits. Therefore, 4.863 when rounded to tenths is 4.9.

EXAMPLE 8 Round 5.4738 to hundredths.

Solution Since we are rounding to hundredths, we identify the number in the thousandths position as 3. Because the number 3 is less than 5, we drop the 3 and all numbers to the right of 3. Therefore, 5.4738 when rounded to hundredths is 5.47.

Percent

The word *percent* means "per hundred." The symbol % means percent. One percent means "one per hundred."

One Percent

$$1\% = \frac{1}{100} \quad \text{or} \quad 1\% = 0.01$$

EXAMPLE 9 Convert 16% to a decimal.

Solution Since 1% = 0.01,

$$16\% = 16(0.01) = 0.16$$

EXAMPLE 10 Convert 4.7% to a decimal.

Solution $4.7\% = 4.7(0.01) = 0.047$

EXAMPLE 11 Convert 1.14 to a percent.

Solution To change a decimal number to a percent, we multiply the number by 100%.

$$1.14 = 1.14 \times 100\% = 114\%$$

Often, you will need to find an amount that is a certain percent of a number. For example, when you purchase an item in a state or county that has a sales tax you must often pay a percent of the item's price as the sales tax. Examples 12 and 13 show how to find a certain percent of a number.

EXAMPLE 12 Find 32% of 300.

Solution To find a percent of a number, use multiplication. Change 32% to a decimal number, then multiply by 300.

$$(0.32)(300) = 96$$

Thus, 32% of 300 is 96.

EXAMPLE 13 Johnson County charges an 8% sales tax.

a) Find the sales tax on a stereo system that cost $580.

b) Find the total cost of the system, including tax.

Solution

a) The sales tax is 8% of 580.

$$(0.08)(580) = 46.40$$

The sales tax is $46.40.

b) The total cost is the purchase price plus the sales tax:

$$\text{total cost} = \$580 + \$46.40 = \$626.40$$

Appendix B Geometry

This appendix introduces or reviews important geometric concepts. **Table B.1** gives the names and descriptions of various types of angles.

Angles

TABLE B.1

Angle	Sketch of Angle
An **acute angle** is an angle whose measure is between 0° and 90°.	
A **right angle** is an angle whose measure is 90°.	

(Continued on next page)

TABLE B.1 ***(Continued)***

Angle	Sketch of Angle
An **obtuse angle** is an angle whose measure is between 90° and 180°.	
A **straight angle** is an angle whose measure is 180°.	
Two angles are **complementary angles** when the sum of their measures is 90°. Each angle is the complement of the other. Angles A and B are complementary angles.	
Two angles are **supplementary angles** when the sum of their measures is 180°. Each angle is the supplement of the other. Angles A and B are supplementary angles.	

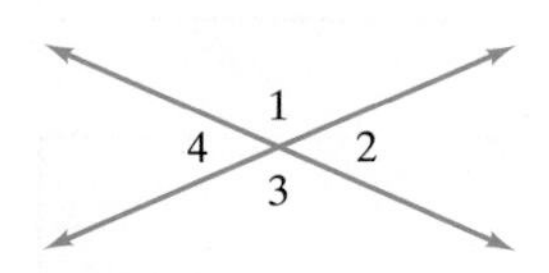

FIGURE B.1

When two lines intersect, four angles are formed as shown in **Figure B.1**. The pair of opposite angles formed by the intersecting lines are called **vertical angles**.

Angles 1 and 3 are vertical angles. Angles 2 and 4 are also vertical angles. *Vertical angles have equal measures*. Thus, angle 1, symbolized by $\angle 1$, is equal to angle 3, symbolized by $\angle 3$. We can write $\angle 1 = \angle 3$. Similarly, $\angle 2 = \angle 4$.

Parallel and Perpendicular Lines

Parallel lines are two lines in the same plane that do not intersect (**Fig. B.2**). **Perpendicular lines** are lines that intersect at right angles (**Fig. B.3**).

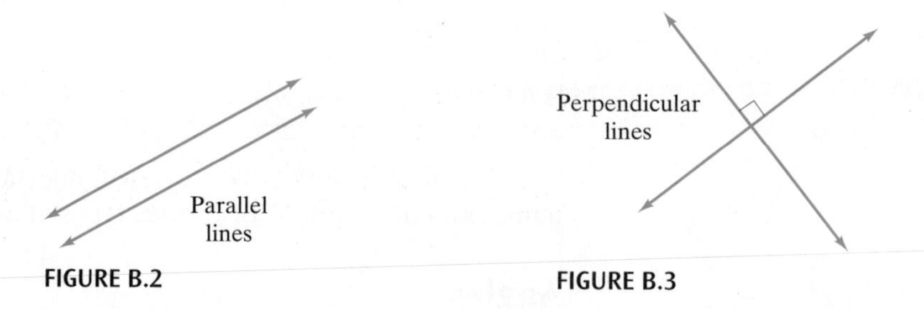

FIGURE B.2 FIGURE B.3

A **transversal** is a line that intersects two or more lines at different points. When a transversal line intersects two other lines, eight angles are formed, as illustrated in **Figure B.4**. Some of these angles are given special names.

FIGURE B.4

Interior angles: 3, 4, 5, 6

Exterior angles: 1, 2, 7, 8

Pairs of corresponding angles: 1 and 5; 2 and 6; 3 and 7; 4 and 8

Pairs of alternate interior angles: 3 and 6; 4 and 5

Pairs of alternate exterior angles: 1 and 8; 2 and 7

Parallel Lines Cut by a Transversal

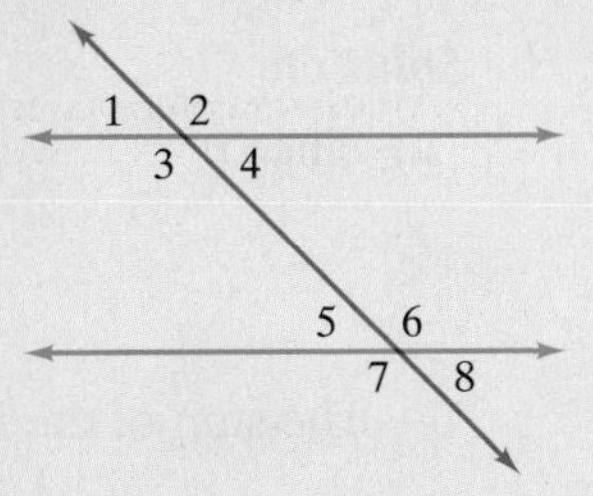

When two parallel lines are cut by a transversal,

1. Corresponding angles are equal $(\angle 1 = \angle 5, \angle 2 = \angle 6, \angle 3 = \angle 7, \angle 4 = \angle 8)$.
2. Alternate interior angles are equal $(\angle 3 = \angle 6, \angle 4 = \angle 5)$.
3. Alternate exterior angles are equal $(\angle 1 = \angle 8, \angle 2 = \angle 7)$.

EXAMPLE 1 If line 1 and line 2 are parallel lines and the measure of angle 1 is $112°$ $(m\angle 1 = 112°)$, find the measure of angles 2 through 8.

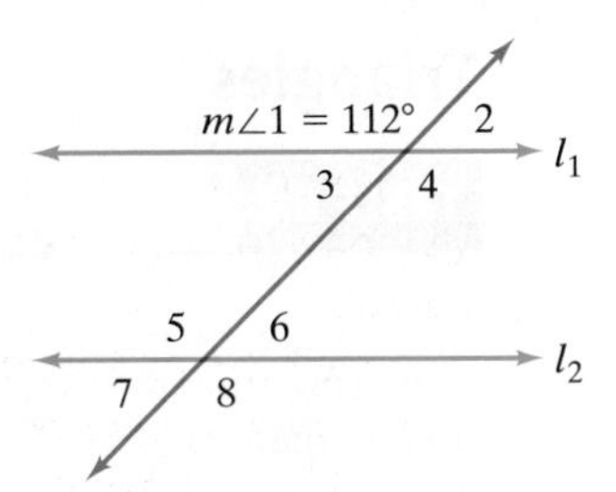

Solution Angles 1 and 2 are supplementary, so $m\angle 2$ is $180° - 112° = 68°$. The measures of angles 1 and 4 are equal since they are vertical angles. Thus, $m\angle 4 = 112°$. Angles 1 and 5 are corresponding angles. Thus, $m\angle 5 = 112°$. It is equal to its vertical angle, $\angle 8$, so $m\angle 8 = 112°$. The measures of angles 2, 3, 6, and 7 are all equal and measure $68°$.

Polygons

A **polygon** is a closed figure in a plane determined by three or more line segments. Some polygons are illustrated in **Figure B.5**.

A **regular polygon** has sides that are all the same length, and interior angles that all have the same measure. In **Figure B.5**, (b) and (d) are regular polygons.

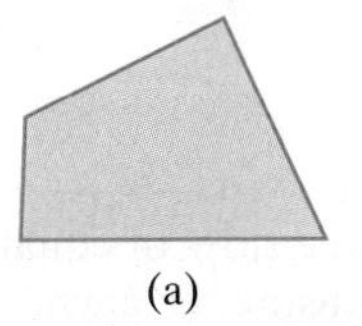
(a)

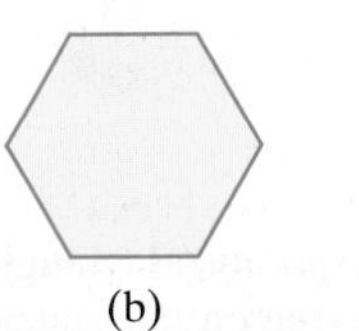
(b)

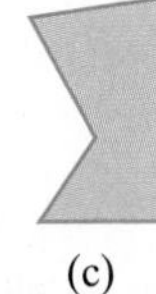
(c)

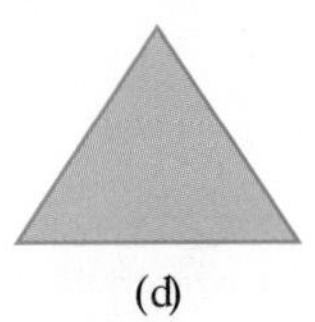
(d)

FIGURE B.5

Sum of the Interior Angles of a Polygon

The sum of the interior angles of a polygon can be found by the formula

$$\text{Sum} = (n - 2)180°$$

where n is the number of sides of the polygon.

EXAMPLE 2 Find the sum of the measures of the interior angles of **a)** a triangle; **b)** a quadrilateral (4 sides); **c)** an octagon (8 sides).

Solution

a) Since $n = 3$, we write

$$\begin{aligned}\text{Sum} &= (n - 2)180^\circ \\ &= (3 - 2)180^\circ = 1(180^\circ) = 180^\circ\end{aligned}$$

The sum of the measures of the interior angles in a triangle is 180°.

b)

$$\begin{aligned}\text{Sum} &= (n - 2)180^\circ \\ &= (4 - 2)180^\circ = 2(180^\circ) = 360^\circ\end{aligned}$$

The sum of the measures of the interior angles in a quadrilateral is 360°.

c)

$$\text{Sum} = (n - 2)(180^\circ) = (8 - 2)180^\circ = 6(180^\circ) = 1080^\circ$$

The sum of the measures of the interior angles in an octagon is 1080°.

Now we will briefly define several types of triangles in **Table B.2**.

Triangles

TABLE B.2

Triangle	Sketch of Triangle
An **acute triangle** is one that has three acute angles (angles of less than 90°).	
An **obtuse triangle** has one obtuse angle (an angle greater than 90°).	
A **right triangle** has one right angle (an angle equal to 90°). The longest side of a right triangle is opposite the right angle and is called the **hypotenuse**. The other two sides are called the **legs**.	
An **isosceles triangle** has two sides of equal length. The angles opposite the equal sides have the same measure.	
An **equilateral triangle** has three sides of equal length. It also has three equal angles that measure 60° each.	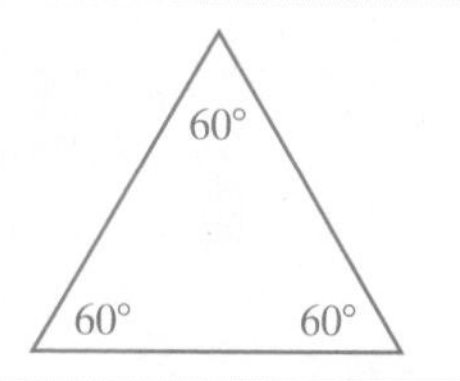

When two sides of a *right triangle* are known, the third side can be found using the **Pythagorean Theorem**, $a^2 + b^2 = c^2$, where a and b are the legs and c is the hypotenuse of the triangle. (See Section 5.7 for examples.)

Congruent and Similar Figures

If two triangles are **congruent**, it means that the two triangles are identical in size and shape. Two congruent triangles would match up exactly if one were placed on the other.

Two Triangles Are Congruent If Any One of the Following Statements Is True

1. Two angles of one triangle are equal to two corresponding angles of the other triangle, and the lengths of the sides between each pair of angles are equal. This method of showing that triangles are congruent is called the *angle, side, angle* method.

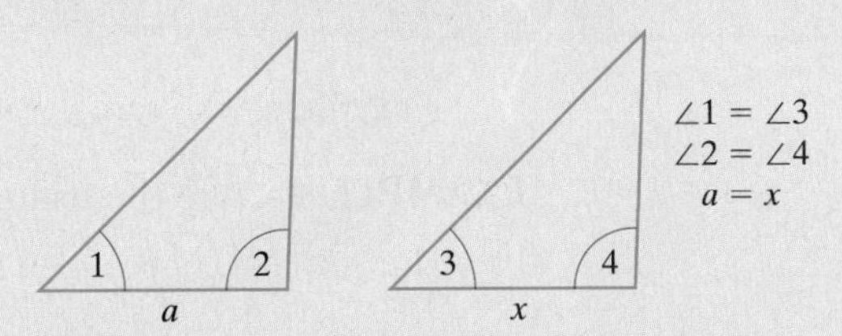

2. Corresponding sides of both triangles are equal. This is called the *side, side, side* method.

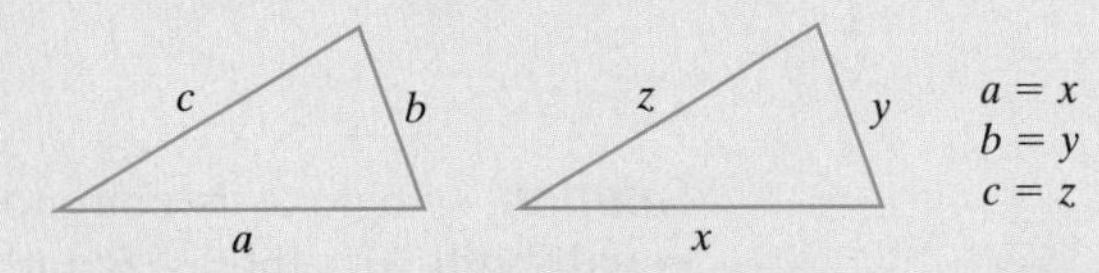

3. Two corresponding pairs of sides are equal, and the angle between them is equal. This is referred to as the *side, angle, side* method.

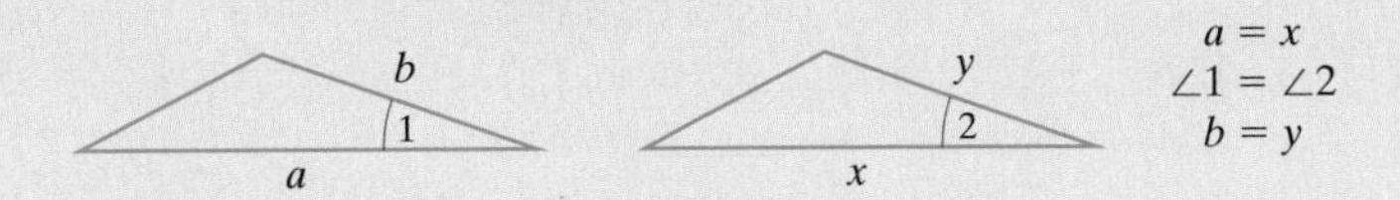

EXAMPLE 3 Determine whether the two triangles are congruent.

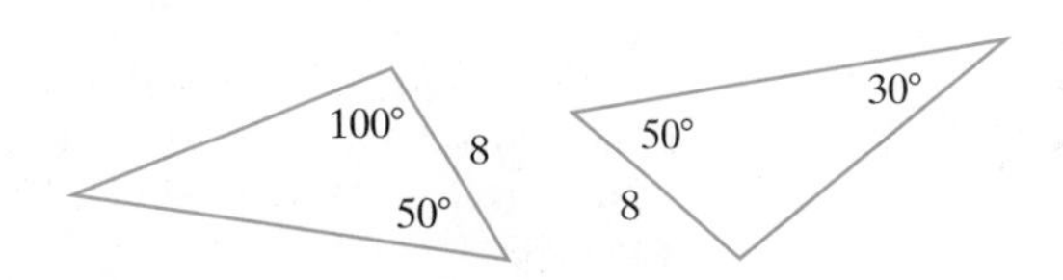

Solution The unknown angle in the figure on the right must measure 100° since the sum of the angles of a triangle is 180°. Both triangles have the same two angles (100° and 50°), with the same length side between them, 8 units. Thus, these two triangles are congruent by the angle, side, angle method.

Two triangles are **similar** if all three pairs of corresponding angles are equal and corresponding sides are in proportion. Similar figures do not have to be the same size but must have the same general shape.

Two Triangles Are Similar If Any One of the Following Statements Is True

1. Two angles of one triangle equal two angles of the other triangle.

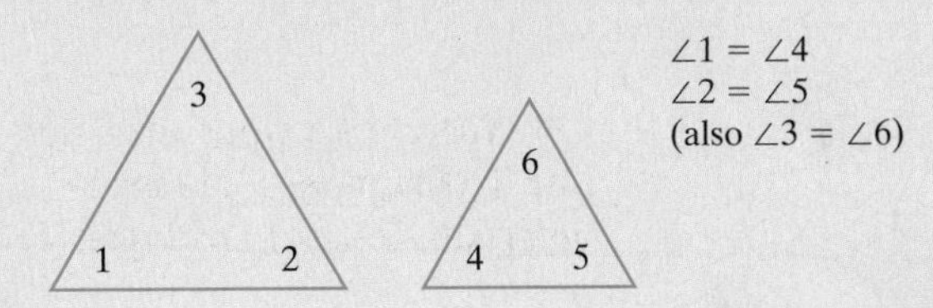

(*Continued on next page*)

2. Corresponding sides of the two triangles are proportional.

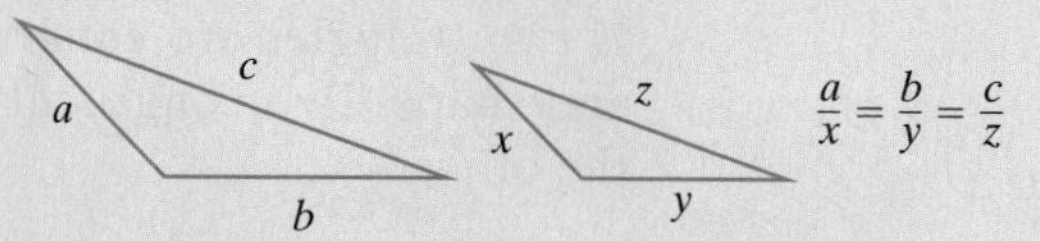

3. Two pairs of corresponding sides are proportional, and the angles between them are equal.

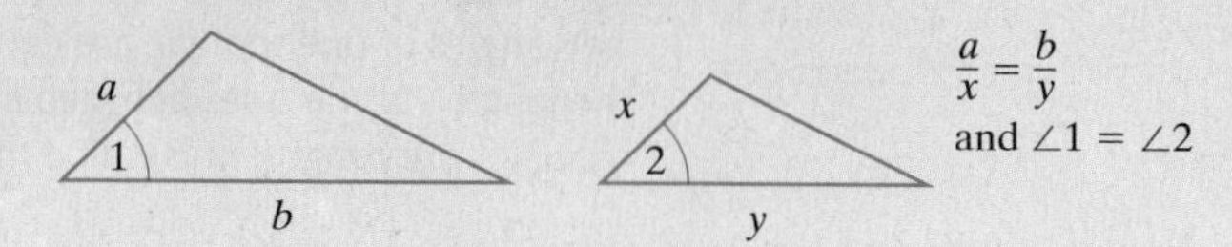

EXAMPLE 4 Are the triangles ABC and $AB'C'$ similar?

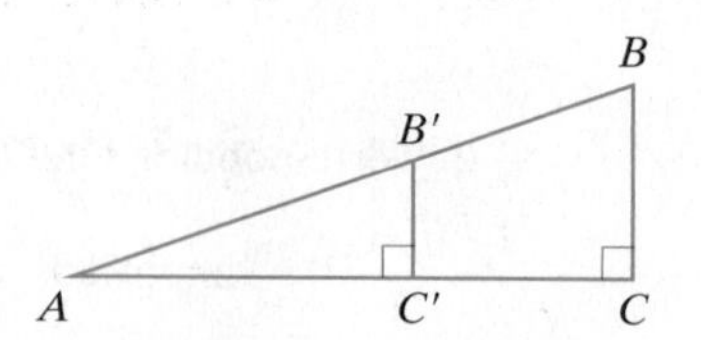

Solution Angle A is common to both triangles. Since angle C and angle C' are equal (both 90°), then $\angle B$ and $\angle B'$ must be equal. Since the three angles of triangle ABC equal the three angles of triangle $AB'C'$, the two triangles are similar.

Answers

Chapter 1

Exercise Set 1.1 **1.** Answers will vary. **3.** Answers will vary. **5.** Answers will vary. **7.** Answers will vary. **9.** Answers will vary. **11.** Do all the homework carefully and completely and preview the new material that is to be covered in class. **13.** At least 2 hours of study and homework time for each hour of class time in generally recommended. **15.** Answers will vary. **17.** Answers will vary.

Exercise Set 1.2 **1.** Median **3.** Approximately **5.** Checking **7.** Circle **9.** Problem **11. a)** 76.4 **b)** 74 **13. a)** \$87.32 **b)** \$86.57 **15. a)** 108.6 thousand **b)** 73.5 thousand **17.** \$470 **19. a)** \$161 **b)** \$2461 **21.** 1,610,000,000 operations **23. a)** 19.375 minutes **b)** 88 minutes **c)** 10 minutes **25.** $\approx$18.49 miles per gallon **27.** \$153 **29.** On the 3 on the right **31. a)** 4106.25 gallons **b)** $\approx$\$21.35 **33. a)** \$193 **b)** \$172 **35. a)** Finland; 540 **b)** Mexico; 420 **c)** 120 **37. a)** 1,200,000 motorcycles and 450,000 motorcycles **b)** 750,000 motorcycles **c)** $\approx$2.67 times greater **39. a)** 1.476 million or 1,476,000 **b)** 0.27 million or 270,000 **c)** 0.054 million or 54,000 **41. a)** 48 **b)** He cannot get a C. **43. a)** $\approx$1.30 times greater **b)** $\approx$1.42 times greater **c)** $\approx$1.30 times greater **45.** One example is 50, 60, 70, 80, 90. **47.** Mean

Exercise Set 1.3 **1.** Added **3.** Variables **5.** $\frac{2}{3}$ **7.** Denominator **9.** $\frac{3}{2}$ **11.** $2\cdot2\cdot3$ **13.** $2\cdot2\cdot3\cdot5$ **15.** $2\cdot3\cdot5\cdot5$ **17.** 6 **19.** 20 **21.** 6 **23.** $\frac{4}{5}$ **25.** $\frac{6}{7}$ **27.** $\frac{9}{19}$ **29.** $\frac{3}{7}$ **31.** Simplified **33.** $\frac{2}{3}$ **35.** $\frac{43}{15}$ **37.** $\frac{23}{3}$ **39.** $\frac{59}{18}$ **41.** $1\frac{3}{4}$ **43.** $3\frac{1}{4}$ **45.** $4\frac{4}{7}$ **47.** $\frac{4}{15}$ **49.** $\frac{1}{9}$ **51.** $\frac{3}{2}$ or $1\frac{1}{2}$ **53.** $\frac{5}{4}$ or $1\frac{1}{4}$ **55.** 6 **57.** $\frac{2}{9}$ **59.** $\frac{43}{10}$ or $4\frac{3}{10}$ **61.** $\frac{8}{13}$ **63.** $\frac{5}{8}$ **65.** $\frac{1}{7}$ **67.** $\frac{6}{5}$ or $1\frac{1}{5}$ **69.** $\frac{10}{17}$ **71.** $\frac{7}{12}$ **73.** $\frac{13}{36}$ **75.** $\frac{23}{100}$ **77.** $\frac{17}{6}$ or $2\frac{5}{6}$ **79.** $\frac{29}{10}$ or $2\frac{9}{10}$ **81.** $\frac{277}{30}$ or $9\frac{7}{30}$ **83.** $\frac{11}{24}$ mile **85.** $\frac{11}{24}$ centimeter **87. a)** $\frac{17}{12}$ or $1\frac{5}{12}$ **b)** $\frac{1}{12}$ **c)** $\frac{1}{2}$ **d)** $\frac{9}{8}$ or $1\frac{1}{8}$ **89. a)** $\frac{85}{18}$ or $4\frac{13}{18}$ **b)** $\frac{9}{2}$ or $4\frac{1}{2}$ **c)** $\frac{17}{10}$ or $1\frac{7}{10}$ **d)** $\frac{7}{6}$ or $1\frac{1}{6}$ **91.** $8\frac{15}{16}$ inches **93.** $\frac{9}{55}$ **95.** $\frac{63}{100}$ **97.** $\frac{7}{16}$ centimeters **99.** $2\frac{3}{10}$ minutes **101.** $1\frac{9}{16}$ inches **103.** 5 milligrams **105.** 40 times **107.** $1\frac{1}{2}$ inches **109.** 6 **111. a)** Yes **b)** $82\frac{3}{8}$ inches **113.** Answers will vary. **115. a)** $\frac{* + ?}{a}$ **b)** $\frac{\odot - \square}{?}$ **c)** $\frac{\triangle + 4}{\square}$ **d)** $\frac{x-2}{3}$ **e)** $\frac{8}{x}$ **117.** 270 pills **119.** Answers will vary. **120.** 16 **121.** 15 **122.** Variables are letters used to represent numbers.

Exercise Set 1.4 **1.** Irrational **3.** Counting **5.** Line **7.** Whole **9.** $\sqrt{3}$ **11.** $\{1, 2, 3, 4, \ldots\}$ **13.** $\{0, 1, 2, 3, \ldots\}$ **15.** $\{\ldots, -3, -2, -1, 0, 1, 2, 3, \ldots\}$ **17.** True **19.** False **21.** False **23.** True **25.** False **27.** True **29.** True **31.** False **33.** True **35.** False **37.** True **39.** True **41.** False **43.** True **45.** False **47.** True **49. a)** 13 **b)** $-2, 13$ **c)** $-2, 13$ **d)** 13 **51. a)** 3, 77 **b)** 0, 3, 77 **c)** 0, -2, 3, 77 **d)** $-\frac{5}{7}, 0, -2, 3, 6\frac{1}{4}, 1.63, 77$ **e)** $\sqrt{7}, -\sqrt{3}$ **f)** $-\frac{5}{7}, 0, -2, 3, 6\frac{1}{4}, \sqrt{7}, -\sqrt{3}, 1.63, 77$ **53.** Answers will vary; three examples are 0, 1, and 2. **55.** Answers will vary; three examples are $-\sqrt{2}, -\sqrt{3}$, and $-\sqrt{7}$. **57.** Answers will vary; three examples are $-\frac{2}{3}, \frac{1}{2}$, and 6.3. **59.** Answers will vary; three examples are $-13, -5$, and -1. **61.** Answers will vary; three examples are $\sqrt{2}, \sqrt{3}$, and $-\sqrt{5}$. **63.** Answers will vary; three examples are -7, 1, and 5. **65.** 87 **67. a)** $\{1, 3, 4, 5, 8\}$ **b)** $\{2, 5, 6, 7, 8\}$ **c)** $\{5, 8\}$ **d)** $\{1, 2, 3, 4, 5, 6, 7, 8\}$ **69. a)** Set *B* continues beyond 4. **b)** 4 **c)** An infinite number of elements **d)** An infinite set **71. a)** An infinite number **b)** An infinite number **73.** $\frac{29}{5}$ **74.** $5\frac{1}{3}$ **75.** $\frac{13}{24}$ **76.** $\frac{4}{45}$

Exercise Set 1.5 **1.** 0 **3.** $|a|$ **5.** True **7.** $|6 - (-4)|$ **9.** Negative **11.** 7 **13.** 15 **15.** 0 **17.** -5 **19.** -26 **21. a)** $<$ **b)** $>$ **23. a)** $>$ **b)** $<$ **25.** $<$ **27.** $>$ **29.** $>$ **31.** $<$ **33.** $<$ **35.** $<$ **37.** $<$ **39.** $>$ **41.** $>$ **43.** $<$ **45.** $>$ **47.** $<$ **49.** $<$ **51.** $<$ **53.** $>$ **55.** $>$ **57.** $>$ **59.** $<$ **61.** $<$ **63.** $<$ **65.** $<$ **67.** $=$ **69.** $=$ **71.** $<$ **73.** $<$ **75.** $-|-1|, \frac{3}{7}, \frac{4}{9}, 0.46, |-5|$ **77.** $\frac{5}{12}, 0.6, \frac{2}{3}, \frac{19}{25}, |-2.6|$ **79.** 4, -4 **81.** Not possible **83.** Answers will vary: one example is $-3, -4$ and -5. **85.** Answers will vary: one example is 4, 5, and 6. **87.** Answers will vary: one example is 3, 4, and 5. **89. a)** Does not include the endpoints **b)** Answers will vary: one example is 4.1, 5, and $5\frac{1}{2}$. **c)** No **d)** Yes **e)** True **91. a)** Dietary fiber and thiamin **b)** Vitamin E, niacin, and riboflavin **93.** Yes **95.** No, $|-4| > |-3|$ but $-4 < -3$. **97.** Greater than **99.** No **102.** $\frac{89}{15}$ or $5\frac{14}{15}$ **103.** $\{\ldots, -3, -2, -1, 0, 1, 2, 3, \ldots\}$ **104.** $\{0, 1, 2, 3, \ldots\}$ **105. a)** 5 **b)** 5, 0 **c)** 5, -2, 0 **d)** $5, -2, 0, \frac{1}{3}, -\frac{5}{9}, 2.3$ **e)** $\sqrt{3}, \pi$ **f)** $5, -2, 0, \frac{1}{3}, \sqrt{3}, -\frac{5}{9}, 2.3, \pi$

Mid-Chapter Test: Sections 1.1–1.5*

1. At least two hours of study and homework for each hour of class time is generally recommended. [1.1] **2. a)** \$80.63 **b)** \$83.81 [1.2] **3.** \$824.59 [1.2] **4. a)** Natwora's **b)** \$24 [1.2] **5.** \$62.35 [1.2] **6.** $\frac{1}{6}$ [1.3] **7.** $\frac{39}{80}$ [1.3] **8.** $\frac{49}{40}$ or $1\frac{9}{40}$ [1.3] **9.** $\frac{61}{20}$ or $3\frac{1}{20}$ [1.3] **10.** $54\frac{1}{3}$ feet [1.3] **11.** False [1.4] **12.** True [1.4] **13.** False [1.4] **14.** True [1.4] **15.** False [1.4] **16.** $-\frac{7}{10}$ [1.5] **17.** $>$ [1.5] **18.** $>$ [1.5] **19.** $<$ [1.5] **20.** $=$ [1.5]

Exercise Set 1.6

1. Negative **3.** Absolute **5.** Sum **7.** −8 **9.** Denominator **11.** Correct **13.** −19 **15.** 28 **17.** 0 **19.** $-\frac{5}{3}$ **21.** $-2\frac{3}{5}$ **23.** −3.72 **25.** 21 **27.** 1 **29.** −6 **31.** 0 **33.** 0 **35.** −10 **37.** −4 **39.** −13 **41.** 0 **43.** −8 **45.** −27 **47.** −64 **49.** 16 **51.** −12 **53.** 3 **55.** −6 **57.** −40 **59.** −31 **61.** −39 **63.** 91 **65.** −9.9 **67.** −144.0 **69.** 4.22 **71.** −107.15 **73.** $\frac{26}{35}$ **75.** $\frac{107}{84}$ or $1\frac{23}{84}$ **77.** $\frac{4}{55}$ **79.** $-\frac{26}{45}$ **81.** $-\frac{31}{30}$ or $-1\frac{1}{30}$ **83.** $-\frac{13}{15}$ **85.** $\frac{3}{10}$ **87.** $\frac{19}{56}$ **89.** $-\frac{43}{60}$ **91.** $-\frac{23}{21}$ or $-1\frac{2}{21}$ **93. a)** Positive **b)** 390 **95. a)** Negative **b)** −373 **97. a)** Negative **b)** −452 **99. a)** Negative **b)** −1300 **101. a)** Negative **b)** −112 **103. a)** Negative **b)** −3880 **105. a)** Positive **b)** 1111 **107. a)** Negative **b)** −2050 **109.** True **111.** True **113.** False **115.** \$277 **117.** 21 yards **119.** 61 feet **121.** 13,796 feet **123. a)** −\$12 thousand **b)** \$32,000, \$36,000, \$26,000; \$94,000 **125.** −26 **127.** 20 **129.** 0 **131.** $\frac{11}{30}$ **133.** 55 **135.** $\frac{19}{14}$ or $1\frac{5}{14}$ **136.** $\frac{43}{16}$ or $2\frac{11}{16}$ **137.** False **138.** $>$ **139.** $<$

Exercise Set 1.7

1. Minuend **3.** Difference **5.** Opposite **7.** Left **9.** $-a + b$ **11.** Correct **13.** 6 **15.** 7 **17.** −1 **19.** 12 **21.** −16 **23.** −9 **25.** 0 **27.** −4 **29.** −3 **31.** 9 **33.** −20 **35.** 9.8 **37.** 0.3 **39.** 37 **41.** 4 **43.** 22 **45.** −11 **47.** −131.0 **49.** −84 **51.** 140 **53.** −7.4 **55.** 0.09 **57.** −11 **59.** 0 **61.** −11 **63.** 6.1 **65.** 18.2 **67.** −11.47 **69.** $\frac{1}{6}$ **71.** $-\frac{7}{10}$ **73.** $-\frac{67}{60}$ or $-1\frac{7}{60}$ **75.** $-\frac{5}{12}$ **77.** $-\frac{17}{24}$ **79.** $\frac{1}{2}$ **81.** $\frac{7}{45}$ **83.** $\frac{13}{16}$ **85.** $-\frac{13}{63}$ **87.** $-\frac{7}{60}$ **89. a)** Positive **b)** 99 **c)** Yes **91. a)** Negative **b)** −619 **c)** Yes **93. a)** Positive **b)** 1588 **c)** Yes **95. a)** Positive **b)** 196 **c)** Yes **97. a)** Negative **b)** −448 **c)** Yes **99. a)** Positive **b)** 116.1 **c)** Yes **101. a)** Negative **b)** −69 **c)** Yes **103. a)** Negative **b)** −1670 **c)** Yes **105. a)** Zero **b)** 0 **c)** Yes **107.** 4 **109.** 4 **111.** −15 **113.** −2 **115.** 13 **117.** −5 **119.** −32 **121.** 9 **123.** −12 **125.** 12 **127.** −18 **129.** 20.9 miles **131.** 14,787 feet **133.** $1\frac{7}{8}$ inches **135.** Dropped 100°F **137. a)** 278 **b)** Stricker's score was 17 strokes more **139.** $x - y = -5, y - x = 5$, and $x - (-y) = 11$ **141.** −5 **143. a)** 8 units **b)** $-3 - (-11)$ **145. a)** 9 feet **b)** −3 feet **146.** $\{1, 2, 3, \ldots\}$ **147.** The set of rational numbers together with the set of irrational numbers form the set of real numbers. **148.** $>$ **149.** $<$ **150.** $-\frac{1}{24}$

Exercise Set 1.8

1. Negative **3.** Zero **5.** $\frac{a}{b}$ **7.** Positive **9.** Product **11.** Negative **13.** Positive **15.** Negative **17.** 24 **19.** −15 **21.** 54 **23.** −21 **25.** −9.6 **27.** −4.67 **29.** 0 **31.** 0 **33.** −84 **35.** −72 **37.** 1400 **39.** 0 **41.** $-\frac{3}{10}$ **43.** $\frac{7}{27}$ **45.** 4 **47.** $-\frac{1}{10}$ **49.** −7 **51.** 4 **53.** 4 **55.** −18 **57.** 9.9 **59.** −10 **61.** −33 **63.** −4 **65.** 6 **67.** 0 **69.** 16.2 **71.** ≈ -5.92 **73.** $-\frac{2}{5}$ **75.** $\frac{5}{36}$ **77.** 1 **79.** $-\frac{144}{5}$ or $-28\frac{4}{5}$ **81.** −32 **83.** 20 **85.** −14 **87.** −9.3 **89.** −20 **91.** 1 **93.** 0 **95.** Undefined **97.** 0 **99.** Undefined **101. a)** Negative **b)** −3496 **c)** Yes **103. a)** Negative **b)** −16 **c)** Yes **105. a)** Negative **b)** −9 **c)** Yes **107. a)** Positive **b)** 6174 **c)** Yes **109. a)** Zero **b)** 0 **c)** Yes **111. a)** Undefined **b)** Undefined **c)** Yes **113. a)** Positive **b)** 3.2 **c)** Yes **115. a)** Positive **b)** 226.8 **c)** Yes **117.** False **119.** False **121.** True **123.** True **125.** False **127.** True **129.** 45 yard loss or −45 yards **131. a)** \$104 **b)** −\$416 **133.** \$143.85 **135. a)** 20 point loss or −20 points **b)** 80 **137. a)** 102 to 128 beats per minute **b)** Answers will vary. **139.** −125 **141.** 1 **143.** Positive, 10 negative numbers **146.** $>$ **147.** $-\frac{41}{60}$ **148.** −2 **149.** −3 **150.** 3

Exercise Set 1.9

1. Left **3.** Grouping **5.** Exponent **7.** 8 **9.** 1 **11.** 5 **13.** 1 **15.** −1 **17.** −9 **19.** 9 **21.** −25 **23.** 36 **25.** −16 **27.** $\frac{9}{16}$ **29.** $\frac{9}{16}$ **31.** $\frac{27}{64}$ **33.** $-\frac{27}{64}$ **35. a)** Negative **b)** −49 **37. a)** Positive **b)** 49 **39. a)** Positive **b)** 49 **41. a)** Negative **b)** −49 **43. a)** Negative **b)** −1.728 **45. a)** Positive **b)** $\frac{25}{64}$ **47.** 21 **49.** 8 **51.** 57 **53.** 0 **55.** −16 **57.** 29 **59.** −19 **61.** −77 **63.** $\frac{83}{100}$ **65.** 2 **67.** 10 **69.** −34 **71.** 103 **73.** 169 **75.** −23 **77.** 36.75 **79.** $\frac{5}{8}$ **81.** $\frac{1}{4}$ **83.** $\frac{49}{30}$ or $1\frac{19}{30}$ **85.** $\frac{5}{27}$ **87.** $\frac{32}{53}$ **89.** 9 **91.** −4 **93.** 1 **95. a)** 25 **b)** −25 **c)** 25 **97. a)** 4 **b)** −4 **c)** 4 **99. a)** 36 **b)** −36 **c)** 36 **101. a)** $\frac{1}{9}$ **b)** $-\frac{1}{9}$ **c)** $\frac{1}{9}$ **103.** 4 **105.** −45 **107.** 3 **109.** −4 **111.** −1 **113.** $\frac{15}{4}$ or $3\frac{3}{4}$ **115.** 994 **117.** −5 **119.** 193

*Numbers in blue brackets after the answer indicates the section where the material was discussed.

121. -25 **123.** $[(6 \cdot 3) - 4] - 2; 12$ **125.** $\{[(10 \cdot 4) + 9] - 6\} \div 7; \frac{43}{7}$ or $6\frac{1}{7}$ **127.** $\left(\frac{4}{5} + \frac{3}{7}\right) \cdot \frac{2}{3}; \frac{86}{105}$ **129.** All real numbers

131. 162.5 miles **133.** 47 feet **135.** \$22,470 **137.** 1.71 inches **139.** $(14 + 6) \div 2 \times 4 = 40$ **144. a)** 3

b)

Dogs	Houses
0	4
1	5
2	3
3	1
4	1

c) 18 **d)** ≈ 1.29 **145.** \$6.40 **146.** $-\frac{5}{36}$ **147.** $\frac{10}{3}$ or $3\frac{1}{3}$

Exercise Set 1.10

1. Associative **3.** 0 **5.** -5 **7.** Distributive **9.** Identity **11. a)** -6 **b)** $\frac{1}{6}$ **13. a)** 3 **b)** $-\frac{1}{3}$ **15. a)** $-x$ **b)** $\frac{1}{x}$ **17. a)** 0 **b)** Does not exist **19. a)** $-\frac{1}{5}$ **b)** 5 **21. a)** $\frac{5}{6}$ **b)** $-\frac{6}{5}$ **23.** Distributive property **25.** Associative property of addition **27.** Commutative property of multiplication **29.** Associative property of multiplication **31.** Inverse property of addition **33.** Identity property of multiplication **35.** Inverse property of multiplication **37.** Commutative property of addition **39.** Identity property of addition **41.** $(-6 \cdot 4) \cdot 2$ **43.** $y \cdot x$ **45.** $3y + 4x$ **47.** 1 **49.** $3x + (4 + 6)$ **51.** $-5x$ **53.** $4x + 4y + 12$ **55.** 0 **57.** $\frac{5}{2}n$ **59.** Yes **61.** Yes **63.** No **65.** Yes **67.** No **69.** No **71.** The $(3 + 4)$ is treated as one value. **73.** Commutative property of addition **75.** No; associative property of addition **77.** $\frac{49}{15}$ or $3\frac{4}{15}$ **78.** $\frac{23}{16}$ or $1\frac{7}{16}$ **79.** -11.2 **80.** $-\frac{7}{8}$

Chapter Review Exercises

1. 66 hot dogs **2.** \$572.45 **3. a)** \$74.25 **b)** \$974.24 **4.** \$100 **5. a)** 78.4 **b)** 79 **6. a)** 11 **b)** 9.5 **7. a)** 29 minutes **b)** 34 minutes **8. a)** 225 **b)** 165 **9.** $\frac{1}{2}$ **10.** $\frac{127}{21}$ or $6\frac{1}{21}$ **11.** $\frac{25}{36}$ **12.** $\frac{7}{6}$ or $1\frac{1}{6}$ **13.** $\frac{23}{12}$ or $1\frac{11}{12}$ **14.** $\frac{177}{10}$ or $17\frac{7}{10}$ **15.** $\{1, 2, 3, \ldots\}$ **16.** $\{0, 1, 2, 3, \ldots\}$ **17.** $\{\ldots, -3, -2, -1, 0, 1, 2, 3, \ldots\}$ **18.** The set of all numbers which can be expressed as the quotient of two integers, denominator not zero **19. a)** 3, 426 **b)** 3, 0, 426 **c)** 3, -5, -12, 0, 426 **d)** 3, -5, -12, 0, $\frac{1}{2}$, -0.62, 426, $-3\frac{1}{4}$ **e)** $\sqrt{7}$ **f)** 3, -5, -12, 0, $\frac{1}{2}$, -0.62, $\sqrt{7}$, 426, $-3\frac{1}{4}$ **20. a)** 1 **b)** 1 **c)** $-8, -9$ **d)** $-8, -9, 1$ **e)** $-2.3, -8, -9, 1\frac{1}{2}, 1, -\frac{3}{17}$ **f)** $\sqrt{2}, -\sqrt{2}$ **g)** $-2.3, -8, -9, 1\frac{1}{2}, \sqrt{2}, -\sqrt{2}, 1, -\frac{3}{17}$ **21.** $<$ **22.** $>$ **23.** $<$ **24.** $<$ **25.** $<$ **26.** $>$ **27.** $=$ **28.** $<$ **29.** -14 **30.** 0 **31.** -3 **32.** -6 **33.** -6 **34.** 2 **35.** 8 **36.** 0 **37.** -5 **38.** 14 **39.** 4 **40.** -12 **41.** $\frac{7}{12}$ **42.** $\frac{11}{10}$ or $1\frac{1}{10}$ **43.** $-\frac{7}{36}$ **44.** $-\frac{19}{56}$ **45.** $-\frac{5}{4}$ or $-1\frac{1}{4}$ **46.** $-\frac{37}{84}$ **47.** $-\frac{7}{90}$ **48.** $\frac{61}{60}$ or $1\frac{1}{60}$ **49.** 14 **50.** -3 **51.** -12 **52.** -7 **53.** 16 **54.** 11 **55.** -63 **56.** 25.42 **57.** -120 **58.** $-\frac{6}{35}$ **59.** $-\frac{6}{11}$ **60.** $\frac{15}{56}$ **61.** 0 **62.** 144 **63.** -15 **64.** -6 **65.** -3.2 **66.** 4.3 **67.** 8 **68.** 9 **69.** $\frac{56}{27}$ or $2\frac{2}{27}$ **70.** $-\frac{35}{9}$ or $-3\frac{8}{9}$ **71.** 0 **72.** 0 **73.** Undefined **74.** Undefined **75.** Undefined **76.** 0 **77.** 25 **78.** -8 **79.** 1 **80.** 3 **81.** -6 **82.** -32 **83.** 6 **84.** -4 **85.** 10 **86.** 1 **87.** 15 **88.** -4 **89.** -36 **90.** 36 **91.** 16 **92.** -27 **93.** -1 **94.** -32 **95.** $\frac{16}{25}$ **96.** $\frac{8}{125}$ **97.** 500 **98.** 4 **99.** 12 **100.** -256 **101.** 9 **102.** 16 **103.** 6.36 **104.** -17 **105.** -39 **106.** -2.3 **107.** 0 **108.** $\frac{9}{7}$ or $1\frac{2}{7}$ **109.** -60 **110.** 10 **111.** 20 **112.** 20 **113.** 14 **114.** 9 **115.** -4 **116.** 50 **117.** 5 **118.** 26 **119.** 45 **120.** 0 **121.** -11 **122.** -3 **123.** -3 **124.** 39 **125.** -215 **126.** 353.6 **127.** -2.88 **128.** 117.8 **129.** 65,536 **130.** -74.088 **131.** Associative property of addition **132.** Distributive property **133.** Commutative property of addition **134.** Commutative property of multiplication **135.** Associative property of multiplication **136.** Inverse property of multiplication **137.** Identity property of multiplication **138.** Inverse property of addition **139.** Identity property of addition **140.** Associative property of addition

Chapter 1 Practice Test

1. a) \$10.65 **b)** \$0.23 **c)** \$10.88 **d)** \$39.12 [1.2] **2.** ≈ 2.49 times greater [1.2] **3. a)** ≈ 13 thousand **b)** During this specific time, half the time KFUN had more than 8.8 thousand listeners and half the time KFUN had less than 8.8 thousand listeners. [1.2] **4. a)** 42 **b)** 42, 0 **c)** -6, 42, 0, -7, -1 **d)** $-6, 42, -3\frac{1}{2}, 0, 6.52, \frac{5}{9}, -7, -1$ **e)** $\sqrt{5}$ **f)** $-6, 42, -3\frac{1}{2}, 0, 6.52, \sqrt{5}, \frac{5}{9}, -7, -1$ [1.4] **5.** $<$ [1.5] **6.** $>$ [1.5] **7.** -15 [1.6] **8.** -11 [1.7] **9.** -14 [1.7] **10.** 8 [1.9] **11.** -24 [1.8] **12.** $\frac{16}{63}$ [1.8] **13.** -3 [1.9] **14.** $-\frac{53}{56}$ [1.7] **15.** 12 [1.9] **16.** $-\frac{32}{243}$ [1.9] **17.** 100 [1.9] **18.** $-x^2$ means $-(x^2)$ and x^2 will always be positive for any nonzero value of x. Therefore, $-x^2$ will always be negative. [1.9] **19.** 37 [1.9] **20.** 11 [1.9] **21.** 10 [1.9] **22.** 1 [1.9] **23.** Commutative property of addition [1.10] **24.** Distributive property [1.10] **25.** Associative property of addition [1.10]

Chapter 2

Exercises Set 2.1 **1.** Constant **3.** Like **5.** Unlike **7.** Terms **9.** $14x$ **11.** $3x + 6$ **13.** $5y + 3$ **15.** $\frac{9}{44}a$ **17.** $-6x + 7t$ **19.** $-5w + 5$ **21.** $-2x$ **23.** 0 **25.** $-2t + 21$ **27.** $-12p - 8$ **29.** $10x^2 - 10y^2 - 7$ **31.** $2x - 8$ **33.** $b + \frac{23}{5}$ **35.** $0.8n + 6.42$ **37.** $-\frac{2}{3}x + \frac{5}{9}y + \frac{1}{9}$ **39.** $14.6x + 8.3$ **41.** $x^2 + y$ **43.** $-3x - 5y$ **45.** $-3n^2 - 2n + 13$ **47.** $21.72x - 7.11$ **49.** $-\frac{23}{20}x - 5$ **51.** $5w^3 + 2w^2 + w + 3$ **53.** $-7z^3 - z^2 + 2z$ **55.** $2x^2 - 3x - 5$ **57.** $2a^3 - 5a^2 - a + 1$ **59.** $5x + 10$ **61.** $5x + 20$ **63.** $3x - 18$ **65.** $-x + 2$ **67.** $x - 4$ **69.** $\frac{4}{5}s - 4$ **71.** $-0.9x^2 - 1.5$ **73.** $-r + 4$ **75.** $1.4x + 0.35$ **77.** $x - y$ **79.** $-2x - 4y + 8$ **81.** $3.41x - 5.72y + 3.08$ **83.** $10x - 45y$ **85.** $r + 3s - 19$ **87.** $3x - 6y - 12$ **89.** $-3x + 1$ **91.** $2x + 1$ **93.** $14x + 18$ **95.** $4x - 2y + 3$ **97.** $5c$ **99.** $7x + 3$ **101.** $\frac{5}{4}x + \frac{1}{3}$ **103.** $\frac{19}{6}x - 2$ **105.** $-4s - 6$ **107.** $2x - 2$ **109.** 0 **111.** $-y - 6$ **113.** $3x - 5$ **115.** $x + 15$ **117.** $0.2x - 4y - 2.8$ **119.** $-6x + 7y$ **121.** $\frac{3}{2}x + \frac{7}{2}$ **123.** $2\square + 3\ominus$ **125.** $2x + 3y + 2\triangle$ **127.** 1, 2, 3, 6, 9, 18 **129. a)** The signs of all terms inside the parentheses change when the parentheses are removed. **b)** $-x + 8$ **131.** $22x^2 - 25y^2 - 4x + 3$ **133.** $9x - 39$ **135.** 7 **136.** -16 **137.** -11 **138.** Answers will vary. **139.** -12

Exercise Set 2.2 **1.** 2 **3.** Isolate **5.** Linear **7.** Equation **9.** Opposite **11.** Yes **13.** No **15.** Yes **17.** Yes **19.** No **21.** Yes **23.** 5 **25.** -7 **27.** -4 **29.** 43 **31.** 15 **33.** 11 **35.** -4 **37.** -5 **39.** -12 **41.** -30 **43.** 0 **45.** -57 **47.** -4 **49.** -65 **51.** 0 **53.** 17 **55.** -26 **57.** 28 **59.** -46.1 **61.** 46.5 **63.** -8.23 **65.** -9.91 **67.** No, the equation is equivalent to $1 = 2$, a false statement. **69.** Use properties that allow us to get the variable by itself on one side of the equation. **71.** $x = \square + \triangle$ **73.** $\square = ☺ - \triangle$ **76.** $\frac{11}{30}$ **77.** $-\frac{31}{24}$ **78.** $2x - 13$ **79.** $7t - 25$

Exercise Set 2.3 **1.** $\frac{x}{3}$ **3.** 3 **5.** Checking **7.** $-\frac{7}{x}$ **9.** 4 **11.** 21 **13.** -3 **15.** -80 **17.** 20 **19.** -3 **21.** $-\frac{7}{3}$ **23.** -13 **25.** 8 **27.** 30 **29.** $-\frac{1}{3}$ **31.** 6 **33.** $\frac{26}{43}$ **35.** 2 **37.** $\frac{1}{5}$ **39.** $-\frac{3}{40}$ **41.** -64 **43.** 240 **45.** -45 **47.** 20 **49.** -50 **51.** 0 **53.** 0 **55.** 22.5 **57.** 6 **59.** -20.2 **61.** 7 **63.** 9 **65. a)** In $5 + x = 10$, 5 is added to the variable, whereas in $5x = 10$, 5 is multiplied by the variable. **b)** $x = 5$ **c)** $x = 2$ **67.** Multiply by $\frac{3}{2}$; 6 **69.** Multiply by $\frac{7}{3}$; $\frac{28}{15}$ **71. a)** $\square$ **b)** Divide both sides of the equation by $\triangle$. **c)** $\square = \frac{☺}{\triangle}$ **73.** -4 **74.** -30 **75.** 6 **76.** Associative property of addition **77.** -57

Exercise Set 2.4 **1.** Nonzero **3.** 10 **5.** Multiplying **7.** 5 **9.** -5 **11.** 2 **13.** $\frac{12}{5}$ **15.** 2 **17.** 7 **19.** $\frac{11}{3}$ **21.** $-\frac{19}{16}$ **23.** 0 **25.** 2 **27.** $-\frac{51}{5}$ **29.** 3 **31.** 6.8 **33.** 4 **35.** 12 **37.** 22 **39.** 60 **41.** -14 **43.** -11 **45.** -2 **47.** $\frac{19}{8}$ **49.** 0 **51.** -10 **53.** -3 **55.** 6 **57.** -21 **59.** $-\frac{19}{7}$ **61.** -9 **63.** 4 **65.** 5 **67.** 0.8 **69.** -1 **71.** $\frac{2}{7}$ **73.** -3.1 **75.** $-\frac{14}{5}$ **77.** 18 **79.** $-\frac{1}{15}$ **81.** $\frac{18}{7}$ **83.** $-\frac{16}{21}$ **85.** 10 **87.** 5 **89.** $-\frac{39}{4}$ **91.** 2 **93.** $\frac{25}{3}$ **95.** $\frac{26}{45}$ **97. a)** You will not have to work with fractions. **b)** 3 **99.** $\frac{35}{6}$ **101.** -4 **105.** False **106.** 64 **107.** Isolate the variable on one side of the equation. **108.** Divide both sides of the equation by -4 to isolate the variable.

Mid-Chapter Test: Sections 2.1–2.4 **1.** $-2x - 5y - 6$ [2.1] **2.** $-\frac{7}{20}x - \frac{15}{2}$ [2.1] **3.** $-8a + 12b - 64$ [2.1] **4.** $3.36x - 5.44y - 8.32$ [2.1] **5.** $2t - 38$ [2.1] **6.** Yes [2.2] **7.** No [2.2] **8.** -4 [2.2] **9.** -160 [2.2] **10.** -23 [2.2] **11.** Multiply both sides by 4. [2.3] **12.** $\frac{1}{3}$ [2.3] **13.** 24 [2.3] **14.** 10 [2.3] **15.** $-\frac{3}{7}$ [2.3] **16.** $\frac{5}{2}$ [2.4] **17.** $-\frac{3}{2}$ [2.4] **18.** $\frac{13}{16}$ [2.4] **19.** -6 [2.4] **20.** $-\frac{20}{9}$ [2.4]

Exercise Set 2.5 **1.** Isolate **3.** Denominator **5.** Identity **7.** Real **9.** Always **11.** $\frac{3}{5}$ **13.** 1 **15.** 2 **17.** 3 **19.** -2 **21.** No solution **23.** 3 **25.** -2 **27.** No solution **29.** $\frac{13}{5}$ **31.** $\frac{7}{9}$ **33.** 6.5 **35.** 3.2 **37.** 5 **39.** 30 **41.** $\frac{3}{2}$ **43.** $\frac{5}{2}$ **45.** 25 **47.** All real numbers **49.** 23 **51.** 0 **53.** All real numbers **55.** $\frac{21}{20}$ **57.** 14 **59.** $-\frac{15}{4}$ **61.** 5 **63.** 4 **65.** 0 **67.** 16 **69.** $-\frac{24}{5}$

71. $-\frac{4}{21}$ **73.** 5 **75.** 30 **77.** 4 **79. a)** One example is $x + x + 1 = x + 2$. **b)** It has a single solution. **c)** For the example given in part **a)**, $x = 1$. **81. a)** One example is $x + x + 1 = 2x + 1$. **b)** Both sides simplify to the same expression. **c)** All real numbers **83. a)** One example is $x + x + 1 = 2x + 2$. **b)** It simplifies to a false statement. **c)** No solution **85.** $-\frac{1}{4}$ **87.** All real numbers **89.** -4 **91. a)** 4 **b)** 7 **c)** 0 **92.** x^{a-b} **93.** Factors are expressions that are multiplied. Terms are expressions that are added. **94.** $7x - 10$ **95.** $\frac{10}{7}$ **96.** -3

Exercise Set 2.6

1. Formula **3.** Quadrilateral **5.** Time **7.** Radius **9.** Cubic **11.** 480 **13.** 96 **15.** 360 **17.** 36 **19.** 314.16 **21.** 36 **23.** 2 **25.** 8 **27.** 6.00 **29.** 77 square yards **31.** 28.27 square meters **33.** 12 square inches **35.** 16.5 square feet **37.** 60 cubic feet **39.** ≈452.39 cubic centimeters **41.** ≈134.04 cubic meters **43.** $C = 10°$ **45.** $F = 77°$ **47.** $l = A/w$ **49.** $t = i/(pr)$ **51.** $b = 2A/h$ **53.** $w = (P - 2l)/2$ **55.** $r = (-n + 3)/2$ **57.** $b = y - mx$ **59.** $b = d - a - c$ **61.** $y = (-ax - c)/b$ **63.** $h = 3V/(\pi r^2)$ **65.** $m = 2A - d$ **67.** $y = -2x + 8$ **69.** $y = x - 6$ **71.** $y = \frac{2}{3}x + \frac{4}{3}$ **73.** $y = \frac{3}{5}x - 2$ **75.** $y = \frac{1}{2}x - \frac{5}{2}$ **77.** $y = \frac{5}{3}x - \frac{11}{3}$ **79.** $y = -\frac{1}{3}x - \frac{5}{3}$ **81.** $y = 2x + \frac{13}{15}$ **83.** \$1440 **85.** \$5000 **87.** 60 miles per hour **89.** 7.632 miles **91.** 48 square inches **93.** 109 inches **95.** ≈75.40 feet **97.** 3 square feet **99.** ≈153.94 square feet **101.** 2700 cubic feet **103.** ≈150.80 cubic feet **105.** 7 square feet **107.** ≈381.7 cubic inches **109. a)** $B = \frac{703w}{h^2}$ **b)** ≈23.91 **111. a)** $V = 18x^3 - 3x^2$ **b)** 6027 cubic centimeters **c)** $S = 54x^2 - 8x$ **d)** 2590 square centimeters **113.** Doubles **115.** 8 times as large **117.** Circle **119.** $\frac{1}{3}$ **120.** -6 **121.** 0 **122.** 8

Exercise Set 2.7

1. Proportion **3.** Means **5.** Similar **7.** Units **9.** Yes **11.** No **13.** 2 : 3 **15.** 1 : 12 **17.** 5 : 12 **19.** 7 : 4 **21.** 1 : 3 **23.** 6 : 1 **25.** 7 : 30 **27.** 8 : 1 **29. a)** 50 : 23 **b)** ≈2.17 : 1 **31. a)** 29 : 19 **b)** ≈1.53 to 1 **33. a)** 33 : 17 **b)** 25 : 19 **35. a)** 40 : 32 or 5 : 4 **b)** 15 : 11 **37.** 12 **39.** 45 **41.** -9 **43.** -2 **45.** -54 **47.** 6 **49.** 32 inches **51.** 15.75 inches **53.** 19.5 inches **55.** 25 loads **57.** 527.5 miles **59.** 1.5 feet **61.** 24 teaspoons **63.** ≈0.43 feet **65.** 3.75 cups **67.** ≈9.49 feet **69.** 0.55 milliliter **71.** 570 minutes or 9 hours 30 minutes **73.** ≈360 children **75.** 6.5 feet **77.** 2.9 square yards **79.** 20 inches **81.** ≈97 hits **83.** \$307 **85.** 2618 pesos **87.** Yes. It is 2.12 : 1. **89.** It must increase. **91.** ≈41,667 miles **93.** 0.625 cubic centimeters **96.** Commutative property of addition **97.** Associative property of multiplication **98.** Distributive property **99.** All real numbers. **100.** $m = (y - b)/x$

Exercise Set 2.8

1. Inequality **3.** Not **5.** Real **7. a)** $28 > -12$ **b)** $\frac{7}{4} > -\frac{3}{4}$ **9. a)** $x > -3$ **b)** $x \geq 4$ **c)** $x < -\frac{7}{3}$ **d)** $x \leq \frac{5}{3}$ **11.** $x > 4$, $(4, \infty)$ **13.** $x \geq -6$, $[-6, \infty)$ **15.** $x > -5$, $(-5, \infty)$ **17.** $r \leq -6$, $(-\infty, -6]$ **19.** $x > -\frac{3}{2}$, $\left(-\frac{3}{2}, \infty\right)$ **21.** $t \leq 1$, $(-\infty, 1]$ **23.** $x < -2$, $(-\infty, -2)$ **25.** $x < \frac{3}{2}$, $\left(-\infty, \frac{3}{2}\right)$ **27.** $x > \frac{35}{9}$, $\left(\frac{35}{9}, \infty\right)$ **29.** $x < -\frac{3}{2}$, $\left(-\infty, -\frac{3}{2}\right)$ **31.** No solution, * **33.** $x \geq -6$, $[-6, \infty)$ **35.** $x < 1$, $(\infty, 1)$ **37.** All real numbers, $(-\infty, \infty)$ **39.** All real numbers, $(-\infty, \infty)$, **41.** $x > \frac{3}{8}$, $\left(\frac{3}{8}, \infty\right)$ **43.** No solution, * **45.** $x \geq -\frac{7}{11}$, $\left[-\frac{7}{11}, \infty\right)$ **47.** $x > 3$, $(3, \infty)$ **49.** $m \geq 2.5$, $[2.5, \infty)$ **51.** $x \geq 30$, $[30, \infty)$ **53.** $t > -\frac{1}{2}$, $\left(-\frac{1}{2}, \infty\right)$ **55.** $r \geq 2$, $[2, \infty)$ **57.** $t \leq -17$, $(-\infty, -17]$ **59. a)** May, September, June, August, and July **b)** January, February, December, March, November, and April **c)** January, February, and December **d)** June, August, and July **61.** ≠ **63.** We do not know that y is positive. If y is negative, we must reverse the sign of the inequality. **65.** $x > 4$ **66.** -9 **67.** -25 **68.** $\frac{14}{5}$ **69.** 500 kilowatt-hours

Chapter 2 Review Exercises

1. $3x + 24$ **2.** $5x - 10$ **3.** $-2x - 8$ **4.** $-x - 2$ **5.** $-m - 8$ **6.** $-16 + 4x$ **7.** $25 - 5p$ **8.** $24x - 30$ **9.** $-25t + 25$ **10.** $-4x + 12$ **11.** $x + 2$ **12.** $-1 - 2y$ **13.** $-x - 2y + z$ **14.** $-6a + 15b - 21$ **15.** $4m$ **16.** $-3y + 8$ **17.** $5x + 1$ **18.** $-3x + 3y$ **19.** $8m + 8n$ **20.** $9x + 3y + 2$ **21.** $4x + 3y + 6$ **22.** 3 **23.** $-12x^2 + 3$ **24.** 0

---

*No interval notation for no solution.

25. $5x + 7$ **26.** $-3b + 2$ **27.** 0 **28.** $4x - 4$ **29.** $22x - 42$ **30.** $6x^2 + 2.03x - 2.11$ **31.** $-\frac{7}{20}d + 7$ **32.** 3 **33.** $\frac{1}{6}x + 2$ **34.** $-\frac{7}{12}n$ **35.** 1 **36.** -13 **37.** 11 **38.** -27 **39.** -15 **40.** $\frac{11}{2}$ **41.** -8 **42.** -3 **43.** 12 **44.** 4 **45.** 2 **46.** -3 **47.** $\frac{3}{2}$ **48.** -3 **49.** $-\frac{1}{2}$ **50.** No solution **51.** All real numbers **52.** 2 **53.** 4 **54.** -35.5 **55.** -1.125 **56.** 0.6 **57.** $-\frac{21}{4}$ **58.** $\frac{78}{7}$ **59.** $-\frac{3}{2}$ **60.** $-\frac{2}{3}$ **61.** 2 **62.** $\frac{10}{7}$ **63.** 0 **64.** -1 **65.** 10 **66.** No solution **67.** All real numbers **68.** -4 **69.** No solution **70.** All real numbers **71.** $\frac{17}{3}$ **72.** $-\frac{20}{7}$ **73.** No solution **74.** 7 **75.** 52 **76.** 32 **77.** 7 **78.** -18 **79.** 3 **80.** 48 **81.** 12 square centimeters **82.** ≈ 33.51 cubic inches **83.** $l = (P - 2w)/2$ **84.** $m = \frac{y - y_1}{x - x_1}$ **85.** $y = \frac{1}{3}x + \frac{2}{3}$ **86.** 308.5 miles **87.** 240 square feet **88.** ≈ 25.13 cubic inches **89.** 3 : 4 **90.** 5 : 12 **91.** 6 : 1 **92.** 2 **93.** 20 **94.** 9 **95.** $\frac{135}{4}$ **96.** -10 **97.** -16 **98.** $\frac{108}{7}$ **99.** 90 **100.** 40 inches **101.** 1 foot **102.** $x \geq 2$, 

, $[2, \infty)$ **103.** $a < 3$, 

, $(-\infty, 3)$ **104.** $r \geq -2$, 

, $[-2, \infty)$ **105.** No solution, 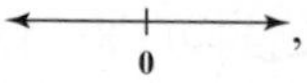

, no interval notation **106.** All real numbers, 

, $(-\infty, \infty)$ **107.** $x < -3$, 

, $(-\infty, -3)$ **108.** $x \leq \frac{19}{4}$, 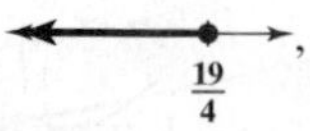

, $\left(-\infty, \frac{19}{4}\right]$ **109.** $y > \frac{8}{5}$, 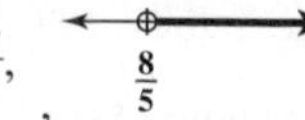

, $\left(\frac{8}{5}, \infty\right)$ **110.** $x > -12$, 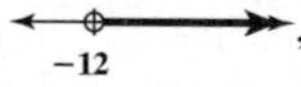

, $(-12, \infty)$ **111.** $t \geq -3$, 

, $[-3, \infty)$ **112.** 6.3 hours **113.** 72 dishes **114.** 440 pages **115.** $6\frac{1}{3}$ inches **116.** 15.75 feet **117.** $\approx \$0.078$ **118.** 192 bottles

Chapter 2 Practice Test

1. $6x - 12$ [2.1] **2.** $-x - 3y + 4$ [2.1] **3.** $-3x + 4$ [2.1] **4.** $-x + 10$ [2.1] **5.** $-5x - y - 6$ [2.1] **6.** $7a - 8b - 3$ [2.1] **7.** $2x^2 + 6x - 1$ [2.1] **8.** 4 [2.4] **9.** 8 [2.5] **10.** $-\frac{1}{7}$ [2.5] **11.** No solution [2.5] **12.** All real numbers [2.5] **13.** $x = \frac{-by - c}{a}$ [2.6] **14.** $y = \frac{6}{5}x - \frac{2}{5}$ [2.6] **15.** 0 [2.5] **16.** -45 [2.7] **17. a)** Conditional equation **b)** Contradiction **c)** Identity [2.5] **18.** $x > -7$, 

, $(-7, \infty)$ [2.8] **19.** $x \leq 12$, 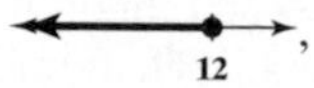

, $(-\infty, 12]$ [2.8] **20.** No solution, 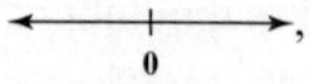

, no interval notation [2.8] **21.** All real numbers, 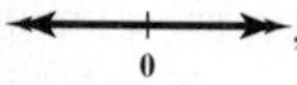

, $(-\infty, \infty)$ [2.5] **22.** $\frac{32}{3}$ feet or $10\frac{2}{3}$ feet [2.7] **23.** 4% [2.6] **24.** ≈ 28.27 inches [2.6] **25.** 175 minutes or 2 hours 55 minutes [2.6]

Cumulative Review Test

1. $\frac{8}{3}$ [1.3] **2.** $\frac{15}{16}$ [1.3] **3.** > [1.5] **4.** 3 [1.7] **5.** -1 [1.7] **6.** 16 [1.9] **7.** 168 [1.9] **8.** 12 [1.9] **9.** Distributive property [1.10] **10.** $12x + y$ [2.1] **11.** $\frac{1}{12}x + 25$ [2.1] **12.** -1 [2.4] **13.** $-\frac{5}{4}$ [2.5] **14.** $\frac{12}{5}$ [2.5] **15.** $b = 3A - a - c$ [2.6] **16.** $\frac{9}{4}$ or 2.25 [2.7] **17.** $x > 10$, 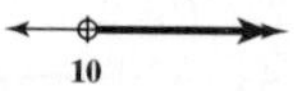

, $(10, \infty)$ [2.8] **18.** $x \geq -12$, 

, $[-12, \infty)$ [2.8] **19.** ≈ 380.13 square feet [2.6] **20.** \$42 [2.7]

Chapter 3

Exercise Set 3.1

1. Increased **3.** Product **5.** $8 - x$ **7.** $y + 0.06y$ **9.** $x + 2$ **11.** $h + 4$ **13.** $a - 5$ **15.** $5h$ **17.** $2d$ **19.** $\frac{1}{2}a$ **21.** $r - 5$ **23.** $12 - m$ **25.** $2w + 8$ **27.** $5a - 4$ **29.** $\frac{1}{3}w - 9$ **31.** x = Sonya's height **33.** x = length of Jones Beach **35.** x = number of medals Finland won **37.** x = cost of the Chevy **39.** x = Teri's grade **41.** x = amount Kristen receives or x = amount Yvonne receives **43.** x = Don's weight or x = Angela's weight **45.** Let c = cost of chair, then $5c$ = cost of table. **47.** Let a = area of the kitchen, then $2a + 20$ = area of the living room. **49.** Let w = width of rectangle, then $6w - 2$ = length of rectangle. **51.** Let w = number of medals won by Sweden, then $20 - w$ = number of medals won by Brazil, or let w = number of medals won by Brazil, then $20 - w$ = number of medals won by Sweden. **53.** Let g = George's age, then $\frac{1}{2}g + 2$ = Mike's age. **55.** Let m = number of miles Jan walked, then $6.4 - m$ = number of miles Edward walked, or let m = number of miles Edward walked, then $6.4 - m$ = number of miles Jan walked. **57.** $n + 8$ **59.** $\frac{1}{2}x$ **61.** $2a - 1$ **63.** $2t - 30$ **65.** $2p - 2.3$ **67.** $80{,}000 - m$ **69.** $2r - 673$ **71.** $10x$ **73.** $600 - a$ **75.** $45 + 0.40x$ **77.** $s + 0.20s$ **79.** $e - 0.12e$ **81.** $c + 0.07c$ **83.** $m - 0.316m$ **85.** $f + (f + 15)$ **87.** $(2l - 1) - l$ **89.** $x - (3x - 40)$ **91.** $w + (2w - 3)$ **93.** $r + (479r + 462)$ **95.** $r + (r + 0.394r)$ **97.** $a - (a + 0.112a)$ **99.** $x + 4x = 20$ **101.** $x + (x + 1) = 41$ **103.** $2x - 8 = 12$ **105.** $\frac{1}{3}(x + 12) = 5$

107. $x + 2(x + 2) = 22$ **109.** $3x + (x + 2) = 14$ **111.** $12.50h = 150$ **113.** $2.99x = 17.94$ **115.** $25q = 175$

117. $a + (2a + 1) = 52$ **119.** $(2s + 300) - s = 420$ **121.** $s + (2s - 4) = 890$ **123.** $m + (3m - 2) = 12.6$
125. $c + 0.002c = 89{,}560$ **127.** $p - 0.019p = 12{,}087$ **129.** $c + 0.07c = 32{,}600$ **131.** $c + 0.15c = 42.50$

133. a) $86{,}400d + 3600h + 60m + s$ **b)** 368,125 sec **136.** 4 **137.** 15 **138.** $y = \frac{3x - 6}{2}$ or $y = \frac{3}{2}x - 3$ **139.** 2.52
140. $x > \frac{7}{2}$, 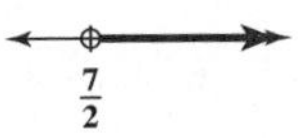 , $\left(\frac{7}{2}, \infty\right)$

Exercise Set 3.2

1. $x + (2x + 2) = 20$ **3.** $x + (x + 2) = 20$ **5.** $(2x + 2) - x = 20$ **7.** 5 **9.** 43, 44 **11.** 47, 49 **13.** 8, 10 **15.** 8, 19 **17.** 25, 42 **19.** 65 cards **21.** 72, 73 **23.** 13 weeks **25.** 9.6 years **27.** December: 22, June: 226 **29.** $15.25 **31.** 5500 pounds **33.** 12.5 gallons **35.** 18,100 copies **37.** 6 movies **39.** 150 miles **41.** 6 years **43.** 4000 pages **45.** 2000 newsletters **47.** $198.82 **49.** $23,230.77 **51.** ≈1.05 million **53.** $25,000 **55.** $39,387.76 **57.** $500 **59.** $7500 **61.** 500 pages **63.** $1071.43 **65.** $120 **67.** $77,777.78 **69.** Russia: 10.54 million bpd; Saudi Arabia: 10.27 million bpd **71.** Answers will vary **73. a)** $80 = \frac{74 + 88 + 76 + x}{4}$ **b)** 82 **75.** 4 **76.** Commutative property of addition **77.** $h = \frac{2A}{b}$ **78.** 12

Mid-Chapter Test: Sections 3.1–3.2

1. $6w$ [3.1] **2.** $3h + 5$ [3.1] **3.** $c + 0.20c$ [3.1] **4.** $60 + 0.95m$ [3.1] **5.** $50n$ [3.1] **6.** $25 - x$ [3.1] **7.** $c - 0.25c$ [3.1] **8.** x = length of Poison Dart Frog [3.1] **9.** Let p = distance Pedro traveled, then $4p + 6$ = distance Mary traveled. [3.1] **10.** $v - (v - 0.18v)$ [3.1] **11.** $p + 0.12p = 38{,}619$ [3.1] **12.** $x + 3(x + 2) = 26$ [3.1] **13.** 46, 47 [3.2] **14.** 4, 11 [3.2] **15.** 18 days [3.2] **16.** 15 hours [3.2] **17.** $700 [3.2] **18.** Betty: 196 clients, Anita: 404 clients [3.2] **19.** 52 miles [3.2] **20.** $5000 [3.2]

Exercise Set 3.3

1. Width **3.** Complementary **5.** 180° **7.** Isosceles **9.** 46°, 46°, 88° **11.** 82 feet, 82 feet, 32 feet **13.** 11.5 inches **15.** $A = 67°$, $B = 23°$ **17.** $A = 47°$, $B = 133°$ **19.** 88° **21.** 50°, 60°, 70° **23.** Length is 14 feet, width is 8 feet **25.** Length is 78 feet, width is 36 feet **27.** Smaller angles are 50°, larger angles are 130° **29.** 30°, 30°, 150°, 150° **31.** 63°, 73°, 140°, 84° **33.** Width is 4 feet, height is 7 feet **35. a)** length: 72 inches, width: 36 inches **b)** 18 inches **c)** length: 36 inches, width: 36 inches **d)** 2592 square inches **37.** Width is 11 feet, length is 16 feet **39.** $l = 12$ feet, $w = 10$ feet **42.** $<$ **43.** $>$ **44.** -10 **45.** $-2x - 5y + 6$ **46.** $y = -2x + 3$

Exercise Set 3.4

1. Subtracting **3.** $d = r \cdot t$ **5.** Adding **7.** 1 hour **9.** 6 miles per hour **11.** 0.7 hour **13.** 2.4 hours **15.** 70 miles per hour **17.** 45 seconds **19.** 35 miles per hour, 40 miles per hour **21.** 0.75 hour **23.** 1.5 miles/day, 2.25 miles/day **25.** *Apollo*: 5 miles per hour, *Pythagoras*: 9 miles per hour **27.** 2.5 hours at 70 miles per hour, 1.5 hours at 50 miles per hour **29.** Dom: 21 kilometers per hour, Sue: 11 kilometers per hour **31. a)** 2.4 miles **b)** 112.0 miles **c)** 26.2 miles **d)** 140.6 miles **e)** 9.23 hours **33.** $10,000 at 7%, $2000 at 5% **35.** $2400 at 6%, $3600 at 4% **37.** $2000 at 4%, $8000 at 5% **39.** She paid 3 months at the lower rate, so the increase occurred in April. **41.** 8 hours at Home Depot, 10 hours at clinic **43.** 1200 adult admissions **45. a)** eBay: 50 shares, Facebook: 250 shares **b)** $150 **47.** 11.25 pounds almonds, 18.75 pounds walnuts **49.** $160: 6 cubic yards, $120: 2 cubic yards **51.** $2.74 per pound **53.** 5 liters **55.** 11.25 pints **57.** 4.4 cups Clorox, 2.6 cups shock treatment **59.** 4 ounces of 5% solution, 2 ounces of 2% solution **61.** 18.96% **63.** 11.2% **65.** 40% pure juice **67.** ≈5.74 hours **70. a)** $\frac{22}{13}$ or $1\frac{9}{13}$ **b)** $\frac{35}{8}$ or $4\frac{3}{8}$ **71.** All real numbers **72.** $\frac{3}{4}$ or 0.75 **73.** $x \le \frac{1}{4}$

Chapter 3 Review Exercises

1. $3n + 7$ **2.** $1.2g$ **3.** $d - 0.25d$ **4.** $16y$ **5.** $200 - x$ **6.** Let d = Dino's age, then $7d + 6$ = Mario's age. **7.** $c - (c - 0.12c)$ **8.** $n - (3n - 24) = 8$ **9.** 33 and 41 **10.** 118 and 119 **11.** 38 and 7 **12.** $21,738.32 **13.** 19 months **14.** $2000 **15.** $618.75 **16.** 10 hours **17.** 2800 square feet **18.** Texas: 269,000 square miles, Alaska: 663,000 square miles **19.** 45°, 55°, 80° **20.** 30°, 40°, 150°, 140° **21.** Width is 15.5 feet, length is 19.5 feet **22.** Width is 50 feet, length is 80 feet **23.** 45°, 45°, 135°, 135° **24.** Height is 2 feet, length is 4 feet **25.** 2 hours **26.** 4 hours **27.** ≈8.8 feet per second **28.** $4000 at 8%, $8000 at $7\frac{1}{4}$% **29.** $700 at 3%, $3300 at 3.5% **30.** 0.5 gallons **31.** 9 smaller and 21 larger **32.** 1.2 liters of 10%, 0.8 liters of 5% **33.** 103 and 105 **34.** $450 **35.** $12,000 **36.** 42°, 50°, 88° **37.** 8 years **38.** 70°, 70°, 110°, 110° **39.** 500 copies **40.** 0.6 miles per hour **41.** 60 pounds of $3.50, 20 pounds of $4.10 **42.** Older brother: 55 miles per hour, younger brother: 60 miles per hour **43.** 0.4 liters **44.** Width is 16 feet, length is 24 feet **45.** 1.5 liters

Chapter 3 Practice Test

1. $500 - n$ [3.1] **2.** $2w + 6000$ [3.1] **3.** $60t$ [3.1] **4.** $c + 0.06c$ [3.1] **5.** Let n = number of packages of orange, then $7n - 105$ = number of packages of peppermint. [3.1] **6.** Let x = number of men, then $600 - x$ = number of women, or let x = number of women, then $600 - x$ = number of men. [3.1] **7.** $(2n - 1) - n$ [3.1] **8.** $n + (n + 18)$ [3.1] **9.** $(c + 0.84c) - c$ [3.1] **10.** 56 and 102 [3.2] **11.** 5 and 7 [3.2] **12.** 9 and 33 [3.2] **13.** $2500 [3.2] **14.** $34.78 [3.2] **15.** Peter: $40,000, Julie: $80,000 [3.2] **16.** 6 times [3.2] **17.** 7450 pages [3.2] **18.** 15 inches, 30 inches, 30 inches [3.3] **19.** Width is 6 feet, length is 8 feet [3.3] **20.** 59°, 59°, 121°, 121° [3.3] **21.** Harlene: 0.3 feet per minute, Ellis: 0.5 feet per minute [3.4] **22.** 6 miles per hour [3.4] **23.** Jelly Belly: ≈1.91 pounds, Kits: ≈1.09 pounds [3.4] **24.** 20 liters [3.4] **25.** 1 liter of 8%, 2 liters of 5% [3.4]

Cumulative Review Test **1.** \$14,600 [1.2] **2. a)** 11.6% **b)** 46 [1.2] **3. a)** 7.2 parts per million **b)** 6 parts per million [1.2] **4.** $\frac{5}{9}$ [1.3] **5.** $\frac{13}{24}$ inch [1.3] **6. a)** $\{1, 2, 3, 4, \dots\}$ **b)** $\{0, 1, 2, 3, \dots\}$ **c)** A rational number is a quotient of two integers where the denominator is not 0. [1.4] **7. a)** 4 **b)** $|-5|$ [1.5] **8.** -34 [1.9] **9.** $2x - 28$ [2.1] **10.** 5 [2.5] **11.** $\frac{1}{5}$ [2.5] **12.** No solution [2.5] **13.** ≈ 113.10 [2.6] **14. a)** $y = -\frac{1}{2}x + 2$ **b)** 4 [2.6] **15.** $w = \frac{P - 2l}{2}$ [2.6] **16.** 9 gallons [2.7] **17.** $x \le 1$,  , $(-\infty, 1]$ [2.8] **18.** 40 minutes [3.2] **19.** 6, 23 [3.2] **20.** 40°, 45°, 90°, 185° [3.3]

Chapter 4

Exercise Set 4.1 **1.** Base **3.** Add **5.** 0 **7.** Factor **9.** a^5b^3 **11.** x^{10} **13.** $-z^5$ **15.** y^5 **17.** 243 **19.** z^8 **21.** 6 **23.** x^7 **25.** 81 **27.** $\frac{1}{y^2}$ **29.** 1 **31.** $\frac{1}{q^6}$ **33.** 1 **35.** 3 **37.** 4 **39.** -9 **41.** $6x^3y^2$ **43.** $-8r$ **45.** x^8 **47.** x^{25} **49.** x^{12} **51.** $9x^2$ **53.** a^6b^3 **55.** $-8w^6$ **57.** $9x^2$ **59.** $64x^9y^6$ **61.** $\frac{x^2}{9}$ **63.** $\frac{y^4}{x^4}$ **65.** $-\frac{216}{x^3}$ **67.** $\frac{8x^3}{y^3}$ **69.** $\frac{16p^2}{25}$ **71.** $\frac{25z^6}{64}$ **73.** $\frac{a^7}{b^3}$ **75.** $\frac{x^{11}}{2y^7}$ **77.** $\frac{6y^4}{z^3}$ **79.** $\frac{7}{3x^5y^3}$ **81.** $-\frac{3y^2}{x^3}$ **83.** $-\frac{2}{x^3y^2z^5}$ **85.** $\frac{8}{x^6}$ **87.** $27y^9$ **89.** 1 **91.** $\frac{x^4}{y^4}$ **93.** $\frac{z^{24}}{16y^{28}}$ **95.** $-\frac{125}{s^6t^9}$ **97.** $9x^2y^8$ **99.** $15x^3y^6$ **101.** $-6x^2y^2$ **103.** $32s^3t^7$ **105.** $-9p^6q^3$ **107.** $49r^6s^4$ **109.** x^2 **111.** x^{12} **113.** $6.25x^6$ **115.** $\frac{x^7}{y^4}$ **117.** $-\frac{m^{12}}{n^9}$ **119.** $-216x^9y^6$ **121.** $w^4x^8y^{12}z^{16}$ **123.** $729r^{12}s^{15}$ **125.** $108x^5y^7$ **127.** $1.69x^4y^8$ **129.** $x^{11}y^{13}$ **131.** x^8z^8 **133.** Cannot be simplified **135.** Cannot be simplified **137.** $6z^2$ **139.** Cannot be simplified **141.** 40 **143.** 1 **145.** The sign will be positive because a negative number with an even exponent will be positive. This is because $(-1)^m = 1$ when m is even. **147.** $8x^2$ **149.** $ab + a^2 + b^2$ **151.** product rule: $x^m \cdot x^n = x^{m+n}$; power rule: $(x^m)^n = x^{m \cdot n}$ **153.** $576y^8z^{13}$ **156.** 13 **157.** $x + 10$ **158.** All real numbers **159. a)** 4 inches, 4 inches, 9 inches, 9 inches **b)** $w = \frac{P - 2l}{2}$

Exercise Set 4.2 **1.** Positive **3.** Multiplying **5.** $\frac{1}{9}$ **7.** $-\frac{1}{9}$ **9.** -25 **11.** $\frac{1}{x^6}$ **13.** $\frac{1}{5}$ **15.** t^3 **17.** a **19.** 36 **21.** $\frac{1}{x^{20}}$ **23.** $\frac{1}{y^{20}}$ **25.** m^{10} **27.** 9 **29.** y^2 **31.** $\frac{1}{x^2}$ **33.** 9 **35.** $\frac{1}{r}$ **37.** p^3 **39.** $\frac{1}{x^4}$ **41.** 27 **43.** $\frac{1}{125}$ **45.** z^9 **47.** p^{24} **49.** y^6 **51.** $\frac{1}{x^4}$ **53.** $\frac{1}{x^{15}}$ **55.** $-\frac{1}{16}$ **57.** $\frac{1}{16}$ **59.** $-\frac{1}{16}$ **61.** 16 **63.** $\frac{1}{x^{10}}$ **65.** n^2 **67.** 1 **69.** 1 **71.** 64 **73.** x^8 **75.** 1 **77.** $\frac{1}{4}$ **79.** $\frac{1}{49}$ **81.** x^3 **83.** $\frac{1}{16}$ **85.** 125 **87.** $\frac{1}{9}$ **89.** 1 **91.** $\frac{1}{36x^4}$ **93.** $\frac{3y^2}{x^2}$ **95.** 4 **97.** $\frac{64}{125}$ **99.** $\frac{d^4}{c^8}$ **101.** $-\frac{s^4}{r^{16}}$ **103.** $-\frac{7}{a^3b^4}$ **105.** $\frac{y^9}{64x^{15}}$ **107.** $\frac{18}{z^9}$ **109.** -8 **111.** $12x^5$ **113.** $-\frac{20z^5}{y}$ **115.** $8d^4$ **117.** $\frac{4}{x^2}$ **119.** $\frac{x^4}{2y^5}$ **121.** $\frac{8x^6}{y^2}$ **123.** $\frac{y^{14}z^6}{25x^8}$ **125.** $\frac{r^{20}t^{48}}{16s^{36}}$ **127.** $\frac{y^{12}}{x^{18}z^6}$ **129.** $\frac{1}{16p^4q^6}$ **131. a)** Yes **b)** No **133.** $16\frac{1}{16}$ **135.** $125\frac{1}{125}$ **137.** $\frac{2}{3}$ **139.** $-\frac{7}{8}$ **141.** $-\frac{5}{6}$ **143.** $\frac{22}{9}$ **145.** -2 **147.** -3 **149.** Answers will vary. **151.** -2 **153.** 2, -3 **155.** $(x + y)^m \neq x^m + y^m$ **157.** 186 miles **158.** 94, 96 **159.** 12 feet by 16 feet **160.** \$3400 at 4%, \$5600 at 3%

Exercise Set 4.3 **1.** Greater **3.** Negative **5.** Nano **7.** Mega **9.** 2×10^4 **11.** 5×10^{-4} **13.** 3.5×10^5 **15.** 7.95×10^3 **17.** 5.3×10^{-2} **19.** 7.26×10^{-4} **21.** 5.26×10^9 **23.** 9.14×10^{-6} **25.** 2.203×10^5 **27.** 5.104×10^{-3} **29.** 43,000 **31.** 0.00000932 **33.** 0.0000213 **35.** 625,000 **37.** 28.91 **39.** 535 **41.** 0.000000773 **43.** 10,000 **45.** 0.18 gram $= 1.8 \times 10^{-1}$ gram **47.** 200,000,000 watts $= 2 \times 10^8$ watts **49.** 0.0004 gram $= 4 \times 10^{-4}$ gram **51.** 5,400,000 meters $= 5.4 \times 10^6$ meters **53.** 90,000,000 **55.** 0.243 **57.** 0.000064 **59.** 2500 **61.** 250,000 **63.** 0.0013 **65.** 5.6×10^{12} **67.** 1.28×10^{-1} **69.** 7×10^2 **71.** 1.75×10^2 **73.** 3.3×10^{-4}, 5.3, 7.3×10^2, 1.75×10^6 **75. a)** $\approx 6.735 \times 10^9$ people **b)** ≈ 22.38 times **77.** 8.64×10^9 cubic feet **79.** 9×10^4 seconds **81. a)** \$$7.61 \times 10^8$, \$$4.61 \times 10^8$ **b)** \$$3.00 \times 10^8$ **c)** ≈ 1.65 times **83.** 5×10^2 seconds **85. a)** 5.97×10^{21} metric tons, 7.35×10^{19} metric tons **b)** ≈ 81.2 times **87. a)** \$$5.0 \times 10^{10}$ **b)** ≈ 5.55 times **89. a)** 2.0×10^7; **b)** 7.6×10^7 **c)** 2.2×10^7 **91.** 17 **93.** A number greater than or equal to 1 and less than 10 multiplied by some power of 10 **95. a)** Less **b)** 1.5×10^{78} **97.** Answers will vary. **99.** 1,000,000 times greater **102.** 0 **103. a)** $\frac{3}{2}$ **b)** 0 **104.** 2 **105.** $-\frac{y^{12}}{64x^9}$

Mid-Chapter Test: Sections 4.1–4.3 **1.** y^{31} [4.1] **2.** x^3 [4.1] **3.** $6x^6y^{13}$ [4.1] **4.** $\frac{2a^5b^6}{3}$ [4.1] **5.** $-64x^6y^{12}$ [4.1] **6.** $\frac{t^6}{4s^4}$ [4.1] **7.** $63x^{11}y^{13}$ [4.1] **8.** $\frac{1}{p^8}$ [4.2] **9.** $\frac{1}{x^{10}}$ [4.2] **10.** 1 [4.2] **11.** $\frac{49}{9}$ [4.2] **12.** $\frac{32x}{y}$ [4.2] **13.** $\frac{3}{m^4n^6}$ [4.2] **14.** $\frac{x^{12}y^{10}}{4z^4}$ [4.2] **15. a)** and **b)** Answers will vary. [4.3] **16.** 6.54×10^9 [4.3] **17.** 0.0000327 [4.3] **18.** 18,900 meters [4.3] **19.** 0.0238 [4.3] **20.** 3.0×10^{-7} [4.3]

Exercise Set 4.4 **1.** Whole **3.** Monomial **5.** Trinomial **7.** Sum **9.** Like **11.** 3 **13.** 1 **15.** 0 **17.** 4 **19.** 5 **21.** 8 **23.** 3 **25.** 9 **27.** Trinomial **29.** Monomial **31.** Binomial **33.** Not a polynomial **35.** Not a polynomial **37.** Polynomial **39.** Not a polynomial **41.** Trinomial **43.** Already in descending order, 0 **45.** $5x + 4$, 1 **47.** $x^2 - 2x - 4$, 2 **49.** Already in descending order, 1 **51.** Already in descending order, 2 **53.** $4x^3 - 3x^2 + x - 4$, 3 **55.** $-2x^4 + 3x^2 + 5x - 6$, 4 **57.** $13x - 9$ **59.** $-x + 11$ **61.** $5m^2 + 4$ **63.** $x^2 + 3x - 3$ **65.** $-2x^3 - 3x^2 + 4x - 3$ **67.** $11x^2 - 7x - y + 5$ **69.** $5x^2y - 3x + 2$ **71.** $x^2 + 6.6x + 0.8$ **73.** $8.2n^2 - 4.8n - 0.6$ **75.** $\frac{1}{2}x + \frac{1}{4}$ **77.** $-3x^2 + x + \frac{17}{2}$ **79.** $11x - 3$ **81.** $7y^2 - 2y + 5$ **83.** $4x^2 + 2x - 4$ **85.** $2x^3 - x^2 + 6x - 2$ **87.** $3n^3 - 11n^2 - n + 2$ **89.** $2x - 6$ **91.** $3x + 4$ **93.** $-3r$ **95.** $8x^2 + x + 4$ **97.** $8x^3 - 7x^2 - 4x - 3$ **99.** $-6y^2 + 1.9y - 12.7$ **101.** $-6.4n^2 + 6n - 7.6$ **103.** $2x^3 - \frac{23}{5}x^2 + 2x - 2$ **105.** $x - 2$ **107.** $2x^2 - 9x + 14$ **109.** $-5c^3 - 4c^2 - 7c + 14$ **111.** $3x + 8$ **113.** $3a^2 - 16a + 28$ **115.** $x^2 - 3x - 3$ **117.** $-4x^2$ **119.** $4x^3 - 7x^2 + x - 2$ **121.** Answers will vary. **123.** Answers will vary. **125.** Sometimes **127.** Sometimes **129.** Answers will vary; one example is: $x^4 - 2x^3 + x$. **131.** $a^2 + 2ab + b^2$ **133.** $4x^2 + 3xy$ **135.** No, all three terms would have to be degree 4 or 1. Therefore at least two of the terms would be like terms. **137.** $-12x + 18$ **139.** $8x^2 + 28x - 24$ **141.** $>$ **142.** True **143.** True **144.** False **145.** False **146.** $\frac{4c^8}{b^8}$

Exercise Set 4.5 **1.** Distributive **3.** Outer **5.** Last **7.** $a^2 + 2ab + b^2$ **9.** $15x^5$ **11.** $30x^5$ **13.** $-20x^6$ **15.** $-24y^8$ **17.** $20x^5y^6$ **19.** $-28x^3y^{15}$ **21.** $3x^6y$ **23.** $\frac{1}{5}x^8y^4$ **25.** $5.94x^8y^3$ **27.** $9x - 54$ **29.** $-6x^2 + 6x$ **31.** $14z^3 - 7z^2 + 21z$ **33.** $-2x^3 + 4x^2 - 10x$ **35.** $5z^3 - 4z^2 + 2z$ **37.** $0.5x^5 - 3x^4 - 0.5x^2$ **39.** $-4m^3 + 2m^2 + m$ **41.** $2n^3 - 2n^2 + \frac{2}{3}n$ **43.** $x^2 + 12x + 35$ **45.** $5x^2 + 18x - 8$ **47.** $6t^2 + 3t - 30$ **49.** $4x^2 - 10x + 4$ **51.** $12k^2 - 30k + 12$ **53.** $x^2 - 4$ **55.** $4x^2 - 12x + 9$ **57.** $4x^2 + 12x + 9$ **59.** $-12x^2 + 32x - 5$ **61.** $-6z^2 + 46z - 28$ **63.** $8x^2 - 50x + 63$ **65.** $xy - 3x + 7y - 21$ **67.** $6x^2 - 5xy - 6y^2$ **69.** $3x^2 - 2x - 8$ **71.** $x^2 + 0.9x + 0.18$ **73.** $x^2 + \frac{7}{2}x - 2$ **75.** $x^2 - 9$ **77.** $x^2 + 8x + 16$ **79.** $x^2 - 12x + 36$ **81.** $4x^2 - 1$ **83.** $49x^2 + 56x + 16$ **85.** $25x^2 - 60x + 36$ **87.** $25x^2 - y^2$ **89.** $\frac{1}{4}x^2 + 4xy + 16y^2$ **91.** $0.09x^2 - 4.2xy + 49y^2$ **93.** $3x^3 + 16x^2 + 15x - 4$ **95.** $16m^3 - 8m^2 + 9m + 18$ **97.** $-14x^3 - 22x^2 + 19x - 3$ **99.** $a^3 + b^3$ **101.** $6t^4 + 5t^3 + 5t^2 + 10t + 4$ **103.** $x^4 - 3x^3 + 5x^2 - 6x$ **105.** $2x^5 + 2x^4 - 23x^3 + x^2 - 12x$ **107.** $b^3 - 3b^2 + 3b - 1$ **109.** $27a^3 - 135a^2 + 225a - 125$ **111.** Yes **113.** No **115.** 6, 3, 1 **117. a)** $(x + 2)(2x + 1)$ or $2x^2 + 5x + 2$ **b)** 54 square feet **c)** 1 foot **119. a)** $3x^2 + 19x + 20$ **b)** $6x^3 + 32x^2 + 2x - 40$ **c)** 864 cubic feet **d)** 864 cubic feet **e)** Yes **121.** $\frac{1}{3}x^2 + \frac{11}{45}x - \frac{4}{15}$ **123.** $x^3 - 6x^2 + 11x - 6$ **125.** No solution **126.** $C = 53°, D = 37°$ **127.** $\frac{x^4}{16y^8}$ **128. a)** -216 **b)** $\frac{1}{216}$ **129.** $-5x^2 - 2x + 14$

Exercise Set 4.6 **1.** Dividend **3.** Quotient **5.** $6x^3 + 8x^2 + 0x + 16$ **7. a)** False **b)** False **c)** True **9. a)** False **b)** False **c)** True **11.** $\frac{x^2 - x - 42}{x - 7} = x + 6$ or $\frac{x^2 - x - 42}{x + 6} = x - 7$ **13.** $\frac{2x^2 + 5x + 3}{2x + 3} = x + 1$ or $\frac{2x^2 + 5x + 3}{x + 1} = 2x + 3$ **15.** $\frac{4x^2 - 9}{2x + 3} = 2x - 3$ or $\frac{4x^2 - 9}{2x - 3} = 2x + 3$ **17.** $t + 2$ **19.** $\frac{3}{7}a + 1$ **21.** $\frac{7}{3}x + 2$ **23.** $-3x + 2$ **25.** $3x + 1$ **27.** $1 - \frac{9}{5y}$ **29.** $\frac{5}{4} - \frac{1}{z}$ **31.** $1 + \frac{2}{x} - \frac{3}{x^2}$ **33.** $-2x^3 + \frac{3}{x} + \frac{4}{x^2}$ **35.** $x^2 + 3x - \frac{3}{x^3}$ **37.** $3x^2 - 2x + 6 - \frac{5}{2x}$ **39.** $-2k^2 - \frac{3}{2}k + \frac{2}{k}$ **41.** $-4x^3 - x^2 + \frac{10}{3} + \frac{3}{x^2}$ **43.** $x + 3$ **45.** $5y + 1$ **47.** $2x + 4$ **49.** $3x + 2$ **51.** $x + 5 - \frac{3}{2x - 3}$ **53.** $2x - 3 + \frac{5}{3x + 5}$ **55.** $3x^2 + 4x + 5 + \frac{4}{3x - 4}$ **57.** $2s^2 - s + 3 - \frac{5}{s + 2}$ **59.** $7x^2 - 5$ **61.** $3s^2 - 3s + 12 - \frac{17}{s + 1}$ **63.** $x^2 - 7x + 1 + \frac{2}{2x - 3}$ **65.** $x^2 + 3x + 9$ **67.** $2x^2 + x - 2 - \frac{2}{2x - 1}$ **69.** $-m^2 - 7m - 5 - \frac{8}{m - 1}$ **71.** $4t^2 - 8t + 15 - \frac{26}{t + 2}$ **73.** No; for example $\frac{x + 2}{x} = 1 + \frac{2}{x}$ which is not a binomial. **75.** $2x^2 + 11x + 16$ **77.** 3 **79.** $4x$ **81.** Since the shaded areas minus 2 must equal 3, 1, 0, and -1, respectively, the shaded areas are 5, 3, 2, and 1, respectively. **83.** $x^2 + \frac{2}{3}x + \frac{4}{9} - \frac{37}{9(3x - 2)}$ **85.** $-3x + 3 + \frac{1}{x + 3}$ **88. a)** 2 **b)** 2, 0 **c)** $2, -5, 0, \frac{2}{5}, -6.3, -\frac{23}{34}$ **d)** $\sqrt{7}, \sqrt{3}$ **e)** $2, -5, 0, \sqrt{7}, \frac{2}{5}, -6.3, \sqrt{3}, -\frac{23}{34}$ **89. a)** 0 **b)** Undefined **90.** Parentheses, exponents, multiplication or division from left to right, addition or subtraction from left to right **91.** $-\frac{2}{3}$ **92.** \$39.50 **93.** x^{13}

Chapter 4 Review Exercises

1. x^8 **2.** x^6 **3.** 243 **4.** 32 **5.** x^3 **6.** 1 **7.** 25 **8.** 64 **9.** $\frac{1}{x^2}$ **10.** y^3 **11.** 1 **12.** 7 **13.** 1 **14.** 1 **15.** $25x^2$ **16.** $27a^3$ **17.** $-27x^3$ **18.** $216s^3$ **19.** $16x^8$ **20.** t^{24} **21.** p^{32} **22.** $\frac{4x^6}{y^2}$ **23.** $\frac{25y^4}{4b^2}$ **24.** $24x^5$ **25.** $\frac{4x}{y}$ **26.** $54x^4y^9$ **27.** $9x^2$ **28.** $24x^7y^7$ **29.** $16x^8y^{11}$ **30.** $6c^6d^3$ **31.** $\frac{27a^6}{b^{15}}$ **32.** $27x^{12}y^3$ **33.** $\frac{1}{b^9}$ **34.** $\frac{1}{27}$ **35.** $\frac{1}{25}$ **36.** z^2 **37.** x^7 **38.** 16 **39.** $\frac{1}{y^3}$ **40.** $\frac{1}{x^5}$ **41.** $\frac{1}{p^2}$ **42.** $\frac{1}{a^5}$ **43.** m^{10} **44.** x^7 **45.** $\frac{1}{x^6}$ **46.** $\frac{1}{9x^8}$ **47.** $\frac{x^9}{64y^3}$ **48.** $\frac{4n^2}{m^6}$ **49.** $12y^2$ **50.** $-\frac{125z^3}{y^9}$ **51.** $\frac{x^4}{16y^6}$ **52.** $\frac{6}{x}$ **53.** $10x^2y^2$ **54.** $\frac{12x^2}{y}$ **55.** $\frac{24y^2}{x^2}$ **56.** $3y^5$ **57.** $\frac{4y}{x^3}$ **58.** $\frac{7x^5}{y^4}$ **59.** $\frac{x}{2y^5}$ **60.** $\frac{4y^{10}}{x}$ **61.** 1.72×10^6 **62.** 1.53×10^{-1} **63.** 7.63×10^{-3} **64.** 4.7×10^4 **65.** 5.76×10^3 **66.** 3.14×10^{-4} **67.** 0.0075 **68.** 0.000652 **69.** 8,900,000 **70.** 51,200 **71.** 0.0000314 **72.** 11,030,000 **73.** 0.092 liter, 9.2×10^{-2} liter **74.** 6,000,000,000 meters, 6.0×10^9 meters **75.** 0.0000128 gram, 1.28×10^{-5} gram **76.** 19,200 grams, 1.92×10^4 grams **77.** 0.085 **78.** 1260 **79.** 245 **80.** 397,000,000 **81.** 0.00003 **82.** 0.0325 **83.** 3.64×10^9 **84.** 2.12×10^1 **85.** 5.0×10^9 **86.** 5.0×10^{-4} **87.** 3.4×10^{-3} **88.** 3.4×10^7 **89.** 50,000 gallons **90. a)** $\approx 6.812 \times 10^9$ people **b)** ≈ 29.62 times **91.** Not a polynomial **92.** Monomial, 0 **93.** $x^2 + 3x - 4$, trinomial, 2 **94.** $4x^2 - x - 3$, trinomial, 2 **95.** Not a polynomial **96.** Binomial, 3 **97.** $-4x^2 + x$, binomial, 2 **98.** Not a polynomial **99.** $2x^3 + 4x^2 - 3x - 7$, polynomial, 3 **100.** $5x - 3$ **101.** $7d + 4$ **102.** $-3x - 5$ **103.** $-4x^2 + 11x - 7$ **104.** $5m^2 - 10$ **105.** $8.1p + 2.8$ **106.** $-3y - 15$ **107.** $4x^2 - 12x - 15$ **108.** $3a^2 - 5a - 21$ **109.** $4x + 2$ **110.** $-5x^2 + 8x - 19$ **111.** $3x^2 + 3x$ **112.** $-15x^2 - 12x$ **113.** $6x^3 - 12x^2 + 21x$ **114.** $-2c^3 + 3c^2 - 5c$ **115.** $28b^3 + 21b^2 + 35b$ **116.** $x^2 + 9x + 20$ **117.** $-12x^2 - 21x + 6$ **118.** $25x^2 - 30x + 9$ **119.** $-6x^2 + 14x + 12$ **120.** $r^2 - 25$ **121.** $3x^3 + x^2 - 10x + 6$ **122.** $3x^3 + 7x^2 + 14x + 4$ **123.** $-12x^3 + 10x^2 - 30x + 14$ **124.** $x + 2$ **125.** $4y + 6$ **126.** $8x + 4$ **127.** $2x^2 + 3x - \frac{4}{3}$ **128.** $2w - \frac{5}{3} + \frac{1}{w}$ **129.** $4x^5 - 2x^4 - \frac{3}{4}x^2 + \frac{1}{4x}$ **130.** $-4m + 2$ **131.** $\frac{5}{2}x + \frac{5}{x} + \frac{1}{x^2}$ **132.** $\frac{5}{3}x - 2 + \frac{5}{x}$ **133.** $x + 4$ **134.** $5x - 2 + \frac{2}{x + 6}$ **135.** $n + 3$ **136.** $2x^2 + 3x - 4$ **137.** $2x - 3$

Chapter 4 Practice Test

1. $15x^6$ [4.1] **2.** $27x^3y^6$ [4.1] **3.** $8p^5$ [4.1] **4.** $\frac{x^3}{8y^6}$ [4.1] **5.** $\frac{y^4}{4x^6}$ [4.2] **6.** 4 [4.1] **7.** $\frac{2x^7y}{3}$ [4.2] **8.** 1.425×10^{10} [4.3] **9.** 2.0×10^{-7} [4.3] **10.** Monomial [4.4] **11.** Binomial [4.4] **12.** Not a polynomial [4.4] **13.** $6x^3 - 2x^2 + 5x - 5$, 3 [4.4] **14.** $2x^2 + x - 7$ [4.4] **15.** $-3y^2 - 2y + 5$ [4.4] **16.** $3x^2 - x + 3$ [4.4] **17.** $15d^2 - 40d$ [4.5] **18.** $15x^2 + 4x - 32$ [4.5] **19.** $-12c^2 + 7c + 45$ [4.5] **20.** $6x^3 + 2x^2 - 35x + 25$ [4.5] **21.** $4x^2 + 2x - 1$ [4.6] **22.** $4x + 2 - \frac{5}{3x}$ [4.6] **23.** $4x + 5$ [4.6] **24.** $3x - 2 - \frac{2}{4x + 5}$ [4.6] **25. a)** 5.73×10^3 **b)** $\approx 7.80 \times 10^5$ [4.3]

Cumulative Review Test

1. 17 [1.9] **2.** $8x + 2$ [2.1] **3.** -25 [1.9] **4. a)** Associative property of addition **b)** Commutative property of multiplication **c)** Commutative property of multiplication [1.10] **5.** $-\frac{13}{3}$ [2.5] **6.** 0 [2.5] **7.** $x > -\frac{9}{2}$,  , $\left(-\frac{9}{2}, \infty\right)$ [2.8] **8.** $y = 3x + 5$ [2.6] **9.** $y = \frac{7x - 21}{3}$, 7 [2.6] **10.** $\frac{25x^6}{y^{16}}$ [4.1] **11.** $-7x^2 - 5x + 2$, 2 [4.4] **12.** $3x^2 + 9x - 2$ [4.4] **13.** $5a^2 + 6a + 5$ [4.4] **14.** $10t^2 - 11t + 3$ [4.5] **15.** $6x^3 - 13x^2 + 9x - 2$ [4.5] **16.** $\frac{5}{2}d + 3 - \frac{2}{d}$ [4.6] **17.** $2x + 5$ [4.6] **18.** \$3.33 [2.7] **19.** Bob: 56.5 miles per hour, Nick: 63.5 miles per hour [3.4] **20.** $l = 10$ feet, $w = 4$ feet [3.3]

Chapter 5

Exercise Set 5.1

1. Product **3.** Prime **5.** Factorization **7.** $1, 2, 4, x, 2x, 4x, x^2, 2x^2, 4x^2$ **9.** $1, 2, 3, 6, x, 2x, 3x, 6x, y, 2y, 3y, 6y, xy, 2xy, 3xy, 6xy$ **11.** $2 \cdot 3^2 \cdot 5$ **13.** $2^3 \cdot 31$ **15.** 8 **17.** 14 **19.** 2 **21.** x^2 **23.** $3x$ **25.** a **27.** qr **29.** x^3y^5 **31.** 1 **33.** $4x^4y^2$ **35.** x **37.** $x - 4$ **39.** $2x - 3$ **41.** $3w + 5$ **43.** $x - 4$ **45.** $x - 1$ **47.** $x - 9$ **49.** $4(x + 3)$ **51.** $3(3x + 1)$ **53.** $5(x - 8)$ **55.** $2(4x - 1)$ **57.** $3x(3x - 4)$ **59.** $3x^2(x^3 - 4)$ **61.** $12x^8(3x^4 + 2)$ **63.** $9y^3(3y^{12} - 1)$ **65.** $y(1 + 6x^3)$ **67.** $a^2(7a^2 + 3)$ **69.** $4xy(4yz + x^2)$ **71.** $4x^2yz^3(20x^3y^2z - 9)$ **73.** $25x^2yz(z^2 + x)$ **75.** $y^2z^3(13y^3 - 11xz^2)$ **77.** $4(2c^2 - c - 8)$ **79.** $3(3x^2 + 6x + 1)$ **81.** $4x(x^2 - 2x + 3)$ **83.** $8(5b^2 - 6c + 3)$ **85.** $x(5x^2 - y^2 + 1)$ **87.** $3a(3a^3 - 2a^2 + b)$ **89.** $xy(8x + 12y + 5)$ **91.** $(x - 7)(x + 6)$ **93.** $(a - 2)(3b - 4)$ **95.** $(2x + 1)(4x + 1)$ **97.** $(2x + 1)(5x + 1)$ **99.** $(6c + 7)(3c - 2)$ **101.** $6\nabla(2 - \nabla)$ **103.** $4\square(3\square^2 - \square + 1)$ **105.** Multiply to obtain the original expression and be sure all common factors have been factored out. **107.** $2x^2(2x + 7)(3x^3 + 2x - 1)$ **109.** $(x + 2)(x + 3)$ **110.** $-3x + 17$ **111.** 2 **112.** $y = \frac{4x - 20}{5}$ or $y = \frac{4}{5}x - 4$ **113.** ≈ 201.06 cubic inches **114.** 14, 27 **115.** $\frac{9y^2}{4x^6}$

Exercise Set 5.2 **1.** Grouping **3.** Positive **5.** Common **7.** $(w+z)(c+d)$ **9.** $(x+3)(x+2)$ **11.** $(w-z)(c+d)$ **13.** $(c-4)(c+7)$ **15.** $(y+z)(w-x)$ **17.** $(3x+4)(2x+5)$ **19.** $(3x+4)(2x-5)$ **21.** $(x+4)(8x+1)$ **23.** $(3t-2)(4t-1)$ **25.** $(x+9)(x-1)$ **27.** $(2p+5)(3p-2)$ **29.** $(x+2y)(x-3y)$ **31.** $(3x+2y)(x-3y)$ **33.** $(5x-6y)(2x-5y)$ **35.** $(x-b)(x-a)$ **37.** $(y+9)(x-5)$ **39.** $(a+3)(a+b)$ **41.** $(y-1)(x+5)$ **43.** $(3+2y)(4-x)$ **45.** $(z+5)(z^2+1)$ **47.** $(x-5)(x^2+8)$ **49.** $5(x-3)(x^2+2)$ **51.** $4(x+2)(x+2)=4(x+2)^2$ **53.** $x(2x+3)(3x-1)$ **55.** $p(p-6q)(p+2q)$ **57.** $(x+5)(y+6)$ **59.** $(a+7)(x-3)$ **61.** $(y+5)(x+5)$ **63.** $(r+6)(s-7)$ **65.** $(c-a)(d+3)$ **67.** No; $xy+2x+5y+10$ is factorable; $xy+10+2x+5y$ is not factorable in this arrangement. **69.** $(\odot+3)(\odot-5)$ **71.** $x^2-2xy-3x+6y$ or $x^2-3x-2xy+6y$ **73. a)** $2x^2-5x-6x+15$ **b)** $(2x-5)(x-3)$ **75. a)** $2x^2-6x-5x+15$ **b)** $(x-3)(2x-5)$ **77. a)** $4x^2+3x-20x-15$ **b)** $(4x+3)(x-5)$ **79.** $(\odot+3)(\bigstar+2)$ **81.** $\frac{6}{5}$ **82.** 30 pounds of jelly beans, 20 pounds of gumdrops **83.** $5x^2-2x-3+\frac{5}{3x}$ **84.** $a-4$

Exercise Set 5.3 **1.** Common **3.** Same **5.** Negative **7.** Descending **9.** Check **11.** $(x+2)(x+4)$ **13.** $(x+3)(x+8)$ **15.** $(y-8)(y-2)$ **17.** $(h-3)(h-1)$ **19.** $(x+8)(x-3)$ **21.** Prime **23.** $(y-12)(y-1)$ **25.** $(a-4)(a+2)$ **27.** $(r-5)(r+3)$ **29.** $(b-9)(b-2)$ **31.** Prime **33.** $(q+9)(q-5)$ **35.** $(x-10)(x+3)$ **37.** $(x+2)^2$ **39.** $(s-4)^2$ **41.** $(p-6)^2$ **43.** $(w-15)(w-3)$ **45.** $(x+13)(x-3)$ **47.** $(x-5)(x+4)$ **49.** $(y+5)(y+8)$ **51.** $(x+16)(x-4)$ **53.** Prime **55.** $(x-16)(x-4)$ **57.** $(a-9)(a-11)$ **59.** $(x+2)(x+1)$ **61.** $(w+9)(w-2)$ **63.** $(x-3y)(x-5y)$ **65.** $(m-3n)^2$ **67.** $(x+6y)(x+2y)$ **69.** $(m+3n)(m-8n)$ **71.** $6(x-4)(x-1)$ **73.** $5(x+3)(x+1)$ **75.** $2(x-4)(x-5)$ **77.** $b(b-5)(b-2)$ **79.** $3z(z-9)(z+2)$ **81.** $x(x+4)^2$ **83.** $3x(x+3y)(x-2y)$ **85.** $3r(r+4t)(r-2t)$ **87.** $x^2(x-7)(x+3)$ **89.** Both negative; one positive and one negative; one positive and one negative; both positive **91.** $x^2-12x+32=(x-8)(x-4)$ **93.** $x^2-2x-35=(x-7)(x+5)$ **95.** The GCF, 2, was not factored out first. **97.** $(x+0.4)(x+0.2)$ **99.** $\left(x+\frac{1}{5}\right)\left(x+\frac{1}{5}\right)=\left(x+\frac{1}{5}\right)^2$ **101.** $(x+8)(x-32)$ **103.** 9 **104.** 19.6% **105.** $2x^3+x^2-16x+12$ **106.** $3x+2-\frac{2}{x-4}$ **107.** $(5x+2)(4x-3)$

Exercise Set 5.4 **1.** Same **3.** Multiplying **5.** $(2x+1)(x+5)$ **7.** $(3x+2)(x+4)$ **9.** $(5x+1)(x-2)$ **11.** $(3r-2)(r+5)$ **13.** $(2z-3)^2$ **15.** $(2z+3)(3z-4)$ **17.** Prime **19.** $(8x+3)(x+2)$ **21.** Prime **23.** $(5y-1)(y-3)$ **25.** $(7x+1)(x+6)$ **27.** $(2x+5)(2x-3)$ **29.** $(7t-1)^2$ **31.** $(5z+4)(z-2)$ **33.** $(4y-3)(y+2)$ **35.** $(5x-1)(2x-5)$ **37.** Prime **39.** $(7m+5)(3m-4)$ **41.** $(7t+3)(t+1)$ **43.** $2(3x+5)(x+1)$ **45.** $4x(3x+1)(x+2)$ **47.** $x(2x+1)(3x-4)$ **49.** $2x(2x+3)(x-2)$ **51.** $8(2c-1)(3c+2)$ **53.** $4(2p+3)(p-1)$ **55.** $(8c+d)(c+5d)$ **57.** $(5x+3y)(3x-2y)$ **59.** $2(2x-y)(3x+4y)$ **61.** $(7p+6q)(p+q)$ **63.** $(3m-2n)(2m+n)$ **65.** $x(4x+3y)(2x+y)$ **67.** $x^2(2x+y)(2x+3y)$ **69.** $3x^2-20x-7$; obtained by multiplying the factors. **71.** $10x^2+35x+15$; obtained by multiplying the factors. **73.** $3t^4+11t^3-4t^2$; obtained by multiplying the factors. **75. a)** Dividing the trinomial by binomial gives the second factor. **b)** $6x+11$ **77.** Factoring trinomials is the reverse process of multiplying binomials. **79.** $(6x-5)(3x+4)$ **81.** $(5x-8)(3x-20)$ **83.** $5(3a-8)(7a+4)$ **85.** $2x+45$, the product of the three first terms must equal $6x^3$, and the product of the constants must equal 2250. **87.** 49 **88.** 140.25 mph **89.** $12xy^2(3x^3y-1+2x^4y^4)$ **90.** $(b+12)(b-8)$

Mid-Chapter Test: Sections 5.1–5.4 **1.** A factoring problem may be checked by multiplying the factors. [5.1] **2.** $3xy^2$ [5.1] **3.** $4a^2b(b^2-6a)$ [5.1] **4.** $(d-6)(5c-3)$ [5.1] **5.** $(2x+9)(7x+1)$ [5.1] **6.** $(x+4)(x+7)$ [5.2] **7.** $(x+5)(x-3)$ [5.2] **8.** $(2a+5b)(3a-b)$ [5.2] **9.** $(5x-2y)(x-9)$ [5.2] **10.** $4x(2x+1)(x-6)$ [5.2] **11.** $(x-3)(x-7)$ [5.3] **12.** $(t+4)(t+5)$ [5.3] **13.** Prime [5.3] **14.** $(x+8)^2$ [5.3] **15.** $(m+5n)(m-9n)$ [5.3] **16.** $(3x+2)(x+5)$ [5.4] **17.** $(4z-3)(z-2)$ [5.4] **18.** Prime [5.4] **19.** $(3x-1)^2$ [5.4] **20.** $3(2a-b)(a+b)$ [5.4]

Exercise Set 5.5 **1.** Difference **3.** a^2-ab+b^2 **5.** Common **7.** Prime **9.** $4(z^2+4)$ **11.** Prime **13.** $(y+5)(y-5)$ **15.** $(x+7)(x-7)$ **17.** $(9+z)(9-z)$ **19.** $(x+y)(x-y)$ **21.** $(4x+3)(4x-3)$ **23.** $(3y+5z)(3y-5z)$ **25.** $(6+7x)(6-7x)$ **27.** $(z^2+9x)(z^2-9x)$ **29.** $(5x^2+7y^2)(5x^2-7y^2)$ **31.** $6(x+5)(x-5)$ **33.** $2(x^2+5y)(x^2-5y)$ **35.** $5(x^2+9)(x+3)(x-3)$ **37.** $(x+y)(x^2-xy+y^2)$ **39.** $(x-y)(x^2+xy+y^2)$ **41.** $(x+4)(x^2-4x+16)$ **43.** $(x-3)(x^2+3x+9)$ **45.** $(a+1)(a^2-a+1)$ **47.** $(3x-1)(9x^2+3x+1)$ **49.** $(3k-5)(9k^2+15k+25)$ **51.** $(3-2y)(9+6y+4y^2)$ **53.** $(4m+3n)(16m^2-12mn+9n^2)$ **55.** $2(2x+3y)(4x^2-6xy+9y^2)$ **57.** $4(t-3)^2$ **59.** $2(5x+2)(5x-3)$ **61.** $3(x+4)(x+3)$ **63.** $5(x-3)(x+1)$ **65.** $5(x+2)(x-2)$ **67.** $2(x+5)(x-5)$ **69.** $2y(x+3)(x-3)$ **71.** $3y^2(x+1)(x^2-x+1)$ **73.** $2(x-2)(x^2+2x+4)$ **75.** $2(3x+5)(3x-5)$ **77.** $3r(2t^2-5t+7)$ **79.** $2(3x-2)(x+4)$ **81.** $2r(s+3)(s-8)$ **83.** $(x+2)(4x-3)$ **85.** $25(b+2)(b-2)$ **87.** $5(x-2)(x^2+3)$ **89.** $5x^2(x+1)^2$ **91.** $x(x^2+25)$ **93.** $(y^2+4)(y+2)(y-2)$ **95.** $2(2m+5)(4m^2-10m+25)$ **97.** $(c+2)(a+b)$ **99.** $3x^2y^2(2x-3y)(6x+5y)$ **101.** (✹ + 2)(◆ + ☺) **103.** 2✹(2◆ − 3)(◆ − 5) **105.** Cannot divide both sides of equation by $a-b$, because it equals 0. **107.** $(x^2+1)(x^4-x^2+1)$ **109.** $(x+y)(x-y)(x^2-xy+y^2)(x^2+xy+y^2)$ **111.** $x\le 1$; 

, $(-\infty,-1]$ **112.** 4 inches **113.** -9 **114.** $\frac{8x^9}{27y^{12}}$ **115.** $\frac{1}{a^{11}}$

Exercise Set 5.6 **1.** Quadratic **3.** Product **5.** Subtracting **7.** $-8, 7$ **9.** $0, 8$ **11.** $-\frac{7}{3}, \frac{11}{2}$ **13.** $0, 12$ **15.** $0, 5$ **17.** $-3, 8$ **19.** $\frac{5}{3}, 7$ **21.** 4 **23.** $4, -4$ **25.** $-2, -10$ **27.** $-2, -10$ **29.** $-3, 4$ **31.** $5, -3$ **33.** $30, -1$ **35.** $-3, -\frac{5}{4}$ **37.** $-\frac{1}{3}, -2$ **39.** $\frac{2}{3}, -5$

41. $-4, 3$ **43.** $-\frac{3}{4}, \frac{1}{2}$ **45.** $10, -10$ **47.** $0, 25$ **49.** $\frac{5}{2}, -\frac{5}{2}$ **51.** $-2, 5$ **53.** $-2, \frac{1}{3}$ **55.** $\frac{3}{2}, 6$ **57.** $x^2 - 2x - 24 = 0$ (other answers are possible) **59.** $x^2 - 6x = 0$ (other answers are possible) **61. a)** $(2x - 1)$ and $(3x + 1)$ **b)** $6x^2 - x - 1 = 0$ **63.** They are constant multiples of one another. **65.** $4, 5$ **67.** $0, 3, -2$ **69.** $\frac{17}{45}$ **70. a)** Identity **b)** Contradiction **71.** ≈ 738 people **72.** $\frac{9}{p^8q^2}$ **73.** Monomial **74.** Binomial **75.** Not a polynomial **76.** Trinomial

Exercise Set 5.7 **1.** Standard **3.** Hypotenuse **5.** $11, 33$ **7.** $-7, -14$ *or* $7, 14$ **9.** $6, 14$ **11.** $7, 8$ **13.** $16, 18$ **15.** $w = 3$ feet, $l = 12$ feet **17.** $l = 15$ inches, $w = 12$ inches **19.** 5 meters **21.** 4 seconds **23.** Yes **25.** No **27.** Yes **29.** No **31.** 3 **33.** 12 **35.** 6 feet, 8 feet, 10 feet **37.** Width: 9 inches, length: 12 inches **39.** $w = 6$ feet, $l = 8$ feet **41.** 30 books **43. a)** 4 **b)** 9 **45.** 12 feet **47.** 24 feet **49.** -9 and 7 **54.** $3x - 7$ **55.** $-x^2 + 7x - 4$ **56.** $6x^3 + x^2 - 10x + 4$ **57.** $2x - 3$ **58.** $2x - 3$

Chapter 5 Review Exercises **1.** y^3 **2.** $3p$ **3.** $6c^2$ **4.** $5x^2y^2$ **5.** 1 **6.** s **7.** $x - 3$ **8.** $x + 5$ **9.** $7(x - 5)$ **10.** $5(7x - 1)$ **11.** $4y(6y - 1)$ **12.** $5p^2(11p - 4)$ **13.** $12ab(5a - 3b)$ **14.** $9xy(1 - 4x^2y)$ **15.** $4x^3y^2(5 + 2x^6y - 4x^2)$ **16.** Prime **17.** Prime **18.** $(5x + 3)(x - 2)$ **19.** $(t - 1)(3t + 4)$ **20.** $(4x - 3)(2x + 1)$ **21.** $(x + 6)(x + 2)$ **22.** $(x - 5)(x + 4)$ **23.** $(y - 6)^2$ **24.** $(y + 1)(3x + 2)$ **25.** $(a - b)(4a - 1)$ **26.** $(x + 6)(2x - 1)$ **27.** $(x + 3)(x - 2y)$ **28.** $(5x - y)(x + 4y)$ **29.** $(x + 3y)(4x - 5y)$ **30.** $(3a - 5b)(2a - b)$ **31.** $(p - 3)(q + 4)$ **32.** $(x - 3y)(3x + 2y)$ **33.** $(3x - 4)(x^2 + 5)$ **34.** $(x - 3)(2x^2 - 7)$ **35.** $(x + 3)(x + 2)$ **36.** Prime **37.** $(x + 2)(x + 9)$ **38.** $(n + 8)(n - 5)$ **39.** $(b + 5)(b - 4)$ **40.** $(x - 8)(x - 7)$ **41.** Prime **42.** Prime **43.** $x(x - 9)(x - 8)$ **44.** $t(t - 9)(t + 4)$ **45.** $(x + 3y)(x - 5y)$ **46.** $4x(x + 5y)(x + 3y)$ **47.** $(2x + 5)(x - 3)$ **48.** $(6x + 1)(x - 5)$ **49.** $(4x - 5)(x - 1)$ **50.** $(5m - 4)(m - 2)$ **51.** $(4y + 3)(4y - 1)$ **52.** $(5x - 2)(x - 6)$ **53.** Prime **54.** $(5x - 3)(x + 8)$ **55.** $(2s + 1)(3s + 5)$ **56.** $(3x - 2)(2x + 5)$ **57.** $2(3x + 2)(2x - 1)$ **58.** $(5x - 3)^2$ **59.** $x(3x - 2)^2$ **60.** $2x(3x + 4)(3x - 2)$ **61.** $(2a - 3b)(2a - 5b)$ **62.** $(8a + b)(2a - 3b)$ **63.** $(x + 10)(x - 10)$ **64.** $(x + 6)(x - 6)$ **65.** $3(x + 4)(x - 4)$ **66.** $9(3x + y)(3x - y)$ **67.** $(9 + a)(9 - a)$ **68.** $(8 + x)(8 - x)$ **69.** $(4x^2 + 7y)(4x^2 - 7y)$ **70.** $(8x^3 + 7y^3)(8x^3 - 7y^3)$ **71.** $(a + b)(a^2 - ab + b^2)$ **72.** $(x - y)(x^2 + xy + y^2)$ **73.** $(x - 1)(x^2 + x + 1)$ **74.** $(x + 2)(x^2 - 2x + 4)$ **75.** $(a + 3)(a^2 - 3a + 9)$ **76.** $(b - 4)(b^2 + 4b + 16)$ **77.** $(5a + b)(25a^2 - 5ab + b^2)$ **78.** $(3 - 2y)(9 + 6y + 4y^2)$ **79.** $3(x - 4y)(x^2 + 4xy + 16y^2)$ **80.** $3(3x^2 + 5y)(3x^2 - 5y)$ **81.** $(x - 6)(x - 8)$ **82.** $3(x - 3)^2$ **83.** $5(q + 1)(q - 1)$ **84.** $8(x + 3)(x - 1)$ **85.** $4(y + 3)(y - 3)$ **86.** $(x - 9)(x + 3)$ **87.** $(3x - 1)^2$ **88.** $(7x - 3)(x + 4)$ **89.** $6(b - 1)(b^2 + b + 1)$ **90.** $y(x - 3)(x^2 + 3x + 9)$ **91.** $b(a + 3)(a - 5)$ **92.** $3x(x + 5)(2x + 3)$ **93.** $(x - 3y)(x - y)$ **94.** $(3m - 4n)(m + 2n)$ **95.** $(2x + 3y)^2$ **96.** $(5a + 7b)(5a - 7b)$ **97.** $(x + 2)(y - 7)$ **98.** $y^5(4 + 5y)(4 - 5y)$ **99.** $(2x - 3y)(3x + 7y)$ **100.** $2x(2x + 5y)(x + 2y)$ **101.** $x^2(4x + 1)(4x - 3)$ **102.** $(d^2 + 4)(d + 2)(d - 2)$ **103.** $0, -9$ **104.** $2, -6$ **105.** $-5, \frac{3}{4}$ **106.** $0, -7$ **107.** $0, -5$ **108.** $0, -3$ **109.** $-3, -6$ **110.** $1, 2$ **111.** $-4, 3$ **112.** $-1, -4$ **113.** $2, 4$ **114.** $3, -5$ **115.** $\frac{1}{4}, -\frac{3}{2}$ **116.** $-\frac{1}{3}, 4$ **117.** $2, -2$ **118.** $\frac{10}{7}, -\frac{10}{7}$ **119.** $\frac{3}{2}, \frac{1}{4}$ **120.** $\frac{3}{2}, \frac{5}{2}$ **121.** $a^2 + b^2 = c^2$ **122.** Hypotenuse **123.** 6 feet **124.** 12 meters **125.** $9, 11$ **126.** $4, 14$ **127.** Width: 12 feet, length: 15 feet **128.** 8 feet, 15 feet, 17 feet **129.** 9 inches **130.** 10 feet **131.** 1 second **132.** 80 dozen

Chapter 5 Practice Test **1.** $3y^3$ [5.1] **2.** $8p^2q^2$ [5.1] **3.** $5x^2y^2(y - 3x^3)$ [5.1] **4.** $4a^2b(2a - 3b + 7)$ [5.1] **5.** $(x - 5)(4x + 1)$ [5.2] **6.** $(a - 4b)(a - 5b)$ [5.2] **7.** $(r + 8)(r - 3)$ [5.3] **8.** $(5a - 3b)(5a + 2b)$ [5.4] **9.** $4(x + 2)(x - 6)$ [5.4] **10.** $y(2y - 3)(y + 1)$ [5.4] **11.** $(3x + 2y)(4x - 3y)$ [5.4] **12.** $(x + 3y)(x - 3y)$ [5.5] **13.** $(x - 4)(x^2 + 4x + 16)$ [5.5] **14.** $\frac{5}{6}, -3$ [5.6] **15.** $0, 6$ [5.6] **16.** $-8, 8$ [5.6] **17.** -9 [5.6] **18.** $3, 4$ [5.6] **19.** $-2, -3$ [5.6] **20.** 5 inches [5.7] **21.** 34 feet [5.7] **22.** $4, 9$ [5.7] **23.** $12, 14$ [5.7] **24.** Length: 6 meters, width: 4 meters [5.7] **25.** 10 seconds [5.7]

Cumulative Review Test **1.** -171 [1.9] **2.** 9 [1.9] **3.** \$90 [3.2] **4. a)** 7 **b)** $-6, -0.2, \frac{3}{5}, 7, 0, -\frac{5}{9}, 1.34$ **c)** $\sqrt{7}, -\sqrt{2}$ **d)** $-6, -0.2, \frac{3}{5}, \sqrt{7}, -\sqrt{2}, 7, 0, -\frac{5}{9}, 1.34$ [1.4] **5.** $|-8|$ [1.5] **6.** 13 [2.5] **7.** 19.2 [2.7] **8.** $x \geq 5$,  , $[5, \infty)$ [2.8] **9.** $y = -\frac{4}{3}x + \frac{7}{3}$ [2.6] **10.** 6 liters [3.4] **11.** $47, 49$ [3.2] **12.** $\frac{1}{4}$ hour [3.4] **13.** $\frac{16x^6}{81y^8}$ [4.1] **14.** $\frac{16y^6}{x^3}$ [4.2] **15.** $-3x^3 + 2x^2 + 6x - 12$ [4.4] **16.** $3x^3 + 13x^2 - 28x + 12$ [4.5] **17.** $x - 5 + \frac{21}{x + 3}$ [4.6] **18.** $(r + 2)(q - 8)$ [5.2] **19.** $(5x + 3)(x - 2)$ [5.4] **20.** $7y(y + 3)(y - 3)$ [5.5]

Chapter 6

Exercise Set 6.1 **1.** Rational **3.** $\frac{2}{3}$ **5.** Reduced **7.** 1 **9.** 3 **11.** All real numbers except $x = 1$ **13.** All real numbers except $x = 0$ **15.** All real numbers except $n = 4$ **17.** All real numbers except $x = 2$ and $x = -2$ **19.** All real numbers except $x = \frac{3}{2}$ and $x = 3$ **21.** All real numbers **23.** All real numbers except $p = \frac{5}{2}$ and $p = -\frac{5}{2}$ **25.** $\frac{x}{3y^4}$ **27.** $\frac{4}{b^5}$ **29.** $\frac{x}{x + 3}$ **31.** 5

33. $\frac{x^2+5x-11}{4x}$ **35.** $r+1$ **37.** $\frac{x}{x+2}$ **39.** $\frac{z-5}{z+5}$ **41.** $\frac{x+1}{x+2}$ **43.** $\frac{x+6}{x-6}$ **45.** $\frac{1}{4m-5}$ **47.** $\frac{x-5}{x+5}$ **49.** $2x-3$ **51.** $\frac{x-4}{x+4}$ **53.** a^2+2a+4 **55.** $3s+4t$ **57.** -1 **59.** $-(x+2)$ **61.** $-\frac{x+6}{2x}$ **63.** $-(x+3)$ **65.** $-\frac{p^2+pq+q^2}{p+q}$ **67.** $\frac{☺}{5}$ **69.** $\frac{\Delta}{2\Delta+9}$ **71.** -1 **73.** $x+2$; $(x+2)(x-3)=x^2-x-6$ **75.** $x^2+9x+20$; $(x+5)(x+4)=x^2+9x+20$ **77.** Set the denominator equal to zero and then solve the resulting equation. **79. a)** $x\neq 0, x\neq -5, x\neq \frac{3}{2}$ **b)** $\frac{1}{x(2x-3)}$ **81.** 1, the numerator and denominator are identical. **84.** $y=x-4z$ **85.** 28°, 58°, and 94° **86.** $\frac{25}{81x^4y^2}$ **87.** $8x^2-10x-19$ **88.** $3(a+4)(a-6)$ **89.** 13 inches

Exercise Set 6.2

1. Denominators **3.** $\frac{ad}{bc}$ **5.** $\frac{3}{19}$ **7.** $-\frac{4}{49}$ **9.** $\frac{5}{16}$ **11.** $\frac{3}{5}$ **13.** $-\frac{13}{48}$ **15.** $\frac{75}{88}$ **17.** $\frac{6}{b}$ **19.** $-\frac{50p^3}{3q^6}$ **21.** $\frac{36x^9y^2}{25z^7}$ **23.** $\frac{-3x+2}{3x+2}$ **25.** 1 **27.** $\frac{1}{(a+b)(a-b)}$ **29.** 1 **31.** $\frac{x+2}{x+3}$ **33.** $x+9$ **35.** $4x^2y$ **37.** $\frac{9z}{x}$ **39.** $-\frac{5y}{2xz^2}$ **41.** $\frac{x+2}{5x}$ **43.** $\frac{x-15}{x+9}$ **45.** $\frac{x-8}{x+2}$ **47.** $\frac{x+3}{x-1}$ **49.** -1 **51.** $\frac{x+1}{x-4}$ **53.** $4x^2y^2$ **55.** $\frac{7c}{5ab^2}$ **57.** $\frac{3y^2}{a^2}$ **59.** $-\frac{9km^3n}{2}$ **61.** $\frac{2(x+3)}{x(x-3)}$ **63.** $\frac{7x}{y}$ **65.** $\frac{4}{m^4n^{11}}$ **67.** $\frac{r+2}{r-3}$ **69.** $\frac{x-6}{x-3}$ **71.** $\frac{2w-7}{w+1}$ **73.** $\frac{q-5}{2q+3}$ **75.** -1 **77.** $\frac{1}{6\Delta^3}$ **79.** $\frac{\Delta+☺}{9(\Delta-☺)}$ **81.** x^2+5x+6 **83.** $x^2-4x-12$ **85.** x^2-3x+2 **87.** Answers will vary. **89.** $\frac{x-1}{x-3}$ **91.** x^2-5x+6, x^2-x-20 **94.** 1 hour **95.** $12x^4y^5z^{11}$ **96.** $2x^2+x-2-\frac{2}{2x-1}$ **97.** $6(x-5)(x+2)$ **98.** $5, -2$

Exercise Set 6.3

1. Denominator **3.** $\frac{a-b}{c}$ **5.** Numerator **7.** $\frac{5}{7}$ **9.** $\frac{1}{4}$ **11.** $\frac{5r-1}{4}$ **13.** $\frac{x+6}{x}$ **15.** $\frac{n+8}{n+1}$ **17.** $\frac{2a-3}{a+1}$ **19.** $\frac{5x+9}{x-3}$ **21.** $\frac{c+7}{c+4}$ **23.** $\frac{t+3}{5t^2}$ **25.** $\frac{1}{x-4}$ **27.** $\frac{1}{m-3}$ **29.** $\frac{p-12}{p-5}$ **31.** $x-3$ **33.** $x-5$ **35.** 1 **37.** 4 **39.** $\frac{5}{x-2}$ **41.** $\frac{3x+2}{x-4}$ **43.** $\frac{x+5}{x^2+3}$ **45.** $\frac{x-3}{x-1}$ **47.** $\frac{6x+1}{x-8}$ **49.** 5 **51.** $9n$ **53.** $3(x+3)$ **55.** p^3 **57.** $3m-4$ **59.** $6t^2$ **61.** $(x+8)(x+4)$ **63.** $18r^4s^7$ **65.** $m(m+2)$ **67.** $x(x+1)$ **69.** $4n-1$ or $1-4n$ **71.** $4k-5r$ or $-4k+5r$ **73.** $18q(q+1)$ **75.** $120x^2y^3$ **77.** $6(x+4)(x+2)$ **79.** $(x+1)(x+8)$ **81.** $(x-8)(x+3)(x+8)$ **83.** $(a-4)^2(a-3)$ **85.** $(x+5)(x+1)(x+3)$ **87.** $(x-3)^2$ **89.** $(x-6)(x-1)$ **91.** $(3t-2)(t+4)(3t-1)$ **93.** $(2x+1)^2(4x+3)$ **95.** $\frac{19}{35}$ **97.** $\frac{35}{36}$ **99.** $\frac{1}{18}$ **101.** x^2+x-9; the sum of the numerators must be $2x^2-5x-6$ **103.** $x^2+9x-10$; the sum of the numerators must be $5x-7$ **105.** 5☺ **107.** $(\Delta+3)(\Delta-3)$ **109.** Answers will vary. **111.** $\frac{-3x^2+12x}{x^2-25}$ **113.** $30x^{12}y^9$ **115.** $(x-4)(x+3)(x-2)$ **117.** $\frac{92}{45}$ or $2\frac{2}{45}$ **118.** $-\frac{1}{5}$ **119.** 2.25 ounces **120.** 70 hours **121.** 4.0×10^{11} **122.** $\frac{3}{2}, -1$

Exercise Set 6.4

1. Denominator **3.** Maintaining **5.** -1 **7.** $\frac{5x+2y}{xy}$ **9.** $\frac{9-4t}{3t^2}$ **11.** $\frac{3x+8}{x}$ **13.** $\frac{3x+10}{5x^2}$ **15.** $\frac{35x-4y}{30x^3y^4}$ **17.** $\frac{4y^2+x}{y}$ **19.** $\frac{9a+7}{6a}$ **21.** $\frac{3x-xy+4y}{xy}$ **23.** $\frac{11p+6}{p(p+3)}$ **25.** $\frac{x+6}{2x^2}$ **27.** $\frac{a^2}{b(a-b)}$ **29.** $\frac{2x+7}{(x-7)(x-4)}$ **31.** $\frac{6}{p-3}$ **33.** $\frac{14}{x+7}$ **35.** $\frac{a+16}{2(a-2)}$ **37.** $\frac{20x}{(x-5)(x+5)}$ **39.** $\frac{5q-18}{2q(q+3)}$ **41.** $\frac{15w+66}{2(w+5)(w+2)}$ **43.** $\frac{2x-7}{(x-5)^2}$ **45.** $\frac{-x+6}{(x+2)(x-2)}$ **47.** $\frac{r+12}{(r-4)(r-6)}$ **49.** $\frac{x^2-17}{(x+4)^2}$ **51.** $\frac{x^2+3x-21}{(x+5)^2}$ **53.** $\frac{-a+16}{(a-8)(a-1)(a+2)}$ **55.** $\frac{9x+17}{(x+3)^2(x-2)}$ **57.** $\frac{3x^2-12x-5}{(2x+1)(3x-2)(x+3)}$ **59.** $\frac{2x^2-3x-4}{(4x+3)(x+2)(2x-1)}$ **61.** $\frac{1}{w-3}$ **63.** $\frac{6}{r-3}$ **65.** $-\frac{x+4}{4x}$ **67.** All real numbers except $x=0$ **69.** All real numbers except $x=4$ and $x=-6$ **71.** $\frac{7}{\Delta-2}$ **73.** All real numbers except $a=-b$ and $a=0$ **75.** 0 **77.** $\frac{2x-3}{2-x}$ **79.** $\frac{6x+5}{(x+2)(x-3)(x+1)}$ **82.** $\approx$1.53 hours **83.** $x>-8$, -8, $(-8,\infty)$ **84.** $4x-3-\frac{6}{2x+3}$ **85.** -1

Mid-Chapter Test

1. All real numbers except $x=\frac{2}{3}$ [6.1] **2.** All real numbers except $x=-2, x=7$ [6.1] **3.** 9 [6.1] **4.** $\frac{2x+3}{3x-1}$ [6.1] **5.** $5r+6t$ [6.1] **6.** $\frac{6y^3}{x^3}$ [6.2] **7.** $-(m+4)$ or $-m-4$ [6.2] **8.** $x-2$ [6.2] **9.** $\frac{x+7}{2(x+1)}$ [6.2] **10.** $\frac{5x+2}{7x+3}$ [6.2]

11. $x - 6$ [6.3] **12.** $x - 3$ [6.3] **13.** $\frac{3t + 2}{4t - 1}$ [6.3] **14.** $3m(2m + 1)$ [6.3] **15.** $(2x + 3)(x - 4)(x - 5)$ [6.3] **16.** $\frac{13x - 1}{10x}$ [6.4] **17.** $-\frac{a^2 + 13a + 23}{(a + 3)(a - 4)}$ [6.4] **18.** $\frac{4x^2 + 17x - 1}{2x^2 + 13x + 6}$ [6.4] **19.** $\frac{x^2 - 7x - 4}{(x + 1)(x + 2)(x - 3)}$ [6.4] **20.** Need common denominator of $x(x + 1)$, $\frac{15x + 8}{x(x + 1)}$ [6.4]

Exercise Set 6.5 **1.** Complex **3.** Reciprocal **5.** $\frac{5}{16}$ **7.** $\frac{57}{32}$ **9.** $\frac{11}{6}$ **11.** $\frac{44a}{b^5}$ **13.** $\frac{8rs}{3p}$ **15.** $\frac{ab - a}{3 + a}$ **17.** $\frac{3}{t}$ **19.** $\frac{5x - 1}{4x - 1}$ **21.** $\frac{m - n}{m}$ **23.** $-\frac{a}{b}$ **25.** -1 **27.** $\frac{t}{s + t}$ **29.** $b - a$ **31.** $\frac{a^2 + b}{b(b + 1)}$ **33.** $\frac{x^2y}{y - x}$ **35.** $\frac{a + b}{a}$ **37. b)-c)** $-\frac{224}{155}$ **39. b)-c)** $\frac{x - y + 6}{2x + 2y - 7}$ **41. a)** $\frac{\frac{5}{12x}}{\frac{8}{x^2} - \frac{4}{3x}}$ **b)** $\frac{5x}{96 - 16x}$ **43.** A complex fraction is a rational expression that contains a rational expression in its numerator or its denominator or in both. **45.** $\frac{y + x}{3xy}$ **47.** $x + y$ **49. a)** $\frac{2}{7}$ **b)** $\frac{4}{13}$ **51.** $\frac{a^3b + a^2b^3 - ab^2}{a^3 - ab^3 + 3b^2}$ **53.** $\frac{17}{2}$ **54.** A polynomial is an expression containing a finite number of terms of the form ax^n where a is a real number and n is a whole number. **55.** $(x - 5)(x - 8)$ **56.** $\frac{x^2 - 9x + 2}{(3x - 1)(x + 6)(x - 3)}$

Exercise Set 6.6 **1.** Rational **3.** Check **5.** Proportions **7.** Equals **9.** 12 **11.** -6 **13.** 12 **15.** -4 **17.** -20 **19.** 30 **21.** 2 **23.** 18 **25.** 11 **27.** 8 **29.** -8 **31.** 3 **33.** 4 **35.** 2 **37.** 5 **39.** No solution **41.** 7 **43.** 4 **45.** No solution **47.** -3 **49.** $\frac{13}{16}$ **51.** $-\frac{1}{3}, 3$ **53.** $6, -1$ **55.** $4, -4$ **57.** $-4, -5$ **59.** No solution **61.** $-\frac{5}{2}$ **63.** $\frac{1}{2}$ **65.** -12 **67.** -12 **69.** -3 **71.** -2 **73.** 5 **75.** 0 **77.** x can be any real number since the sum on the left is also $\frac{2x - 4}{3}$. **79.** 15 centimeters **81.** -4 **83.** No, it is impossible for both sides of the equation to be equal. **85.** More than $6\frac{1}{3}$ hours **86.** 150 minutes **87.** 40°, 140° **88.** 6.8×10^8

Exercise Set 6.7 **1.** $\frac{\text{distance}}{\text{rate}}$ **3.** $\frac{1}{5}$ **5.** length = 14 meters, width = 3 meters **7.** base = 12 centimeters, height = 7 centimeters **9.** 4 feet **11.** $\frac{8}{9}, 8$ **13.** 7 **15.** 9 miles per hour **17.** 15 miles per hour **19.** 4 miles per hour, 16 miles per hour **21.** 150 miles per hour, 600 miles per hour **23.** 11 miles **25.** 1200 feet **27.** 24 minutes **29.** $3\frac{3}{7}$ hours **31.** 120 minutes or 2 hours. **33.** 60 hours **35.** 60 minutes or 1 hour **37.** 100 minutes or 1 hour 40 minutes **39.** $8\frac{3}{4}$ days **41.** $\frac{6}{5}$ hours or 1 hour, 12 minutes **43.** 3 or $\frac{1}{2}$ **45.** 8 pints **47.** $-\frac{3}{2}x - \frac{27}{2}$ **48.** $(y + 6)(y - 1)$ **49.** 1 **50.** $\frac{3x^2 - 9x - 15}{(2x + 3)(3x - 5)(3x + 1)}$

Exercise Set 6.8 **1.** Decrease **3.** Constant **5.** Direct **7.** Inverse **9.** Inverse **11.** Direct **13.** Inverse **15.** 440 **17.** $\frac{1}{5}$ **19.** 75 **21.** 2.5 **23.** 54 **25.** 3.5 **27.** 36 **29.** 80 **31.** 110 miles **33.** 0.5 hour **35.** $57 **37.** 4.2 hours **39.** 1000 people **41.** 6400 cubic centimeters **43.** $\approx$452.16 square inches **45.** 14.4 inches **47.** 1296 horsepower **49.** 0.6 footcandle **51.** It will be doubled. **53.** It will be halved. **55.** Pounds per inch **57. a)** $x = kyz$ **b)** 216 **59.** $2x - 3 + \frac{6}{4x + 9}$ **60.** $(z - 2)(y + 8)$ **61.** $-4, 2$ **62.** $x + 8$

Chapter 6 Review Exercises **1.** All real numbers except $x = 19$ **2.** All real numbers except $x = 3$ and $x = 5$ **3.** All real numbers except $x = \frac{1}{5}$ and $x = -1$ **4.** $\frac{1}{x - 8}$ **5.** $x^2 + 5x + 12$ **6.** $3x + y$ **7.** $x + 4$ **8.** $a + 9$ **9.** $-(2x + 1)$ **10.** $\frac{b - 2}{b + 2}$ **11.** $\frac{x - 3}{x - 2}$ **12.** $\frac{x - 8}{2x + 3}$ **13.** $\frac{5}{12b^2}$ **14.** $12xz^2$ **15.** $\frac{8b^3c^4}{a^2}$ **16.** $-\frac{2}{9}$ **17.** $-\frac{2}{3}$ **18.** 1 **19.** $\frac{36x^2}{y}$ **20.** $\frac{20z}{x^3}$ **21.** $\frac{6}{a - b}$ **22.** $\frac{1}{8(a + 3)}$ **23.** 1 **24.** $6y(x - y)$ **25.** $\frac{n - 2}{n + 5}$ **26.** 4 **27.** 5 **28.** $\frac{4}{x + 10}$ **29.** $4h - 3$ **30.** $3x + 4$ **31.** 24 **32.** $x + 3$ **33.** $20x^2y^3$ **34.** $x(x - 3)$ **35.** $(n + 5)(n - 4)$ **36.** $x(x + 2)$ **37.** $(r + s)(r - s)$ **38.** $x - 9$ **39.** $(x + 7)(x - 5)(x + 2)$ **40.** $\frac{3y^2 + 10}{6y^2}$ **41.** $\frac{12x + y}{4xy}$ **42.** $\frac{5x^2 - 18y}{3x^2y}$ **43.** $\frac{7x + 12}{x + 2}$ **44.** $\frac{x^2 - 2xy - y^2}{xy}$ **45.** $\frac{9x + 8}{x(x + 4)}$ **46.** $\frac{-x - 4}{3x(x - 2)}$ **47.** $\frac{z + 14}{(z + 5)^2}$ **48.** $\frac{2x - 8}{(x - 3)(x - 5)}$ **49.** $\frac{5x + 38}{(x + 6)(x + 2)}$ **50.** $\frac{3x - 8}{x - 4}$ **51.** $\frac{5ab + 10b}{a - 2}$ **52.** $\frac{3x - 1}{(x + 3)(x - 3)}$ **53.** $\frac{6p^3}{q}$ **54.** $\frac{2x + 2}{(x + 2)(x - 3)(x - 2)}$

55. $\dfrac{8x-29}{(x+2)(x-7)(x+7)}$ **56.** $\dfrac{x}{x+y}$ **57.** $\dfrac{3(x+3y)}{5(x-3y)}$ **58.** $a-3$ **59.** $\dfrac{3a^2-8a+3}{(a+1)(a-1)(3a-5)}$ **60.** $\dfrac{4x-8}{x}$ **61.** $\dfrac{64}{9}$ **62.** $\dfrac{2}{3}$

63. $\dfrac{bc}{3}$ **64.** $\dfrac{8x^3z^2}{y^3}$ **65.** $\dfrac{ab-a}{a+1}$ **66.** $\dfrac{r^2s+7}{s^3}$ **67.** $\dfrac{3x+2}{x(5x-1)}$ **68.** $\dfrac{4}{x}$ **69.** x **70.** $\dfrac{3a+1}{4}$ **71.** $\dfrac{-x+1}{x+1}$ **72.** $\dfrac{8x^2-x^2y}{y(y-x)}$ **73.** 13 **74.** 6

75. 12, −4 **76.** 20, −5 **77.** −16 **78.** $\dfrac{1}{2}$ **79.** −6 **80.** 28 **81.** No solution **82.** 2.4 hours **83.** $16\frac{4}{5}$ hours **84.** $\dfrac{1}{6}$, 1 **85.** Robert: 2.1 miles per hour, Tran: 5.6 miles per hour **86.** 273 milligrams **87.** 12 cubic inches

Chapter 6 Practice Test

1. 1 [6.1] **2.** $\dfrac{x^2+x+1}{x+1}$ [6.1] **3.** $\dfrac{8x^2z}{y}$ [6.2] **4.** $a+3$ [6.2] **5.** $\dfrac{x^2-6x+9}{(x+3)(x+2)}$ [6.2] **6.** −1 [6.2] **7.** $\dfrac{x-2y}{5}$ [6.2] **8.** $\dfrac{3}{y+5}$ [6.2] **9.** $-\dfrac{m+6}{m-5}$ [6.2] **10.** $\dfrac{3x-1}{4y}$ [6.3] **11.** $\dfrac{7x^2-6x-13}{x+3}$ [6.3] **12.** $\dfrac{2y^2-8}{xy^3}$ [6.4] **13.** $-\dfrac{2z+15}{z-5}$ [6.4] **14.** $\dfrac{-1}{(x+4)(x-4)}$ [6.4] **15.** $\dfrac{25}{28}$ [6.5] **16.** $\dfrac{x^2+x^2y}{7y}$ [6.5] **17.** $\dfrac{4x+3}{9-5x}$ [6.5] **18.** 2 [6.6] **19.** $-\dfrac{12}{7}$ [6.6] **20.** 12 [6.6] **21.** 6 hours [6.7] **22.** 1 [6.7] **23.** Base: 6 inches, height: 10 inches [6.7] **24.** 2 miles [6.7] **25.** ≈1.13 feet [6.8]

Cumulative Review Test

1. 121 [1.9] **2.** $\dfrac{17}{8}$ [2.5] **3.** $\dfrac{125x^3}{y^6}$ [4.1] **4.** $c+(c+0.018c)$ [3.1] **5.** $8x^2+5x+14$ [4.4] **6.** $6n^3-23n^2+26n-15$ [4.5] **7.** $(a-1)(8a-5)$ [5.2] **8.** $13(x+3)(x-1)$ [5.3] **9.** 36 [1.9] **10.** $x \le -4$, [number line, closed dot at −4, shaded left], $(-\infty, -4]$ [2.8] **11.** $\dfrac{1}{2}x - \dfrac{19}{4}$ [4.6] **12.** 4, $\dfrac{3}{2}$ [5.6] **13.** $\dfrac{x+4}{2x+1}$ [6.2] **14.** $\dfrac{r^2-8r-6}{(r+2)(r-5)}$ [6.4] **15.** $\dfrac{10x-18}{(x-5)(x+2)(x+3)}$ [6.4] **16.** $-\dfrac{3}{2}$ [6.6] **17.** No solution [6.6] **18.** \$3000 [3.2] **19.** 20 pounds sunflower seed, 30 pounds premixed assorted seed mix; [3.4] **20.** First leg: 3.25 miles, second leg: 9.5 miles; [3.4]

Chapter 7

Exercise Set 7.1

1. x-coordinate **3.** x-axis **5.** Linear **7.** Solution **9.** II **11.** IV **13.** I **15.** III **17.** I **19.** II

21. $A(3,1)$; $B(-3,0)$; $C(1,-3)$; $D(-2,-3)$; $E(0,3)$; $F\left(\frac{3}{2},-1\right)$ **23.**

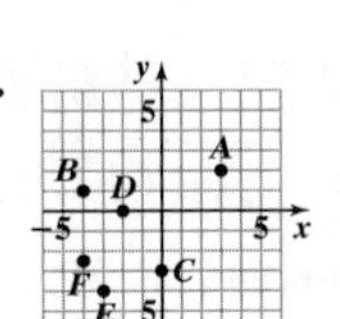

25. 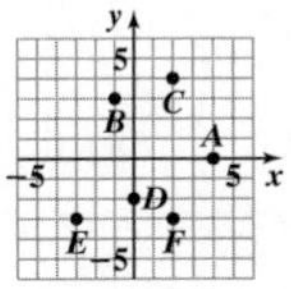 **27.** The points are collinear.

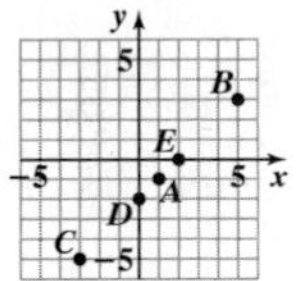

29. $(-5,-3)$ is not on the line 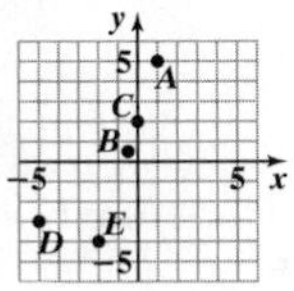 **31. a)** Point c) does not satisfy the equation. **b)**

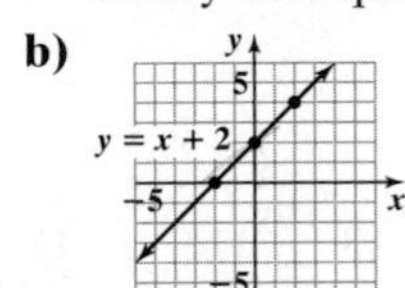

33. a) Point a) does not satisfy the equation. **b)**

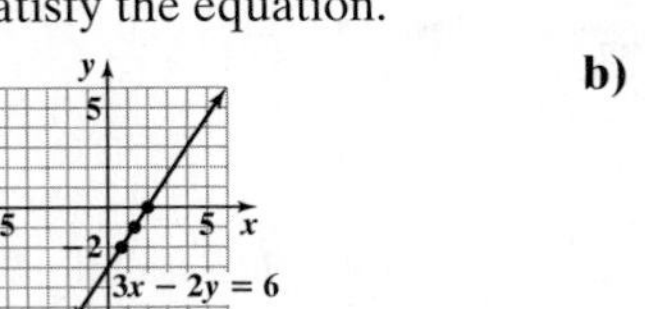

35. a) Point a) does not satisfy the equation. **b)**

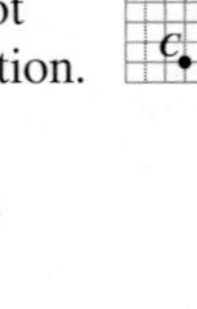

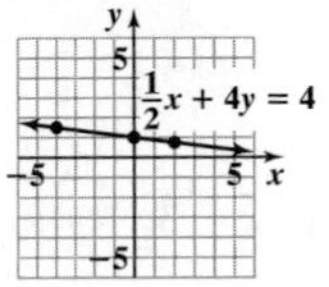

37. 2 **39.** −4 **41.** 6 **43.** $\dfrac{1}{2}$ **45. a)** Latitude, 16°N; Longitude, 56°W **b)** Latitude, 29°N; Longitude, 90.5°W **c)** Latitude 26°N; Longitude, 80.5°W **d)** Answers will vary. **47.** 0 **49.** To show that the line extends in both directions **55.** 21 **56.** $y = \dfrac{3x-4}{2} = \dfrac{3}{2}x - 2$ **57.** $16x^{12}$ **58.** $(x+3)(x-9)$ **59.** 0, 7 **60.** $\dfrac{5x+18}{3x^2}$

Exercise Set 7.2

1. x-axis **3.** y-intercept **5.** $(x, 0)$ **7.** Horizontal **9.** 3 **11.** 5 **13.** $\dfrac{21}{2}$ **15.** 2 **17.** $\dfrac{8}{3}$ **19.** 3

21. **23.**

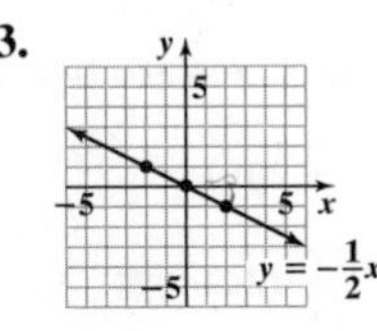

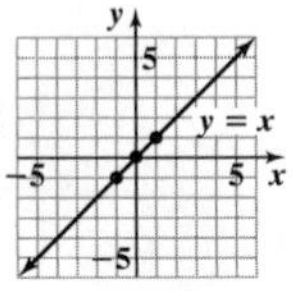

25. 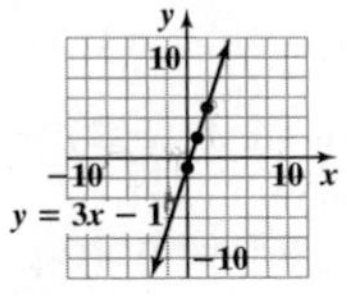 **27.** 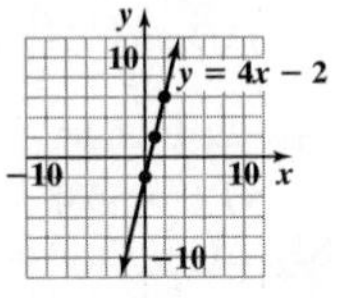 **29.**

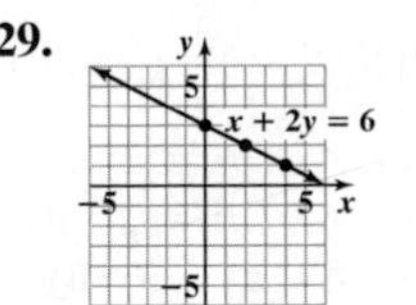

31.

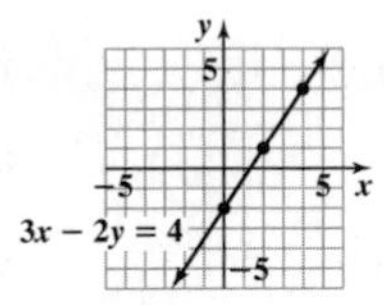

33. 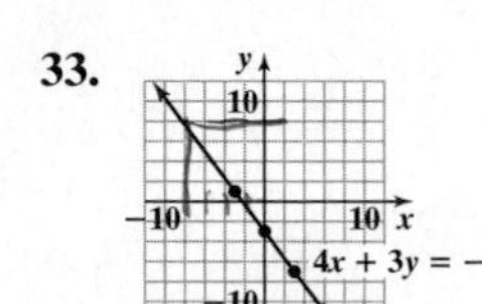

35. 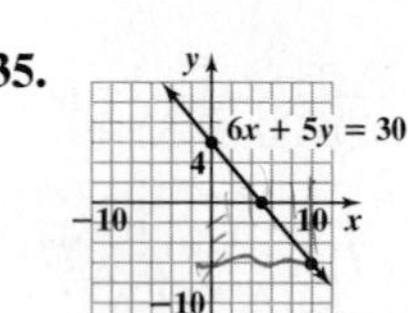

37. 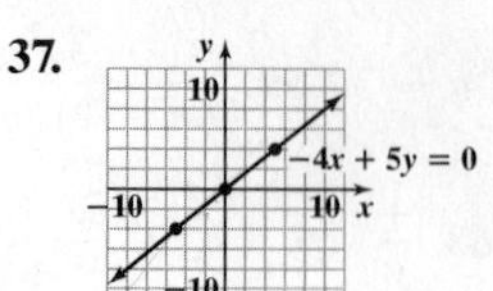

39. 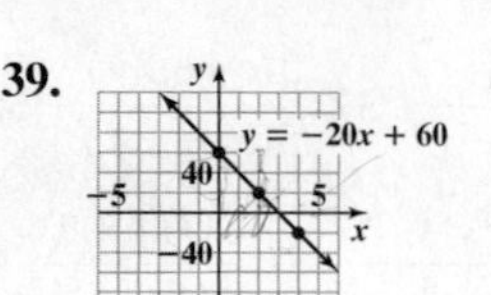

41. 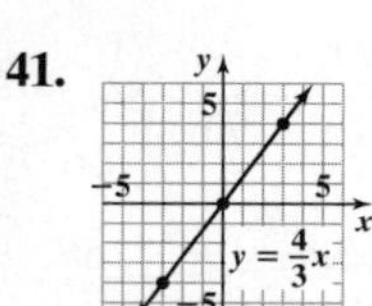

43. 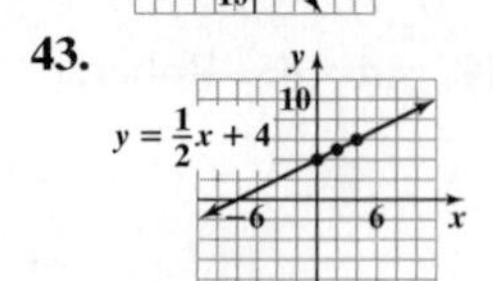

45. 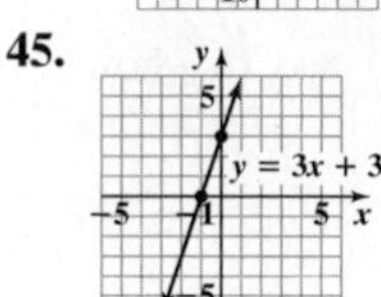

47. 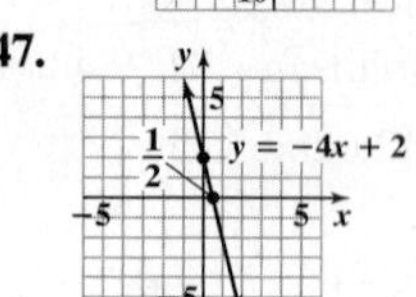

49. 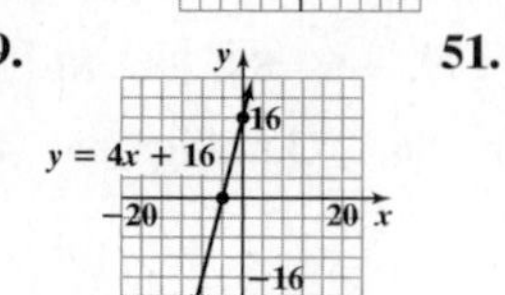

51. 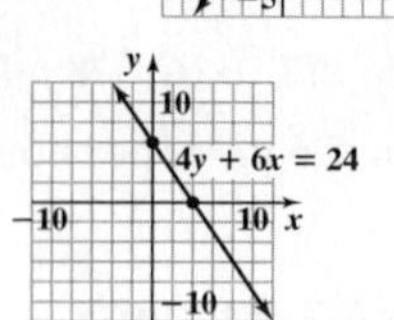

53. 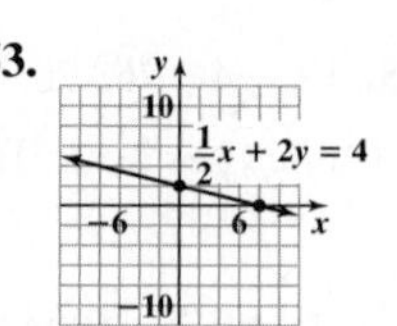

55. 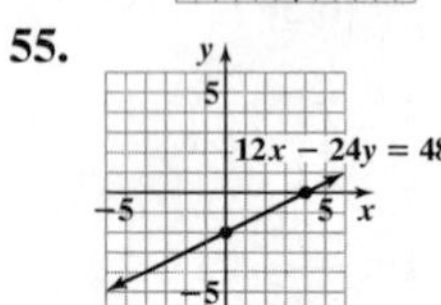

57. 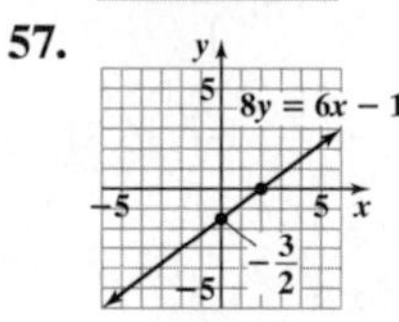

59. 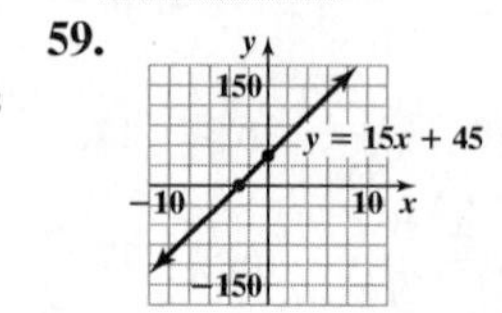

61. 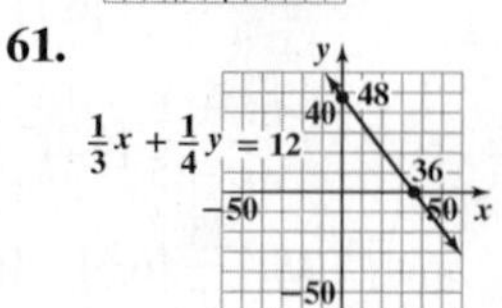

63. 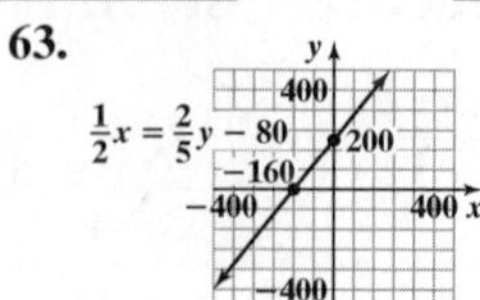

65. 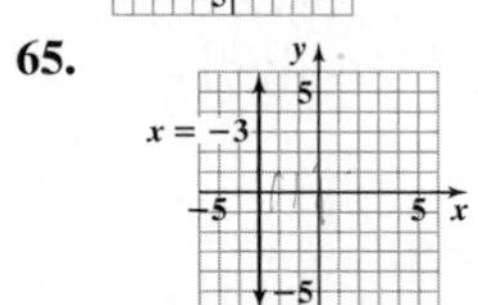

67. 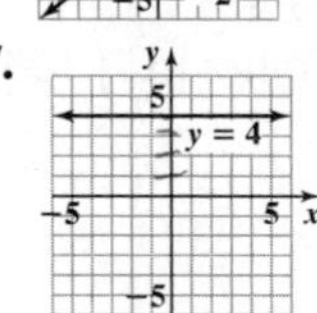

69. $x = 3$ **71.** $y = 3$ **73.** Yes

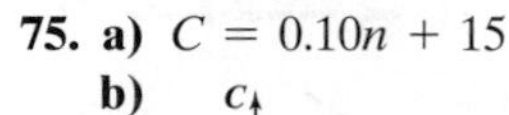

75. a) $C = 0.10n + 15$

b) 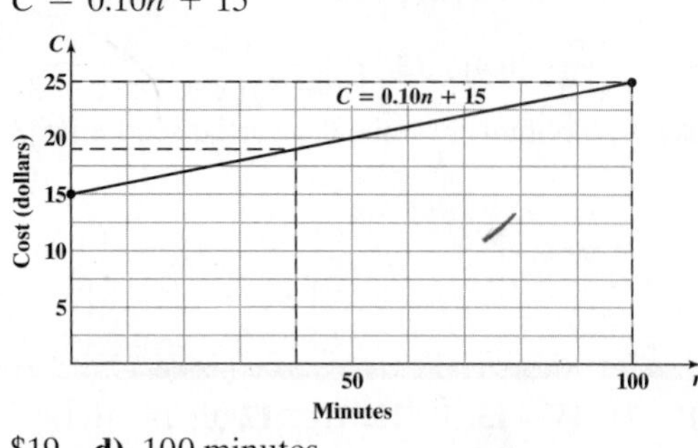

c) \$19 **d)** 100 minutes

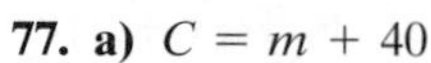

77. a) $C = m + 40$

b) 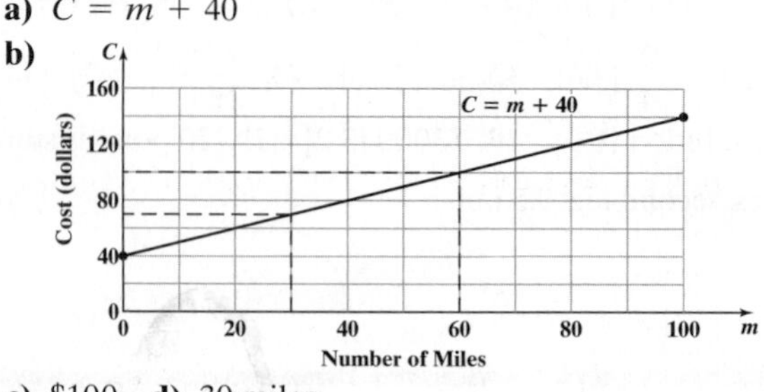

c) \$100 **d)** 30 miles

79. a) 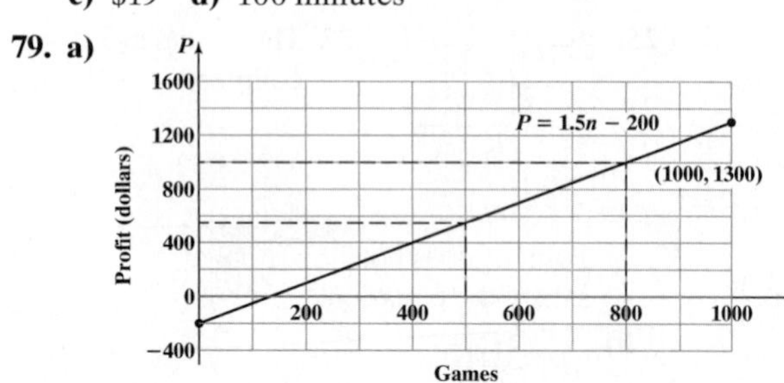

b) \$550 **c)** 800 games **81.** 5 **83.** 2 **85.** 3, 2 **87.** 6, 4

89. a) 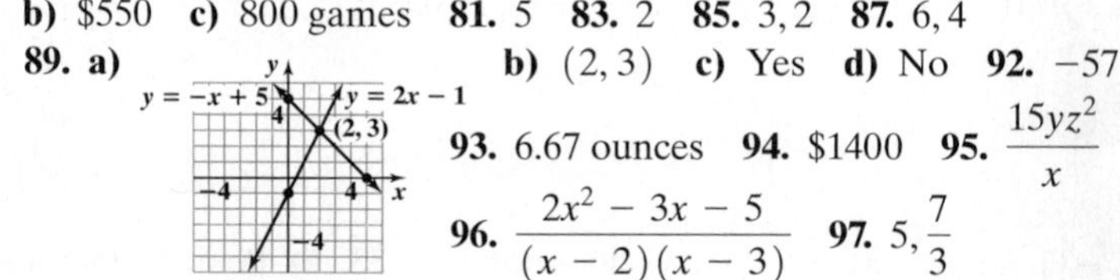

b) $(2, 3)$ **c)** Yes **d)** No **92.** -57 **93.** 6.67 ounces **94.** \$1400 **95.** $\dfrac{15yz^2}{x}$ **96.** $\dfrac{2x^2 - 3x - 5}{(x - 2)(x - 3)}$ **97.** $5, \dfrac{7}{3}$

Exercise Set 7.3

1. Rise **3.** Ratio **5.** Horizontal **7.** Parallel **9.** Equal **11.** 2 **13.** $\frac{1}{2}$ **15.** 0 **17.** 1 **19.** Undefined **21.** $-\frac{3}{8}$ **23.** $\frac{2}{3}$ **25.** $m = 2$ **27.** $m = -\frac{2}{5}$ **29.** $m = -\frac{4}{7}$ **31.** $m = \frac{7}{4}$ **33.** $m = 0$ **35.** $m = -\frac{2}{3}$ **37.** Undefined

39. 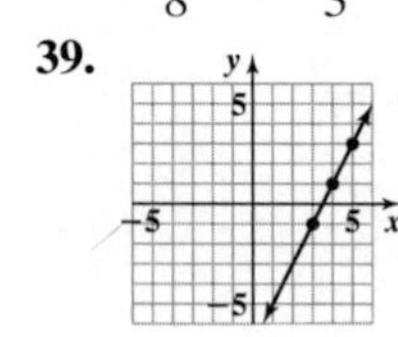

41. 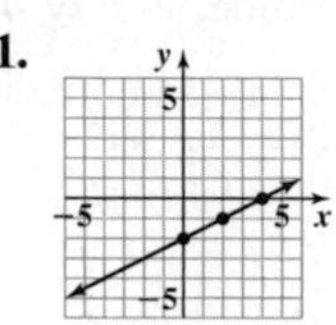

43.

45.

47.

49. Parallel **51.** Neither **53.** Perpendicular **55.** Neither **57.** Neither **59.** Perpendicular **61.** Parallel **63.** Perpendicular **65.** 3 **67.** $\frac{1}{4}$ **69.** First **71. a)** 60 **b)** 75 **73.** -4 **75.** $-\frac{1}{8}$ **77.** 3

79. a) 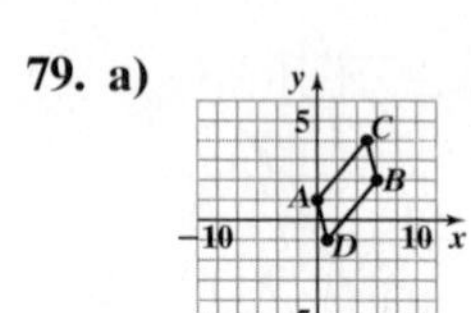

b) $AC, m = \frac{3}{5}$; $CB, m = -2$; $DB, m = \frac{3}{5}$; $AD, m = -2$ **c)** Yes, opposite sides are parallel.

81. a) $AB, m = 4$; $BC, m = -2$; $CD, m = 4$ **b)** $[4 + (-2) + 4]/3 = 2$ **c)** $AD, m = 2$ **d)** Yes **e)** Answers will vary.

83. 2 **84. a)** -3 **b)** 0 **85.** $2x + 16$ **86.** -3 **87.** x-intercept: $(6, 0)$; y-intercept: $(0, -10)$

Mid-Chapter Test

1. IV [7.1] **2.**

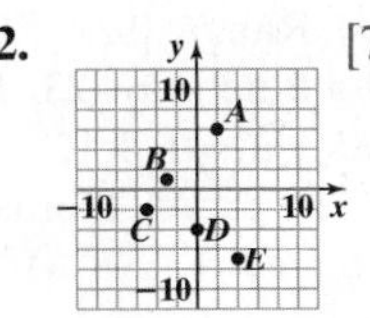

[7.1] **3. b)** (0, 2) does not satisfy the equation. [7.1] **4.** −4 [7.1] **5.** 3 [7.1] **6.** A graph of an equation in two variables is an illustration of a set of points whose coordinates satisfy the equation. [7.1]

7.

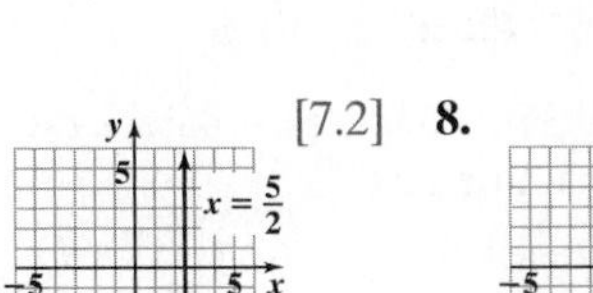

[7.2] **8.** 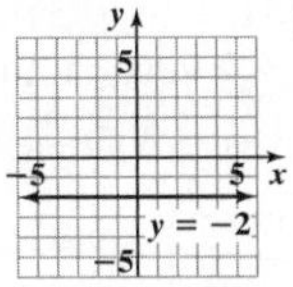 [7.2] **9.**

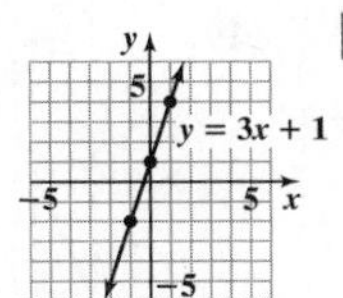

[7.2] **10.**

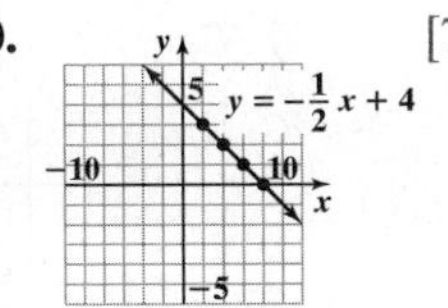

[7.2] **11.**

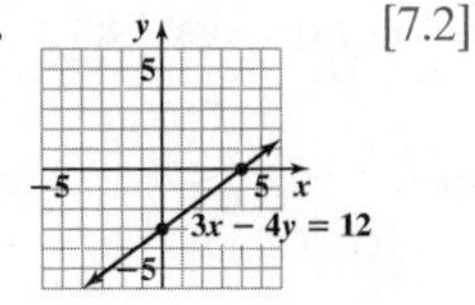

[7.2]

12.

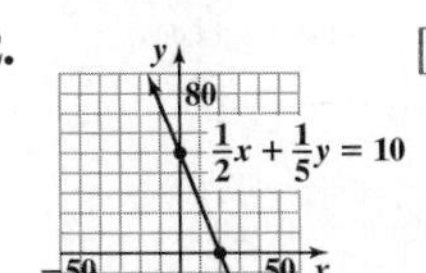

[7.2]

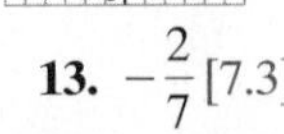

13. $-\frac{2}{7}$ [7.3]

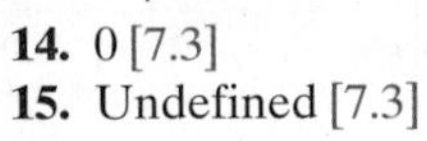

14. 0 [7.3] **15.** Undefined [7.3]

16.

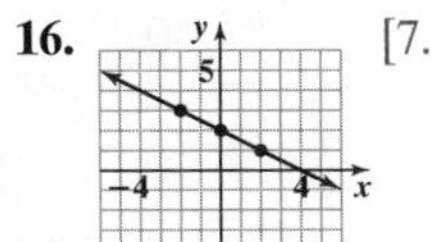

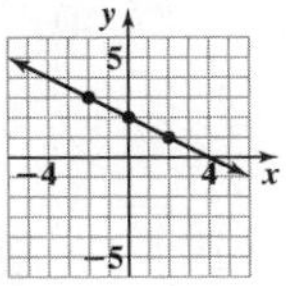

[7.3] **17.**

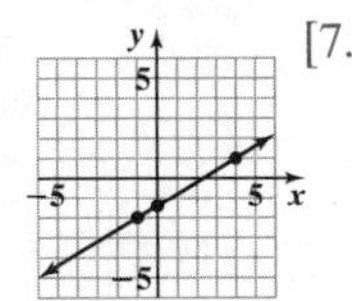

[7.3] **18.** Neither [7.3] **19.** Perpendicular [7.3] **20.** 10 [7.3]

Exercise Set 7.4

1. Point-slope **3.** Slope-intercept **5.** *y*-intercept **7.** Slope **9.** 3, (0, 1) **11.** $\frac{4}{3}$, (0, −7)

13. 1, (0, −3)

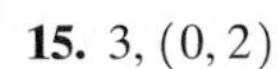

15. 3, (0, 2)

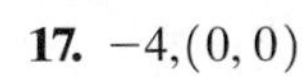

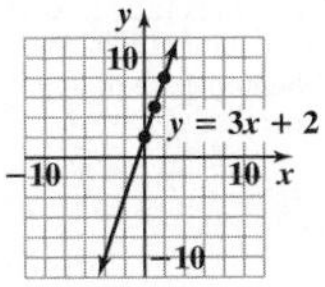

17. −4, (0, 0) **19.** 2, (0, −3)

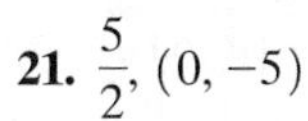

21. $\frac{5}{2}$, (0, −5)

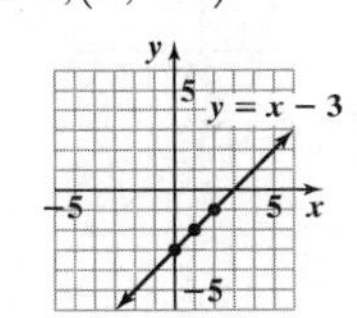

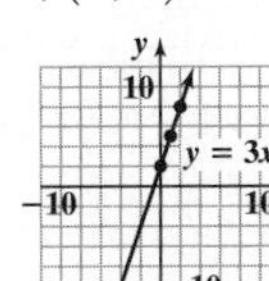

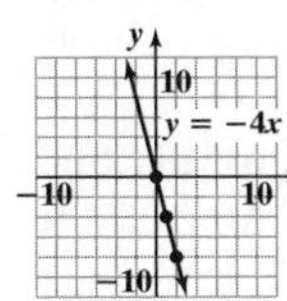

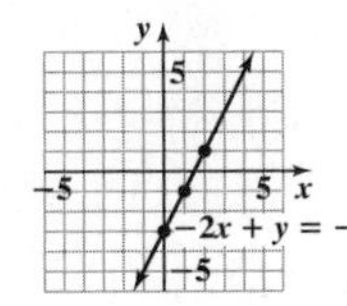

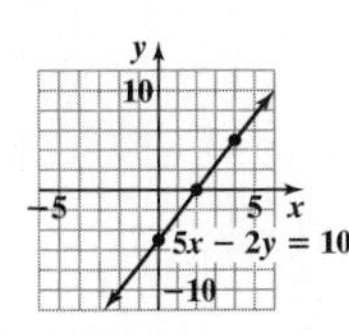

23. $-\frac{1}{2}, \left(0, \frac{3}{2}\right)$ **25.** 3, (0, 4) **27.** $\frac{3}{2}$, (0, 2) 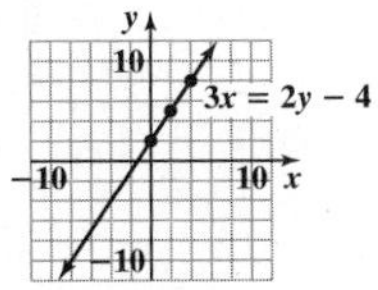 **29.** $y = x - 2$ **31.** $y = -\frac{1}{3}x + 2$ **33.** $y = -3x - 5$

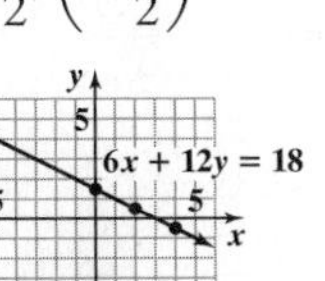

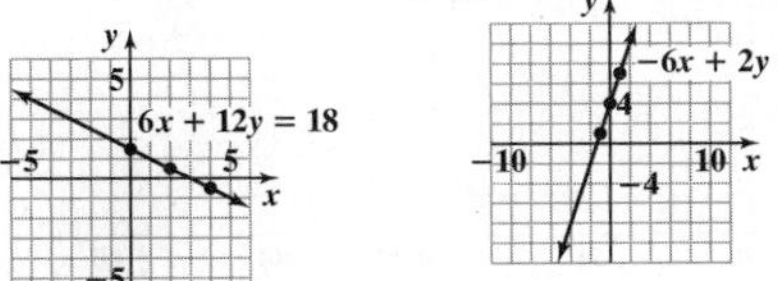

35. $y = \frac{1}{3}x + 5$ **37.** Parallel **39.** Perpendicular **41.** Parallel **43.** Neither **45.** Perpendicular **47.** $y = 3x + 2$ **49.** $y = \frac{2}{3}x + 6$ **51.** $y = -3x - 7$

53. $y = -\frac{5}{6}x + 11$ **55.** $y = \frac{1}{2}x - \frac{5}{2}$ **57.** $y = \frac{1}{2}x - 2$ **59.** $y = 2x - 1$ **61.** $y = 3x + 10$ **63.** $y = -\frac{3}{2}x$ **65.** $y = -\frac{6}{7}x + \frac{32}{7}$ **67. a)** $y = 5x + 60$ **b)** $210 **69. a)** No **b)** $y + 4 = 2(x + 5)$ **c)** $y - 12 = 2(x - 3)$ **d)** $y = 2x + 6$ **e)** $y = 2x + 6$ **f)** Yes **71. a)** 1.465 **b)** $f = 1.465m$ **c)** ≈205.5 feet per second **d)** 150 feet per second **e)** 55 miles per hour **73.** $y = -2x + 5$ **75.** $y = \frac{3}{4}x + 5$ **78.** < **79.** $r = \frac{i}{pt}$ **80.** $x \le -4$, (number line, closed point at −4, shaded left), $(-\infty, -4]$ **81.** $(x - 4y)(x + 3y)$ **82.** −5

Exercise Set 7.5

1. Solid **3.** Solutions

5.

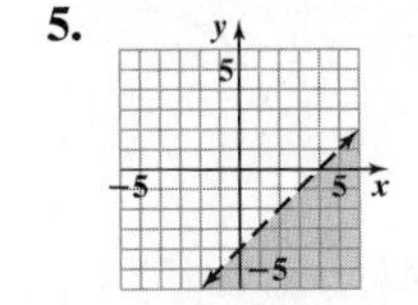

7.

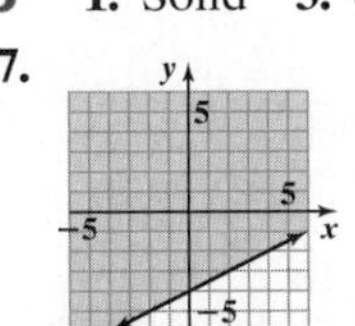

9. 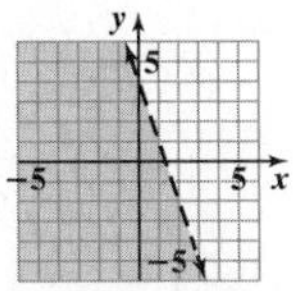 **11.**

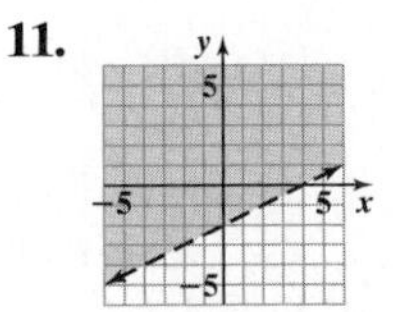

13.

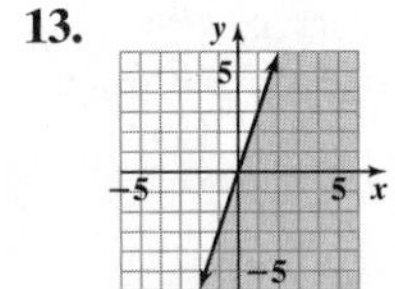

15.

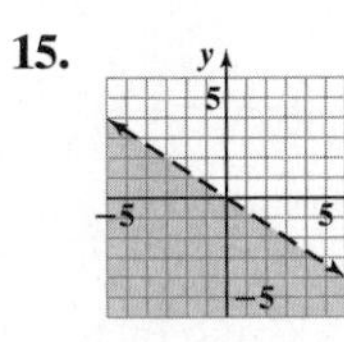

17.

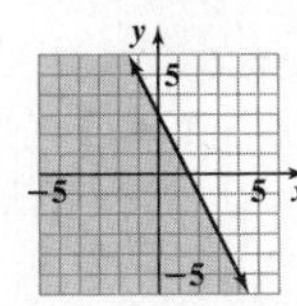

19.

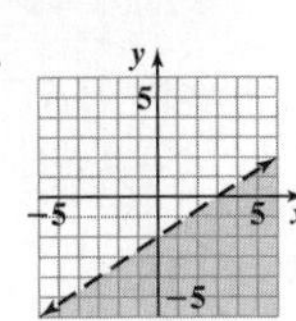

21.

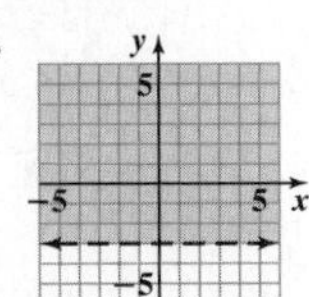

23.

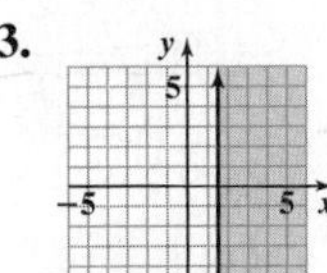

25. a) No **b)** No **c)** Yes **d)** Yes **27. a)** Less than or equal to **b)** Greater than or equal to **c)** Less than or equal to **d)** Greater than or equal to

29. Points on the line satisfy the = part of the inequality. **31.** No, it could be a solution to $ax + by = c$. **33.** No, the location of an ordered pair which satisfies the first inequality lies on one side of the line while an ordered pair which satisfies the other inequality lies either on the line or on the other side of the line. **35.** Shading on opposite sides of the boundary line **37. a)** 2 **b)** 2, 0 **c)** $2, -5, 0, \frac{2}{5}, -6.3, -\frac{23}{34}$ **d)** $\sqrt{7}, \sqrt{3}$ **e)** $2, -5, 0, \sqrt{7}, \frac{2}{5}, -6.3, \sqrt{3}, -\frac{23}{34}$ **38.** 2 **39.** $2x - 3 + \frac{5}{x}$ **40.** $\frac{1}{x + 2y}$

Exercise Set 7.6

1. Domain **3.** Relation **5.** Vertical **7.** *x*-component **9.** Function, Domain: {1, 2, 3, 4, 5}, Range: {1, 2, 3, 4, 5} **11.** Not a function, Domain: {1, 2, 3, 5, 7}, Range: {−2, 0, 2, 4, 5} **13.** Function, Domain: {0, 1, 3, 4, 5}, Range: {−4, −1, 0, 1, 2}

15. Function, Domain: $\{0, 1, 2, 3, 4\}$, Range: $\{3\}$ **17.** Not a function, Domain: $\{3\}$, Range: $\{0, 1, 2, 3, 4\}$ **19. a)** $\{(1, 4), (2, 5), (3, 5), (4, 7)\}$ **b)** Function **21. a)** $\{(-5, 4), (0, 7), (6, 9), (6, 3)\}$ **b)** Not a function **23.** Function, Domain: $\{-2, -1, 2, 3\}$, Range: $\{-3, -1, 1, 2\}$ **25.** Not a function, Domain: $\{-2, 0, 1, 2\}$, Range: $\{-3, -2, -1, 1, 2\}$ **27.** Function, Domain: $(-\infty, \infty)$, Range: $(-\infty, \infty)$ **29.** Function, Domain: $(-\infty, \infty)$, Range: $[0, \infty)$ **31.** Not a Function, Domain: $(-\infty, 1]$, Range: $(-\infty, \infty)$ **33.** Function, Domain: $(-\infty, \infty)$, Range: $[-2, 2]$ **35. a)** 14 **b)** -2 **37. a)** 8 **b)** 8 **39. a)** 4 **b)** 6 **41. a)** 3 **b)** 8

43.

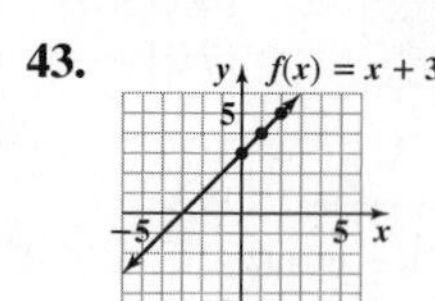

45.

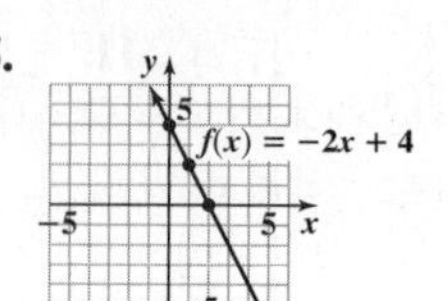

47.

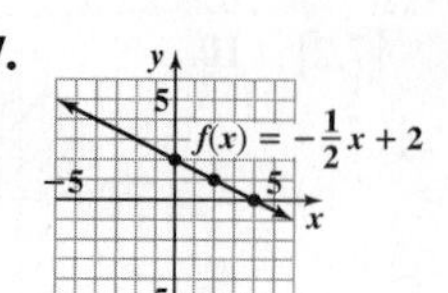

49.

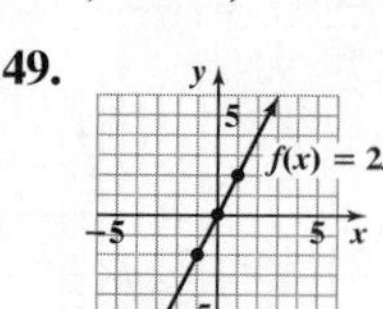

51. $c = 0.35n$ **53.** Yes, it passes the vertical line test.

55. a)

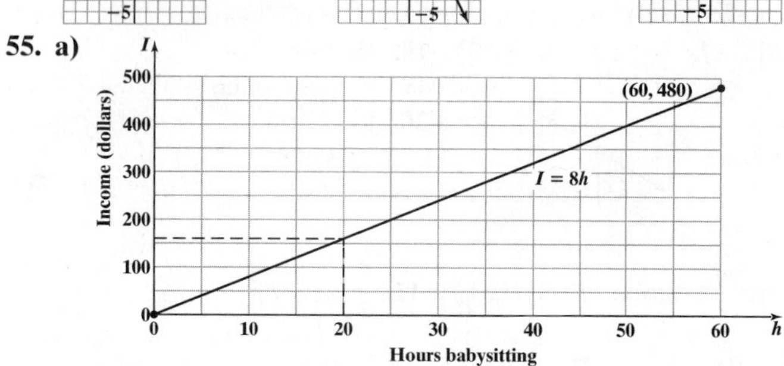

b) \$160

57. a)

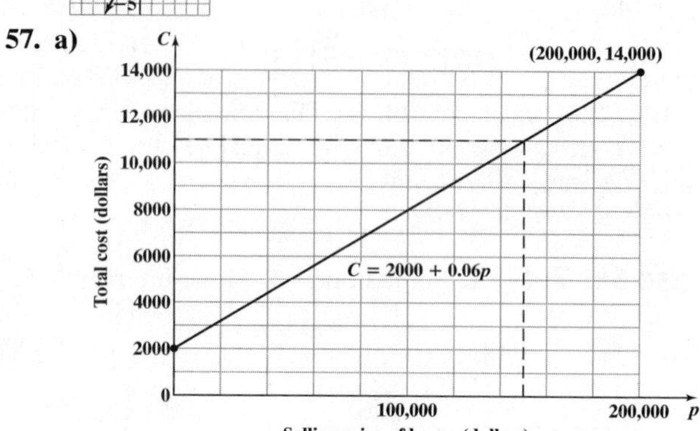

b) \$11,000

59. a)

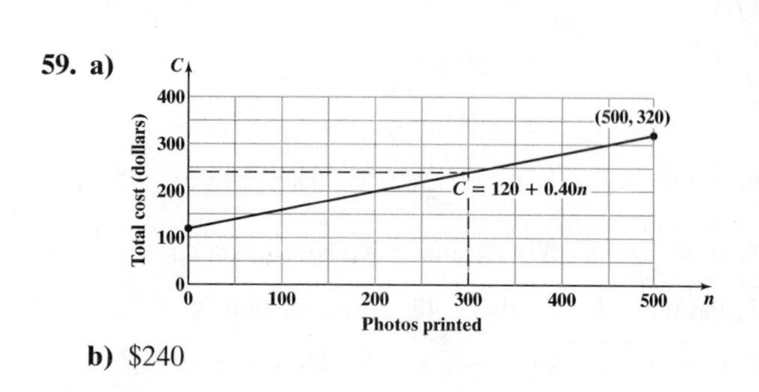

b) \$240

61. a)

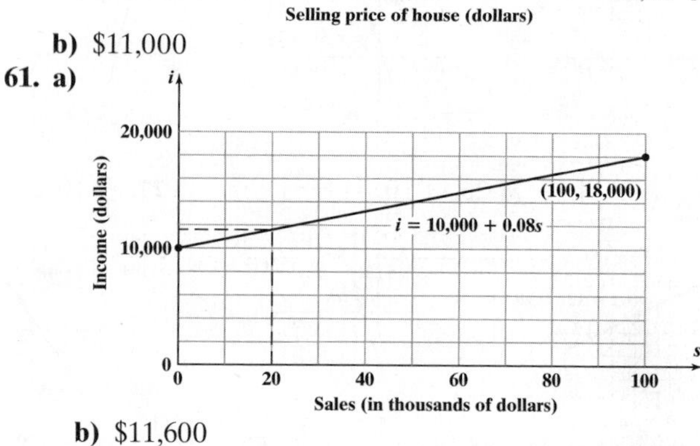

b) \$11,600

63. a) No; each x must have exactly one y. **b)** No; a y may correspond to more than one x. **65.** Yes, it passes the vertical line test. **67.** No, the vertical line $x = 1$ intersects the graph at more than one point. **69. a)** $\frac{29}{8}$ **b)** $\frac{29}{9}$ **c)** 3.88 **73.** $\frac{1}{63}$ **74.** -14 **75.** 13 miles **76.** $(5x + 7y)(5x - 7y)$ **77.** $\frac{4x^2}{y}$ **78.** A graph is an illustration of a set of points whose coordinates satisfy an equation.

Chapter 7 Review Exercises

1.

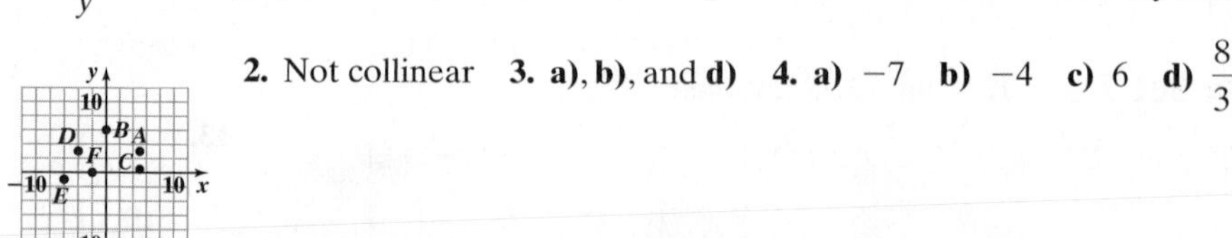

2. Not collinear **3. a), b),** and **d)** **4. a)** -7 **b)** -4 **c)** 6 **d)** $\frac{8}{3}$

5.

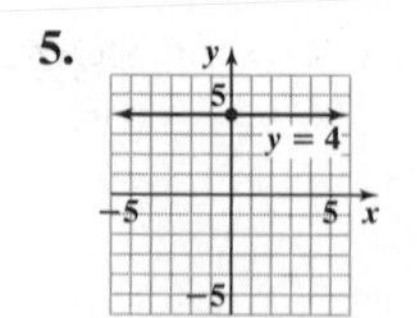

6.

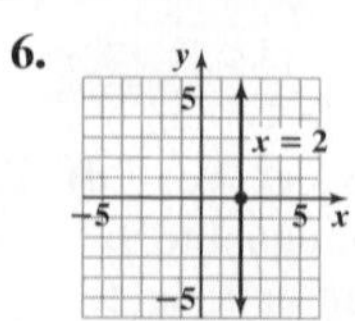

7.

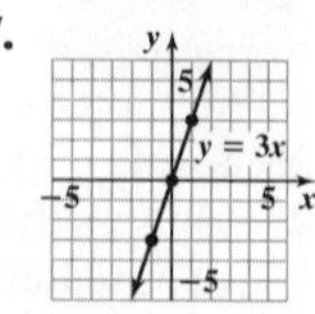

8.

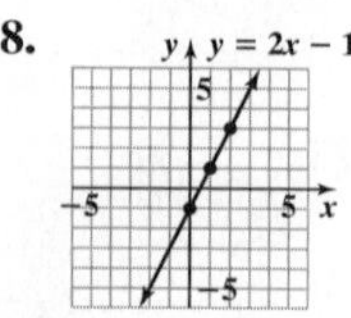

9.

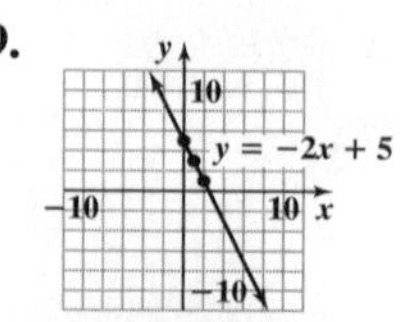

10.

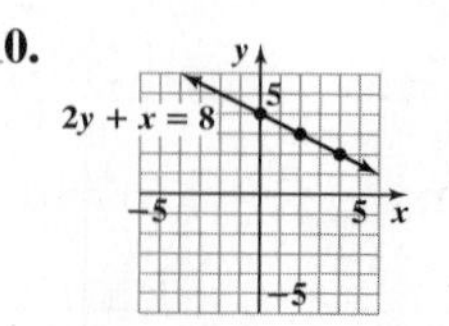

11.

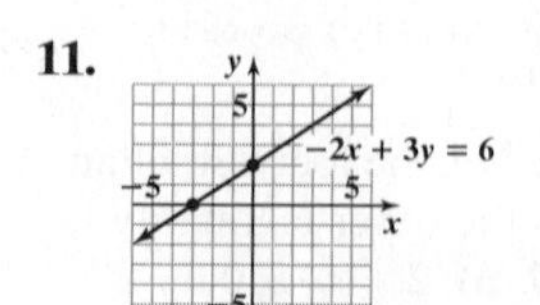

12.

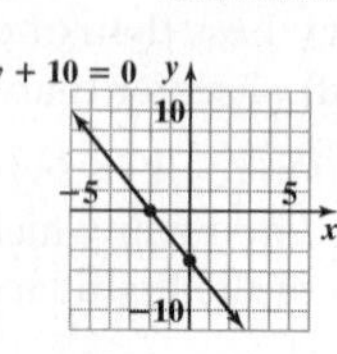

13.

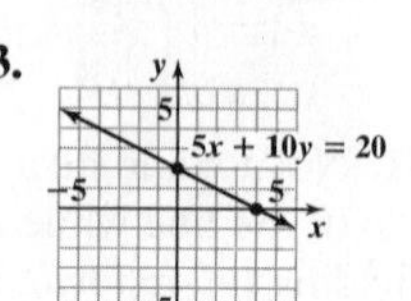

14.

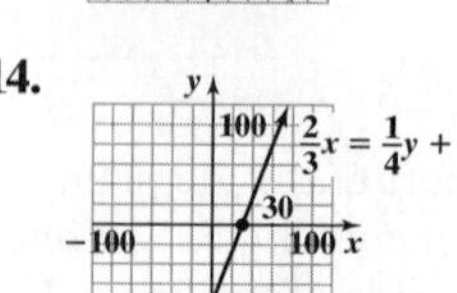

15. $-\frac{9}{5}$ **16.** $-\frac{1}{12}$ **17.** -2 **18.** 0 **19.** Undefined

20. The slope of a line is the ratio of the vertical change to the horizontal change between any two points on the line. **21.** -2 **22.** $\frac{1}{4}$ **23.** Neither **24.** Perpendicular **25. a)** 18 **b)** 10 **26.** $m = -\frac{6}{7}$, $(0, 3)$ **27.** Slope is undefined, no y-intercept **28.** $m = 0$, $(0, -3)$ **29.** $y = 3x - 3$ **30.** $y = -\frac{1}{2}x + 2$ **31.** Parallel **32.** Perpendicular **33.** $y = 3x + 1$ **34.** $y = -\frac{2}{3}x + 4$

35. $y = 2$ **36.** $x = 4$ **37.** $y = -\frac{7}{2}x - 3$ **38.** $x = -5$

39. **40.** **41.** **42.**

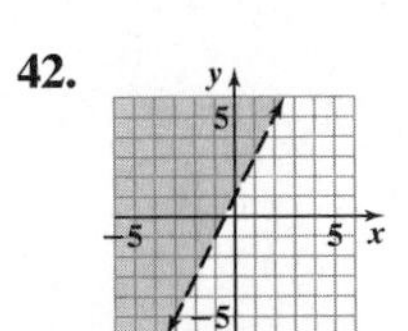

43.

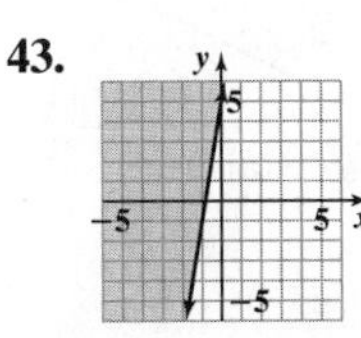

44.

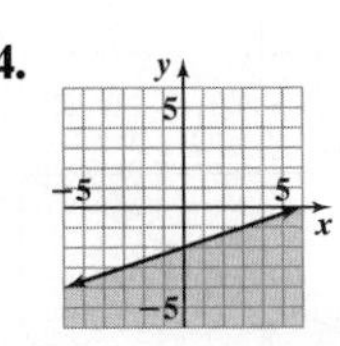

45. Function, Domain: {1, 2, 3, 4, 6}, Range: $\{-3, -1, 2, 4, 5\}$ **46.** Not a function, Domain: {3, 4, 6, 7}, Range: {0, 1, 2, 5} **47.** Not a function, Domain: {3, 4, 5, 6}, Range: $\{-3, 1, 2\}$ **48.** Function, Domain: $\{-2, 3, 4, 5, 9\}$, Range: $\{-2\}$ **49. a)** $\{(1, 3), (4, 5), (7, 2), (9, 2)\}$ **b)** Function **50. a)** $\{(4, 1)(6, 3), (6, 5), (8, 7)\}$ **b)** Not a function **51. a)** Domain: {Mary, Pete, George, Carlos}, Range: {Apple, Orange, Grape} **b)** Not a function **52. a)** Domain: {Sarah, Jacob, Kristen, Erin}, Range: {Seat 1, Seat 2, Seat 3, Seat 4} **b)** Not a function **53. a)** Domain: {Blue, Green, Yellow}, Range: {Paul, Maria, Lalo, Duc} **b)** Function **54. a)** Domain: {1, 2, 3, 4}, Range: {A, B, C} **b)** Function **55.** Function, Domain: $(-\infty, \infty)$, Range: $(-\infty, \infty)$ **56.** Function, Domain: $(-\infty, \infty)$, Range: $(-\infty, 3]$ **57.** Not a function, Domain: $[-2, 4]$, Range: $[-1, 3]$ **58.** Function, Domain: $(-\infty, \infty)$, Range: $[-3, \infty)$ **59. a)** 5 **b)** -31 **60. a)** 9 **b)** -39 **61. a)** -4 **b)** -8 **62. a)** 12 **b)** 76 **63.** Yes, it passes the vertical line test. **64.** Yes, each year corresponds to exactly one y-value. **65.**

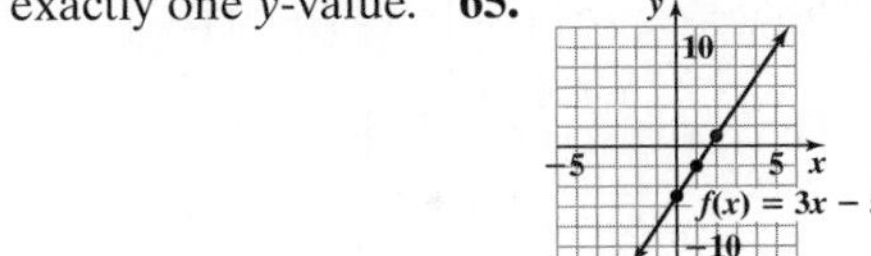

66.

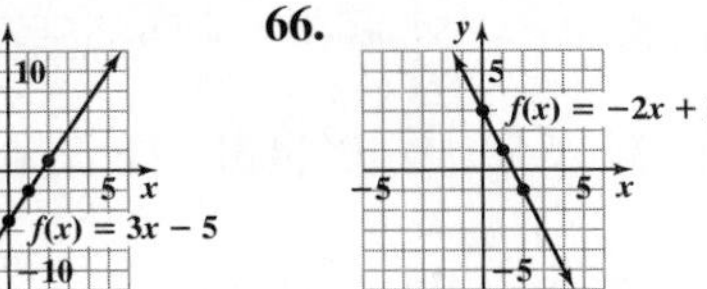

67. a)

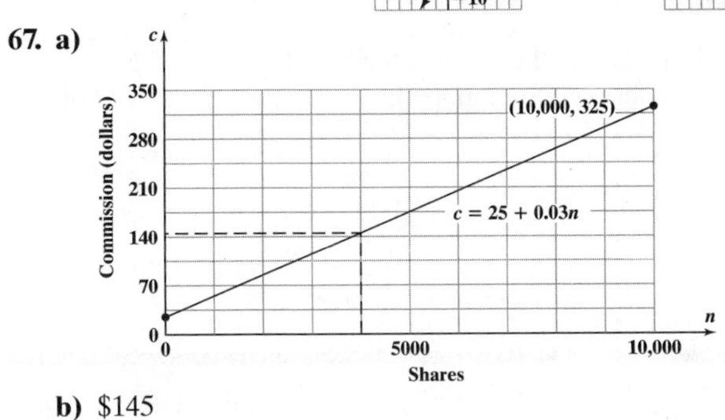

b) \$145

68. a)

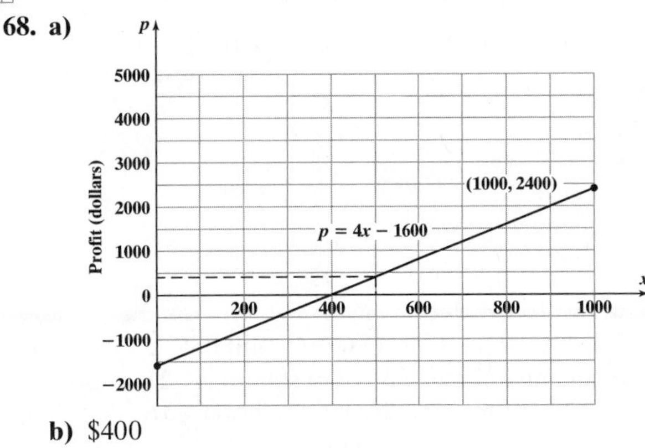

b) \$400

Chapter 7 Practice Test

1. A graph is an illustration of the set of points whose coordinates satisfy an equation. [7.1] **2. a)** IV **b)** II [7.1] **3. a)** $ax + by = c$ **b)** $y = mx + b$ **c)** $y - y_1 = m(x - x_1)$ [7.1–7.4] **4. b)** and **d)** [7.1] **5.** $-\frac{4}{3}$ [7.3] **6.** $\frac{4}{9}, \left(0, -\frac{5}{3}\right)$ [7.4] **7.** $y = -x - 1$ [7.4]

8.

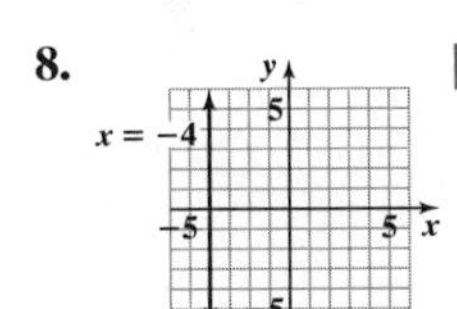

[7.2] **9.**

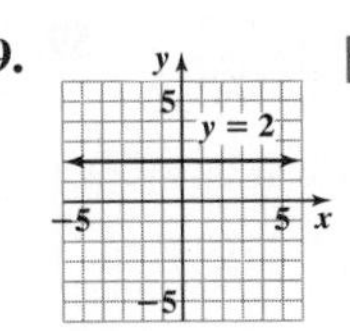

[7.2] **10.**

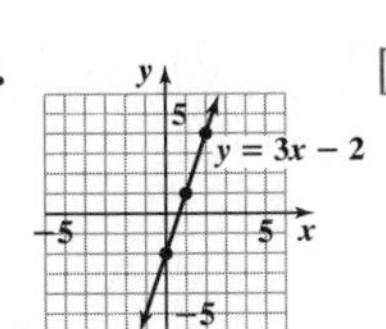

[7.2] **11. a)** $y = \frac{1}{2}x - 2$ **b)**

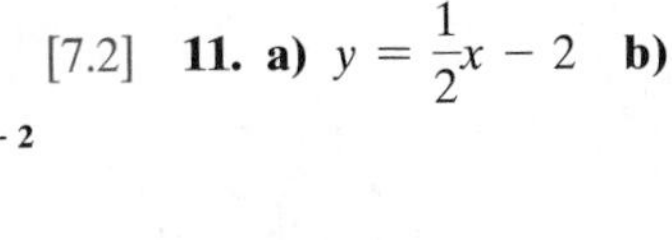

[7.2]

12.

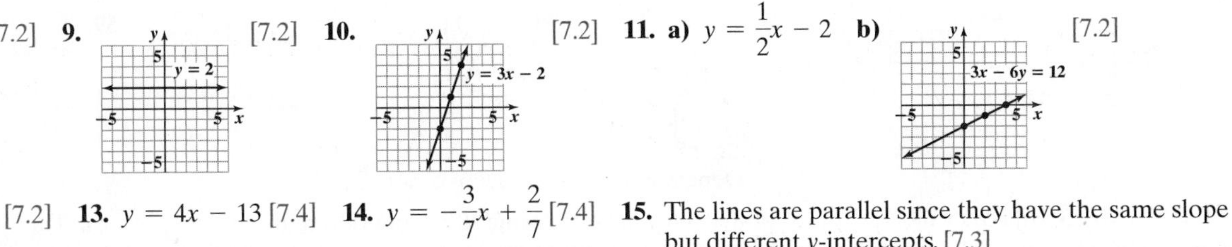

[7.2] **13.** $y = 4x - 13$ [7.4] **14.** $y = -\frac{3}{7}x + \frac{2}{7}$ [7.4] **15.** The lines are parallel since they have the same slope but different y-intercepts. [7.3]

16. [7.4] **17.**

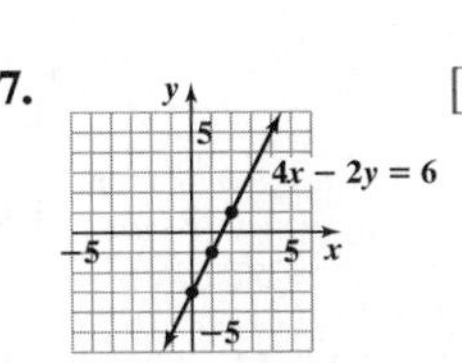

[7.4] **18.** A set of ordered pairs in which each first component corresponds to exactly one second component. [7.4]

19. a) Not a function; 3, a first component, is paired with more than one value

b) Domain: {1, 3 5, 6}, Range: $\{-4, 0, 2, 3, 5\}$ [7.6]

20. a) Not a function, Domain: $[-3, \infty)$, Range: $(-\infty, \infty)$ **b)** Function, Domain: $(-\infty, \infty)$, Range: $(-\infty, 3]$ [7.6]

21. a) 15 **b)** 10 [7.6] **22.** [7.6] **23.**

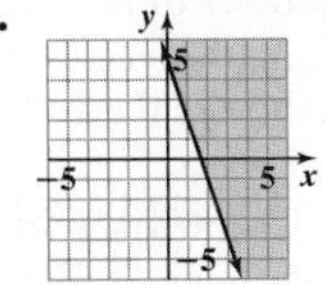

[7.5] **24.**

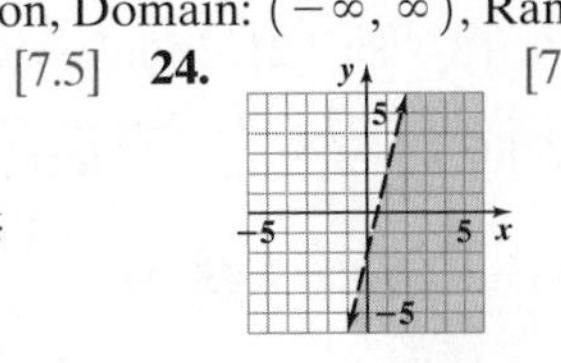

[7.5]

25. a)

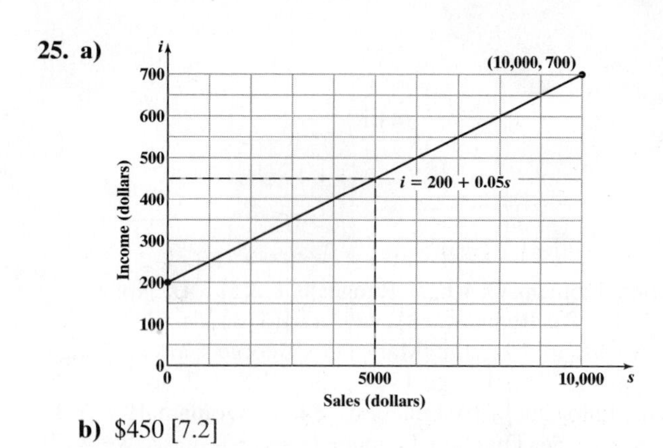

b) \$450 [7.2]

Cumulative Review Test

1. a) $\{1, 2, 3, 4, \ldots\}$ **b)** $\{0, 1, 2, 3, 4, \ldots\}$ [1.4] **2. a)** Distributive property **b)** Commutative property of addition [1.10] **3.** 20 [2.5] **4.** All real numbers [2.5] **5.** $x < -5$, (number line, open circle at −5, shaded left), $(-\infty, -5)$ [2.8] **6.** \$4.00 [2.7] **7.** width: 5 feet, length: 13 feet [3.3] **8.** 2 hours [3.4] **9.** $\frac{1}{x^{15}}$ [4.2] **10.** 6.523×10^2 [4.3] **11.** $2(x - 5)(x - 1)$ [5.3] **12.** $(2a - 5)(2a + 7)$ [5.4] **13.** 0, 7 [5.6] **14.** $-\frac{1}{2}$ [6.1] **15.** $\frac{8x}{3x + 7}$ [6.2] **16.** -6 [6.6] **17.** [7.2] **18.** [7.4]

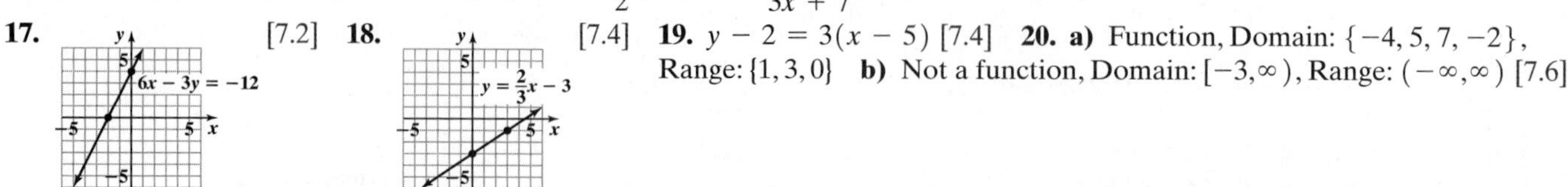

19. $y - 2 = 3(x - 5)$ [7.4] **20. a)** Function, Domain: $\{-4, 5, 7, -2\}$, Range: $\{1, 3, 0\}$ **b)** Not a function, Domain: $[-3, \infty)$, Range: $(-\infty, \infty)$ [7.6]

Chapter 8

Exercise Set 8.1

1. System **3.** One **5.** Infinite **7.** c) **9.** a) **11.** None **13.** a), b), c) **15.** c) **17.** Consistent; one solution **19.** Dependent; infinite number of solutions **21.** Consistent; one solution **23.** Inconsistent; no solution **25.** One solution **27.** No solution **29.** One solution **31.** No solution **33.** Infinite number of solutions **35.** One solution

37. (0, 3) **39.** (3, 3) **41.** (−1, 2) **43.** (2, 3) **45.** Inconsistent **47.** (−2, 5)

49. Dependent **51.** (3, 3) **53.** Dependent **55.** Inconsistent **57.** (2, −3) **59.** (0, 0)

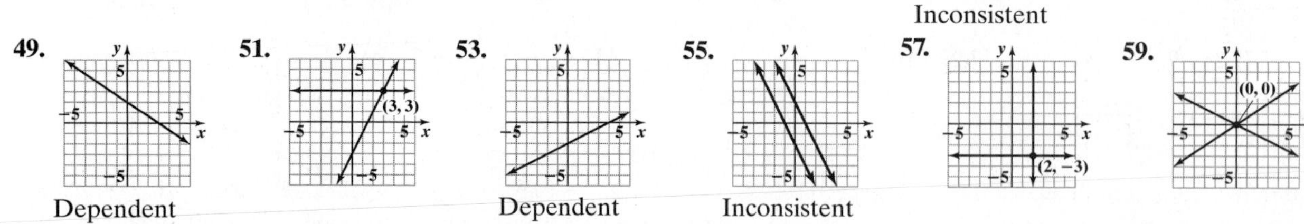

61. 6 years **63.** 7 hours **65.** Lines are parallel; they have the same slope and different y-intercepts. **67.** The system has an infinite number of solutions. If the two lines have two points in common then they must be the same line. **69.** The system has no solution. Parallel lines do not intersect. **71.** One; (5, 3) **79.** $2x + 18$ **80.** 1 **81.** $-(x - 7)$ **82.** $\frac{1}{3}, 6$ **83. a)** $(6, 0), (0, 4)$ **b)**

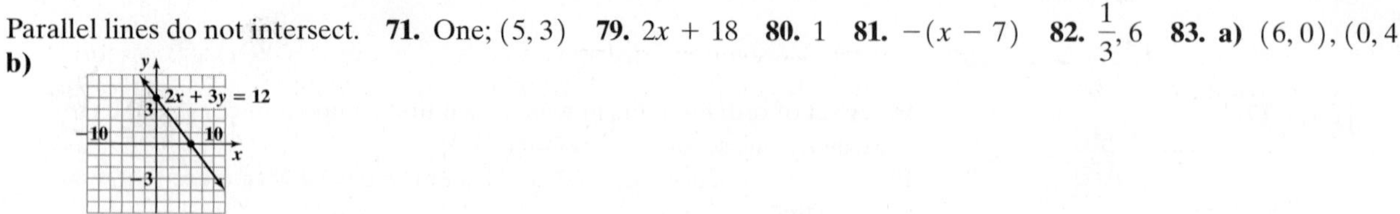

Exercise Set 8.2

1. Coefficient **3.** Solution **5.** $(-2, -3)$ **7.** $(2, 1)$ **9.** $(4, -1)$ **11.** $(2, 1)$ **13.** No solution **15.** Infinite number of solutions **17.** $(7, 2)$ **19.** $(3, -8)$ **21.** Infinite number of solutions **23.** No solution **25.** $(-2, -3)$ **27.** $\left(3, \frac{1}{2}\right)$ **29.** $\left(-\frac{12}{7}, \frac{6}{35}\right)$ **31.** 42, 21 **33.** Width: 8 feet, length: 17 feet **35.** 200 adults, 350 students **37.** \$30,000 **39. a)** 16 months **b)** Yes **41. a)** 1.25 hours **b)** 155 mile marker **43. a)** $T = 180 - 10t$ **b)** $T = 20 + 6t$ **c)** 10 minutes **d)** 80°F **45.** 23 years

46. $18x^2 + 9x - 14$ **47.**

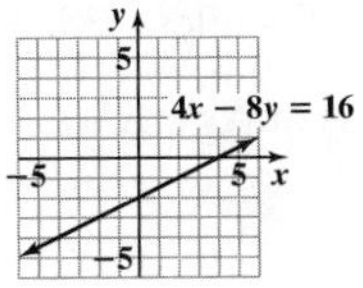

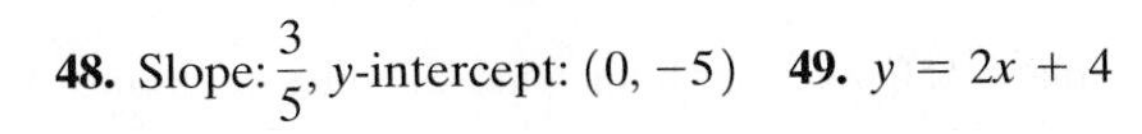

48. Slope: $\frac{3}{5}$, y-intercept: $(0, -5)$ **49.** $y = 2x + 4$

Exercise Set 8.3

1. Eliminate **3.** Infinite **5.** $(5, 1)$ **7.** $(-2, -1)$ **9.** $(-8, -26)$ **11.** $(3, -1)$ **13.** $(-2, 3)$ **15.** $(-1, 4)$ **17.** No solution **19.** $\left(-\frac{3}{2}, -3\right)$ **21.** Infinite number of solutions **23.** $(-3, 1)$ **25.** $(1, 2)$ **27.** $(2, -1)$ **29.** No solution **31.** $(-6, -12)$ **33.** Infinite number of solutions **35.** $\left(10, \frac{15}{2}\right)$ **37.** $\left(\frac{20}{39}, -\frac{16}{39}\right)$ **39.** $\left(\frac{14}{5}, -\frac{12}{5}\right)$ **41.** First number: 49, second number: 11 **43.** First number: 6, second number: 4 **45.** Width: 3 inches, length: 6 inches **47.** Width: 8 inches, length: 10 inches **49.** Answers will vary. **51. a)** $(200, 100)$ **b)** Same solution; dividing both sides of an equation by a nonzero number does not change the solution. **53.** $(8, -1)$ **55.** $(1, 2, 3)$ **57.** 125 **58.** 7 **59.** $2x^2y - 9xy + 8y$ **60.** $36a^6b^9c^5$ **61.** $(y + c)(x - a)$ **62.** 5

Mid-Chapter Test: Sections 8.1–8.3

1. b) [8.1] **2.** a) [8.1] **3.** One solution [8.1] **4.** Infinite number of solutions [8.1] **5.** No solution [8.1] **6.**

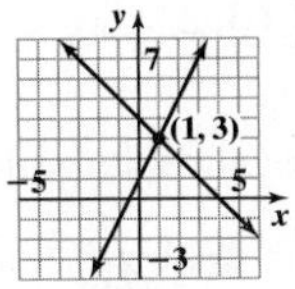

[8.1] **7.**

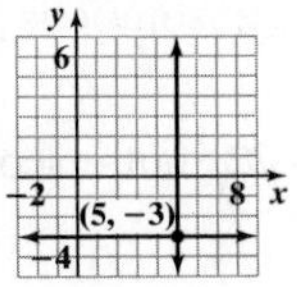

[8.1] **8.** $(-2, 4)$ [8.2] **9.** $\left(1, -\frac{1}{3}\right)$ [8.2] **10.** No solution [8.2] **11.** Length: 15 feet, width: 7 feet [8.2] **12.** $(4, -1)$ [8.3] **13.** $\left(-\frac{1}{2}, 2\right)$ [8.3] **14.** Infinite number of solutions [8.3] **15.** Solution must have two values, one for x and one for y. The solution is $x = 3$ and $y = 5$ or $(3, 5)$ [8.1–8.3]

Exercise Set 8.4

1. 35, 42 **3.** 12, 32 **5.** $A = 36°$; $B = 54°$ **7.** $A = 112°$, $B = 68°$ **9.** Width: 390 feet, length: 740 feet **11.** 30 at 46 cents, 40 at 66 cents **13.** kayak: 4.05 miles per hour, current: 0.65 miles per hour **15.** adults': 12, children's: 15 **17. a)** 1400 copies **b)** Office Copier Depot **19.** \$2500 at 4%; \$5500 at 5% **21. a)** 375 pages **b)** University Bookstore **23.** Dave: 57 miles per hour, Alice: 72 miles per hour **25.** Melissa: 30 miles per hour, Elizabeth: 34 miles per hour **27.** 1 hour **29.** 0.75 hour or 45 minutes **31.** 8 liters of 15%, 12 liters of 40% **33.** 3.2 oz juice, 4.8 oz drink **35.** 140 gallons 5%, 60 gallons skim **37.** Soybean: 50 pounds, Cornmeal: 100 pounds **39.** hot dog: \$1.50, sunflower seeds: \$0.75 **41.** 1.125 miles **43.** 45 minutes **45. a)** Commutative property of addition **b)** Associative property of multiplication **c)** Distributive property **46.** Width: 3 feet, length: 8 feet **47.** −2, 2 **48.** A graph is an illustration of the set of points whose ordered pairs satisfy an equation.

Exercise Set 8.5

1. Dashed **3.** Solutions **5.**

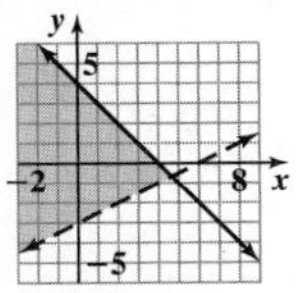

7.

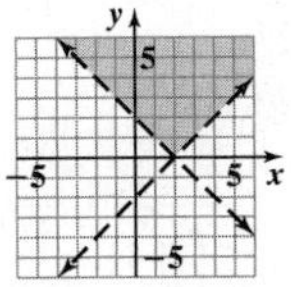

9.

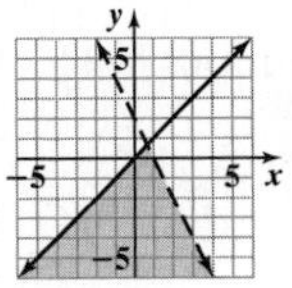

11.

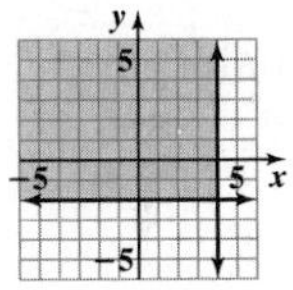

13.

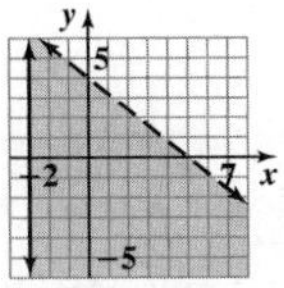

15.

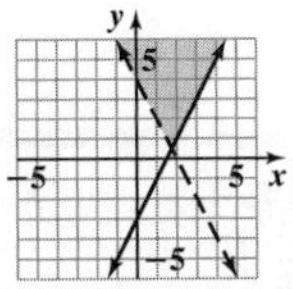

17.

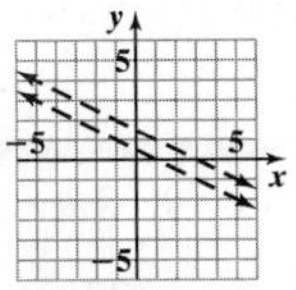

19.

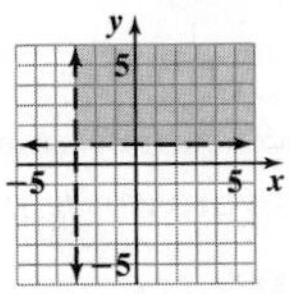

21.

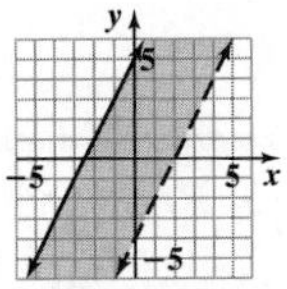

23.

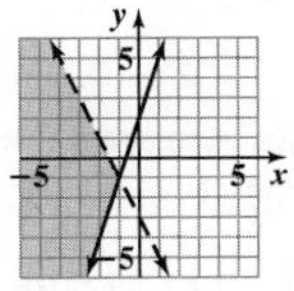

25. Yes, the solution to a system of linear inequalities contains all of the ordered pairs which satisfy both inequalities. **27.** Yes, when the lines are parallel. One possible system is $x + y > 2$, $x + y < 1$. **29.** No, the system can have no solution or infinitely many solutions. **33.** all real numbers;

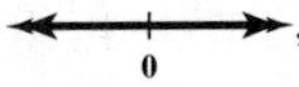

$(-\infty, \infty)$ **34.** $y = \frac{3}{5}x - \frac{6}{5}$ **35.** $-\frac{1}{4}, 3$ **36.** $\frac{y^7}{x^9}$

Chapter 8 Review Exercises

1. c) **2.** a) **3.** Consistent, one solution **4.** Inconsistent, no solution **5.** Dependent, infinite number of solutions **6.** Consistent, one solution **7.** No solution **8.** One solution **9.** Infinite number of solutions **10.** One solution

11.

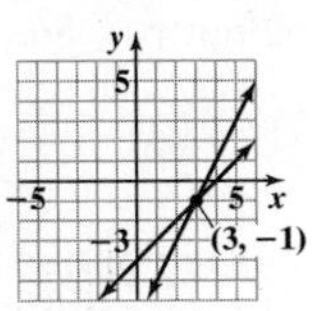

12.

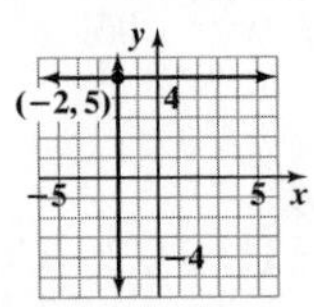

13.

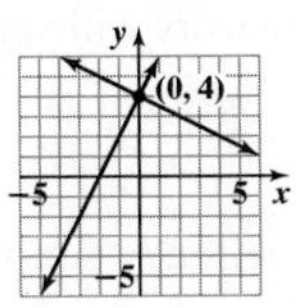

14.

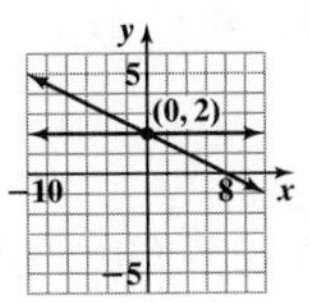

15.

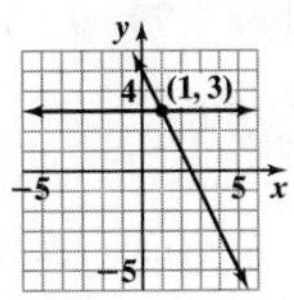

16.

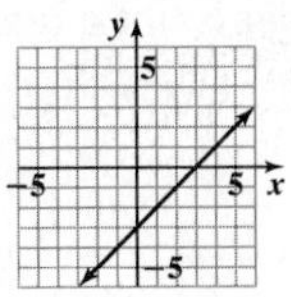

17.

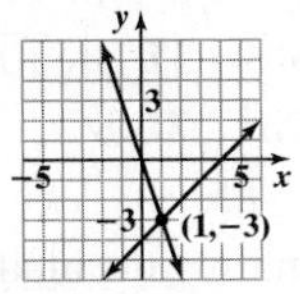

18.

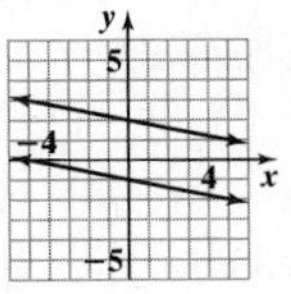

19. $(2, 1)$ **20.** $(-3, 2)$ **21.** $(4, 1)$ **22.** $(-15, 5)$ **23.** No solution **24.** Infinite number of solutions **25.** $\left(\frac{5}{2}, -1\right)$ **26.** $\left(-\frac{2}{7}, \frac{29}{7}\right)$ **27.** $(-6, -2)$ **28.** $(1, -2)$ **29.** $(-7, 19)$ **30.** $(2, 2)$ **31.** $\left(-1, \frac{13}{3}\right)$ **32.** No solution **33.** Infinite number of solutions

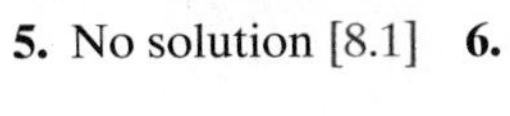

34. $\left(-\frac{78}{7}, -\frac{48}{7}\right)$ **35.** 16 and 24 **36.** plane: 550 miles per hour, wind: 30 miles per hour **37.** 50 miles **38.** \$10,000 at 4%, \$6000 at 6% **39.** Liz: 57 miles per hour, Mary: 63 miles per hour **40.** 20 pounds of Green Turf's, 30 pounds of Agway's **41.** 5 liters of each

42.
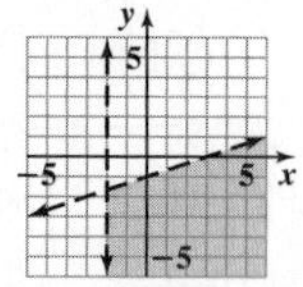
43.
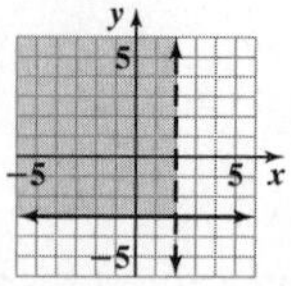
44.
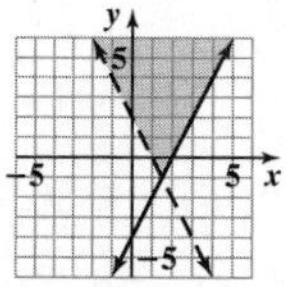
45.
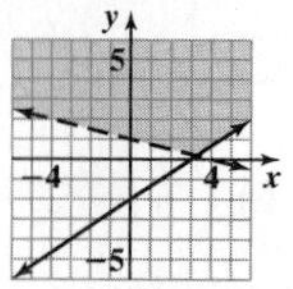

Chapter 8 Practice Test

1. b) [8.1] **2.** Consistent, one solution [8.1] **3.** Inconsistent, no solution [8.1] **4.** Dependent, infinite number of solutions [8.1] **5.** No solution [8.1] **6.** One solution [8.1] **7.** Infinite number of solutions [8.1] **8. a)** You will obtain a false statement, such as $6 = 0$. **b)** You will obtain a true statement, such as $0 = 0$. [8.1]

9.
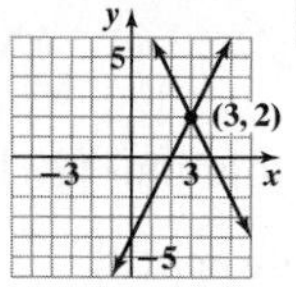

[8.1] **10.**
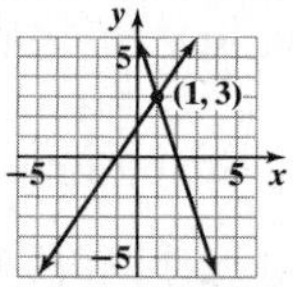

[8.1] **11.**
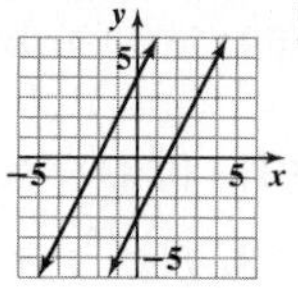
No solution

[8.1] **12.** (6, 21) [8.2] **13.** $\left(\frac{7}{2}, -\frac{5}{2}\right)$ [8.2] **14.** (4, 5) [8.2] **15.** $(-2, 2)$ [8.3] **16.** $\left(\frac{1}{2}, -6\right)$ [8.3] **17.** Infinite number of solutions [8.3] **18.** (3, 2) [8.1–8.3]

19. $(-5, -4)$ [8.1–8.3] **20.** $\left(-\frac{40}{7}, \frac{52}{7}\right)$ [8.1–8.3] **21.** 100 miles [8.4] **22.** butterscotch: $13\frac{1}{3}$ pounds, lemon: $6\frac{2}{3}$ pounds [8.4] **23.** Deja's boat: 20 miles per hour, Dante's boat: 24 miles per hour [8.4] **24.**
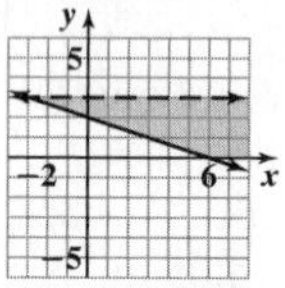
[8.5] **25.**
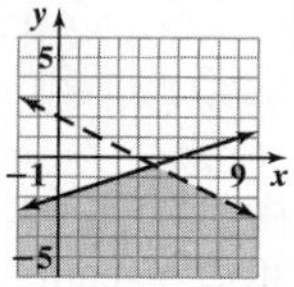
[8.5]

Cumulative Review Test

1. a) Air Force, Army **b)** Air Force, Army, Navy [1.2] **2. a)** 7 **b)** $-5, -0.6, \frac{3}{5}, 7, 0, -\frac{5}{9}, 1.34$ **c)** $\sqrt{10}, -\sqrt{2}$ **d)** $-5, -0.6, \frac{3}{5}, \sqrt{10}, -\sqrt{2}, 7, 0, -\frac{5}{9}, 1.34$ [1.4] **3.** -244 [1.7] **4.** $a + 24$ [2.1] **5.** -2 [2.5]
6. $x \le 5$;
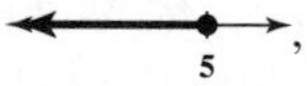
, $(-\infty, 5]$ [2.8] **7.** $39 - y$ [3.1] **8.** \$5800 [3.2] **9.** 15°, 45°, 120° [3.3] **10.** $64x^{11}y^7$ [4.1] **11.** $x^2 - 4xy + 4y^2$ [4.5] **12.** $(2n + 3)(n - 4)$ [5.4] **13.** $-6, 9$ [5.6] **14.** $\frac{x + 2}{7}$ [6.2] **15.** 16 [6.6] **16.**
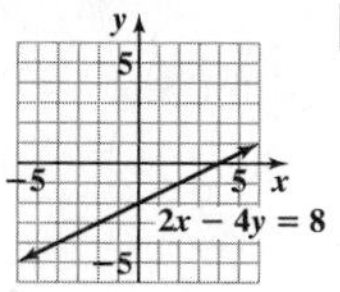

[7.2] **17.**
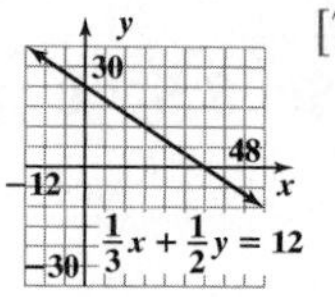

[7.2]

18. An infinite number of solutions [8.1] **19.**
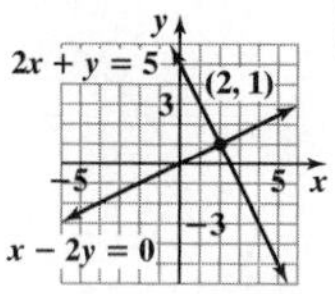

[8.1] **20.** $\left(\frac{11}{3}, -1\right)$ [8.3]

Chapter 9

Exercise Set 9.1

1. Radical **3.** Radicand **5.** Negative **7.** Imaginary **9.** Rational **11.** Exponential **13.** 0 **15.** 2 **17.** -7 **19.** 20 **21.** -4 **23.** 12 **25.** 13 **27.** -1 **29.** -12 **31.** -11 **33.** $\frac{1}{2}$ **35.** $-\frac{5}{6}$ **37.** $\frac{5}{4}$ **39.** $-\frac{10}{3}$ **41.** True **43.** True **45.** False **47.** True **49.** False **51.** True **53.** $7^{1/2}$ **55.** $(29)^{1/2}$ **57.** $(8x)^{1/2}$ **59.** $(12x^2)^{1/2}$ **61.** $(21ab^2)^{1/2}$ **63.** $(38n^3)^{1/2}$ **65.** Rational: $9.83, \frac{3}{5}, 0.333\ldots, 5, \sqrt{\frac{4}{49}}, \frac{3}{7}, -\sqrt{9}$; Irrational: $\sqrt{\frac{5}{16}}$; Imaginary: $\sqrt{-4}, -\sqrt{-16}$ **67.** 6 and 7 **69. a)** Square 4.6 and compare the result to 20. **b)** 4.6 **71.** $-\sqrt{9}, -\sqrt{7}, -\frac{1}{2}, 2.5, \sqrt{16}, 4.01, 5, 12$ **73.** $\sqrt{4} = 2$; $(36)^{1/2} = 6$; $-\sqrt{9} = -3$; $-(25)^{1/2} = -5$; $(81)^{1/2} = 9$; $(-4)^{1/2}$, imaginary number **75. a)** Yes **b)** No **c)** No **d)** Yes **e)** No **77.** It is nonnegative. The square root of a negative number is not a real number. **79. a)** Yes **b)** Yes **c)** a **d)** Answers will vary. **81.** $x^{3/2}$ **83.** x^3 **85.** 52 jumps **86.** -6 **87.** -1 **88.** $\frac{5}{11}$ **89.** 6

Exercise Set 9.2

1. Factors **3.** Real **5.** Squares **7.** One-half **9.** Already simplified **11.** $2\sqrt{2}$ **13.** Already simplified **15.** $2\sqrt{7}$ **17.** $4\sqrt{2}$ **19.** $2\sqrt{5}$ **21.** $4\sqrt{10}$ **23.** $2\sqrt{11}$ **25.** $6\sqrt{2}$ **27.** $2\sqrt{21}$ **29.** $5\sqrt{5}$ **31.** x^3 **33.** $z^3\sqrt{z}$ **35.** xy^4 **37.** $a^6b^4\sqrt{b}$ **39.** $ab^2\sqrt{c}$ **41.** $n\sqrt{6n}$ **43.** $6ab\sqrt{3a}$ **45.** $9xy^2\sqrt{3x}$ **47.** $8z^3\sqrt{xyz}$ **49.** $8ab^4\sqrt{3bc}$ **51.** $6rs^5t^2\sqrt{5rt}$ **53.** 7 **55.** $2\sqrt{3}$ **57.** $3\sqrt{10}$ **59.** $x\sqrt{14}$ **61.** $4ab\sqrt{3a}$ **63.** $6xy^2\sqrt{2x}$ **65.** $3r^5s^6\sqrt{11}$ **67.** $3xy^3\sqrt{10yz}$ **69.** $3a^3b^5\sqrt{6}$ **71.** $5x$ **73.** $13x^4y^6$ **75.** $42a^2$ **77.** 8 **79.** Exponent on x: 12, on y: 5 **81.** Coefficient: 8, exponent on x: 12, exponent on y: 7 **83. a)** $7x^4$ **b)** $7x^4$ **c)** Yes **85. a)** Answers will vary. **b)** $2\sqrt{5}$ **87.** Answers will vary. **89.** $x^{1/6}$ **91.** $2x^{2/5}$ **93.** Rational, since $\sqrt{6.25} = 2.5$ and 2.5 is a terminating decimal number **95. a)** 4 feet **b)** No; the side length increased $\sqrt{2}$ or ≈ 1.414 times **c)** 4 times **97. a)** Yes **b)** No; for example $\sqrt{2} \cdot \sqrt{2} = 2$. **102.** 148.5 square inches **103.** ≈ 25.13 cubic feet **104.** $\frac{3x+1}{x-1}$ **105.** $y = -\frac{1}{2}x + \frac{3}{2}, m = -\frac{1}{2}, \left(0, \frac{3}{2}\right)$

106.

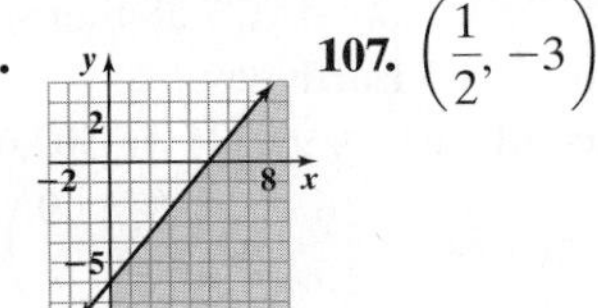

107. $\left(\frac{1}{2}, -3\right)$

Exercise Set 9.3

1. Index **3.** Distributive **5.** $15\sqrt{2}$ **7.** $-2\sqrt{3}$ **9.** $2\sqrt{7} + 5$ **11.** $6\sqrt{x}$ **13.** $-4\sqrt{y}$ **15.** $-4\sqrt{t} - 19$ **17.** $\sqrt{x} + 8\sqrt{y} + x$ **19.** $-2\sqrt{7} - 9\sqrt{x}$ **21.** $11 - 2\sqrt{m} + 5m$ **23.** $2\sqrt{2} - 3\sqrt{3}$ **25.** $8\sqrt{3}$ **27.** $11\sqrt{3}$ **29.** $16\sqrt{2}$ **31.** $-15\sqrt{5} + 20\sqrt{3}$ **33.** $24\sqrt{10}$ **35.** $32 - 4\sqrt{3}$ **37.** $\sqrt{3} + 3$ **39.** $4\sqrt{x} - 4\sqrt{5}$ **41.** $y\sqrt{y} + y^2$ **43.** $3\sqrt{2} + 2\sqrt{3}$ **45.** $x + \sqrt{3x}$ **47.** $9\sqrt{a} - a\sqrt{2}$ **49.** $x^2 + 3x\sqrt{y}$ **51.** $12x^2 - 9x\sqrt{x}$ **53.** $6 - 6\sqrt{2} + \sqrt{5} - \sqrt{10}$ **55.** $\sqrt{30} + 3\sqrt{5} - 4\sqrt{6} - 12$ **57.** $76 - 28\sqrt{7}$ **59.** $64 + 48\sqrt{k} + 9k$ **61.** $z\sqrt{15} + \sqrt{3z} - 4\sqrt{5z} - 4$ **63.** $2r^2 + 9r\sqrt{s} - 18s$ **65.** $6x^2 - 8x\sqrt{2y} + 4y$ **67.** $4p^2 + 6p\sqrt{3q} - 12q$ **69.** -2 **71.** 4 **73.** $x - 4$ **75.** $5 - m^2$ **77.** $5x - y$ **79.** $12w - 80z$ **81. a)** No **b)** $2\sqrt{5}$ **83.** $x + 3$ **85.** Yes **87.** 343 **89.** Perimeter: $4(\sqrt{2} + \sqrt{3})$ units, area: $5 + 2\sqrt{6}$ square units **91.** Perimeter: $\sqrt{7} + \sqrt{15} + \sqrt{22}$ units, area: $\frac{1}{2}\sqrt{105}$ square units **93.** 1 **95.** $\sqrt{216} = 6\sqrt{6}$ **97.** $\sqrt{72} = 6\sqrt{2}$ **99.** Square roots having the same radicand; one example is $\sqrt{3}, 5\sqrt{3}$ **101.** Only like square roots can be added or subtracted. **103.** 48 **105.** 0.05 hour or 3 minutes **106.** $3(x+4)(x-8)$ **107.** $\frac{x+1}{x^2+x+1}$ **108.** $6, -4$ **109.**

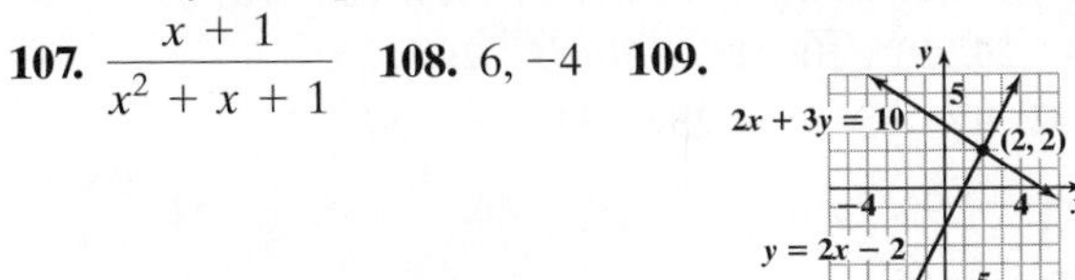

Exercise Set 9.4

1. Square **3.** Denominator **5.** Rational **7.** Conjugates **9.** 3 **11.** 5 **13.** 3 **15.** 6 **17.** $\frac{3}{11}$ **19.** $\frac{1}{10}$ **21.** $\sqrt{3}$ **23.** $\frac{\sqrt{13}}{5}$ **25.** $5x$ **27.** $\frac{4y}{9x}$ **29.** $2x\sqrt{6}$ **31.** $\frac{3\sqrt{5}}{4y^2}$ **33.** $2n^2$ **35.** $\frac{b\sqrt{5}}{c}$ **37.** $5a^2b^3$ **39.** $\frac{3x\sqrt{2x}}{yz^4}$ **41.** $\frac{\sqrt{7}}{7}$ **43.** $4\sqrt{3}$ **45.** $\sqrt{3}$ **47.** $\frac{\sqrt{15}}{5}$ **49.** $\frac{\sqrt{3}}{3}$ **51.** $\frac{\sqrt{10}}{4}$ **53.** $\frac{\sqrt{6x}}{x}$ **55.** $\frac{\sqrt{ab}}{b}$ **57.** $\frac{\sqrt{cd}}{d}$ **59.** $\frac{\sqrt{15p}}{6}$ **61.** $\frac{x\sqrt{5}}{5}$ **63.** $\frac{z\sqrt{2z}}{4}$ **65.** $\frac{t^2\sqrt{6t}}{6}$ **67.** $\frac{a^4\sqrt{14b}}{14b}$ **69.** $\frac{\sqrt{5d}}{5d^2}$ **71.** $\frac{5\sqrt{3}}{6x^2y^2z^4}$ **73.** $\frac{3\sqrt{5x}}{xy^2}$ **75.** 33 **77.** -8 **79.** $x - y^2$ **81.** $a - b$ **83.** $3\sqrt{5} - 6$ **85.** $\frac{4\sqrt{6}+4}{5}$ **87.** $-6\sqrt{3} + 6\sqrt{5}$ **89.** $\frac{-8\sqrt{5} - 16\sqrt{2}}{3}$ **91.** $\frac{4\sqrt{y} - 12}{y - 9}$ **93.** $\frac{28 + 7\sqrt{y}}{16 - y}$ **95.** $\frac{16\sqrt{y} - 16x}{y - x^2}$ **97.** $\frac{5\sqrt{x} - 5\sqrt{y}}{x - y}$ **99.** $\frac{5 + \sqrt{5n}}{5 - n}$ **101.** $\frac{6\sqrt{x} + x}{36 - x}$ **103.** Yes **105.** 5.40 **107.** 0.71 **109.** 7.23 **111.** 0.58 **113.** $\frac{24(4 - \sqrt{3})}{13}$ **115.** Cannot be simplified **117.** Yes, $x\sqrt{3}$ **119.** Cannot be simplified **121.** No perfect square factors in any radicand, no radicand contains a fraction, no square roots in any denominator **123.** 2 **125.** $64x^{10}$ **127.** $\frac{-\sqrt{x} - \sqrt{3x}}{2}$ **130.** $3x - 8 + \frac{6}{x+4}$ **131.** $\frac{9}{2}, -4$ **132.** $\frac{-3x - 5}{(x+2)(x-2)}$ **133.** 30 minutes

Mid-Chapter Test: Sections 9.1–9.4

1. 7 [9.1] **2.** -11 [9.1] **3.** $\frac{13}{9}$ [9.1] **4. a)** Irrational **b)** Rational **c)** Imaginary **d)** Rational [9.1] **5.** $(59xy^2)^{1/2}$ [9.1] **6.** $2\sqrt{10}$ [9.2] **7.** $3\sqrt{7}$ [9.2] **8.** $ab^2c^4\sqrt{b}$ [9.2] **9.** $8x^4y^7\sqrt{2x}$ [9.2] **10.** $3x^2y^4\sqrt{6xy}$ [9.2] **11.** $12\sqrt{x} + 6\sqrt{5} + 17$ [9.3] **12.** $8\sqrt{3}$ [9.3] **13.** $35 + 5\sqrt{3} - 7\sqrt{2} - \sqrt{6}$ [9.3] **14.** $x + z\sqrt{x} - 20z^2$ [9.3] **15.** $3a - 2b$ [9.3] **16.** Can add only like radicals; $\sqrt{8}$ and $\sqrt{32}$ are not like radicals. $\sqrt{8} + \sqrt{32} = 2\sqrt{2} + 4\sqrt{2} = 6\sqrt{2}$ [9.3] **17.** $\frac{1}{2}$ [9.4] **18.** $\frac{3x^2}{y}$ [9.4] **19.** $\frac{7\sqrt{6} + 14}{2}$ [9.4] **20.** $\frac{4x - 20\sqrt{x}}{x - 25}$ [9.4]

Exercise Set 9.5

1. Variable **3.** Extraneous **5.** Isolate **7.** Yes **9.** No **11.** Yes **13.** No **15.** Yes **17.** 25 **19.** No solution **21.** 4 **23.** No solution **25.** 25 **27.** 36 **29.** 6 **31.** 3 **33.** 2 **35.** 5 **37.** 7 **39.** 20 **41.** 1 **43.** 1, 9 **45.** 7, 8 **47.** 7 **49.** 3, 5 **51.** 3 **53.** 3 **55.** 2, 10 **57.** 7 **59.** 11 **61.** $\sqrt{x+4} = 7, 45$ **63.** $\sqrt{x-2} = \sqrt{2x-9}, 7$ **65. a)** $x - 9 = 40$ **b)** 49 **67. a)** $35 + 2\sqrt{x} - x = 35$ **b)** 0, 4 **69.** 1 **71.** 25 **73.** 9 **77.**

78. (2, 0) **79.** (2, 0) **80.** Ferryboat: 16 mph, current: 2 mph

Exercise Set 9.6 **1.** Pythagorean **3.** Leg **5.** Distance **7.** $\sqrt{25} = 5$ **9.** $\sqrt{73} \approx 8.54$ **11.** $\sqrt{33} \approx 5.74$ **13.** $\sqrt{67} \approx 8.19$ **15.** $\sqrt{108} \approx 10.39$ **17.** $\sqrt{202} \approx 14.21$ **19.** $\sqrt{180} \approx 13.42$ **21.** $\sqrt{34} \approx 5.83$ **23.** $\sqrt{193} \approx 13.89$ **25.** $\sqrt{17{,}240.89} \approx 131.30$ yards **27.** $\sqrt{60} \approx 7.75$ meters **29.** 16 feet **31.** $\approx\sqrt{625.80} \approx 25.02$ feet **33.** $\approx\sqrt{64.01} \approx 8.00$ inches **35.** $\sqrt{16{,}200} \approx 127.28$ feet **37.** $\sqrt{914} \approx 30.23$ inches **39.** $6.28\sqrt{0.11} \approx 2.08$ seconds **41.** $6.28\sqrt{1.925} \approx 8.71$ seconds **43.** About 365 Earth days **45.** 33 miles per hour **47.** ≈ 52.2 miles per hour **49.** $\sqrt{19.62(6{,}370{,}000)} \approx 11{,}179.42$ meters per second **51.** $\sqrt{91{,}584} \approx 302.63$ feet per second **53.** $\approx 2.06\text{ m}^2$ **55.** No **57.** 9 inches by 12 inches **59.** ≈ 3.25 feet **60.** $x > 6$ **61.** $\frac{x^4}{y^3}$ **62.** $-\frac{16}{15}$ **63.** $\left(7, -\frac{9}{4}\right)$

Exercise Set 9.7 **1.** Cube **3.** Positive **5.** Real **7.** Index **9.** 2 **11.** -3 **13.** 2 **15.** 6 **17.** 3 **19.** 4 **21.** $2\sqrt[3]{5}$ **23.** $2\sqrt[3]{2}$ **25.** $4\sqrt[3]{2}$ **27.** $2\sqrt[4]{3}$ **29.** $2\sqrt[5]{3}$ **31.** $x^{7/3}$ **33.** $a^{3/5}$ **35.** $y^{15/4}$ **37.** $y^{8/3}$ **39.** x **41.** x^5 **43.** x **45.** a^6 **47.** k **49.** $\sqrt{m}$ **51.** $\sqrt{c}$ **53.** $\sqrt[3]{x}$ **55.** $\sqrt[7]{k}$ **57.** $\sqrt[3]{x^2}$ **59.** $\sqrt[4]{z^3}$ **61.** 32 **63.** 27 **65.** 16 **67.** 16 **69.** $\frac{1}{2}$ **71.** $\frac{1}{81}$ **73.** $\frac{1}{25}$ **75.** $\frac{1}{64}$ **77.** 3 **79.** 2 **81.** 8 **83.** 49 **85.** 27 **87.** 36 **89.** x **91.** $x^{7/12}$ **93.** r^8 **95.** x^6 **97.** both equal 4 **99.** $\sqrt[3]{3^2}$ **101.** $\sqrt[4]{6^2}$ **103.** $\sqrt[4]{7^3}$ **105.** 3 **107.** 1 **109.** xy **111.** $\sqrt[4]{2}$ **113. a)** $\sqrt[3]{2^3} = 2$ **b)** $\frac{\sqrt[3]{4}}{2}$ **114.** -44 **115.** $(3x - 4)(x - 8)$ **116.** slope: $\frac{2}{3}$, y-intercept: $\left(0, -\frac{4}{3}\right)$ **117.** $\frac{3y^3\sqrt{2xy}}{x}$

Chapter 9 Review Exercises **1.** 9 **2.** 12 **3.** -10 **4.** $(21)^{1/2}$ **5.** $(43z)^{1/2}$ **6.** $(58x^2y)^{1/2}$ **7.** $7\sqrt{2}$ **8.** $2\sqrt{14}$ **9.** $4x^3y^2\sqrt{3xy}$ **10.** $5x^2y^4\sqrt{5}$ **11.** $2b^2c^2\sqrt{13ab}$ **12.** $6abc^3\sqrt{2c}$ **13.** $8\sqrt{10}$ **14.** $7a$ **15.** $8x\sqrt{y}$ **16.** $5xy\sqrt{3}$ **17.** $6ab^4\sqrt{a}$ **18.** $10ab^3\sqrt{b}$ **19.** $4\sqrt{3}$ **20.** $-2\sqrt{5}$ **21.** $16\sqrt{x}$ **22.** $-3\sqrt{k}$ **23.** $9\sqrt{2} - 3\sqrt{3}$ **24.** $13\sqrt{10}$ **25.** $-10\sqrt{2}$ **26.** $25\sqrt{2}$ **27.** $2\sqrt{7} + 7$ **28.** $3 + 11\sqrt{3}$ **29.** $x\sqrt{y} - 9y$ **30.** $12a^2 + 4a\sqrt{2a}$ **31.** -57 **32.** 95 **33.** $x^2 - 25y$ **34.** $c - 16d$ **35.** $m^2 - 3m\sqrt{r} - 10r$ **36.** $3t + 5s\sqrt{t} - 2s^2$ **37.** $15m + 17\sqrt{5mn} - 6n$ **38.** $14 - 9\sqrt{7p} + 9p$ **39.** 4 **40.** $\frac{1}{7}$ **41.** $\frac{1}{5}$ **42.** $\frac{2\sqrt{3}}{3}$ **43.** $\frac{\sqrt{11n}}{11}$ **44.** $\frac{\sqrt{15a}}{6}$ **45.** $\frac{x\sqrt{7}}{7}$ **46.** $\frac{z^3\sqrt{2}}{4}$ **47.** $y^2\sqrt{7}$ **48.** $\frac{x\sqrt{2y}}{y^2}$ **49.** $\frac{2\sqrt{5a}}{3a^2b}$ **50.** $\frac{c\sqrt{14a}}{7a}$ **51.** $\frac{-2 - 2\sqrt{6}}{5}$ **52.** $\frac{16 + 4\sqrt{5}}{9}$ **53.** $\frac{2\sqrt{3} - \sqrt{3y}}{4 - y}$ **54.** $\frac{3\sqrt{x} + 24}{x - 64}$ **55.** $\frac{\sqrt{10x} - \sqrt{30}}{x - 3}$ **56.** $\frac{\sqrt{55} + x\sqrt{11}}{5 - x^2}$ **57.** 81 **58.** No solution **59.** 17 **60.** 7 **61.** 2 **62.** 4 **63.** 3 **64.** 7 **65.** 2 **66.** 26 **67.** $\sqrt{125} \approx 11.18$ **68.** $\sqrt{12} \approx 3.46$ **69.** $\sqrt{61} \approx 7.81$ **70.** $\sqrt{108} \approx 10.39$ feet **71.** $\sqrt{325} \approx 18.03$ inches **72.** $\sqrt{109} \approx 10.44$ **73.** $\sqrt{153} \approx 12.37$ **74.** ≈ 561.18 square inches **75.** $\sqrt{60} \approx 7.75$ miles **76.** ≈ 646 feet **77.** 2 **78.** -4 **79.** 2 **80.** 3 **81.** $2\sqrt[3]{2}$ **82.** $2\sqrt[3]{4}$ **83.** $2\sqrt[3]{7}$ **84.** $2\sqrt[4]{2}$ **85.** $3\sqrt[4]{2}$ **86.** $2\sqrt[4]{5}$ **87.** x^7 **88.** s^{12} **89.** 9 **90.** 128 **91.** $\frac{1}{9}$ **92.** 256 **93.** $\frac{1}{625}$ **94.** 243 **95.** $x^{5/2}$ **96.** $a^{7/2}$ **97.** $z^{11/3}$ **98.** $a^{13/3}$ **99.** $b^{17/4}$ **100.** $m^{6/4} = m^{3/2}$ **101.** x **102.** $\sqrt[3]{x^2}$ **103.** a^4 **104.** x^2 **105.** q^3 **106.** b^2 **107.** x^6 **108.** x^{10}

Chapter 9 Practice Test **1.** $(5x)^{1/2}$ [9.1] **2.** $\sqrt[4]{x^3}$ [9.1] **3.** 13 [9.1] **4.** $3\sqrt{10}$ [9.2] **5.** $2x\sqrt{3}$ [9.2] **6.** $5x^3y\sqrt{3xy}$ [9.2] **7.** $4xy\sqrt{5x}$ [9.3] **8.** $5x^2y^2\sqrt{2y}$ [9.3] **9.** $\frac{1}{5}$ [9.4] **10.** $\frac{c^2}{d}$ [9.4] **11.** $\frac{\sqrt{5}}{5}$ [9.4] **12.** $\frac{3\sqrt{5r}}{5}$ [9.4] **13.** $\frac{2\sqrt{30x}}{3xy}$ [9.4] **14.** $-4 - 2\sqrt{7}$ [9.4] **15.** $\frac{7\sqrt{x} + 21}{x - 9}$ [9.4] **16.** $16\sqrt{3}$ [9.3] **17.** $5\sqrt{y}$ [9.3] **18.** 26 [9.5] **19.** 4, 8 [9.5] **20.** $\sqrt{106} \approx 10.30$ [9.6] **21.** $\sqrt{58} \approx 7.62$ [9.6] **22.** $\frac{1}{81}$ [9.7] **23.** x^3 [9.6] **24.** 13 meters [9.6] **25.** $\sqrt{640} \approx 25.30$ feet per second [9.6]

Cumulative Review Test **1. a)** $-5, 735, 4$ **b)** $735, 4$ **c)** $-5, 735, 0.5, 4, \frac{1}{2}$ **d)** $\sqrt{12}$ **e)** $-5, \sqrt{12}, 735, 0.5, 4, \frac{1}{2}$ [1.4] **2.** 29 [1.9] **3.** $\frac{29}{6}$ [2.5] **4.** $x > \frac{27}{11}$,

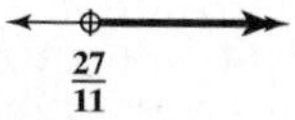

$\left(\frac{27}{11}, \infty\right)$ [2.8] **5.** $(4x + 1)(x^2 + 3)$ [5.2] **6.** $(2x - 3)(x - 9)$ [5.4] **7.** 0, 8 [5.6] **8.** $\frac{1}{5a^5b^6}$ [4.2] **9.**

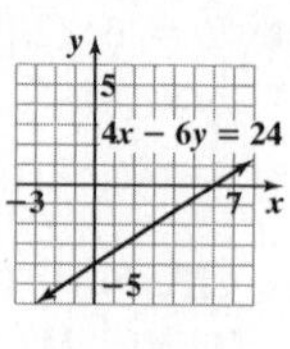

[7.2] **10.** $y = 3x - 2$ [7.4] **11.** $y = \frac{2}{5}x + 4$ [7.4] **12.** $(0, 2)$ [8.3] **13.** $\frac{3y - 5}{8}$ [6.3] **14.** $\frac{5}{7}$ [6.6] **15.** $7\sqrt{11}$ [9.3] **16.** $\frac{\sqrt{21yz}}{14y^3}$ [9.4] **17.** 26 [9.5] **18.** $2\frac{8}{11}$ cups [2.7] **19.** \$125 [3.2] **20.** ≈ 111.1 miles [6.7]

Chapter 10

Exercise Set 10.1 **1.** Quadratic **3.** Plus **5.** 9, -9 **7.** 13, -13 **9.** $5\sqrt{3}, -5\sqrt{3}$ **11.** 8, -8 **13.** 10, -10 **15.** $2\sqrt{5}, -2\sqrt{5}$ **17.** 2, -2 **19.** $\sqrt{17}, -\sqrt{17}$ **21.** 3, -3 **23.** $\frac{\sqrt{15}}{3}, -\frac{\sqrt{15}}{3}$ **25.** $2\sqrt{7}, -2\sqrt{7}$ **27.** 3, 5 **29.** 3, -9 **31.** $-7, 3$ **33.** $-8 + 4\sqrt{2}, -8 - 4\sqrt{2}$ **35.** $-1 + 3\sqrt{5}, -1 - 3\sqrt{5}$ **37.** 0, 16 **39.** $-\frac{14}{3}, \frac{4}{3}$ **41.** $\frac{4}{3}, 4$ **43.** $\frac{-3 \pm 3\sqrt{2}}{2}$ **45.** $\frac{7 + 3\sqrt{2}}{2}, \frac{7 - 3\sqrt{2}}{2}$

47. 4 and 17 **49.** width = 23 inches, length = 28.75 inches **51.** length ≈ 56.58 feet, width ≈ 35.36 feet **53.** $x^2 = 49$ **55.** 19 **57. a)** $3x^2 - 9x + 6 = 0$ **b)** $x^2 - 3x + 2 = 0$ **59. a)** Blue: x^2, red: $(x + 3)^2$ **b)** 6 inches **c)** $\sqrt{50} \approx 7.07$ inches **d)** 9 inches **e)** $\sqrt{92} \approx 9.59$ inches **61.** $s = \sqrt{A}$ **63.** $r = \sqrt{\dfrac{A}{\pi}}$ **65.** $d = \sqrt{\dfrac{k}{I}}$ **67.** $3(2x + 3)(x - 4)$ **68.** $\dfrac{5y - 1}{6y - 1}$ **69.** $y = 4x - 1$ **70.** $\dfrac{3\sqrt{5a}}{ab^3}$

Exercise Set 10.2

1. Perfect **3.** Coefficient **5.** Root **7.** $-4, -6$ **9.** $7, 1$ **11.** $5, -3$ **13.** $1, -3$ **15.** $-5, -2$ **17.** $4, -1$ **19.** $-2, -3$ **21.** $6, -1$ **23.** -3 **25.** $2 + \sqrt{2}, 2 - \sqrt{2}$ **27.** $-3 + \sqrt{6}, -3 - \sqrt{6}$ **29.** $-2 + \sqrt{13}, -2 - \sqrt{13}$ **31.** $\dfrac{-7 + \sqrt{41}}{2}, \dfrac{-7 - \sqrt{41}}{2}$ **33.** $\dfrac{-9 + \sqrt{73}}{2}, \dfrac{-9 - \sqrt{73}}{2}$ **35.** $4, -\dfrac{1}{3}$ **37.** $\dfrac{1}{3}, -\dfrac{3}{2}$ **39.** $\dfrac{-1 + \sqrt{7}}{3}, \dfrac{-1 - \sqrt{7}}{3}$ **41.** $8, 0$ **43.** $9, 0$ **45.** $1, -4$ **47.** $3, 7$ **49.** 24 feet **51.** 10 **53.** 3 seconds and 5 seconds **55. a)** $x^2 + 10x + 25$ **b)** Answers will vary. **57.** $+18x$ or $-18x$ **59.** $7 + 5\sqrt{2}, 7 - 5\sqrt{2}$ **61.** $\dfrac{-3 + \sqrt{59}}{10}, \dfrac{-3 - \sqrt{59}}{10}$ **63.** $\dfrac{-1 + \sqrt{193}}{12}, \dfrac{-1 - \sqrt{193}}{12}$ **65.** $0.75 + \sqrt{3.5625}, 0.75 - \sqrt{3.5625}$ **66.** $\dfrac{4}{(x + 2)(x - 3)}$ **67.** If the slopes are the same and the y-intercepts are different, the equations represent parallel lines. **68.** $(-3, 4)$ **69.** 3

Exercise Set 10.3

1. Standard **3. a)** 5 **b)** -3 **c)** -7 **5.** Discriminant **7.** One **9.** Two distinct real number solutions **11.** No real number solution **13.** One real number solution **15.** Two distinct real number solutions **17.** No real number solution **19.** One real number solution **21.** No real number solution **23.** Two distinct real number solutions **25.** One real number solution **27.** $-3, -6$ **29.** $2, -4$ **31.** $4, 6$ **33.** $8, -1$ **35.** $-\dfrac{1}{2}, -3$ **37.** $\dfrac{1}{3}, -\dfrac{1}{2}$ **39.** No real number solution **41.** $\dfrac{7}{2}, -1$ **43.** $\dfrac{-7 + \sqrt{37}}{2}, \dfrac{-7 - \sqrt{37}}{2}$ **45.** $\dfrac{-1 + \sqrt{41}}{2}, \dfrac{-1 - \sqrt{41}}{2}$ **47.** $2 + \sqrt{2}, 2 - \sqrt{2}$ **49.** $3 + 2\sqrt{3}, 3 - 2\sqrt{3}$ **51.** No real number solution **53.** $\dfrac{7 + \sqrt{17}}{4}, \dfrac{7 - \sqrt{17}}{4}$ **55.** $7, -4$ **57.** $\dfrac{\sqrt{21}}{7}, -\dfrac{\sqrt{21}}{7}$ **59.** Width = 4 feet, length = 5 feet **61.** 2 feet **63.** ≈4.14 feet **65.** 7 sides **67.** 20 flags **69.** ≈1.11 seconds and ≈4.52 seconds **71. a)** $c < 16$ **b)** $c = 16$ **c)** $c > 16$ **73. a)** $c > -3$ **b)** $c = -3$ **c)** $c < -3$ **76.** $10, 4$ **77.** $\dfrac{5}{3}, -\dfrac{7}{2}$ **78.** Cannot be solved by factoring; $\dfrac{-3 + \sqrt{41}}{4}, \dfrac{-3 - \sqrt{41}}{4}$ **79.** $4, -4$ **80.** $\dfrac{x^2 - 9x + 2}{(2x - 1)(x + 4)(x - 5)}$

Mid-Chapter Test

1. $7, -7$ [10.1] **2.** $\sqrt{21}, -\sqrt{21}$ [10.1] **3.** $\dfrac{\sqrt{15}}{4}, -\dfrac{\sqrt{15}}{4}$ [10.1] **4.** $1, 5$ [10.1] **5.** $3, -15$ [10.1] **6.** $7 + 2\sqrt{6}, 7 - 2\sqrt{6}$ [10.1] **7.** $4, 10$ [10.1] **8.** $2, -4$ [10.2] **9.** $3, -5$ [10.2] **10.** $-2, -9$ [10.2] **11.** $0, 8$ [10.2] **12.** $-1 + \sqrt{7}, -1 - \sqrt{7}$ [10.2] **13.** $\dfrac{9 + \sqrt{77}}{2}, \dfrac{9 - \sqrt{77}}{2}$ [10.2] **14.** $8, -5$ [10.3] **15.** $-7, -6$ [10.3] **16.** $\dfrac{5 + \sqrt{33}}{2}, \dfrac{5 - \sqrt{33}}{2}$ [10.3] **17. a)** $b^2 - 4ac > 0$ **b)** $b^2 - 4ac = 0$ **c)** $b^2 - 4ac < 0$ [10.3] **18.** Two distinct real number solutions [10.3] **19.** No real number solution [10.3] **20.** Length is 11 inches, width is 8 inches [10.3]

Exercise Set 10.4

1. Parabola **3.** Downward **5.** x-coordinate **7.** Symmetry **9.** $x = -1, (-1, -8)$, upward **11.** $x = -1, (-1, -1)$, upward **13.** $x = \dfrac{1}{3}, \left(\dfrac{1}{3}, \dfrac{4}{3}\right)$, downward **15.** $x = \dfrac{3}{2}, \left(\dfrac{3}{2}, -\dfrac{7}{4}\right)$, downward **17.** $x = -\dfrac{3}{4}, \left(-\dfrac{3}{4}, \dfrac{7}{8}\right)$, upward **19.** $x = \dfrac{1}{2}, \left(\dfrac{1}{2}, \dfrac{33}{4}\right)$, downward

21.
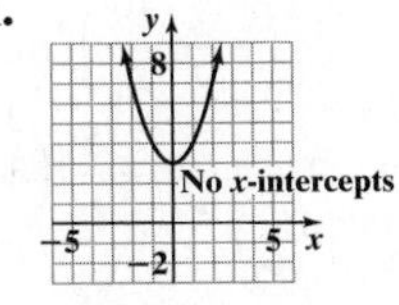

23.

25.
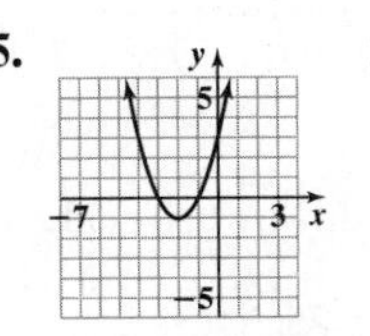

27.
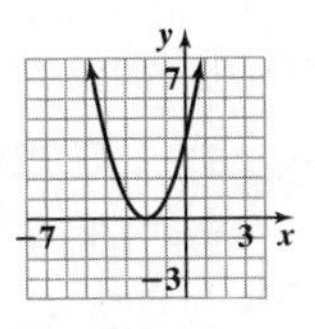

29.
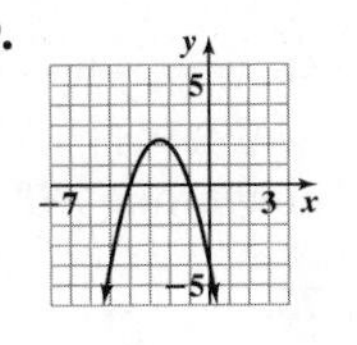

31.

33.
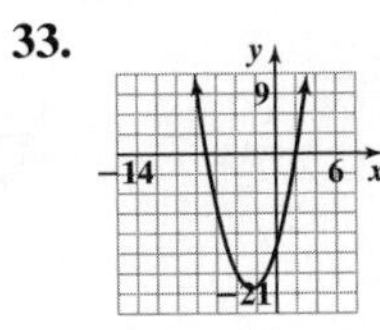

35.
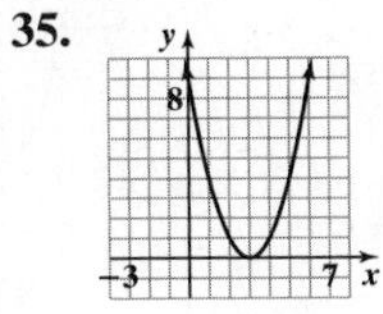

37.
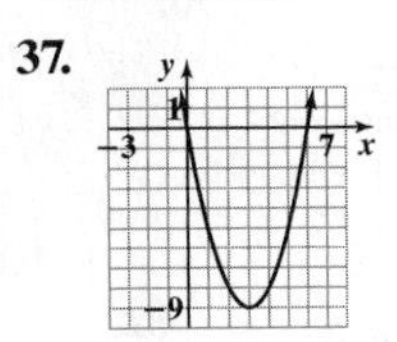

39.
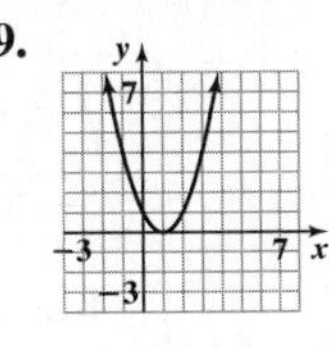

41.
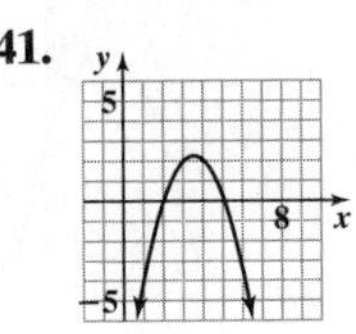

43. 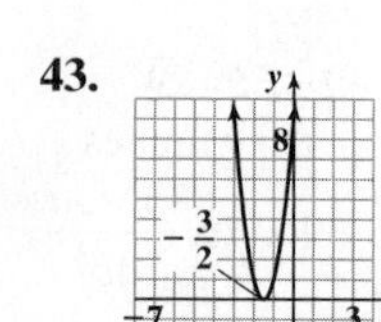

45. 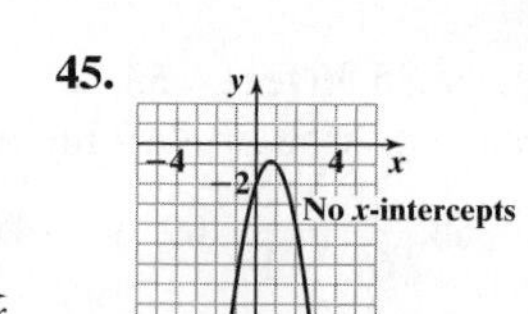

47.

49. Two **51.** One **53.** None **55.** Two **57.** None; the vertex is below the x-axis and the parabola opens downward **59.** One; the vertex of the parabola is on the x-axis **61. a)** ≈ 255 feet **b)** 4 seconds **c)** 8 seconds **d)** ≈ 190 feet, ≈ 240 feet

63. a) 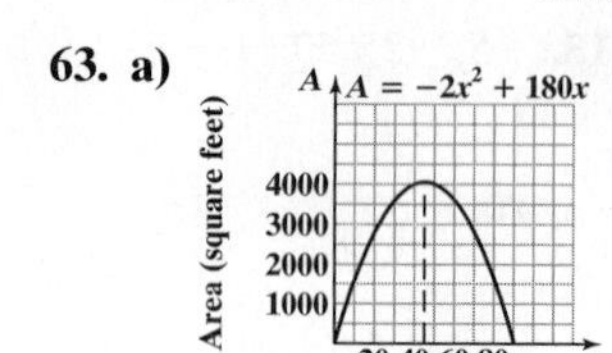

b) 45 feet **c)** 4050 square feet

65. Yes; if y is set to 0, both equations have the same solutions, 5 and -3 **67.** Answers will vary. **69. a)** The x-intercepts are where the graph crosses the x-axis. **b)** The x-intercepts are found by setting $y = 0$ and solving for x. **71.** $x = -\frac{b}{2a}$

73. a) 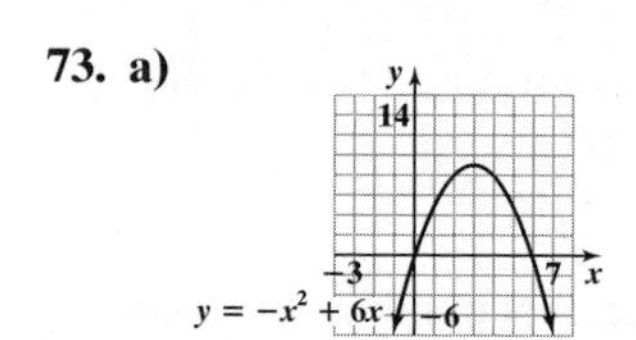

b) 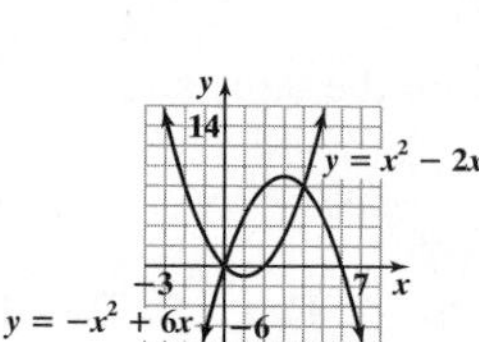

c) $(0, 0), (4, 8)$ **75.** $\frac{-x^2 + 4x - 14}{(x + 3)(x - 4)}$ **76.** $\frac{27}{7}$ **77.** $(6, -5)$ **78.** 7

Exercise Set 10.5

1. Negative **3.** Complex **5.** Imaginary **7.** $4i$ **9.** $10i$ **11.** $i\sqrt{23}$ **13.** $i\sqrt{10}$ **15.** $4i\sqrt{2}$ **17.** $3i\sqrt{5}$ **19.** $8 + 2i$ **21.** $10 + 4i$ **23.** $-3 - 3i$ **25.** $-9 - i\sqrt{15}$ **27.** $5.2 + 4i\sqrt{3}$ **29.** $\frac{1}{6} + 5i\sqrt{3}$ **31.** $10 + 3i$ **33.** $13 - 11i$ **35.** $11 + 5i$ **37.** $-2 - 7i$ **39.** $-13 + i$ **41.** $-9 - 4i$ **43.** $4i, -4i$ **45.** $i\sqrt{30}, -i\sqrt{30}$ **47.** $\frac{-3 + i\sqrt{31}}{4}, \frac{-3 - i\sqrt{31}}{4}$ **49.** $2 + i, 2 - i$ **51.** $\frac{-2 + i\sqrt{10}}{2}, \frac{-2 - i\sqrt{10}}{2}$ **53.** $\frac{1 + i\sqrt{14}}{5}, \frac{1 - i\sqrt{14}}{5}$ **55.** When $c < 0$ **57.** When $b^2 - 4ac < 0$ **59.** 5 **60.** $-\frac{5}{3}$ **61.** 2, 9 **62.** 13

Chapter 10 Review Exercises

1. $5, -5$ **2.** $12, -12$ **3.** $\sqrt{5}, -\sqrt{5}$ **4.** $\sqrt{6}, -\sqrt{6}$ **5.** $\sqrt{7}, -\sqrt{7}$ **6.** $2\sqrt{5}, -2\sqrt{5}$ **7.** $2\sqrt{2}, -2\sqrt{2}$ **8.** $5 + 2\sqrt{6}, 5 - 2\sqrt{6}$ **9.** $\frac{9 + 5\sqrt{2}}{4}, \frac{9 - 5\sqrt{2}}{4}$ **10.** $\frac{-4 + \sqrt{30}}{3}, \frac{-4 - \sqrt{30}}{3}$ **11.** 5, 2 **12.** 7, 4 **13.** $-14, -1$ **14.** $-10, -8$ **15.** $9, -5$ **16.** $1, -6$ **17.** $8, -4$ **18.** $5, -3$ **19.** $-1 + \sqrt{10}, -1 - \sqrt{10}$ **20.** $\frac{5 + \sqrt{33}}{2}, \frac{5 - \sqrt{33}}{2}$ **21.** $\frac{2}{3}, -1$ **22.** $-\frac{4}{3}, -\frac{1}{4}$ **23.** No real solution **24.** No real solution **25.** One solution **26.** Two solutions **27.** Two solutions **28.** No real solution **29.** Two solutions **30.** Two solutions **31.** 2, 8 **32.** 3, 9 **33.** $11, -4$ **34.** $1, \frac{5}{4}$ **35.** $5, -7$ **36.** No real solution **37.** $\frac{3}{2}, -\frac{4}{5}$ **38.** $\frac{3 + \sqrt{57}}{4}, \frac{3 - \sqrt{57}}{4}$ **39.** $\frac{-2 + \sqrt{10}}{2}, \frac{-2 - \sqrt{10}}{2}$ **40.** $3 + \sqrt{6}, 3 - \sqrt{6}$ **41.** No real solution **42.** $\frac{3 + \sqrt{33}}{2}, \frac{3 - \sqrt{33}}{2}$ **43.** $0, \frac{5}{7}$ **44.** $0, \frac{5}{6}$ **45.** 4, 7 **46.** 2, 12 **47.** $10, -6$ **48.** $-9, 3$ **49.** $-4, -11$ **50.** $-5, -8$ **51.** $3, -12$ **52.** 0, 8 **53.** $9, -9$ **54.** $14, -14$ **55.** $\frac{5}{3}, 2$ **56.** $\frac{2}{3}, -\frac{3}{2}$ **57.** $0, \frac{11}{5}$ **58.** $0, -\frac{13}{3}$ **59.** $1, -\frac{8}{3}$ **60.** $2, \frac{5}{3}$ **61.** $\frac{3 + 3\sqrt{3}}{2}, \frac{3 - 3\sqrt{3}}{2}$ **62.** $\frac{-3 + \sqrt{33}}{2}, \frac{-3 - \sqrt{33}}{2}$ **63.** $x = 2$, $(2, -9)$, upward **64.** $x = 6$, $(6, -14)$, upward **65.** $x = \frac{3}{2}, \left(\frac{3}{2}, \frac{19}{4}\right)$, upward **66.** $x = -1$, $(-1, 10)$, downward **67.** $x = -\frac{7}{6}, \left(-\frac{7}{6}, -\frac{13}{12}\right)$, upward **68.** $x = -\frac{7}{2}, \left(-\frac{7}{2}, \frac{49}{4}\right)$, downward **69.** $x = 0$, $(0, -12)$, downward **70.** $x = -\frac{1}{4}\left(-\frac{1}{4}, \frac{121}{8}\right)$, downward **71.** $x = 1$, $(1, 13)$, downward **72.** $x = -\frac{5}{6}, \left(-\frac{5}{6}, -\frac{97}{12}\right)$, upward

73. 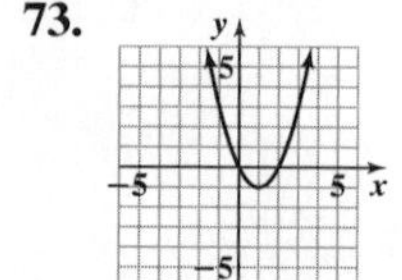

74. 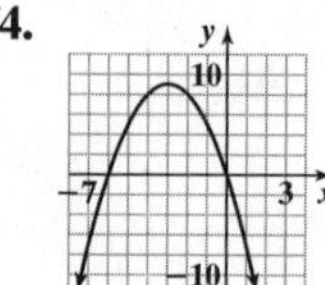

75. 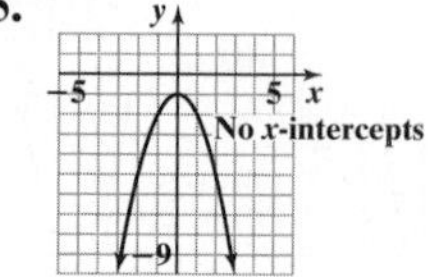

76. 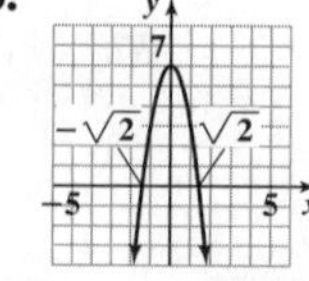

77. 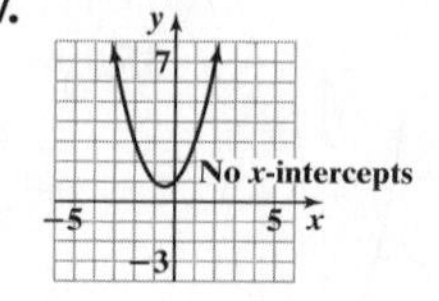

78. 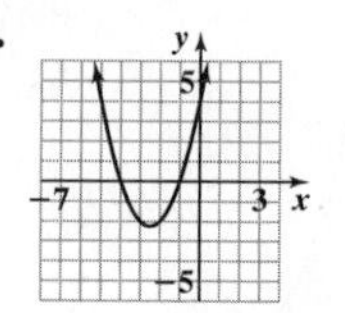

79. 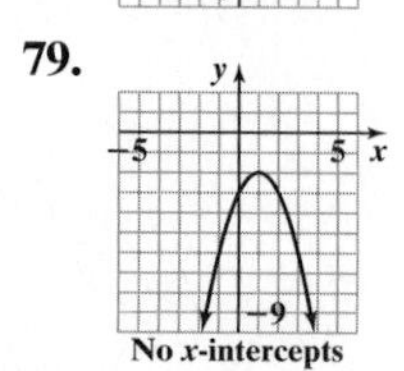

80. 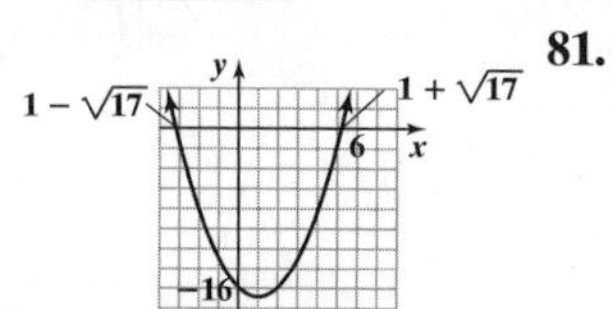

81. 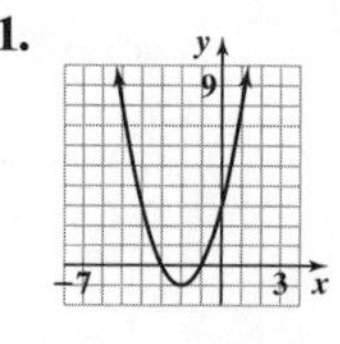

82. 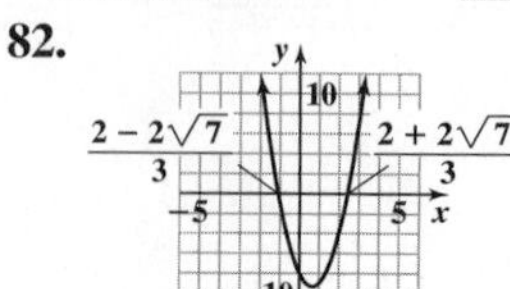

83. 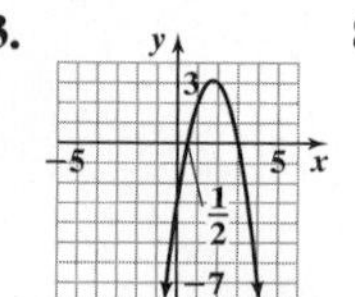

84. 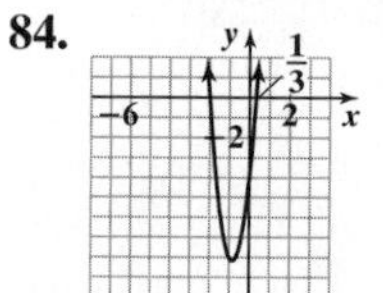

85. 9, 11 **86.** 6, 13 **87.** Width = 20 inches, length = 46 inches **88.** Width = 24 inches, length = 40 inches **89.** $3i$ **90.** $i\sqrt{26}$
91. $8 - 5i$ **92.** $9 - 2i\sqrt{15}$ **93.** $12 - 9i$ **94.** $3 + 8i$ **95.** $4i, -4i$ **96.** $i\sqrt{6}, -i\sqrt{6}$ **97.** $\frac{5 + i\sqrt{23}}{4}, \frac{5 - i\sqrt{23}}{4}$
98. $\frac{2 + i\sqrt{5}}{2}, \frac{2 - i\sqrt{5}}{2}$

Chapter 10 Practice Test

1. $4\sqrt{2}, -4\sqrt{2}$ [10.1] **2.** $\frac{3 + \sqrt{13}}{2}, \frac{3 - \sqrt{13}}{2}$ [10.1] **3.** 10, −5 [10.2] **4.** 3, −11 [10.2]
5. 6, 7 [10.3] **6.** $\frac{-4 + \sqrt{6}}{2}, \frac{-4 - \sqrt{6}}{2}$ [10.3] **7.** $\frac{5}{4}, -\frac{5}{4}$ [10.1–10.3] **8.** $x = \frac{-b \pm \sqrt{b^2 - 4ac}}{2a}$ [10.3] **9.** Answers will vary. [10.2]
10. Two real solutions [10.3] **11.** One real solution [10.3] **12.** $x = -4$ [10.4] **13.** $x = 2$ [10.4] **14.** Downward; $a < 0$ [10.4]
15. Upward; $a > 0$ [10.4] **16.** The vertex is the lowest point on a parabola that opens upward; highest point on a parabola that opens downward [10.4]
17. $(-2, -8)$ [10.4] **18.** $\left(\frac{4}{3}, \frac{11}{3}\right)$ [10.4] **19.**

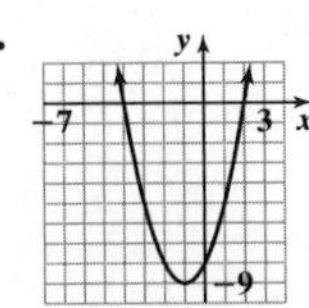

[10.4] **20.** [10.4] **21.** [10.4]

22. Width = 3 feet, length = 11 feet [10.2–10.3] **23.** 7 [10.2–10.3] **24.** 11 [10.2–10.3] **25.** $\frac{1 + i\sqrt{17}}{3}, \frac{1 - i\sqrt{17}}{3}$ [10.5]

Cumulative Review Test

1. 255 [1.9] **2.** $\frac{14}{27}$ [2.5] **3.** $5\frac{1}{3}$ inches [2.7] **4.** $x \geq -\frac{1}{4}$, $\left[-\frac{1}{4}, \infty\right)$ [2.8]
5. $P = 2A - m - n$ [2.6] **6.** $72a^{16}b^{25}$ [4.1] **7.** $x + 4 - \frac{15}{x + 2}$ [4.6] **8.** $(2x - 3y)(x - 2y)$ [5.2] **9.** $3(2x + 3)(x - 6)$ [5.4]
10. $\frac{5a - 4}{(a + 4)(a - 4)^2}$ [6.4] **11.** 3, 8 [6.6] **12.** [7.2] **13.** $(-3, -9)$ [8.3] **14.** $\frac{y\sqrt{6xy}}{6}$ [9.4] **15.** $4\sqrt{7}$ [9.3]
16. 4 [9.5] **17.** $\frac{-3 + \sqrt{73}}{4}, \frac{-3 - \sqrt{73}}{4}$ [10.3]

18. 25.6 pounds [2.7] **19.** Width = 7 feet, length = 25 feet [3.3] **20.** Walks: 3 miles per hour, jogs: 6 miles per hour [6.7]

Applications Index

Housing/construction

Insurance

Manufacturing

Medical/medications

Miscellaneous

Miscellaneous (*Cont.*)

Retail

Salary

Scientific/astronomical

Sports/exercise

Tax

Travel/transportation. *See also* Automotive/motor vehicle; Aviation/aeronautical

Weather/temperature

Subject Index

Q

R

U

V

W

X

Y

Z

Photo Credits

Chapter 1

Page 1 Fotolia; **p. 3** Fotolia; **p. 4** Yuri Arcurs/Fotolia; **p. 5** Fotolia; **p. 8** Fotolia; **p. 16 (left)** Fotolia; **p. 16 (right)** Photos.com; **p. 18** Fotolia; **p. 31** Steve Byland/Fotolia; **p. 37** Allen R. Angel; **p. 43** Allen R. Angel; **p. 51** Fotolia; **p. 61** Byron Moore/Fotolia; **p. 69** Fotolia; **p. 78 (top)** Fotolia; **p. 78 (bottom)** Doug Menuez/Photodisc/Getty Images; **p. 85** Christophe Fouquin/Fotolia; **p. 91** Fotolia

Chapter 2

Page 95 Fotolia; **p. 126** Allen R. Angel; **p. 136** Fotolia; **p. 140** Allen R. Angel; **p. 142** Dan Bannister/Rough Guides/Dorling Kindersley, Ltd.; **p. 147 (left)** Chris Lofty/Fotolia; **p. 147 (right)** Kittycat/Dreamstime; **p. 148** Sophie Bluy/Pearson Education, Inc.; **p. 149** Allen R. Angel; **p. 150** Eric Isselée/Fotolia; **p. 153** Kurhan/Fotolia; **p. 154** Rosawolke/Fotolia; **p. 160 (top, left)** Fotolia; **p. 160 (bottom, left)** Allen R. Angel; **p. 160 (right)** Allen R. Angel; **p. 161 (top)** Steffen Niclas/Fotolia; **p. 161 (bottom)** Barbara Helgason/Fotolia; **p. 176** Fotolia; **p. 177** Allen R. Angel

Chapter 3

Page 179 William Casey/Fotolia; **p. 182** Fotolia; **p. 183** Allen R. Angel; **p. 184** Fotolia; **p. 187** Allen R. Angel; **p. 189** Allen R. Angel; **p. 190 (left)** Fotolia; **p. 190 (top, right and bottom, right)** Allen R. Angel; **p. 191 (left)** Allen R. Angel; **p. 191 (top, right)** Fotolia; **p. 191 (bottom, right)** Allen R. Angel; **p. 192 (top and bottom)** Allen R. Angel; **p. 193 (left and right)** Allen R. Angel; **p. 195** MIXA Co., Ltd./Dorling Kindersley, Ltd.; **p. 196** Allen R. Angel; **p. 1978 (top)** Fotolia; **p. 197 (bottom)** Allen R. Angel; **p. 198** Allen R. Angel; **p. 199** Allen R. Angel; **p. 201 (top, left)** Fotolia; **p. 201 (bottom, left)** Igor Mojzes/Fotolia; **p. 201 (right)** Tupungato/Fotolia; **p. 202 (left)** Ljupco Smokovski/Fotolia; **p. 202 (right)** Allen R. Angel; **p. 203 (top, left)** Fotolia; **p. 203 (bottom, left)** Allen R. Angel; **p. 203 (right)** Sarapinas Valery/Fotolia; **p. 204 (top)** Gene Chutka/iStockPhoto; **p. 204 (middle and bottom)** Allen R. Angel; **p. 205** Fotolia; **p. 208** Allen R. Angel; **p. 213** Allen R. Angel; **p. 219 (top, left)** Natalia Bratslavsky/Fotolia; **p. 219 (bottom, left and right)** Allen R. Angel; **p. 220 (top)** EyeWire Collection/Photodisc/Getty Images; **p. 220 (bottom)** Ivonne Wierink/Fotolia; **p. 221 (top and bottom)** Allen R. Angel; **p. 222 (left)** Fotolia; **p. 22 (right)** Ulga/Fotolia; **p. 223** Gina Smith/Fotolia; **p. 226 (left)** Eduard Stelmakh/Fotolia; **p. 226 (right)** Allen R. Angel; **p. 227 (left)** SeanPavone/Fotolia; **p. 227 (right)** Comstock; **p. 228 (left)** Allen R. Angel; **p. 228 (right)** MIXA Co., Ltd./Dorling Kindersley, Ltd.

Chapter 4

Page 231 Allen R. Angel; **p. 251** Allen R. Angel; **p. 254** Anke van Wyk/Fotolia; **p. 255** Allen R. Angel; **p. 256** Courtesy of Oak Ridge National Laboratory; **p. 257** Allen R. Angel; **p. 258** Canada Bureau of Travel; **p. 259** Mihail Glushakov/Fotolia; **p. 260** NASA; **p. 285** Allen R. Angel; **p. 288** Fotolia; **p. 290 (top and bottom)** Allen R. Angel

Chapter 5

Page 291 Allen R. Angel; **p. 304** Daniel Wiedemann/Fotolia; **p. 312** Michaeljung/Fotolia; **p. 322** Allen R. Angel; **p. 335** Allen R. Angel; **p. 340** Jaimie Duplass/Fotolia; **p. 341 (top and bottom)** Allen R. Angel; **p. 342** Allen R. Angel; **p. 346** Jaimie Duplass/Fotolia; **p. 347 (top)** Allen R. Angel; **p. 347 (bottom)** Fotolia

Chapter 6

Page 348 Konstantin Shevtsov/Fotolia; **p. 363** Oleg Zabielin/Fotolia; **p. 370** Fotolia; **p. 377** Allen R. Angel; **p. 389** Gareth Boden/Pearson Education Ltd.; **p. 390** Mcininch/Dreamstime; **p. 392** Shchipkova Elena/Fotolia; **p. 393** Diggkat/Fotolia; **p. 394** Doug Menuez/Photodisc/Getty Images; **p. 397 (left)** Allen R. Angel; **p. 397 (right)** Nikola Spasenoski/Fotolia; **p. 398 (top)** Mark Atkins/Fotolia; **p. 398 (bottom)** Llandrea/Fotolia; **p. 399 (top, left and bottom, left)** Allen R. Angel; **p. 399 (right)** Igor Mojzes/Fotolia; **p. 400 (top)** Marina/Fotolia; **p. 400 (bottom)** Fotolia; **p. 403 (top)** Bugsy/Dreamstime; **p. 403 (bottom)** Allen R. Angel; **p. 404** Stephen Coburn/Fotolia; **p. 405 (top, left)** Allen R. Angel; **p. 405 (bottom, left)** Fotolia; **p. 405 (right)** Rpernell/Dreamstime; **p. 406** Allen R. Angel; **p. 411** Allen R. Angel; **p. 412 (left)** Nathan Allred/Fotolia; **p. 412 (top, right)** Monkey Business/Fotolia; **p. 412 (bottom, right)** Allen R. Angel

Chapter 7

Page 413 Thinkstock; **p. 414** Library of Congress Prints and Photographs Division [LC-USZ62-61365]; **p. 426** JanMika/Fotolia; **p. 429 (top)** Yuri Arcurs/Fotolia; **p. 429 (bottom)** Thinkstock; **p. 433** Andrey Armyagov/Fotolia; **p. 439** Allen R. Angel; **p. 440** Dell/Fotolia; **p. 444** Allen R. Angel; **p. 450** John Narewski/U.S. Navy; **p. 462** Morane/Fotolia; **p. 465** Michael Harbison/Fotolia; **p. 466 (left)** Corbis; **p. 466 (right)** Allen R. Angel

Chapter 8

Page 475 Thinkstock; **p. 483** Magnus Rew/Dorling Kindersley, Ltd.; **p. 489 (left)** MIXA Co., Ltd./Dorling Kindersley, Ltd.; **p. 489 (right)** Jakob Radlgruber/Fotolia; **p. 497** Allen R. Angel; **p. 500** Allen R. Angel; **p. 501** Allen R. Angel; **p. 506 (left)** NASA; **p. 506 (top, right)** Allen R. Angel; **p. 506 (bottom, right)** Brenda Carson/Fotolia; **p. 507 (top)** Fotolia; **p. 507 (bottom)** Allen R. Angel; **p. 508 (top, left)** Alexey Fursov/Fotolia; **p. 508 (bottom, left)** Comstock; **p. 508 (right)** Allen R. Angel; **p. 515** Robert Pernell/Fotolia

Chapter 9

Page 518 Jorge Felix Costa/Shutterstock; **p. 525** Allen R. Angel; **p. 538** Allen R. Angel; **p. 546** Thinkstock; **p. 553** Allen R. Angel; **p. 555** Dan Race/Fotolia; **p. 556 (top)** Allen R. Angel; **p. 556 (bottom)** Antonio Gravante/Fotolia; **p. 558** Thinkstock; **p. 559** Allen R. Angel; **p. 570 (left)** Mark Ross/Fotolia; **p. 570 (right)** Allen R. Angel

Chapter 10

Page 573 Jana Lumley/Fotolia; **p. 577** Monkey Business/Fotolia; **p. 593 (left)** Tudor Photography/Pearson Education, Inc.; **p. 593 (right)** Kathy Libby/Fotolia; **p. 594** Matjaz Boncina/iStockphoto; **p. 599 (left)** Fotolia; **p. 599 (right)** Allen R. Angel; **p. 611** Monkey Business/Fotolia; **p. 612** Fotolia; **p. 613** Krawczyk/Fotolia

Chapter 1 Real Numbers

Fractions

Addition

$$\frac{a}{c} + \frac{b}{c} = \frac{a+b}{c}$$

Subtraction

$$\frac{a}{c} - \frac{b}{c} = \frac{a-b}{c}$$

Multiplication

$$\frac{a}{b} \cdot \frac{c}{d} = \frac{a \cdot c}{b \cdot d}$$

Division

$$\frac{a}{b} \div \frac{c}{d} = \frac{a}{b} \cdot \frac{d}{c} = \frac{a \cdot d}{b \cdot c}$$

Natural numbers $\{1,2,3,4,\ldots\}$

Whole numbers $\{0,1,2,3,\ldots\}$

Integers $\{\ldots,-3,-2,-1,0,1,2,3,\ldots\}$

Rational numbers {quotient of two integers, denominator not 0}

The sum of two positive numbers will be a positive number.
The sum of two negative numbers will be a negative number.
The sum of a positive number and a negative number can be either a positive or negative number.

The product (or quotient) of two numbers with like signs will be a positive number.
The product (or quotient) of two numbers with unlike signs will be a negative number.

$a - b$ means $a + (-b)$

$$\frac{a}{-b} = \frac{-a}{b} = -\frac{a}{b}$$

$$b^n = \underbrace{b \cdot b \cdot b \cdot \cdots \cdot b}_{n \text{ factors of } b}$$

Order of Operations

1. Evaluate expressions within parentheses.
2. Evaluate expressions with exponents.
3. Perform multiplications or divisions moving from left to right.
4. Perform additions or subtractions moving from left to right.

Properties of the Real Numbers

Commutative: $a + b = b + a, a \cdot b = b \cdot a$

Associative: $(a + b) + c = a + (b + c), (a \cdot b) \cdot c = a \cdot (b \cdot c)$

Distributive: $a(b + c) = a \cdot b + a \cdot c$

Identity: $a + 0 = 0 + a = a, 1 \cdot a = a \cdot 1 = a$

Inverse: $a + (-a) = -a + a = 0, a \cdot \frac{1}{a} = \frac{1}{a} \cdot a = 1$ for $a \neq 0$

Chapter 2 Solving Linear Equations and Inequalities

Addition property of equality: If $a = b$, then $a + c = b + c$ for any real numbers a, b, and c.

Multiplication property of equality: If $a = b$, then $a \cdot c = b \cdot c$ for any real numbers a, b, and c.

Linear equation: $ax + b = c$, for real numbers a, b, and c.

To Solve Linear Equations with the Variable on Both Sides of the Equation

1. If the equation contains fractions, multiply both sides of the equation by the LCD.
2. Use the distributive property to remove parentheses.
3. Combine like terms on the same side of the equation.
4. Use the addition property to rewrite the equation with all terms containing the variable on one side of the equation and all terms not containing the variable on the other side of the equation. Repeated use of the addition property will eventually result in an equation of the form $ax = b$.
5. Use the multiplication property to isolate the variable. This will give a solution of the form $x =$ some number.
6. Check the solution in the original equation.

Simple interest formula: $i = prt$

Distance formula: $d = rt$

Geometric formulas: See Section 2.6 and Appendix B.

Cross multiplication: If $\frac{a}{b} = \frac{c}{d}$, then $ad = bc$.

Inequalities

If $a > b$, then $a + c > b + c$.
If $a > b$, then $a - c > b - c$.
If $a > b$ and $c > 0$, then $ac > bc$.
If $a > b$ and $c > 0$, then $\frac{a}{c} > \frac{b}{c}$.
If $a > b$ and $c < 0$, then $ac < bc$.
If $a > b$ and $c < 0$, then $\frac{a}{c} < \frac{b}{c}$.

Chapter 3 Formulas and Applications of Algebra

Problem-Solving Procedure for Solving Applications Problems

1. **Understand the problem.**
   Identify the quantity or quantities you are being asked to find.
2. **Translate the problem into mathematical language (express the problem as an equation).**
   a) Choose a variable to represent one quantity, *and write down exactly what it represents.* Represent any other quantity to be found in terms of this variable.
   b) Using the information from step a) write an equation that represents the application.
3. **Carry out the mathematical calculations (solve the equation).**
4. **Check the answer (using the original application).**
5. **Answer the question asked.**

Chapter 4 Exponents and Polynomials

Rules of Exponents

1. $x^m \cdot x^n = x^{m+n}$ **product rule**
2. $\dfrac{x^m}{x^n} = x^{m-n}, x \neq 0$ **quotient rule**
3. $(x^m)^n = x^{m \cdot n}$ **power rule**
4. $x^0 = 1, x \neq 0$ **zero exponent rule**
5. $x^{-m} = \dfrac{1}{x^m}, x \neq 0$ **negative exponent rule**
6. $\left(\dfrac{ax}{by}\right)^m = \dfrac{a^m x^m}{b^m y^m}, b \neq 0, y \neq 0$ **expanded power rule**
7. $\left(\dfrac{a}{b}\right)^{-m} = \left(\dfrac{b}{a}\right)^m, a \neq 0, b \neq 0$ **a fraction raised to a negative exponent rule**

FOIL method (*F*irst, *O*uter, *I*nner, *L*ast) of multiplying binomials: $(a+b)(c+d) = ac + ad + bc + bd$

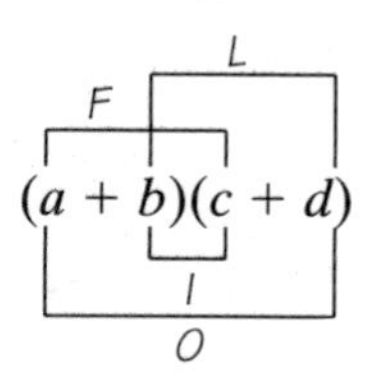

Product of the sum and difference of the same two terms:
$(a+b)(a-b) = a^2 - b^2$

Squares of binomials:
$(a+b)^2 = a^2 + 2ab + b^2$
$(a-b)^2 = a^2 - 2ab + b^2$

Chapter 5 Factoring

If $a \cdot b = c$, then a and b are **factors** of c.
Difference of two squares: $a^2 - b^2 = (a+b)(a-b)$
Sum of two cubes: $a^3 + b^3 = (a+b)(a^2 - ab + b^2)$
Difference of two cubes: $a^3 - b^3 = (a-b)(a^2 + ab + b^2)$

To Factor a Polynomial

1. If all the terms of the polynomial have a greatest common factor other than 1, factor it out.
2. If the polynomial has two terms, determine if it is a difference of two squares or a sum or a difference of two cubes. If so, factor using the appropriate formula.
3. If the polynomial has three terms, factor the trinomial using one of the procedures discussed.
4. If the polynomial has more than three terms, try factoring by grouping.
5. As a final step, examine your factored polynomial to see if the terms in any factors listed have a common factor. If you find a common factor, factor it out at this point.

Quadratic equation: $ax^2 + bx + c = 0, a \neq 0$
Zero-factor property: If $ab = 0$, then $a = 0$ or $b = 0$.

To Solve a Quadratic Equation by Factoring

1. Write the equation in standard form with the squared term positive. This will result in one side of the equation being 0.
2. Factor the side of the equation that is not 0.
3. Set each factor containing a variable equal zero and solve each equation.
4. Check the solution found in step 3 in the original equation.

Pythagorean Theorem: $a^2 + b^2 = c^2$

Chapter 6 Rational Expressions and Equations

To Simplify Rational Expressions

1. Factor both the numerator and denominator as completely as possible.
2. Divide out any factors common to both the numerator and denominator.

To Multiply Rational Expressions

1. Factor all numerators and denominators completely.
2. Divide out common factors.
3. Multiply the numerators together and multiply the denominators together.

To Add or Subtract Two Rational Expressions

1. Determine the least common denominator (LCD).
2. Rewrite each fraction as an equivalent fraction with the LCD.
3. Add or subtract numerators while maintaining the LCD.
4. When possible, factor the remaining numerator and simplify the fraction.

To Solve Rational Equations

1. Determine the LCD of all fractions in the equation.
2. Multiply both sides of the equation by the LCD. This will result in every term in the equation being multiplied by the LCD.
3. Remove any parentheses and combine like terms on each side of the equation.
4. Solve the equation.
5. Check your solution in the original equation.

Variation

Direct Variation: $y = kx$
Inverse Variation: $y = \dfrac{k}{x}$

Chapter 7 Graphing Linear Equations

Linear equation in two variables: $a + by = c$

A **graph** is an illustration of the set of points whose coordinates satisfy the equation.
Every **linear equation** of the form $ax + by = c$ will be a straight line when graphed.
To find the *y*-intercept (where the graph crosses the y-axis) set $x = 0$ and solve for $\boldsymbol{y}$.
To find the *x*-intercept (where the graph crosses the x-axis) set $y = 0$ and solve for $\boldsymbol{x}$.

$$\text{slope } (m) = \frac{\text{rise}}{\text{run}} = \frac{\text{change in } y}{\text{change in } x} = \frac{y_2 - y_1}{x_2 - x_1}$$

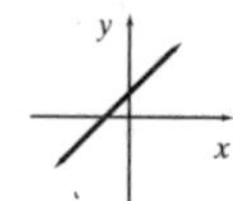

Positive slope (rises to right)

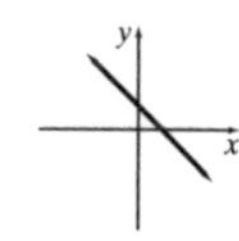

Negative slope (falls to right)

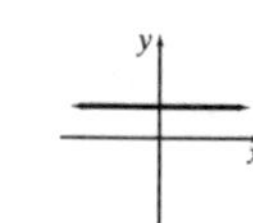

Slope is 0. (horizontal line)

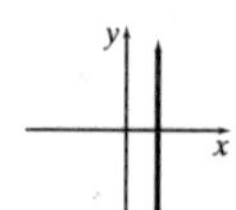

Slope is undefined. (vertical line)